Chemistry and Biochemistry of Plant Pigments

Chemistry and Biochemistry of Plant Pigments

Edited by

T. W. GOODWIN

Department of Biochemistry,
University of Liverpool, England

Second Edition

Volume 1

1976

ACADEMIC PRESS LONDON NEW YORK SAN FRANCISCO

A Subsidiary of Harcourt Brace Jovanovich, Publishers

ACADEMIC PRESS INC. (LONDON) LTD.
24/28 Oval Road,
London NW1

United States Edition published by
ACADEMIC PRESS INC.
111 Fifth Avenue
New York, New York 10003

Library of Congress Catalog Card Number: 75-19642
ISBN: 0-12-289901-6

Printed in Great Britain by
William Clowes & Sons Limited
London, Colchester and Beccles

List of Contributors

L. BOGORAD *The Biological Laboratories, Harvard University, 16 Divinity Avenue, Cambridge, Massachusetts, 02138, U.S.A.* (Vol. 1, p. 64.)

J. H. BURNETT *Department of Agricultural Science, Oxford University, Oxford, U.K.* (Vol. 1, p. 655.)

G. BRITTON *Department of Biochemistry, University of Liverpool, P.O. Box 147, Liverpool L69 3BX, U.K.* (Vol. 1, p. 262.)

C. O. CHICHESTER *Department of Food and Resource Chemistry, University of Rhode Island, Kingston, R.I. 02881, U.S.A.* (Vol. 1, p. 779.)

B. H. DAVIES *Department of Biochemistry and Agricultural Biochemistry, Institute of Rural Science, University College of Wales, Penglais, Aberystwyth, U.K.* (Vol. 2, p. 38.)

A. L. GALSTON *Department of Biology, Yale University, New Haven, Connecticut 06520, U.S.A.* (Vol. 1, p. 680.)

T. W. GOODWIN *Department of Biochemistry, University of Liverpool, P.O. Box 147, Liverpool L69 3BX, U.K.* (Vol. 1, p. 225.)

J. B. HARBORNE *Department of Botany, Plant Science Laboratories, The University, Whiteknights, Reading RG6 2AS, U.K.* (Vol. 1, p. 736.)

M. HOLDEN *Biochemistry Department, Rothamsted Experimental Station, Harpenden, Hertfordshire AL5 2JQ, U.K.* (Vol. 2, p. 1.)

A. H. JACKSON *Department of Chemistry, University College, Cardiff, P.O. Box 78, Cardiff CF1 1XL, U.K.* (Vol. 1, p. 1.)

W. JUNGE *Max-Vomer Institut für Physikalische Chemie und Molekularbiologie, Technische Universität Berlin, Berlin 12, Strasse des 17 Juni 135, Berlin, Germany.* (Vol. 2, p. 233.)

R. E. KENDRICK *Department of Plant Biology, The University, Newcastle Upon Tyne NE1 7RU, U.K.* (Vol. 1, p. 377; Vol. 2, p. 334.)

T.-C. LEE *Department of Food and Resource Chemistry, University of Rhode Island, Kingston, R.I. 02881, U.S.A.* (Vol. 1, p. 779.)

G. P. MOSS *Department of Chemistry, Queen Mary College, University of London, Mile End Road, London E1 4NS, U.K.* (Vol. 1, p. 149.)

P. Ó CARRA *Department of Biochemistry, University College, Galway, Ireland.* (Vol. 1, p. 328.)

C. Ó hEOCHA *Department of Biochemistry, University College, Galway, Ireland.* (Vol. 1, p. 328.)

M. PIATELLI *Istituto di Chimica Organica, Università di Catania, Catania, 95125, Viale A. Doria, Italy.* (Vol. 1, p. 560.)

D. B. RODRIGUEZ *Department of Food and Resource Chemistry, University of Rhode Island, Kingston, R.I. 02881, U.S.A.* (Vol. 1, p. 779.)

R. L. SATTER *Department of Biology, Yale University, New Haven, Connecticut 06520, U.S.A.* (Vol. 1, p. 680.)

K. L. SIMPSON *Department of Food and Resource Chemistry, University of Rhode Island, Kingston, R.I. 02881, U.S.A.* (Vol. 1, p. 779.)

H. SMITH *Department of Physiology and Environmental Studies, University of Nottingham, School of Agriculture, Sutton Bonington, Loughborough, Leicestershire LE12 5RD, U.K.* (Vol. 1, p. 377; Vol. 2, p. 334.)

T. SWAIN *ARC Laboratory of Biochemical Systematics, Royal Botanic Gardens, Kew, Richmond, Surrey, U.K.* (Vol. 1, p. 425; Vol. 2, p. 166.)

R. H. THOMSON *Department of Chemistry, University of Aberdeen, Meston Walk, Old Aberdeen AB9 2UE, U.K.* (Vol. 1, p. 527, p. 597; Vol. 2, p. 207.)

B. C. L. WEEDON *Department of Chemistry, Queen Mary College, University of London, Mile End Road, London, E1 4NS, U.K.* (Vol. 1, p. 149.)

C. P. WHITTINGHAM *Botany Department, Rothamsted Experimental Station, Harpenden, Hertfordshire AL5 2JQ, U.K.* (Vol. 1, p. 624.)

E. WONG *Applied Biochemistry Division, D.S.I.R., Palmerston North, New Zealand.* (Vol. 1, p. 464.)

Preface to First Edition

In August 1962 a colloquium on the "Biochemistry of Plant Pigments" was held by the Biochemical Society in Aberystwyth. It was successful in bringing together interested scientists from a number of countries to discuss critically the latest developments in this field. However, it also clearly demonstrated the need for collecting in one volume the scattered information discussed, and by early 1963 the editor had planned the book and persuaded the various authors to take part in the venture and to prepare their contributions by early 1964. The coverage of the colloquium was extended to include the chemistry of the plant pigments and an analytical section was also planned. All the contributors have international reputations and the fact that they all willingly agreed to undertake the work re-emphasizes the need for this volume. A few manuscripts inevitably failed to be produced on time, but apart from this the co-operation of the contributors with the editor has been excellent, and no editor could have had an easier assignment. Furthermore, the expertise of the publishers in dealing with late manuscripts has resulted in only a minimum delay in the appearance of the book. Owing to unforeseen circumstances it has not been possible to include a projected chapter on plant cytochromes.

The plan of the book is simple; it is divided into four main sections, the first dealing with the chemistry and biosynthesis of the various pigments; the second and third with their function and metabolism respectively; and the fourth with analytical procedures used in dealing with plant pigments. All the contributors are still actively concerned in research in the speciality on which they write and this has resulted in critically written, forward-looking chapters in which not only is the well-established information thoroughly reviewed but the gaps in our knowledge are clearly indicated, and sufficient suggestions for future investigations are made, either implicitly or explicitly, to keep an army of research workers busy for a long time. The detailed analytical chapters, considered in association with Parts I, II and III, should be of particular value.

Biochemical investigations on higher plants have been relatively few compared with those on animals and protista. During the past few years, however, more and more biochemists have become interested in this field and, in spite of the many difficulties in dealing with higher plants, the number of investigations reported in the journals is increasing rapidly each year. This book should be of background value to those who are contemplating entry into the field of plant biochemistry via plant pigments; it should also be useful to senior undergraduate students in biochemistry who are finding that more and more plant biochemistry is appearing in their courses every year. Their instructors may also find useful suggestions for laboratory exercises in the analytical sections. Finally, the "Chemistry and Biochemistry of Plant Pigments" should appeal to the increasing number of botanists who are keen to specialize in the more biochemical aspects of plant physiology.

The final stages of the editing of this volume were carried out whilst the editor was holder of a National Science Foundation Senior Foreign Scientist Fellowship at the University of California, Davis.

Dr B. H. Davies kindly prepared the subject index.

T. W. GOODWIN

Department of Biochemistry and Agricultural Biochemistry
University College of Wales
Aberystwyth

November, 1964

Preface to Second Edition

The success of the First Edition of "Chemistry and Biochemistry of Plant Pigments" and the great advances in the subject during the last decade clearly warranted a new edition. The editor has been extremely lucky to have been able to have persuaded many of the original contributors to provide "second editions" of their first contributions; where this has not been possible he has been equally lucky to have found other eminent authors to fill the gaps. A comparison with the first edition will soon make it clear that the second edition is in fact a new book: all the chapters have been completely rewritten and, such is the speed of development, many bear very little resemblance to their original counterparts. In addition new topics have been included: these are the Betalains and Flash Kinetic Spectrophotometry; sections on analytical methods for quinones and phytochrome are also new.

Once again I must express my thanks to all contributors for the understanding help given to me during the preparation of this volume: I trust that it will be worthy of their excellent individual contributions.

It was considered that the need for an author index was outweighed by the need for economy but it was considered important to include a specific name index. I am most grateful to my wife for preparing the subject index.

August, 1975 T. W. GOODWIN

Contents of Volume 1

Contents of Volume 2

Part IV. Analytical Methods

Part I. NATURE, DISTRIBUTION AND BIOSYNTHESIS

Chapter 1

Structure, Properties and Distribution of Chlorophylls

A. H. JACKSON

Department of Chemistry, University College, Cardiff, Wales

I Introduction

Chlorophyll is the generally accepted name for the green pigments in organisms capable of photosynthesis. It was first used by Pelletier and Caventou (1818) to describe the pigment complex responsible for the green colour of leaves, but some considerable time elapsed before Stokes (1864) demonstrated that it was a mixture of two green pigments and several yellow pigments; the green pigments were later isolated by Sorby (1873). Tswett (1906) separated the two chlorophylls and the carotenoids by column chromatography on sugar, and his experiments with these pigments are now traditionally regarded as the classical beginnings of chromatography. These chlorophylls were originally known as α and β, but subsequently the terms *a* and *b* came into general use. Since that time other closely related chlorophylls have been isolated from algae, and *bacteriochlorophyll* and the *Chlorobium chlorophylls* have been isolated from photosynthetic bacteria (Table I).

In higher plants and all algae except the blue-greens chlorophyll is found in organelles known as chloroplasts, whilst in the blue-green algae and photosynthetic bacteria the chlorophylls are located on intracellular lamellae. The function of chlorophylls is to absorb energy from sunlight and convert this energy into chemical energy. The overall process involves the production of electrons which effectively carry out the reduction of carbon dioxide and water to carbohydrates, and the energy trapped is then released in oxidative

TABLE I

Distribution of chlorophylls among photosynthetic organisms[a]

Organism	Pigment				
	Chlorophyll				
	a	*b*	*c*	*d*	*e*
Higher plants, ferns and mosses	+	+	−	−	−
Algae[b]					
Chlorophyta	+	+	−	−	−
Chrysophyta					
Xanthophyceae	+	−	−	−	+[c]
Chrysophyceae	+	−	±	−	−
Bacillariophyceae	+	−	+	−	−
Euglenophyta	+	+	−	−	−
Pyrrophyta					
Cryptophyceae	+	−	+	−	−
Dinophyceae	+	−	+	−	−
Phaeophyta	+	−	+	−	−
Rhodophyta	+	−	−	+	−
Cyanophyta	+	−	−	−	−
	Bacteriochlorophyll			*Chlorobium* chlorophylls	
	a	*b*		650	660
Bacteria					
Thio- and Athiorhodaceae	+ or	+		−	−
Chlorobacteriaceae	+	−		+ or	+

[a] See Fischer and Stern (1940), Strain (1958), Haxo and Fork (1959), Allen *et al.* (1960), Stanier and Smith (1960) and Eimhjellen *et al.* (1963).
[b] Taxonomic classification according to Prescott (1962).
[c] Chlorophyll *e* has been isolated from only one organism; it seems to resemble chlorophyllide *c* (Strain, 1951).

processes in plants, and ultimately in animals, to provide the chemical and mechanical energy necessary to sustain life.

In this article it has been convenient to consider firstly the various aspects of the physical and chemical properties of chlorophylls *a* and *b*, which, because of their ubiquitous nature in land plants, have received much more attention than the marine and bacterial chlorophylls. The chemistry of the latter is then discussed in relation to chlorophylls *a* and *b*.

II Nomenclature

The chlorophylls are all magnesium complexes of compounds derived from porphin, and its di- and tetrahydro- derivatives (Fig. 1).

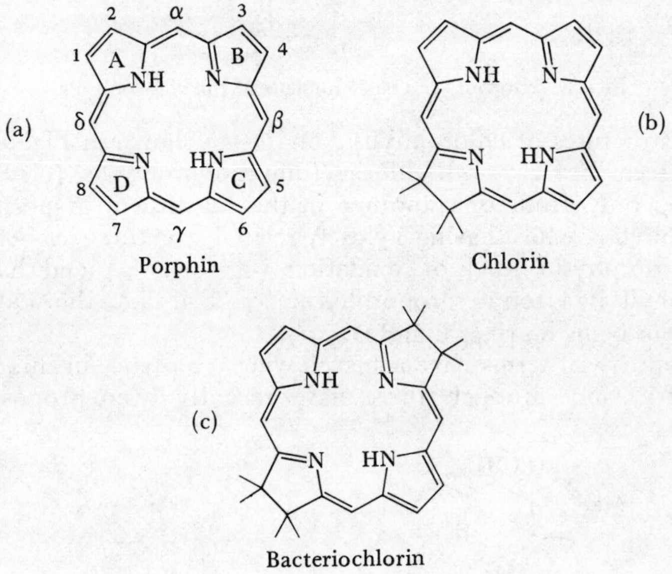

FIG. 1. (a) Porphin, (b) Dihydroporphin (Chlorin), (c) Tetrahydroporphin (Bacteriochlorin).

The peripheral carbon atoms are numbered 1–8 and the methine (or *meso*) bridge carbon atoms are designated α–δ in the Fischer system (Fischer and Stern, 1940). An alternative numbering system has been proposed by IUPAC (1960) to allow for inclusion of the chemically and biosynthetically related vitamin B_{12} and its derivatives; however, this has not yet been widely accepted in the porphyrin series, and to preserve continuity with the older literature the Fischer system will be used in this article. Unlike the haems which are iron complexes of porphyrins, the chlorophylls all contain another ring in addition to the four pyrrole rings A–D, or I–IV (Fig. 2). This is often referred to as the *isocyclic* ring (or as ring V) and it is derived biosynthetically by oxidation and cyclization of a C-6 propionic acid onto the γ bridge atom (Cox *et al.*, 1969).

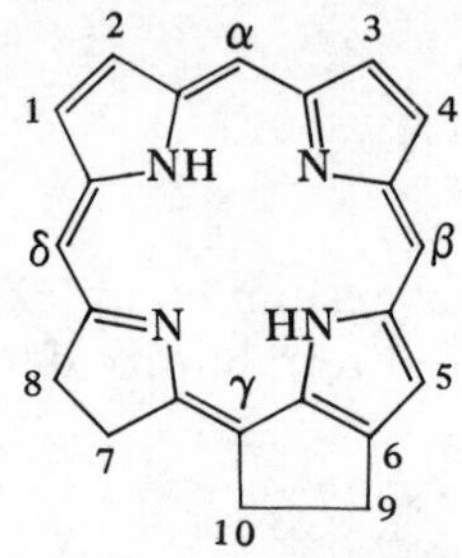

FIG. 2. Phorbin—the basic nucleus of the chlorophylls.

The structures of chlorophylls *a* and *b* are shown in Fig. 3 and as can be seen they are both chlorins (dihydroporphyrins) (cf. Fig. 1b) differing only from one another in the substituent at position 2. Chlorophyll *c*, whilst having a closely related structure (see below), is at the porphyrin level of oxidation (cf. Fig. 1a) and bacteriochlorophyll is a tetrahydroporphyrin (cf. Fig. 1c), the additional hydrogens being on rings B and D.

Certain trivial terms have achieved wide acceptance in chlorophyll chemistry, and although there have recently been proposals for

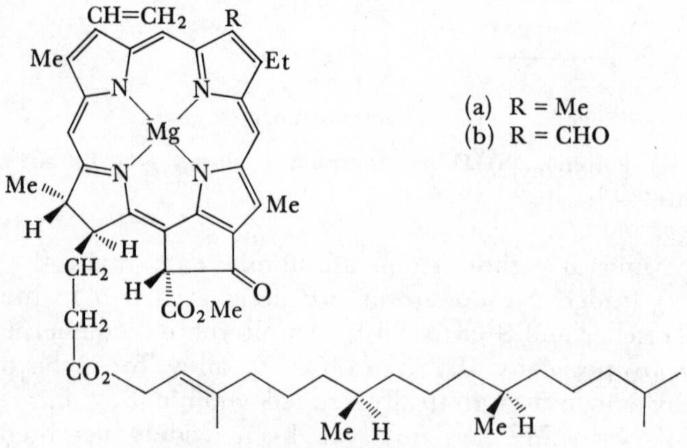

FIG. 3. Chlorophylls *a* and *b*.

rationalizing and reducing their number, no definite agreement has yet been reached (cf. Bonnett, 1973). Some of the most important of these trivial names are as follows (cf. Fischer and Stern, 1940).

Phyllins: chlorophyll derivatives containing magnesium.
Phaeophytins: the magnesium-free derivatives of the chlorophylls.
Chlorophyllide: the acid derivative resulting from enzymic or chemical hydrolysis of the C_7 propionate ester.
Chlorophyllase: the enzyme present in leaves which catalyses hydrolysis of the C_7 propionate ester.
Phaeophorbides: the products containing a C_7 propionic acid resulting from removal of magnesium and hydrolysis of the phytyl ester. The corresponding 7-propionate methyl (or ethyl) ester is, however, somewhat unsystematically named as methyl (or ethyl) phaeophorbide.
"Meso" compounds: derivatives in which the C-2 vinyl group has been reduced to ethyl.
"Pyro" compounds: derivatives in which the C-10 carbomethoxy group has been replaced by hydrogen.
Chlorins e: derivatives of phaeophorbide *a* resulting from cleavage of the isocyclic ring; these are usually given a subscript number, e.g. chlorine e_6 specifies a product with six oxygen atoms (in three ester groups).
Rhodins g: the corresponding derivatives from phaeophorbide *b*.

TABLE II
Hydrochloric acid numbers of porphyrins and chlorophyll derivatives

Compound	Acid number	
Phaeophytin	*a* 29	*b* 35
Phaeophorbide	*a* 15	*b* 19·5
Methyl phaeophorbide	*a* 16	*b* 21
Chlorin e_6 trimethyl ester	8	
Rhodin g_7 trimethyl ester	12 – 13	
Rhodoporphyrin	4·0	
Pyrroporphyrin	1·3	
Phylloporphyrin	0·35	
Phaeophorbide *c*	12	
Methyl phaeophorbide *d*	16·5	
Bacteriophaephytin *a*	>36	
Methyl bacterio phaeophorbide *a*	<22	
Chlorobium phaeophytins	*–660*, 18	*–650*, 22
Chlorobium phaeophorbides	*–650*, 7 – 11	*–650*, 9 – 13

Purpurins: derivatives of chlorins in which C-10 has been oxidized.
Porphyrinogens: hexahydro derivatives of porphyrins in which four pyrrole rings are joined through saturated methylene bridges.
Hydrochloric acid number: the % (w/w) hydrochloric acid which extracts two thirds of a metal-free derivative from an equal volume of ether (Table II).

The interrelationships between some of these terms are indicated schematically in Fig. 4.

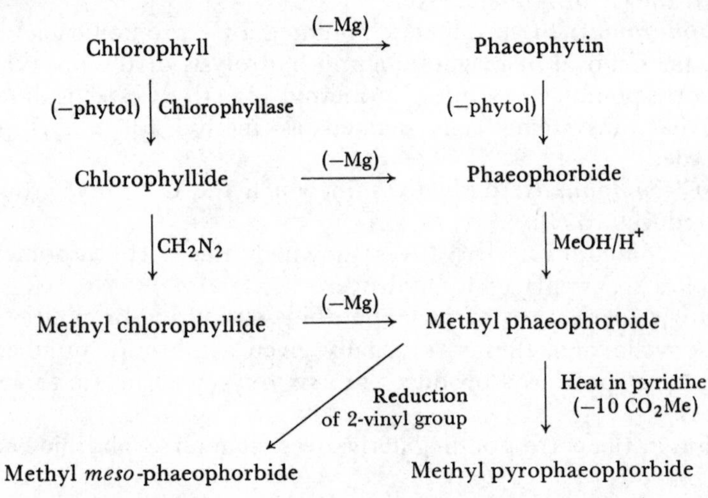

FIG. 4. Interrelationships between chlorophyll derivatives.

III Chlorophylls *a* and *b*

A STRUCTURE

Chlorophylls *a* and *b* are generally found associated with each other in the leaves of green plants and the ratio is usually (but not invariably) 2 : 1/*a* : *b*.

The elucidation of the complete structures of chlorophylls *a* and *b* involved many investigators over a period stretching from the end of the nineteenth century until the late 1960s. Most of this work has been reviewed extensively elsewhere (Willstätter and Stoll, 1913; Fischer and Stern, 1940; Rothemund, 1944; Stoll and Wiedemann, 1952; Aronoff, 1960, 1966; Holt, 1966; Seely, 1966; Strain and Svec, 1966), and only the recent developments are discussed in detail below.

The earliest systematic investigations of chlorophyll chemistry were carried out by Willstätter and Stoll (1913), and the essential structures of chlorophylls *a* and *b*, as well as bacteriochlorophyll, were established by the work of Hans Fischer's school in Munich during the 1920s and 1930s. His monumental, analytical, degradative, and synthetic investigations of both the blood and plant pigments will forever remain a classic in structural studies of natural products (Fischer and Orth, 1937; Fischer and Stern, 1940). Valuable contributions to this work were made particularly by Conant, and after the war Linstead's school in London provided essential confirmatory evidence of the dihydro nature of the D ring of chlorophylls *a* and *b* (Eisner and Linstead, 1955), and the *trans* disposition of the peripheral methyl and propionate ester groups (Ficken *et al.*, 1956).

Synthetic evidence (cf. Jackson and Smith, 1973) for the structure of chlorophyll *a* was first provided by Fischer's work in Munich, in the later stages of which he completed the synthesis of the related porphyrin derivative, phaeoporphyrin a_5 (Fig. 5). Strell and Kalojanoff (1962) extended Fischer's approach to the synthesis of methyl phaeophorbide *a* (Fig. 6a) itself, although the method used has subsequently been criticized (Inhoffen, 1964; Inhoffen *et al.*, 1968).

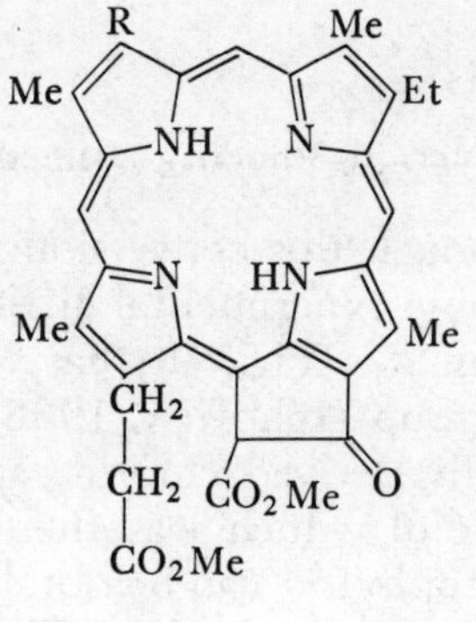

(a) R = Et
(b) R = $CH{=}CH_2$

FIG. 5. (a) Phaeoporphyrin a_5 and (b) 2-vinyl phaeoporphyrin a_5 dimethyl esters.

The Harvard synthesis of methyl phaeophorbide *a*, also completed in 1960 (Woodward *et al.*, 1960), however, utilized an ingenious and novel method of joining two dipyromethane units together in a completely specific manner; the porphyrin produced was then modified to give chlorin e_6 trimethyl ester (Fig. 7). As the latter had been previously converted into chlorophyll *a* this therefore formally

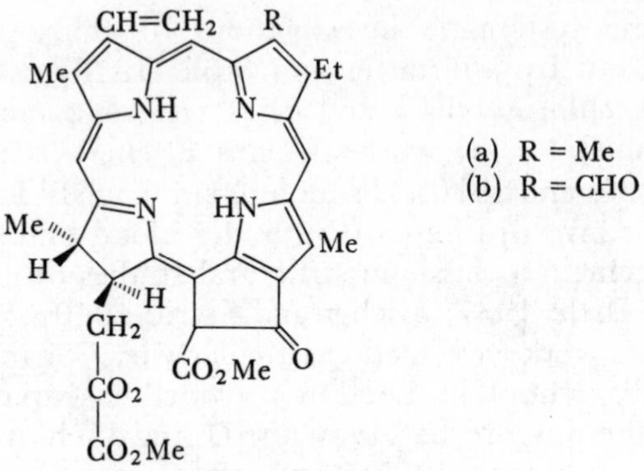

(a) R = Me
(b) R = CHO

FIG. 6. Methylphaeophorbides *a* and *b*.

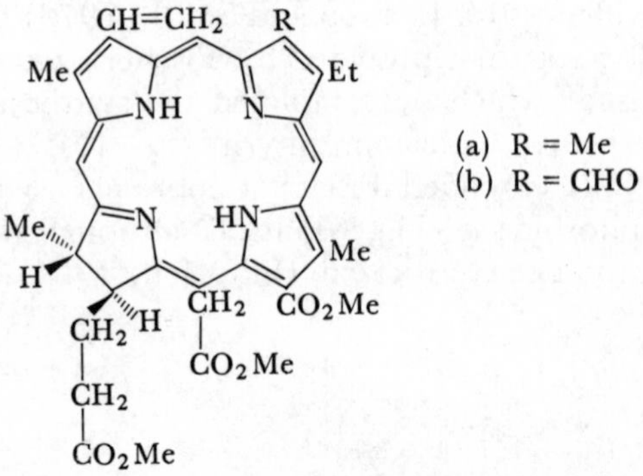

(a) R = Me
(b) R = CHO

FIG. 7. (a) Chlorin e_6 trimethyl ester. (b) Rhodin g_7 trimethyl ester.

completed its total synthesis, although this conversion has not been repeated, and it presents considerable experimental difficulties.

The partial synthesis of rhodin g_7 (Fig. 7b) has recently been described by the Braunschweig group (Inhoffen, 1968; Inhoffen *et al.*, 1968). Electrolysis of naturally derived chlorin e_6 followed by photo-oxidation gave a 3,4-*trans*-diol, which was then transformed into rhodin g_7 (Fig. 7b). As chlorophyll *b* can be obtained from the latter, and as chlorin e_7 had been synthesized by Woodward, this formally constitutes a total synthesis of chlorophyll *b*. The synthesis of phaeoporphyrin a_5 derivatives by oxidative ring closure of a porphyrin 6-β- keto ester has also been described recently; these syntheses are closely akin to the natural biosynthetic process for formation of the isocyclic ring (Cox *et al.*, 1969; Kenner *et al.*, 1972).

It was not until 1959 that the phytyl group of the chlorophylls was shown to be 2′-*trans* (7′R: 11′R) (cf. Fig. 3) (Burrell *et al.*, 1959,

1966; Crabbe *et al.*, 1959) and although Woodward in the course of his work on chlorophyll synthesis had pointed out that it was very likely on steric grounds that the C-10 methoxy carbonyl was *trans* to the C_7 propionate group, this was not finally proven correct until 1967 (Wolf *et al.*, 1967; see also Closs *et al.*, 1963). The absolute stereochemistry of chlorophyll *a* (7S : 8S : 1OR : 7′R : 11′R) was finally shown by Fleming (1967, 1968) by means of a chemical correlation of a degradation product derived from the D ring of phaephorbide *a* with a transformation product of the terpenoid lactone (−)-α-santonin, whose absolute configuration had been thoroughly established previously. Brockmann (1968, 1971) has also described a direct correlation of chlorophyll *a* itself with the degradation product from (−)-α-santonin. Fleming and Brockmann's work also showed the absolute configuration of chlorophyll *b*, as it had previously been shown to have the same configuration as chlorophyll *a* (Fischer and Gibian, 1942).

In earlier studies of chlorophylls *a* and *b*, Strain and Manning (1942) chromatographically separated small quantities of two minor green pigments designated *a*′ and *b*′, which were almost indistinguishable spectroscopically from the main pigments. Equilibrium between chlorophylls *a* and *a*′, and between *b* and *b*′, could be established on heating in alcohol, or pyridine, solution, but the *a*′ and *b*′ series were only formed to the extent of about 20%. The minor pigments were, therefore, considered to be isomers of chlorophylls *a* and *b*, and it was also found that methyl and ethyl chlorophyllides (which lack the phytyl ester group) behaved in a similar fashion; thus the reversible isomerization of the chlorophylls could not involve a *cis–trans*-type isomerization of the phytyl double bond. Chlorophylls which have been subject to "allomerization" (oxidation at C-10) do not form isomers and it was later suggested that the isomers were due to epimerization at C-10 via keto–enol tautomerism in the isocyclic ring (Strain, 1954; Aronoff, 1960).

On the basis largely of proton magnetic resonance data it has been proposed that chlorophylls *a*′ and *b*′ are the 10-*epi*-chlorophylls *a* and *b* (Katz *et al.*, 1968a). The chemical shifts of the two C-10 protons of pyrochlorophyll *a*, have slightly different chemical shifts, because of their differing magnetic environments, due to the chiral centres at C-7 (and to a lesser extent C-8) (Pennington *et al.*, 1964), and similar differences have been observed in the chemical shifts of the two isomeric 10-methoxymethyl pyrophaeophorbides *a* (Wolf *et al.*, 1967). Katz *et al.* (1968a) observed small satellite peaks at slightly higher field of the C-10 proton resonance of chlorophylls *a* and *b* and concluded, on the basis of the foregoing observations, that these were due to the small amount of the epiforms present at equilibrium. Kinetic studies of the equilibration of freshly prepared samples of chlorophylls *a*, *a*′ and *b* in solvents such as tetrahydrofuran and pyridine, showed that the half-life for chlorophyll *a* to reach equilibrium is about 2 h at 30°C, and isomerization is probably facilitated by basic solvents which catalyse enolization, for in benzene equilibration takes at least 24 h; isomerization of chlorophyll *a*′ is much faster and occurs rapidly at 10°C, and the spectrum must be observed at low temperatures.

Assignments of the C-10 protons were confirmed by exchange with perdeuteriomethanol which is known to occur rapidly (Dougherty *et al.*, 1965). Further proof was obtained by methanol titrations of carbon tetrachloride solutions of perdeuteriochlorophyll (Closs *et al.*, 1963) containing hydrogen at C-10 (in which there is no possibility of confusion of the C-10 proton resonances with part of the C-2 vinyl resonances); disaggregation of chlorophyll dimers then occurs, and the *two* C-10 resonances undergo shifts of the order of 2 p.p.m. (Katz *et al.*, 1968a).

More recently the preparation of the complete series of stereoisomers of the C-10 methoxy (and -trideuteriomethoxy) derivatives, and the related C-9 hydroxy derivatives (prepared by borohydride reduction of the isocyclic ring carbonyl group) has been described. Their n.m.r. optical rotatory dispersion and circular dichroism spectra were studied in detail (Wolf and Scheer, 1971, 1973).

Very recently Katz's conclusions regarding the precise nature of chlorophylls *a*′ and *b*′ have been challenged by Hynninen (1973) who has suggested that they are in fact the chelated enol forms of the chlorophylls. Evidence cited in favour of this view includes: (i) the much higher solubility of the *a*′ and *b*′ isomers in non-polar solvents; (ii) the marked difference in visible absorption between chlorophylls *a*′ and *a*; (iii) chlorophylls *a*′ and *b*′ are not detected in counter-current separations because the chelated enol is unstable in the highly polar lower phase (cf. Hynninen and Ellfolk, 1973). Hynninen (1973) suggests that the chelated enol forms are favoured by solvents (such as pyridine or tetrahydrofuran) which are polar enough to prevent aggregation, but which do not possess strong hydrogen bonding capacity. The satellite resonance peak (at 3·9τ) observed by Katz *et al.* (1963a) is not likely to be due to a chelated enolic hydroxyl group as suggested by Hynninen (1973), because it is at much too high a field (that of acetylactone is at *c.* -5τ).

Further work on this phenomenon is clearly desirable to provide a definitive solution, for chlorophylls *a*′ and *b*′ are not found in carefully prepared plant extracts; their possible presence must, however, be taken into account in studying the spectral properties of chlorophylls (especially n.m.r.), and could be of importance in the formation of ordered aggregates (Katz *et al.* 1968a, b).

B CHEMICAL PROPERTIES

Much of the detailed chemistry of the chlorophylls has been reviewed in earlier publications (e.g. Willstätter and Stoll, 1913; Fischer and Stern, 1940; Aronoff, 1960, 1966; Seely, 1966; Marks, 1969), but some of the important features will be described here to set the scene for discussions of the more recent work, particularly in relation to the structural studies of chlorophyll *c*, and the *Chlorobium* chlorophylls. Visible spectroscopy and later nuclear magnetic resonance spectroscopy played a vital role in following the course of many of these reactions, and this is discussed in detail below.

(*i*) *Magnesium removal and insertion*

Magnesium is removed immediately from the chlorophylls by the action of dilute mineral acids as can be seen from the consequential change in the visible absorption spectrum. It can be reintroduced by treatment with a Grignard reagent, or its alkoxy derivative, but this reaction must be carried out in an inert atmosphere to avoid aerial oxidation ("allomerization"); furthermore, transesterification of the side-chain esters may also occur (M. T. Cox, A. H. Jackson, K. M. Smith and G. W. Kenner, unpublished work). Tracer experiments with ^{28}Mg *in vivo* or in aqueous acetone solution show that the magnesium of chlorophyll does not exchange (Becker and Sheline, 1955; Aronoff, 1962).

Other metals can be introduced into the macrocyclic nucleus using metal acetates in acetic acid solution, or metal acetates or chlorides in methanolic solutions containing a small amount of pyridine, e.g. zinc, copper, iron etc. (Falk, 1964).

(*ii*) *Hydrolysis or transesterification of the phytyl ester group*

The phytyl ester on the C-7 propionate group is removed enzymically by chlorophyllase, to give phaeophytin. It can also be hydrolysed by acid or alkali under mild conditions, without affecting the C-10 methoxycarbonyl group, but alkaline hydrolysis must be carried out in an inert atmosphere to avoid oxidation at C-10. Acid hydrolysis also effects removal of the magnesium, whilst treatment with alcoholic acid results in transesterification with formation of the corresponding ester.

(*iii*) *Vinyl groups*

The vinyl group at position 2 in the chlorophylls may be reduced to ethyl by hydriodic acid, hydrazine hydrate or catalytically by palladium and hydrogen. Oxidation by permanganate in acetone affords a mixture of products including the corresponding glycol and the formyl and carboxyl cleavage products.

Under mildly acidic conditions (hydrobromic acid in acetic acid or dilute hydrochloric acid) the vinyl group can be hydrated to the hydroxyethyl derivative, which can readily be oxidized to an acetyl group, or dehydrated back to vinyl. Dilute hydriodic acid in acetic acid, on the other hand, converts a vinyl group into an acetyl group, whilst phaeophorbides, chlorins and purpurins may undergo simultaneous conversion into porphyrins.

Diazoacetic ester adds to the vinyl group to form the related cyclopropane carboxylate group, and this reaction helped to establish the presence of a vinyl group in both chlorophylls *a* and *b*, i.e.

$$-CH{=}CH_2 \xrightarrow{N_2CHCO_2Et} -\underset{\diagdown\ \diagup \atop CHCO_2Et}{CH{-}CH_2}$$

(*iv*) *Carbonyl groups*

These form the normal types of derivatives such as oximes, semicarbazones, acetals etc. and can be reduced to the alcohols by sodium borohydride or to methylene by the Wolff–Kishner method. The formyl group of phaeophytin *b* also reacts with cysteine (Tyray, 1944) and with Girard reagents; the latter reaction has recently been shown to be very valuable in effecting large-scale separations of phaeophytin *b* from phaeophytin *a* (Wetherall and Hendrickson, 1959; Kenner *et al.*, 1973).

$$R{-}CHO \rightarrow RCH{=}N.NH\,COCH_2N^+Me_3Cl^-$$

(*v*) *Conversion of di- and tetrahydroporphyrins to porphyrins*

Hydriodic acid in warm acetic acid, or other solvents, catalytic hydrogenation, or powdered iron in formic acid, all effect reduction of di- and tetrahydroporphyrins to the colourless porphyrinogens, which can then be reoxidized aerobically to the corresponding porphyrins. Vinyl groups, however, are often reduced to ethyl groups in the process although this occurs to a lesser extent when powdered iron in formic acid is used. Other methods involving the use of acidic or basic reagents at high temperatures, both in the presence and absence of air, have also been described, and these may involve disproportionation processes (Fischer and Stern, 1940). Transformation of vinyl into ethyl may also occur as well as degradation of the isocyclic ring, and a classic example is the conversion of phaeophorbide *a* into rhodo-, phyllo- and pyrroporphyrins (Willstätter and Stoll, 1913) (Fig. 8).

Direct dehydrogenation of ring D in simple chlorins has been effected by silver acetate–acetic acid, or by oxygen catalysed by copper acetate in acetic acid. However, the most efficient method of carrying out the conversion of chlorins into porphyrins is the use of *p*-benzoquinone and its derivatives, e.g. "DDQ" (2,3-dichloro-5,6-dicyanobenzoquinone). The use of these reagents was studied in detail by Linstead and his colleagues (cf. Eisner and Linstead, 1955; Eisner *et al.*, 1957) and their work demonstrated the stoicheiometry of the reaction, i.e. that one mole of quinone was needed to oxidize one mole of chlorin to porphyrin thus showing directly that the chlorophylls and other chlorins were dihydroporphyrins (see also Kenner *et al.*, 1973).

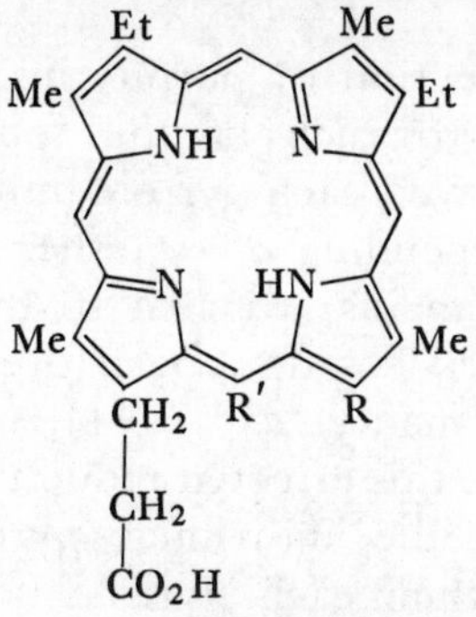

(a) $R = CO_2H$, $R' = H$
(b) $R = H$, $R' = Me$
(c) $R = R' = H$

FIG. 8. Rhodo-, phyllo- and pyrroporphyrin XV.

(*vi*) *Degradation of the macrocyclic nucleus*

The close chemical relationship between haem and chlorophylls was shown at the turn of the century not only by the ready conversion of chlorophylls into porphyrins but also by the fact that both pigments gave mixtures of maleic imides on chromic acid oxidation, and mixtures of simple pyrroles on hydriodic acid reduction (Falk, 1964).

Chromic acid oxidation of porphyrins affords maleic imides corresponding to the four pyrrolic units of the original macrocycle, but rings with unstable side chains (e.g. vinyl, hydroxyethyl, formyl or acetyl) usually decompose, so that some preliminary treatment (e.g. reduction) is required. The partly reduced D ring of pyrophaeophorbide *a* affords, under carefully controlled conditions, either dihydrohaematinic acid or its imide (Fig. 9); the *trans* disposition of the peripheral methyl and propionate groups was shown by chemical comparisons with synthetic material (Ficken *et al.*, 1956) and the absolute configuration was deduced later (Fleming, 1967, 1968; Brockman, 1968) (see above).

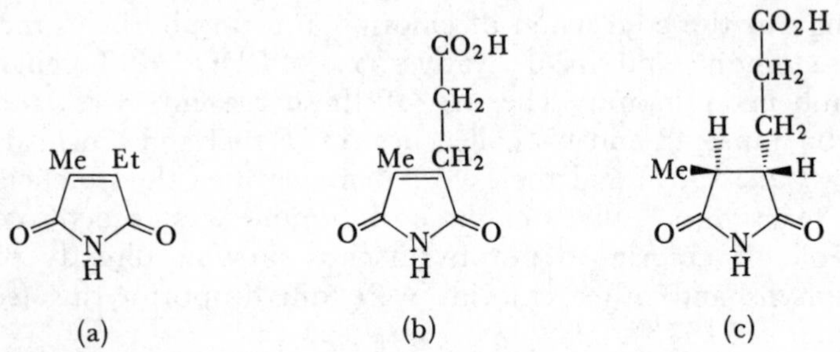

CO_2H CH_2 CO_2H CH_2 H CH_2 Me Et Me CH_2 Me H O N O O N O O N O H H H

(a) (b) (c)

FIG. 9. Chromic acid degradation products of chlorophylls: (a) methyl ethyl maleimide; (b) haematinic imide; (c) *trans*-dihydrohaematinic imide.

Hydriodic acid in acetic acid reduction of porphyrins affords a mixture of pyrroles corresponding to each of the four original pyrrole units of the macrocycle. However, each pyrrole unit may give rise to as many as four products depending on whether either, or both, of the bridge carbon atoms remains attached to the pyrrole ring or not (Fig. 10). This drawback has now been overcome by MacDonald's group in Canada (Chapman *et al.*, 1971), who added formaldehyde to the reaction medium; this effected reductive methylation of any vacant positions in the individual pyrrole units produced in the degradation, and then each pyrrole unit of the original porphyrin gave only one pyrrole, e.g. 2,5-dimethylpyrroles from porphyrins with unsubstituted bridge carbon atoms (Fig. 10). *Meso* substituents are retained in the reductive alkylations, and chlorophyll derivatives (after pre-reduction of the isocyclic carbonyl function) afford cyclopenteno pyrroles from the C ring. Vinyl and hydroxy ethyl substituents are reduced to ethyl in the course of these reductive degradations.

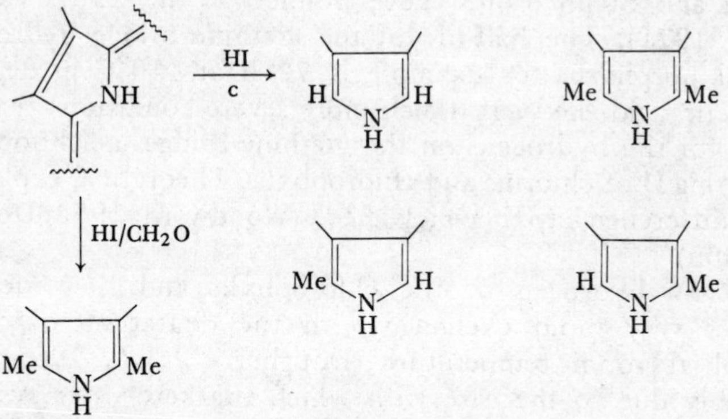

NH HI c H N H H Me N H Me HI/CH_2O Me N H H H N H Me Me N H Me

FIG. 10. Products from hydrogen iodide, or hydrogen iodide and formaldehyde reductions of porphyrins.

The sensitivity of these degradative methods for determining porphyrin and chlorin structures has recently been dramatically improved by use of thin-layer chromatography (Rudiger, 1968) and gas–liquid chromatography (Hughes and Holt, 1962; Holt *et al.*, 1966; Chapman *et al.*, 1971) so that fractions of a milligram can be studied routinely. This work has been extended recently by use of gas–liquid chromatography in combination with mass spectrometry (g.l.c./m.s.); the reductive alkylation technique may now be applied to 100 μg samples of porphyrins, whereas the chromic acid oxidation to maleic imides is even more sensitive and with g.l.c./m.s. can be carried out with as little as 5–10 μg porphyrin (Games *et al.*, 1974a, b; Stoll *et al.*, 1973). In favourable cases the methyl vinyl maleimide has been observed in low yield (from a pyrrole unit bearing methyl and vinyl groups), but *meso* substituents are lost in the oxidative degradation.

(*vii*) *Hydrogen exchange reactions of chlorins*

During the course of his work on the synthesis of chlorophyll *a* Woodward discovered that the δ-methine bridge hydrogen atom of chlorins such as phaeophorbide *a* and chlorin e_6 is replaced by deuterium on heating in hot deuterioacetic acid, whereas the α- and β-methine bridge hydrogen atoms are inert under these conditions (Woodward and Skaric, 1961; Woodward, 1962). These observations have since been amply confirmed, and systematic kinetic comparisons of exchange reactions at the *meso* positions of porphyrins and chlorins and their metal complexes have been undertaken (Bonnett and Stephenson, 1964; Bonnett *et al.*, 1967; Paine and Dolphin, 1971). The half-life of the methine bridge hydrogens of octaethyl porphyrin for example is 72 h at 110°C in deuterotrifluoroacetic acid, i.e. very much more severe conditions than those required for the hydrogens on the methine bridges neighbouring the reduced ring D of chlorins and chlorophylls. Theoretical explanations for this difference have been advanced (Woodward, 1962; Dougherty *et al.*, 1965).

The C-10 hydrogen of the chlorophylls and their derivatives undergoes very rapid exchange with the deuterium of deuteriomethanol at room temperature (Dougherty *et al.*, 1965); this is presumably due to the ease with which the keto ester system can undergo enolization and equilibration. Exchange at C-10 in chlorophyll *a* is some two orders of magnitude greater than at the δ position whereas in phaeophytin *a* the rate difference is *c*. 10^6. This has been attributed to the effect of the magnesium atom in potentiating electrophilic substitution. Recent work has confirmed this view because exchange of the *meso* protons of magnesium porphyrin complexes occurs readily in hot deuteriomethanol (Cavaleiro *et al.*, 1973).

In the presence of base the hydrogen atoms of the 6-methyl group will also undergo exchange with deuterium due to the activating effect of the isocyclic ring carbonyl group.

(*viii*) *Reactions of the isocyclic ring*

Solutions of chlorophyll derivatives containing the isocyclic pentenone ring and bearing a methoxycarbonyl group at C-10 undergo the well-known Molisch "phase" test; on treatment with alkali a colour change to reddish brown occurs, but the original green colour is rapidly regenerated on shaking in acid or oxygen. The "pyro" compounds (i.e. those which have lost the C-10 methoxycarbonyl group) do not give the phase test nor do the phorbides (which have lost the magnesium). Solutions of chlorophylls that fail to give the phase test are said to be "allomerized". Such allomerization occurs slowly when solutions in suitable solvents (e.g. methanol or ethanol) are left to stand in air and it is now clear that this involves oxidation of small amounts of the intermediate enolate anion generated in solution by traces of alkali. The 10-methoxy or 10-hydroxy derivative can be isolated under appropriate conditions as well as a 10-methoxy or 10-hydroxy lactone (Fig. 11) (Weller, 1954; Holt, 1958, 1963; Seely, 1966; Pennington *et al.*, 1967; Hynninen and Assandri, 1973). 10-Hydroxy chlorophyll *a* is also formed by enzymic oxidation (Pennington *et al.*, 1967) whilst the main allomerization product from chlorophyll *b* is the 10-methoxy lactone derivative (Hynninen and Assandri, 1973) other minor products are also formed, and the hydroxy lactone may undergo further transformations to purpurins (Fig. 11).

The initial intermediates in these complex processes have not been isolated but it seems likely that a peroxide may be the precursor of the lactones (Kenner *et al.*, 1973), whilst the 10-hydroxy and 10-methoxy derivatives may arise through formation of a radical species at C-10 which reacts with the solvent (cf. Cox *et al.*, 1969) (Fig. 11).

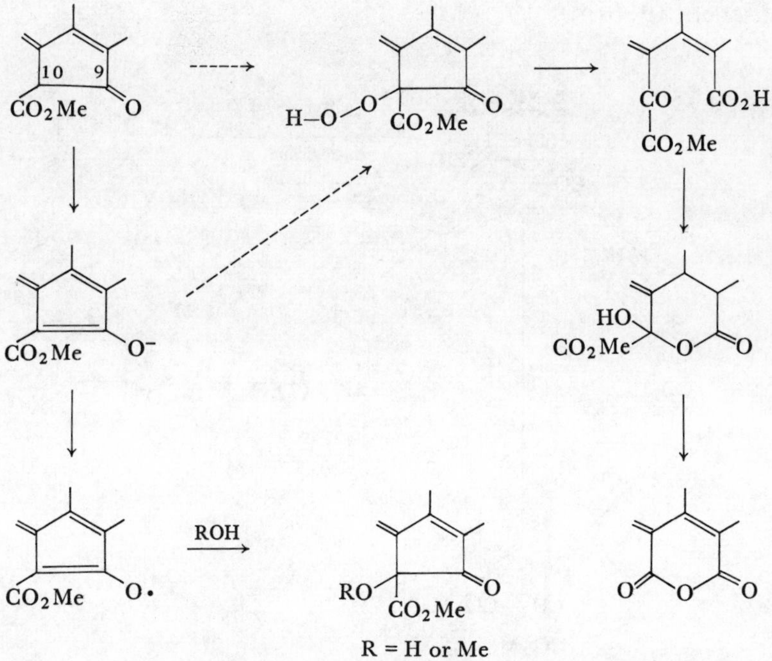

FIG. 11. Allomerization reactions of chlorophylls.

Treatment of phaeophytin *a* with hot strong alkali hydrolyses the phytyl ester side chain, and also cleaves the isocyclic β keto ester ring to form chlorin e_6 (Fig. 12), which can be methylated to the trimethyl ester with diazomethane. Chlorin e_6 trimethyl ester can also be obtained directly by use of boiling dilute methanolic alkali. Cleavage of the isocyclic ring in this fashion does not occur with allomerized pigment. Amines will also effect cleavage of the isocyclic ring of phaeophorbides to form C-6 amides derived from chlorin e_6 (Pennington *et al.*, 1967; Hynninen, 1973b).

On heating phaeophorbide derivatives in refluxing pyridine, or by heating to higher temperatures in absence of solvent the C-10 methoxycarbonyl group is cleaved to form the "pyro" analogues; the use of boiling collidine has recently been reported to give nearly quantitative conversions of methyl phaeophorbides *a* and *b* to the corresponding "pyro" derivatives (Kenner *et ai.*, 1973). Whilst the latter do not give a positive phase test, methyl *meso* pyrophaeophorbide *a* can be oxidized to a diketone using quinones such as DDQ (Kenner *et al.*, 1973); a similar oxidation has been utilized in the structure determination of the *Chlorobium* chlorophylls (see below), but using air with alkaline dimethyl formamide solutions (Holt *et al.*, 1966).

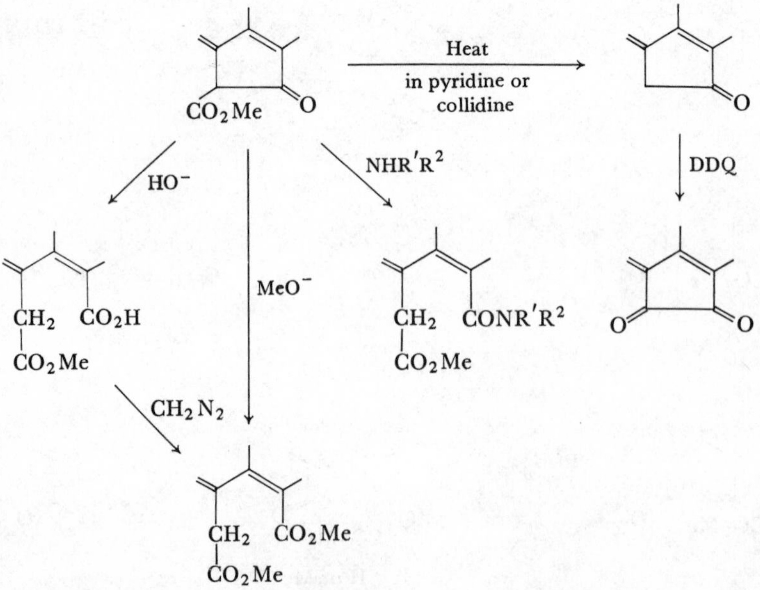

FIG. 12. Cleavage and oxidation of the isocyclic ring of phaeophorbides.

C ISOLATION AND PHYSICAL PROPERTIES

Chlorophylls are usually extracted from plant material by use of mixtures of methanol and petroleum ether (Smith and Benitez, 1955; Strain and Svec, 1966). The chlorophylls are finally transferred into the petroleum ether solution and then purified from other pigments and separated by chromatography. In order to prevent hydrolysis of the phytyl side chain by chlorophyllase the plant material is treated briefly with boiling water before extraction.

Chlorophylls *a* and *b* in which all the hydrogen was exchanged by deuterium have been prepared from green algae (*Chlorella vulgaris* or *Scenedesmus obliquus*) grown in heavy water (Strain *et al.*, 1971); these have played a very important role in n.m.r. spectroscopic studies.

Chlorophylls *a* and *b* were first separated from other leaf pigments and from each other by chromatography on weak adsorbents such as sugar, cellulose or starch. Many of the earlier methods such as column chromatography have been thoroughly reviewed recently (Strain and Svec, 1966, 1969). In the last few years analytical and small scale preparations on thin-layer chromatograms have become very popular because of their speed and efficiency. There is no doubt that the recently introduced technique of high pressure liquid chromatography (h.p.l.c.) will also come into widespread use because of its high resolving power and the quantitation possible by monitoring the effluent from the columns using a u.v. detector. Preliminary studies in the author's laboratory, and elsewhere, clearly show the power of h.p.l.c. on the one hand for the extreme sensitivity possible, and on the other hand for small-scale preparative work with larger columns. Other recent developments in the chromatographic separation of chlorophylls have been the use of reverse phase methods (Sherma, 1970); a recent patent describes their separation from carotenoids by gel permeation on a polystyrene gel (Zwolenik, 1970), and the use of Sephadex LH20 has also been described (Shimizu, 1971). Counter-current distribution techniques have also been used for the separation of chlorophyll derivatives, but to a lesser extent than for porphyrins derived from the blood pigments (Falk, 1964). A solvent system developed by the author and his colleagues (Burbridge, Jackson and Kenner, unpublished work) has been used to separate phaeophorbides *a* and *b* by partition between methyl isobutyl ketone/benzene and hydrochloric acid. Hynninen has recently described counter-current separations of chlorophylls *a* and *b* using petroleum-formamide and similar solvent systems

(Hynninen and Ellfolk, 1973). Partition chromatography on columns has also been used for preparative separation of small amounts of chlorophylls and their derivatives, and this has been particularly useful for the *Chlorobium* chlorophylls (see below).

Among the problems which may arise during these chromatographic procedures is the relative instability of chlorophylls and their derivatives, e.g. the slow decomposition which occurs even on the mildest adsorbents. This is due to the facile loss of magnesium in presence of traces of acid, the possibility of aerial oxidation especially in presence of traces of bases (allomerization), and the light sensitivity of chlorophylls. Unexpected minor additional zones may thus be due to alteration products and must be carefully investigated to avoid misinterpretation. Furthermore, the visible spectral characteristics of the various products eluted may be very dependent on the nature of the solvent, and the presence of minor contaminants (see below). For these reasons the precise estimation of the composition of mixtures of chlorophylls cannot be determined very accurately; the separation of derivatives of chlorophylls *a* and *b* has also been achieved preparatively via the Girard complex of the *b* series.

Chlorophylls, phaeophytin, and phaeophorbide esters are usually soluble in organic solvents such as ether, acetone, methanol, chloroform and especially pyridine, but relatively insoluble in hydrocarbons (when pure). The phaeophorbide free acids are less soluble than the esters, and both the phaeophorbides and phaeophytins are soluble in warm acetic and formic acids. Porphyrins and chlorins are usually soluble in aqueous hydrochloric acid and their *acid number* (which depends on the relative partition ratios with ether) has often been used as an indication of structure as well as for determining the conditions for separation by solvent partition. Table II shows the acid numbers of a representative selection of chlorophyll derivatives, and it can be seen that the acid number is lower the shorter the ester side chain on the C-7 propionate group. Cleavage of the isocyclic ring (as in chlorin e_6) or reduction of the isocyclic ring carbonyl group also leads to lower acid numbers. Pigments of the *b* series have a higher acid number than the *a* series presumably because the electron withdrawing character of the 2-formyl group makes them less basic. In general, chlorins also have higher acid numbers than the corresponding porphyrins, again presumably because of their lower basicity.

Chlorophylls *a* and *b* are usually obtained as waxy solids but can be crystallized from ether solutions of chromatographically pure material in the presence of water when they form their lancet-shaped leaflets, or thin triangular platelets. Chlorophyll *a* was reported to melt at 117–120°C, whilst chlorophyll *b* sintered between 86 and 92°C. Such preparations do not show sharp diffraction patterns (Hanson, 1939) and other X-ray data were later reported by Jacobs *et al.* (1953, 1954). Donnay (1959) calculated that there were two molecules per unit cell. A full X-ray structure has not yet been described, because of the disordered structures shown by the chlorophylls and their derivatives.

However, the structures of three chlorophyll derivatives have now been described, phyllochlorin ester (Hoppe *et al.*, 1969), vanadyl deoxyphylloerythroaetioporphyrin (Pettersen, 1969) and methyl phaeophorbide *a* (Gassman *et al.*, 1971; Fischer *et al.*, 1972). The structure of the latter is particularly interesting in relation to the way in which chlorophyll molecules may be arranged in chloroplasts (see below).

Esters of chlorophyll derivatives usually melt at lower temperatures than the free acids. This reduces the possibility of decomposition and mixed melting points have often been used for establishing identity, especially in the porphyrin series, but polymorphism may occasionally be encountered.

D SPECTROSCOPIC PROPERTIES

In recent years the application of physical methods has led to great advances in our knowledge of structure and function in chemistry. Visible spectra have long played a very important role in structural studies of haems and chlorophylls, and it is interesting that their use in this way foreshadowed by many years the widespread post-war developments in the use of other types of spectroscopy in analytical and structural chemistry.

The recent introduction of these newer techniques (i.r., n.m.r., and mass spectrometry, X-ray crystallography etc.) has been of great importance in structural studies especially as they can be applied to the very small amounts of materials which are often all that is available from natural sources. Fundamental theoretical and chemical studies of chlorophyll derivatives are also made possible by these modern physical methods, and moreover they are now beginning to provide an insight into the nature of the chemical and physical processes occurring in photosynthesis.

(*i*) *Electronic absorption spectra*

The colours exhibited by porphyrins, chlorins and related compounds have long had an immense aesthetic appeal, and especially because of the relative sharpness of the ultraviolet and visible absorption bands. These features have been very helpful in providing information both on the gross structure of the macrocycle (i.e. whether porphyrin, chlorin, tetrahydroporphyrin etc.) and on the nature of groups conjugated with the macrocycle.

Measurements of the positions of the various absorption bands were originally made visually with hand spectroscopes, but these have now been largely superseded by automatic recording spectrophotometers. Porphyrin free bases usually exhibit four main bands in the visible region of the spectrum between 500 and 700 nm and a much more intense band at around 400 nm (the Soret band) in the ultraviolet. Stern and Wenderlein (1936) distinguished three types of porphyrin spectra (Fig. 13). The "aetio"-type spectra (Fig. 13a) which are typical of octaalkyl substituted porphyrins have four bands which increase in intensity from long to short wavelength, and are sometimes referred to as "ladder"-type spectra. In "rhodo"-type spectra (Fig. 13b) band III is more intense than bands II and IV and

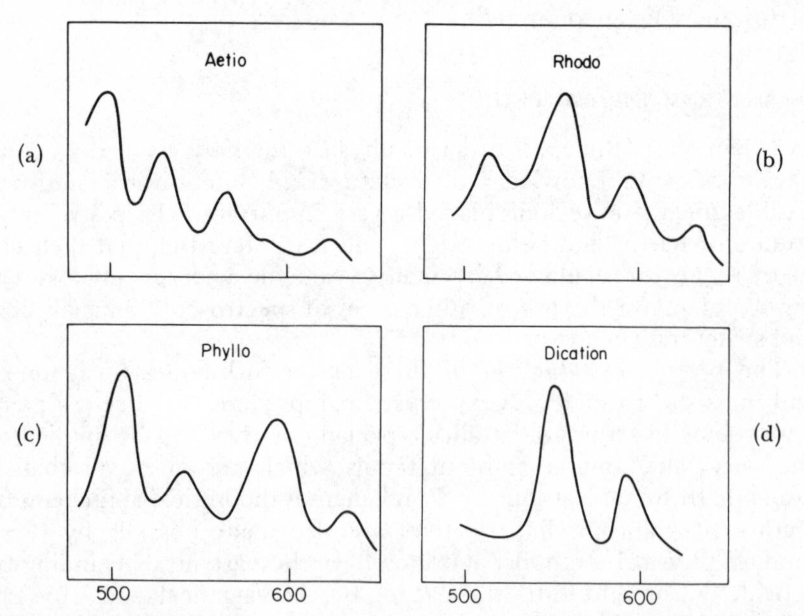

FIG. 13. Porphyrin visible spectra: (a) aetio type; (b) rhodo type; (c) phyllo type; (d) porphyrin di-cations (formed in acidic media).

this is typical of porphyrins bearing a carbonyl group in one of the pyrrole rings, whilst in "phyllo" spectra (Fig. 13c) band III is less intense than bands II and IV and this is characteristic of *meso*-alkyl substituted porphyrins.

Aetio-type spectra are named after aetioporphyrin which is a degradation product of haem bearing one methyl and one ethyl group on each pyrrole ring, whilst rhodo- and phyllo-type spectra are named after the chlorophyll degradation products rhodoporphyrin and phylloporphyrin (Fig. 8). A vinyl group has a small "rhodofying" effect and causes a slight shift to longer wavelengths of all the peaks of an aetio-type spectrum. If, however, two carbonyl groups occur in neighbouring rings, or both a carbonyl group and a *meso*-alkyl group are present, or an acetyl or formyl group is converted to its oxime, or a carbonyl group is converted to a salt, then reversion to aetio-type spectra occurs. On the other hand the rhodo spectrum is retained on conversion of the isocyclic ring carbonyl group (of phaeoporphyrins) to its oxime. If the C-10 methoxycarbonyl group is also present in the isocyclic ring, or two carbonyl groups are present in opposite rings then band II is more intense than band IV, and affords an "oxorhodo"-type spectrum. In the absence of the C-10 methoxycarbonyl group, as in phylloerythrin, bands II and IV absorb equally.

Marked changes in the spectra of porphyrins occur in acidic solution when the violet porphyrin di-cations are formed; these exhibit two bands in the visible region of the spectrum, typically around 550 and 590 nm, and the Soret band usually shows a large increase in intensity and a slight shift to longer wavelength (Fig. 13d). Similar two-banded visible spectra are exhibited by porphyrin–metal complexes, or by porphyrin di-anions which are formed in presence of strong bases. The differences between the four-banded and two-banded visible spectra have been attributed on theoretical grounds essentially to changes in symmetry; the free bases are less symmetrical, having only two nitrogens substituted by hydrogen.

Chlorins (dihydroporphyrins) are green in colour owing to the characteristic strong absorption at the red end of the spectrum (Fig. 14a,b); they also exhibit other, weaker, bands in the visible region but like porphyrins have an intense Soret band in the near ultraviolet. Two types of spectra are characteristic in the chlorophyll series: (i) the "chlorin" type as shown by phaeophytin *a* or phaeophorbide *a* (Fig. 14a); and (ii) the so-called "rhodin" type of phaeophytin *b* (Fig. 14b). Reduction of the 3-formyl group in the *b* series causes reversion to a "chlorin"-type spectrum with a bathochromic shift of the long wavelength band by 15–20 nm; conversely reduction of the 2 vinyl group or the isocyclic ring carbonyl group causes the "red" maximum to move hypsochromically by 8–10 nm. Oxidation of the vinyl group to carboxyl, acetyl or formyl results in a long wavelength shift of the "red" maximum, but cleavage of the isocyclic ring (e.g. to chlorin e_6 or rhodin g_7) has relatively little effect on the spectra of the free bases, although the long wavelength maxima of the magnesium complexes are shifted some 20 nm to shorter wavelengths.

Tetrahydroporphyrins, of the bacteriochlorophyll type, also show a strong band at the red end of the spectrum, but this is located in the infrared near 800 nm, i.e. some 150 nm longer wavelength than the chlorin owing to the reinforcing effect of a second dihydro ring opposite the first (Fig. 14).

The introduction of magnesium into either the di- or tetrahydroporphyrins, or the formation of salts, has relatively little effect on the spectral type other than causing small shifts in wavelength (and intensity) as compared with the dramatic effects observed with porphyrins. This is due to the inherently less symmetrical nature of the hydroporphyrins whether in the free base, salt or metal complex form.

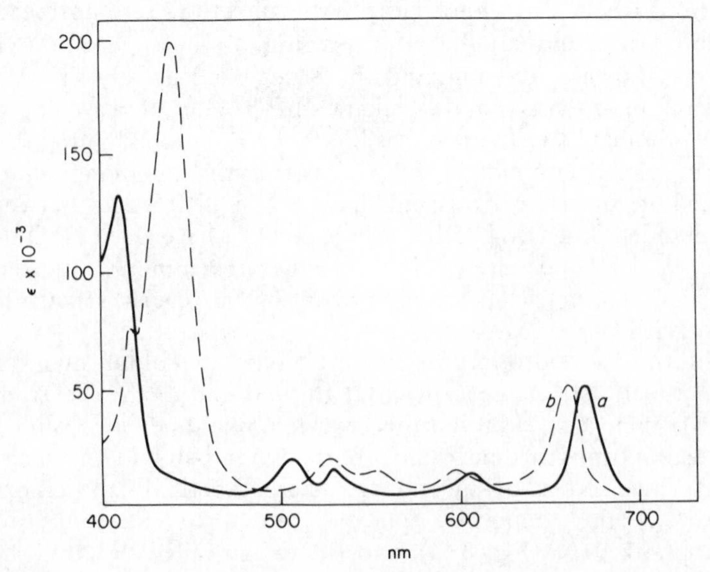

FIG. 14. Absorption spectra of chlorins: (a) phaeophytin *a*; (b) phaeophytin *b*.

A general feature of both porphyrins, and dihydroporphyrins, is that two "rhodofying" (i.e. conjugating) substituents on opposite rings tend to reinforce each other in their effects on the spectra, whereas if such substituents are present on neighbouring rings then the effects tend to cancel out, e.g. in porphyrins reversion to aetio-type occurs.

(*ii*) *Infrared spectra*

The infrared spectra of the chlorophylls and their derivatives have been measured as mulls (in mineral oil), in potassium bromide discs, and in solution. The waxy properties of chlorophylls, however, tend to make the preparation of mulls or discs difficult, and moreover the solid-state spectra are less well resolved than the solution spectra.

Functional groups present in the side-chain substituents of chlorophylls give rise to infrared absorption bands in the expected regions of the spectrum, e.g. hydroxyl stretching frequencies 3600 cm^{-1},

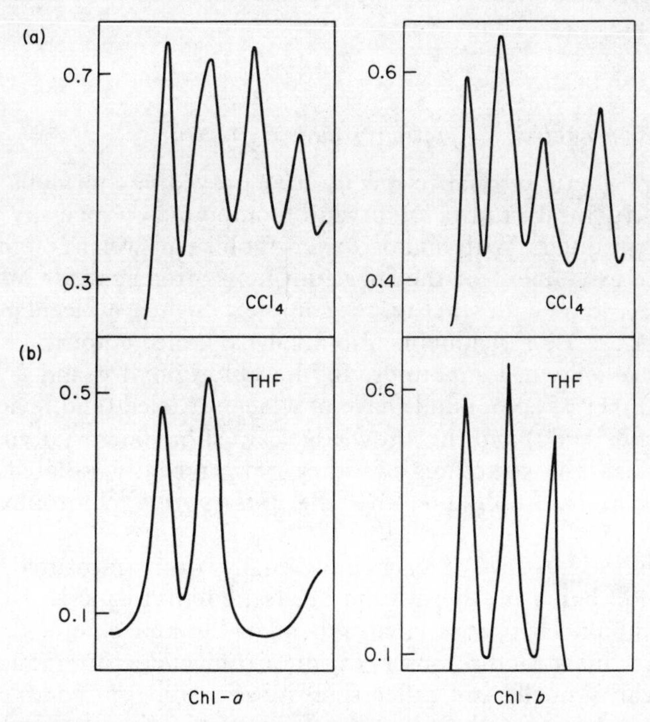

FIG. 15. Infrared spectra of chlorophylls *a* and *b* in the carbonyl region: (a) in CCl_4; (b) in tetrahydrofuran. (Katz *et al.*, 1963a.)

and carbonyl frequencies in the range 1650–1750 cm^{-1} (Falk and Willis, 1951; Holt and Jacobs, 1955). The infrared spectra of chlorophylls *a* and *b* in non-polar solvents such as carbon tetrachloride are strikingly similar, even though there is an additional carbonyl group in chlorophyll *b*; however, in more polar solvents such as tetrahydrofuran, the two chlorophylls have very different spectra (Katz *et al.*, 1963). This phenomenon is particularly marked in the carbonyl region of the spectrum (Fig. 15) and moreover it had been known for some time that the spectra in non-polar solvents were concentration dependent. Holt and Jacob (1955) suggested that this concentration dependence might be due to keto–enol type tautomerism, but Katz *et al.* (1963) later provided strong evidence that the observed effects could be attributed to intramolecular aggregation in non-polar solvents, on the basis of detailed infrared spectral studies, and osmometric determination of molecular weights. Further work on the far infrared spectra (667–160 cm^{-1}) of chlorophylls, where metal–ligand vibrations are observed, has confirmed the deductions made from studies of the carbonyl region (Boucher *et al.*, 1966; Katz *et al.*, 1966).

(*iii*) *Proton nuclear magnetic resonance spectra*

Proton magnetic resonance spectra have proved of enormous value in recent structural studies of organic compounds, especially of new natural products. With many lower molecular weight compounds complete assignment of the spectrum may often be made on general principles, but with the more complex higher molecular weight materials precise assignments also involve detailed empirical comparisons with known compounds. In the chlorophyll *a* and *b* series a large number of compounds have now been studied (and reviewed by Katz *et al.*, 1966) and the knowledge gained has been vital in helping to establish the structures of other more recently isolated chlorophylls, e.g. chlorophyll *c* and the *Chlorobium* chlorophylls (see below).

The chemical shift of each individual proton (measured in parts per million based on the τ scale, i.e. tetramethylsilane as 10) is very dependent on its precise environment in the molecule, and furthermore the fine structure observed (due to magnetic interactions with other nearby nuclei and called spin–spin coupling) provides evidence of the numbers and stereochemical relationships of hydrogens attached to neighbouring carbon atoms.

The proton magnetic resonance spectra of porphyrins and chlorins exhibit several striking features, especially the wide range of the chemical shifts observed, and the marked concentration dependence of the spectra in non-acidic solvents. The wide range is due to the "ring current" in the conjugated macrocyclic ring (Becker and Bradley, 1959; Ellis *et al.*, 1960; Abraham, 1961) so that the *meso* protons of porphyrins resonate around 0τ (compared to $2{\cdot}73\tau$ for the protons of benzene) and the NH protons resonate in the 14–15τ range (Table III). Because of the wide spread in the resonances assignments to particular types of substituent are relatively easy, but assignments of specific protons are not so straightforward and require much more detailed investigations of related compounds, taking into account the concentration dependence. The remarkable concentration dependence of the free base spectra in, for example, deuteriochloroform (Table IV) has been clearly demonstrated to be due to aggregation (Abraham *et al.*, 1966). Because of the magnitude of the concentration shifts (Table IV) it is essential in recording n.m.r. spectra of porphyrins and chlorins to report the concentration of the solution, or to give values extrapolated to infinite dilution.

Proton magnetic resonance spectra of porphyrins and chlorins in trifluoroacetic acid (in which the di-cationic species are formed) are virtually independent of concentration (Ellis *et al.*, 1960; Abraham *et al.*, 1961). The effects of β substituents on neighbouring *meso* proton chemical shifts in these spectra are additive, and conversely a

TABLE III

Proton nuclear magnetic resonance spectra of porphyrins and chlorins: chemical shifts of some typical substituent groups[a]

	Porphyrins	Chlorins
Methine $-H$	0·0–0·4	0·3–1·5
Peripheral $-H$	~1·0	5·5–5·9[b]
$-CHO$	~–0·5	–
Peripheral $-CH_3$	6·2–6·5	~8·2[b]
$-CH_2CH_3$	5·7–6·1, 8·4–8·6	–
$-CH_2CH_2CO_2CH_3$	5·7–6·0	~7·5[b]
	6·5–7·0	~7·5
	6·3–6·4	6·3–6·4
$-CH{=}CH_2$	~2·2 ~4·0	–
N–H	13·8–14·3	11·7–13·0

[a] Approximate values (measured on the τ scale relative to tetramethylsilane as 10) in $CDCl_3$ solution.
[b] Substituents associated with the reduced ring.

TABLE IV
Concentration effects in the proton n.m.r. spectra of coproporphyrin-III tetramethyl ester ($CDCl_3$ solution)

	0·15 M	Extrapolated to infinite dilution
NH	14·30	13·64
Methine-H	0·21 and 0·33	−0·14 and −0·17
Peripheral-CH_3	6·50, 6·59, 6·67	6·33, 6·31, 6·29
Ester-CH_3	6·37	6·32

meso-alkyl substituent causes the resonance of the neighbouring β-substituents to move to higher field; this latter effect has been attributed to distortion of the macrocycle by steric strain, thus reducing the effective ring current.

Complete assignment of the p.m.r. spectra of chlorophylls *a* and *b* was complicated owing to the marked concentration dependence due to molecular aggregation, and depended to a large extent on assignment of the resonances in methyl phaeophorbides *a* and *b* (Closs *et al.*, 1963). These were also concentration dependent but the assignments made were based on several considerations as well as on information available from the earlier studies of porphyrin n.m.r. spectra, and from previously reported partial assignments of the p.m.r. spectrum of chlorin e_6 trimethyl ester (Woodward and Skaric, 1961). Full details are given in the original paper (Closs *et al.*, 1963) but Fig. 16 shows the data for methyl phaeophorbide *a*. The three *meso* protons resonances, for example, are at lowest field, and the highest of these three peaks was assigned to the δ proton, because it would be the most shielded, being next to the partially reduced ring; the lowest *meso* proton resonance was assigned to the β proton as this would be expected to be somewhat more deshielded by the isocyclic ring carbonyl group than the α proton. As already mentioned above the δ proton undergoes exchange with deuterium in hot deuterioacetic acid whilst the C-10 proton exchanges even in deuteriomethanol at room temperature. The spectrum of methyl phaeophorbide *b* is very similar to that of methyl phaeophorbide *a*, except for the substitution of the formyl proton for one of the methyl resonances, and because of the additional deshielding effect of the 2-formyl group the α proton resonates at lower yield than the β proton. The phaeophytins have similar spectra to the chlorophylls but with the addition of the phytyl resonances (Closs *et al.*, 1963), and the spectrum of phytol itself has also been described (Rapaport and Hamlow, 1961).

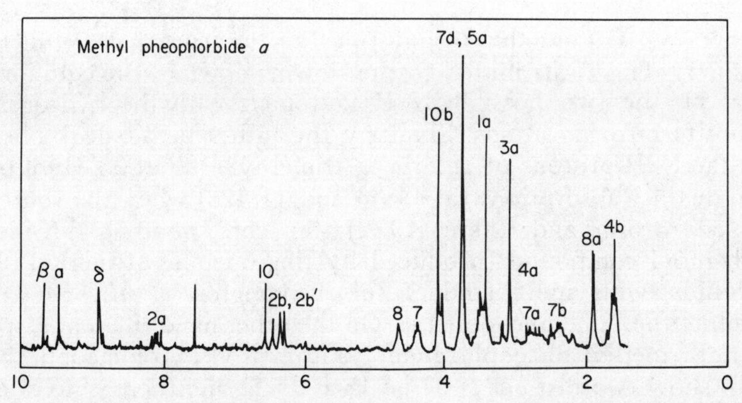

FIG. 16. Proton n.m.r. spectrum of methyl phaeophorbide *a* (0·08M) in $CDCl_3$ at 220 MHz. (Katz and Janson, 1973.)

The p.m.r. spectra of the chlorophylls themselves in alcohol-free chloroform show a series of broad poorly defined peaks due to aggregation. However, on addition of alcohol (which causes disaggregation) or in more polar electron-donating solvents (such as acetone d_6 or tetrahydrofuran d_8) well resolved spectra are obtained which are very similar to those of the metal-free methyl phaeophorbides and assignments can then be readily made for all the protons. Similar features have also been described in the p.m.r. of pyrochlorophyll *a*, methyl pyrochlorophyllide *a*, pyrophaeophytin *a*, and methyl pyrophaeophorbide *a* (the products derived by thermal removal of the 10-methoxycarbonyl group from the corresponding chlorophyll derivatives) (Pennington *et al.*, 1964).

Exchange reactions of the C-10 and *meso* protons, which were studied mainly by n.m.r., have been described in detail above.

A long-standing question in porphyrin and chlorin chemistry has been whether the central hydrogen atoms are bound to specific nitrogen atoms, are in tautomeric equilibrium, or even perhaps in a bridged resonance form. Arguments for and against the various possibilities have been adduced from studies of infrared, visible spectra and from X-ray crystallographic studies, and these have been reviewed recently by Storm *et al.* (1973). A solution to this problem has now been provided by proton magnetic resonance studies at low temperatures (Storm and Teklu, 1972). Whereas relatively symmetrical porphyrins like the coproporphyrin tetramethyl esters do not show any evidence of the existence of N–H tautomers even at –50°C (Abraham *et al.*, 1966), the low temperature spectrum of deuteroporphyrin dimethyl ester shows two broad N–H resonances below −70°C (and the 2- and 4-proton resonances also split into a doublet). This is attributed to the slowing down of the equilibration between the two possible N–H tautomers with the hydrogens on opposite nitrogen atoms. Similarly the rather broad peak observed for the N–H protons of chlorin e_6 trimethyl ester at 35°C splits into two distinct absorptions (at 11·35 and 11·42τ) when the solution is cooled (Storm and Teklu, 1972); the corresponding 6-N-methyl-carbamoyl compound (produced by ring opening of methyl phaeophorbide with methylamine) (cf. Pennington *et al.*, 1967) also exhibits the same phenomenon. On the other hand the n.m.r. spectra of both methyl phaeophorbide *a* and methyl pyrophaeophorbide *a* both show two distinct N–H peaks at 35°C, and there is no evidence from lowering the temperature that any exchange is taking place between −60 and 60°C; the higher field peak in each case is assigned to a hydrogen attached to N_a and the low field peak to a hydrogen attached to N_c (Storm *et al.*, 1973). Presumably exchange would take place at higher temperatures but this has not yet been tested. It is interesting to note that *meso*-tetraphenylporphin, which is symmetrical, but severely distorted from planarity owing to the steric effects of the *meso* substituents, also shows evidence of the existence of two forms at low temperatures (Storm *et al.*, 1973).

The conclusion to be drawn from this evidence is that the existence of N–H tautomers in porphyrins has clearly been demonstrated; the rate of exchange between the tautomeric forms (and whether they can be identified discretely by spectroscopic methods) must depend on such factors as the degree of symmetry and overall planarity of the ring, and on the nature of the substituents attached to each pyrrole ring; in general it would seem that the more symmetrical and the more nearly planar the macrocyclic ring the more facile the intromolecular exchange will be.

(*iv*) *Carbon-13 nuclear magnetic resonance spectra*

The recent development of Fourier transform methods for determining the carbon-13 magnetic resonance spectra of organic compounds at natural abundance (only about 1% of carbon-12) has made possible some very exciting advances both in the structural chemistry (including configurational and conformational studies) and biosynthesis of a variety of natural products. Sensitivity is still a problem for large molecules like the porphyrins, and the spectra of chlorophylls which have recently been described were determined with enriched material prepared by growing algae in an atmosphere of $^{13}CO_2$ (Katz *et al.*, 1972a; Strouse *et al.*, 1972; Katz and Janson, 1973; Matwiyoff and Burnham, 1973).

Assignments of some of the ^{13}C signals from chlorophyll *a* were made with highly enriched material ($^{13}C \sim 90\%$) at 25 MHz (Strouse *et al.*, 1972; Matwiyoff and Burnham, 1973), but high enrichments make interpretation less straightforward because of the complex spin–spin interactions between neighbouring carbon atoms. Katz and his colleagues, however, worked with chlorophyll containing 15% distributed ^{13}C so that the spectra were essentially first order because most of the ^{13}C atoms in any one molecule would be next to ^{12}C; furthermore, the spectra were collected at 55 MHz which further facilitated assignments.

Correlation of ^{13}C chemical shifts with structural features of an organic molecule has not so far achieved the same precision, either experimentally or theoretically, as proton chemical shifts. However, the observed shifts are at least an order of magnitude greater than for protons, and with partly enriched material and proton decoupling (with off-resonance white noise) ^{13}C spectra appear as a series of discrete singlets (Fig. 17a), even with highly enriched samples the wide range of the chemical shifts, and the deceptively simple multiplet patterns allow a straightforward assignment of the resonances in many regions of the spectrum. Further assistance in the assignment of specific resonances can also be made by use of single frequency off-resonance decoupling in which a particular proton resonance is selectively irradiated (Ernst, 1966). Figure 17b shows the effect of irradiating the δ proton of chlorophyll *a* when the doublet collapses to a singlet and simultaneously the coupling of the other α- and β-methine protons also falls slightly. The ^{13}C n.m.r. spectrum of chlorophyll *a* is shown in Fig. 17a and assignments which have been made are given in Table V; detailed discussions of the manner in which these assignments have been made are given in a recent paper by Katz and Janson (1973). As with the p.m.r. spectra the c.m.r. spectra are profoundly affected by complexing in non-polar solvents and the spectra of chlorophylls *a* and *b* change on addition of Lewis bases such as tetrahydrofuran or pyridine which bring about disaggregation. Studies of the c.m.r. spectra of methyl phaeophorbide *a* helped considerably in the interpretation of the spectra of the chlorophylls, especially because the former lacks the side-chain phytyl group the resonances of which overlap those of the chlorophyll macrocycle (Katz and Janson, 1973).

This work on the c.m.r. spectra of chlorophylls is clearly of great importance both for structural studies of newly discovered chloro-

TABLE V

^{13}C n.m.r. chemical shifts for chlorophyll *a*

Peak position[a]	Coupled multiplicity[b]	Single-frequency decoupling (Hz)[c]	Assignment
189·3	S		C-9
172·7	S		C-7c or C-10b
170·9	S		C-7c or C-10b
167·4	S		C–N(8)
161·5	S		
155·8	S		
154·0	S		
151·5	S		
148·0	S		
147·7	S		
146·1	S		
144·1	S		C–C(6)
142·4	S		
139·1	S		
135·5	S		
134·2	S		
134·0	S		
131·9	S		C-phy-3
131·5	D	220,011,625	C-2a
119·5	D	220,011,125	C-phy-2
118·9	T	220,011,200	C-2b
107·1	D	220,011,950	C-β
106·2	S		C-γ
100·1	D	220,011,905	C-α
92·8	D	220,011,705	C-δ
67·8	T	—	Solvent THF
66·4	D		C-10
61·3	T	220,010,825	C-phy-1
52·1	D		C-7
52·0	Q		C-10b
50·2	D		C-8
40·5	Aliphatic carbons from chlorophyll *a* and phytyl resonances		
40·2			
38·1			
37·5			
33·6			
31·2			
30·8			
28·9			
27·3			
26·1	T	—	Solvent THF

TABLE V—*continued*

Peak position[a]	Coupled multiplicity[b]	Single-frequency decoupling (Hz)[c]	Assignment
25·6	Aliphatic carbons from chlorophyll *a* and phytyl resonances		
23·9			
23·0			
20·2			
17·9			
16·2			
12·6	Q		C-1a, C-5a
11·2	Q		C-3a

[a] In p.p.m. from "internal" TMS (TMS–THF/upfield resonance separation was measured as 26·11 p.p.m.).
[b] S, singlet; D, doublet; T, triplet; Q, quartet.
[c] ± 25 Hz.

phyll and in studies of the aggregation of chlorophylls which is already proving of value in relation to the way in which chlorophyll is organized in the cells of living organisms and hence sheds very important light on the primary processes of photosynthesis (see below).

A recent detailed study of the natural abundance ^{13}C n.m.r. spectra of chlorophyll *a*, phytol and phytyl acetate has enabled complete assignments of all the phytyl carbon resonances in the spectrum of chlorophyll *a* (Goodman *et al.*, 1973).

(*v*) *Mass spectra*

Electron impact mass spectrometric studies of porphyrins have proved to be very useful in recent structural studies of porphyrins of both synthetic and natural origin. The molecular ions are usually the base peaks in the spectra, and fragmentations of the various side chains occur in fairly predictable fashion, but there is little evidence of fragmentation of the aromatic macrocycle (Jackson *et al.*, 1965); substantial doubly charged ions (up to 30% of the base peak) are also a common feature of porphyrin mass spectra. Porphyrins and chlorins with acidic side chains must usually be converted into their methyl esters as otherwise they may be too involatile for mass spectrometry; even with the esters source temperatures in the 200–300°C range are usually necessary. Whilst the mass spectrum of an unknown porphyrin or chlorin may yield a good deal of information about the nature of the side chains (from the molecular ion and fragmentation patterns, and the use of high resolution to

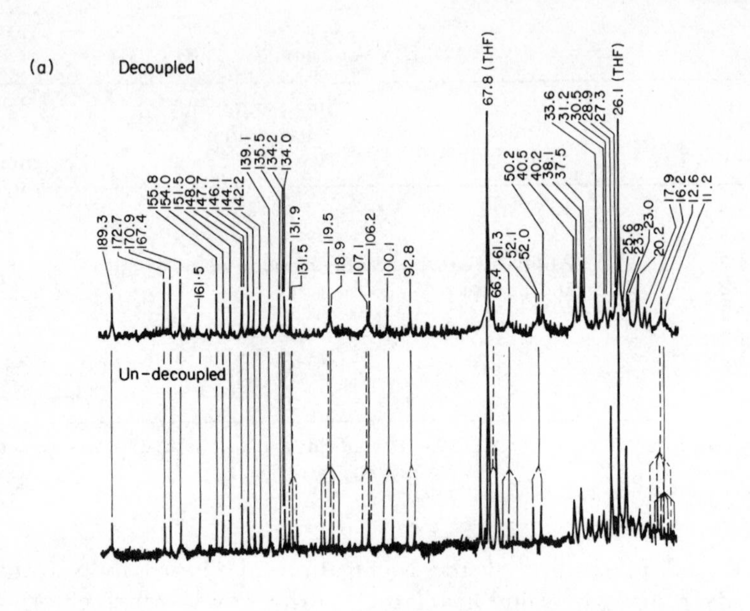

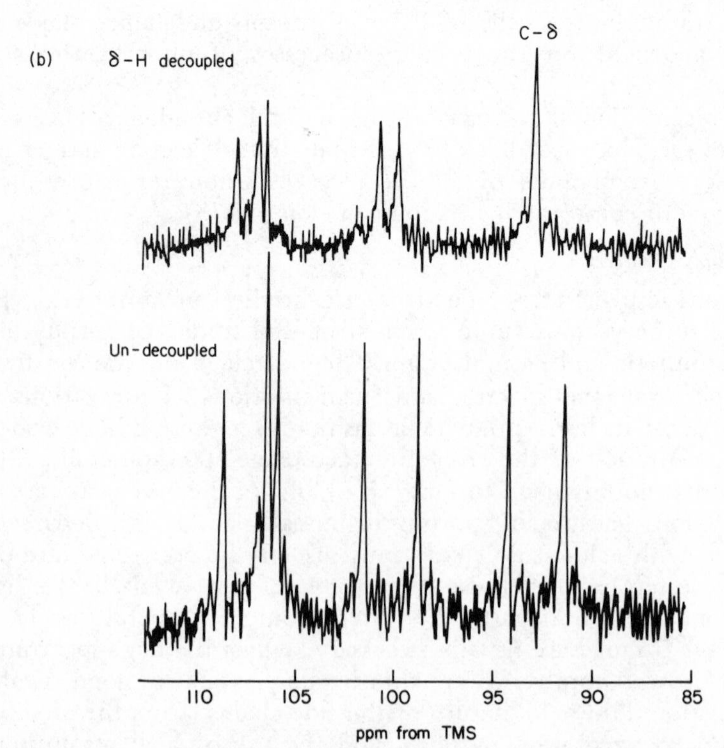

FIG. 17. (a) ^{13}C N.m.r. spectrum of chlorophyll *a* at 55 MHz. (b) Effects of decoupling the δ-H of chlorophyll *a* on the *meso*-^{13}C resonances. (Katz and Janson, 1973.)

obtain molecular compositions of the various ions) it is not possible to determine the order of the side chains around the periphery of the molecule; this often still rests on biogenetic arguments in the first instance, and must be proven ultimately by synthesis.

The spectrum of methyl phaeophorbide *a* is shown in Fig. 18 together with a partial analysis of the fragmentation pattern (Jackson *et al.*, 1965). The spectrum of the chlorophylls themselves have not yet been determined owing to their involatility (presumably a consequence of their strong donor–acceptor properties (see below)); phaeophytins *a* and *b* do not give very well defined mass spectra, although very weak molecular ions have been observed. Two newer methods of ionization offer some hope for the future in dealing with the less volatile compounds, namely field ionization and field desorption mass spectrometry (cf. Beckey and Schulten, 1973). The field ionization mass spectra of porphyrins are extremely simple and generally show only the molecular ion, and the corresponding doubly charged species (Games and Jackson, unpublished work); however, the porphyrin must still be volatilized before ionization. In field desorption ionization takes place directly from the solid state (from carbon fibres grown on a tiny filament through which a current is passed); spectra can be obtained of many involatile materials at room

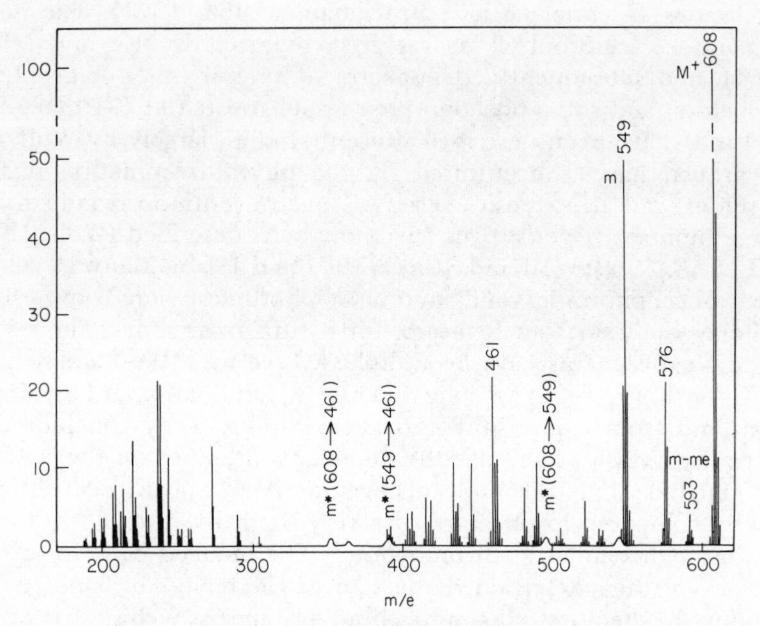

FIG. 18. Mass spectrum of methylphaeophorbide *a*. (Jackson *et al.*, 1965.)

temperature or very little above, and for example spectra of porphyrin free acids and of haemin itself have already been obtained (Games and Jackson, unpublished work). Chlorophyll has so far proved intransigent, but there is no doubt of the potential of field desorption for use with the metal-free pigments and free acids.

(*vi*) *Optical rotatory dispersion and circular dichroism*

In contrast to haem and other naturally occurring porphyrins, the chlorophylls have three chiral centres at C-7, C-8 and C-10. The first observations of optical activity in chlorophyll *a* were made by Stoll and Wiedmann (1933) and confirmed by Fischer and Stern (1935), and it was known at that time that chlorophyll possessed a chiral centre at C-10 (Fischer and Siebel, 1932). As discussed above recognition of the *trans* disposition of the groups at C-7 and C-8 followed from the isolation of *trans*-dihydro haematinic acid after oxidative degradation (Ficken *et al.*, 1956). The relative configuration was also deduced from spin–spin coupling constants of the 7- and 8-hydrogen atoms (Closs *et al.*, 1963), and from n.m.r. and optical rotatory dispersion measurements (Wolf *et al.*, 1967); subsequently the absolute configuration was determined by chemical correlations (Fleming, 1967; Brockmann, 1968, 1971). The o.r.d. spectrum of chlorophyll *a* was first reported by Ke and Miller (1964), and subsequently the spectra of several other chlorophylls (see below) and some phaeoporphyrins substituted at C-10 (Wolf *et al.*, 1968) have been described. Recent studies, largely by Wolf and co-workers, have concentrated on the phytol-free methyl phaeophorbides, and an extensive series of spectra (both o.r.d and c.d.) of a large number of derivatives have now been described (Wolf, 1966; Wolf *et al.*, 1969; Wolf and Scheer, 1971 and 1973). Many 10-substituted phaeophorbides and pyrophaeophorbides were synthesized including the 7,8-*cis* analogues of the natural *trans* series. The results of these experiments have been discussed recently (Wolf and Scheer, 1973) both in regard to experimental determinations of configuration, and from a more theoretical standpoint. They conclude that the configuration at C-7 has the dominant influence on the spectra, and conclude that the molecular geometry of substituted chlorins may be compared to one turn of a very flat helix (thus affording an inherently disymmetric chromophore). The induced chirality of the macrocycle then determined the sign of the main c.d. band (corresponding to the Soret maximum about 410 nm) as well as that of the long wavelength Cotton effect.

Magnetic circular dichroism studies of porphyrins are now beginning to attract increasing interest because of the possibility of analytical applications and for their theoretical interest. A recent paper gives spectra for a number of porphyrin di-cations including that of the chlorophyll degradation product, phylloerythrin methyl ester (Barth *et al.*, 1973) and this also gives pertinent references to recent studies of this relatively new spectral technique (cf. also Djerassi *et al.*, 1971).

E AGGREGATION OF CHLOROPHYLLS AND PHOTOSYNTHESIS

The chlorophylls have long been recognized as the primary light acceptors in plants and they are invariably present in every organism which carries out photosynthesis with absorption of carbon dioxide and evolution of molecular oxygen. Chlorophyll is organized in the plant in small photosynthetic units containing a few hundred molecules, and the function of most of the chlorophyll ("antenna" chlorophyll) appears to be to gather light, whilst a small proportion acts as the primary reaction centre where light conversion occurs. The antenna chlorophyll absorbs at about 680 nm, i.e. somewhat further towards the infrared than chlorophyll solutions in polar solvents which have their long wavelength absorption at about 665 nm.

It has long been recognized that the spectral characteristics of chlorophylls and their derivatives are highly dependent on the nature of the solvent and upon the presence of polar or hydrogen bonding type impurities. A large variety of spectral bands have been described for chlorophyll *in vivo* (French *et al.*, 1968) and attempts have been made to analyse these in terms of various species by curve resolving techniques. Furthermore, it is well known that solutions of chlorophylls in polar solvents are intensely fluorescent whereas in pure non-polar solvents they are virtually non-fluorescent (Livingston *et al.*, 1949); chloroform or benzene solutions of chlorophyll also show a long wavelength shoulder on the side of the 665 nm band observed in both polar and non-polar solvents. Recent studies of the infrared and n.m.r. spectra of chlorophylls and their derivatives have now, however, provided the key to the explanation of this behaviour, as demonstrated by a very elegant and thorough series of investigations carried out over the last few years by Katz and his colleagues at Argonne (cf. Katz and Janson, 1973).

As described above they very early realized that the anomalous behaviour observed in the infrared and n.m.r. spectra of chlorophylls could be accounted for by aggregation in solution in non-polar solvents (Katz *et al.*, 1963a; Boucher and Katz, 1967; see also Anderson and Calvin, 1964; Closs *et al.*, 1963; Katz *et al.*, 1966). The results of these experiments were interpreted in terms of electron donor and acceptor properties present in the chlorophyll *a* molecules giving rise to formation of dimers (Chl_2) and oligomers $(Chl_2)_n$. Katz suggested that the co-ordination number of the magnesium in chlorophyll is not 4 as implied by the usual structural formula but is preferably 5, and may often be 6 (in neat solvents such as pyridine, or in very concentrated solution); the fifth ligand is supplied by the 9–C=O function in the isocyclic ring (in the absence of extraneous nucleophiles or a coordinating solvent) as shown by the infrared and n.m.r. studies. In presence of bifunctional ligands (e.g. dioxane, 1,4, diazabicyclo-(2,2,2)-octane, or water) cross linking may occur with formation of polymeric species (Chl–L–Chl–L–. . .–Chl–L–Chl). Water acts both as a hydrogen bonding agent and as an electron donor through the oxygen atom, so that in hydrocarbon solvents large polymers $(Chl{-}H_2O)_n$ (where n may be greater than 100) may be formed (Katz and Ballschmitter, 1968; Katz *et al.*, 1968c). Osmometric studies of chlorophyll-chlorophyll interactions in hexane provide evidence for the existence of species containing over twenty individual molecules (Ballschmitter *et al.*, 1969). The visible spectra of various chlorophyll species in presence of various ligands and in non-polar solvents are summarized in Table VI. The properties of chlorophyll oligomers which exist in concentrated solution in non-polar solvents such as hexane correspond very closely to those of antenna chlorophyll *in vivo*, and thus the nature of the complexing mechanism in non-polar solvents provides a very good model for the natural system.

The proton magnetic resonance studies gave a fairly good if indirect indication of the role of the 9-carbonyl function in co-ordinating with the magnesium of an adjacent molecule, but more direct evidence has recently been obtained from studies of the ^{13}C n.m.r. spectra (Katz *et al.*, 1972a; Katz and Janson, 1973). In the earlier studies of the p.m.r. spectra of chlorophyll *a* in non-polar solvents (e.g. $CDCl_3$) successive additions of small amounts of Lewis bases such as fully deuteriated tetrahydrofuran, or pyridine, caused increasing disaggregation until the monomeric species was obtained, and the resulting changes in the chemical shifts of the various protons were studied. As can be seen from Fig. 19 the C-10 proton showed the greatest low field shift, and a so-called aggregation map was deduced (Fig. 20). Similar studies were carried out with ^{13}C

TABLE VI

Visible spectra of chlorophylls in presence of various ligands[a]

Interaction	Species		
	Type	Structure	λ_{max} (nm)
Exogamous			
Monofunctional ligands	Chl L_1, and/or Chl L_2		
Methanol		Chl CH_3OH	663
Water		Chl H_2O	663
Tetrahydrofuran		Chl THF	665
Pyridine		Chl py	671
Bifunctional ligands	(–Chl–L–Chl) and/or (–Chl.Chl–L–Chl.Chl–)$_n$		
Pyrazine		$(Chl_2\ pyr)_n$	690
Bipyrimidine		$(Chl\ bipyr)_n$	715
Pheophytin + H_2O		$(Chl\ pheo\ H_2O)_n$	715–720
Water		$(Chl\ H_2O)_n$	743
Endogamous (non-polar solvents, films)	$(Chl_2)_n$		
CCl_4, Benzene, $n = 1$		(Chl_2)	665, 678
Aliphatic hydrocarbons, $n = 20$		$(Chl_2)_n$	665, 678

[a] Katz and Janson, 1973.

spectra in carbon tetrachloride by titration with tetrahydrofuran (Katz *et al.*, 1972) but the C-9 resonance moves upfield rather than downfield on disaggregation (Fig. 21); Katz suggests that this is because the electronic effects of co-ordination (which may amount to 10 p.p.m. or more) far outweigh ring-current effects (which are the same for ^{13}C as for ^{1}H) and thus disaggregation leads to a net upfield shift as co-ordination at the oxygen ceases.

Schematic drawings of the structures proposed for the chlorophyll dimers (and oligomers) and chlorophyll–water adducts are shown in Fig. 22; in the dimers and oligomers the macrocyclic rings cannot be coplanar, whereas they are nearly parallel in the water adducts (as is consistent with the long wavelength shifts from 665 to 740 nm observed in these species. Similar conclusions about the structure of chlorophyll *a* dimer have also been drawn from circular dichroism studies (Houssier and Sauer, 1970).

The aggregation behaviour of chlorophyll *b* and pyrochlorophyll *a* has also been studied (Closs et al., 1963; Boucher *et al.*, 1966; Ballschmitter and Katz, 1969), but the data available are not yet as

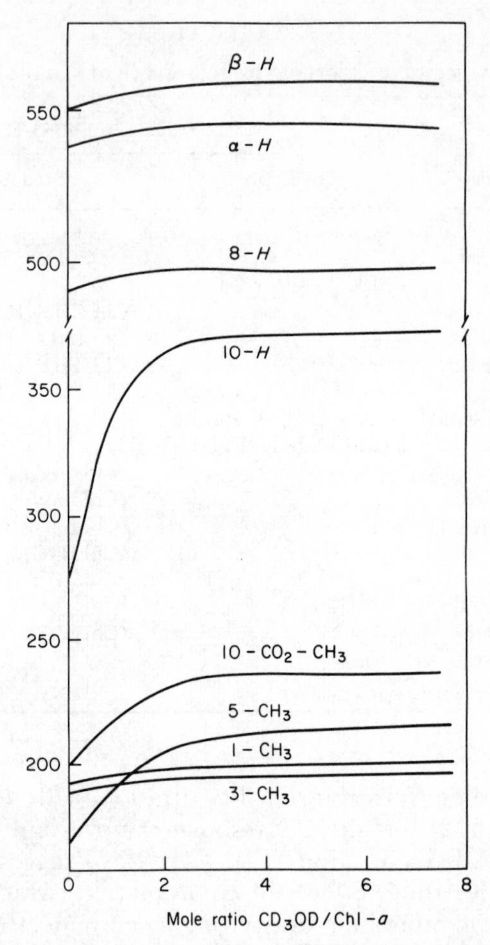

FIG. 19. Disaggregation of chlorophyll *a* (0·15 M) in $CDCl_3$ solution on addition of CD_3OD as shown by changes in proton n.m.r. chemical shifts. (Closs *et al.*, 1963.)

definitive as for chlorophyll *a*. It appears that the fundamental unit with dry chlorophyll *b* is the trimer, and that both the aldehyde and isocyclic ring carbonyl groups are involved in C=O Mg interactions. Chlorophyll *b* oligomer, or water micelle formation appears to be stronger than is the case with chlorophyll *a* as judged from the red shifts in the visible spectrum in carbon tetrachloride and from i.r. spectra.

X-ray structure determinations have not so far been successful with chlorophylls, because of the problems resulting from association with traces of water, or other ligands, but the complete structure

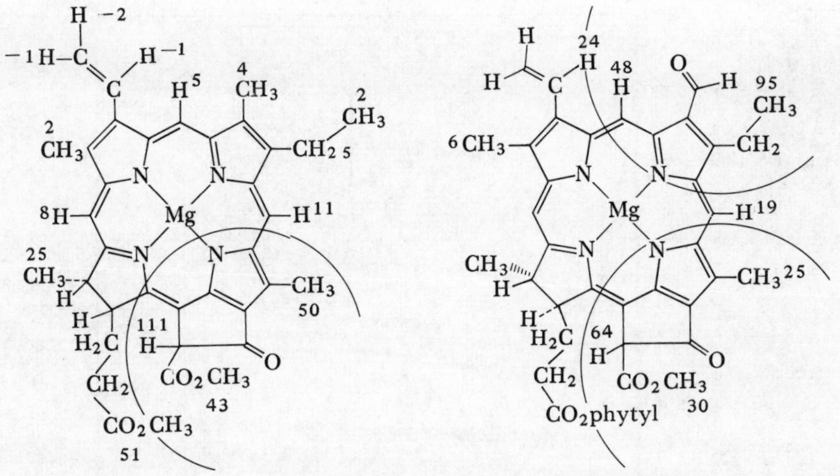

FIG. 20. "Aggregation maps" of methyl chlorophyllide *a* and chlorophyll *b* from proton n.m.r. chemical shift differences (Hz) between $CHCl_3$ and $CDCl_3$-CD_3OD solution for indicated protons at 100 MHz. (Closs *et al.*, 1963.)

determination of methyl phaeophorbide *a* has recently been reported by two groups (Gassmann *et al.*, 1971; Fischer *et al.*, 1972). Based on this work Calvin and his colleagues (Fischer *et al.*, 1972) have proposed a model for chlorophyll *a* in which the photosynthetic lamellae are related by a 2_1 screw axis, and the chlorophyll molecules are linked by water molecules in such a way that the planes of each pair of rings are inclined at a slight angle to each other.

In vivo as is well known, chlorophyll is associated with proteins in the photosynthetic system of chloroplasts (Goedheer, 1966; Park, 1966). The spectral and crystallographic investigations described above, although carried out largely with *in vitro* systems, have been of great importance in connection with studies of the mechanisms of photosynthesis *in vivo*. The nature of the interactions between water and chlorophyll (cf. Rabinowitch, 1945, 1951, 1956) is of considerable interest because the key event in photosynthesis is the photolysis of water, and for example Witt (1968) has shown that a long wavelength form of chlorophyll is involved. The primary effects of light on aggregates of chlorophylls appear to be the formation of excited states through oxidoreduction reactions; two types of reaction centre have been recognized in green plants, the so-called systems I and II (Clayton, 1965, 1966; Seely, 1966; Vernon and Ke, 1966). System I forms a weak oxidant and a strong reductant (which can reduce pyridine nucleotides) whilst system II forms a weak

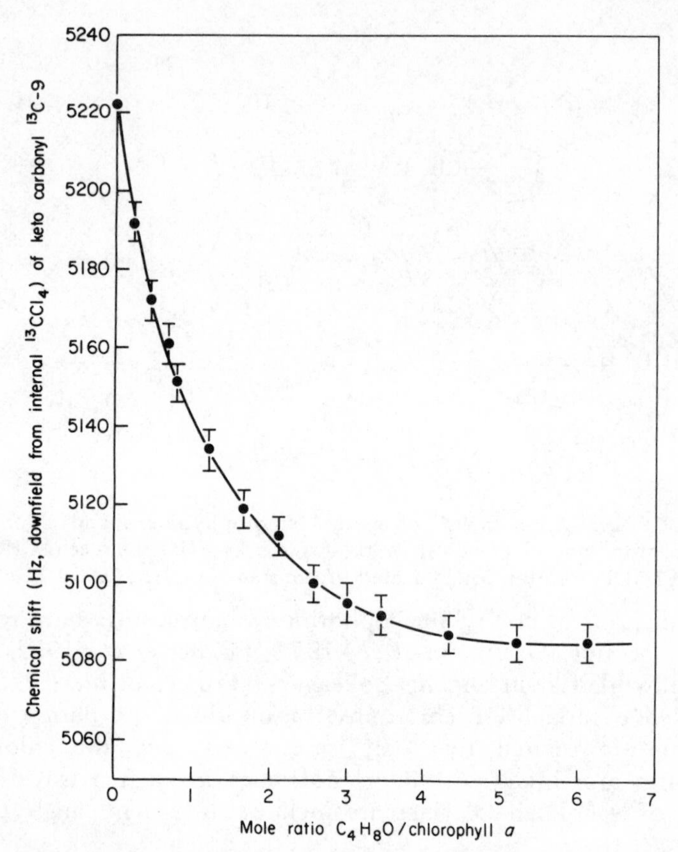

FIG. 21. Disaggregation of chlorophyll *a* in CCl_4 solution on addition of tetrahydrofuran as shown by the ^{13}C n.m.r. chemical shifts of the C-9 carbonyl carbon resonance. (Katz *et al.*, 1972.)

reductant and a strong oxidant which leads to oxygen evolution. These reactions are now thought to involve electron addition or removal from the macrocycle (Fuhrhop and Mauzerall, 1968, 1969; Felton *et al.*, 1969); radicals are readily produced by chemical oxidations or in polarographic processes and their electron spin resonance (e.s.r.) and optical spectra have been studied (Druyan *et al.*, 1973). It has long been known that e.s.r. signals can be observed in photosynthesizing systems, but it has now been shown that *dry* oxygen-free chlorophyll systems *in vitro* do not give rise to such signals on irradiation (Garcia-Morin *et al.*, 1969). On the other hand chlorophyll–water micelles yield a narrow, reversible, photo-induced e.s.r. signal on irradiation with red light, and it has now been suggested that the primary light conversion step is the formation of

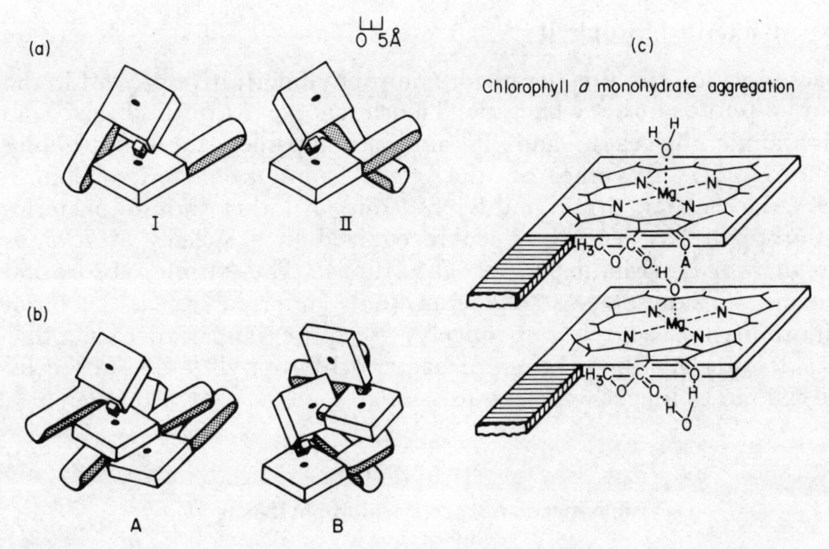

FIG. 22. Structures proposed for (a) chlorophyll dimers, (b) oligomers, (c) chlorophyll–water adducts. (Ballschmitter and Katz, 1969.)

chlorophyll radical cations and anions which provide the reducing and oxidizing functions of photosynthesis (Garcia-Morin *et al.*, 1969). Borg *et al.* (1970) have concluded that the first photoproduct ("oxidized P700") of system I in green plants contains a cation radical by comparison of its e.s.r. and optical properties with those of the product of electrochemical oxidation of chlorophyll *a* (cf. Dolphin and Felton, 1974). The e.s.r. signal associated with the primary step of photosynthesis ("signal I") has been compared with the free radicals generated *in vitro* by various chlorophyll species, and the line-width and spectral properties of reaction-centre chlorophyll were found to be consistent with spin delocalization over an entity (Chl–H_2O–Chl) containing two chlorophyll molecules (Norris *et al.*, 1971). These results have led to a proposal for a mode of the photosynthetic unit in which the bulk of the chlorophyll is present as an oligomer $(Chl_2)_n$ absorbing near 680 nm (the light-gathering or antenna chlorophyll) whilst the active centre is Chl–H_2O–Chl. The latter makes two molecules of chlorophyll available for simultaneous removal and replacement of the electron, whilst the integrity of the antenna chlorophyll oligomer is maintained by a highly water-free environment. This picture agrees very well with that deduced from infrared and visible spectral studies, and described in detail above (Ballschmitter and Katz, 1969).

IV Bacteriochlorophyll

Bacteriochlorophyll is the major chlorophyll derivative present in the purple photosynthetic bacteria (Thiorhodaceae, Athiorhodaceae, and Hyphomicrobiaceae), and it also accompanies the *Chlorobium* chlorophylls in some of the green sulphur bacteria (Chlorobacteriaceae) (cf. Strain and Svec, 1966). Another form of bacteriochlorophyll has recently been recognized in a species of *Rhodopseudomonas* (Eimhjellen *et al.*, 1963). The visible absorption spectra of these two pigments (now referred to as bacteriochlorophylls *a* and *b* respectively) are quite similar to each other (Table VII) but the maxima of bacteriochlorophyll *b* are shifted by 10–20 nm to longer wavelengths.

TABLE VII

Visible spectra of bacteriochlorophylls *a* and *b* in acetone[a]

a λ_{max} (nm)	*b* λ_{max} (nm)
358	368
390 sh	407
(575)	(580)
700 sh	720 sh
771	794

[a] Eimhjellen *et al.*, 1963.

Bacteriochlorophyll *a* (Fig. 23) was shown by Schneider (1934) to contain magnesium and to give a positive phase test; it also yielded a phaeophorbide which could be converted into porphyrins by oxidation. Photo-oxidation also converted bacteriochlorophyll into a product with an absorption band near that of chlorophyll *a*. The close relationship with chlorophyll *a* was confirmed by treatment with hydriodic acid which converted bacteriochlorophyll *a* into "oxophaeoporphyrin a_5" (the 2-acetyl-2-desvinyl analogue of phaeoporphyrin a_5) which can also be obtained direct from phaeoporphyrin a_5 by mild treatment with the same reagent. Bacteriochlorophyll *a* was also shown to contain a phytyl ester side-chain.

Elemental analysis showed that bacteriochlorophyll *a* was probably a tetrahydroporphyrin derivative, and a tentative structure was proposed (Fischer *et al.*, 1940). The position of the extra

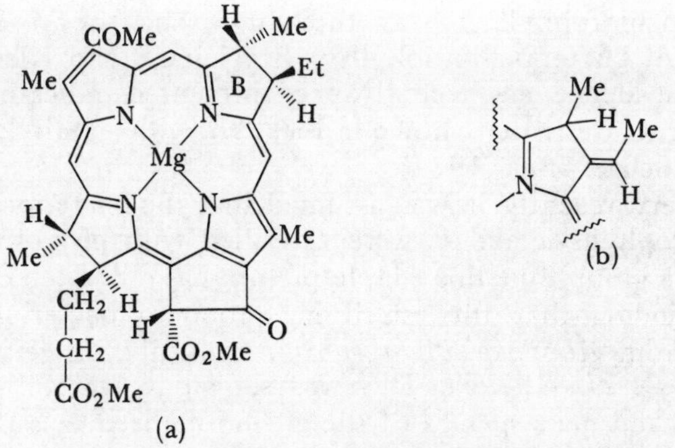

FIG. 23. Bacteriochlorophylls *a* and *b*.

hydrogen atoms has, however, only been confirmed relatively recently. Bacteriochlorophyll *a* from *Chromatium* was converted into bacteriochlorin e_6 trimethyl ester, and this was oxidized to give a mixture of maleimides and succinimides (Golden *et al.*, 1958) when methyl ethyl succinimide was formed in addition to the dihydrohaematinicimide which is found in similar degradations of chlorophyll *a* (cf. Ficken *et al* 1956). This confirmed the location of the two additional hydrogen atoms in the nucleus of bacteriochlorophyll *a* and moreover the methyl ethyl succinimide was shown to be *trans*. Oxidation of bacteriochlorophyll *a* with quinones affords 2-desvinyl-2-acetyl chlorophyll *a*.

Bacteriochlorophyll *a* also occurs in green sulphur bacteria in about one-twentieth the ratio of the major pigments the *Chlorobium* chlorophylls. It was first observed as a minor peak in organic extracts (Katz and Wassink, 1939; Larsen, 1953) and isolated as a protein-pigment complex by Olsen and Romano (1962). Its properties were studied by Holt *et al.* (1963), who showed that it contained a phytyl ester, was oxidizable to 2-acetyl-2-desvinyl chlorophyll *a* and afforded oxophylloerythrin on degradation; the latter was identical with a sample obtained from pyrophaeophorbide by the "oxo reaction".

The absolute configuration and *trans* relationship of the 3- and 4-substituents of bacteriochlorophyll was finally proved by comparison of the di-*p*-bromophenacyl esters of the 2-methyl-3-ethyl succinic acid (derived from the chromic acid oxidations) and synthetic (+)-*threo*-methyl-3-ethylsuccinic acid (Brockmann, 1968); the substituents in ring B were thus shown to have the absolute configuration 3R, 4R (Fig. 23a).

Bacteriochlorophyll *b* was thought to be the *cis*-3,4-dihydro analogue of bacteriochlorophyll *a* (Brockmann and Kleber, 1970), but new evidence has recently been presented indicating that the correct structure is that shown in Fig. 23b, with an ethylidene group in ring-B (Schaer *et al.*, 1974).

Until very recently it was assumed that the bacteriochlorophylls like chlorophylls *a* and *b* were esterified with phytol on the C-7 propionate group, but this simple picture has now been complicated by the finding that the esterifying group of bacteriochlorophyll isolated from *Rhodospirillum rubrum* was all-*trans*-geranylgeraniol (see Katz *et al.*, 1972b). This was shown by careful analysis of the n.m.r. and mass spectra of the alcohol derived by saponification of the bacteriochlorophyll, and in the light of the knowledge that the esterifying alcohol of the *Chlorobium* chlorophylls is farnesol (see below) Katz and his colleagues suggest that other polyprenyl esters of chlorophylls may also exist, albeit in small quantities. They point out that their findings raise obvious implications regarding photosynthesis and polyprenoid biosynthesis, as well as taxonomic questions. Brockmann and Knobloch (1972) have described evidence that *R. rubrum* bacteriochlorophyll is a farnesyl ester, and it seems possible that the esterifying alcohol may be strain dependent.

As with chlorophylls *a* and *b* the p.m.r. spectrum of bacteriochlorophyll *a* in tetrahydrofuran shows the presence of an isomeric form, in somewhat less than 10% at equilibrium. This isomeric form has not been isolated chromatographically but it may be the C-10 epimer (Katz *et al.*, 1968b) the alternative possibility that it is an enol chelate must, however, now also be considered in view of Hynninen's (1973) arguments concerning the analogous chlorophyll *a*′ and *b*′.

Bacteriochlorophyll *a* also forms dimers and oligomers, as well as complexes with water in the same manner as chlorophylls *a* and *b* (Ballschmitter and Katz, 1969), but the 2-acetyl group surprisingly does not appear to be involved to any significant extent in coordination with the magnesium, in either case. The visible spectra of bacteriochlorophylls *in vivo* have been reviewed recently (Olsen and Stanton, 1966).

The photosynthetic function of the bacterial chlorophylls in purple bacteria is probably exercised in a similar manner to chlorophyll in green plants and algae, i.e. by photo-oxidation of the bacteriochlorophyll–water complexes (BChl–H_2O–BChl) to form cation radicals (Borg *et al.*, 1970; Fajer *et al.*, 1970; Norris *et al.*, 1971; McElroy *et al.*, 1972). Recent studies of the e.s.r. and visible spectrum of the *anion* radical of bacteriochlorophyll have now led to the conclusion that it is not the primary acceptor in photosynthesis (Fajer *et al.*, 1973). Electron nuclear double resonance ("endor") studies of bacteriochlorophyll and e.s.r. studies are consistent with the suggestion that the 9-G e.s.r. signal associated with photosynthesis in purple bacteria is due to delocalization of the unpaired electron over two "active-centre" bacteriochlorophyll molecules (Norris *et al.*, 1973); similar conclusions were reached concerning chlorophyll *a* free radicals in algae.

V Chlorophyll *c*

This pigment was first discovered in extracts of "olive-coloured seaweeds" (Stokes, 1864) and later named chlorofuscine by Sorby (1873), who also determined its visible spectrum with a visual spectroscope. The possibility that it might be an artefact (Willstätter and Page, 1914) was discounted by later work (Strain and Manning, 1942; Strain *et al.*, 1943) and chlorophyll *c* has now been found in association with chlorophyll *a* in a variety of marine algae, dinoflagellates and marine diatoms (cf. Dougherty *et al.*, 1966, 1970).

Some of earlier investigations showed that the pigment was a mixture of free acids (Strain and Manning, 1942; Smith and Benitez, 1955), and the magnesium-free derivative showed an "oxorhodo"-type spectrum like that of 2-vinyl-2-desethyl phaeoporphyrin a_5 (the magnesium-free porphyrin analogue of phaeophorbide *a*). Chlorophyll *c* also gave a positive phase test, and on treatment with methanolic hydrochloric acid it underwent a transformation to give a product with an "aetio-type" visible spectrum (Granick, 1949). These early findings showed that the pigment was a magnesium derivative of a *porphyrin*-free acid, containing a cyclopentanone ring and a C-10 ester group like chlorophylls *a* and *b*.

These findings have been confirmed by more recent studies in which chlorophyll *c* (λ_{max} 447, 579, 627 nm in diethyl ether) was isolated from the marine diatom (*Nitzschia closterium*) and shown to be a mixture of magnesium hexadehydro- and tetradehydrophaeoporphyrin a_5 (Fig. 24) (Dougherty *et al.*, 1970). Crystallization of chlorophyll *c* from tetrahydrofuran and petroleum ether (Strain and Svec, 1966) gave the bis-tetrahydrofuranate complex as shown by the high field resonance of the tetrahydrofuran (at 8τ) in the n.m.r. spectrum, and by elemental analysis. Crystalline chlorophyll *c* proved to be infusible, and moreover as it was relatively insoluble in organic solvents the n.m.r. spectrum was determined in trifluoroacetic acid.

Assignment of the various resonances present in the spectrum (Fig. 25) followed in part from n.m.r. studies of other porphyrins and chlorins (see above); the variability in the relative areas of certain of the resonances (e.g. the high field methyl triplet at 8·3 and the vinyl resonances) confirmed that a mixture of species was present. Reduction of chlorophyll *c* with hydriodic acid afforded phaeoporphyrin a_5 monomethyl ester which was identical with material derived from chlorophyll *a*.

Neither chlorophyll *c* nor its magnesium-free derivative gave molecular ions in their mass spectra because of their involatility due to the free carboxylic acid group. Attempts to esterify the free acid group with diazomethane led to a complex mixture of products, whilst prolonged treatment with methanolic hydrogen chloride afforded a rearrangement product "isochloroporphyrin-c_6 trimethyl ester" (Fig. 24) arising from opening of the isocyclic ring and reclosure onto the acrylic acid residue at C-7. This new material, unlike chlorophyll *c*, gave a good mass spectrum showing two molecular ions at *m/e* 632 and 634. If the chlorophyll *c* was reduced in trifluoroacetic acid solution with hydrogen over palladium on carbon prior to methylation with methanolic hydrogen chloride, another product, *meso*-isochloroporphyrin trimethyl ester, was formed; this showed a molecular ion at *m/e* 636 corresponding to the addition of two or four hydrogen atoms to the former product, and confirmed the presence of one and two vinyl groups in the two

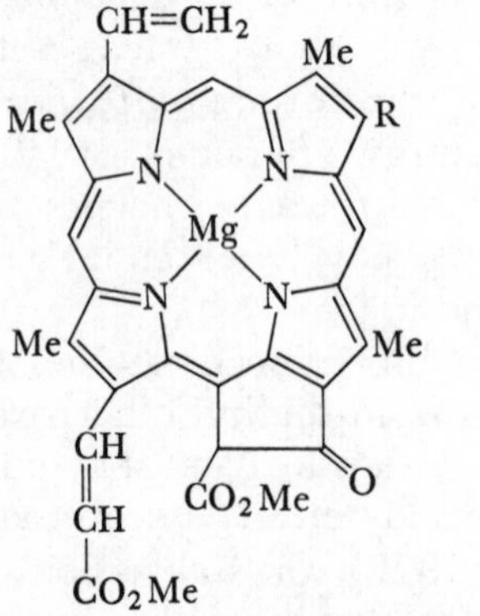

Chlorophyll c_1 R = Et
Chlorophyll c_2 R = CH=CH$_2$

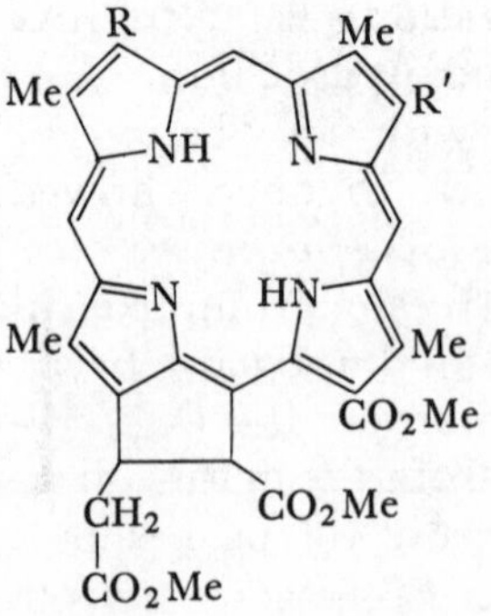

Isochloroporphyrin c_6 trimethyl ester
(R = CH=CH$_2$, R′ = Et and CH=CH$_2$)
Meso-isochloroporphyrin c_6 trimethyl ester
(R = R′ = Et)

FIG. 24. Chlorophylls c_1 and c_2, isochloroporphyrin methyl ester and *meso*-isochloroporphyrin methyl ester.

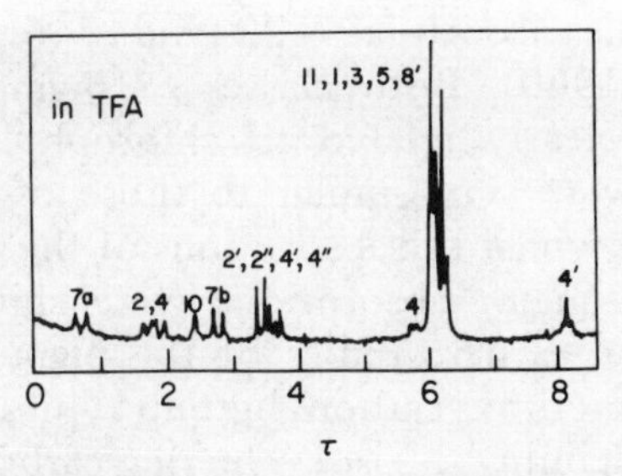

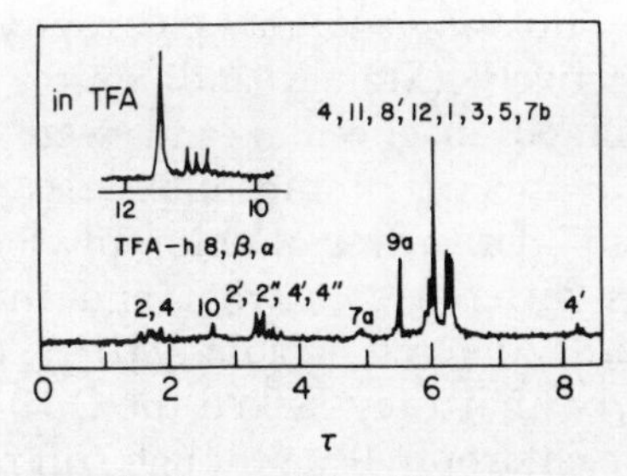

FIG. 25. Proton n.m.r. spectrum of chlorophylls *c* and isochloroporphyrin.

components of chlorophyll *c*. Detailed analyses of the mass and n.m.r. spectra of the isochloroporphyrins are given in the original paper (Dougherty *et al.*, 1970), and these together with biosynthetic considerations and the relationship established with phaeoporphyrin a_5 clearly showed the structure of the two components of chlorophyll *c* (Fig. 24).

Subsequently the two components of the mixture have been separated by chromatography on polyethylene and the ratio of c_1 and c_2 was found to be about 0·6 (Strain *et al.*, 1971; Jeffrey, 1972). The wide distribution of chlorophyll *c* in marine diatoms, dinoflagellates and brown algae has been emphasized by Dougherty *et al.* (1970), who have pointed out that these organisms constitute a major part of the world's carbon dioxide fixation capacity. The chlorophyll *c* contained in a number of other organisms has now also been shown to have the structure given in Fig. 24 (Croft and Howden, 1970; De-Greefe and Caubergs, 1970; Wasley *et al.*, 1970; Budzikiewicz and Taraz, 1971).

VI *Chlorobium* chlorophylls

The *Chlorobium* chlorophylls are the principal pigments of the green sulphur bacteria (Chlorobacteriaceae), and as has already been mentioned they are sometimes accompanied by small amounts of bacteriochlorophyll. Most of the published work on their structure has been carried out with various strains of *Chlorobium thiosulphatophilum*, but other green bacteria have also been investigated, e.g. *Chlorobium limicola* and *Chloropseudomonas ethylicum*. Two types of *Chlorobium* chlorophylls were obtained from different strains of *Chlorobium thiosulphatophilum* and designated as the "650" and "660" series depending on the position of their long wavelength absorption in diethyl ether solution (Holt and Morley, 1960a,b; Stanier and Smith, 1960; Smith and French, 1960). *In vivo* the 650 and 660 *Chlorobium* chlorophylls absorb at 725 and 746 nm respectively (Stanier and Smith, 1960). Both classes of pigments contained magnesium and were clearly related to chlorophyll *a*; visible spectra of the 650 series were very similar to those of the "meso" derivatives of chlorophyll *a* whilst the absorptions of the 660 series differed somewhat and the major absorptions were shifted towards the red (Table VIII). The earlier studies of the pigments showed that they lacked the C-10 methoxycarbonyl group typical of other chlorophylls, although still retaining the isocyclic ring carbonyl group (Morley *et al.*, 1959; Holt and Hughes, 1961; Morley and Holt, 1961; Holt *et al.*, 1962). Furthermore, they contained farnesol in place of phytol as shown by alkaline hydrolysis of the magnesium-free pigments (Rapoport and Hamlow, 1961; Holt *et al.*, 1963).

Each of the pigments was separated by partition chromatography into six fractions (numbered 1–6 in order of their decreasing acid number) (Hughes and Holt, 1962; Holt *et al.*, 1962, 1963), but all the pigments of both series (Fig. 26) contained an α-hydroxyalklyl group (at C-2) as shown by acetylation, oxidation to a keto group, and dehydration to an alkenyl group (Holt and Hughes, 1961). After conversion into a porphyrin by hydriodic acid in acetic acid (at 65°C) the 650 series pigments gave a rhodo-type spectrum, whereas the 660 series pigments gave an aetio-type spectrum. The oxime of the 650 series retained sufficient rhodo-type character (band III = band IV in intensity) to indicate the presence of a cyclopentanone ring, whilst the oxime of the 660 series retained an aetio-type spectrum.

Chromic acid oxidation of the phaeophorbides of both series afforded dihydrohaematinic acid imide which was identical with the *trans* product synthesized by Ficken *et al.* (1956), whilst methyl

TABLE VIII

Visible spectra of *Chlorobium* chlorophylls 650 and 660 in ether

650 λ_{max} (nm)	660 λ_{max} (nm)
410	410
425	430
(575)	
(605)	(625)
(613)	
650	660

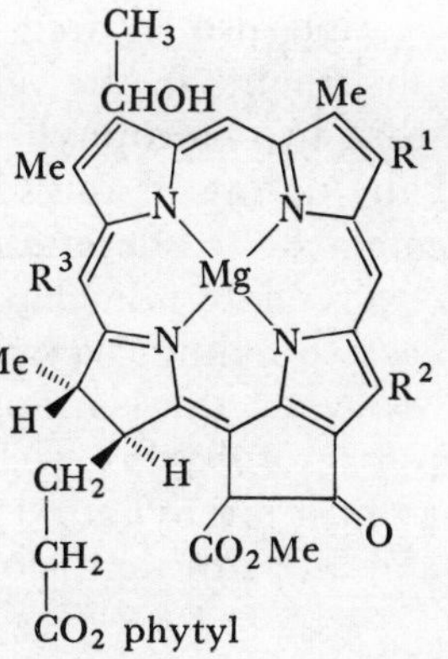

650 Series	R^1	R^2	R^3
Fraction 1	isobutyl	ethyl	H
2	*n*-propyl	ethyl	H
3	isobutyl	methyl	H
4	ethyl	ethyl	H
5	*n*-propyl	methyl	H
6	ethyl	methyl	H

660 Series	R^1	R^2	R^3
Fraction 1	isobutyl	ethyl	methyl
2	isobutyl	ethyl	ethyl?
3	*n*-propyl	ethyl	methyl
4	*n*-propyl	ethyl	ethyl?
5	ethyl	ethyl	methyl
6	ethyl	methyl	methyl

FIG. 26. Structures of *Chlorobium* chlorophylls 650 and 660.

ethyl maleimide, methyl *n*-propylmaleimide and methyl isobutylmaleimide were also obtained depending on which fraction was oxidized. The imides were identified by gas–liquid chromatographic comparisons with synthetic materials (Morley and Holt, 1961; Hughes and Holt, 1962; Holt *et al.*, 1966). In the 650 series the precise nature of the substituents in ring A was determined by converting the hydroxyalkyl group into an alkyl group; oxidation of those fractions which previously yielded only methyl *n*-propyl maleimide, then also afforded methyl ethyl maleimide, thus showing that the C-2 substituent was hydroxyethyl. This was confirmed by dehydration of fraction 6 which gave a product identical with pyrophaeophorbide *a*, and also showed that the arrangement of the peripheral methyl substituents in the various fractions was probably also identical with that in chlorophyll *a*. The information thus far obtained concerned the nature of the substituents in rings A, B and D and it only remained to determine the nature of the C-5 substituents. These were determined by conversion of the various fractions to the corresponding "pyrroporphyrins" by oxidative degradation of the isocyclic ring as shown in Fig. 27; chromic acid oxidation of the pyrroporphyrins yielded ethyl maleimide from fractions 1, 2 and 4, whilst methyl ethyl maleimide was also obtained from fractions 1, 2 and 3 (Purdie and Holt, 1965).

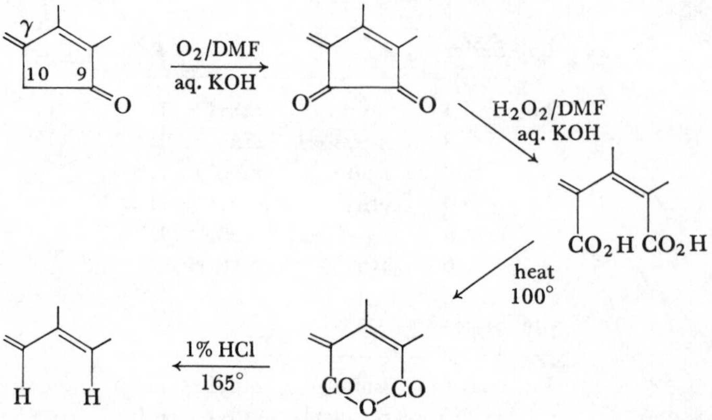

FIG. 27. Oxidative degradation of the isocyclic ring of the *Chlorobium* chlorophylls.

In the 660 series similar results were obtained (Holt *et al.*, 1966) but the porphyrin obtained on oxidative degradation of the isocyclic ring (cf. Fig. 26) had a phyllo-type spectrum in contrast to the aetio-type observed for the pyrroporphyrins. This indicated the presence of a *meso*-alkyl substituent which was assigned to the δ-position by comparison of the proton magnetic resonance spectra of the methyl esters of the "meso" derivative of fraction 5 (the major fraction), mesopyrophaeophorbide *a* (i.e. 2-ethyl 2-desvinylpyrophaeophorbide *a*) and δ-chloromesopyrophaeophorbide *a*; the signal ($\tau = 1{\cdot}53$) assigned to the δ-proton of mesopyrophaeophorbide *a* (cf. Woodward and Skaric, 1961) was missing from the spectra of fraction 5 and the δ-chloromesophaeophorbide *a*. Rapaport and his colleagues (Mathewson *et al.*, 1963a,b) have also investigated the major fraction of the *Chlorobium* chlorophylls 660 and concluded on the basis of n.m.r.

studies that the structure proposed (Holt *et al.*, 1962) was correct, except that the *meso*-alkyl substituent was assigned to the α or β position; this assignment was made because treatment of fraction 5 with warm deuterioacetic acid effected exchange of one of the two low field *meso* proton peaks (that at $\tau = 0{\cdot}28$) and Woodward and Skaric (1961) had shown that the δ proton of mesopyrophaeophorbide *a* was the only *meso* proton to exchange with deuterioacetic acid. However, the latter is the *highest* field proton of mesopyrophaeophorbide *a* whereas the proton which exchanges in fraction 5 of the 660 series is at *lowest* field; moreover, as the *meso* substituents are likely to arise biogenetically by electrophilic alkylation, Holt's original assignment seems more likely because electrophilic substitution is most likely to occur at the δ position of a phaeophorbide (cf. Woodward and Skaric, 1961; Woodward, 1962). The reasons for the unusual ease of exchange of the α (or β?) proton of fraction 5 of the 660 series must await the synthesis of suitable pyrophaeophorbides for definitive n.m.r. experiments, but the distorting effect of the δ-alkyl substituent may be responsible.

The analytical and degradative evidence described above did not completely prove the structures of the pigments of the 650 and 660 series for it only revealed the nature of the pairs of substituents on the four rings, and their order was assumed because of their relationship to chlorophyll *a*. Moreover there was no direct evidence concerning the nature of the C-5 substituents in fractions 3, 5 and 6 of the 650 series, as citraconimide (methyl maleimide) had not been detected among the oxidation products; in the 660 series improved analytical techniques showed that fractions 1–5 yielded ethylmaleimide, and fraction 6, citraconimide, from ring C (Holt *et al.*, 1966).

The structures of the *Chlorobium* chlorophylls 650 fractions 1–5 were finally proven by synthesis of the pyroporphyrins from fractions 1–4 and of the desoxophylloerythrin from fraction 5 (Archibald *et al.*, 1963, 1966); the structure of fraction 6 followed from its relationship with pyrophaeophorbide as described above. The position in the 660 series is a good deal less satisfactory, and conclusive proof is only available for the structures of fractions 5 and 6 through synthesis of the corresponding δ-phylloporphyrins (Archibald *et al.*, 1966); the synthetic phylloporphyrins which would correspond with the structures assigned to fractions 3 and 4 were not well characterized nor could they be identified with the naturally derived material, whilst insufficient of fractions 1 and 2 was available to attempt degradation (Archibald *et al.*, 1966). However, both fractions 3 and 4 also gave one pyrroporphyrin, which was identical with synthetic pyrroporphyrin corresponding to the 650 series fraction 2.

The original structures assigned to the various fractions in the 660 series (Holt, 1965) were revised (Holt *et al.*, 1966) in the light of mass spectrometric determinations of molecular weights of the methyl esters of the phaeophorbides derived from fractions 2, 4, 5 and 6 (A. H. Jackson, G. W. Kenner and K. M. Smith, unpublished work, 1965); the values found for fractions 5 and 6 corresponded to those deduced from the analytical and synthetic evidence, whilst those for fractions 2 and 4 corresponded to δ-methyl compounds (whereas it had originally been suggested that δ-ethyl groups were present in these fractions). The assignments for fractions 1 and 2, and 3 and 4 were therefore reversed. However, there is no real evidence for the presence of an ethyl group, and recent synthetic work has cast further doubt on this suggestion (Cox *et al.*, 1971); it may well be that fractions 1 and 2 are identical with each other, and similarly that fractions 3 and 4 are also identical, but a test of this possibility must await the outcome of a reinvestigation of the nature of the *Chlorobium* chlorophylls 660. Confirmatory evidence for the nature of the substituents on the pyrrole rings has recently also been obtained by gas chromatographic studies of the alkyl pyrroles produced by reductive alkylation (Chapman *et al.*, 1971).

VII Other chlorophylls

A CHLOROPHYLL *d*

Chlorophyll *d* was discovered as a minor constituent accompanying chlorophyll *a* in various extracts of Rhodophyceae (Manning and Strain, 1943). Chromatographically pure samples of the phyllin and corresponding phaeophytin were later prepared and their spectra determined (Smith and Benitez, 1955); the long wavelength shifts relative to chlorophyll *a* indicated that another conjugating substituent was probably present. Holt and Morley (1959) showed that the visible absorption spectra were identical with the permanganate oxidation product of chlorophyll *a*, the 2-formyl-2-desvinyl derivative, and the corresponding phaeophytin. Holt (1961) also showed that chlorophyll *d* obtained from three species of red algae had similar chemical and chromatographic properties to those of the product obtained by oxidation of chlorophyll *a*, e.g. identical spectra, changing in the same manner on reduction of the formyl group with borohydride, and both gave a red coloured phase-test

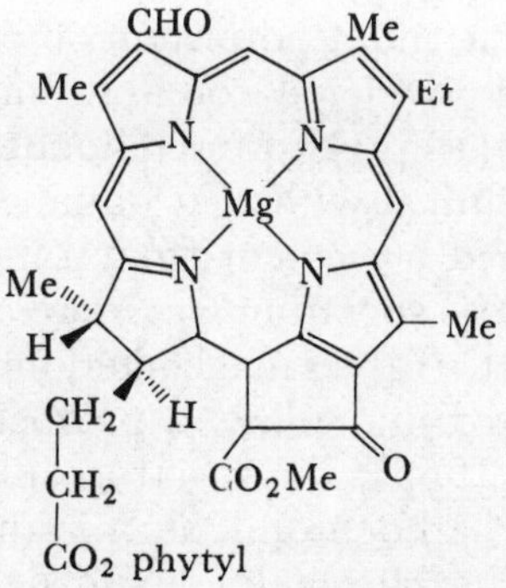

FIG. 28. Proposed structure for chlorophyll *d*.

intermediate. The formyl group is readily converted into its dimethyl acetal (Holt and Morley, 1959) which was identical with a product previously obtained by Manning and Strain (1943) and referred to as an "iso" derivative.

It is still uncertain whether chlorophyll *d* exists *in vivo* or whether it is an artefact; some thalli of *Gigartina pupillata* afforded appreciable amounts of chlorophyll *d* whilst others gave none. Sagromsky (1960) also isolated chlorophyll *d*, and claimed that sunlight, or standing in dilute methanolic potassium hydroxide, potentiated its formation; she reported a shift of the long wavelength absorption from 667 to 695 nm on standing in the alkaline medium, but did not report any evidence for the existence of a formyl group, so that her product may well have arisen from allomerization (Holt, 1958). Trace amounts of a chlorophyll *d*-like pigment have also been described in *Chlorella* extracts (Michel-Wolwertz *et al.*, 1965).

B PROTOCHLOROPHYLL

Protochlorophyll is a minor green pigment which has long been known to be present in the leaves of etiolated plants (Boardman, 1966), and Smith and Young (1956) have shown that it is a precursor of chlorophyll *a*. Fischer *et al.* (1939) concluded that protochlorophyll was probably the magnesium complex of the phytyl ester of 2-vinyl-2-desethyl phaeoporphyrin a_5 i.e. 7,8-dehydrochlorophyll *a* (Fig. 29a). The visible absorption spectrum in ether has its main long wavelength absorption at 625 nm, whilst the magnesium-free derivative has an oxorhodo-type spectrum. *In vivo* protochlorophyll absorbs at 638 nm (Smith and Coomber, 1959). A similar pigment had also been isolated from cucurbit seed coats by Noack and Kiessling (1930) and they showed a close relationship between it and a porphyrin obtained by treating methyl phaeophorbide *a* with powdered iron in formic acid.

Stanier and Smith (1959) expressed doubts as to whether the seed coat pigment was identical with that obtained from etiolated leaves, and they also obtained a pigment from a mutant of *Rhodopseudomonas spheroides* whose spectrum was more closely similar to that of the seed coat pigment. If the same bacterium is grown in presence of the inhibitor, 8-hydroxyquinoline, a protochlorophyll-like material accumulates (Jones, 1963); the spectroscopic properties, formation of a mono-oxime, the bathochromic shift in the spectrum on catalytic hydrogenation, and its chromatographic behaviour led to its formulation as magnesium 2,4-desethyl-2,4-divinylphaeoporphyrin a_5 monomethyl ester (Fig. 29b). It was further concluded that this bacterioprotochlorophyll was probably an intermediate in plant and bacterial chlorophyll biosynthesis. Another pigment isolated from the same system was shown to be 2-hydroxyethyl-2-desvinylphaeophorbide *a* by spectroscopic and chromatographic comparisons with authentic material (Jones, 1964); chemical evidence for the hydroxylated side chain was also obtained, e.g. dehydration to a product spectroscopically identical with phaeophorbide *a* and oxidation to the acetyl derivative.

Further studies of seed coat "protochlorophyll" have now shown that the pigment from *Cucurbita pepo* can be chromatographically separated into two fractions (Jones, 1965, 1966); one pigment was identical with true protochlorophyll, and the other was identified as 2,4-divinylphaeoporphyrin a_5 (Fig. 29b), the "bacterial proto-

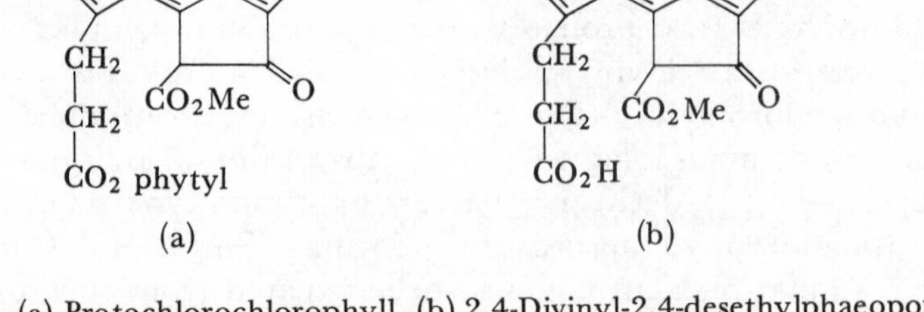

FIG. 29. (a) Protochlorochlorophyll. (b) 2,4-Divinyl-2,4-desethylphaeoporphyrin a_5 monomethyl ester.

chlorophyll" isolated from a mutant of *R. spheroides*, or from *R. spheroides* treated with 8-hydroxyquinoline. Protophaeophytin *a* (the magnesium-free derivative of protochlorophyll) has also been isolated recently from pumpkin seed coats (Houssier and Sauer, 1969).

The spectral properties of the protochlorophylls and the corresponding phaeophytins have been reported (Houssier and Sauer, 1969) and an extensive discussion of their circular dichroism and magnetic circular dichroism spectra in relation to those of other chlorophylls has also appeared (Houssier and Sauer, 1970); the magnesium complexes form dimers in non-polar solvents such as carbon tetrachloride but the structures of the dimers differ significantly from those of the chlorophylls and bacteriochlorophyll.

In recent studies of chlorophyll biosynthesis carried out with *Chlorella* mutants a number of intermediates between magnesium protoporphyrin monomethyl ester and magnesium 2-vinyl-2-desethylphaeoporphyrin-2_5 have also been isolated and characterized (Ellsworth and Arnoff, 1968, 1969).

REFERENCES

Abraham, R. J. (1961). *Molec. Phys.* **4**, 145.

Abraham, R. J., Burbidge, P., Jackson, A. H. and Kenner, G. S. (1963). *Proc. chem. Soc.* 134.

Abraham, R. J., Burbidge, P., Jackson, A. H. and MacDonald, D. B. (1966). *J. chem. Soc.* **B** 620.

Abraham, R. J., Jackson, A. H. and Kenner, G. W. (1961). *J. chem. Soc.* 3468.

Allen, M. B., French, C. S. and Brown J. S. (1960). *In* "Comparative Biochemistry of Photoreactive Systems" (M. B. Allen, ed.), pp. 33-52. Academic Press, New York and London.

Anderson, A. F. H. and Calvin, M. (1964). *Archs Biochem. Biophys.* **107**, 251.

Archibald, J. L., MacDonald, S. F. and Shaw, K. B. (1963). *J. Am. chem. Soc.* **85**, 644.

Archibald, J. L., Walker, D. M., Shaw, K. B., Markovac, A. and MacDonald, S. F. (1966). *Can. J. Chem.* **44**, 345.

Aronoff, S. (1960). *In* "Handbuch Der Pflanzenphysiologie" (W. Ruhland, ed.), vol. 5, part 1, pp. 234-251. Springer-Verlag, Berlin.

Aronoff, S. (1962). *Biochim. biophys. Acta* **60**, 193.

Aronoff, S. (1966). *In* "The Chlorophylls" (L. P. Vernon and G. R. Seely, eds), p. 3. Academic Press, New York and London.

Ballschmitter, K. and Katz, J. J. (1969). *J. Am. chem. Soc.* **91**, 2661.

Ballschmitter, K., Truesdell, K. and Katz, J. J. (1969). *Biochim. biophys. Acta* **184**, 604.

Barth, G., Linder, R. E., Bunnenberg, E. and Djerassi, C. (1973). *Ann. N.Y. Acad. Sci.* **206**, 223.

Becker, E. D. and Bradley, R. B. (1959). *J. chem. Phys.* **31**, 1413.
Becker, R. S. and Sheline, R. K. (1955). *Archs Biochem. Biophys.* **54**, 259.
Beckey, H. D. and Schulten, H.-R. (1973). *Org. Mass Spectrometry* 7, 861.
Boardman, N. K. (1966). *In* "The Chlorophylls" (L. P. Vernon and G. R. Seely, eds), p. 437. Academic Press, New York and London.
Bonnett, R. (1973). *Ann. N.Y. Acad. Sci.* **206**, 745.
Bonnett, R. and Stephenson, G. F. (1964). *J. chem. Soc.* 291.
Bonnett, R., Gale, I. A. D. and Stephenson, G. S. (1967). *J. chem. Soc. C* 1168.
Borg, D. C., Fajer, J., Felton, R. H. and Dolphin, D. (1970). *Proc. natn. Acad. Sci. U.S.A.* **67**, 813.
Boucher, L. J. and Katz, J. J. (1967). *J. Am. chem. Soc.* **89**, 1340, 4103.
Boucher, L. J., Strain, H. H. and Katz, J. J. (1966). *J. Am. chem. Soc.* **88**, 1341.
Brockmann, H. H. (1968). *Angew. Chem. int. Edn* 7, 221; *Angew. Chem.* **80**, 233.
Brockmann, H. Jr. (1971). *Justus Liebig's Annln Chem.* **754**, 139.
Brockmann, H. and Kleber, I. (1970). *Tetrahedron Lett.* 2195.
Brockmann, H. and Knobloch, G. (1972). *Arch. Mikrobiol.* **85**, 123.
Budzikiewicz, H. and Taraz, K. (1971). *Tetrahedron* 27, 1447.
Burrell, J. W. K., Jackman, L. M. and Weedon, B. C. L. (1959). *Proc. chem. Soc.* 263.
Burrell, J. W. K., Garwood, R. F., Jackman, L. M., Oskay, E. and Weedon, B. C. L. (1966). *J. chem. Soc. C* 2144.
Cavaleiro, J. A. S., Kenner, G. W. and Smith K. M. (1973). *J. chem. Soc.* 2478.
Chapman, R. A., Roomi, M. W., Morton, T. C., Krajcarski, D. T. and MacDonald, S. F. (1971). *Can. J. Chem.* **49**, 3544.
Clayton, R. K. (1965). *Science, N.Y.* **149**, 1346.
Clayton, R. K. (1966). *In* "The Chlorophylls" (L. P. Vernon and G. R. Seely, eds), p. 610. Academic Press, New York.
Closs, G. L., Katz, J. J., Pennington, F. C., Thomas, M. R. and Strain, H. H. (1963). *J. Am. chem. Soc.* **85**, 3809.
Cox, M. T., Haworth, T. T., Jackson, A. H. and Kenner, G. W. (1969). *J. Am. chem. Soc.* **91**, 1232.
Cox, M. T., Jackson, A. H. and Kenner, G. W. (1971). *J. chem. Soc. C* 1974.
Crabbé, P., Djerassi, C., Eisenbraun, E. J. and Liu, S. (1959). *Proc. chem. Soc.* 264.
Croft, J. A. and Howden, M. E. H. (1970). *Phytochemistry* 9, 901.
De-Greefe, J. A. and Caubergs, R. (1970). *Naturwissenschaften* **57**, 673.
Djerassi, C., Brunnenberg, E. and Elder, D. (1971). *Pure appl. Chem.* **25**, 57.
Dolphin, D. and Felton, R. H. (1974). *Accts. chem. Res.* 7, 26.
Donnay, G. (1959). *Archs Biochem. Biophys.* **80**, 80.
Dougherty, R. C., Strain, H. H. and Katz, J. J. (1965). *J. Am. chem. Soc.* **87**, 104.
Dougherty, R. C., Strain, H. H., Svec, W. A., Uphams, R. A. and Katz, J. J. (1966). *J. Am. chem. Soc.* **88**, 5037.
Dougherty, R. C., Strain, H. H., Svec, W. A., Uphams, R. A. and Katz, J. J. (1970). *J. Am. chem. Soc.* 92, 2826.
Druyan, M. F., Norris, J. R. and Katz, J. J. (1973). *J. Am. chem. Soc.* 95, 1682.
Eimhjellen, K. E., Aasmundrud, O. and Jensen, A. (1963). *Biochem. biophys. Res. Commun.* 10, 232.
Eisner, U. and Linstead, R. P. (1955). *J. chem. Soc.* 3749.

Eisner, U., Lichtarowicz, A. and Linstead, R. P. (1957). *J. chem. Soc.* 733.

Ellis, J., Jackson, A. H., Kenner, G. W. and Lee, J. (1960). *Tetrahedron Lett.* no. 2, 23.

Ellsworth, R. K. and Aronoff, S. (1968). *Archs Biochem. Biophys.* **125**, 269, 358.

Ellsworth, R. K. and Aronoff, S. (1969). *Archs Biochem. Biophys.* **130**, 374.

Fajer, J., Borg, D. C., Forman, A., Dolphin, D. and Felton, R. H. (1970). *J. Am. chem. Soc.* **92**, 3451.

Fajer, J., Borg, D. C., Forman, A., Dolphin, D. and Felton, R. H. (1973). *J. Am. chem. Soc.* **95**, 2739.

Falk, J. E. (1964). "Porphyrins and Metalloporphyrins". Elsevier, Amsterdam.

Falk, J. E. and Willis, J. B. (1951). *Aust. J. sci. Res.*, **A4**, 579.

Felton, R. H., Dolphin, D., Borg, D. C. and Fajer, J. (1969). *J. Am. chem. Soc.* **91**, 196.

Ficken, G. E., Johns, R. B. and Linstead, R. P. (1956). *J. chem. Soc.* 2272.

Fischer, H. and Gibian, H. (1942). *Justus Liebig's Annln Chem.* **552**, 153.

Fischer, H. and Orth, H. (1937). "Die Chemie des Pyrrols", vol. 2 (i), Akademische Verlag, Leipzig.

Fischer, H. and Siebel, H. (1932). *Justus Liebig's Annln Chem.* **499**, 84.

Fischer, H. and Stern, H. (1935). *Justus Liebig's Annln Chem.* **519**, 58; **520**, 88.

Fischer, H. and Stern, A. (1940). "Die Chemie des Pyrrols", vol. 2 (ii), Akademische Verlagsgesellschaft, Leipzig.

Fischer, H., Mittenzwei, H. and Oestreicher, A. (1939). *Hoppe-Seyler's Z. physiol. Chem.* **257**, iv-vii.

Fischer, H., Mittenzwei, H. and Hever, D. B. (1940). *Justus Liebig's Annln Chem.* **545**, 154.

Fischer, M. S., Templeton, D. H., Zalkin, A. and Calvin, M. (1972). *J. Am. chem. Soc.* **94**, 3613.

Fleming, I. (1967). *Nature, Lond.* **216**, 151.

Fleming, I. (1968). *J. chem. Soc. C* 2765.

French, C. S., Michel-Wolwertz, M. R., Michel, J. M., Brown, J. S. and Prager, V. K. (1968). *In* "Porphyrins and Related Compounds" (T. W. Goodwin, ed.), p. 147. Academic Press, London and New York.

Fuhrhop, J.-H. and Mauzerall, D. (1968). *J. Am. chem. Soc.* **90**, 3875.

Fuhrhop, J.-H. and Mauzerall, D. (1969). *J. Am. chem. Soc.* **91**, 4174.

Games, D. E., Jackson, A. H. and Millington, D. S. (1974a). *In* "Proc. Int. Symp. on Mass Spectrometry in Biochemistry and Medicine", p. 257. Raven Press, New York. (1974b). *In* "Proc. 6th Int. Mass Spect. Conf., Edinburgh", p. 215.

Garcia-Morin, M., Uphaus, R. A., Norris, J. R. and Katz, J. J. (1969). *J. phys. Chem.* **73**, 1066.

Gassman, J., Strell, I., Brandl, F., Sturm, M. and Hoppe, W. (1971). *Tetrahedron Lett.* 4609.

Goedheer, J. C. (1966). *In* "The Chlorophylls" (L. P. Vernon and G. R. Seely, eds), p. 147. Academic Press, New York and London.

Golden, J. H., Linstead, R. P. and Whitham, G. H. (1958). *J. chem. Soc.* 1725.

Goodman, A., Oldfield, E. and Allerhand, A. (1973). *J. Am. chem. Soc.* **95**, 7553.

Granick, S. (1949). *J. biol. Chem.* **179**, 505.

Hanson, E. A. (1939). *Recl. Trav. Bot. neerl.* **36**, 180.

Haxo, F. T. and Fork, D. C. (1959). *Nature, Lond.* **184**, 1051.
Holt, A. S. (1958). *Can. J. Biochem. Physiol.* **36**, 439.
Holt, A. S. (1961). *Can. J. Bot.* **39**, 327.
Holt, A. S. (1963). *In* "Mechanism of Photosynthesis" (H. Tamiya, ed.), vol. 6, pp. 59–63. Pergamon Press, Oxford.
Holt, A. S. (1965). "Chemistry and Biochemistry of Plant Pigments" (T. W. Goodwin, ed.), 1st edn, pp. 3–28. Academic Press, London and New York.
Holt, A. S. (1966). *In* "The Chlorophylls" (C. P. Vernon and G. R. Seely, eds), p. 111. Academic Press, New York and London.
Holt, A. S. and Hughes, D. W. (1961). *J. Am. chem. Soc.* **83**, 499.
Holt, A. S. and Jacobs, E. E. (1955). *Pl. Physiol., Lancaster* **30**, 553.
Holt, A. S. and Morley, H. V. (1959). *Can. J. Chem.* **37**, 507.
Holt, A. S. and Morley, H. V. (1960a). *J. Am. chem. Soc.* **82**, 500.
Holt, A. S. and Morley, H. V. (1960b). *In* "Comparative Biochemistry of Photoreactive Systems" (M. B. Allen, ed.), pp. 169–179. Academic Press, New York and London.
Holt, A. S., Hughes, D. W., Kende, H. J. and Purdie, J. W. (1962). *J. Am. chem. Soc.* **84**, 2835.
Holt, A. S., Hughes, D. W., Kende, H. J. and Purdie, J. W. (1963). *Pl. Cell Physiol.* **4**, 49.
Holt, A. S., Purdie, J. W. and Wasley, J. W. F. (1966). *Can. J. Chem.* **44**, 88.
Hoppe, W., Will, G., Gassman, H. and Weichselgartrier, H. (1969). *Z. Kristallogr,* **128**, 18.
Houssier C. and Sauer, K. (1969). *Biochim. biophys. Acta* **172**, 476, 492.
Houssier, C. and Sauer, K. (1970). *J. Am. chem. Soc.* **92**, 779.
Hughes, D. W. and Holt, A. S. (1962). *Can. J. Chem.* **40**, 171.
Hynninen, P. H. (1973). *Acta chem. scand.* **27**, 1487, 1771.
Hynninen, P. H. and Assandri, S. (1973). *Acta chem. scand.* **27**, 1478.
Hynninen, P. H. and Ellfolk, N. (1973). *Acta chem. scand.* **27**, 1463.
Inhoffen, H. H. (1964). *Angew. Chem. int. Edn* **3**, 322.
Inhoffen, H. H. (1968). *Pure appl. Chem.* **17**, 443.
Inhoffen, H. H., Buchler, J. W. and Jager, P. (1968). *Fortschr. Chem. org. NatStoffe* **26**, 284.
IUPAC Rules for Nomenclature, (1960). *J. Am. chem. Soc.* **82**, 5582.
Jackson, A. H. and Smith, K. M. (1973). *In* "The Total Synthesis of Natural Products" (J. Ap Simon, ed.), p. 144. Wiley, New York.
Jackson, A. H., Kenner, G. W., Smith, K. M., Aplin, R. T., Budzikiewicz, H. and Djerassi, C. (1965). *Tetrahedron* **21**, 2913.
Jacobs, E. E., Vatter, A. E. and Holt, A. S. (1953). *J. chem. Phys.* **21**, 2246.
Jacobs, E. E., Vatter, A. E. and Holt, A. S. (1954). *Archs Biochem. Biophys.* **53**, 228.
Jeffrey, S. W. (1972). *Biochim. biophys. Acta* **279**, 15.
Jones, O. T. G. (1963). *Biochem. J.* **88**, 182, 335.
Jones, O. T. G. (1964). *Biochem. J.* **91**, 572.
Jones, O. T. G. (1965). *Biochem. J.* **96**, 6P.
Jones, O. T. G. (1966). *Biochem. J.* **101**, 153.
Katz, E. and Wassink, E. C. (1939). *Enzymologia* **7**, 97.
Katz, J. J. and Ballschmitter, K. (1968). *Angew Chem. Int. Edn.* **7**, 286.
Katz, J. J. and Janson, T. R. (1973). *Ann. N.Y. Acad. Sci.* **206**, 579.
Katz, J. J., Closs, G. L., Pennington, F. C., Thomas, M. R. and Strain, H. H. (1963a). *J. Am. chem. Soc.* **85**, 3801.

Katz, J. J., Dougherty, R. C., Pennington, F. C., Strain, H. H. and Closs, G. L. (1963b). *J. Am. chem. Soc.* **85**, 4049.

Katz, J. J., Dougherty, R. C. and Boucher, L. J. (1966). *In* "The Chlorophylls", (L. P. Vernon and G. R. Seely, eds), Chap. 7. Academic Press, New York and London.

Katz, J. J., Strain, H. H., Lenssing, D. L. and Dougherty, R. C. (1968a). *J. Am. chem. Soc.* **90**, 784.

Katz, J. J., Norman, G. D., Svec, W. A. and Strain, H. H. (1968b). *J. Am. chem. Soc.* **90**, 6841.

Katz, J. J., Ballschmitter, K., Garcia-Morin, M., Strain, H. H. and Uphaus, R. A. (1968c). *Proc. natn. Acad. Sci. U.S.A.* **60**, 100.

Katz, J. J., Janson, T. R., Kostka, A. G., Uphaus, R. A. and Closs, G. L. (1972a). *J. Am. chem. Soc.* **94**, 2883.

Katz, J. J., Strain, H. H., Harkness, A. L., Studier, M. H., Svec, W. A., Janson, T. R. and Cope, B. T. (1972b). *J. Am. chem. Soc.* **94**, 7938.

Ke, B. and Miller, R. M. (1964). *Naturwissenschaften* **51**, 436.

Kenner, G. W., McCombie, S. W. and Smith, K. M. (1972). *Chem. Commun.* 844.

Kenner, G. W., McCombie, S. W. and Smith, K. M. (1973). *J. chem. Soc. Perkin I.* 2517.

Larsen, H. (1953). *K. norske Vidensk. Selsk. Forh.* 1-205.

Livingston, R., Watson, W. F. and McArdle, J. (1949). *J. Am. chem. Soc.* **71**, 1542.

McElroy, J. C., Feher, G. and Mauzerall, D. (1972). *Biochim. biophys. Acta* **153**, 248.

Manning, W. M. and Strain, H. H. (1943). *J. biol. Chem.*, **151**, 1.

Marks, G. S. (1969). "Heme and Chlorophyll". Van Nostrand, London.

Mathewson, J. W., Richards, W. R. and Rapaport, H. (1963a), *Biochem. biophys. Res. Commun.* **13**, 1.

Mathewson, J. W., Richards, W. R. and Rapaport, H. (1963b). *J. Am. chem. Soc.* **85**, 364.

Matwiyoff, N. A. and Burnham, B. F. (1973). *Ann. N.Y. Acad. Sci.* **206**, 365.

Michel-Wolwertz, M. R., Sironval, C. and Goedheer, J. C. (1965). *Biochim. biophys. Acta* **94**, 584.

Morley, H. V. and Holt, A. S. (1961). *Can. J. Chem.* **39**, 755.

Morley, H. V., Cooper, F. P. and Holt, A. S. (1959). *Chemy Ind.* **32**, 1018.

Noack, K. and Kiessling, W. (1930). *Hoppe-Seyler's Z. physiol. Chem.* **193**, 97.

Norris, J. R., Uphaus, R. A., Crespi, H. L. and Katz, J. J. (1971). *Proc. natn. Acad. Sci. U.S.A.* **68**, 625.

Norris, J. R., Druyan, M. F. and Katz, J. J. (1973). *J. Am. chem. Soc.* **95**, 1680.

Olsen, J. M. and Romano, C. A. (1962). *Biochim. biophys. Acta* **59**, 726.

Olsen, J. M. and Stanton, E. K. (1966). *In* "The Chlorophylls" (L. P. Vernon and G. R. Seely, eds), p. 381. Academic Press, New York and London.

Paine, J. B. and Dolphin, D. (1971). *J. Am. chem. Soc.* **93**, 4080.

Park, R. B. (1966). *In* "The Chlorophylls" (L. P. Vernon and G. R. Seely, eds), p. 283. Academic Press, New York and London.

Pelletier, P. J. and Caventou, J. B. (1818). *Ann. Chim. et Phys.* **9**, 194.

Pennington, F. C., Strain, H. H., Svec, W. A. and Katz, J. J. (1964), *J. Am. chem. Soc.* **86**, 1418.

Pennington, F. C., Boyd, S. D., Horton, H., Taylor, S. W., Wulf, D. G., Katz, J. J. and Strain, H. H. (1967). *J. Am. chem. Soc.* **89**, 3871.

R. C. Pettersen, (1969). *Acta cryst.* **B25**, 2527.
Purdie, J. W. and Holt, A. S. (1965). *Can. J. Chem.* **43**, 3347.
Prescott, G. W. (1962). "Algae of the Western Great Lakes Area". W. C. Brown Co., Duduque, Iowa.
Rabinowitch, E. I. (1945, 1951, 1956). "Photosynthesis and Related Processes", vol. 1, pp. 1-599; vol. 2, part 1, pp. 603-1208; vol. 2, part 2, pp. 1211-2088. Interscience, New York.
Rapaport, H. and Hamlow, H. P. (1961). *Biochem. biophys. Res. Commun.* **6**, 134.
Rothemund, P. (1944). *In* Medical Physics" (O. Glasser, ed.), pp. 154-180. Year Book Pub., Chicago.
Rudiger, W. (1968). *In* "Porphyrins and Related Compounds" (T. W. Goodwin, ed.), p. 121. Academic Press, London and New York.
Sagromsky, H. (1960). *Ber. dt. bot. Ges.* **73**, 3, 358.
Schaer, H., Svec, W. A., Cope, B. T., Studier, M. H., Scott, R. G. and Katz, J. J. (1974). *J. Am. chem. Soc.* **96**, 3714.
Schneider, E. (1934). *Hoppe-Seyler's Z. physiol. Chem.* **226**, 221.
Seely, G. R. (1966), *In* "The Chlorophylls" (L. P. Vernon and G. R. Seely, eds), pp. 67, 523. Academic Press, New York and London.
Sherma, J. (1970). *J. Chromat.* **52**, 177.
Shimizu, S. (1971). *J. Chromat.* **59**, 440.
Smith, J. H. C. and Benitez, A. (1954). *Yearb. Carneg. Instn Wash.* **53**, 168.
Smith, J. H. C. and Benitez, (1955). "Modern Methods of Plant Analysis" (K. Peach and M. V. Tracey, eds), vol. 4, pp. 142-196. Springer-Verlag, Berlin.
Smith, J. H. C. and Coomber, J. (1959). *Yearb. Carneg. Instn Wash.* **58**, 331.
Smith, J. H. C. and French, C. S. (1960). *Ann. Rev. Pl. Physiol.* **14**, 181.
Smith, J. H. C. and Young, V. M. K. (1956). *In* "Radiation Biology" (A. Hollaender, ed.), vol. 3, pp. 393-442. McGraw-Hill, New York.
Sorby, H. C. (1873). *Proc. R. Soc.* **21**, 442.
Stanier, R. Y. and Smith, J. H. C. (1959). *Yearb. Carneg. Instn Wash.* **58**, 336.
Stanier, R. Y. and Smith, J. H. C. (1960). *Biochim. biophys. Acta* **41**, 478.
Stern, A. and Wenderlein, H. (1936). *Z. phys. Chem. A.* **176**, 81.
Stokes, G. G. (1864). *Proc. R. Soc.* **13**, 144.
Stoll, A. and Wiedemann, E. (1933). *Helv. chim. Acta* **16**, 307.
Stoll, A. and E. Wiedemann, (1952). *Fortschr. der Chem. org. NatStoffe.* **2**, 538.
Stoll, M. S., Elder, G. H., Games, D. E., O'Hanlon, P. J., Millington, D. S. and Jackson, A. H. (1973). *Biochem. J.* **131**, 429.
Storm, C. B. and Teklu, Y. (1972). *J. Am. chem. Soc.* **94**, 1745.
Storm, C. B., Teklu, Y. and Sokoloski, E. A. (1973). *Ann. N.Y. Acad. Sci.* **206**, 631.
Strain, H. H. (1951). *In* "Manual of Phycology" (G. M. Smith, ed.), pp. 243-262. Chronica Botanica Co., Waltham, Mass.
Strain, H. H. (1954). *J. Agric. Fd Chem.* **2**, 1222.
Strain, H. H. and Manning, W. M. (1942). *J. biol. Chem.* **144**, 625.
Strain, H. H. and Svec, W. A. (1966). *In* "The Chlorophylls" (C. P. Vernon and G. R. Seely, eds), p. 22. Academic Press, New York.
Strain, H. H. and Svec, W. A. (1969). *Adv. Chromat.* **8**, 119.
Strain, H. W., Manning, W. M. and Hardin, G. (1943). *J. biol. Chem.* **148**, 655.

Strain, H. H., Cope, B. T., McDonald, G. N., Svec, W. A. and Katz, J. J. (1971). *Photochemistry* **10**, 1109.

Strouse, C. E., Kollman, V. A. and Maturijoff, N. A. (1972). *Biochem. biophys. Res. Commun.* **46**, 328.

Tswett, M. (1906). *Ber. dt. bot. Ges.* **24**, 384.

Tyray, E. (1944). *Justus Liebig's Annln Chem.* **556**, 171.

Vernon, L. P. and Ke, B. (1966). *In* "The Chlorophylls" (L. P. Vernon and G. R. Seely, eds), p. 569. Academic Press, New York and London.

Wasley, J. W. F., Scott, W. T. and Holt, A. S. (1970). *Can. J. Biochem.* **48**, 376.

Weller, A. (1954). *J. Am. chem.Soc.* **76**, 5819.

Wetherell, H. R. and Hendrickson, M. J. (1959). *J. org. Chem.* **24**, 710.

Willstätter, R. and Page, H. J. (1914). *Justus Liebig's Annln Chem.* **404**, 237.

Willstätter, R. and Stoll, A. (1913). "Untersuchungen über Chlorophyll". J. Springer, Berlin.

Witt, H. T. (1968). *Naturwissenschaften* **55**, 219.

Wolf, H. (1966). *Justus Liebig's Annln Chem.* **695**, 98.

Wolf, H. and Scheer, H. (1971). *Justus Liebig's Annln Chem.* **745**, 87.

Wolf, H. and Scheer, H. (1973). *Ann. N.Y. Acad. Sci.* **206**, 549.

Wolf, H., Brockmann, H., Biere, H. and Inhoffen, H. H. (1967). *Justus Liebig's Annln Chem.* **704**, 208.

Wolf, H., Brockmann, H. Jr., Richter, I., Mengler, C. D. and Inhoffen, H. H. (1968). *Justus Liebig's Annln Chem.* **718**, 162.

Wolf, H., Richter, I. and Inhoffen, H. H. (1969). *Justus Liebig's Annln Chem.* **725**, 177.

Woodward, R. B., Ayer, W. A., Beaton, J. M., Bickelhaupt, F., Bonnet, R., Buchschacher, P., Closs, G. L., Dutler, H., Hannah, J., Hauck, F. P., Ito, S., Langemann, A., LeGoff, E., Leimgruber, W., Lwowski, W., Sauer, J., Valenta, Z. and Volz, H. (1960). *J. Am. chem. Soc.* **82**, 3800.

Woodward, R. B. and Skaric, V. (1961). *J. Am. chem. Soc.* **83**, 4676.

Woodward, R. B. (1962). *Ind. Chem. Belge* **15**.

Zwolenik, J. J. (1970). U.S. Patent, 3 513 467, *Chem. Absr.* **73**, 32263m.

Chapter 2

Chlorophyll Biosynthesis*

LAWRENCE BOGORAD

The Biological Laboratories, Harvard University, Cambridge, Massachusetts, U.S.A.

I Introduction

The biosynthesis of porphyrins, including the chlorophylls, is discussed in this chapter. Evidence for the chemical structures of the chlorophylls and their natural distribution has been described in the preceding chapter; the role of the chlorophylls in photosynthesis is examined in a later chapter.

It seems almost unnecessary to call the attention of any reader of this book to the ubiquity of chlorophylls among photosynthetic

* Preparation supported in part by a research grant from the National Science Foundation.

organisms or to the distribution of iron-porphyrin-proteins, haemoproteins, among organisms.

Most of the intermediates in the biosynthesis of chlorophyll *a* have been identified; this knowledge restricts speculations on the nature of some "missing" ones. In common with much of the rest of current biochemistry, understanding of the enzymic mechanisms involved lags far behind what has quickly become descriptive biochemistry, but a great deal of work is being done on the question of modes of control of porphyrin biogenesis.

The discussion of chlorophyll biosynthesis will be prefaced by a brief outline of pyrrole porphyrin chemistry, as a supplement to Chapter 1, and will be followed by considerations of demonstrated as well as imagined mechanisms by which biogenesis is controlled

II Pyrroles, porphyrinogens, and porphyrins

Porphyrins are cyclic tetrapyrroles. The fundamental pyrrole heterocycle is shown in Fig. 1. It is apparent that each of the four carbon atoms in the pyrrole ring is capable of forming a bond with another carbon atom. The two ring carbon atoms adjacent to the nitrogen are designated α and the other two carbon atoms β.

Also shown in Fig. 1 is a porphyrinogen. This compound is formed when four pyrroles are joined to one another through their α carbon atoms by saturated carbon bridges. A porphyrinogen of the sort shown in Fig. 1 is readily oxidized by air, by iodine, or by other mild oxidizing agents to form a porphyrin. The porphyrin, unlike the porphyrinogen, is a planar molecule capable of chelating any number of metals. It is highly coloured. All these new properties are a consequence of the oxidation to this new state.

It is apparent that in the porphyrinogen and the porphyrin shown in Fig. 1 the two β carbons of each pyrrole ring can be substituted.

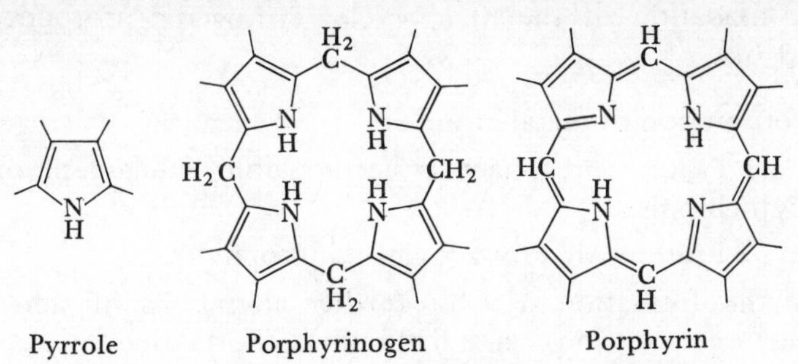

Pyrrole Porphyrinogen Porphyrin

FIG. 1. Pyrrole and cyclic tetrapyrrole skeletons.

Porphyrins can be distinguished from one another by the nature of the substituents on these eight β-carbon atoms as well as by the absence or presence of a metal held in the centre of this cyclic tetrapyrrole. Only magnesium and iron have so far been found in biologically active porphyrins.

Much of the discussion of porphyrin biosynthesis dwells on modifications of side chains. Falk (1964) and Phillips (1963) have reviewed chemical and physiochemical properties of biologically important porphyrins.

III The biosynthesis of porphyrins

The probable biosynthetic pathway for chlorophyll *a* (Chl *a*) is shown in Fig. 2. The reactions can be grouped as follows:

(1) The diversion of carbon and nitrogen atoms from the general metabolism of the cell into porphyrin formation via δ-aminolaevulinic acid (ALA).

(2) The formation of the precursor pyrrole: 2ALA → 1 porphobilinogen (PBG).

(3) The formation of a cyclic tetrapyrrole: 4PBG → uroporphyrinogen (Urogen) + 4 NH_3.

(4) The modifications of porphyrinogen side chains (i.e. substituents on β-carbon atoms):

(a) Decarboxylation of acetic side chains: Urogen → coproporphyrinogen (Coprogen) + 4 CO_2.

(b) Oxidative decarboxylation of propionic acid side chains on rings A and B: Coprogen → protoporphyrinogen (Protogen) + 2 CO_2 + 4H.

(5) The oxidation of the macrocycle: Protogen → protoporphyrin (Proto) + 6H.

(6) Incorporation of metal atoms:

(a) Fe + Proto → protohaem → haemoglobin, catalase, peroxidase, cytochromes.

(b) Mg + Proto → Mg Proto → Chl (chlorophyll).

(7) For the formation of Chl: further alterations of side chains including esterifications and reductions, formation of a cyclopantanone ring, and the reduction of one (Chl) or two pyrrole rings (bacteriochlorophyll).

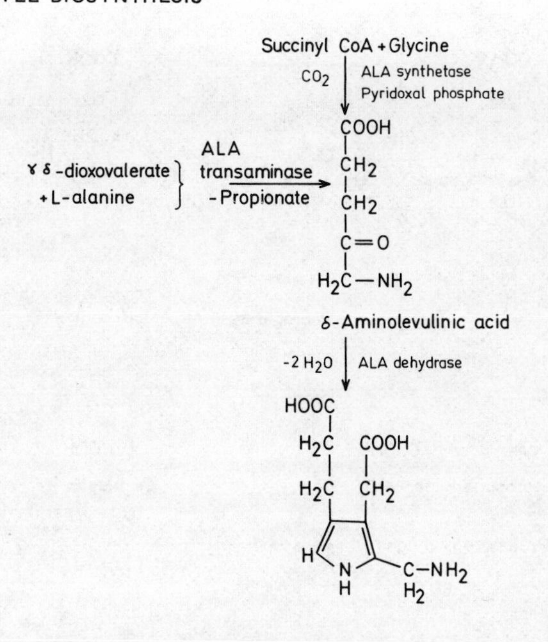
Succinyl CoA + Glycine
CO_2
ALA synthetase
Pyridoxal phosphate
COOH
CH_2
CH_2
C=O
H_2C-NH_2
γδ-dioxovalerate
+L-alanine
ALA
transaminase
-Propionate
δ-Aminolevulinic acid
$-2 H_2O$
ALA dehydrase
HOOC
H_2C
COOH
H_2C
CH_2
H
N
H
$C-NH_2$
H_2

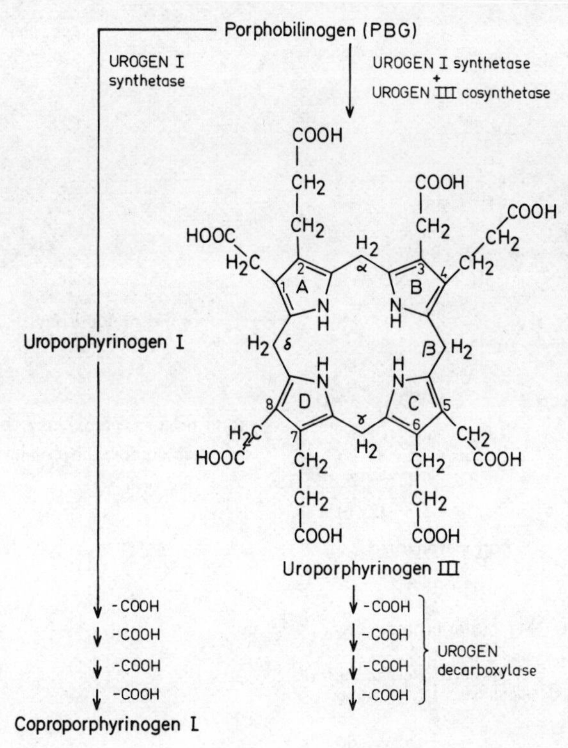
Porphobilinogen (PBG)
UROGEN I
synthetase
UROGEN I synthetase
+
UROGEN III cosynthetase
Uroporphyrinogen I
Uroporphyrinogen III
-COOH
-COOH
-COOH
-COOH
UROGEN
decarboxylase
Coproporphyrinogen I

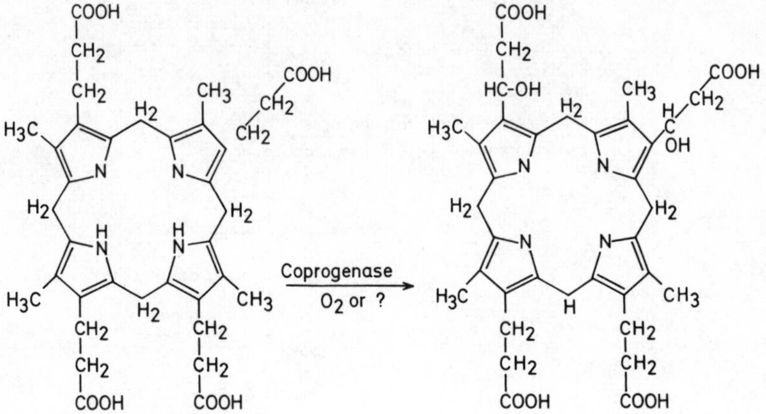

Coproporphyrinogen III

Coprogenase
O_2 or ?

2,4-bis-(β-hydroxy propionic) deutero porphyrinogen IX

Protoporphyrinogen IX

-6H

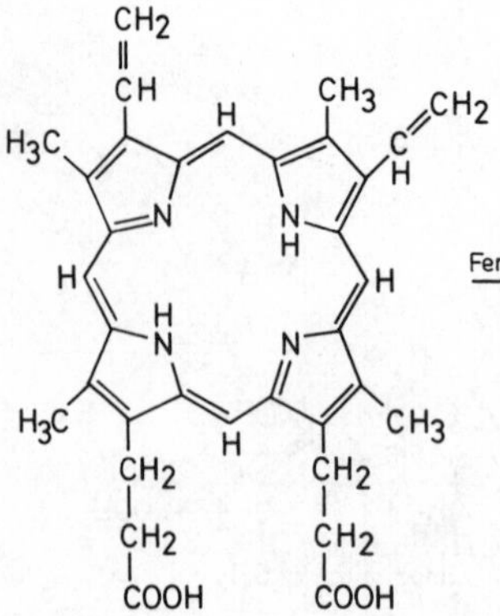

Protoporphyrin IX

Ferrochelatase
$+Fe^{++}$

Fe protoporphyrin IX

Cytochromes, peroxidase, catalase, hemoglobin, phycobilins

+Mg

Mg protoporphyrin IX

+ S-adenosyl methionine

Mg proto methyl transferase

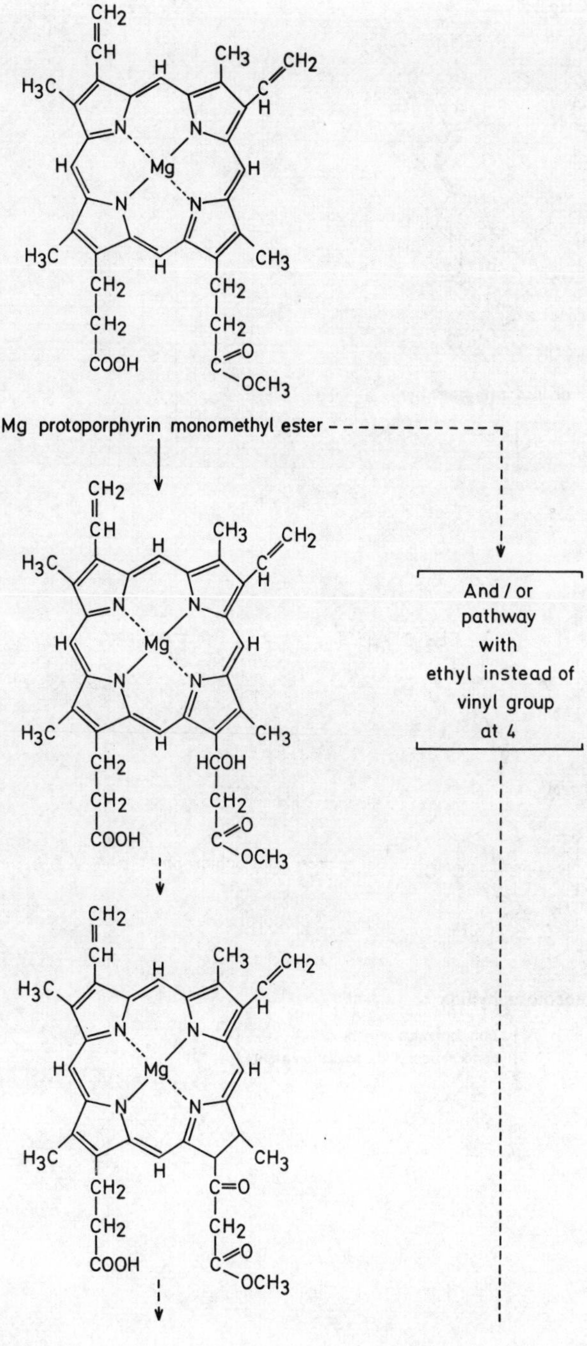

Mg protoporphyrin monomethyl ester

And / or
pathway
with
ethyl instead of
vinyl group
at 4

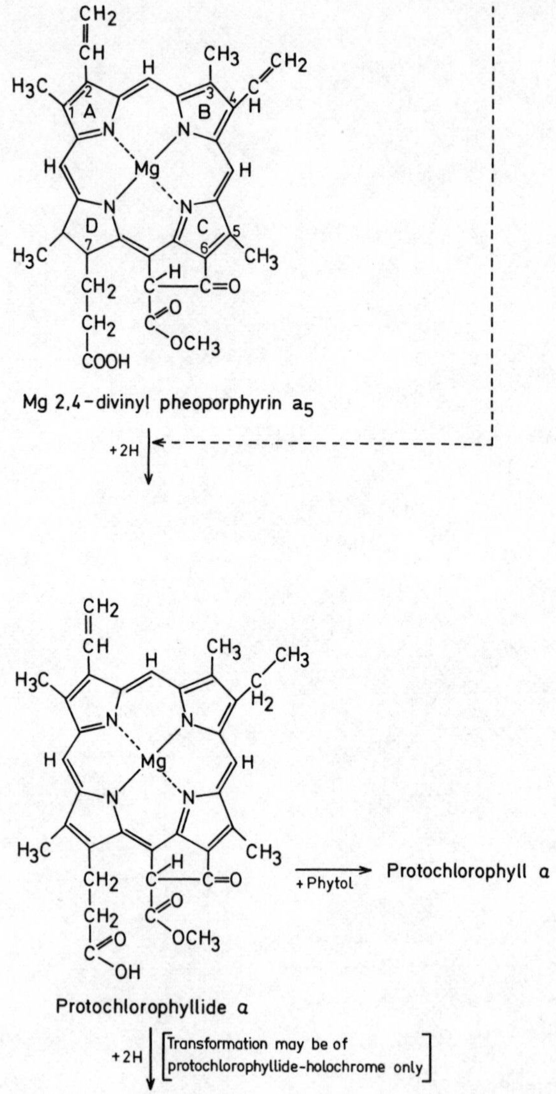

Mg 2,4-divinyl pheoporphyrin a_5

+2H

+Phytol → Protochlorophyll *a*

Protochlorophyllide *a*

+2H [Transformation may be of protochlorophyllide-holochrome only]

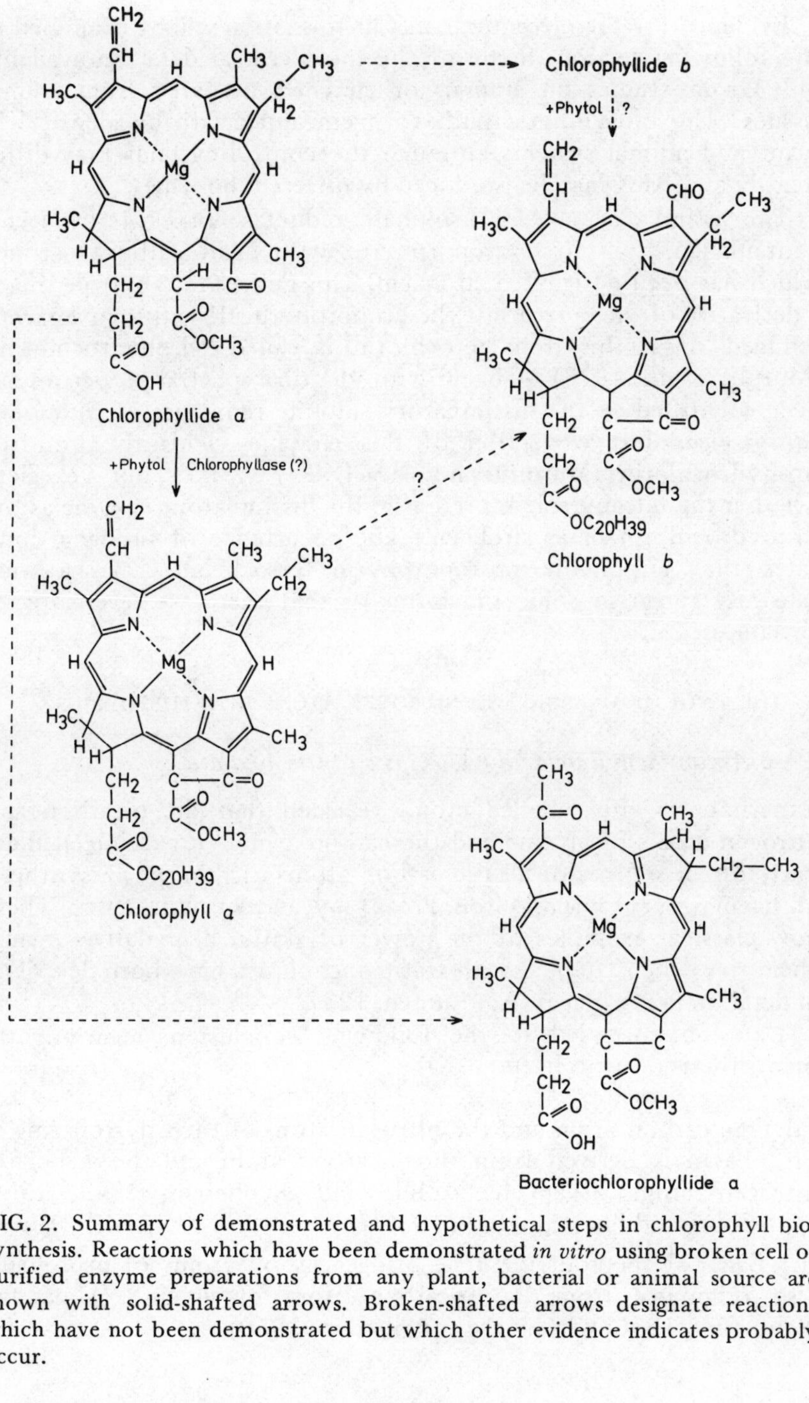

FIG. 2. Summary of demonstrated and hypothetical steps in chlorophyll biosynthesis. Reactions which have been demonstrated *in vitro* using broken cell or purified enzyme preparations from any plant, bacterial or animal source are shown with solid-shafted arrows. Broken-shafted arrows designate reactions which have not been demonstrated but which other evidence indicates probably occur.

Evidence for these reactions in Chl formation will be examined in the following section. Some of the most crucial data are available only from studies on animals or enzymes prepared from animal tissues. The biosynthetic path to haem appears to be identical in plant and animal systems although the control systems may differ sharply and ALA may be produced by different enzymes.

The assimilatory NADPH-sulphate reductase of *Escherichia coli* contains an iron-tetrahydroporphyrin with eight carboxyl groups which has been designated sirohaem. The Fe-free tetrapyrrole ring is a derivative of the octacarboxylic uroporphyrin III with two adjacent reduced rings; this isobacterochlorin is called "sirohydrochlorin" (Murphy *et al.*, 1973). A haem with identical spectral properties has been identified in the dissimilatory sulphite reductase of *Desulphovibrio gigas* but extraction of this enzyme yields the iron-free sirohydrochlorin (Murphy and Siegel, 1973). It is not yet clear whether the tetrapyrrole is present in the dissimilatory enzyme as the sirohydrochlorin or as sirohaem. The occurrence of sirohaem indicates that either the protoporphyrin ferrochelatase has a wide substrate range in some organisms or that there are several ferrochelatases.

A THE PATH OF CARBON AND NITROGEN ATOMS INTO PORPHYRINS

1 δ-Aminolaevulinic acid (ALA) is an intermediate

Experiments with labelled atoms revealed that the α-carbon and nitrogen atoms of glycine and the carbon atoms of succinic acid are fairly direct sources of all the carbon atoms used in the biosynthesis of haem (= protohaem = iron Proto) by avian erythrocytes. These now classical examples of the power of skilful degradative organic chemistry when coupled with tracer techniques have been described in detail in many places (e.g. Kamen, 1957).

Data obtained led to the following conclusions regarding the biosynthesis of protohaem:

(a) One carbon atom and the nitrogen atom of each pyrrole ring of protohaem is derived from the α-carbon atom and the associated nitrogen atom of glycine (Muir and Neuberger, 1949, 1950; Wittenberg and Shemin, 1949, 1950).

(b) Each of the four methene bridge carbon atoms of protohaem also originates from the α-carbon atom of glycine (Muir and Neuberger, 1950; Wittenberg and Shemin, 1950).

(c) The remaining carbon atoms of the pyrrole rings as well as of the side chains are derived from acetate (Shemin and Wittenberg, 1951) or, more directly, from succinate (Shemin and Kumin, 1952). This observation led Shemin to suggest that the carbon atoms of succinate might enter porphyrin metabolism as succinyl coenzyme A.

(d) Finally, contrary to some ideas current at that time (e.g. Lemberg and Legge, 1949), strong support was provided for the growing conviction that all four pyrrole rings of protohaem arise from a common precursor pyrrole.

Investigations of the course of utilization of succinate and the α-carbon plus nitrogen of glycine for protohaem biosynthesis were brought to a logical and highly satisfactory conclusion in Shemin's laboratory (Shemin and Russell, 1953). It was shown that δ-aminolaevulinic acid (ALA) can serve as the sole source of all the carbon and nitrogen atoms for the synthesis of protohaem by suspensions of avian erythrocytes (Fig. 2). This discovery, which was quickly confirmed by Neuberger and Scott (1953) immediately led to a search for enzymes which on the one hand catalyse the formation of ALA and on the other hand mediate the utilization of this compound for porphyrin formation.

2 *Enzymes which synthesize ALA in animals and bacteria*

a. ALA synthetase. The enzyme ALA synthetase catalyses the synthesis of ALA from succinyl coenzyme A and glycine. It was first prepared as freeze-dried particles from chicken red cells (Gibson *et al.*, 1958) and in higher states of purity as fractionated cell-free extracts of *Rhodopseudomonas spheroides.* Kikuchi *et al.* (1958) obtained an eightyfold and Burnham (1962) a twentyfold purified preparation of ALA synthetase from this bacterium. The ALA synthetase preparations of Kikuchi *et al.* (1958) and Gibson *et al.* (1958) required pyridoxal phosphate as a cofactor for the condensation of glycine and succinyl coenzyme A, the substrates for ALA synthesis. Neuberger (1961) has discussed a possible mechanism for the condensation.

The expected product of the condensation of glycine and succinyl coenzyme A is δ-amino-β-ketoadipic acid (Shemin and Russell, 1953) but this compound has not been detected in reaction mixtures, probably because it is rapidly and spontaneously decarboxylated to δ-aminolaevulinic acid (Laver *et al.*, 1959). The half-life of δ-amino-β-ketoadipic acid at pH 7 in aqueous solution is estimated to be less than 1 min. The ALA synthetase reaction is inhibited by substances which complex with the aldehydic group of pyridoxal phosphate (e.g. L-cysteine, penicillamine, and cyanide) and also by aminomalonate (Gibson *et al.*, 1958, 1961).

At the time of the discovery of the role of δ-aminolaevulinic acid in porphyrin metabolism Shemin and Russell (1953) proposed another possible metabolic role for the aminoketone. A modification of their succinate–glycine cycle, for which there is some evidence from studies with intact rats (Nemeth *et al.*, 1957) is shown in Fig. 3. Reactions involving aminoketones other than ALA have also been studied. Mitochondria of guinea-pig liver contain an enzyme which seems to catalyse preferentially the formation of aminoketones other than ALA: it catalyses the condensation of glycine with acetyl coenzyme A, malonyl coenzyme A, or propionyl coenzyme A; however, only traces of ALA were formed with succinyl coenzyme A (Urata and Granick, 1963). The reaction mechanism has been considered by Shemin and Kikuchi (1958) and by Neuberger (1961), who has also discussed the apparent lack of specificity of aminoketone synthetase systems examined in his laboratory. Gibson *et al.* (1961) have suggested the possibility that Chl synthesis might be controlled by competition for glycine and for the aminoketone synthetase capacity of the organism by acyl coenzyme As other than succinyl coenzyme A. If plant as well as animal cells possess specific aminoketone synthetases of the kinds described by Urata and Granick the value of the suggestion of Gibson *et al.* would be partly reduced and even now, as they point out, there is no evidence to support it.

ALA synthetases have been prepared in highly purified forms from *Rhodopseudomonas spheroides* (Tuboi *et al.*, 1970a; Warnick and Burnham, 1971) from rabbit reticulocytes (Aoki *et al.*, 1971) and from *Micrococcus denitrificans* (Tait, 1973).

The rabbit reticulocyte enzyme has been purified 4400-fold; it is homogeneous on polyacrylamide gel electrophoresis. The isoelectric point of this enzyme is 5·9. Its pH optimum is at 7·6; the optima for activity of enzymes from the three sources ranged from pH 7·0 to 7·8. The K_ms for these three ALA synthetases are: for succinyl CoA about 5–60 μM; for glycine about 10 mM; and for pyridoxal phosphate the range is 1–11 μM.

The molecular weights reported for ALA synthetases range from about 50 000 to 600 000 daltons. The *R. spheroides* enzyme is estimated to be about 57 000–60 000 daltons (Tuboi *et al.*, 1970a; Warnick and Burnham, 1971) based on SDS polyacrylamide gel

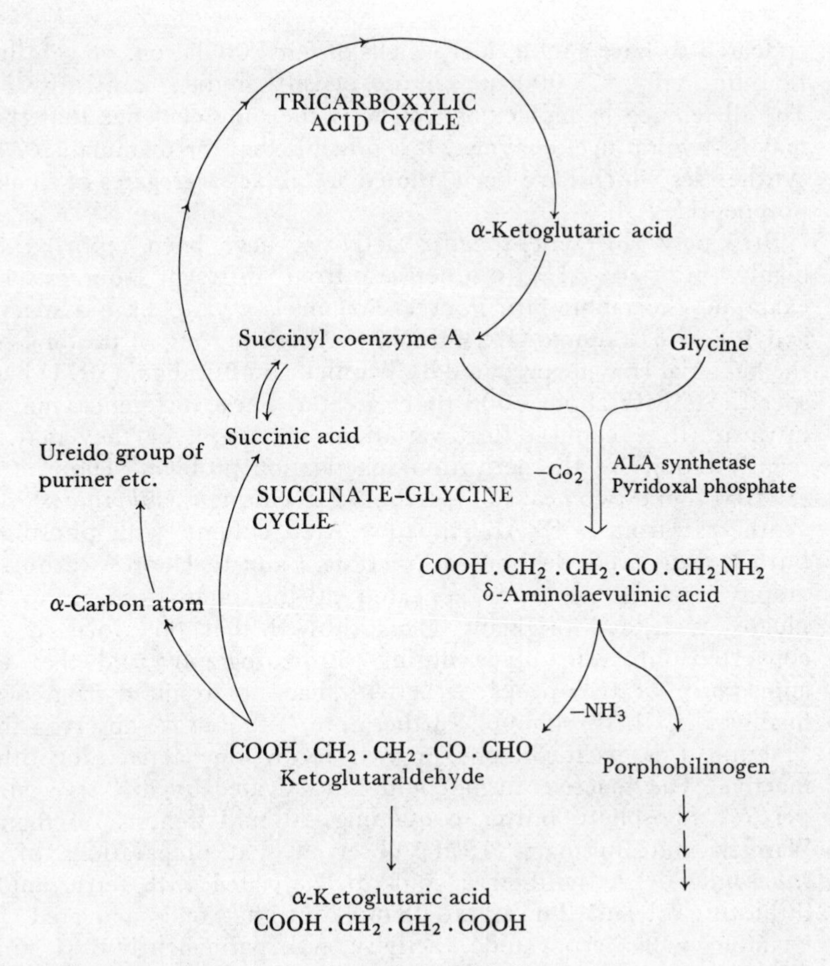

FIG. 3. The succinate-glycine cycle and its relationship to the tricarboxylic acid cycle (Shemin and Russell, 1953).

electrophoresis and on gel filtration. *Micrococcus denitrificans* ALA synthetase is in the same range; it has a molecular weight of about 68 000 daltons (Tait, 1973).

Much higher values have been published for rabbit reticulocyte and rat liver ALA synthetases. According to Aoki *et al.* (1971) rabbit reticulocyte ALA synthetase is 200 000 daltons based on elution from Sephadex. Hayashi *et al.* (1970) have described a rat liver mitochondrial enzyme with a molecular weight of about 113 000–115 000 daltons determined by sucrose density gradient centrifugation as well as by gel filtration. A soluble enzyme from the same source appeared to have a molecular weight of 600 000 daltons on gel filtration but only 178 000 on sucrose density gradient centrifugation. The difference in molecular weight is the only evidence that there may be two rat liver enzymes. It is possible that the mammalian ALA synthetases which have been studied are in fact aggregates of smaller polypeptides.

Strikingly different specific activities have been reported for highly purified ALA synthetases from different sources. For example, the rabbit reticulocyte enzyme is said to have a specific activity of 312 nmol ALA formed per hour per mg of protein, but the bacterial enzyme prepared by Warnick and Burnham (1971) has a specific activity about 4000 times greater. These differences may be intrinsic, they could reflect variations in purity, or they may be manifestations of the activation–inactivation problem. Tuboi *et al.* (1970a) found two peaks of activity when eluting *R. spheroides* ALA synthetase from a DEAE-Sephadex A-25 column with phosphate buffer; these were designated Fractions I and II. Upon rechromatography each active form appeared at the same position in the eluate as it had originally. This showed that one form is not converted into the other during chromatography and that the appearance of two peaks of activity had not resulted from overloading the DEAE column. Furthermore, Tuboi *et al.* observed that Fraction I is itself present in two forms—one active, the other inactive. The inactive form could be activated by dialysis against pH 7·4 phosphate buffer containing 10 mM 2-mercaptoethanol. Warnick and Burnham (1971) observed that preparations of *R. spheroides* ALA synthetase could be activated with ferricyanide, dithiothreitol and 2-mercaptoethanol but that once activated, for example with ferricyanide, activity was permanently lost more rapidly on storage. It is possible that the great differences in specific activities of ALA synthetases can be accounted for by variations in the amount of inactive enzyme present.

(*i*) *Inhibitors*

Natural inhibitors and inhibition–activation systems in R. spheroides. As already noted, ALA synthetase from *R. spheroides* has been separated into two fractions by Kikuchi and his co-workers (Tuboi *et al.*, 1970a). Fraction I, one of the two forms separated on DEAE-Sephadex, occurs in inactive and active forms; the inactive form has been purified about 100-fold.

The inactive Fraction I enzyme was not activated by chromatography on Dowex 1. This treatment removes inhibitors found in ALA synthetases prepared from some other sources. However, Tuboi and Hayasaka (1971) could activate the enzyme by incubation with rat liver mitochondria or a protein fraction from acetone-dried rat livers.

Based on filtration on Sephadex G-100, the inactive form was judged to have a molecular weight of 100 000 while the active form appeared to be 80 000 daltons.

Tuboi and Hayasaka (1972a,b) later found that the inactive form of ALA synthetase Fraction I in relatively crude preparations could be activated by some naturally occurring sulphur-containing compounds, by L-cystine or by any one of a group of other sulphydryl-containing compounds. Half maximal activation with L-cystine occurred at 8×10^{-6} M. However, the 100-fold purified inactive enzyme could not be activated by sulphydryl-containing compounds alone. Activation also required a protein different from ALA synthetase. The "activating enzyme" was purified about seventyfold by Tuboi and Hayasaka (1972b). Tuboi and Hayasaka (1972a) have shown that ^{14}C from ^{14}C-cystine is not incorporated into the enzyme if this disulphide is used as an activator.

On the basis of filtration on Sephadex G-100 the inactive form was judged to have a molecular weight of 100 000 while the active form had an apparent molecular weight of 80 000. Tuboi and Hayasaka (1971) thought this difference could not be accounted for by association and dissociation of relatively large polypeptides but might be accounted for by a change in shape of the molecule or by the removal of a segment of a polypeptide.

Davies *et al.* (1973) and Neuberger *et al.* (1973a,b) have also investigated the control of the activity of ALA synthetase in *R. spheroides.* They suggest that a trisulphide of cystine or some other sulphur-containing amino acid may serve as or with the activator.

Haem. Haemin has long been known to be a potent inhibitor of ALA synthetase (Burnham and Lascelles, 1963). Porra and Irving (1972) found that incubation of haemin with imidazole before addition of the iron porphyrin to a crude extract of *R. spheroides* diminished the inhibitory effect of the iron porphyrin. The lower effectiveness of the bis-imidazole chelate than free haemin suggested to Porra and Irving that haemin may inhibit ALA synthetase by forming a co-ordination complex with the enzyme through the iron of haem. Other possible modes of interaction of haem and ALA synthetase are not explored.

(ii) How many ALA synthetases in the cell?

Hayashi *et al.* (1970) have reported, as already noted, that there are two forms of ALA synthetase in rat liver; a smaller mitochondrial form and a larger soluble form. They found it "tempting to speculate that ALA synthetase in the soluble fraction is modified in some way to a smaller protein before or after being transferred to the mitochondria". Kurashima *et al.* (1970) found that after injection of haemin into male Wistar rats the ALA synthetase level in the mitochondria dropped and the level in the extramitochondrial fraction increased. The result was no net difference as compared with the controls. This does not seem to support the suggestion made by Hayashi *et al.* (1970) but does not eliminate the possibility that there is more than one ALA synthetase in liver cells. Tuboi *et al.* (1970b) found that ALA synthetase in both Fractions I and II are repressed in *R. spheroides* under high partial pressures of oxygen. However, activity in Fraction I increases after the oxygen tension in the environment of the cells is reduced; activity in Fraction II is enhanced only if *R. spheroides* cells are illuminated as well. Tuboi *et al.* believed that Fractions I and II contain different ALA synthetases and they advanced the idea that the formation or activities of these two enzymes are controlled independently. On the other hand Lascelles and Altschuler (1969) had concluded earlier from the analysis of some ALA-requiring mutant strains of this bacterium that there is only one ALA synthetase in *R. spheroides.* Recent successes in purifying the *R. spheroides* ALA synthetase may permit a clear resolution of this question.

b. ALA transaminase. ALA can also be formed by the amination of α-oxogluturate (i.e. γ,δ-dioxovalerate = DOVA). ALA transaminase catalyses the transfer of an amino group, preferably from L-alanine, to DOVA. The equilibrium for the reaction catalysed by the *R. spheroides* enzyme is toward the formation of ALA (Bagdasarian, 1958; Kowalski *et al.*, 1959; Gibson *et al.*, 1961; Neuberger and Turner, 1963; Gassman *et al.*, 1968).

(3) ALA synthesis in algae and higher green plants

a. ALA transaminase. This enzyme has been demonstrated in broken cell preparations of *Chlorella* by Gassman *et al.* (1968b). Indications that the normal path for ALA production in green leaves may include an ALA transaminase-catalysed step has been obtained by Beale and Castelfranco (1974b). Radioactive amino acids were supplied to greening cucumber cotyledons and red kidney bean leaves in which utilization of ALA was partly blocked by the administration of laevulinic acid, a competitive inhibitor of ALA dehydrase. The newly formed ALA was collected and its radioactivity determined. The analysis of the radioactivity in ALA using the techniques of Beale and Castelfranco (*vide infra*) gives much more direct and easier-to-interpret results than would analyses of the chlorophylls produced from the labelled amino acids. The most effective labelled precursors of ALA were glutamate, glutamine and α-keto glutarate. [^{14}C]Glycine and succinate were relatively poor precursors and, most strikingly, the carboxyl and α-carbon atoms of glycine were incorporated into ALA about equally. The latter would not be expected if the ALA had been produced only via the reaction catalysed by ALA synthetase. These experiments are discussed in the next section.

b. ALA synthetase. Ramaswamy and Nair (1973) have reported detecting ALA synthetase in preparations from green potato peelings and Wider de Xifra *et al.* (1971) have reported detecting this enzyme in preparations of soyabean callus. *Chorella* incorporates the α-carbon of glycine into compounds which copurify with ALA through ion exchange, paper and thin-layer chromatography (Gassman, 1967). This could also occur in other organisms. Because of previously experienced difficulties, the identification of the product as ALA must be especially rigorous. The published reports of the work with the potato peel enzyme and the soyabean callus enzyme do not show beyond all doubt that ALA was produced. In addition, the potato peel ALA synthetase is reported to work quite well in the presence of succinate but neither succinyl CoA nor succinyl thiokinase is required. The reaction mixtures for assaying the soyabean callus enzyme include all the components used to assay ALA synthetase from other sources but there is no experimental evidence presented that any of these components are needed for the reaction which is being measured. The product of the enzymic reaction purporting to demonstrate the synthesis of ALA should be convertible into porphyrins by an enzyme system. Beale and Castelfranco (1974b) have described a gas–liquid chromatographic separation procedure for assaying the methylated pyrrole which can be derived from ALA. This technique may permit more positive identification of small amounts of ALA.

Attempts to analyse the path of synthesis of chlorophylls from amino acids *in vivo* by feeding labelled precursors were first reported by della Rossa *et al.* (1953). They found that both the α- and the carboxyl-carbons of glycine were incorporated into Chl by *Chlorella.*

If ALA for chlorophyll synthesis were formed exclusively via the ALA synthetase reaction, only the α-carbon should be incorporated; CO_2 liberated from the carboxyl carbon could have been taken up photosynthetically and incorporated into Chl indirectly. Experiments of this kind are also difficult because of great technical problems in bringing Chls and their derivatives to radiopurity (Roberts and Perkins, 1966). It is comparatively easy to determine whether labelled carbon atoms are in the porphyrin ring system or in the esterifying methyl or phytyl groups and this is the first question which should be resolved.

Wellburn and Wellburn (1971) reported that radioactivity from glycine 2-^{14}C was incorporated into Chl by chloroplast preparations which were also capable of incorporating ALA into Chl. Unfortunately, the radioactivity was not shown to reside in the tetrapyrrole ring nor was it shown whether the carboxyl- as well as the methyl-carbon of glycine, was incorporated into the product.

Beale and Castelfranco (1974a,b) have designed experiments which eliminate or at least reduce some of the experimental problems. First, they established conditions to measure incorporation of ^{14}C from precursors into ALA instead of Chl by feeding laevulinic acid. Secondly, they have developed a gas–liquid chromatographic procedure for purification of the methyl pyrrole which is easily produced from ALA. This should reduce the chance of carrying along radio-impurities.

Laevulinic acid is a specific competitive inhibitor of ALA dehydrase, the enzyme responsible for the consumption of ALA in the synthesis of porphyrins (Nandi and Shemin, 1968b; Beale, 1970). Chl production is partly arrested in cucumber cotyledons treated with 100 mM laevulinic acid but these cotyledons accumulate ALA in an amount corresponding to the decrease in Chl production when compared with control plants. This quantitative relationship suggests that there are no other major disturbances in the metabolism of the cotyledons during the experimental period as a result of treatment with laevulinic acid.

Glutamate labelled in carbon 1 or at carbons 3 and 4, glycine labelled at carbon 1 or at carbon 2, and a variety of other labelled organic acids were administered to greening leaves of barley or red kidney bean or to greening cucumber cotyledons. The radioactivity in the ALA which accumulated was measured. Of the compounds administered, some carbons from glutamate, glutamine, and α-ketoglutarate were used best; glycine and succinate were relatively poor precursors. Most significantly, both carbons of glycine were incorporated into ALA and the carboxyl carbon of glutamate was incorporated about as well as carbons 3 and 4 of this compound. If the ALA were formed by ALA synthetase, as already pointed out, the carboxyl carbon of glycine should not be in the ALA. Even if the glycine were metabolized, and the carboxyl carbon were taken up photosynthetically the radioactivity from the carboxyl carbon should be very much diluted compared with the carboxyl carbon; this was not the case.

Some carbons of α-keto glutarate or glutamate might find their way into ALA via succinyl CoA. This conversion occurs through the loss of the carboxyl carbon 1; the internal carbons of glutamate or α-keto glutarate would remain in succinyl CoA. The data obtained by Beale and Castelfranco (1974b) suggest that the carbon atoms from these dicarboxylic acids are not introduced into ALA via the ALA synthetase reaction for the carboxyl- as well as the internal carbons are found in ALA. As a control, experiments of the same type were performed on turkey blood. The ratio of glycine carbons, C-2/C-1, in ALA was 8 in the case of turkey blood, the direction expected if the ALA were made by the ALA synthetase reaction, but was only 1 for cucumber cotyledons, and 2·4 for greening bean or barley leaves. Incorporation of the carboxyl-carbon glycine into ALA was not entirely excluded even in the turkey blood but the ratio was strikingly different. No radioactivity from C-1 of glutamate was incorporated into turkey blood ALA, again in line with expectation if synthesis were by ALA synthetase, whereas it was incorporated, although not to a great extent, into the ALA formed by the three plants.

Beale and Castelfranco concluded that their experiments confirmed the role of ALA synthetase in the biosynthesis of avian haem but raised serious doubts about the importance of this enzyme in the biosynthesis of Chl. Their data are not inconsistent, however, with the occurrence of the ALA transaminase step or the cyclization of glutamate in the biosynthetic sequence.

Tracer experiments of the sort carried out by Beale and Castelfranco can be misleading if there are differences in the uptake of the various compounds supplied to the organism. Beale and Castelfranco have studied the formation of $^{14}CO_2$ by some of the tissues they have studied from the labelled organic and amino acids supplied as possible precursors. All of the precursors were metabolized, indicating that they had been incorporated into the plants to some extent. Another problem is the possible existence of different sized pools of potential substrates. Neither of these two possible difficulties is significant in experiments where the ratio of incorporation of different carbon atoms from a single precursor is being determined. Finally, although these experiments may not be entirely free of some possible errors in the analysis of radioactivity of the product, more powerful methods than employed previously have been used to isolate a derivative of ALA and thus to ensure that the radioactivity being measured is in ALA.

(4) Summary

ALA synthetase has been purified and characterized from *Rhodopseudomonas spheriodes, Micrococcus denitrificans* and some animal tissues. The bacterial enzymes have molecular masses of 57 000-68 000 daltons. The size of rat liver and rabbit reticulocyte ALA synthetases are not so well established. A number of natural inhibitors of ALA synthetase have been studied. Some are described in this section, but a more extensive discussion of this problem is found in Section IVA.

Although there are several reports that ALA synthetase plays a role in the formation of chlorophyll in green plants (Miller and Teng, 1967; Wellburn and Wellburn, 1971; Wider de Xifra *et al.*, 1971; Ramaswamy and Nair, 1973), none of these provides unequivocal evidence. It has sometimes been assumed (e.g. Schneider, 1973) that the accumulation of ALA after the administration of laevulinic acid demonstrates the involvement of ALA synthetase. Such a conclusion may be proved correct ultimately, but at present it is uncertain.

The only other known biosynthetic route to ALA is catalysed by ALA transaminase. This enzyme catalyses the transfer of an amino group from an amino acid (e.g. L-alanine or L-glutamate) to γ,δ-dioxylvalerate. Among photosynthetic organisms, ALA transaminase has been demonstrated in extracts of *R. spheroides* (Gibson *et al.*, 1961) and *Chlorella vulgaris* (Gassman *et al.*, 1968b). Recent experiments by Beale and Castelfranco (1974a,b) raise additional doubts that chlorophyll synthesis proceeds via ALA synthetase in a number of higher green plants; their data can be taken to support the view that ALA is synthesized by ALA transaminase or some as yet undiscovered enzyme in these organisms.

B PYRROLE FORMATION, THE BIOSYNTHESIS OF PORPHOBILINOGEN

The isolation and crystallization of porphobilinogen (PBG) from the urine of acute porphyrin sufferers by Westall (1952), the characterization of this compound as a monopyrrole (Cookson and Rimington,

1953; Granick and Bogorad, 1953), and the demonstration of its utilization in the enzymic synthesis of porphyrins (Bogorad and Granick, 1953b; Falk *et al.*, 1953) coincided with the experiments by Shemin and Russell (1953) which established the role of ALA in the biogenesis of protohaem.

ALA dehydrase catalyses the condensation of two molecules of ALA to form one of PBG; two molecules of water are eliminated. This enzyme was first purified to various extents from avian erythrocytes (Granick, 1954; Schmid and Shemin, 1955), rabbit reticulocytes (Granick and Mauzerall, 1958) and liver from various animals (Gibson *et al.*, 1955; Iodice *et al.*, 1958). Granick (1954) demonstrated the conversion of ALA into porphyrin by extracts of *Chlorella* and of spinach.

The ALA dehydrase from *R. spheroides* has been studied more intensively than any other. The subunit molecular weight as determined on SDS-polyacrylamide gel electrophoresis is about 40 000. It normally occurs as aggregates of about 240 000 daltons, as judged by gel filtration. Similar, although not always identical, values have been obtained for ALA dehydrase from livers of mice, rats and cows as well as from *R. capsulata* (Nandi *et al.*, 1968; Doyle, 1971; Gurba *et al.*, 1972; Nandi and Shemin, 1973).

Burnham and Lascelles (1963) found that the activity of ALA dehydrase is dependent upon sulphydryl-containing compounds and that potassium ions further activate the enzyme. Shemin and his co-workers reinvestigated the effects of potassium ions and observed that they also promote the aggregation of the enzyme (Nandi *et al.*, 1968; Nandi and Shemin, 1968a,b; Shemin 1968; van Heyningen and Shemin, 1971). In the presence of potassium ions (50 mM potassium was used in these experiments) aggregates of about 480 000 and 720 000 daltons were observed in addition to a 240 000 dalton species. The latter can be reversibly dissociated into 120 000 dalton units by 1M urea which also inhibits the activity of the enzyme. In addition, potassium ions at concentrations of 10–30 mM bring about an up to twentyfold increase in the activity of the enzyme. Sodium and magnesium ions also activate the enzyme to about 50% and 80%, respectively, as much as potassium; but neither sodium nor magnesium ions influence the aggregation of the enzyme. The higher states of aggregation are apparently not required for enhanced activity of *R. spheroides* ALA dehydrase. Enzymes from some other sources require no activation by metals.

Rhodopseudomonas spheroides ALA dehydrase is inhibited by haemin (Burnham and Lascelles, 1963) and by Proto (Nandi *et al.*,

1968). The enzyme is completely inhibited by 3·3 mM haemin when incubated with dithiothreitol-activated enzyme before the addition of substrate.

ALA dehydrases from mouse, rat, and cow livers require no metal activation, but they are inhibited very strongly by metal chelators such as ethylenediamine tetraacetate and 8-quinolinol. Thus, these enzymes differ from the *R. spheroides* dehydrase in having no requirement for metal activation and in being strongly inhibited by metal chelators.

The ALA dehydrase from *R. capusulata* has still another set of characteristics (Nandi and Shemin, 1973). It is about the same aggregate weight as the enzyme from *R. spheroides* but differs from the latter in two ways. It does not require metal ions for activation at low substrate concentrations and it is not inhibited by haemin or protoporphyrin. It resembles *R. spheroides* dehydrase but differs from the various liver enzymes in being insensitive to metal chelators (Burnham and Lascelles, 1963; Nandi and Shemin, 1973).

Even the limited amount of work which has been done with ALA dehydrase of higher plants shows it to be a widely distributed, and relatively easily detected enzyme. ALA dehydrase has been purified about eightyfold from tobacco leaves by Shetty and Miller (1969). Less extensive purification from soyabean callus tissue has been reported by Tigier *et al.* (1968, 1970) and from wheat leaves by Nandi and Waygood (1967). Some properties of an ALA dehydrase in *Phaseolus vulgaris* have been studied by Slutiers-Scholten *et al.* (1973). The enzyme increases in activity during greening of bean leaves and tissue cultures (Steer and Gibbs, 1969; Stobart and Thomas, 1968).

The tobacco leaf enzyme has an optimum at about pH 7·4, considerably lower than that of the photosynthetic bacterial enzymes. Like the liver enzymes, EDTA is inhibitory and potassium ions are not required for activation (Nandi and Waygood, 1967; Shetty and Miller, 1969). However, ferrous iron is inhibitory.

Shemin and his co-workers have obtained evidence for the formation of a Schiff base between an amino group in the active site of the enzyme and ALA (Nandi and Shemin, 1968b; Shemin, 1968).

ALA dehydrase is competitively inhibited by laevulinic acid (Nandi and Shemin, 1968b). Laevulinic acid and its ethyl ester, like ALA, form Schiff bases with the enzyme. The enzyme can catalyse the condensation of laevulinic acid with ALA. Based on these observations Shemin has proposed a detailed outline of a mechanism for PBG synthesis from two molecules of ALA (Shemin, 1968). The inhibition of ALA dehydrase in plants by laevulinic acid has been very successfully exploited by Beale (e.g. 1971) in studies on Chl formation described elsewhere in this discussion.

C TETRAPYRROLE FORMATION

Uroporphyrinogens (Urogens) are cyclic tetrapyrroles with one acetic acid and one propionic acid residue on each of the pyrrole rings. Four isomers are possible. Only isomers I and III, or their products, are found in biological systems.

Urogen III is the first cyclic tetrapyrrole on the biosynthetic path to chlorophylls, haems, and other biologically useful porphyrins. It is produced from PBG and is transformed in several steps into protoporphyrin IX, the porphyrin moiety of haemoglobin, cytochrome *b*, catalase and peroxidases, and a precursor of chlorophyll by broken-cell preparations of *Chlorella* (Bogorad and Granick, 1953b; Bogorad, 1955, 1957a, 1958c) and by haemolysed red cells (Neve *et al.*, 1956; Mauzerall and Granick, 1958). Uroporphyrin I (Uro I) and its derivatives are best known as constituents of the urine of humans or other animals afflicted with certain hereditary or acquired porphyrias (for discussion and review of porphyria diseases see, e.g. Schmid, 1960; Bogorad, 1963).

The head-to-tail condensation of four molecules of PBG would yield the tetrapyrrole Urogen I. Urogen III differs from isomer I in the arrangement of the acetic and propionic acid side chains on ring D (Fig. 4).

In the next steps in the biosynthesis of porphyrins the acetic acid side chains of Urogen III are decarboxylated to methyl groups. Later, during the formation of protoporphyrin IX, the propionic acid residues on rings A and B are converted into vinyl groups. If these subsequent alterations are taken into account, the close relationship between Urogen III and Proto IX are apparent. A comparison of the positions of the propionic and acetic side chains on ring D in Urogen I and Urogen III and Proto IX demonstrates that Proto IX is more closely related to Urogen III than to Urogen I.

Chl *a* is derived from Proto IX. The propionic acid side chain on ring C of Proto IX forms the cyclopentanone ring of Chl *a*; the vinyl group on ring B of Proto IX is converted into the ethyl group at the same position of Chl *a*; the carboxyl groups of the two propionic acid residues (rings C and D) of Proto IX are esterified in Chl *a*. Thus Chl *a* is a close structural relative of Proto IX, which, in turn, is derived from Urogen III. Our current knowledge of steps in the

A = $-CH_2-COOH$
P = $-CH_2-CH_2-COOH$
V = $CH{=}CH_2$
E = $-CH_2-CH_3$

M = CH_3

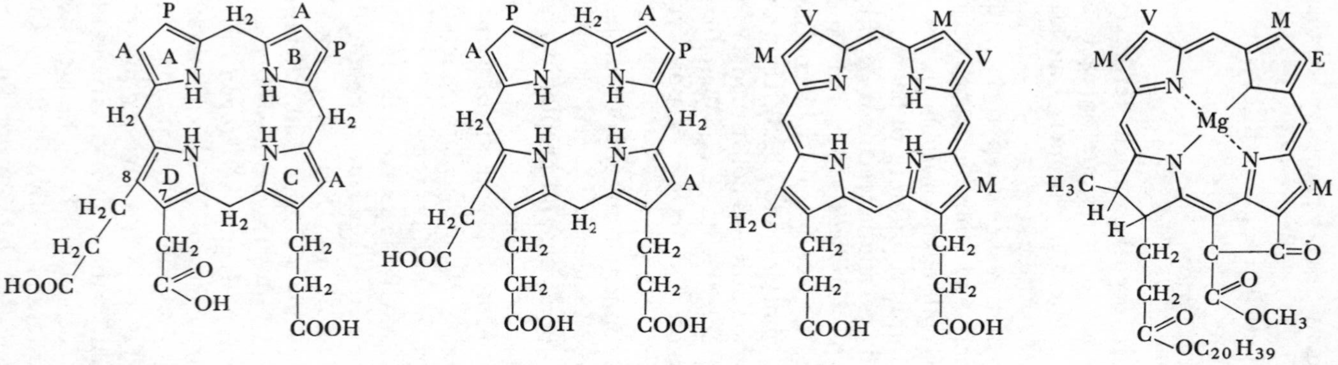

Uroporphyrinogen I

Uroporphyrinogen III

Protoporphyrin IX

Chlorophyll *a*

FIG. 4. Structural relationships among some cyclic tetrapyrroles. Compare the arrangement of substituents on positions 7 and 8 (ring D).

formation of Chl from PBG will be discussed. The precursor–product relationship of Urogens, Proto IX and the various chlorophylls should become clear in the course of the ensuing discussions of steps in porphyrin biosynthesis.

Frozen and thawed preparations of the green alga *Chlorella* catalyse the consumption of PBG and the production of a number of porphyrins including Proto IX, Uro III and Uro I (presumably some of the Urogen formed is oxidized during the incubation while some of the Urogen III is used for Proto IX synthesis; oxidized intermediates between Urogen III and Proto IX are also detected). If such preparations are heated at 55°C for 30 min prior to incubation with substrate, their capacity to consume PBG is essentially unaltered quantitatively, i.e. the rate of utilization of PBG is unaffected, but only Uro I is made from PBG (Bogorad and Granick, 1953b). (Roughly similar observations have been made with haemolysed avian and human erythrocytes by Booij and Rimington (1957) and by Rimington and Booij (1957).)

The observation that the rate of utilization of PBG was unaffected by heating at 55°C was taken to indicate that an enzyme which plays the major role in the utilization of PBG for tetrapyrrole synthesis is relatively heat stable. But the most striking effect of the heat treatment was the change in the nature of the tetrapyrrole isomer produced. Only Urogen I was formed, not a mixture of isomers I and III. Heating the enzyme preparation prior to incubation with PBG eliminated the capacity to produce Urogen III—it seemed reasonable to assume that an enzyme which is inactivated at 55°C is required for the synthesis of Urogen III from PBG. Furthermore, the relatively heat-stable enzyme when incubated alone with PBG catalyses the formation of Urogen I. To summarize this: (a) the crude preparations could be assumed to contain two enzymes—both required for the synthesis of Urogen III from PBG; (b) an enzyme which is required for the formation of isomer III is easily destroyed by heating the preparation to 55°C. Because heating had virtually no effect on the rate of PBG consumption, the enzyme which survives heating should be the one primarily responsible for the removal of PBG from the reaction mixture but this surviving enzyme produces only Urogen I (Bogorad and Granick, 1953b; Bogorad, 1958a,b,c).

The hypothesis that two enzymes are involved in the synthesis of Urogen III (Bogorad and Granick, 1953b) was confirmed by direct demonstration of their existence (Bogorad, 1958a,b). Urogen I synthase, the enzyme which catalyses the consumption of PBG and the formation of Urogen I, was first demonstrated and partly purified from spinach leaf tissue (Bogorad, 1955, 1958a). The enzyme was easily assayed by following the rate of consumption of PBG. Urogen III cosynthase, the other enzyme involved in the synthesis of Urogen III, was first prepared free from Urogen I synthase activity in a fraction of an extract of wheat germ (Bogorad, 1958b).

The original experiments in which *Chlorella* extracts were heated to 55°C showed that there was little, if any, change in the rate of PBG consumption as a consequence of this treatment. Thus, if there were a second enzyme involved in the biosynthesis of Urogen III one would not expect to be able to detect it by assaying PBG-consuming activity. However, experiments in which spinach Urogen I synthase was combined with wheat germ fractions, which alone did not catalyse the consumption of PBG, permitted the identification of a fraction containing an enzyme which was required together with Urogen I synthase for the synthesis of Urogen III.

After a discussion of the properties of each of the two enzymes of Urogen biosynthesis some possible modes of interaction of the two enzymes will be discussed.

1 Uroporphyrinogen I synthase

Urogen I synthase (formerly designated PBG deaminase) has been partly purified from spinach leaf tissue (Bogorad, 1955, 1958a) and from *Rhodopseudomonas spheroides* (Heath and Hoare, 1959; Hoare and Heath, 1959). The enzyme from spinach is only slightly affected by heating at 70°C at pH 8·2 but is inactivated at 100°C at this pH value and at lower temperature in more acid or alkaline solutions.

More highly purified Urogen I synthases have been prepared from spinach (Frydman and Frydman, 1970; Higuchi and Bogorad, 1975), wheat germ (Higuchi and Bogorad, 1975), *R. spheroides* (Davies and Neuberger, 1973; Jordan and Shemin, 1973), and soyabean callus tissue (Llambias and Batlle, 1971a) as well as from a number of animal sources (Levin and Coleman, 1967; Stevens *et al.*, 1968; Sancovich *et al.*, 1969; Llambias and Batlle, 1971b).

Wheat germ Urogen I synthase has a molecular weight of about 40 000 daltons (Higuchi and Bogorad, 1975). The soyabean callus synthase is also reported to have a molecular weight of about 40 000, judged from gel filtration, but particularly at low concentrations of protein some dissociation into about 10 000 dalton fractions was observed (Llambias and Batlle, 1971a). Further work must be done to determine whether the soyabean enzyme really is one-fourth the size of that from wheat germ or spinach leaves. The molecular weight of the spinach enzyme was determined by several methods including SDS-polyacrylamide gel electrophoresis which would reveal the molecular weights of small subunits if they occurred. The same molecular weight was obtained by gel filtration and by sucrose density gradient centrifugation. This indicates that the 40 000 dalton polypeptides do not aggregate. The isoelectric point for the spinach and wheat germ enzymes is 4·2–4·5.

Urogen I synthase from *R. spheroides* has a molecular weight of about 35 000–39 000 based on gel filtration, sucrose gradient centrifugation, and SDS-polyacrylamide gel electrophoresis (Jordan and Shemin, 1973; Davies and Neuberger, 1973). This enzyme has an isoelectric point and a pH optimum similar to that of the Urogen I synthases from spinach and wheat germ.

Since the product of the action of Urogen I synthase on PBG is Urogen I in yields close to 100%, and this cyclic tetrapyrrole can be visualized as arising by the sequential addition of four PBG molecules to one another with the elimination of ammonia to form a linear tetrapyrrole which cyclizes, the simplest role to assign Urogen I synthase is the catalysis of the condensation of PBGs to form first a dipyrrole, then a tripyrrole and finally a linear tetrapyrrole. The last of these would have a very high probability of cyclizing to form Urogen I perhaps non-enzymically; it has been known since the work of Waldenström and Vahlquist in 1939, when PBG was known only by its products, that uroporphyrins (Uro) are formed in the absence of enzymes upon heating aqueous solutions of PBG.

Three classes of inhibitors of Urogen I synthase are known:

(a) Non-competitive inhibitors such as formaldehyde, *p*-chloromercuribenzoate, silver and mercuric ions (Bogorad, 1958a). Inhibition by the mercurials is reversed by sulphydryl-containing compounds.

(b) Competitive inhibitors including PBG analogues such as opsopyrrole dicarboxylic acid (Bogorad, 1957b, 1960; Carpenter and Scott, 1959) and isoporphobilinogen (Carpenter and Scott, 1961).

(c) A group of inhibitors whose effect, unlike that of the two types already described, is not primarily on the rate of substrate consumption, although this is affected to some extent, but rather upon the nature of the reaction. In the presence of sufficiently high concentrations of ammonium ions or hydroxylamine the rate of PBG consumption is not matched by that of Urogen I formation—compounds which appear to be linear polypyrroles accumulate in the reaction mixture. Under some conditions these accumulated compounds can form additional Urogen I thus they are intermediates in Urogen I biosynthesis (Bogorad, 1961, 1963; Pluscec and Bogorad, 1970; Radmer and Bogorad, 1972). Similar pyrrolic intermediates can be detected on paper or thin-layer electrophoresis if the reaction is run at high concentrations of PBG or in the presence of dipyrrylmethane 1 (Fig. 5.).

One of the intermediates which accumulates is dipyrrane 2 (Pluscec and Bogorad, 1970), which has also been synthesized by Osgerby *et al.* (1972). Pairs of dipyrrylmethane 2 are not condensed to Urogen I by Urogen I synthase. However, if PBG is added to a reaction mixture containing Urogen I synthase and dipyrrane 2 the latter is incorporated into Urogen I. This shows that PBG molecules are not all condensed to dipyrroles, as the first step in the Urogen I synthase reaction, and that pairs of dipyrroles are not subsequently condensed to produce Urogen I. The enzyme thus appears to work by adding 1 PBG to another then a third PBG to the dipyrrole, and presumably a fourth PBG to a tripyrrole (Pluscec and Bogorad, 1970). Frydman *et al.* (1973) have synthesized the same dipyrrylmethane and confirmed the observations of Pluscec and Bogorad (1970) using wheat germ rather than spinach Urogen I synthase.

Experiments described by Battersby (1971) indicate that the enzymic condensation of dipyrranes may occur under the conditions he employed. However, the possibility that this condensation is spontaneous and not enzymic is not completely excluded by the information provided.

An open-chain tetrapyrrole accumulates in reaction mixtures of PBG and spinach leaf Urogen I synthase in the presence of ammon-

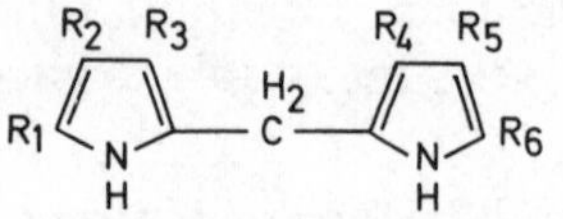

R_1 R_2 R_3 — CH_2 — R_4 R_5 R_6 (two NH pyrrole rings)

Ac = $-CH_2-COOH$
Pr = $-CH_2-CH_2-COOH$
AM = $-CH_2NH_2$

	R_1	R_2	R_3	R_4	R_5	R_6
DPM-1	AM	Ac	Pr	Pr	Ac	AM
DPM-2	AM	Ac	Pr	Ac	Pr	H
DPM-3	AM	Pr	A	Ac	Pr	H

FIG. 5. The structures of some dipyrranes mentioned in the text.

ium ions (Bogorad, 1963; Radmer and Bogorad, 1972). Urogen I synthase of *R. spheroides* accumulates polypyrroles in the presence of hydroxylamine, methoxyamine or ammonia (Davies and Neuberger, 1973). In the latter experiments it could be shown that radioactivity from methoxyamine was incorporated into the pyrrane which was formed. Hydroxylamine was released during the non-enzymic cyclization of the tetrapyrrole.

Llambias and Batlle (1970b) and Stella *et al.* (1971) have reported the presence of presumptive polypyrranes in studies employing enzymes from soyabean callus.

Is one enzyme responsible for the polymerization of PBG molecules and the cyclization of the tetrapyrrole to form Urogen I? The question has been raised repeatedly (e.g. Bogorad, 1958a). Frydman and Frydman (1970) concluded that their data favoured the conclusion that several enzymes are involved in the transformation of PBG into a cyclic tetrapyrrole. The purification to homogeneity of Urogen I synthase from several sources indicates that only one polypeptide chain is necessary to carry the reaction from PBG to Urogen I.

In the normal course of the enzymic reaction there is no detectable free tetrapyrrole unless very large amounts of substrate are used in the reaction. Thus in the presence of a moderate amount of substrate the reaction goes completely to the cyclized product. The uncyclized immediate precursor of Urogen I has been obtained by Radmer and Bogorad (1972) from incubation mixtures of PBG, spinach leaf Urogen I synthase and ammonia. Under these circumstances the reaction does not go to completion but the uncyclized tetrapyrrole accumulates. On continued incubation it cyclizes to make Urogen I. This demonstrates that the lifetime of the uncyclized tetrapyrrole is much shorter on the surface of the enzyme than free in solution.

Thus the enzyme appears to catalyse the cyclization. However, the addition of Urogen I synthase to solutions of the uncyclized tetrapyrrole do not accelerate its cyclization. Why are these observations not in direct conflict with one another? The tetrapyrrane can assume an infinite number of spatial configurations in solution. Suppose that the two ends of the molecule fit onto the surface of the enzyme in only one of these configurations—perhaps a planar one. The two terminal PBG residues are estimated to be as close to one another as they would be in a 0·4 M solution when the tetrapyrrane is planar (Mauzerall, 1960). At this position the probability that two free ends of the tetrapyrrole will condense may be as high as the probability that the whole molecule will first go on to the enzyme, especially since the enzyme is present at a concentration much below 0·4 M.

In view of these considerations it seems likely that the addition of a fourth PBG molecule to the tripyrrane on the enzyme would normally be likely also to result in the almost immediate condensation of the two free ends of the newly formed tetrapyrrane. The two terminal residues would be very close together on the enzyme surface at that time.

These experiments and interpretations describe characteristics of the system which complicate work with free polypyrranes. In its native form the active centre of the enzyme, and indeed its entire surface, is relatively limited in the shapes it can take. In contrast, a polypyrrane can assume many different forms in solution but only one or few forms when it is associated with the surface of the enzyme. If the polypyrrane leaves the enzyme surface it is more likely to pass randomly through many forms; in most of these shapes it has a low probability of binding to the enzyme again. What bearing has this discussion on Urogen I biosynthesis?

How many active sites does each Urogen I synthase bear? If there is only one active site for the condensation of PBGs with the elimination of ammonia, after the first dipyrrylmethane is made it would leave the enzyme surface, move over, and reassociate with the enzyme in a position which would permit the addition of another PBG to the chain of two etc. Up to a point the farther the pyrrane moves away from the enzyme the lower the probability of re-association. On the other hand, the principal argument against there being four active sites and a "mould" for a cyclic tetrapyrrole, perhaps with charged groups to bind the carboxyls of the propionic and acetic acid residues, is the failure of pairs of DPM-2 to be condensed enzymically. The issue is not resolved.

2 *Uroporphyrinogen III cosynthase*

The III rather than the I isomer of Urogen is the precursor of protohaem IX and of the chlorophylls. How is the III isomer formed? Some aspects of the phenomenon are known but the mechanism is not.

Urogen III is formed entirely from PBG by crude enzyme preparations (e.g. preparations from: *Chlorella*—Bogorad and Granick, 1953b; haemolysed chicken erythrocytes—Falk *et al.*, 1953, 1956: Dresel and Falk, 1956; rabbit reticulocytes—Granick and Mauzerall, 1958;

haemolysed chicken erythrocytes—Lockwood and Benson, 1960; *Rhodopseudomonas spheroides*—Heath and Hoare, 1959). It can also be synthesized enzymically *in vitro* by the action of two enzymes: Urogen I synthase and Urogen III cosynthase (previously designated uroporphyrinogen isomerase). The former has been discussed above; the latter enzyme was originally partly purified from aqueous extracts of wheat germ (Bogorad, 1958b). Urogen III cosynthase has been purified more extensively from wheat germ, mouse spleen, and human erythrocytes (Levin, 1968; Stevens and Frydman, 1968; Stevens *et al.*, 1968; Higuchi and Bogorad, 1975).

The wheat germ enzyme has a molecular mass of about 62 000 daltons and its pI is 6·1–6·7 (Higuchi and Bogorad, 1975). Llambias and Batlle (1971a) estimated the Urogen III cosynthase from soyabean callus to be 210 000 daltons. However, they pointed out that other proteins might still have been bound to the enzyme.

Urogen III cosynthase does not catalyse the condensation of PBG molecules when this enzyme and the pyrrole are incubated together, but when included in a reaction mixture with Urogen I synthase and PBG it brings about the formation of Urogen III (Bogorad, 1958b). The most obvious possibility for the formation of Urogen III by the two-enzyme mixture is that Urogen I synthase produces Urogen I from PBG and Urogen III cosynthase rearranges the cyclic tetrapyrrole. This does not occur (Bogorad, 1958b; Granick and Mauzerall, 1958).

In reaction mixtures of Urogen I synthase, Urogen III cosynthase and PBG the rate of utilization of the monopyrrole is governed chiefly by the concentration of Urogen I synthase, in accord with the earlier observations on the crude *Chlorella* system (Bogorad and Granick, 1953b). However, kinetic studies reveal that the maximum velocity of the reaction is higher in the presence of both Urogen III cosynthase and Urogen I synthase than when the former is omitted—despite its failure to catalyse any measurable consumption of PBG when alone (Bogorad, 1958b). (Levin (1968) has developed an Urogen III cosynthase assay based on the latter system.)

Two categories of possible mechanisms of action of Urogen III cosynthase and its relationship to Urogen I synthase can be postulated: (a) a product–substrate relationship and (b) a co-operative relationship between the two enzymes.

One of the possible product–substrate relationships has already been examined. Urogen I formed by Urogen I synthase is not converted to Urogen III by Urogen III cosynthase. Experiments dealing with the possible utilization of polypyrranes formed from PBG by Urogen I synthase as substrates for Urogen III cosynthase are described in the ensuing paragraphs.

A product–substrate relationship in which Urogen I synthase uses a product of the other enzyme, can also be considered. Urogen III can be considered as three molecules of PBG condensed head-to-tail (rings A, B and C) plus one molecule of *iso*-PBG (ring D). Perhaps Urogen III cosynthase converts some PBG to *iso*-PBG, and Urogen I synthase condenses those PBG molecules and one *iso*-PBG? However, no altered PBG can be detected after incubation with Urogen III cosynthase. It is conceivable, however, that in the absence of Urogen I synthase the amount of altered PBG is undetectably small because, for example, the product is a feedback inhibitor of Urogen III cosynthase. In any event, preincubation of the very heat-labile Urogen III cosynthase with PBG, destruction of the enzyme by mild heating, and addition of Urogen I synthase results in the synthesis of only Urogen I, thus arguing against the Urogen III cosynthase-catalysed step occurring first (Bogorad, 1957a,b) and *iso*-PBG is an inhibitor of Urogen I synthase (Carpenter and Scott, 1961).

The other type of possibility is that a cooperative relationship may exist between the two enzymes. That is, the two proteins might interact to form a complex with novel enzymatic properties. The first dipyrrole and all of the intermediate pyrranes formed from PBG by this complex would be different from the product of Urogen I synthase acting alone on PBG. This possibility is also examined in more detail below.

DPM-2, the dipyrrylmethane which can be formed by the head-to-tail condensation of two molecules of PBG, i.e. [PA]–C–[PA]–CH_2NH_2, is incorporated enzymically into Urogen I (Pluscec and Bogorad, 1970; Frydman *et al.*, 1971) but, according to Frydman *et al.* (1971, 1973), not into Urogen III. Unfortunately, the dipyrrane undergoes relatively rapid non-enzymic dimerization and rearrangment under the incubation conditions employed by Frydman *et al.*; in the absence of PBG at 37°C after 60 min 9% of 14 nmol of dipyrrane was converted to an equal mixture of Urogens I and III. The corrections for blanks are quite large and the rearrangements create doubts but there is some reassurance from the report that the dipyrrane [PA]–C–[AP]–CH_2NH_2 [DPM-3] is incorporated enzymically into Urogen III although some is incorporated into Urogen I as well (Frydman *et al.*, 1972). On the other hand, Battersby (1971) has reported that DPM-2 is incorporated into Proto IX, presumably via Urogen III.

The data of the Frydmans and co-workers can be taken to support the possibility that the dipyrrolic precursor in the enzymic synthesis of Urogen III is different from the one from which Urogen I forms; that is, to support the proposal that the first condensation steps are different in the biosynthetic pathways leading to the I and III isomers. Battersby's data do not support this view. The chemical, non-enzymic, rearrangements which Urogens can undergo complicate the experiments and their evaluation.

The mechanism of Urogen III biosynthesis can be studied also by considering possible physical interactions between Urogen I synthase and Urogen III cosynthase. First, a "product–substrate" relationship between the two enzymes would not require that they interact physically but if the product of one enzyme cannot diffuse to the other one easily or in an unaltered form the transfer may be effective only if physical contact occurs. Evidence that a polypyrrane once detached from a protein surface may have difficulty reassociating with the same enzyme before it reacts as another way has been discussed in the section on the biosynthesis of Urogen I. The contact between enzymes could be momentary and occur only late in the reaction after a polypyrrane had been built up. If, on the other hand, the two enzymes co-operate to form a "starting dipyrrane" of a different type (e.g. DPM-3 instead of -2) from that which Urogen I synthase would form alone, one should expect the two enzymes to be together at least at the start of the reaction if not throughout the synthesis of Urogen III.

Is there any evidence for the physical interaction of these two enzymes? Attempts to separate them from one another were not always successful, depending on the source. Some workers continue to use the term "porphobilinogenase" to designate an unresolved mixture of the two enzymes. The difficulties of separating these two enzymes led some to conclude, without valid evidence, that they form a tight complex. Subsequent experiments showed that the methods being used then were simply inadequate to permit a good separation.

Urogen I synthase has a molecular weight of about 40 000 and Urogen III cosynthase is about 60 000 (Higuchi and Bogorad, 1975). An aggregate of one of each of these two molecules should be about 100 000 daltons. The detection of Urogen I synthase by measuring PBG consumption is simple and accurate. The possible association of the enzymes with one another can be studied by determining whether Urogen III cosynthase influences the sedimentation behaviour of the easy-to-measure Urogen I synthase during centrifugation in a density gradient. PBG alone has no effect on the sedimentation of Urogen I synthase. This indicates that the enzyme does not aggregate while it is catalysing the formation of Urogen I. Urogen I synthase does sediment more rapidly if Urogen I cosynthase and PBG are present but not if PBG is omitted from the gradient (Higuchi and Bogorad, 1975); both Urogen III synthase and PBG are required for complex formation. Although these experiments show that Urogen III cosynthase may interact with Urogen I synthase at least long enough to affect the latter's sedimentation behaviour, it is not possible to judge from these data whether such a complex lasts throughout the synthesis of Urogen III or whether it is ordinarily very short lived—e.g. lasting only long enough to transfer some precursor of Urogen III from one protein to the other. Further experiments along these lines should help us understand more about this complex set of interactions and the functional interactions of these proteins.

A number of possible mechanisms for the formation of Urogen III have been advanced in the years following the discovery of participation of two enzymes in this reaction. Many of the schemes proposed prior to 1960 have been discussed extensively (Bogorad, 1960). A scheme based on the alteration by Urogen III cosynthase of a linear tetrapyrrole formed from PBG by the action of Urogen I synthase was advanced by Mathewson and Corwin (1961) and has received a good deal of attention. A system with many features similar to that of the Mathewson-Corwin proposal but involving the exchange of pyrrole residues between a pair of linear tetrapyrroles was proposed by Wittenberg (1959). The mechanism of synthesis remains unresolved at this time.

As noted above, the PBG in aqueous solution condenses non-enzymatically. The Uro isomer (or isomers) produced is affected by the acidity or alkalinity of the solution (e.g. Cookson and Rimington, 1954; Mauzerall, 1960). Current biological practice appears to differ greatly from the mechanisms of non-enzymatic formation of Uro III but the latter is of considerable interest in speculations on the origin of life.

D THE DECARBOXYLATION OF UROGENS. THE FORMATION OF COPROGENS

Mammals afflicted with hereditary or acquired porphyrias excrete varying amounts of Uro and coproporphyrins (Copro) of the I and III isomeric series. How the biological occurrence of these porphyrins and of Proto IX could be rationalized was the subject of extensive speculation before the biogenetic relationships among these compounds became clear. Lemberg and Legge (1949) discussed some of the then current ideas about the interrelationships of Copros, Uros, and Protos.

By the early 1950s most students of porphyrin biosynthesis had become convinced that Uro III and Copro III were closely related to Proto IX or precursors of it. At that time not only the tetracarboxylic coproporphyrins and octacarboxylic uropophyrins, but porphyrins with seven, six and five carboxyl groups per molecule had been identified in the urine of patients with congenital porphyria (McSwiney *et al.*, 1950; Rimington and Miles, 1951); a somewhat similar array of porphyrins had been identified in a Chl-less mutant strain of *Chlorella* (Bogorad and Granick, 1953a) and, most important, Wittenberg, and Shemin's (1949, 1950) degradative analyses of protohaem synthesized from position-labelled acetate supplied to cell preparations strongly suggested, first, that the methyl substituents on each of the four pyrrole rings of protohaem had a common origin and, second, that the vinyl substituents on rings A and B were probably derived from propionic acid residues formed initially by the same mechanism as those on rings C and D. Numerous attempts, generally not published, were made to demonstrate the conversion of Uro into Copro or Proto in biological systems shown to be capable of making Proto from ALA or PBG. Then, from a few experiments it became clear that Urogens, the unoxidized precursors of uroporphyrins, were the substrates for the synthesis of Coprogens. In turn, Coprogen III rather than Copro III was shown to be the precursor of Protogen IX (Bogorad, 1955, 1957a, 1958c: Neve *et al.*, 1956; Mauzerall and Granick, 1958).

Urogen decarboxylase catalyses the removal of the carboxyl group from each of the four acetic acid side chains of Urogen; the methyl residues of the corresponding Coprogen isomers remain. The enzyme attacks not only the biologically occurring isomers, Urogens I and III, but also isomers II and IV. The reaction rate varies with the isomeric configuration of the substrate in the following order: Urogen III > IV > II > I. Porphyrins with seven, six, or five carboxyl groups per molecule, i.e. with four propionic acid residues plus three, two, or one not yet decarboxylated acetic acid substituents, have been identified in oxidized reaction mixtures of Urogen and its decarboxylase (Mauzerall and Granick, 1958). Although it is clear that this enzyme is present in tissues of higher plants and in broken cell preparations of *Chlorella* which can catalyse the synthesis of Proto from PBG or Urogen III and of Coprogen I from Urogen I (Bogorad and Granick, 1953b; Bogorad, 1958c) the most detailed investigations have been conducted on an enzyme prepared from rabbit red cells (Granick and Mauzerall, 1958; Mauzerall and Granick, 1958).

Conditions affecting the production of Coprogen by cultures of

Rhodopseudomonas capsulata have been studied by Cooper (1963); Cohen-Bazire *et al.* (1957) have investigated this problem in *Rhodopseudomonas spheroides.* Heath and Hoare (1959) reported the presence of a heat-stable factor for Copro formation in broken-cell preparations of *R. spheroides*; this factor has not been identified further.

E THE OXIDATIVE DECARBOXYLATION OF COPROGEN III. THE FORMATION OF PROTOGEN IX

The conversion of Coprogen III into Protogen IX requires the formation of vinyl groups from propionic acid groups on rings A and B. That this decarboxylation and dehydrogenation might require aerobic conditions was indicated by the observations of Bénard *et al.* (1951) that blood from anaemic rabbits formed Proto when incubated with glycine aerobically, but predominantly Copro when incubated under nitrogen. This observation of an oxygen requirement was extended by Falk *et al.* (1953) in studies on the formation of porphyrins from porphyrinogen by haemolysed chicken erythrocytes.

In work with about twentyfold purified Coprogen oxidative decarboxylase (Coprogenase) from guinea-pig liver mitochondria (Sano and Granick, 1961) or with acetone dried powders of ox liver mitochondria (Porra and Falk, 1964) no oxidant other than O_2 was used by the enzyme. Sano and Granick reported trying flavin mononucleotide, 1,4-naphthoquinone, and hydrogen peroxide without success. Porra and Falk found the product of the reaction to be Protogen IX.

The guinea-pig mitochondrial enzyme studied by Sano and Granick does not act on Copro, mesoporphyrinogen, hematoporphyrinogen, Urogen III, monopropionic monoacrylic deutero porphyrinogen, 2,4-diacrylic Proto or the corresponding porphyrinogen. Sano and Granick report that this enzyme is also inactive against Coprogen isomers other than III. However, Porra and Falk (1964) found the substrate specificity not to be absolute; in their experiments Coprogen IV was used by the enzyme they obtained from ox liver mitochondria. Sano and Granick reported that some Proto IX was produced from Coprogen I; they believe that the Proto formed was derived from Urogen III which contaminated the sample of the isomer I they used. Small amounts of Proto IX were also detected upon incubation of Urogen I with broken *Chlorella* cells; it is not clear whether there was indeed porphyrin formation by the *Chlorella* enzyme from isomers other than those of the III series in these experiments (Bogorad, 1958c).

Coprogenase was purified about sixtyfold from rat liver mitochondria by Batlle *et al.* (1965). The isoelectric point of the enzyme was found to be about pH 5·0 and the molecular weight to be about 80 000 daltons. The enzyme has also been extracted and purified about seventyfold from tobacco by Hsu and Miller (1970); the activity was located mainly in the mitochondria. The K_m of the tobacco enzyme is $3{\cdot}6 \times 10^{-5}$ M, in fairly close agreement with the value obtained with the rat liver enzyme. The rat liver enzyme was unaffected by metal-chelating agents but tobacco Coprogenase is sharply inhibited by *o*-phenanthroline and EDTA, each at 1 mM. The activity of the tobacco enzyme was enhanced by increasing the Fe^{2+} concentration up to 0·5 mM. Co^{2+} and Mn^{2+} also activate the enzyme at low concentrations (0·2 mM). Batlle *et al.* (1965), on the other hand, found that the activity of the rat enzyme was not affected by Mg^{2+}, Ca^{2+}, K^{+}, or Na^{+}. The effect of Fe^{2+} was not tested by Batlle *et al.* (1965). All of the enzymes studied to date exhibit maximal activity at about pH 7·4.

Sano (1966) has prepared 2,4-bis-(β-hydroxypropionic acid) deuteroporphyrinogen IX, i.e. a compound which is like Coprogen III and Protogen IX with respect to pyrrole rings C and D but with a methyl and a β-hydroxypropionic acid side chain on each of the other two rings. This compound was converted by a mitochondrial enzyme into Protogen IX under anaerobic conditions. Oxygen is required when Coprogen III is the substrate, as was mentioned earlier. Sano (1966) has suggested that the β carbons of the propionic acid side chains on rings A and B are oxidized to hydroxyl groups by an oxygenase-type of reaction. Then that the β-hydroxypropionic acid side chains are dehydrated and decarboxylated to yield Protogen IX. As demonstrated in Sano's experiments, the latter step can proceed anaerobically. Support for the idea that the reaction goes via a β-hydroxypropionic acid side chain rather than β-keto propionate has been provided by Battersby (1971), Battersby *et al.* (1972), and by Zaman *et al.* (1972). Only one of the two hydrogen atoms on the β-carbon atom on the propionic acid side chains is lost during the conversion of Coprogen III into Protogen IX and the hydrogen atoms associated with the α-carbon atom are not involved in the formation of the vinyl side chains.

The absolute requirement for oxygen by Coprogenase has generated interest in Proto IX formation by anaerobic organisms including photosynthetic bacteria. For example, the green sulphur bacterium *Chromatium* as well as *Rhodopseudomonas spheroides* form chlorophyll when grown anaerobically.

Mori and Sano (1968) showed that extracts of *Chromatium* cells can form Proto IX from Coprogen III, but only when the extract was incubated with substrate aerobically. The reaction did not occur under anaerobic conditions. Ehteshamuddin (1968), on the contrary, has reported that enzyme preparations from *Pseudomonas*, an anaerobe, form Proto IX under either anaerobic or aerobic conditions. The production under anaerobic conditions was about half that under aerobic conditions but the formation of any Proto *in vitro* under anaerobic conditions makes these two reports conflicting. Mori and Sano (1972) have purified Coprogenase of *Chromatium* about twentythree-fold. It is a heat-tolerant enzyme: it survives 60°C. Chelating agents such as *o*-phenanthroline and α,α′-dipyridyl did not inhibit the enzyme; Fe^{2+} did not stimulate it. In most of these respects the enzyme was similar to the mitochondrial enzyme studied previously by Batlle *et al.*

The problem of the aerobic versus anaerobic formation of Proto IX by *R. spheroides* and *Chromatium* has been examined more extensively by Tait (1969, 1972). When *R. spheroides* cells are grown under reduced O_2 tension in the light they form some chlorophyll. Extracts from such cells convert Coprogen into Proto when incubated anaerobically in the dark with Mg^{2+}, ATP and methionine. All these components are essential for activity. *S*-Adenosyl methionine can replace methionine but not ATP in the reaction. The role of *S*-adenosyl methionine in this reaction is not known but Davies *et al.* (1973) have suggested that the positively charged sulphur atom of *S*-adenosyl methionine facilitates the removal of a hydride ion from the β-carbon atom of the propionic acid side chain thus initiating the concerted dehydrogenation and decarboxylation which may occur. Ethionine, an inhibitor of bacteriochlorophyll production by *R. spheroides* (Gibson *et al.*, 1962b), inhibits Proto formation in the presence of methionine. By using an oxygen electrode Tait established that the reaction mixture became completely anaerobic after about 2 min; incubation times for the reaction ranged from 1 to 3 h.

On the other hand, extracts prepared from *R. spheroides* grown in air or O_2 in the dark, and consequently poorly pigmented or lacking pigment entirely, failed to form Proto under anaerobic conditions even if the three required compounds were present. Proto could be formed from Coprogen by these preparations if O_2 were available. The anaerobic reaction can also be distinguished from the aerobic one in its sensitivity to inhibition by the metal chelators *o*-phenanthroline and α,α′-dipyridyl, and by flavins, 2,4-dinitrophenol and 1,4-naphthoquinone. The inhibition by low concentrations of flavins suggests that a flavoprotein might be involved; the inhibition by both *o*-phenanthroline and α,α'-dipyridyl suggests that a metal, possibly iron, is involved; the requirement for ATP and the inhibitory effect of 2,4-dinitrophenol, both suggest that a substantial amount of energy is required. On the basis of these observations plus the finding that nicotinamide nucleotides had to be added for activity of more highly purified enzymes, Tait suggests that the hydrogen removed from propionic acid side chains of Coprogen "is accepted by a component of the electron-transport chain, perhaps a non-haem iron-protein, and that it is then transferred first to a flavoprotein and then to NAD(P) in an energy-requiring step. Energy-dependent reverse electron transport has been shown to occur in mitochondria and in autotrophic bacteria."

Tait (1972) was also able to detect activity in extracts of *Chromatium* under anaerobic conditions. He suggests that there are two enzymes: one requires aerobic conditions and the other works anaerobically. The enzyme from *Chromatium* responsible for aerobic activity has been purified about sixtyfold. Its K_m is about the same as those of other Coprogenases studied and it is estimated to have a molecular weight of about 44 000 based on Sephadex G-100 filtration. Tait concludes that a single polypeptide can account for all the aerobic activity. In contrast, the anaerobic activity appears to require at least two proteins because the activity which is detected under anaerobic conditions in the crude extract can be detected neither in the supernatant nor in the pellet after centrifugation. However, if the supernatant and pellet are recombined 75% of the activity is recovered. One of the required proteins is thought to be in the supernatant and the other in the pellet; anaerobic Coprogenase activity is exhibited only when the two are in the same solution.

The two kinds of activity suggest to Tait (1972) that there may be two different mechanisms for the conversion of propionic acid side chains to vinyl side chains.

F METAL CHELATION. THE FORMATION OF IRON AND MAGNESIUM PROTOPORPHYRINS

Protogen, like other porphyrinogens, cannot chelate metal ions. Haems and chlorophylls are iron- and magnesium-porphyrins respectively. The step after Protogen IX synthesis must be oxidation to Proto for metal insertion is required for further utilization. Evidence has been presented from time to time for biological systems which stimulate the oxidation of porphyrinogens but none of the data is compelling (Bogorad, 1958a; Sano and Granick, 1961).

Proto IX and several other porphyrins can serve as substrates for porphyrin-iron chelating enzymes, ferrochelatases, which have been prepared from various organisms. The first of these enzymes was prepared from rat liver mitochondria by Labbe and his coworkers (Labbe, 1959; Nishida and Labbe, 1959; Labbe and Hubbard, 1960, 1961; Labbe *et al.*, 1963; Porra *et al.*, 1967). Ferrochelatases were studied at about the same time from duck and chicken erythrocytes (Schwartz *et al.*, 1959; Oyami *et al.*, 1961), from pig liver and from a number of micro-organisms including *Chromatium* (Porra and Jones, 1963a,b).

Labbe and co-workers found that the rat liver enzyme catalysed the incorporation of ferrous or cobaltous but neither ferric nor stannous ions into proto-, deutero-, haemato-, *meso*-, or 2,4-dibromo-deuteroporphyrins. Neither Uro nor Copro served as substrates. The pig liver enzyme of Porra and Jones (1963b) was generally similar with regard to porphyrin specificity.

Schwartz *et al.* (1959) separated chicken erythrocyte ferrochelatase into two heat-labile non-dialysable components; no activity was shown for either component alone. Porra and Jones (1963b) found that the porphyrin substrate range for fresh ferrochelatase preparations from *Thiobacillus X* was broader than for aged preparations. They suggested that more than one type of iron-porphyrin chelatase might be present in a single organism.

Jones and Jones (1970) have obtained a soluble and a particulate form of ferrochelatase from preparations of sonicated *R. spheroides.* The soluble form catalyses the incorporation of Fe^{2+}, Co^{2+} or Zn^{2+} into deutero- and haematoporphyrins. The particulate form incorporates these ions into proto-, *meso*-, deutero- or haemato-porphyrins. Neither enzyme promotes the incorporation of magnesium into porphyrins.

Ferrochelatase is located principally in the membrane fraction of broken *Spirillum itersonii* preparations; less than 20% of the activity appeared to be in the cytoplasm, i.e. was found as a soluble enzyme (Dailey and Lascelles, 1974). The membrane-bound enzyme incorporates iron more effectively than zinc or cobalt into proto-, deutero-, or *meso*-porphyrins in increasing order of activity but like the soluble enzyme from *R. spheroides*, the soluble *S. itersonii* ferrochelatase is inactive with Proto. An interesting haemin-requiring mutant (H9) lacks the membrane-bound ferrochelatase but has a soluble enzyme like the wild-type.

Do each of these micro-organisms have at least two different ferrochelatases or just one type which may be altered during preparation? Dailey and Lascelles (1974) have solubilized the membrane-bound ferrochelatase; this enzyme has several properties which still set it apart from the "cytoplasmic" ferrochelatase. Under these conditions, solubilization alone is not enough to change the properties of the membrane-bound enzyme to those of the "cytoplasmic" one.

A ferrochelatase has been isolated from spinach leaves (Jones, 1967) and a zinc- or iron-incorporating enzyme has been prepared from barley leaves (Goldin and Little, 1969) following on the work of Little and Kelsey (1964). A zinc protoporphyrin chelatase from *R. spheroides* chromatophores was detected and studied by Neuberger and Tait in 1964; this enzyme appeared to be specific for Zn^{2+}; ions of Fe^{2+} competitively inhibit zinc protoporphyrin production. The bacterial enzyme preparation did, however, show ferrochelatase activity after incubation with ascorbate. Neuberger and Tait (1964) suggested that two ferrochelatases are present in the *R. spheroides* extracts—a zinc chelatase and an iron chelatase.

It is not known for certain that there are separate zinc- and iron-porphyrin chelating enzymes. The two activities have not been completely separated from one another. Furthermore, porphyrins take up zinc easily and spontaneously in neutral or alkaline solutions. The incorporation of iron and magnesium is chemically much more difficult.

Neither zinc- nor iron-porphyrin chelatase preparations catalyse the insertion of magnesium into protoporphyrin. Gorchein (1972) has studied the formation by cells of *R. spheroides* of magnesium Proto monomethyl ester produced from Proto supplied in a lipid emulsion. Only cells grown under conditions favouring bacteriochlorophyll formation, e.g. low O_2 tension (5% O_2 in N_2) in the light, were active. The reaction is best in cells incubated with EDTA or some related chelators. (These were added to remove contaminating zinc but this may not be their sole effect.) Because (a) the product is the methyl ester of Mg Proto, (b) dialysed cells require *S*-adenosylmethionine or methionine and ATP, and (c) ethionine and inorganic phosphate inhibit the reaction, Gorchein suggests that a multienzyme complex catalyses two reactions and that the insertion of Mg is obligatorily linked to methylation. The methylation of Mg Proto has been studied *in vitro* independent of Mg incorporation and a *Chlorella* mutant which produces Mg Proto is known (Granick, 1948b); however, the situation could be different in *R. spheroides*.

G THE ESTERIFICATION OF MAGNESIUM PROTOPORPHYRIN: MAGNESIUM PROTOPORPHYRIN METHYLTRANSFERASE

In 1948 Granick reported on a series of mutants of *Chlorella* which formed no Chl but accumulated large amounts of other porphyrins. One strain accumulated Proto IX, another magnesium Proto and later (Granick, 1950) a third mutant was found which accumulated magnesium vinyl phaeoporphyrin a_5. These compounds, in the order listed, then appeared to establish a few landmarks along the path between Proto IX and Chl (see Fig. 2).

The first step in the further utilization of magnesium Proto appears to be the esterification of the carboxyl group of the propionic acid residue on ring C to make magnesium Proto IX monomethyl ester. Such a compound has been isolated from *Chlorella vulgaris* mutant no. 60A and from etiolated barley leaves supplied with δ-aminolaevulinic acid and α,α′-dipyridyl (Granick, 1961); from cultures of *Rhodopseudomonas spheroides* (Jones, 1963a) and from cultures of *R. capsulata* (Cooper, 1963). Tait and Gibson (1961) and Gibson *et al.* (1963) found that chromatophores from *R. spheroides* when incubated with magnesium protoporphyrin and *S*-adenosyl methionine catalyse the formation of magnesium Proto monomethylester. They found that while Proto IX could be esterified by this enzyme the rate of this reaction was only about one-fifteenth that of the esterification of magnesium Proto. This appears to support the contention that magnesium Proto methyl ester is derived from magnesium Proto which in turn is formed from Proto.

Radmer and Bogorad (1967) demonstrated the transferase in *Zea mays* and studied its specificity. Ebbon and Tait (1969) have found *S*-adenosyl methionine-Mg Proto methyltransferase to be localized principally but not solely in the chloroplast fraction of light-grown *Euglena gracilis.* This enzyme was solubilized by Tween 80 and partially purified.

H STEPS IN THE FORMATION OF PROTOCHLOROPHYLLIDE (PCHLIDE) *a* FROM MAGNESIUM PROTOPORPHYRIN IX METHYL ESTER

In 1948 Smith showed a mole-for-mole correspondence between the disappearance of Pchlide (then thought to be protochlorophyll (Pchl) *a*) and the appearance of Chlide *a* (thought at the time to be Chl *a*) during brief illumination of etiolated barley leaves. This established the terminal step in Chlide production as the reduction of Pchlide. This step is discussed in greater detail below.

Pchlide *a* differs from magnesium Proto IX monomethyl ester in the presence of an ethyl rather than a vinyl group on carbon 4 (ring B), and of the cyclopentanone ring formed from the esterified propionic acid residue on carbon 6 (ring C).

Jones (1963b) isolated Mg 2,4-divinyl pheoporphyrin a_5 monomethyl ester from cultures of *R. spheroides* grown in the presence of 8-hydroxyquinoline. This suggested that the cyclopentanone ring is formed before the vinyl group on ring B is reduced. Aronoff and Ellsworth (1968) have summarized their extensive analyses of porphyrins accumulated by mutants of *Chlorella* blocked between Mg Proto IX methyl ester and Pchlide (Mg vinyl pheoporphyrin a_5). In their investigations (Ellsworth and Aronoff, 1968a,b,c, 1969) both monovinyl (on ring A) and divinyl (on rings A and B) forms were judged to occur of (a) Mg Proto IX monomethyl ester, and of (b) three derivatives of this compound in which the propionic acid methyl ester side chains on ring C are at different oxidation levels. The latter are judged to constitute the members of this biosynthetic sequence: an acrylic methyl ester, a derivative in which the β carbon is hydroxylated, and a compound in which the β carbon is oxidized to a ketone. Pchlide, a monovinyl compound in which the oxidized side chain is condensed with the α-bridge carbon atom to form the cyclopentanone ring, was also detected together with a possible divinyl precursor of this compound. Two parallel β oxidation pathways are suggested; one with a series of divinyl intermediates, the other with monovinyl, monoethyl intermediates. The compounds were identified primarily by the judicious use of mass spectrometry of oxidation products in addition to infrared and visible spectrometry. It elegantly closes a gap in knowledge about the biosynthesis of Pchlide *a* although the reality of the parallel pathways in normal cells remains to be determined. This issue is likely to be resolved when the specificity of enzymes which catalyse the β oxidation of the propionic residue on ring C can be studied.

To summarize: The propionic acid residue on ring C of Mg Proto IX is first esterified with a methyl group transferred from *S*-adenosyl methionine. Then the β-carbon atom is oxidized to a carbonyl. The activated α carbon then condenses with the γ-bridge carbon to yield the cyclopentanone, or isocyclic, ring. In the *Chlorella* mutants studied by Ellsworth and Aronoff, each of the compounds described occurs in a divinyl and a monovinyl form. The vinyl side chain on ring B of Proto has been reduced to an ethyl group. It is suggested by Aronoff and Ellsworth that the vinyl group oxidizing enzyme is not entirely specific some traces of diethyl (*meso*-porphyrin type) compounds were also detected. It is also implied that the propionic acid β-carbon-oxidizing enzymes are not specific with regard to the presence of a vinyl or ethyl group on ring B. Whether these enzymes are specific for this and other features remains to be discovered.

Cox *et al.* (1969) reported the formation *in vitro* of a substituted form of an isocyclic ring like that in Chl (and other phaeoporphyrins) by treating a magnesium porphyrin bearing a $-CH_2-CO-COCH_3$ substituent with iodine in 98% methanol containing sodium carbonate at 20°C. Kenner *et al.* (1972) have also investigated the conversion of porphyrin β-keto esters into phaeoporphyrins *in vitro*. Carr and Cox (see Cox *et al.*, 1969) are reported to have shown that isolated chloroplasts can convert Mg Proto IX into an isocyclic ring-bearing porphyrin. Kenner *et al.* (1972) reported that Jackson, Kenner and co-workers have converted Mg porphyrin β-keto acids to phaeoporphyrins enzymatically.

I CONVERSION OF PCHLIDE *a* INTO CHLIDE *a*

Many, though not all, algae and some higher plants can form chlorophylls in darkness. Other plants, including leaves of all flowering plants examined, form Chl only upon illumination. Plants which have grown in total darkness from seed are etiolated, i.e. they are pale yellow in colour and are markedly different morphologically from light-grown plants of the same age. Etiolated dicotyledonous plants such as red kidney beans characteristically have unexpanded leaves and an elongated hypocotyl; etiolated monocots such as barley or maize bear long thin leaves. Yellow, etiolated plants form Chl when illuminated. Much work on the final steps in the formation of Chls has been done with illuminated leaves from etiolated plants.

Interest in the final steps in the formation of Chls, including curiosity about the role of Pchl, antedates most of the research already described in this chapter on other aspects of Chl biosynthesis.

The discovery of Pchl dates to Monteverde's (1893) discovery in an alcoholic extract of etiolated leaves of a compound with absorption properties closer to those of Chl than anything then known. As might be expected, for a pigment which was found not only in etiolated leaves but also in inner seed coats of cucurbits such as squash and pumpkin, two opposing positions regarding the role of Pchl were taken: Lubimenko's that Pchl is a by-product of Chl synthesis and for example, Noack and Kiessling's (1930) that Pchl is the immediate precursor of Chl. Pchl was shown to be 7,8-dehydrochlorophyll *a* by Noack and Kiessling (1929), Fischer *et al*, (1939) and Fischer and Stern (1940) and thus it became clear that if Pchl were to be converted into Chl *a* reduction of pyrrole ring D would be required.

In 1946 Sylvia Frank reported that light of wavelengths 445 and 645 nm most effectively promoted the greening of oat seedlings. Minor maxima in the action spectrum occurred at 575 and 545 nm. This action spectrum was in general accord with the absorption spectrum as then known for Pchl dissolved in organic solvents. Not long afterwards J. H. C. Smith began a series of studies which established that Pchl is the immediate precursor of Chl *a*. Koski and Smith (1948) began by isolating sufficient quantities of Pchl from etiolated barley to permit the accurate determination of the absolute absorption spectrum in ether solution. Then, since similar data were available for Chl *a*, it was possible to estimate the concentration of both Chl *a* and Pchl *a* in extracts of leaves. (See Smith and Benitez (1955) for discussion of procedures.) The Pchl present in an etiolated leaf is of the order of 1/100 or 1/200 to 1/1000 of the maximum amount of Chl which the same leaf may contain after prolonged illumination. As will be discussed in greater detail later, a small amount of Chl is formed immediately upon illumination of an etiolated leaf; much larger amounts are produced later, sometimes after a lag of several hours. Smith (1948), in order to deal only or primarily with photochemical conversions, cooled etiolated barley leaves to 0–7°C, took samples for determination of Chl and Pchl, and then, still at lowered temperatures, illuminated the remaining leaves for various periods of time before harvest. He was able to demonstrate a mole-for-mole correspondence between the decrease in Pchl and the increase in Chl during the experiment. Similar results were obtained by Koski (1950) using dark-grown maize seedlings illuminated at room temperature but for short periods ranging from 10 s to 16 min.

(Pchl *a* and its phytolless acid, Pchlide *a*, have identical absorption maxima; the same is true for Chlide *a* and Chl *a*, the phytol ester of Chlide *a*. The biosynthesis sequence was later shown to be protochlorophyllide *a* → chlorophyllide *a* → chlorophyll *a*. The phytol esters were not distinguished from the free acids in the original work.)

Koski *et al.* (1951) studied the action spectrum for Chl formation by etiolated normal and albino maize seedlings exposed to short periods of illumination. Since the albino strain was almost free from carotenoids possible screening by these pigments in the blue region of the spectrum was eliminated. In general agreement with Frank's data on oats, action maxima at 445 and 650 nm were observed for Chl formation by both albino and normal maize plants. However, while the response to 650 nm light was about equal in the two strains of maize, 445 nm light was almost three times more effective in evoking Chl formation by albino than normal plants in accord with expectations of the screening effect of carotenoids.

The data enumerated provided firm support for the proposal that Chl *a* is formed upon illumination of Pchl *a* in seed plants. However, the action spectrum maxima at 445 nm do not precisely coincide with the absorption spectrum maxima for Pchl dissolved in organic solvents (434 and 629 nm in methanol; 432 and 623 nm in diethyl ether (Koski *et al.*, 1951)). The differences between the action spectrum for Chl formation and the absorption spectrum of Pchlide *a* in organic solvents in a general way resembled those between the absorption maxima of an extract of Chl *a* and *b* from a plant and the wavelengths of maximum photosynthesis; the differences were attributed to the arrangement of Pchl molecules within the cell or their attachment to some cell component. Shibata (1957), using his opal glass technique for studying absorption spectra of opaque, i.e. highly light-scattering materials such as leaves (for full discussion of this method, see Shibata, 1959) with commercial spectrophotometers, followed spectroscopic changes in intact etiolated leaves during greening. (Investigations of this sort have generally been confined to the red region of the spectrum because the carotenoids do not absorb there.) He observed that unilluminated etiolated leaves absorb light of 650 nm strongly, i.e. the absorption maximum for Pchl in the red region of the spectrum *in vivo* is at 650 nm, in excellent agreement with the maximum for effectiveness of Chl formation.

Shibata (1957) also observed a complex series of absorbance changes following the illumination of leaves of dark-grown bean or maize. He observed, first, that as the absorption at 650 nm attributed to Pchlide 650 (protochlorophyllide) declined, a new absorption band at 684 nm was seen. The 684 nm absorbing compound is chlorophyllide *a*, this species will be designated Chlide 684. Thus the appearance of the 684 nm absorption band shows that ring D of Pchlide has been reduced. At least one intermediate between Pchlide and Chlide 684 has been detected (Gassman *et al.*, 1968a; Bonner, 1969; Sironval and Brouers, 1970; Thorne, 1971a,b). There is general agreement on the occurrence of a Chlide 678 (fluorescence emission maximum reported as 687–688 nm) as a short-lived spectral intermediate between Pchlide 650 and Chlide 684. Chlide 678 appears after etiolated leaves have been illuminated for 1 s at room temperature (Gassman *et al.*, 1968a). Thorne (1971a) has, in addition, detected Chlide 668 as an intermediate in the conversion of Pchlide 650 to Chlide 678 *in vivo*. He believes that two photosteps, each requiring a single photon, are needed for the photoconversion of Pchlide 650 into Chlide 678. Inasmuch as the 668 nm absorption band is attributed to Chlide the second photon must be needed for some other process if two are indeed required. (The question of the possible nature of the 668 nm absorbing species is taken up later.) Shibata's Chlide 684 (or 683) has been detected at room temperature about 30 s after the appearance of Chlide 678; this transformation is independent of light.

Shibata also observed that at room temperature the 684 nm absorption band decreased in intensity while the absorbance at 671–673 nm increased. After a further 2 h in either light or darkness a shift back to about 677 nm was observed.

To summarize: All the absorbance changes discussed above in an etiolated leaf before and after illumination are shown diagrammatically on p. 110. Not all have been studied equally thoroughly. Only the first change is a result of the reduction of some of the Pchlide; the others reflect changes in the environment of the Chlide or Chl. There is a growing body of evidence (e.g. Kahn *et al.*, 1970; Thorne, 1971a, b; Mathis and Sauer, 1972; Sironval, 1972; Schultz and Sauer, 1972) that Pchlide molecules occur *in vivo* as dimers or larger aggregates and photochemical interactions may occur among as many as twenty molecules. Thorne (1971a) studied the reduction of Pchlide using subsaturating light and only partly converting one spectral species to another. He also observed that only a small fraction of the pigment passed through the 668 nm form which he designated as "Intermediate 668". The circular dichroism analyses of Mathis and Sauer (1972) can be taken to suggest that "Intermediate 668" is composed of a mixed conversion unit–i.e. one containing both Pchlide and Chlide. The requirement for two photons to go from Pchlide 650 to Chlide 678 may show that pairs of pigment molecules are associated.

In addition to the absorption maximum at 650 nm which changes upon illumination, Shibata (1957) observed a band at 632–636 nm (designated for convenience as 632 nm). The latter was especially prominent in older etiolated leaves. This band was unaltered by brief

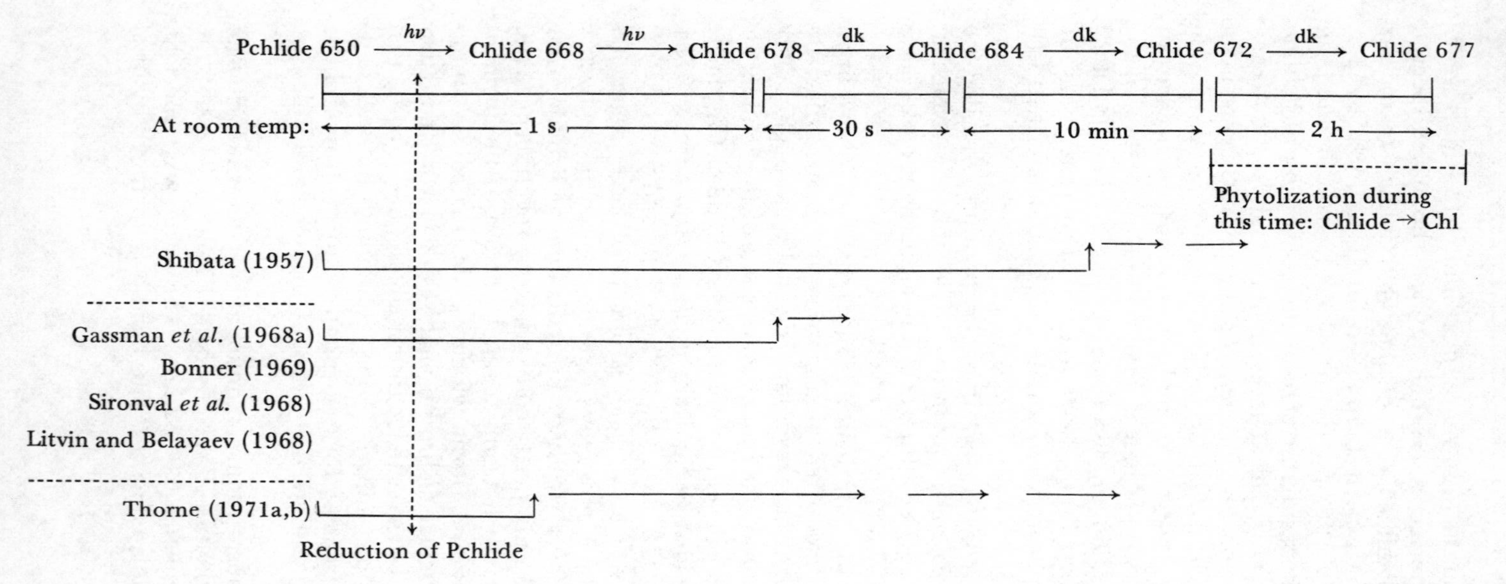

Pchlide 650 —hν→ Chlide 668 —hν→ Chlide 678 —dk→ Chlide 684 —dk→ Chlide 672 —dk→ Chlide 677
At room temp:
1 s
30 s
10 min
2 h
Phytolization during this time: Chlide → Chl
Shibata (1957)
Gassman *et al.* (1968a)
Bonner (1969)
Sironval *et al.* (1968)
Litvin and Belayaev (1968)
Thorne (1971a,b)
Reduction of Pchlide

illumination. Shibata attributed the 632 nm absorption to an inactive, i.e. non-transformable, type of Pchl (later determined to largely or entirely Pchlide). Etiolated leaves utilize exogenous ALA and accumulate enough Pchlide 636 to make them green to the eye (Granick, 1959). Although Pchlide 632 is not converted during a brief illumination period it is utilized after the first Pchlide 650 has been reduced (Gassman and Bogorad, 1967b; Sundqvist, 1969; Granick and Gassman, 1970; Gassman, 1973).

Granick (1950) isolated a *Chlorella* mutant which fails to form Chls in darkness but accumulates Pchlide. *Chlorella* mutant C-10 isolated by Bryan and Bogorad (1963) also forms Pchlide in darkness. The absorption spectrum of intact C-10 cells shows a maximum at 632 nm but the action spectrum for Chl formation peaks at 647 nm; this in a way resembles ALA-fed etiolated leaves in which the main absorption band is at about 632 but the 650 nm form of Pchlide is the absorber for its own photoreduction.

Pchlide *a* and Pchl *a* have identical absorption maxima but are considerably different in solubility in non-polar organic solvents such as light petroleum. The presence of the C-20 isoprenoid phytol side chain of Pchl confers upon it solubility in light petroleum or iso-octane which Chlide lacks. An analogous set of similarities in spectral property and differences in solubility mark the relationship between Chlide *a* and Chl *a*. Wolff and Price (1957) took advantage of the ability to separate easily "-phylls" from "-phyllides" in aqueous acetone extracts of etiolated or illuminated bean leaves. In etiolated leaves they found about 20–25% Pchl *a*; 75–80% Pchlide *a*. After illumination for 10 s the Pchl *a* level was unchanged, i.e. the Pchl *a* was not reduced, but the Pchlide was converted to Chl *a*.

The esterification of Chlide *a* with phytol starts after illumination, following a sigmoid pattern with a lag phase extending from about 10 min to about 40 min from the termination of the irradiation, it is completed only after about 50–60 min; this can all occur in darkness. The period of phytolation covers almost the entire time span of the spectral shifts observed by Shibata (1957). This leads to the conjecture that the last of the spectral shifts—673 to 677 nm—*may* coincide with the phytolation of Chlide *a*; however, the data of Smith *et al.* (1959) on spectral shifts and phytolation in a group of maize mutants illustrates the difficulty of strongly supporting such a guess.

Godnev *et al.* (1963) determined the amounts of Pchl and Pchlide in etiolated plants of seventeen species; they report the percentage of Pchlide to range from 12 to 90%. Smith *et al.* (1959) examined a number of maize mutants in this respect.

In green leaves which are steadily accumulating Chl *a* the near terminal step is the photoreduction of a pigment with the fluorescence spectrum of Pchl or Pchlide *a* (Litvin *et al.*, 1959). Shlyk *et al.* (1960) have shown that in green as well as etiolated leaves protochlorophyllide *a* is the pigment which is converted.

1 *Protochlorophyllide holochrome*

Before Shibata had carried out his detailed studies of the spectral shifts which accompany the conversion of Pchlide *a* into Chlide *a in vivo*, the transformation has been observed in glycerol homogenates or buffered aqueous extracts of etiolated leaves. The unit of conversion *in vitro*, and presumably *in vivo* as well, is not Pchlide alone but a Pchlide–protein complex termed the "Pchl-holochrome" (now more accurately: Pchlide-holochrome) by Smith (1952). (See also Smith and Young, 1956; Smith, 1960.)

Smith (1952) found that the opalescent supernatant fluid from centrifuged glycerine homogenates of dark-grown barley leaves contained a complex of Pchlide and protein which absorbed at about 650 nm; upon illumination as the 650 nm absorption maximum disappeared a new absorption band, presumably that of chlorophyllide holochrome, appeared at 680 nm. Similar phototransformable Pchlide holochromes, sometimes with slightly different absorption maxima, can also be prepared from bean leaves and from squash cotyledons (Smith *et al.*, 1957). The discovery by Krasnovsky and Kosobutskaya (1952) that soluble phototransformable Pchlide holochrome could be prepared by grinding etiolated bean leaves in buffered aqueous solution opened the way for extensive purification of the homochrome.

Smith has summarized the findings of his laboratory on purified bean leaf Pchl holochrome: it is estimated to have one Pchlide molecule per protein of molecular weight about $0{\cdot}5–1{\cdot}4 \times 10^6$; a volume of $2{\cdot}05 \times 10^{-18}$ cm^3; a density, as determined on a sucrose gradient, of 1·16; and in electron micrographs it appears as an oblate spheroid particle with axial diameters of 218, 193, and 93 Å (Smith, 1960; Smith and Coomber, 1961). Boardman (1962a) obtained slightly different values from studies of the isolated Pchlide protein particle: an average of one Pchlide molecule per protein of molecular weight 600 000 ± 50 000; a sedimentation coefficient of 18S; a density of 1·37. Boardman reported the complex to be a sphere of 100–110 Å diameter when viewed with the electron microscope; this is a major difference compared with reports from Smith's laboratory.

Based on studies of Pchlide *a* conversion *in situ* in etiolated leaves and *in vitro*, as a component of the holochrome, the following conclusions have been reached: the quantum yield for the reaction is about 0·6 or about one quantum required per molecule being transformed; the reaction proceeds at −70°C but not at −195°C and the reaction rate is also directly proportional to light intensity, suggesting the possibility that the reaction has a bimolecular aspect; however, the rate of conversion is independent of the initial concentration of complex and of the viscosity of the medium (Smith and Benitez, 1954; Smith, 1960; Boardman, 1962b). In agreement with work by Smith's group, Boardman (1962b) found that only the native Pchlide complex is required for conversion; no evidence for the involvement of a dialysable cofactor could be obtained. Boardman's data led him to "suggest that the transformation does not involve a collision process either between independent protein molecules or between a protein molecule and a hydrogen donor molecule" and he concluded that "the phototransformation of protochlorophyll to chlorophyll *a* can be explained by a restricted collision process within the protochlorophyll–protein complex. . ."

Pchlide holochrome extracted from etiolated bean leaves in dilute Triton X-100 was purified much more extensively by Schopfer and Siegelman (1968). After a series of enrichment steps they identified 300 000 and 550 000 dalton types during agarose gel chromatography. These two were thought to be different-sized aggregates of some smaller units and at least two chromophore molecules were judged to be associated with each 550 000 dalton unit. Henningsen and Kahn (1971) prepared soluble Pchlide holochrome from barley using the natural detergent saponin. Prior to this Pchlide holochrome had been obtained in solution only from bean leaves. The barley saponin holochrome was partially purified by fraction ammonium sulphate precipitation. Then the molecular weight was estimated to be about 63 000 by gel filtration through Sephadex G-100 (filtration through other-sized Sephadexes gave values of 51–83 000 daltons). Earlier Kahn *et al.* (1970) had measured energy transfer between Pchlide molecules in bean holochrome prepared without saponin; they concluded that the "observed efficiencies of energy transfer are consistent with a model of protochlorophyll holochrome containing at least four chromophores". Henningsen and Kahn (1971) observed that there was no energy transfer in their 63 000 dalton barley saponin Pchlide holochrome preparations. They concluded from this that each of these units bears a single chromophore. However, the effect of saponin on energy transfer in this system has not been determined nor would this analysis reveal whether every 63 000 dalton holochrome unit in their preparation carries one molecule of Pchlide. Bean holochrome prepared with saponin had an apparent molecule weight of 100–150 000 daltons.

Bean Pchlide holochrome extracted with dilute Triton X-100 and purified until free of ribulose diphosphate carboxylase and homogenous in a non-denaturing polyacrylamide gel electrophoresis system contains both 37 000 and 54 000 dalton polypeptides (Rosinski and Bogorad, in prep.). It is not yet known whether one or both of these polypeptides bears a chromophore(s). It is also possible that one of these polypeptides is a contamination although the purification methods employed included molecular sizing. Both polypeptides would have had to be in the same larger aggregate or would have had to be in two different but about equally large aggregates.

Guignery *et al.* (1974) administered [^{3}H]ALA to etiolated maize and then prepared a 1500 x *g* plastid-containing pellet. After washing, the pellet was dissolved in a solution of SDS. Polyacrylamide gel electrophoresis in SDS revealed that the radioactivity, entirely attributed to Pchlide, was associated with 21 000 and 29 000 molecular weight bands. The association of non-covalently bound lipophilic molecules, such as Chl, Chlide, Pchlide and Pchl, with proteins in the presence of strong detergents like SDS is not yet well understood; transfer of pigment molecules from protein to protein seems easily possible in the presence of 1–2% SDS. In these particular experiments most of the radioactivity associated with the pellet appears to run free of the two polypeptide bands. Radioactivity from [^{3}H]ALA was shown to be in Pchlide but a small amount could be in other compounds, e.g. amino acids or haems. The approach is interesting but it was not shown here that the radioactivity migrating with the two polypeptide bands is entirely in Pchlide.

It seems likely that the holochrome protein is a photo-enzyme—a reductase (Gassman and Bogorad, 1967b; Sundqvist, 1969; Granick and Gassman, 1970; Gassman, 1973; Sizer and Sauer, 1971; see also Thorne, 1971b). The holochrome protein will be designated "Pchlide reductase" in the ensuing discussions. As described above, Pchlide 650 is regenerated *in vivo* with Pchlide formed from newly made or from exogenous ALA. Pchlide 650 can also be regenerated from non-transformable Pchlide 632 *in vivo*. Some features of the regeneration indicate that the protein, i.e. Pchlide reductase and its substrate, Pchlide, i.e. the two portions of the holochrome complex, are not formed concomitantly. It appears that the holochrome protein can be reused; it can be recharged with new Pchlide after the porphyrin originally present is reduced to Chlide and moves from the active site of the reductase. (Bogorad *et al.* (1968) have considered some possible relationships between the protein and the Chl precursors in relation to the enzymic step and the incorporation of the porphyrin into the photosynthetic membrane. A detailed examination of the latter problem is beyond the scope of this discussion.)

Earlier data (Smith and Benitez, 1954; Smith, 1960; Boardman, 1962b) indicated that, to quote Boardman again "... The transformation [of Pchlide to Chlide] does not involve a collision process either between independent protein molecules or between a protein molecule and a hydrogen donor molecule." In other words, the Pchlide reductase molecule is the immediate source of the hydrogen atoms; it is the reductant.

If a single holochrome protein molecule is reused, as appears to be the case, and it is oxidized during the reduction of its substrate, there must be a mechanism to reduce it again before or after it has picked up a new Pchlide molecule. There should be a Pchlide reduction cycle with a sequence something like this: illumination of the enzyme–substrate complex brings about the transfer of a pair of electrons (or H atoms) from the reductase to the Pchlide, the enzyme is then discharged, reduced, recharged with a new Pchlide molecule, photocatalyses the reduction of the Pchlide etc. A hypothetical cycle is shown in Fig. 6. (Some related schemes have been suggested by Bogorad *et al.* (1968).) According to this view, there are two transfers of a pair of electrons during the cycle: (a) from some unidentified donor to the reductase, and (b) from the latter to Pchlide. Pchlide on the reductase (i.e. on the holochrome protein) can be photo-reduced *in vitro* but there are no reports of the recharging of the enzyme. The possibility that oxidized Pchlide reductase is reduced by a specific transhydrogenase keyed to some enzyme or step in ALA synthesis is suggested as a mechanism for regulating ALA synthesis in Section IV, "The control of chlorophyll metabolism". The successes in purifying the Pchlide holochrome will be used as the basis for exploring the entire Pchlide reduction cycle.

Pchl holochrome *in vivo* absorbs at 650 nm. *In vitro*, for example, in the careful preparation of Schopfer and Siegelman (1968), it absorbs at 639 nm and converts into Chlide 678 (it is not known whether this is equivalent to the *in vivo* Chlide 678). It is obvious that the environment of the Pchlide has changed, perhaps by some diaggregation.

The occurrence of one or two broad red-fluorescing regions in

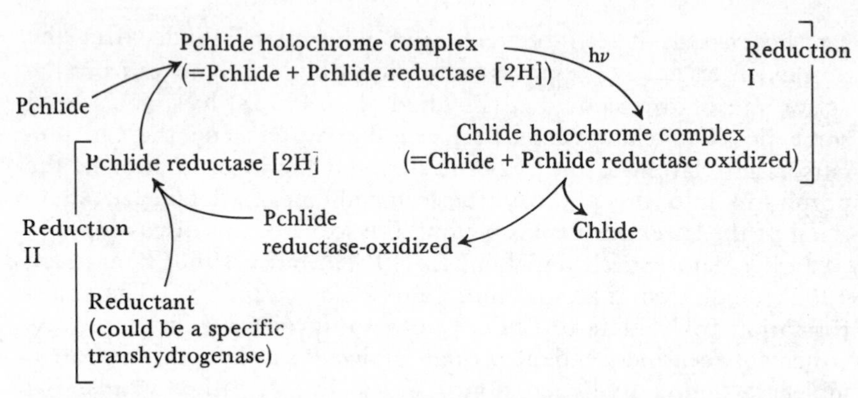

FIG. 6. A hypothetical light-driven Pchlide reductase cycle. As an alternative, Reduction II could occur after the formation of a complex between Pchlide and Pchlide reductase-oxidized instead of where shown here (see discussion of Chl formation in darkness).

etioplasts together with the electron microscopic observation of one or two prolamellar bodies per plastid has led to the conclusion that the pigment is in the prolamellar body (see Boardman, 1966). This view is strengthened by the apparent insolubility of the holochrome but there is no unequivocal evidence to indicate whether: (a) the localization of the Pchlide holochrome coincides with the prolamellar body; (b) even if it does, whether the complex is precipitated onto or is an integral normal component of the membrane. Some possibilities have been reviewed by Bogorad *et al.* (1968), Schultz and Sauer (1972), and by Mathis and Sauer (1972) among others.

2 *Chlorophyll accumulation in darkness*

Those algae and flowering plants which produce Chl in total darkness apparently contain enzyme systems capable of reducing Pchl (or Pchlide). There are, however, numerous species or strains of algae which require light for Chl formation; among these are: *Euglena gracilis*, which has been widely used in studies of Chl formation and plastid development (e.g. Wolken, 1961; Schiff and Epstein, 1965; Holowinsky and Schiff, 1970); a wild-type *Cyanidium caldarium* which requires light for the formation of Chl *a*, phycocyanin, and allophycocyanin (Allen, 1959; Nichols and Bogorad, 1962); a strain of *Chlorella vulgaris* which requires light for growth and greening even on a nutrient medium containing 1% glucose (Finkle *et al.*, 1950); and *Chlorella vulgaris* mutants which grow and accumulate Pchlide in darkness but when grown in the light closely resemble wild-type cells in Chl production and content (Granick, 1950; Bryan and Bogorad, 1963). There is no evidence upon which to decide whether the substrate for enzymic (dark) reduction is Pchl or Pchlide or whether a protochlorophyll(ide) holochrome is involved. However, as discussed earlier, a *Chlorella* mutant which accumulates Pchlide which absorbs at 632 nm when grown in darkness (Bryan and Bogorad, 1963) produces maximal amounts of Chl when illuminated with light of 647 nm the wavelength at which a protein–protochlorophyllide complex would be expected to absorb.

Traces of Chl have been detected in dark-grown oats (Goodwin and Owens, 1947) and in a variety of angiosperms by Godnev *et al.* (1959). However, among higher plants, Chl production in darkness by conifer seedlings is of particular interest (Schmidt, 1924; Bogorad, 1950; Schou, 1951; also see general discussion by Egle, 1960) because large amounts of Chl are formed and two distinct tissues are involved. Besides the various seed coats, a conifer seed consists of a massive haploid megagametophyte tissue in which is embedded the sporophyte, a diploid embryo complete with a number of long thin cotyledons. During germination the cotyledons elongate, pushing the radicle out of the enveloping megagametophyte first. As the cotyledons enlarge the megagametophyte is digested away, but almost to its last remnant it remains in contact with the tips of the expanding cotyledons. Even in total darkness the cotyledons become green, forming Chls *a* and *b* at about the same rate as those of illuminated excised seedlings. However, if the embryos of *Pinus jeffreyi* are excized before germination and grown in darkness on a medium containing sugar, little (Schou, 1951) or no (Bogorad, 1950) Chl is reported to be formed. On the other hand, if seedlings are separated from megagametophytes after germination then after a few days growth in darkness they continue to produce some Chl although at a lower rate than do intact seedlings; the rate of Chl production by isolated seedlings depends in part on the duration of previous contact with megagametophyte tissue. It appears as though the megagametophyte may provide a specific compound which promotes greening without light but the possibility of a general growth promoting effect of digested megagametophyte cannot be excluded.

Bogdanovic (1973) has shown that Chl is formed in darkness by wheat embryos transplanted to the megagametophytes of black pine (*Pinus nigra*).

The spectral response curve for greening of *Chlorella* strain C-10

(Bryan, 1962) indicates that Pchlide reductase plays a role. If we assume that it also functions in the reduction of Pchlide in darkness, as in wild-type *C. vulgaris* and in pine seedlings, how could the dark Pchlide reductase cycle differ from that proposed for light-requiring plants? Certainly the need for light-energy for Reduction I (Fig. 6) is eliminated. Especially if a complex can be formed between Pchlide and Pchlide reductase-oxidized, the use of a higher potential reductant could eliminate Reduction II, i.e. the transfer of $2e^-$ to Pchlide reductase. Pchlide on the enzyme would be reduced directly. Another alternative is that some mechanism for activating Pchlide in the dark or for transferring the hydrogen atoms from the reductase coupled with some energy-yielding step occurs. The capacity to reduce Pchlide in darkness is lost relatively easily by mutation. Such mutants may be useful to help unravel permutations in the Pchlide reductase cycle.

J CHLOROPHYLLASE: AN ENZYME FOR THE PHYTOLATION OF CHLIDE *a*?

From current data it seems likely that the final step in Chl *a* synthesis is the esterification with phytol of the propionic acid residue at position 7 (ring D) of Chlide *a*.

The hydrolysis of Chls to Chlides and phytol by the action of chlorophyllase has been known indirectly since the work of Borodin, who, in 1881, found crystals of "chlorophyll" in and around alcohol-treated leaf sections. The "chlorophyll" crystals were later found to be ethyl Chlide; the enzyme chlorophyllase which is present in the leaf apparently catalyses the hydrolytic removal of the phytol from Chl and esterification of the resulting Chlide with the ethanol in which the leaf is bathed. It is probably more appropriate (although less lyrical) to ascribe the discovery of the enzyme to Willstätter and his co-workers. Chapters on the "Action of chlorophyllase," the "Application of the enzyme of chlorophyllase for making preparations: the chlorophyllides" are included in an extensive summary of investigations on Chl by Willstätter and Stoll published in 1913.

From the first reinterpretations of Borodin's experiments it was clear that chlorophyllase is an exceptional enzyme: although it is destroyed if the leaves are boiled in water the enzyme is active in the presence of high concentrations of alcohol or acetone.

Chlorophyllase appears normally to be intimately associated with lipids or, if it has a removable lipid segment, this portion is not essential for activity. Upon finding large amounts of the enzyme in their preparatons of spinach chloroplastin, a chloroplast–lipoprotein complex obtained by digitonin treatment of chloroplasts, Ardao and Vennesland (1960) suggested that all the chlorophyllase of a mature spinach leaf may be present in a Chl-lipoprotein complex. Klein and Vishniac (1961) prepared 500-fold purified chlorophyllase (or a sodium deoxycholate-chlorophyllase complex) by extracting iso-butanol-treated etiolated rye seedlings with sodium deoxycholate. On the other hand, Holden (1961, 1963) obtained a 500-600-fold purified water-soluble enzyme preparation by cellulose ion-exchange chromatography of an alkaline buffer extract of sugar beet leaf acetone powder. Shimizu and Tamaki (1962) prepared soluble chlorophyllase from tobacco chloroplasts by treatment with *n*-butanol, heating, and acetone and ammonium sulphate fractionation.

There is strong but not obligatory dependence upon organic solvents for enzyme activity; the chlorophyllase in Ardao and Vennesland's chloroplastin did not act upon the Chl of the complex until an organic solvent suitable for dissociating the pigment had been introduced but Klein and Vishniac devised an active aqueous system in which the substrate was solubilized by the detergent Triton X-100. However, in a solution containing 0·2% of this detergent purified sugar beet chlorophyllase hydrolyses Chl at about one-third the rate observed in 40% acetone (Holden, 1963).

Chlorophyllase acts upon Chls including bacteriochlorophyll and *Chlorobium* Chl "650" (chlorophyll $\leftrightarrow$ chlorophyllide + phytol) but not protochlorophyll (Mayer, 1930; Sudyina, 1963). The activity of the enzyme is limited to porphyrins with a carboxymethyl group at carbon 10 and hydrogens at positions 7 and 8—i.e. cyclic tetrapyrroles with ring D reduced as in chlorophyll (Fischer and Stern, 1940). These interesting data leave unexplained the presence of Pchl in etiolated tissues—perhaps the specificity of chlorophyllase is not absolute although other explanations could be concocted readily.

Chlorophyllase catalyses a reversible reaction; thus by manipulating the concentrations of reactants the net accumulation of either Chl or Chlide can be effected. But what is the normal role of the enzyme *in vivo*? Does it participate in Chl synthesis? Knowledge of the specificity of the enzyme and the view that Chl *a* synthesis proceeds via the sequence Pchlide *a* $\rightarrow$ Chlide *a* $\rightarrow$ Chl *a* does not eliminate the possibility that chlorophyllase acts synthetically. Indeed, indirect evidence favours assigning chlorophyllase a synthetic as well as a degradative role. Holden (1963) reported that etiolated tissues have considerably lower chlorophyllase activity than green leaves, that when etiolated leaves are transferred to the light chlorophyllase activity increases, and that this increase parallels the accumulation of Chl. Furthermore, Sudyina (1963) found that chlorophyllase activity increased sharply during the first few minutes of illumination of etiolated plants, a more rapid response than observed by Holden. Finally, Shimizu and Tamaki (1962, 1963) studied longer term changes in chlorophyllase activity. They found that chlorophyllase activity in tobacco was highest in 51-day-old plants and began to decline even before the maximum Chl content was reached after 91 days of growth.

Among more recent studies on chlorophyllase are those by Chiba *et al.* (1967), McFeeters *et al.* (1971) and Ogura (1972). In the former a water-soluble preparation was obtained by *n*-butanol treatment of *Chlorella protothecoides*; partial purification was by acetone fractionation. McFeeters *et al.* (1971) solubilized the enzyme from acetone powders of *Ailanthus altissima* by 0·5% Triton X-100 (which slightly inhibited the enzyme) and purified it about sixtythree-fold by combined heat treatment, ultracentrifugation, gel filtration and DEAE chromatography.

Granick (1967) has suggested that phytol might be added via phytol pyrophosphate rather than via the chlorophyllase reaction. Another possibility is esterification with, for example, a methyl group followed quickly by transesterification with phytol; no methylated intermediate has been detected.

Assuming that chlorophyllase does participate in the synthetic reaction, future work on this problem may reveal whether the substrate is newly formed Chlide holochrome or may relate the action of this enzyme to final stages in situating Chl at its functional site in the chloroplast either indirectly, by conferring lipophilic properties upon it, or directly by the enzymes being situated in the lamellar system.

K THE ORIGIN OF CHLOROPHYLL *b*

It might be simpler to understand the origin of Chl *b* if Seybold and Egle's (1937) observations on the existence of two chromatographically separable Pchls in pumpkin seed coats had been confirmed and extended to leaf tissue. There is, however, no assurance of the existence of a Pchlide *b* (Seybold 1949; Egle, 1960) and other possible precursors of Chl *b* must be considered.

Shlyk and Godnev (1958) have summarized some logical possible interrelationships between the biosynthesis of Chls *a* and *b*: "(1)

Each pigment is formed as a result of a separate biosynthetic chain; (2) Both pigments are formed in parallel from a common precursor. (3) Chl *b* is produced from previously formed Chl *a* . . .. (4) Chl *a* is produced from previously formed Chl *b*. . . ." The choice among these possibilities is made mainly by default.

Proposal 4 cannot be embraced warmly because many plants make Chl *a* but no Chl *b*. Among these are mutants of strains which normally make both Chl *a* and *b* (e.g. Highkin, 1950; Allen, 1958). Besides this indirect evidence, the photoconversion *in vitro* of Pchlide *a* holochrome into Chlide *a* holochrome clearly identifies the precursor of the latter.

Propositions 1 and 2 are related, depending only upon whether the common precursor is taken to be glycine or Pchlide *a*. The former seems extraordinarily unlikely; the latter is possible but not strongly supported by presently available data.

Much of the evidence which has accumulated, some of it contradictory, has been interpreted to support the concept that Chl *b* is formed from *a*. The most direct experiment is that of Godnev *et al.* (1959, 1963) in which the tops were trimmed off onion leaves which were just starting to become green and [^{14}C] Chl *a* dissolved in sunflower oil was introduced into the hollow centre of the remaining leaf. Two days later the leaves were extracted with acetone and, after purification of the extract, [^{14}C] Chl *b* was found. No radioactivity was found in Chl *a* from a leaf similarly exposed to Chl *b*. Most other pertinent data are from experiments in which changes in the specific radioactivity of each Chl was found after the administration of a pulse or during continuous exposure to $^{14}CO_2$ or to [^{14}C] glycine, [^{14}C] acetate, [^{14}C] δ-aminolaevulinic acid or some other appropriate precursor of Chl of green leaves, a growing algal culture, or etiolated leaves subsequently illuminated.

Shlyk and Godnev (1958) found, for a number of plants, the specific radioactivity of Chl *a* to be about 2·5–4·7 times higher than that of Chl *b* in the same leaf. Perkins and Roberts (1960) obtained similar data with wheat leaves. In a 10-day-long experiment with *Ceratophyllum demersum* L., Shlyk and Godnev (1958) found that the ratio "specific activity chlorophyll *a* specific activity chlorophyll *b*" rose to a maximum of 4·7 on the third day but then declined to 2·8 on the tenth day. However, there are further complexities: an effect of day length on the ratio of specific activities of the two Chls has been observed by Sironval *et al.* (1961). Furthermore, Duranton *et al.* (1958) found that Chl *a* from tobacco supplied with [^{14}C] δ-aminolaevulinic acid had at first three times the specific activity of the

Chl *b* but in the course of the experiment the activity of *b* remained constant while that of *a* declined.

The data from some studies of algal cultures, *Chlorella* (Becker and Sheline, 1955; Blass *et al.*, 1959; Michel-Wolwertz, 1963; Shlyk *et al.*, 1963b) and *Scenedesmus* (Blass *et al.*, 1959) are similar to those from investigations on green leaves, i.e. the specific activity of Chl *a* remains 2·5–3 times higher than that of *b*. On the other hand, under different conditions the two Chls can become equally radioactive in *Chlorella* (della Rosa *et al.*, 1953; Shlyk *et al.*, 1963a) and in one case Chl *b* of specific radioactivity more than 1·6 times that of Chl *a* was isolated from *Chlorella vulgaris* cells which had been incubated with [2-^{14}C]glycine for 6 h (Brzeski and Rucker, 1960). Michel-Wolwertz's (1963) observations on the effect of light intensity upon the relative specific activities of the two Chls in *Chlorella* may help explain such variations: "The specific activity of chlorophyll *b* increases progressively with the light intensity and very rapidly with the strongest intensities. On the other hand, chlorophyll *a*, so long as there is light, reveals a fairly constant specific activity. . . ." All these observations may not argue very strongly for the conversion of Chl *a* into *b* but they do not preclude the possibility of such a transformation. The complexities of interpretation of changes in relative specific activities are also apparent; they are undoubtedly compounded by any Chl turnover which may occur.

Much more satisfactory are labelling studies with etiolated plants which convert their Pchlide *a* to Chl *a* immediately upon illumination, later produce additional "new" Chl *a*, and sometimes at about the same time (e.g. maize–Koski, 1950), sometimes considerably later (oats–Goodwin and Owens, 1947), begin to make Chl *b*(also see Seybold and Egle, 1938). Michel-Wolwertz (1963) supplied [^{14}C] δ-aminolaevulinic acid to etiolated barley leaves and later illuminated them. After 4 and 6 h of illumination the Chls were extracted and their specific activities were found to be equal. According to Michel-Wolwertz this argues for a sequential synthesis of Chl *b* from *a*; Smith and French (1963) point out that it could equally support the concept of the production of the two pigments from a close common precursor.

In another type of experiment Shlyk *et al.* (1963a) supplied illuminated green barley plants with $^{14}CO_2$ for 20–30 min, transferred the plants to darkness, and over the next 8 days periodically took samples and determined the specific activities of Chl *a* and *b*. During this interval the specific activity of the Chl *b* rose, while that of *a* dropped. Shlyk believes that these and other data demonstrate the origin of Chl *b* from *a*.

It is not known whether there is normally a light-requiring step specific for Chl *b* synthesis but Allen (1958) has isolated a *Chlorella* mutant which forms only Chl *a* in darkness but both Chls *a* and *b* when illuminated. Such an organism might be useful in studies of possible Chl interconversions.

The acceptance and accommodation of conflicting data from tracer experiments is complicated by the great difficulty of bringing Chls to radiopurity; a problem recognized at some time in the writing of almost every worker in this area. The complications introduced by differences in the rate of labelling with carbon of the tetrapyrrole nucleus and the esterifying groups of Chls are fairly generally recognized if not always taken into account.

Smith (1960) has argued that Chls *a* and *b* might arise from a common precursor (Chlide *a*?) because the rates of formation of *a* and *b* are equal once Chl *b* production has begun in greening etiolated barley. He also cites the experiments of della Rosa *et al.* (1953) and Duranton *et al.* (1958) as evidence *against* the formation of Chl *b* from *a*.

In trying to interpret the effects of light intensity on the relative radioactivity of Chls *a* and *b* in *Chlorella* supplied with [^{14}C]acetate Michel-Wolwertz (1963) suggests that two kinds of Chl *a* molecules might be recognized, old ones which are more readily photoxidized than young ones, the latter remaining more available for transformation to Chl *b*. This could account for some of the observations, but it does bring up other important issues. Where might Chlide *a* fit in this biosynthetic scheme (chlorophyllase can act on Chlide *b* as well as *a*)? Where is the enzyme for the oxidation of the methyl group at position 3? Is it in the lamellar structure? And the major question: When and how are the Chls set into their functional places?

To summarize: the two most likely possible immediate precursors of Chl *b* are (1) Chl *a* or (2) a compound which can be transformed into either Chl *a* or *b*.

In an excellent review Shlyk (1971) has reconsidered the problem and concluded that "on the strength of all the evidence, a verdict is now reached in favour of the sequential scheme of Chl *b* from Chl *a* formation".

L CHLOROPHYLLS *c* AND *d*

Chls c_1 and c_2 have been separated from one another and prepared in crystalline form by Jeffrey (1972) from brown seaweeds (*Sargassum* sp.) and from diatom cultures (*Phaeodactylum tricornutum*). Chl c_2 was also prepared from symbiotic dinoflagellates which grow in the mantle of the rock-boring clam *Tridacna croecea* which lack Chl c_1. The amounts of these Chls and the ratio $c_1:c_2$ were determined for a much larger number and variety of algae.

Chl c_1 in acetone has absorption maxima in the red region at 633 and 694 nm; Chl c_2 at 635 and 696 nm. The absorption maxima and relative band intensities in dioxane are: Chl c_1–450·6, 584·2, 630·3 (10·22:0·88:1·00); Chl c_2–454·2, 584·9, 631·4 (14·37:1·34:1·00) (Jeffrey, 1972). Chl c_1 is Mg tetrahydropheoporphyrin a_5 monomethyl ester (a monovinyl compound); Chl c_2 is Mg hexahydropheoporphyrin a_5 monomethyl ester (it has a vinyl group on ring A and one on ring B like Proto). An acrylate residue is present in both of these Chls on ring D (in lieu of the propionate residue of Pchlide) (Dougherty *et al.*, 1966, 1970; Wasley *et al.*, 1970; Budzikiewicz and Taraz, 1971; Strain *et al.*, 1971).

Chl c_1 is identical to Pchlide except for the acrylate residue on ring D; it could be derived from Pchlide. Chl c_2, the divinyl compound, again except for the acrylate residue, is similarly related to Mg 2,4-divinyl phaeoporphyrin a_5 monomethyl ester. The latter is believed to be the immediate precursor of Pchlide; it has been recovered from *R. spheroides* cultured in 8-hydroxy quinoline (Jones, 1963b), several unicellar flagellates by Ricketts (1966), and earlier from *R. spheroides* mutants (Stanier and Smith, 1959; Griffiths, 1962).

Chl *d*, 2-desvinyl,2-formyl Chl *a* (Holt, 1960, 1961) could arise by oxidation of Chl *a* but its precursor has not been identified.

M THE CHLOROPHYLLS OF BACTERIA

Early steps in the formation of Chlide *a* and of bacteriochlorophyll *a* (Bchl *a*) are the same. Stanier and Smith (see Smith, 1960) isolated a pigment which closely resembles Pchlide spectroscopically from a "tan" mutant of *R. spheroides.* Sistrom *et al.* (1956) observed that another mutant of this bacterium excreted pheophorbide *a* (magnesium-less derivative of Chlide *a*). Griffiths (1962) also found a pigment which spectroscopically resembles Pchlide *a* (or Pchl *a*) as well as another pigment similar to Chlide *a* in mutants of *R. spheroides* which lack Bchl. And, again, Drews *et al.* (1971) discovered a mutant of *Rhodopseudomonas capsulata* which produces Pchlide and its Mg-free derivative as well as Bchl *a*. All these observations show that bacteria which normally form Bchl *a* can, in mutant forms, accumulate Chlide *a* and its immediate precursor

Pchlide *a*. In confirmation of this Jones (1963c) grew *R. spheroides* in the presence of 8-hydroxyquinoline and found that they excreted Chlide and its Mg-free derivative. Magnesium is easily removed from magnesium porphyrins in solutions which are slightly acid. It is likely that most of the Mg-free derivatives are secondary although some observations of the presence of bacteriopheophytin are present in the reaction centres of *R. spheroides*.

Rhodopseudomonas spheroides growing in the presence of 8-hydroxyquinoline also excreted the Mg-free derivative of 2-devinyl, 2-hydroxyethyl, Chlide *a* (Jones, 1964). Lascelles (1966) found that some *R. spheroides* mutants accumulate the same compound but containing Mg. None of these compounds has ring 2 reduced. They are at the same oxidation level as Chlide *a*. This suggests that the acetyl group on position 2 on Bchl is formed by the hydration of the vinyl residue at that position on Chl *a* and that the hydroxyethyl group is then oxidized to an acetyl group.

The next step is then the reduction of ring 2 to form Bchlide *a*. Some mutants of *R. spheroides* studied by Lascelles and Altschuler (1967) accumulate this pigment.

The final step in this sequence should then be the esterification of the propionic acid residue at position 7 on ring D. Other mutants of *R. spheroides* and *R. rubrum* which accumulate 2-devinyl,2-hydroxyethyl Chlide or its Mg-free derivative have been studied by Oelze and Drews (1970), by Oelze *et al.* (1970), and by Richards and Lascelles (1969). Brockmann and Knobloch (1972) have found that *Rhodospirillum rubrum* and *Rhodopseudomonas palustris* contain not only Bch a_P ("P" to designate the phytyl ester) but also Bchl a_F, a Bchl *a* which is a farnesyl rather than a phytyl ester. *Chlorobium* Chls 650 and 660 (sometimes designated Bchl-*c* and -*d*) are also esterified with farnesol rather than phytol (Pfennig, 1967).

Chlorobium Chls 650 and 660, like higher plant Chls, have only ring D reduced. Consequently, their red absorption bands are in the 650 nm region rather than much farther into the infrared. In both of these Chls there is a hydroxyethyl ($-CHOH-CH_3$) substituent at position 2 (ring A) rather than a vinyl group as in Chls *a* or *b* or an acetyl group as in Bchl *a*. Thus the *Chlorobium* Chls are at an intermediate level of oxidation. The carboxymethyl group is missing from the cyclopentanone ring and in *Chlorobium* Chl 660 (Bchl *c*), but not the 650 form, the δ-bridge carbon is substituted with a methyl group (Pfennig, 1967).

If the biosynthetic chain leading toward the *Chlorobium* Chls resembles that in *Rhodopseudomonas* and *Rhodospirillum* species the pathways to these chlorophylls and to Bchl *a* is probably common up to the formation of the 2-devinyl,2-hydroxyethyl Chlide *a*. Removal of the carboxyethyl group attached to the cyclopentanone ring would lead directly to the formation of *Chlorobium* Chl 650. This group would probably be removed easily after hydrolysis of the methyl ester. Finally, or prior to the removal of the carboxymethyl group, the propionic acid residue on ring 4 would be esterified with farnesol.

Richards and Rapaport (1967) have observed the synthesis of Mg Proto monomethyl ester, Bchl *a*, and *Chlorobium* phaeoporphyrins by *Chlorobium thiosulphatophilum*-660.

In summary, the sequence discussed is:

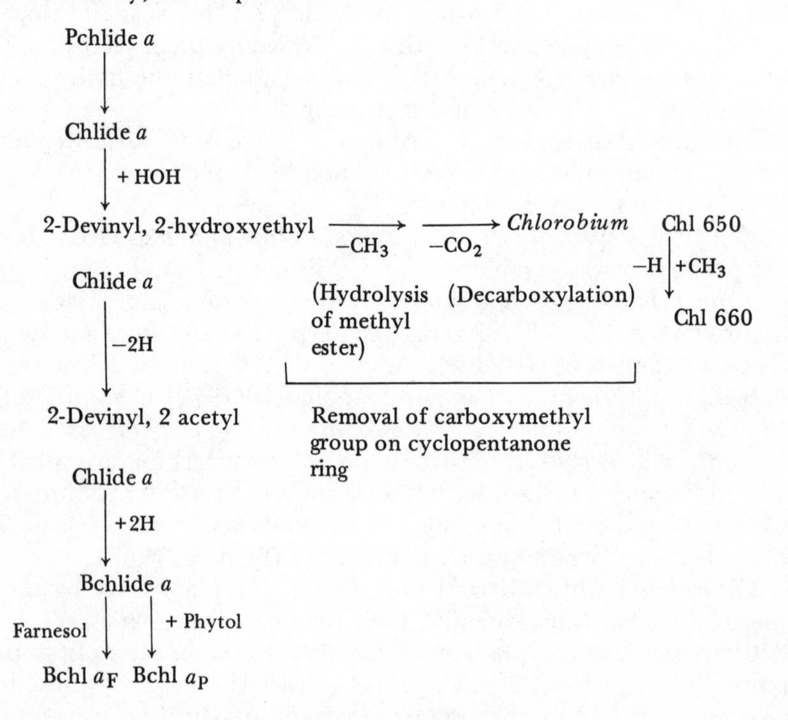

IV The control of chlorophyll metabolism

The control of Chl formation has been studied principally in two situations: (a) greening of plants which fail to form Chls unless illuminated, e.g. dark-grown seedlings and some algae which can be grown heterotrophically in darkness but do not accumulate Chls under these conditions; and (b) formation of tetrapyrrolic pigments upon transfer from aerobic to anaerobic conditions of *Rhodopseudomonas spheroides, Rhodospirillum rubrum*, and other photosynthetic bacteria which lack Chls when grown aerobically, particularly aerobically in darkness, but form these pigments when grown anaerobically, particularly anaerobically in the light.

Three types of control mechanisms have been proposed: (a) feedback inhibition by an intermediate or a derivative of an intermediate in Chl biosynthesis of an enzyme acting early in the chain, e.g. feedback inhibition of the synthesis of ALA by haem; (b) repression–derepression control over the synthesis of the enzyme which catalyses the synthesis of ALA; and (c) activation–inactivation control of a more or less constant amount of a controlling enzyme of ALA synthesis.

A BACTERIOCHLOROPHYLL BIOSYNTHESIS IN ATHIORHODACEAE

Rhodopseudomonas spheroides forms bacteriochlorophyll when grown anaerobically in light of moderate intensity but only traces of the pigment are found in thoroughly aerated cultures maintained in darkness (van Niel, 1944; Cohen-Bazire *et al.*, 1957). The virtual absence of ALA, PBG or porphyrin precursors of bacteriochlorophyll show that the biosynthetic path is blocked prior to the synthesis of any of these compounds in aerobically dark-grown cells.

Small amounts of pigment are accumulated by unilluminated cells grown under reduced oxygen tension (Cohen-Bazire *et al.*, 1957; Lascelles, 1959). Lascelles (1959) found that ALA synthetase and ALA dehydrase are five times more active in organisms grown photosynthetically, i.e. anaerobically in the light, than in aerated dark-grown cells which are bacteriochlorophyll poor. In general, the level of bacteriochlorophyll parallels, and is presumably at least in part regulated by, the activity of one of these enzymes.

How is the activity of ALA synthetase regulated in response to O_2 tension and illumination? Inhibitors of protein synthesis (DL-*p*-fluorophenylalanine and chloramphenicol—Sistrom, 1962; Bull and Lascelles, 1963) and of nucleic acid synthesis (5-fluorouracil—Bull and Lascelles, 1963) block the increase in ALA synthetase activity and bacteriochlorophyll formation which normally occur during anaerobic growth in the light. The simplest and most direct explanation would be: the production of ALA synthetase, and of mRNA for this protein are promoted by light and anaerobiosis; inhibitors of nucleic acid or protein synthesis block the formation of the enzyme and so there is no rise in activity. The most cautious users of transcription and translation inhibitors have usually pointed out that the data show only that the synthesis of some RNA and/or *some* protein is necessary. Others have guessed that the effect of the inhibitors was directly upon the production of the enzyme or the mRNA which specifies it. The latter view now seems difficult to defend. Neuberger *et al.* (1973a, b), Davies *et al.* (1973), Marriott (1968), Marriott *et al.* (1969) have shown that ALA synthetase activity is the same in freshly prepared extracts of *R. spheroides* cells grown anaerobically in the light as in similar extracts from cells oxygenated for 1 h before harvest. The activity was low in both extracts initially but during storage at 4°C for 1 h in the presence of air ALA synthetase activity in the extract of light-anaerobically-grown cells increased eightfold. It appears, as discussed in the section on ALA synthetase, that an activating system is present in light-anaerobic cells—or at least extracts of these cells, but is absent from oxygenated cells—or their extracts. Neuberger *et al.* (1973b) obtained from extracts of unoxygenated cells material which brought about six- to eightfold increase (activation) of ALA synthetase in extracts of oxygenated cells. This showed that the ALA synthetase is not quickly destroyed on oxygenation of the culture but that the enzyme becomes inactive and, unlike the enzyme from anaerobic cells, it is not reactivated. The maintenance, build-up or destruction of an activating system could be the control point. This moves the control problem back a step but the change in activity of ALA synthetase is not the only consequence of oxygenation. For example, the activities of some other enzymes of porphyrin biosynthesis also change and the rapid accumulation (or increase in activity) of ribulose 1,5-diphosphate carboxylase under light-anaerobic conditions stops abruptly upon vigorous aeration and resumes upon flushing with N_2–CO_2 (Lascelles, 1960). Some more central change in the cell which affects several metabolic processes including ALA synthesis simultaneously may occur.

An ALA synthetase activating enzyme has been studied by Tuboi and Hayasaka (1972a). Davies *et al.* (1973) have suggested that the activity of ALA synthetase depends on the concentration of a trisulphide component of the activating system. The concentration of this activating component could be affected by the oxidation–reduction balance of the cell, i.e. the activity of its electron-transport chain, depending in turn upon the degree of oxygenation of the cells.

ALA synthetase of *R. spheroides* is strongly inhibited by low concentrations of haem and haemin (Burnham and Lascelles, 1963; Gibson *et al.*, 1961; Warnick and Burnham, 1971). Feedback inhibition of ALA synthetase by haemin in *R. spheroides* had been suggested as a control mechanism earlier (Lascelles, 1968).

Lascelles (1968) has pointed out that protein synthesis inhibitors stop bacteriochlorophyll synthesis in *R. spheroides* quickly but that haem synthesis continues for several hours. The following sequence leading to termination of ALA synthetase action is suggested: (a) A block in protein synthesis by puromycin stops bacteriochlorophyll synthesis (for some unknown reason) but haem synthesis continues unimpaired for a few hours longer. No other porphyrins accumulate. (b) The haem acts as a feedback inhibitor of ALA synthesis. Haem synthesis almost stops about 3 h after addition of puromycin. If *o*-phenanthroline, a metal chelator, is added together with puromycin, large amounts of Coprogen accumulates for at least 10 h. The *o*-phenanthroline effect is taken to show that if haem formation is arrested there is no effect of puromycin on bacteriochlorophyll formation.

The possibility that feedback inhibition by haem plays an important role in the control of ALA synthetase in *R. spheroides* has been strengthened by the conclusion that the insertion of magnesium into Proto is the rate-limiting step in the formation of bacteriochlorophyll from Proto and is also the only step at which oxygen inhibits the conversion of Proto to bacteriochlorophyll (Lascelles and Altschuler, 1959; Davies *et al.*, 1973). The accumulated Proto or the haem formed from it could block ALA synthetase activity.

Extracts of cells of *R. spheroides* grown anaerobically in the light, compared with preparations from aerobically light-grown cells, in addition to differences in ALA synthetase activation also have much higher levels of activity of ALA dehydrase (Lascelles, 1959) and of *S*-adenosyl methionine Mg Proto transferase (Davies *et al.*, 1973). They incorporate Mg into Proto much more actively (Gorchein, 1973) as well.

Do these changes in activity result from the formation of additional enzyme molecules, from activation or inactivation of a fixed number of enzyme molecules, or a combination of these? These questions can be answered only after each enzyme has been purified completely and specific activities can be determined. The recent observations on an activation system for ALA synthetase show how estimates of changes in gross activity, as were made earlier, can sometimes provide misleading information about regulatory mechanisms. A similar case may be developed with regard to apparent changes in ALA synthetase activity in animal mitochondria during chemically induced porphyria. Patton and Beattie (1973) report that equally high activities can be detected in sonicated preparations of mitochondria from livers of treated and untreated animals under some assay conditions; other assays show less activity in control samples.

In summary: ALA synthetase activity in *Rhodopseudomonas spheroides* can appear to be higher in extracts of cells grown anaerobically in the light than in extracts of cells cultured in the dark with air or oxygen. The increase in activity which normally follows quickly upon transfer of cells to anaerobic conditions in the light is blocked by various inhibitors of protein and RNA synthesis. Yet, under some assay conditions it can be seen that the potential ALA synthetase activity is the same in extracts from the two types of cells. The enzyme can exist in either an active or an inactive form; control of ALA synthesis may be via an ALA synthetase-activating system and thus probably the sum of active and inactive enzyme molecules is the same in aerobic and anaerobic cells. It seems likely

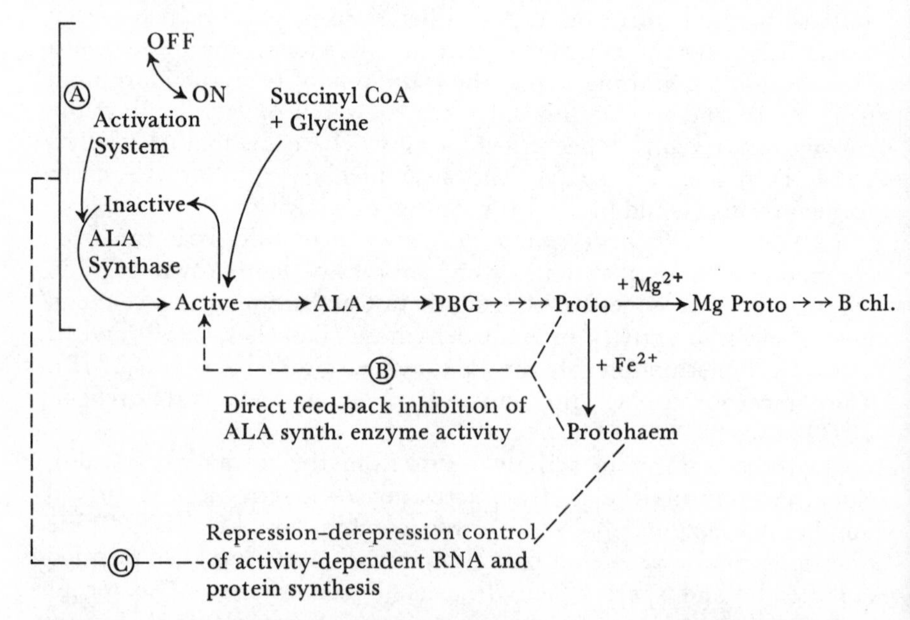

FIG. 7. A summary of possible mechanisms for the control of bacteriochlorophyll synthesis in *Rhodopseudomonas spheroides.* A: An activation–inactivation system working on ALA synthetase. B: Direct feedback inhibition of ALA synthetase by Proto and/or protohaem. C: Repression–derepression control of the synthesis of RNA and protein required to sustain ALA synthetase activity. May be direct control of ALA synthetase production in some circumstances.

that such a system, or the conditions which promote its action, may also regulate other parts of the porphyrin biosynthetic chain, the formation or activation of enzymes of photosynthetic carbon fixation, and the formation of photosynthetic membranes (Takemoto and Lascelles, 1973) in *R. spheroides*. In addition, ALA synthetase activity may be moderated via feedback inhibition by haem or haemin which could accumulate when some step in bacteriochlorophyll formation beyond Proto is blocked. There are no data now available from which to judge whether feedback inhibition, activation–inactivation or both play important roles *in vivo*. It is also uncertain now whether ALA synthetase itself (as opposed to some component of the ALA synthetase control system) is an inducible enzyme in this organism. The activation system will need to be understood and the amount ALA synthetase rather than only its activity will need to be measured (Fig. 7).

In considering the control of ALA synthesis in chick embryo liver cells by steroid hormones and other agents, Granick (1966) and Kappas and Granick (1968) proposed that ALA synthetase formation is controlled by a protein apo-repressor and haem, acting as a co-repressor. The steroids or porphyria-producing drugs serve as inducers in this working hypothesis by displacing haem from the apo-repressor. The repressor system is thus inactivated and more ALA synthetase is produced. Granick (1966) found the activity of rat liver mitochondrial ALA synthetase to be insensitive to haem.

B CHLOROPHYLL BIOSYNTHESIS IN GREEN PLANTS

When etiolated leaves are illuminated their Pchlide is converted to Chlide *a*. If such leaves are returned to darkness after the brief exposure to light required for the conversion of the pigment they accumulate Pchlide again. Madsen (1963) reported that 12–15 min after being returned to darkness etiolated wheat and barley leaves contained as much Pchlide as before they were illuminated; from about this point Pchlide production continues at a constantly decreasing rate and finally ceases after a total of about 1 h in darkness. While new Pchlide is being formed, the Chlide produced during illumination is being esterified with phytol. By repeatedly illuminating leaves with light flashes as short as 1/10000 s a considerable amount of Chl can be accumulated if adequate (10–15 min) dark periods are interposed (Madsen, 1963).

The situation is more complex in leaves under continuous light. The production of additional Pchlide *a* apparently occurs just as in darkness (as has already been discussed); the Pchlide *a* is converted into Chlide *a*; and finally Chl is formed. However, the rate of Chl production generally increases drastically during illumination.

During the early stages of illumination new Chl is formed in continuously illuminated leaves at about the same rate as Pchlide accumulates in leaves returned to darkness (Virgin, 1955). This is termed the "lag phase" because soon afterward the rate of Chl synthesis increases sharply and continues at a constant rapid rate until net Chl synthesis ceases in the fully-greened leaf. For example, 12-day-old etiolated bean leaves illuminated with 30 μW cm^2 of red (600–700 nm) light formed less than 50 μg of Chl per leaf during the 2 h lag period but about 400 μg per leaf during the next hour, i.e. the first hour of the period of rapid greening (Sisler and Klein 1963). Thus when etiolated leaves are illuminated continuously the Pchlide present is rapidly converted to Chlide, new Pchlide is formed (and converted into Chlide) first slowly and then, accelerating to a constant rapid level, as much as ten times more rapidly than during the "lag phase" until the leaf is completely green. Phytolation and the development of lamellae in the chloroplast occur during greening (Virgin *et al.*, 1963).

The lag phase varies from being imperceptible to lasting about 5 h depending upon the age of the leaf, the species or variety of plant, light intensity, temperature, and some other conditions (Virgin, 1955; Sisler and Klein, 1963). Etiolated beans 2, 3 or 4 days old formed Chl at a constant slow rate throughout the 3 h period of illumination studied by Sisler and Klein (1963), i.e. a period of rapid greening never developed or, according to their view, there was no lag phase. However, in leaves of the 5-day-old etiolated Black Valentine bean plants Chl formed rapidly from the very beginning of illumination. Leaves from 7-, 12-, and 21-day-old plants exhibited lag phases of 1·5–2 h duration.

During the rapid linear phase of Chl formation in greening dark-grown leaves, Chl accumulation stops abruptly when the leaves are returned to darkness (Gassman and Bogorad, 1967a). No additional Chl or Pchlide is detectable but the limit of accumulation seems likely to be equivalent to the pigment formed after return to darkness of briefly illuminated dark-grown leaves. Such amounts would be undetectable in rapidly greening leaves; they would constitute a small increment against a large background.

In both the case of etiolated tissue and greening plants returned to darkness no precursors of Pchlide can be detected. The biosynthetic chain of Chl is either on or off. The failure to detect precursors of Pchlide demonstrates that the synthetic system is turned off and on at the very first more or less irreversible step in porphyrin biosynthesis—ALA synthesis. To emphasize this point, the very limited amount of Pchlide in etiolated leaves is augmented at least tenfold by the administration of ALA in darkness, as already mentioned (Granick, 1959) and the arrest of Chl-Pchlide production in greening leaves by the administration of chloramphenicol, which mimics the effect of darkness, can also be partly overcome by ALA feeding (Gassman and Bogorad, 1967a). The conclusion that all the enzymes from ALA dehydrase on required for Pchlide formation are constitutive fits with these observations; the enzyme for ALA synthesis or for the production of its substrate or some activator of one of these enzymes is photoinducible. The production of the (or a) controlling material by greening leaves requires continued RNA and protein synthesis (Margulies, 1962; Bogorad and Jacobson, 1964; Gassman and Bogorad, 1967a,b). What is the controlling component? How ALA is synthesized in green plants, including algae, is not known. We do not know what enzymic reaction in the biosynthesis of porphyrins could be regulated. The complete answer cannot be given now but some guesses about controls can be made.

Some algae (e.g. *Chlorella vulgaris* and *Chlamydomonas rheinhardi*) form Chls equally well when growing heterotrophically in darkness or purely photosynthetically. Some mutant strains of *C. vulgaris* and *C. rheinhardi* as well as wild-type cells of, for example, *Euglena gracilis* and *Cyanidium caldarium*, resemble angiosperms in requiring light for Chl production.

Illumination of dark-grown seedlings or algae which require light for greening has two prompt effects as described above. Holochrome-associated Pchlide *a* is immediately reduced to Chlide *a* and ALA synthesis, which begins within minutes, provides the substrate for making more Pchlide. Are these two events related?

A little later, complex metabolic changes occur including: an increase in the activity of plastid RNA polymerase (Bogorad, 1967); increases in the activity of some enzymes of the photosynthetic carbon reduction cycle notably including ribulose 1-5-diphosphate carboxylase (Chen *et al.*, 1967; Keller and Huffaker, 1967) build-up of thylakoid membranes by the formation and/or incorporation of some lipids and previously lacking polypeptides (e.g. Forger and Bogorad, 1973). Are these changes related to the early changes in porphyrin metabolism or are they under separate control?

The latter question is beyond the scope of the present discussion but the possible mechanisms by which the synthesis of ALA is regulated will be explored.

Three kinds of observations suggest that the accumulation of Pchlide or Chlide may directly or indirectly block ALA synthesis:

1. There are two classes of Chl-less mutants of *Chlorella vulgaris* with regard to accumulation of Chl precursors or their derivatives. One class is exemplified by strain C-10 of Bryan *et al.* (Bryan, 1962; Bogorad, 1963; Bryan *et al.*, 1967). These cells accumulate very little total porphyrin in darkness, almost all is Pchlide of which some is associated with the holochrome, judging from the action spectrum for Chl production.

Members of the other class accumulate large amounts of porphyrins—about equal to the Chl in wild-type cells, i.e. about 5% of the dry weight of the cells. The porphyrins are Mg Proto, Proto and some of their precursors; no Pchlide. This class is exemplified by a number of porphyrin-accumulating mutants isolated by Granick (1948a,b, 1949, 1950) and by Bogorad and Granick (1953a). In these strains the pigments are produced in darkness and, just as with wild-type *C. vulgaris*, the amount of porphyrin is the same whether cells are grown in light or darkness.

The total amount of Chl-related porphyrins formed is small because of limited ALA synthesis when Pchlide accumulates, as in strain C-10, but the total is as large as in wild-type cells in several cases where the block in biosynthesis is before Pchlide as in *Chlorella* strains which accumulate Proto or Mg Proto.

2. Cotyledons of pine seedlings germinated in darkness contain the same concentrations of Chls *a* and *b* as those germinated in the light (Schmidt, 1924; Bogorad, 1950). Greening of the cotyledons in darkness depends upon continued contact with the megagametophyte tissue during germination. When the contact is broken Chl soon stops forming (Bogorad, 1950). The megagametophyte provides something which permits the reduction of Pchlide to Chlide in the cotyledons without light, just as it must occur in wild-type *C. vulgaris* for example.

These data can be interpreted as follows: Pchlide accumulates (which could lead to the arrest of ALA synthesis) but its reduction to Chlide occurs through the action of something from the pine megagametophyte and ALA and consequently more Pchlide is formed.

Bogdanovic (1973) has found that primary leaves of dark-grown wheat (*Triticum vulgare*) embryos form Chl in darkness when transplanted with a piece of scutellum onto black pine (*Pinus nigra*

Arnold) megagametophytes. Not all of the Pchlide is reduced. The data provided do not make it clear whether, like pine cotyledons, the leaves of transplanted wheat form a great deal more Chl plus Pchlide than the controls. They do make some extra pigment.

3. Pchlide conversion by light is soon followed by ALA synthesis. If etiolated plants are returned to darkness they continue to accumulate Pchlide to about the same level (or twice the level) as they contained before they were first illuminated. This suggests that there may be some stoichiometric relationship between the level of Pchlide or Pchlide holochrome and the activity of the ALA-synthesizing system.

All three of these points might also be explained by the backing up of the biosynthetic chain leading to the accumulation of Proto which could be converted into haem. If this system is inhibited by haem, ALA formation would be blocked (The effect of haem on ALA synthesis in plants is not known.) However, the behaviour of the *Chlorella* mutants, especially those which accumulate very high concentrations of Proto and Mg Proto argue against pile-up back-up effects up to Mg Proto or Mg Proto monomethyl ester.*

The effects of iron on Chl formation are complex. Iron-deficient plants are chlorotic. Marsh *et al.* (1963a) found that chlorotic leaves from iron-deficient cowpea plants formed protoporphyrin when incubated with ALA in darkness indicating that porphyrin biosynthesis in these leaves was blocked by their failure to synthesize ALA and to incorporate magnesium into protoporphyrin. They found, however, that illuminated chlorotic leaf discs can incorporate radioactivity from [^{14}C]ALA into chlorophyll at the same rate as normal tissues (Marsh *et al.*, 1963b). Differences in the effect of iron

* Gough (1972) carefully analysed data on mutants of algae and photosynthetic bacteria which accumulate Chl precursors and concluded that Pchlide levels may be involved in the regulation of ALA synthesis. He also conducted an interesting series of experiments in which ALA was administered to an albina mutant and to a number of *xantha* strains of barley and then determined which porphyrins accumulate. The objective was to determine the site of the lesion in the biosynthetic chain. In some mutants 80–97% of the product of ALA metabolism was Proto. Others accumulated smaller fractions as Uro, Mg Proto, or Pchlide. Gough concluded that Uro, Proto, Mg Proto or any tetrapyrrolic intermediate in Pchlide biosynthesis can directly or indirectly inhibit ALA formation in higher plants. He also pointed out that none of the mutants is absolutely blocked in Pchlide synthesis. They all made some Pchlide. The ALA-feeding is, however, working as though the synthesis of ALA has not been stopped; that is why enough intermediates accumulate for Gough to measure. Since each mutant makes some Pchlide, this pigment could act in the regulating system to limit ALA synthesis although some of the other porphyrins would accumulate at low levels. Then upon illumination the Pchlide would be converted but the other porphyrins would act as photopoisons. The data do not appear to rule out the possibility that Pchlide plays a controlling role because all of the mutants do accumulate some of this pigment.

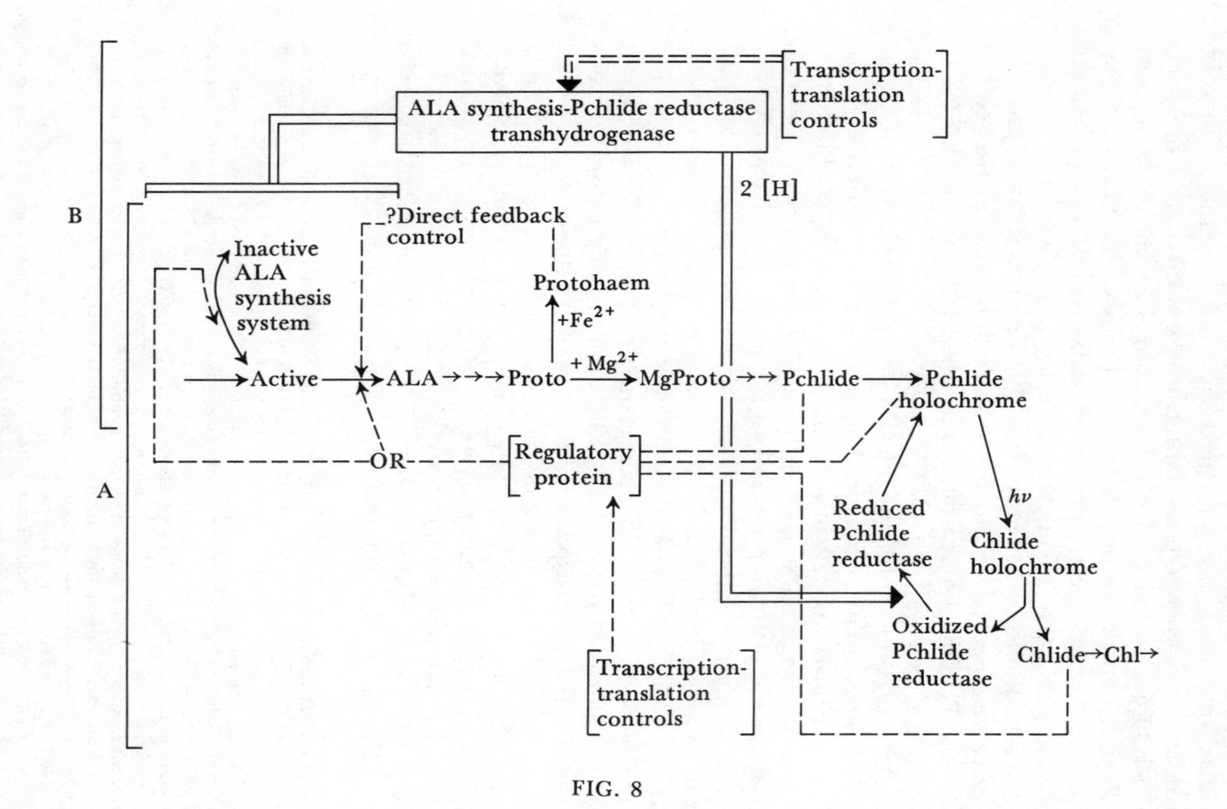

Transcription-translation controls
ALA synthesis-Pchlide reductase transhydrogenase
2 [H]
B
?Direct feedback control
Inactive ALA synthesis system
Protohaem
$+Fe^{2+}$
$+Mg^{2+}$
Active
ALA
Proto
MgProto
Pchlide
Pchlide holochrome
OR
Regulatory protein
A
$h\nu$
Reduced Pchlide reductase
Chlide holochrome
Oxidized Pchlide reductase
Chlide→Chl→
Transcription-translation controls

FIG. 8

deficiency in darkness and light have also been observed in a *Chlorella* mutant (Bryan and Bogorad, 1963).

Depending upon the degree of chlorosis, discs from iron-deficient leaves of Swiss chard accumulate either protoporphyrin or protochlorophyllide when incubated with ALA in darkness (T. Treffry and L. Bogorad, unpublished data). Machold and co-workers (Machold and Stephan, 1969; Stephan and Machold, 1969; Machold and Scholz, 1969) conclude from studies with a wild-type and a *xantha* mutant strain of *Lycopersicon esculentum* that iron metabolism affects the conversion on Coprogen into Proto. Hsu and Miller (1965) came to a similar conclusion; they found that Copro accumulated in iron-deficient *Nicotiana tabacum* supplied with ALA.

Iron chelators such as *o*-phenanthroline overcome the inhibitory effect of puromycin on tetrapyrrole biosynthesis in *R. spheroides* (Lascelles, 1968). An interpretation of this effect has been examined above. Duggan and Gassman (1974) have now shown that the administration of α,α'-dipyridyl, *o*-phenthroline and certain other metal chelators to etiolated bean leaves results in their forming large amounts of Mg Proto monomethyl ester and less Mg Proto, Proto and Pchlide. The amount of porphyrin accumulated is comparable to that found in leaves after feeding ALA. The chelators stimulate ALA synthesis in some way. Furthermore, the normal photostimulation of

---

FIG. 8. Hypothetical schemes for the regulation of Chl formation in green plants which require light for massive porphyrin synthesis. A: It is suggested that a protein required to keep ALA synthesis active becomes inoperative in the presence of excess Pchlide holochrome (or Pchlide or free Chlide or Chl) which perhaps might bind the protein. The activity factor would need to be able to move freely in the compartments of ALA synthesis and Pchlide reduction. The abundance of the activity factor to maintain ALA synthesis might also be reduced by interference with its synthesis. Metal chelators might release the inhibition of ALA synthetase in darkness in this scheme by substituting for the activity factor or by altering conditions in the compartment (unidentified) in which ALA is synthesized for Chl production so as to permit the synthesizing system to function; the chelators would thus operationally substitute for the activity factor. B: Here is shown the possible coupling between Reduction II of Pchlide reductase (Fig. 6) and some unspecified step in ALA synthesis. The abundance of the transhydrogenase would be under the same sort of transcription–translation control as the regulatory protein in Scheme A. In this scheme the metal chelators could disconnect the two parts of the system or introduce a bypass around the transhydrogenase.

The possibility of some direct feedback inhibition of ALA synthesis by a more or less soluble intermediate such as haem seems unlikely from the existence of Proto and Mg Proto accumulating mutants of *Chlorella* as discussed in the text.

ALA synthesis in etiolated leaves is strongly inhibited by chloramphenicol but this antibiotic does not block chelator-promoted porphyrin production. Unless chelators prevent chloramphenicol from blocking protein synthesis (this point has not been examined), these experiments show that the enzyme for ALA synthesis is present but inactive in etiolated leaves. Chelators of iron somehow directly or indirectly activate the enzyme.

The chelators do not convert Pchlide into Chlide nor do they influence this process indirectly. If ALA synthesis is controlled by Pchlide levels, the chelators work elsewhere later in the control process. Suppose we assume that if chloramphenicol works here by stopping protein synthesis it blocks photostimulated synthesis of ALA not by preventing the production of the enzyme for ALA synthesis but by stopping the synthesis of some protein in the control reaction chain between Pchlide and ALA synthesis. The chelator effect should be beyond the synthesis of this activator or derepressor (Fig. 8).

Foster *et al.* (1971) have described a number of *tigrina* mutants of barley (*Hordeum vulgare* L.) which accumulate much more Pchlide in darkness than normal plants. They behave like wild-type plants which have been supplied with α,α'-dipyridyl. They produce larger than normal amounts of Chl precursor porphyrins without reducing their Pchlide.

We have considered feedback control via Pchlide, Pchlide reductase (the free holochrome), or free (i.e. not inserted into the thylakoid membrane) Chlide acting through a protein which activates the synthetic system (Fig. 8). The idea of a "communicating protein" is especially attractive because the final production of Chlide occurs on, in or near the thylakoid membrane but ALA synthesis probably does not. Some long-distance signal is needed.

Feedback inhibition via haem has also been considered indirectly. A back-up of intermediates along the biosynthetic path would raise the level of Proto. Insertion of Fe would produce haem. The Fe chelators would interfere with haem formation. The *tigrina* mutant could have a defective ferrochelatase. Against this view is the behaviour of Proto and Mg Proto accumulating mutants of *Chlorella.* They accumulate very large amounts of porphyrins but ALA synthesis does not seem to be blocked.

A coupled oxido-reduction system between Pchlide reductase and a step in ALA synthesis has not yet been discussed. The immediate source of the reductant for Pchlide is probably the holochrome protein, i.e. Pchlide reductase, but what reduces the enzyme for the next cycle of Pchlide reduction? A tight coupling between Pchlide reduction and ALA synthesis could be provided if an oxidative step in ALA synthesis or in the maintenance of an enzyme (not entirely unlike some proposals for ALA synthetase activation in *R. spheroides* but by reduction) were the specific source of electrons for the reduction of Pchlide. A carrier keyed to both the oxidative and reductive ends of the couple would serve as an electron acceptor and part of the ALA synthesis activating system at one end and a donor of electrons for the Pchlide photo-reduction system at the other.

The immediately preceding discussion has treated the control of ALA and Pchlide synthesis as isolated from the rest of the maturation of the photosynthetic machinery. It may well be just one manifestation of this process. The role of phytochrome as the photoreceptor for many de-etiolation phenomena is well documented (see Chapter 15). For example, the accumulation of proteins in plastids after illumination of etiolated leaves is under phytochrome control (Mego and Jagendorf, 1961). The quick shut-off features of ALA synthesis, as when greening leaves are returned to darkness, could be a simpler feedback loop super-imposed on a more complicated generalized plastic maturation turn-on system. Yet inhibitors of nucleic acid and protein synthesis shut off Chl synthesis.

The full resolution of this intriguing problem must await more information about several enzymatic reactions at or near the start and the end of the chain, as well as further knowledge of the mode of interaction of nuclear and organellar genomes in eukaryotes (Bogorad *et al.*, 1974; Bogorad, 1975).

REFERENCES

Allen, M. B. (1958). *Brookhaven Symp. Biol.* **11**, 339.
Allen, M. B. (1959). *Arch. Mikrobiol.* **32**, 270.
Aoki, Y., Wada, O., Urata, G., Takakku, F. and Nakao, K. (1971). *Biochem. biophys. Res. Commun.* **42**, 568.
Ardao, C. and Vennesland, B. (1960). *Pl. Physiol., Lancaster* **35**, 368.
Aronoff, S. and Ellsworth, R. K. (1968). *Photosynthetica* **2**, 288.
Bagdasarian, M. (1958). *Nature, Lond.* **181**, 1399.
Batlle, A. M. del C., Benson, A. and Rimington, C. (1965). *Biochem. J.* **97**, 731.
Battersby, A. R. (1971). *Int. Congr. Int. Un. pure appl. Chem.* **23**,(5), 1.
Battersby, A. R., Baldas, J., Collins, J., Grayson, D. H., James, K. J. and McDonald, E. (1972). *Chem. Commun.* **23**, 1265.
Beale, S. I. (1970). *Pl. Physiol., Lancaster* **45**, 504–506.
Beale, S. I. (1971). *Pl. Physiol., Lancaster* **48**, 316.
Beale, S. I. and Castelfranco, P. A. (1974a). *Pl. Physiol., Lancaster* **53**, 291.
Beale, S. I. and Castelfranco, P. A. (1974b). *Pl. Physiol., Lancaster* **53**, 297.

Becker, R. S. and Sheline, R. K. (1955). *Archs Biochem. Biophys.* **54**, 259.
Bénard, H., Gadjos, A. and Gadjos-Török, M. (1951), *C.r. Séanc. Soc. Bio.* **145**, 538.
Blass, U., Anderson, J. M., and Calvin, M. (1959). *Pl Physiol., Lancaster* **34**, 329.
Boardman, N. K. (1962a). *Biochim. biophys. Acta* **62**, 63.
Boardman, N. K. (1962b). *Biochim. biophys. Acta* **64**, 279.
Boardman, N. K. (1966). *In* "The Chlorophylls" (L. P. Vernon and G. R. Seely, eds), p. 437. Academic Press, New York and London.
Bogdanovic, M. (1973). *Physiologia Pl.* **29**, 19.
Bogorad, L. (1950). *Bot. Gaz.* **111**, 221.
Bogorad, L. (1955). *Science N.Y.* **121**, 878.
Bogorad, L. (1957a). *In* "Research in Photosynthesis" (H. Gaffron, ed.), p. 475. Interscience, New York.
Bogorad, L. (1957b). *Pl. Physiol., Lancaster* **32**, xli.
Bogorad, L. (1958a). *J. biol. Chem.* **233**, 501.
Bogorad, L. (1958b). *J. biol. Chem.* **233**, 510.
Bogorad, L. (1958c). *J. biol. Chem.* **233**, 516.
Bogorad, L. (1960). *In* "Comparative Biochemistry of Photoreactive Systems" (M. B. Allen, ed.), p. 227. Academic Press, New York and London.
Bogorad, L. (1961). *Fifth int. Congr. Biochem.* p. 158. Pergamon Press, New York and Oxford.
Bogorad, L. (1963). *Ann. N.Y. Acad. Sci.* **104**, 676.
Bogorad, L. (1967). *In* "Control Mechanisms in Developmental Processes." (M. Locke, ed.), pp. 1-31. Academic Press, New York and London. (Developmental Biology Suppl. 1, 1-31).
Bogorad, L. (1975). *Science, N.Y.* **188**, 891.
Bogorad, L. and Granick, S. (1953a). *J. biol. Chem.* **202**, 793.
Bogorad, L. and Granick, S. (1953b). *Proc. natn. Acad. Sci. U.S.A.* **30**, 1176.
Bogorad, L. and Jacobson, A. B. (1964). *Biochem. biophys. Res. Commun.* **14**, 113.
Bogorad, L., Laber, L. and Gassman, M. (1968). *In* "Comparative Biochemistry and Biophysics of Photosynthesis" (K. Shibata, A. Takamiya, A. T. Jagendorf and R. C. Fuller, eds), p. 299. University Park Press State College, Pa.
Bogorad, L., Mets, L. J., Mullinix, K. P., Smith, H. J. and Strain, G. C. (1974). *Biochem. Soc. Symp.* **38**, 17.
Bonner, B. A. (1969). *Pl. Physiol., Lancaster* **44**, 739.
Booij, H. L. and Rimington, C. (1957). *Biochem. J.* **65**, 4P.
Brockmann, H. Jr. and Knobloch, G. (1972). *Arch. Mikrobiol.* **85**, 123.
Bryan, G. W. (1962). Protochlorophyllide *a* Metabolism and Chlorophyllide *a* Production by a Chlorella mutant in Response to Iron and Spectral Exposures. Ph.D. thesis, University of Chicago.
Bryan, G. W. and Bogorad, L. (1963). "Microalgae and Photosynthetic Bacteria" (a special supplement to *Pl. Cell Physiol.*), p. 399.
Bryan, G. W., Zadylak, A. H. and Ehret, C. F. (1967). *J. Cell Sci.* **2**, 513.
Brzeski, W. and Rucker, W. (1960). *Nature, Lond.* **185**, 922.
Budzikiewicz, H. and Taraz, K. (1971). *Tetrahedron* **27**, 1447.
Bull, M. J. and Lascelles, J. (1963). *Biochem. J.* **87**, 15.
Burnham, B. F. (1962). *Biochem. biophys. Res. Commun.* **7**, 351.
Burnham, B. F. and Lascelles, J. (1963). *Biochem. J.* **87**, 462.
Carpenter, A. T. and Scott, J. J. (1959). *Biochem. J.* **71**, 325.

Carpenter, A. T. and Scott, J. J. (1961). *Biochim. biophys. Acta* **52**, 195.
Chen, S. D., McMahon, D. and Bogorad, L. (1967), *Pl. Physiol., Lancaster* **42**, 1.
Chiba, Y., Aiga, I., Indemori, M., Satoh, Y., Matsushita, K. and Sasa, T. (1967). *Pl. Cell Physiol.* **8**, 623.
Cohen-Bazire, G., Sistrom, W. R. and Stanier, R. Y. (1957). *J. cell. comp. Physiol.* **43**, 25.
Cookson, G. H. and Rimington, C. (1953). *Nature, Lond.* **171**, 375.
Cookson, G. H. and Rimington, C. (1954). *Biochem. J.* **57**, 476.
Cooper, R. (1963). *Biochem. J.* **89**, 100.
Cox, M. T., Howarth, T. T., Jackson, A. H. and Kenner, G. W. (1969). *J. Am. chem. Soc.* **91**, 1232.
Dailey, H. A. Jr. and Lascelles, J. (1974). *Archs Biochem. Biophys.* **160**, 523.
Davies, R. C., Gorchein, A., Neuberger, A., Sandy, J. D. and Tait, G. H. (1973). *Nature, Lond.* **245**, 15.
Davies, R. C. and Neuberger, A. (1973). *Biochem. J.* **133**, 471.
della Rosa, R. J., Altman, K. I. and Saloman, K. (1953). *J. biol. Chem.* **202**, 771.
Dougherty, R. C., Strain, H. H., Sveč, W. A., Uphaus, R. A. and Katz, J. J. (1966). *J. Am. chem. Soc.* **88**, 5037.
Dougherty, R. C., Strain, H. H., Svec, W. A., Uphaus, R. A. and Katz, J. J. (1970). *J. Am. chem. Soc.* **92**, 2826.
Doyle, G. (1971). *J. biol. Chem.* **246**, 4965.
Dresel, E. I. B. and Falk, J. E. (1956). *Biochem. J.* **63**, 80.
Drews, G., Leutiger, I. and Ludwig, R. (1971). *Arch. Mikrobiol.* **76**, 349.
Duggan, J. and Gassman, M. (1974). *Pl. Physiol., Lancaster* **53**, 206.
Duranton, J., Galmiche, J. M. and Roux, E. (1958). *C.r. hebd. Séanc. Acad. Sci., Paris* **246**, 992.
Ebbon, J. G. and Tait, G. H. (1969). *Biochem. J.* **111**, 573.
Egle, K. (1960). *In* "Encyclopedia of Plant Physiology" (W. Ruhland, ed.), vol. 5. Part 1, p. 323. Springer Verlag, Berlin.
Ehtesthamuddin, A. F. M. (1968). *Biochem. J.* **107**, 446.
Ellsworth, R. K. and Aronoff, S. (1968a). *Archs Biochem. Biophys.* **124**, 358.
Ellsworth, R. K. and Aronoff, S. (1968b). *Archs Biochem. Biophys.* **125**, 35.
Ellsworth, R. K. and Aronoff, S. (1968c). *Archs Biochem. Biophys.* **125**, 269.
Ellsworth, R. K. and Aronoff, S. (1969). *Archs Biochem. Biophys.* **130**, 374.
Falk, J. E. (1964). "Porphyrins and Metalloporphyrins". Elsevier, New York.
Falk, J. E., Drexel, E. I. B. and Rimington, C. (1953). *Nature, Lond.* **172**, 292.
Falk, J. E., Dresel, E. I. B., Benson, A. and Knight, B. C. (1956). *Biochem. J.* **63**, 87.
Finkle, B. J., Appleman, D. and Fleischer, F. K. (1950). *Science, N.Y.* **111**, 309.
Fischer, H. and Stern, A. (1940). "Die Chemie des Pyrrols", vol. 2, part 2. Akademische VerlagsgesellSchaft M. B. II., Leipzig.
Fischer, H., Mittenzwei, H. and Ocstreicher, A. (1939). *Hoppe-Seyler's Z. Physiol. Chem.* **257**, IV–VII.
Forger, J. M., III and Bogorad, L. (1973). *Pl. Physiol., Lancaster* **52**, 491.
Foster, R. J., Fibbons, G. C., Gough, S., Henningsen, K. W., Kahn, A., Nielsen, O. F. and von Wettstein, D. (1971). *In* "Proc. First European Biophysics Congress", vol. 4, p. 137. Verl. Weiner Mediz. Akad., Wien, Austria.
Frank, S. R. (1946). *J. gen. Physiol.* **29**, 157.
Frydman, R. B. and Frydman, B. (1970). *Archs Biochem. Biophys.* **136**, 193.

Frydman, R. B., Reil, S., Valasinas, A., Frydman, B. and Rapaport, H. (1971). *J. Am. chem. Soc.* **93** 2739.
Frydman, R. B., Valasinas, A., Rapaport, H. and Frydman, B. (1972). *FEBS Lett.* **25**, 309.
Frydman, R. B., Tomaro, M. L., Anschelbaum, A., Anderson, J. and Frydman, B. (1973). *Biochemistry, N.Y.* **12**, 5253.
Frydman, R. B., Valasinas, A. and Frydman, B. (1973). *Biochemistry, N.Y.* **12**, 80.
Gassman, M. L. (1967). "Studies on the control of chlorophyll synthesis". Ph.D. thesis, University of Chicago, Chicago.
Gassman, M. L. (1973). *Pl. Physiol., Lancaster* **52**, 590.
Gassman, M. and Bogorad, L. (1967a). *Pl. Physiol., Lancaster* **42**, 774.
Gassman, M. and Bogorad, L. (1967b). *Pl. Physiol., Lancaster* **42**, 781.
Gassman, M., Granick, S. and Mauzerall, D. (1968a). *Biochem. biophys. Res. Commun.* **32**, 295.
Gassman, M., Plauscec, J. and Bogorad, L. (1968b). *Pl. Physiol., Lancaster* **43**, 1411.
Gibson, K. D., Neuberger, A. and Scott, J. J. (1955). *Biochem. J.* **61**, 618.
Gibson, K., Laver, W. and Neuberger, A. (1958). *Biochem. J.* **70**, 71.
Gibson, K. D., Matthew, M., Neuberger, A. and Tait, G. H. (1961). *Nature, Lond.* **192**, 204.
Gibson, K. D., Neuberger, A. and Tait, G. H. (1962a). *Biochem. J.* **83**, 539.
Gibson, K. D., Neuberger, A. and Tait, G. H. (1962b). *Biochem. J.* **83**, 550.
Gibson, K. D., Neuberger, A. and Tait, G. H. (1963). *Biochem. J.* **88**, 325.
Godnev, T. N., Shlyk, A. A. and Rotfarb, R. M. (1959). *Fiziol. Rostenii* **6**, 36. (Eng. Trans; *Soviet Pl. Physiol.* **33**, 38.)
Godnev, T. N., Rotfarb, R. M. and Akulovich, N. K. (1963). *Photochem. Photobiol.* **2**, 119.
Goldin, B. R. and Little, H. N. (1969). *Biochim. biophys. Acta* **171**, 321.
Goodwin, R. H. and Owens, O. H. (1947). *Pl. Physiol., Lancaster* **22**, 197.
Gorchein, A. (1972). *Biochem. J.* **127**, 97.
Gorchein, A. (1973). *Biochem. J.* **134**, 833.
Gough, S. (1972). *Biochim. biophys. Acta* **286**, 36.
Granick, S. (1948a). *J. biol. Chem.* **172**, 717.
Granick, S. (1948b). *J. biol. Chem.* **175**, 333.
Granick, S. (1949). *J. biol. Chem.* **179**, 505.
Granick, S. (1950). *J. biol. Chem.* **183**, 713.
Granick, S. (1954). *Science, N.Y.* **120**, 105.
Granick, S. (1959). *Pl. Physiol., Lancaster* **34**, xviii.
Granick, S. (1961). *J. biol. Chem.* **236**, 1168.
Granick, S. (1966). *J. biol. Chem.* **241**, 1359.
Granick, S. (1967). *In* "Biochemistry of Chloroplasts" (T. W. Goodwin, ed.), vol. 2, p. 400. Academic Press, London and New York.
Granick, S. and Bogorad, L. (1953). *J. Am. chem. Soc.* **75**, 3610.
Granick, S. and Gassman, M. (1970). *Pl. Physiol., Lancaster* **45**, 201.
Granick, S. and Mauzerall, D. (1958). *J. biol. Chem.* **232**, 1119.
Griffiths, M. (1962). *J. gen. Microbiol.* **27**, 427.
Guignery, G., Luzzati, A. and Duranton, J. (1974). *Planta* **115**, 227.
Gurba, P. E., Sennett, R. E., and Kobes, R. D. (1972). *Archs Biochem. Biophys.* **150**, 130.
Hayashi, N., Yoda, B. and Kikuchi, G. (1970). *J. Biochem., Tokyo* **67**, 859.

Heath, H. and Hoare, D. S. (1959). *Biochem. J.* 72, 14.
Henningsen, K. W. and Kahn, A. (1971). *Pl. Physiol., Lancaster* 47, 685.
Highkin, H. R. (1950). *Pl. Physiol., Lancaster* 25, 291.
Higuchi, M. and Bogorad, L. (1975). *Ann. N.Y. Acad. Sci.* 224, 401.
Hoare, D. S. and Heath, H. (1959). *Biochem. J.* 73, 379.
Holden, M. (1961). *Biochem. J.* 78, 359.
Holden, M. (1963). *Photochem. Photobiol.* 2, 175.
Holowinsky, A. W. and Schiff, J. A. (1970). *Pl. Physiol., Lancaster* 45, 339.
Holt, A. S. (1960). *In* "Comparative Biochemistry of Photoreactive Systems" (M.B. Allen, ed.), p. 169. Academic Press, New York and London.
Holt, A. S. (1961). *Can. J. Bot.* 39, 327.
Hsu, W. P. and Miller, G. W. (1965). *Biochim. biophys. Acta* 111, 393.
Hsu, W. P. and Miller, G. W. (1970). *Biochem. J.* 117, 215.
Iodice, A. A., Richert, D. A. and Schulman, M. P. (1958). *Fedn Proc. Fedn Am. Socs exp. Biol.* 17, 248.
Jeffrey, S. W. (1972). *Biochim. biophys. Acta* 279, 15.
Jones, O. T. G. (1963a). *Biochem. J.* 86, 429.
Jones, O. T. G. (1963b). *Biochem. J.* 89, 182.
Jones, O. T. G. (1963c). *Biochem. J.* 88, 335.
Jones, O. T. G. (1964). *Biochem. J.* 91, 572.
Jones, O. T. G. (1967). *Biochem. biophys. Res. Commun.* 28, 671.
Jones, M. S. and Jones, O. T. G. (1970). *Biochem. J.* 119, 453.
Jordan, P. M. and Shemin, D. (1973). *J. biol. Chem.* 248, 1019.
Kahn, A., Boardman, N. K. and Thorne, S. W. (1970). *J. molec. Biol.* 48, 85.
Kamen, M. (1957). "Isotopic Tracers in Biology". Academic Press, New York and London.
Kappas, A. and Granick, S. (1968). *J. biol. Chem.* 243, 346.
Keller, C. J. and Huffaker, R. C. (1967). *Pl. Physiol., Lancaster* 42, 1277.
Kenner, G. W., McCombie, S. W. and Smith, K. M. (1972). *Chem. Commun.* 844-845.
Kikuchi, G., Kumer, A., Talmadge, P. and Shemin, D. (1958). *J. biol. Chem.* 233, 1214.
Klein, A. and Vishniac, W. (1961). *J. biol. Chem.* 236, 2544.
Koski, V. M. (1950). *Archs Biochem. Biophys.* 29, 339.
Koski, V. M. and Smith, J. H. C. (1948). *J. Am. chem. Soc.* 70, 3558.
Koski, Violet M., French, C. S. and Smith, J. H. C. (1951). *Archs Biochem. Biophys.* 31, 1.
Kowalski, E., Dancewicz, A. M., Szot, Z., Lipinski, B. and Rosiek, O. (1959). *Acta biochem. Polon.* 6, 257.
Krasnovsky, A. A. and Kosobutskaya, L. M. (1952). *C.r. Acad. Sci., U.R.S.S.* 85, 177.
Kurashima, Y., Hayashi, N. and Kikuchi, G. (1970). *J. Biochem., Tokyo* 67, 863.
Labbe, R. F. (1959). *Biochim. biophys. Acta* 31, 589.
Labbe, R. F. and Hubbard, N. (1960). *Biochim. biophys. Acta* 41, 185.
Labbe, R. F. and Hubbard, N. (1961). *Fedn Proc. Fedn Am. Socs exp. Biol.* 20, 376.
Labbe, R. F., Hubbard, N. and Caughey, W. (1963). *Biochemistry, N.Y.* 2, 372.
Lascelles, J. (1959). *Biochem. J.* 72, 508.
Lascelles, J. (1960). *J. gen. Microbiol.* 23, 499.
Lascelles, J. (1966). *Biochem. J.* 100, 175.

Lascelles, J. (1968). *Biochem. Soc. Symp.* **28**, 49.
Lascelles, J. and Altshuler, T. (1967). *Arch. Mikrobiol.* **59**, 204.
Lascelles, J. and Altschuler, T. (1969). *J. Bact.* **98**, 721.
Laver, W. G., Neuberger, A. and Scott, J. J. (1959). *J. Chem. Soc.* **1483**.
Lemberg, R. and Legge, J. W. (1949). "Haematin Compounds and Bile Pigments". Interscience, New York.
Levin, E. Y. (1968). *Biochemistry, N.Y.* **7**, 3781.
Levin, E. Y. and Coleman, D. L. (1967). *J. biol. Chem.* **242**, 4248.
Little, H. N. and Kelsey, M. I. (1964). *Fedn Proc. Fedn Am. Socs exp. Biol.* **23**, 223.
Litvin, F. F. and Belayaev, O. B. (1968). *Biokhimiya* **33**, 928.
Litvin, F. F., Krasnovsky, A. A. and Rikhireva, G. T. (1959). *C.r. Acad. Sci., U.R.S.S.* **127**, 699.
Llambias, E. B. C. and Batlle, A. M. del. C. (1970a). *Biochim biophys. Acta* **220**, 552.
Llambias, E. B. C. and Batlle, A. M. del C. (1970b). *FEBS Lett.* **6**, 285.
Llambias, E. B. C. and Batlle, A. M. del C. (1971a). *Biochem. J.* **121**, 327.
Llambias, E. B. C. and Batlle, A. M. del C. (1971b). *Biochem. biophys. Acta* **227**, 180.
Lockwood, W. H. and Benson, A. (1960). *Biochem. J.* **75**, 372.
Machold, O. and Stephan, U. W. (1969). *Phytochemistry* **8**, 2189.
Machold, O. and Scholz, G. (1969). *Naturwissenschaften* **56**, 447.
Madsen, A. (1963). *Photochem. Photobiol.* **2**, 93.
Margulies, M. M. (1962). *Pl. Physiol, Lancaster* **37**, 473.
Marriott, J. (1968). *Biochem. Soc. Symp.* **28**, 61.
Marriott, J., Neuberger, A. and Tait, G. H. (1969). *Biochem. J.* **111**, 385.
Marsh, H. V. Jr., Evans, H. J. and Matrone, G. (1963a). *Pl. Physiol., Lancaster* **38**, 632.
Marsh, H. V. Jr., Evans, H. J. and Matrone, G. (1963b). *Pl. Physiol., Lancaster* **38**, 638.
Mathewson, J. H. and Corwin, A. H. (1961). *J. Am. chem. Soc.* **83**, 135.
Mathis, P. and Sauer, K. (1972). *Biochim. biophys. Acta* **267**, 498.
Mauzerall, D. (1960). *J. Am. chem. Soc.* **82**, 2601.
Mauzerall, D. and Granick, S. (1958). *J. biol. Chem.* **232**, 1141.
Mayer, H. (1930). *Planta* **11**, 294.
McFeeters, R. F., Chichester, C. O. and Whitaker, J. R. (1971). *Pl. Physiol., Lancaster* **47**, 609.
McSwiney, R. R., Nicholas, R. E. H. and Prunty, F. T. G. (1950). *Biochem. J.* **46**, 147.
Mego, J. L. and Jagendorf, A. (1961). *Biochim. biophys. Acta* **53** 237.
Michel-Wolwertz, M. R. (1963). *Photochem. Photobiol.* **2**, 149.
Miller, J. W. and Teng, D. (1967). *Seventh int. Congr. Biochem.* p. 1059.
Monteverde, N. A. (1893–94). *Acta Horti petropolitani* **13**, 201.
Mori, M. and Sano, S. (1968). *Biochem. biophys. Res. Commun.* **32**, 610.
Mori, M. and Sano, S. (1972). *Biochim. biophys. Acta* **264**, 252.
Muir, H. M., and Neuberger, A. (1949). *Biochem. J.* **45**, 163.
Muir, H. M. and Neuberger, A. (1950). *Biochem. J.* **47**, 97.
Murphy, M. J. and Siegel, L. M. (1973). *J. biol Chem.* **248**, 6911.
Murphy, M. J., Siegel, L. M., Kamin, H. and Rosenthal, D. (1973). *J. biol. Chem.* **248**, 2801.

Nandi, D. L. and Shemin, D. (1968a). *J. biol. Chem.* **243**, 1231.
Nandi, D. L. and Shemin, D. (1968b). *J. biol. Chem.* **243**, 1236.
Nandi, D. L. and Shemin, D. (1973). *Archs Biochem. Biophys.* **158**, 305.
Nandi, D. L. and Waygood, R. (1967). *Can. J. Biochem.* **45**, 327.
Nandi, D. L., Baker-Cohen, K. F. and Shemin, D. (1968). *J. biol. Chem.* **243**, 1224.
Nemeth, A. M., Russell, C. S. and Shemin, D. (1957). *J. biol. Chem.* **229**, 415.
Neuberger, A. (1961). *Biochem. J.* **78**, 1.
Neuberger, A. and Scott, J. J. (1953). *Nature, Lond.* **172**, 1093.
Neuberger, A. and Tait, G. (1964). *Biochem. J.* **90**, 607.
Neuberger, A., Sandy, J. D. and Tait, G. H. (1973a). *Biochem. J.* **136**, 477.
Neuberger, A., Sandy, J. D. and Tait, G. H. (1973b). *Biochem. J.* **136**, 491.
Neuberger, A. and Turner, J. M. (1963). *Biochim. biophys. Acta* **67**, 345.
Neve, R. A., Labbe, R. F. and Aldrich, R. A. (1956). *J. Am. Chem. Soc.* **78**, 691.
Nichols, K. E. and Bogorad, L. (1962). *Bot. Gaz.* **124**, 85.
Nishida, G. and Labbe, R. F. (1959). *Biochim. biophys. Acta* **31**, 519.
Noack, K. and Kiessling, W. (1929). *Hoppe-Seyler's Z. physiol. Chem.* **182**, 13.
Noack, K. and Kiessling, W. (1930). *Hoppe-Seyler's Z. physiol. Chem.* **193**, 97.
Oelze, J. and Drews, G. (1970). *Arch. Mikrobiol.* **73**, 19.
Oelze, J., Schroeder, J. and Drews, G. (1970). *J. Bact.* **101**, 669.
Ogura, N. (1972). *Pl. Cell Physiol.* **13**, 971.
Osgerby, J. M., Pluscec, J., Kim, Y. C., Boyer, F., Stojanac, N., Mah, H. D. and Macdonald, S. F. (1972). *Can. J. Chem.* **50**, 2652.
Oyama, H., Sugita, Y., Yoneyama, Y. and Yoshikaya, H. (1961). *Biochim. biophys. Acta* **47**, 413.
Patton, G. M. and Beattie, D. S. (1973). *J. biol. Chem.* **248**, 4467.
Perkins, H. J. and Roberts, D. W. A. (1960). *Biochim. biophys. Acta* **45**, 613.
Pfennig, N. (1967). *A. Rev. Microbiol.* **21**, 285.
Phillips, J. N. (1963). *In* "Comprehensive Biochemistry" (M. Florkin and E. H. Stotz, eds), vol. 9, p. 34. Elsevier, Amsterdam.
Pluscec, J. and Bogorad, L. (1970). *Biochemistry, N.Y.* **9**, 4736.
Porra, R. J. and Falk, J. E. (1964). *Biochem. J.* **90**, 69.
Porra, R. J. and Irving, E. A. (1972). *Archs Biochem. Biophys.* **148**, 37.
Porra, R. J. and Jones, O. T. G. (1963a). *Biochem. J.* **87**, 181.
Porra, R. J. and Jones, O. T. G. (1963b). *Biochem. J.* **87**, 186.
Porra, R. J., Vitols, K. S., Labbe, R. F. and Newton, N. A. (1967). *Biochem. J.* **104**, 321.
Radmer, R. J. and Bogorad, L. (1967). *Pl. Physiol., Lancaster* **42**, 463.
Radmer, R. J. and Bogorad, L. (1967). *Biochemistry, N.Y.* **11**, 904.
Ramaswamy, N. K. and Nair, P. M. (1973). *Biochim. biophys. Acta* **293**, 269.
Richards, W. R. and Lascelles, J. (1969). *Biochemistry*, **8**, 3473.
Richards, W. R. and Rapaport, H. (1967). *J. Am. chem. Soc.* **6**, 3830.
Ricketts, T. R. (1966). *Phytochemistry* **5**, 223.
Rimington, C. and Booij, H. L. (1957). *Biochem. J.* **65**, 3P.
Rimington, D. and Miles, P. A. (1951). *Biochem. J.* **50**, 202.
Roberts, D. W. A. and Perkins, H. J. (1966). *Biochim. biophys. Acta* **127**, 42–46.
Rosinski, J. and Bogorad, L. (1976). In preparation.
Sancovich, H. A., Battle, A. M. del C. and Grinestein, M. (1969). *Biochim. biophys. Acta* **191**, 130.

Sano, S. (1966). *J. biol. Chem.* **44**, 5276.
Sano, S. and Granick, S. (1961). *J. biol. Chem.* **236**, 1173.
Schiff, J. A. and Epstein, H. T. (1965). *In* "Reproduction: Molecular, Subcellular, and Cellular" (M. Locke, ed.), pp. 131–189. Academic Press, New York and London.
Schmid, R. (1960). *In* "The Metabolic Basis of Inherited Diseases" (J. B. Stanbury, ed.), p. 939. McGraw-Hill, New York.
Schmid, R., and Shemin, D. (1955). *J. Am. chem. Soc.* **77**, 506.
Schmidt, A. (1924). *Bot. Arch.* **5**, 260.
Schneider, H. A. W. (1973). *Z. PflPhysiol.* **69**, 68.
Schopfer, P. and Siegelman, H. W. (1968). *Pl. Physiol., Lancaster* **43**, 990.
Schou, L. (1951). *Physiologia. Pl.* **4**, 617.
Schultz, A. and Sauer, K. (1972). *Biochim, biophys. Acta* **267**, 320.
Schwartz, H. C., Hill, R. L., Cartwright, C. E. and Wintrobe, M. M. (1959). *Fedn Proc. Fedn Am. Socs. exp. Biol.* **18**, 545.
Seybold, A. (1949). *Planta* **36**, 371.
Seybold, A. and Egle, K. (1937). *Planta* **26**, 491.
Seybold, A. and Egle, K. (1938). *Planta* **28**, 87.
Shemin, D. (1968). *Biochem. Soc. Symp.* **28**, 75.
Shemin, D. and Kikuchi, G. (1958). *Ann. N.Y. Acad. Sci.* **75**, 122.
Shemin, D. and Kumin, S. (1952). *J. biol. Chem.* **198**, 827.
Shemin, D. and Russell, C. S. (1953). *J. Am. chem. Soc.* **76**, 4873.
Shemin, D. and Wittenberg, J. (1951). *J. biol. Chem.* **192**, 315.
Shetty, A. S. and Miller, G. W. (1969). *Biochim. biophys. Acta* **185**, 485.
Shibata, K. (1957). *J. Biochem., Tokyo* **44**, 147.
Shibata, K. (1959). *In* "Methods of Biochemical Analysis" (D. Glick, ed.), vol. 7, p. 77. Interscience, New York.
Shimizu, S. and Tamaki, E. (1962). *Bot. Mag., Tokyo* **75**, 462.
Shimizu, S. and Tamaki, E. (1963). *Archs Biochem. Biophys.* **102**, 152.
Shlyk, A. A. (1971). *A. Rev. Pl. Physiol.* **22**, 169.
Shlyk, A. A. and Godnev, T. N. (1958). *In* "First International Conference on Radioisotopes in Scientific Research," vol. 4, p. 479. Pergamon Press, London.
Shlyk, A. A., Kaler, V. L. and Podchufavova, G. M. (1960). *C.r. Acad. Sci., U.R.S.S.* **133**, 1472.
Shlyk, A. A., Kaler, V. L., Vlasenok, L. I. and Gaponenko, V. I. (1963a). *Photochem. Photobiol.* **2**, 129.
Shlyk, A. A., Mikhailova, S. A., Gaponenko, V. I. and Kukhtenko, T. V. (1963b). *Fiziol. Rastenii* **10**, 275. (Engl. trans. *Soviet Pl. Physiol.* **10**, 227.)
Sironval, C. (1972). *Photosynthetica* **6**, 375.
Sironval, C. and Brouers, M. (1970). *Photosynthetica* **4**, 38.
Sironval, C., Brouers, M., Michel, J. M. and Kniper, Y. (1968). *Photosynthetica* **2**, 268.
Sironval, C., Verly, W. G., and Marcelle, R. (1961). *Physiologia Pl.* **14**, 303.
Sisler, E. G., and Klein, W. H. (1963). *Physiologia Pl.* 315.
Sistrom, W. R. (1962). *J. gen. Microbiol.* **28**, 599.
Sistrom, W. R., Griffiths, M. and Stanier, R. Y. (1956). *J. cell. comp. Physiol.* **48**, 459.
Sizer, S. and Sauer, K. (1971). *Pl. Physiol., Lancaster* **48**, 60.
Sluiters-Scholten, C. M. Th., van den Berg, F. M. and Stegwee, D. (1973). *Z. PflPhysiol.* **69**, 217.

Smith, J. H. C. (1948). *Archs Biochem. Biophys.* **19**, 449.
Smith, J. H. C. (1952). *Yearb. Carneg. Instn. Wash.* **51**, 151.
Smith, J. H. C. (1960). In "Comparative Biochemistry of Photoreactive Systems" (M. B. Allen, ed.), p. 257. Academic Press, New York and London.
Smith, J. H. C. and Benitez, A. (1954). *Pl. Physiol., Lancaster* **29**, 135.
Smith, J. H. C. and Benitez, A. (1955). *In* "Modern Methods of Plant Analysis" (K. Paech and M. V. Tracey, eds), vol. 4, p. 142. Springer Verlag, Berlin.
Smith, J. H. C. and Coomber, J. C. (1961). *Yearb. Carneg. Instn Wash.* **60**, 371.
Smith, J. H. C. and French, C. S. (1963). *A. Rev. Pl. Physiol.* **14**, 181.
Smith, J. H. C. and Young, V. M. K. (1956). *In* "Radiation Biology" (A. Hollaender, ed.), vol. 3, p. 393. McGraw-Hill, New York.
Smith, J. H. C., Kupke, D. W., Loeffler, J. E., Benitez, A., Ahrue, I. and Giese, A. T. (1957). *In* "Research in Photosynthesis" (H. Gaffron *et al.*, eds), p. 464. Interscience, New York.
Smith, J. H. C., Durham, L. J. and Wurster, C. F. (1959). *Pl. Physiol., Lancaster* **34**, 340.
Stanier, R. Y. and Smith, J. H. C. (1959). *Yearb. Carneg. Instn Wash.* **58**, 336.
Steer, B. T. and Gibbs, M. (1969). *Pl. Physiol., Lancaster* **44**, 781.
Stella, A. M., Parera, V. E., Llambias, E. B. C. and Batlle, A. M. del C. (1971). *Biochim. biophys. Acta* **252**, 481.
Stephan, U. W. and Machold, O. (1969). *Z. PflPhysiol.* **61**, 98.
Stevens, E. and Frydman, B. (1968). *Biochim. biophys. Acta* **151**, 429.
Stevens, E., Frydman, R. B., and Frydman, B. (1968). *Biochim. biophys. Acta* **158**, 496.
Stobart, A. K. and Thomas, D. R. (1968). *Phytochemistry* **7**, 1313.
Strain, H. H., Cope, B. T. Jr., McDonald, G. N., Svec̆, W. A. and Katz, J. J. (1971). *Phytochemistry* **10**, 1109.
Sudyina, E. G. (1963). *Photochem. Photobiol.* **2**, 181.
Sundqvist, C. (1969). *Physiologia Pl.* **22**, 147.
Tait, G. H. (1969). *Biochem. biophys. Res. Commun.* **37**, 116.
Tait, G. H. (1972). *Biochem. J.* **128**, 1159.
Tait, G. H. (1973). *Biochem. J.* **131**, 389.
Tait, G. H. and Gibson, K. D. (1961). *Biochim. biophys. Acta* **52**, 614.
Takemoto, J. and Lascelles, J. (1973). *Proc. natn. Acad. Sci. U.S.A.* **70**, 799.
Thorne, S. W. (1971a). *Biochim. biophys. Acta* **226**, 113.
Thorne, S. W. (1971b). *Biochim. biophys. Acta* **226**, 128.
Tigier, H. A., Batlle, A. M. del C. and Locascio, G. A. (1968). *Biochim. biophys. Acta* **151**, 300.
Tigier, H. A., Batlle, A. M. del C. and Locascio, G. (1970). *Enzymol. Acta Biocatal.* **38**, 43.
Tuboi, S., Kim, H. J. and Kikuchi, G. (1970a). *Archs Biochem. Biophys.* **138**, 147.
Tuboi, S., Kim, H. J., and Kikuchi, G. (1970b). *Archs Biochem. Biophys.* **138**, 155.
Tuboi, S. and Hayasaka, S. (1971). *Archs Biochem. Biophys.* **146**, 282.
Tuboi, S. and Hayasaka, S. (1972a). *Archs Biochem. Biophys.* **150**, 690.
Tuboi, S. and Hayasaka, S. (1972b). *J. Biochem., Tokyo* **72**, 219.
Urata, G. and Granick, S. (1963). *J. Biol. Chem.* **238**, 811.
Van Heyningen, S. and Shemin, D. (1971). *Biochemistry, N.Y.* **10**, 4676.
van Niel, C. B. (1944). *Bact. Rev.* **8**, 1.
Virgin, H. I. (1955). *Physiologia. Pl.* **8**, 630.

Virgin, H. I., Kahn, A. and von Wettstein, D. (1963). *Photochem. Photobiol.* 2, 83.

Waldenström, J. and Vahlquist, B. (1939). *Hoppe-Seyler's Z. physiol. Chem.* **260**, 189.

Warnick, G. R. and Burnham, B. F. (1971). *J. biol. Chem.* **246**, 6880.

Wasley, J. W. F., Scott, W. T. and Holt, A. S. (1970). *Can. J. Biochem.* **48**, 376.

Wellburn, F. A. M. and Wellburn, A. R. (1971). *J. Cell Sci.* 9, 271.

Westall, R. G. (1952). *Nature, Lond.* **170**, 614.

Wider, de Xifra, E. A., Batlle, A. M. del C. and Tigier, H. A. (1971). *Biochim. biophys. Acta* **235**, 511.

Willstätter, R. and Stoll, A. (1913). "Investigations on Chlorophyll." English translation: (1928) by F. M. Schertz and G. R. Mertz. Science Press, Lancaster, Pa.

Wittenberg, J. B. (1959). *Nature, Lond.* **184**, 876.

Wittenberg, J. and Shemin, D. (1949). *J. biol. Chem.* **178**, 47.

Wittenberg, J. and Shemin, D. (1950). *J. biol. Chem.* **185**, 103.

Wolff, J. B. and Price, L. (1957). *Archs Biochem. Biophys.* 72, 293.

Wolken, J. J. (1961). "*Euglena*; an Experimental Organism for Biochemical and Biophysical Studies". Institute of Microbiology, Rutgers, New Brunswick, New Jersey.

Zaman, Z., Abboud, M. N. and Akhtar, M. (1972). *Chem. Commun.* 23, 1263.

Chapter 3

Chemistry of the Carotenoids

G. P. MOSS and B. C. L. WEEDON

Queen Mary College, Mile End Road, London, England

I Introduction

Although the occurrence in nature of red and yellow pigments, now known to be carotenoids, has been reported for well over a century, the chemistry of these substances can, for most practical purposes, be regarded as starting in the late 1920s. In the succeeding decade the structures of some of the more readily available representatives were determined, largely through the classical work of the Karrer school, and other important studies were undertaken by Kuhn, Zechmeister, Heilbron, and their respective collaborators.

Following these pioneer structural studies, much effort was devoted to the synthesis of carotenoids and related polyenes. A crude product which apparently contained traces of β-carotene (I) was prepared from vitamin A (IV) by Karrer and Schwyzer in 1948,

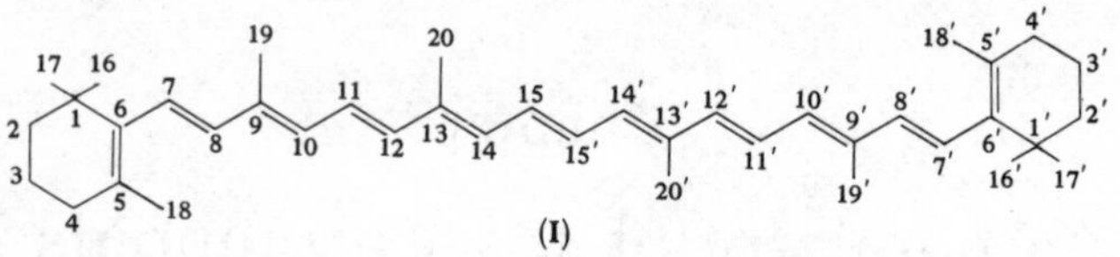

(I)

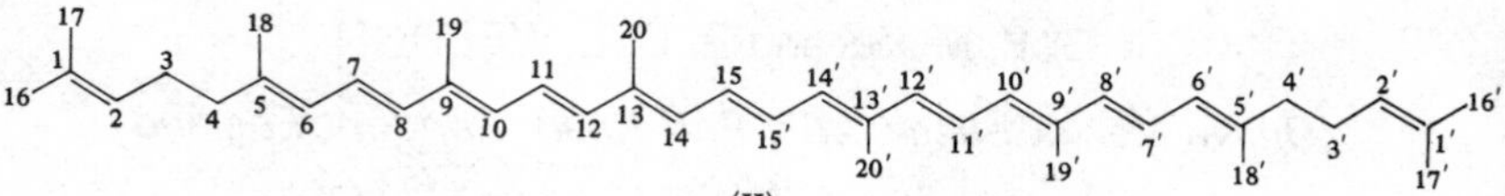

(II)

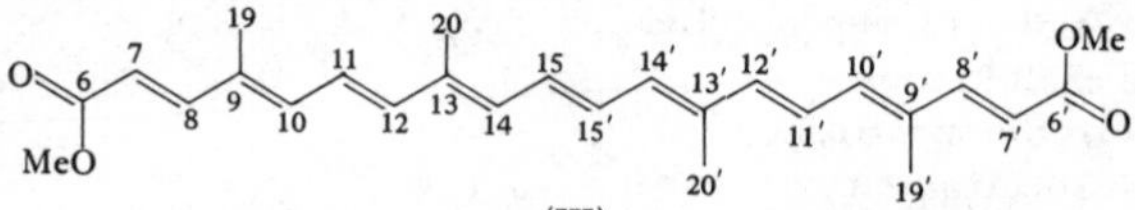

(III)

but the unambiguous total synthesis of carotenoids in isolatable amounts was first achieved with the hydrocarbons β-carotene (I) and lycopene (II) in 1950 (Inhoffen *et al.*, b,c; Karrer and Eugster; Milas *et al.*) and with the oxygenated carotenoid all-*trans*-methylbixin (III) in 1952 (Ahmad and Weedon), all by the same basic approach which was developed independently in several laboratories. The range of synthetic methods in the carotenoid field has since been greatly extended, principally by the groups of Karrer, Inhoffen, Weedon, Isler, and Pommer (cf. Mayer and Isler, 1971). As a result, many natural carotenoids have now been synthesized, some even on an industrial scale.

Since 1950 attention has also been directed to the intriguing question of the way in which carotenoids are formed in nature, and the main biosynthetic outlines have now been revealed by the work of Goodwin, Grob, Davies, Porter, Britton and others (see Chapter 5).

In the last 10–15 years the widespread adoption of spectroscopic techniques, notably nuclear magnetic resonance (n.m.r.) and mass spectrometry (m.s.), has revolutionized structural studies in this field. As a result many new carotenoids, often representing novel structural types, have been recognized, and the number continues to grow steadily. Another feature of the same period is the advance that has been made with the aid of other spectroscopic techniques such as optical rotatory dispersion (o.r.d.) and circular dichroism (c.d.), and with X-ray crystallography, in elucidating the absolute configuration of natural carotenoids, about half of which are believed to be chiral.

Although at different periods different interests, such as isolation, structure, properties, and synthesis, have predominated, these have always tended to overlap. This broad interest is scarcely surprising in view of the growing realization of the importance and variety of these widespread natural pigments. The emphasis throughout this chapter is on the methods currently used for elucidating their structures.

II Isolation and general properties

With the exception of some glycosides such as crocin, the digentiobioside of crocetin (80-80)* and some protein complexes, such as the astaxanthin (30-30) derivative ovoverdin, the carotenoids are fat-soluble pigments which can be isolated by extraction with a suitable solvent, e.g. benzene, petrol, (peroxide-free) ether, carbon disulphide, ethanol, methanol, or (acid-free) chloroform. Evaporation of the extract then gives a crude material which occasionally need only be crystallized to yield the desired product directly (e.g. in the isolation of lycopene from tomatoes). Usually, however, it is necessary to carry out a preliminary purification. This is often best done after treatment with 10% methanolic potassium hydroxide to hydrolyse any lipids or esters of hydroxylated carotenoids. (In some isolations the destruction of lipids is most conveniently achieved by reaction with lithium aluminium hydride.) A rough separation can then be obtained by partitioning the crude mixture between two immiscible solvents. Petrol and 90% methanol are commonly used, the former dissolving the non-hydroxylated carotenoids ("epiphasic fraction") and the latter the carotenoids with two or more hydroxyl groups ("hypophasic fraction"); the monohydroxy carotenoids tend to be distributed between both phases. This simple partition procedure can, of course, be greatly improved by the use of a Craig machine (Curl, 1960, 1963; cf. Krinsky, 1963).

Despite the procedures mentioned above, the great success which has been achieved in isolating carotenoids, and in separating the

* Formulae: For brevity most structures are referred to in the text by symbols such as (9-9) for β-carotene, and (53-63) for lutein. The figures refer to the two "halves" of the molecule listed in Table I, and which are attached to one another to give the 15,15′ single or multiple bond.

The conventional numbering of the positions in carotenoid structures is shown on formulae I–III.

Proposals have been made by the International Union of Pure and Applied Chemistry and the International Union of Biochemistry for a semisystematic nomenclature for carotenoids (Klyne *et al.*, 1972).

TABLE I
Formulae of half carotenoids

1

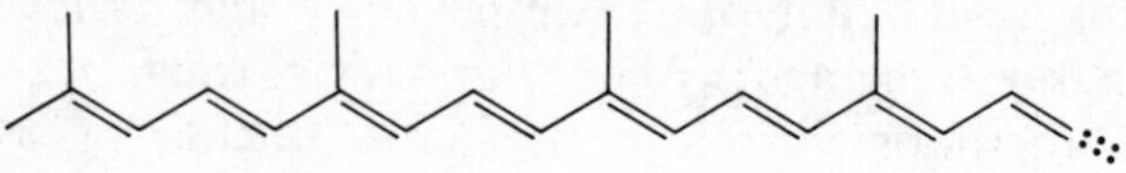

2

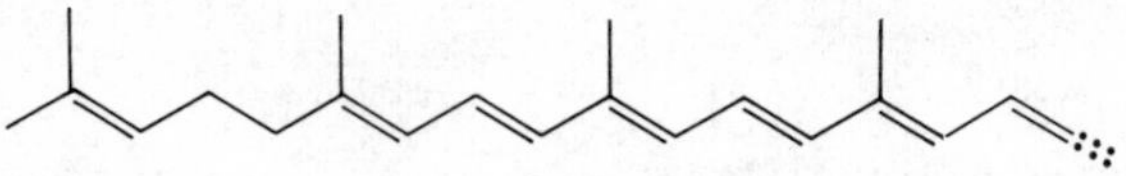

3

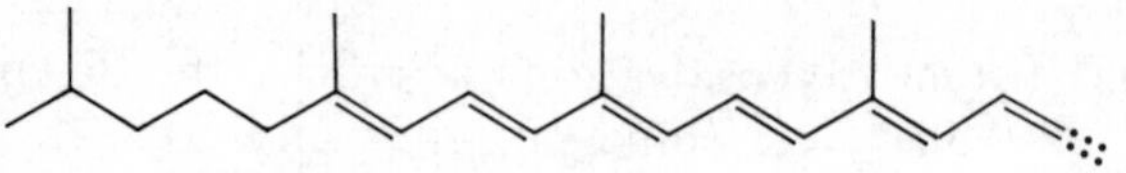

4

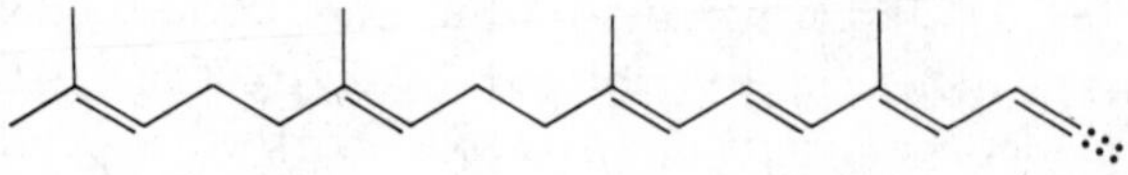

5

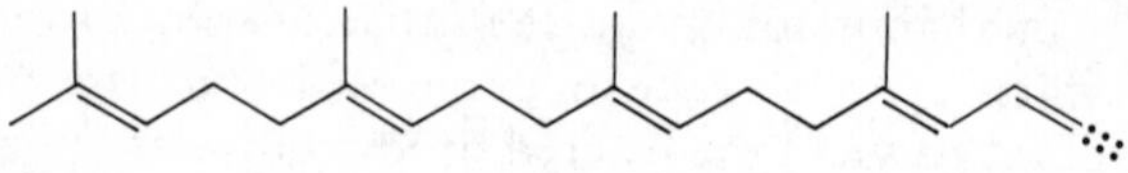

6

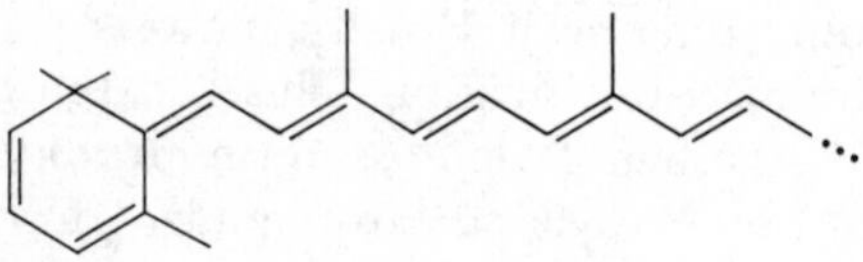

7

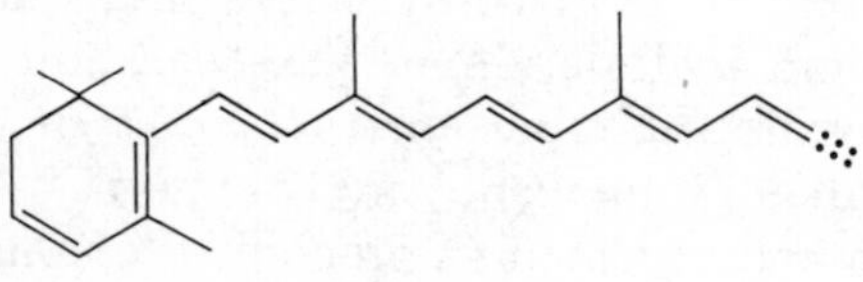

8

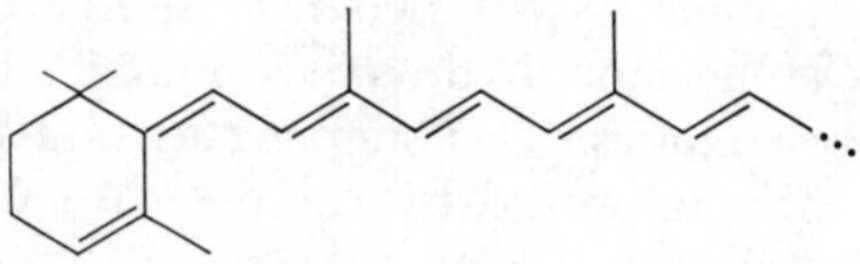

9

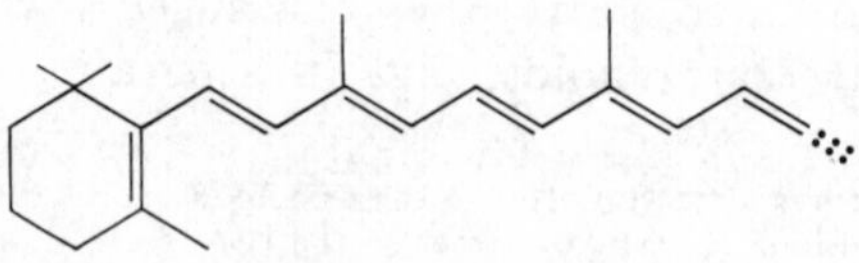

10

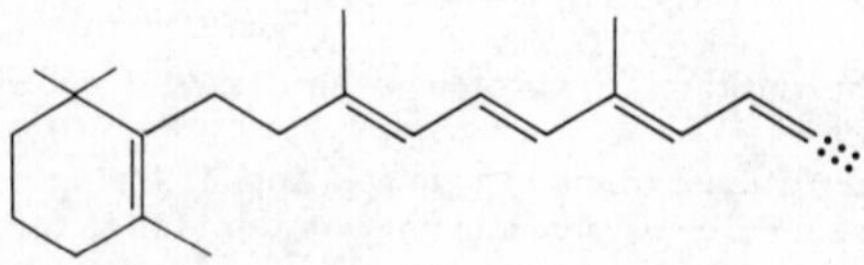

TABLE I—*continued*

11

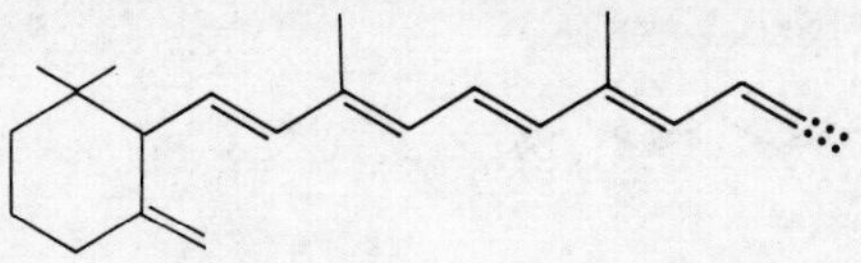

H

12

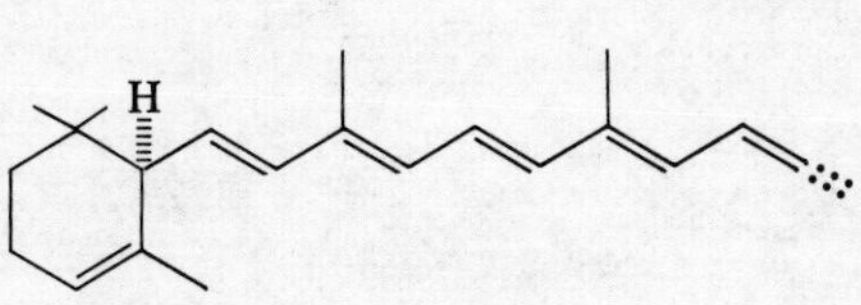

13

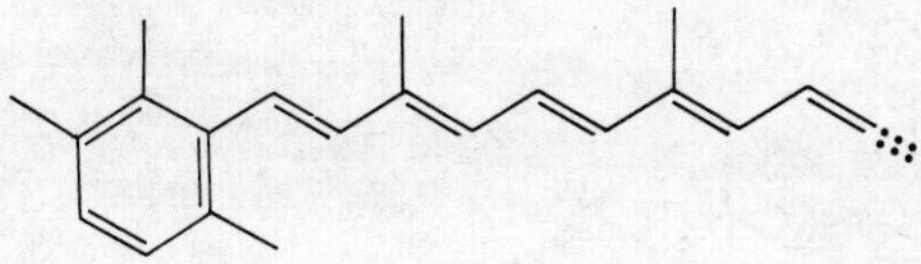

14

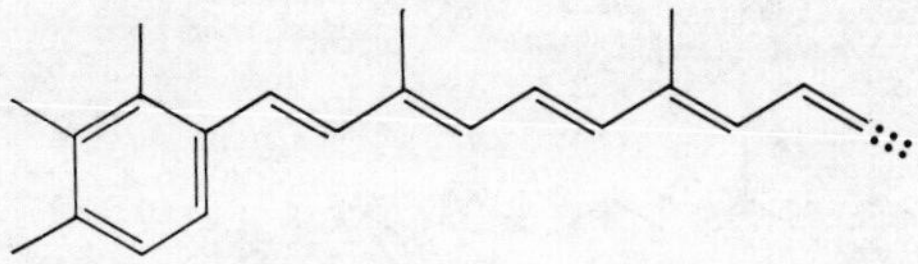

O

15

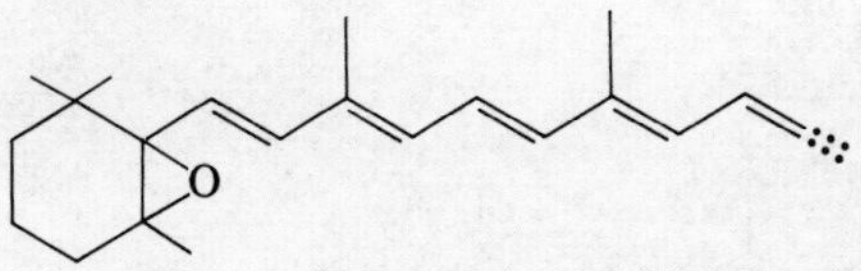

O

16

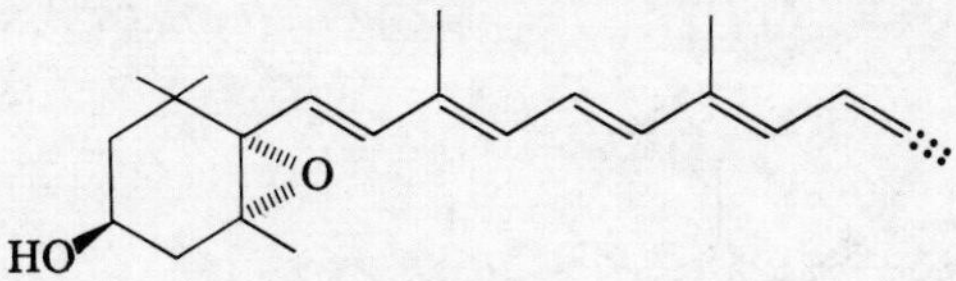

HO

O

17

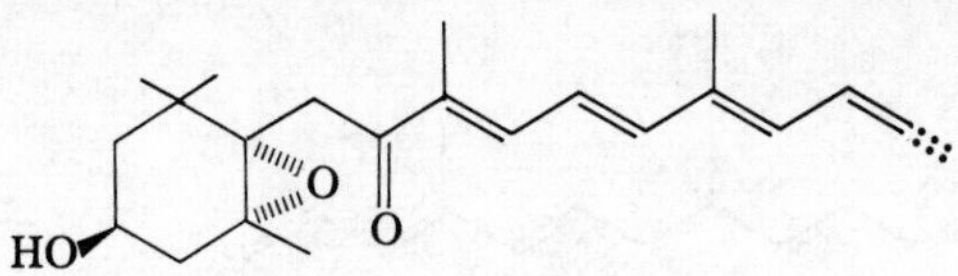

O

HO

O

O

O

18

HO

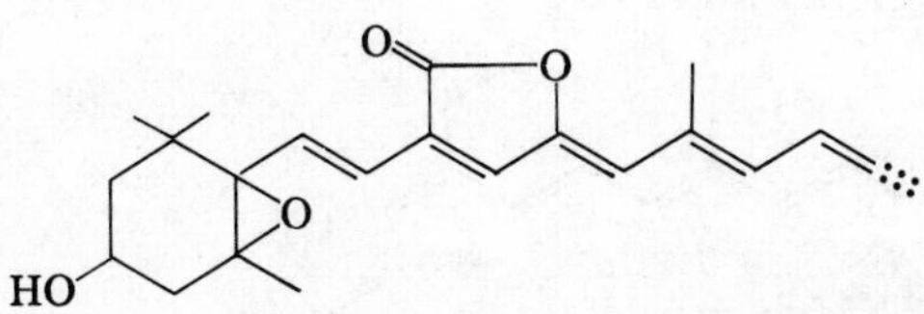

TABLE I—*continued*

19 HO OH O *

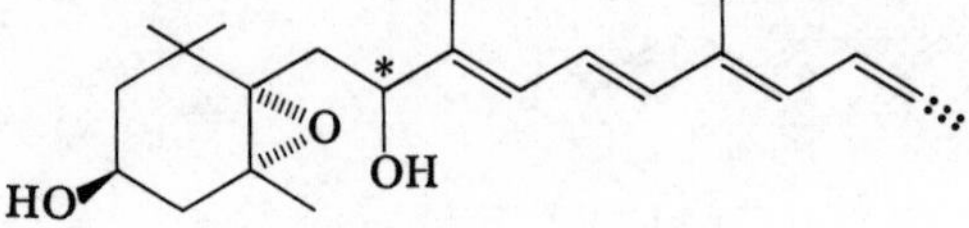

20 O *

21 HO O *

22 O OMe

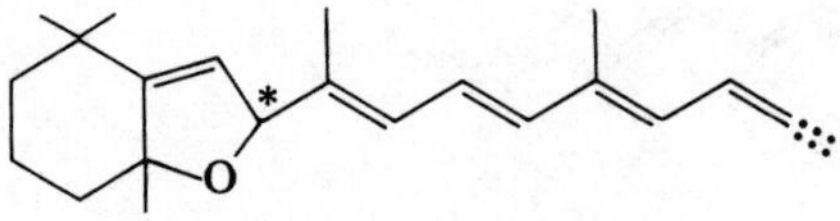

23 HO O

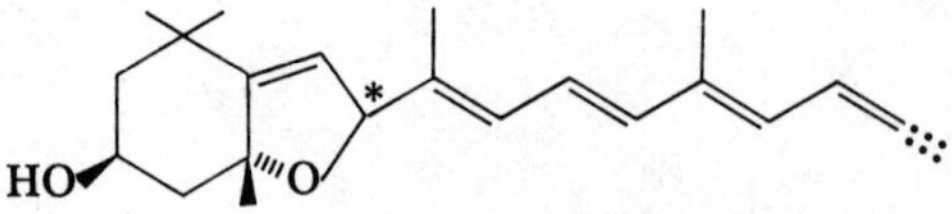

24 MeO O

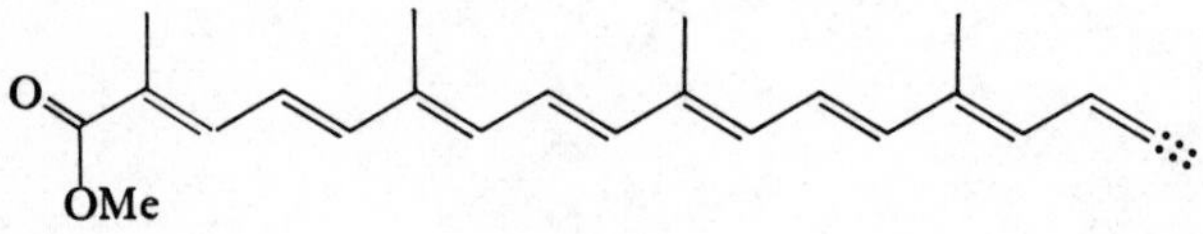

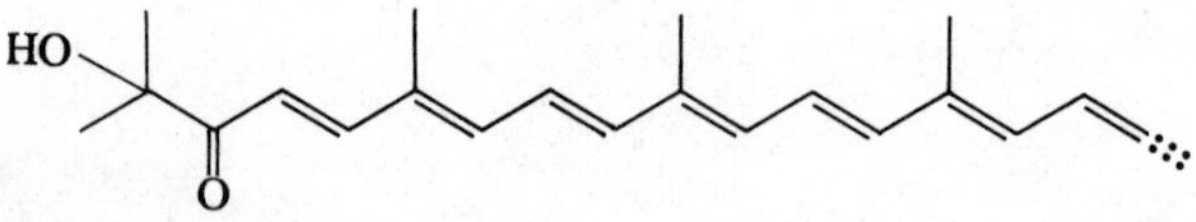

25 MeO O

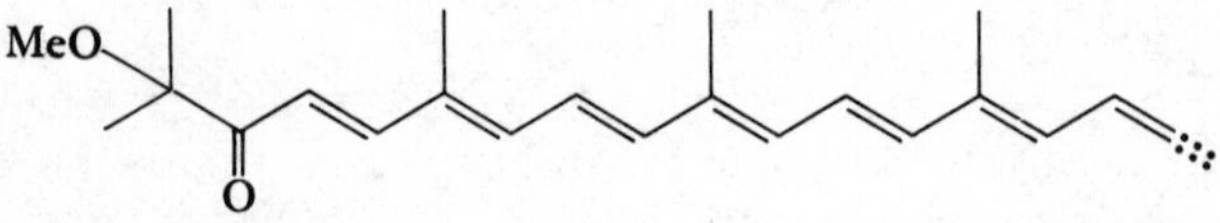

26 HO O

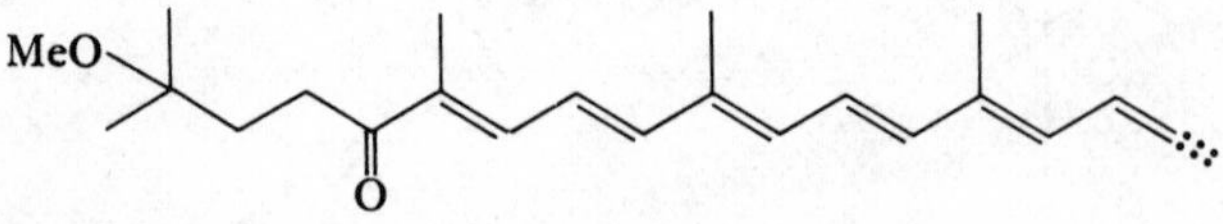

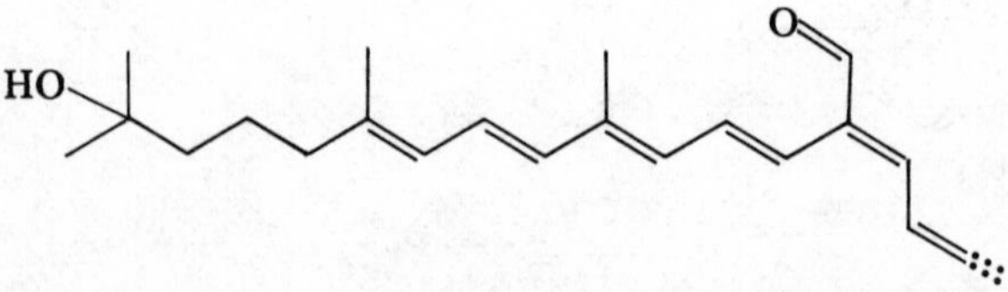

TABLE I—*continued*

27

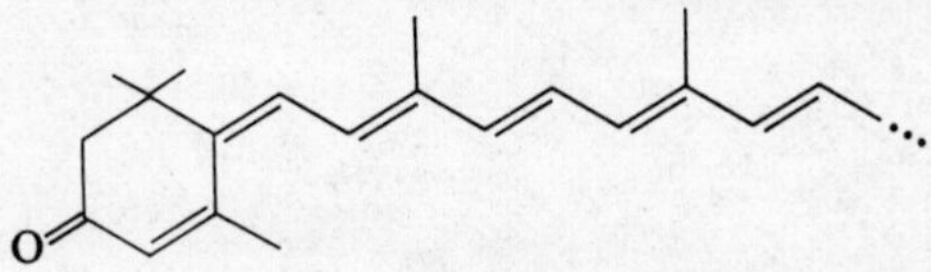

O

28

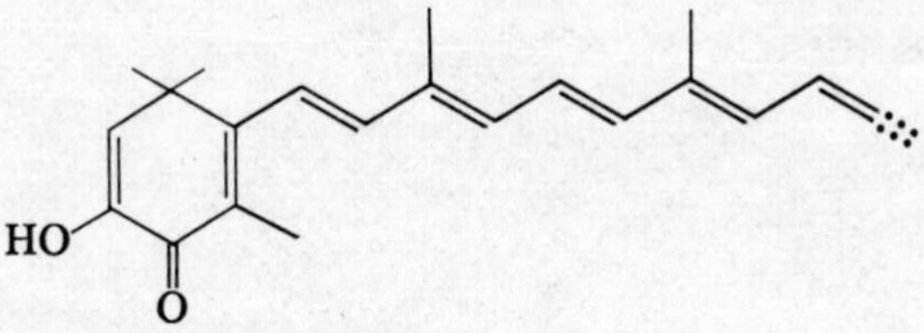

HO

O

29

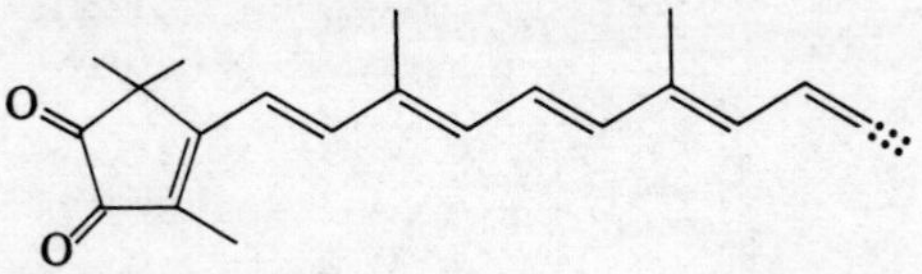

O

O

30

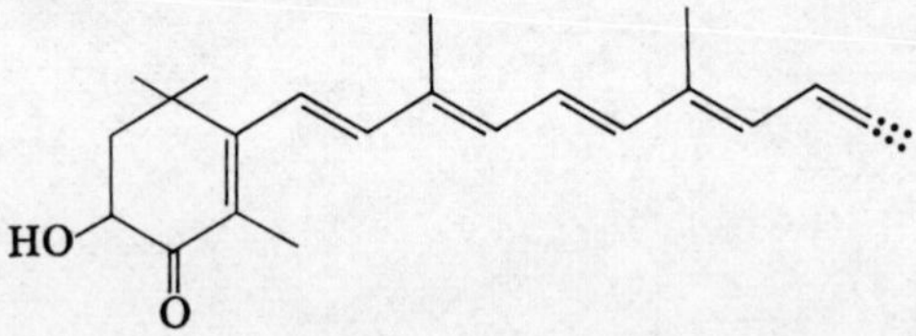

HO

O

31

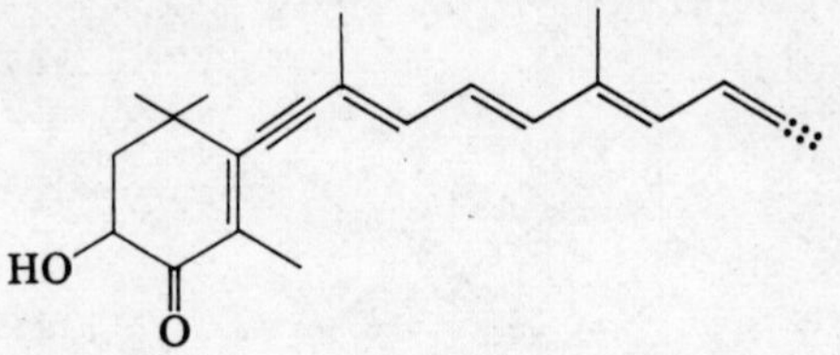

HO

O

32

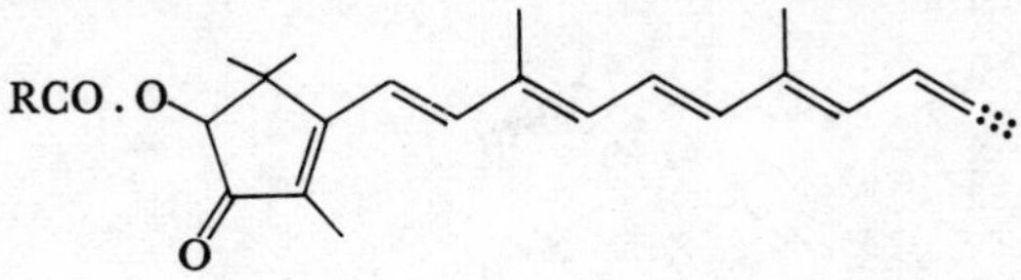

RCO.O

O

33

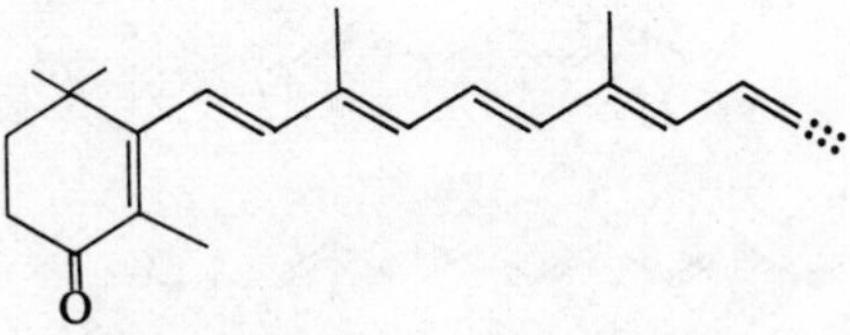

O

TABLE I—*continued*

34

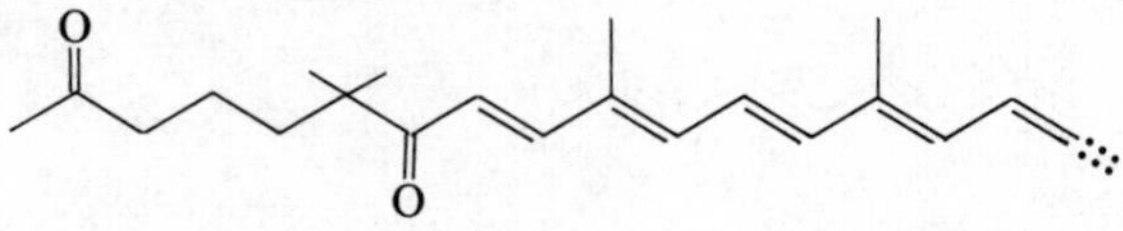

35

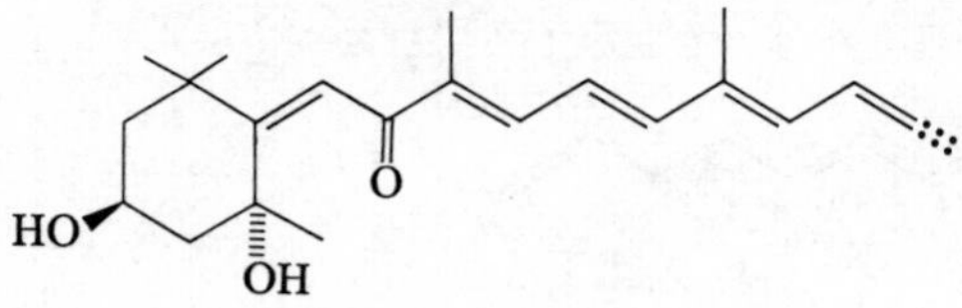

HO OH

36

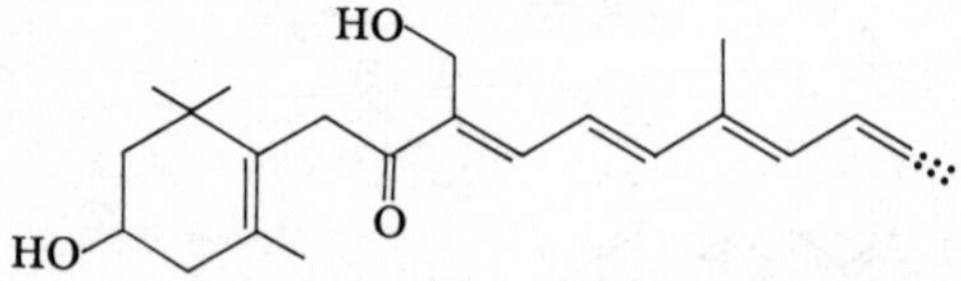

HO HO

37

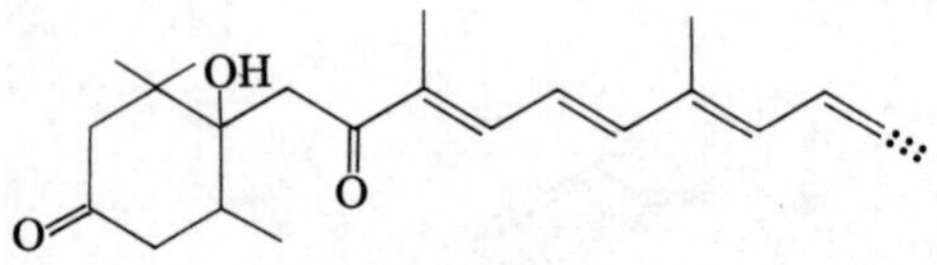

OH O

38

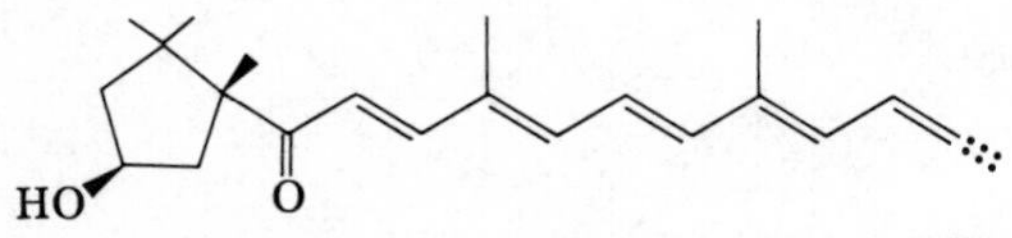

HO O

39

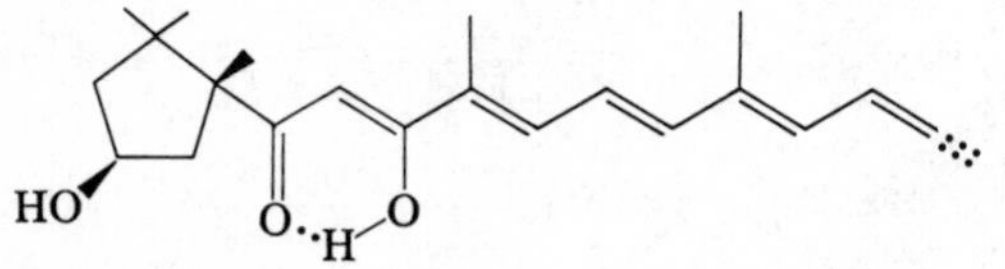

HO O··H O

40

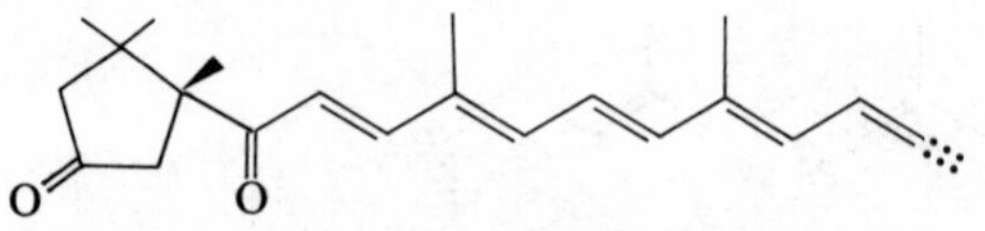

O O

41

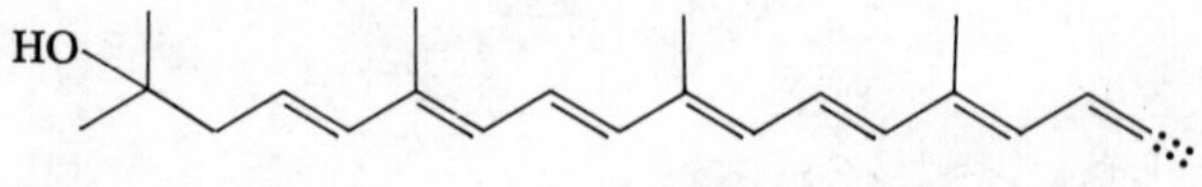

HO

42

MeO

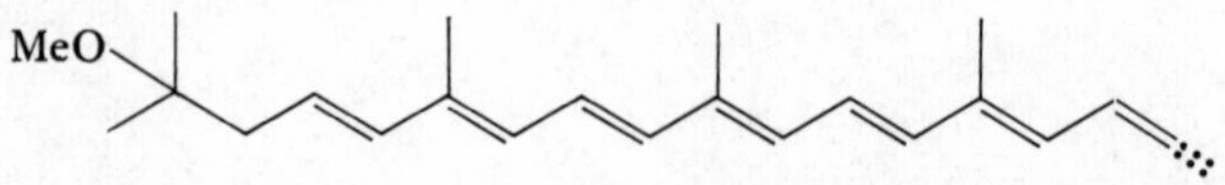

TABLE I—*continued*

43 HO OH

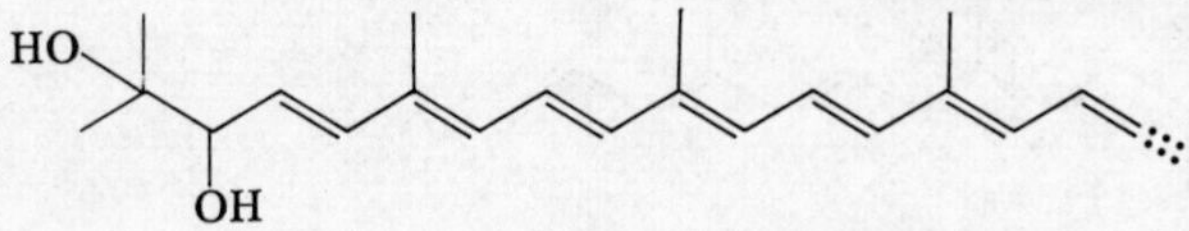

44 HO OGly

45 HO

46 OH

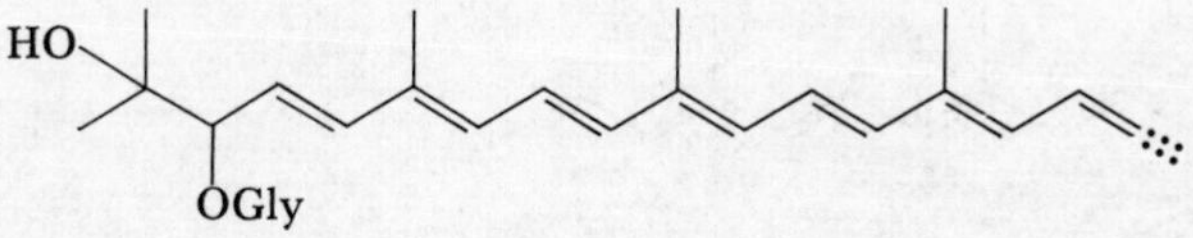

47 HO

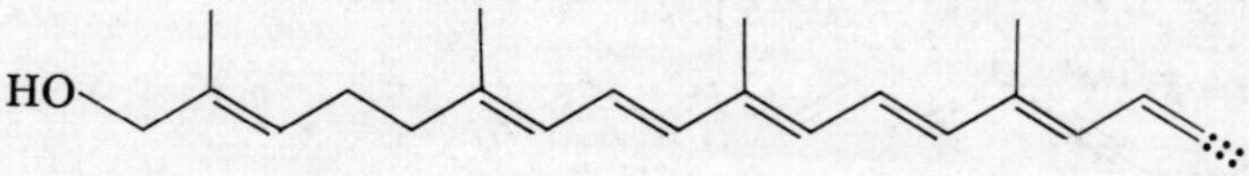

48 MeO

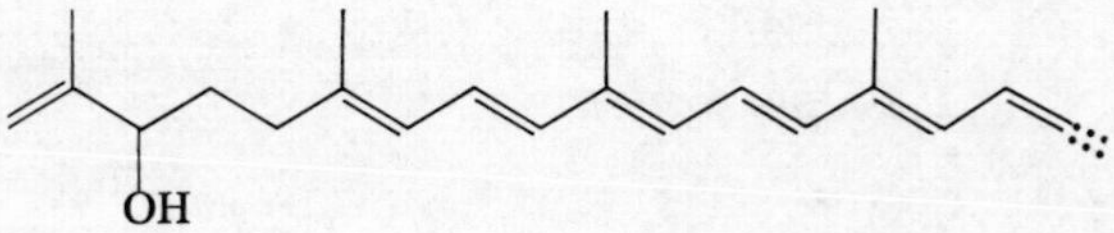

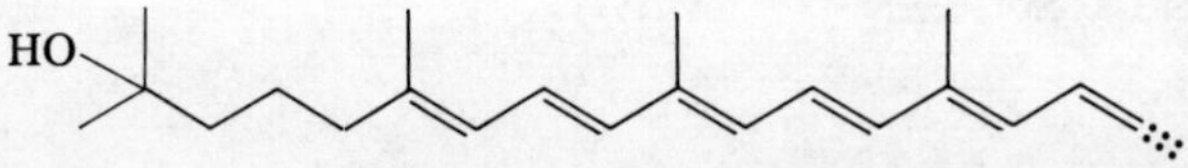

49 HO HO

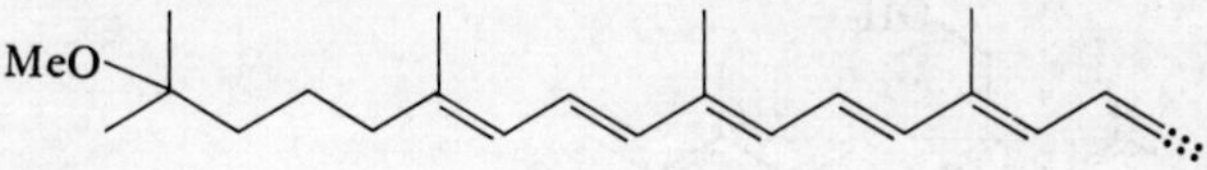

50 HO

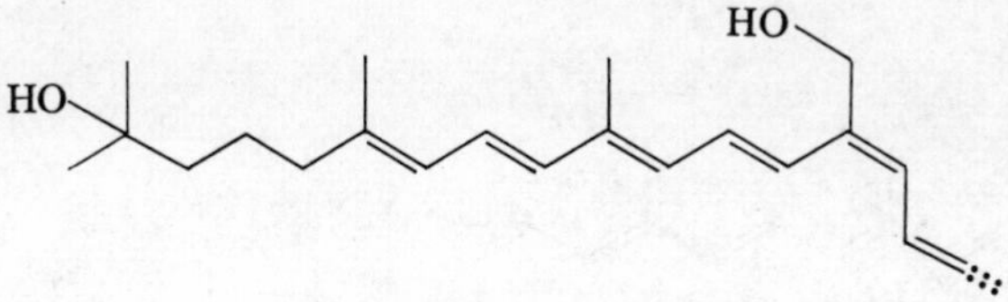

51 HO

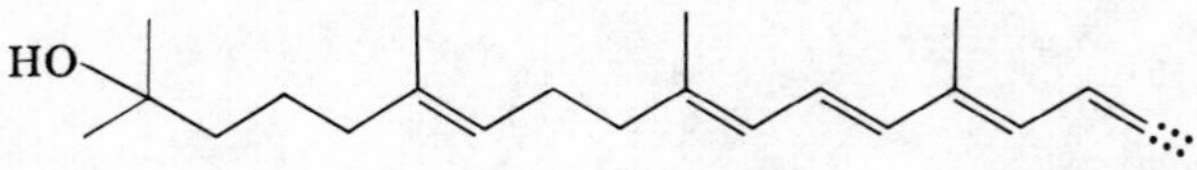

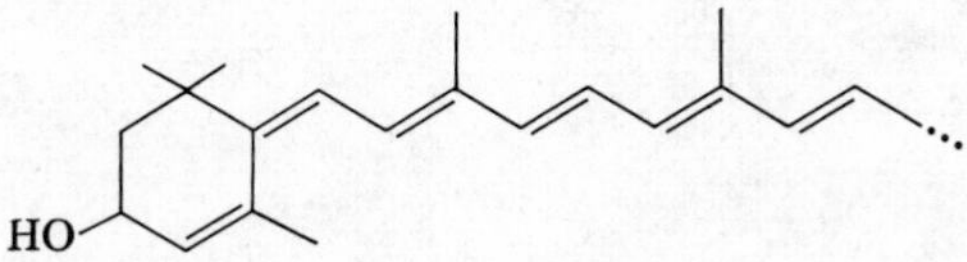

TABLE I—*continued*

52 HO

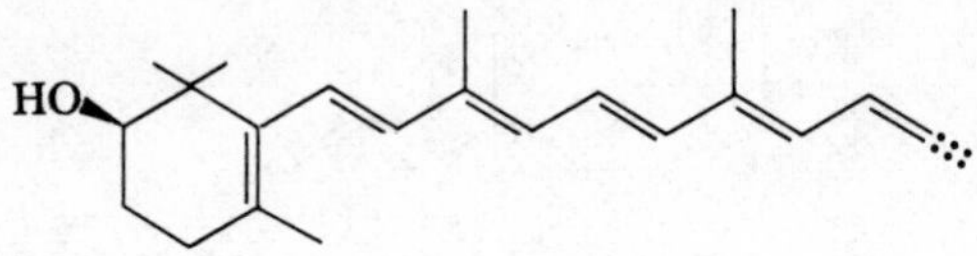

53 HO

54 HO

55 OH

56 OH HO

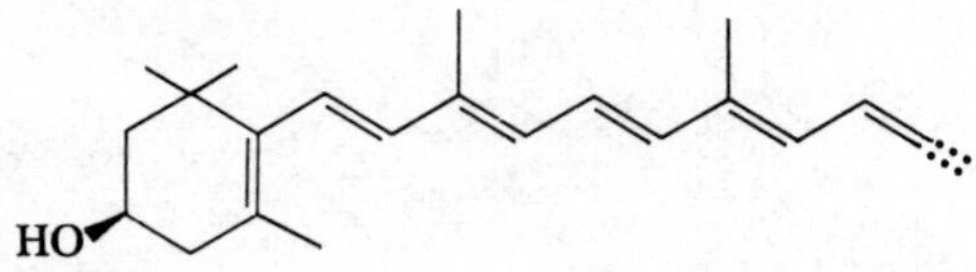

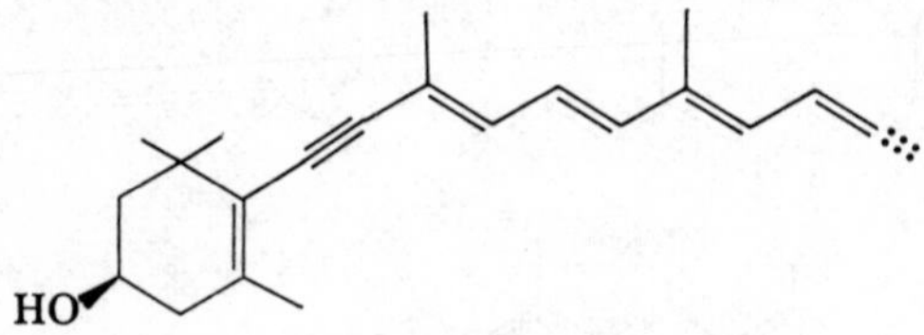

57 OH HO

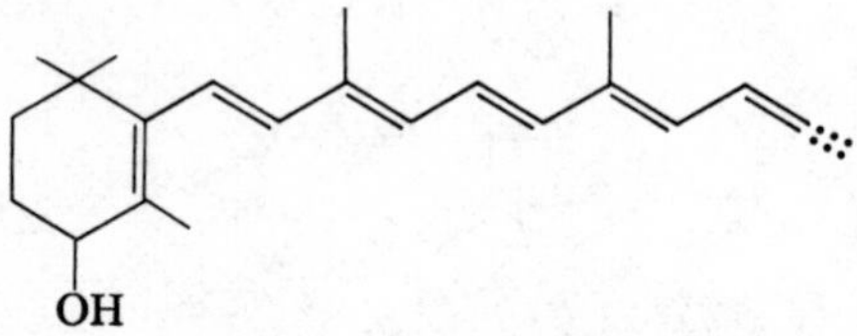

58 H HO OH

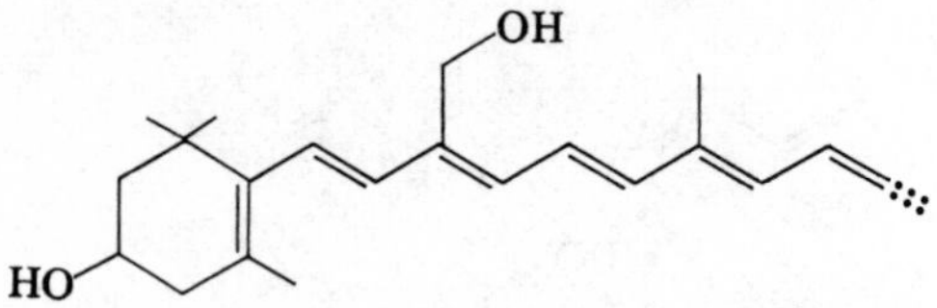

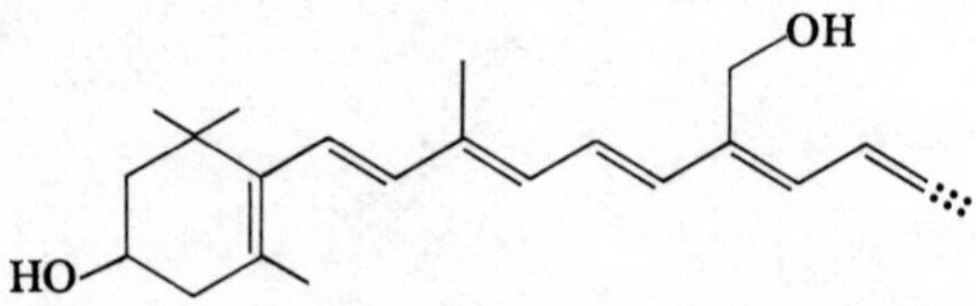

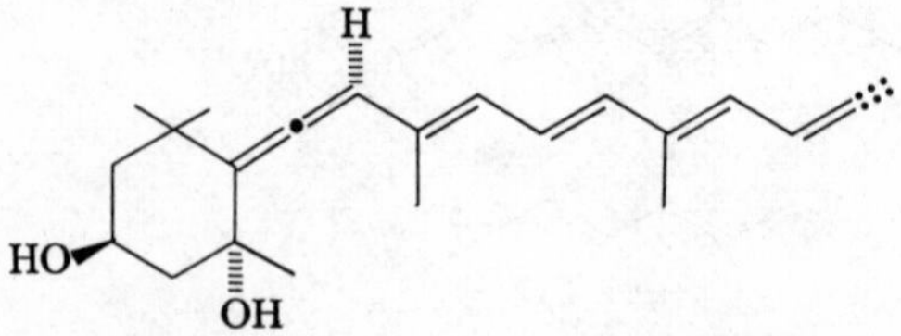

TABLE I—*continued*

59
MeCO.O
OH
H

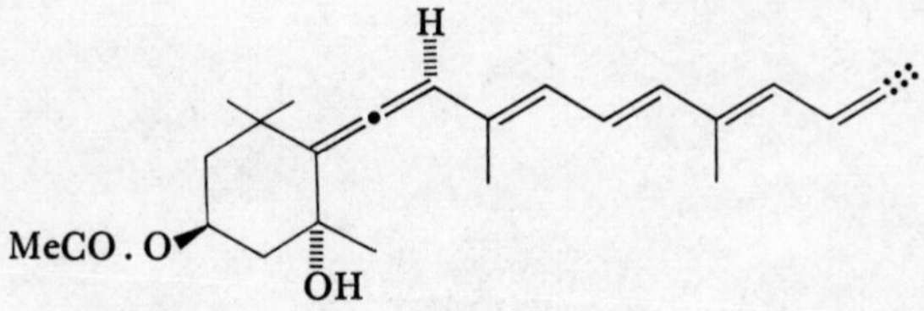

60
MeCO.O
OH
H

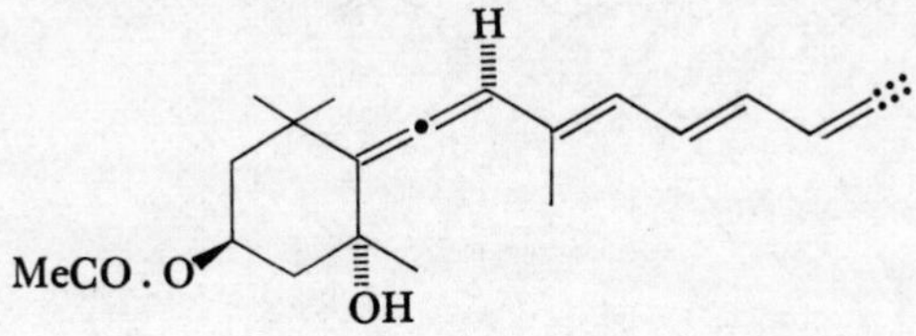

61
OH
OH

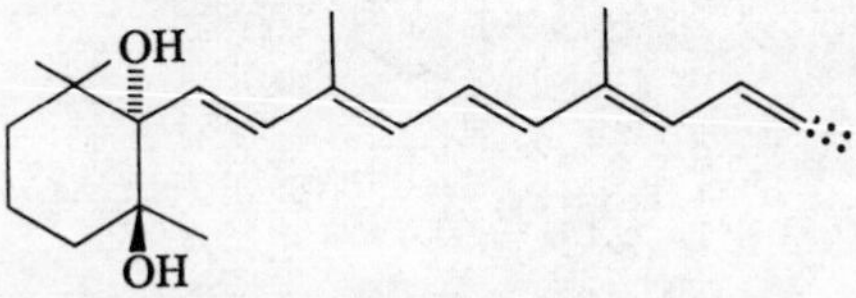

62
OH
HO
OH

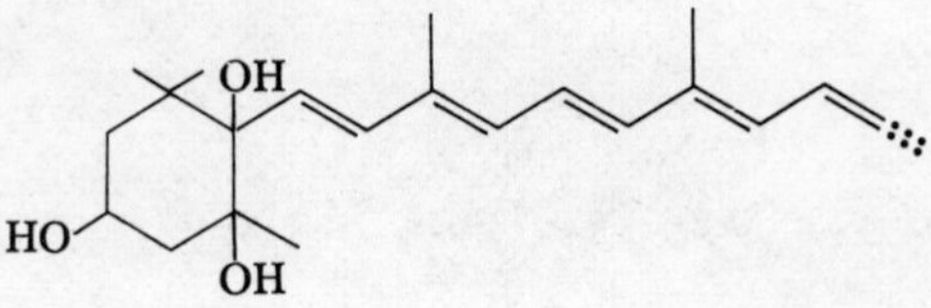

63
H
HO

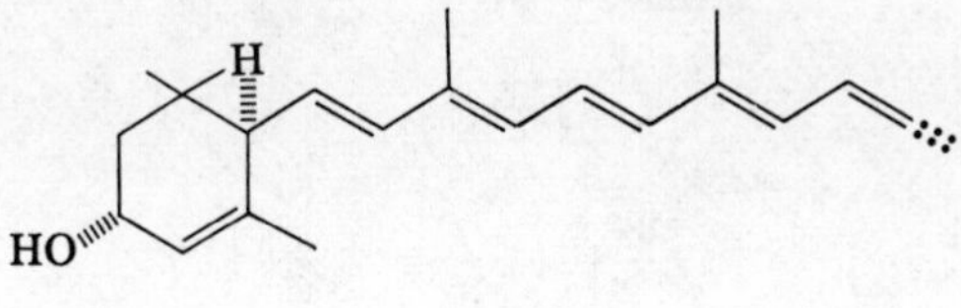

64
*
HO
OH

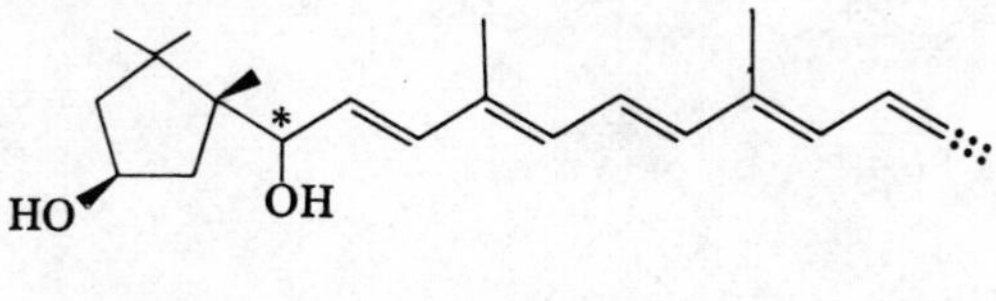

65
HO

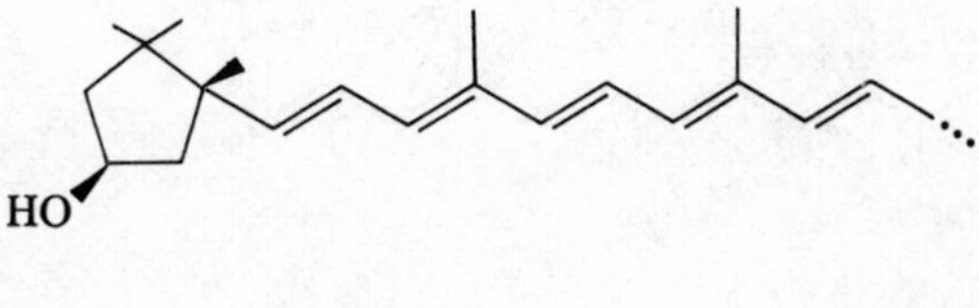

TABLE I—*continued*

66

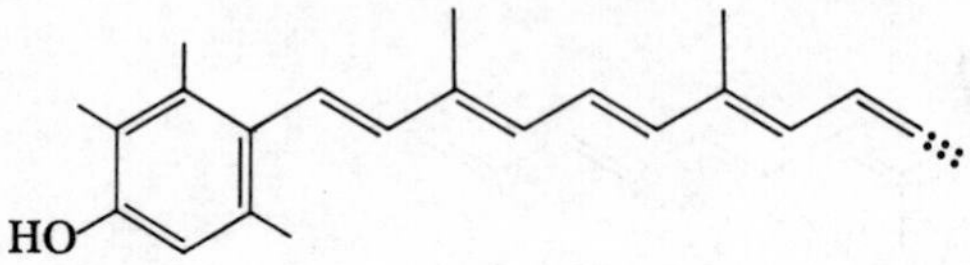

HO

67

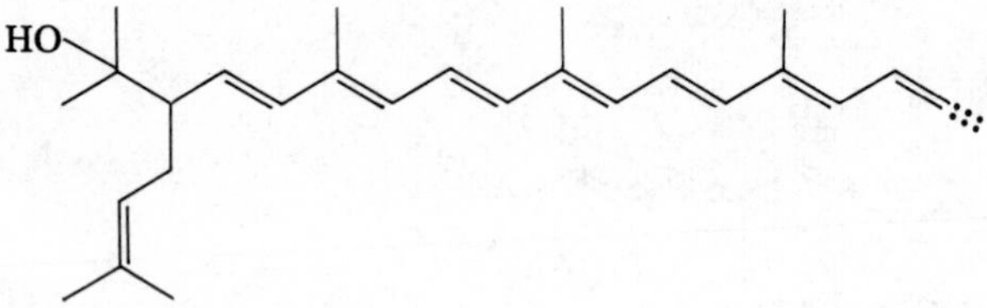

HO

68

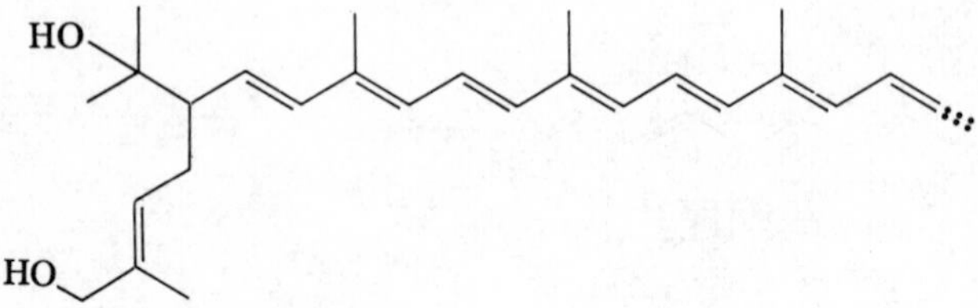

HO

HO

69

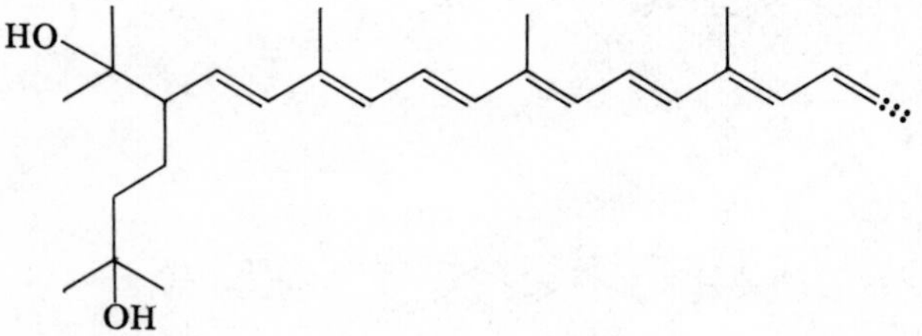

HO

OH

70

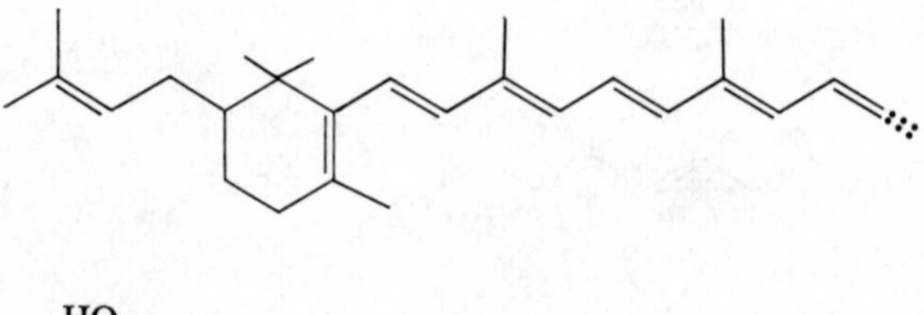

71

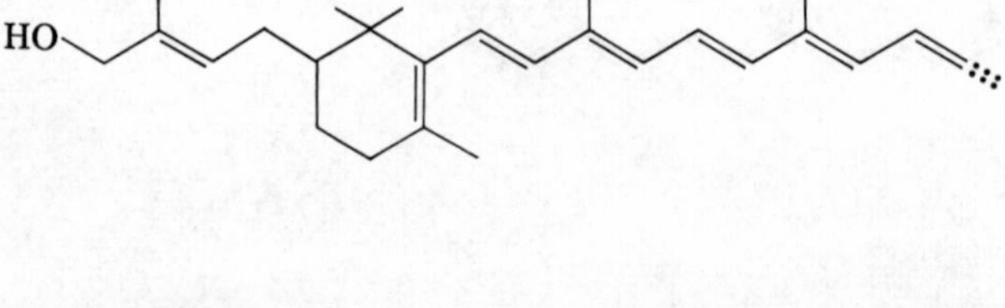

HO

HO

TABLE I—*continued*

72 OH

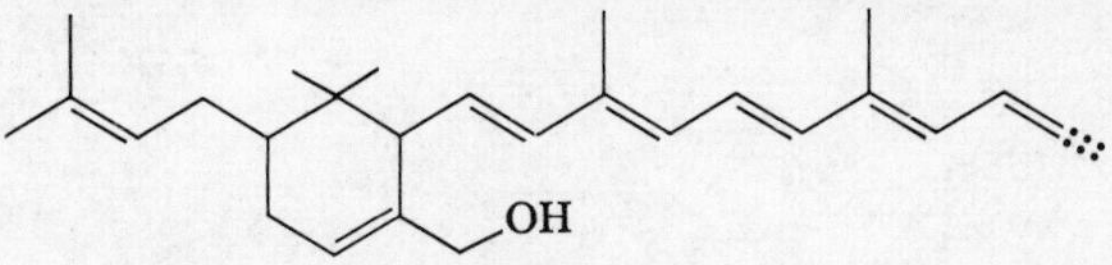

73 HO H H

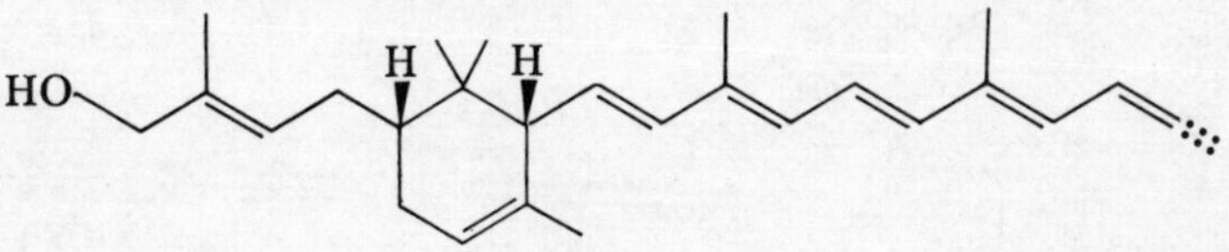

73a HO

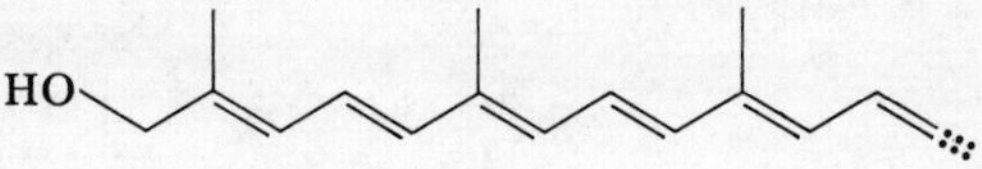

74 O OMe

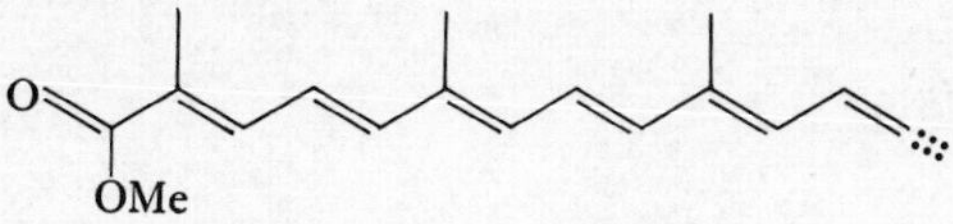

74a O

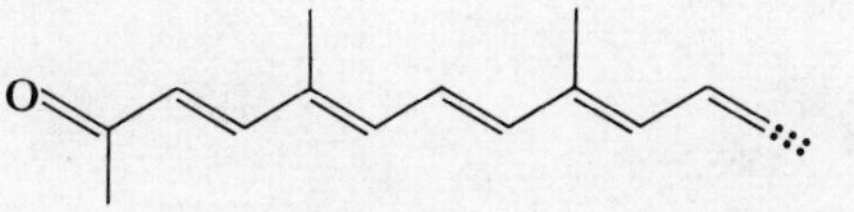

75 O

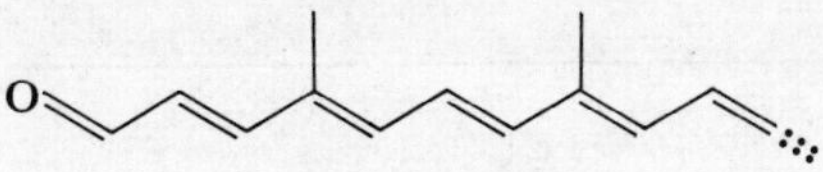

76 O OH

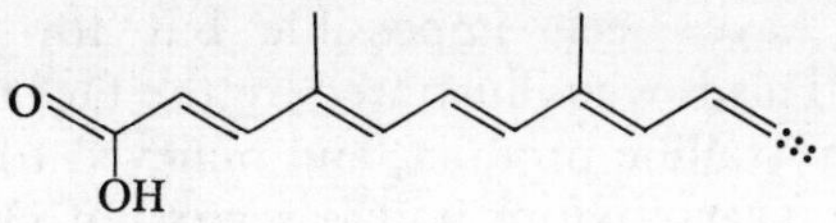

77 O OMe

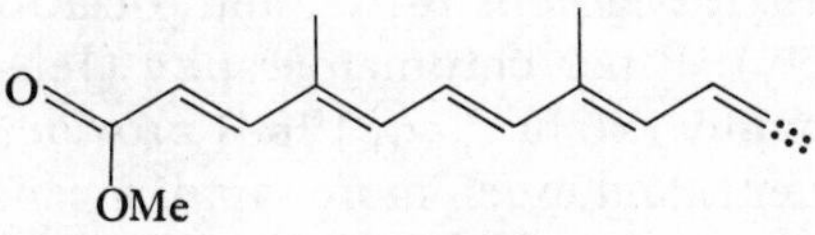

78 O

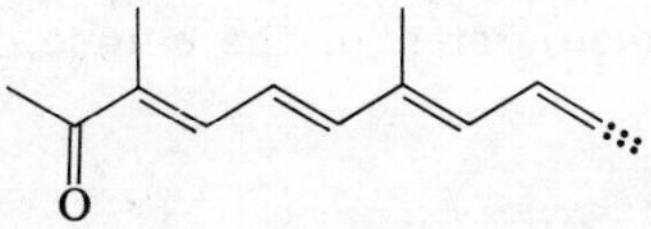

TABLE I—*continued*

79 O

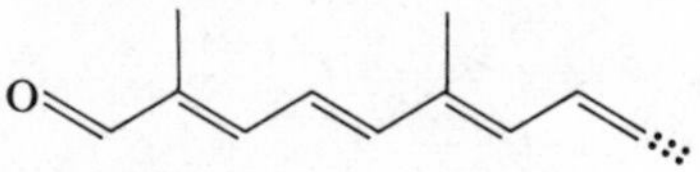

80 O OH

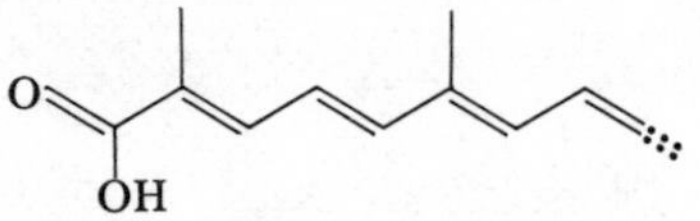

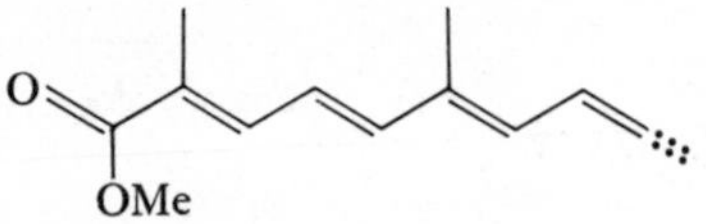

81 O OMe

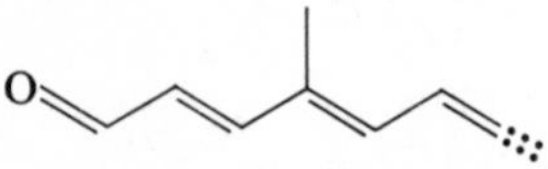

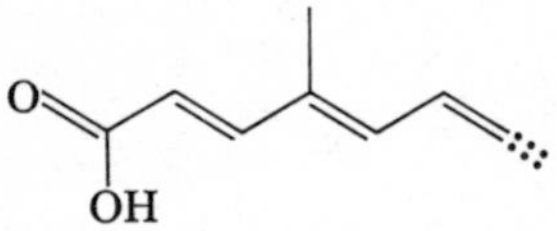

82 O

83 O OH

84 O

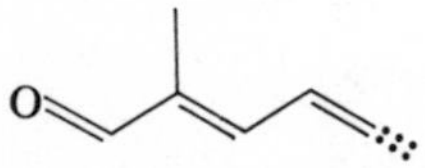

complex mixtures of these pigments which are frequently encountered in nature, would have been impossible but for the technique of chromatography. This is well illustrated by the fact that "carotene" was known as a crystalline product, and believed to be homogeneous, for nearly 100 years before it was separated chromatographically into the three hydrocarbons α-, β-, and γ-carotene (9–12, 9–9, and 2–9 respectively). Paper chromatography (Jensen, 1963) and thin-layer chromatography (Stahl *et al.*, 1963) provide, on an analytical scale, a more efficient, and much more rapid, means of separation, and these techniques are now widely used to monitor the more conventional (liquid–solid) chromatograms, or the course of a reaction.*

* For further details see Chapter 19.

Although many carotenoids appear to have sharp melting points, these may vary greatly with the rate of heating (decomposition, stereomutation?) and the way in which the determination is made (Kofler block, evacuated capillary, open tube). Some compounds may also exhibit polymorphism. In practice visible, infrared, n.m.r. and mass spectra afford the best means of characterization, and a mixed chromatogram with an authentic specimen (preferably using thin-layer or paper chromatography) provides a far more reliable proof of identity than the classical mixed melting point. Nevertheless it should always be remembered that, by itself, failure to observe a separation in a mixed chromatogram constitutes premissive, and not conclusive, evidence of identity. A micromethod for the comparison of carotenoids, based on mixed chromatograms and stereomutation studies, has been developed by Liaaen-Jensen (1962), and is particularly valuable when the supplies of pigment are inadequate for full spectral studies.

Some carotenoids, notably the various epoxides such as violaxanthin (16-16) and antheraxanthin (16-53), are sensitive to acids, even to the trace amounts of hydrogen chloride in "aged" chloroform, whilst fucoxanthin (17-59) and peridinin (18-60) are labile to dilute alkalis.

A few carotenoids (e.g. bixin (9′-*cis* 76-77) and crocetindial (79-79)) are surprisingly stable to air, and others (e.g. β-carotene (9-9), and astacene (28-28) reasonably stable provided that they are pure and well crystallized. Lycopene (2-2) is noticeably less stable than β-carotene (9-9) and ζ-carotene (4-4), a tetrahydrolycopene, is notoriously unstable in air. However, phytoene (5-5), an octahydro lycopene, is comparatively stable. Compounds such as spirilloxanthin (42-42), lacking the 1,2 double bond, but containing an additional double bond in the 3,4 position, are considerably less stable than lycopene. Compounds containing an additional conjugated double bond in the cyclic end group also seem to possess lowered stability. Thus vitamin A_2 (V) is noticeably less stable than vitamin A_1 (IV), and bis-dehydro-β-carotene (7-7) than β-carotene. Astaxanthin (30-30), a bis-acyloin, is rapidly oxidized in alkaline solution by air to the bis-diosphenol, astacene (28-28) (Kuhn and Sörensen, 1938).

Because of the instability of many carotenoids, and their tendency to undergo stereomutation (see Section VA), all operations involving these compounds should, whenever possible, be carried out in an inert atmosphere (e.g. nitrogen), solvents should be freshly purified, and solutions should not be heated more than necessary nor exposed to bright light.

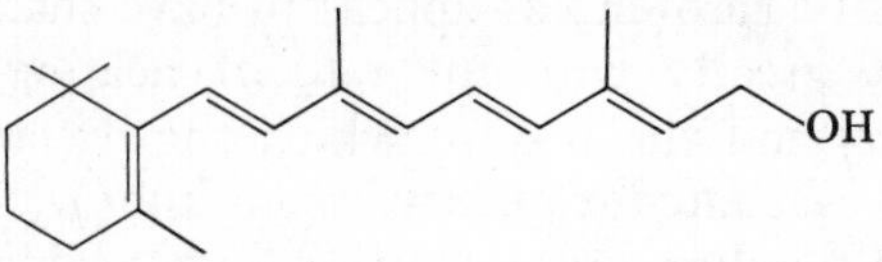

OH

(IV)

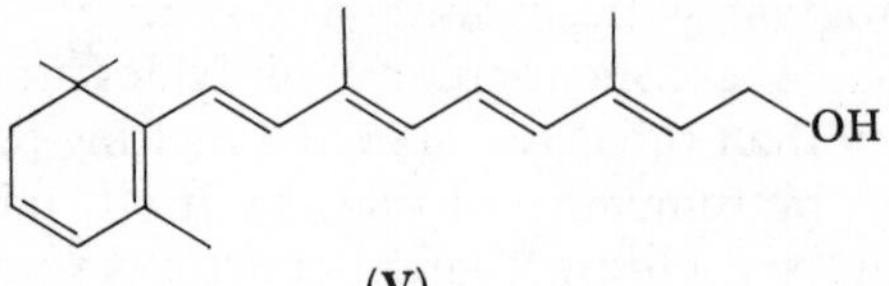

OH

(V)

III Physical examination

This section deals with the examination of carotenoids by a range of non-destructive techniques and by mass spectrometry. Much structural information can be obtained from a combination of these methods. Thus the mass spectrum will indicate the molecular formula and possibly indicate some structural features from the fragmentation pattern. The number of double bond equivalents present may be deduced from the molecular formula. The electronic spectrum will indicate the type of chromophore present. The nature of any oxygen substituents may be deduced from the infrared spectrum, while the proton magnetic resonance spectrum will characterize at least the methyl substituents. After the gross structure has been deduced, chiroptical and/or X-ray techniques may be used to derive the absolute configuration of asymmetric units. These various techniques are discussed below in the order in which they are commonly invoked.

A ELECTRONIC SPECTROSCOPY

The ultraviolet or visible light absorption properties of a polyene provide the best indication of the chromophoric system present. The wavelengths of the light absorption maxima increase with the number of conjugated double bonds, but the actual positions of the maxima are dependent on various structural features and on the solvent used (cf. Table II). The principal light absorption maxima of some standard polyenes in the carotenoid series are given in Table III and Fig. 1. The wavelengths of the maxima are longer than those observed with the corresponding polyenes devoid of methyl side

TABLE II
Effects of substituents and solvents on the main absorption maximum of carotenoids

Double bond	Chain	+7 to +35 nm
Double bond	Ring	+5 to +9 nm
First carbonyl	Chain	+28 nm
Second carbonyl	Chain	+1 to +7 nm
First carbonyl	Ring	+7 nm
Second carbonyl	Ring	+5 to +9 nm
First carbonyl	Ring *retro*	+9 nm
Second carbonyl	Ring *retro*	+5 nm
First or second	5,6-Epoxide	−8 nm
First or second	5,8-Epoxide	−20 nm
Trans → *cis*		−4 nm
Normal → *retro*		−10 nm
Solvent effect:	Ethanol, hexane, petrol, ether	Small effect
	Benzene, chloroform	+15 nm
	Carbon disulphide	+35 nm

From Vetter *et al.*, 1971.

TABLE III
Visible light absorption properties of some isoprenoid polyene hydrocarbons (acylic chromophore)[a]

Polyene	Number of conjugated double bonds	Principal light absorption maxima (nm)			Reference
Phytoene (5–5)	3	298	286	276	1
Photofluene (4–5)	5	366	347	331	1
ζ-Carotene (4–4)	7	425	401	380	1
α-Zeacarotene (4–12)	8	449	421	398	2
Neurosporene (2–4)	9	470	440	416	1
ϵ-Carotene (12–12)	9	471	440	418	3
δ-Carotene (2–12)	10	487	456	431	3
Lycopene (2–2)	11	504	472	443	4
1,2-Dihydro-3,4-dehydrolycopene	12	518	483	457	5
3,4-Dehydrolycopene (1–2)	13	535	500	468	6
3,4,3′,4′-bisDehydrolycopene (1–1)	15	540	510	480	7

References
1. Davis *et al.*, 1961.
2. Petzold *et al.*, 1959.
3. Manchand *et al.*, 1965.
4. Karrer and Jucker, 1948.
5. Malhotra *et al.*, 1970.
6. Winterstein *et al.*, 1960.
7. Surmatis and Ofner, 1963.

Note:
[a] Determined on dilute solutions of the all-*trans* isomers in light petroleum or hexane.

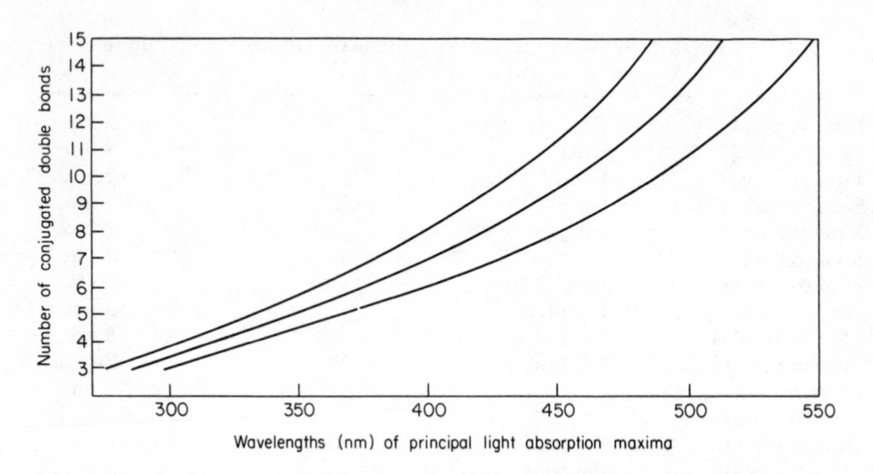

FIG. 1. Positions of principal light absorption maxima of polyenes in the carotenoid series with acyclic chromophores (plotted from data given in Table III).

chains (cf. Dale, 1954, 1957). A comparison of β-carotene with 13,13′-bis-desmethyl-β-carotene (Inhoffen *et al.*, 1950a), and of crocetin dimethyl ester (81-81) with less highly methylated analogues (Mildner and Weedon, 1953), suggests that each methyl substituent on the polyene chain produces a bathochromic shift comparable to that (*c.* 5 nm*) observed in simple dienes (cf. Braude, 1945).

Many carotenoids exhibit light absorption properties which do not correspond exactly with the data given in Fig. 1 (see Table III). These departures, which are often of great help in making structural assignments, are due to the presence of cyclic end groups of the type found in β-carotene, to conjugation of the polyene system with another chromophore (–CHO, $-CO_2H$, $-CO_2Me$, >C=O, aryl), or to the fact that one or more double bonds in the polyene chain has the *cis* configuration. The first two of these causes for deviation will be discussed briefly here, and the third deferred to Section VA.

Although lycopene (2-2), γ-carotene (2-9) and β-carotene (9-9) formally possess the same chromophore, the positions of their light absorption maxima differ significantly (Table IV). A study of molecular models reveals that the presence of the cyclic end groups leads to steric hindrance between the ring methyls and the polyene side chain:

* The units nm were formerly given as mμ.

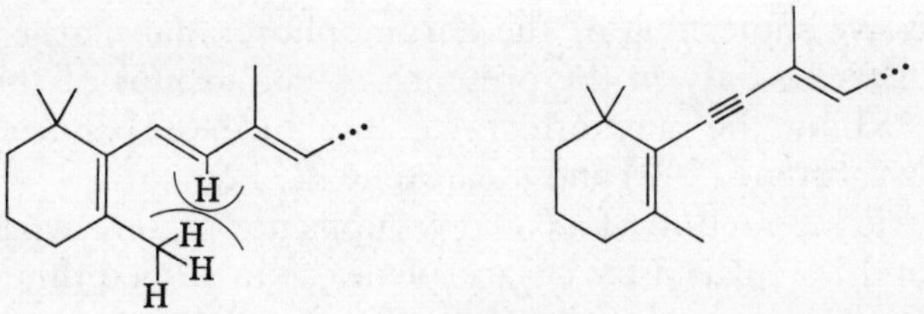

TABLE IV
Visible light absorption spectra of some carotenoids with eleven conjugated multiple bonds (wavelength in nm for a petroleum ether solution)

Carotenoid	Maxima			Reference
Lycopene (2–2)	507	476	448	Manchand *et al.*, 1965
γ-Carotene (2–9)	490	459	433	Manchand *et al.*, 1965
β-Carotene (9–9)	477	449	427	Unpublished results
Isozeaxanthin (55–55)	475	448	427	Unpublished results
Zeaxanthin (53–53)	479	451	429	Unpublished results
Alloxanthin (54–54)	478	450		Unpublished results
15,15′-Dehydro-β-carotene	457	431		Isler *et al.*, 1956
11,12,11′,12′-Tetradehydro-β-carotene		405		Isler *et al.*, 1957b
4′-Apo-β-carotenol (9–73a)	488	459	436	Rüegg *et al.*, 1959
Citranaxanthin (9–74a)	495	463		Yokoyama and White, 1965
6′-Apo-β-carotenal (9–75)		473		Rüegg *et al.*, 1959
Methyl 6′-apo-β-carotenoate (9–77)	491	464		Isler *et al.*, 1959
6′-Apo-β-carotenoic acid (9–76)	495	458		Isler *et al.*, 1959
6′-Apo-lycopenal (2–75)	515	479	462	Kjøsen and Liaaen-Jensen, 1969
Methyl 6′-apo-lycopenoate (2–77)	503	471	448	Kjøsen and Liaaen-Jensen, 1969
Trans-methyl bixin (77-77)	490	456	432	Isler *et al.*, 1957a

The resulting lack of planarity in the molecule limits overlap of the π orbitals associated with the ring double bond and the polyene chain, and produces the observed shift of the light absorption maxima to shorter wavelengths (*c.* 12 nm for each end which has been cyclized) together with a partial loss of fine structure. Similar changes are also observed in the spectra of some of the aryl carotenoids (Yamaguchi, 1958; Cooper *et al.*, 1963).

Despite the "incomplete conjugation" of the double bond in β end groups, the difference between the light absorption maxima of β-carotene (9–9), α-carotene (9–12), and ε-carotene (12–12) indicates the progressive shortening of the chromophore. Many other pairs of isomers, differing only in the presence of end groups of the "α-" or "β-type", exhibit the same effect, e.g. lutein (53-63) and zeaxanthin (53-53); γ-carotene (2-9) and δ-carotene (2-12).

This steric interaction of a β end group is not present when the 7,8 double bond is replaced by an acetylene, as in alloxanthin (54-54). The principal maximum of alloxanthin is virtually the same as that of zeaxanthin (53-53). This is the result of two opposing effects. Due to the acetylene the chromophore may be essentially planar across the 5=6—7≡8—9=10 system giving full conjugation. However, when an acetylene replaces an unhindered double bond the principal maxima occurs at a shorter wavelength (see Table IV).

The presence of an allene as in neoxanthin (16-58) will have no significant effect on the spectrum, compared with that of a carotenoid such as lutein epoxide (16-63) in which chromophore ends at C-7′. This is due to the plane of the 6,7 double bond being perpendicular to that of the 7,8 double bond.

The conjugation of the polyene chain with another chromophore will, of course, result in a shift of the light absorption maxima to longer wavelengths (cf. Table V), and this shift is usually accompanied by loss of fine structure. The presence of a carbonyl group conjugated with the main chromophore is readily detected by treatment of the solution in the u.v. cell with sodium borohydride (or lithium aluminium hydride, in which case a non-hydroxylic solvent is necessary). Reduction of the carbonyl group shortens the chromophore, resulting in a spectrum with more fine structure and a lower wavelength principal maximum.

TABLE V
Wavelength of maximal light absorption of β-carotene and its keto derivatives[a]

Polyene	λ_{max} (nm)
β-Carotene (9-9)	466
Echinenone (9-33)	480
Canthaxanthin (33-33)	486·5
3-Ketoechinenone (9-28)	490·5
3-Ketocanthaxanthin (28-33)	494·5
Astacene (33-33)	499·5

[a] Determined on chloroform solutions; C. W. Price and B. C. L. Weedon, unpublished results.

Tests may also be applied on a u.v. scale for the presence of a 5,6-epoxide (by acid) or a carboxylic acid (by base). Further details of these reactions are given in Section IV.

A number of methods have been developed to calculate the principal maxima of carotenoids (see, for example, Dale, 1954, 1957; Hirayama, 1955; Arpin *et al.*, 1969–70; Schwieter *et al.*, 1969). However, suitable model compounds are usually available which will give a far better estimate of the expected spectrum.

B MASS SPECTROMETRY

With care, mass spectrometry is an extremely powerful tool for the determination of carotenoid structures. Very small quantities of material are required, usually limited only by handling problems (i.e. 100 μg or less of a pure sample). With a modern double focusing spectrometer it is possible to measure the mass of the individual ions produced to an accuracy of at least 2 p.p.m. This allows direct calculation of the molecular formula using the known exact isotopic masses of the elements present (^{12}C = 12·000000, ^{1}H = 1·007825, ^{16}O = 15·994915). Thus many of the inherent limitations of microanalysis are circumvented.

A particularly good example of this is the ready distinction between astaxanthin (30–30) and its 7,8-dehydro analogues (30–31) and (31–31). On a microscale the mass spectrum clearly shows the molecular ion at two or four mass units less (Francis *et al.*, 1970). On this scale the infrared acetylenic absorption is not measurable and there is no significant difference between the ultraviolet spectra of these three compounds.

The limitations of the mass spectrometric technique should be appreciated. In the presence of more volatile impurities the observed spectrum may not represent the major component of the sample; changes in the spectrum on increasing the probe temperature may indicate this problem. Not all carotenoids will exhibit a molecular ion due to rapid loss of a fragment such as water. However, in the case of a hydroxyl group its presence is readily demonstrated by its absorption in the infrared spectrum.

Different instruments, ionization or insertion conditions, and probe temperatures may each give rise to marked variation in the intensity and relative abundance of the parent and daughter ions. Thus while many reported significant fragment ions are listed in Table VI there is no guarantee on their abundance, or even that they will all be detected in any one spectrum.

TABL

^{1}H-Nuclear magnetic resonance, infrar

(n.m.r. data in p.p.m. from Me_4Si

	Structure of half carotenoid[a]								
Formula	Skeleton	Epoxy	Hydroxy and Deriv.	Oxo	Dehydro-	Hydro-	Notes	1,1 Me's	5 Me
				Hydrocarbon					
1	ψ				3,4			1·82, 1·82	1·96
2	ψ							1·61(c), 1·68(t)	1·81
3	ψ					1,2		0·88(6H, d, J = 6)	1·81
4	ψ					7,8		1·60(c), 1·67(t)	1·60
5	ψ					7,8,11,12	15,15′-*cis*	1·61(c), 1·66(t)	1·58
7	β				3,4			1·03(6H)	1·86
8	β				4		*retro*	1·29(6H)	1·94
9	β							1·03(6H)	1·72
10	β					7,8		1·01(6H)	1·61
11	γ							0·83, 0·91	–
12	ϵ							0·82, 0·90	1·58
13	φ							2·22*, 2·27*	2·27*
14	χ							2·19*, 2·28*	2·28*
				Epoxides					
15	β	5,6				5,6		0·95, 1·10*	1·14*
16	β	5,6α	3β			5,6		0·95, 1·12*	1·16*
17	β	5,6α	3β	8		5,6,7,8		0·96, 1·04	1·22
18	β	5,6	3,11 ←	19		5,6	lactone	0·97(ax), 1·20(eq)	1·20
20	β	5,8						1·11(ax), 1·15(eq)	1·43
20*	β	5,8epi						1·11(ax), 1·18(eq)	1·46
21	β	5α,8	3β					1·33(ax), 1·18(eq)	1·62
21*	β	5α,8epi	3β					1·34(ax), 1·19(eq)	1·68
				Ketones and esters (see also 17, 18, 74–84)					
22	ψ			16(CO_2Me)	3,4			1·94, –	1·94
23	ψ		1	2	3,4	1,2		1·43(6H)	2·00
24	ψ		1(OMe)	2	3,4	1,2		1·35(6H)	1·98
25	ψ		1(OMe)	4		1,2		1·17(6H)	1·93
26	ψ		1	20			13-*cis*	1·22(6H)	1·81
27	β			3	4		*retro*	1·37(6H)	2·18
28	β		3	4	2,3		diosphenol	1·30(6H)	2·10
29[d]	β			3,4			2-nor	1·41(6H)	2·07
30	β		3	4				1·21, 1·32	1·94
31	β		3	4	7,8		acetylene		
32	β		3(OCOR)	4			2-nor	1·21, 1·42	1·92

d mass spectral data of half carotenoids
•Cl_3 ; i.r. in cm^{-1} ; m.s. in m/e)

ı.r.[b]					
Me	13 Me	Other signals	i.r.	m.s.	References[c]
6	1·96			M-69, M-82, M-122, M-135, M-148, M-175	
7	1·97	see Table VIII		M-69, 69	
7	1·97			M-71, 43	
0	1·96			M-69, M-137, 69	
8.	1·74		film: 766(*cis* CH=CH)	M-69, M-137, M-205	(1)
8	1·98				
1	1·91				
8	1·98	see Table VIII		M-137	
3	1·93				(2)
7	1·97	4·57, 4·75(5-CH_2), 2·51(6-H, d, J = 8)		M-123, M-135, M-176, M-188, M-190 M-203, M-215, M-255, M-268, 268	
1	1·95	see Table VIII		M-56, M-123	
7	1·98		CS_2: 810(ArH)	M-133, 133	
4	1·97		KBr: 813(ArH)	M-133, 133	
5	1·92	see Table VIII		M-16, M-80, 205, 165	
8	1·92	3·85(3α-H)	$CHCl_3$: 3597(O—H)	M-18, M-80, 221, 181	
5	1·99	3·81(3α-H), see Table VIII	$CHCl_3$: 3615(O—H), 1660(C=O)	M-18, M-155, M-197	
	2·20	3·8(3-H), see Table VIII	KBr: 3440 (O—H), *c.* 1745(C=O)	M-18, M-44, 234, 212, 207, 181	(10)
5	1·94	5·16(7,8-H), see Table VIII		M-16, M-80, 205, 165	(3)
0	1·94	5·23(7-H), 5·06(8-H), $J_{7,8}$ = 2		M-16, M-80, 205, 165	(3)
2	1·94	5·25(7-H), 5·17(8-H), 4·23(3α-H)	$CHCl_3$: 3595(O—H)	M-16, M-18, M-80, 221, 181	(3)
0	1·94	5·31(7-H), 5·07(8-H), $J_{7,8}$ = 2, 4·23(3α-H)	$CHCl_3$: 3595(O—H)	M-16, M-18, M-80, 221, 181	(3)
4	1·94	3·72(OMe)	KBr: 1705(C=O)	M-31, M-59, M-113	
0	2·00		KBr: 3460(O—H), 1667(C=O)	M-16, M-18, M-59, M-87, M-88, 59	
8	1·98	3·22(OMe)	CCl_4: 1665(C=O)	M-30, M-32, M-60, M-73, M-101, 73	
8	1·98	3·16(OMe)	$CHCl_3$: 1648(C=O)	M-32, M-101, M-129, 73	
8	–	9·5(CHO)	KBr: 3410(O—H), 1670(C=O)	M-106(C_7H_6O), M-120, M-172, M-195, M-208, M-221	
2	1·98	2 36(2-H)	KBr: 1665(C=O)	M-56, M-137, M-150, M-163, M-190, M-203, M-216	
2	2·02		$CHCl_3$: 3410(O—H), 1617(C=O)	M-16, M-154, 203, 152, 137	
7	2·02		$CHCl_3$: 1750(C=O), 1678(C=O)	M-28, M-164, M-190, M-204, M-216, M-256	
)	2·00	4·28(3-H)	$CHCl_3$: 3500(O—H), 1660(C=O)	M-16, M-18, M-154, M-167, M-207, M-219, M-233	(6)
:	1·98		KBr: 3400(O—H), 2150(C≡C), 1660(C=O)	M-16, M-87, M-153, M-205, M-231	
2	2·02	5·18(3-H)	$CHCl_3$: 1750(CO_2R), 1710(C=O)	depends on ester	

TAB

Formula	Skeleton	Epoxy	Hydroxy and Deriv.	Oxo	Dehydro-	Hydro-	Notes	1,1 Me's	5 Me
Structure of half carotenoid[a]									
Ketones and esters (see also 17, 18, 74–84)									
33	β			4				1·19(6H)	1·87
34	β			5,6			5,6-seco	1·16(6H)	2·10
35	β		3β,5α	8		5,8		1·07, 1·37	1·55
36	β		3,19	8		7,8		0·99, 0·99	1·48
37	β		6	3,8		5,6,7,8		1·09, 1·26	0·98 (d, J = 6
38	κ		3β	6				0·84, 1·36	1·20
39	κ		3β,8	6			β-enol	0·85, 1·34	1·19
40	κ			3,6				0·98, 1·23	1·35
Remaining alcohols and derivatives									
41	ψ		1		3,4	1,2		1·24(6H)	1·92
42	ψ		1(OMe)		3,4	1,2		1·13(6H)	1·90
43	ψ		1,2		3,4	1,2		1·18, 1·25	1·93
44	ψ		1,2(OGly)		3,4	1,2		1·20, 1·28	1·90
45	ψ		16					1·69, –	1·81
46	ψ		2		1,16	1,2		1·74, –	1·85
47	ψ		1			1,2		1·21(6H)	1·81
48	ψ		1(OMe)			1,2		1·14(6H)	1·81
49	ψ		1,20			1,2	13-*cis*	1·17(6H)	1·80
50	ψ		1			1,2,7,8		1·21(6H)	1·62
51	β		3		4		*retro*	1·24, 1·44	1·93
52	β		2β					1·04, 1·09	1·72
53	β		3β					1·07, 1·07	1·74
54	β		3β		7,8		acetylene	1·14, 1·20	1·91
55	β		4					1·02, 1·04	1·82
56	β		3,19					1·02, 1·02	1·68
57	β		3,20					1·08, 1·08	1·74
58	β		3β,5α		6,7	5,6	allene	1·07, 1·34	1·34
59	β		3β(OAc),5α		6,7	5,6	allene	1·08, 1·35*	1·39*
60	β		3β(OAc),5α		6,7	5,6	13,14,20-tris nor[e], allene	1·07, 1·35*	1·37*
61	β		5,6(*threo*)					0·84, 1·14*	1·19*
62	β		3,5,6					0·82, 1·10*	1·13*
63	ε		3α					0·85, 1·00	1·62
66	φ		3					2·26–2·30[f]	

VI—*continued*

n.m.r.[b]					
9 Me	13 Me	Other signals	i.r.	m.s.	References[c]
·99	2·00	see Table VIII	$CHCl_3$: 1655(C=O)	M-28, M-56, M-138, M-151, M-191, M-203, M-204, M-217	
·99	1·99	see Table VIII	$CHCl_3$: 1715(C=O), 1660(C=O)	M-155, 127, 109, 69, 43	
·98	1·98		$CHCl_3$: 3600(O—H), 1628(C=O)	M-18, M-36	
—	1·96	4·46(19-CH_2O), 3·48(7-H)	$CHCl_3$: (O—H), 1655(C=O)	M-18, M-36, M-44, M-153, M-171, M-181, M-199	(4)
·94	1·98		$CHCl_3$: 3610(O—H), 1710(C=O), 1640(C=O)	M-154, M-169, M-197, 197, 169, 155	(5)
·95	1·98	4·48(3α-H), 2·93(dd, J = 8, 14)	$CHCl_3$: 3600(O—H), 1664(C=O)	M-18, M-155, 127, 109	
·97	1·97	4·49(3α-H), 2·85(dd, J = 8, 14), 5·82(7-H), 16·25(8-OH)	$CHCl_3$: 3610(O—H), 1605(C=O)	M-197, 197, 127, 109	(5)
·97	1·97		$CHCl_3$: 1739(C=O), 1664(C=O)	125, 83	
·97	1·97	2·29(2-H, d, J = 6·5)	KBr: 3390(O—H)	M-18, M-58, 59	
·96	1·96	2·31(2-H, d, J = 6·5), 3·21(OMe)		M-32, M-73, 73	
·98	1·98		KBr: 3380(O—H)	M-16, M-18, M-58, M-59, M-60, M-88, M-90	
·97	1·97		KBr: 3325(O—H)	M-18, M-58, M-146, M-147, M-164, M-180, M-236	
·97	1·97	5·42(2-H), 4·00(16-CH_2O)	KBr: 3400(O—H)	M-16, M-18, M-85, M-153	
·97	1·97	4·97, 4·88(16CH_2=), 4·07(2-H, t, J = 6)	KBr: 3300(O—H)	M-18, M-26, M-42, M-43, M-57, M-71, M-79, M-85	(7)
·97	1·97		$CHCl_3$: 3600(O—H)	M-18, 59	
·97	1·97	3·13(OMe)		M-30, M-32, M-88, 73	
·95	—	(20-CH_2O)	KBr: 3430(O—H)	M-18, M-108, M-122, M-174, M-194, M-206, M-219	
·83	1·98		KBr: 3410(O—H)	M-18	
		4·35(3-H), 5·75(4-H)	KBr: 3400(O—H)		
·97	1·97	3·55(2-H, dd, J = 4·5, 7)	KBr: 3400(O—H)	M-18, M-153	(8)
·97	1·97	4·00(3α-H), see Table VIII	$CHCl_3$: 3597(O—H)	M-18, M-153, M-193	
·00	1·95	3·95(3α-H)	$CHCl_3$: 3610(O—H), 2167(C≡C)	M-18	
·97	1·97	4·03(4-H)	CS_2: 3605(O—H)	M-18	
—	1·89	4·46(19-CH_2O)	KBr: 3380(O—H)	M-18, M-36, M-122	
·96	—	4·54(20-CH_2O)	KBr: 3450(O—H)	M-18, M-36, 299	
·81	1·98		$CHCl_3$: 3597(O—H), 1923(C=C=C)	M-18, M-36, M-101	
·82	1·99	5·36(3α-H), 2·02(Ac), 6·05(8-H), see Table VIII	CCl_4: 3615(O—H), 1927(C=C=C), 1740(OAc)	M-18, M-60, M-78, M-143	
·81	—	5·4(3α-H), 2·01(Ac), 6·05(8-H)	KBr: 3440, (O—H), 1930 (C=C=C), *c.* 1745 (OAc)	M-18, M-60	(11)
·98†	2·00†	see Table VIII	CCl_4: 3497(O—H)	M-18, M-36, M-98, M-127, M-155, 109	
·92	1·92		CS_2: 3610(O—H)	M-143, M-161, M-173, M-186, M-199, M-226	
·91	1·95	4·22(3β-H)	$CHCl_3$: 3615(O—H)	M-18, M-56, M-138, M-153, M-193, M-206, M-219	
·06[e]	1·96[e]	2·06[e](Ac)	KBr: 3400(O—H)	149	

TABL

For- mula	Skele- ton	Epoxy	Hydroxy and Deriv.	Oxo	Dehydro-	Hydro-	Notes	1,1 Me's	5 Me
Structure of half carotenoid[a]									
				Isoprenylated (all with 2-(3′-methylbutyl) group)					
67	ψ		1		3,4,2′,3′	1,2		1·19, 1·21	1·92
68	ψ		1,4′(t)		3,4,2′,3′	1,2		1·22, 1·22	1·92
69	ψ		1,3′		3,4	1,2		1·23, 1·23	1·91
70	β				2′,3′			0·95, 1·08	1·70
71	β		4′,4″		2′,3′			0·94, 1·08	1·69
72	ε		18		2′,3′			0·75, 0·95	–
73	ε		4′(t)		2′,3′			0·75, 0·95	1·53
				Apo					
74	ψ			4(CO_2Me)			4-apo	– –	1·97
75	ψ			6			6-apo	– –	–
77	ψ			6(CO_2Me)			6-apo	– –	–
78	β			8		7,8	7-apo	– –	–
79	β			8			8-apo	– –	–
81	β			8(CO_2Me)			8-apo	– –	–
82	β			10			10-apo	– –	–
83	β			10(CO_2H)			10-apo	– –	–
84	β			12			12-apo	– –	–

[a] The physical data of natural carotenoids are associated with the appropriate halves separated by the 15,15′-bon The structures have been divided into six groups—hydrocarbons, epoxides, ketones and esters, remaining alcohols an derivatives, isoprenylated, and apo. Systematic nomenclature is used as outlined in Section II.

[b] Signals which may have their assignments reversed are marked with an asterisk or dagger. J. values are quoted in Hz.

[c] References are only given to papers not referred to in the main reviews (Weedon, 1969; Enzell *et al.*, 1969; Enze 1969; Budzikiewicz *et al.*, 1970; Foppen, 1971; Vetter *et al.*, 1971) or the list of natural carotenoids (Straub, 197 augmented where necessary by unpublished results from Queen Mary College.

[d] Artefact commonly isolated instead of 32.

[e] Or 9,10,19-trisnor.

[f] Data for the acetate.

I—*continued*

m.r.[b]					
Me	13 Me	Other signals	i.r.	m.s.	References[c]
98	1·98	1·67(c, 4″-Me), 1·61(t, 4′-Me)	KBr: 3400(O–H)	M-18, M-58, M-69, M-87	
98	1·98	1·70(4″-Me), 4·04(4′-CH_2O)	KBr: 3400(O–H)	M-18, M-58, M-87, M-182	
97	1·97	1·18(6H, 4′-Me)	KBr: 3320(O–H)	M-18, M-36, M-58, M-76	
98	1·98	1·68(c, 4″-Me), 1·61(t, 4′-Me)		M-69, 69	
98	1·98	4·03(4H, CH_2O)	KBr: 3400(O–H)	M-18, M-128	
96	1·96	4·02(18-CH_2O), 1·68(c, 4″-Me), 1·59(t, 4′-Me)	KBr: 3330(O–H)	M-124, M-207, M-261, M-274	(9)
94	1·98	1·68(4″-Me), 4·02(4′-CH_2O)	KBr: 3500(O–H)	M-16, M-18, M-140	
97	1·97	3·74(OMe)	(C=O)	M-31, M-59, M-113	
99	2·01	9·58(6-CHO, d, J = 8)	(C=O)	M-28, M-29	
97	1·97	7·44(8-H, d, J = 15·5), 5·90(7-H, d, J = 15·5), 3·76(OMe)	KBr: 1705(C=O)	M-31, M-59, M-113	
94	1·98	2·35(7-Me)	CCl_4: 1660(C=O)	M-43, 43	
90	2·01	9·44(8-CHO), see Table VIII	$CHCl_3$: 1667(C=O)	M-28, M-29	
00	2·00	3·76(OMe)	KBr: 1685(C=O)	M-31, M-59, M-113	
-	1·96	9·57(10-CHO)	CS_2: 1675(C=O)	M-28, M-29	
-			(C=O)	M-44, M-45, M-99	
-	1·87	9·45(12-CHO), see Table VIII	CCl_4: 1675(C=O)	M-28, M-29	

ferences
) Aung Than *et al.*, 1972.
) Andrewes and Liaaen-Jensen, 1973.
) Goodfellow *et al.*, 1973.
) Ricketts, 1971.
) Khare *et al.*, 1973a.
) Khare *et al.*, 1973b.
) Kjøsen and Liaaen-Jensen, 1973; Arpin *et al.*, 1973b.
) Kjøsen *et al.*, 1972.
) Arpin *et al.*, 1973a.
) Kjøsen, H., thesis, 1972.

One of the most characteristic fragmentation patterns observed with conjugated polyene carotenoids is the loss of C_7H_8 (toluene, M-92), C_8H_{10} (xylene, M-106), and $C_{12}H_{14}$ (M-154). The M-154 ion is often weak or absent. With about eight or more double bonds the M-92 and M-106 ions are usually observed. Their relative intensity is influenced by the length of the polyene chain and other substituents (Francis *et al.*, 1974). A cyclic mechanism for their loss has been suggested:

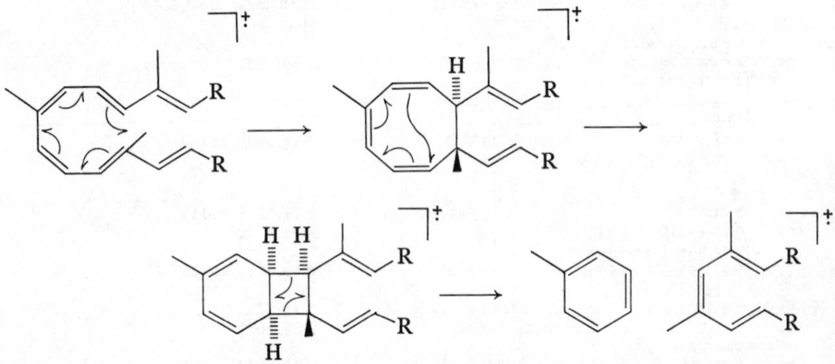

Substitution of the in-chain methyl groups by an oxygen function is readily shown by the corresponding loss of fragments analogous to toluene and xylene. With an aldehyde such as rhodopinal (2-26) the loss of the fragment analogous to toluene is C_7H_6O (M-106). This loss is only recognized by accurate mass determination; however the M-120 ion is characteristic of this aldehyde (Francis and Liaaen-Jensen, 1970).

Fragment ions are frequently encountered at M-2 and M-79. Although typical of carotenoids the process involved in their formation is not understood.

Typical fragmentations of carotenoid end groups have been extensively reviewed (Budzikiewicz *et al.*, 1970; Enzell, 1969; Enzell *et al.*, 1969; Vetter *et al.*, 1971; Weedon, 1969). Loss of water, M-18, is usually observed with alcohols. This may be confirmed by conversion of the carotenoid into the corresponding trimethylsilyl derivative (McCormick and Liaaen-Jensen, 1966) when the increase in the molecular weight by 72 mass units per trimethylsilyl ether group is a good measure of the number of hydroxyl groups originally present. Primary or secondary alcohols may be acetylated to give products with their own characteristic mass spectra in which the loss of acetic acid, M-60, is usually significant. Ketene (42 mass units) is lost when a McLafferty type elimination of acetic acid is not favourable:

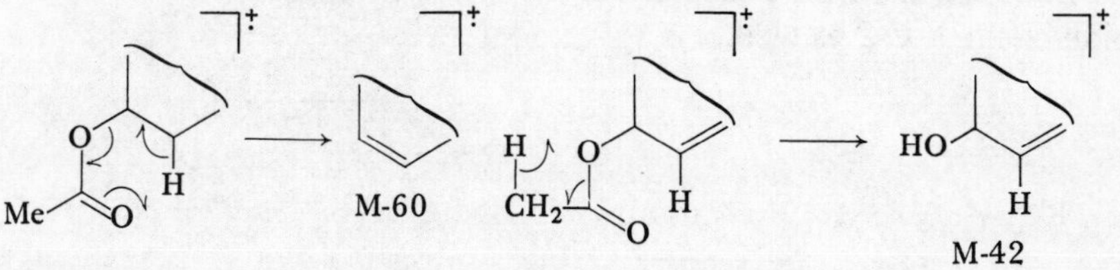

Me O O H → M-60 H O CH_2 O H → HO H M-42

Tertiary alcohols cannot normally be acetylated, but they may be detected in the presence of secondary alcohols by first acetylating the latter groups and then observing the loss of water from the resulting (tertiary) hydroxy acetates. Acyclic tertiary alcohols or their derivatives exhibit cleavage of the 1,2 bond:

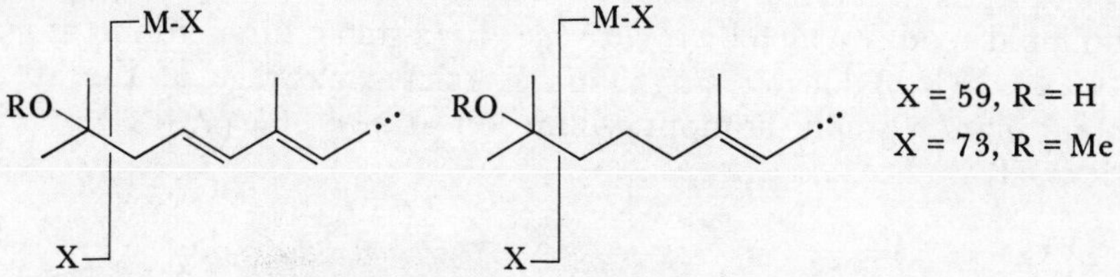

M-X RO X M-X RO X

X = 59, R = H
X = 73, R = Me

Some alcohols, as well as showing loss of water, also show a loss of an oxygen atom. The mechanism of this process is not known.

Carbonyl containing carotenoids frequently undergo α cleavage. For example β-carotenone (34-34) exhibits fragmentation of three of the α bonds:

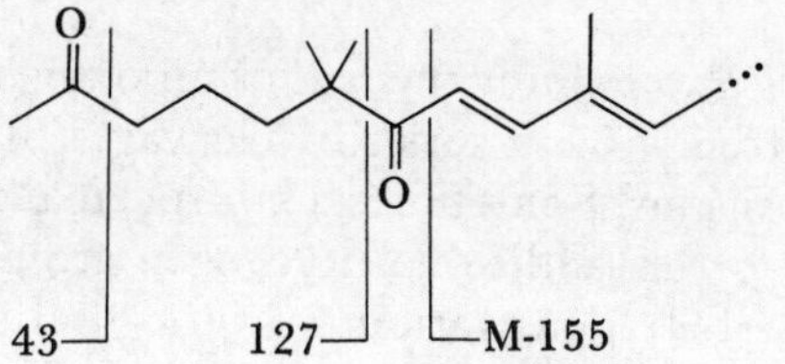

O O 43 127 M-155

The bisallylic single bond of lycopene (2-2) readily fragments with loss of 69 mass units. With several such bonds, as in phytoene (5-5), their position is clearly indicated by such fragmentations:

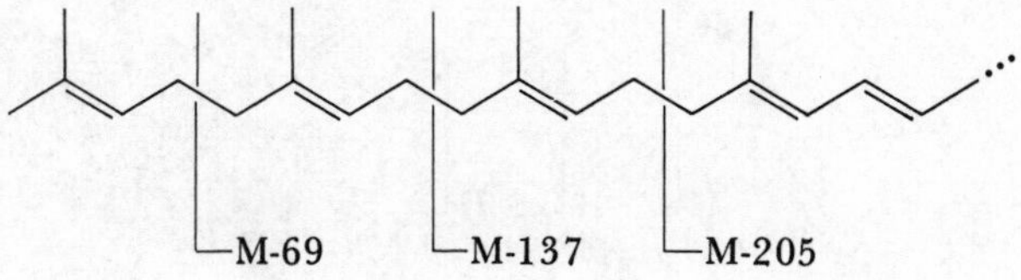

M-69 M-137 M-205

The ε ring system is characterized by a *retro* Diels Alder fragmentation with loss of 56 mass units. The C_{45} or C_{50} isoprenylated carotenoids with an ε ring show the corresponding loss with the C_5 side chain at C-2 as well:

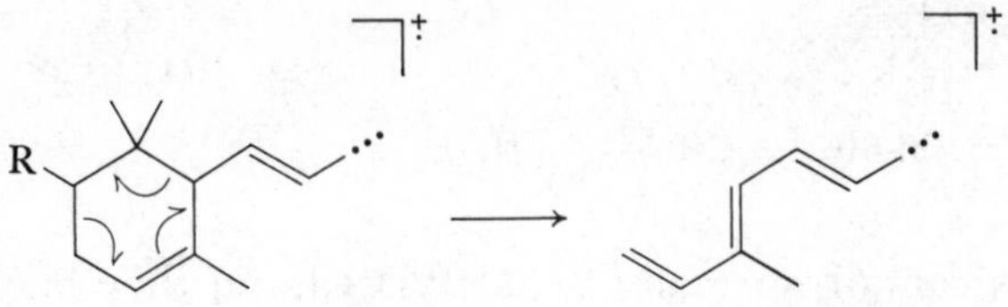

M-56, R = H
M-140, R = $HOCH_2.CMe{=}CH.CH_2$.

Both 5,6- and 5,8-epoxides fragment with cleavage of the 8,9 bond. It is suggested that 5,6-epoxides thermally rearrange to the furanoid oxides which by α cleavage give a stable furanoid ion at *m/e* 165 or 181 (Baldas *et al.*, 1966). A related cleavage of the 10,11 bond gives a stable homopyrylium ion at *m/e* 205 (R = H) or 221 (R=OH):

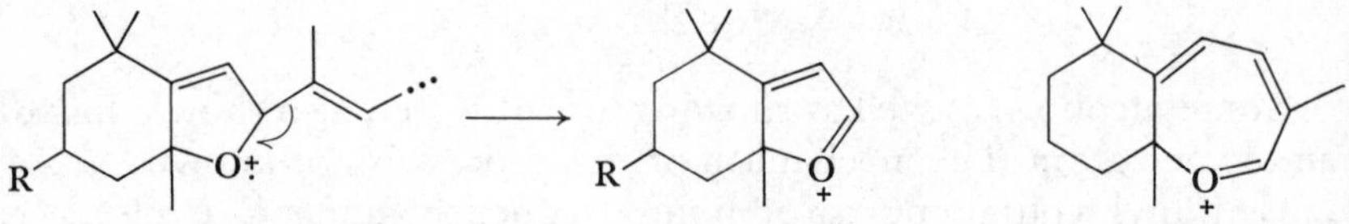

m/e 165, R = H
m/e 181, R = OH

m/e 205, R = H
m/e 221, R = OH

Many epoxides also exhibit the loss of an oxygen atom.

Carotenoids often show a series of cleavages of the polyene chain at most bond positions. Sometimes a hydrogen atom is transferred to the radical lost, or an additional hydrogen atom is retained by the ion. This type of fragmentation is very useful for limiting the possible positions of the oxygen atoms.

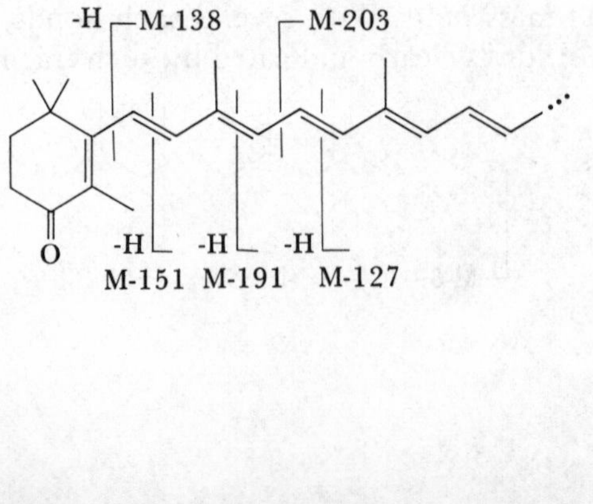

-H M-138 M-203

-H -H -H
M-151 M-191 M-127

C NUCLEAR MAGNETIC RESONANCE SPECTROSCOPY

Until recently carotenoid end groups have been identified most clearly by proton magnetic resonance spectroscopy (p.m.r.) (cf. Fig. 2). New developments in ^{13}C-magnetic resonance (c.m.r.) suggest that this technique will prove an even more powerful method for structural studies.

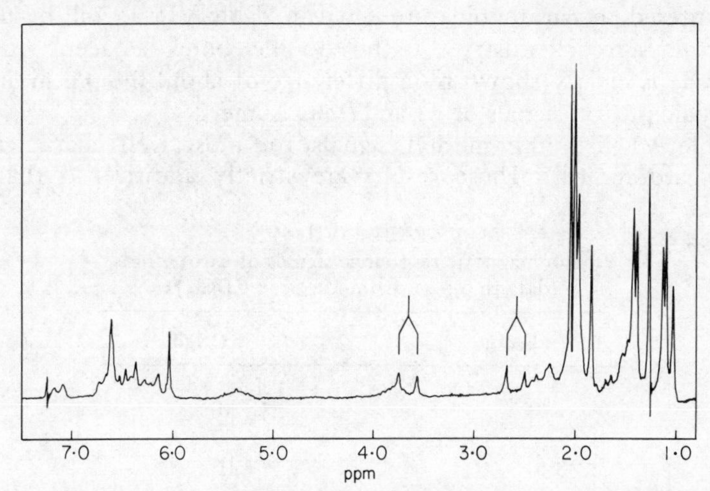

FIG. 2. P.m.r. spectrum of fucoxanthin (17–59) in $CDCl_3$ at 100 MHz (data in p.p.m. from Me_4Si, signal at δ 7·24 due to $CHCl_3$ in solvent) (cf. Table VIII).

(i) *P.m.r.*

Modern instrumentation is capable of measuring high resolution p.m.r. spectra on as little as 1 mg, though with the development of pulsed Fourier transform techniques this requirement should be readily reduced by an order of magnitude or more. Unlike other spectroscopic techniques p.m.r. gives quantitative data even when the structure is not known. Although other solvents such as pyridine, benzene, carbon disulphide or carbon tetrachloride have been used only data measured in deuteriochloroform and referenced to tetramethyl silane (δ = 0) will be considered in this section.*

In the spectra of most carotenoids run on a 60 or 100 MHz instrument the only clearly recognizable signals are the methyl singlets in the region δ 0·7–3·8. Occasionally spin–spin coupling gives rise to a methyl doublet. Some other signals such as those due to protons on a carbon α to an oxygen function may also be readily identified, e.g. –OMe, –CO_2Me and –CHO. In general methylene

* Much of the literature cites data in τ ($\tau = 10 - \delta$).

and vinyl signals are too complex to analyse due to spin–spin coupling, unless these signals are examined at 220 MHz (see Table VIII; Patel, 1969; Schwieter *et al.*, 1969; Vetter *et al.*, 1971), or by the use of lanthanide shift reagents (Strain *et al.*, 1971; Kjøsen and Liaaen-Jensen, 1972).

Typical chemical shift values for the different types of protons encountered in carotenoids are listed in Table VII. It will be noted that the stereochemistry of the double bond adjacent to the aldehyde is clearly shown by a difference of about 0·8 p.p.m. in the aldehydic proton signals of *cis* and *trans* isomers.

Table VI lists the methyl signals for most well characterized "half carotenoids" These results are strictly additive so that the

TABLE VII
Proton magnetic resonance signals of carotenoids
(data in p.p.m. from Me_4Si in $CDCl_3$)

Range	Origin
0·7–1·7	$-\overset{\vert}{\underset{\vert}{C}}-Me$
1·2–3·5	$-\overset{\vert}{\underset{\vert}{C}}-H$
1·6–2·3	$=\overset{\vert}{C}-Me$
2·0–2·1	O—CO—Me
3·1–3·3	$-\overset{\vert}{\underset{\vert}{C}}-O-Me$
3·7–3·8	CO—O—Me
3·5–4·5	$-\overset{\vert}{C}H-OH$
5·2–5·4	$-\overset{\vert}{C}H-OCOR$
4·5–7·0	$=\overset{\vert}{C}-H$
9·4–9·6	*trans*-C=C—CHO
10·0–10·4	*cis*-C=C—CHO
10·0–12·0 (very broad)	CO_2H
~16	$CO.\overset{\vert}{C}=\overset{\vert}{C}-OH$

TABLE VIII

Proton magnetic resonance spectra of some half carotenoids (220 MHz spectra in $CDCl_3$; data in p.p.m. from Me_4Si)[a]

Formula	H-2	H-3	H-4	H-6	H-7	H-8	H-10	H-11	H-12	H-14	H-15	1,1-Me's	5-Me	9-Me	13-Me
2	5·10	2·12	2·12	5·95	6·48	6·24	6·17	6·62	6·34	6·24	6·62	1·61, 1·68	1·81	1·97	1·97
9	1·45	1·60	2·02	—	6·14	6·14	6·14	6·65	6·34	6·24	6·63	1·03(6H)	1·72	1·98	1·98
12	1·18, 1·45	2·00	5·40	2·17	5·51	6·10	6·11	6·60	6·33	6·23	6·62	0·82, 0·90	1·58	1·91	1·95
15				—	5·90	6·30	6·19	6·72	6·36	6·25	6·78	0·95, 1·10[b]	1·14[b]	1·95	1·92
17		3·81		—	2·59, 3·68	—	7·16	6·56	6·71	6·40	6·64	0·96, 1·04	1·22	1·95	1·99
18		3·80		—	4·35	7·18	7·01	—	5·75			0·97, 1·20	1·20	—	2·20
20				—	5·18	5·18	6·20	6·57	6·31	6·23	6·78	1·10, 1·15	1·42	1·75	1·97
33	1·85	2·51	—	—	6·21	6·35	6·26	6·63	6·41	6·26	6·65	1·19(6H)	1·87	1·99	2·00
34			2·40	—	6·54	7·39	6·51			6·35	6·70	1·17(6H)	2·10	1·99	1·99
53	1·47, 1·70	4·00	2·04, 2·39	—	6·11	6·11	6·14	6·63	6·35	6·24	6·62	1·07, 1·07	1·74	1·97	1·97
59		5·36		—	—	6·05	6·12	6·75	6·35	6·25	6·64	1·08, 1·35	1·39	1·82	1·99
61				—	6·22	6·40	6·22	6·70	6·36	6·27	6·79	0·84, 1·14	1·19	1·98	2·00
79	—	—	—	—	—	9·44	6·94	6·62	6·73	6·43	6·63	— —	—	1·90	2·01
84	—	—	—	—	—	—	—	—	9·45	6·94	6·66	— —	—	—	1·87

[a] Data from Vetter *et al.*, 1971.

[b] Assignment may be reversed.

spectrum corresponding to any pair of "half carotenoids" will be the sum of the signals in Table VI. Most carotenoids have at least ten methyl groups which, even with four in-chain methyl signals at about 1·97, still leaves six to characterize the end groups. Additional confirmatory evidence in many cases may be derived from other signals listed in Table VI, for example the 7-H and 8-H of a furanoid oxide (Goodfellow *et al.*, 1973). The methyl signals in this example were used to show that the 3-hydroxy was axial and *trans* to the oxygen at C-5.

The additivity of the methyl and related signals does not apply to the conjugated vinyl signals. Table VIII lists some 220 MHz data which clearly show the effect of a carbonyl group. The signals due to protons at positions α, γ, ϵ etc. to the carbonyl group are shifted to higher field, whereas those due to protons β, δ, ζ etc. are moved to lower field.

Spin–spin coupling has in general proved of limited value for structure determination. Vicinal coupling across a *trans* di-substituted double bond is 14–15 Hz, rising to about 16 Hz when $\alpha\beta$ to a carbonyl group. Further values are tabulated in Table IX. A

TABLE IX
Carotenoid spin–spin coupling constants

System	Coupling constant (Hz)
trans-CH=CH–	14–15
cis-CH=CH–	*c.* 10·5
–C(–)=CH–CH=C(–)–	10–11
–C(–)=CH–CHO	*c.* 8
–CH(–)–CH(–)– diaxial	*c.* 8
–CH(–)–CH(–)– axial-equatorial	*c.* 4·5
$J_{6,7}$ ϵ –CH(–)–CH=C	*c.* 9·5
$J_{6,7}$ γ –CH(–)–CH=C	*c.* 8
$J_{2,3}$ ψ –CH–CH=C	*c.* 6·5
$J_{7,8}$ (5,8 oxide) –CH(–)–CH=C	2 or *c.* 1
>CH_2	14–18
>C=CH_2	*c.* 2

notable example of the use of spin–spin coupling data is the determination of the stereochemistry of phytoene (Khatoon *et al.*, 1972; Granger *et al.*, 1973). A second-order analysis of the AA′BB′ systems of vinyl protons shows that the central double bond of the most common naturally occurring phytoene is *cis* (J = 10·6 Hz whereas in the *trans* isomer J = 14·6 Hz). In conjunction with the chemical shift data the *trans* (E) stereochemistry of the 13,14 double bond was also demonstrated.

The characteristic positions of the methyl signals of most carotenoids readily defines the end group involved. However, in appropriate cases the assignment can be further confirmed by examining the shifts induced by additions of a lanthanide complex such as $Eu(dpm)_3$ (Kjøsen and Liaaen-Jensen, 1972; see also Kjøsen *et al.*, 1972).

(ii) C.m.r.

The recent development of pulsed Fourier transform techniques for nuclear magnetic resonance spectroscopy permits the routine measurement of ^{13}C spectra at natural abundance (1·1%). Broad band irradiation of the ^{1}H region removes ^{1}H-^{13}C spin–spin coupling and improves the sensitivity. Owing to the low probability of one ^{13}C atom being next to another no ^{13}C–^{13}C spin–spin coupling is observed. Thus a series of singlets are recorded corresponding to each type of carbon atom present in the molecule. The tremendous gain over ^{1}H magnetic resonance is that the signals occur over a range of more than 200 p.p.m. Data are again referenced to tetramethyl silane.

Applications in the carotenoid field have so far been quite limited (Strain *et al.*, 1971; Granger *et al.*, 1973). Full spectral details have been published of β-carotene (9–9) and its 15,15′-*cis* isomer and 15,15′-dehydro analogue (Jautelat *et al.*, 1970). While the assignment of the sp^3 hybridized carbon atoms seems to be correct there is no doubt that at least some of the other published assignments are incorrect (Vetter *et al.*, 1971 (N.B. the δ values quoted in this reference should be multiplied by 0·97656); unpublished results from Queen Mary College).

Comparison of the spectrum of β-carotene with those of model compounds such as vitamin A acetate and β-ionone again show some of the sp^2 signals need revision (Bremser *et al.*, 1971; Hollenstein and von Philipsborn, 1972). Further caution is needed in some cases following the observation of significant concentration and solvent

TABLE X

C.m.r. spectra of some typical carotenoids (22·63 MHz spectra in $CDCl_3$; data in p.p.m. from Me_4Si)[a]

Carotenoid	1,1-Me's	5-Me	9-Me	13-Me	1	2	3	4	5	6	7	8
β-Carotene (9–9)	29·0	21·7	12·7	12·7	34·3	39·7	19·3	33·2	129·3	138·0	126·7	137·8
Zeaxanthin (53–53)	28·7, 30·2	21·7	12·8	12·8	37·1	48·2	65·1	42·4	125·5	142·3	126·1	138·5
Lutein (53–63)	28·7, 30·2	21·6	12·7	12·7	37·1	48·4		42·5	125·6	n.a.	126·2	138·5
	24·3*, 29·5*	22·8*	13·2	12·7	34·0	44·7	65·9	129·6	137·8	55·0	128·6	137·8
Alloxanthin (54–54)	28·9, 30·6	22·5	18·1	12·8	37·0	46·8	65·0	41·6	138·2	124·5	99·0*	89·3*
Isomytiloxanthin (37–54)[b]	28·9, 30·7	22·8	18·2	12·9	37·0	46·9	65·0	41·7	138·2	124·6	n.d.	n.d.
	29·3#, 31.2#	23·7#	12·1#	14·9#	41·7#	50·8#	208·5	48·3#	40·7#	84·0	n.a.	201·0
Isozeaxanthin (55–55)	28·6, 29·1	27·6	12·7	12·7	34·9	34·9	18·6	70·3	129·8	142·0	125·6	138·9
Canthaxanthin (33–33)	27·7	13·7	12·7	12·5	35·7	37·7*	34·3*	198·7	129·9	160·9	124·2	141·1
Violaxanthin (16–16)	94·7, 29·7	20·0	13·1	12·8	35·4	47·2	64·1	40·9	70·5*	67·3*	123·9	137·2
Fucoxanthin (17–59)[c]	25·0●, 29·2●	21·5●	11·9▼	14·0▼	35·2+	47·3△	64·2	40·9△	67·2*	72·6*	41·8△	197·9
	31·3●, 32·2●	28·2●	13·0▼	12·7▼	35·8+	45·4△	68·2	45·5△	66·3*	117·6	202·4	103·4
trans Methyl bixin (77–77)[d]	— —	—	12·6	12·6	—	—	—	—	—	167·9	115·9	149·0
Crocetindial (79–79)	— —	—	9·7	12·8	—	—	—	—	—	—	—	194·2
cis Phytoene (5–5)[e]	17·8(c), 25·8(t)	16·2	16·2	16·7	131·4	124·2*	27·0	39·9	135·1+	124·5*	27·0	39·9

n.a. = Not assigned. n.d. = Not detected. * + ● ▼ △ Assignments may be reversed. # Tentative assignment from data derived from models, but without SFOR data.

Notes:

[a] Unpublished data from Queen Mary College, unless otherwise stated.

[b] See Weedon, 1973.

[c] Acetate signals at 21·2 (Me) 170·5 (CO) (cf. peridinin bromobenzoate 145·5 (C—11′), 164·5* (benzoate C=O) 168·5* (C—19) 171·5 (acetate C=O) 202·5 (C—7′), Strain *et al.*, 1971).

[d] Methyl ester signal 51·5 (OMe).

[e] See Granger *et al.*, 1973. Other signals are C–9, 135·5+; C–10, 124·7*; C–11, 27·0; C–12, 40·7; C–13, 123·6; C–14, 120·5; C–15, 139·6.

effects with β-ionone and other compounds (unpublished results from Queen Mary College, in collaboration with Professor J. A. Elvidge, University of Surrey). The use of these effects for assignment purposes is currently being developed.

Unpublished results from the authors' laboratories are partially tabulated in Table X. While in most cases the assignment of the sp^3 signals is reasonably certain there still remain many uncertainties with the sp^2 signals. Considerable assistance in tackling this problem is obtained by the study of β-apo-carotenoic esters where the polarization effect of the carbonyl group is clearly seen (cf. ^1H-magnetic resonance).

The potential of c.m.r. is well illustrated by the spectrum of fucoxanthin (17-59) where forty-one separate signals are observed for this C_{42} carotenoid (Fig. 3). Single frequency off-resonance (SFOR) gives residual coupling from one bond ^1H–^{13}C spin-spin

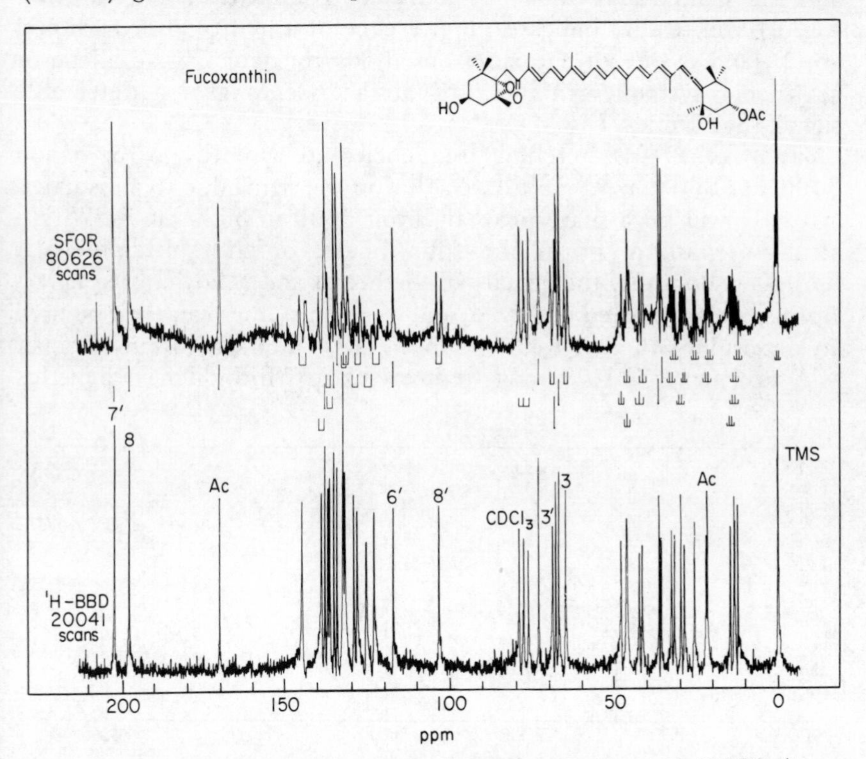

FIG. 3. C.m.r. spectrum of fucoxanthin (17-59) in $CDCl_3$ at 22·63 MHz (lower spectrum with broad band decoupling by irradiation of the proton frequencies, upper spectrum with single frequency off-resonance to show residual one-bond couplings) (cf. Table 10).

coupling. Thus signals can be readily divided into groups by virtue of the number of hydrogen atoms attached, with methyl groups resulting in a quartet etc. Knowing the typical δ values for most sp^3 carbons (0–80 p.p.m.), and sp^2 carbons (100–165 p.p.m), as well as the characteristic position for a carbonyl group (165–220 p.p.m.), allene sp (200–210 p.p.m.) and acetylene (70–100 p.p.m.), the assignment of many signals is straightforward. Full assignment of all signals requires the study of suitable model compounds.

D INFRARED SPECTROSCOPY

With the advances in other spectroscopic techniques infrared spectroscopy has become less widely used for structural studies in the carotenoid field. However, a sample as small as 0.2 mg may be used and the sample may then be recovered. The infrared spectrum (cf. Fig. 4) will readily indicate the presence of a hydroxyl or carbonyl group, and is one of the surest methods for detecting an allene or acetylenic system, although not all acetylenes give a detectable signal. (See Table VI.)

Hydroxyl O–H stretching frequencies are typically in the region 3400 to 3700 cm^{-1}.* In dilute solution the signal due to an isolated free OH will be a sharp peak at about 3600 to 3620 cm^{-1}. With a stronger solution, or in the solid phase, or with intramolecular hydrogen bonding, the signal will be broad and at lower frequency. Both signals are often observed, and dilution studies can then be used to determine whether hydrogen bonding is inter- or intramolecular.

Acetylene C≡C stretching frequencies are "forbidden" when this

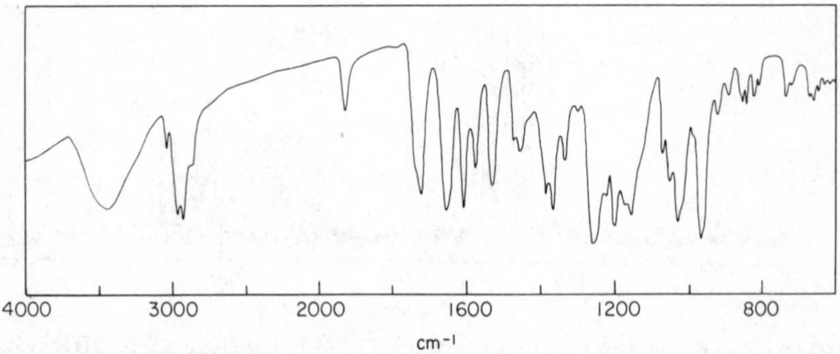

FIG. 4. Infrared spectrum of fucoxanthin (17–59) in KBr disc (cf. Table VI).

* Some of the literature cites data in μm, formerly $\mu (cm^{-1} = 10^4/\mu m)$.

bond is symmetrically substituted. Thus a 15,15′-dehydro carotenoid may not show this signal, but the natural 7,8-dehydro carotenoids such as alloxanthin (54–54) exhibit a weak characteristic peak at 2170 cm^{-1}. Additional conjugation with a carbonyl group at C-4 (31-31) gives a slight shift to 2150 cm^{-1} and the polarization effect of the carbonyl group results in a more intense signal.

Allenic C=C=C stretching frequencies are also very characteristic, occurring in a region of the spectrum normally transparent. Fucoxanthin (17–59) and neoxanthin (16–58) exhibit this infrared absorption in the region 1920–1930 cm^{-1}.

Carbonyl C=O stretching frequencies are the only other infrared absorption frequencies which occur in a region of the spectrum (1600–1740 cm^{-1}) where confusion with other signals is unlikely. The position of this band is strongly correlated with the nature of the group. Conjugation and/or intramolecular hydrogen bonding lowers the frequency, whereas ring strain, such as with a cyclopentanone structure, raises the frequency of the carbonyl stretching (see Table XI). In most cases the carbonyl absorption bands are among the strongest signals in the infrared spectrum. In the region near 1600 cm^{-1} there may be confusion with C=C stretching frequencies. Isolated double bonds only give very weak bands, but conjugation, especially with a carbonyl group, intensifies this signal.

The C—H out-of-plane deformation vibrations of a double bond or aromatic ring are usually recognizable. Normally the *trans* disubstituted double bonds will give a signal at about 965 cm^{-1} whereas the *cis* isomer absorbs at about 780 cm^{-1}. (In *cis*-phytoene (5–5) it is as low as 758 cm^{-1}.) Aromatic carotenoids such as isorenieratene (13–13) or renierapurpurin (14–14) exhibit a band in this region at about 810 cm^{-1} owing to two adjacent aromatic protons. Further substitution will give an isolated aromatic hydrogen atom which will absorb at about 860 cm^{-1}. Splitting of the 965 cm^{-1} band is observed with *retro* carotenoids such as eschscholtzxanthin (51-51) as well as with some carbonyl conjugated and *cis* double bond systems. (Lunde and Zechmeister, 1955; Nicoară *et al.*, 1966.)

A C—O stretching frequency (1000-1150 cm^{-1}) is often recognizable, although many other skeletal vibrations occur in this fingerprint region. Secondary alcohols have a band in the region 1000-1045 cm^{-1} whereas tertiary alcohols have a weaker absorption at 1130-1150 cm^{-1} (Liaaen-Jensen and Jensen, 1965). Methoxy groups such as in spirilloxanthin (42–42) show a C—O stretching frequency at 1070-1080 cm^{-1}.

TABLE XI
Infrared absorption bands of carotenoids

Origin	Frequency (cm^{-1})	Notes
Free O—H stretch	3595–3650	Sharp band especially in dilute solution, characteristic
H-bonded O—H stretch	3300–3510	Broad band
=C—H stretch	3000–3050	Usually weak bands, present in all carotenoids
—C—H stretch	2850–3000	Normally strong bands, present in all carotenoids
—C≡C— stretch	2150–2170	Usually weak band, characteristic
C=C=C stretch	1920–1930	Medium band, characteristic
C=O stretch	1740–1750	Strong band, non-conjugated ester, or 5-ring ketone
	c. 1715	Strong band, acyclic non-conjugated ketone
	c. 1710	Strong band, 6-ring non-conjugated ketone, or 5-ring conjugated ketone
	c. 1705	Strong band, conjugated ester
	1660–1670	Strong band, conjugated aldehyde (sometimes C—H stretches at *c.* 2700 cm^{-1})
	1640–1665	Strong band, conjugated ketone, or 6-ring conjugated ketone
	c. 1630	Strong band, cross conjugated ketone
	c. 1620	Strong band, diosphenol
	c. 1605	Strong band, enolic β-diketone
C=C stretch	*c.* 1600	Usual weak band present in all carotenoids
Various	900–1600	Fingerprint region, limited value for structure determination
C—O stretch	1130–1150	Weak band, tertiary alcohol, limited value
	1070–1080	Medium band, tertiary methoxy group, limited value
	1000–1050	Medium band, secondary alcohol, limited value
=C—H out-of-plane	950–980	Strong band, *trans* di-substituted double bond, present in most carotenoids
	c. 830	Weak band, tri-substituted double bond, limited value
	800–820	Medium band, two adjacent aromatic hydrogen atoms, limited value
	c. 780	Medium band, *cis* di-substituted double bond, fairly characteristic

E OPTICAL ROTATORY DISPERSION AND CIRCULAR DICHROISM

The strong light absorption of carotenoids has until recently limited studies on their optical rotatory and circular dichroism properties to the ultraviolet region. Circular dichrographs now developed permit

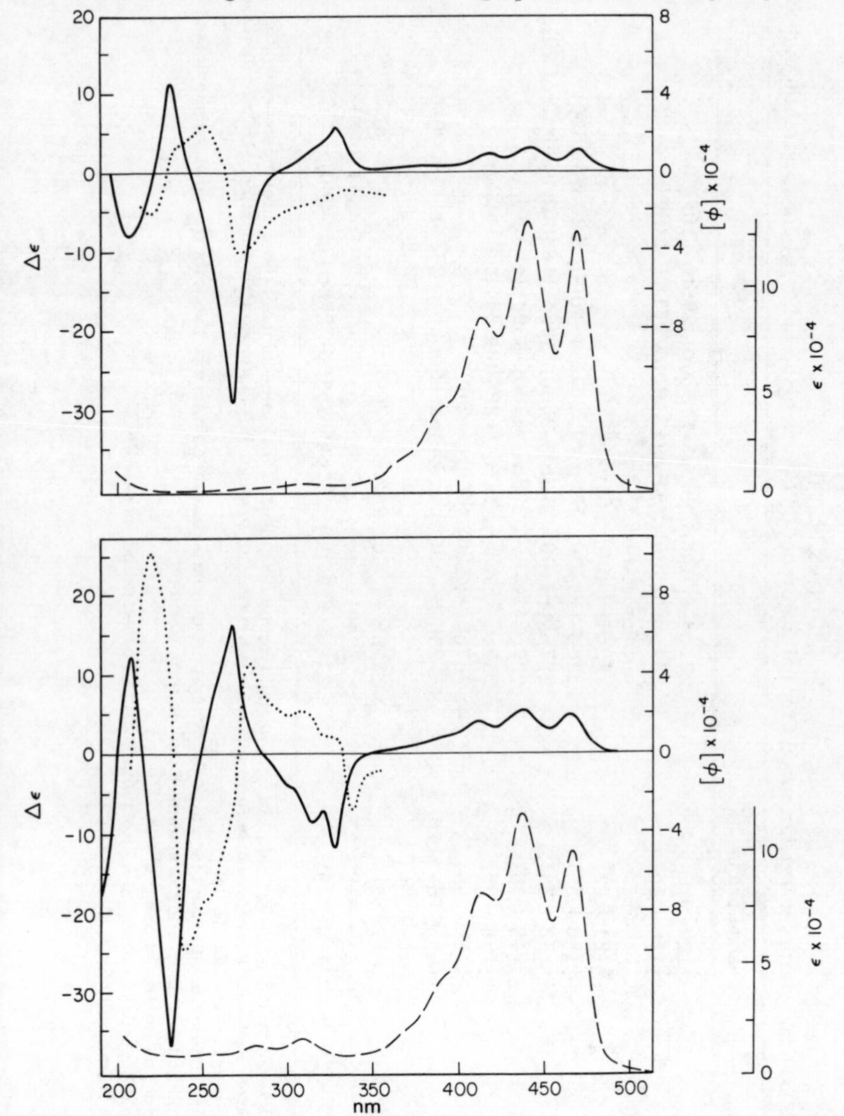

FIG. 5. O.r.d. in dioxan (····), c.d. in methanol (——) and electronic spectra in methanol (- - - -) of violaxanthin (16–16) (above) and violeoxanthin (9-*cis* 16–16) (below). (From Moss *et al.*, 1975.)

TABLE XII

Optical rotatory dispersion spectra for half carotenoids[a] (wavelength in nm, [ϕ] for spectra run in dioxan)

Formula	Data origin	Spectrum
12	(12 + 9)	227tr(−37000), 260pk(+15300), 296tr(−10700)
16	$\frac{1}{2}$(16 + 16)	234pk(+7900), 251pk(+11800), 271tr(−20750)
17 + 59		not detectable[b]
21[c]	$\frac{1}{2}$(21 + 21)	232tr(−5400), 265pk(+6700)
30[d]	$\frac{1}{2}$(30 + 30)	238pk(+3100), 252tr(0), 281pk(+4000), 330tr(−1300)
38	$\frac{1}{2}$(38 + 38)	227pk(+4100), 278tr(−2500), 306pk(+8300), 328inf(+5000), 376tr(−1400)
40	$\frac{1}{2}$(40 + 40)	227pk(+4700), 278tr(−4700), 309pk(+5500)
53	$\frac{1}{2}$(53 + 53)	232tr(−19200), 263pk(+22800), 297tr(−16000)
54	$\frac{1}{2}$(54 + 54)	267pk(−2500), 292tr(−4800)
58[e]	(58 + 16) − 16	*c*. 236tr(−4300), *c*. 249pk(+5000), *c*. 272pk(+10000)
59	See above	
61	(61 + 83)	231pk(+1300), 252tr(−5700)
63	(63 + 53) − 53	*c*. 234pk(−400), *c*. 263tr(−16200), *c*. 291pk(+11000)
73	$\frac{1}{2}$(73 + 73)	231sh(+15700), 255pk(+25300), 276tr(−29000)

[a] Data based on half of the [ϕ] values of homodichiral carotenoids, all of the [ϕ] values of monochiral carotenoids or by difference for a heterodichiral carotenoid using the other entry from this table (pk = peak, tr = trough, inf = inflection, sh = shoulder). Unless otherwise stated the data are taken from Bartlett *et al*., 1969.

[b] Antia (1965) claimed to have measured the o.r.d. spectrum of fucoxanthin. This is rejected on the basis that the sample contained *cis* isomers from the quoted u.v. curve, and calculations based on the c.d. curve.

[c] Mixture of both enantiomers at C-8.

[d] Probably partially racemized on isolation. See also Buchwald and Jencks, 1968.

[e] Data from Cholnoky *et al*., 1969.

measurement of good reproducible spectra even in the region of the main absorption maxima (Fig. 5).

Although there is not an adequate theory to explain the observed spectra of carotenoids, they do present a useful method of correlating absolute configurations. An extensive study of optical rotatory dispersion spectra shows that they are additive within experimental limitations (Bartlett *et al.*, 1969). Thus a monochiral carotenoid such as β-cryptoxanthin (9–53) has a very similar spectrum to that of rubixanthin (2–53) with approximately half of the amplitude of the spectrum of the homodichiral zeaxanthin (53–53). Similar correlations also show that the heterodichiral carotenoid capsanthin (39–53) has a spectrum close to that predicted from the sum of "half zeaxanthin" and "half capsorubin" (39–39). The circular dichroism spectra of these carotenoids confirms that this additivity also applies here.

Tables XII and XIII list the data for "half carotenoids" based as far as possible on the figures for homodichiral carotenoids. Monochiral data are included when necessary and are preferred to heterodichiral data (where the result can only be obtained by subtraction). The application of these studies to the determination of absolute stereochemistry is discussed in Section VC.

F X-RAY CRYSTALLOGRAPHY

Determination of the absolute stereochemistry of a carotenoid must rely on the X-ray crystallographic examination of a crystal of a related substance containing a heavy atom (cf. Section VC). Some configurational problems are best solved by similar studies, even without a heavy atom. Thus the examination of a suitable model has established that the 5,6-diol system of azafrin (61–83) is *threo* (unpublished results).

In addition X-ray studies can provide much information on the conformation of the molecule in the solid state (cf. Section VB). With caution these results may be used to suggest the conformation of the molecule in solution. Frequently, a major problem with crystallography is the preparation of a suitable crystal. Thus it is often necessary to use a derivative, or a model compound, which may be related unambiguously to the carotenoid under investigation.

An example of the use of X-ray crystallography to the determination of absolute configuration is provided by studies on the capsorubin end group (38). The stereochemistry at C-5 has been related by degradation and synthesis to C-1 of (+)-camphor (cf. p. 217). An

TABLE XIII

Circular dichroism spectra of half carotenoids (wavelength in nm with $\Delta\epsilon$ values, see note [a] Table XII)

Formula	Data origin	Solvent[a]	Circular dichroism spectra	Reference
12	(12 + 9)	EPA	240pk(+5·50), 254tr(+3·70), 263pk(+4·50), 293tr(+0·15), 340pk(+2·10), 373tr(+0·30)	1
16	½(16 + 16)	EtOH	205tr(–4·02), 230pk(+5·8), 266(–14·7), 316pk(+1·79), 328pk(+2·90), 420pk(+1·07), 444pk(+1·70), 471pk(+1·34)	2
17 + 59	(17 + 59)	EtOH	210tr(–0·44), 225pk(+0·44), 250tr(–0·64), 263tr(–0·55), 313pk(+0·40), 328pk(+0·29), *c.* 468pk(+0·94)	2
21[b]	½(21 + 21)	EtOH	199pk(+7·52), 220tr(+3·12), 235pk(+4·74), 297tr(–0·31), 381pk(+0·50), 410pk(+0·58), 432pk(+0·83)	2
21*[b]	(21 + 21*) – 21	EtOH	212tr(–8·7), 228inf(–6·9)	2
30	½(30 + 30)	dioxan	223pk(+8·6), 246tr(–6·5), 260pk(+3·9), 279pk(+7·3), [315tr(–10)[4]]	3
37 + 54	(37 + 54)	dioxan	214(–1·82), 228(–1·14)	2
38	½(38 + 38)	dioxan	221pk(+1·10), 250tr(–0·49), 303pk(+3·30), 374tr(–0·98)	5
39	(39 + 54) – 54	dioxan	*c.* 216pk(+1·72), *c.* 254tr(–1·08), *c.* 298pk(+3·37)	2
46	(9 + 46)	EPA	not significant	11
52	½(52 + 52)	EPA	223pk(+1·98), 250tr(–1·48), 281pk(+2·47), 345tr(–0·40)	6
53	½(53 + 53)	EtOH	205pk(+2·42), 220tr(–5·35), 246pk(+4·93), 283tr(–7·75), 341pk(+1·75), 432pk(+0·27), 459pk(+0·34), 485pk(+0·40)	7
54	½(54 + 54)	dioxan	210tr(–3·43), 258pk(–0·63), 289tr(–2·52)	2

58	(58 + 16) – 16	EtOH	204pk(+5·5), 230tr(–6·8), 265pk(+8·2), 328(–2·6)	2
59	See above			
61	(61 + 83)	dioxan	246tr(–3·64)	2
63	(63 + 7)	dioxan	235pk(+5·3), 263pk(+7·2), *c*. 290pk(+2·1)	8
67	$\frac{1}{2}$(67 + 67)	EPA	illustration only	9
69	$\frac{1}{2}$(69 + 69)	EPA	illustration only	9
73	$\frac{1}{2}$(73 + 73)	MeOH	204tr(–2·4), 224pk(+1·7), 264tr(–10·8), 325pk(+1·8), 414pk(+0·49), 445pk(+0·73), 473pk(+0·65)	10

References:

1. Kjøsen *et al.*, 1972; see also Buchecker *et al.*, 1974 and references therein.
2. Unpublished results from Queen Mary College in collaboration with Westfield College, London.
3. Khare *et al.*, 1973b.
4. Estimated from an unpublished spectrum in ethanol on partially racemized material.
5. Weedon, 1973.
6. Kjøsen *et al.*, 1972; see also Buchecker *et al.*, 1973a,b.
7. Unpublished results; see also Buchecker *et al.*, 1974; Andrewes *et al.*, 1974 and references therein.
8. Buchecker *et al.*, 1974.
9. Borch *et al.*, 1972.
10. Unpublished result; see also Borch *et al.*, 1972.
11. Arpin *et al.* 1973b.

Notes:

[a] EPA is ether : isopropanol : ethanol (5:5:2).

[b] 21 is the epoxide end group of chrysanthemaxanthin, 21* of flavoxanthin.

X-ray study of 3-bromo camphor proved conclusively (Allen and Rogers, 1971) that the absolute stereochemistry of capsorubin is as shown below. Further details of related carotenoid applications are given in Section VB (p. 213).

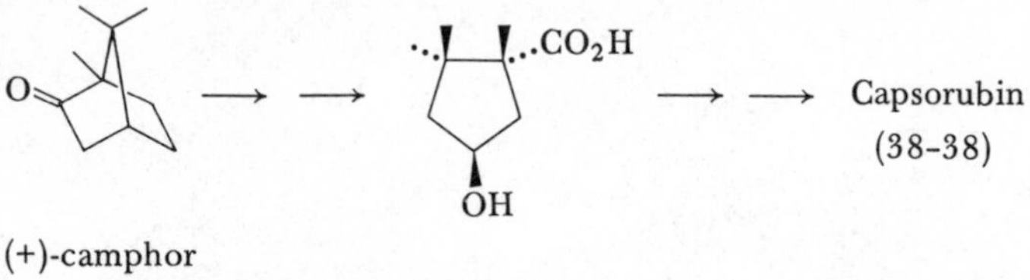

$\longrightarrow \longrightarrow$ CO_2H … OH $\longrightarrow \longrightarrow$ Capsorubin (38–38)

(+)-camphor

Only three natural carotenoids have been successfully studied directly by X-ray crystallography. Each of them is (chemically) symmetric about the 15,15′ bond. The conformation of β-carotene (Sterling, 1964), canthaxanthin (Bart and McGillavry, 1968) and crocetindial (Hjortås, 1972) are very similar to those found for model compounds which are discussed in Section VB.

IV Chemical examination

This section deals mainly with those reactions of carotenoids which are commonly used for structural studies. Of particular importance in this connection are the reactions which can be carried out on a "spectroscopic scale", i.e. which require no more material than is necessary to record the main visible light absorption maxima of the initial solution and of that of the product. These methods allow much valuable information to be obtained even when the quantities available do not permit full spectroscopic characterization of starting material and/or product.

Processes which formerly played an important part in the structural elucidation of a carotenoid, but which have now been superseded by the newer spectroscopic methods reviewed in the previous section, will not be discussed here. To this category belong such procedures as quantitative microhydrogenation for determining the number of double bonds, the Kuhn–Roth method for estimating *C*-methyl, ozonolysis for the detection of isopropylidene groups and 1,5-diene structures, the detection of α and β end groups by oxidation to isogeronic and geronic acid respectively, and the location of a hydroxyl substituent in one of these end groups by analysis of the dicarboxylic acids formed on drastic oxidation. Information on these and many other reactions of carotenoids may be obtained from the extensive review published recently by Liaaen-Jensen (1971).

A HYDRIDE REDUCTION

Lack of fine structure in the visible light absorption curve of a carotenoid is often the result of conjugation of the polyene chain with a carbonyl group (aldehyde, ketone, carboxylic acid or carboxylate). Even on a spectroscopic scale, confirmation of the presence of such a group can readily be obtained by treatment of the carotenoid in ether or tetrahydrofuran with lithium aluminium hydride and observing both the shift (20-30 nm) in the light absorption maximum to shorter wavelengths and the increase in fine structure. Examples of the application of this technique are to be found in the studies on capsanthin (38-53) (Barber *et al.*, 1961b), spheroidenone (4-24) (Barber *et al.*, 1966), and okenone (14-25) (Liaaen-Jensen, 1967). With isomytiloxanthin (37-54) a tetrahydro derivative is formed by reduction of both the conjugated and the unconjugated ketone group (Khare *et al.*, 1973a). With fucoxanthin (17-59) "reductive hydrolysis" of the acetate function accompanies the reduction of the keto group resulting in a mixture of fucoxanthols (19-58) differing in configuration at C-8 (Bonnett *et al.*, 1969).

Reduction with sodium or potassium borohydride in methanol or ethanol is an obvious modification of the above test (cf. Fig. 6). This has the advantage that it allows a distinction to be made between aldehydes and ketones, which are reduced, and the carotenoic acids and their esters which are not normally reduced.

Since compounds such as capsorubin (38-38) and canthaxanthin (33-33) with a conjugated keto group at both ends of the polyene chain have light absorption properties which do not differ greatly from those of the corresponding monoketones, the hydride reduction test does not distinguish clearly between the two classes. However, large shifts in λ_{max} accompany the reduction of the enolic β-diketone function in mytiloxanthin (39-54; 42 nm) (Khare *et al.*, 1973a), the enolic α-diketone functions in astacene (28-28; 33 nm), and the diketone groupings in violerythrin (29-29; ~100 nm) (Hertzberg *et al.*, 1969).

When quantities of material permit, confirmation of the results of hydride reduction can be obtained by isolating the resulting allylic alcohol, determining its composition by mass spectrometry, and then submitting it either to the dehydration test discussed in Section IVB, or to oxidation with a reagent which exhibits a high selectivity for allylic alcohols (MnO_2, NiO_2, dichlorodicyanobenzoquinone). Such an oxidation should regenerate the original chromophore, though not necessarily the original carotenoid; thus selective oxidation of the

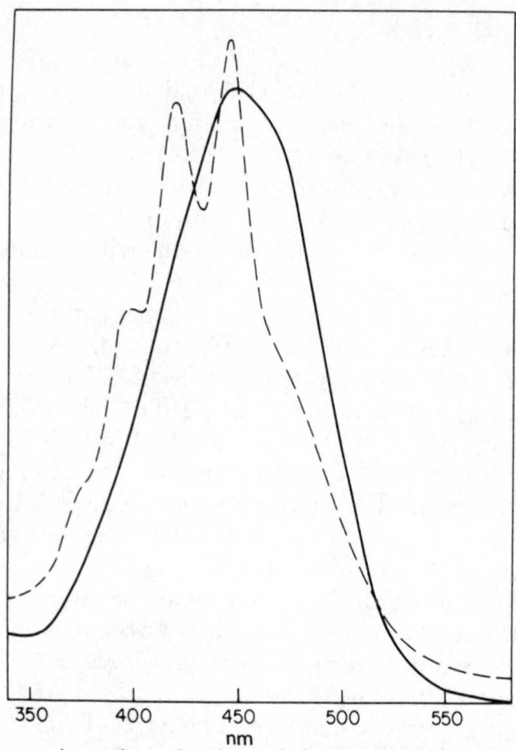

FIG. 6. Spectroscopic scale reduction of fucoxanthin (17–59) to the allylic alcohol (19–59) in ethanol by sodium borohydride (6 h) (——— before, - - - - after).

fucoxanthols (19–58) from fucoxanthin (17–59) leads to fucoxanthinol (17–58) (Bonnett *et al.*, 1969).

Although the 5,6-epoxide group is not attached when fucoxanthin (17–59) is treated with lithium aluminium hydride under mild conditions (Bonnett *et al.*, 1969), examples have been reported of the hydrogenolysis of both 5,6- and 5,8-epoxides with this reagent to give 5-hydroxy compounds. Related to these processes is the hydrogenolysis of allylic glycosides (cf. Liaaen-Jensen, 1971). However, of greater value for structural studies is the observation that, under appropriate conditions with lithium aluminium hydride, both 5,6- and 5,8-epoxides undergo de-epoxidation to give the "parent" carotenoid. Thus violaxanthin (16–16) yields zeaxanthin (53–53) (Cholnoky *et al.*, 1967). Under these conditions neochrome (21–58) also gives zeaxanthin (Bonnett *et al.*, 1969; Cholnoky *et al.*, 1969); the transformation of the allenic end group probably involves initial reduction of the α-hydroxy allene followed by β elimination.

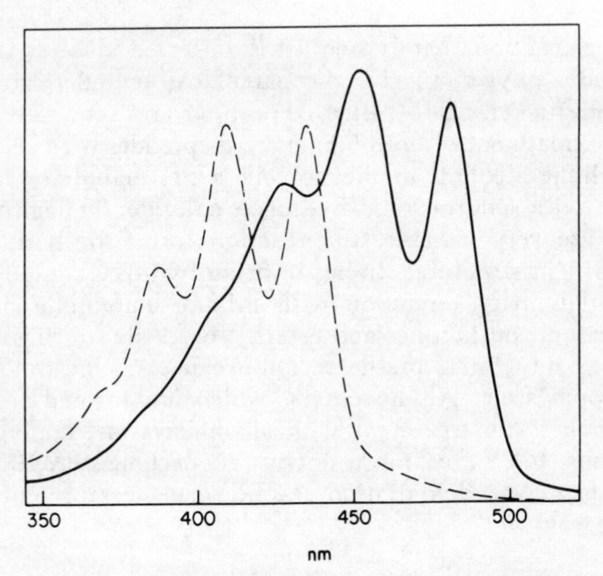

FIG. 7. Spectroscopic scale rearrangement of violaxanthin (16–16) to auroxanthin (21–21) in benzene by hydrogen chloride (—— before, - - - - after).

B REACTIONS PROMOTED BY ACID

Violaxanthin (16–16), and related carotenoids which possess one end group of the violaxanthin type, e.g. lutein epoxide (16–63) and neoxanthin (16–58), comprise an important group of pigments which are widely distributed in algae and higher plants. In the presence of mineral acids (aged chloroform contains sufficient free HCl!) these 5,6-epoxides are rapidly converted into a pair of 5,8-epoxides (the so-called furanoid oxides) which differ in configuration at C-8.

Thus violaxanthin gives a mixture of auroxanthins (21–21). Similar rearrangements are observed with the semisynthetic 5,6-epoxides which are obtained on treating β-carotene, zeaxanthin (acetate), and related xanthophylls with monoperphthalic acid (Karrer, 1948; Mayer and Isler, 1971). However, the principal semisynthetic epoxides in the 3-hydroxy series differ from the natural epoxides in their configuration at C-5 and C-6, and hence give rise to a pair of 5,8-epoxides which differ in the configuration at C-5 from those obtained from the natural 5,6-epoxides (Goodfellow *et al.*, 1973).

Since these 5,6- to 5,8-epoxide rearrangements involve a shortening of the chromophore, they too provide the basis of a test which can conveniently be carried out on a spectroscopic scale (Fig. 8). A minor side reaction that is frequently observed involves the loss of the epoxidic oxygen to give the parent carotenoid (Karrer, 1948; Karrer and Jucker, 1948). Blue colorations are frequently observed during the treatment of both 5,6- and 5,8-epoxides with acids.

Most allylic alcohols in the polyene series dehydrate rapidly on treatment with chloroformic hydrogen chloride. Though the yields obtained are very variable, the reaction forms the basis of a test widely used in structural studies both on hydroxy compounds and, after a preliminary reduction with lithium aluminium hydride or similar reagent, on ketones and esters. Very little starting material is required as the course of the reaction can easily be followed on a spectroscopic scale. Allylic ethers, which are formed from many allylic alcohols by treatment with alcoholic acids, undergo similar eliminations, but rather more slowly (cf. Zechmeister, 1958). In its simplest form the dehydration reaction can be represented in the following way:

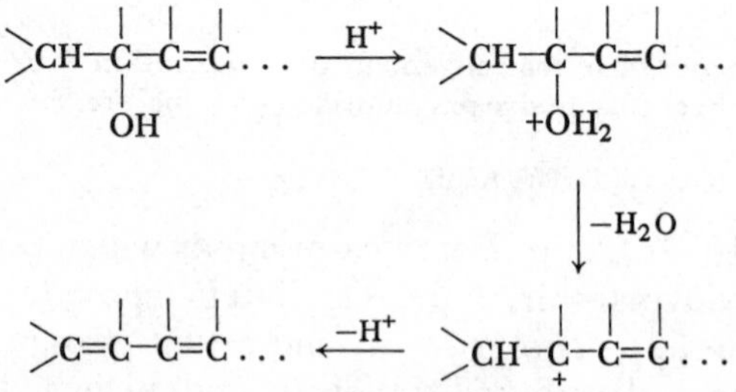

$$>CH{-}\underset{OH}{C}{-}C{=}C\ldots \xrightarrow{H^+} >CH{-}\underset{+OH_2}{C}{-}C{=}C\ldots$$

$$\downarrow -H_2O$$

$$>C{=}C{-}C{=}C\ldots \xleftarrow{-H^+} >CH{-}\overset{}{\underset{+}{C}}{-}C{=}C\ldots$$

Thus eschscholtzxanthin (51-51) gives the bis-anhydro derivative (6-6) (Karrer and Leumann, 1951). Frequently, however, dehydration involves loss of a proton not from the adjacent carbon atom, but from the vinylogous position at the other end of the polyene chain. Thus cryptocapsol (9-64), prepared by reduction of cryptocapsin (9-38) with lithium aluminium hydride, gives the anhydro compound (8-65) (Cholnoky *et al.*, 1963), and the reduction product (9-55) from echinenone gives the anhydro compound (8-8) (Petracek and Zechmeister, 1956) (Fig. 8). These reactions are obviously analogous to the formation of anhydro vitamin A (VI) on treatment of vitamin A with alcoholic hydrogen chloride (Shantz *et al.*, 1943).

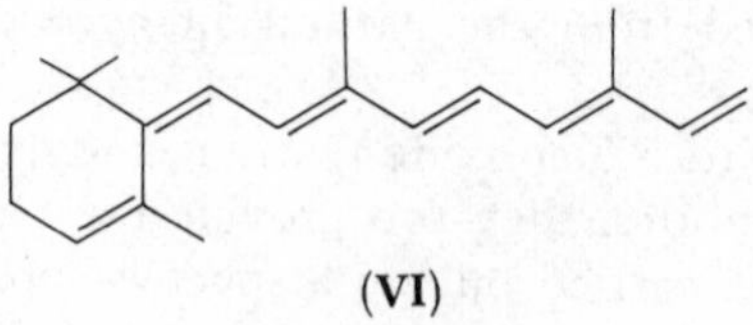

(VI)

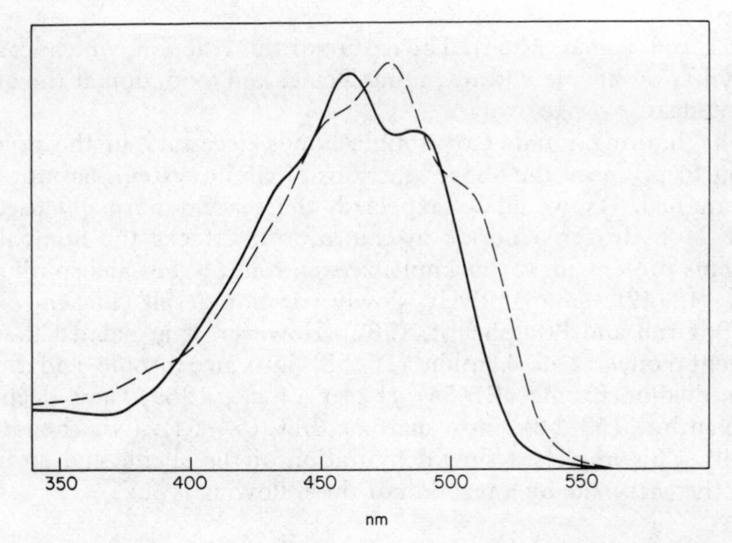

FIG. 8. Spectroscopic scale dehydration of isocryptoxanthin (5–55) to the hydrocarbon (8–8) in chloroform by hydrogen chloride (1 min) (—— before, - - - - after).

Limitations to the chloroformic hydrogen chloride reaction are known. Thus the substitution pattern present in capsorubinol (64–64), the reduction product of capsorubin (38–38), precludes dehydration (Cholnoky and Szabolcs, 1957).

The elimination of an allylic function may produce another allylic system, which then reacts further. Thus capsanthol (53–64), prepared from capsanthin (38–53), yields the bis-anhydro derivative (6–65) (Cholnoky and Szabolcs, 1957), and the reduction product of spheroidenone (4–24) gives 3,4-dehydrolycopene (1–2) (Goodwin *et al.*, 1956; Barber *et al.*, 1966), the chromophores being lengthened by two and three double bonds respectively. With the fucoxanthols (19–58), allylic dehydration is followed by normal acid catalysed rearrangement of the 5,6-epoxide to give the furanoid oxides (21–58) (Bonnett *et al.*, 1969; Goodfellow *et al.*, 1973). Though this transformation occurs rapidly, it is unusual in that there is no accompanying change in the visible light absorption maxima and has therefore to be monitored by t.l.c.

Isozeaxanthin (55–55), formed on reduction of canthaxanthin (33–33), exhibits an unexpected reaction on treatment with chloroformic hydrogen chloride. In addition to a mixture of hydrocarbons, some of which seem to have chromophores of the *retro* type, 3′,4′-dehydroechinenone is produced (Petracek and Zechmeister, 1956; C. W. Price and B. C. L. Weedon, unpublished results; cf.

Bodea and Tămas, 1961). The nature of this reaction, which leads to dehydration at one end of the molecule, and oxidation at the other, is not clear.

The limiting number of double bonds necessary in the polyene chain to promote the above reactions of allylic systems has not been determined. As would be expected, the reagent normally used (*c.* 0·01 M hydrogen chloride in chloroform) attacks the homoallylic systems present in, for example, zeaxanthin (53–53) and spirilloxanthin (42–42), comparatively slowly, or not at all (Liaaen-Jensen, 1959; Grob and Pflugshaupt, 1962). However, it is claimed that the reagent converts neoxanthin (16–58) into neochrome and thence into diadinochrome (21–54) (Egger *et al.*, 1969), and deepoxyneoxanthin (53–58) into diatoxanthin (53–54) (Nitsche *et al.*, 1969). This implies a slow dehydration of the allenic end group to give the acetylene by a reaction of the following type:

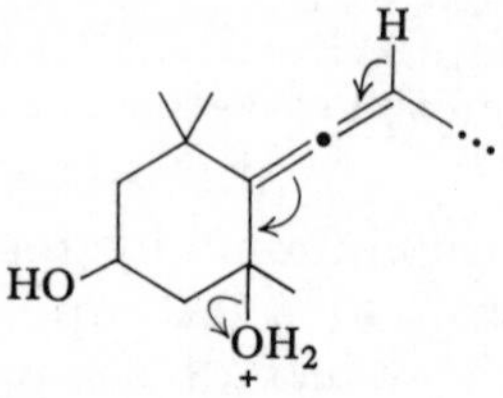

Finally it should be mentioned that acid hydrolysis (e.g. HCl/MeOH) is frequently used in the identification of carotenoid glycosides (Liaaen-Jensen, 1971).

C REACTIONS PROMOTED BY BASE

In the presence of base, astaxanthin (30–30) is rapidly oxidized by air to astacene (28–28). Unless special precautions are taken, this oxidation is likely to occur during attempts to isolate astaxanthin (30–30). With carotenoids such as flexixanthin (30–41), which contain only one end group of the astaxanthin type, the reaction is slower (Liaaen-Jenson, 1971). Alkaline hydrolysis of actinioerythrin (32–32) is accompanied by a similar autoxidation to give the bis-α-diketone violerythrin (29–29) and subsequently a variety of other products (Hertzberg *et al.*, 1969).

Peridinin (18–60) (Liaaen-Jensen, 1971) and fucoxanthin (17–59) are unstable to alkali even in the absence of air. On basic alumina fucoxanthin is slowly isomerized into isofucoxanthin (35–59) by opening of the epoxide ring.

Polyene aldehydes, and particularly dialdehydes, are surprisingly stable to dilute alkali, even in the presence of oxygen. (They are not readily oxidized to the corresponding acids by treatment with alkaline silver oxide.) This stability can be put to good use in identifying the two classes when only trace amounts of material are available (for instance from a partial oxidative degradation). Addition of base to an acetone solution of the polyene leads to an aldol condensation of any aldehyde function. Facile elimination of water from the β-hydroxy ketones formed results in an extension of the chromophore by one or two carbon–carbon double bonds. These changes are readily detected by the change in the light absorption properties of the solution (Bonnett *et al.*, 1969) (Fig. 9). When handling polyene aldehydes such as rhodopinal (2–26) in the presence of acetone, or similar ketones, care is necessary to ensure that such condensations are not carried out inadvertently (Liaaen-Jensen, 1973). An intramolecular aldol condensation is observed when semi-β-carotenone (9–34), and related compounds possessing 1,6-diketone systems, are treated with base (Kuhn and Brockmann, 1935).

Both carotenoic esters and carotenol esters may be hydrolysed by treatment with (1–5%) methanolic potassium hydroxide at room temperature. The conversion of a carotenoic acid into the correspond-

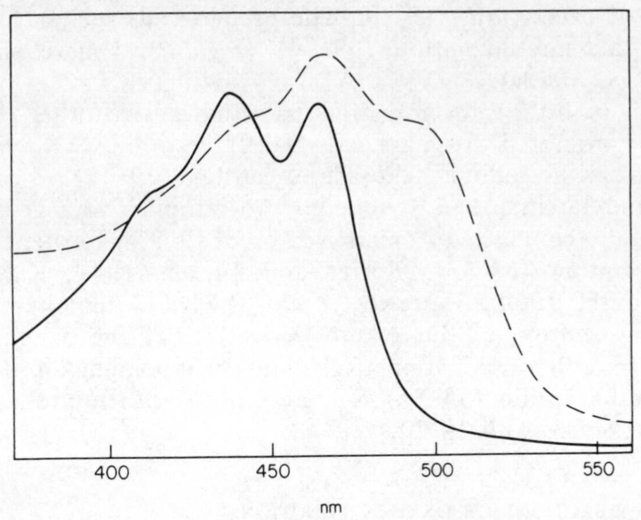

FIG. 9. Spectroscopic scale condensation of crocetindial (79–79) to the diketone (74a–74a) in acetone on addition of 3% potassium hydroxide in ethanol (1 min) (—— before, - - - - after).

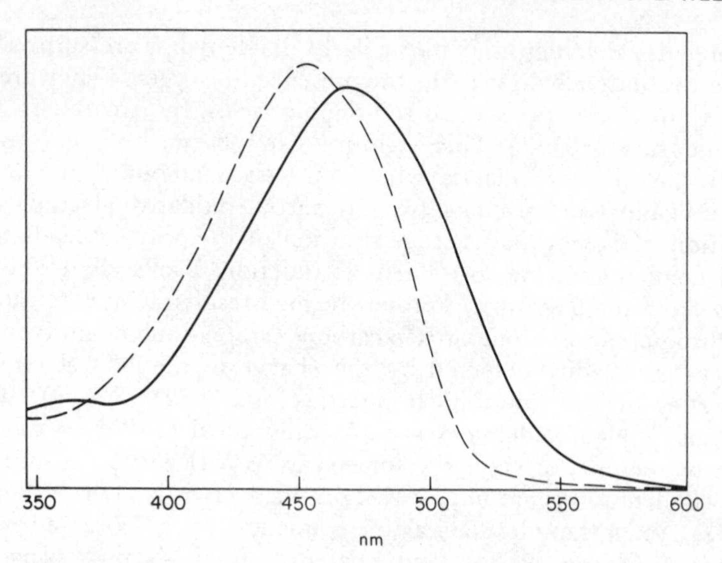

FIG. 10. Spectroscopic scale treatment of mytiloxanthin (39–54) with ethanolic potassium hydroxide (5 min) to give an enolate (—— before, - - - - after).

ing carboxylate which occurs on addition of alkali to an alcoholic solution is normally accompanied by a shift in the visible light absorption maxima. The change is, of course, reversible by the addition of excess mineral acid. The pseudo-acids such as astacene (28–28) and mytiloxanthin (39–54) (Fig. 10) behave similarly (Khare *et al.*, 1973a).

Though of little value for structural studies, two other types of reactions promoted by base are worth recording. On vigorous treatment with sodium alkoxide α-carotene (9–12), δ-carotene (2–12) and lutein (53–63) undergo prototropic rearrangement to give the fully conjugated isomers β-carotene (9–9), γ-carotene (2–9), and zeaxanthin (53–53) (Karrer and Jucker, 1947; Kargl and Quackenbush, 1960; Andrewes *et al.*, 1974). In the presence of potassium t-butoxide, the 4-keto derivatives of the β end group undergo smooth autoxidation to give the corresponding diosphenols. Thus canthaxanthin (33–33) is converted (*c.* 90%) into astacene (28–28) (Cooper *et al.*, 1975).

D OTHER REACTIONS OF OXYGEN FUNCTIONS

Primary and secondary hydroxy groups are acetylated by treatment with acetic anhydride (or acetyl chloride) in pyridine at room temperature. Determination of the composition of the product by mass spectrometry then reveals the total number of primary and secondary hydroxyl groups in the starting material.

Tertiary hydroxy groups do not react under the above conditions. However, should the presence of a tertiary hydroxyl group be suspected from the i.r. and m.s. spectra of the acetate, this can be confirmed by dehydration with phosphorus oxychloride in pyridine (Surmatis and Ofner, 1963; Bonnett *et al.*, 1969), or by trimethylsilylation. All types of hydroxyl are susceptible to the latter reaction, though azafrin (61-83) only gives a mono derivative due to steric hindrance at the 6 position (McCormick and Liaaen-Jensen, 1966).

Methylation of hydroxy groups can be achieved under suitable conditions. Thus treatment of lutein with methyl iodide in the presence of potassium t-butoxide gives the dimethyl ether (Buchecker *et al.*, 1974). Carotenoic acids are best converted into their methyl esters by reaction in pyridine with diazomethane. Rather surprisingly neither the enolic hydroxyls such as those in astacene (28-28), nor the phenolic carotenoids such as 3-hydroxyisorenieratene (13-66), react with diazomethane (Liaaen-Jensen, 1971).

Reference has already been made to the oxidation of allylic hydroxy compounds (Section IVA). Thus the allylic hydroxyl group in lutein (53-63) is readily oxidized with nickel dioxide (Liaaen-Jensen and Herzberg, 1966). Non-allylic hydroxyls are not readily oxidized. Vigorous treatment of zeaxanthin (53-53) with manganese dioxide gives rhodoxanthin (27-27) (Entschel and Karrer, 1959), but no oxidation has been observed using the Oppenauer method. Under the latter conditions, oxidation at the 3 position is believed to occur in fucoxanthin (17-59) (Hora *et al.*, 1970), and oxidation at the 3 position of the five-membered rings in capsanthin (38-53) and capsorubin (38-38) has been clearly demonstrated (Entschel and Karrer, 1960; Barber *et al.*, 1961b; Cholnoky *et al.*, 1963). Astaxanthin (30-30) is readily oxidized to astacene (28-28) by manganese dioxide or nickel dioxide; ring contraction occurs subsequently to give violerythrin (29-29) (Holzel *et al.*, 1969). The α-glycol structure present in azafrin (61-83) is oxidized by lead tetraacetate to give the corresponding diketone (Kuhn and Brockmann, 1935).

E PARTIAL DEGRADATION

In the early structural studies on carotenoids much valuable information was obtained by partial degradation. By carefully controlled chromic acid oxidation of β-carotene (9–9) Kuhn and Brockmann (1935) succeeded in isolating a whole series of partial oxidation products including β-carotenone (34–34) and the C_{27}-aldehyde (34–82) which not only forged a link between β-carotene and azafrin (61–83), but also provided valuable confirmation of the structures assigned to these two carotenoids. Similar, though less detailed, studies have also been carried out with α-carotene (9–12), capsanthin (38–53), physalien (the dipalmitate of zeaxanthin) (53–53) and lycopene (2–2). The latter yielded, among other products, bixin dialdehyde (75–75), which was transformed via the dioxime and dinitrile into norbixin (76–76) (Kuhn and Grundmann, 1932; cf. Karrer and Jucker, 1948).

By partial permanganate oxidation of α-carotene, β-carotene, azafrin, and lycopene, Karrer *et al.* obtained polyene aldehydes which were different from those reported from the chromic acid oxidations; thus β-carotene gave apo-8′- and apo-12′-carotenal (9–79 and 9–84) (Karrer and Solmssen, 1937), and lycopene gave apo-8′-lycopenal (2–79) (Karrer and Jaffé, 1939). With the cyclic carotenoids, permanganate degrades the polyene chain leaving one ring intact. With chromic acid, however, degradation of the chain seems to follow the initial opening of the β end group (9 → 34).

It is noteworthy that in both permanganate and chromic acid oxidations, α-carotene degrades from the β end to yield products isomeric with those obtained from β-carotene. The same difference is observed in the permanganate oxidation of lutein (53–63) and zeaxanthin (53–53) which give mainly α-citraurin (63–79) and β-citraurin (53–79) respectively (cf. Karrer and Jucker, 1948).

Partial oxidation also provided much helpful information in studies on bixin (76–77) (Karrer and Solmssen, 1937), spirilloxanthin (42–42) (Karrer and Koenig, 1940; Barber *et al.*, 1966), renieratene (13–14) and isorenieratene (13–13) (Yamaguchi, 1957, 1958).

Clearly oxidative degradations of the type referred to above require substantial amounts of starting material, and frequently these are either not available, or can nowadays be used more effectively in other ways. However, the method retains its usefulness in special instances. Even in the "spectroscopic era" identification of (17–82) and (**XI**) after oxidation of fucoxanthin (17–59) with zinc permanganate played a major part in elucidating the structure of this carotenoid (Bonnett *et al.*, 1969). The isolation of (**XI**) after ozonolysis was similarly important in establishing the structure of peridinin (18–60) (Strain *et al.*, 1971). Recently the partial oxidation with nickel dioxide of lutein dimethyl ether, and of acetates of the 2-hydroxy xanthophylls such as (9–52), was used very effectively in stereochemical studies on these carotenoids (Buchecker *et al.*, 1973a,b, 1974).

V Stereochemistry

A GEOMETRICAL ISOMERISM

The transformation of bixin (*cis* 76–77) into a higher melting form provided the first example of *cis–trans* isomerism in a natural polyene (Herzig and Faltis, 1923; Karrer *et al.*, 1929). Later a *cis* form of crocetin (80–80) was reported (Kuhn and Winterstein, 1933), but it was the pioneer observations of Gillam and the subsequent extensive studies of Zechmeister and his collaborators that revealed the full extent and importance of geometrical isomerism in the carotenoid field (cf. Zechmeister, 1962). It was shown that all carotenoids may be converted into mixtures of stereoisomers under appropriate conditions. A number of *cis* and poly-*cis* isomers of carotenoids have also been isolated from natural sources, though a few of the compounds containing one or two *cis* bonds may conceivably be artefacts formed by stereomutation of the more familiar *trans* forms during isolation. Phytoene is a natural 15-*cis* carotenoid. The natural occurrence of these *cis* compounds, and the realization that the *cis–trans* isomerism of retinal, the aldehyde of vitamin A_1 (**IV**), plays an all-important role in the chemistry of the visual process, emphasizes the need for careful consideration to be given to the stereochemical aspects of the problem when considering the biosynthesis and functions of carotenoids.

A detailed review of the *cis–trans* isomerism of carotenoids, vitamins A, and related polyenes, was published by Zechmeister in 1962. The account given below is therefore confined to the more important general aspects of the subject, and to recent developments.

1 *Number and types of* cis-*carotenoids*

By methods outlined below, any carotenoid can be converted into a quasi-equilibrium mixture of *cis–trans* isomers. Once this had been appreciated, calculation of the numbers of isomers theoretically possible gave some rather startling figures. For a conjugated system with n non-cyclic double bonds, the number of stereoisomers N is given by the expressions:

$N = 2^n$ for unsymmetrical systems
$N = 2^{(n-1)/2} . (2^{(n-1)/2} + 1)$ for symmetrical systems, n odd
$N = 2^{(n/2)-1} . (2^{n/2} + 1)$ for symmetrical systems, n even.

Thus lycopene, with a symmetrical carbon skeleton and eleven conjugated double bonds, should be capable of existing in 1056 different forms! Of these only about forty have so far been encountered.

Fortunately the situation is not, in practice, as complex as has just been implied. As was first pointed out by Pauling (1939), the double bonds in the acyclic polyene chain of carotenoids are of two types: those for which the adoption of a *cis* configuration involves very little steric hindrance (between two hydrogen atoms), and those for which a *cis* configuration leads to a serious clash between a methyl group and a hydrogen atom (cf. Fig. 11). The methyl substituted double bonds come into the first category, as do the 15,15′ double bonds corresponding to the centre of the carbon skeleton in most carotenoids. All other acyclic di-substituted double bonds are adjacent to a methyl side chain, and therefore fall into the second category. This classification is helpful in indicating the double bonds about which stereomutation is most likely to take place. A further calculation reveals that the number of "sterically unhindered" isomers of lycopene is only 72. It must not, however, be concluded, as was widely done at one time, that carotenoids with a *cis* configuration about a double bond of the second category are incapable of existence. A number of these "sterically hindered" isomers have been prepared indirectly by partial hydrogenation of acetylenic analogues; moreover, it is known that retinal, the aldehyde of vitamin A, (IV), undergoes stereomutation about the 11,12 double bond before it combines with opsin in the visual cycle (cf. Heilbron and Weedon, 1958; Zechmeister, 1962). However, in

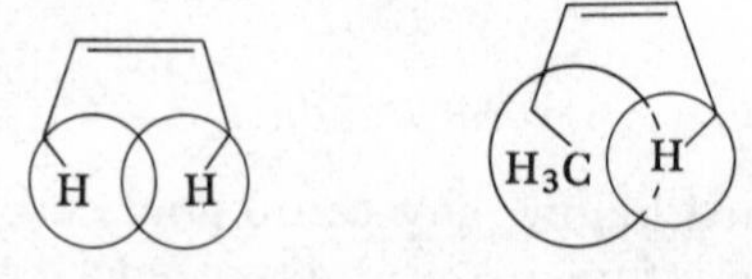

FIG. 11. Overlapping of hydrogen atoms in –CH–CH=CH–CH–, and of hydrogen and methyl in –CH–CH=CH–CMe– with *cis* configuration; according to Pauling (from *Fortschr. Chem. org. Naturstoffe* 3, 203 (1939)).

practice, relatively few isomers constitute the bulk of the mixtures produced by stereomutation, as would be expected on purely statistical grounds (Zechmeister, 1962).

2 *Stereomutation*

Stereomutation of a carotenoid occurs under a variety of conditions but the rate of reaction depends greatly on the structure and configuration of the starting material. Moreover, different mixtures of products may be formed using different procedures.

Normally the most stable isomer of a carotenoid is the all-*trans* form. Important exceptions to this generalization are provided by the natural acetylenic carotenoids. Thus with crocoxanthin (12-54) the main constituent of the equilibrium mixture is the 9-*cis* isomer, and with the diacetylene alloxanthin (54-54) the 9,9′-di-*cis* isomer is the thermodynamically most stable form. It has been suggested that, in contrast to their polyene analogues, these *cis* isomers in the acetylene series are somewhat less sterically hindered than their all-*trans* forms (Weedon, 1970). Further exceptions are provided by compounds of the rhodopinal (2-26) series. Here cross-conjugation with the side-chain aldehyde group favours the 13-*cis* structure, and it is only the allylic alcohol rhodopinol (2-49) without this group that can be partly isomerized to the all-*trans* form (Liaaen-Jensen, 1969).

In general 15-*cis* and "hindered" *cis* isomers are particularly labile. The central *cis* bond in phytoene (5-5) is, however, comparatively stable.

The principal methods of isomerizing carotenoids are given below.*

a. Thermal methods. Stereomutation of a carotenoid begins immediately on solution. The process is usually slow at room temperature. Thus in benzene or light petroleum solution, in diffuse daylight, only *c.* 1-2% of α-, β-, and γ-carotene undergo stereomutation in 24 h. However, with all-*trans* lycopene (2-2) under similar conditions the proportion is 10%, with spirilloxanthin (42-42) 23%, and with α-bacterioruberin (69-69) as much as 42% (Liaaen-Jensen, 1962; Zechmeister, 1962).

Stereomutation is more rapid at elevated temperatures, and with all-*trans* carotenoids in boiling benzene or hexane a quasi-equilibrium mixture of geometrical isomers is produced within 10-60 min. Since

* See also Chapter 19.

many of these mixtures contain several isomers, it is obvious that heating or storage of carotenoid solutions should be avoided as far as possible.

Most *cis* isomers of carotenoids exhibit marked thermal lability, but some poly-*cis* carotenoids are as stable as the corresponding all-*trans* forms (Zechmeister, 1962), and (*cis*) methylbixin yields a di-*cis* isomer as the main product (Barber *et al.*, 1961a).

As would be expected, stereomutation also occurs when a carotenoid is melted.

b. Photochemical methods (*without catalyst*). All carotenoids undergo *cis–trans* isomerization on irradiation in solution, light of wavelengths corresponding to the main absorption band being most effective. A similar process probably operates in nature in the conversion of *cis* and poly-*cis* isomers into the more familiar all-*trans* forms, and reference has already been made to the formation of a "hindered" *cis* isomer on irradiation of (all-*trans*) retinal.

c. Iodine catalysis, in light. The most commonly used method of stereomutation is to expose a solution of the carotenoid, containing catalytic amounts of iodine, to light. The quasi-equilibrium mixture of isomers, containing as a rule one-third to one-half of the pigment in *cis* configurations, is formed rapidly. Each isomer when isolated and submitted again to rearrangement yields a mixture of approximately the same composition.

3 Synthesis of cis *isomers*

Many total syntheses of carotenoids involve the partial reduction of an acetylenic analogue over a palladium or nickel catalyst. Such hydrogenations occur by *cis* addition, and under appropriate conditions the *cis* carotenoids formed initially may be isolated before rearrangement to the all-*trans* forms. In this way several central (15-) *cis* isomers have been prepared, and a few compounds with "hindered" *cis* configurations (cf. Zechmeister, 1962; Mayer and Isler, 1971).

The synthesis of these and other polyene isomers of known configuration has necessitated revision of the structures tentatively assigned to some stereomutation products, but has confirmed many of the general principles used in drawing structural conclusions from spectral properties (cf. Akhtar *et al.*, 1959; Zechmeister, 1962).

Extensive studies have been made on the controlled synthesis of *cis* isomers in the vitamin A_1 and vitamin A_2 series; all "unhindered" forms, and some "hindered" forms of the vitamins and the related retinals have been prepared (cf. Heilbron and Weedon, 1958; Mayer and Isler, 1971; Ramamurthy and Liu, 1975). The 9-*cis* structure assigned to the dimethyl ester derived from natural bixin (*cis* 76–77) has been confirmed by stereoselective synthesis (Pattenden *et al.*, 1970).

4 *Differences between* cis–trans *isomers*

a. General properties. The carotenoids so far studied conform to the general principle that the all-*trans* compound is the isomer of lowest solubility and highest melting point. Because of stereomutation on fusion, some *cis* isomers exhibit the phenomenon of a double melting point.

The *cis* forms of optically active carotenoids exhibit rotation which may differ markedly from that of the all-*trans* isomer, both in size and sometimes even in sign. This is strikingly illustrated by the c.d. curves for violaxanthin (16–16) and its 9-*cis* isomer violeoxanthin (Fig. 5) (Moss *et al.*, 1975).

The geometrical configuration of a carotenoid has a profound influence on its adsorption affinity. This means that even complex mixtures of isomers can usually be separated by chromatography. However, some *cis* isomers are more, and others less, strongly adsorbed than the corresponding all-*trans* compound.

b. Visible and ultraviolet light absorption. In the carotenoid series most isomers with one "unhindered" *cis* double bond absorb at wavelengths which are 2–5 nm shorter than those of the corresponding all-*trans* compounds (Fig. 12). The only exception so far reported is gazaniaxanthin, the 5′-*cis* isomer of rubixanthin (2–53). Here the isomer with the *cis* configuration about the terminal double bond in the (acyclic) polyene chain absorbs at virtually the same wavelengths as the all-*trans* form. With the terminal (13-) *cis* isomers of vitamin A_1 (IV) and vitamin A_2 (V) maximal absorption occurs at slightly (2–3 nm) *longer* wavelengths than with the all-*trans* forms (Hubbard, 1956).

As would be expected on theoretical grounds, the central-*cis* isomer has less intense absorption than the other mono- ("unhindered") *cis* forms. With di- "unhindered" *cis* isomers the "λ shift" seems to be *c.* 10 nm.

The poly-*cis* carotenoids, such as the natural prolycopene and proneurosporene, exhibit light absorption at much shorter wavelengths, and of much lower intensity, than either the all-*trans* or the mono- "unhindered"-*cis* forms. Their spectra also show little fine structure. In many respects these spectra resemble the "degraded" spectra seen with mono-*cis* isomers if the *cis* bond is of the

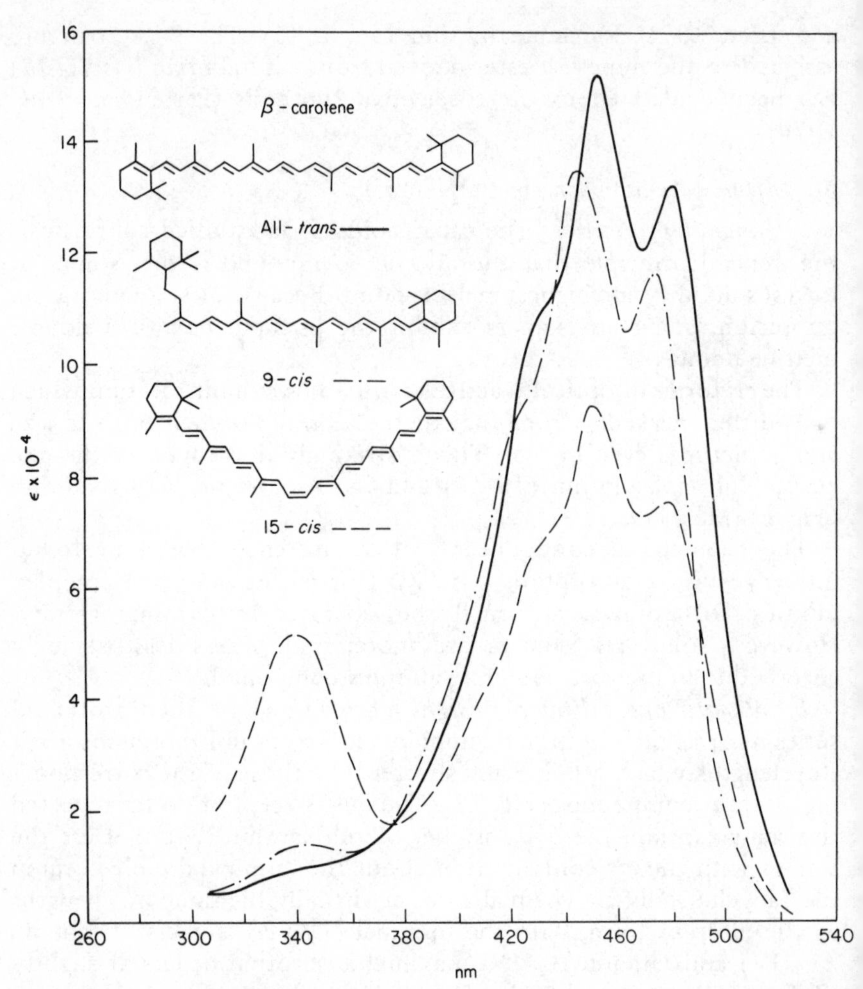

FIG. 12. Light absorption spectra of three stereoisomeric β-carotenes. (From "Theory and Applications of Ultraviolet Spectroscopy" (p. 233) by H. H. Jaffé and M. Orchin, Wiley, New York (1962).)

"hindered" type. With both classes of carotenoids there is a spectacular change in light absorption properties on iodine-catalysed isomerization (Fig. 13).

One of the most noticeable features of the spectra of mono-*cis* carotenoids is the appearance of a subsidiary peak in the near ultraviolet region (Fig. 12). The wavelength difference between the "*cis* peak" and the longest wavelength maximum of the all-*trans* compound is practically a constant (142 ± 2 nm for C_{40} carotenoids

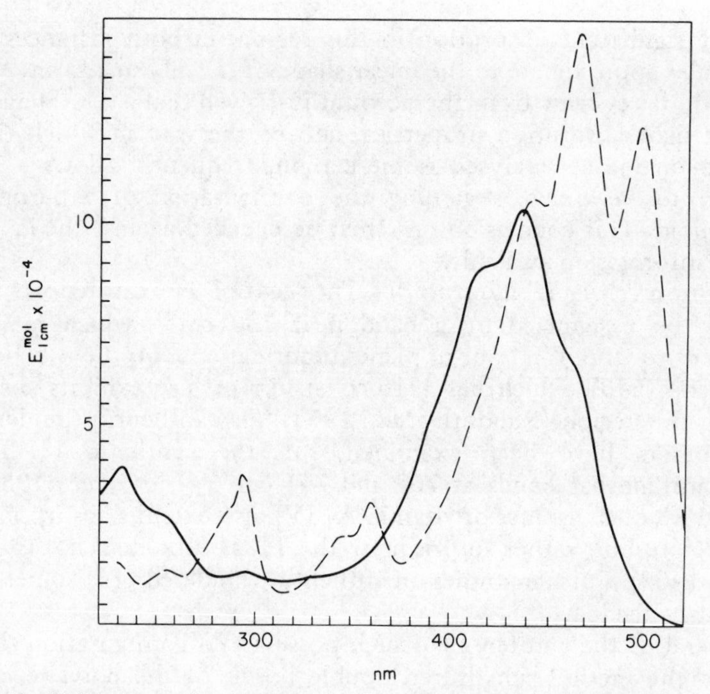

FIG. 13. Molecular extinction curves of prolycopene, in hexane: ——, fresh solutions, and - - - -, mixture of stereoisomers after catalysis by iodine; according to Zechmeister and Pinckard. (From *Fortschr. Chem. org. Naturstoffe* **18**, 284 (1960)).

possessing 10–11 conjugated double bonds). Although the evidence is not conclusive, mono- "hindered" *cis* isomers also seems to exhibit *cis* peaks, but the band appears to be masked with the *cis* forms of carotenoids possessing chromophores of the *retro* type (Zechmeister, 1962).

On theoretical grounds, it has been predicted that the intensity of the *cis* peak should be roughly proportional to the square of the distance between the centre of the conjugated system and the midpoint of a straight line joining its two ends (Zechmeister *et al.*, 1943) (i.e. zero in the case of the all-*trans* and terminal *cis* isomers). In agreement with this prediction the central-*cis* isomers which have been prepared in a number of series by unambiguous total synthesis have more intense *cis* peaks than any of the other isomers that have been observed. Since the intensity of the *cis* peak seems to depend primarily on the overall shape of the chromophore, it is scarcely surprising that neither the poly-*cis* nor the authentic di-*cis* isomers exhibit significant absorption in this region; in both instances the molecules approximate to the linear shape of the all-*trans* form.

It will be evident from the account just given that a consideration of the light absorption properties, and of the way in which these change on iodine-catalysed isomerization, frequently allows a conclusion to be drawn regarding the configuration of a particular carotenoid. This conclusion can then be checked against the i.r. and n.m.r. information available.

c. Infrared light absorption. The central *cis* carotenoids can readily be recognized by a band near 780 cm^{-1} which may be ascribed to the C-H out-of-plane deformations of the *cis* bond. Phytoene (5–5), which has a 15-*cis* structure, also exhibits absorption in this region (Rabourn *et al.*, 1954). Few authentic "hindered" *cis* isomers have been examined, but the synthetic 11,11′-di-*cis*-β-carotene has bands at 762 and 741 cm^{-1} (Isler *et al.*, 1957b). The 11,13-di-*cis* isomer of vitamin A_2 (V) also exhibits absorption in this region, but, rather surprisingly, the 11-*cis* form does not (Planta *et al.*, 1962). Further studies on authentic "hindered" *cis* isomers are obviously needed.

Several of the carotenoids which possess a *cis* configuration about one of the methyl substituted double bonds of the polyene chain have a band at 1380 cm^{-1}, ascribable to the C—H in-plane deformation of the *cis* bond (Lunde and Zechmeister, 1955). However, this band is not always easily resolved, and is absent from the spectrum of methyl natural bixin (*cis*- 76–77) though this pigment is known to have a *cis* bond of the type in question (Barber *et al.*, 1961a).

In the spectra of some *cis* isomers, the characteristic band at about 966 cm^{-1} due to C—H out-of-plane deformations of the *trans* di-substituted double bonds, is split into a doublet (Lunde and Zechmeister, 1955).

d. N.m.r. spectra. Methyl groups which form part of unconjugated tri-substituted, double bonds give rise to bands at positions which differ slightly according to whether the methyl group is *cis* or *trans* to the olefinic hydrogen atom (1·66 and 1·60 respectively) (Davis *et al.*, 1961). This difference is no doubt due to long-range shielding. Differences are also observed between *cis* and *trans* isomers in the β-methyl bands of allylic systems; these are particularly marked when the functional substituent contains a carbonyl group (e.g. —CHO, —CO.Me, —CO_2Me) (Burrell *et al.*, 1966). Thus the C-13 methyl in all-*trans* vitamin A_1 acid methyl ester, and in the A_2 analogue, gives a band (2·16 and 2·42 respectively) at lower fields than the corresponding group in the 13-*cis* isomers (1·88 and 2·14 respectively) (Planta *et al.*, 1962). Such effects are attributed to the greater deshielding of the β-methyl protons in the *trans* isomers owing to the proximity of the carbonyl and β-methyl groups. No significant differences have been reported between geometrical isomers in the bands associated with methyl groups attached to the polyene chain, except for the 11,13-di-*cis* isomer of vitamin A acid and some related di-*cis* polyenes (Pattenden and Weedon, 1968; Ramamurthy and Liu, 1975). It may be significant that a similar difference in shielding has been observed in two of the methyls attached to the polyene chain in prolycopene (Weedon, 1971).

In both the violaxanthin (16-16) and neoxanthin (16-58) series a switch from the all-*trans* to the 9-*cis* configuration is accompanied by slight, but detectable, differences in shielding of the methyls attached to fully substituted carbon atoms in the adjacent ring (Cholnoky *et al.*, 1969; Moss *et al.*, 1975).

Little use has yet been made of differences in the olefinic proton bands to determine stereochemistry. However, this method has been used to elucidate the 9′-*cis* configuration of natural bixin (*cis* 76-77) (Barber *et al.*, 1961a), and to confirm the *trans*, *cis*, *trans* configuration of the central triene unit in phytoene (5-5) (Khatoon *et al.*, 1972). In both instances the conditions were particularly favourable in that the appropriate signals were resolved at 60 and 100 MHz respectively. It is possible to extend these procedures to other cases by taking advantage of the increase in chemical shifts that can be achieved with the 220 MHz spectrometers.

The significant changes in the n.m.r. spectra of carotenoids with a *cis* double bond, compared with the corresponding all-*trans* isomer, are summarized below (based on data from Vetter *et al.*, 1971, changes ⩾ 0·1 p.p.m. quoted)

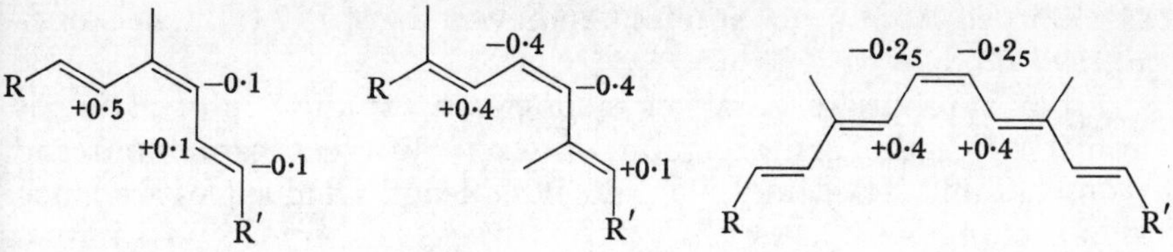

9- or 13-*cis*

11-*cis*

15-*cis*

B CONFORMATION

The determination of molecular structure by crystallographic analysis gives detailed information on the conformation of the molecule in the crystalline lattice. Attempts to determine the conformations in solution suggest that these resemble that in the solid state. A p.m.r. study of β-ionone (VII) by Honig *et al.* (1971) showed from the long-range spin-spin coupling constants and from the nuclear Overhauser effect that the diene is s-*cis* and non-planar (by about 30° from J values, or 70° from NOE results). This conclusion was supported by a study of the low resolution microwave spectrum (Steinmetz, 1974).

Steinmetz also examined α-ionone (VIII) and both *cis* (IX) and *trans* α-irones (X). He found that the C-2 and C-6 substituents in *cis* α-irone are both equatorial as expected, and that the other two compounds exist as a mixture of two main conformations. His conclusion that *trans* α-irone is predominantly in the conformation with a pseudo-axial enone side chain is supported by a study of c.m.r. substituent effects (unpublished results from Queen Mary College).

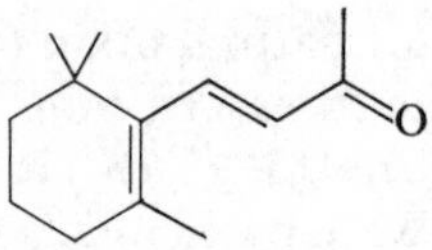

(VII)

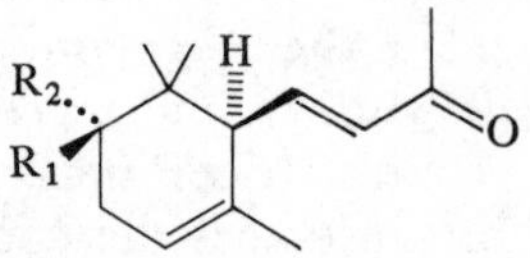

(VIII) $R_1 = R_2 = H$
(IX) $R_1 = Me, R_2 = H$
(X) $R_1 = H, R_2 = Me$

A study of c.m.r. substituent effects with lutein (55-63) and related model compounds shows that at the α-end the 3-hydroxyl group and the polyene chain are both psuedo-equatorial (unpublished results from Queen Mary College). This implies a *trans* relationship, as proposed from other studies (Andrewes *et al.*, 1974; Buchecker *et al.*, 1974).

From appropriate X-ray crystallographic studies, information is available on the conformation of seven different carotenoid end groups (Table XIV). In general the bond lengths and angles are much as expected.

The cyclohexane ring of 11-apoazafrinol is slightly flattened in the heavily substituted region of C-6. Thus the dihedral angles associated with C-6 are significantly less than for cyclohexane (55·9°; Geise *et al.*, 1971). This X-ray study also confirms the *threo* relationship of the 5,6-diol systems and the predicted preferred conformation with both hydroxyl groups axial.

TABLE XIV
Conformation of carotenoid end groups from X-ray crystallographic studies

Related end group[a]	Dihedral angle 1–2–3–4	2–3–4–5	3–4–5–6	4–5–6–1	5–6–1–2	6–1–2–3	Side chain dihedral angle	6–7–8 angle	Reference
(61)	57·4	−57·6	54·5	−49·6	48·6	−53·3	O–6–7–8 = −1·3	126·1	1
(59)	66·8	−65·3	51·2	−47·9	48·0	−56·1	5–6–8–9 = −83·3	178·8	2
(9)	61·3	−43·1	12·0	2·7	13·6	−45·3	5–6–7–8 = 58·0	123·2	3
(54)	60·7	−48·1	17·9	0·9	7·9	−40·8	5–6–9–10 = −130·2	178·7	4
(33)	50·6	−26·8	2·4	−1·3	24·3	−48·5	5–6–7–8 = −42·7	126·0	5
(53)	76·7	−55·8	18·7	9·6	7·9	−51·8	5–6–7–8 = −176·6	144·1	6
(38)	−14·8	−9·6	33·4[b]	−47·0[b]	37·8[b]	−173·8[b]	5–6–7–8 = 177·7	143·8	6

References:

1. 11-Apoazafrinol; Flo and Hursthouse, unpublished data. See also Schwieter *et al.*, 1971.
2. Allenic ketone (**XI**); DeVille *et al.*, 1969a; and unpublished data. See also DeVille *et al.*, 1969b, 1970.
3. Vitamin A acetate; Oberhänsli *et al.*, 1974. See also Braun, 1971; Sundaralingam and Beddell, 1972; Hamanaka *et al.*, 1972; Stam, 1972; Koch, 1972; Gilardi *et al.*, 1972.
4. 9-*Cis*-C_{15}-Wittig salt; Flo and Hursthouse, unpublished data.
5. 15,15′-Dehydrocanthaxanthin; Bart and MacGillavry, 1968. See also Sundaralingam and Beddell, 1972.
6. Capsanthin bis-*p*-bromobenzoate (molecule B); Ueda and Nowacki, 1974.

Notes:

[a] The compounds used in the X-ray studies cited are given in references 1-5.

[b] Dihedral angles 3–4–5–1, 4–5–1–2, 5–1–2–3, 4–5–6–7 respectively.

Conversion of C-6 into an sp^2 hybrid as in the allenic ketone (**XI**) does not change the ring dihedral angles associated with C-6 significantly from the previous case. However, there does seem to be an effect of the equatorial 3-OH group on the dihedral angles associated with C-3, suggesting a slight movement of C-3 towards the axial substituents at C-1 and C-5 (DeVille *et al.*, 1969a; see also DeVille *et al.*, 1969b, 1970).

The most studied end group is that of β-carotene and related compounds. The results show that the cyclohexene ring takes up a typical half-chair conformation. There is only a slight change in the 9-*cis*-C_{15}-Wittig salt related to alloxanthin (54–54) (unpublished results). Oxidation of β-carotene to canthaxanthin (33–33) results in the expected flatter ring which favours conjugation between the ketone and the 5,6 double bond.

As mentioned in Section IIIA, the cyclohexene double bond of a β end group is not fully conjugated with the polyene chromophore. The reason for this is the non-planarity of the 5=6—7=8 system. In the solid phase most examples of this chromophore so far studied have an s-*cis* conformation with a dihedral angle of 40–60°. Isolated examples of the s-*trans* conformation have been found. In these structures the interaction between the hydrogen atom at C-8 and the C-1 methyl groups is minimized with a relatively small dihedral angle 1-6-7-8 (*c.* 15°). In solution it is probable that the s-*cis* conformation is preferred (Langlet *et al.*, 1970). The non-planarity is then caused by interaction between the hydrogen atom at C-8 and the C-5 methyl group. This is partly relieved by an increase in the 6-7-8 angle. Crystallographic data confirm this predicted inverse relationship between the 6-7-8 angle and the 5-6-7-8 dihedral angle (Oberhänsli *et al.*, 1974).

Bond alternation along the polyene chain is clearly demonstrated by the bond lengths of about $1{\cdot}4_5$ Å for the single bonds and $1{\cdot}3_5$ Å for the double bonds. The conjugated system is essentially planar but shows a marked in-plane bending caused by steric interference between hydrogen atoms and methyl groups on the polyene chain (Oberhänsli *et al.*, 1974).

The presence of a *cis* double bond in the polyene chain results in distortion of the planarity. A crystallographic study (Gilardi *et al.*, 1972) of 11-*cis* retinal revealed that the 12–13 bond of the polyene is s-*cis* with a dihedral angle of 38·7°. This conformation is supported by the calculations of Rowan III *et al.* (1974) (see also Schaffer *et al.*, 1974), and by the p.m.r. studies of Patel (1969).

C ABSOLUTE CONFIGURATION

Though carotenoids have been the subject of study for many years, information on their absolute configuration is of comparatively recent origin. However, a number of correlations have now been established between various carotenoids by chemical, n.m.r., and optical (o.r.d. or c.d.) means (Bartlett *et al.*, 1969), and these correlations, together with success in determining the configurations of five key end groups, now permits the assignment of absolute configuration to a substantial number of carotenoids (Weedon, 1971). No instance has yet been found of two natural carotenoids having an end group with the same gross structure but with different absolute stereochemistry. For this reason absolute configuration is indicated where possible in the list of "half structures" in Table I. However, further study may reveal examples in which these assignments are inappropriate.

Faigle and Karrer (1961) degraded capsanthin (38–53) and capsorubin (38–38) to a derivative of natural camphor, viz. (−)-camphoronic acid, and thus revealed the stereochemistry at the C-5 position in the five-membered rings of these pigments. Subsequently Cooper *et al.* (1962) proved that the two oxygen substituents on these end groups are *trans* to one another, thus establishing the 3*S*,5*R* configuration as shown in the "half structure" (38). Recently these conclusions were confirmed by synthesizing 3*S*,5*R*,3′*S*,5′*R*-capsorubin (38–38) from (+)-camphor, and showing that it had the same c.d. properties as the natural pigment (Weedon, 1973).

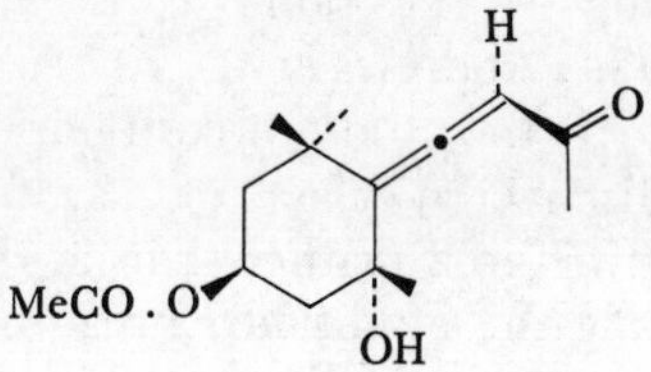

H
O
MeCO.O
OH
(XI)

The configuration of the common 3-hydroxy end groups in carotenoids was revealed as the result of structural studies on fucoxanthin (17–59). The key degradation product (**XI**) was converted from the acetate into the corresponding *p*-bromobenzoate. X-ray crystallographic analysis of this derivative established that the allenic end group in fucoxanthin (17–59), and that in neoxanthin (16–58), must have the 3*S*,5*R*,6*R* configuration given in the "half structures" (59) and (58) respectively (DeVille *et al.*, 1969a). Moreover, since a correlation had previously been established between fucoxanthin, zeaxanthin (53–53), violaxanthin (16–16), and alloxanthin (54–54), this X-ray analysis also indicated the 3*R*,3′*R* configuration of zeaxanthin, and the corresponding configurations at C-3 and C-3′ in all the carotenoids cited (including C-3 in the epoxide end group of fucoxanthin).

The configuration of the epoxide end group of violaxanthin (16–16), lutein epoxide (16–63), and neoxanthin (16–58) was elucidated by studies on the derived furanoid oxides (cf. p. 182). This rearrangement was shown to occur with retention of configuration at C-5 and with the formation of both possible epimers at the new chiral centre, C-8. The n.m.r. properties of the products leave no doubt that the 3-hydroxy-5,6-epoxide end groups of violaxanthin and related natural carotenoids have the 3*S*,5*R*,6*S* configuration shown in (16). Extension of this work shows the epoxide end group in fucoxanthin is also 3*S*,5*R*,6*S* (Goodfellow *et al.*, 1973).

By judicious chemical correlations with manool and ambrein, it has been established that α-carotene (9–12), and several related carotenoids with the α end group, have the 6′*R* configuration (12) (Buchecker *et al.*, 1973a). This also applies to the 6′ position in the common xanthophyll, lutein (53–63) (Goodfellow *et al.*, 1970; Buchecker *et al.*, 1974). By further studies on lutein degradation products it has been found that the 3′-position to which the allylic hydroxyl is attached has the *R* configuration shown in (63), and that the other end group is the same as those in zeaxanthin (53) (Buchecker *et al.*, 1974; Andrews *et al.*, 1974).

Another approach to the study of configuration has been adopted for the recently discovered class of carotenoids which have a hydroxy substituent in the 2 (rather than at the usual 3) position of the β end group. Using the modified Horeau method these pigments have been shown to possess the *R* stereochemistry (52) (Buchecker *et al.*, 1973b).

As yet, no absolute configuration has been determined by X-ray crystallographic analysis of an intact chiral carotenoid, or even of one of its derivatives which preserves the entire carbon skeleton. However, developments along these lines are to be expected in the future. The recent study of capsanthin bis-*p*-bromobenzoate (Ueda and Nowacki, 1974) confirmed the above correlations, but was unsuitable for direct determination of absolute configuration.

REFERENCES

Ahmad, R. and Weedon, B. C. L. (1952). *Chemy Ind.* 882.

Akhtar, M., Richards, T. A. and Weedon, B. C. L. (1959). *J. chem. Soc.* 933-940.

Allen, F. H. and Rogers, D. (1971). *J. chem. Soc. B* 632-636.

Andrewes, A. G. and Liaaen-Jensen, S. (1973). *Acta chem. scand.* 27, 1401-1409.

Andrewes, A. G., Borch, G. and Liaaen-Jensen, S. (1974). *Acta chem. scand.* **B28**, 139-140.

Antia, N. J. (1965). *Can. J. Chem.* 43, 302-303.

Arpin, N., Fiasson, J. L. and Lebreton, P. (1969-70). *Prod. prob. pharm.* 24, 630-644; 25, 21-34.

Arpin, N., Norgård, S., Francis, G. W. and Liaaen-Jensen, S. (1973a). *Acta chem. scand.* 27, 2321-2334.

Arpin, N., Kjøsen, H., Francis, G. W. and Liaaen-Jensen, S. (1973b). *Phytochemistry* 12, 2751-2758.

Aung Than, Bramley, P. M., Davies, B. H. and Rees, A. F. (1972). *Phytochemistry* 11, 3187-3192.

Baldas, J., Porter, Q. N., Cholnoky, L., Szabolcs, J. and Weedon, B. C. L. (1966). *Chem. Commun.* 852-854.

Barber, M. S., Hardisson, A., Jackman, L. M. and Weedon, B. C. L. (1961a). *J. chem. Soc.* 1625-1630.

Barber, M. S., Jackman, L. M., Warren, C. K. and Weedon, B. C. L. (1961b). *J. chem. Soc.* 4019-4024.

Barber, M. S., Jackman, L. M., Manchand, P. S. and Weedon, B. C. L. (1966). *J. chem. Soc. C* 2166-2176.

Bart, J. C. J. and MacGillavry, C. H. (1968). *Acta cryst.* **B24**, 1569-1587, 1587-1606.

Bartlett, L., Klyne, W., Mose, W. P., Scopes, P. M., Galasko, G., Mallams, A. K., Weedon, B. C. L., Szabolcs, J. and Tóth, Gy. (1969). *J. chem. Soc. C* 2527-2544.

Bodea, C. and Tămas, V. (1961). *Angew. Chem.* 73, 532.

Bonnett, R., Mallams, A. K., Spark, A. A., Tee, J. L., Weedon, B. C. L. and McCormick, A. (1969). *J. chem. Soc. C* 429-454.

Borch, G., Norgård, S. and Liaaen-Jensen, S. (1972). *Acta chem. scand.* 26, 402-403.

Braude, E. A. (1945). *Rep. Prog. Chem.* 42, 105-130.

Braun, P. B., Hornstra, J. and Leenhouts, J. I. (1971). *Acta cryst.* **B27**, 90-95.

Bremser, W., Hill, H. D. W. and Freeman, R. (1971). *Messtechnik* 79, 14-21.

Buchecker, R., Egli, R., Regel-Wild, H., Tscharner, C., Eugster, C. H., Uhde, G. and Ohloff, G. (1973a). *Helv. chim. Acta* 56, 2548-2563.

Buchecker, R., Eugster, C. H., Kjøsen, H. and Liaaen-Jensen, S. (1973b). *Helv. chim. Acta* 56, 2899-2901.

Buchecker, R., Hamm, P. and Eugster, C. H. (1974). *Helv. chim. Acta* 57, 631-656.

Buchwald, M. and Jencks, W. P. (1969). *Biochemistry, N.Y.* 7, 834-843, 844-859.

Budzikiewicz, H., Brzezinka, H. and Johannes, B. (1970). *Monatsh. Chem.* 101, 579-609.

Burrell, J. W. K., Garwood, R. F., Jackman, L. M., Oskay, E. and Weedon, B. C. L. (1966). *J. chem. Soc. C* 2144–2154.

Cholnoky, L. and Szabolcs, J. (1957). *Naturwissenschaften* **44**, 513.

Cholnoky, L., Szabolcs, J., Cooper, R. D. G. and Weedon, B. C. L. (1963). *Tetrahedron Lett.* 1257–1259.

Cholnoky, L., Szabolcs, J. and Tóth, Gy. (1967). *Justus Liebig's Annln Chem.* **708**, 218–223.

Cholnoky, K., Györgyfy, K. D., Rónai, A., Szabolcs, J., Tóth, Gy., Galasko, G., Mallams, A. K., Waight, E. S. and Weedon, B. C. L. (1969). *J. chem. Soc. C* 1256–1263.

Cooper, R. D. G., Jackmann, L. M. and Weedon, B. C. L. (1962). *Proc. chem. Soc.* 215.

Cooper, R. D. G., Davis, J. B. and Weedon, B. C. L. (1963). *J. chem. Soc.* 5637–5641.

Cooper, R. D. G., Davis, J. B., Leftwick, A. P., Price, C. and Weedon, B. C. L. (1975). *J. chem. Soc., Perkin I* in press.

Curl, A. L. (1960). *J. agric. Fd Chem.* **8**, 356–358.

Curl, A. L. (1963). *J. Fd. Sci.* **28**, 623–627.

Dale, J. (1954). *Acta chem. scand.* **8**, 1235–1256.

Dale, J. (1957). *Acta chem. scand.* **11**, 265–274.

Davis, J. B., Jackman, L. M., Siddons, P. T. and Weedon, B. C. L. (1961). *Proc. chem. Soc.* 261–263.

DeVille, T. E., Hursthouse, M. B., Russell. S. W. and Weedon, B. C. L. (1969a). *Chem. Commun.* 1311–1312.

DeVille, T. E., Hursthouse, M. B., Russell, S. W. and Weedon, B. C. L. (1969b). *Chem. Commun.* 754–755.

DeVille, T. E., Hora, J., Hursthouse, M. B., Toube, T. P. and Weedon, B. C. L. (1970). *Chem. Commun.* 1231–1232.

Egger, K., Dabbagh, A. G. and Nitsche, H. (1969). *Tetrahedron Lett.* 2995–2998.

Entschel, R. and Karrer, P. (1959). *Helv. chim. Acta* **42**, 466–472.

Entschel, R. and Karrer, P. (1960). *Helv. chim. Acta* **43**, 89–94.

Enzell, C. R. (1969). *Pure appl. Chem.* **20**, 497–515.

Enzell, C. R., Francis, G. W. and Liaaen-Jensen, S. (1969). *Acta chem. scand.* **23**, 727–750.

Faigle, H. and Karrer, P. (1961). *Helv. chim. Acta* **44**, 1904–1907.

Foppen, F. H. (1971). *Chromatog. Rev.* **14**, 133–298.

Francis, G. W. and Liaaen-Jensen, S. (1970). *Acta chem. scand.* **24**, 2705–2712.

Francis, G. W., Upadhyay, R. R. and Liaaen-Jensen, S. (1970). *Acta chem. scand.* **24**, 3050–3087.

Francis, G. W., Norgård, S. and Liaaen-Jensen, S. (1974). *Acta chem. scand.* **B28**, 244–248.

Geise, H. J., Buys, H. R. and Mijlhoff, F. C. (1971). *J. molec. Struct.* **9**, 447–454.

Gilardi, R. D., Karle, I. L. and Karle, J. (1972). *Acta cryst.* **B28**, 2605–2612.

Goodfellow, D., Moss, G. P. and Weedon, B. C. L. (1970). *Chem. Commun.* 1578.

Goodfellow, D., Moss, G. P., Szabolcs, J., Tóth, Gy. and Weedon, B. C. L. (1973). *Tetrahedron Lett.* 3925–3928.

Goodwin, T. W., Land, D. G. and Sissins, M. E. (1956). *Biochem. J.* **64**, 486-492.

Granger, P., Maudinas, B., Herber, R. and Villoutreix, J. (1973). *J. mag. Res.* **10**, 43-50.

Grob, E. C. and Pflugshaupt, R. P. (1962). *Helv. chim. Acta* **45**, 1592-1598.

Hamanaka, T., Mitsui, T., Ashida, T. and Kakudo, M. (1972). *Acta cryst.* **B28**, 214-222.

Heilbron, I. M., and Weedon, B. C. L. (1958). *Bull. soc. chim. France* 83-97.

Hertzberg, S., Liaaen-Jensen, S., Enzell, C. R. and Francis, G. W. (1969). *Acta chem. scand.* **23**, 3290-3312.

Herzig, J. and Faltis, F. (1923). *Justus Liebig's Annln Chem.* **431**, 40-70.

Hirayama, K. (1955). *J. Am. chem. Soc.* **77**, 373-381, 382-383, 383-384.

Hjortås, J. (1972). *Acta cryst.* **B28**, 2252-2259.

Hollenstein, R. and Philipsborn, W. v. (1972). *Helv. chim. Acta* **55**, 2030-2044.

Holzel, R., Leftwick, A. P. and Weedon, B. C. L. (1969). *Chem. Commun.* 128-129.

Honig, B., Hudson, B., Sykes, B. D. and Karplus, M. (1971). *Proc. natn. Acad. Sci. U.S.A.* **68**, 1289-1293.

Hora, J., Toube, T. P. and Weedon, B. C. L. (1970). *J. chem. Soc. C* 241-242.

Hubbard, R. (1956). *J. Am. chem. Soc.* **78**, 4662-4667.

Inhoffen, H. H., Bohlmann, F. and Rummert, G. (1950a). *Justus Liebig's Annln Chem.* **569**, 226-237.

Inhoffen, H. H., Pommer, H. and Bohlmann, F. (1950b). *Justus Liebig's Annln Chem.* **569** 237-246.

Inhoffen, H. H., Pommer, H. and Westphal, F. (1950c). *Justus Liebig's Annln Chem.* **570**, 69-72.

Isler, O., Lindlar, H., Montavon, M., Rüegg, R. and Zeller, P. (1956). *Helv. chim. Acta* **39**, 249-259.

Isler, O., Gutmann, H., Montavon, M., Rüegg, R., Ryser, G. and Zeller, P. (1957a). *Helv. chim. Acta* **40**, 1242-1249.

Isler, O., Chopard-dit-Jean, L. H., Montavon, M., Rüegg, R. and Zeller, P. (1957b). *Helv. chim. Acta* **40**, 1256-1262.

Isler, O., Guex, W., Rüegg, R., Ryser, G., Saucy, G., Schwieter, U., Walter, M. and Winterstein, A. (1959). *Helv. chim. Acta* **42**, 864-871.

Jautelat, M., Grutzner, J. B. and Roberts, J. D. (1970). *Proc. natn. Acad. Sci. U.S.A.* **65**, 288-292.

Jensen, A. (1963). "Carotine und Carotinoide", paper 5. Dietrich Steinkopff Verlag, Darmstadt.

Jensen, S. L., see Liaaen-Jensen, S.

Kargl, T. E. and Quackenbush, F. W. (1960). *Archs Biochem. Biophys.* **88**, 59-63.

Karrer, P. (1948). *Fortschr. Chem. org. Naturstoffe* **5**, 1-19.

Karrer, P. and Eugster, C. H. (1950). *Helv. chim. Acta* **33**, 1172-1174.

Karrer, P. and Jaffé, W. (1939). *Helv. chim. Acta* **22**, 69-71.

Karrer, P. and Jucker, E. (1947). *Helv. chim. Acta* **30**, 266-267.

Karrer, P. and Jucker, E. (1948). "Carotinoide" Birkhäuser, Basle, English translation by Braude, E. A. (1950), Elsevier, New York.

Karrer, P. and Koenig, H. (1940). *Helv. chim. Acta* **23**, 460-463.

Karrer, P. and Leumann, E. (1951). *Helv. chim. Acta* **34**, 445-453.

Karrer, P. and Schwyzer, A. (1948). *Helv. chim. Acta* 31, 1055-1062.
Karrer, P. and Solmssen, U. (1937). *Helv. chim. Acta* 20, 1396-1407.
Karrer, P., Helfenstein, A., Widmer, R. and van Itallie, Th.B. (1929). *Helv. chim. Acta* 12, 741-756.
Khare, A., Moss, G. P. and Weedon, B. C. L. (1973a). *Tetrahedron Lett.* 3921-3924.
Khare, A., Moss, G. P., Weedon, B. C. L. and Matthews, A. D. (1973b). *Comp. Biochem. Physiol.* **45B**, 971-973.
Khatoon, N., Loeber, D. E., Toube, T. P. and Weedon, B. C. L. (1972). *Chem. Commun.* 996-997.
Kjøsen, H. and Liaaen-Jensen, S. (1969). *Phytochemistry* **8**, 483-491.
Kjøsen, H. and Liaaen-Jensen, S. (1972). *Acta chem. scand.* 26, 2185-2193.
Kjøsen, H. and Liaaen-Jensen, S. (1973). *Acta chem. scand.* 27, 2495-2502.
Kjøsen, H., Arpin, N. and Liaaen-Jensen, S. (1972). *Acta chem. scand.* 26, 3053-3067.
Klyne, W., Veibel, S., Bodea, C., Goodwin, T. W., Haxo, F., Isler, O., Liaaen-Jensen, S., Powell, W., Quackenbush, W., Sørensen, N. A. and Weedon, B. C. L. (1972). *Inf. Bull. Int. Un. pure appl. Chem.* no. 19. See also: (1971). *Biochemistry, N.Y.* **10**, 4827-4837; (1972). *Biochim. biophys. Acta* 286, 217-242; (1972). *Biochem. J.* 127, 741-751; (1972). *J. biol. Chem.* 247, 2633-2643; (1971). *In* "Carotenoids" (O. Isler, ed.), pp. 851-864, Birkhäuser, Basle; *Pure appl. Chem.* in press.
Koch, B. (1972). *Acta cryst.* **B28**, 1151-1159.
Krinsky, N. I. (1963). *Analyt. Biochem.* 6, 293-302.
Kuhn, R. and Brockmann, H. (1935). *Justus Liebig's Annln Chem.* 516, 95-143.
Kuhn, R. and Grundmann, C. (1932). *Ber. deut. chem. Ges.* 65, 898-902, 1880-1889.
Kuhn, R., and Sörensen, N. A. (1938). *Ber. dt. chem. Ges.* 71, 1879-1888.
Kuhn, R., and Winterstein, A. (1933). *Ber. dt. chem. Ges.* 66, 209-214.
Langlet, J., Pullman, B. and Bothod, H. (1970). *J. molec. Struct.* 6, 139-144.
Liaaen-Jensen, S. (1959). *Acta chem. scand.* 13, 381-383.
Liaaen-Jensen, S. (1962). *K. norske Vidensk. Selsk. Forh.* no. 8.
Liaaen-Jensen, S. (1967). *Acta chem. scand.* 21, 961-969.
Liaaen-Jensen, S. (1969). *Pure appl. Chem.* 20, 421-448.
Liaaen-Jensen, S. (1971). *In* "Carotenoids" (O. Isler, ed.), pp. 61-188. Birkhäuser, Basle.
Liaaen-Jensen, S. (1973). *Pure appl. Chem.* 35, 81-112.
Liaaen-Jensen, S. and Hertzberg, S. (1966). *Acta chem. scand.* 20, 1703-1709.
Liaaen-Jensen, S. and Jensen, A. (1965). *Prog. Chem. Fats* 8, 133-205.
Lunde, K. and Zechmeister, L. (1955). *J. Am. chem. Soc.* 77, 1647-1653.
McCormick, A. and Liaaen-Jensen, S. (1966). *Acta chem. scand.* 20, 1989-1991.
Malhotra, H. C., Britton, G. and Goodwin, T. W. (1970). *Chem. Commun.* 127-128.
Manchand, P. S., Rüegg, R., Schwieter, U., Siddons, P. T. and Weedon, B. C. L. (1965). *J. chem. Soc.* 2019-2026.
Mayer, H. and Isler, O. (1971). *In* "Carotenoids" (O. Isler, ed.), pp. 325-575. Birkhäuser, Basle.

Milas, N. A., Davis, P., Belič, I. and Fleš, D. A. (1950). *J. Am. chem. Soc.* 72, 4844.

Mildner, P. and Weedon, B. C. L. (1953). *J. chem. Soc.* 3294–3298.

Moss, G. P., Szabolcs, J., Tóth, Gy. and Weedon, B. C. L. (1975). *Acta chim. Acad. Sci. hung.* in press.

Nicoară, E., Tămaş, V., Neamţu, G. and Bodea, C. (1966). *Justus Liebig's Annln Chem.* 697, 201–203.

Nitsche, H., Egger, K. and Dabbagh, A. G. (1969). *Tetrahedron Lett.* 2999–3002. 2999–3002.

Oberhänsli, W. E., Wagner, H. P. and Isler, O. (1947). *Acta cryst.* B30, 161–166.

Patel, D. J. (1969). *Nature, Lond.* 221, 825–828.

Pattenden, G. and Weedon, B. C. L. (1968). *J. chem. Soc. C* 1984–1997.

Pattenden, G., Way, J. E. and Weedon, B. C. L. (1970). *J. chem. Soc. C* 235–241.

Pauling, L. (1939). *Fortschr. Chem. org. Naturstoffe* 3, 203–235.

Petracek, F. J. and Zechmeister, L. (1956). *J. Am. chem. Soc.* 78, 1427–1434.

Petzold, E. N., Quackenbush, F. W. and McQuistan, M. (1959). *Archs Biochem. Biophys.* 82, 117–124.

Planta, C. v., Schwieter, U., Chopard-dit-Jean, L., Rüegg, R., Kofler, M. and Isler, O. (1962). *Helv. chim. Acta* 45, 548–561.

Rabourn, W. J., Quackenbush, F. W. and Porter, J. W. (1954). *Archs Biochem. Biophys.* 48, 267–274.

Ramamurthy, V. and Liu, R. S. H. (1975). *Tetrahedron* 31, 201–206.

Ricketts, T. R. (1971). *Phytochemistry* 10, 155–160.

Rowan III, R., Warshel, A., Sykes, B. D. and Karplus, M. (1974). *Biochemistry, N.Y.* 13, 970–981.

Rüegg, R., Montavon, M., Ryser, G., Saucy, G., Schwieter, U. and Isler, O. (1959). *Helv. chim. Acta* 42, 854–864.

Schaffer, A. M., Waddell, W. H. and Becker, R. S. (1974). *J. Am. chem. Soc.* 96, 2063–2068.

Schwieter, U., Englert, G., Rigassi, N. and Vetter, W. (1969). *Pure appl. Chem.* 20, 365–420.

Schwieter, U., Arnold, W., Oberhänsli, W. E., Rigassi, N. and Vetter, W. (1971). *Helv. chim. Acta* 54, 2447–2459.

Schantz, E. M., Cawley, J. D. and Embree, N. D. (1943). *J. Am. chem. Soc.* 65, 901–906.

Stahl, E., Bolliger, H. R. and Lehnert, L. (1963). "Carotine und Carotinoide", paper 6. Dietrich Steinkopff Verlag, Darmstadt.

Stam, C. H. (1972). *Acta cryst.* B28, 2936–2945.

Steinmetz, W. E. (1974). *J. Am. chem. Soc.* 96, 685–692.

Sterling, C. (1964). *Acta cryst.* 17, 1224–1228.

Strain, H. H., Svec, W. A., Aitzetmüller, K., Grandolfo, M. C., Katz, J. J., Kjøsen, H., Norgård, S., Liaaen-Jensen, S., Haxo, F. T., Wegfahrt, P. and Rapaport, H. (1971). *J. Am. chem. Soc.* 93, 1823–1825.

Straub, O. (1971). *In* "Carotenoids" (O. Isler, ed.), pp. 771–850. Birkhäuser, Basle.

Sundaralingam, M. and Beddel, C. (1972). *Proc. natn. Acad. Sci. U.S.A.* 69, 1569–1573.

Surmatis, J. D. and Ofner, A. (1963). *J. org. Chem.* 28, 2735–2739.

Vetter, W., Englert, G., Rigassi, N. and Schwieter, U. (1971). *In* "Carotenoids" (O. Isler, ed.), pp. 189–266. Birkhäuser, Basle.
Weedon, B. C. L. (1969). *Fortschr. chem. org. Naturstoffe* **27**, 81–130.
Weedon, B. C. L. (1970). *Rev. pure appl. Chem., Aust.* **20**, 51–66.
Weedon, B. C. L. (1971). *In* "Carotenoids" (O. Isler, ed.), pp. 29–59. Birkhäuser, Basle.
Weedon, B. C. L. (1973). *Pure appl. Chem.* **35**, 113–130.
Winterstein, A., Studer, A. and Rüegg, R. (1960). *Chem. Ber.* **93**, 2951–2965.
Yamaguchi, M. (1957). *Bull. chem. Soc. Japan* **30**, 111–114; 979–983.
Yamaguchi, M. (1958). *Bull. chem. Soc. Japan* **31**, 51–55, 739–742.
Yokoyama, H. and White, M. J. (1965). *J. org. Chem.* **30**, 2481–2482.
Zechmeister, L. (1958). *Fortschr. chem. org. Naturstoffe* **15**, 31–82.
Zechmeister, L. (1962). "*Cis–Trans* Isomeric Carotenoids, Vitamin A, and Arylpolyenes". Springer-Verlag, Vienna.
Zechmeister, L., Le Rosen, A. L., Schroeder, W. A. Polgár, A. and Pauling, L. (1943). *J. Am. chem. Soc.* **65**, 1940–1951.

Chapter 4
Distribution of Carotenoids

T. W. GOODWIN

Department of Biochemistry, The University of Liverpool, England

I Introduction

As far as is known carotenoids are the only naturally occurring tetraterpenes and are widely distributed throughout the living world. However, they appear to be synthesized *de novo* only by higher plants, spore-bearing vascular plants, algae and photosynthetic bacteria. The carotenoids which have been isolated exclusively from animal tissues originated from plant or bacterial sources and have been altered, mainly by oxidation, by animal enzymes.

The general distribution of carotenoids in plants has been discussed in detail in a recent review (Goodwin, 1973a) and more restricted reviews have concentrated on algae (Goodwin, 1971), fungi (Goodwin, 1972) and photosynthetic bacteria (Andrewes and Liaaen-Jensen, 1972). In this chapter an attempt will be made to give an overall picture of carotenoid distribution and the reader is referred to these specialized reviews for further detailed information. For the earlier literature Goodwin (1952) should be consulted. Photosynthetic bacteria are outside the scope of this volume and will not be considered further.

II Photosynthetic tissues

A HIGHER PLANTS

1 *Qualitative distribution*

All green tissues of higher plants generally contain the same major carotenoids which are located, probably exclusively, in the chloroplasts. These pigments are β-carotene (I), lutein (II), violaxanthin (III) and neoxanthin (IV). Pigments which appear frequently but not constantly in smaller amounts are α-carotene (V), β-cryptoxanthin (VI), zeaxanthin (VII) and antheraxanthin (VIII). The partly saturated biosynthetic precursors phytoene (IX) and phytofluene (X) occur only in trace amounts, for example in *Vicia sativa* their concentration is about one two-hundredth that of β-carotene (Mercer *et al*., 1963). The xanthophylls normally occur unesterified in the chloroplast but during senescence when chloroplasts disintegrate the xanthophylls released into the cytoplasm are esterified (Goodwin, 1958) (see also Chapter 17).

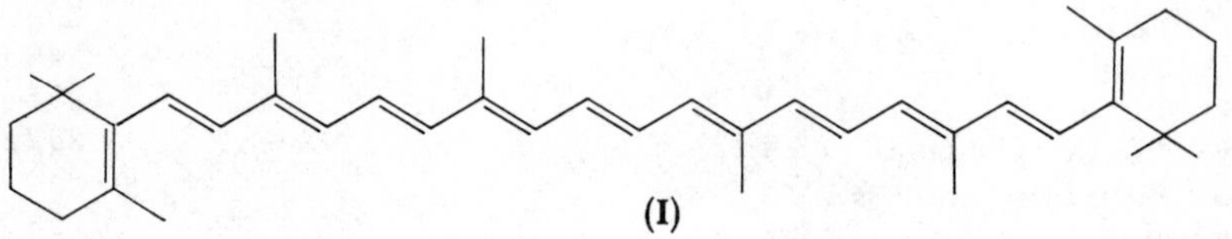

(I)

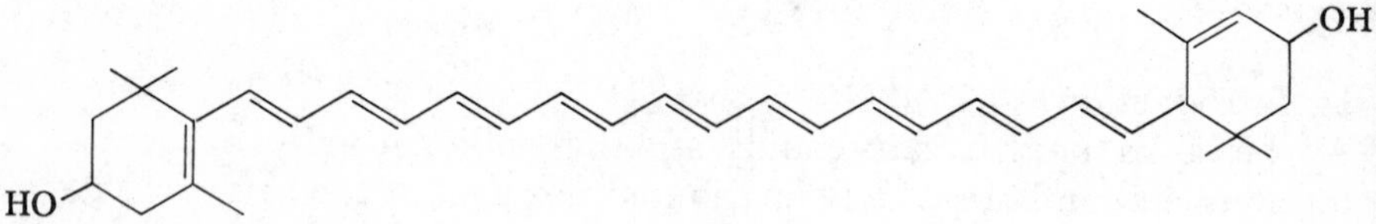

(II)

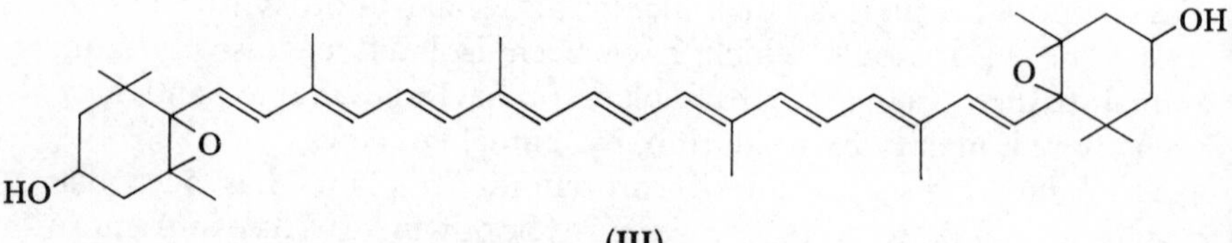

(III)

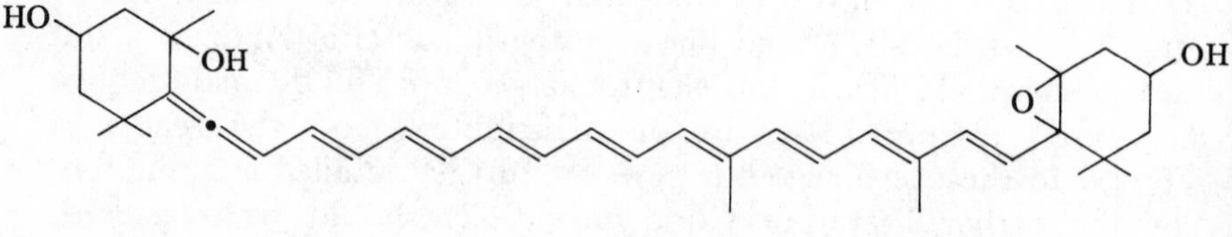

(IV)

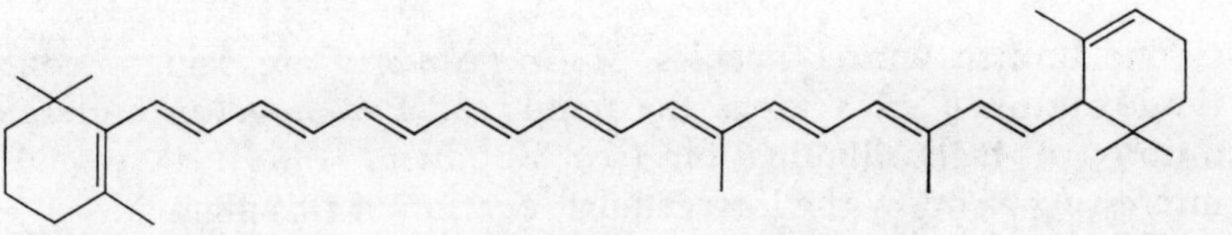

HO

(V)

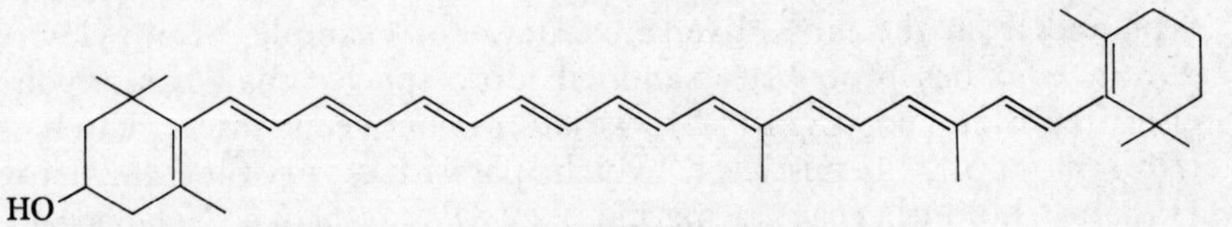

HO

(VI)

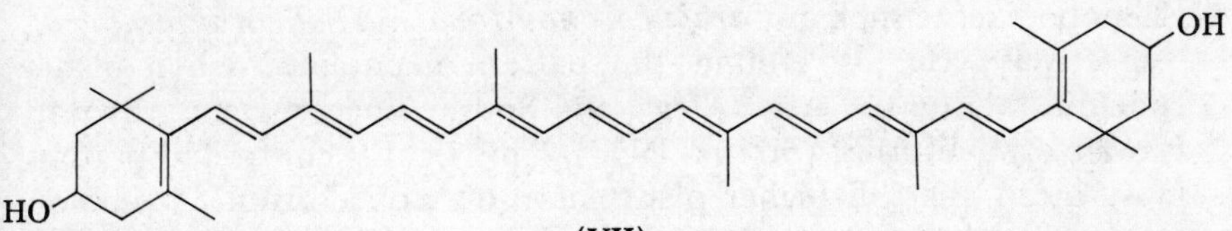

HO

OH

(VII)

HO

O

OH

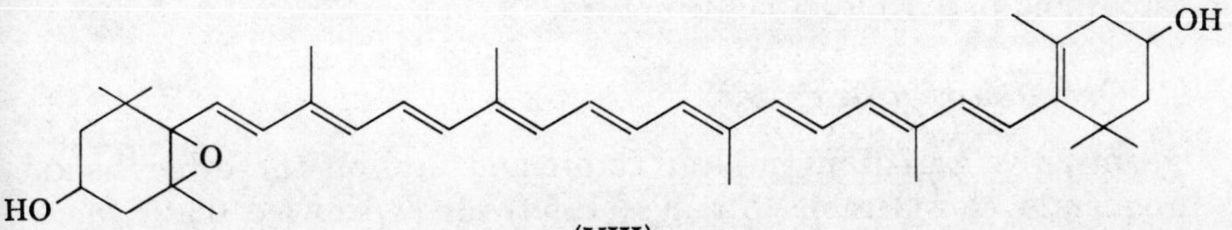

(VIII)

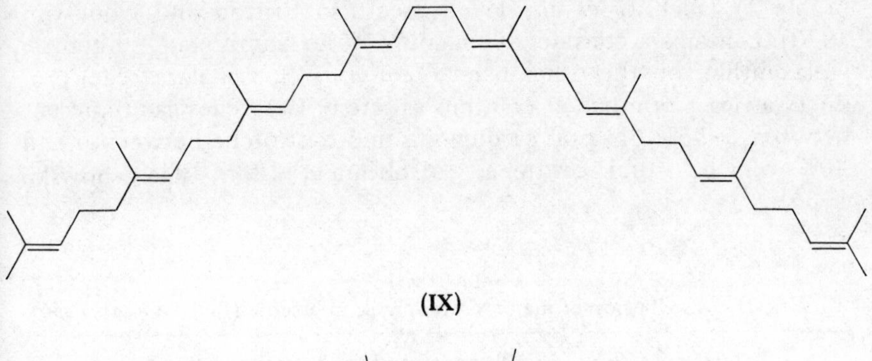

(IX)

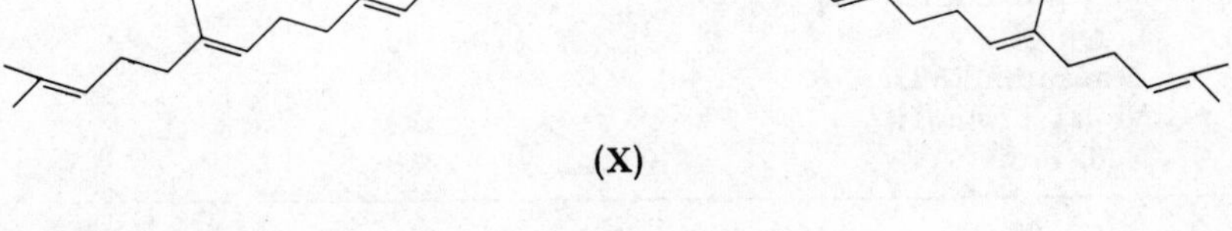

(X)

The bronze winter needles of *Cryptomeria* are said to contain rhodoxanthin (**XI**) as does the pondweed *Potamogeton* under conditions of high illumination (see Goodwin, 1959). It would be interesting to know the intracellular location of this pigment.

The consistency of the qualitative distribution in the leaves of higher plants is remarkable and applies to plants of the most diverse type and from the most diverse habitat. For example, Strain (1958, 1966), who has probably examined more species than has anyone else, reported no significant variations between insect catchers (*Drosera* spp.), a mistletoe which parasitizes another mistletoe (*Viscum*: Loranthaceae), a marine plant (*Phyllaspadix*: Nejadaceae), a salt marsh mangrove, and members of the Compositae and the Chenopodiaceae such as *Cotula coronopifolia* and *Salicornia ambigua*, respectively. He also found the pattern maintained when plants endemic to Europe, Asia, Africa and South America were grown in Hawaii or California (Strain, 1958). This tightly controlled pattern must mean that all higher plants have developed from a common ancestor and that any mutation which significantly changes the carotenoid pattern is lethal (Goodwin, 1973a).

2 *Quantitative distribution*

Quantitative variations in leaf carotenoid are, on the other hand, frequently considerable. Strain's (1958) survey showed that *Syngonium* (Araceae), compared with a typical forage plant (lucerne) (Table I) (Bickoff *et al.*, 1954) (see also Britton and Goodwin, 1973), contains a considerable amount of neoxanthin and only little violaxanthin; on the other hand *Fremontia* (Sterculalareaceae) produces at least as much zeaxanthin as lutein. Carotenes can represent between 5–35% of total carotenoids and α-carotene between 0 and 40% of the total carotenes (MacKinney, 1935; Wierzchowski, 1965).

TABLE I
Quantitative distribution of major xanthophylls in lucerne (Bickoff *et al.* 1954)

Pigment	Percentage of total xanthophylls present
β-Cryptoxanthin (**VI**)	4
Lutein (**II**)	40
Zeaxanthin (**VII**)	2
Violaxanthin (**III**)	34
Neoxanthin (**IV**)	19

Seedlings germinated in the dark do not produce chloroplasts and thus, as would be expected, contain only small amounts of carotenoids, mainly xanthophylls (Goodwin and Phagpolngarm, 1960; Valadon and Mummery, 1969). These pigments are found either in the etioplasts which contain small yellowish granules or outside the plastids (Granick, 1961). Illumination of etiolated seedlings rapidly results in the development of chloroplasts and the simultaneous synthesis of the characteristic green leaf pigments (Goodwin, 1958; Wieçkowski, 1961; Wolf, 1963; Valadon and Mummery, 1969).

3 *Location*

The pigments are probably located in the chloroplast grana in the form of chromoproteins. Two chromoproteins, complex I and complex II which were isolated by Bailey *et al.* (1966), now appear to correspond to photosytems I and II in photosynthesis (Tevini and Lichtenthaler, 1970) (see Chapter 13). There is a differential distribution of carotenoids in these complexes; I contains mainly β-carotene with a trace of lutein, whilst II contains more xanthophylls than β-carotene. In another preparation of system I the β-carotene to xanthophyll ratio was about 1·4 : 1 (Suzuki *et al.*, 1970). This differential distribution may be connected with the observation that in spinach stroma lamellae and grana stacks contain the same amount of β-carotene but the latter contain considerably more xanthophylls (Trosper and Allen, 1973).

4 *Mutants*

Mutants in coloured carotenoids have been described but such mutation is lethal and the plants can only be grown heterotrophically in the dark (see Chapter 14). Well-characterized mutants which have been obtained from maize include W-3, which accumulates phytoene (**IX**) (Anderson and Robertson, 1960), and a group which accumulates lycopene (**XII**) and ζ-carotene (**XIII**) (Faludi-Daniel, 1965).

B BRYOPHYTES AND PTERIDOPHYTES

The limited information available suggests that the carotenoids present are the same as those in green plants (e.g. in the moss *Fontinalis antipyretica*, Bendz *et al.*, 1968) although two, *Equisetum* and *Selaginella*, are reported to contain rhodoxanthin (**XI**) (see Goodwin, 1955). The gametes and young sporophytes of *Lophocolea heterophylla* contain the usual chloroplast carotenoids (Taylor *et al.*, 1972).

C LICHENS

Lichens would be expected to contain carotenoids characteristic of their algal symbionts together with carotenoids made by their fungal symbionts if the fungi happened to be carotenogenic. The first lichen to be examined in detail, *Ramelia reticulata*, has a carotenoid composition similar to that of green leaves (Strain, 1951). β-Carotene is present in *Roccella fuciformis* (Huneck *et al.*, 1967). *Trebouxia decolorans* the phycobiont of the lichen *Xanthoria parietina* also contains carotenoids characteristic of green leaves (de Nicola and Tomaselli, 1961; de Nicola and di Benedetto, 1962). This is not surprising as it is a green alga (see next section).

D ALGAE

1 General

Carotenoids are normally found only in the chloroplasts of algae but under certain adverse cultural conditions they can appear in the cytoplasm (see p. 232). They can also accumulate in the reproductive areas of colonial algae and in the eye spots of certain *Euglena* species (Goodwin, 1971).

In marked contrast to the higher plants the chloroplast carotenoids of the various classes of algae show considerable qualitative differences and these have been used in attempts to help solve the problems of algal taxonomy and evolution.

2 Chlorophyta

The general distribution of carotenoids in both unicellular and colonial green algae except the Siphonales and Prasinophyceae is similar to that observed in green leaves, viz. β-carotene (**I**), lutein (**II**), violaxanthin (**III**) and neoxanthin (**IV**) (Strain, 1958, 1966; Goodwin, 1971a), although zeaxanthin (**VII**) would also appear to be a constant component (Hager and Stransky, 1970). The pigments originally described as zeaxanthin and lutein in *Trentepohlia iolithus* have been characterized as 2-hydroxy-β-carotene and 2-hydroxy-α-carotene respectively; they are accompanied by 2,2′-dihydroxy-β-carotene (Kjøsen *et al.*, 1972). This is the first report of naturally occurring 2-hydroxy carotenoids. As in higher plants α-carotene (**V**) is sporadically encountered in small amounts but in *Chlamydomonas algae formis* Strain (1958) found much α-carotene.

Certain Chlorophyta (Table II) contain small amounts of loroxanthin (**XIV**), a carotenoid not yet observed in higher plants; it is characterized by the fact that the in-chain methyl group at C-19 has been oxidized to a hydroxymethyl group (Aitzetmüller *et al.*, 1969).

Pyrenoxanthin, isomeric with loroxanthin (C_{20} hydroxyl instead of C_{19} hydroxyl), is present in *Chlorella pyrenoidosa* (Yokoyama *et al.*, 1972); the possibility exists that it is identical with loroxanthin.

Siphonaxanthin (XV) (Strain *et al.*, 1970; Walton *et al.*, 1970; Rickets, 1971a) is known to exist as a characteristic pigment of the members of Bryopsidophyceae, particularly in the Siphonales, Derbesidales Codiales and Cauleropales (Strain, 1966; Kleinig, 1969). It also

TABLE II
Green algae known to synthesize loroxanthin
(Aitzetmüller *et al.*, 1969)

Chlorella vulgaris	*Scenedesmus obliquus*
Cladophora trichotomata	*Ulva rigida*
C. ovoidea	

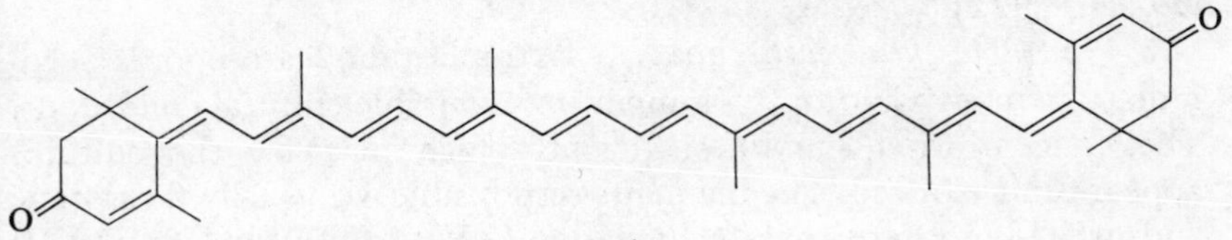

(XI)

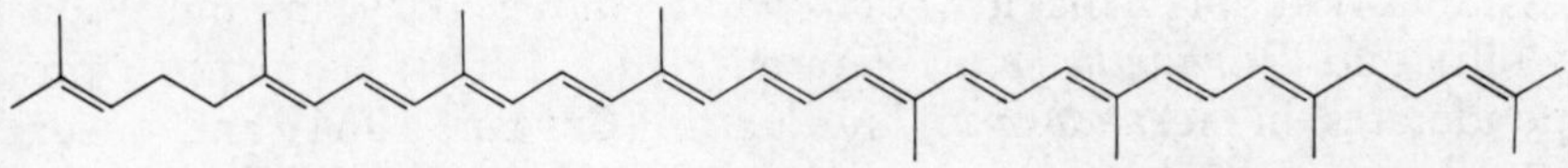

(XII)

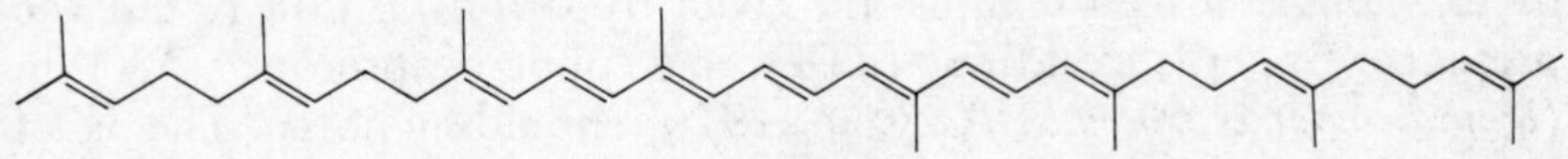

(XIII)

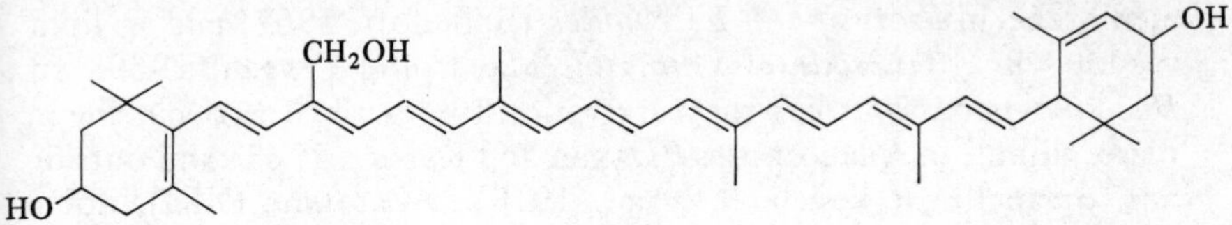

(XIV)

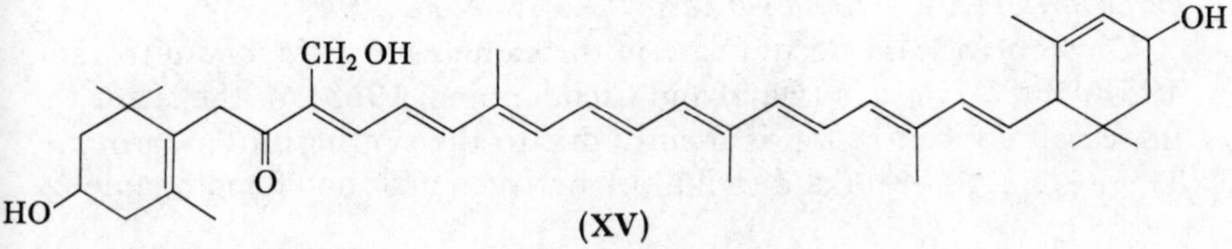

(XV)

frequently exists in the esterified form siphonein) and this represents a rare occurrence in the chloroplast pigments. The usual Chlorophyta carotenoids exist alongside siphonaxanthin in the Bryopsidophyceae but another characteristic is that with one exception (*Dichotomosiphon tuberosus*) α-carotene preponderates over β-carotene (Strain, 1965). However, siphonaxanthin and siphonein are now known to be widely but sporadically distributed amongst the Cladophorales and are constant constituents in all Derbesidales, Codiales and Cauleropales examined by Kleinig (1969). He also found that some but not all Siphonocladales contained siphonaxanthin but not siphonein, whilst siphonein but not siphonaxanthin was present in some Dichotomosiphonales. Both pigments are also present in a number of Prasinophyceae (Ricketts, 1970, 1971b) and in one member of the order Ulotrichales, *Microthamnion knetzingianum* (Weber and Czygan, 1972).

a. Extraplastidic carotenoids. Extraplastidic carotenoids accumulate in many Chlorophyta under unfavourable cultural conditions. The concentration can be sufficiently high to make the cultures appear orange or red. The pigments responsible are usually β-carotene and/or its keto derivatives echinenone (**XVI**), canthaxanthin (**XVII**) and astaxanthin (**XVIII**). For example β-carotene is the major extraplastidic pigment in *Trentepohlia aurea* (Czygan and Kalb, 1966) and *Dunaliella salina* (Aasen *et al.*, 1969); echinenone preponderates in *Scenedesmus brasiliensis* (Czygan, 1964) and occurs together with canthaxanthin and astaxanthin in *Protosiphon botryoides* (Kleinig and Czygan, 1969). Further details of the distribution of extraplastidic carotenoids are given by Goodwin (1971) but the presence of crustaxanthin (**XIX**) and phoenicopterone (**XX**) in *Haematococcus pluvialis* (Czygan, 1970) should be noted. The exact location of extraplastidic carotenoids appears to vary with species; they occur in intracytoplasmic deposits which have no limiting membrane in *Protosiphon botryoides* (Berkaloff, 1967) and in lipid vacuoles in *Ankistrodesmus braunii* (Mayer and Czygan, 1969). In *Haematococcus pluvialis* they are variously described as occurring in plastoglobuli in aplanospores (Czygan and Kessler, 1967) and outside any organelle or vesicle (Czygan, 1970). γ-Carotene (**XXI**), not a normal chloroplast carotenoid, accumulates in the antherida of *Chara ceratophylla* and *Nitella syncarpa* (Karrer *et al.*, 1943).

Carotenoids also accumulate in the gametes of *Ulva lobata* (Strain, 1951) and *U. lactuca* (Haxo and Clendenning, 1953); in the latter the increased concentration is mainly due to the synthesis of γ-carotene. There is no differential distribution between male and female gametes.

b. Mutants. Mutants in which carotenoid synthesis has been interrupted have been obtained from *Chlorella vulgaris* (Claes, 1954, 1956) and *C. pyrenoidosa* (Kessler and Czygan, 1966).

3 *Rhodophyta*

The general pattern of carotenoid distribution in Rhodophyta is usually very simple: α- and β-carotenes, lutein and zeaxanthin (Strain, 1958, 1966; Allen *et al.*, 1964). In some algae it is made even simpler by the occasional absence of lutein, e.g. from *Porphyridium aerugineum* (Chapman, 1966) and *P. cruentum* (Stransky and Hager, 1970), the frequent absence of α-carotene and, in one unexpected case, *Phycodrys sinuosa*, the absence of β-carotene (Larsen and Haug, 1956). Monohydroxy carotenes are occasionally found; β-cryptoxanthin (VI) was reported in two algae, *Acantophora spicifera* and *Gracilaria lichenoides*, collected in Hawaii (Aihara and Yamamoto, 1968) and α-cryptoxanthin (XXI) occurs in *Lenormandia prolifera* (Saenger and Rowan, 1968).

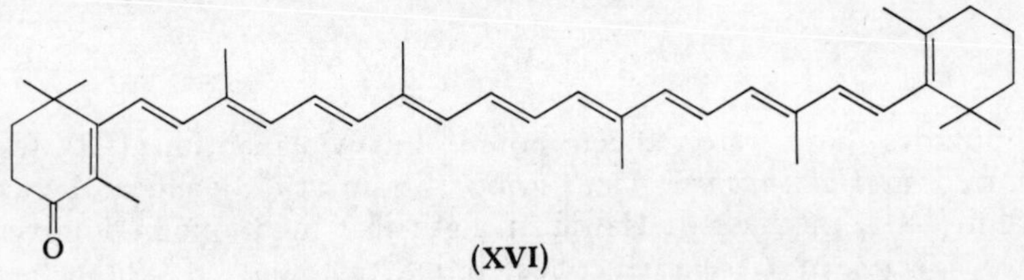

O

(XVI)

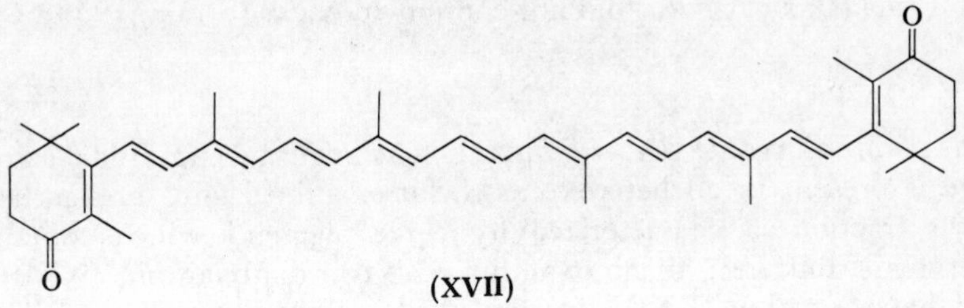

O

O

(XVII)

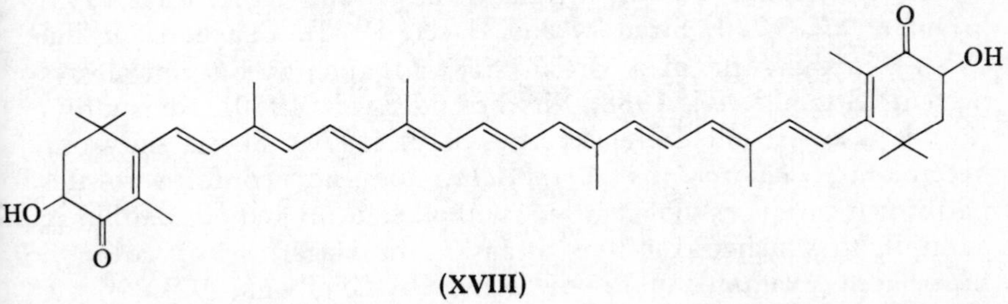

O

OH

HO

O

(XVIII)

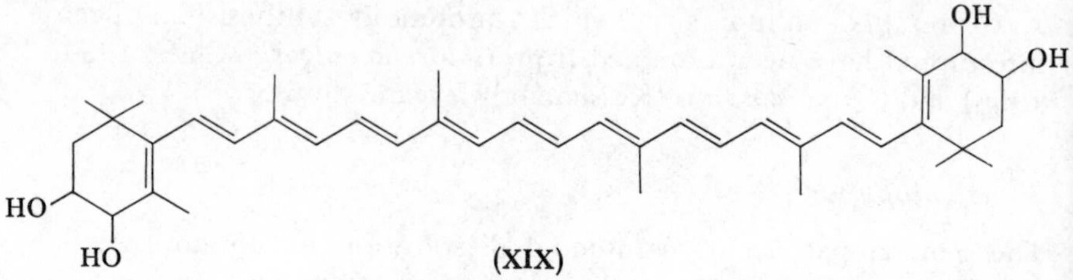

OH OH

HO HO

(XIX)

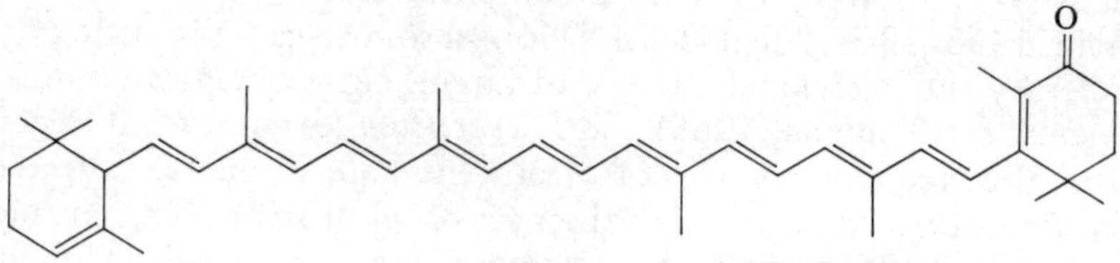

O

(XX)

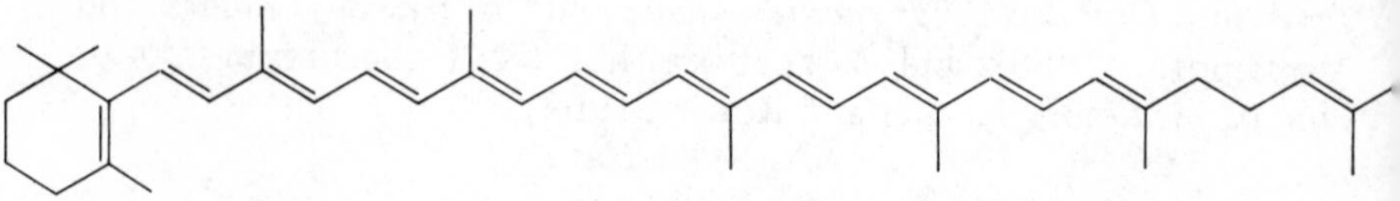

(XXI)

Epoxides have rarely been noted, but violaxanthin (**III**) was reported in *Halosaccion glandiforma* (Strain, 1958) and antheraxanthin (**VIII**) in the two Hawaiian algae just mentioned. However, other species of *Gracilaria* collected in Australia (*G. edulis*) and California (*G. sjoestedtii*) did not contain epoxides (Strain, 1958).

4 *Chrysophyta*

a. Xanthophyceae (*Heterokontae*). As might be expected β-carotene is present in all heterokonts so far examined, but the xanthophyll fraction is characterized by three pigments which contain acetylenic linkages: diadinoxanthin (**XXII**), diatoxanthin (**XXIII**) and heteroxanthin (**XXIV**) (Strain *et al.*, 1968; Egger *et al.*, 1969; Strain *et al.*, 1970; Stransky and Hager, 1970). Vaucheriaxanthin (**XXV**), in the form of a partial ester, is frequently but not always present (Strain *et al.*, 1968; Nitsche and Egger, 1970). Neoxanthin (**IV**) is also present (Stransky and Hager, 1970). So far only one heterokont, *Pleurochloris commutata*, does not conform to this pattern; it produces violaxanthin, antheraxanthin and zeaxanthin in addition to vaucheriaxanthin (Stransky and Hager, 1970). A list of heterokonts examined up to now is given by Goodwin (1971).

b. Chrysophyceae. β-Carotene is either the only carotene present (Allen *et al.*, 1960) or the major component of a hydrocarbon fraction which included traces of α- and γ-carotenes (Dales, 1960). The principal xanthophyll is fucoxanthin (**XXVI**) (Allen *et al.*, 1960; Dales 1960; Jeffrey, 1961; Parsons, 1961; Bunt, 1964; Jeffrey and Allen, 1964; Riley and Wilson, 1965), which is usually accompanied by diatoxanthin, diadinoxanthin and dinoxanthin (neoxanthin 3-acetate) (Jeffrey, 1961; Hager and Stransky, 1970). It is interesting that in *Hymenomonas huxleyii* the pigments appear to be localized in a finely coiled lamellar system which is distinct from the chloroplast (Olson *et al.*, 1967).

c. Haptophyceae. Only three members have been examined. *Pavlova* sp. contains β-carotene (**V**), diadinoxanthin (**XXII**) whilst a *Hymenomonas* sp. and *Coccolithus huxleyi* accumulate β-carotene, diadinoxanthin and a pigment which is probably not fucoxanthin but a closely related pigment (Jeffrey and Allen, 1964; Norgård *et al.*, 1974).

d. Eustagmaphyceae. Many members of this new class were originally grouped under Xanthophyceae; they resemble this class in not synthesizing fucoxanthin (**XXVI**). The pigments are rather varied, from β-carotene (**V**), vaucheriaxanthin ester (**XXV**), neoxanthin (**IV**), diadinoxanthin (**XXII**) and diatoxanthin (**XXIII**) in *Botyridiopsus alpina* (Thomas and Goodwin, 1965; Stransky and Hager, 1970) to the non-acetylenic pigments β-carotene, neoxanthin and violaxanthin in *Pleurochloris magna* (Whittle and Casselton, 1969).

e. Bacillariophyceae (diatoms). The distribution is in general very similar to that in the Phaeophyceae, viz. β-carotene (**I**), diatoxanthin, (**XXIII**), diadinoxanthin (**XXII**) and fucoxanthin (**XXVI**) (Strain, 1951, 1958, 1966; Jeffrey, 1961; Bunt, 1964). The rare ϵ-carotene (**XXVII**) is present in traces in *Nitzschia closterium* (Strain *et al.*, 1944) and in *Navicula pelliculosa* (Hager and Stransky, 1970).

5 *Phaeophyta*

The main pigments encountered in an extensive survey of the Phaeophyta were β-carotene (**I**), violaxanthin (**III**) and fucoxanthin (**XXVI**) (Jensen, 1966). No α-carotene (**V**) or lutein (**II**) was detected (Jensen, 1966; Strain, 1966) although traces of diatoxanthin (**XXIII**) and diadinoxanthin (**XXII**) were occasionally present (Jensen, 1966).

f. Extraplastidic pigments. β-Carotene accumulates in the male gametes of *Fucus* spp. and *Ascophyllum nodosum* whilst fucoxanthin (**XXVI**), together with chlorophylls, accumulates in the ova (Carter *et al.*, 1948).

6 *Pyrrophyta* (*Dinophyceae*)

The characteristic carotenoid of this class is peridinin (**XXVIII**) (Strain *et al.*, 1971); it is accompanied by smaller amounts of diadinoxanthin (**XXII**) and dinoxanthin as well as β-carotene (Pinckard *et al.*, 1953; Sweeney *et al.*, 1959; Jeffrey, 1961; Parsons and Strickland, 1963; Bunt, 1964; Jeffrey and Haxo, 1968). However, *Gymnodinium veneficum* (Riley and Wilson, 1965) and *Glenodinium foliaceum* (Mandelli, 1968) are said to make fucoxanthin (**XXVI**) rather than peridinin. In contrast an Australian *Gymnodinium* sp. synthesizes peridinin (Jeffrey, 1961).

7 *Cryptophyta*

The Cryptophyta are characterized by the preponderance of α-carotene over β-carotene (Allen *et al.*, 1964; Chapman and Haxo, 1963) and by the presence of the acetylenic carotenoid alloxanthin (**XXIX**) (Chapman, 1966a; Mallams *et al.*, 1967). ϵ-Carotene (**XXVII**), crocoxanthin (**XXX**) and monadoxanthin (**XXXI**) were noted in *Cryptomonas ovata* (Chapman and Haxo, 1966).

8 *Euglenophyta*

The chloroplast pigments of various *Euglena* species are β-carotene (**I**), zeaxanthin (**VII**), neoxanthin (**IV**) and, as major xanthophyll, diadinoxanthin (**XXII**) (Aitzetmüller *et al.*, 1968). Dinoxanthin (Hager and Stransky, 1970; Johannes *et al.*, 1971) and heteroxanthin (**XXIV**) (Nitsche, 1973) have also been reported as well as a pigment considered to be deepoxyneoxanthin (**XXXII**) (Nitsche *et al.*, 1969). Traces of keto carotenoids such as echinenone (**XVI**), 3-hydroxyechinenone and canthaxanthin (**XVII**) (euglenanone) are always encountered (Goodwin and Gross, 1958; Krinsky and Goldsmith, 1960); they are considered to be localized in the eye spot of these organisms. Astaxanthin (**XVIII**) is the main pigment of *E. helicorubescens* (Tischer, 1941).

9 *Cyanophyta: blue-green algae*

The characteristic carotenoids of blue-green algae are β-carotene (**I**), echinenone (**XVI**) and zeaxanthin (**VII**) myxoxanthin, aphanin

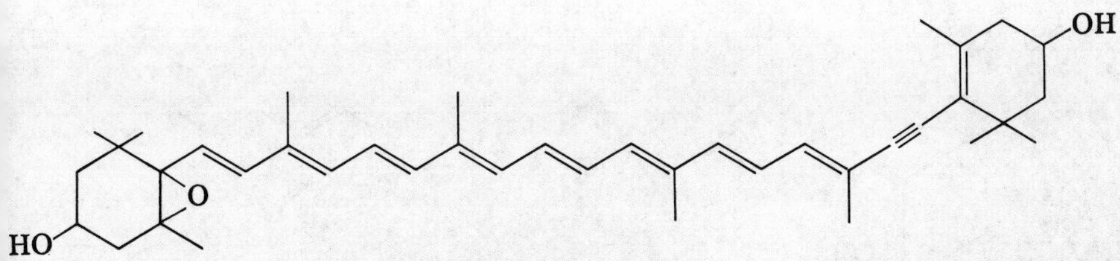

OH
O
HO

(XXII)

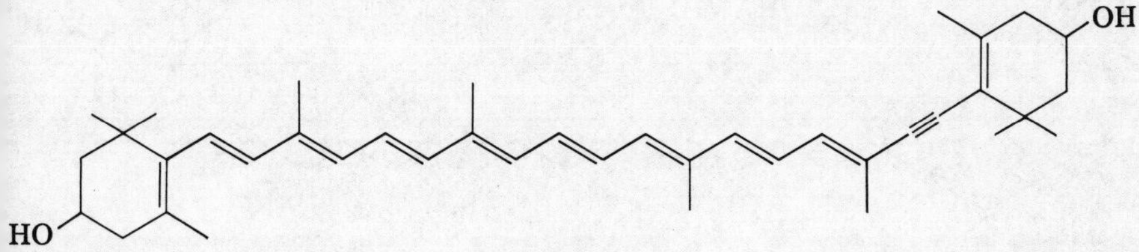

OH
HO

(XXIII)

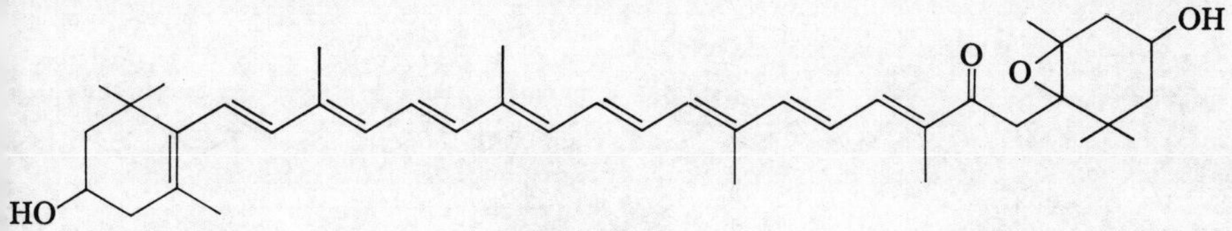

OH
O
O
HO

(XXIV)

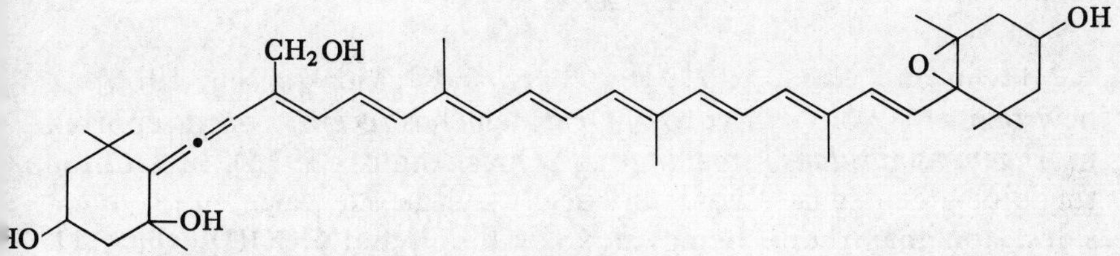

OH
CH_2OH
O
OH
HO

(XXV)

O
O
HO
$OCOCH_3$

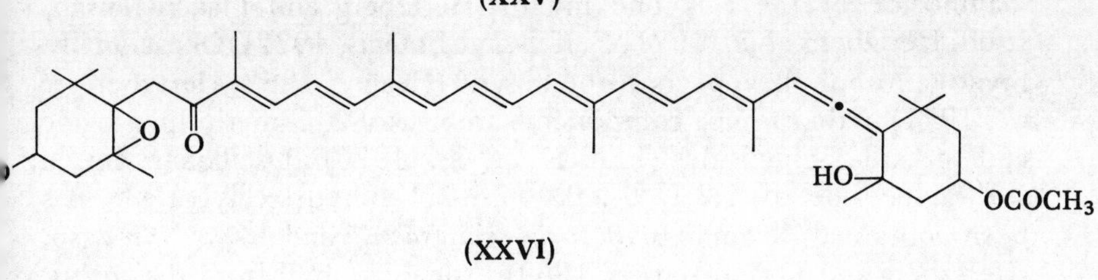

(XXVI)

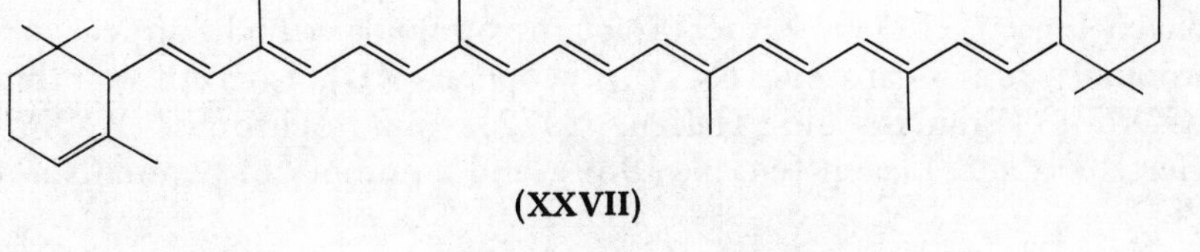

(XXVII)

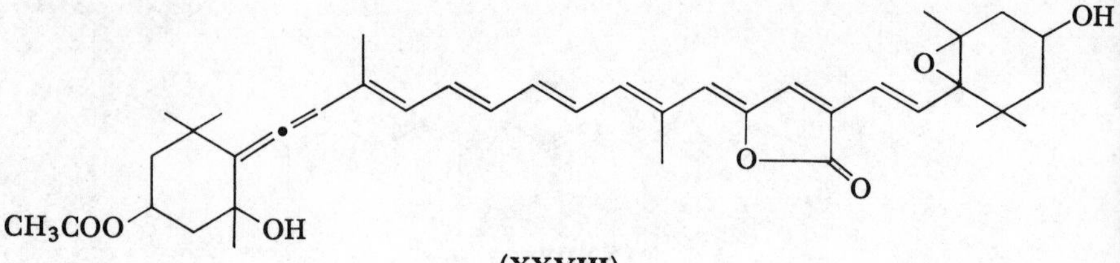

(XXVIII)

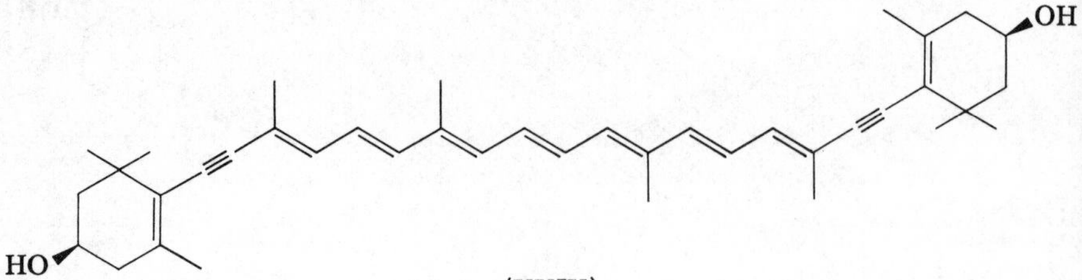

(XXIX)

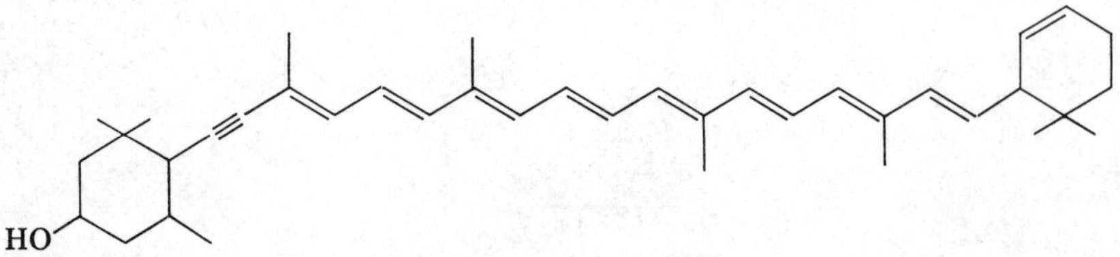

(XXX)

(Goodwin and Taha, 1951; Hertzberg and Liaaen-Jensen, 1966). A new pigment 2,3,3′-trihydroxy-β-carotene has recently been reported in *Anacystis nidulans* (Smallidge and Quackenbush, 1973). In addition the blue-green algae characteristically synthesize carotenoid glycosides, the major one being myxoxanthophyll (**XXXIII**), the 2-*O*-rhamnoside of the aglycone myxol (Hertzberg and Liaaen-Jensen, 1969; Hertzberg *et al.*, 1971; Halfen and Francis, 1972). Occasionally myxoxanthophyll appears to be absent (Healey, 1968; Hertzberg *et al.*, 1971). Other minor components are myxol-2′-*O*-methylpentoside and myxol-2′-glucoside (Francis *et al.*, 1970). Oscillaxanthin, a 1,1′-diglucoside of 1,2,1′,2′-tetrahydro-1,1′-dihydroxylycopene, has been obtained from *Oscillatoria rubescens* and *Athrospira* sp. (Hertzberg and Liaaen-Jensen, 1969). Aphanizophyll from *Aphanizomenon flos-aquae* is 4-hydroxymyxoxanthophyll (Hertzberg and Liaaen-Jensen, 1971). Other trace carotenoids which appear occasionally are γ-carotene (**XXI**), lycopene (**XII**), isocryptoxanthin (**XXXIV**) (Francis and Halfen, 1972), mutatochrome (**XXXV**) (Hertzberg and Liaaen-Jensen, 1969), and a number of pigments not yet fully characterized, such as the allenic xanthophylls caloxanthin and nostoxanthin (Stransky and Hager, 1970; Hertzberg *et al.*, 1971).

Two symbiotic blue-green algae which co-exist with the colourless algae *Cyanophora paradoxa* and *Glaucocystis nostochinearum* synthesize only β-carotene and zeaxanthin; no echinenone nor myxoxanthophyll was noted (Chapman, 1966b).

The distribution in the vegetative cells of *Anabaena cylindrica* is the same as that in the heterocysts (Winkenbach *et al.*, 1972).

It should be emphasized that blue-green algae contain no α-carotene derivatives.

10 Chloromonadophyta

Only two members of this order have been examined in any detail and they appear to contain the same pigments as the heterokonts (Chapman and Haxo, 1966).

11 Colourless algae

Polytoma uvella, a heterotrophic phytoflagellate, synthesizes polytomaxanthin, an unknown keto carotenoid (Links *et al.*, 1960), and *Astasia ocellata* also synthesizes a keto carotenoid, probably phoenicopterone (**XX**) (Thomas *et al.*, 1967).

Flexibacteria are sometimes considered as non-photosynthetic blue-green algae and they produce carotenoids closely related to

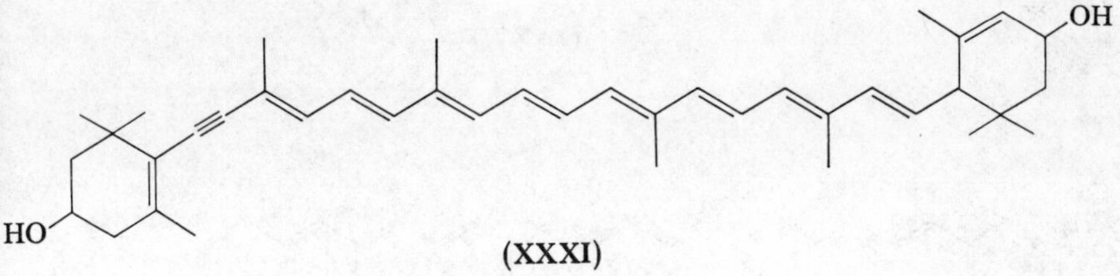

(XXXI)

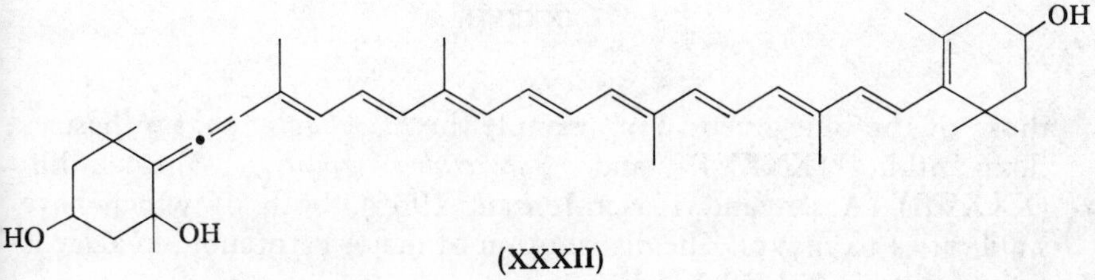

(XXXII)

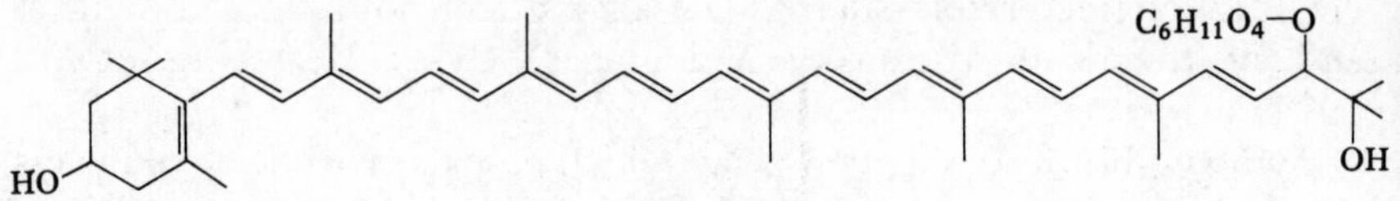

(XXXIII)

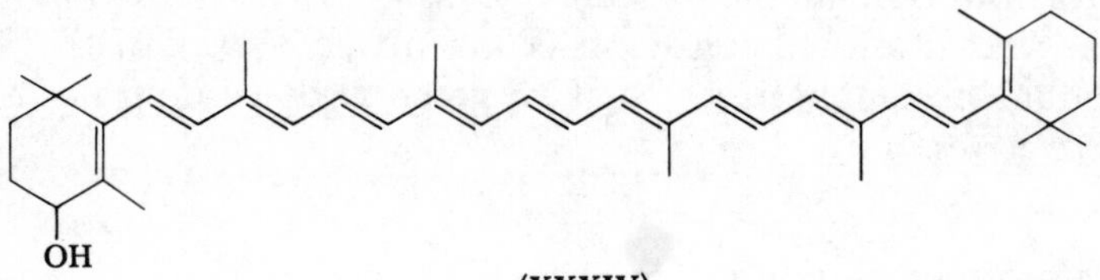

(XXXIV)

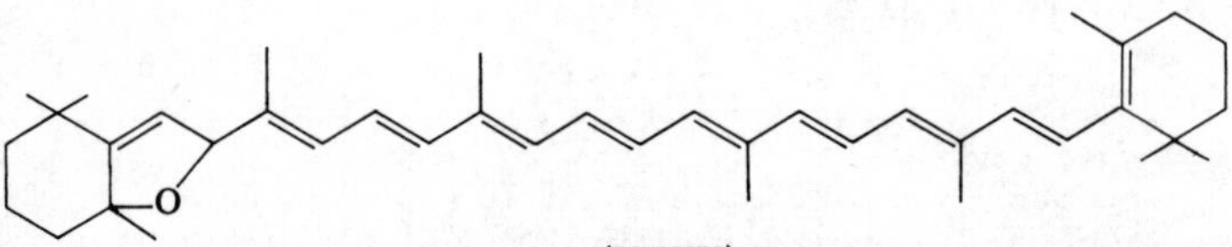

(XXXV)

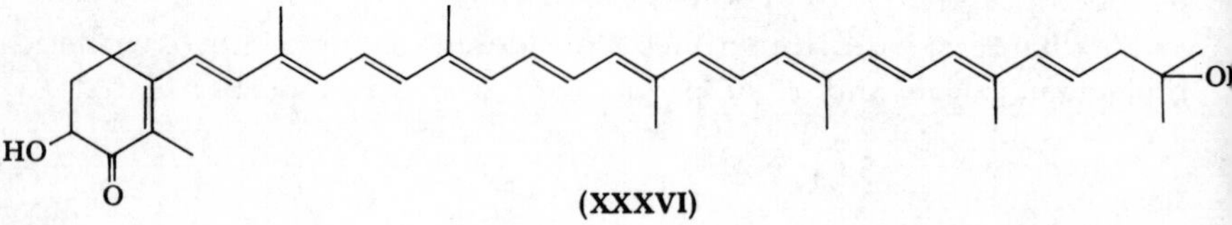

(XXXVI)

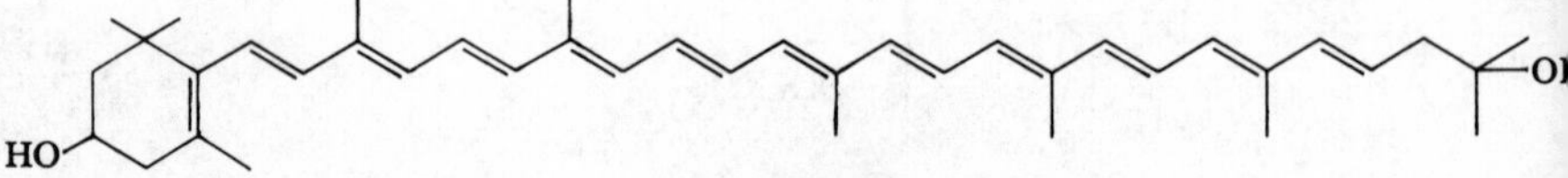

(XXXVII)

those of the blue-greens. For example the *Flexibacter* sp. synthesizes flexixanthin (XXXVI) and *Saprospira grandis* saproxanthin (XXXVII) (Aasen and Liaaen-Jensen, 1966), both of which have similarities to myxol. The distribution of major carotenoid in algae is summarized in Table III.

TABLE III
Distribution of major carotenoids in algae[a]

Pigment	Chlorophyta[b]	Rhodophyta	Chrysophyta: Xanthophyceae[d]	Chrysophyta: Chrysophyceae	Chrysophyta: Haptophyceae	Chrysophyta: Eustigmaphyceae	Chrysophyta: Bacillariophyceae[d]	Phaeophyta	Pyrrophyta	Euglenaphyta	Cyanophyta	Cryptophyta	Chloromondadophyta
α-Carotene (V)	+	+										+	
β-Carotene (I)	+	+	+	+	+	+[f]	+	+	+	+	+	+	+[j]
Echinenone (XVI)										+[h]	+[h]		
Lutein (II)	+	+											
Zeaxanthin (VII)	+	+	+							+	+		
Antheraxanthin (VIII)		+[c]											
Violaxanthin (III)	+	+[c]						+					
Neoxanthin (IV)	+		+							+			
Fucoxanthin (XXVI)				+	?[e]		+	+	+[g]				
Diatoxanthin (XXIII)			+	+			+			+			
Peridinin (XXVIII)									+				
Alloxanthin (XXIX)												+[i]	
Myxoxanthophyll (XXXIII)											+		
Oscillaxanthin (p. 238)											+		

[a] For variations from the general pattern see the text.
[b] Siphonaxanthin and siphonein are widely present in the Bryopsidophyceae.
[c] Rarely encountered.
[d] Also vaucheriaxanthin (XXV) and heteroxanthin (XXIV).
[e] Probably a pigment closely related to but not fucoxanthin.
[f] Xanthophyll distribution rather varied (see text).
[g] Only rarely encountered—peridinin is usually the main pigment.
[h] Other keto carotenoids also encountered.
[i] Other acetylenic carotenoids also encountered.
[j] Chapman and Haxo, 1966.

III Non-photosynthetic tissues of higher plants

A REPRODUCTIVE TISSUES

Carotenoids occur in the anthers and pollen of some but by no means all higher plants. The known distribution is given in Tables IV and V. Sporopollenin, the highly resistant macromolecule of the outer wall of pollen is considered to be a carotenoid polymer (Brooks and Shaw, 1968).

TABLE IV
Carotenoids present in anthers

Species	Pigments	Reference
Allium spp.[a]	1	Genchev (1970)
Colchicum autumnale	1,2,3	Karrer *et al.* (1950)
Dahlia spp.	4	Karrer *et al.* (1950)
Delonia regia[b]	5	Jungalwala and Cama (1962)
Lilium maxime	1,6	Karrer *et al.* (1950)
L. regale	1,6	Karrer *et al.* (1950)
L. tigrinum	6,7,8	Karrer and Oswald (1935); Karrer and Jucker (1945)
L. umbellatum	1,8	Karrer *et al.* (1950)
L. willomottiae unicolor	1,8	Karrer *et al.* (1950)
Narcissus spp.[c]	1,9,10	Valadon and Mummery (1968)
Olivia miniata	1,2,9	Karrer and Krause-Voith (1948)
Rosa spp.[d]	1,11,12,13	Valadon and Mummery (1968)
Tulipa spp.	1,10	von Euler and Klussman (1932)

Pigments
1. β-Carotene (**I**)
2. Lutein-5,6-epoxide
3. Lutein(esters) (**II**)
4. Lycopene (**XII**)
5. Zeaxanthin (**VII**)
6. *Cis*-Antheraxanthin (**VIII**)
7. Antheraxanthin (**VIII**)
8. Capsanthin (**XLI**)
9. α-Carotene (**V**)
10. Lutein (**II**)
11. Auroxanthin (**XXXVIII**)
12. β-Carotene-5,6,5-6-diepoxide
13. Flavoxanthin and chrysanthemaxanthin (**XXXIX**)

Notes:
[a] Carotene in sterile forms lower than in normal forms.
[b] A complex mixture with zeaxanthin as major component.
[c] Also unspecified epoxides.
[d] Pigments cited are common to all species examined.

B FLOWERS

Not all flowers contain carotenoids but those that do can be divided into three main groups: (1) those that synthesize mainly higher oxidized pigments such as auroxanthin (**XXXVIII**) and flavoxanthin

TABLE V
Carotenoid composition of pollen

Species	Pigments	References
Acacia dealbata v. Le Gaulois	1,2,3,4,5,6	Tappi (1949–50)
Helianthus annuus	2,7	Karrer *et al.* (1950)
H. tuberosus	2,7	Karrer *et al.* (1950)
Zea mays	2,7	Karrer *et al.* (1950)

Pigments
1. α-Carotene (**V**)
2. β-Carotene (**I**)
3. α-Carotene-5,6-epoxide
4. Lutein-5,6-epoxide (previously known as taraxanthin) (Tóth and Szabolcs, 1970)
5. Flavoxanthin (**XXXIX**)
6. Lutein (**II**)
7. Lutein esters

(**XXXIX**); (2) those which synthesize large amounts of carotene such as β-carotene and lycopene; and (3) those that produce highly species-specific pigments such as eschscholtzxanthin (**XL**) and the diapocarotenoid acid crocetin (**XLII**). It is not possible to give here a full list of carotenoid-containing flowers (Goodwin, 1974) but

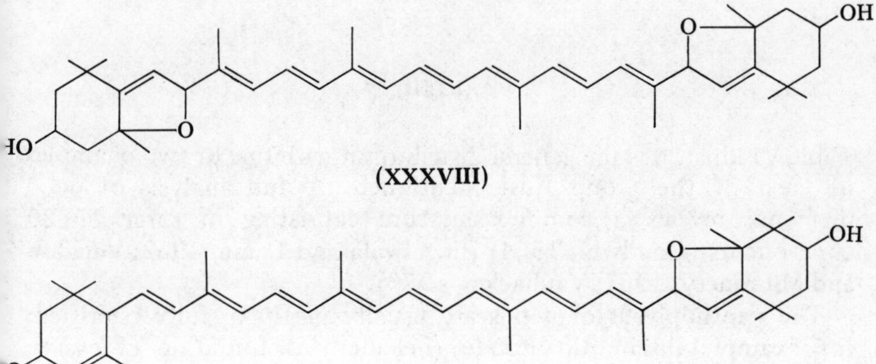

(XXXVIII)

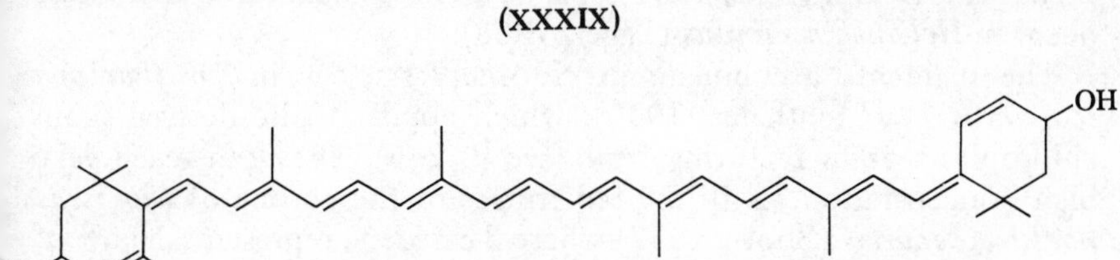

(XXXIX)

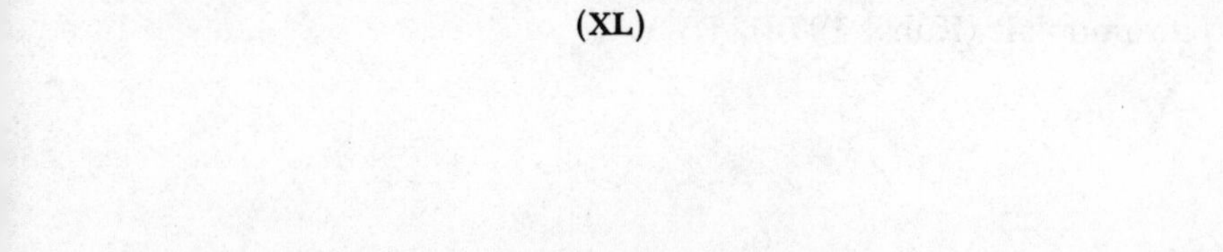

(XL)

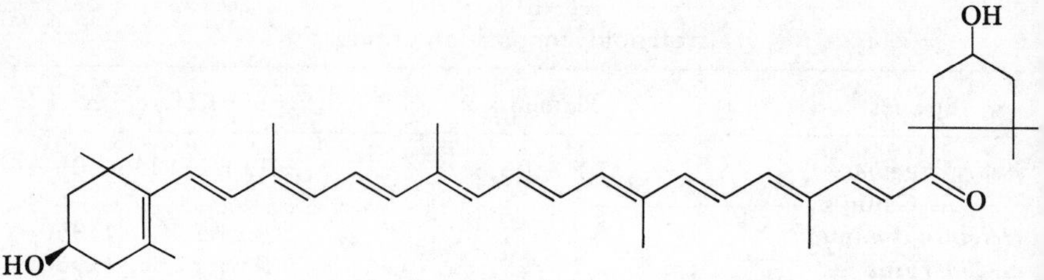

(XLI)

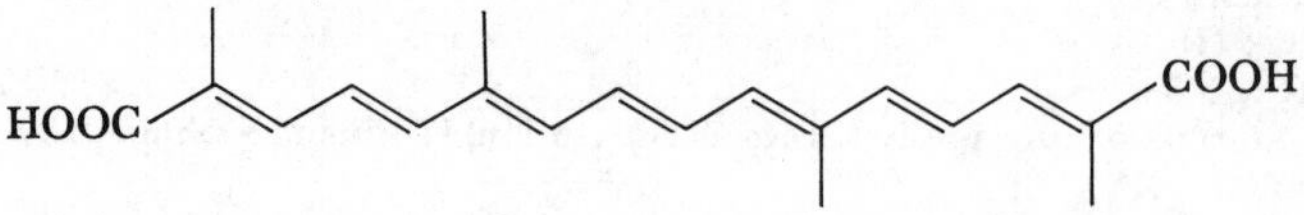

(XLII)

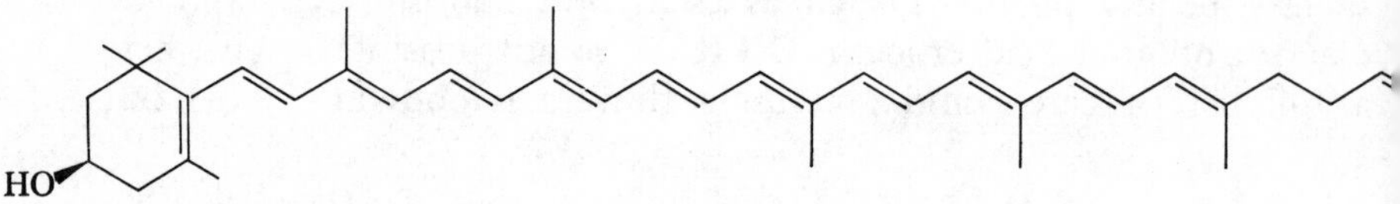

(XLIII)

Table VI illustrates the general distribution with one or two examples of each of the groups just mentioned. A full analysis of petal pigments reveals a complex mixture consisting of some 15–20 components (Goodwin, 1954; Jungalwala and Cama, 1962; Valadon and Mummery, 1967; Vanhaelen, 1973).

The xanthophylls of petals are usually partly or fully esterified; for example lutein dipalmitate (helenien) is found in *Helenium autumnale* (Kuhn *et al.*, 1931) and other long-chain fatty acid esters occur in *Helianthus annuus* (Egger, 1968).

The pigments accumulate in chromoplasts, e.g. in *Sarothamnus scopamus* (Lichtenthaler, 1970), which appear to be derived from chloroplasts as in maturing fruit (see p. 246). They can reach very high concentration, as in the red fringes of the corona of *Narcissus poeticus recurvis* (Booth, 1957) where β-carotene represents 16·5% of the dry weight of the tissue and actually exists as crystals within the chromoplast (Kuhn, 1970).

TABLE VI
Major carotenoid distribution in representative groups of flowers

Group no.	Main pigments	Reference
I Highly oxidized pigments		
Calendula officinalis (yellow)	1,2,3	Goodwin (1954)
Chrysanthemum coronarium	1,2,4	Valadon and Mummery (1971)
II Hydrocarbons		
Calendula officinalis (orange)	5	Goodwin (1954)
Delonia regia	6	Jungalwala and Cama (1962)
Mimulus cupreus	6	Goodwin and Thomas (1964)
III Highly species-specific pigments		
Gazania rigens	7	Zechmeister and Schroeder (1943)
Eschscholtzia californica	8	Strain (1938); Karrer and Leumann (1951)
Crocus sativa	9	(see Karrer and Jucker, 1950)

Pigments
1. Mutatochrome (β-carotene-5,8-epoxide)
2. Flavoxanthin } (**XXXIX**)
3. Chrysanthemaxanthin } (**XXXIX**)
4. β-Carotene-5,6-epoxide
5. Lycopene (**XII**)
6. β-Carotene (**I**)
7. Gazanixanthin (5′-*cis*-rubixanthin, **XLIII**)
8. Eschscholtzxanthin (**XL**)
9. Crocetin (**XLII**)

C SEEDS

1 *Cereals*

Carotenoids are present in traces in many seeds but maize has much more than most and has been examined in detail. It contains mainly β-carotene (**I**), β-cryptoxanthin (**VI**) and zeaxanthin (**VII**), with as minor components α-carotene (**V**), ζ-carotene (**XIII**), γ-carotene (**XXI**) (White *et al.*, 1942), β-zeacarotene (**XLIV**) (Petzold *et al.*, 1959) and zeinoxanthin (**XLV**) (Petzold and Quackenbush, 1960), which has been identified with α-cryptoxanthin (Szabolcs and Ronai, 1969).

2 *Legumes*

As would be expected green beans and peas contain "chloroplast carotenoids" (Goodwin, 1952). The pigments of etiolated cotyledons of French beans consist mainly of auroxanthin (**XXXVIII**) and chrysanthemaxanthin (**XXXIX**) (Goodwin and Phagpolngarm, 1960) whilst lutein is the major pigment in etiolated cucumber cotyledons (Rebeiz, 1968).

3 *Bixa orellana*

The seeds of this plant produce the unique diapocarotenoid, bixin (**LII**) (see Karrer and Jucker, 1950).

D FRUIT

1 *General distribution*

It is not possible here to summarize the extensive literature on fruit carotenoids which has been recently reviewed in detail by Goad and Goodwin (1970). A consideration of this information allows plants to be divided into eight main groups in which one distribution pattern predominates in the ripe fruit: (1) those which produce insignificant amounts of carotenoids; (2) those which produce mainly carotenoids characteristic of chloroplasts; (3) those in which there is a marked synthesis of lycopene and its precursors; (4) those in which β-carotene and its derivatives are synthesized; (5) those which synthesize unusually large amounts of epoxides; (6) those that synthesize pigments which are either completely or almost species specific; (7) those which synthesize mainly poly-*cis* carotenoids; and (8) those that synthesize mainly apocarotenoids. Typical examples are given in Table VII. Although one pattern predominates in any given fruit other patterns are sometimes discernible. For example in the avocado pear (*Persea americana*) the chloroplast pattern predominates but more oxidized pigments are also present (Gross *et al*., 1973).

Recently acyclic epoxides, such as phytoene-1,2-epoxide (**LIII**) and lycopene-5,6-epoxide have been isolated from tomatoes (Ben-Aziz *et al*., 1973).

2 *Taxonomic significance*

Attempts have been made to assess the taxonomic significance of carotenoid distribution in fruit (Goodwin, 1966) but many more data need to be collected before authoritative statements can be made. For example in one genus, *Rosa*, the qualitative distribution in

TABLE VII
Carotenoid distribution in representative groups of fruit

Group no.	Main pigments	References
I Traces		
Pyracantha rogersiana	—	Goodwin (1956)
II Chloroplast carotenoids		
Sambucus nigra	1,2,3,4	Goodwin (1956)
Cucurbita maxima	1,2,5,6	Zechmeister and Tuzson (1936)
III Lycopene series		
Diospyros kaki[a]	1,2,4,5,6,7,8 9, 10,11,12,13,14	Brossard and MacKinney (1963)
Hippophae rhamnoides	1,2,6,8,9,10,15	Goodwin (1956)
IV β-Carotene derivatives		
Physalis alkekengi	1,2,5,6,11,12,16	Bodea and Nicoara (1959)
Mangifera indica	1,2,6,7,10,12,13, 14,17	Jungalwala and Cama (1963)
V Epoxides		
Crataegus pratensis	1,12,17	Goodwin (1956)
VI Species-specific pigments[b]		
Capsicum annuum (red)[c]	18,19	(see Goad and Goodwin, 1970)
Rosa spp.	20	Goodwin (1956)
Lycopersicon esculentum	21,22	(see Goad and Goodwin, 1970)
VII Poly-*cis*-derivatives		
Pyracantha angustifolia	23,24	Zechmeister and Pinckard (1947)
Arum maculatum	23	Goodwin (1956)
VIII Seco- and apocarotenoids[b]		
Murraya exotica	25	Yokoyama and White (1968)
Sheperdia canadensis[d]	26	Kjøsen and Liaaen-Jensen (1969)

Pigments

1. β-Carotene (I)
2. Lutein (II)
3. Flavoxanthin (XXXIX)
4. Neoxanthin (IV)
5. α-Carotene (V)
6. Zeaxanthin (VII)
7. Phytofluene (X)
8. ζ-Carotene (XIII)
9. Lycopene (XII)
10. γ-Carotene (XIII)
11. α-Cryptoxanthin (XLV)
12. β-Cryptoxanthin (VI)
13. Antheraxanthin (VIII)
14. Violaxanthin (III)
15. Aurochrome (5,8,5,8-diepoxy-β-carotene)
16. Phytoene (IX)
17. Mutatochrome (XXXV)
18. Capsanthin (XLI)
19. Capsorubin (XLVI)
20. Rubixanthin (XLIII)
21. Lycoxanthin (XLVIII)
22. Lycophyll (XLIX)
23. Prolycopenes } Poly-*cis* compounds
24. Pro-γ-carotene } Poly-*cis* compounds
25. Semi-β-carotenone (L)
26. Apo-6′-lycopenoic acid (LI)

Notes:
[a] Note the appearance also of type IV pattern.
[b] Only the specific pigments quoted.
[c] See also Table IX.
[d] Occurs as the methyl ester.

a number of species examined is very similar, in particular rubixanthin (**XLIV**) is a major pigment (Goodwin, 1956). This is not so in other genera, e.g. *Pyracantha*, of the family Rosaceae. Indeed *Pyracantha* is a good example of interspecies variation in carotenoid accumulation: *P. rogersiana* only produces very small amounts of carotenoids, whilst *P. flava* synthesizes large amounts of β-carotene derivatives (Goodwin, 1956) and *P. angustifolia* produces poly-*cis* carotenes (Magoon and Zechmeister, 1957). As in the case with flower petals fruit carotenoids accumulate in chromoplasts which have developed from chloroplasts, e.g. tomato and peppers. However, in some cases chromoplasts can be formed from colourless proplastids as in *Convallaria majalis* (Steffen, 1964).

3 *Mutants and crosses*

As might be expected tomato fruit have been examined in great detail and the carotenoids in a number of mutants have been clearly characterized. Table VIII outlines the general distribution in the best-known mutants. Mutants of *Capsicum annum*, *Carica papaya* and water melons have also been examined (see Goad and Goodwin, 1970).

In the case of *Capsicum annuum* it is clear that the ripe fruit of the yellow variety synthesize essentially chloroplast pigments whilst orange and red fruit contain the characteristic pigments capsanthin (**XLI**) and capsorubin (**XLVI**) (Table IX) (Davies *et al.*, 1970).

A number of citrus hybrids with deep red flavedos have been examined, for example the Sinton citrangequat, a trigeneric hybrid of the oval kumquat (*Fortunella margarita*) with the Rusk citrange (*Poncirus trifoliata Citrus sinensis*). The colour is due mainly to the presence of very large amounts of apocarotenoids, in particular β-citraurin (**LVII**) and β-apo-8-carotenal (**LVIII**) (Stewart and Wheaton, 1973). The large amounts of reticulataxanthin (**LIX**) and citranaxanthin (**LX**) previously reported (Yokoyama and White, 1966; Yokoyama *et al.*, 1972) are artefacts produced in the presence of acetone and alkali used in the extraction procedure (Stewart and Wheaton, 1973).

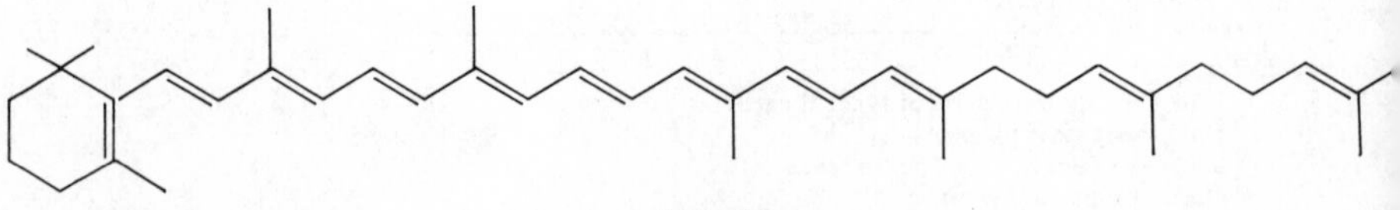

(XLIV)

TABLE VIII

Carotenoid distribution in various tomato mutants

Genotypes	Pigments and colourless polyenes	Total concentration (μg/g fresh wt)		References
		Total polyenes	Pigments only	
Red (normal)	1,2,3,4,5	87	50	Tomes (1963)
High pigment (*hp*)	1,2,3,4,5	88	69	
Tangerine (r^+r^+tt)	2,3,4,5,6	158	90	
Yellow (*rrtt*)	1,2,4,5	4	2	Mackinney and Jenkins (1952)
Yellow–Tangerine (*rrtt*)	1,2,4,5,7	24	14	Jenkins and MacKinney (1955)
Apricot (*atat*)	1 2,4,5	13	11	MacKinney and Jenkins (1952)
Yellow–Apricot (*ttrratat*)	2,3,5 (trace)	2	1	Jenkins and MacKinney (1955)
Tangerine–Apricot (*ttatat*)	1,2,4,5,6,7	29	16	Jenkins and MacKinney (1955)
Yellow–Tangerine–Apricot (*rrtatat*)	4,5	10	0	Jenkins and MacKinney (1955)
Ghost (*ghgh*)	1,4,5	295	1	MacKinney *et al.* (1956)
Intermediate β (*BB*)	1,2,3,4,5	50	46	Tomes (1963)
High β ($BB^{mo^+}B^{mo^+}B$)	1,2,3,4,5	80	70	
Delta (*deldel*)	1,2,3,4,5,7,8,9,10	84	68	
High δ (*deldelhphp*)	1,2,3,4,5,8,9,10	55	46	

Key:

1. Lycopene (**XII**)
2. β-Carotene (**I**)
3. γ-Carotene (**XXI**)
4. Phytoene (**IX**)
5. Phytofluene (**X**)
6. Prolycopenes (poly-*cis*-lycopene)
7. ζ-Carotene (**XIII**)
8. α-Carotene (**V**)
9. δ-Carotene (**LIV**)
10. Neurosporene (**LV**)

TABLE IX
Carotenoid distribution in ripe fruit of colour variants of *Capsicum annuum*
(Adapted from Davies *et al.*, 1970)
(Expressed as percentage of total pigment recovered)

Pigment	Red	Orange	Yellow
β-Carotene (**I**)	12·3	—	1·0
β-Carotene 5,6-epoxide	3·1	—	—
Hydroxy-α-carotene (?)	—	—	5·7
β-Cryptoxanthin (**VI**)	7·8	—	—
Cryptocapsin (**XLVII**)	5·0	8·0	—
Lutein (**II**)	—	—	28·1
Zeaxanthin (**VII**)	6·5	7·3	15·9
Antheraxanthin (**VIII**)	9·2	10·6	31·3
Violaxanthin (**III**)	9·8	14·9	—
Capsanthin (**XLI**)	31·7	35·4	—
Capsanthin-5,6-epoxide	4·2	6·1	—
Capsorubin (**XLVI**)	7·5	14·3	—
Neoxanthin (**IV**)	2·0	2·7	20·0

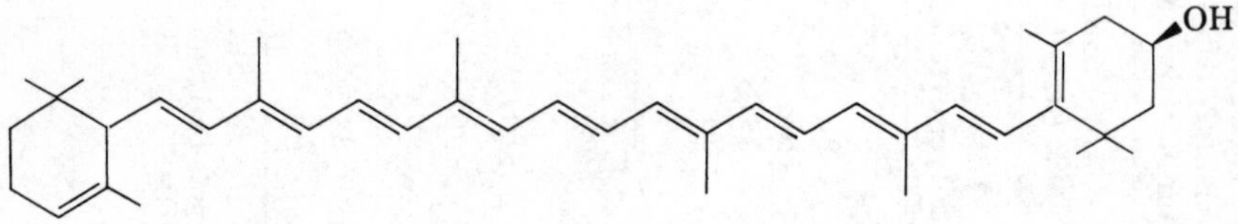

OH

(XLV)

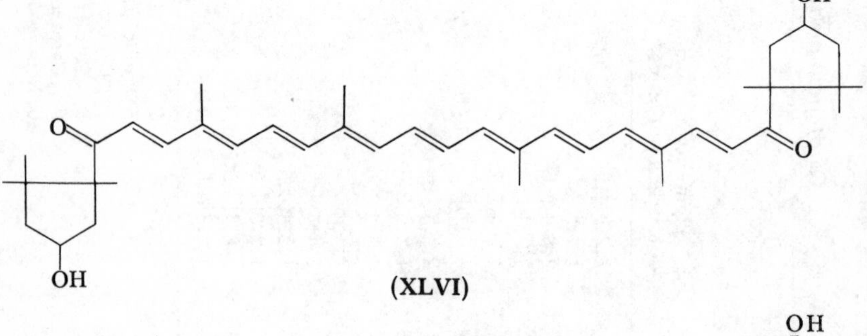

OH
O
O
OH

(XLVI)

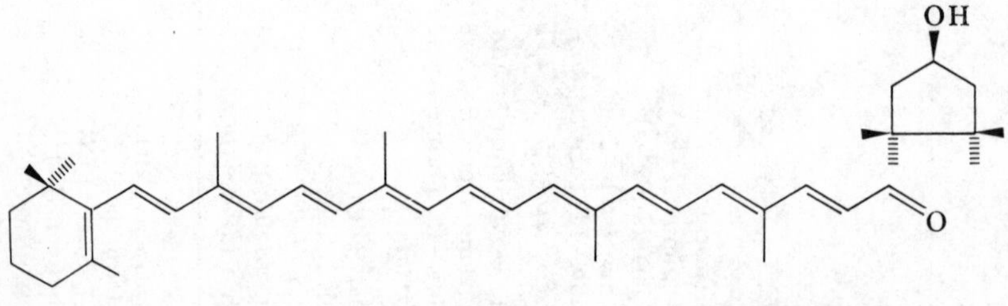

OH
O

(XLVII)

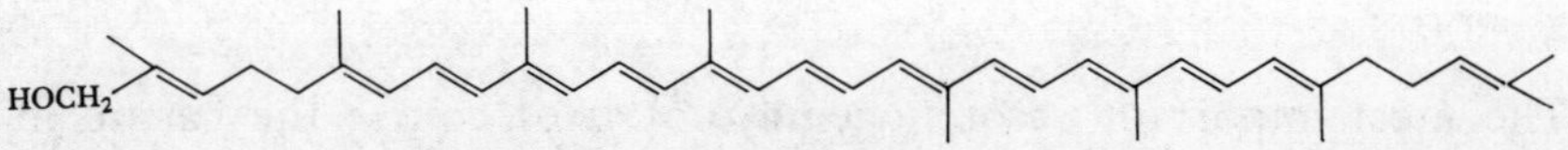

HOCH$_2$

(XLVIII)

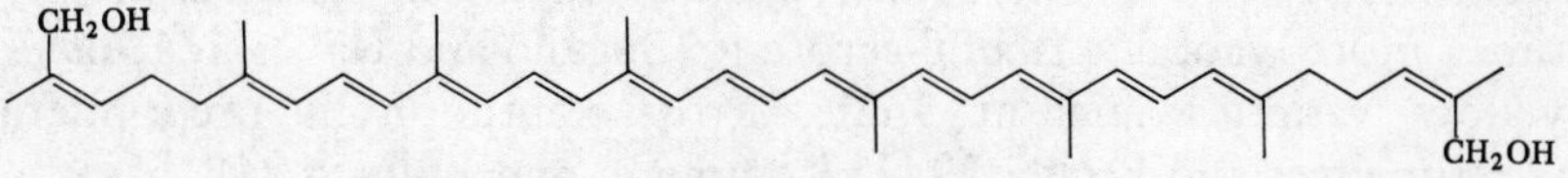

CH$_2$OH

CH$_2$OH

(XLIX)

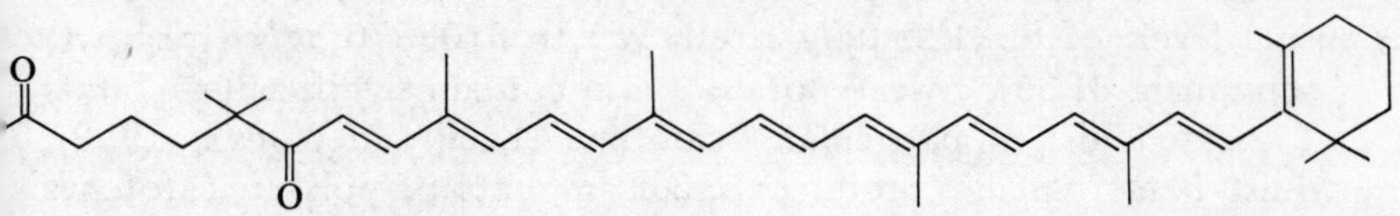

O

O

(L)

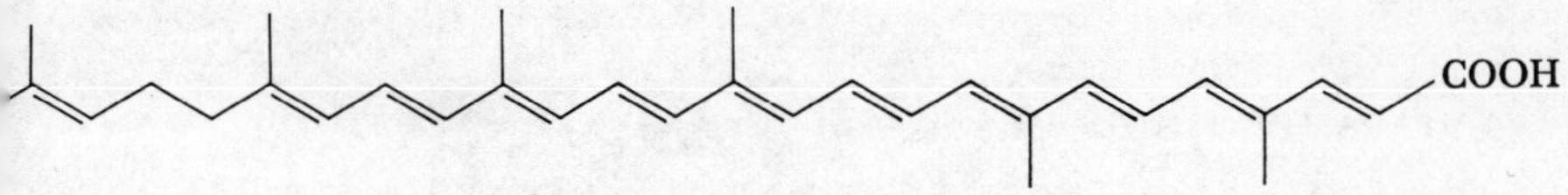

COOH

(LI)

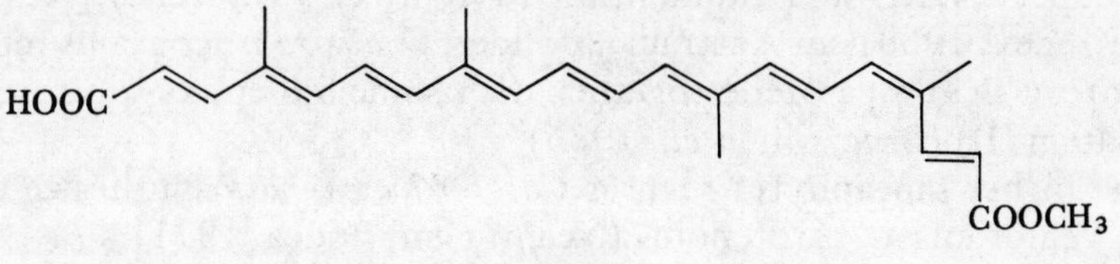

HOOC

COOCH$_3$

(LII)

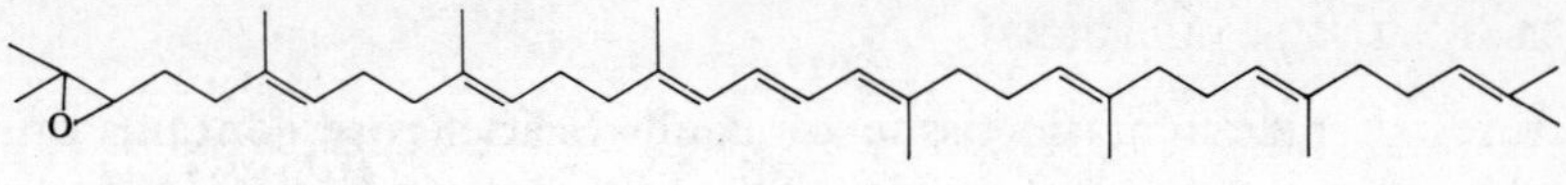

O

(LIII)

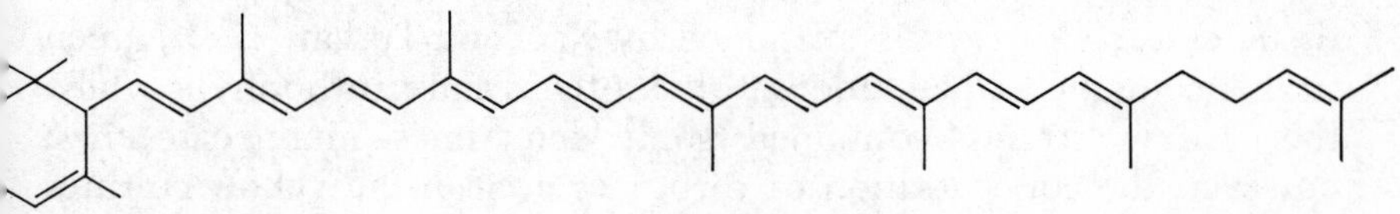

(LIV)

E ROOTS

The most important carotenogenic root is of course the carrot, in which the main pigment in commercial varieties is β-carotene, with xanthophylls only representing 5–10% of the total pigments (see Goodwin, 1973). One strain is reported which synthesizes some three times more lycopene than β-carotene (Sugano and Hayashi, 1967). In yellow varieties and in wild carrots xanthophylls preponderate (Zechmeister and Escue, 1941; Kemmerer and Fraps, 1945).

Inheritance of pigments in carrots has been examined in detail (Imam and Gableman, 1968) and strain selection can increase the usual levels of 60–120 μg/g (fresh wt) to 310–370 μg/g (fresh wt) (Schuphan, 1965). Sweet potatoes also contain significant amounts of carotenoids, in particular β-carotene (Ezell and Wilcox, 1952), whilst *Beta rutabaga* produces small amounts of poly-*cis* carotenes (Joyce, 1954). The unique apocarotenoid, azafrin (**LXI**) is the major pigment in the roots of *Escobedia scabrifolia* (see Karrer and Jucker, 1950).

F PARASITIC PLANTS

The yellow dodders *Cuscuta salina* and *C. subinclusa* contain mainly γ-carotene (**XXI**) with smaller amounts of β-carotene (**I**) and traces of lycopene (**XII**) and rubixanthin (3-hydroxy-γ-carotene) (**XLIV**) (MacKinney, 1935); an Australian dodder, *C. australis,* contains more β-carotene than γ-carotene together with some 5,6-epoxy-α-carotene and lutein (**II**) (Baccarini *et al.*, 1965).

The higher saprophytes such as *Corallorhiza innata* synthesize the usual "chloroplast" carotenoids (Neamtu and Bodea, 1971).

G PLANT TISSUE CULTURES

Cultures of meristematic tissue of Paul's Scarlet rose contains only small amounts of carotenoids but with violaxanthin (**III**), auroxanthin (**XXXVIII**) and neoxanthin (**IV**) predominating; β-carotene (**I**) is present in traces (Williams and Goodwin, 1965). In cambial tissue cultures of carrots examined by Naef and Turian (1963) green strains contained equal amount of carotenes and xanthophylls whilst the yellow strains contained some ten times more carotenes; however, the concentration of carotenes even in the yellow cultures was about five times less than in the roots themselves. On the other hand, in one strain which was derived from roots synthesizing both β-carotene and lycopene, lycopene only was synthesized; the concentration of lycopene was similar in both root and tissue culture (Sugano *et al.*, 1971).

IV Fungi

1 *Distribution*

Carotenoids are widely but not ubiquitously distributed in fungi and the known distribution has recently been listed in detail (Goodwin, 1972, 1973b). Some generalities which can be extracted from this list will be summarized here: (i) many fungi, in particular the Phycomycetes, synthesize only carotenes; (ii) β-carotene (**I**) is not a constant component but is well distributed and frequently quantitatively the major pigment; it is absent from some Chytridrales and Blastocladiales, where it is replaced by γ-carotene (**XXI**) (e.g. in *Cladochytrium replicatum*, Fuller and Tavares, 1960); indeed γ-carotene is more widely distributed in fungi than is β-carotene; (iii) unique carotenes appear in which an additional double bond appears at position 3,4; for example torulene (3,4-dehydro-γ-carotene) (**LXII**), which is characteristic of red yeasts and certain Ascomycetes but is rarely formed by Basidiomycetes; 3,4-dehydrolycopene (**LXIII**), which is found, *inter alia*, in *Neurospora* spp. (Aasen and Liaaen-Jensen, 1965); Pigments with a 3,4 double bond are rarely if ever found in higher plants and algae but are characteristic of photosynthetic bacteria (Andrewes and Liaaen-Jensen, 1972), and βγ-carotene (**LXIV**) from *Caloscypha fulgens* (Discomycetales) (Arpin *et al.*, 1971): (iv) no carotenoids with an α-ionone residue has been unequivocally identified in fungi: (v) similarly no leaf xanthophylls have been clearly identified, these include 3-hydroxyderivatives (e.g. β-cryptoxanthin, **VI**), 5,6-epoxides and 5,8-epoxides (although β-carotene 5,8-epoxide is present as a degradation product in old cultures of *Phycomyces blakesleeanus*) and allenes (e.g. neoxanthin, **IV**), one characteristic group of xanthophylls are carboxylic acids, for example, rhodotorulin (**LXV**) in which one gem dimethyl group is oxidized, or neurosporaxanthin, which is an apocarotenoid acid. Similar pigments with oxidized methyl groups appear in some fruit and algae (e.g. lycoxanthin, **XLVII**, p. 247, and loroxanthin, **XIV**, p. 231) but they are never oxidized beyond the alcohol level. Apocarotenals but not apocarotenoid acids appear in citrus fruit (p. 248), but such acids do appear in other plants such as the seeds of

Bixa orellana (bixin, **LII**, p. 246), the roots of *Escobedia sabrifolia* (azafrin, **LXI**) and the flowers of *Crocus sativus* (crocetin, **XLII**): (VIII) many Discomycetes are characterized by unique xanthophylls and their derivatives; for example esterified phillipsiaxanthin (**LXVI**) occurs in *Phillipsia carminea* (Pezizates) (Arpin and Liaaen-Jensen, 1967a), plectaniaxanthin (**LXVII**) in *Plectania coccinea* (Pezizales) (Arpin and Liaaen-Jensen, 1967b) and esterified aleuriaxanthin (**XLVIII**) in *Aleuria aurantia* (Pezizales) (Arpin *et al.*, 1973). The first two closely resemble the characteristic pigments in the photosynthetic bacteria in having an additional double bond at C-3, the usual double bond at C-1 hydrated and a keto group at C-2. Aleuriaxanthin, on the other hand, has an entirely novel end group (IX). Keto carotenoids which are occasionally encountered in algae and higher plants are also occasionally found in fungi; e.g. rhodoxanthin (**XI**) in *Epicoccum nigrum* (Deuteromycetes) (Gribanovski-Sassu and Foppen, 1967), echinenone (**XVI**) and astaxanthin (**XVIII**) in *Peniophora aurantiaca* (Aphyllophorales) Arpin *et al.*, 1966), and canthaxanthin (**XVII**) in *Cantharellus cinnabarinus* (Aphyllophorales), the original source (Haxo, 1950), (X) acetylenic carotenoids, widespread in algae, have not yet been detected in fungi.

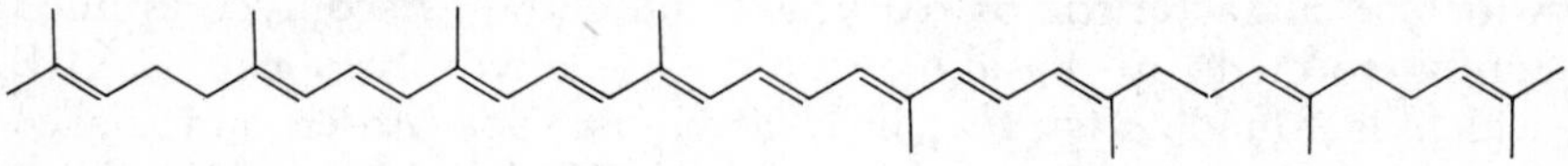

(LV)

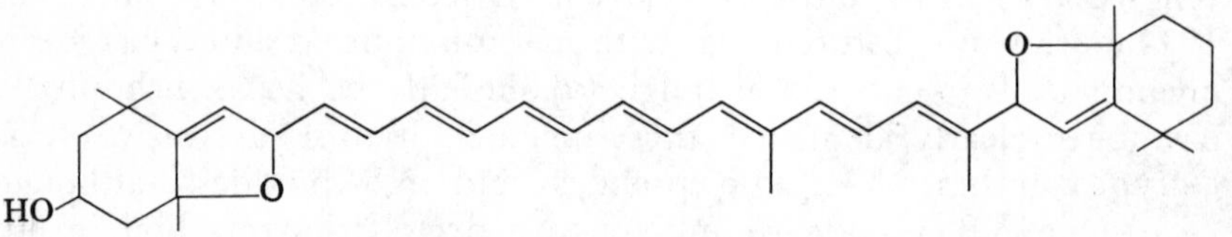

HO O O

(LVI)

HO CHO

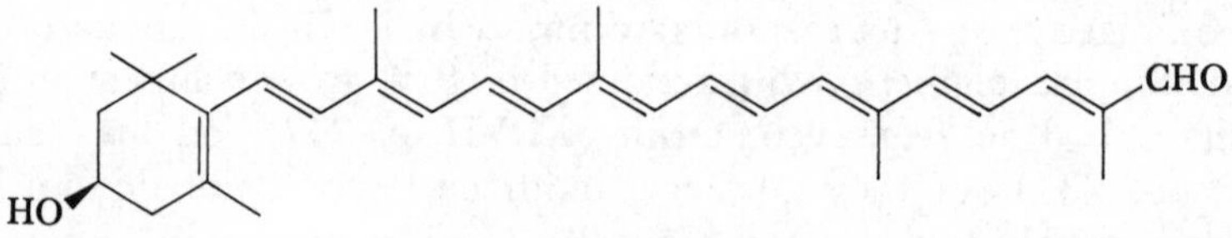

(LVII)

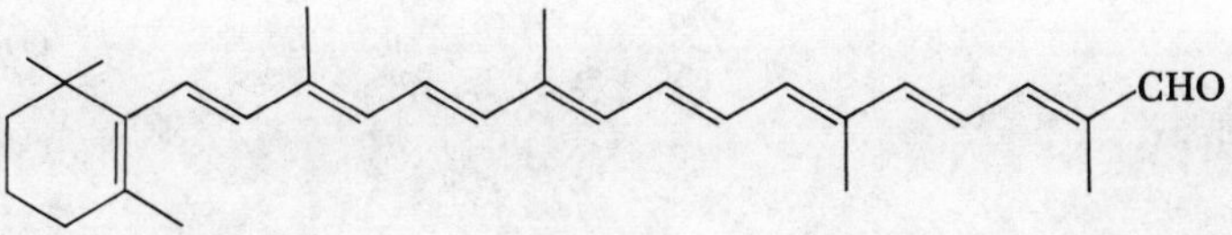

(LVIII)

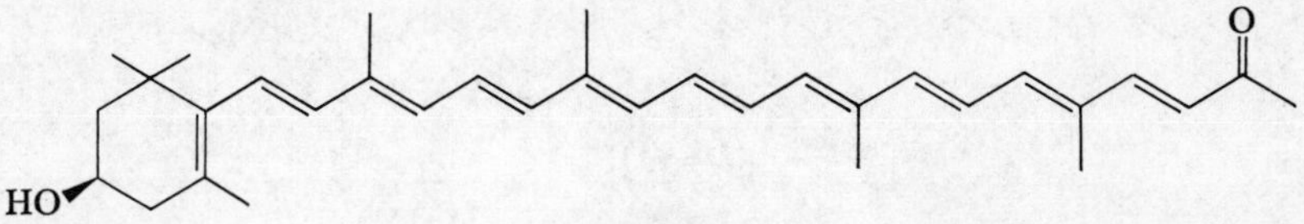

(LIX)

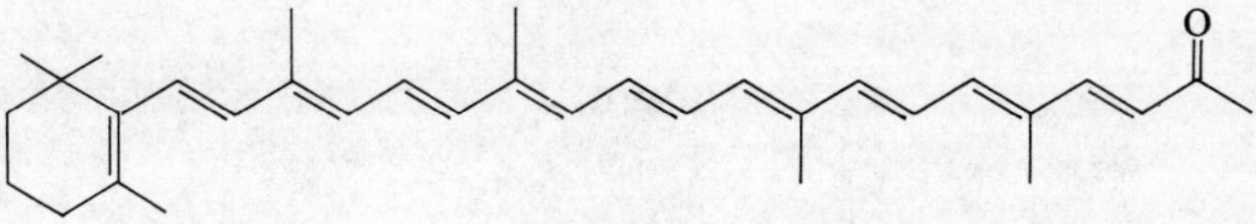

(LX)

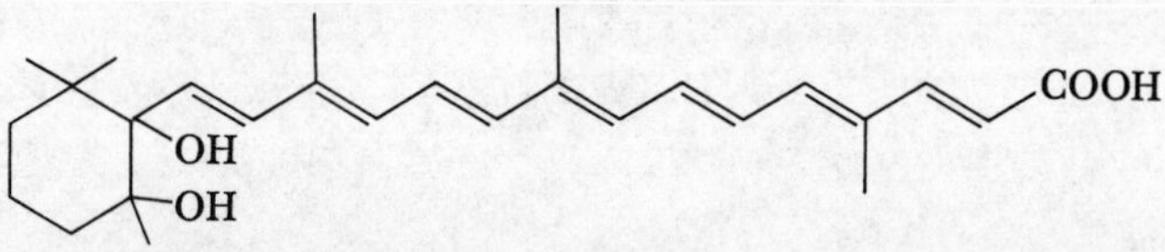

(LXI)

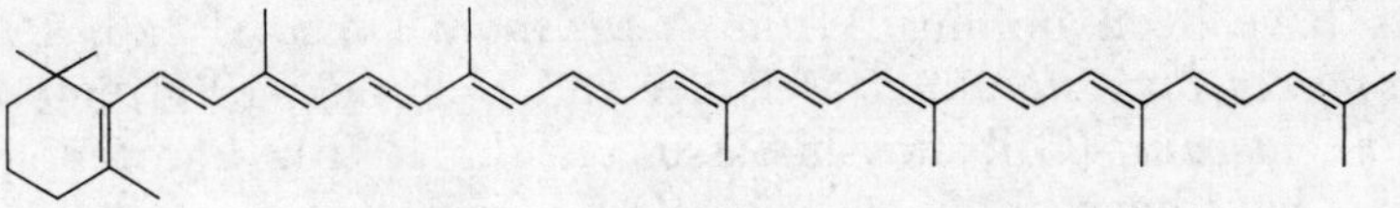

(LXII)

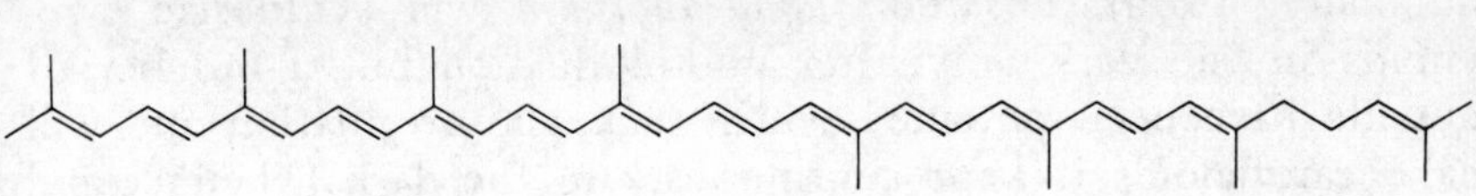

(LXIII)

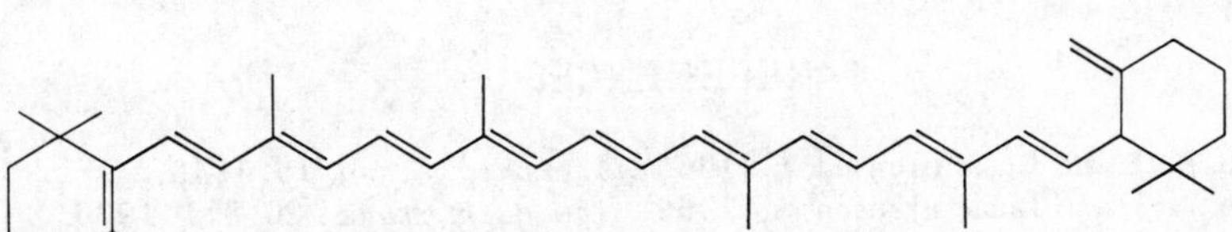

(LXIV)

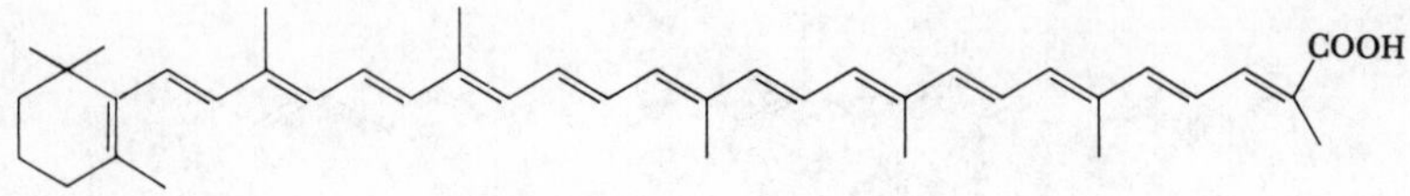

(LXV)

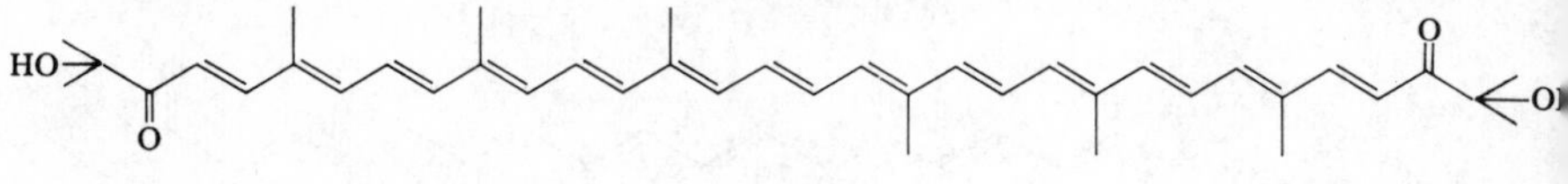

(LXVI)

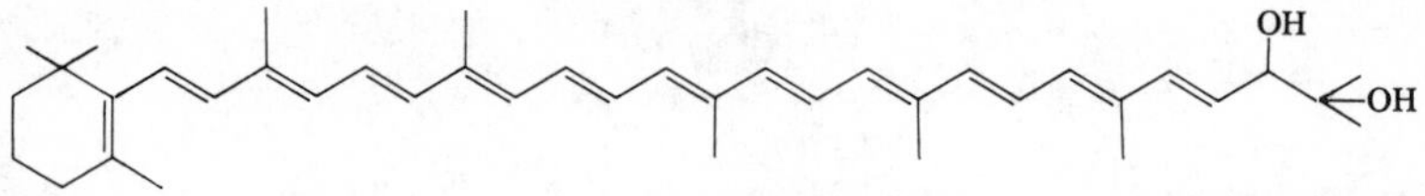

(LXVII)

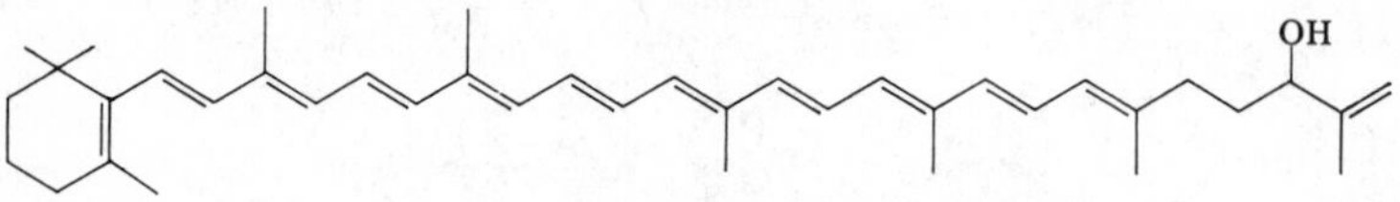

(LXVIII)

2 *Mutants*

Mutants in which carotenoid synthesis has been blocked at various points have been obtained from *Neurospora crassa* (Haxo, 1956), *Phycomyces blakesleeanus* (Meissner and Delbruck, 1968) and *Epicoccum nigrum* (Gribanovski-Sassu *et al.*, 1970). Carotenogenic mutants have been obtained from *Verticillium albo-atrum*, which in its native form does not synthesize carotenoids (Valadon and Mummery, 1966). *Fusarium aquaeductuum* will synthesize carotenoids in the dark only after photoinduction but a number of mutants have been obtained which lack photoregulation and will make carotenoids if kept continuously in the dark (Theimer and Rau, 1969).

REFERENCES

Aasen, A. J. and Liaaen-Jensen, S. (1965). *Acta chem. scand.* 19, 1843.
Aasen, A. J. and Liaaen-Jensen, S. (1966). *Acta chem. scand.*, 20, 811, 1970.
Aasen, A. J., Eimhjellen, K. E. and Liaaen-Jensen, S. (1969). *Acta chem. Scand.* 23, 2544–2545.

Aihara, M. S. and Yamamoto, H. Y. (1968). *Phytochemistry* 7, 497-499.
Aitzetmüller, K., Svec, W. A., Katz, J. J. and Strain, H. H. (1968). *Chem. Commun.* 32-33.
Aitzetmüller, K., Strain, H. H., Svec, W. A., Grandolfo, M. and Katz, J. J. (1969) *Phytochemistry* **8**, 1761-1770.
Allen, M. B., Goodwin, T. W. and Phagpolngarm, S. (1960). *J. gen. Microbiol.* **23**, 93-103.
Allen, M. B., Fries, L., Goodwin, T. W. and Thomas, D. M. (1964). *J. gen. Microbiol.* **34**, 259-267.
Anderson, I. C. and Robertson, D. S. (1960). *Pl. Physiol, Lancaster* **35**, 531-534.
Andrewes, A. G. and Liaaen-Jensen, S. (1972). *A. Rev. Microbiol.* **26**, 225-248.
Arpin, N. and Liaaen-Jensen, S. (1967a). *Bull. Soc. Chim. biol.* **49**, 527.
Arpin, N. and Liaaen-Jensen, S. (1967b). *Phytochemistry* **6**, 995-1005.
Arpin, N., Fiasson, J. L., Bouchez-Dangye-Caye, M. P., Francis, G. W. and Liaaen-Jensen, S. (1971). *Phytochemistry* **10**, 1595.
Arpin, N., Kjøsen, H., Francis, G. W. and Liaaen-Jensen, S. (1973). *Phytochemistry,* **12**, 2751-2758.
Arpin, N., Lebreton, P. and Fiasson, J. L. (1966). *Bull. Soc. mycol. Fr.* **82**, 450.
Baccarini, A., Bertossi, F. and Bagni, (1965). *Phytochemistry* **4**, 349-351.
Bailey, J. L., Thornber, J. P. and Whyborn, A. G.. (1966). *In* "Biochemistry of Chloroplasts" (T. W. Goodwin, ed.), vol. 1, pp. 275-284. Academic Press, London and New York.
Ben-Aziz, A., Britton, G. and Goodwin, T. W. (1973). *Phytochemistry* **12**, 2759-2784.
Bendz, G., Loof, L. G. and Martensson, O. (1968). *Acta chem. scand.* **22**, 2215.
Berkaloff, C. (1967). *J. Microscopie* **6**, 839-850.
Bickoff, E. M., Livingston, A. L., Bailey, G. F. and Thompson, C. R. (1954). *Agric. Fd Chem.* **2**, 563-567.
Bodea, C. and Nicoara, E. (1959). *Justus Liebig's Annln Chem.* **622**, 188.
Booth, V. H. (1957). *Biochem. J.* **65**, 660-663.
Britton, G. and Goodwin, T. W. (1973). *In* "Chemistry and Biochemistry of Herbage" (G. W. Butler and R. W. Bailey, eds), vol. 1, pp. 477-510. Academic Press, London and New York.
Brooks, J. and Shaw, G. (1968). *Nature, Lond.* **220**, 678-679.
Brossard, J. and MacKinney, G. (1963). *Agric. Fd Chem.* **11**, 501.
Bunt, J. S. (1964). *Nature, Lond.* **203**, 1261-1263.
Carter, P. W., Cross, L. C., Heilbron, I. M. and Jones, E. R. H. (1948). *Biochem. J.* **43**, 349-352.
Chapman, D. J. (1966a). *Phytochemistry* **5**, 1331-1333.
Chapman, D. J. (1966b). *Arch. Mikrobiol.* **55**, 19.
Chapman, D. J. and Haxo, F. T. (1963). *Pl. Cell Physiol.* **4**, 57-63.
Chapman, D. J. and Haxo, F. T. (1966). *J. Phycol.* **2**, 89.
Claes, H. (1954). *Z. Naturf.* **9b**, 462-469.
Class, H. (1956). *Z. Naturf.* **11b**, 260-266.
Czygan, F. C. (1964). *Experientia* **20**, 573.
Czygan, F. C. (1970). *Flora,* Abt. A *Physiol-Biochem.* **3**, 25.
Czygan, F. C. and Kalb, K. (1966). *Z. PflPhysiol.* **55**, 59-64.
Czygan, F. C. and Kessler, E. (1967). *Z. Naturf.* **22b**, 1085.
Dales, R. P. (1960). *J. mar. biol. Ass. U.K.* **39**, 693.

Davies, B. H., Matthews, S. and Kirk, J. T. O. (1970). *Phytochemistry* 9, 797-805.
de Nicola, M. and di Benedetto, G. (1962). *Boll. Ist. Bot. Univ. Catania* 3, 22.
de Nicola, M. and Tomaselli, R. (1961). *Boll. Ist. Bot. Univ. Catania* 2, 22.
Egger, K. (1968). *Z. Naturf.* **23b**, 733-735.
Egger, K., Nitsche, H. and Kleinig, H. (1969). *Phytochemistry* 8, 1583-1585.
Euler, H. von and Klussman, E. (1932). *Svensk Rem. Tidskr.* **44**, 198.
Ezell, B. D. and Wilcox, M. S. (1952). *Pl. Physiol, Lancaster* 27, 81.
Faludi-Daniel, A. (1965). *Acta agron. hung.* 14, 203.
Francis, G. W. and Halfen, L. N. (1972). *Phytochemistry* 11, 2347-2348.
Francis, G. W., Hertzberg, S., Andersen, K. and Liaaen-Jensen, S. (1970). *Phytochemistry,* 9, 629-635.
Fuller, M. and Tavares, J. E. (1960). *Biochim. biophys. Acta* **44**, 589.
Genchev, S. (1970). *Dokl. Akad. Sel'skokhoz Nauk Bolg.* 3, 25.
Goad, L. J. and Goodwin, T. W. (1970). *In* "The Biochemistry of Fruits and Their Products" (A. C. Hulme, ed.), vol. 1, pp. 305-368. Academic Press, London and New York.
Goodwin, T. W. (1954). *Biochem. J.* **58**, 90.
Goodwin, T. W. (1952). *In* "The Comparative Biochemistry of the Carotenoids". Chapman and Hall, London.
Goodwin, T. W. (1955). *A. Rev. Biochem.* 24, 497-522.
Goodwin, T. W. (1956). *Biochem. J.* 62, 346.
Goodwin, T. W. (1958). *Biochem. J.* **68**, 503-511.
Goodwin, T. W. (1959). *In* "Encyclopaedia of Plant Physiology" (W. Ruhland, ed.), vol. 5, pp. 394-443. Springer-Verlag, Heidelberg.
Goodwin, T. W. (1966). *In* "Comparative Phytochemistry" (T. Swain, ed.), pp. 121-137. Academic Press, London and New York.
Goodwin, T. W. (1971). *In* "Aspects of Terpenoid Chemistry and Biochemistry (T. W. Goodwin, ed.), pp. 315-356. Academic Press, London and New York.
Goodwin, T. W. (1972). *In* "Progress in Industrial Microbiology" (D. J. D. Hockenhull, ed.), vol. II, pp. 29-88. Churchill Livingstone, Edinburgh.
Goodwin, T. W. (1973a). *In* "Phytochemistry" (L. P. Miller, ed.), vol. 1, pp. 112-142. Van Nostrand Reinhold, New York.
Goodwin, T. W. (1973b). *In* "Handbook of Microbiology" (A. I. Laskin and H. A. LeChevalier, eds), vol. 3, pp. 75-83. CRC Press, Cleveland.
Goodwin, T. W. (1974). "Comparative Biochemistry of Carotenoids", 2nd edn. Chapman and Hall, London.
Goodwin, T. W. and Gross, J. A. (1958). *J. Protozool.* 5, 292-296.
Goodwin, T. W. and Phagpolngarm, S. (1960). *Biochem. J.* 76, 197-199.
Goodwin, T. W. and Taha, M. M. (1951). *Biochem. J.* 47, 513-514.
Goodwin, T. W. and Thomas, D. M. (1964). *Phytochemistry* 3, 47-50.
Granick, S. (1961). *Proc. 5th int. Congr. Biochem.* p. 276.
Gribanovski-Sassu, O. and Foppen, F. H. (1967). *Phytochemistry* 6, 907.
Gribanovski-Sassu, O., Tuttobello, L. and Foppen, F. H. (1970). *Arch. Mikrobiol.* 73, 344.
Hager, A. and Stransky, H. (1970). *Arch. Mikrobiol.* 72, 68; 73, 77.
Halfen, L. N. and Francis, G. W. (1972). *Arch. Mikrobiol.* **81**, 25.
Haxo, F. (1950). *Bot. Gaz.* 112, 228.
Haxo, F. (1956). *Fortschr. Chem. org. NatStoffe* 12, 169-197.
Haxo, F. T. and Clendenning, K. A. (1953). *Biol. Bull. mar. biol. Lab., Woods Hole* 105, 103-114.

Healey, F. P. (1968). *J. Phycol.* 4, 126.
Hertzberg, S. and Liaaen-Jensen, S. (1966). *Phytochemistry* 5, 557-563, 565-570.
Hertzberg, S. and Liaaen-Jensen, S. (1969). *Phytochemistry* 8, 1259-1280, 1281-1293.
Hertzberg, S. and Liaaen-Jensen, S. (1971). *Phytochemistry* 10, 3251-3252.
Hertzberg, S., Liaaen-Jensen, S. and Siegelman, H. W. (1971). *Phytochemistry* 10, 3121-3127.
Huneck, S., Mathey, A. and Trotet, G. (1967). *Z. Naturf.* 22b, 1367.
Imam, M. K. and Gabelman, W. H. (1968). *Proc. Am. Soc. hort. Sci.* 93, 419.
Jeffrey, S. W. (1961). *Biochem. J.* 80, 336-342.
Jeffrey, S. W. and Allen, M. B. (1964). *J. gen. Microbiol.* 36, 277-288.
Jeffrey, S. W. and Haxo, F. T. (1968). *Biol. Bull. mar. biol. Lab., Woods Hole* 135, 149-165.
Jenkins, J. A. and MacKinney, G. (1955). *Genetics, Princeton* 40, 715-720.
Jensen, A. (1966). Rep. norw. Inst. Seaweed Res. no. 31.
Johannes, B., Brzezinka, H. and Budziekiewicz, H. (1971). *Z. Naturf.* 26B, 377-378.
Joyce, A. E. (1954). *Nature, Lond.* 173, 311.
Jungalwala, F. B. and Cama, H. R. (1962). *Biochem. J.*, 85, 1-8.
Jungalwala, F. B. and Cama, H. R. (1963). *Indian J. Chem.* 1, 36.
Karrer, P. and Jucker, E. (1954). *Helv. chim. Acta* 28, 300-315.
Karrer, P. and Jucker, E. (1950). "Carotenoids" (E. A. Braude, Trans.). Birkhäuser, Basle.
Karrer, P. and Krause-Voith, E. (1948). *Helv. chim. Acta* 31, 802.
Karrer, P. and Leumann, E. (1951). *Helv. chim. Acta* 34, 445.
Karrer, P. and Oswald, A. (1935). *Helv. chim. Acta* 18, 1303.
Karrer, P., Eugster, C. H. and Faust, M. (1950). *Helv. chim. Acta* 33, 300.
Karrer, P., Fatzer, W., Favarger, M. and Jucker, E. (1943). *Helv. chim. Acta* 26, 2121-2122.
Kemmerer, A. R. and Fraps, G. S. (1945). *Fd Res.* 10, 459.
Kessler, E. and Czygan, F.-C. (1966). *Arch. Mikrobiol.* 54, 37-45.
Kjøsen, H. and Liaaen-Jensen, S. (1969). *Phytochemistry* 8, 483-491.
Kjøsen, H., Arpin, N. and Liaaen-Jensen, S. (1972). *Acta chem. scand.* 26, 3053-3067.
Kleinig, H. (1969). *J. Phycol.* 5, 281-284.
Kleinig, H. and Czygan, F.-C. (1969). *Z. Naturf.* 24b, 927.
Krinsky, N. I. and Goldsmith, T. H. (1960). *Archs Biochem. Biophys.* 91, 271.
Kuhn, H. (1970). *J. Ultrastruct. Res.* 33, 332.
Kuhn, R., Winterstein, A. and Lederer, E. (1931). *Hoppe-Seyler's Z. Physiol. Chem.* 197, 141-160.
Larsen, B. and Haug, A. (1956). *Acta chem. scand.* 10, 470-472.
Lichtenthaler, H. K. (1970). *Planta* 90, 142-152.
Links, J., Verloop, A. and Havinga, E. (1960). *Arch. Mikrobiol.* 36, 306.
MacKinney, G. (1935). *J. Biol. Chem.* 112, 421-424.
MacKinney, G. and Jenkins, J. A. (1952). *Proc. natn. Acad. Sci. U.S.A.* 38, 48.
MacKinney, G., Rick, C. M. and Jenkins, J. A. (1956). *Proc. natn. Acad. Sci. U.S.A.* 42, 404.
Magoon, E. F. and Zechmeister, L. (1957). *Archs Biochem. Biophys.* 68, 263.
Mallams, A. K., Waight, E. S., Weedon, B. C. L., Chapman, D. J., Haxo, F. T., Goodwin, T. W. and Thomas, D. M. (1967). *Proc. chem. Soc.* 301-302.

Mandelli, E. F. (1968). *J. Phycol.* 4, 347.
Mayer, F. and Czygan, F. C. (1969). *Planta* 86, 175.
Meissner, G. and Delbrück, M. (1968). *Pl. Physiol., Lancaster* 43, 1279-1283.
Mercer, E. I., Davies, B. H. and Goodwin, T. W. (1963). *Biochem. J.* 87, 317-325.
Naef, J. and Turian, G. (1963). *Phytochemistry* 2, 173-177.
Neamtu, G. and Bodea, C. (1971). *Revue roum. biochim.* 8, 129-133.
Nitsche, H. (1973). *Arch. Mikrobiol.* 90, 151.
Nitsche, H. and Egger, K. (1970). *Tetrahedron Lett.* 1435-1438.
Nitsche, H., Egger, K. and Dabbagh, A. G. (1969). *Tetrahedron Lett.* 2999-3002.
Norgård, S., Svec, W. A., Liaaen-Jensen, S., Jensen, A. and Guillard, R. R. L. (1974). *Biochem. Systematics Ecol.* 2, 3-6.
Olson, R. A., Jennings, W. R. and Allen, M. B. (1967). *J. cell. comp. Physiol.* 70, 133.
Parsons, T. R. (1961). *J. Fish. Res. Bd Can.* 18, 1017-1025.
Parsons, T. R. and Strickland, J. D. H. (1963). *J. mar. Res.* 21, 155-163.
Petzold, E. N. and Quackenbush, F. W. (1960). *Archs Biochem. Biophys.* 86, 163-165.
Petzold, E. N., Quackenbush, F. W. and McQuistan, M. (1959). *Archs Biochem. Biophys.* 82, 117-124.
Pinckard, J. H., Kittredge, J. S., Fox, D. L., Haxo, F. T. and Zechmeister, L. (1953). *Archs. Biochem. Biophys.* 44, 189-199.
Rebeiz, C. A. (1968). *Inst. Rech. Agron. Liban.* pub. no. 21.
Ricketts, T. R. (1970). *Phytochemistry* 9, 1835-1842.
Ricketts, T. R. (1971a). *Phytochemistry* 10, 155-160.
Ricketts, T. R. (1971b). *Phytochemistry* 10, 161-164.
Riley, J. P. and Wilson, T. R. S. (1967) *J. mar. biol. Ass. U.K.* 45, 583-591.
Saenger, P. and Rowan, K. S. (1968). *Helgoloendes wiss. Meeresuntin* 18, 549.
Schuphan, W. (1965). "Nutritional Values in Crops and Plants". Faber, London.
Smallidge, R. L. and Quackenbush, F. W. (1973). *Phytochemistry* 12, 2481-2485.
Sprey, B. (1970). *Protoplasma* 71, 235-250.
Steffen, K. (1964). *Planta* 60, 506.
Stewart, I. and Wheaton, T. A. (1973). *Phytochemistry*, 12, 2947.
Strain, H. H. (1938). *J. biol. Chem.* 123, 425.
Strain, H. H. (1951). *In* "Manual of Phycology" (G. M. Smith, ed.), 243-262. Chronica Botanica, Waltham, Mass.
Strain, H. H. (1958). "Chloroplast Pigments and Chromatographic Analysis, p. 180. 32nd Annual Priestley Lectures, Penn. State University, University Park, Pa, U.S.A.
Strain, H. H. (1965). *Biol. Bull. mar. biol. Lab., Woods Hole,* 129, 366.
Strain, H. H. (1966). *In* "Biochemistry of Chloroplasts" (T. W. Goodwin, ed.), vol. 1, pp. 387-406. Academic Press, London and New York.
Strain, H. H., Manning, W. M. and Hardin, G. J. (1944). *Biol. Bull. mar. biol. Lab. Woods Hole* 86, 169-191.
Strain, H. H., Svec, W. A., Aitzetmüller, K., Grandolfo, M. and Katz, J. J. (1969). *Phytochemistry* 7, 1417-1418.
Strain, H. H., Benton, F. L., Grandolfo, M. C., Aitzetmüller, K., Svec, W. A. and Katz, J. J. (1970). *Phytochemistry* 9, 2561-2565.

Strain, H. H., Svec, W. A., Aitzetmüller, K., Grandolfo, M. C., Katz, J. J., Kjøsen, H., Norgård, S., Liaaen-Jensen, S., Haxo, F. T., Wegfahrt, P. and Rapaport, H. (1971). *J. Am. chem. Soc.* **93**, 1823.
Stransky, H. and Hager, A. (1970). *Arch. Mikrobiol.* **71**, 164; **72**, 84.
Sugano, N. and Hayashi, K. (1967). *Bot. Mag., Tokyo* **80**, 440.
Sugano, N., Miya, S. and Nishi, A. (1971). *Pl. Cell Physiol.* **12**, 525.
Suzuki, K., Ishu, T. and Amano, H. (1970). *Scient. Rep. Saitama Univ. Ser. B.* **5**, 169.
Sweeney, B. M., Haxo, F. T. and Hastings, F. W. (1959). *J. gen. Physiol.* **43**, 285.
Szabolcs, J. and Ronai, A. (1969). *Acta chim. hung.* **61**, 309.
Tappi, G. (1949-50). *Atti Accad. Sci. Torino* **84**, 97.
Taylor, J., Thomas, R. J. and Otero, J. G. (1972). *Bryologist* **75**, 36.
Tevini, M. and Lichtenthaler, H. K. (1970). *Z. PflPhysiol.* **62**, 17.
Theimer, R. R. and Rau, W. (1969). *Biochim. biophys. Acta* **177**, 180.
Thomas, D. M. and Goodwin, T. W. (1965). *J. Phycol.* **1**, 118-121.
Thomas, D. M., Goodwin, T. W. and Ryley, J. F. (1967). *J. Protozool.* **14**, 654-657.
Tischer, J. (1941). *Hoppe-Seyler's Z. physiol. Chem.* **240**, 191.
Tóth, G. and Szabolcs, J. (1970). *Acta chim. hung.* **64**, 393.
Tomes, M. L. (1963). *Bot. Gaz.* **124**, 180-185.
Trosper, T. and Allen, C. F. (1973). *Pl. Physiol., Lancaster* **51**, 584-585.
Valadon, L. R. G. and Mummery, R. S. (1966). *J. gen. Microbiol.* **45**, 531.
Valadon, L. R. S. and Mummery, R. S. (1967). *Ann. Bot.* **31**, 494.
Valadon, L. R. G. and Mummery, R. S. (1968). *Nature, Lond.* **217**, 1066.
Valadon, L. R. G. and Mummery, R. S. (1969). *J. exp. Bot.* **20**, 732-742.
Valadon, L. R. G. and Mummery, R. S. (1971). *Phytochemistry* **10**, 2349-2353.
Vanhaelen, M. (1973). *Planta med.* **23**, 301.
Walton, T. J., Britton, G., Goodwin, T. W., Diner, B. and Moshier, S. (1970). *Phytochemistry* **9**, 2545-2552.
Weber, A. and Czygan, F. C. (1972). *Arch. Mikrobiol.* **84**, 243.
White, J. W., Zscheile, F. P. and Brunson, A. M. (1942). *J. Am. chem. Soc.* **64**, 2693.
Whittle, S. J. and Casselton, P. J. (1969). *Br. Phycol. J.* **4**, 55.
Więckowski, S. (1961). *Bull. Acad. pol. Sci. Cl. II Ser. Sci. biol.* **9**, 325-332.
Wierzchowski, Z. (1965). *Z. Proto Post. Nauk Pol.* **53**, 305.
Williams, B. L. and Goodwin, T. W. (1965). *Phytochemistry* **4**, 81-88.
Winkenbach, F., Wolk, C. P. and Jost, M. (1972). *Planta* **107**, 69.
Wolf, F. T. (1963). *Pl. Physiol., Lancaster* **38**, 649-652.
Yamamoto, H., Yokoyama, H. and Boettger, H. (1969). *J. org. Chem.*, **34**, 4207-4208.
Yokoyama, H. and White, M. J. (1966). *Phytochemistry*, **5**, 1159-1173.
Yokoyama, H. and White, M. J. (1968). *Phytochemistry* **7**, 1031-1034.
Yokoyama, H., Guerrero, H. C. and Boettger, H. (1972). *In* "Chemistry of Plant Pigments" (C. O. Chichester, ed.). Academic Press, New York and London.
Zechmeister, L. and Escue, R. B. (1941). *Proc. natn. Acad. Sci. U.S.A.* **27**, 528.
Zechmeister, L. and Pinckard, J. C. (1947). *J. Am. chem. Soc.* **69**, 1930.
Zechmeister, L. and Schroeder, W. A. (1943). *J. Am. chem. Soc.* **65**, 1535.
Zechmeister, L. and Tuzson, P. (1936). *Hoppe-Seyler's Z. physiol. Chem.* **240**, 191.

Chapter 5

Biosynthesis of Carotenoids

GEORGE BRITTON

Department of Biochemistry, University of Liverpool, England

I Introduction

Although C_{30}, C_{45} and C_{50} carotenoids are produced by some non-photosynthetic bacteria, most naturally occurring carotenoids, in particular those of higher plants, are C_{40} tetraterpenes, biosynthesized by the well-established terpenoid pathway. In general, terpenoid compounds are built up from C_5 "isoprene units", the carbon skeleton of which is readily seen in the important intermediate isopentenyl pyrophosphate (IPP, I). A series of intermediates with carbon chains composed of multiples of five carbon atoms may be built up by successive addition of C_5 units. In carotenoid biosynthesis, the terpenoid chain is built up to the C_{20} level, and two C_{20} units condense to give the typical C_{40} carotenoid skeleton. Figure 1 presents a general scheme which illustrates the biosynthetic relationship between carotenoids and other major terpenoid classes.

The early steps in terpenoid biosynthesis, up to and including prenyl pyrophosphate production, are considered to be identical in the biosynthesis of all terpenoids. In describing details of these steps, information obtained in studies of the biosynthesis of terpenoids

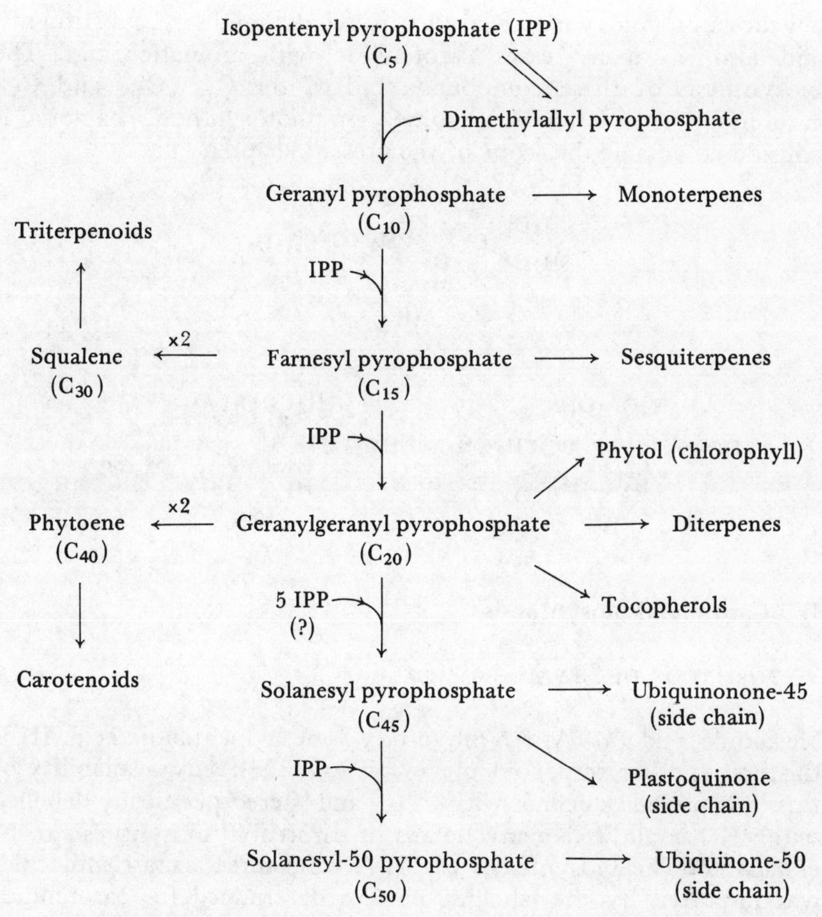

FIG. 1. General scheme for biosynthesis of terpenoids.

other than carotenoids, and in many cases concerned with steroid biosynthesis in animal systems, will be used when this gives an insight into the mechanism or control of an enzyme, when this information is not available from work directly concerned with carotenoids. Evidence for the involvement of each intermediate and each enzyme in carotenoid biosynthesis will be reviewed.

The main emphasis, however, will be placed on the later stages in the pathway, those unique to carotenoids. These discussions will concentrate, as far as possible, on work with higher plants, algae and fungi, but the biosynthesis of bacterial carotenoids will be used to illustrate points when insufficient information is available from plant and fungal studies. Photosynthetic bacteria produce a wide range of acyclic carotenoids with tertiary hydroxyl and methoxyl substituents, and also, in many cases carotenoids with aromatic rings. The biosynthesis of these compounds and of the C_{30}, C_{45} and C_{50} carotenoids encountered in some non-photosynthetic bacteria is considered outside the scope of the present chapter.

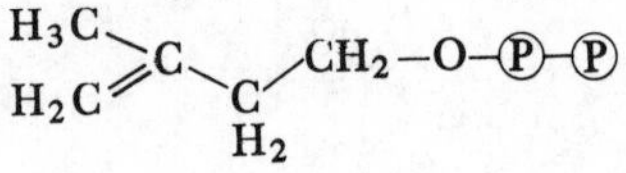

$H_3C(H_2C{=})C{-}CH_2{-}CH_2{-}O{-}Ⓟ{-}Ⓟ$

(I)

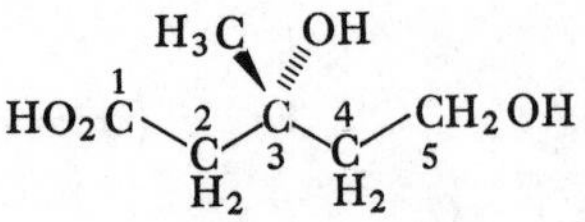

$HO_2\overset{1}{C}{-}\overset{2}{C}H_2{-}\overset{3}{C}(CH_3)(OH){-}\overset{4}{C}H_2{-}\overset{5}{C}H_2OH$

(II)

$HO_2C{-}CH_2{-}C(CH_3)(OH){-}CH_2{-}C({=}O){-}S{-}CoA$

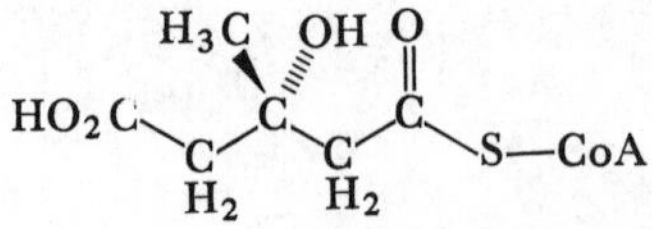

(III)

II Carotenoid biosynthesis

A FORMATION OF MEVALONIC ACID

Mevalonic acid (MVA, 3,5-dihydroxy-3-methylpentanoic acid, **II**) is the first specific terpenoid precursor, and the ready availability of this compound labelled with ^{14}C, and stereospecifically labelled with ^{3}H has allowed many details of carotenoid biosynthesis to be elucidated. The biosynthesis of MVA from three acetate molecules was indicated by the labelling pattern determined for carotenoids and other terpenoids biosynthesized from [1-^{14}C]- and [2-^{14}C]acetate (Grob and Bütler, 1955, 1956; Lotspeich *et al.*, 1959; Braithwaite and Goodwin, 1960; Steele and Gurin, 1960).

1 3-Hydroxy-3-methylglutaryl coenzyme A (HMG-CoA)

It is now established that in the main pathway of MVA biosynthesis, acetyl-CoA and acetoacetyl-CoA are condensed by the enzyme HMG-CoA synthetase (HMG-CoA: acetoacetyl-CoA lyase, EC 4.1.3.5) to form the C_6 compound (3*S*)-3-hydroxy-3-methylglutaryl-CoA (HMG-CoA, **III**), the CoA moiety of acetoacetyl-CoA being retained in the HMG-CoA (Fig. 2a) (Rudney, 1957; Stewart and Rudney, 1966). The involvement of malonyl-CoA in HMG-CoA production (Fig. 2b) has been proposed (Brodie *et al.*,

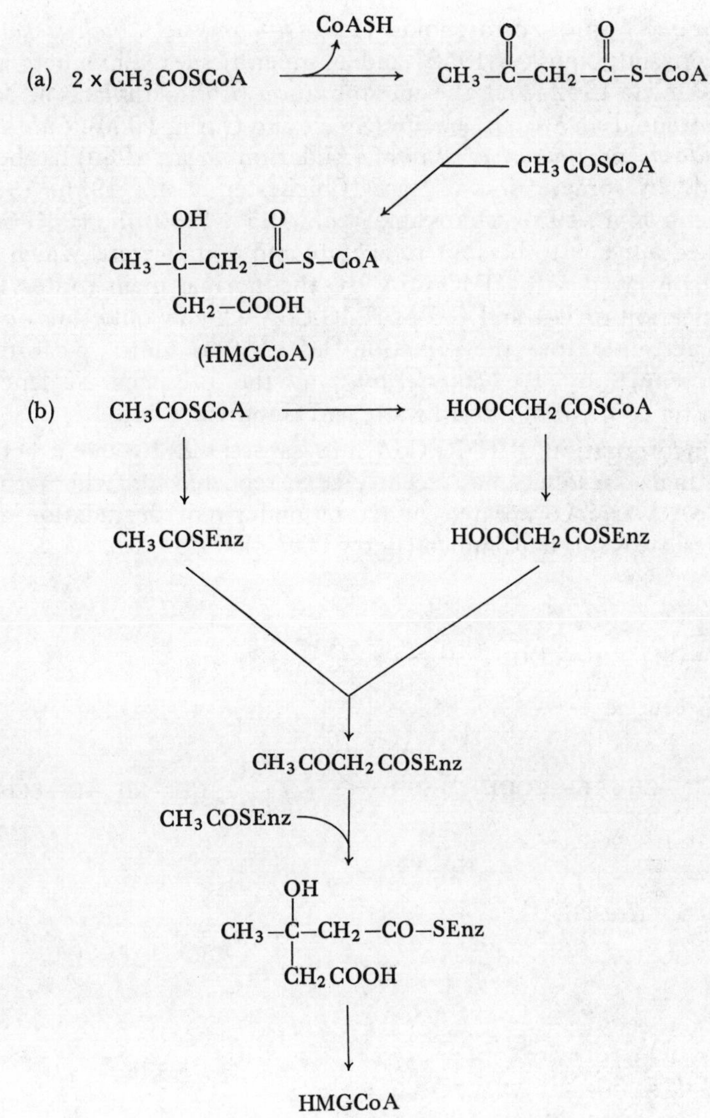

FIG. 2. Alternative pathways for formation of HMG-CoA.

1963), but malonyl-CoA has now been shown not to be involved in HMG-CoA synthesis in liver, yeast and higher plant systems (Higgins and Kekwick, 1973).

HMG-CoA can also be formed from the amino acids leucine and valine, by a process (Fig. 3) involving a biotin-dependent CO_2 fixation (Coon *et al.*, 1959). Leucine and valine are efficiently incorporated into carotenoids in *Phycomyces blakesleeanus* (Goodwin and Lijinsky, 1952) and in green tissues (Ehrenberg and Faludi-Dániel, 1962) and the incorporation of dimethylacrylic acid into carotenoids in *Euglena gracilis* (Steele and Gurin, 1960), *Chlorella pyrenoidosa* and *Blakeslea trispora* (Anderson *et al.*, 1960) has been reported. In some cases at least (Chichester *et al.*, 1955, 1959; Wuhrmann *et al.*, 1957; Yokoyama *et al.*, 1957, 1960) the HMG-CoA so formed appears to be split to acetate and acetoacetate, which are then reconverted into HMG-CoA by the normal main route. The randomization of C-1 and C-5 of HMG-CoA which would thus occur could account for the fixation of $^{14}CO_2$ into β-carotene (β,β-carotene)* by *P. blakesleeanus* in the presence of leucine (Chichester *et al.*, 1959; Braithwaite and Goodwin, 1960).

The incorporation of HMG-CoA into carotenoids by an extract of excised maize seedlings has recently been reported but whether the HMG-CoA was incorporated intact or underwent degradation and resynthesis was not determined (Berry *et al.*, 1972).

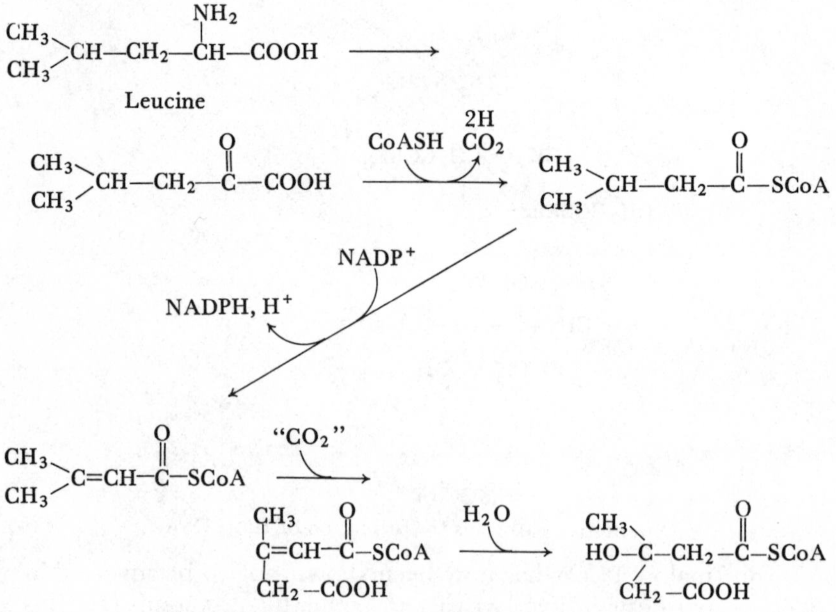

$$(CH_3)_2CH{-}CH_2{-}CH(NH_2){-}COOH \longrightarrow$$

Leucine

$$(CH_3)_2CH{-}CH_2{-}\overset{O}{\overset{\|}{C}}{-}COOH \xrightarrow[\text{CoASH} \quad 2H,\ CO_2]{} (CH_3)_2CH{-}CH_2{-}\overset{O}{\overset{\|}{C}}{-}SCoA$$

$$\xrightarrow{NADP^+ \quad NADPH,\ H^+}$$

$$(CH_3)_2C{=}CH{-}\overset{O}{\overset{\|}{C}}{-}SCoA \xrightarrow{\text{“}CO_2\text{”}} CH_3{-}C(CH_2{-}COOH){=}CH{-}\overset{O}{\overset{\|}{C}}{-}SCoA \xrightarrow{H_2O} HO{-}C(CH_3)(CH_2{-}COOH){-}CH_2{-}\overset{O}{\overset{\|}{C}}{-}SCoA$$

FIG. 3. Formation of HMG-CoA from leucine.

* In general, well-accepted trivial names of carotenoids will be used in this article, but the semisystematic name of each carotenoid according to the IUPAC-IUB Commission will be given, in parentheses, at the first mention.

2 *Mevalonic acid*

The conversion of HMG-CoA into MVA in yeast and liver has been thoroughly studied, but only in a *Hevea* latex system has this reaction been demonstrated in plants (Hepper and Audley, 1969).

(3*S*)HMG-CoA undergoes an essentially irreversible two-step reduction to give (3*R*)MVA (**IV**), the reduction being catalysed by HMG-CoA reductase (mevalonate: NADP oxidoreductase (acylating CoA) (EC 1.1.1.34)) and requiring two molecules of NADPH (Knappe *et al.*, 1959; Brodie and Porter, 1960; Durr and Rudney, 1960) (Fig. 4). Each step is stereospecific for the 4-*pro-R* hydrogen atom of NADPH (Blattmann and Rétey, 1970, 1971; Dugan and Porter, 1971; Beedle *et al.*, 1972). The first step is thought to produce the CoA hemithioacetal (**V**) of mevaldic acid (**VI**) (Rétey *et al.*, 1970) and the hydrogen atom introduced is that which ultimately becomes the 5-*pro-R* hydrogen of MVA. In the second step, the 5-*pro-S* hydrogen is introduced (Blattmann and Rétey, 1970, 1971). Free mevaldate appears not to be involved in this reaction (Brodie and Porter, 1960), although in liver an enzyme, mevaldate reductase (mevalonate: NAD or NADP oxidoreductase, EC 1.1.1.32 or 33), exists which will reduce mevaldate to MVA (Nakamura and Greenberg, 1961; Schlesinger and Coon, 1961; Knauss *et al.*, 1962).

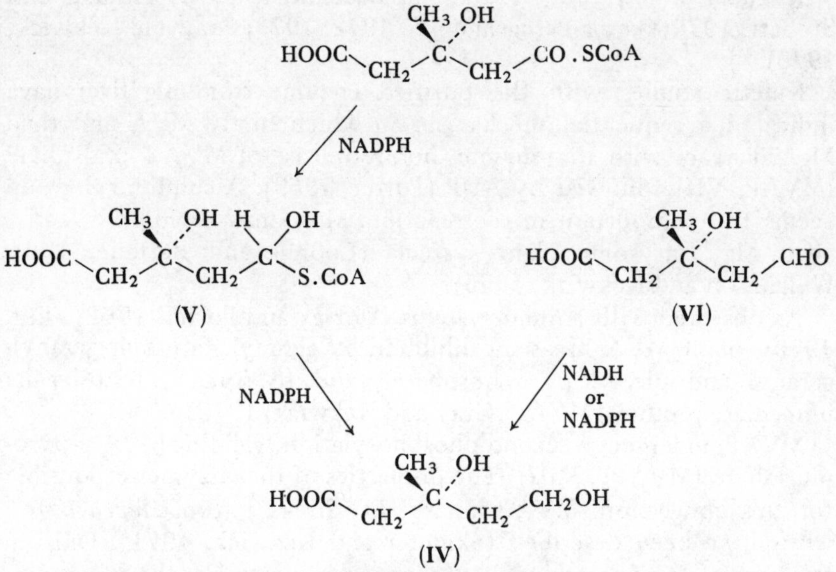

HOOC–CH_2–C(CH_3)(OH)–CH_2–CO.SCoA

↓ NADPH

HOOC–CH_2–C(CH_3)(OH)–CH_2–C(H)(OH)–S.CoA (**V**)   HOOC–CH_2–C(CH_3)(OH)–CH_2–CHO (**VI**)

(**V**) → NADPH; (**VI**) → NADH or NADPH

HOOC–CH_2–C(CH_3)(OH)–CH_2–CH_2OH (**IV**)

FIG. 4. Formation of MVA from HMG-CoA or mevaldate.

In this process, the hydrogen introduced becomes the 5-*pro-R* hydrogen atom of MVA (Donninger and Popják, 1966).

It is now clear that in sterol-synthesizing systems HMG-CoA reductase is a key control point. In rat liver, cholesterol exerts end-product control on this enzyme and HMG-CoA synthase (Siperstein and Fagan, 1966; White and Rudney, 1970; Shepherd and Booth, 1971). No such regulatory mechanisms have yet been demonstrated in carotenogenic systems.

B FORMATION OF GERANYLGERANYL PYROPHOSPHATE

1 *MVA-5-pyrophosphate*

The incorporation of MVA into carotenoids has been accomplished in very many carotenogenic systems, and has become a routine laboratory technique. The presence of the first enzyme in the pathway from MVA to carotenoids, MVA kinase (ATP-mevalonate 5-phosphotransferase, EC 2.7.1.36) has been demonstrated in many plant tissues, and pH optima ranging from 5·5 to 7·5 have been reported (Archer *et al.*, 1963; Loomis and Battaile, 1963; Rogers *et al.*, 1965; Williamson and Kekwick, 1965; Pollard *et al.*, 1966; Valenzuela *et al.*, 1966; Potty and Bruemmer, 1970; Thomas and Stobart, 1970; García-Peregrín *et al.*, 1972, 1973; Gray and Kekwick, 1973).

Kinetic studies with the purified enzyme from pig liver have indicated a sequential mechanism in which firstly MVA and then MgATP react with the enzyme before release of MVA 5-phosphate (MVAP, **VII**) followed by ADP (Porter, 1969). A sulphydryl group seems to be important in the reaction. Mn^{2+} may be more effective than Mg^{2+} in some plant systems (Loomis and Battaile, 1963; Williamson and Kekwick, 1965).

As observed with animal systems (Dorsey and Porter, 1968; Flint, 1970), plant MVA kinase is inhibited by geranyl, farnesyl, geranylgeranyl and phytyl pyrophosphates, and to some extent by its immediate product, MVAP (Gray and Kekwick, 1972).

MVAP undergoes a second phosphorylation, yielding MVA 5-pyrophosphate (MVAPP, **VIII**). The properties of the enzyme responsible for this conversion, MVAP kinase (EC 2.7.4.2) from *Hevea brasiliensis* have been described (Skilleter and Kekwick, 1971). Only in the case of the bacterium *Staphylococcus aureus* has the incorporation of MVAP into carotenoids been reported (Ohnoki *et al.*, 1962).

2 *Isopentenyl pyrophosphate*

In the presence of ATP, MVAPP is converted into the extremely important isopentenyl pyrophosphate (IPP, IX). This reaction is catalysed by the enzyme MVAPP anhydrodecarboxylase (EC 4.1.1.33), and a mechanism has been proposed in which MVAPP is phosphorylated at C-3, and the resultant MVA-3-phosphate-5-pyrophosphate is dephosphorylated and loses CO_2 to give IPP (Lindberg *et al.*, 1962). However, no MVA derivative with a phosphate group at C-3 has ever been detected, and a concerted mechanism is thought more likely (Cornforth *et al.*, 1966b).

Although the involvement of MVAPP in carotenoid biosynthesis has been indicated only in a bacterial system (Suzue, 1964), the incorporation of IPP into carotenoids has been demonstrated in many systems, notably the spinach chloroplast and tomato plastid preparations of Porter (Suzue and Porter, 1969; Subbarayan *et al.*, 1970).

The pathway of IPP formation from MVA is illustrated in Fig. 5.

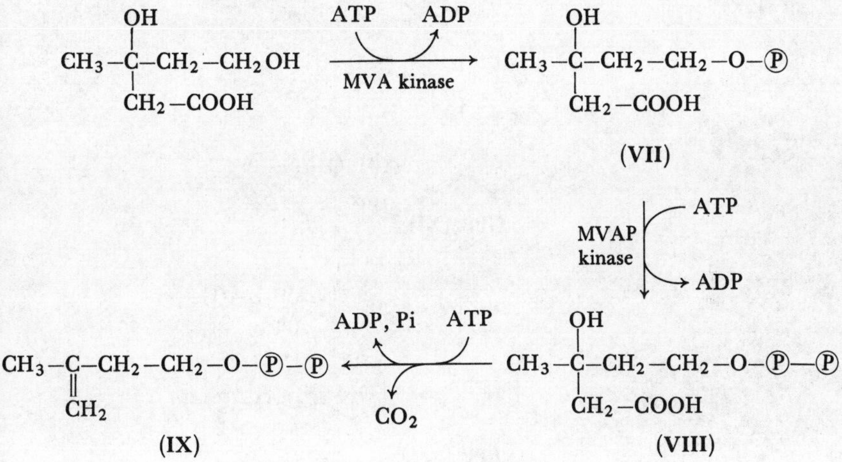

$CH_3-C(OH)(CH_2-COOH)-CH_2-CH_2OH$ —(ATP → ADP; MVA kinase)→ $CH_3-C(OH)(CH_2-COOH)-CH_2-CH_2-O-Ⓟ$ **(VII)**

(VII) —(ATP → ADP; MVAP kinase)→ $CH_3-C(OH)(CH_2-COOH)-CH_2-CH_2-O-Ⓟ-Ⓟ$ **(VIII)**

(VIII) —(ATP → ADP, Pi; CO_2)→ $CH_3-C(=CH_2)-CH_2-CH_2-O-Ⓟ-Ⓟ$ **(IX)**

FIG. 5. Formation of IPP from MVA.

3 *Geranylgeranyl pyrophosphate*

IPP is the fundamental C_5 biosynthetic unit from which the carotenoids, and indeed all terpenoids, are constructed. However, an isomerization of IPP into dimethylallyl pyrophosphate (DMAPP, X) must occur before chain elongation can begin (Agranoff *et al.*, 1960; Shah *et al.*, 1965; Holloway and Popják, 1968). The DMAPP thus formed is the substrate of the prenyl transferase enzyme and condenses with a molecule of IPP to give the C_{10} compound geranyl pyrophosphate (GPP, **XI**) (Agranoff *et al.*, 1959, 1960; Dorsey *et al.*, 1966; Holloway and Popják, 1967). This then adds two further IPP units to produce successively farnesyl pyrophosphate (FPP, **XII**), the C_{15} precursor of sterols and triterpenes (Lynen *et al.*, 1958, 1959; Popják, 1959; Goodman and Popják, 1960), and the C_{20} geranylgeranyl pyrophosphate (GGPP, **XIII**) (Grob *et al.*, 1961; Kandutsch *et al.*, 1964; Nandi and Porter, 1964; Jungalwala and Porter, 1967). The purification and some properties of GGPP synthetase from *Micrococcus lysodeikticus* (Kandutsch *et al.*, 1964), tomato plastids (Porter, 1969) and pumpkin fruit (*Cucurbita pepo*)

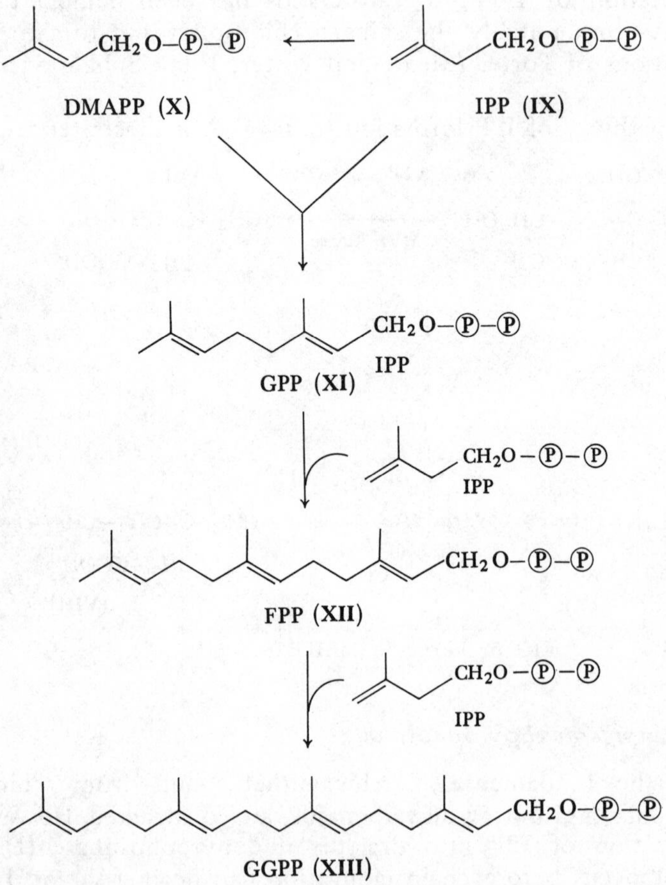

CH_2O–Ⓟ–Ⓟ ← CH_2O–Ⓟ–Ⓟ

DMAPP (X) IPP (IX)

CH_2O–Ⓟ–Ⓟ

GPP (XI) IPP

CH_2O–Ⓟ–Ⓟ
IPP

CH_2O–Ⓟ–Ⓟ

FPP (XII)

CH_2O–Ⓟ–Ⓟ
IPP

CH_2O–Ⓟ–Ⓟ

GGPP (XIII)

FIG. 6. Formation of GGPP from IPP.

(Ogura *et al.* 1972) have been described. The biosynthesis of GGPP from IPP is outlined in Fig. 6.

The presence of IPP isomerase (EC 5.3.3.2) in carotenogenic tomato plastid preparations has been demonstrated and the production of GGPP from IPP and DMAPP by a tomato plastid enzyme system has been studied in some detail (Porter, 1969), but the direct incorporation of DMAPP into carotenoids has not been reported. No demonstration of the incorporation of GPP into carotenoids has been achieved, but the reported incorporation of free geraniol into β-carotene by carrot root slices (Varma and Chichester, 1962) provides an example of the apparent ability of higher plants to pyrophosphorylate higher terpenoid alcohols (Battersby, 1970). FPP has been incorporated into phytoene by a tomato plastid enzyme system (Jungalwala and Porter, 1967) and into β-carotene by a *Phycomyces* preparation (Yamamoto *et al.*, 1961). It was necessary to add MVA or IPP to these systems, thus providing the extra C_5 unit necessary for GGPP production. GGPP has been shown to be a direct precursor of carotenoids in a *Phycomyces* preparation (Lee and Chichester, 1969), a tomato plastid enzyme system (Shah *et al.*, 1968) and in bean leaf chloroplasts (Buggy *et al.*, 1974).

4 *Stereochemistry*

The absolute configuration of HMG-CoA (3*S*) is known (Stewart and Rudney, 1966) and natural MVA has been shown to be the 3*R* isomer (Eberle and Arigoni, 1960); only this isomer serves as substrate for MVA kinase (Cornforth *et al.*, 1962). In addition to this one chiral centre, MVA also contains three prochiral centres, C-2, C-4 and C-5. The preparation of MVA with the hydrogen atoms of the methylene groups at C-2, C-4 and C-5 stereospecifically labelled with deuterium or tritium (Cornforth *et al.*, 1966a,b; Donninger and Popják, 1966; Blattmann and Rétey, 1970; Cornforth and Ross, 1970; Scott *et al.*, 1970) has enabled details of the stereochemistry of many of the steps in terpenoid biosynthesis to be determined. In particular, the stereochemistry of formation of IPP from MVAPP, of the isomerization of IPP to DMAPP, and of the prenyl transferase reaction by which the terpenoid chain is lengthened has been established. In this work, enzymes obtained from sterol-synthesizing tissues were used, but the stereochemistry of the same reactions in carotenoid-synthesizing systems is unlikely to be different.

In the formation of IPP from MVAPP (Fig. 7), *trans* elimination of

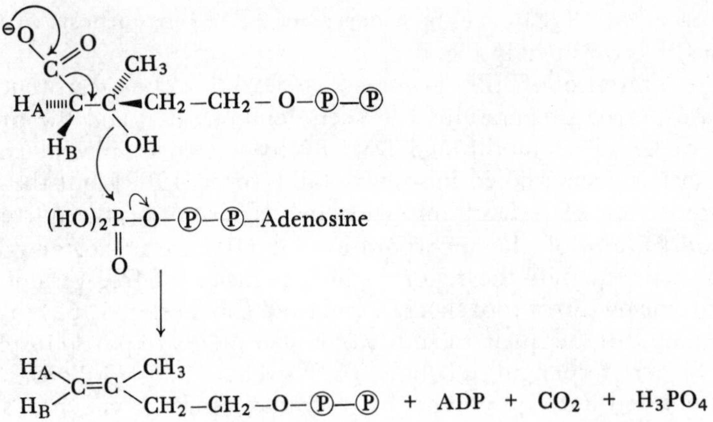

FIG. 7. Stereochemistry of IPP formation from MVAPP.

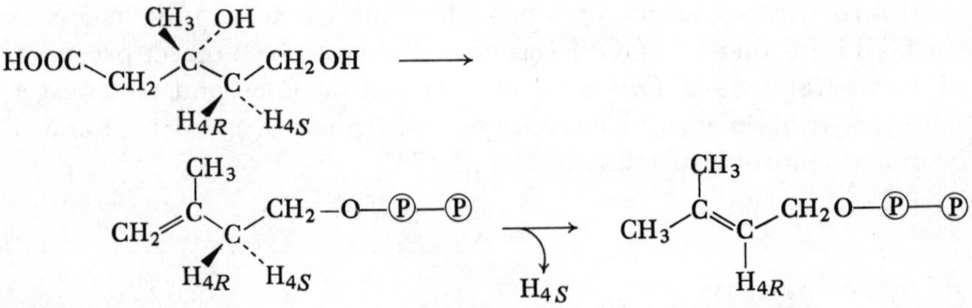

FIG. 8. Stereochemistry of hydrogen loss in the isomerization of IPP into DMAPP.

the hydroxyl group at C-3 and the carboxyl group occurs, and the stereochemical results are consistent with a concerted mechanism, as shown (Cornforth *et al.*, 1966b).

In the isomerization of IPP to DMAPP, proton loss from C-2 of IPP (originally C-4 of MVA) is stereospecific, the hydrogen lost being that which was originally the 4-*pro-S* hydrogen atom of MVA (Fig. 8) (Cornforth *et al.*, 1966a). It has recently been established (Cornforth *et al.*, 1972) that the proton added to the vinylic carbon atom is added to the *re,re* side of the double bond. The experimental results are in agreement with the concerted mechanism shown in Fig. 9.

It is important to note that the reaction catalysed by IPP isomerase is reversible, the proportions at equilibrium of IPP and DMAPP being approximately 1:9 (Agranoff *et al.*, 1960; Shah *et al.*, 1965; Holloway and Popják, 1968). Thus although the olefinic protons of IPP are stereochemically distinct, conversion of IPP into

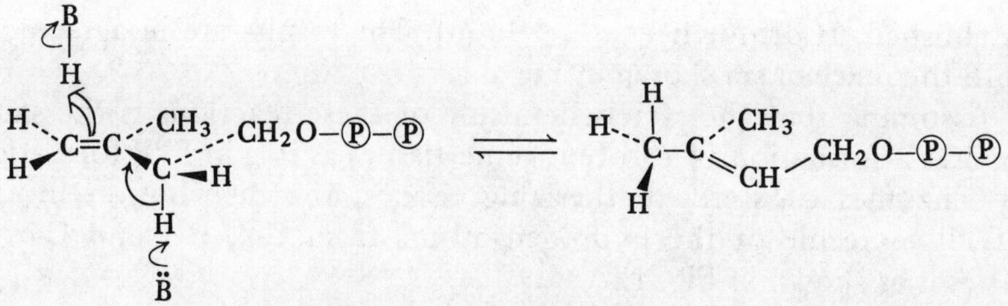

FIG. 9. Mechanism of the IPP isomerase reaction.

DMAPP destroys this stereospecificity. Isomerization of DMAPP can then give any one of three species of IPP, including an equal mixture of the two labelled species (**XIV**) and (**XV**) (Fig. 10). The overall stereospecificity of labelling of IPP is thus lost. This means that great care must be exercised in interpreting the results of experiments with (2*R*)-[2-$^3H_1$]- and (2*S*)-[2-$^3H_1$]MVA as substrates, in tissues where prenyl transferase may be rather sluggish compared with IPP isomerase. Green and Baisted (1972) have shown that in germinating peas, IPP isomerase is indeed more active than prenyl transferase.

The prenyl transferase reaction also involves loss from C-2 of IPP of the hydrogen atom that was originally the 4-*pro-S* hydrogen of MVA (Cornforth *et al.*, 1966a). The behaviour in this reaction of the hydrogen atoms arising from C-2 and C-5 of MVA has been

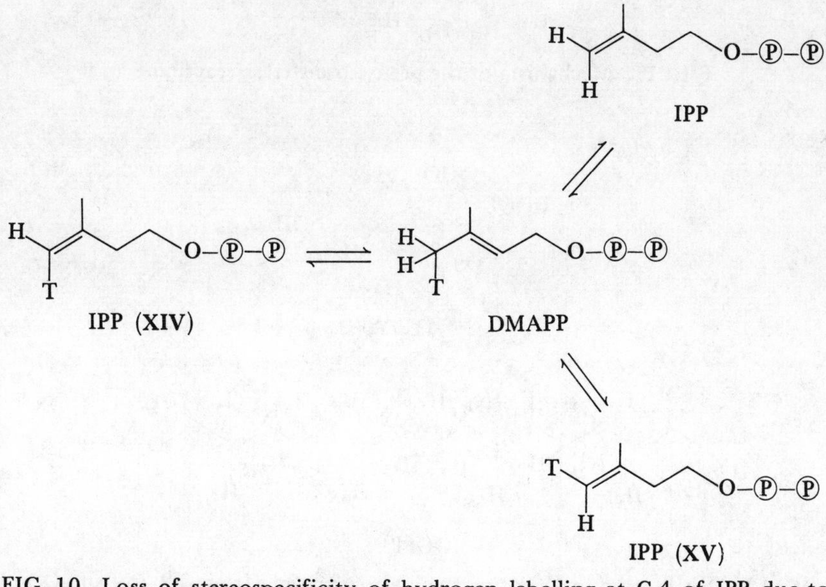

FIG. 10. Loss of stereospecificity of hydrogen labelling at C-4 of IPP due to isomerization to DMAPP.

established. (Cornforth *et al.*, 1966b). The results are in agreement with the mechanism shown in Fig. 11.

Assuming that the stereochemistry of these reactions is the same in GGPP formation in carotenogenic tissues as it is in FPP formation by enzymes of sterol-synthesizing tissues, the distribution in the GGPP molecule of the hydrogen atoms from C-2, C-4 and C-5 of MVA is as shown in Fig. 12.

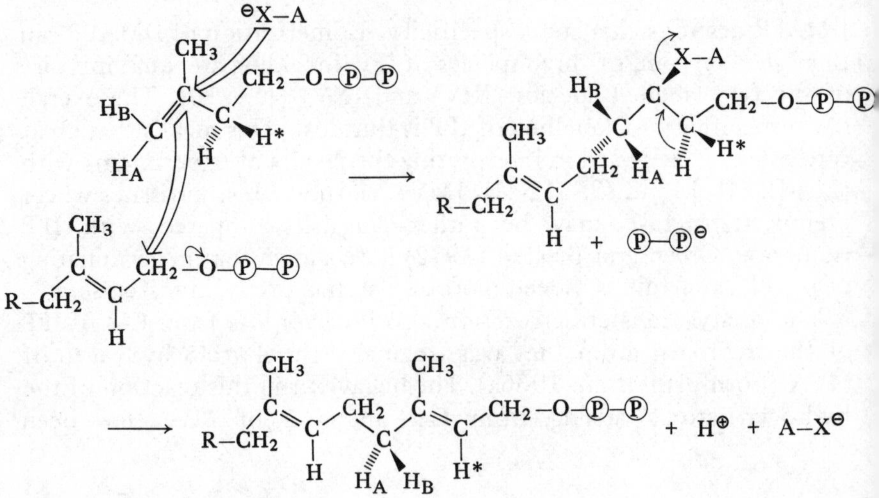

$^{\ominus}$X—A
CH_3
C
CH_2—O—Ⓟ—Ⓟ
H_B
C
C
H*
H
H_A
CH_3
C
CH_2—O—Ⓟ—Ⓟ
R—CH_2
C
H
⟶
X—A
H_B
C
CH_2—O—Ⓟ—Ⓟ
C
C
CH_3
H*
C
CH_2
H_A
H
R—CH_2
C
H + Ⓟ—Ⓟ$^{\ominus}$

⟶
CH_3
CH_3
R—CH_2
C
C
CH_2
C
C
C
CH_2—O—Ⓟ—Ⓟ + $H^{\oplus}$ + A—$X^{\ominus}$
H
H_A
H_B
H*

FIG. 11. Mechanism of the prenyl transferase reaction.

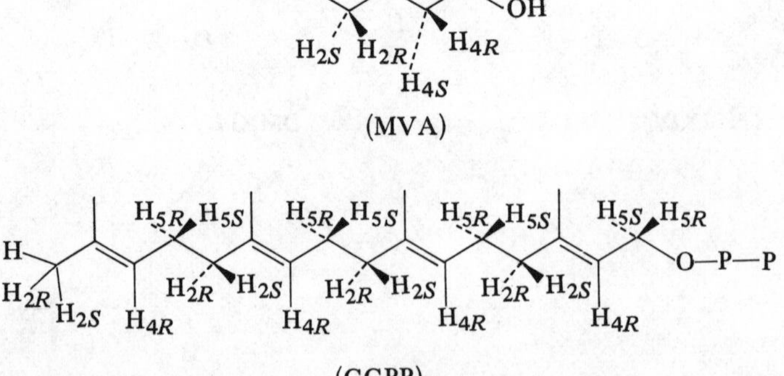

H_{5S}
HO
CH_3
H_{5R}
HOOC
OH
H_{2S}
H_{2R}
H_{4R}
H_{4S}
(MVA)

H_{5R} H_{5S} H_{5R} H_{5S} H_{5R} H_{5S} H_{5S} H_{5R}
H
O—P—P
H_{2R} H_{2R} H_{2S} H_{2R} H_{2S} H_{2R} H_{2S}
H_{2S} H_{4R} H_{4R} H_{4R} H_{4R}

(GGPP)

FIG. 12. Distribution in GGPP of the hydrogen atoms from C-2, C-4 and C-5 of MVA.

The absence of tritium from carotenoids biosynthesized from $[2\text{-}^{14}C,(4S)\text{-}4\text{-}^{3}H_1]$ MVA confirms that the stereochemistry of hydrogen loss occurring in the IPP isomerase and prenyl transferase reactions is the same in carotenogenic systems as described above (Goodwin and Williams, 1966).

C FORMATION OF PHYTOENE

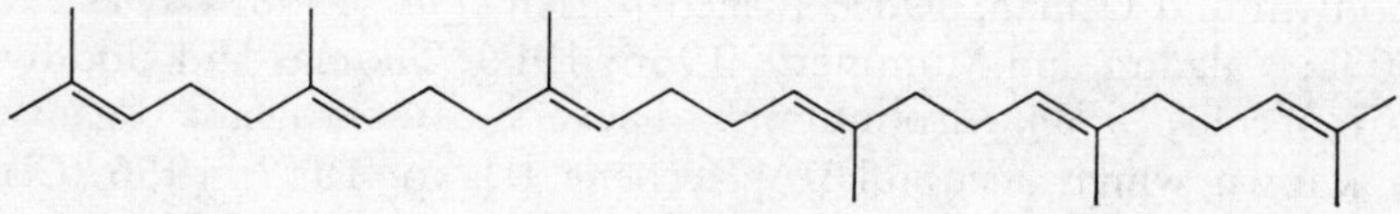

Squalene (XVI)

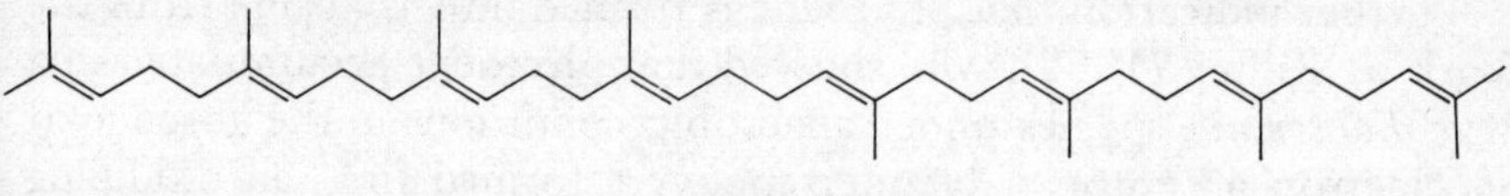

Lycopersene (XVII)

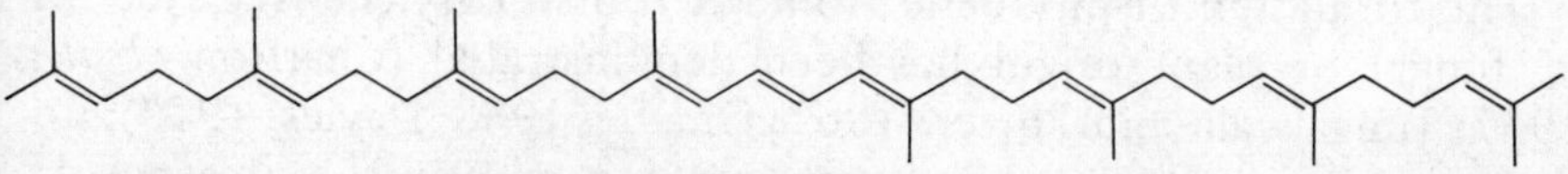

Phytoene (XVIII)

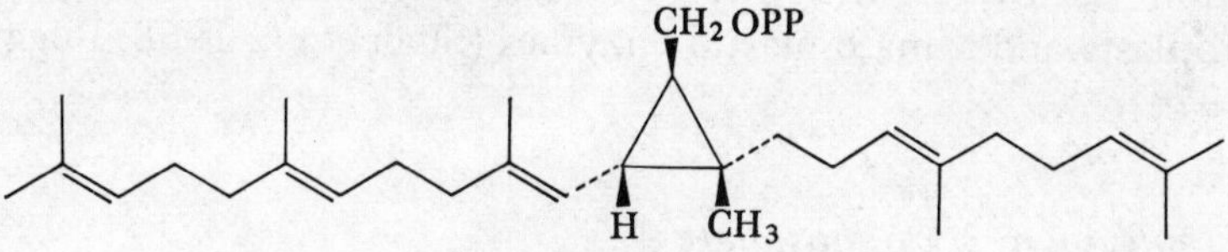

CH_2OPP

H CH_3

Presqualene pyrophosphate (XIX)

In sterol biosynthesis, the condensation of two molecules of the C_{15} FPP in the presence of NADPH gives the triterpene squalene (**XVI**). An analogous reaction in carotenoid biosynthesis would yield lycopersene (7,8,11,12,15,7′,8′,11′,12′,15′-decahydro-ψ,ψ-carotene, **XVII**) from two molecules of GGPP. Early work did suggest the presence of lycopersene in *Neurospora crassa* (Grob and Boschetti, 1962) and some carrot strains (Nusbaum-Cassuto and Villoutreix, 1965), and the formation of lycopersene from GGPP by *N. crassa* extracts was reported (Grob *et al.*, 1961). Many later investigations, however, failed to reveal the presence of lycopersene (Anderson and

Porter, 1962; Pennock *et al.*, 1962; Beeler *et al.*, 1963; Davies *et al.*, 1963a; Mercer *et al.*, 1963), and evidence accumulated to support the idea that phytoene (7,8,11,12,7′,8′,11′,12′-octahydro-ψ,ψ-carotene, **XVII**), rather than lycopersene, is the first C_{40} hydrocarbon produced. Thus, in a number of micro-organisms, diphenylamine (DPA) inhibits carotene production and causes an accumulation of large amounts of phytoene (Turian, 1950, 1957; Turian and Haxo, 1952; Goodwin, 1953; Goodwin *et al.*, 1953; Goodwin and Osman, 1954; Liaaen-Jensen *et al.*, 1958; Davies *et al.*, 1963a; Valadon and Mummery, 1966; 1969; Thomas and Goodwin, 1967; Weeks, 1971). Also mutants of several carotenogenic organisms are known which accumulate phytoene (Haxo, 1952, 1956; Claes, 1954; Griffiths and Stanier, 1956; Kessler and Czygan, 1966). In none of these cases was lycopersene detected.

Further indication that phytoene is formed directly came from the work of Rilling (1962), who showed that phytoene accumulated in a *Mycobacterium* species under anaerobic conditions in the absence of an electron acceptor; if lycopersene were formed first, an oxidizing step would be necessary to convert this into phytoene.

The formation of phytoene from MVA or IPP by cell-free systems of fungal or plant origin has been demonstrated (Charlton *et al.*, 1967; Jungalwala and Porter, 1967; Graebe, 1968; Davies, 1973); no lycopersene was detected in these systems, even in the presence of NADPH, which stimulated squalene production. The direct incorporation of GGPP into phytoene has also been achieved, with chloroplasts and tomato plastid enzymes (Shah *et al.*, 1968; Buggy *et al.*, 1974).

1 *Prephytoene pyrophosphate*

An intermediate between FPP and squalene, first detected some years ago (Rilling, 1966), has now been isolated and characterized as presqualene pyrophosphate (**XIX**) (Epstein and Rilling, 1970; Edmond *et al.*, 1971; Rilling *et al.*, 1971). This prompted the search for a similar intermediate between GGPP and phytoene in carotenoid biosynthesis. Such a compound, termed prephytoene pyrophosphate (**XX**), was isolated from a *Mycobacterium* preparation, and the natural and a synthetic sample were each converted into phytoene by a cell-free system from *Mycobacterium* (Altman *et al.*, 1972). Porter and co-workers (Qureshi *et al.*, 1972, 1973) have shown that incubation of yeast squalene synthetase with GGPP in the absence of NADPH produces prephytoene pyrophosphate (which they call prelycopersene pyrophosphate); in the presence of NADPH, lycopersene was produced. A similar result was obtained with a tomato plastid enzyme system (Qureshi *et al.*, 1972; Barnes *et al.*, 1973). In this case, however, incubation with GGPP in the presence of NADPH and $NADP^+$ resulted in the formation of phytoene. With NADPH alone, lycopersene was produced. The presence in the preparation of squalene synthetase, acting on GGPP, could account for the production of lycopersene, though not for the apparent incorporation of lycopersene into phytoene by the same system in the presence of $NADP^+$ and FAD.

2 *Stereochemistry*

Any mechanism proposed for phytoene formation must take into account two details of stereochemistry, the stereochemistry of hydrogen loss, and the configuration of the phytoene produced.

The situation regarding the nature of the phytoene produced is somewhat confusing. Phytoene isolated from tomatoes, carrot oil and a *Chlorella vulgaris* mutant has the 15-*cis* configuration (Rabourn *et al.*, 1954; Rabourn and Quackenbush, 1956; Davis *et al.*, 1961; Jungalwala and Porter, 1965); this, and the *trans* nature of the C-13 and C-13′ double bonds, have been confirmed by modern methods (Khatoon *et al.*, 1972). Recent studies have confirmed these findings, and show that the 15-*cis* isomer is the main component of the phytoene isolated from fungi (*Phycomyces blakesleeanus, Neurospora crassa* and *Mucor hiemalis*) (Aung Than *et al.*, 1972; Herber *et al.*, 1972), but two bacteria, *Flavobacterium dehydrogenans* and *Mycobacterium* species, produce almost entirely all-*trans* phytoene (Weeks, 1971; Gregonis and Rilling, 1973).

In the formation of phytoene, one hydrogen atom is lost from each of the two molecules of GGPP. Incubations of tomato slices and isolated bean chloropasts with [2-^{14}C,(5*R*)-5-$^{3}H_1$]MVA and [2-^{14}C,5-$^{3}H_2$]MVA have shown that the hydrogen lost in each case is the 1-*pro-S* hydrogen atom of GGPP, originally the 5-*pro-S* hydrogen of MVA (Williams *et al.*, 1967b; Buggy *et al.*, 1969). As indicated above, all samples of higher plant phytoene so far examined have been predominantly the 15-*cis* isomer. Very recent work by Gregonis and Rilling (1974) has confirmed the stereochemistry of hydrogen loss in the formation of *cis*-phytoene. Incubation of a *Phycomyces blakesleeanus* preparation with (1*S*)-[1-$^{3}H_1$,4-^{14}C]GGPP gave 15-*cis* phytoene which retained no tritium. In contrast to this, all-*trans* phytoene isolated from a

Mycobacterium preparation after incubation with the same substrate retained one labelled hydrogen atom. The stereochemistry of hydrogen loss from GGPP thus appears different in the formation of *cis*- and *trans*-phytoene.

3 *Mechanism*

The formation of prephytoene pyrophosphate and phytoene from GGPP is likely to involve a mechanism similar to that of the formation of presqualene pyrophosphate and squalene from FPP. Several mechanisms have been proposed for this (Epstein and Rilling, 1970; Altman *et al.*, 1971; Edmond *et al.*, 1971; Rilling *et al.*, 1971; van Tamelen and Schwartz, 1971; Beytia *et al.*, 1973).

a. Prephytoene pyrophosphate. Figure 13 illustrates a scheme for the formation of prephytoene pyrophosphate from GGPP, consistent with the available data, including kinetic data (Beytia *et al.*, 1973) for presqualene biosynthesis from FPP. This mechanism, rather than that proposed by the same authors (Qureshi *et al.*, 1973) for prelycopersene pyrophosphate, is quoted because it gives rise to prephytoene pyrophosphate with the same absolute configuration (1*R*, 2*R*, 3*R*) as has been established for presqualene pyrophosphate (Popják *et al.*, 1973); consideration of optical properties indicates that the stereochemistry of the two intermediates is the same (Qureshi *et al.*, 1973).

b. Phytoene via lycopersene. The formation of lycopersene from prelycopersene pyrophosphate (= prephytoene pyrophosphate) would be entirely analogous to the formation of squalene from presqualene pyrophosphate, and a hydrogen atom from NADPH would be added as the final step. Figures 14(a) and 15(a) illustrate two mechanisms for lycopersene formation, based upon two mechanisms proposed for squalene synthesis (Epstein and Rilling, 1970; Edmond *et al.*, 1971).

Detailed analysis of the stereochemistry of hydrogen loss in the formation of phytoene, in conjunction with the distribution of hydrogen atoms at the two central carbon atoms of lycopersene (assumed to be the same as in squalene (Popják and Cornforth, 1966)) is considered to indicate that it is very unlikely that phytoene could be formed via lycopersene (Gregonis and Rilling, 1974).

c. Direct formation of phytoene. Only minor modification to the scheme proposed for lycopersene biosynthesis is required to account for the direct formation of phytoene from prephytoene pyrophosphate. The final step in the reaction would involve the stereo-

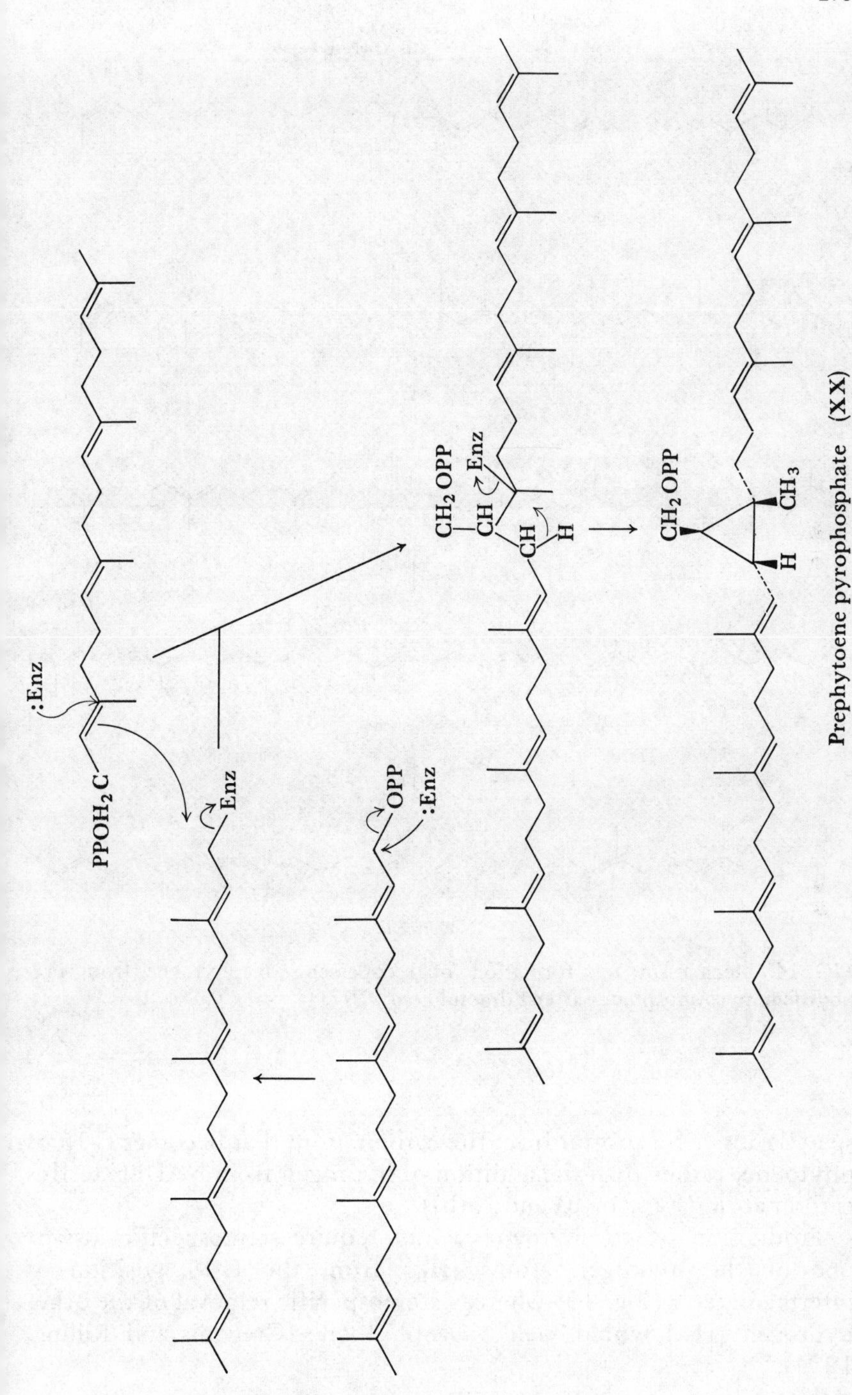

FIG. 13. Formation of prephytoene pyrophosphate.

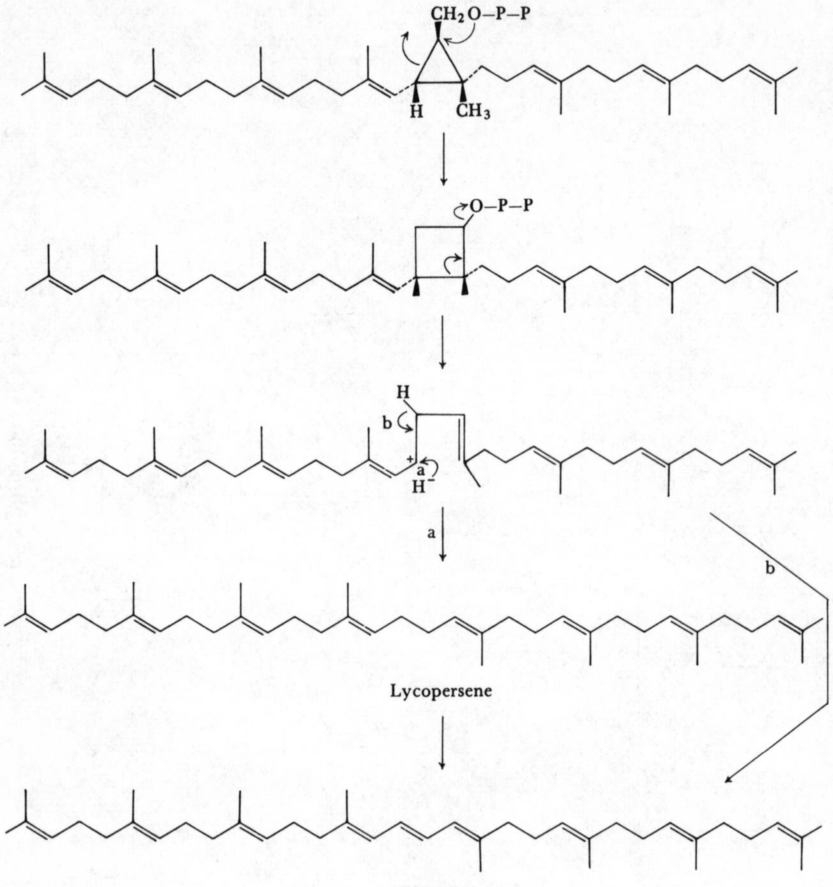

CH₂O–P–P

H CH₃

O–P–P

H

b

a

H⁻

a

b

Lycopersene

Phytoene

FIG. 14. Mechanism for formation of lycopersene or phytoene from pre-phytoene pyrophosphate (after Edmond *et al.*, 1971).

specific loss of a proton from the carbon atom that becomes C-15 of phytoene, rather than the addition of hydrogen from NADPH to this carbon atom (Figs. 14(b) and 15(b)).

Production of *cis*-phytoene would require stereospecific loss of one of the hydrogen atoms (H_B) from the C-15 position of intermediate A (Fig. 16) whereas stereospecific removal of the other hydrogen (H_A) would yield *trans*-phytoene (Gregonis and Rilling, 1974).

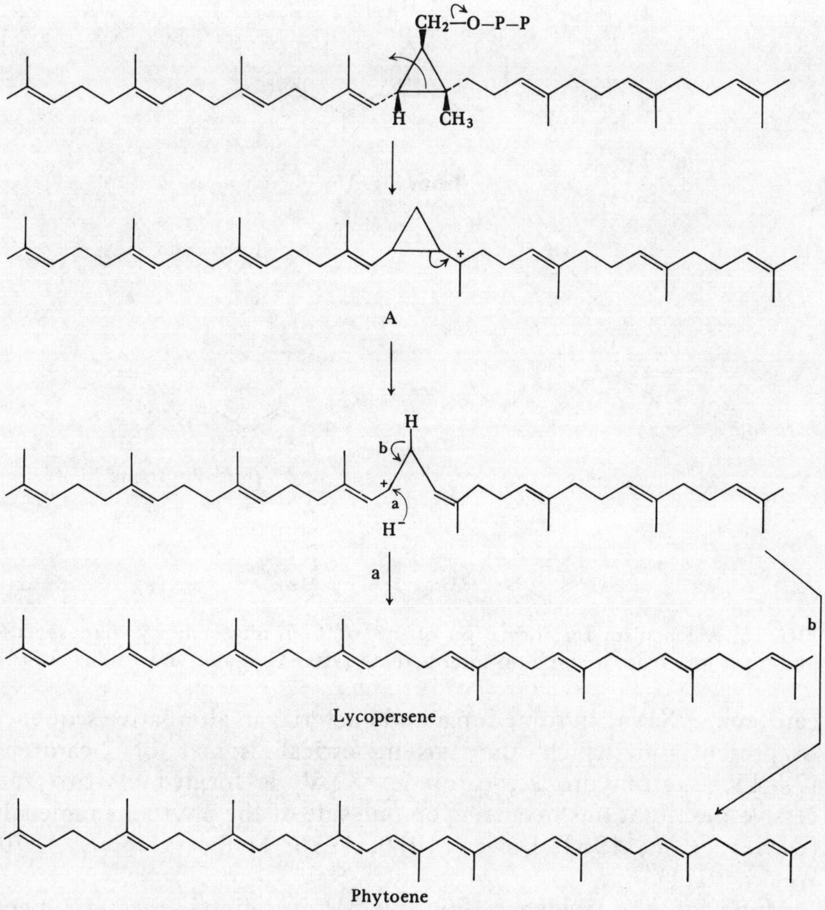

FIG. 15. Mechanism for formation of lycopersene or phytoene from prephytoene pyrophosphate (after Epstein and Rilling, 1970).

D DESATURATION

Even before the structures of the intermediates had been determined, a scheme was proposed for the stepwise conversion of a saturated precursor into the more unsaturated carotenes (Porter and Lincoln, 1950). The scheme now accepted for this desaturation series is shown in Fig. 17. This sequence involves the removal of pairs of hydrogen atoms alternately from the two sides of the molecule, producing phytofluene (7,8,11,12,7′,8′-hexahydro-ψ,ψ-carotene, **XXI**), ζ-carotene (7,8,7′,8′-tetrahydro-ψ,ψ-carotene, **XXII**) neurosporene (7,8-dihydro-ψ,ψ-carotene, **XXIII**) and finally lycopene (ψ,ψ-

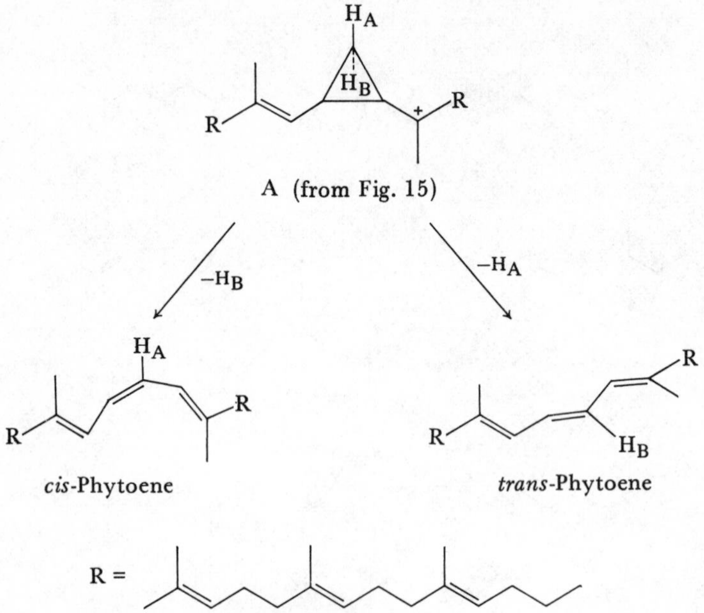

H_A

H_B

R R

A (from Fig. 15)

$-H_B$ $-H_A$

H_A R

R R

R H_B

cis-Phytoene *trans*-Phytoene

R =

FIG. 16. Mechanism for formation of *cis*- or *trans*-phytoene by stereospecific hydrogen loss from a common intermediate (after Gregonis and Rilling, 1974).

carotene, **XXIV**). In some fungi and bacteria, an alternative sequence is present, in which the unsymmetrical isomer of ζ-carotene (7,8,11,12-tetrahydro-ψ,ψ-carotene, **XXV**) is formed by two successive desaturations occurring on one side of the phytoene molecule (Davies *et al.*, 1969; Davies, 1970, 1973; Malhotra *et al.*, 1970; Weeks, 1971).

Much of the evidence for this desaturation series has been obtained from genetic studies and from experiments with the inhibitor DPA. As previously mentioned, DPA inhibits desaturation, so that phytoene and other carotenoid intermediates accumulate. Removal of the DPA allows synthesis of the normal unsaturated carotenoids to proceed, in some cases apparently at the expense of the accumulated saturated precursors (Goodwin and Osman, 1954; Liaaen-Jensen *et al.*, 1958; Braithwaite and Goodwin, 1960; Villoutreix, 1960); all the accumulated phytoene, however, did not appear to be metabolized.

Many other experiments have been performed in which the time course of pigment synthesis following removal of an inhibitor or change in nutrient conditions or the time course of incorporation of radioactive substrates into carotenoids has been followed. Several

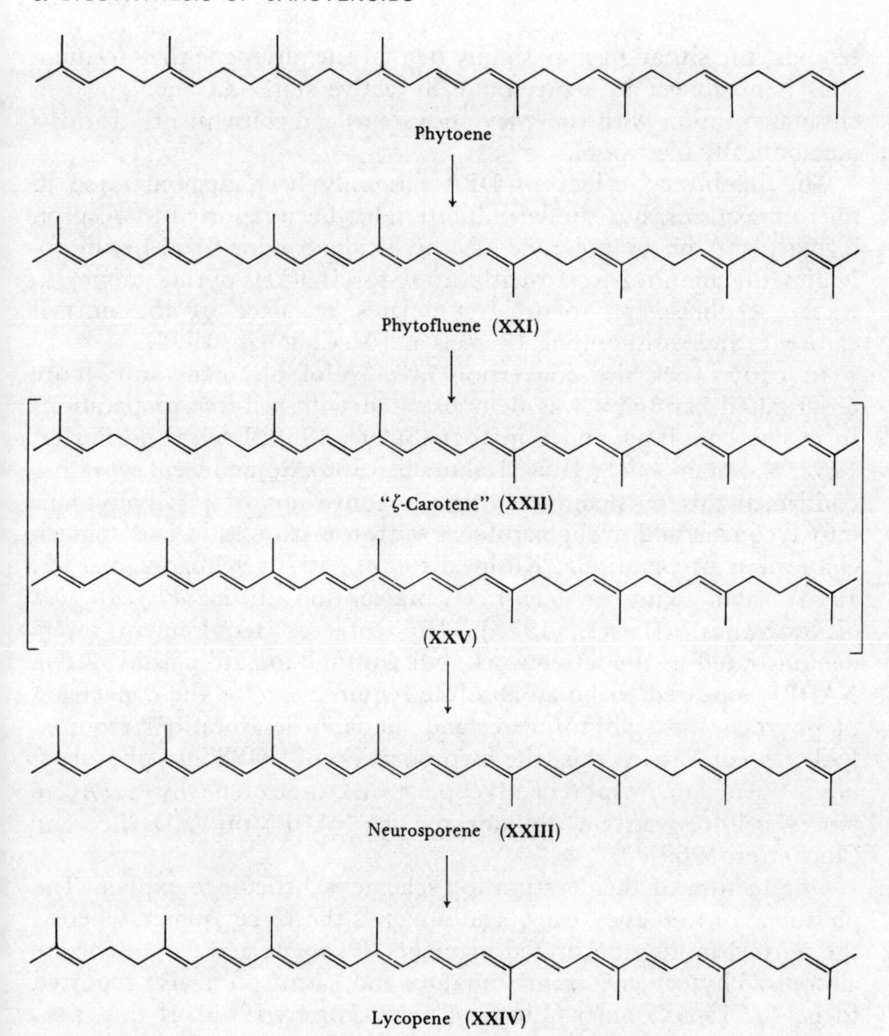

FIG. 17. Scheme for the stepwise desaturation of phytoene to lycopene.

workers have concluded that phytoene did not behave as a precursor of other carotenoids (Purcell *et al.*, 1959; Krzeminski and Quackenbush, 1960; Yokoyama *et al.*, 1961; Purcell, 1964; Karunakaran *et al.*, 1966); other workers (Beeler and Porter, 1963; Davies *et al.*, 1963b; Nusbaum-Cassuto *et al.*, 1967; Harding *et al.*, 1969) have deduced that phytoene is a precursor of other carotenoids.

Although some of these findings have been used as an argument against phytoene being a precursor of the more unsaturated carotenoids, the situation may simply be that the phytoene that accumulates is no longer in a metabolically active state (e.g. no longer in close association with the enzyme system) and consequently forms a metabolically inert pool.

The inhibitory effect of DPA has only been demonstrated in micro-organisms, but similar inhibition has been reported to occur in higher plants in the presence of a herbicide, Sandoz 6706 [4-chloro-5-(dimethylamino)-2-(α,α,α-trifluoro-*m*-tolyl)-3(2H)-pyridazinone]. In wheat seedlings, phytoene accumulates in place of the normal carotenes and xanthophylls (Bartels and McCullough, 1972).

In early work the conversion *in vitro* of phytoene into more unsaturated carotenes was demonstrated with cell-free preparations from bacteria, fungi and tomatoes (Suzue, 1960; Beeler and Porter, 1962; Kakutani *et al.*, 1964; Kakutani, 1966). Some recent work has confirmed this by demonstrating the conversion of [^{14}C] phytoene into lycopene and cyclic carotenes with tomato plastid and spinach chloroplast preparations (Kushwaha *et al.*, 1970; Subbarayan *et al.*, 1970) and with a cell-free preparation from *Phycomyces blakesleeanus* (Davies, 1973). No cofactor requirements were demonstrated in the latter work, but with the tomato plastid system $NADP^+$ appeared to be an absolute requirement for the conversion of phytoene into phytofluene, and the later desaturations required FAD. In contrast to this, the incorporation of GGPP into phytoene, phytofluene, neurosporene, lycopene and β-carotene by a *Phycomyces* cell-free system did not require $NADP^+$ or FAD (Lee and Chichester, 1969).

One feature of the desaturation scheme is difficult to explain. The phytoene of tomatoes, carrots and fungi is the 15-*cis* isomer, whereas the normal more unsaturated pigments, lycopene and β-carotene are all-*trans*. Phytofluene from tomatoes and carrot oil is also reported to be the 15-*cis* isomer (Jungalwala and Porter, 1965). If this is so, then the conversion of phytofluene into ζ-carotene (normally all-*trans*) would involve an isomerization as well as a desaturation (Fig. 18). Recent work by Davies, however (Aung Than *et al.*, 1972; Davies, 1973), has shown that the phytoene of tomato, carrot and fungal origin, although predominantly 15-*cis*, contains a small amount of the all-*trans* isomer. In the case of a *Phycomyces* mutant grown in the presence of DPA, synthesis of lycopene on removal of the inhibitor is accompanied by a decrease in the amount of *trans*- rather than *cis*-phytoene. Davies thus suggests that *trans*-phytoene (possibly produced from 15-*cis*-phytoene first synthesized) is the isomer involved in the desaturation series, with the *cis* isomer

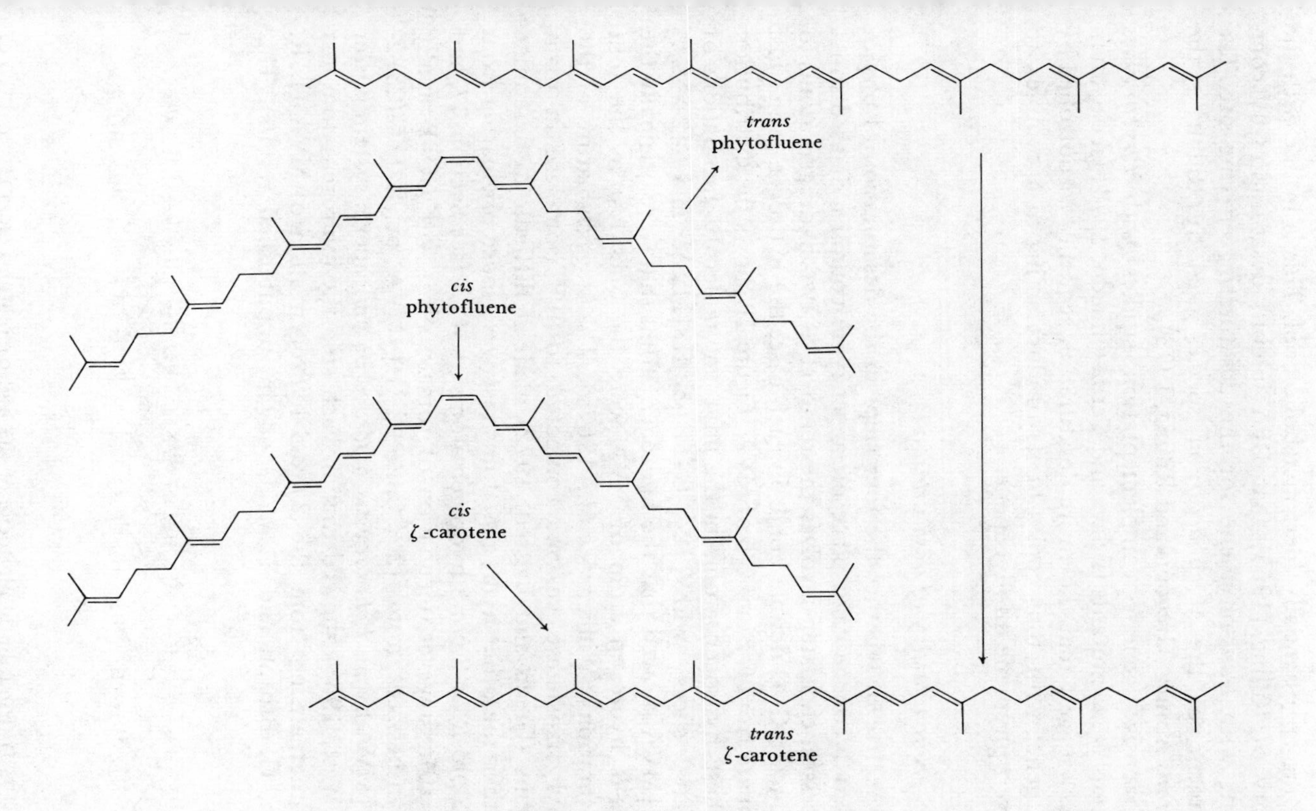

FIG. 18. Formation of *trans*-ζ-carotene from *cis*-phytofluene.

forming an essentially inert metabolic pool. This is in line with the theory of Rilling (1965) that DPA inhibits desaturation by competing with *trans*-phytoene for the desaturating enzyme site. The phytoene of the *Mycobacterium* species used by Rilling is the all-*trans* isomer (Gregonis and Rilling, 1973).

Extensive genetic complementation studies with *Phycomyces blakesleeanus* mutants (Eslava and Cerdá-Olmedo, 1974) (see Section IIIA*1*) suggest that the entire desaturation sequence is controlled by one gene, with four copies of its product acting in a complex to convert phytoene into lycopene.

1 Stereochemistry of desaturation

Irrespective of whether the four steps in the desaturation of phytoene to lycopene take place on an enzyme complex or as distinct processes, each step involves the removal of two hydrogen atoms to produce a C–C double bond. In each case, the hydrogen atoms lost originate from C-2 and C-5 of MVA. Figure 19 shows the distribution of these hydrogen atoms in the phytoene molecule. Incubations of tomato slices with [2-^{14}C,(5*R*)-5-$^{3}H_1$]MVA and [2-^{14}C,5-$^{3}H_2$]MVA showed that the hydrogen atoms that were originally the 5-*pro-R* hydrogen atom of MVA were the ones lost in the four desaturations (Williams *et al.*, 1967b). The sluggish nature of the prenyl transferase enzyme compared to IPP isomerase in plant systems (Green and Baisted, 1972) made it difficult to determine whether the other hydrogen atoms lost were those originating from the 2-*pro-R* or 2-*pro-S* hydrogen atoms of MVA (see Section B*4*, and Fig. 10). However, incubation of tomato slices with GGPP rapidly biosynthesized from [2-^{14}C,(2*R*)-2-$^{3}H_1$]MVA or [2-^{14}C,(2*S*)-2-$^{3}H_1$]MVA by an *Echinocystis macrocarpa* endosperm system (Oster and West, 1968) showed that in each step the hydrogen atom lost was that arising from the 2-*pro-S* hydrogen atom of MVA (J. R. Vose, G. Britton and T. W. Goodwin, unpublished results). This

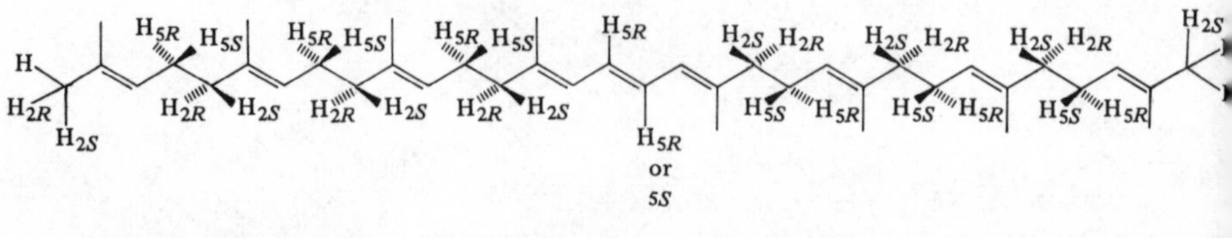

FIG. 19. Distribution in phytoene of the hydrogen atoms from C-2 and C-5 of MVA.

has now been confirmed with a very efficient carotenoid synthesizing preparation from a *Flavobacterium* species (McDermott *et al.*, 1973b).

Each desaturation step thus involves a *trans* elimination of two hydrogen atoms (Fig. 20).

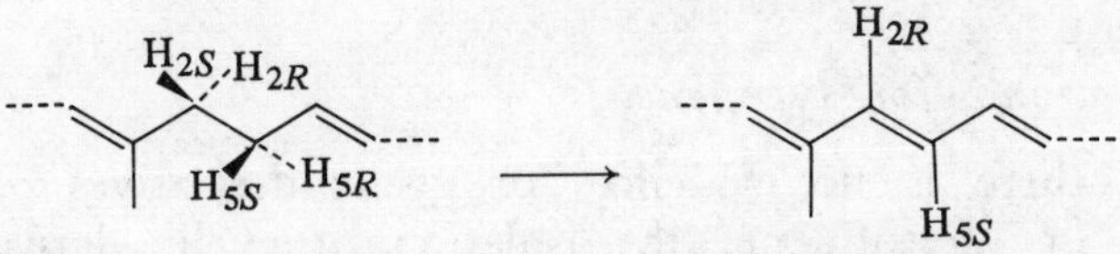

FIG. 20. *Trans* elimination of hydrogen in the desaturation reaction.

E CYCLIZATION

1 *Cyclization of lycopene*

In recent years, a considerable amount of evidence has been obtained to support the view first proposed by Porter and Lincoln (1950) that lycopene is the immediate acyclic precursor of the cyclic carotenoids.

Three compounds have been found to inhibit the production of cyclic carotenoids; in all cases lycopene accumulates. Thus lycopene accumulates in pumpkin cotyledons in the presence of cycocel [(2-chloroethyl) trimethylammonium chloride] (Knypl, 1969), and in apricot, peach, tomato and citrus fruits, carrot and sweet potato roots, and the mycelia of *Phycomyces blakesleeanus* and *Blakeslea trispora* in the presence of CPTA [2-(*p*-chlorophenylthio) triethylammonium chloride] (Coggins *et al.*, 1970; Yokoyama *et al.*, 1971, 1972; Hsu *et al.*, 1972; Elahi *et al.*, 1973). Nicotine also inhibits the cyclization reaction in several micro-organisms, and in particular causes an accumulation of lycopene in a *Mycobacterium* species that normally produces β-carotene (Howes and Batra, 1970; Batra *et al.*, 1973) and in a *Flavobacterium* species normally producing zeaxanthin (β,β-carotene-3,3′-diol) (McDermott *et al.*, 1973a,c, 1974). In these cases, removal of the nicotine by washing allows formation of cyclic carotenoids to proceed at the expense of the accumulated lycopene.

Several demonstrations of the *in vitro* incorporation of labelled lycopene into β-carotene and other carotenes have been reported Decker and Uehleke (1961) and Hill *et al.* (1971) have achieved conversions of lycopene into cyclic carotenoids in bean leaf chloroplast preparations, and the tomato plastid and spinach chloroplast enzyme systems of Porter (Kushwaha *et al.*, 1969, 1970) have also been used to demonstrate the conversion of labelled lycopene into α- and β-carotene and monocyclic carotenes. In this work, a requirement for light, $NADP^+$ and FAD was noted. The conversion of lycopene into β-carotene has also been achieved with a *Phycomyces blakesleeanus* cell-free preparation (Davies, 1973).

2 *Cyclization before lycopene*

Although there is no evidence for cyclic derivatives of GGPP, phytoene or phytofluene, the isolation of cyclic derivatives of ζ-carotene and neurosporene has been reported. A "cyclic ζ-carotene", tetrahydro-γ-carotene (7′,8′,11′,12′-tetrahydro-β,ψ-carotene, **XXVI**) has been found in DPA-inhibited cultures of *Phycomyces blakesleeanus* (Davies and Rees, 1973) and of the two "cyclic neurosporenes", α-zeacarotene (7′,8′-dihydro-ε,ψ-carotene, **XXVII**) has been found in maize (Mase *et al.*, 1957; Petzold *et al.*, 1959) and in the Delta tomato strain (Williams *et al.*, 1967a), and β-zeacarotene (7′,8′-dihydro-β,ψ-carotene, **XXVIII**) is of fairly common occurrence. In particular, β-zeacarotene is accumulated by inhibited cultures of *Rhodotorula* and *Phycomyces blakesleeanus* under conditions where lycopene does not accumulate (Davies *et al.*, 1963b; Simpson *et al.*, 1964b).

Much of the older evidence that has been used against lycopene and in favour of neurosporene as the precyclization intermediate is open to other interpretations. Tomatoes ripening above 30°C do not synthesize the normal main pigment, lycopene, although β-carotene production continues to the normal extent (Goodwin and Jamikorn, 1952; Tomes *et al.*, 1956), suggesting at first sight that lycopene cannot be an intermediate in β-carotene formation. It is likely, however, that two pools and sites of biosynthesis are involved, with the main system that produces lycopene being inhibited (probably at the GGPP-phytoene step (Porter, 1969)) by the higher temperature, and the minor, β-carotene-producing system not being affected. Dimethyl sulphoxide (DMSO) produces similar effects (Raymundo *et al.*, 1967).

The fungus *Rhizophlyctis rosea* synthesizes lycopene during the early stages of growth and γ-carotene (β,ψ-carotene, **XXIX**) in later stages. Addition of [2-^{14}C]MVA to the medium at the time of inoculation strongly labels the lycopene, but if the MVA is removed when the lycopene level is maximal, the γ-carotene later produced is essentially unlabelled. If the [2-^{14}C]MVA is added after the lycopene level has reached its peak, the γ-carotene becomes heavily labelled, but the lycopene is only slightly labelled (Davies, 1961). Although at first sight these results appear to be strong evidence against lycopene being a precursor of γ-carotene, the situation could also be explained by invoking compartmentation of the two phases of synthesis.

The direct incorporation of labelled neurosporene into β-carotene has been demonstrated only in the crude *Phycomyces* cell-free system of Davies (1973). Even in this case, however, the neurosporene could be desaturated to lycopene before cyclizing. Unlabelled lycopene or β-zeacarotene added to such incubations at the same time as the labelled neurosporene substrate dilute out the incorporation of radioactivity into β-carotene to an approximately equal extent; the radioactivity appears in the carotenoid added. It was thus concluded that two pathways of β-carotene formation, via lycopene or via neurosporene, were of equal significance.

3 *Mechanism of cyclization*

In the mechanism currently accepted for the cyclization reaction (Fig. 21), the C-1,2 double bond of the acyclic precursor is folded into position near the C-5,6 double bond, and proton attack at C-2 initiates cyclization to yield a "carbonium ion" intermediate (**XXX**). This intermediate, by loss of a proton from C-4 or C-6 respectively, would give rise to an ϵ ring (α ring) or a β ring. (Although usually depicted as involving the carbonium ion intermediate, the reaction is stereospecific, and therefore likely either to be a concerted process

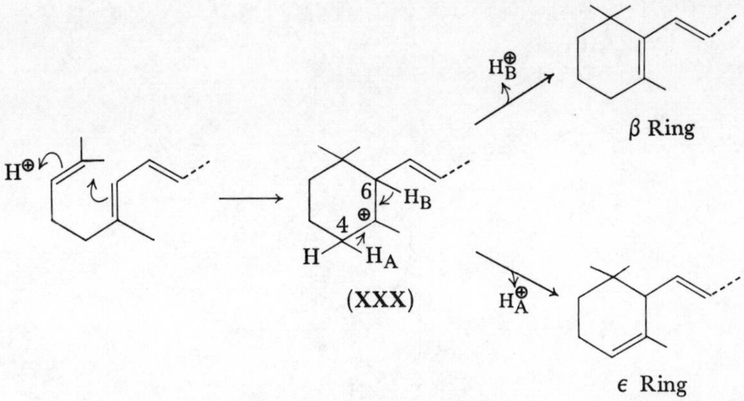

FIG. 21. General mechanism of cyclization.

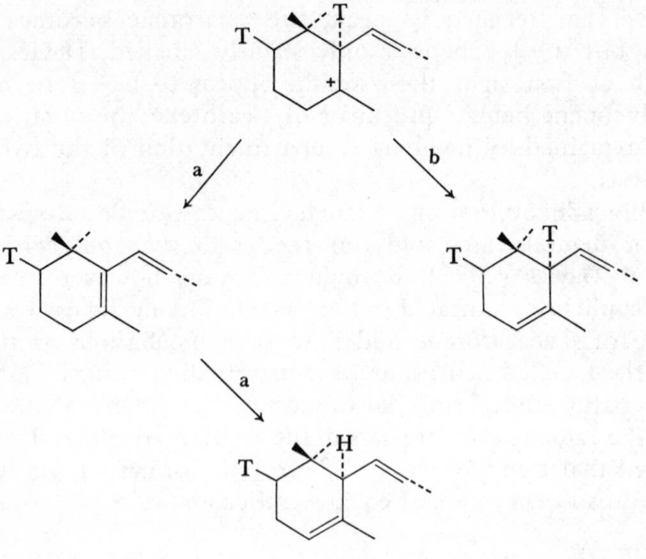

FIG. 22a,b. Distribution of tritium from $[2\text{-}^{14}C,(4R)\text{-}4\text{-}^{3}H_1]$MVA in ϵ and β rings.

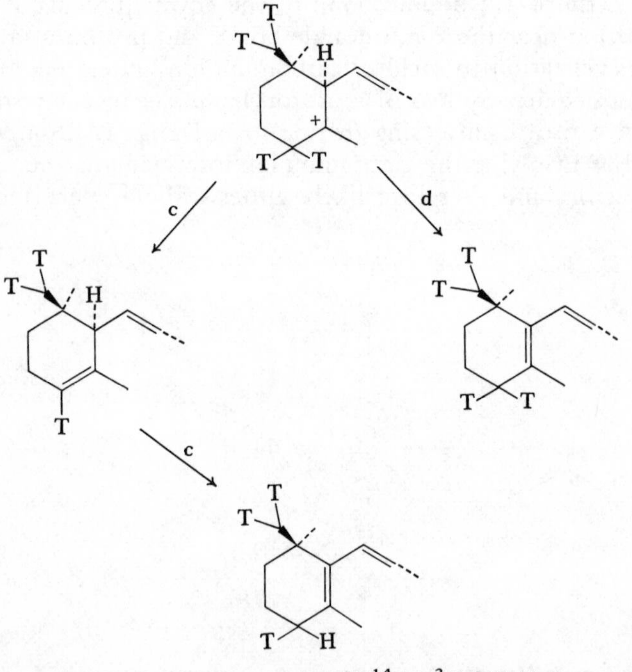

FIG. 22c,d. Distribution of tritium from $[2\text{-}^{14}C,2\text{-}^{3}H_2]$MVA in ϵ and β rings.

or to involve stabilization of the electron-deficient site in the intermediate by an enzyme group.) Genetic evidence is in agreement with this mechanism which indicates the separate formation of the ϵ and β rings; the ability to form the two ring types seems to be inherited independently (Tomes, 1967).

Incubations with doubly labelled species of MVA have confirmed the independent formation of the two ring types (Goodwin and Williams, 1965a,b; Williams *et al.*, 1967a). Figure 22a shows the positions occupied by tritium in the rings of β-carotene biosynthesized from [2-^{14}C,(4*R*)-4-^{3}H$_1$]MVA. α-Carotene (β,ϵ-carotene, **XXXI**), if formed from β-carotene, would have the same distribution of tritium. α-Carotene formed directly by removal of a proton from C-4 of the "carbonium ion intermediate" would, however, retain tritium at C-6 of the ϵ ring (Fig. 22b). The latter situation, retention of tritium at C-6 of the ϵ ring of α-carotene and related compounds, has been demonstrated in experiments with carrots and with tomato strains.

Similarly, formation of α-carotene from [2-^{14}C, 2-^{3}H$_2$]MVA would involve the loss of one labelled hydrogen atom from C-4; this labelled hydrogen atom would be absent from β-carotene, if this were formed from α-carotene (Fig. 22c). β-Carotene formed directly from the "carbonium-ion intermediate", however, would retain both labelled hydrogen atoms at C-4 (Fig. 22d). Again the latter situation has been demonstrated in several plant systems. The two ring types are therefore formed independently, though possibily from the same intermediate.

4 *Stereochemistry*

The cyclization is obviously a stereospecific process; the ϵ ring produced has a chiral carbon atom at C-6, and in all natural ϵ ring carotenoids this has been found to have the (*R*) configuration (Eugster *et al.*, 1969; Buchecker and Eugster, 1973). The loss of hydrogen from C-4 is also stereospecific, the hydrogen atom lost being that which was originally the 2-*pro-S* hydrogen atom of MVA (Fig. 23) (J. R. Vose, G. Britton and T. W. Goodwin, unpublished results). That the formation of the β ring is also a stereospecific process is indicated by work with *Blakeslea trispora* (Bu'lock *et al.*, 1970). This organism produces a carotenogenesis-stimulating compound, trisporic acid, with the absolute configuration shown (**XXXII**). The formation of this metabolite from β-carotene has been demonstrated (Austin *et al.*, 1970), and it has been shown that the

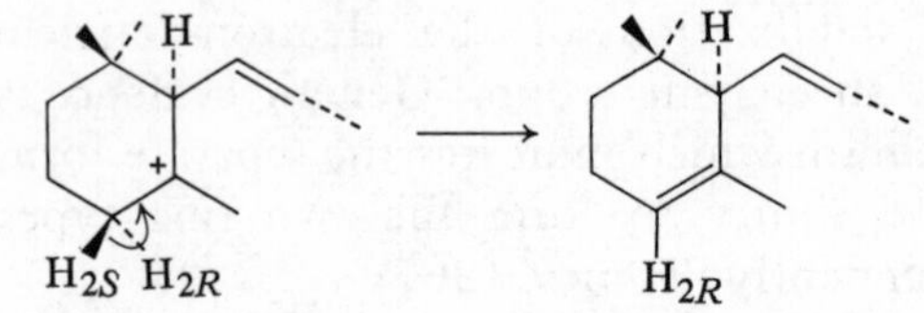

FIG. 23. Stereochemistry of hydrogen loss in formation of the (6*R*) ϵ ring.

label from [2-^{14}C]MVA is located in the methyl substituent at C-1, no activity being detected in the carboxyl group. The two C-1 methyl groups thus retain their individuality throughout the biosynthetic process (Fig. 24). A series of schemes has been produced to interpret these known results in terms of chair and boat folding of the acyclic precursor, and of which face of the C-5,6 double bond is approached by the C-1,2 double bond during the cyclization process (Britton, 1971; Goodwin, 1971a). More unknowns need to be determined before it can be decided which of these schemes is closest to the true pattern.

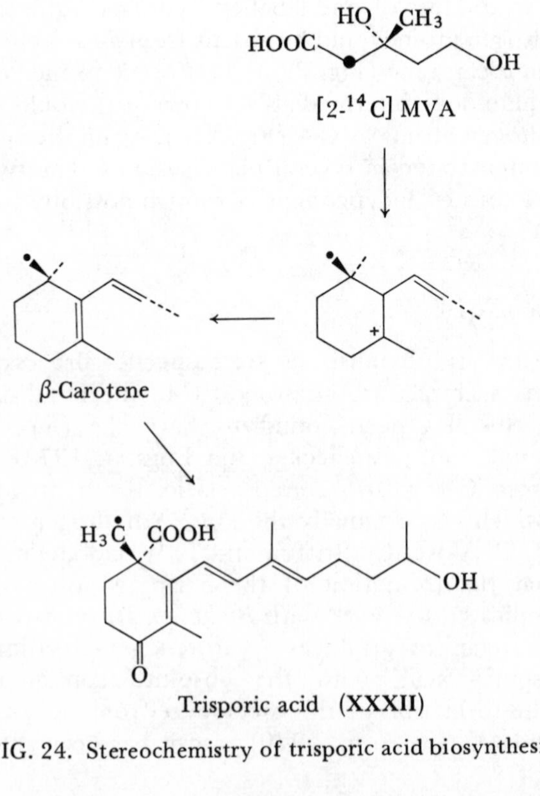

Trisporic acid (XXXII)

FIG. 24. Stereochemistry of trisporic acid biosynthesis.

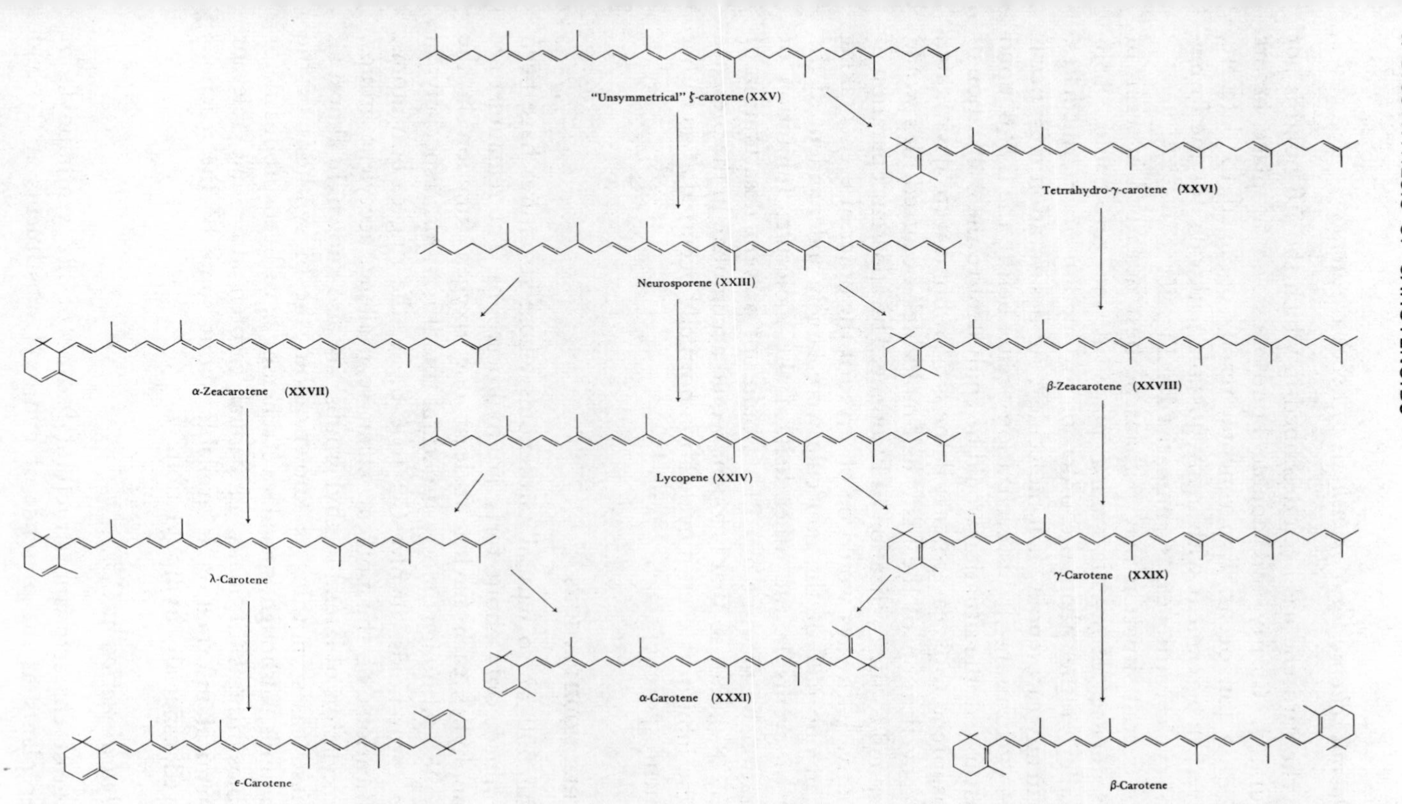

FIG. 25. Overall scheme for biosynthesis of cyclic carotenes.

5 *Overall scheme for biosynthesis of cyclic carotenes*

No cyclic carotene has been described in which the 7,8 bond is not desaturated. Clearly this double bond must be in place before cyclization can occur. Experimental results obtained to date are, however, in agreement with many possible pathways for the formation of cyclic carotenes, as summarized in Fig. 25.

The overall situation may perhaps be best described in terms of the sequence of events taking place in the carotenoid "half-molecule", rather than the order of production of intermediates. Thus in any carotenoid "half-molecule", 7,8 desaturation must occur before cyclization. If this cyclization occurs before the C-7′,8′ bond is desaturated, the situation would be equivalent to the cyclization of neurosporene to β-zeacarotene; if the C-7′,8′ bond were desaturated before cyclization of the first "half-molecule" occurred, this would correspond to the cyclization of lycopene. The apparent operation of alternative pathways would be observed if the order of events taking place in one half of the molecule were largely independent of the order of events in the other half of the molecule. Inhibition of cyclization by nicotine or CPTA would not prevent desaturation of the C-7′,8′ bond, so that lycopene would accumulate in the presence of these inhibitors even if cyclization normally occurred at an earlier (e.g. neurosporene) stage.

F OTHER MODIFICATIONS

Of the 300 or so natural carotenoids whose structures have been determined, only about forty are hydrocarbons. The remainder are xanthophylls, each having at least one oxygen function in the molecule. Although most structural modifications, especially in cyclic carotenoids, involve C-1 to C-6, i.e. the ring carbon atoms, modifications of the polyene chain (e.g. allene, acetylene groups, hydroxylation of lateral methyl groups) are also known. In almost all of these cases, nothing is known about the biosynthesis of the compounds, although speculative schemes have been proposed for the biosynthesis of many individual carotenoids. Some relevant biochemical information is available about some of the reactions involved, especially hydroxylation.

1 *Hydroxylation at C-3*

The most characteristic structural feature of the xanthophylls of higher plants, is the presence of hydroxyl substituents at C-3 and

C-3′, e.g. lutein (β,ε-carotene-3,3′-diol, **XXXIII**) and zeaxanthin (β,β-carotene-3,3′-diol, **XXXIV**).

Hydroxyl groups are also present at C-3 and C-3′ of the other chloroplast xanthophylls, such as violaxanthin (5,6,5′,6′-diepoxy-5,6,5′,6′-tetrahydro-β,β-carotene-3,3′-diol, **XXXV**) and neoxanthin (5′,6′-epoxy-6,7-didehydro-5,6,5′,6′-tetrahydro-β,β-carotene-3,5,3′-triol, **XXXVI**).

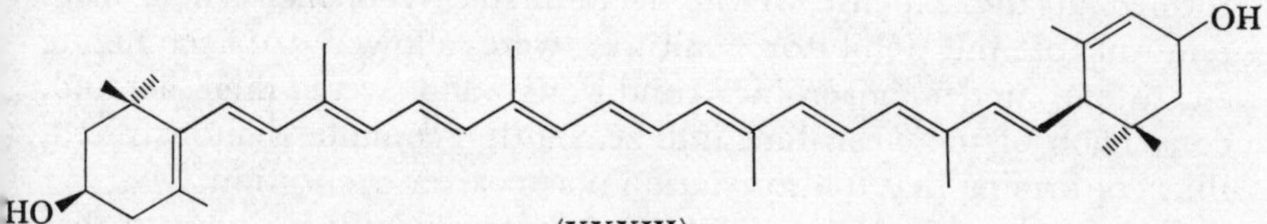

(XXXIII)

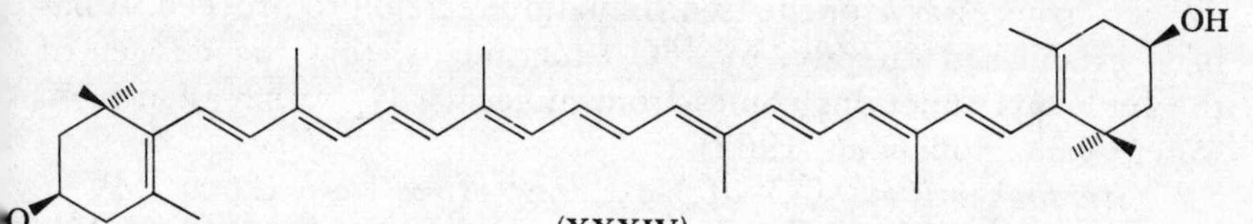

(XXXIV)

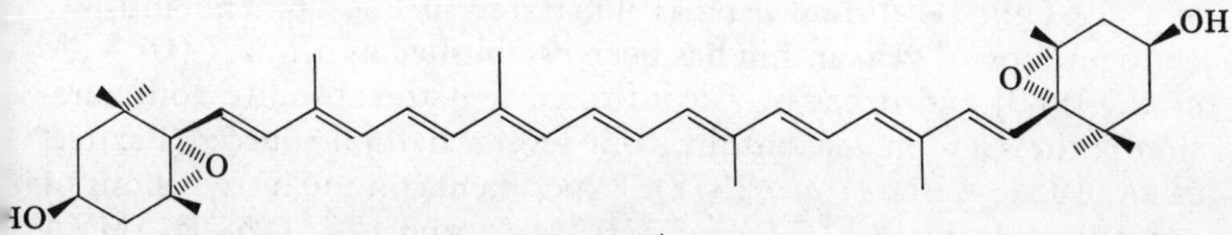

(XXXV)

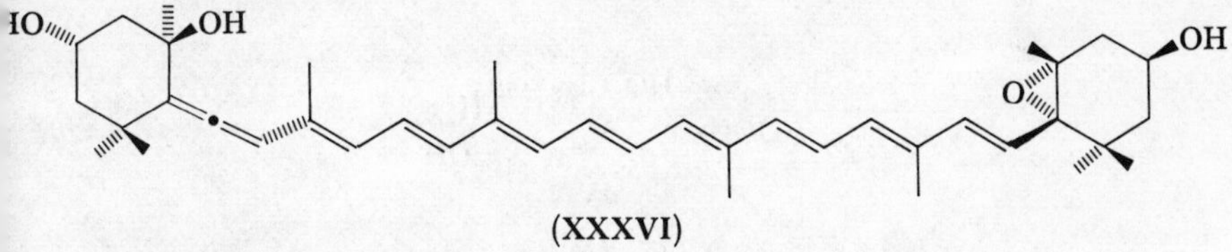

(XXXVI)

Claes (1959) showed that when a *Chlorella* mutant was cultured heterotrophically in the dark, and then illuminated in the absence of oxygen, carotene hydrocarbons accumulated; when these cultures were placed in the dark and allowed access to oxygen, the carotenes were apparently replaced by xanthophylls. Largely as a result of this work, it has generally been accepted that the introduction of hydroxyl groups occurs at a late stage in the biosynthetic pathway, i.e. lutein and zeaxanthin are generally considered to be formed by hydroxylation of α- and β-carotene respectively. This has now been confirmed by a series of inhibitor studies with a *Flavobacterium* species (McDermott *et al.*, 1973a,c; 1974). When this organism, which normally produces large amounts of zeaxanthin, is grown in the presence of nicotine, cyclization is inhibited, and in place of zeaxanthin, lycopene accumulates rather than the hydroxylycopene that would be expected if hydroxylation occurred before cyclization. Removal of the inhibitor allows cyclization to proceed, and β-carotene (anaerobic conditions) and zeaxanthin (aerobic conditions) are formed at the expense of the accumulated lycopene. When, after removal of the inhibitor, cultures were allowed to accumulate β-carotene under anaerobic conditions, and were then aerated, conversion of the β-carotene into zeaxanthin could be demonstrated, thus confirming that hydroxylation occurs after cyclization.

This work of McDermott *et al.*, and that of Claes, indicate that oxygen is necessary for the hydroxylation reaction to proceed. It has been established directly, by ^{18}O experiments, that the oxygen of the hydroxyl functions comes from molecular O_2 rather than from water (Yamamota *et al.*, 1962).

a. Stereochemistry. C-3 of β-carotene arises from C-5 of MVA, and the location of the 5-*pro-R* and 5-*pro-S* hydrogen atoms of MVA at C-3 of the β-carotene ring is illustrated in Fig. 26. The absolute configuration of zeaxanthin has been established as 3*R*, 3′*R* (De Ville *et al.*, 1969) and it has been confirmed that the absolute configuration is the same in zeaxanthin from several natural sources (Bartlett *et al.*, 1969; Aasen *et al.*, 1972). Experiments on the biosynthesis of zeaxanthin from [2-^{14}C,(5*R*)-5-$^3H_1$]MVA and [2-^{14}C,5-$^3H_2$]MVA by higher plant and bacterial systems (Goodwin *et al.*, 1968; T. J.

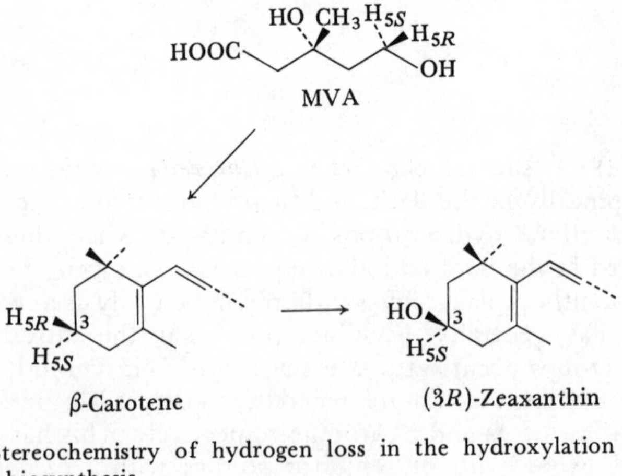

HO CH$_3$ H$_{5S}$ H$_{5R}$
HOOC OH
MVA

H$_{5R}$ 3 H$_{5S}$
β-Carotene

HO 3 H$_{5S}$
(3*R*)-Zeaxanthin

FIG. 26. Stereochemistry of hydrogen loss in the hydroxylation reaction of zeaxanthin biosynthesis.

Walton, J. C. B. McDermott, F. J. Leuenberger, G. Britton and T. W. Goodwin, unpublished results) have shown that the hydrogen lost from C-3 during hydroxylation is that which was originally the 5-*pro-R* hydrogen atom of MVA, the 5-*pro-S* hydrogen atom being retained (Fig. 26). These results clearly demonstrate that no ketonic intermediate is involved in the hydroxylation, and also that hydroxylation occurs with retention of configuration, a situation typical of mixed-function oxidase hydroxylations (Hayano, 1962).

The results of preliminary experiments on the incorporation of $[2\text{-}^{14}C,(5R)\text{-}5\text{-}^{3}H_1]$MVA and $[2\text{-}^{14}C,5\text{-}^{3}H_2]$MVA into lutein in leaves indicate that the 5-*pro-R* hydrogen atom of MVA is lost in the introduction of the hydroxyl group at C-3′ (Goodwin *et al.*, 1968; T. J. Walton, G. Britton and T. W. Goodwin, unpublished results). The absolute stereochemistry at C-3′ of lutein has recently been shown to be *R* (**XXXVII**), i.e. opposite to that at C-3′ of zeaxanthin (Buchecker *et al.*, 1972), but the significance of this in the light of the known stereochemistry of hydrogen loss is not clear.

No enzyme work on the hydroxylation reaction has been reported.

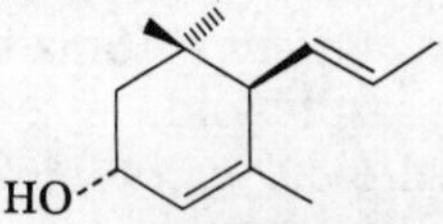

HO

(XXXVII)

2 *Other hydroxylations*

a. At C-1. Tertiary hydroxyl groups at C-1 are characteristic structural features of the acyclic carotenoids of the photosynthetic bacteria and other bacteria and fungi, and are also present in myxoxanthophyll [2′-(β-L-rhamnopyranosyloxy)-3′,4′-didehydro-1′,2′-dihydro-β,ψ-carotene-3,1′-diol, **XXXVIII**] and oscillaxanthin [2,2′-bis(β-L-rhamnopyranosyloxy)-3,4,3′,4′-tetradehydro-1,2,1′,2′-tetrahydro-ψ,ψ-carotene-1,1′-diol, **XXXIX**], carotenoids of blue-green algae (Hertzberg and Liaaen-Jensen, 1969a,b).

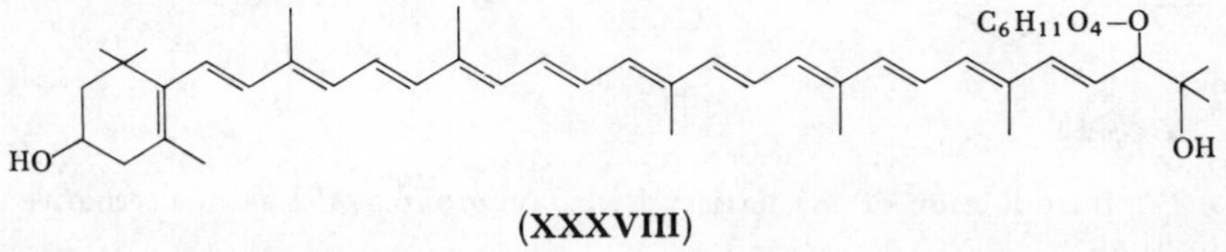

$C_6H_{11}O_4-O$

HO OH

(XXXVIII)

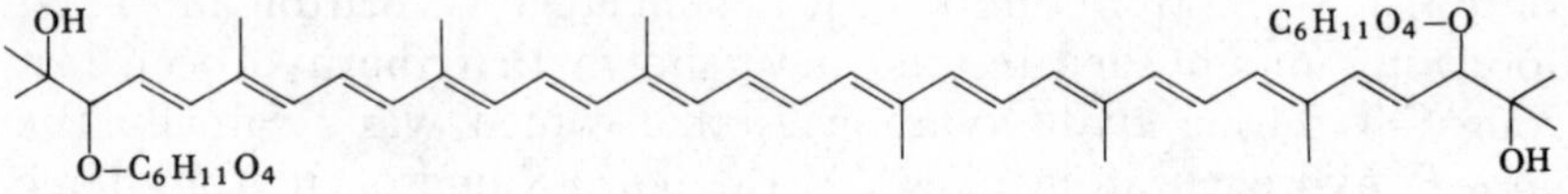

(XXXIX)

At least in the photosynthetic bacteria, the introduction of the C-1 hydroxyl groups is an anaerobic process, probably a hydration reaction occurring as an alternative to cyclization (Fig. 27) and inhibited by the same inhibitors, nicotine and CPTA (McDermott *et al.*, 1973a; Singh *et al.*, 1973). However, in addition to the hydroxyl group at C-1, the algal carotenoids also have a glycosylated hydroxyl group at the adjacent C-2 position, and in this case a different mechanism, such as hydrolysis of an epoxide group could be involved.

b. At C-2. Four cyclic carotenoids from algae have been shown to have hydroxyl substituents at C-2 of a β ring (XL). The green alga *Trentepohlia iolithus* produces large amounts of β,β-caroten-2-ol, β,ϵ-caroten-2-ol and β,β-carotene-2,2$'$-diol, as fatty acid esters (Kjøsen *et al.*, 1972), whereas β,β-carotene-2,3,3$'$-triol is produced by the blue-green alga *Anacystis nidulans* (Smallidge and Quackenbush, 1973). A 2-hydroxycarotenoid, 2-hydroxyplectaniaxanthin (3$'$,4$'$-didehydro-1$'$,2$'$-dihydro-β,ψ-carotene-2,1$'$,2$'$-triol) has been isolated from a red yeast, *Rhodotorula aurantiaca* (Liu *et al.*, 1973).

The absolute configuration (R) at C-2 of the *Trentepohlia* carotenoids has been determined, and their formation by cyclization of

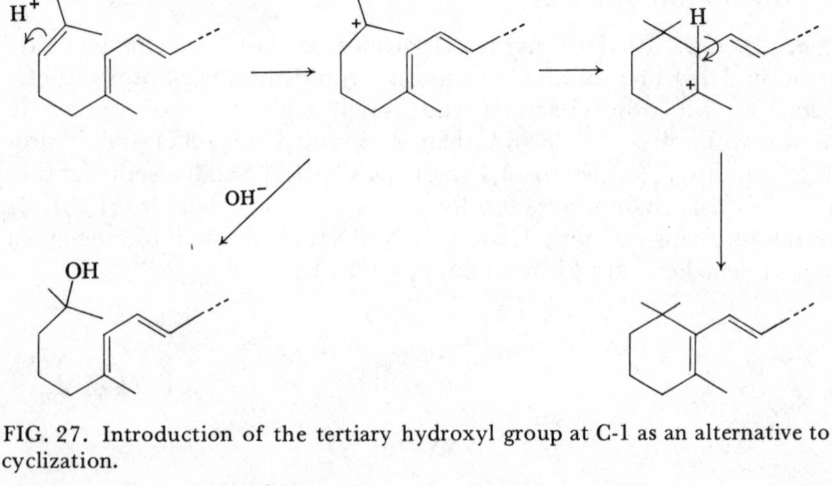

FIG. 27. Introduction of the tertiary hydroxyl group at C-1 as an alternative to cyclization.

an acyclic 1,2-epoxide, in a manner similar to the first step of squalene epoxide cyclization in triterpene biosynthesis has been suggested (Fig. 28) (Kjøsen *et al.*, 1972).

c. At C-4. Carotenoids with an oxo group at C-4 of a β ring are produced as "secondary carotenoids" by many algae, especially when grown heterotrophically or under unusual nutritional conditions (see Goodwin, 1971b). The green alga *Dictyococcus cinnabarinus*, when grown in submerged culture on a glucose medium, produces echinenone (β,β-caroten-4-one, **XLII**) and canthaxanthin (β,β-carotene-4,4′-dione, **XLIV**) as its main carotenoids. The inclusion of low concentrations of DPA in the medium inhibits the production of these oxocarotenoids, and in their place the corresponding 4-hydroxy-carotenoids, isocryptoxanthin (β,β-caroten-4-ol, **XLI**) and isozeaxanthin (β,β-carotene-4,4′-diol, **XLIII**) are accumulated (Gribanovski-Sassu, 1972). It thus appears that the introduction of oxo groups at C-4 is a two-stage process, via the corresponding hydroxy intermediate (Fig. 29). No details of the mechanism or stereochemistry of the hydroxylation are available.

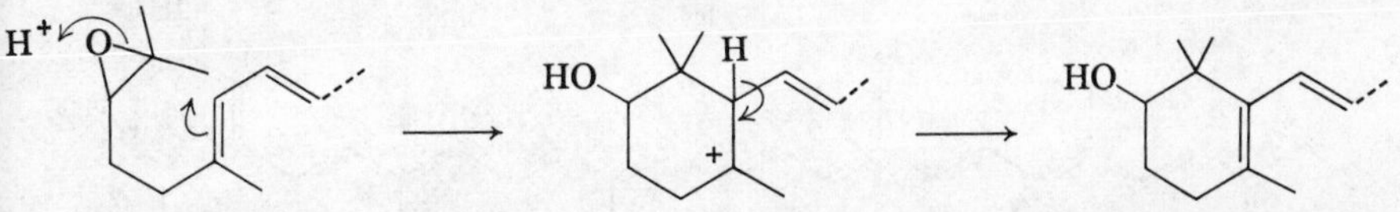

H^+ HO H + HO

(XL)

FIG. 28. Mechanism for formation of the 2-hydroxy β ring.

d. Hydroxylation of lateral methyl groups. Several cases are now known of carotenoids in which lateral methyl groups are hydroxylated. In higher plants, the fruit of *Solanum dulcamara* contain lycoxanthin (ψ,ψ-caroten-16-ol, **XLV**), in which one of the geminal methyl groups at C-1 is hydroxylated, and the dihydroxy analogue, lycophyll (ψ,ψ-carotene-16,16′-diol, **XLVI**) (Markham and Liaaen-

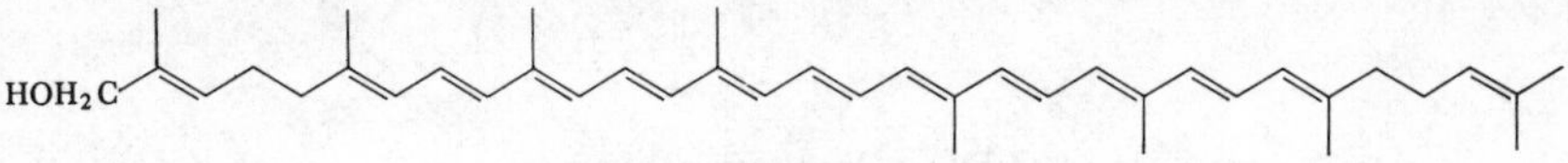

HOH_2C

(XLV)

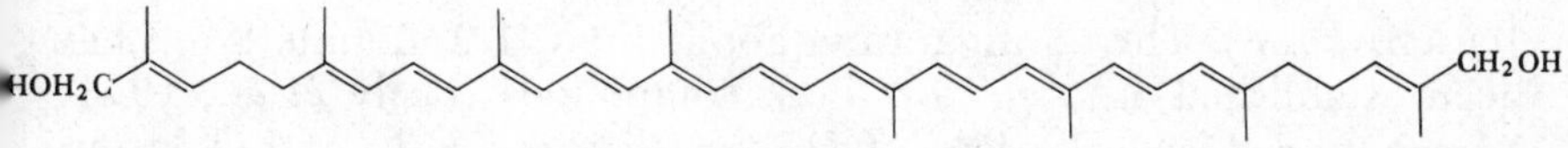

HOH_2C CH_2OH

(XLVI)

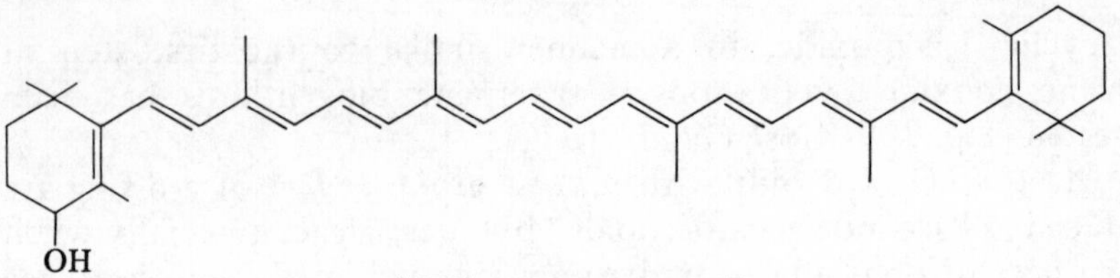

OH

(XLI) Isocryptoxanthin

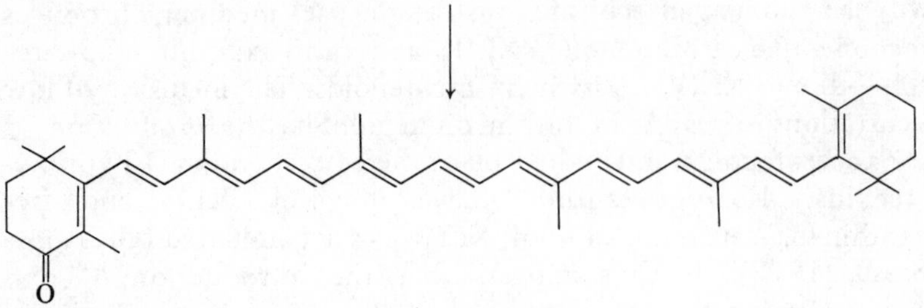

O

(XLII) Echinenone

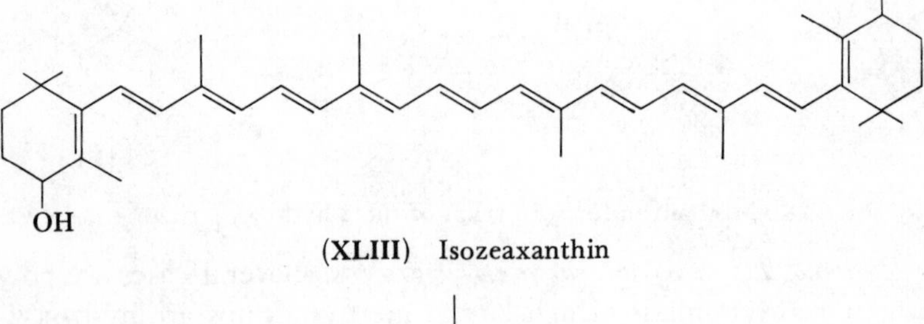

(XLIII) Isozeaxanthin

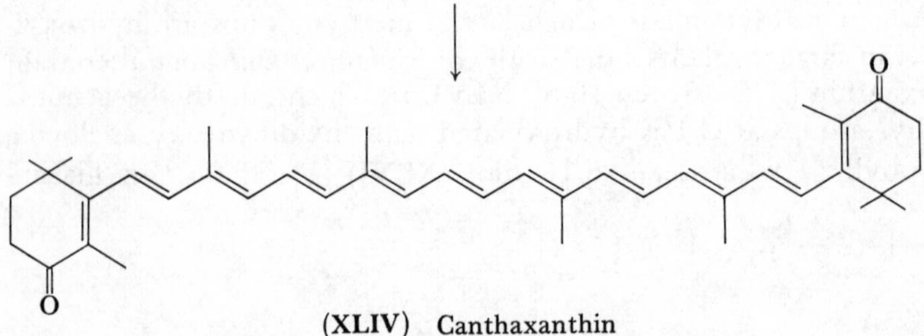

(XLIV) Canthaxanthin

FIG. 29. Introduction of oxo groups at C-4 and C-4′.

Jensen, 1968). The configuration about the C-1,2, double bond has been established as *trans* for these compounds (Kelly *et al.*, 1971; Kjøsen and Liaaen-Jensen, 1971), so that the C-1 methyl group hydroxylated should be that arising from C-2 of MVA rather than from the MVA methyl group. This has not, however, been demonstrated. The occurrence of similar hydroxy derivatives of phytofluene in *S. dulcamara* (S. F. Dunkeyson, G. Britton and T. W. Goodwin, unpublished results) indicates that the hydroxylation can occur at an earlier stage of desaturation, but the hydroxyphytofluene is not necessarily an intermediate in lycoxanthin biosynthesis.

Several algal carotenoids, e.g. loroxanthin (β,ε-carotene-3,19,3′-triol, **XLVII**) from green algae including *Chlorella vulgaris, Scenedesmus obliquus* (Aitzetmüller *et al.*, 1969; Walton *et al.*, 1970), siphonaxanthin (3,19,3′-trihydroxy-7,8-dihydro-β,ε-caroten-8-one, **XLVIII**), from siphonalean green algae including *Codium fragile* (Kleinig *et al.*, 1969; Walton *et al.*, 1970; Ricketts, 1971), and vaucheriaxanthin (5′,6′-epoxy-6,7-didehydro-5,6,5′,6′-tetrahydro-β,β-carotene-3,5,3′,19′-tetrol, **XLIX**) from *Vaucheria* spp. (Heterokontae) (Nitsche and Egger, 1970; Strain *et al.*, 1968), have hydroxyl groups at C-19. No biochemical information is available about the introduction of these hydroxyl functions.

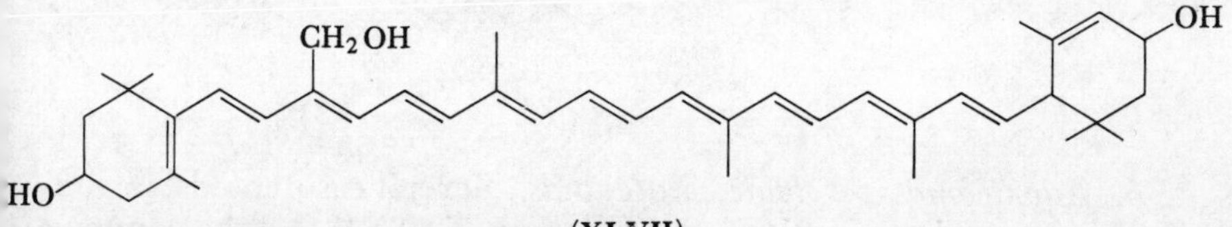

(XLVII)

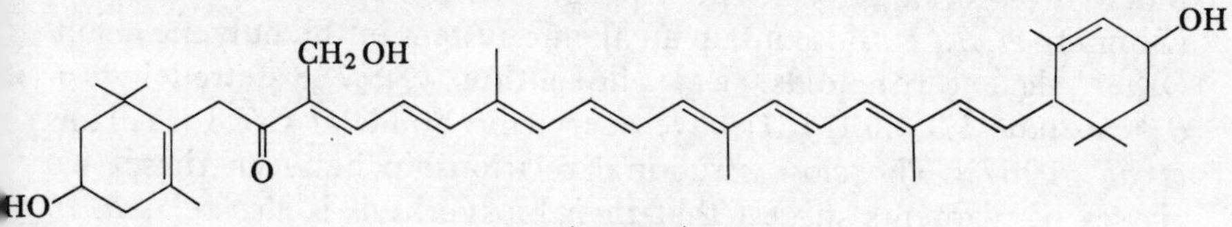

(XLVIII)

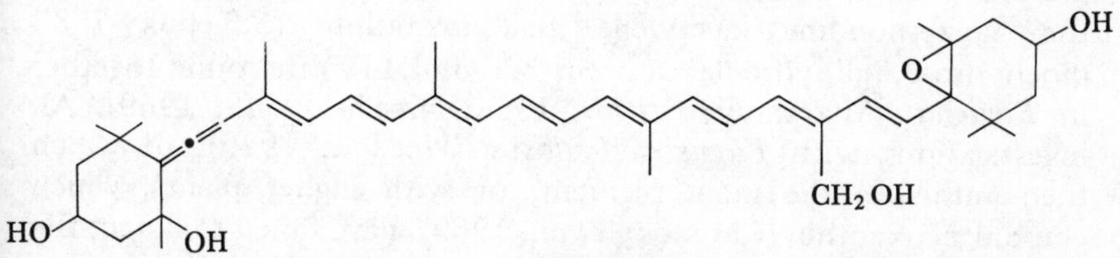

(XLIX)

3 *Miscellaneous*

a. Epoxides. Epoxycarotenoids, especially 5,6-epoxides of cyclic carotenoids are of widespread distribution, and include the chloroplast constituents violaxanthin (**XXXV**) and neoxanthin (**XXXVI**). The epoxide groups are introduced stereospecifically—the 5*R*, 6*S* configuration has been established for natural violaxanthin (Bartlett *et al.*, 1969)—indicating that an enzymic reaction is involved. Experiments with ^{18}O have confirmed that the epoxide oxygen comes from molecular oxygen (Yamamoto and Chichester, 1965) and not from water as was first thought (Yamamoto *et al.*, 1962; Sapozhnikov *et al.*, 1964).

A series of acyclic carotene epoxides, e.g. phytoene 1,2-epoxide (1,2-epoxy-1,2,7,8,11,12,7′,8′,11′,12′-decahydro-ψ,ψ-carotene, **L**) has been isolated from tomato fruit (Britton and Goodwin, 1969; Ben-Aziz *et al.*, 1973). It is not known whether their formation is analogous to that of squalene-2,3-epoxide, or whether they have any significance in the biosynthesis of other carotenoids.

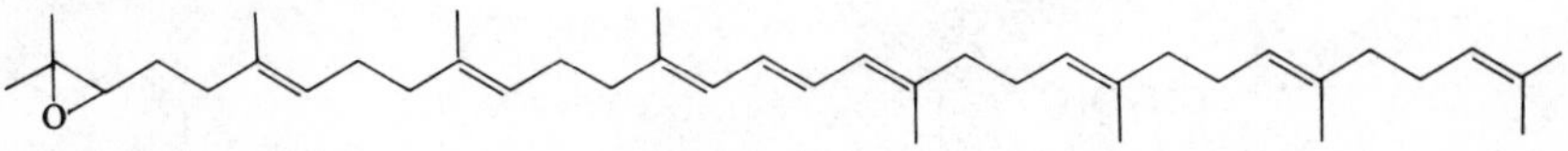

O

(L)

b. Allenic and acetylenic carotenoids. Several carotenoids, including the chloroplast constituent neoxanthin (**XXXVI**) and the common algal product fucoxanthin (5,6-epoxy-3,3′,5′-trihydroxy-6′,7′-didehydro-5,6,7,8,5′,6′-hexahydro-β,β-caroten-8-one 3-acetate, **LI**) (Bonnett *et al.*, 1969) contain an allenic system in the polyene chain. Other algal carotenoids, e.g. alloxanthin (7,8,7′,8′-tetradehydro-β,β-carotene-3,3′-diol, **LII**) have acetylenic bonds at C-7,8 (Mallams *et al.*, 1967). The close structural relationship between these two classes of pigments suggest that their biosynthesis is also related. In addition, the allene, fucoxanthin and the acetylene, diatoxanthin (7,8-didehydro-β,β-carotene-3,3′-diol, **LIII**), occur together in some diatoms (Strain *et al.*, 1944; Goodwin, 1971b) and neoxanthin and the corresponding acetylene diadinoxanthin (5,6-epoxy-7′,8′-didehydro-5,6-dihydro-β,β-carotene-3,3′-diol, **LIV**) are found together in *Euglena* (Aitzetmüller *et al.*, 1968; Nitsche *et al.*, 1969). All investigations with *Fucus vesiculosus* (Weedon, 1970), of which fucoxanthin is the main pigment, or with higher plants, which contain neoxanthin (Cholnoky *et al.*, 1969), have failed to detect the acetylenic carotenoids that could be intermediates in the conversion of zeaxanthin or lutein into fucoxanthin or neoxanthin. It therefore seems possible that the allenic carotenoids are formed directly from the normal polyene carotenoids with few, if any, intermediates that are liberated from the sites at which the conversion occurs.

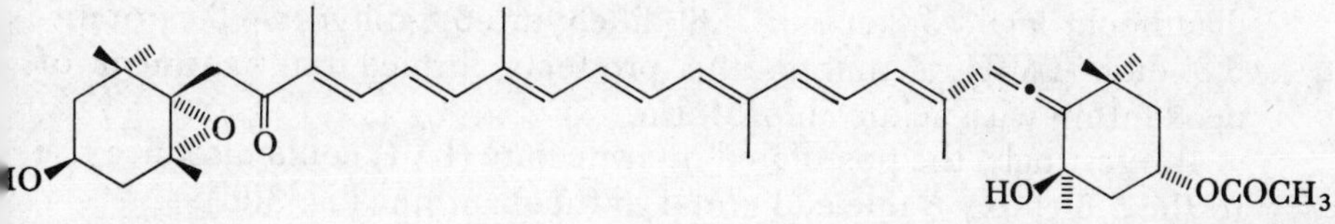

(LI)

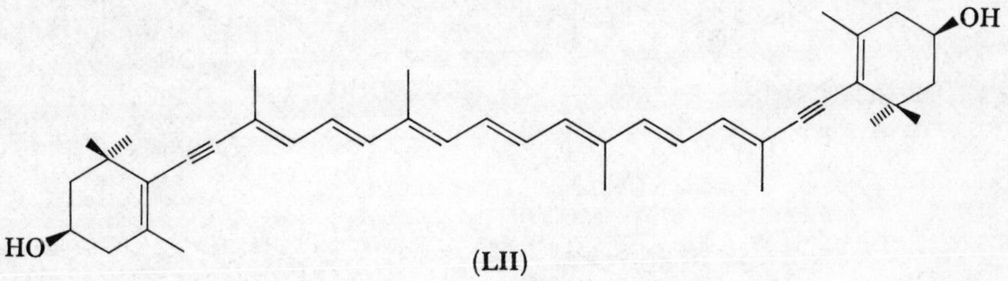

(LII)

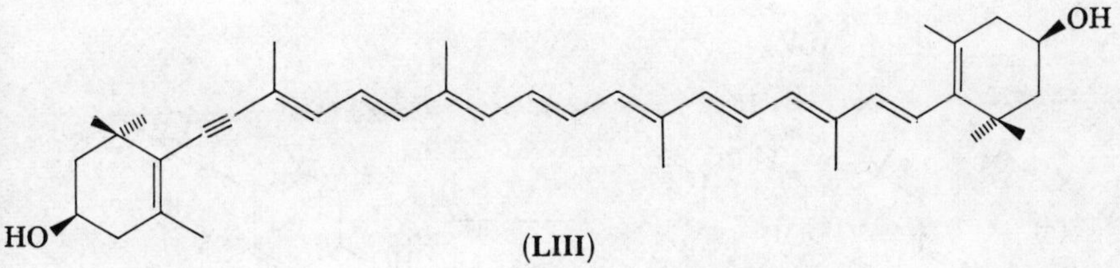

(LIII)

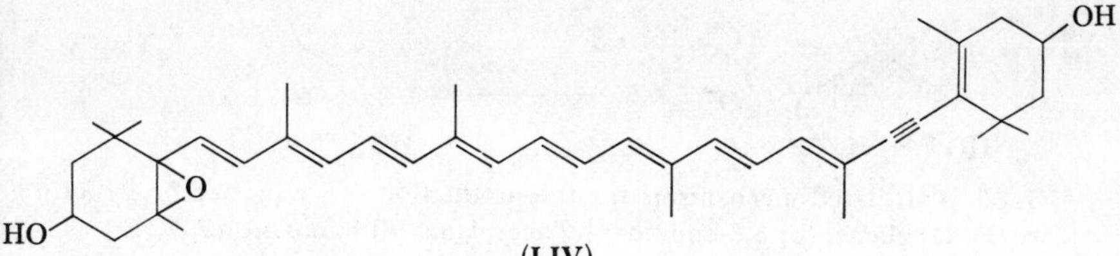

(LIV)

Weedon (1970) has suggested a possible mechanism for the formation of the allenic end groups in carotenoids by oxidation of zeaxanthin and related pigments with the formation of intermediates such as **(LV)** or **(LVI)**. A rearrangement of the type illustrated in Fig. 30a would lead to an allene with the natural stereochemistry. Simple loss of an oxygen atom might result in the formation of the epoxy group of neoxanthin and violaxanthin (Fig. 30b). Mechanisms were also proposed for the conversion of allenic carotenoids into the related acetylenes (Fig. 30c). The feasibility of such a scheme is supported by a recent report (Egger *et al.*, 1969) that the acetylene diadinochrome (5,8-epoxy-7′,8′-didehydro-5,8-dihydro-β,β-carotene-3,3′-diol, **LVII**) is among the products formed on treatment of neoxanthin with acidic chloroform.

Interestingly the postulated intermediate (**LVI**) could also give rise to the 5,6-epoxy-8-one end group in fucoxanthin (Fig. 30d).

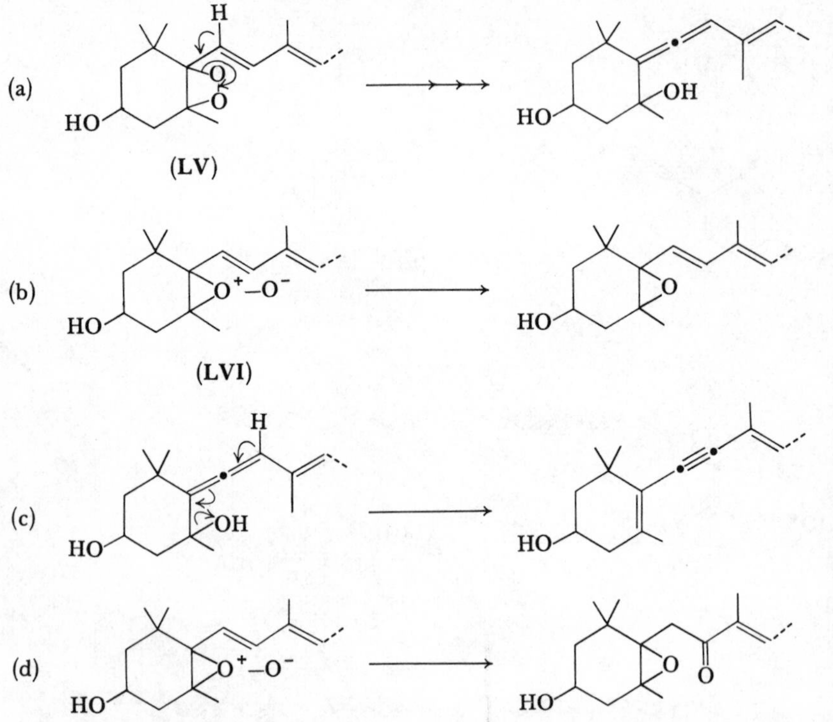

(a) HO H O O (**LV**) → HO OH

(b) HO $O^+–O^-$ (**LVI**) → HO O

(c) HO H OH → HO

(d) HO $O^+–O^-$ → HO O O

FIG. 30. Postulated mechanisms for the production of some specific structural features: (a) allene; (b) 5,6-epoxide; (c) acetylene; (d) 8-oxo group.

c. Cyclopentane carotenoids, e.g. capsanthin. The stereochemical correlation between capsanthin (3,3′-dihydroxy-β,κ-caroten-6′-one, **LVIII**) and zeaxanthin, and the co-occurrence of the two pigments in *Capsicum* spp. has suggested a close biogenetic relationship. A scheme for the formation of capsanthin from zeaxanthin via antheraxanthin (5,6-epoxy-5,6-dihydro-β,β-carotene-3,3′-diol) has been proposed (Fig. 31) (Cholnoky and Szabolcs, 1960; Entschel and Karrer, 1960; Barber *et al.*, 1961; Weedon, 1967). Alternatively a peroxide derivative of zeaxanthin, such as (LV), could be involved.

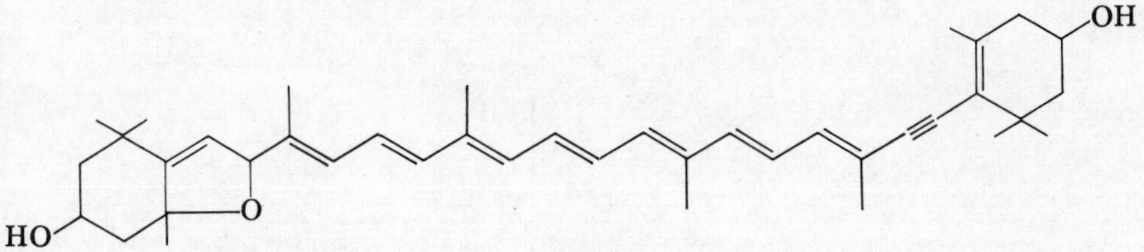

(LVII)

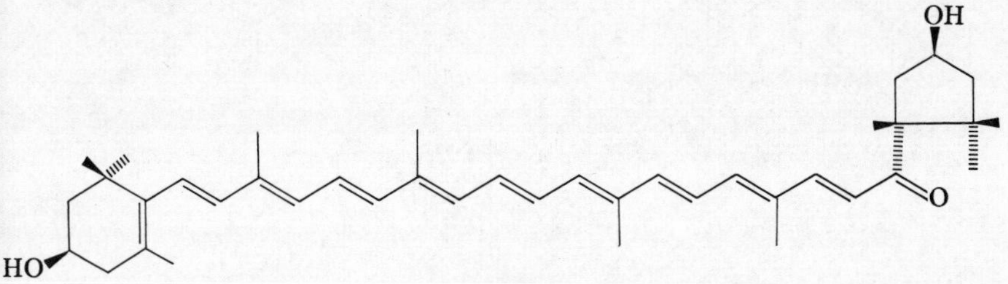

(LVIII)

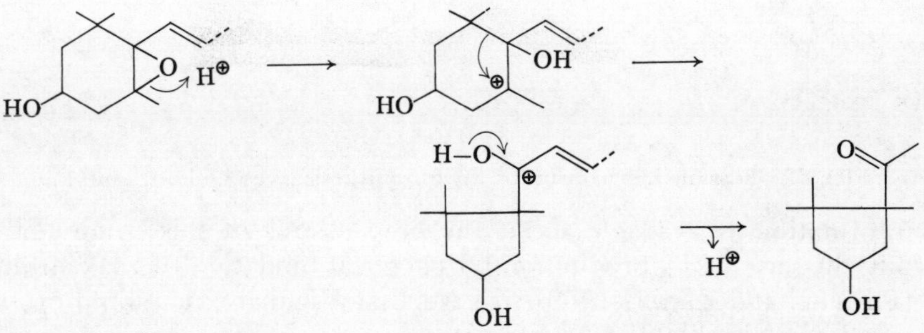

FIG. 31. Possible mechanism for formation of capsanthin from antheraxanthin.

d. Retro-*carotenoids.* Two well-known carotenoids have the *retro* arrangement of double bonds. The biosynthesis of eschscholtz-xanthin (4′,5′-didehydro-4,5′-*retro*-β,β-carotene-3,3′-diol, **LIX**), the characteristic pigment of the Californian poppy *Eschscholtzia californica*, from [2-^{14}C,(4*R*)-4-^{3}H$_1$]MVA has been studied, and a scheme (Fig. 32) for its biosynthesis has been proposed (Williams *et al.*, 1966). Again a peroxide derivative could be involved rather than an epoxide.

Rhodoxanthin (4′,5′-didehydro-4,5′-*retro*-β,β-carotene-3,3′-dione, LX) is the main pigment of yew berries (*Taxus baccata*). No

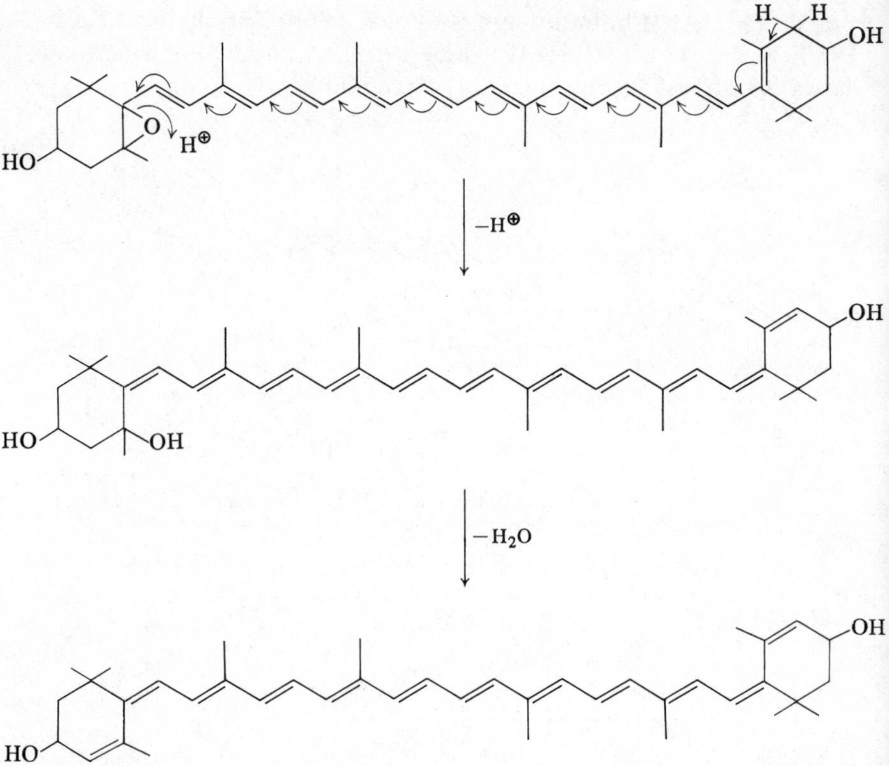

$-H^{\oplus}$

$-H_2O$

(LIX)

FIG. 32. Possible mechanism for the biosynthesis of eschscholtzxanthin.

information is available about the biosynthesis of this compound, but the ease of its production by chemical oxidation of zeaxanthin (Entschel and Karrer, 1959) suggests that a similar oxidation process may be involved in its biosynthesis.

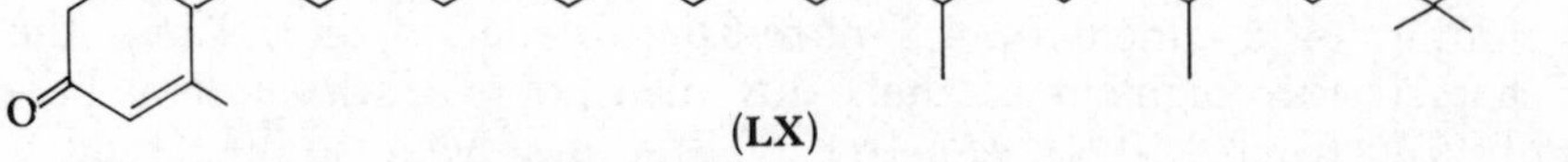

(LX)

e. Seco-carotenoids. The seco-carotenoids, e.g. semi-β-carotenone (5,6-seco-β,β-carotene-5,6-dione, **LXI**) found in some *Citrus* relatives (Yokoyama and White, 1968; Yokoyama *et al.*, 1970; Yokoyama and Guerrero, 1970) are identical to products of partial chemical oxidation of β-carotene and other carotenoids (Kuhn and

Brockmann, 1933, 1935). Their biosynthesis by rearrangement of a peroxide derivative of β-carotene (Fig. 33) has been suggested (Weedon, 1970).

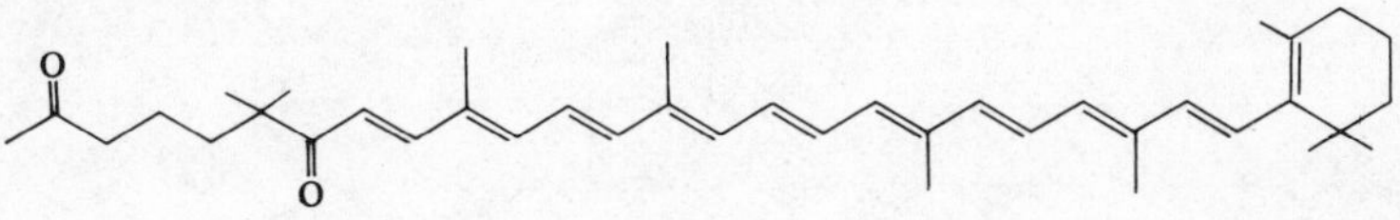

(LXI)

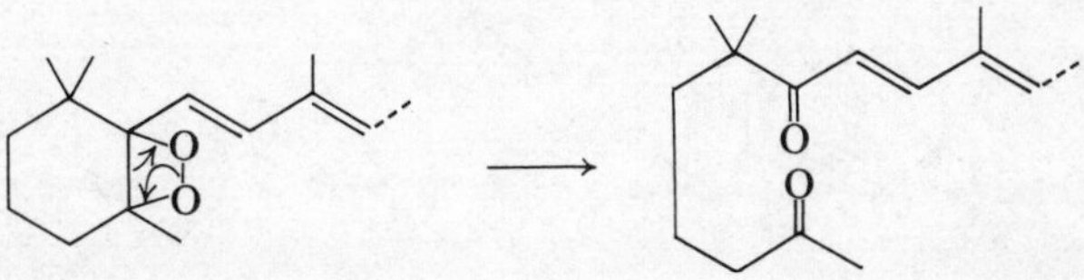

FIG. 33. Possible mechanism for the formation of seco-carotenoids.

f. Torularhodin. The carboxylic acid torularhodin (3′,4′-didehydro-β,ψ-caroten-16′-oic acid, **LXII**) together with the corresponding hydrocarbon torulene (3′,4′-didehydro-β,ψ-carotene, **LXIII**) are the characteristic pigments of the red yeasts *Rhodotorula* spp. A scheme (Fig. 34) for the biosynthesis of torularhodin, via β-zeacarotene, has been proposed (Kayser and Villoutreix, 1961; Simpson *et al.*, 1964a). Work with $^{18}O_2$ indicates that one oxygen atom from molecular oxygen is incorporated into the carboxyl group of torularhodin, in agreement with the hydroxylation of $-CH_3$ to $-CH_2OH$ by a mixed-function oxygenase, followed by dehydrogenation reactions to –CHO and –COOH (Simpson *et al.*, 1964a). The intermediates in such a pathway, torularhodin alcohol (3′,4′-didehydro-β,ψ-caroten-16′-ol, **LXIV**) and torularhodin aldehyde (3′,4′-didehydro-β,ψ-caroten-16′,-al, **LXV**) have been detected (Bonaly.and Malengé, 1968).

Experiments with [2-^{14}C]MVA and [2-^{14}C,2-$^{3}H_2$]MVA have shown that the carboxyl group of torularhodin arises entirely from C-2 of MVA, i.e. the geminal methyl groups at C-1 retain their individuality (Tefft *et al.*, 1970).

g. γ Rings. The presence in the discomycete *Caloscypha fulgens* of a carotene (β,γ-carotene, **LXVI**) with a methylene group substituent at C-5′, as in γ-ionone, has recently been reported (Arpin *et al.*, 1971). The biosynthesis of this new ring type (**LXVII**, designated γ ring) presumably involves the "carbonium ion" cyclization intermediate (**XXX**) being stabilized by loss of a proton from the

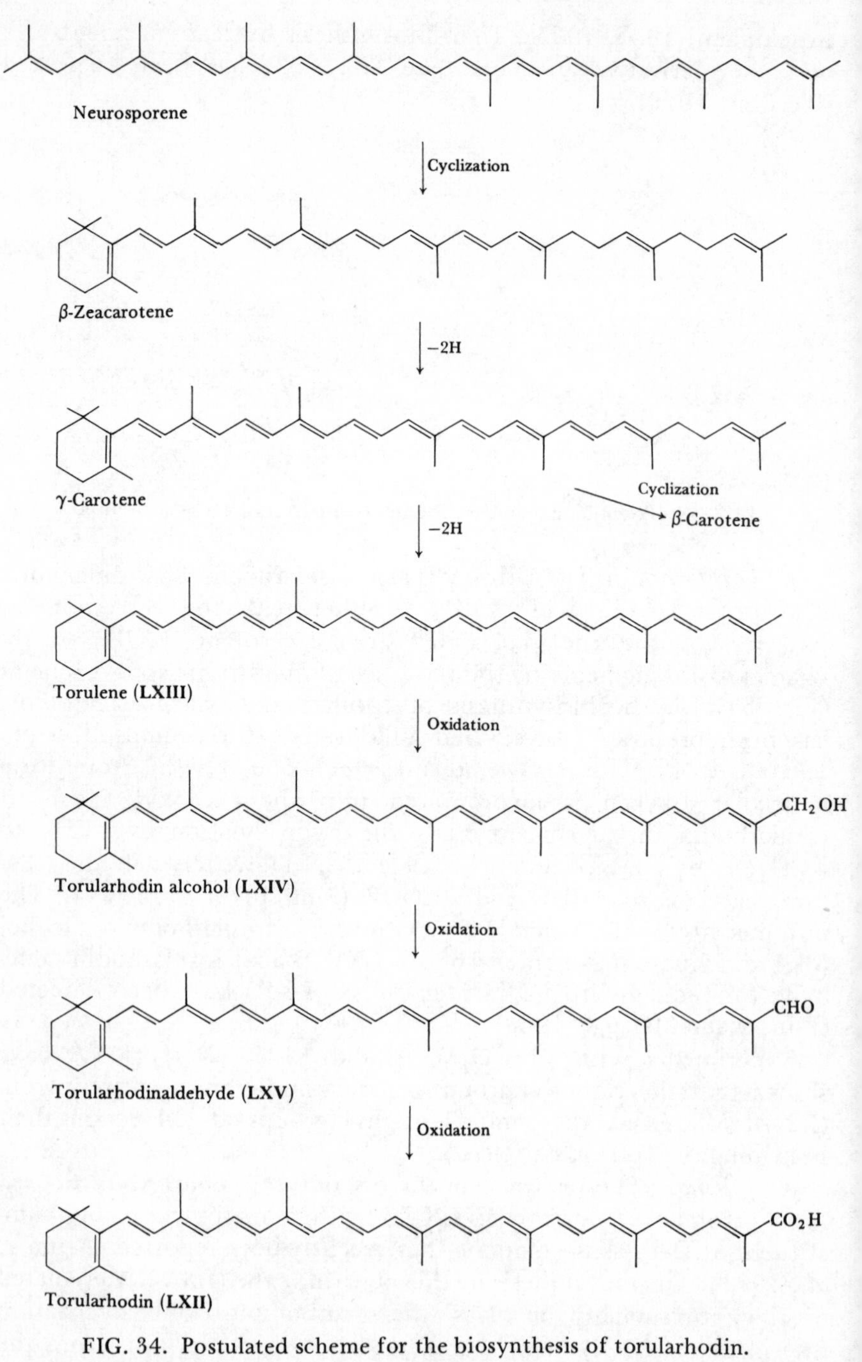

Neurosporene

Cyclization

β-Zeacarotene

−2H

γ-Carotene

Cyclization → β-Carotene

−2H

Torulene (**LXIII**)

Oxidation

CH_2OH

Torularhodin alcohol (**LXIV**)

Oxidation

CHO

Torularhodinaldehyde (**LXV**)

Oxidation

CO_2H

Torularhodin (**LXII**)

FIG. 34. Postulated scheme for the biosynthesis of torularhodin.

C-18 methyl group (Fig. 35) rather than the usual loss from C-4 or C-6 as observed in the formation of ε and β rings.

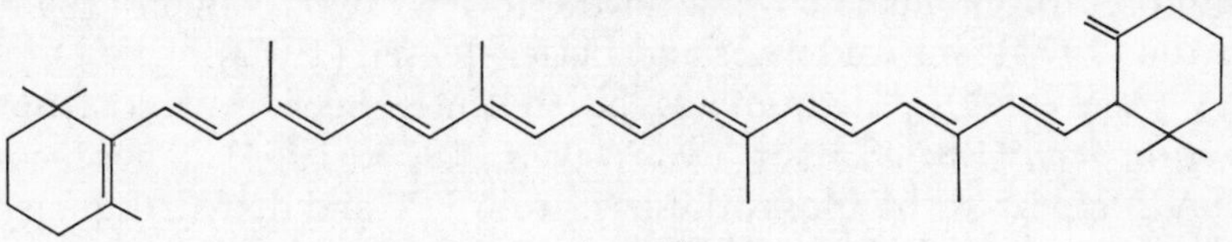

(LXVI)

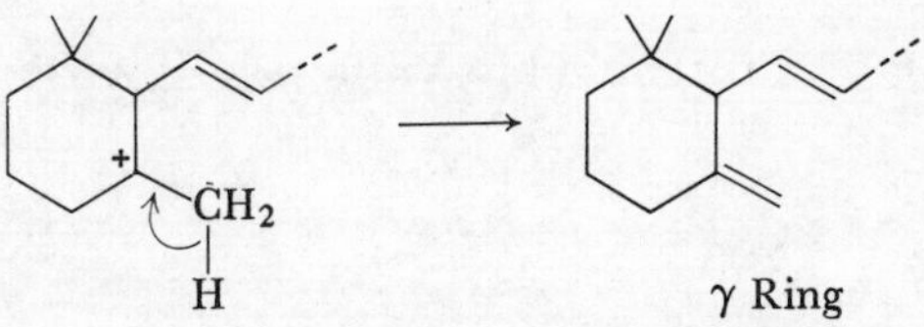

(XXX)

γ Ring

FIG. 35. A mechanism for biosynthesis of γ rings.

h. Poly-cis *carotenoids: prolycopene.* A poly-*cis* isomer, prolycopene, replaces all-*trans* lycopene as the characteristic pigment of the Tangerine tomato mutant (Le Rosen and Zechmeister, 1942), and has also been reported to occur in a dark-grown *Chlorella* mutant (Claes, 1954). Little is known of the biosynthesis or significance of this carotenoid except that the stereochemistry of hydrogen loss from C-4 and C-5 of MVA in the biosynthesis appears to be the same as in the formation of all-*trans* lycopene (Williams *et al.*, 1967a; R. J. H. Williams, G. Britton and T. W. Goodwin, unpublished results). The significance of these results cannot be assessed until the full stereochemical structure of prolycopene is determined. Smaller amounts of poly-*cis* isomers of neurosporene (proneurosporene) and ζ-carotene are also found in Tangerine tomatoes (Le Rosen and Zechmeister, 1942; Raymundo and Simpson, 1972) but the significance of these compounds in prolycopene biosynthesis is not known.

j. Others. Several other structural features are found in natural carotenoids. Particularly noteworthy are carotenoids with aromatic rings, found in several bacteria, particularly photosynthetic bacteria, and C_{45} and C_{50} carotenoids, found quite commonly in non-photosynthetic bacteria. The detailed biosynthesis of these carotenoid types, as well as the tertiary hydroxy- and methoxycarotenoids characteristic of photosynthetic bacteria, is considered outside the scope of this chapter, but information on the biosynthesis of aromatic carotenoids is given by Moshier and Chapman (1973), the biosynthesis of C_{45} and C_{50} carotenoids has been reviewed by Weeks (1971) and the biosynthesis of the acyclic hydroxy- and methoxycarotenoids is covered in the reviews by Goodwin (1971a), Britton (1971) and Andrewes and Liaaen-Jensen (1972).

A series of C_{30} carotenoids has recently been isolated from a *Streptococcus* sp. (Taylor and Davies, 1973, 1974). These compounds appear to be biosynthesized from FPP and dehydrosqualene, rather than from GGPP and phytoene.

III Regulation of carotenoid biosynthesis

A FUNGI

1 *Genetic control*

Several carotenoid mutants of *Phycomyces blakesleeanus* have been used extensively in studies of carotenoid biosynthesis. Three main types of these mutants are available, those that accumulate lycopene, those accumulating phytoene and those that are unable to synthesize carotenoids. Extensive complementation studies have revealed that each type corresponds to mutations in a single cistron, termed *car R*, *car B* and *car A*, respectively (Ootaki *et al.*, 1973). Cyclization (of lycopene?) is carried out by the product of gene *car R*; two copies of its product (i.e. two cyclase enzymes) in an enzyme complex are considered to be concerned in β-carotene formation (de la Guardia *et al.*, 1971). Similarly, four copies of the product of gene *car B* are considered to act in a dehydrogenase complex which carries out the four successive desaturations required to convert phytoene into lycopene (Eslava and Cerdá-Olmedo, 1974).

2 *Nutritional control*

a. Carbon and nitrogen sources. The amount of β-carotene synthesized by *Phycomyces blakesleeanus* during the initial period of active synthesis leading to maximal concentration depends on the availability of excess carbohydrate after growth has been completed (Goodwin and Willmer, 1952). Maltose and glucose are equally effective for carotenoid synthesis, but fructose and xylose, although supporting growth, are much less carotenogenic, and lactose and glycerol support neither growth nor carotenogenesis (Garton *et al.*, 1951). Acetate will support growth and carotenogenesis (Friend *et al.*, 1955).

Replacement of the normal nitrogen source, asparagine, by leucine or valine causes a considerable stimulation of β-carotene synthesis; these amino acids can act as precursors of HMGCoA (see Section IIA*1*). With $(NH_4)_2SO_4$ as nitrogen source, β-carotene levels are normal only in the presence of succinate or acetate (Friend *et al.*, 1955).

An interesting situation is found with *Blastocladiella*. This organism is normally non-carotenogenic, but in the presence of high levels of bicarbonate it synthesizes γ-carotene (Cantino and Hyatt, 1953).

b. pH and temperature. From an initial pH of 6·2, a fall to 2·6–3·0 occurs during growth of *Phycomyces* on the normal glucose-asparagine medium. If the pH change is prevented by buffering the medium, carotenogenesis is almost completely inhibited, although growth is unaffected (Goodwin and Willmer, 1952).

No qualitative changes in carotenoid composition occur in *Phycomyces* cultures grown over the temperature range 5–25°C, although the amount of carotenoid produced is less at the lower temperatures (Friend and Goodwin, 1954). A similar situation has been found with the yeasts *Rhodotorula rubra* and *R. penaus* (Nakayama *et al.*, 1954). *Rhodotorula glutinis*, however, produces mainly β- and γ-carotenes when cultured at 5–8°C, but at 25°C the levels of these carotenes are reduced and torulene and torularhodin accumulate (Nakayama *et al.*, 1954; Simpson *et al.*, 1964b).

3 *Photoinduction*

Light stimulates additional carotenoid synthesis in fungi which normally form reasonable amounts in the dark, e.g. *Phycomyces blakesleeanus* (Garton *et al.*, 1951; Bergman *et al.*, 1973). However, in many other fungi carotenoid synthesis occurs to a very limited extent or not at all in the dark, but can be initiated by photoinduction, involving a short simultaneous exposure to light and oxygen. After photoinduction, there is usually a time lag (about 4 h) before carotenogenesis begins (Rau, 1967).

The process of photoinduction has been studied in detail in *Fusarium aquaeductuum*, and appears to involve the production of a fairly stable "induction factor" which stimulates synthesis of carotenogenic enzymes. The enzyme synthesis can occur under anaerobic conditions, but access to oxygen within 48 hours is necessary for pigment synthesis to take place (Rau *et al.*, 1968; Rau, 1969, 1971; Theimer and Rau, 1969; Lang and Rau, 1972). Inhibitor studies have implicated ribosomal RNA in the photoinduction of carotenogenesis in *Verticillium albo-atrum* (Mummery and Valadon, 1973).

Hydrogen peroxide and *p*-chloromercuribenzoate simulate the photoinduction of carotenogenesis in *F. aquaeductuum* in that they induce synthesis in the dark. The locus of this effect, however, is different from that of light (Rau *et al.*, 1967; Theimer and Rau, 1969, 1970, 1972).

The photoinduction of carotenogenesis in micro-organisms has recently been reviewed by Batra (1971) and by Weeks *et al.* (1973).

4 *Chemical bioinduction*

When (+) and (−) strains of the heterothallic *Choanephora cucurbita* are cultured together, some twenty times more carotenoid is synthesized than by the (+) or (−) strains cultured separately (Barnett *et al.*, 1956). The same phenomenon is observed with (+) and (−) strains of *Blakeslea trispora* (Plempel, 1963; Prieto *et al.*, 1964; Sebek and Jäger, 1964; Ciegler, 1965; Thomas and Goodwin, 1967). The β factor responsible for this stimulation consists of a series of acids, of which trisporic acid C (**XXXII**) is the major component (Caglioti *et al.*, 1964; 1966; Van den Ende, 1968). Trisporic acid stimulates carotenogenesis in the (−) strain but not in the (+) strain of *B. trispora* (Thomas and Goodwin, 1967; Sutter and Rafelson, 1968; Yuldasheva *et al.*, 1972), and it appears to be the (−) strain which synthesizes the factor from precursors made by the (+) strain (Sutter, 1970; Van den Ende *et al.*, 1972). Physical contact between the (+) and (−) strains is probably not necessary for trisporic acid production, which is stimulated by a diffusible "progamone" formed and released by the (−) strain (Van den Ende, 1968).

Trisporic acid is not incorporated into β-carotene (Sebek and Jäger, 1964), but is thought to be formed from β-carotene via retinaldehyde (Austin *et al.*, 1970).

The effect of trisporic acid on carotenogenesis is inhibited by actidione (Thomas *et al.*, 1967), indicating that it probably acts by derepressing synthesis of an enzyme concerned with carotene synthesis. Trisporic acid also stimulates the synthesis of enormous amounts of phytoene by *B. trispora* in the presence of DPA, suggesting that the enzyme involved must be concerned with steps of carotenoid biosynthesis before phytoene formation (Thomas and Goodwin, 1967).

β-Ionone also has a stimulatory effect on carotenogenesis in *Phycomyces blakesleeanus* (Mackinney *et al.*, 1952; Engel *et al.*, 1953; Chichester *et al.*, 1954) and in mated (±) cultures of *B. trispora* (Ciegler *et al.*, 1959; Reyes *et al.*, 1964). The mechanism of action of β-ionone is different from that of trisporic acid; induction of enzyme synthesis may not be involved (Reyes *et al.*, 1964).

B ALGAE

1 *Nutritional control*

Under unfavourable cultural conditions, such as nitrogen deficiency, many green algae become yellow or red, owing to the formation of large amounts of extraplastidic carotenoids ("secondary carotenoids"), especially β-carotene, echinenone, canthaxanthin and astaxanthin (3,3′-dihydroxy-β,β-carotene-4,4′-dione, **LXVII**) (Lwoff and Lwoff, 1930; Goodwin and Jamikorn, 1954a; Dentice di Accadia *et al.*, 1966). The mechanism involved in this phenomenon is not understood.

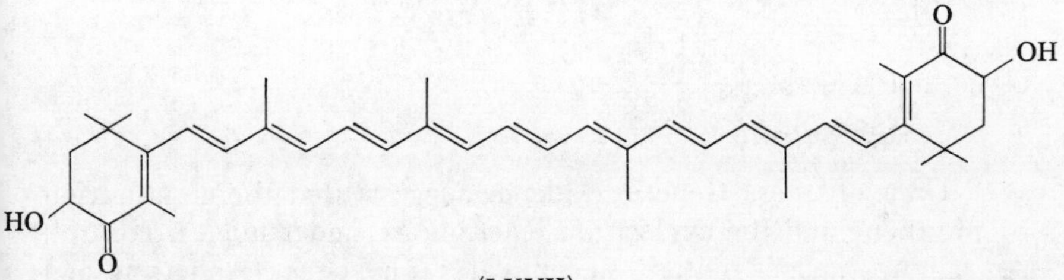

(LXVII)

In *Ankistrodesmus* spp. deficiency of PO_4^{3-}, SO_4^{2-}, Ca^{2+} or Fe^{3+} stimulates carotenogenesis, whilst Mg^{2+} and Mn^{2+} deficiency have no noticeable effect (Dersch, 1960). No effect of deficiencies in these minerals on carotenogenesis in *Trentepohlia aurea* was observed (Czygan and Kalb, 1966). A low phosphate medium stimulates carotenogenesis threefold in a bleached strain of *Euglena* (Blum and Bégin-Heick, 1967). *Dunaliella salina* produces α- and β-carotenes, lutein, violaxanthin and neoxanthin, but in salinities above 5% NaCl a large increase in β-carotene synthesis is observed (Loeblich, 1969).

2 *Effect of light*

The carotenoid levels of *Euglena* spp. are greatly reduced when the algae are cultured heterotrophically in the dark (Goodwin and

Jamikorn, 1954b; Wolken *et al.*, 1955); if the cells are continually subcultured in the dark, the carotenoid levels become extremely low because chloroplasts cease to be produced. Some algae, however, e.g. *Chlorella*, when grown in the dark produce chloroplasts and thus also the usual plastid carotenoids, although the amount in young dark-grown cultures is relatively less than in light-grown cultures (Goodwin, 1954).

Under natural autotrophic conditions, changes in light intensity appear to have little effect on carotenoid production (Dutton and Juday, 1944), but there are some early reports of qualitative differences in carotenoid composition between algae grown in the light or dark (Strain and Manning, 1943; Strain *et al.*, 1944).

Claes (1966) has reported that red light (670 nm) stimulates formation of cyclic carotenes in a *Chlorella* mutant that accumulates acyclic carotenoid precursors in the dark.

Synthesis of extraplastidic carotenoids appears to be independent of light intensity (Czygan and Kalb, 1966; Czygan, 1968).

A consideration of the factors controlling carotenoid synthesis in algae is included in a general review of algal carotenoids by Goodwin (1971b).

C HIGHER PLANTS

1 *Genetic control*

a. Green tissues. Genetic evidence suggests that the desaturation of phytoene and the cyclization reaction are under nuclear control. An albino maize mutant accumulates phytoene (Anderson and Robertson, 1960; Treharne *et al.*, 1966), and two other maize mutants accumulate ζ-carotene and lycopene respectively (Faludi-Dániel *et al.*, 1966). These mutations are inherited in normal Mendelian fashion.

A yellow mutant of *Helianthus annuus* accumulates xanthophylls rather than carotene (Wallace and Schwarting, 1954; Wallace and Habermann, 1959; Habermann, 1960). This might suggest that the nucleus controls a step in carotene synthesis which is not involved in xanthophyll synthesis, or it may reflect a biosynthetic compartmentation, so that xanthophyll biosynthesis, although it must take place via carotene intermediates, is a separate process from the formation of the carotene that accumulates. This characteristic also shows Mendelian inheritance, and is controlled by a single recessive factor.

b. Fruits. The most important genetic studies have been carried out with tomatoes. Normal (red) tomatoes possess the dominant allele r^+ while yellow tomatoes are homozygous for the recessive allele *r*. The r^+/r gene thus controls the total amount of pigment formed; the *rr* genotype synthesizes only about 5% of that formed by the r^+r^+ genotype (Le Rosen *et al.*, 1941). Apricot-coloured fruit homozygous for the recessive allele *at* have the synthesis of the acyclic series considerably decreased, while β-carotene synthesis is unaffected (Jenkins and Mackinney, 1955). A third recessive gene, *hp*, tends to increase the content of all components by about 100% (Tomes *et al.*, 1958).

Qualitative changes are controlled by another series of genes. Tangerine tomatoes are homozygous for the recessive *t*, while red fruit carry the dominant allele t^+ (MacArthur, 1934). Poly-*cis* isomers, e.g. prolycopene, accumulate at the expense of lycopene in Tangerine tomatoes (Zechmeister and Went, 1948). An orange phenotype obtained by back-crossing a *Lycopersicom esculentum* and a *L. hirsutum* hybrid has β-carotene and not lycopene as its major carotenoid (Lincoln and Porter, 1950). These "high-β" fruit carry the dominant allele b^+, whereas normal red fruit are homozygous for the recessive *b*. The expression of the b^+/b gene is partly regulated by an independently inherited modifier $mo_b{}^+/mo_b$ (Tomes *et al.*, 1954). Tomatoes of the Delta strain have a high concentration of δ-carotene replacing lycopene as the main pigment (Kargl *et al.*, 1960). These fruit possess the high delta allele del^+. A ghost phenotype appears spontaneously in tomato lines carrying red fruit. Grafting is necessary to obtain "ghost" fruit, which are white-yellow and contain phytoene in amounts comparable to the lycopene level in red fruit. The leaves also accumulate phytoene in place of the normal carotenoids (Mackinney *et al.*, 1956).

Tentative proposals can be made about the sites of action of these genes. Several (r^+/r, at^+/at and hp^+/hp) act before the phytoene stage and control quantitatively the flow of precursor into phytoene, b^+/b controls the cyclization of lycopene to β-carotene, del^+/del appears to regulate the cyclization to produce ϵ- rather than β-rings, and t^+/t determines the stereochemistry of the final molecule.

Genetic information related to carotenoids in fruit has been discussed in more detail by Goodwin (1971a) and Goodwin and Goad (1970).

2 *Carotenoid synthesis in ripening fruit*

In many fruits, ripening is accompanied by massive synthesis of carotenoids, as chloroplasts change into chromoplasts.

These changes can take place in fruit that have been removed from the plant, e.g. a stored tomato will ripen, and can synthesize over 1 mg lycopene per day (Sadana and Ahmad, 1948). It has also been shown that isolated pericarp discs of *Capsicum annuum* will ripen and form the characteristic red carotenoids such as capsanthin, when cultured aseptically in an aerated liquid medium (Von Abrams and Pratt, 1964).

The ripening process frequently involves the synthesis of large amounts of non-chloroplast pigments such as lycopene (tomato), capsanthin (red pepper) or increased synthesis of one possible chloroplast carotenoid without concomitant synthesis of the others (e.g. zeaxanthin in *Physalis alkekengi*). It is likely that a physically separate pathway of synthesis arises in developing fruit, and this is superimposed on the chloroplast pathway, which may continue or gradually disappear. For example, in normal red tomatoes, lycopene synthesis is inhibited if the fruit are held at temperatures above 30°C, while synthesis of the chloroplast pigment, β-carotene, is not (Goodwin and Jamikorn, 1952; Tomes *et al.*, 1956; Czygan and Willuhn, 1967). On the other hand, in high-β tomatoes, in which lycopene is replaced by β-carotene, the synthesis of this extra β-carotene is also temperature sensitive (Tomes *et al.*, 1956; Tomes, 1963). Light apparently has no significant effect on the ripening process or on its associated carotenogenesis (Smith and Smith, 1931).

The factors governing carotenoid formation in ripening fruits are discussed more fully by Goodwin and Goad (1970).

3 *Regulation of synthesis in green tissues*

In photosynthetic tissues carotenoids exist in the chloroplasts, and it is likely that they are synthesized in these organelles. Etiolated seedlings produce only small amounts of carotenoids, mainly xanthophylls (Kay and Phinney, 1956; Goodwin, 1958; Goodwin and Phagpolngarm, 1960), but in response to light, "chloroplast carotenoids", including considerable amounts of β-carotene, are synthesized as functional chloroplasts are formed (Goodwin, 1958; Valadon and Mummery, 1969). Although chloroplasts exhibit a considerable degree of autonomy, there is no evidence that they can carry out the steps of glycolysis leading to pyruvate and eventually acetyl-CoA, the primary carotenoid precursor. Evidence has been obtained for the existence of a pathway for the formation of pyruvate from glycolate (Breidenbach *et al.*, 1968; Shah and Rogers, 1969; Lord and Merrett, 1970; Tait, 1970), but there is no evidence that the chloroplast can convert pyruvate into acetyl CoA (Bassham, 1965) or convert acetyl CoA into MVA.

A compartmentation of biosynthesis has been suggested as a mechanism whereby formation of different classes of terpenoid compounds is regulated (Goodwin and Mercer, 1963; Goodwin, 1967). If excised etiolated seedlings are placed in [2-^{14}C] MVA and allowed to green up, then little radioactivity is recovered in the chloroplast terpenoids, e.g. carotenoids and the side chains of the chlorophylls, plastoquinones, tocopherols and phylloquinone, although the extraplastidic terpenoids such as squalene, sterols and the side chain of ubiquinone are very strongly labelled. On the other hand, if the seedlings are allowed to green in the presence of $^{14}CO_2$, the situation is reversed; the chloroplast terpenoids are highly labelled and the extraplastidic terpenoids only slightly labelled. Thus regulation of terpenoid synthesis in developing seedlings may be achieved by compartmentation with two separate sites of terpenoid synthesis from MVA onwards, one inside and the other outside the chloroplast. The basic reactions of terpenoid biosynthesis (i.e. formation of prenyl pyrophosphates) can occur at both sites, but the specific reactions involved in the biosynthesis of a particular group of terpenoids can only occur at one site, in the case of carotenoids inside the chloroplast.

REFERENCES

Aasen, A. J., Liaaen-Jensen, S. and Borsch, G. (1972). *Acta chem. scand.* **26**, 404–405.

Abrams, G. J. von and Pratt, H. K. (1964). *Pl. Physiol., Lancaster* **39** (suppl.), 65.

Agranoff, B. W., Eggerer, H., Henning, U. and Lynen, F. (1959). *J. Am. chem. Soc.* **81**, 1254–1255.

Agranoff, B. W., Eggerer, H., Henning, U. and Lynen, F. (1960). *J. biol. Chem.* **235**, 326–332.

Aitzetmüller, K., Svec, W. A., Katz, J. J. and Strain, H. H. (1968). *Chem. Commun.* 32–33.

Aitzetmüller, K., Strain, H. H., Svec, W. A., Grandolfo, M. and Katz, J. J. (1969). *Phytochemistry* **8**, 1761–1770.

Altman, L. J., Kowerski, R. C. and Rilling, H. C. (1971). *J. Am. chem. Soc.* **93**, 1782–1783.

Altman, L. J., Ash, L., Kowerski, R. C., Epstein, W. W., Larsen, B. R., Rilling, H. C., Muscio, F. and Gregonis, D. E. (1972). *J. Am. chem. Soc.* **94**, 3257–3259.

Anderson, D. G. and Porter, J. W. (1962). *Archs Biochem. Biophys.* **97**, 509–519.

Anderson, D. G., Norgård, D. W. and Porter, J. W. (1960). *Archs Biochem. Biophys.* **88**, 68–77.

Anderson, I. C. and Robertson, D. S. (1960). *Pl. Physiol., Lancaster* **35**, 531–534.

Andrewes, A. G. and Liaaen-Jensen, S. (1972). *A. Rev. Microbiol.* **26**, 225-248.

Archer, B. L., Audley, B. G., Cockbain, E. G. and McSweeney, G. P. (1963). *Biochem. J.* **89**, 565-574.

Arpin, N., Fiasson, J.-L., Dangye-Caye, M. P., Francis, G. W. and Liaaen-Jensen, S. (1971). *Phytochemistry* **10**, 1595-1601.

Aung Than, Bramley, P. M., Davies, B. H. and Rees, A. F. (1972). *Phytochemistry* **11**, 3187-3192.

Austin, D. J., Bu'Lock, J. D. and Drake, D. (1970). *Experientia* **26**, 348-349.

Barber, M. S., Jackman, L. M., Warren, C. K. and Weedon, B. C. L. (1961). *J. chem. Soc.* 4019-4024.

Barnes, F. J., Qureshi, A. A., Semmler, E. J. and Porter, J. W. (1973). *J. biol. Chem.* **248**, 2768-2773.

Barnett, H. L., Lilly, V. G. and Krause, R. F. (1956). *Science, N.Y.* **123**, 141.

Bartels, P. G. and McCullough, C. (1972). *Biochem. biophys. Res. Commun.* **48**, 16-22.

Bartlett, L., Klyne, W., Mose, W. P., Scopes, P. M., Galasko, G., Mallams, A. K., Weedon, B. C. L., Szabolcs, J. and Tóth, G. (1969). *J. chem. Soc. C* 2527-2543.

Bassham, J. A. (1965). *In* "Plant Biochemistry" (J. Bonner and J. E. Varner, eds), pp. 875-902. Academic Press, New York and London.

Batra, P. P. (1971). *In* "Photophysiology" (A. C. Giese, ed.), vol. 6, pp. 47-76. Academic Press, New York and London.

Batra, P. P., Gleason, R. M., Jr. and Louda, J. W. (1973). *Phytochemistry* **12**, 1309-1313.

Battersby, A. R. (1970). *In* "Substances Formed Biologically from Mevalonic Acid" (T. W. Goodwin, ed.), pp. 157-168. Academic Press, London and New York.

Beedle, A. S., Munday, K. A. and Wilton, D. C. (1972). *Eur. J. Biochem.* **28**, 151-155.

Beeler, D. A. and Porter, J. W. (1962). *Biochem. biophys. Res. Commun.* **8**, 367-371.

Beeler, D. A. and Porter, J. W. (1963). *Archs Biochem. Biophys.* **100**, 167-170.

Beeler, D. A., Anderson, D. G. and Porter, J. W. (1963). *Archs Biochem. Biophys.* **102**, 26-32.

Ben-Aziz, A., Britton, G. and Goodwin, T. W. (1973). *Phytochemistry* **12**, 2759-2764.

Bergman, K., Eslava, A. P. and Cerdá-Olmedo, E. (1973). *Molec. gen. Genet.* **123**, 1-16.

Berry, D. L., Singh, B. and Salunkhe, D. K. (1972). *Pl. Cell Physiol.* **13**, 157-165.

Beytia, E., Qureshi, A. A. and Porter, J. W. (1973). *J. biol. Chem.* **248**, 1856-1867.

Blattmann, P. and Rétey, J. (1970). *Chem. Commun.* 1394.

Blattmann, P. and Rétey, J. (1971). *Hoppe-Seyler's Z. physiol. Chem.* **352**, 367-376.

Blum, J. J. and Bégin-Heick, N. (1967). *Biochem. J.* **105**, 821-829.

Bonaly, R. and Malengé, J. P. (1968). *Biochim. biophys. Acta* **164**, 306.

Bonnett, R., Mallams, A. K., Spark, A. A., Tee, J. L., Weedon, B. C. L. and McCormick, A. (1969). *J. chem. Soc. C* 429-454.

Braithwaite, G. D. and Goodwin, T. W. (1960). *Biochem. J.* **76**, 1-5, 5-10, 192-197.

Breidenbach, R. W., Kahn, A. and Beevers, H. (1968). *Pl. Physiol, Lancaster* **43**, 705–713.

Britton, G. (1971). *In* "Aspects of Terpenoid Chemistry and Biochemistry" (T. W. Goodwin, ed.), pp. 255–289. Academic Press, London and New York.

Britton, G. and Goodwin, T. W. (1969). *Phytochemistry* **8**, 2257–2258.

Brodie, J. D. and Porter, J. W. (1960). *Biochem. biophys. Res. Commun.* **3**, 173–177.

Brodie, J. D., Wasson, G. and Porter, J. W. (1963). *J. biol. Chem.* **238**, 1294–1301.

Buchecker, R. and Eugster, C. H. (1973). *Helv. chim. Acta* **56**, 1124–1128.

Buchecker, R., Hamm, P. and Eugster, C. H. (1972). *Chimia* **26**, 134–136.

Buggy, M. J., Britton, G. and Goodwin, T. W. (1969). *Biochem. J.* **114**, 641–643.

Buggy, M. J., Britton, G. and Goodwin, T. W. (1974). *Phytochemistry* **13**, 125–129.

Bu'Lock, J. D., Austin, D. J., Snatzke, G. and Hruban, L. (1970). *Chem. Commun.* 255–256.

Caglioti, L., Cainelli, G., Camerino, B., Mondelli, R., Prieto, A., Quilico, A., Salvatori, T. and Selva, A. (1964). *Chim. Ind., Milan* **46**, 961.

Caglioti, L., Cainelli, G., Camerino, B., Mondelli, R., Prieto, A., Quilico, A., Salvatori, T. and Selva, A. (1966). *Tetrahedron suppl.* **7**, 175–187.

Cantino, E. C. and Hyatt, M. T. (1953). *Am. J. Bot.* **40**, 688–694.

Charlton, J. M., Treharne, K. J. and Goodwin, T. W. (1967). *Biochem. J.* **105**, 205–212.

Chichester, C. O., Wong, P. S. and Mackinney, G. (1954). *Pl. Physiol., Lancaster* **29**, 238–241.

Chichester, C. O., Nakayama, T. O. M., Mackinney, G. and Goodwin, T. W. (1955). *J. biol. Chem.* **214**, 515–517.

Chichester, C. O., Yokoyama, H., Nakayama, T. O. M., Lukton, A. and Mackinney, G. (1959). *J. biol. Chem.* **234**, 598–602.

Cholnoky, L. and Szabolcs, J. (1960). *Experientia* **16**, 483–484.

Cholnoky, L., Györgyfy, K., Rónai, A., Szabolcs, J., Tóth, G., Galasko, G., Mallams, A. K., Waight, E. S. and Weedon, B. C. L. (1969). *J. chem. Soc. C* 1256–1263.

Ciegler, A. (1965). *Adv. appl. Microbiol.* **7**, 1–34.

Ciegler, A., Arnold, M. and Anderson, R. F. (1959). *Appl. Microbiol.* **7**, 94.

Claes, H. (1954). *Z. Naturf. B* **9**, 461–469.

Claes, H. (1959). *Z. Naturf. B* **14**, 4–17.

Claes, H. (1966). *In* "Biochemistry of Chloroplasts" (T. W. Goodwin, ed.), vol. 2, pp. 441–444. Academic Press, London and New York.

Coggins, C. W. Jr., Henning, G. L. and Yokoyama, H. (1970). *Science, N.Y.* **168**, 1589–1590.

Coon, M. J., Kupiecki, F. P., Dekker, E. E., Schlesinger, M. J. and del Campillo, A. (1959). *In* "Biosynthesis of Terpenes and Sterols" (G. E. W. Wolstenholme and M. O'Connor, eds), p. 62. Churchill, London.

Cornforth, J. W. and Ross, F. P. (1970). *Chem. Commun.* 1395–1396.

Cornforth, J. W., Cornforth, R. H. and Popják, G. (1962). *Tetrahedron* **18**, 1351–1354.

Cornforth, J. W., Cornforth, R. H., Donninger, C. and Popják, G. (1966a). *Proc. R. Soc. B* **163**, 492–514.

Cornforth, J. W., Cornforth, R. H., Popják, G. and Yengoyan, L. (1966b). *J. biol. Chem.* **241**, 3970–3987.
Cornforth, J. W., Clifford, K., Mallaby, R. and Phillips, G. T. (1972), *Proc. R. Soc. B* **182**, 277–295.
Czygan, F.-C. (1968). *Arch. Mikrobiol.* **62**, 209–236.
Czygan, F.-C. and Kalb, K. (1966). *Z. PflPhysiol.* **55**, 59–64.
Czygan, F.-C. and Willuhn, G. (1967). *Planta med.* **15**, 404.
Davies, B. H. (1961). *Biochem. J.* **80**, 48P.
Davies, B. H. (1970). *Biochem. J.* **116**, 93–99.
Davies, B. H. (1973). *Pure appl. Chem.* **35**, 1–28.
Davies, B. H. and Rees, A. F. (1973). *Phytochemistry* **12**, 2745–2750.
Davies, B. H., Jones, D. and Goodwin, T. W. (1963a). *Biochem. J.* **87**, 326–329.
Davies, B. H., Villoutreix, J., Williams, R. J. H. and Goodwin, T. W. (1963b). *Biochem. J.* **89**, 96P.
Davies, B. H., Holmes, E. A., Loeber, D. E., Toube, T. P. and Weedon, B. C. L. (1969). *J. chem. Soc. C* 1266–1268.
Davis, J. B., Jackman, L. M., Siddons, P. T. and Weedon, B. C. L. (1961). *Proc. chem. Soc.* 261–263.
Decker, K. and Uehleke, H. (1961). *Hoppe-Seyler's Z. physiol. Chem.* **323**, 61–76.
de la Guardia, M. D., Aragon, C. M. G., Murillo, F. J. and Cerdá-Olmedo, E. (1971). *Proc. natn. Acad. Sci. U.S.A.* **68**, 2012–2015.
Dentice di Accadia, F., Gribanovski-Sassu, O., Romagnoli, A. and Tuttobello, L. (1966). *Biochem. J.* **101**, 735–740.
Dersch, G. (1960). *Flora, Jena* **149**, 566–603.
De Ville, T. E., Hursthouse, M. B., Russell, S. W. and Weedon, B. C. L. (1969). *Chem. Commun.* 1311–1312.
Donninger, C. and Popják, G. (1966). *Proc. R. Soc. B* **163**, 465–491.
Dorsey, J. K. and Porter, J. W. (1968). *J. biol. Chem.* **243**, 4667–4670.
Dorsey, J. K., Dorsey, J. A. and Porter, J. W. (1966). *J. biol. Chem.* **241**, 5353–5360.
Dugan, R. E. and Porter, J. W. (1971). *J. biol. Chem.* **246**, 5361–5364.
Durr, I. F. and Rudney, H. (1960). *J. biol. Chem.* **235**, 2572–2578.
Dutton, H. J. and Juday, C. (1944). *Ecology* **25**, 273–283.
Eberle, M. and Arigoni, D. (1960). *Helv. chim. Acta* **43**, 1508–1513.
Edmond, J., Popják, G., Wong, S.-M. and Williams, V. P. (1971). *J. biol. Chem.* **246**, 6254–6271.
Egger, K., Dabbagh, A. G. and Nitsche, H. (1969). *Tetrahedron Lett.* 2995–2998.
Ehrenberg, L. and Faludi-Dániel, A. (1962). *Acta chem. scand.* **16**, 1523–1527.
Elahi, M., Lee, T. H., Simpson, K. L. and Chichester, C. O. (1973). *Phytochemistry* **12**, 1633–1639.
Engel, B. G., Würsch, J. and Zimmermann, M. (1953). *Helv. chim. Acta* **36**, 1771–1776.
Entschel, R. and Karrer, P. (1959). *Helv. chim. Acta* **42**, 466–472.
Entschel, R. and Karrer, P. (1960). *Helv. chim. Acta* **43**, 89–94.
Epstein, W. W. and Rilling, H. C. (1970). *J. biol. Chem.* **245**, 4597–4605.
Eslava, A. P. and Cerdá-Olmedo, E. (1974). *Plant Sci. Lett.* **2**, 9–14.
Eugster, C. H., Buchecker, R., Tscharner, C., Uhde, G. and Ohloff, G. (1969). *Helv. chim. Acta* **52**, 1729–1731.
Faludi-Dániel, A., Láng, F. and Fradkin, L. I. (1966). *In* "Biochemistry of

Chloroplasts" (T. W. Goodwin, ed.), vol. 1, pp. 269-274. Academic Press, London and New York.
Flint, A. P. F. (1970). *Biochem. J.* 120, 145-150.
Friend, J. and Goodwin, T. W. (1954). *Biochem. J.* 57, 434-437.
Friend, J., Goodwin, T. W. and Griffiths, L. A. (1955). *Biochem. J.* **60**, 649-655.
García-Peregrín, E., Suárez, M. D., Aragón, M. C. and Mayor, F. (1972). *Phytochemistry* 11, 2495-2498.
García-Peregrín, E., Suárez, M. D. and Mayor, F. (1973). *FEBS Lett.* **30**, 15-17.
Garton, G. A., Goodwin, T. W. and Lijinsky, W. (1951). *Biochem. J.* 48, 154-163.
Goodman, DeW. S. and Popják, G. (1960). *J. Lipid Res.* 1, 286-300.
Goodwin, T. W. (1953). *J. Sci. Fd Agric.* 5, 209-220.
Goodwin, T. W. (1954). *Experientia* 10, 213-214.
Goodwin, T. W. (1958). *Biochem. J.* 70, 612-617.
Goodwin, T. W. (1967). *In* "The Biochemistry of Chloroplasts" (T. W. Goodwin, ed.), vol. 2, pp. 721-733. Academic Press, London and New York.
Goodwin, T. W. (1971a). *In* "Carotenoids" (O. Isler, ed.), pp. 577-636. Birkhäuser Verlag, Basle.
Goodwin, T. W. (1971b). *In* "Aspects of Terpenoid Chemistry and Biochemistry" (T. W. Goodwin, ed.), pp. 315-356. Academic Press, London and New York.
Goodwin, T. W. and Goad, L. J. (1970). *In* "The Biochemistry of Fruits and their Products" (A. C. Hulme, ed.), vol. 1, pp. 305-368. Academic Press, London and New York.
Goodwin, T. W. and Jamikorn, M. (1952). *Nature, Lond.* 170, 104-105.
Goodwin, T. W. and Jamikorn, M. (1954a). *Biochem. J.* 57, 376-381.
Goodwin, T. W. and Jamikorn, M. (1954b). *J. Protozool.* 1, 216-219.
Goodwin, T. W. and Lijinsky, W. (1952). *Biochem. J.* 50, 268-273.
Goodwin, T. W. and Mercer, E. I. (1963). *Biochem. Soc. Symp.*, 24, 37-41.
Goodwin, T. W. and Osman, H. G. (1954). *Biochem. J.* 56, 222-230.
Goodwin, T. W. and Phagpolngarm, S. (1960). *Biochem. J.* 76, 197-199.
Goodwin, T. W. and Williams, R. J. H. (1965a). *Biochem. J.* 94, 5C-7C.
Goodwin, T. W. and Williams, R. J. H. (1965b). *Biochem. J.* 97, 28C-32C.
Goodwin, T. W. and Williams, R. J. H. (1966). *Proc. R. Soc. B.* 163, 515-518.
Goodwin, T. W. and Willmer, J. S. (1952). *Biochem. J.* 51, 213-217.
Goodwin, T. W., Jamikorn, M. and Willmer, J. S. (1953). *Biochem. J.* 53, 531-538.
Goodwin, T. W., Britton, G. and Walton, T. J. (1968). *Pl. Physiol, Lancaster* 43, (Suppl.) S-46.
Graebe, J. E. (1968). *Phytochemistry* 7, 2003-2020.
Gray J. C. and Kekwick, R. G. O. (1972). *Biochim. biophys. Acta* 279, 290-296.
Gray, J. C. and Kekwick, R. G. O. (1973). *Biochem. J.* 133, 335-347.
Green, T. R. and Baisted, D. J. (1972). *Biochem. J.* 130. 983-995.
Gregonis, D. E. and Rilling, H. C. (1973). *Biochem. biophys. Res. Commun.* 54, 449-454.
Gregonis, D. E. and Rilling, H. C. (1974). *Biochemistry* 13, 1538.
Gribanovski-Sassu, O. (1972). *Phytochemistry* 11, 3195-3198.
Griffiths, M. and Stainer, R. Y. (1956). *J. gen. Microbiol.* 14, 698-715.

Grob, E. C. and Boschetti, A. (1962). *Chimia* **16**, 15-16.
Grob, E. C. and Bütler, R. (1955). *Helv. chim. Acta* **38**, 1313-1316.
Grob, E. C. and Bütler, R. (1956). *Helv. chim. Acta* **39**, 1975-1980.
Grob, E. C., Kirschner, K. and Lynen, F. (1961). *Chimia* **15**, 308-310.
Habermann, H. M. (1960). *Physiologia Pl.* **13**, 718-725.
Harding, R. W., Huang, P. C. and Mitchell, H. K. (1969). *Archs Biochem. Biophys.* **129**, 696-707.
Haxo, F. T. (1952). *Biol. Bull.* **103**, 286.
Haxo, F. T. (1956). *Fortschr. Chem. org. NatStoffe* **12**, 169-197.
Hayano, M. (1962). *In* "Oxygenases" (O. Hayaishi, ed.), pp. 181-240. Academic Press, London and New York.
Hepper, C. M. and Audley, B. G. (1969). *Biochem. J.* **114**, 379-386.
Herber, R., Maudinas, B., Villoutreix, J. and Granger, P. (1972). *Biochim. biophys. Acta* **280**, 194-202.
Hertzberg, S. and Liaaen-Jensen, S. (1969a). *Phytochemistry* **8**, 1259-1280.
Hertzberg, S. and Liaaen-Jensen, S. (1969b). *Phytochemistry* **8**, 1281-1292.
Higgins, M. J. P. and Kekwick, R. G. O. (1973). *Biochem. J.* **134**, 295-310.
Hill, H. M., Calderwood, S. K. and Rogers, L. J. (1971). *Phytochemistry* **10**, 2051-2058.
Holloway, P. W. and Popják, G. (1967). *Biochem. J.* **104**, 57-70.
Holloway, P. W. and Popják, G. (1968). *Biochem. J.* **106**, 835-840.
Howes, C. D. and Batra, P. P. (1970). *Biochim. biophys. Acta* **222**, 174-178.
Hsu, W. J., Yokoyama, H. and Coggins, C. W., Jr. (1972). *Phytochemistry* **11**, 2985-2990.
Jenkins, J. A. and Mackinney, G. (1955). *Genetics, Princeton* **40**, 715-720.
Jungalwala, F. B. and Porter, J. W. (1965). *Archs Biochem. Biophys.* **110**, 291-299.
Jungalwala, F. B. and Porter, J. W. (1967). *Archs Biochem. Biophys.* **119**, 209-219.
Kakutani, Y. (1966). *J. Biochem., Tokyo* **59**, 135-138.
Kakutani, Y., Suzue, G. and Tanaka, S. (1964). *J. Biochem., Tokyo* **56**, 195-196.
Kandutsch, A. A., Paulus, H., Levin, E. and Bloch, K. (1964). *J. biol. Chem.* **239**, 2507-2515.
Kargl, T. E., Quackenbush, F. W. and Tomes, M. L. (1960). *Proc. Am. Soc. hort. Sci.* **75**, 574.
Karunakaran, A., Karunakaran, M. and Quackenbush, F. W. (1966). *Archs Biochem. Biophys.* **114**, 326-330.
Kay, R. E. and Phinney, B. (1956). *Pl. Physiol., Lancaster* **31**, 226-231.
Kayser, F. and Villoutriex, J. (1961). *C.r. Séanc. Soc. Biol.* **155**, 1094-1095.
Kelly, M., Andresen, S. A. and Liaaen-Jensen, S. (1971). *Acta chem. scand.* **25**, 1607-1614.
Kessler, E. and Czygan, F.-C. (1966). *Arch. Mikrobiol.* **54**, 37-45.
Khatoon, N., Loeber, D. E., Toube, T. P. and Weedon, B. C. L. (1972). *Chem. Commun.* 996-997.
Kjøsen, H. and Liaaen-Jensen, S. (1971). *Acta chem. scand.* **25**, 1500-1502.
Kjøsen, H., Arpin, N. and Liaaen-Jensen, S. (1972). *Acta chem. scand.* **26**, 3053-3067.
Kleinig, H., Nitsche, H. and Egger, K. (1969). *Tetrahedron Lett.* 5139-5142.
Knappe, J., Ringelmann, E. and Lynen, F. (1959). *Biochem. Z.* **332**, 195-213.

Knauss, H., Brodie, J. D. and Porter, J. W. (1962). *J. Lipid Res.* 3, 197–206.
Knypl, J. S. (1969). *Naturwissenschaften* **56**, 572.
Krzeminski, L. F. and Quackenbush, F. W. (1960). *Archs Biochem. Biophys.* **88**, 287–293.
Kuhn, R. and Brockmann, H. (1933). *Chem. Ber.* **66**, 1319–1326.
Kuhn, R. and Brockmann, H. (1935). *Justus Liebig's Annln Chem.* **516**, 95.
Kushwaha, S. C., Subbarayan, C., Beeler, D. A. and Porter, J. W. (1969). *J. biol. Chem.* **244**, 3635–3642.
Kushwaha, S. C., Suzue, G., Subbarayan, C. and Porter, J. W. (1970). *J. biol. Chem.* **245**, 4708–4717.
Lang, W. and Rau, W. (1972). *Planta* **106**, 345–354.
Lee, T.-C. and Chichester, C. O. (1969). *Phytochemistry*, **8**, 603–609.
Le Rosen, A. L. and Zechmeister, L. (1942). *J. Am. chem. Soc.* **64**, 1075–1079.
Le Rosen, A. L., Went, F. W. and Zechmeister, L. (1941). *Proc. natn. Acad. Sci. U.S.A.* 27, 236–242.
Liaaen-Jensen, S., Cohen-Bazire, G., Nakayama, T. O. M. and Stanier, R. Y. (1958). *Biochim. biophys. Acta* 29, 477–498.
Lincoln, R. E. and Porter, J. W. (1950). *Genetics, Princeton* 35, 206–211.
Lindberg, M., Yuan, C., de Waard, A. and Bloch, K. (1962). *Biochemistry, N.Y.* 1, 182–188.
Liu, I.-S., Lee, T. H., Yokoyama, H., Simpson, K. L. and Chichester, C. O. (1973). *Phytochemistry* 12, 2953–2956.
Loeblich, L. A. (1969). *J. Protozool.* 16 (suppl.), 22–23.
Loomis, W. D. and Battaile, J. (1963). *Biochim. biophys. Acta* 67, 54–63.
Lord, J. M. and Merrett, M. J. (1970). *Biochem. J.* 117, 929–937.
Lotspeich, F. J., Krause, R. F., Lilly, V. G. and Barnett, H. L. (1959). *J. biol. Chem.* 234, 3109–3110.
Lwoff, M. and Lwoff, A. (1930). *C.r. Séanc. Soc. Biol.* 105, 454–456.
Lynen, F., Eggerer, H., Henning, U. and Kessel, I. (1958). *Angew. Chem.* 70, 738–742.
Lynen, F., Agranoff, B. W., Eggerer, H., Henning, U. and Moslein, E. M. (1959). *Angew Chem.* 71, 657–663.
MacArthur, J. W. (1934). *J. Genet.* 29, 123–133.
McDermott, J. C. B., Ben-Aziz, A., Singh, R. K., Britton, G. and Goodwin, T. W. (1973a). *Pure appl. Chem.* 35, 29–45.
McDermott, J. C. B., Britton, G. and Goodwin, T. W. (1973b). *Biochem. J.* 134, 1115–1117.
McDermott, J. C. B., Britton, G. and Goodwin, T. W. (1973c). *J. gen. Microbiol.* 77, 161–171.
McDermott, J. C. B., Brown, D. J., Britton, G. and Goodwin, T. W. (1974). *Biochem. J.* 144, 231–243.
Mackinney, G., Nakayama, T. O. M., Buss, C. D. and Chichester, C. O. (1952). *J. Am. chem. Soc.* 74, 3456–3457.
Mackinney, G., Rick, C. M. and Jenkins, J. A. (1956). *Proc. natn. Acad. Sci. U.S.A.* 42, 404–408.
Malhotra, H. C., Britton, G. and Goodwin, T. W. (1970). *Int. J. vit. Res.* 40, 315–322.
Mallams, A. K., Waight, E. S., Weedon, B. C. L., Chapman, D. J., Haxo, F. T., Goodwin, T. W. and Thomas, D. M. (1967). *Chem. Commun.* 301–302.
Markham, M. C. and Liaaen-Jensen, S. (1968). *Phytochemistry*, 7, 839–844.

Mase, Y., Rabourn, W. J. and Quackenbush, F. W. (1957). *Archs Biochem. Biophys.* **68**, 150–156.
Mercer, E. I., Davies, B. H. and Goodwin, T. W. (1963). *Biochem. J.* **87**, 317–325.
Moshier, S. E. and Chapman, D. J. (1973). *Biochem. J.* **136**, 395–404.
Mummery, R. S. and Valadon, L. R. G. (1973). *Physiologia Pl.* **28**, 254–258.
Nakamura, H. and Greenberg, G. M. (1961). *Archs Biochem. Biophys.* **93**, 153–156.
Nakayama, T. O. M., Mackinney, G. and Phaff, H. J. (1954). *Antonie van Leeuwenhoek* **20**, 217–228.
Nandi, D. L. and Porter, J. W. (1964). *Archs Biochem. Biophys.* **105**, 7–19.
Nitsche, H. and Egger, K. (1970). *Tetrahedron Lett.* 1435–1438.
Nitsche, H., Egger, K. and Dabbagh, A. G. (1969). *Tetrahedron Lett.* 2999–3002.
Nusbaum-Cassuto, E. and Villoutreix, J. (1965). *C.r hebd. Séanc. Acad. Sci., Paris* **260**, 1013–1015.
Nusbaum-Cassuto, E., Villoutreix, J. and Malengé, J.-P. (1967). *Biochim. biophys. Acta* **136**, 459.
Ogura, K., Shinka, T. and Seto, S. (1972). *J. Biochem., Tokyo* **72**, 1101–1108.
Ohnoki, S., Suzue, G. and Tanaka, S. (1962). *J. Biochem., Tokyo* **52**, 423–427.
Ootaki, T., Lighty, A. C., Delbrück, M. and Hsu, W.-J. (1973). *Molec. gen. Genet.* **121**, 57–70.
Oster, M. O. and West, C. A. (1968). *Archs Biochem. Biophys.* **127**, 112–123.
Pennock, J. F., Hemming, F. W. and Morton, R. A. (1962). *Biochem. J.* **82**, 11P.
Petzold, E. N., Quackenbush, F. W. and McQuistan, M. (1959). *Archs Biochem. Biophys.* **82**, 117–124.
Plempel, M. (1963). *Naturwissenschaften* **50**, 226.
Pollard, C. J., Bonner, J., Haagen-Smit, A. J. and Nimmo, C. C. (1966). *Pl. Physiol., Lancaster* **41**, 66–70.
Popják, G. (1959). *Tetrahedron Lett.* no. 19, pp. 19–28.
Popják, G. and Cornforth, J. W. (1966). *Biochem. J.* **101**, 553–568.
Popják, G., Edmond, J. and Wong, S.-M. (1973). *J. Am. chem. Soc.* **95**, 2713–2714.
Porter, J. W. (1969). *Pure appl. Chem.* **20**, 449–481.
Porter, J. W. and Lincoln, R. E. (1950). *Archs Biochem. Biophys.* **27**, 390–403.
Potty, V. H. and Bruemmer, J. H. (1970). *Phytochemistry* **9**, 99–105.
Prieto, A., Spalla, C., Bianchi, M. and Biffi, G. (1964). *Chemy Ind.* 551.
Purcell, A. E. (1964). *Archs Biochem. Biophys.* **105**, 606–611.
Purcell, A. E., Thompson, G. A. Jr. and Bonner, J. (1959). *J. biol. Chem.* **234**, 1081–1084.
Qureshi, A. A., Barnes, F. J. and Porter, J. W. (1972). *J. biol. Chem.* **247**, 6730–6732.
Qureshi, A. A., Barnes, F. J., Semmler, E. J. and Porter, J. W. (1973). *J. biol. Chem.* **248**, 2755–2767.
Rabourn, W. J. and Quackenbush, F. W. (1956). *Archs Biochem. Biophys.* **61**, 111–118.
Rabourn, W. J., Quackenbush, F. W. and Porter, J. W. (1954). *Archs Biochem. Biophys.* **48**, 267–274.
Rau, W. (1967). *Planta* **74**, 263–277.
Rau, W. (1969). *Planta* **84**, 30–42.

Rau, W. (1971). *Planta* **101**, 251-264.
Rau, W., Feuser, A. and Rau-Hund, A. (1967). *Biochim. biophys. Acta* **136**, 589.
Rau, W. Lindemann, I. and Rau-Hund, A. (1968). *Planta* **80**, 309-316.
Raymundo, L. C. and Simpson, K. L. (1972). *Phytochemistry* **11**, 397-400.
Raymundo, L. C., Griffiths, A. E. and Simpson, K. L. (1967). *Phytochemistry* **6**, 1527-1532.
Rétey, J., Stetten, E. von, Coy, U. and Lynen, F. (1970). *Eur. J. Biochem.* **15**, 72-76.
Reyes, P., Nakayama, T. O. M. and Chichester, C. O. (1964). *Biochim. biophys. Acta* **90**, 578-592.
Ricketts, T. R. (1971). *Phytochemistry*, **10**, 155-164.
Rilling, H. C. (1962). *Biochim. biophys. Acta* **65**, 156-158.
Rilling, H. C. (1965). *Archs Biochem. Biophys.* **110**, 39-46.
Rilling, H. C. (1966). *J. biol. Chem.* **241**, 3233-3236.
Rilling, H. C., Poulter, C. D., Epstein, W. W. and Larsen, B. R. (1971). *J. Am. chem. Soc.* **93**, 1783-1785.
Rogers, L. J., Shah, S. P. J. and Goodwin, T. W. (1965). *Biochem. J.* **96**, 7P-8P.
Rudney, H. (1957). *J. biol. Chem.* **227**, 363-377.
Sadana, J. C. and Ahmad, B. (1948). *J. scient. ind. Res.* **78**, 172.
Sapozhnikov, D. I., Alkhazov, D. G., Eidel'man, Z. M., Bazhanova, I. V., Lemberg, I. K., Maslova, T. G., Girshin, A. B., Popova, I. A., Saakov, B. C., Povova, O. F. and Shiryaeva, G. A. (1964). *Dokl. Akad. Nauk SSSR* **154**, 974-977.
Schlesinger, M. J. and Coon, M. J. (1961). *J. biol. Chem.* **236**, 2421-2424.
Scott, A. I., Phillips, T., Reichardt, P. B. and Sweeny, J. G. (1970). *Chem. Commun.* 1396-1397.
Sebek, O. and Jäger, H. (1964). *Abstr. 148th Meet. Am. chem. Soc.* 9Q.
Shah, D. H., Cleland, W. W. and Porter, J. W. (1965). *J. biol. Chem.* **240**, 1946-1956.
Shah, D. V., Feldbruegge, D. H., Houser, A. R. and Porter, J. W. (1968). *Archs Biochem. Biophys.* **127**, 124-131.
Shah, S. P. J. and Rogers, L. J. (1969). *Biochem. J.* **114**, 395-405.
Shepherd, K. S. and Booth, R. (1971). *Biochem. J.* **125**, 39P-40P.
Simpson, K. L., Nakayama, T. O. M. and Chichester, C. O. (1964a). *Biochem. J.* **92**, 508-510.
Simpson, K. L., Nakayama, T. O. M. and Chichester, C. O. (1964b). *J. Bacteriol.* **88**, 1688-1694.
Singh, R. K., Ben-Aziz, A., Britton, G. and Goodwin, T. W. (1973). *Biochem. J.* **132**, 649-652.
Siperstein, M. D. and Fagan, V. M. (1966). *J. biol. Chem.* **241**, 602-609.
Skilleter, D. N. and Kekwick, R. G. O. (1971). *Biochem. J.* **124**, 407-417.
Smallidge, R. L. and Quackenbush, F. W. (1973). *Phytochemistry* **12**, 2481-2482.
Smith, L. W. and Smith, O. (1931). *Pl. Physiol., Lancaster* **6**, 265.
Steele, W. J. and Gurin, S. (1960). *J. biol. Chem.* **235**, 2778-2785.
Stewart, P. R. and Rudney, H. (1966). *J. biol. Chem.* **241**, 1212-1221.
Strain, H. H. and Manning, W. M. (1943). *J. Am. chem. Soc.* **65**, 2258-2259.
Strain, H. H., Manning, W. M. and Hardin, G. J. (1944). *Biol. Bull.* **86**, 169-192.
Strain, H. H., Svec, W. A., Aitzetmüller, K., Grandolfo, M. and Katz, J. J. (1968). *Phytochemistry* **7**, 1417-1418.

Subbarayan, C., Kushwaha, S. C., Suzue, G. and Porter, J. W. (1970). *Archs Biochem. Biophys.* **137**, 547-557.
Sutter, R. P. (1970). *Science, N.Y.* **168**, 1590-1592.
Sutter, R. P. and Rafelson, M. E., Jr. (1968). *J. Bacteriol.* **95**, 426-432.
Suzue, G. (1960). *Biochim. biophys. Acta* **45**, 616-617.
Suzue, G. (1964). *Bull. chem. Soc. Japan* **37**, 613-616.
Suzue, G. and Porter, J. W. (1969). *Biochim. biophys. Acta* **176**, 653-656.
Tait, G. H. (1970). *Biochem. J.* **118**, 819-830.
Taylor, R. F. and Davies, B. H. (1973). *Trans. biochem. Soc.* **1**, 1091-1092.
Taylor, R. F. and Davies, B. H. (1974). *Biochem. J.* **139**, 751-760, 761-769.
Tefft, R. E., Goodwin, T. W. and Simpson, K. L. (1970). *Biochem. J.* **117**, 921-927.
Theimer, R. R. and Rau, W. (1969). *Biochim. biophys. Acta* **177**, 180-181.
Theimer, R. R. and Rau, W. (1970). *Planta* **92**, 129-137.
Theimer, R. R. and Rau, W. (1972). *Planta* **106**, 331-343.
Thomas, D. M. and Goodwin, T. W. (1967). *Phytochemistry* **6**, 355-360.
Thomas, D. M., Harris, R. C., Kirk, J. T. O. and Goodwin, T. W. (1967). *Phytochemistry* **6**, 361-366.
Thomas, D. R. and Stobart, A. K. (1970). *Phytochemistry* **9**, 1443-1451.
Tomes, M. L. (1963). *Bot. Gaz.* **124**, 180-185.
Tomes, M. L. (1967). *Genetics, Princeton* **56**, 227-232.
Tomes, M. L., Quackenbush, F. W. and McQuistan, M. (1954). *Genetics, Princeton* **39**, 810-817.
Tomes, M. L., Quackenbush, F. W. and Kargl, T. E. (1956). *Bot. Gaz.* **117**, 248-253.
Tomes, M. L., Quackenbush, F. W. and Kargl, T. E. (1958). *Bot. Gaz.* **119**, 250-253.
Treharne, K. J., Mercer, E. I. and Goodwin, T. W. (1966). *Phytochemistry* **5**, 581-587.
Turian, G. (1950). *Helv. chim. Acta* **33**, 1988-1993.
Turian, G. (1957). *Physiologia Pl.* **10**, 667-674.
Turian, G. and Haxo, F. T. (1952). *J. Bacteriol.* **63**, 690-691.
Valadon, L. R. G. and Mummery, R. S. (1966). *J. gen. Microbiol.* **45**, 531-540.
Valadon, L. R. G. and Mummery, R. S. (1969a). *J. exp. Bot.* **20**, 732.
Valadon, L. R. G. and Mummery, R. S. (1969b). *Microbios.* **1**, 3.
Valenzuela, P., Beytia, E., Cori, O. and Yudelevich, A. (1966). *Archs Biochem. Biophys.* **113**, 536-539.
Van den Ende, H. (1968). *J. Bacteriol.* **96**, 1298-1303.
Van den Ende, H., Werkman, B. A. and Van den Briel, M. L. (1972). *Arch. Mikrobiol.* **86**, 175-184.
Van Tamelen, E. and Schwartz, M. A. (1971). *J. Am. chem. Soc.* **93**, 1780-1782.
Varma, T. N. R. and Chichester, C. O. (1962). *Archs Biochem. Biophys.* **96**, 419-422.
Villoutreix, J. (1960). *Biochim. biophys. Acta* **40**, 434-441.
Wallace, R. H. and Habermann, H. M. (1959). *Am. J. Bot.* **46**, 157-162.
Wallace, R. H. and Schwarting, A. E. (1954). *Pl. Physiol., Lancaster* **29**, 431-436.
Walton, T. J., Britton, G. and Goodwin, T. W. (1970). *Phytochemistry* **9**, 2545-2552.

Weedon, B. C. L. (1967). *Chem. in Britain* 3, 424-432.
Weedon, B. C. L. (1970). *Rev. pure appl. Chem.* 20, 51-66.
Weeks, O. B. (1971). *In* "Aspects of Terpenoid Chemistry and Biochemistry" (T. W. Goodwin, ed.), pp. 291-313. Academic Press, London and New York.
Weeks, O. B., Saleh, F. K., Wirahadikusumah, M. and Berry, R. A. (1973). *Pure appl. Chem.* 35, 63-80.
White, L. W. and Rudney, H. (1970). *Biochemistry, N.Y.* 9, 2725-2731.
Williams, R. J. H., Britton, G. and Goodwin, T. W. (1966). *Biochim. biophys. Acta* 124, 200-203.
Williams, R. J. H., Britton, G. and Goodwin, T. W. (1967a). *Biochem. J.* **105**, 99-105.
Williams, R. J. H., Charlton, J. M., Britton, G. and Goodwin, T. W. (1967b). *Biochem. J.* **104**, 767-777.
Williamson, I. P. and Kekwick, R. G. O. (1965). *Biochem. J.* **96**, 862-871.
Wolken, J. J., Mellon, A. D. and Greenblatt, C. L. (1955). *J. Protozool.* 2, 89-96.
Wuhrmann, J. J., Yokoyama, H. and Chichester, C. O. (1957). *J. Am. chem. Soc.* 79, 4569-4570.
Yamamoto, H. Y. and Chichester, C. O. (1965). *Biochim. biophys. Acta* **109**, 303-305.
Yamamoto, H. Y., Yokoyama, H., Simpson, K. L., Nakayama, T. O. M. and Chichester, C. O. (1961). *Nature, Lond.* **191**, 1299-1300.
Yamamoto, H. Y., Chichester, C. O. and Nakayama, T. O. M. (1962). *Archs Biochem. Biophys.* **96**, 645-649.
Yokoyama, H. and White, M. J. (1968). *Phytochemistry* 7, 1031-1034.
Yokoyama, H. and Guerrero, H. C. (1970). *Phytochemistry* 9, 231-232.
Yokoyama, H., Chichester, C. O., Nakayama, T. O. M., Lukton, A. and Mackinney, G. (1957). *J. Am. chem. Soc.* 79, 2029-2030.
Yokoyama, H., Chichester, C. O. and Mackinney, G. (1960). *Nature, Lond.* **185**, 687-688
Yokoyama, H., Nakayama, T. O. M. and Chichester, C. O. (1961). *J. biol. Chem.* 237, 681-686.
Yokoyama, H., Guerrero, H. C. and Boettger, H. (1970). *J. org. Chem.* 35, 2080-2082.
Yokoyama, H., Coggins, C. W. Jr., and Henning, G. L. (1971). *Phytochemistry* 10, 1831-1834.
Yokoyama, H., DeBenedict, C., Coggins, C. W. Jr. and Henning, G. L. (1972). *Phytochemistry* 11, 1721-1724.
Yuldasheva, L. S., Feofilova, E. P., Samokhvalov, G. I. and Bekhtereva, M. N. (1972). *Mikrobiologiya* 41, 430.
Zechmeister, L. and Went, F. W. (1948). *Nature, Lond.* **162**, 847-848.

Chapter 6

Algal Biliproteins and Phycobilins

PÁDRAIG Ó CARRA and COLM Ó hEOCHA

Department of Biochemistry, University College, Galway, Ireland

I General properties, composition and distribution

Bilins are open tetrapyrroles containing the skeleton ring system of structure I. They are widely, though irregularly, distributed in the plant and animal kingdoms. In animals, bilins appear mostly as functionless catabolites of haem although in some species they play marginal roles, principally as skin or eggshell pigments (see e.g. Rüdiger, 1971). Plant bilins on the other hand perform important functional roles as light harvesters and light sensors. They also differ from the generality of animal bilins in being covalently attached to apoproteins, forming conjugates known as biliproteins.

The algal biliproteins (also termed phycobiliproteins) are a group of pigments which occur in the divisions Rhodophyta (red algae), Cyanophyta (blue-green algae) and Cryptophyta (cryptomonad algae), where they function as the principal absorbers of light for the

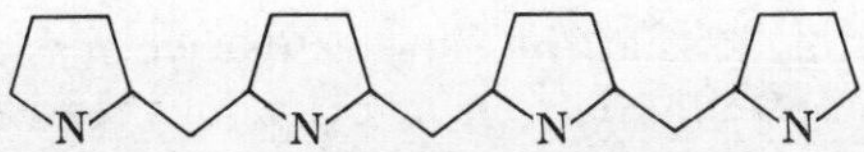

(I)

photosynthetic apparatus. There has been a tendency to regard these pigments as specialized evolutionary developments, but their seemingly universal distribution in these primitive algal divisions suggests that they may represent an ancestral type of photoreceptor which was later lost from the photosynthetic apparatus of other algae and higher plants.

Phytochrome, another plant photoreceptor, functions in photomorphogenesis and is seemingly universally distributed in the plant kingdom (see Chapters 7 and 18). The demonstration that this pigment is also a biliprotein, with many points of close similarity to the algal biliproteins, lends some support to the view that the latter are more than isolated evolutionary curiosities.

Scientific accounts of the algal biliproteins date back to 1836, when von Esenbeck described the extraction of the water-soluble blue pigment from a blue-green alga. The brilliant and attractive coloration and fluorescence of biliprotein solutions have been repeatedly remarked upon, and this coloration was clearly a major factor in the attraction of the pigments for many early scientific workers. Apart from their aesthetic attraction, their intense coloration allows easy visual monitoring even at low concentrations. This, in conjunction with the fact that these pigments are well defined and relatively stable globular proteins, led to their use by workers such as Svedberg and Tiselius in the development of many now classical physical techniques for protein separation and characterization.

Like most globular proteins, the biliproteins may be readily extracted into water or dilute salt solution from ruptured algal cells. Individual biliproteins may be separated from one another and from other proteins by a combination of ammonium sulphate fractionation, adsorption or ion exchange chromatography and gel filtration, followed in most cases by crystallization from ammonium sulphate solution; however, some biliproteins, notably those from cryptomonad algae, have as yet resisted all efforts at crystallization. For a selection of purification procedures see Haxo *et al.* (1955), Ó hEocha *et al.* (1964), Ó Carra (1965), Siegelman *et al.* (1968), Bennett and Bogorad (1971).

Algal biliproteins have isoelectric points in the range 4·3 to 4·9, and crystallization is most readily achieved within this pH range. It was originally thought that their molecular weights all fell within the range 250 000 to 300 000 daltons at pH values close to the isoelectric points, with some biliproteins tending to dissociate into half molecules at neutral or alkaline pH values (see Eriksson-Quensel, 1938; Ó hEocha, 1965b). However, much more complex aggregation-disaggregation phenomena have since been revealed for certain biliproteins, and molecular weights in the range 30 000–45 000 daltons have been reported for some. Cryptomonad biliproteins in particular seem to exist, at least in solution, as such low molecular weight forms (see Section IVB).

The algal biliproteins are differentiated into two main groups on the basis of their coloration. *Phycoerythrins* are a clear red by transmitted light and emit a brilliant orange-yellow fluorescence. The *phycocyanins* are blue with a strong red fluorescence.

Most species of red, blue-green and cryptomonad algae contain both a phycoerythrin and one or more phycocyanins, although a single biliprotein usually predominates. Generally, phycoerythrin dominates in red algae and phycocyanin in blue-green algae, resulting in the characterisitic coloration from which these algal divisions derive their names. However, this is not a universal rule; in several blue-green algal species phycoerythrin is dominant, resulting paradoxically in red-coloured "blue-green algae", and at least one member of the red algae—the unicellular *Porphyridium aerugineum*—is coloured blue owing to its containing phycocyanin as its only biliprotein (Haxo, 1960). The ratio of phycoerythrin to phycocyanin in most species is variable and dependent on the quality of the incident light and other growth conditions (see Section VI). The overall biliprotein content of algae is also dependent on such variables. Thus considerable seasonal variation is observed in the biliprotein content of free-growing macroscopic red algae; values in the region of 2% on a dry weight basis seem to be maximal (see Rabinowitch, 1945). The biliprotein content of cultured blue-green algae is commonly in the region of 15% on a dry weight basis but values as high as 24% have been reported (Myers and Kratz, 1955; Hattori and Fujita, 1959a,b; Allen, 1968).

Several types of phycoerythrins and phycocyanins are distinguishable on the basis of characteristic differences in absorption spectra. Many of these types are denoted by prefixed letters as indicated in Table I, which summarizes the characteristic visible absorbance maxima of individual biliproteins. Visible and ultraviolet spectra of representative examples of the various biliprotein types are illustrated in Figs 1 to 5. Relatively minor spectral variations may be

TABLE I

Spectral characteristics of algal biliproteins in the visible region. The most intense absorption maximum is in bold-face type; minor maxima or absorption shoulders are in parentheses[a]

Biliprotein	Absorption maxima (nm) in visible region	$A^{1\%}_{1\,cm}$ at max[b]	Algal source (B: blue-green alga R: red alga)
C-phycoerythrin	**565**	126	*Tolypothrix tenuis* (B)
	562	125	*Phormidium persicium* (B)
R-phycoerythrin	498, 540, **568**	80·2	*Ceramium rubrum* (R)
	498, (540), **564**	81·5	*Porphyra umbilicalis* (R)
B-phycoerythrin	(498), **546**, (565)	82·3	*Rhodochorton floridulum* (R)
C-phycocyanin	**615**	65	*Nostoc muscorum* (B)
	620		*Fremyella diplosiphon* (B)
R-phycocyanin	553, **615**	66	*Ceramium rubrum* (R)
	553, **618**	64·5	*Porphyra umbilicalis* (R)[c]
Allophycocyanin	(620), **650**	63·5	*Nostoc muscorum* (B)
Cryptomonad biliproteins			*Ceramium rubrum* (R)
Phycoerythrin type I	**544**, (565)		*Plagioselmis prolonga*
type II	**556**		*Hemiselmis rufescens*
type III	**565**		*Cryptomonas ovata*
Phycocyanin type I	585, **615**		*Hemiselmis virescens*
type II	585, (620) **650**		*Chroomonas* sp.
type III	585, **625–630**		*Cryptomonas cyanomagna*

[a] Taken from: Allen *et al.* (1959); Hattori and Fujita (1959a,b); Fujita and Hattori (1960a,b); Ó Carra (1962); Ó hEocha *et al.* (1964); Ó Carra (1965); Bennett and Bogorad (1971).

[b] $A^{1\%}_{1cm}$ is the absorbance of a 1% solution of the biliprotein in a 1 cm lightpath optical cell at the major absorption maximum (that in bold-face type).

[c] *Porphyra umbilicalis* was formerly known as *P. laciniata.*

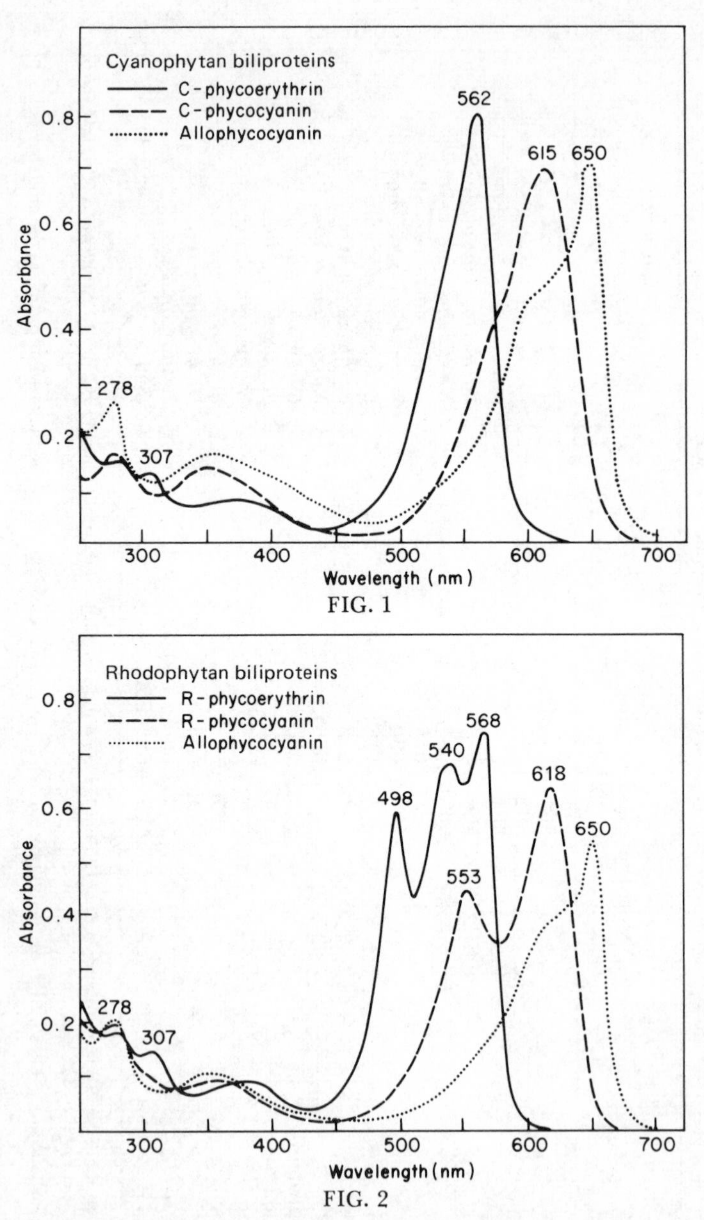

FIG. 1

FIG. 2

FIGS. 1–5. Absorption spectra of biliproteins in dilute phosphate buffer, pH 6·5–7·0. (Taken from Ó Carra, 1962.)

FIG. 1. Biliproteins from blue-green algae: ——, C-phycoerythrin from *Phormidium persicinum*; – – –, C-phycocyanin from *Nostoc muscorum*; · · · · · ·, allophycocyanin from *Nostoc muscorum*.

FIG. 2. Biliproteins from the red alga *Ceramium rubrum*: ——, R-phycoerythrin; – – –, R-phycocyanin; · · · · · ·, allophycocyanin.

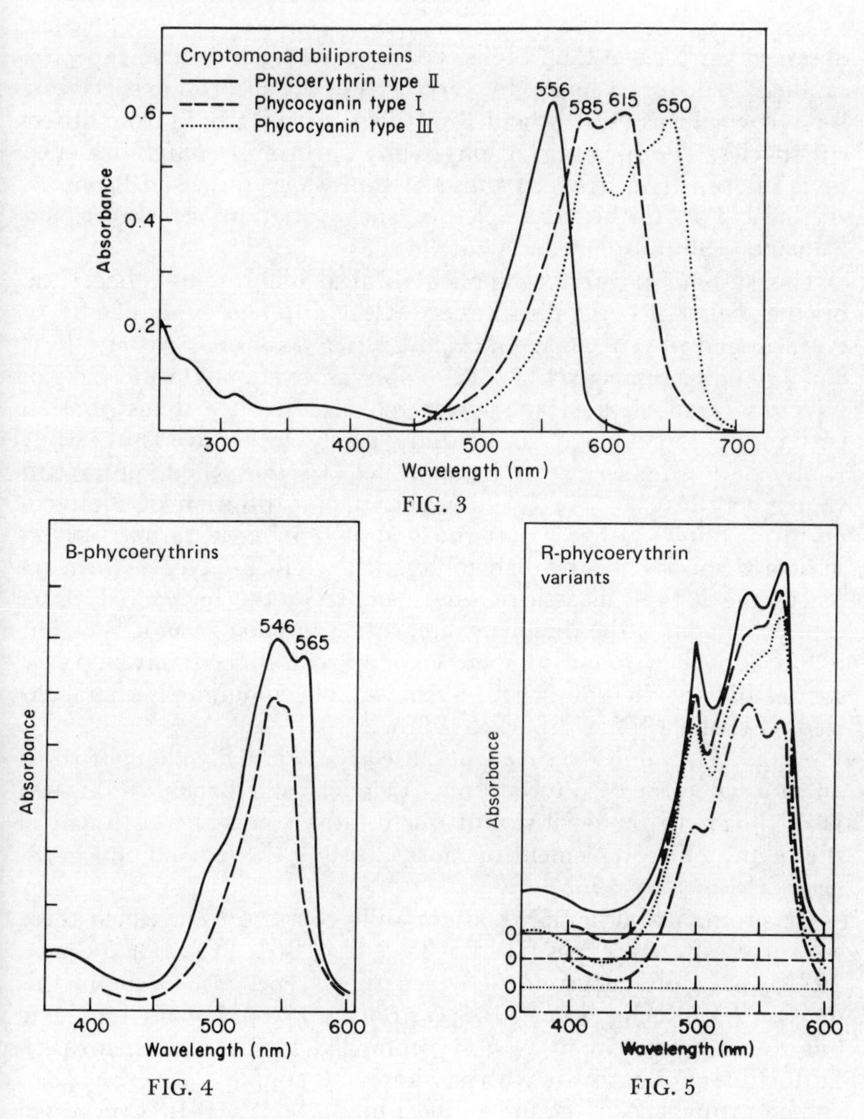

FIG. 3

FIG. 4

FIG. 5

FIG. 3. Some cryptomonad biliproteins: ——, phycoerythrin from *Hemiselmis fugescens*; – – –, phycocyanin from *Hemiselmis virescens*; · · · · · ·, phycocyanin from *Croomonas* sp.

FIG. 4. Spectral variants of B-phycoerythrin: ——, B-phycoerythrin from *Rhodochorton floridulum*; – – –, B-phycoerythrin from *Porphyridium cruentum.*

FIG. 5. Spectral variants of R-phycoerythrin: ——, R-phycoerythrin from *Ceramium rubrum*; – – –, R-phycoerythrin from *Rhodymenia palmata*; · · · · ·, R-phycoerythrin from *Porphyra umbilicalis* (formerly known as *P. laciniata*); —·—·—·, phycoerythrin from *Helminthora diverricata*.

observed for a given biliprotein type isolated from different species, as illustrated, for example, by variations in the absorption spectra of B-phycoerythrins (Fig. 4) and R-phycoerythrins (Fig. 5) from different species. The spectrum of phycoerythrin from *Helminthora* (Fig. 5) is intermediate between those of B-phycoerythrin and R-phycoerythrin. For further examples of such variation see Hirose and Kumano (1966), Boney and White (1968).

The system of prefixes applied to algal biliproteins reflects the original belief that the distribution of the different types of phycoerythrin and phycocyanin followed a strict taxonomic pattern, R, C and B standing respectively for Rhodophyta, Cyanophyta and Bangiophyceae (a subclass of the lower red algae). Such a distribution pattern is now known to be only partly followed. Thus, while R-phycoerythrin seems to be confined to the Rhodophyta and C-phycoerythrin to the Cyanophyta, the distribution of B-phycoerythrin follows no clear pattern; R-, not B-phycoerythrin, is found in many species of the Bangiophyceae, while B-phycoerythrin replaces the R-type in at least one member of the higher red algae. C-phycocyanin is the dominant biliprotein in most Cyanophyta, but it has also been found in some Rhodophyta. Indeed only in a few species of that division has R-phycocyanin been definitely characterized (Ó Carra, 1962; Ó hEocha, 1966).

Allophycocyanin, so-called because it was initially thought to be an unusual form of phycocyanin (Lemberg and Bader, 1933), was later shown to be a native but quantitatively minor constituent of the biliprotein complement of most, possibly all, red and blue-green algae (Haxo *et al.*, 1955).

Cryptomonad algae may contain a phycoerythrin, of which there are three different types (Ó hEocha *et al.*, 1964; Brooks and Gantt, 1973), or a phycocyanin, of which three types also occur in this division (Ó hEocha *et al.*, 1964; Ó hEocha, 1966). Usually, only one biliprotein is present in each cryptomonad species and their distribution does not correlate with any known taxonomic pattern.

The proteinaceous nature of algal biliproteins was first suggested by von Esenbeck (1836) and was definitely established towards the end of the century by Molish (1894, 1895). However, the structural relationship between the coloured prosthetic groups and the animal bilins (bile pigments) remained to be established by Lemberg (1930, 1933), who introduced the generic term *phycobilin* for these chromophores.

The phycobilins are attached to the apoproteins by strong covalent linkages. In this they differ from other photosynthetic pigments, chlorophylls and carotenoids, which are readily released from algal cells by extraction into organic solvents. The cleavage of the phycobilin-apoprotein linkages requires relatively severe hydrolytic conditions, and the lability of the bilins during such treatment has severely hampered and confused structural studies. From spectral examination of the biliproteins themselves and of coloured products derived by partial acid hydrolysis, Lemberg deduced that phycoerythrins and phycocyanins contain chemically distinct phycobilins for which he coined the terms *phycoerythrobilin* and *phycocyanobilin,* respectively. Later, the picture was seen not to be as simple as this nomenclature would suggest.

Thus Lemberg himself (1933) suggested that R-phycocyanin contains both phycoerythrobilin and phycocyanobilin, and this has been confirmed by later workers (Ó hEocha, 1960; Ó Carra, 1962; Chapman *et al.*, 1967b). Evidence for the existence of other phycobilins has also accumulated and R-phycoerythrin has been reported to contain, in addition to phycoerythrobilin, a pigment spectrally similar to urobilins of animal origin (Ó hEocha and Ó Carra, 1961). Proportionately smaller quantities of this urobilinoid chromophore, for which the name *phycourobilin* was suggested, have been shown to occur in B-phycoerythrin (Ó Carra, 1962; Ó Carra *et al.*, 1964). Spectral studies of the cryptomonad phycocyanins suggest that these contain another chromophore additional to phycocyanobilin (Ó Carra, 1962; Ó hEocha *et al.*, 1964). Because of its resemblance to the class of bilins known as violins the name *phycoviolin* is proposed for this pigment. The distribution of these four phycobilins among the algal biliproteins is summarized in Table II.

Purified algal biliproteins do not contain significant levels of metal ions and the spectral evidence indicates that the phycobilin chromophores are not complexed with metals in the native chromoproteins. Estimates of the phycobilin content of various biliproteins on a

TABLE II
Distribution of phycobilins among algal biliproteins

Biliprotein	Phycobilin(s)
C-phycoerythrin	Phycoerythrobilin
C-phycocyanin	Phycocyanobilin
Allophycocyanin	Phycocyanobilin
R-phycocyanin	Phycocyanobilin and phycoerythrobilin
R-phycoerythrin	Phycoerythrobilin and phycourobilin
Cryptomonad phycoerythrins	Phycoerythrobilin
Cryptomonad phycocyanins	Phycocyanobilin and phycoviolin

percentage weight basis have ranged from about 4% for C-phycocyanin (Clendenning, 1954; Ó Carra, 1962; Troxler and Lester, 1968; Schram and Kroes, 1971) to about 7% for R-phycoerythrin (Ó Carra, 1962).

The amino acid composition of algal biliproteins is not specially noteworthy. Most recent analyses account for 85–90% of the mass of the chromoproteins as amino acid residues. When the estimated contribution of the phycobilin residues is added there remains an apparent discrepancy of 5 to 10% of total biliprotein mass unaccounted for. However, amino acid analyses on an absolute quantitative basis rarely account for the full mass even of simple proteins. A suggestion that carbohydrate may account for the discrepancy, particularly in R-phycoerythrin (Fujiwara, 1961), has not been investigated in detail.

II Structures of the phycobilins and covalent linkages to the apoproteins

A PHYCOERYTHROBILIN AND PHYCOCYANOBILIN

The first clue to the chemical nature of the phycobilins was provided by Lemberg (1928), who found that treatment of the biliproteins with hot alkali released their chromophores in the form of a green, chloroform-soluble, pigment which was identified as mesobiliverdin (Scheme 1). The particular arrangement of side chains shown in Scheme 1 is that of the IXα isomer. This has long been assumed to be the isomer of mesobiliverdin formed from the phycobilins, although this was not established unequivocally until recently (Ó Carra and Colleran, 1970).

The alkali-released mesobiliverdin was clearly an artefact. Lemberg (1930, 1933) found that hot concentrated hydrochloric acid released pigments whose colours approximated more closely to those of the parent chromoproteins. Ó hEocha (1958) later showed that these preparations were also artefacts but that milder acid treatment (brief treatment with concentrated hydrochloric acid in the cold) released low yields of chloroform-soluble pigments which, it was suggested, were the native protein-free phycobilins. This was supported by the demonstration that the spectral properties of these pigment preparations corresponded very closely with those of the parent biliproteins when these are unfolded by denaturation to eliminate the distorting effects of the native protein environments on the spectra of the phycobilins (Ó hEocha and Ó Carra, 1961; Ó Carra, 1962; Ó Carra *et al.*, 1964). Lemberg intended the terms

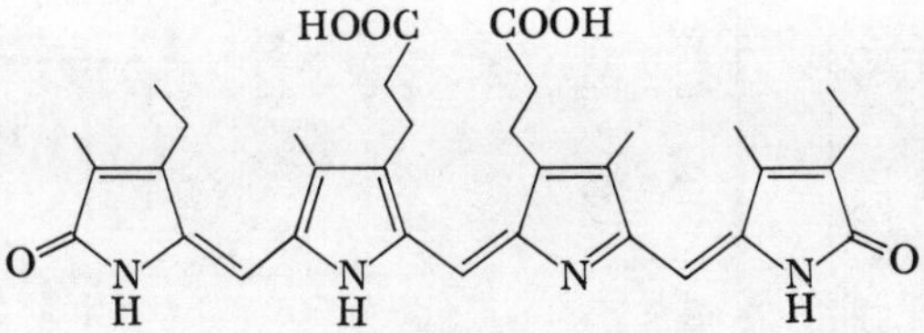

(II)

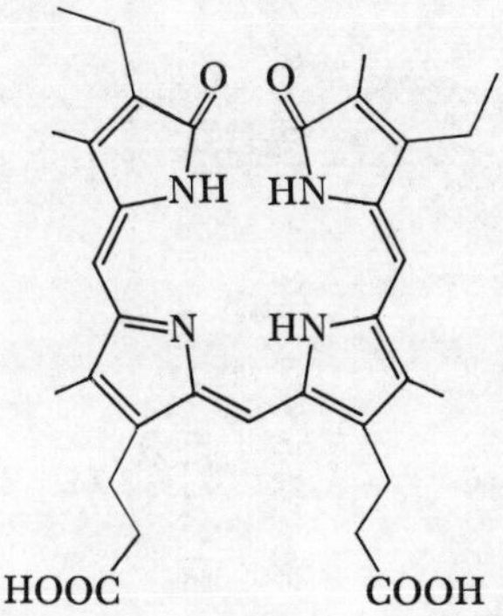

(III)

SCHEME 1. Mesobiliverdin IXα. Structure **II** is the conventional linear representation of the bilin structure but structure **III** more correctly represents the true shape of the molecule and emphasizes its relationship with the porphyrins (see also Scheme 5). Bilins are generally represented in the linear form for convenience and this convention is followed in this chapter.

phycoerythrobilin and phycocyanobilin to apply to the native chromophores of biliproteins and, consequently, these names were attributed by Ó hEocha to the pigments released by mild acid hydrolysis of phycoerythrins and phycocyanins, respectively. The spectral properties of these phycobilins are summarized in Table III.

The released phycobilins are ampholytes which are soluble in alkali and in mineral acids, as well as organic solvents. Their carboxyl groups may be esterified. A pyrrole nitrogen in the conjugated system of phycobilins is sufficiently basic to become protonated in acid solution and the resulting salts have sharper and more characteristic absorption spectra than the unprotonated pigments. Phycobilins form fluorescent complexes with certain divalent metal ions such as Zn^{2+}, and the spectra of such complexes are also sharp and distinctive.

The phycobilins released from biliproteins by acid treatment are

TABLE III

Comparative spectral properties of phycobilins, related bilins and denatured biliproteins. (Major maximum in bold-face type, minor maxima or absorption shoulders in parentheses)[a]

Bilin or denatured biliprotein	Chromophore as hydrochloride: In 0·1 M aqueous methanolic HCl	Chromophore as hydrochloride: In acidified $CHCl_3$	Chromophore as zinc complex in 8 M urea or $CHCl_3$[b]
	Absorption maxima (nm)		
"Purple pigment" derivative of phycoerythrobilin	328, **590**	329, **602**	326, 340, 560, **605**
Phycoerythrobilin (acid-released)	307, **556**	312, **576**	320, (540), **583**
Denatured C-phycoerythrin	278,[c] 307, **556**	insoluble	278,[c] 321, 542 **586**
Mesobilirhodin[d]	307, **555**	312, **576**	319 (543) **583**
"Blue pigment" derivative of phycocyanobilin	374, **684**	**680**	378 **665**
Phycocyanobilin (acid-released)	*c*.360, 655	357, **630**	375 **630**
Denatured C-phycocyanin	278,[c] *c*.350, **655–660**	insoluble	278,[c] 375 **630**

[a] Data from: Ó Carra *et al.* (1964); Ó Carra and Ó hEocha (1966); Killilea (1971); Ó Carra and Killilea (1971).

[b] Denatured biliproteins in 8M urea containing 1% zinc acetate; protein-free bilins in chloroform containing a drop of ethanolic zinc acetate and a trace of pyridine to neutralize protons.

[c] 278 nm maximum attributable to aromatic amino acid residues.

[d] Obtained by isomerization of *i*-urobilin (Ó Carra and Killilea, 1970; Rüdiger *et al.*, 1970).

remarkably labile and readily undergo a variety of isomerizations and oxidations (Ó hEocha, 1963; Ó Carra *et al.*, 1964). As a result, they can be obtained only in very poor yield and efforts to purify and crystallize them have generally resulted in the accumulation of artefacts.

Ó Carra and Ó hEocha (1966) described an alternative method for releasing the chromophores by incubating biliproteins in neutral methanol at about 60°C; highest yields were obtained from finely divided biliprotein preparations (see also Rüdiger and Ó Carra, 1969). It was also shown that the bilins extracted into hot methanol from red and blue-green algae were not free biosynthetic precursors of the phycobilins, as had been postulated by Fujita and Hattori (1962a, 1963b), but artefact derivatives of the phycobilins released from the biliproteins of the algal cells by the hot methanol. The "purple pigment" and "blue pigment" of Fujita and Hattori were thus shown to be altered forms of phycoerythrobilin and phycocyanobilin, respectively*.

These methanol-released "purple and blue pigments" are more stable and more readily purified than the acid-released phycobilins, and they have been obtained in the form of highly purified, crystalline preparations (Chapman *et al.*, 1967a; Cole *et al.*, 1967a; Siegelman *et al.*, 1968). Consequently, most recent chemical studies have tended to concentrate on the methanol-released pigments.

The structures illustrated in Scheme 2 have been established for these pigments by two independent approaches, one involving n.m.r. and mass spectrometric techniques (Chapman *et al.*, 1967a; Cole *et al.*, 1967a, 1968; Crespi *et al.*, 1967, 1968), the other using chemical, mainly microdegradative techniques (Rüdiger *et al.*, 1967; Rüdiger and Ó Carra, 1969). The structure elucidated for "blue pigment" (Scheme 2) has recently been proved by chemical synthesis (Gössauer and Hirsch, 1973). It will be seen that "blue pigment" and "purple pigment" are isomeric, and differ only in the position of one double bond, which constitutes the double bond of the vinyl side chain of ring D in "purple pigment" but occupies the bridge position connecting ring D with the rest of the conjugated system in "blue pigment". The unusual ethylidene side chain on ring A of both pigments is notable.

* "Blue pigment" seems to be identical with phycobilin-655, an artefact form of phycocyanobilin studied by Ó hEocha (1963), while phycobilin-630, described by the same author, is identifiable as acid-released phycocyanobilin (Murphy, 1968; Schram and Kroes, 1971). Rüdiger (1971) has applied the terms phycobiliviolin and phycobilicyanin respectively to "purple pigment" and "blue pigment".

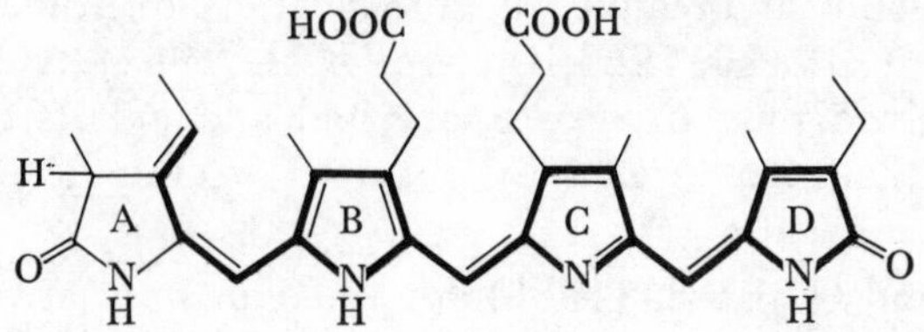

HOOC COOH

(IV)

"Blue pigment" derivative of phycocyanobilin

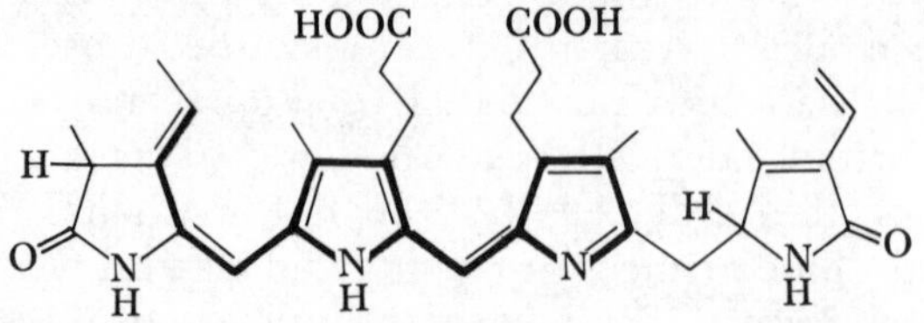

HOOC COOH

(V)

"Purple pigment" derivative of phycoerythrobilin

acid ⇌ hot MeOH

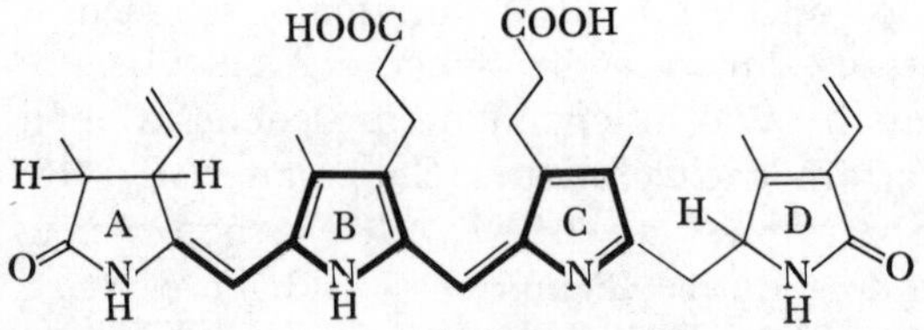

HOOC COOH

(VI)

Suggested structure for acid-released phycoerythrobilin

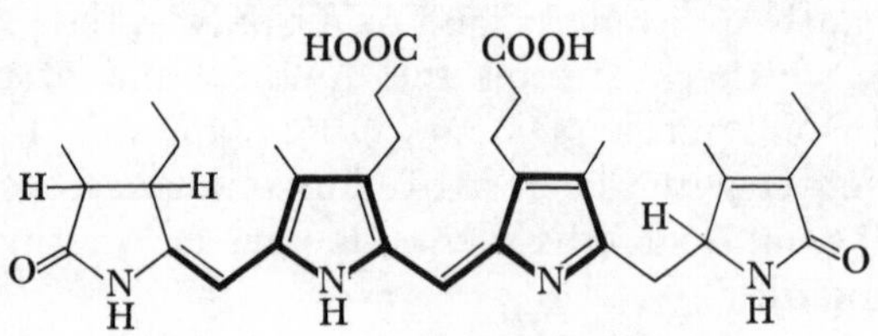

HOOC COOH

(VII)

Mesobilirhodin

SCHEME 2. Structures of protein-free phycobilins and their derivatives. (Chromophore conjugated systems are emphasized by heavy lines.)

While many workers refer to the methanol-released pigments as native phycoerythrobilin and phycocyanobilin (for example Cole *et al.*, 1967a, 1968; Chapman *et al.*, 1967a; Siegelman *et al.*, 1968; Gössauer and Hirsch, 1973), they are clearly very different from the native protein-bound phycobilins which, as mentioned above, resemble instead the acid-released phycobilins (Ó Carra and Ó hEocha 1966). The spectral properties summarized in Table III demonstrate this.

Owing largely to the lability of the acid-released phycobilins, the techniques used in elucidating the structure of "purple pigment" and "blue pigment" have not revealed the exact structural distinction between the two sets of pigments. However, their ready interconvertibility strongly suggests an isomeric relationship (Ó Carra and Ó hEocha, 1966; Rüdiger and Ó Carra, 1969). Acid-released phycoerythrobilin is also related to the semisynthetic mesobilirhodin which was recently shown to have structure **VII** (Scheme 2) by a combination of mass spectrometric and chemical studies (Ó Carra and Killilea, 1970; Rüdiger *et al.*, 1970), later confirmed by n.m.r. methods (Chapman *et al.* 1973). As can be seen from Table III, the very characteristic spectral properties of unreleased and acid-released phycoerythrobilin are almost identical with those of mesobilirhodin, indicating that these pigments contain the same conjugated system. This suggests that the isomerization of phycoerythrobilin to "purple pigment" in methanol involves the shifting of a double bond into the conjugated system to form the ethylidine grouping of "purple pigment". Structure **VI** (Scheme 2), which is isomeric with that elucidated for "purple pigment", satisfies these and other requirements and has been proposed for phycoerythrobilin (Killilea and Ó Carra, in prep.).

The same relationship is thought to exist between "blue pigment" and acid-released phycocyanobilin, the latter containing a vinyl and not an ethylidene side chain on its reduced A ring* (Killilea and Ó Carra, 1974).

* The imides produced by chromic oxidation of the acid-released phycobilins are identical with those produced from the purple and blue pigments (Rüdiger and Ó Carra, 1969). This would seem to contradict the proposed difference in the side chains on ring A of the pigments. However, the imide that might be expected from ring A of the phycoerythrobilin structure **VI** (methyl-vinyl-succinimide) has never been synthesized successfully nor isolated from any other source and it seems likely that this imide, with its reduced ring and vinyl side chain, isomerizes quickly to the ethylidene-containing imide which is isolated after oxidation. Thus Rüdiger (1971) has shown that the corresponding hydrated imide (having a hydroxyethyl side chain) is spontaneously converted to the ethylidene-bearing imide under the conditions of the chromic acid oxidation.

Rüdiger (1971) has proposed alternative structures for the acid-released phycobilins in which the vinyl group on ring A (Scheme 2) is hydrated. However, this must be considered a doubtful proposition as the acid-released phycobilins do not differ significantly from the mechanol-released pigments in solvent partition and adsorption properties which are drastically altered in other bilins by the introduction of a side-chain hydroxyl group (Killilea, 1971).

A suggestion by Schram (1970) and Schram and Kroes (1971) that acid-released phycocyanobilin contains two hydrogen atoms more than phycobilin-655 ("blue pigment") is based on mass spectrometric evidence which is almost certainly complicated by redox disproportionation phenomena, well known in the field of bilin chemistry (see Jackson *et al.*, 1966; Lightner *et al.*, 1969). A similar phenomenon was found to complicate the mass spectrometric studies of Crespi *et al.* (1967) on "blue pigment" and similarly led initially to the proposal of a structure containing two hydrogen atoms too many.

B COVALENT LINKAGES TO THE APOPROTEINS

The microdegradative method used in the elucidation of phycobilin structures was applied by Rüdiger and Ó Carra (1969) to an investigation of the covalent chromophophore–protein linkages. This work indicated that the phycobilins are each linked to the apoproteins through ring A and ring C.

Approaches to the identification of the amino acid residues linked to the chromophores have mainly involved enzymic hydrolysis of the biliproteins with mixtures of endopeptidases and exopeptidases with the aim of completely hydrolysing all peptide bonds in the polypeptide chain, leaving only the directly-linked amino acid residues attached to the chromophores. So far, however, this ideal has not been achieved in practice. The smallest chromopeptides that have been obtained contain significant levels of at least four or five amino acids, including residues such as alanine and glycine whose side chains could hardly be involved in the phycobilin linkages. The resistance of such chromopeptides to further "trimming" by peptidases is probably due to steric interference by the bulky attached phycobilin groups, as has been found in the case of chromopeptides from certain cytochromes (see for example, Dus *et al.*, 1962). Further complexity has undoubtedly been introduced in some studies by the tendency of the phycobilins to form additional artefact linkages with the protein components under certain conditions (see Ó Carra *et al.*, 1964; Rüdiger and Ó Carra, 1969) and by the ability of certain protease preparations to catalyse alteration and splitting of the phycobilin-protein linkages (Siegelman *et al.*, 1967; Murphy, 1968; Killilea, 1971).

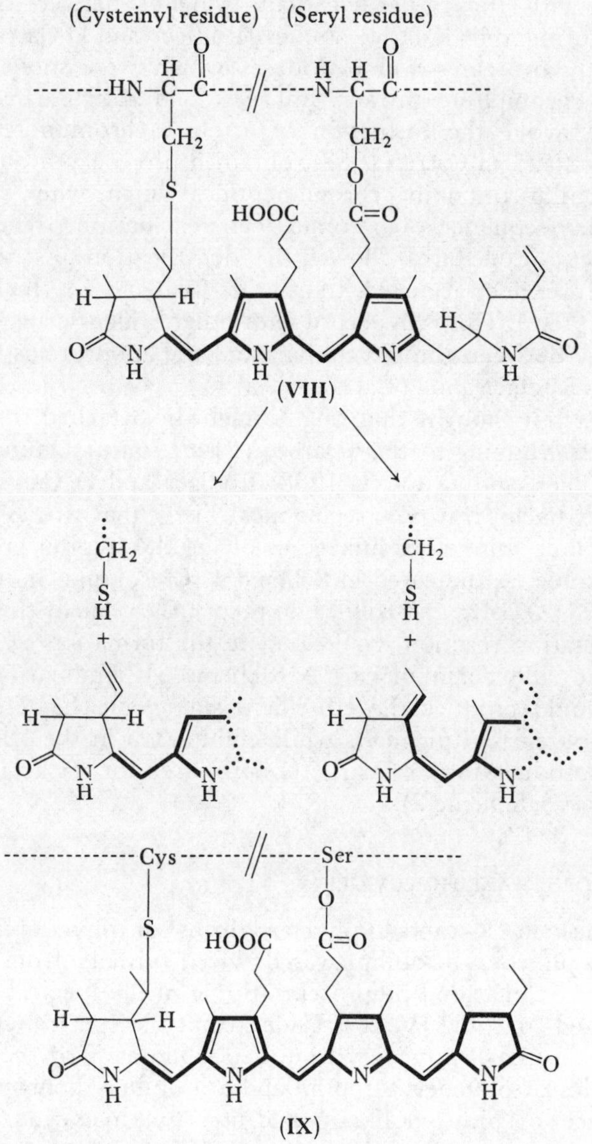

(Cysteinyl residue) (Seryl residue)

(VIII)

(IX)

SCHEME 3. Proposed structure of protein-bound phycoerythrobilin (**VIII**) and phycocyanobilin (**IX**). (Chromophore conjugated systems are emphasized by heavy lines.)

Such complications may be partly responsible for results which led Crespi and Smith (1970), Byfield and Zuber (1972) and Brooks and Chapman (1972) to propose modes of linkage which are not compatible with the rather clear-cut evidence derived from the chromic acid microdegradative studies (Rüdiger and Ó Carra, 1969).

Using only proteases and peptidases which were shown not to affect the chromphore–protein linkages, and taking special precautions to avoid the formation of artefact chromphore–protein links, Killilea and Ó Carra (1968), obtained phycoerythrobilin- and phycocyanobilin-containing chromopeptides which, when subjected to a complex sequence of chromic acid degradation, fractionation and hydrolytic procedures, allowed the identification of serine as the amino acid residue attached to ring C (Scheme 3) (Killilea and Ó Carra, in prep.). This is in accord with other evidence suggesting an ester linkage between the phycobilins and the apoproteins (Ó Carra *et al.*, 1964; Rüdiger and Ó Carra 1969).

It was at first thought that ring A might be attached through its ring nitrogen grouping to the γ-carboxyl function of a glutamic acid residue (Killilea and Ó Carra, 1968; Rüdiger and Ó Carra, 1969). However, on balance, it now seems most likely that ring A is linked through a labile thio–ether linkage involving the sulphur group of a cysteine residue, as indicated in Scheme 4 (Crespi and Smith, 1970; Killilea, 1971; Ó Carra and Killilea, in prep.). Cleavage of this linkage by an elimination reaction would lead to the formation of a double bond in the side chain of ring A (Scheme 3); elimination in one direction would produce the ethylidene side chain characteristic of the methanol-released pigments while elimination in the other direction would produce the vinyl structure proposed for the acid-released phycobilins (see Scheme 2).

C PHYCOUROBILIN AND PHYCOVIOLIN

The presence in R- and B-phycoerythrins of phycourobilin, in addition to phycoerythrobilin, was deduced initially from spectral analysis of the denatured biliproteins. Some of the spectral evidence is summarized in Table IV. It is evident that the spectral characteristics of denatured R-phycoerythrin are a composite of the spectral characteristics of phycoerythrobilin and a urobilinoid chromophore. The presence of phycoviolin, in addition to phycocyanobilin, in cryptomonad phycocyanins can be demonstrated on the basis of similar evidence.

Chapman *et al.* (1968) have disputed the existence of these

TABLE IV

Comparison of the spectral properties of denatured R-phycoerythrin with those of phycoerythrobilin and *i*-urobilin[a]

Pigment	Chromophores as hydrochlorides[b]	Chromophores as zinc complexes[c]
	Absorption maxima (nm)[d]	
Denatured R-phycoerythrin	278,[e] 307, 498, 556	278,[e] 321, 510, (545), 586
Phycoerythrobilin (acid-released)	307, 556	320, (548) 583
i-Urobilin	494	510

[a] Taken from: Ó Carra (1962); Ó Carra *et al.* (1964).
[b] Solvent: 0·1 M aqueous HCl (R-phycoerythrin) or 0·1 M methanolic HCl (bilins).
[c] Solvent: 8 M urea containing 1% zinc acetate (R-phycoerythrin) or 1% ethanolic zinc acetate (bilins).
[d] Diffuse absorbancy in the 370–380 nm region is omitted.
[e] 278 nm absorbance maximum attributable chiefly to aromatic amino acid residue.

additional phycobilins, mainly on the grounds that hot methanol treatment of R-phycoerythrin and of a cryptomonad phycocyanin released only the pigments derived from phycoerythrobilin and phycocyanobilin respectively. However, it had already been shown that phycourobilin is much more firmly attached to the apoprotein than is phycoerythrobilin (Ó Carra *et al.*, 1964). The situation is further complicated by the fact that an artefact urobilinoid pigment is readily formed from phycoerythrobilin, particularly during acid treatment. This artefact urobilin material has been extensively studied by Vaughan (1963, 1964) and by Cole *et al.* (1967b).

Native phycourobilin can be separated from phycoerythrobilin by first digesting R-phycoerythrin with trypsin and then fractionating the typtic peptides under carefully controlled conditions designed to avoid the formation of phycobilin artefacts. By these means, a seemingly homogeneous phycourobilin-containing peptide was cleanly separated from phycoerythrobilin-containing peptide material (Killilea, 1971; Killilea and Ó Carra, in prep.). Degradative and other chemical studies on this chromopeptide suggest that the structure and apoprotein linkages of phycourobilin are as indicated in Scheme 4 (Killilea and Ó Carra, in prep.). An extra linkage through ring D, as revealed by chromic acid degradative studies, distinguishes the attachment of phycourobilin from that of phycoerythrobilin. The positioning of the double bond in ring A of phycourobilin (marked with an asterisk) is based on the observation that the linked

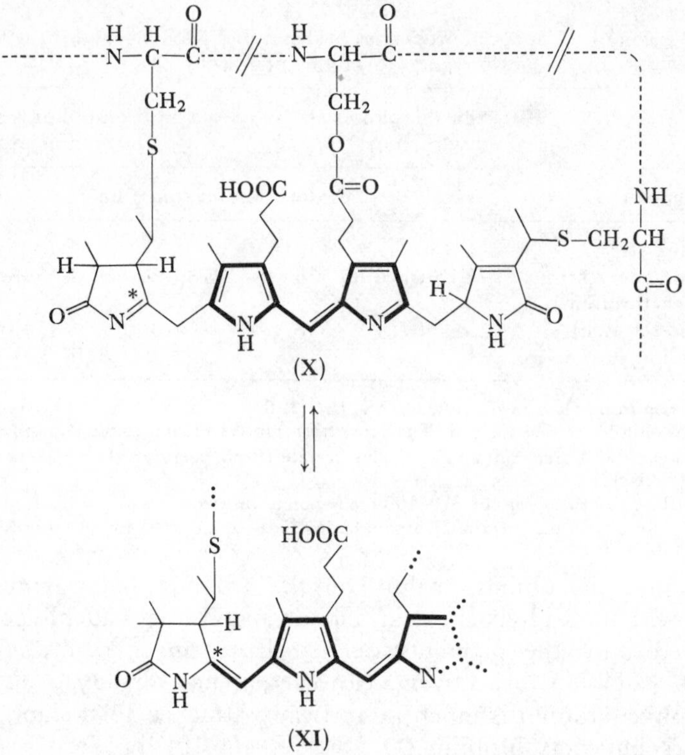

SCHEME 4. Proposed structure of protein-bound phycourobilin (X) and its erythrobilinoid isomeride (XI).

phycourobilin readily and reversibly isomerizes to a chromophore with phycoerythrobilin/mesobilirhodin spectral characteristics, the mechanism of the isomerization being envisaged as shown in Scheme 4. The thio-ether linkage to ring D is much more stable than the labile linkage to ring A. This accounts for the much greater resistance of the phycourobilin-apoprotein linkage to the hydrolytic methods used to release phycoerythrobilin.

III Non-covalent phycobilin-protein interactions and spectral properties of native biliproteins

The visible spectral properties of the biliproteins in the denatured state correspond closely with the spectra of the constituent phycobilins (Table IV). However, in the native biliproteins these spectral properties are considerably altered by the environments in which the chromophores are positioned by the specific folding of the polypeptide chains. Diversity of these native protein environments makes it possible for a given phycobilin to possess different spectral properties depending on the environment in which it is embedded. This accounts for the spectral diversity of the biliproteins over and above that due simply to differences in phycobilin complement (see, for example, Ó hEocha and Ó Carra, 1961; Ó Carra, 1962; Ó hEocha, 1963; Dale and Teale, 1966, 1970; Macdowall *et al.*, 1968; Teale and Dale, 1970).

The attribution of the visible absorbance peaks of native biliproteins to constituent phycobilins is summarized in Table V. The indicated positions of the spectral maxima in this table and in the discussion below are average or typical values because, although the spectral properties of a biliprotein from a given species are well defined and constant, species variation in the exact spectral properties of a given biliprotein type are common, presumably because of variation in the exact structure of conformation of the native protein environment resulting from differences in polypeptide sequence.

Besides modulating the position and shape of the absorption spectra of the phycobilins, the native protein environments in most cases also increase the intensity of light absorption very considerably, thus contributing to the intensity of biliprotein coloration. Certain of the modes of interaction with the native apoproteins also cause the modulated phycobilins to fluoresce brilliantly, producing the distinctive fluorescence properties of the native biliproteins (see Table V). These fluorescence properties are lost on denaturation of the proteins and even comparatively minor conformational perturbations may result in drastic quenching or complete loss of fluorescence.

At least two types of modulated phycoerythrobilin may be distinguished in native phycoerythrins. One is characterized by an absorbance peak at about 565 nm (the exact location of the maximum varies from 562 to 568 nm in phycoerythrins from different sources), the other by a peak at 540 or 545 nm (see Table V and Figs 1–5). The 565 nm type is present in B-, C- and R-phycoerythrins. In C-phycoerythrin it is the sole chromophore contributing to the visible spectrum, but it represents only a quantitatively minor proportion of the chromophore complement of B-phycoerythrin, where the 545 nm type predominates. In R-phycoerythrins the 565 nm type contributes the most intense absorbance maximum and the proportion of the 540 nm type varies according to the species of origin, being much less in R-phycoerythrin from

TABLE V

Correlation of spectral properties and phycobilin complements of native biliproteins. (Bold-face type indicates the most intense maximum; minor peaks or shoulders are in parentheses)[a]

Biliprotein type	Constituent phycobilins; absorbance maxima (nm)				Fluorescence max. (nm)
	Phycourobilin	Phycoerythrobilin	Phycoviolin	Phycocyanobilin	
Phycoerythrins:					
C–		**565**			575
B–	(498)	**546**, (565)			578
R–	498	540, **565**			578
Phycocyanins:					
R–		553		**615**	565,[b] 635
C–				**615**	647
allo				(620), **650**	660
Cryptomonad, type I			585	**615**	637
Cryptomonad, type II			585	(620), **650**	660
Cryptomonad, type III			585	**630**	not available

[a] The indicated maxima are typical or average values and are based on data derived mainly from: French *et al.* (1956); Hattori and Fujita (1959a); Ó hEocha and Ó Carra (1961); Ó Carra (1962); Ó hEocha *et al.* (1964); Macdowell *et al.* (1968); Teale and Dale (1970).

[b] This 565 nm fluorescence peak was reported by Ó Carra (1962) but not by Gantt and Lipschultz (1973).

Porphyra umbilicalis, for example, than in that from *Ceramium rubrum* (see Figs 2 and 5).

The fluorescence spectra of C-, B- and R-phycoerythrins are almost identical, consisting of a single sharp emission band at 575–578 nm, evidently emitted solely by the 565 nm phycoerythrobilin type. However, the fluorescence activation spectrum of each biliprotein closely follows the entire visible absorption spectrum, showing that in R- and B-phycoerythrins light quanta absorbed by phycourobilin and by the 540 nm type phycoerythrobilin transfer very efficiently to the 565 nm type phycoerythrobilin residues (Ó hEocha and Ó Carra, 1961; Ó Carra, 1962; Ó hEocha *et al.*, 1964; Eriksson and Halldal, 1965; Teale and Dale, 1966, 1970; Macdowall *et al.*, 1968). Teale and Dale (1970) designated the 540 and 565 nm type phycoerythrobilins in the native biliprotein as sensitizing ("s") and fluorescing ("f"), respectively.

The native protein environments also mask the phycobilins from the external medium, as evidenced by sluggishness or complete inhibition of the characteristic complexing reactions of the tetrapyrrole systems with divalent ions such as Zn^{2+}. This masking effect is abolished when the proteins are denatured and unfolded (Ó hEocha and Ó Carra, 1961).

The "primary" phycobilins, phycoerythrobilin and phycocyanobilin, are much more heavily masked and spectrally modulated by the native protein environments than are the "auxiliary" chromophores, phycourobilin and phycoviolin. Thus zinc ions, when added to native R-phycoerythrin, complex sluggishly with the urobilin residues (with a corresponding shift of the absorbance peak at 498 nm to 510–512 nm) but do not react at all with the erythrobilin residues. Comparatively mild perturbing media (e.g. 2 M urea) eliminates the sluggishness of the urobilin reaction, but much more strongly denaturing conditions (e.g. 8 M urea) are required to allow the complexing of Zn^{2+} with the erythrobilin residues (Ó hEocha and Ó Carra, 1961; Ó Carra, 1962; G. Downey and P. Ó Carra, unpublished results). The absorbance peak of the phycourobilin residues in native R-phycoerythrin is characteristic of the protonated form of the pigment and it remains in this form even at pH values well above the p*K* range of urobilins (p*K* about 7·5). This might possibly be due to ion pairing with an acidic amino acid side chain maintained in close proximity to the urobilin group by the polypeptide conformation. Such a situation would stabilize the protonated form of the chromophore and could also account for the sluggishness of the reaction with zinc ions.

The phycoerythrobilin present in R-phycocyanin constitutes a further type of protein-modulated phycoerythrobilin. The apparent absorbance peak lies at 533 nm but when the phycocyanobilin absorbancy (λ_{max} 615 nm) is subtracted, the true maximum of the phycoerythrobilin peak is found to lie at 549 nm (Ó Carra and Ó hEocha, 1965).

The phycocyanobilin in the various native phycocyanins may also be classified into a number of protein-modulated types. One type, characterized by an absorbance peak at about 615 nm, determines the absorption spectrum of C-phycocyanin in the visible region, and it is found also in R-phycocyanin and in one of the types of cryptomonad phycocyanin (Table V). Another type of modulated phycocyanobilin, with an absorbance peak at about 650 nm, is characteristic of allophycocyanin (Table V). It is not clear whether the shoulder at about 620 nm in the spectrum of this biliprotein is an integral feature of the spectrum of the 650 nm species or derives from a proportion of phycocyanobilin residues of the 615 nm type. The modulation of the phycocyanobilin residues in the second type of cryptomonad phycocyanin (the "650 nm type") seems to be similar to that in allophycocyanin (Table V). A third type of cryptomonad phycocyanin is characterized by a 630 nm absorption maximum (Table V). This presumably represents a further or intermediate type of modulation of phycocyanobilin residues.

The precise nature of the interactions with the protein environments that cause these spectral modulations of bilin residues is not well understood. The modulations are characterized by considerable bathochromic spectral shifts and in most cases by the induction of fluorescence. These effects are reminiscent of those that arise from complexing of the ring nitrogens of free phycobilins with ions such as Zn^{2+} or Cd^{2+}. Such ions are not present in biliproteins but it has been suggested that some type of complex hydrogen bonding of the ring nitrogens of phycoerythrobilin and phycocyanobilin might result in similar spectral effects and may constitute part of the interactions with the native protein environments (Ó Carra and Ó hEocha, 1961). It has also been suggested that the environments of these phycobilins are of a largely hydrophobic character (Ó Carra, 1962; Ó hEocha, 1963).

Besides the transfer of absorbed quanta between the spectrally distinct phycobilin types in the more complex biliproteins, considerable degree of energy transfer among like chromophores has been demonstrated by fluorescence polarization experiments (see, for example, Goedheer and Birnie, 1965; Teale and Dale, 1966; Vernotte, 1971).

IV Protein structure and ultrastructure

A CHEMICAL STUDIES

Some typical examples of the numerous amino acid analyses of biliproteins now available are given in Table VI. Acidic predominate over basic amino acid residues and levels of residues with hydrophobic side chains are comparatively high. The histidine content is

TABLE VI

Amino acid composition of biliproteins from *Porphyra umbilicalis*.[a]
(Calculated on the basis of moles amino acid residue per equivalent weight of biliprotein primary structure[b])

	moles amino acid residue		
Biliprotein Estimated equivalent weight[b]	R-phycoerythrin 39 000 daltons ($\alpha + \beta$ polypeptides)	R-phycocyanin 39 000 daltons ($\alpha + \beta$ polypeptides)	Allophycocyanin 15 500
Lysine	13	9	6
Histidine	2	1	0
Arginine	21	19	8
Aspartic acid	43	35	14
Threonine	13	16	7
Serine	30	32	9
Glutamic acid	24	35	14
Proline	10	10	5
Glycine	25	27	14
Alanine	59	55	16
Half-cystine[c]	6	3	2
Valine	30	27	11
Methionine	7	8	3
Isoleucine	14	13	10
Leucine	22	33	14
Tyrosine	14	15	8
Phenylalanine	7	7	3
Tryptophan[d]	0[d]	0[d]	0[d]
Total	340	345	144

[a] Based on unpublished data of D. Nolan and P. Ó Carra determined with a Joel amino acid analyser. *Porphyra umbilicalis* was formerly known as *Porphyra laciniata*.

[b] The equivalent weight of biliprotein primary structure was taken as the estimated polypeptide size in the case of allophycocyanin and as the sum of the two different polypeptide subunits in R-phycoerythrin and R-phycocyanin (see Table VII). The estimated weight of the phycobilin prosthetic groups was subtracted before calculating the molar proportions of the various amino acid residues.

[c] Estimated as cysteic acid after performate oxidation of samples of the biliproteins.

[d] Absence of tryptophan indicated by negative reaction of tryptic peptide maps to Ehrlich's stain.

generally low and this amino acid is absent from allophycocyanin (Glazer and Cohen-Bazire, 1971; Ó Carra *et al.*, in prep.).

The presence of the phycobilin chromophores complicates the determination of tryptophan. It has been reported as absent from some biliproteins (e.g. Raftery and Ó hEocha, 1965), but has recently been reported present in C-phycocyanin from certain species (Glazer and Fang, 1973a,b; Binder *et al.*, 1972). Cysteine/cystine is low in phycocyanins but higher levels are present in phycoerythrins.

Methionine is N-terminal in all phycoerythrins so far investigated (Ó Carra and Ó hEocha, 1962; Ó Carra, 1965). Quantitative N-terminal analyses indicate a polypeptide size of about 20 000 daltons in R- and B-phycoerythrins and a somewhat larger value for C-phycoerythrin (Ó Carra, 1965). Methionine is also N-terminal in C-, R- and allophycocyanins (Cope *et al.*, 1967; Ó Cara, 1970; Ó Carra *et al.*, in prep.), although threonine may be an additional N-terminal residue in C-phycocyanin from some species (Ó Carra *et al.*, in prep.), while Binder *et al.* (1972) have reported both alanine and valine as N-terminal residues in C-phycocyanin from *Mastigocladus laminosus.* The earlier results suggesting a large polypeptide chain with N-terminal threonine in phycocyanins (Ó Carra, 1965) seem to be attributable to some peculiar conformational masking effect which is eliminated when the protein is alkylated or subjected to performate oxidation (Cope *et al.*, 1967; Ó Carra, 1970). Raftery and Ó hEocha (1965) report that serine is C-terminal in C-phycocyanin and alanine is C-terminal in R-phycoerythrin.

To date, few attempts to elucidate the amino acid sequences of algal biliproteins have been reported in the literature. Most attempts in this direction have concentrated on the isolation and structural investigation of chromopeptide fragments released by proteolytic digestion. Such studies have been greatly hampered, however, by the instability and reactivity of the chromophores, and the complexity of the covalent chromophore-protein attachments.

The presence of two different types of polypeptide subunit in most of the biliproteins (see below) is a further complication revealed only comparatively recently and this also partly accounts for the slow progress of the sequencing work. This complication is well illustrated by the 13-cycle Edman degradation of the N-terminal sequence of C-phycocyanin described by Torjesen and Sletten (1972). No unique sequence could be deduced from the complex results owing, presumably to differences in the sequence of the two polypeptide subunits. The development of methods for the separation of the polypeptide chains of the biliproteins on a preparative scale (Kobayashi *et al.*, 1972; Glazer and Fang, 1973a,b; Ó Carra *et al.*, in prep.) should help to overcome this problem and should greatly facilitate sequence studies.

B CHARACTERISTICS OF THE POLYPEPTIDE SUBUNITS

The study of the subunit structures of the biliproteins has received considerable impetus in recent years from the introduction of new empirical methods for investigating polypeptide subunits, in particular the technique of electrophoresis in polyacrylamide gels in the presence of sodium dodecyl sulphate ("SDS electrophoresis"). Earlier knowledge of the subunit structures was derived mainly from the N- and C-terminal studies, from calculations of minimal molecular weights based on histidine or half-cystine contents and on extrapolations from ultracentrifugal data. All of these approaches were beset by considerable uncertainties and sources of error and led to disagreement and controversy, particularly in the case of C-phycocyanin, the most intensively studied biliprotein in this regard (see, for example, Hattori *et al.* 1965; Berns and Scott, 1966; Kao and Berns, 1968; Bloomfield and Jennings, 1969; Berns and Kao, 1969). The application of newer empirical techniques indicated that the size of the polypeptide subunits in most of the biliprotein types lay in the range 18 000 to 22 000 daltons (Ó Carra, 1970), and, by and large, similar results have been reported by other workers. A selection of these findings is summarized in Table VII.

Extensive studies confirm that each type of algal biliprotein, with the probable exception of allophycocyanin, is composed of two different polypeptide subunits in a 1:1 stoicheiometric ratio. All the polypeptide subunits contain attached phycobilin(s) and there are no covalent cross-links between subunits (Ó Carra, 1970; Bennett and Bogorad, 1971; Glazer and Cohen-Bazire, 1971; Killilea, 1971; Ó Carra and Killilea, 1971; Ó Carra *et al.*, in prep.).

The polypeptide sizes indicated in Table VII for the phycoerythrins are in good agreement with those derived from terminal analyses and minimal molecular weight calculations (see Ó hEocha, 1965a, 1966). Cryptomonad phycoerythrins have also been reported to contain two types of polypeptide subunit whose molecular weights have been estimated as 19 000 and 11 800 by Glazer *et al.* (1971a) and 17 700 and 11 000 by Brooks and Gantt (1973).

The polypeptide sizes indicated for C-phycocyanin (Table VII) are in conflict with most of the earlier values proposed on the basis of

TABLE VII
Constituent polypeptide chains of algal biliproteins

Biliprotein	Proposed subunit nomenclature[a]	Polypeptide size	Colour of subunit	References
C-phycoerythrin	α	19 700	Red	1,2,3,4
	β	22 000	Red	
C-phycocyanin	α	18 500[b]	Blue	1,2
	β	20 500[b]	Blue	
R-phycocyanin	α	18 500	Blue	1,5,7
	β	20 500	Red	
R-phycoerythrin	α	19 500	Red	1,5,6,7
	β	19 500	Orange	
B-phycoerythrin	α	19 500	Red	1,7
	β	19 500	Red	
Allophycocyanin	—	15 500	Blue	3,7

References
1. Ó Carra (1970).
2. Ó Carra and Killilea (1971).
3. Bennett and Bogorad (1971).
4. Glazer and Cohen-Bazire (1971).
5. Killilea (1971).
6. van der Velde (1973).
7. Ó Carra *et al.* (in prep.).

[a] As used in references 4 and 7.

[b] Smaller values have been reported by some investigators—see text.

ultracentrifugal and chemical studies. This conflict is discussed by Ó Carra and Killilea (1971). The 20 500 and 18 500 dalton polypeptide sizes correlate well with the polypeptide sizes of the other major biliproteins, particularly those of R-phycocyanin (Table VII), and gel filtration in 8 M urea independently confirms an approximate size of 20 000 daltons (Ó Carra, 1970; Ó Carra and Killilea, 1971).

However, a number of other workers using the SDS electrophoretic method have since reported somewhat smaller polypeptide sizes for C-phycocyanin. In some cases the differences are comparatively small, possibly within the margin of error of the techniques (Glazer and Cohen-Bazire, 1971; Bennett and Bogorad, 1971) but in other cases the discrepancies are considerable. For example, Kobayashi *et al.* (1972) reported polypeptide sizes of 18 500 and 11 900 daltons, while Binder *et al.* (1972) estimated that both subunits in C-phycocyanin had a size of 14 000 daltons. The possibility of species variation of such magnitude seems remote and, moreover, Ó Carra and Killilea (1971) found polypeptide sizes identical with those given in Table VII for C-phycocyanins from several different species of blue-green algae.

The danger of proteolytic modification of C-phycocyanin during extraction and purification has been indicated in some previous studies (Ó Carra, 1965; Murphy and Ó Carra, 1970) and could be responsible for some of the apparent variation in polypeptide size. C-phycoerythrin has also been shown to undergo proteolytic modification if it is not extracted and purified rapidly (Ó hEocha and Curley, 1961; Nolan and Ó hEocha, 1967). In this case the subunits are both degraded to 17 000 dalton polypeptides (D. Nolan, unpublished data).

Studies on allophycocyanin both from blue-green algae (Bennett and Bogorad, 1971; Rice and Briggs, 1973a) and from red algae (Ó Carra *et al.*, 1974) indicate that, in contrast to the other biliproteins, it contains a single type of polypeptide subunit of 15 000 to 16 000 daltons. However, somewhat more complex results have been reported by Glazer and Cohen-Bazire (1971) for allophycocyanin from blue-green algal species.

The identically sized polypeptide subunits of R-phycoerythrin are separable by electrophoresis in 8 M urea, a technique which also separates the polypeptides of the other biliproteins and can be applied on a preparative scale (e.g. Glazer and Cohen-Bazire, 1971; Ó Carra *et al.*, in prep.). However, ion-exchange chromatography in 8 M urea shows more promise for preparative-scale separations (Binder *et al.*, 1972; Glazer and Fang, 1973a,b; Ó Carra *et al.*, in prep.).

The possibility that the light polypeptide chain in C-phycocyanin is merely a shortened version of the heavy one has been ruled out by two-dimensional peptide mapping of tryptic digests of the separated subunits which reveals that the two polypeptide chains must differ considerably from one another in amino acid sequence (Ó Carra *et al.*, in prep.). The amino acid composition of the two chains also differs considerably, a particularly significant difference being the presence of one histidine residue in the light chain and none in the heavy chain, again indicating that the light chain could not be derived by degradation of the heavy chain (Glazer and Fang 1973a; Ó Carra *et al.*, in prep.). Analogous evidence applies to the polypeptide chains of R-phycocyanin; the tryptic peptide maps and amino acid analyses of the two subunits are quite different, and again histidine is found only in the lighter subunit (Ó Carra *et al.*, in prep.). Peptide mapping has also shown that the two polypeptide chains in each type of phycoerythrin must differ considerably in amino acid sequence (Ó Carra *et al.*, in prep.).

TABLE VIII
Phycobilin complements of the polypeptide subunits

Biliprotein	Subunit (and descriptive designations)	Phycobilin complement[a]	References
C-phycoerythrin	α (Light)	1 PEB	2,3
	β (Heavy)	2 PEB	
C-phycocyanin	α (Light)	1 PCB	1,2,3
	β (Heavy)	2 PCB	
R-phycocyanin	α (Blue, light)	1 or 2 PCB	3
	β (Red, heavy)	1 PEB + PCB	
R-phycoerythrin	α (Red)	2 or 3 PEB + 1 PUB	3
	β (Orange)	1 PEB + 1 PUB	
Allophycocyanin	—	1 PCB	2,3

References
1. Bennett and Bogorad (1971).
2. Glazer and Fang (1973).
3. Ó Carra *et al.* (in prep.).

[a] Abbreviations: PEB, phycoerythrobilin; PCB, phycocyanobilin; PUB, phycourobilin.

In the case of R-phycocyanin and R-phycoerythrin the subunits also differ qualitatively as regards attached phycobilins and, as a result, the two polypeptide chains in each of these biliproteins also differ characteristically in colour (Tables VII and VIII).

The blue subunit of R-phycocyanin contains only phycocyanobilin (Table VIII). It was thought initially that the red subunit of R-phycocyanin contained only phycoerythrobilin (Ó Carra, 1970) but subsequent work on the separated subunits showed that this subunit also contains phycocyanobilin, the colour of which is not apparent visually owing to the more intense absorbancy of the phycoerythrobilin residues (Ó Carra *et al.*, in prep.). The two polypeptide subunits of R-phycoerythrin contain both phycoerythrobilin and phycourobilin, but the red subunit contains two or three times as much phycoerythrobilin as the orange one (Table VIII).

Available quantitative estimates of the phycobilin complements of the various polypeptide subunits are summarized in Table VIII. In general these agree well with previous estimates of the phycobilin contents of the biliproteins.

C SUBUNIT AGGREGATION AND QUATERNARY STRUCTURE

Some recent estimates of the molecular weights of native algal biliproteins are summarized in Table IX together with the subunit

TABLE IX
Molecular weights and probable subunit structures of native biliproteins

Biliprotein	Apparent molecular weight(s) (daltons)	Probable subunit structure	References
C-phycoerythrin	44 000[a]	$\alpha\beta$	1,2,3
	226 000[a]	$(\alpha\beta)_6$	
B-phycoerythrin	268 000	$(\alpha\beta)_6$	3
R-phycoerythrin	268 000	$(\alpha\beta)_6$	3,4
R-phycocyanin	134 000	$(\alpha\beta)_3$	3
	262 000	$[(\alpha\beta)_3]_2$	
C-phycocyanin	134 000	$(\alpha\beta)_3$	3,5
	262 000[b]	$[(\alpha\beta)_3]_2$	
Allophycocyanin	c.100 000[c]	$(\alpha)_6$	3,7
Cryptomonad phycocyanin	37 000	?	6
Cryptomonad phycoerythrins	27 800 35 000	$\alpha\beta$	6,7,8

References
1. Ó Carra and Killilea (1971).
2. Bennett and Bogorad (1971).
3. Killilea *et al.* (in prep.).
4. van der Velde (1973).
5. Lee and Berns (1968b).
6. Nolan and Ó hEocha (1967).
7. Glazer *et al.* (1971).
8. Brooks and Gantt (1973).

[a] The C-phycoerythrin from certain species disaggregates spontaneously and essentially completely to the 45 000 dalton species (see text).

[b] Larger metastable aggregates of C-phycocyanin are detectable in fresh extracts from many species (see text).

[c] Values ranging from 96 000 to 134 000 daltons have been reported (see text).

structures or aggregation patterns that seem most likely on the basis of current evidence.

Some of the biliproteins exist in a number of distinct aggregation states depending on concentration, pH, ionic strength and other factors. In the case of the cyanophytan biliproteins, C-phycoerythrin and C-phycocyanin, particularly, such behaviour seems to vary somewhat with species of origin. Thus, while C-phycoerythrin from *Phormidium persicinum* dissociates spontaneously and completely to a form with a molecular weight of about 44 000 daltons, the same biliprotein from some other species tends to stabilize as a more highly aggregated form about six times this size (Table IX; Bennett and Bogorad, 1971; Ó Carra and Killilea, 1971; Killilea *et al.*, 1974).

By contrast, native R-phycoerythrin exists as a stable high molecular weight form which seems to be a polypeptide dodecamer of about 268 000 daltons. (Earlier estimates of the molecular weight of this protein–about 290 000 daltons–are probably too high (van der Velde, 1973; Killilea *et al.*, 1974).) B-phycoerythrin is similar in size and subunit structure (Table IX). Electron microscopy of R- and B-phycoerythrins show their shapes to approximate to squat cylinders measuring about 100 by 50 Å (Vaughan, 1964; Gantt, 1969). The structures appear quite solid and no subunit or modular boundaries are visible in the electron micrographs.

Traces of a smaller form of phycoerythrin with a molecular weight of about 45 000 daltons were reported in extracts of some red algae whose main biliprotein is R-phycoerythrin (Nolan and Ó hEocha, 1967). Studies on the subunits of this "small phycoerythrin" suggest that it is a dissociated form of R-phycoerythin representing a polypeptide dimer (van der Velde, 1973; Ó Carra *et al.,* in prep.). The dissociation appears to be irreversible.

R-phycocyanin exists in two well-defined aggregation states depending on pH. Above pH 6·5 it exists as an apparent polypeptide hexamer having a molecular weight of about 134 000 daltons. This aggregate dimerizes below pH 6·5 to form a species of about 262 000 daltons (Table IX), apparently a polypeptide dodecamer.

C-phycocyanin appears to follow an aggregation pattern similar to that of R-phycocyanin, but at low protein concentrations it tends to dissociate into a low molecular weight species which seems to represent the polypeptide dimer (Killilea, 1971; Killilea *et al.,* in prep.). Fresh preparations of C-phycocyanin from some species contain higher aggregates which tend to dissociate irreversibly. These higher aggregates may represent fragmentary and metastable survivals from the phycobilisomes, the large granules into which this biliprotein is aggregated *in vivo* (see Section VIA below and Kessel *et al.*, 1973).

C-phycocyanin is by far the most extensively studied biliprotein as regards aggregation-disaggregation phenomena, the studies of Berns and co-workers (for example, Berns and Scott, 1966; Lee and Burns, 1968a,b; Berns and Kao, 1969; Kao *et al.*, 1971), and of Hattori *et al.* (1965), Neufeld and Riggs (1969) and Vernotte (1971) being noteworthy. Unfortunately, disagreement concerning the nature and size of the "monomer subunit" makes some of this literature difficult to correlate and interpret quantitatively. However, most of the association-dissociation data reported seem to be qualitatively compatible with the aggregation scheme outlined above if it is accepted that the "monomer subunit", rather than being a single polypeptide chain of 30 000 daltons as originally proposed (Kao and Berns, 1968; Kao *et al.*, 1971) corresponds instead to the $\alpha\beta$ polypeptide dimer. This seems likely on the basis of data presented above to have a molecular weight closer to 40 000 than to 30 000, but some variability is possible as discussed in Section IVB.

Berns and Edwards (1965) have published electron micrographs of C-phycocyanin which they interpret in terms of an annular hexameric structure about 130 Å in diameter.

In contrast to C- and R-phycocyanins, allophycocyanin appears to remain in the same state of aggregation over a range of pH values. Its exact molecular weight is somewhat uncertain. Earlier reports indicated a molecular weight of about 134 000 (Hattori and Fujita, 1959a; Nolan and Ó hEocha, 1967) but more recent investigations suggest a value nearer to 100 000 daltons (Glazer and Cohen-Bazire, 1971; MacColl *et al.*, 1974), which would indicate a subunit structure consisting of a polypeptide hexamer.

All the cryptomonad biliproteins so far investigated have molecular weights smaller than 40 000 daltons, corresponding to simple polypeptide dimer structures (Table IX).

Dobler *et al.* (1972), in the first published X-ray crystallographic studies on a biliprotein, describe the crystallographic parameters of C-phycocyanin from *Mastiglocladus laminosus* and hold out the promise of data at a molecular level.

D CONFORMATIONAL STABILITY AND REVERSIBLE DENATURATION

The visible spectral properties and fluorescence of the biliproteins are closely dependent on the conformational integrity of the apoproteins and these properties, therefore, provide sensitive and convenient indicators for monitoring conformational unfolding and its reversal (see, for example, Ó hEocha and Ó Carra, 1961; Lavorel and Moniot, 1962; Murphy and Ó Carra, 1970). Complete unfolding or denaturation causes a very substantial decrease in visible light absorbance and also abolishes the visible fluorescence. Less drastic conformational perturbations are also usually accompanied by considerable spectral changes together with partial quenching of fluorescence.

Changes in subunit aggregation may be accompanied by comparatively minor spectral changes which may nevertheless be useful in monitoring alterations in quaternary structure (see, for example, Hattori *et al.*, 1965; Neufield and Riggs, 1969).

Spectral and viscometric studies indicate that phycoerythrins as a class are considerably more stable towards denaturation than phycocyanins. For example, all phycocyanins (C- R-, allo- and cryptomonad types) are rapidly denatured in 4 M urea while phycoerythrins are not (Ó Carra, 1962; Ó Carra and Ó hEocha, 1965; Killilea, 1971). Among the phycoerythrins, the R and B types are considerably more stable than the C and cryptomonad types; R- and B-phycoerythrins are denatured only slowly in neutral 8 M urea and retain their fluorescence and spectral properties at temperatures as high as 70° (Ó hEocha and Ó Carra, 1961; Ó Carra, 1962).

Differences in conformational stability between biliproteins seem to correlate in a general way with difference in the content of cysteine/cystine, and it is possible that intrachain disulphide links are responsible for the greater conformation stability of the phycoerythrins. Thus R- and C-phycocyanins seem to contain only enough half-cystine to account for the covalent chromophore–protein links, while the levels in R- and B-phycoerythrins are higher (D. Nolan, unpublished results).

Although C-phycocyanin is easily denatured, the process can be readily reversed to yield preparations identical with the native protein (Murphy and Ó Carra, 1970; Killilea, 1971). In some elegant experiments, Glazer and Fang (1973b) have taken advantage of this ease of renaturation to produce interspecific hybrid forms of C-phycocyanin.

Denatured C-phycoerythrin may also be renatured with high efficiency from 8 M urea if the denatured protein is not left standing in solution for more than a few minutes. If it is incubated in 8 M urea the proportion of the protein which is renaturable decreases progressively with time. The integrity of cysteine thiol group(s) in this biliprotein seem to be necessary for correct resumption of the native conformational structure and the progressive irreversibility of the denaturation may be due to oxidative modification of such thiol group(s) (Killilea, 1971).

Fully denatured R-phycoerythrin reverts only to a "partially renatured" form having a molecular weight of about 45 000 daltons (D. Nolan and P. Ó Carra, unpublished data). This form closely resembles the "small phycoerythrin" which accompanies R-phycoerythrin in small quantities in extracts of many red algae (see Section IVC). Fujimori and Pecci (1967) described the formation of a similar low molecular weight form of B-phycoerythrin on treatment of the protein with *p*-chloromercuribenzoate. Denatured R-phycocyanin has resisted all attempts at renaturation.

V Functional aspects

A INTRACELLULAR LOCALIZATION AND ORGANIZATION

Electron microscopic studies of red and blue-green algae have shown that the biliproteins in these algae are organized intracellularly into large aggregates or granules measuring upwards of 300 Å in diameter. These aggregates, termed phycobilisomes, are located on the outer (stromal) surface of the photosynthetic membranes (Gantt and Conti, 1966a,b, 1969; Bourdu and Lefort, 1967; Edwards *et al.*, 1968; Lichtlé and Giraud, 1970; Edwards and Gantt, 1971; Gantt and Lipschultz, 1972). In cryptomonad algae, on the other hand, the biliproteins seem to be packed evenly between paired photosynthetic membranes (Gantt *et al.*, 1971). It remains to be established whether this difference is in any way related to the relatively small size of isolated cryptomonad biliproteins (see Section IVB).

In cells of red and blue-green algae phycobilisomes seem to be arranged in a series of parallel rows or chains which, in transverse section, assume the appearance of long solid rods or cylinders running along the lamellar surface (see Bourdu and Lefort, 1967; Lichtlé and Giraud, 1970; Edwards and Gantt, 1971). According to Edwards and Gantt (1971) the long cylindrical structures in the blue-green *Synechococcus lividus* may be visualized in terms of longitudinal stacks in which many individual disc-shaped phycobilisomes are standing on edge and stacked serially face-to-face or face-to-tail. These longitudinal arrays seem to disintegrate spontaneously into individual phycobilisomes when the membranes are disrupted during extraction of the algae.

The phycobilisomes in turn dissociate readily into their constituent biliprotein molecules*. Thus extraction of the algae with distilled water or dilute buffers generally yields the biliproteins as molecular solutions rather than supermolecular aggregates. However, careful extraction of *Porphyridium cruentum* in 0·5 M phosphate buffer containing Triton X-100 enabled Gantt and Lipschultz (1972) to isolate relatively intact phycobilisomes, while Kessel *et al.* (1973) reported that fresh, rapidly prepared extracts of two blue-green algae

* In terms of phycobilisome structure the individual biliprotein molecules are sometimes referred to as "monomers" (see, for example, Gantt, 1969). However, the term "biliprotein monomer" is also commonly applied at the submolecular level with reference both to the modular subunits which combine to make up the biliprotein molecules, and also with reference to the polypeptide chains which combine to make up the modular subunits. In this field it is therefore advisable to determine carefully the exact meaning with which the term "monomer" is used in any particular context.

contained a proportion of phycocyanin in the form of large unstable aggregates reminiscent of phycobilisomes.

Phycobilisomes seem to contain the entire cellular complement of photosynthetic biliproteins (Gantt and Lipschultz, 1972, 1973). The variation with regard to biliprotein complement from species to species, and the variations in the proportions of the different biliproteins which are observed even within a given species under different conditions of growth indicate that the composition and structure of the phycobilisomes must also be variable. Such variation has indeed been demonstrated by Gantt *et al.* (1968) for two species of *Porphyridium*, one of which (*P. cruentum*) contains mainly B-phycoerythrin and has spherical or oblate phycobilisomes, while the other (*P. aerugineum*) contains mainly C-phycocyanin and has disc-shaped phycobilisomes.

The phycobilisomes of *Synechococcus lividus,* a blue-green algal species containing only C-phycocyanin and allophycocyanin, are visualized by Edwards and Gantt (1971) as two-layered discs, each layer consisting of a hexagonal ring of biliprotein molecules around a central molecule. Kessel *et al.* (1973) propose a somewhat different pattern, but again featuring a ring of phycocyanin units arranged around a central unit.

Gantt and Lipschultz (1972, 1973) have studied the phycobilisomes isolated from *Porphyridium cruentum* in some detail. B-phycoerythrin constitutes about 80% of the mass of the particles, with R-phycocyanin and allophycocyanin in about equal proportions making up most of the remainder. A trace of chlorophyll is also present but it is not clear whether this is a native constituent of the phycobilisomes or merely a contaminant adsorbed during the isolation procedure. Gantt and Lipschultz (1972, 1973) postulate that these phycobilisomes consist of a small core of allophycocyanin and R-phycocyanin surrounded by a shell of B-phycoerythrin molecules. They also postulate that the allophycocyanin core is in direct contact with the photosynthetic lamella in the intact cell.

B FUNCTION AND MECHANISM OF ACTION

The algal biliproteins act as accessory photosynthetic pigments absorbing light in the spectral regions in which chlorophyll absorbs only weakly or not at all. The absorbed quanta are transferred to chlorophyll *a* which transmits the energy to the next phase of the photosynthetic process. In spite of their accessory role, however, the biliproteins are far more effective light harvesters than chlorophyll itself in red, blue-green and cryptomonad algae, as revealed by photosynthetic action spectra (e.g. Emerson and Lewis, 1942; Haxo and Blinks, 1950; Blinks, 1954, 1960; Haxo and Fork, 1959; Haxo, 1960; Jones and Myers 1964; Fork and Amesz, 1969). In these algae chlorophyll assumes a major role in direct light harvesting (as distinct from energy transfer) only under conditions where the biliprotein complement is greatly depleted (Lemasson *et al.*, 1973).

The advantage conferred on the algae by this dominance of the biliproteins in light harvesting is not entirely clear. Differential scattering and absorption of both the blue-violet and red ends of the spectrum by sea-water results in a greater penetration through deep water of light in the mid-spectral region, which phycoerythrin absorbs. This appears to confer an ecological advantage on deep-growing marine red algae, which, in general, are relatively richer in phycoerythrin than those growing further up the littoral zone; however, this cannot be regarded as a clear-cut distinction and many species of red algae from shallow water are also rich in phycoerythrin. Certain laboratory-grown and also some wild algae have been shown to adapt the content and proportions of their biliproteins in response to variations in light quality and intensity, though others remain comparatively unaffected (see Section VI).

The photosynthetic efficiency of the biliproteins implies very efficient transfer of quanta from the biliproteins to the chlorophyll *a* molecules located at the photochemical reaction centres of the photosynthetic apparatus. Such highly efficient energy transfer *in vivo* was demonstrated by the fluorescence studies of Duysens (1951), French and Young (1952), Goedheer (1965) and others. It is a prerequisite for even moderately efficient energy transfer that the donor and acceptor chromophores should be less than 50 Å apart (see Forster, 1959). Arnold and Oppenheimer (1950) calculated that the phycobilin chromophores must lie within 7–40 Å of the chlorophyll *a* molecules to permit the necessary degree of transfer in blue-green algae. Since the chlorophyll molecules lie within the photosynthetic lamella while the phycobilin chromophores lie buried within the relatively bulky biliprotein molecules it seems likely that this spatial relationship could only be met by biliproteins directly attached to the lamella. However, owing to the size and geometry of the phycobilisome complexes in red and blue-green algae only a small proportion of their constituent biliprotein molecules could be in such direct contact with the photosynthetic membrane, and it seems clear that these directly attached molecules must act as intermediate acceptors through which the quanta absorbed by the outer biliprotein molecules are chanelled to the lamellar chlorophyll.

In line with this concept, efficient energy transfer among the biliprotein molecules of intact phycobilisomes isolated from *Porphyridium cruentrum* has been demonstrated by Gantt and Lipschultz (1973), whose data indicate that the light energy absorbed by phycoerythrin, the major biliprotein in these phycobilisomes, is efficiently transferred to the small quantity of allophycocyanin present. These authors therefore postulate that allophycocyanin acts as the "energy funnel" between the phycobilisome and the lamellar chlorophyll. Lemasson *et al.* (1973), working with blue-green algae, have produced further evidence in support of this concept. Gantt and Lipschultz (1973) suggest that the allophycocyanin must be in direct contact with the photosynthetic membrane and must also be located centrally as a "core" in the phycobilisome so as to permit efficient transfer of quanta from all the phycocyanin and phycoerythrin molecules in the particle. The spectral properties of allophycocyanin, lying as they do between those of the other biliproteins on the one hand and those of chlorophyll *a* on the other, suit it ideally for its postulated energy-funnelling role.

The bright fluorescence of the isolated biliproteins (and that of chlorophyll) is largely quenched *in vivo* since the light quanta are shuttled away by very rapid resonance energy transfer along the chain of intermediate acceptors to the photochemical reaction centres. Any disruption of the phycobilisome-lamellar arrangement which interferes with the energy transfer mechanism results in increased leakage of quanta as fluorescence. Thus careful detachment of the phycobilisomes from the lamella results in the appearance of strong allophycocyanin fluorescence, which is shifted from 660 nm to 675–680 nm in the aggregated state of the phycobilisomes (Gantt and Lipschultz, 1973). The phycoerythrins and R- and C-phycocyanins fluoresce strongly only when the integrity of the phycobilisomes themselves is disrupted.

It will be recalled that extensive and efficient energy transfer also takes place between the phycobilin groups within the individual biliprotein molecules themselves (cf. Section III). The degree and complexity of excitation energy transfer involving the biliproteins is thus very considerable.

Studies of excitation energy transfer in artificial model complexes of bilins and biliproteins with chlorophyll derivatives have also been reported (Frąckowiak and Grabowski, 1970, 1973; Miedziejko and Frąckowiak, 1969).

Many workers have searched for a more direct involvement of phycoerythrins and phycocyanins in photochemical reactions in addition to their indirect, light-harvesting role. A number of reports have described the promotion of certain photoreductions and other light-mediated reactions by these biliproteins (see, for example, Vernon, 1961a,b; Evstigneev and Gavrilova, 1964; Bellin and Gergel, 1969), but it remains doubtful whether any of these activities are of real biological significance (see also Fujita and Tsuji, 1968).

VI Biogenesis and metabolism

It has long been established that animal bilins are derived from haem by oxidative elimination of one of the carbon bridges followed by loss of the iron atom (Scheme 5). The initial bilin product is biliverdin, which in mammals undergoes subsequent redox transformations.

Most suggested schemes of phycobilin biosynthesis involve an analogous oxidative haem-cleavage process in conjunction with redox and prototropic transformations of the conjugated system. (For reviews of the earlier literature, see: Bogorad, 1965; Bogorad and Troxler, 1967; Ó hEocha, 1968.) A considerable body of data now supports this point of view, although conclusive evidence has remained elusive.

Two of the most characteristic features of haem-cleavage in animals have been shown to apply equally to phycobilin formation in algae. These features are, firstly, the isomeric nature of the bilin carbon skeleton and, secondly, the metabolic fate of the eliminated carbon bridge. In animals, cleavage of haem takes place almost exclusively at the α-carbon bridge, giving rise specifically to only one of the four possible isomeric bilin carbon skeletons (Scheme 5). This particular isomeric form is denoted by the suffix IXα, named after the parent haem IX and the α bridge which is specifically eliminated. The carbon skeleton of the phycobilins has been similarly established as being entirely of the IXα type (Cole *et al.*, 1968; Ó Carra and Colleran, 1970). An even more characteristic feature of animal haem-cleavage is the carbon monoxide formed from the eliminated carbon bridge (Scheme 5). This is not further metabolized and is excreted via the lungs in vertebrates. Troxler and co-workers have established that red and blue-green algae also produce carbon monoxide; in a series of elegant experiments they have shown that phycobilin synthesis correlates stoicheiometrically with carbon monoxide formation (Troxler *et al.*, 1970; Troxler 1972; Troxler and Dokos, 1973).

Direct demonstration of the precursor role of haem by means of radioisotope tracer experiments has been hampered by technical difficulties, particularly the slow uptake of haem and porphyrins by

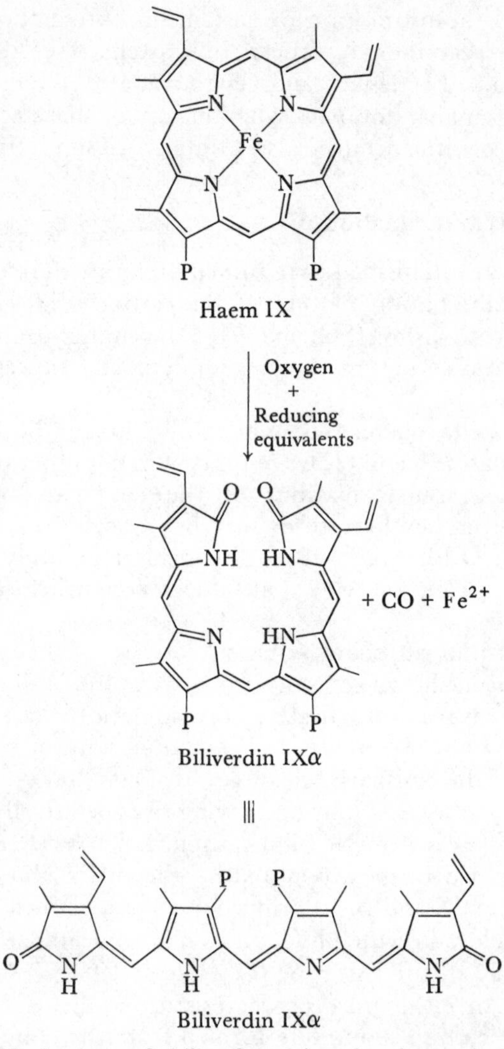

Haem IX

Oxygen + Reducing equivalents

Biliverdin IXα + CO + Fe^{2+}

Biliverdin IXα

SCHEME 5. Mechanism of bilin formation by haem-cleavage in animals. (P: propionic acid side chain.)

algal cells. Such problems do not apply to the simpler precursors, however, and the isotopic tracer experiments of Troxler and Lester (1967, 1968) establish δ-aminolaevulinic acid as a precursor of phycocyanobilin. While this work confirms that the early stages of phycobilin synthesis follow the same pattern as porphyrin synthesis, it does not distinguish between a sequential as against a parallel biosynthetic relationship.

Godnev *et al.* (1966) claimed that addition of protoporphyrin or haem to the growth medium of the blue-green alga *Anacystis nidulans* resulted in an increase in phycocyanobilin synthesis and also diluted the incorporation of ^{14}C acetate into phycocyanobilin. The reported effects, however, are far greater than could be accounted for even by quantitative incorporation of the added porphyrin into phycocyanobilin. Colleran and Ó Carra (see Ó Carra, 1970) failed in their attempts to duplicate these experiments, but in direct feeding experiments, in which ^{14}C haemin was fed to cultures of *Anacystis*, these latter workers were able to demonstrate a low but apparently significant level of isotope incorporation into phycocyanobilin.

The α-specific cleavage of haem in animals (producing functionless catabolites) is considered by some authors to result from metabolically incidental side-effects of the functioning of certain haemoproteins (Ó Carra and Colleran, 1969; Colleran and Ó Carra, 1970), while others (e.g. Tenhunen *et al.*, 1969, 1972) consider such cleavage to involve a functionally-specific enzyme system. In view of the functional importance of the phycobilins in algae, the mechanism of their production seems more likely to involve functionally-specific or purposeful system(s).

Little or no information is available concerning the supplementary transformations which create the characteristic structural details of the phycobilins. If, as seems likely, protohaem is a precursor, these transformations must include reductions and shifts of double bonds. However, efforts to detect systems catalysing such bilin transformations in algal extracts have been unsuccessful to date (M. J. Murphy and C. Ó hEocha, unpublished data). These investigations are complicated by the possibility that the substrates for such systems may be phycobilin precursors already attached to apoproteins, rather than free bilins, since it is not clear at what stage the phycobilins or their precursors become attached to the apoproteins.

Fujita and Hattori (1962a, 1963b) reported that blue-green algae contained a significant pool of protein-free precursors. However, subsequent work (Ó Carra and Ó hEocha, 1966) showed that these supposed precursors were artefact forms of the phycobilins which were released from the biliproteins under the conditions of extraction used (see Section IIA). Fujita and Hattori (1960a,b, 1962b, 1963) considered that chromatic adaptation in algae involved interconversion of pre-existing pools of phycobilins or their immediate precursors, but the data on which they based this view now seem equally interpretable on the basis of *de novo* synthesis of the phycobilins rather than their interconversion. In general, the indications are that red and blue-green algae do not contain significant pools of protein-free bilins.

The double labelling experiments of Troxler and Brown (1970), in which ^{14}C-labelled δ-aminolaevulinic acid (incorporated into phycocyanobilin) and [^{3}H-] leucine (incorporated into the phycocyanin apoprotein) were fed simultaneously to *Cyanidum caldarium*, indicate a strict stoicheometry between phycobilin and apoprotein formation. While this is compatible with the idea that both phycobilin and apoprotein components may be derived by modification of preformed haem–protein conjugates (see, for example, Bogorad, 1965) it is also compatible with strictly coordinated regulation of separate phycobilin and apoprotein synthesis.

As mentioned in Section VB, the intensity and spectral quality of the incident light influences the biliprotein levels in most red and blue-green algae. In general, light of low or moderate intensity favours synthesis of biliproteins relative to that of chlorophyll *a*. Phycoerythrin formation in many red algae seems to be especially stimulated by low-intensity green light. Since this is also the region of the spectrum absorbed by the phycoerythrins, such adaptation presumably allows red algae to maximize their absorption of the dim greenish light that penetrates through deep sea-water.

The ratio of phycoerythrin to phycocyanin has also been found to vary in some algae in response to seasonal or artifically-imposed variations in the quality of the incident light. Fujita and Hattori (1960a,b, 1962b, 1963a) have extensively investigated such chromatic adaptation in the blue-green alga *Tolypothrix tenuis*, which produces a high level of phycoerythrin in addition to phycocyanin when grown under white fluorescent light but produces only phycocyanin when grown under filtered red light. *Fremyella diplosiphon*, another blue-green alga, behaves similarly (Bennett and Bogorad, 1971).

Nichols and Bogorad (1962) studied the effect of varying wavelengths of incident light on phycocyanin formation in a chlorophyll-less mutant form of *Cyanidium caldarium*. The action spectrum of phycocyanin formation had maxima at about 420 nm and 550–595 nm, and these workers drew attention to the resemblance between this action spectrum and the absorption spectra of many haemoproteins. They suggested that a haemoprotein might either be acting as a photoreceptor involved in the control of biliprotein synthesis or, alternatively, that a haemoprotein precursor of phycocyanin might be sensitized, and its conversion to phycocyanin potentiated, by direct light absorption (see also Bogorad, 1965). Neither of these hypotheses has, as yet, been supported by further experimental evidence.

High light intensity, besides inhibiting biliprotein formation, also causes photodestruction of existing biliproteins, a phenomenon that leads to the bleaching observed in some exposed marine red algae during summer months. It has been suggested that similar bleaching of blue-green algae results from destruction of phycocyanin by peroxides produced in the algal cells under conditions of high light intensity (Abeliovich and Shilo, 1972). The production of biliproteins by algae grown in culture is limited by deficiencies in such elements as nitrogen, iron, molybdenum and sodium. Biliproteins are metabolically unstable when nitrogen is limiting. Allen and Smith (1969) have shown that nitrogen-deficient blue-green algae contain normal levels of chlorophyll *a* and carotenoids but not of phycocyanin, and similar conditions lead to loss of phycoerythrin in the red alga *Porphyridium cruentum* (Gantt and Conti, 1966). The biliprotein level begins to decrease when algal growth ceases due to limitation of nitrate and is complete after several hours. The process can be reversed by addition of nitrate to the deficient cultures; cell division does not recommence until normal levels of biliproteins are restored.

Kinetin was shown to increase the phycoerythrin content of the red alga *Hypnea musciformis* (Jennings *et al.*, 1972). These authors suggest that the primary effect of kinetin may be exercised through enhanced synthesis of phycoerythrobilin, although a decreased rate of phycoerythrin turnover was not ruled out.

VII Interrelationships of the algal biliproteins and phytochrome

The available chemical and physical evidence indicates that the different classes of biliproteins (phycoerythrins, phycocyanins, allophycocyanin) differ considerably in their polypeptide structures which, it would appear, could not be metabolically derived from one another (Section IV). However key similarities—in particular with regard to the phycobilins and their mode of attachment—suggest a common, though presumably remote, evolutionary origin for these biliprotein classes.

Immunochemical studies reveal no significant degree of immunological cross-reactivity between the phycoerythrins and the phycocyanins (R and C types as a group) or between either of these classes of biliprotein and allophycocyanin (Vaughan, 1964; Bogorad, 1965; Glazer *et al.*, 1971b). Such negative immunochemical evidence does not necessarily indicate a lack of evolutionary relatedness among these biliprotein classes, however, since the antigenic determinants of native proteins are largely determined by the conformational and quaternary structures and do not necessarily reflect the degree of structural homology at the primary amino acid sequence level. (Thus, for example the M and H type isoenzymes of lactate dehydrogenase display little immunological cross-reactivity in spite of considerable sequence homology.)

Positive immunological cross-reactivity, on the other hand, can usually be taken as a fairly reliable indication of true structural homology, and it is therefore particularly interesting that the homologous biliproteins from the red and blue-green algae cross-react strongly (Vaughan, 1964; Bogorad, 1965; Glazer *et al.*, 1971b). That is, the various types of phycoerythrins from red and blue-green algae all cross-react, as do R- and C-phycocyanins as a group and the allophycocyanin from both algal divisions as a group. The strength of the cross-reaction is remarkable in view of the very remote evolutionary divergence of the procaryotic blue-green algae and the eucaryotic red algae. The possibility that the structural homology responsible for the cross-reaction might be confined to the phycobilin prosthetic groups and their attachment points must be considered unlikely in view of the evidence that these groups are folded into the interior of the protein molecules (see Section III) and therefore inaccessible as antigenic determinants in the native proteins. Further evidence with regard to this point is discussed by Vaughan (1964) and Glazer *et al.* (1971).

The very close structural homology of C-phycocyanins isolated from distantly related blue-green algae, indicated by the immunochemical studies of Berns (1967) and Glazer *et al.* (1971), has been elegantly confirmed by Glazer and Fang (1973b), who showed that the separated α and β polypeptide subunits of C-phycocyanins from distantly related species hybridize efficiently to yield interspecific hybrid phycocyanins.

The immunological cross-reactivity of the homologous biliproteins from red and blue-green algae does not extend to those from cryptomonad algae. Glazer *et al.* (1971b) report that the cryptomonad biliproteins show no detectable cross-reaction with their rhodophytan and cyanophytan counterparts. Again this negative immunological evidence may largely reflect differences in conformational and quaternary structures. As mentioned in Section IVB, the physical and chemical evidence shows that the subunit aggregation patterns in the cryptomonad biliproteins are much simpler than in the biliproteins from the other algal divisions.

The earlier work on phytochrome suggested the possibility of a close relationship with the photosynthetic algal biliproteins, in particular with allophycocyanin. However, more recent studies suggest that the relationship may be confined to a structural similarity between the chromophores.

The precise structure of the phytochrome chromophore, phytobilin, remains to be established, but the available evidence (Siegelman *et al.*, 1966, 1968; Rüdiger, 1971; see also Chapter 7.) suggests a structure very similar to that of phycocyanobilin or its "blue pigment" derivative (Section II and Scheme 2). The work of Rüdiger and Corell (1969) indicates further that phytobilin is covalently attached to the phytochrome apoprotein through ring C and probably also through ring A (see Rüdiger, 1971), in a manner reminiscent of the apoprotein attachments of the phycobilins (Section II).

The spectral properties of phytochrome in the P_r form closely resemble those of allophycocyanin, a resemblance that provided the first clue to the biliprotein nature of phytochrome. However, while allophycocyanin is a photostable light harvester which quickly transfers or re-emits its absorbed quanta, phytochrome is photochemically transformed by the quanta it absorbs to the spectrally distinct P_{fr} form (see Chapters 7 and 18.) This transformation is reversible and, interestingly, the phycobilins undergo a similar reversible spectral shift in response to alkaline pH values (Rüdiger and Ó Carra, 1969; Ó Carra, 1970).

Despite the similarities in spectral properties, the more recent chemical and physical studies on phytochrome from higher plants reveal little or no resemblance between its protein component and that of allophycocyanin (Rice and Briggs, 1973a). Earlier indications of a partial immunological cross-reaction between oat phytochrome and phycocyanin (Berns, 1967) have not been confirmed by later more detailed studies which reveal no cross-reactivity with either C-phycocyanin or allophycocyanin (Rice and Briggs, 1973b).

REFERENCES

Abeliovich, A. and Shilo, M. (1972). *Biochim. biophys. Acta* **283**, 483–491.
Allen, M. B., Dougherty, E. C. and McLaughlin, J. J. A. (1959). *Nature, Lond.* **184**, 1047–1052.
Allen, M. M. (1968). *J. Bact.* **96**, 836–841.
Allen, M. M. and Smith, A. J. (1969). *Arch. Mikrobiol.* **69**, 114–120.
Arnold, W. and Oppenheimer, J. R. (1950). *J. gen. Physiol.* **33**, 423–435.
Bellin, J. S. and Gergel, C. A. (1969). *Photochem. Photobiol.* **10**, 427–439.
Bennett, A. and Bogorad, L. (1971). *Biochemistry, N.Y.* **10**, 3625–3634.
Berns, D. S. (1967). *Pl. Physiol., Lancaster* **42**, 1569–1586.

Berns, D. S. and Edwards, M. R. (1965). *Archs Biochem. Biophys.* **110**, 511-516.
Berns, D. S. and Kao, O. (1969). *Biopolymers* **8**, 293-295.
Berns, D. S. and Scott, E. (1966). *Biochemistry, N.Y.* **5**, 1528-1533.
Binder, A., Wilson, K. and Zuber, H. (1972). *FEBS Lett.* **20**, 111-116.
Blinks, L. R. (1954). *A. Rev. Pl. Physiol.* **5**, 93-114.
Blinks, L. R. (1960). *Proc. natn. Acad. Sci. U.S.A.* **46**, 327-333.
Bloomfield, V. A. and Jennings, B. R. (1969). *Biopolymers* **8**, 297-299.
Bogorad, L. (1965). *Rec. chem. Prog.* **26**, 1-12.
Bogorad, L. and Troxler, R. F. (1967). *In* "Biogenesis of Natural Compounds" (P. Bernfeld, ed.), 2nd edn, pp. 247-313. Pergamon Press, New York and Oxford.
Boney, A. D. and White, E. B. (1968). *Nature, Lond.* **218**, 1068-1069.
Bourdu, R. and Lefort, M. (1967). *C.r. hebd. Seanc. Acad. Sci., Ser. D.* **265**, 37-40.
Brooks, C. and Chapman, D. J. (1972). *Phytochemistry* **11**, 2663-2670.
Brooks, C. and Gantt, E. (1973). *Arch. Mikrobiol.* **88**, 193-204.
Byfield, P. G. H. and Zuber, H. (1972). *FEBS Lett.* **28**, 36-40.
Chapman, D. J., Cole, W. J. and Siegelman, H. W. (1967a). *J. Am. chem. Soc.* **89**, 5976-5977.
Chapman, D. J., Cole, W. J. and Siegelman, H. W. (1967b). *Biochem. J.* **105**, 903-905.
Chapman, D. J., Cole, W. J. and Siegelman, H. W. (1968). *Phytochemistry* **7**, 1831-1835.
Chapman, D. J., Budzikiewicz, H. and Siegelman, H. W. (1973). *Experientia* **28**, 876-878.
Clendenning, K. A. (1954). *Eighth Int. bot. Congr. Rapt. Commun. Sect.* **11**, 21-35.
Cole, W. J., Chapman, D. J. and Siegelman, H. W. (1967a). *J. Am. chem. Soc.* **89**, 3643-3645.
Cole, W. J., Ó hEocha, C., Moscowitz, A. and Krueger, W. R. (1967b). *Eur. J. Biochem.* **3**, 202-207.
Cole, W. J., Chapman, D. J. and Siegelman, H. W. (1968). *Biochemistry, N.Y.* **7**, 2929-2935.
Colleran, E. and Ó Carra, P. (1970). *Biochem. J.* **119**, 905-911.
Cope, B. T., Smith, U., Crespi, H. L. and Katz, J. J. (1967). *Biochim. biophys. Acta* **133**, 446-453.
Crespi, H. L. and Katz, J. J. (1969). *Phytochemistry* **8**, 759-761.
Crespi, H. L. and Smith, U. H. (1970). *Phytochemistry* **9**, 205-212.
Crespi, H. L., Boucher, L. J., Norman, G. D., Katz, J. J. and Dougherty, R. C. (1967). *J. Am. chem. Soc.* **89**, 3642-3643.
Crespi, H. L., Smith, U. and Katz, J. J. (1968). *Biochemistry, N.Y.* **7**, 2232-2242.
Dale, R. E. and Teale, F. W. J. (1966). *Proc. 2nd Western Eur. Conf. Photosyn.* 169-175.
Dale, R. E. and Teale, F. W. J. (1970). *Photochem. Photobiol.* **12**, 99-117.
Dobler, M., Dover, S. D., Laves, K., Binder, A. and Zuber, H. (1972). *J. molec. Biol.* **71**, 785-787.
Dus, K., Bartsch, R. G. and Kamen, M. D. (1962). *J. biol. Chem.* **237**, 3083-3093.
Duysens, L. N. M. (1951). *Nature, Lond.* **168**, 548-550.
Edwards, M. R. and Gantt, E. (1971). *J. Cell. Biol.* **50**, 896-900.

Edwards, M. R., Berns, D. S., Ghiorse, W. C. and Holt, S. C. (1968). *J. Phycol.* 4, 283-298.
Emerson R. and Lewis, C. M. (1942). *J. gen. Physiol.* **25**, 579-595.
Eriksson, C. E. A. and Halldal, P. (1965). *Physiologia. Pl.* **18**, 146-152.
Eriksson-Quensel, I.-B. (1938). *Biochem. J.* 32, 585-589.
Esenbeck, N. von (1836). *Justus Liebig's Annin Chem.* 17, 75-82.
Evstigneev, V. B. and Gavrilova, V. A. (1964). *Biofisika* 9, 739; also *Chem. Abstr.* (1965) **64**, 5489c.
Fork, D. C. and Amesz, J. (1969). *A. Rev. Pl. Physiol.* 20, 305-328.
Forster, T. (1959). *Discussions Faraday Soc.* 27, 7.
Fraçkowiak, D. and Grabowski, J. (1970). *Photosynthetica* 4, 236-242.
Fraçkowiak, D. and Grabowski, J. (1973). *Photosynthetica* 7, 402-404.
French, C. S. and Young, V. K. (1952). *J. gen. Physiol.* 35, 873-890.
French, C. S., Smith, J. H. C., Virgin, H. I. and Airth, R. L. (1956). *Pl. Physiol., Lancaster* 31, 369-374.
Fujimori, E. and Pecci, J. (1967). *Archs Biochem. Biophys.* **118**, 448-455.
Fujita, Y, and Hattori, A. (1960a). *Pl. Cell Physiol.* **1**, 281-292.
Fujita, Y, and Hattori, A. (1960b). *Pl. Cell Physiol.* **1**, 293-303.
Fujita, Y. and Hattori, A. (1962a). *J. Biochem., Tokyo* **51**, 89-91.
Fujita, Y. and Hattori, A. (1962b). *Pl. Cell Physiol.* 3, 209-220.
Fujita, Y. and Hattori, A. (1963a). *In* "Studies on microalgae and photosynthetic bacteria", p. 431. Japanese Soc. of Plant Physiologists, Tokyo.
Fujita, Y. and Hattori, A. (1963b). *J. gen. appl. Microbiol., Tokyo* 9, 253-256.
Fujita, Y. and Tsuji, T. (1968). *Nature, Lond.* **219**, 1270-1271.
Fujiwara, T. (1961). *J. Biochem., Tokyo* **49**, 361-367.
Gantt, E. (1969). *Pl. Physiol., Lancaster* **44**, 1629-1638.
Gantt, E. and Conti, S. F. (1965). *J. Cell Biol.* 26, 365-381.
Gantt, E. and Conti, S. F. (1966a). *J. Cell Biol.* 29, 423-434.
Gantt, E. and Conti, S. F. (1966b). *Brookhaven Symp. Biol.* 19, 393-405.
Gantt, E. and Conti, S. F. (1969). *J. Bacteriol.* 97, 1486-1493.
Gantt, E. and Lipschultz, C. A. (1972). *J. Cell Biol.* 54, 313-324.
Gantt, E. and Lipschultz, C. A. (1973). *Biochim. biophys. Acta*, 292, 858-861.
Gantt, E., Edwards, M. R. and Conti, S. F. (1968). *J. Phycol.* 4, 65-71.
Gantt. E., Edwards, M. R. and Provasoli, L. (1971). *J. Cell Biol.* 48, 280-290.
Glazer, A. N. and Cohen-Bazire, G. (1971). *Proc. natn. Acad. Sci. U.S.A.* **68**, 1398-1401
Glazer, A. N. and Fang, S. (1973a). *J. biol. Chem.* **248**, 659-662.
Glazer, A. N. and Fang, S. (1973b). *J. biol. Chem.* **248**, 663-671.
Glazer, A. N., Cohen-Bazire, G. and Stanier, R. Y. (1971a). *Arch. Mikrobiol.* **80**, 1-18.
Glazer, A. N., Cohen-Bazire, G. and Stanier, R. Y. (1971b). *Proc. natn. Acad. Sci. U.S.A.* 68, 3005-3008.
Godnev, T. N., Rotfarb, R. M., Guardiyan, V. N. (1966). *Dokl. Akad. Nauk SSSR* 169, 1191.
Goedheer, J. C. (1965). *Biochim. biophys. Acta* **102**, 75-89.
Goedheer, J. C. and Birnie, F. (1965). *Biochim. biophys. Acta* **94**, 579-581.
Gössauer and Hirsch (1973). *Tetrahedron Lett.* 1451-1454.
Hattori, A. and Fujita, Y. (1959a). *J. Biochem., Tokyo* **46**, 633-644.
Hattori, A. and Fujita, Y. (1959b). *J. Biochem., Tokyo* **46**, 903-909.
Hattori, A., Crespi, H. L. and Katz, J. J. (1965). *Biochemistry, N.Y.* **4**, 1225-1238.

Haxo, F. T. (1960). *In* "Comparative Biochemistry of Photoreactive Systems" (M. B. Allen, ed.), pp. 339-360. Academic Press, New York and London.
Haxo, F. T. and Blinks, L. R. (1950). *J. gen. Physiol.* 33, 389-422.
Haxo, F. T. and Fork, D. C. (1959). *Nature, Lond.* **184**, 1051-1052.
Haxo, F. T. and Ó hEocha, C. (1960). *In* "Handbuch der Pflanzenphysiologie" (Ruhland, W. ed.), vol. 5, pp. 497-510.
Haxo, F. T., Ó hEocha, C. and Norris, P. S. (1955). *Archs Biochem. Biophys.* **54**, 162-173.
Hirose, H. and Kumano, S. (1966). *Bot. Mag., Tokyo* 79, 105-113.
Jackson, A. H., Smith, K. M., Gray, C. H. and Nicholson, D. C. (1966). *Nature, Lond.* **209**, 581-583.
Jennings, R. C., Broughton, W. J. and McComb, A. J. (1972). *Phytochemistry* **11**, 1937-1943.
Jones, L. W. and Myers, J. (1964). *Pl. Physiol., Lancaster* 39, 938-946.
Kao, O. and Berns, D. S. (1968). *Biochem. biophys. Res. Commun.* 33, 457-462.
Kao, O., Berns, D. S. and MacColl, R. (1971). *Eur. J. Biochem.* 19, 595-599.
Kessel, M., MacColl, R., Berns, D. S. and Edwards, M. R. (1973). *Can. J. Microbiol.* 19, 831-836.
Killilea, S. D. (1971). Doctoral Thesis, National University of Ireland, Galway.
Killilea, S. D. and Ó Carra, P. (1968). *Biochem. J.* **110**, 14-15P.
Kobayashi, Y., Siegelman, H. W. and Hirs, C. H. W. (1972). *Archs Biochem. Biophys.* **152**, 187-198.
Lavorel, J. and Moniot, C. (1962). *J. chem. Phys.* pp. 1007-1012.
Lee, J. J. and Berns, D. S. (1968a). *Biochem. J.* **110**, 457-464.
Lee, J. J. and Berns, D. S. (1968b). *Biochem. J.* **110**, 465-470.
Lemasson, C., Tandeau de Marsac, N. and Cohen-Bazire, G. (1973). *Proc. natn. Acad. Sci. U.S.A.* 70, 3130-3133.
Lemberg, R. (1928). *Justus Liebig's Annln Chem.* **461**, 46-89.
Lemberg, R. (1930). *Justus Liebig's Annln Chem.* **477**, 195-245.
Lemberg, R. and Bader, G. (1933). *Justus Liebig's Annln Chem.* **505**, 151-177.
Lemberg, R. and Legge, J. W. (1949). "Haematin Compounds and Bile Pigments". Interscience, New York.
Lichtlé, C. and Giraud, G. (1970). *J. Phycol.* 6, 281-289.
Lightner, D. A., Moscowitz, A., Petryka, Z. J., Jones, S., Weimer, M., Davis, E., Beach, N. A. and Watson, C. J. (1969). *Archs Biochem. Biophys.* **131**, 566-576.
MacColl, R., Edwards, M. R., Mulks, M. H. and Berns, D. S. (1974). *Biochem. J.* 141, 419-425.
Macdowall, F. D. H., Bednar, T. and Rosenberg, A. (1968). *Proc. natn. Acad. Sci. U.S.A.* 59, 1356-1363.
Miedziejko, E. and Fraçkowiak, D. (1969). *Photochem. Photobiol.* 10, 97-108.
Molisch, H. (1894). *Bot. Ztg* 52, 177-189.
Molisch, H. (1895). *Bot. Ztg* 53, 131-135.
Murphy, R. F. (1968). Doctoral Thesis, National University of Ireland, Galway.
Murphy, R. F. and Ó Carra, P. (1970). *Biochim. biophys. Acta* 214, 371-373.
Myers, J. and Kratz, W. A. (1955). *J. gen. Physiol.* 39, 11-22.
Neufeld, G. J. and Riggs, A. F. (1969). *Biochim. biophys. Acta* **181**, 234-243.
Nichols, K. E. and Bogorad, L. (1962). *Bot. Gaz.* 124, 85-93.
Nolan, D. N. and Ó hEocha, C. (1967). *Biochem. J.* 103, 39-40P.

Ó Carra, P. (1962). Doctoral Thesis, National University of Ireland, Galway.
Ó Carra, P. (1965). *Biochem. J.* **94**, 171-174.
Ó Carra, P. (1970). *Biochem. J.* **119**, 2-3P.
Ó Carra, P. and Colleran, E. (1969). *FEBS Lett.* **5**, 295-298.
Ó Carra, P. and Colleran, E. (1970). *J. Chromat.* **50**, 458-468.
Ó Carra, P. and Killilea, S. D. (1970). *Tetrahedron Lett.* 4211-4214.
Ó Carra, P. and Killilea, S. D. (1971). *Biochem. biophys. Res. Commun.* **45**, 1192-1197.
Ó Carra, P. and Ó hEocha, C. (1962). *Nature, Lond.* **195**, 173-174.
Ó Carra, P. and Ó hEocha, C. (1965). *Phytochemistry* **4**, 635-638.
Ó Carra, P. and Ó hEocha, C. (1966). *Phytochemistry.* **5**, 993-997.
Ó Carra, P., Ó hEocha, C. and Carroll, D. M. (1964). *Biochemistry, N.Y.* **3**, 1343-1350.
Ó Carra, P., Killilea, S. D. and Nolan, D. (1974). In preparation.
Ó hEocha, C. (1958). *Archs Biochem. Biophys.* **73**, 207-219.
Ó hEocha, C. (1960). In "Comparative Biochemistry of Photoreactive Systems" (M. B. Allen, ed.), pp. 181-203. Academic Press, New York and London.
Ó hEocha, C. (1962). In "Physiology and Biochemistry of Algae" (R. A. Lewin, ed.), pp. 421-435. Academic Press, New York and London.
Ó hEocha, C. (1963). *Biochemistry, N.Y.* **2**, 375-382.
Ó hEocha, C. (1965a). *A. Rev. Pl. Physiol.* **16**, 415-434.
Ó hEocha, C. (1965b). In "Chemistry and Biochemistry of Plant Pigments" (T. W. Goodwin, ed.), 1st edn., pp. 175-196. Academic Press, London and New York.
Ó hEocha, C. (1966). In "Biochemistry of Chloroplasts" (T. W. Goodwin, ed.), vol. 1, pp. 407-421. Academic Press, London and New York.
Ó hEocha, C. (1968). In "Porphyrins and Related Compounds" (T. W. Goodwin, ed.), pp. 91-105. Academic Press, London and New York.
Ó hEocha, C. and Curley, D. (1961). *Abst. 5th int. Congr. Biochem.* Sec. 22-20, p. 447.
Ó hEocha, C. and Ó Carra, P. (1961). *J. Am. chem. Soc.* **83**, 1091-1093.
Ó hEocha, C. and Raftery, M. (1959). *Nature, Lond.* **184**, 1049-1051.
Ó hEocha, C., Ó Carra, P. and Mitchell, D. (1964). *Proc. R. Ir. Acad.* **63(B)**, 191-200.
Rabinowitch, E. I. (1945). "Photosynthesis", vol. 1. Interscience, New York.
Raftery, M. A. and Ó hEocha, C. (1965). *Biochem. J.* **94**, 166-170.
Rice, H. V. and Briggs, W. R. (1973a). *Pl. Physiol., Lancaster* **51**, 927-938.
Rice, H. V. and Briggs, W. R. (1973b). *Pl. Physiol., Lancaster* **51**, 939-945.
Rüdiger, W. (1971). *Fortschr. Chem. org. NatStoffe* **29**, 61-139.
Rüdiger, W. and Correll, D. L. (1969). *Justus Liebig's Annln Chem.* **723**, 208-212.
Rüdiger, W. and Ó Carra, P. (1969). *Eur. J. Biochem.* **7**, 509-516.
Rüdiger, W., Ó Carra, P. and Ó hEocha, C. (1967). *Nature, Lond.* **215**, 1477-1478.
Rüdiger, W., Kost, H. P., Budzikiewicz, H. and Kramer, V. (1970). *Justus Liebig's Annln Chem.* **738**, 197-201.
Schram, B. L. (1970). *Biochem. J.* **119**, 15P.
Schram, B. L. and Kroes, H. H. (1971). *Eur. J. Biochem.* **19**, 581-594.
Siegelman, H. W., Turner, B. C. and Hendricks, S. B. (1966). *Pl. Physiol., Lancaster* **41**, 1289-1292.

Siegelman, H. W., Chapman, D. J. and Cole, W. J. (1967). *Archs Biochem. Biophys.* 122, 261.
Siegelman, H. W., Chapman, D. J. and Cole, W. J. (1968). In "Porphyrins and Related Compounds" (T. W. Goodwin, ed.), pp. 107-120. Academic Press, London and New York.
Teale, F. W. J. and Dale, R. E. (1966). *Proc. 2nd Western Eur. Conf. Photosyn.* 165-168.
Teale, F. W. J. and Dale, R. E. (1970). *Biochem. J.* 116, 161-169.
Tenhunen, R., Marver, H. S. and Schmid, R. (1969). *J. biol. Chem.* 244, 6388-6394.
Tenhunen, R., Marver, H., Pimstone, N. R., Trager, W. F., Cooper, D. Y. and Schmid, R. (1972). *Biochemistry, N.Y.* 11, 1716-1720.
Torjesen, P. A. and Sletten, K. (1972). *Biochim. biophys. Acta* 263, 259-271.
Troxler, R. F. (1972). *Biochemistry, N.Y.* 11, 4235-4242.
Troxler, R. F. and Brown, A. (1970). *Biochim. biophys. Acta* 215, 503-511.
Troxler, R. F. and Dokos, J. M. (1973). *Pl. Physiol., Lancaster* 51, 72-75.
Troxler, R. F. and Lester, R. (1967). *Biochemistry, N.Y.* 6, 3840-3846.
Troxler, R. F. and Lester, R. (1968). *Pl. Physiol., Lancaster* 43, 1937-1939.
Troxler, R. F., Brown, A., Lester, R. and White, P. (1970). *Science, N.Y.* 167, 192-193.
Vaughan, M. H. (1963). *Fedn. Proc. Fedn. Am. Socs exp. Biol.* 22, 681.
Vaughan, M. H. (1964). Doctoral Thesis, Massachusetts Institute of Technology, Cambridge, Mass.
van der Velde, H. H. (1973). *Biochim. biophys. Acta* 303, 246-257.
Vernon, L. P. (1961a). *Acta chem. scand.* 15, 1639-1650.
Vernon, L. P. (1961b). *Acta chem. scand.* 15, 1651-1659.
Vernotte, C. and Moya, I. (1973). *Photochem. Photobiol.* 17, 245-254.
Vernotte, C. (1971). *Photochem. Photobiol.* 14, 163-173.

Chapter 7

The Structure and Properties of Phytochrome

HARRY SMITH

Department of Physiology and Environmental Studies, University of Nottingham School of Agriculture, Loughborough, Leicestershire, England

and RICHARD E. KENDRICK

Department of Plant Biology, University of Newcastle-upon-Tyne, Newcastle, England

I Introduction

Phytochrome is a blue-green photochromic pigment which controls a wide range of developmental and metabolic processes in green plants. First demonstrated *in vitro* in 1959 (Butler *et al.*, 1959), phytochrome is a protein with a linear tetrapyrrole chromophore which exists in two forms: P_r, which absorbs maximally in the red region of the spectrum; and P_{fr}, which absorbs maximally in the far-red. These two forms are interconvertible upon absorption of radiant energy and it is assumed that the P_{fr} form is responsible for the initiation of the metabolic and developmental changes. The structure and properties of phytochrome have been comprehensively reviewed in recent years (Briggs and Rice, 1972; Mitrakos and Shropshire, 1972; Shropshire, 1972; Briggs, 1974). However, much of the information obtained before 1972 is now known to relate to partly purified fragments of phytochrome, formed proteolytically during the extraction process. It is now possible to isolate and purify non-degraded phytochrome, at least from certain plant sources (see Chapter 23), and thus this review concentrates on the structure and properties of the native molecule, as far as is possible.

II The protein moiety

The discovery by Briggs and co-workers that most pure phytochrome preparations previously investigated were in fact collections of polypeptide fragments still retaining photoreversibility has enabled a major rationalization of the available data to be made (see Briggs and Rice, 1972). It is useful to chart, briefly, the sequence of events leading to this rationalization. In 1968, Briggs, Zollinger and Platz observed two photoreversible fractions on gel exclusion chromatography of freshly prepared oat phytochrome. The gels were inadequately calibrated, but estimated molecular weights were 80 and 180 x 10^3 daltons. Following up this work, Gardner *et al.* (1971) investigated the possibility that the large phytochrome was due to aggregation of native low molecular weight molecules. However, alterations of pH, salt concentration or phytochrome concentration did not result in the formation of any large phytochrome from the low molecular weight fraction. Nevertheless, the basic observation of the presence of two fractions of differing molecular weight was confirmed; the larger fraction had a sedimentation coefficient of 9*s* whilst that for the smaller fraction was 4·5*s*.

Meanwhile, Pringle (1970) had shown that yeast malate dehydrogenase could retain catalytic activity even after proteolytic attack which resulted in modification of the physical properties of the protein. With this in mind Gardner *et al.* (1971) proceeded to show that the larger fraction of phytochrome is, in fact, the native molecule, the smaller fraction being formed during extraction and purification by proteolysis. The conversion of large into small phytochrome occurred during storage at 4°C, and was almost quantitative. Breakdown occurred more rapidly in preparations of oat phytochrome than in similar preparations from rye. The phytochrome breakdown could be inhibited by phenylmethanesulphonylfluoride (PMSF), a known inhibitor of proteolysis. Furthermore, a wide range of commercial endopeptidases (including trypsin) were capable of converting purified large molecular weight phytochrome to small phytochrome.

Subsequently Pike and Briggs (1972) purified 600-fold a neutral endoprotease from etiolated oat shoots. The estimated molecular weight of the protease is 61 500 daltons. The enzyme is apparently dependent on reduced sulphydryl groups for activity. The uncharged sulphydryl reductants 2-mercaptoethanol and dithiothreitol caused strong enhancement of activity (e.g. 10^{-2}-10^{-1} M 2-mercaptoethanol stimulated activity about threefold) whereas the charged reagents glutathione, sodium metabisulphite and sodium dithionite inhibited activity at between 5 and 20 mM. High ionic strength reduces activity, as do both $HgCl_2$ and PMSF. The conditions which stimulate or, at least, allow, protease activity are generally similar to those normally used in the isolation of phytochrome. Thus proteolysis is to be expected and attempts should be made to reduce its extent (see Chapter 23). In practice it is simpler to use, as starting material, a tissue low in protease. One such is etiolated rye shoots, and much of the following discussion relates to the properties of the native large phytochrome from rye as elucidated, principally, by the elegant experiments of Briggs' group.

A MOLECULAR WEIGHT

Prior to 1972, a wide range of phytochrome molecular weights had been reported (see Table I). With the exception of the results of Walker and Bailey (1970a,b), however, none of the oat phytochrome determinations were less than 55 000–60 000 daltons. The findings of Gardner *et al.* (1971) and Pike and Briggs (1972) suggest that all of these measurements were of proteolytically produced fragments of the native molecule. The phytochrome preparations investigated by Walker and Bailey (1970a,b) must have been even more extensively degraded, although they still retained photoreversibility. If they represent, as seems likely, small polypeptide fragments attached to photoreversible chromophores, they would repay more detailed analysis.

TABLE I
Some molecular weight estimates of phytochrome published prior to 1972

Plant source	Method used	Molecular weight $\times 10^3$ daltons	Reference
Oat	Velocity sedimentation	90–150	Siegelman and Firer (1964)
Oat	Gel exclusion chromatography	55	Mumford and Jenner (1966)
	Equilibrium ultracentrifugation	60	
Oat	Gel exclusion chromatography	80, 180	Briggs *et al.* (1968)
Oat	Gel exclusion chromatography	55	Kroes (1970)
Oat	Velocity sedimentation	18, 23, 130	Walker and Bailey (1970a)
Oat	Velocity sedimentation	26, 79, 127	Walker and Bailey (1970b)
Oat	Gel exclusion chromatography	70, 110	Roux (1971)
Oat	Equilibrium ultracentrifugation	60	Hopkins (1971)
	Velocity sedimentation	60	
	SDS-polyacrylamide gel electrophoresis	69	
Oat	Velocity sedimentation	60, 120	Gardner *et al.* (1971)
Rye	Velocity sedimentation; Equilibrium ultracentrifugation	42, 180	Correll *et al.* (1968)
Pea	Velocity sedimentation	113, 265	Walker and Bailey (1970b)

Rice *et al.* (1973) purified oat and rye phytochrome to homogeneity. The two preparations were significantly different in elution from DEAE-cellulose columns and in mobility on molecular sieve gels. The oat preparation gave a molecular weight on calibrated SDS-polyacrylamide gels of 62 000 daltons, whilst the rye phytochrome ran at 120 000 daltons using the same technique. Oat phytochrome had a partition coefficient (σ_{200}) on Sephadex G200 of 0·350 with an estimated molecular weight of 62 000 daltons. Rye phytochrome, on the other hand, had a σ_{200} of 0·085, with an estimated molecular weight of 375 000 daltons.

The molecular weight discrepancy for the rye phytochrome suggests a multimer structure for the non-denatured protein. Gardner *et al.* (1971), however, using sucrose density gradient centrifugation reported a multimer with an *s* value of 9, corresponding to an estimated molecular weight of 180 000 daltons. Rice and Briggs (1973) also reported preliminary equilibrium centrifugation studies by Gardner (as yet unpublished) which indicate a molecular weight of roughly 240 000 daltons. The resolution of these anomalies may lie in the suggestion of Rice and Briggs (1973) that rye phytochrome does not behave as a globular protein, and thus the observed molecular weights will depend on whether or not the method used is shape dependent.

Table II summarizes the molecular weights of purified native rye phytochrome as determined by Briggs' group using the different methods. Thus the values of 180 000 and 375 000 daltons may be artefacts of shape-dependent methods. SDS-gel electrophoresis has proved an extremely reliable technique for the estimation of the molecular weight of polypeptide chains (Weber and Osborn, 1969). The 240 000 value obtained by equilibrium ultracentrifugation may represent a favoured dimerization of the basic 120 000 mol. wt subunit. As pointed out by Rice and Briggs (1973) the major remaining discrepancy relates to the 42 000 mol. wt value observed by Correll *et al.* (1968) for rye phytochrome. It seems impossible to reconcile this with the 120 000 mol. wt observed by Rice and Briggs unless it is assumed that extensive proteolysis occurred. Rice and Briggs, in fact, observed a 42 000 mol. wt fraction after treatment of 120 000 mol. wt rye phytochrome with either trypsin or the oat protease.

Thus the most likely molecular weight for native, undegraded rye and oat phytochrome is 120 000 daltons. The native molecule is unlikely to be globular, although it is interesting that the photoreversible fragments of about 60 000 daltons produced by proteo-

TABLE II
The molecular weight of highly purified rye phytochrome as measured by different procedures

Method		Mol. wt (daltons)
Velocity sedimentation	Shape dependent	180 000
Molecular exclusion chromatography	Shape dependent	375 000
Equilibrium ultracentrifugation	Shape independent	240 000
SDS-gel electrophoresis	Shape independent	120 000

lysis seem to be globular. Mumford and Jenner (1966), Hopkins (1971) and Rice and Briggs (1973) obtained consistent values of molecular weight for oat phytochrome using both shape-dependent and shape-independent techniques. Thus a non-globular protein yields a globular polypeptide upon proteolysis. Amongst many possible interpretations of this behaviour is an intriguing idea of Briggs (personal communication) in which native phytochrome is thought to exist as a dumb-bell shaped molecule. Electron micrographs of dried high molecular weight phytochrome prepared by J. M. Mackenzie (in Briggs' group) supporting this concept of the molecular structure of phytochrome can be seen in the book by Smith (1975).

At present, it is not possible to state with certainty whether or not there is more than one kind of 120 000 mol. wt phytochrome subunit. Rice and Briggs (1973) carried out N-terminal amino acid analysis by the Sanger (1945) procedure and obtained both glutamate and aspartate. At face value, this suggests the existence of two different polypeptides, but Rice and Briggs urge caution in interpretation partly because of the possibility of slight proteolysis, but mainly because of problems experienced by other workers in using the Sanger procedure on purified phycocyanin (Cope *et al.*, 1967). In this case, glutamate and aspartate were indicated as N-terminal amino acids, whereas the cyanate technique of Stark and Smyth (1963) yielded only methionine.

B AMINO ACID ANALYSIS

The amino acid analysis of 120 000 mol. wt rye phytochrome obtained by Rice and Briggs (1973) is given in Table III. It is significantly different from that previously published for rye phyto-

TABLE III
Amino acid composition of oat phytochrome (62 000 mol. wt) and rye phytochrome (120 000 mol. wt). (Data from Rice and Briggs, 1973)

Amino acid	Residues per molecule	
	Oat	Rye
Lysine	41	58
Histidine	19	28
Arginine	26	47
Aspartic acid	50	104
Threonine	45	46
Serine	37	75
Glutamic acid	63	128
Proline	30	88
Glycine	27	77
Alanine	43	110
Half-cystine	10	26
Valine	42	89
Methionine	31	32
Isoleucine	27	54
Leucine	57	111
Tyrosine	15	23
Phenylalanine	25	43
Total residues	588	1139

chrome by Correll *et al.* (1968). Briggs and Rice (1972) suggest that the reason for the disagreement may be the presence of contaminants in the Correll *et al.* preparations. There appears to be considerable unanimity on the amino acid analysis of 60 000 mol. wt oat phytochrome as determined by Mumford and Jenner (1966), Roux (1971) and Rice and Briggs (1973) (Table III).

The differences between the large rye and small oat amino acid analyses may be related to the loss of protein upon proteolysis. On the other hand, some general similarities emerge: both acidic and basic amino acids are present in high amounts; there are at least 11 cystines per 60 000 mol. wt; and the threonine and serine contents are relatively high. Thus phytochrome is potentially a highly reactive protein.

From the amino acid analyses Rice and Briggs (1973) calculated the partial specific volumes of oat 62 000 mol. wt phytochrome and rye 120 000 mol. wt phytochrome to be 0·736 cm^3/g and 0·728 cm^3/g respectively.

C PROTEIN CONFORMATION

Tobin and Briggs (1973) have recently analysed the circular dichroism spectra of purified 120 000 mol. wt rye phytochrome in order to obtain information on the conformation of the protein. The method used was that of Greenfield and Fasman (1969), which is based on computed circular dichroism curves in the 190–250 nm region for combinations of experimentally determined curves for polylysine in the α helix, random coil and pleated sheet (β structure) conformations. On this method, rye phytochrome has about 17–20% α helix, 30% β structure and 50% random coil. Small rye phytochrome had about 10–13% α helix.

D CARBOHYDRATE CONTENT OF PHYTOCHROME

Evidence that phytochrome is a glycoprotein has been obtained by Roux (1971) (cited in Briggs and Rice, 1972). The total amount of carbohydrate covalently bound to highly purified oat phytochrome (*c.* 60 000 mol. wt) was between 3 and 3·5%. Estimation of sugar content was based on the phenol–sulphuric acid method (Dubois *et al.*, 1956) using a washed TCA precipitate of 90–95% pure phytochrome (A_{280}/A_{667} = 1·15). This value of 3·5%, which could represent up to 23 sugar residues, is probably an underestimate, bearing in mind the high content of sugar-degrading enzymes in crude plant extracts, and the proteolysis of the native phytochrome that had occurred during extraction.

A rough estimate of the number of sugar chains can be made by counting the number of glucosamines and galactosamines present in the protein. Analysis has shown 2 glucosamines and 1 galactosamine per 120 000 mol. wt phytochrome subunit, indicating the presence of 3 carbohydrate chains (S. J. Roux, personal communication). Since these estimations were calculated from data on 60 000 mol. wt preparations, it would be interesting to attempt confirmation by analysis of native rye phytochrome. Briggs and Rice (1972) point out that the presence of significant amounts of carbohydrate can effect protein mobility, and thus molecular weight estimation, on SDS–acrylamide gels. They conclude, however, that the relatively small carbohydrate content of 3·5% is unlikely to have serious effects on the estimation of molecular weight.

E IMMUNOCHEMISTRY OF PHYTOCHROME

Antisera to 60 000 mol. wt oat phytochrome have been raised by several groups (Hopkins and Butler, 1970; Pratt and Coleman, 1971;

Pratt, 1973; Rice and Briggs, 1973b); antisera to high molecular weight phytochrome have only been reported once to date (Cundiff and Pratt, 1973). Hopkins and Butler (1970) reported differences between P_r and P_{fr} on microcomplement fixation, but none of the latter reports confirm this finding (see Section VII).

Using antibody to small oat phytochrome, both Pratt (1973) and Rice and Briggs (1973b) observed lines of identity between oat and rye phytochromes on double diffusion plates (See Chapter 23, Section IIG for technique.) Pratt (1973) also showed antigenic similarity between barley and oat phytochromes. Both groups, however, observed antigenic differences between pea and oat phytochromes.

Cundiff and Pratt (1973) used antisera to large and small oat phytochrome preparations to study the differences between the two preparations. They showed that large phytochrome contains at least two antigenically identifiable moieties in addition to those in the 60 000 mol. wt chromophore-containing polypeptides. By following the antigenic behaviour of large oat phytochrome during proteolysis either by the endogenous protease or by trypsin, a number of polypeptides could be recognized, each of which showed antigenic similarity to large phytochrome, but only one showing antigenic similarity to small phytochrome. One of the fractions, not identical to small phytochrome, was seen to elute from Bio-Gel P-200 ahead of small phytochrome and had no spectral photoreversibility. Thus at least one of the proteolytically derived fragments of native phytochrome appears to be larger than 60 000 mol. wt, but does not contain the chromophore. These findings reinforce the conclusions of Gardner *et al.* (1971) that proteolysis of phytochrome leads to the production of a family of polypeptides, one of which has a molecular weight of 60 000 daltons, contains the chromophore, and is relatively stable. It is this breakdown product which has been assiduously purified in the past.

III The phytochrome chromophore

The first proposal concerning the chemical identity of the phytochrome chromophore came from the Belstville group in 1950, nine years before the first isolation of the pigment (Borthwick *et al.*, 1950; Parker *et al.*, 1950). Based on the similarities between the action spectra of the physiological responses and the absorption spectra of the algal photosynthetic pigments, phycocyanin and allophycocyanin, they suggested that phytochrome was probably a linear tetrapyrrole, a prediction which has turned out to be remarkably accurate.

A EVIDENCE FROM CHROMOPHORE ISOLATION

As yet it has proved extremely difficult to separate the phytochrome chromophore from its associated protein. Siegelman *et al.* (1966) denatured the protein with trichloroacetic acid (TCA) and refluxed for 3–4 hours in acidic methanol to obtain at most 10% of the chromophore in a free form. Methanolic denaturation bleached the pigment and thus was unsatisfactory. Hydrolysis of the TCA-denatured pigment with alkali or 12 M-HCl was unsuccessful, although such methods are effective with algal biliproteins. In a later study, Kroes (1970) was similarly unable to obtain reasonable yields of chromophore. Observations with phycocyanin as a model compound had shown that the algal chromophore was bound to the apoprotein via a serine residue. Such linkages are susceptible to hydrolysis with hydrogen bromide in trifluoroacetic acid solution and it had been hoped that phytochrome would be similarly hydrolysable. The fact that HBr was not effective suggests that the phytochrome chromophore is not coupled to the apoprotein via a serine or threonine residue. Kroes, starting with a purified phytochrome preparation containing 300 mg protein, obtained only 20 μg of chromophore after TCA denaturation, refluxing in methanol, washing in chloroform and chromatography on silica gel. After methanolic extraction, the protein residue was still pale green, indicating that significant amounts of chromophore had not been cleaved.

Siegelman *et al.* (1966) showed that the absorption spectra of the chromophores of phytochrome, phycocyanin and allophycocyanin were similar but not identical; the phytochrome chromophore had absorption maxima at 380 and 690 nm, compared to 375 and 685 nm for the algal chromophores, both of which were later shown to be phycocyanobilin (Siegelman *et al.*, 1968). Thin-layer chromatography of esterified and partially esterified chromophores also showed similarity, but not identity, between phytochrome and the algal pigments. Chromatographic separation of partly esterified phytochrome chromophore with lutidine-H_2O-ammonia as solvent, which is known to separate porphyrin pigments according to the number of carboxyl groups present, was consistent with the presence of two carboxyl groups as free acids, mono- and dimethyl-esters. These workers concluded that the phytochrome chromophore was a bilitriene (i.e. a linear tetrapyrrole with three methyne bridges) and was probably similar to, but not identical with, the algal chromophores. In a later paper Siegelman *et al.* (1968), using the above information and a comparison with detailed studies of phycocyanobilin, proposed a model for the complete structure of the phytochrome chromophore, both in the P_r and the P_{fr} form. These structures are shown in Fig. 1, although discussion of the proposed mechanism of phototransformation is left till Section VII.

Kroes (1970) used the same methods but with rather purer phytochrome as a starting point, and confirmed the results of Siegelman *et al.* (1966). Walker and Bailey (1968), on the other hand, reported differences in chromophore absorption spectra depending on whether the phytochrome was in the P_r or the P_{fr} form upon denaturation and extraction. Kroes (1970), however, discounted this claim on the grounds that any differences observed were probably due to impurities in the raw methanol extracts which could be removed by purification on a silica gel column.

The lack of an effective hydrolysis technique has prevented the further chemical analysis of free chromophore and thus the proposed structure in Fig. 1 can only be considered tentative.

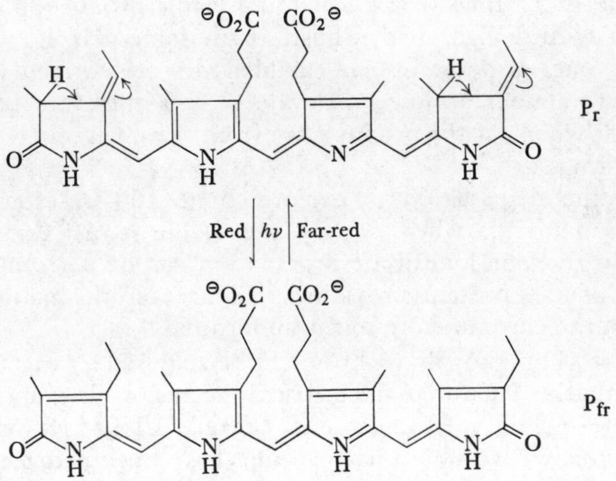

FIG. 1. Structure of the phytochrome chromophore as proposed by Siegelman *et al.* (1968).

B CHROMOPHORE DEGRADATION *IN SITU*

A method used successfully in the elucidation of the chromophoric structure of other bile pigments is oxidation of the chromophore *in situ* followed by analysis of the products. In this technique, the purified biliprotein is treated with chromic acid, whereupon the pyrroles are oxidized to maleimides with unchanged β substituents:

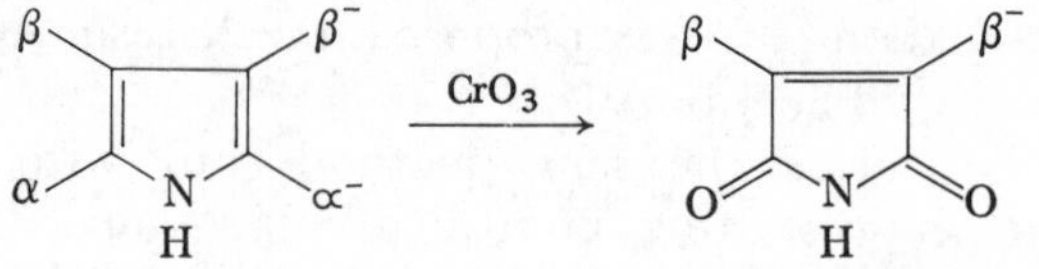

$$\beta, \beta^-, \alpha, \alpha^- \text{ (pyrrole, N–H)} \xrightarrow{CrO_3} \beta, \beta^- \text{ (maleimide, O, N–H, O)}$$

In addition to circumventing the problem of the cleavage of the chromophore from the apoprotein, this technique has several important advantages for the study of phytochrome:

(1) At pH 1·5–1·7, bile pigments are quickly degraded whilst porphyrins and chlorophylls (i.e. ringed tetrapyrroles) are relatively stable; this enables an unknown pyrrole pigment to be characterized as a bile pigment simply by its degradation kinetics.

(2) The outer rings of a linear tetrapyrrole produce imides upon oxidation with a high yield, whilst, at the same pH, the yield from the inner rings is poor; instead, dialdehydes are formed which are oxidized to imides at lower pH values. It is therefore possible to distinguish which of the products are from the inner and which from the outer rings.

(3) If the degradation is performed at 100°C, ester linkages become saponified, whereas they are stable at 20°C; thus both hydrolytic and non-hydrolytic degradation can be performed in the presence of the protein providing evidence on the nature of the binding of the chromophore to the apoprotein.

The first sample of phytochrome to be analysed by this technique was apparently denatured since it was greenish-yellow and had lost its photoreversibility (Rüdiger and Correll, 1969). However, later work on freely reversible phytochrome has confirmed most of the details (Rüdiger, 1972). The denatured greenish-yellow phytochrome used by Rüdiger and Correll (1969) could be converted into a blue form by cold acid and reconverted into the greenish-yellow form by alkali. On the basis of these absorbance changes it was assumed that the greenish-yellow form corresponded to P_{fr} and the blue form to P_r, although clearly it could not be shown that the chromophores were unchanged from the native pigment.

Both forms yielded identical products from three of the four pyrrole rings as shown in Fig. 2:

(i) Ring IV yielded an unidentified imide at 20°C from the blue and the yellow denatured forms, suggesting it was free from the protein.

(ii) Ring III yielded haematinic acid imide only upon exhaustive oxidation at 100°C, suggesting that it was covalently bounded to the protein.

(iii) Ring II yielded the dialydehyde of haematinic acid at 20°C, indicating that it was free from the protein.

(iv) Only ring I showed different products from the blue and yellow forms of the denatured pigment. The blue form gave methylethylidene-succinimide at 20°C, and must therefore contain an exocyclic double bond and be free from the protein. In the yellow form, ring I gave methylethylmaleimide with an endocyclic double bond, but only under hydrolytic conditions, suggesting that ring I is covalently bound to the protein.

Briggs and Rice (1972) have questioned the designation of ring III as the binding point with the apoprotein, pointing out that ring II and ring III both have the same β substituents (methyl and propionic acid) and could form dialdehydes or imides upon oxidation. However, Rüdiger and Correll (1969) do not state that the point has been settled. Ring III is the preferred choice by analogy with the case of phycoerythrin in which the linkage has been shown to be at ring III. Certainly, the chromic acid oxidation technique could not distinguish between ring II and ring III in symmetrical bilins as in phytochrome and phycocyanin (W. Rüdiger, personal communication).

Subsequently, Rüdiger (1972) was able to repeat his analysis using freely reversible native phytochrome. The products of these oxidations are also shown in Fig. 2. The differences between the denatured and the native forms are restricted to rings I and IV. P_r and P_{fr} both gave methylvinylmaleimide at 20°C from ring IV whereas the denatured forms had yielded an unknown imide. No degradation products have yet been identified from ring I of P_r and P_{fr}, suggesting that there may be a linkage to the protein which is not cleaved by the hydrolytic degradation. Brief alkali treatment, however, sufficiently denatures the pigment to allow the chromic acid oxidation to take place, and the products are identical to those formed from the original denatured preparation. A further alkali treatment cleaved the chromophore from the protein, during which process an endocyclic double bond is formed.

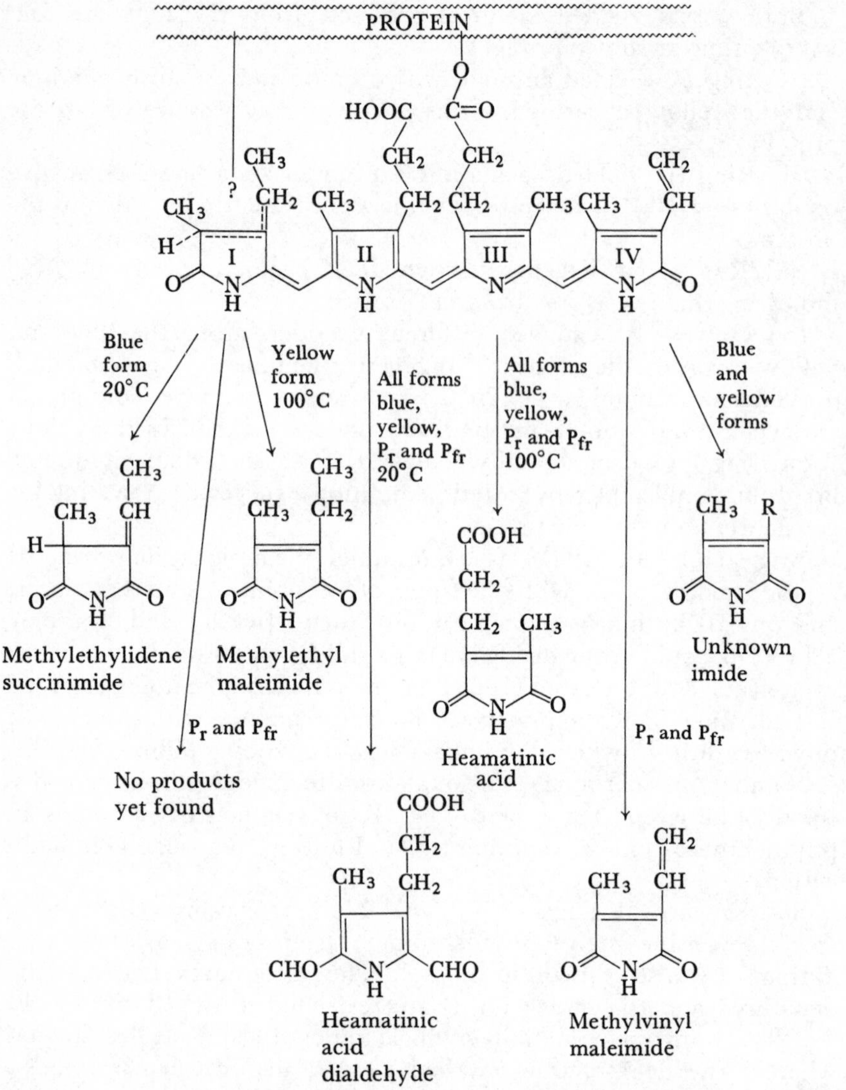

FIG. 2. The chromic acid oxidation of the phytochrome chromophore *in situ* as carried out by Rüdiger and Correll (1969) and Rüdiger (1972).

On the basis of this elegant work, Rüdiger has proposed a complete structure for the chromophores of the blue and greenish-yellow denatured forms, which are presumably analogous to P_r and P_{fr} (Fig. 3) (Rüdiger and Correll, 1969; Rüdiger, 1972). The major difference between this model and that of Siegelman *et al.* (1968) is

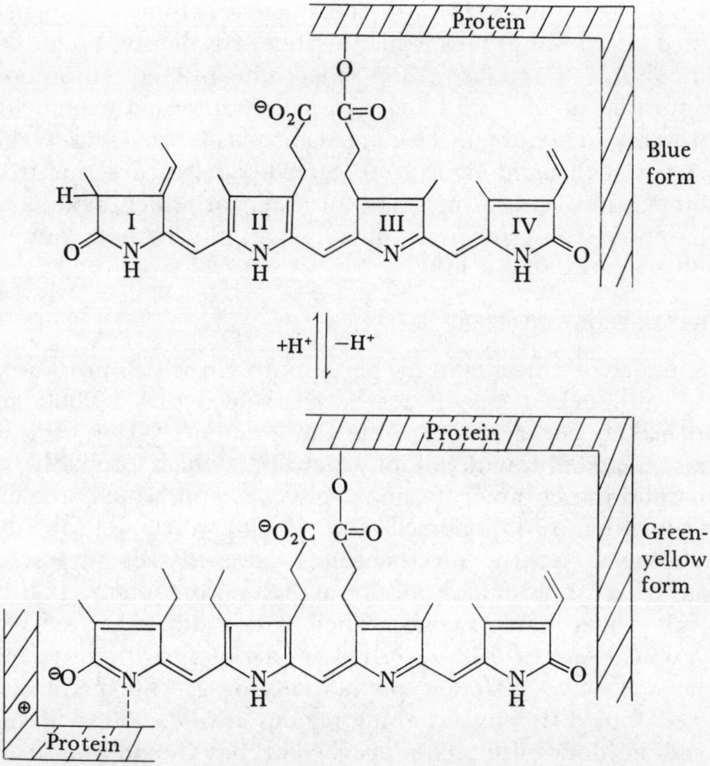

FIG. 3. Structure of the phytochrome chromophore as proposed by Rüdiger (1972).

the presence of the vinyl, as opposed to the ethylidene substituent on ring IV. The only question remaining to be answered is the nature of the binding of ring I to the protein. A thioether of the ethylidene group would be analogous to the corresponding linkage in phycobiliproteins, but there is a difference in stability of this linkage between phytochrome and the phycobiliproteins (W. Rüdiger, personal communication). Either this means that phytochrome contains an additional linkage at ring I, or that the bond is not a thioether, but a more stable bond instead.

C AMINO ACID SEQUENCE AT CHROMOPHORE BINDING SITE

The exact nature of the binding of the chromophore is as yet unknown, but attempts have been made to determine the amino acid sequence of peptide fragments containing the chromophore. Fry and

Mumford (1971) digested purified oat phytochrome with pepsin and obtained a non-photoreversible peptide covalently bound to the chromophore. The absorption spectrum of the chromopeptide showed peaks at 650, 292 and 370 nm. Amino acid sequencing gave the primary structure as: leu-arg-ala-pro-his-(ser,cys)-his-leu-glu-tyr. Treatment with aminopeptidase M readily released serine from the chromopeptide, indicating that an ester or ether linkage of the chromophore to serine was unlikely and the authors preferred a thioether linkage to cystine.

D CHROMOPHORE NUMBER

The number of chromophores per apoprotein is still not known for certain, although it has always been assumed that 60 000 mol. wt phytochrome contains only one chromophore. If so, this would contrast markedly with allophycocyanin, which probably has 12 chromophores per molecule, and C-phycocyanin, which probably has 22 (Ó hEocha, 1965). Correll *et al.* (1968), working probably with 120 000 mol. wt rye phytochrome, have in fact suggested that several different chromophores, or at least chromophores in different molecular environments, exist. Their data indicate the existence of two populations of P_{fr}, reverting in darkness with differing rate constants to two different populations of P_r. Furthermore, they detected four different absorbing regions at 580, 660, 670 and 730 nm; sodium dodecylsulphate bleached all but the 580 nm band and destroyed photoreversibility, whereas glutaraldehyde removed the 730 nm band without affecting photoreversibility.

From this information Correll *et al.* (1968) constructed a rather complex model with four spectrally distinct chromophoric species which can interact by coupled oxidation-reduction reactions between the 580–660 nm and 670–730 nm pairs of chromophores.

IV Spectrophotometric properties

A ABSORPTION SPECTRUM

The absorption spectrum of native 120 000 mol. wt phytochrome extracted from rye is shown in Fig. 4. (Rice *et al.*, 1973). The absorption maximum of P_r is at 666 nm and that of P_{fr} at 730 nm, with minor absorption maxima at 365 nm and 495 nm respectively. As mentioned above, much of the early work was conducted on the degraded form of phytochrome (*c.* 60 000 mol. wt), which makes comparisons of spectral characteristics reported in the literature

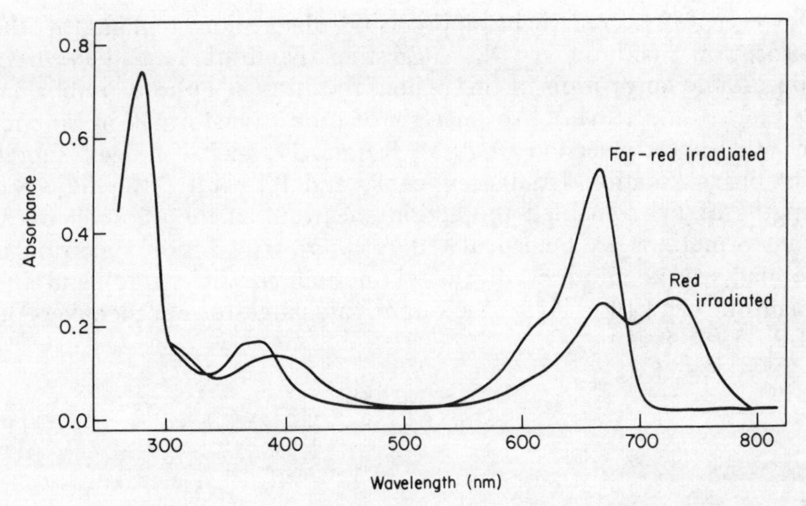

FIG. 4. Absorption spectra of large rye phytochrome after a saturating exposure to red and far-red light. (Rice *et al.*, 1973.)

difficult. An example is the shift of the absorption maximum of P_{fr} from 730 nm to 725 nm on proteolytic degradation in the case of rye (Rice *et al.*, 1973; Rice and Briggs, 1973a). Fortunately, species differences do not appear to be great in angiosperms so far studied, although they may be considerable in algae and bryophytes (Taylor and Bonner, 1967) and in gymnosperms (Grill and Spruit, 1972), where the absorption maxima of P_r and P_{fr} are at lower wavelengths.

Another difficulty is the comparison of *in vitro* and *in vivo* spectral characteristics. The absorption maximum of P_r, and particularly of P_{fr}, occurs at longer wavelengths *in vivo* than *in vitro*. In the case of oats this amounts to a 10 nm shift for P_{fr} (Everett and Briggs, 1970). The maximum in the absorption spectrum at 280 nm (Fig. 4) corresponds to the amino acid residues in the protein moiety. With knowledge of the molecular weight, the molar extinction coefficients (ϵ) of large rye phytochrome have been calculated (Tobin and Briggs, 1973):

$$\epsilon_{280}^{\mathrm{Pr}} = \epsilon_{280}^{\mathrm{Pfr}} = 9 \times 10^4 \ \mathrm{l\ mol^{-1}\ cm^{-1}}$$
$$\epsilon_{665}^{\mathrm{Pfr}} = 7 \times 10^4 \ \mathrm{l\ mol^{-1}\ cm^{-1}}$$
$$\epsilon_{730}^{\mathrm{Pfr}} = 4 \times 10^4 \ \mathrm{l\ mol^{-1}\ cm^{-1}}$$

B ACTION SPECTRUM

The absorption spectrum of phytochrome (Fig. 4) indicates that after far-red irradiation P_{fr} is driven strongly towards the P_r form.

However, after red light appreciable absorption remains at the absorption maximum of P_r, suggesting that both P_r and P_{fr} have appreciable absorption in this region, resulting in a photoequilibrium (Butler *et al.*, 1964a). An analysis of phototransformation kinetics showed that conversion of P_r to P_{fr} and P_{fr} to P_r followed simple first-order kinetics. Irradiation of P_r and P_{fr} with different wavelengths of light enabled the action spectrum of the photochemical transformations to be calculated (Fig. 5). The action spectrum is defined as the product of the extinction coefficient (ϵ) and the quantum yield (Φ). This value can be calculated for photoconversion of P_r

$$\epsilon_\lambda^{\mathrm{Pr}} \Phi_r = \frac{k_\lambda}{I_\lambda} \frac{[P_{fr_\infty}]_\lambda}{[P]} \tag{1}$$

and similarly for P_{fr}

$$\epsilon_\lambda^{\mathrm{Pfr}} \Phi_{fr} = \frac{k_\lambda}{I_\lambda} \frac{[P_{r_\infty}]_\lambda}{[P]} \tag{2}$$

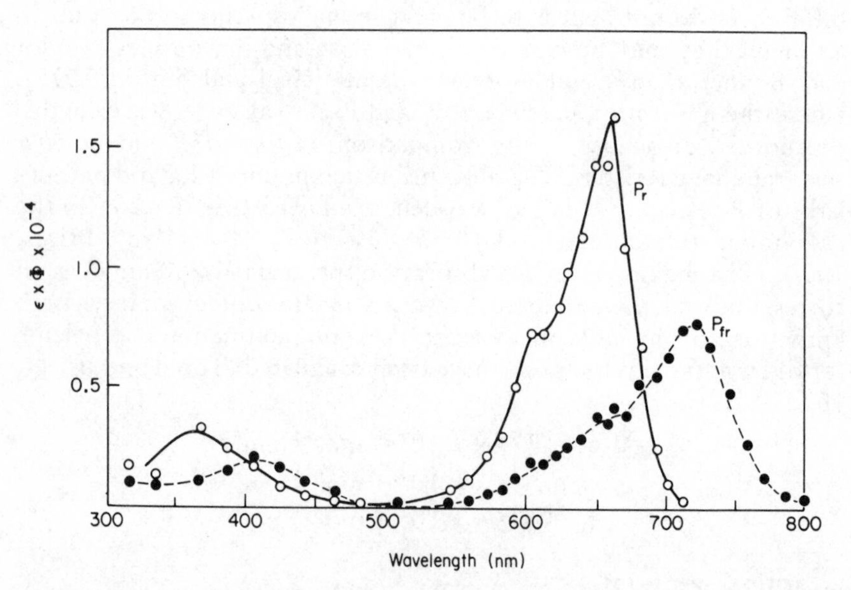

FIG. 5. Action spectrum of photochemical transformations of P_r and P_{fr} in a purified sample from oat. ϵ is extinction coefficient in 1 mol^{-1} cm^{-1}, and Φ is the quantum yield in mol Einstein^{-1}. (Butler *et al.*, 1964a.)

where ϵ_λ^{Pr} and ϵ_λ^{Pfr} are the extinction coefficients; Φ_r and Φ_{fr} are the quantum yields for transformation of P_r and P_{fr}; k_λ is the first-order rate constant of photoconversion with intensity (I) of wavelength (λ); and $[P_{r\infty}]_\lambda/[P]$ and $[P_{fr\infty}]_\lambda/[P]$ are the fractions of the P_r and P_{fr} present at photoequilibrium under wavelength λ.

In order to solve these equations for different wavelengths knowledge is needed of the photoequilibrium under red light. Butler derived expressions for the ratio of the quantum yields Φ_r/Φ_{fr} and the proportion of phytochrome present as P_r at equilibrium $[P_{r\infty}]_{660}/[P]$ using experimentally determinable parameters (Butler *et al.*, 1964a; Butler, 1972).

$$\frac{\Phi_r}{\Phi_{fr}} = -\frac{I_{730}A_{730\,max}(dA_{730}/dt)}{I_{660}A_{660\,max}(dA_{730}/dt)} \qquad \begin{matrix} 660(t=0) \\ 730(t=0) \end{matrix} \tag{3}$$

where I_{730} and I_{660} = intensity of 730 nm and 660 nm light. $A_{730\,max}$ and $A_{660\,max}$ = maximum absorbance at equilibrium of 730 nm and 660 nm of intensity I_{730} and I_{660}. dA_{730}/dt = rate of change of absorbance at 730 nm after 660 nm ($t = 0$) or 730 nm ($t = 0$). Using this value of Φ_r/Φ_{fr} the proportion of phytochrome as P_r at photoequilibrium under red light is given by

$$\frac{[P_{r_\infty}]_{660}}{[P]} = \frac{1}{1 + (\Phi_r/\Phi_{fr})} \frac{A_{660\,min}}{A_{660\,max}} \tag{4}$$

where

$A_{660\,min}$ = absorbance at 660 nm after saturating 660 nm light
$A_{660\,max}$ = absorbance at 660 nm after saturating 730 nm light.

Using these expressions the data from a purified oat sample Butler *et al.* (1964a) showed that red light maintains a photostationary equilibrium P_{fr}/P_{total} (ϕ) = 0·81 (i.e. 19% P_r and 81% P_{fr}) and that the ratio of the quantum yields Φ_r/Φ_{fr} = 1·5. A similar value for the ratio of the quantum yield was later obtained for 280 nm light, presumed to be only absorbed by the protein (Pratt and Butler, 1970b). This apparent energy transfer from protein to chromophore was calculated to be 30% efficient. However, the assumption that the chromophore has no absorbance at 280 nm may be incorrect (Tobin and Briggs, 1973).

Using the assumption that red light also maintains 80% of the phytochrome as P_{fr} at equilibrium *in vivo*, Pratt and Briggs (1966) calculated the action spectra for the phototransformation of P_r and P_{fr} in corn coleoptiles which are essentially similar to those in Fig. 5. Note that blue and near ultraviolet light are capable of converting P_r into P_{fr}, but are less effective than red light.

Accurate information of the molecular weight of large rye phytochrome has enabled calculation of the quantum yields of the photoconversions (Gardner and Briggs, 1974):

$$\Phi_r = 0{\cdot}28 \text{ mol Einstein}^{-1}$$

$$\Phi_{fr} = 0{\cdot}20 \text{ mol Einstein}^{-1}$$

Therefore

$$\frac{\Phi_r}{\Phi_{fr}} = 1{\cdot}4$$

It is obvious from equations (1) and (2) that the time required to reach photoequilibrium is irradiance dependent, whereas the photoequilibrium (ϕ) is irradiance independent.

C PHOTOEQUILIBRIA

The value of photoequilibria established by different wavelengths of light have been calculated for phytochrome *in vitro* (Butler *et al.*,

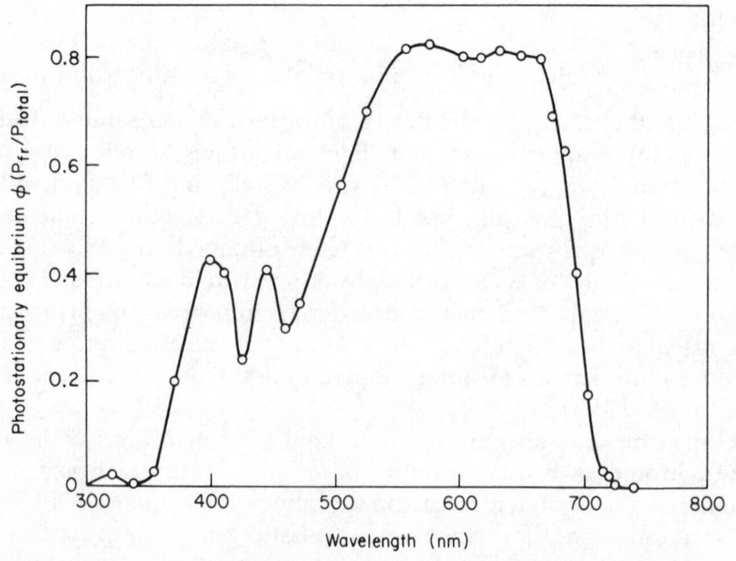

FIG. 6. Photostationary equilibrium (P_{fr}/P_{total}) maintained by different wavelengths of light in mustard hypocotyl hooks, 25°C. (Hanke *et al.*, 1969.)

1964; Hartmann, 1966) and *in vivo* (Pratt and Briggs, 1966; Hanke *et al.*, 1969) (Fig. 6). All these calculations of the photoequilibrium at different wavelengths *in vivo* depend on the assumption that red light maintains the same photoequilibrium as *in vitro*, i.e. $\phi_{red} = 0{\cdot}8$. The practical difficulties and methods used for measuring photoequilibria, particularly under far-red light, are discussed in Chapter 23, Section IIF. Note that blue light maintains a photoequilibrium of about $\phi_{blue} = 0{\cdot}35$, whereas in far-red light $\phi_{far\text{-}red} = c.\ 0{\cdot}05$.

D PHOTOTRANSFORMATION KINETICS

When first investigated, phototransformation of P_r and P_{fr} *in vitro* followed first-order photoreaction kinetics (Butler *et al.*, 1964a). This observation is now known to be correct, although it generated a controversy of sizeable proportions which allowed a number of self-cancelling papers to be published. In 1968 Purves and Briggs presented evidence for phototransformation kinetics *in vitro* and *in vivo* which were not linear when the fraction of pigment converted was plotted on a logarithmic scale against time. The paper was subsequently retracted, however, on the discovery that the derivation from first-order phototransformation kinetics was the result of a non-linear response of the spectrophotometer (Everett *et al.*, 1970). Unfortunately this did not settle the matter and in 1971 non-linear kinetics were again demonstrated, this time in *in vivo* samples of pumpkin and mustard seedlings using several different spectrophotometers (Boisard *et al.*, 1971; Schäfer *et al.*, 1971). It was concluded from this evidence that at least two different pools of phytochrome having different phototransformation kinetics existed. The saga continued when Spruit and Kendrick (1972) presented an alternative explanation on the basis of the optically dense nature of the *in vivo* samples. They calculated theoretical curves on the basis of an exponential light intensity gradient through the cuvette and showed them to fit well with the measured phototransformation curves. Schmidt *et al.* (1973) determined parameters of the light intensity gradients in *in vivo* samples and agreed with this interpretation. They demonstrated first-order phototransformation kinetics *in vivo*, in cases where steps were taken to minimize the light intensity gradient by using a single oat coleoptile sample sandwiched between layers of $CaCO_3$. Further *in vitro* work with large rye phytochrome (Gardner and Briggs, 1974) has shown without doubt the log linearity of phototransformation of P_r to P_{fr} and vice versa.

E FLUORESCENCE

Several reports have indicated that phytochrome can fluoresce. Hendricks *et al.* (1962) reported fluorescence in *in vivo* and *in vitro* extracts. They measured emission above 730 nm and detected an excitation maximum for P_r at 670 nm, but failed to observe any fluorescence of P_{fr}. A more detailed study by Correll *et al.* (1968b) on purified rye phytochrome showed fluorescence at 340 nm, with a characteristic protein excitation spectrum at 290 nm. Also excitation of P_r at 370 nm yielded an emission spectrum maximum at 670 nm. The detailed excitation spectrum for this fluorescence showed a shoulder at 290–300 nm (Fig. 7) and suggests a possible energy transfer between protein and chromophore. There was no evidence of P_{fr} fluorescence since on photoconversion with red light the excitation and emission maxima were only reduced. More recently large rye phytochrome protein fluorescence has been examined and shows excitation and emission maxima at 288 nm and 331 nm respectively, these being identical for both P_r and P_{fr} (Tobin and Briggs, 1973). Although no long-wavelength fluorescence was seen with large rye phytochrome at room temperature, Song *et al.* (1973) have demonstrated fluorescence at low temperature (−259°C) with maxima at 440 and 675 nm on excitation with 380 nm light.

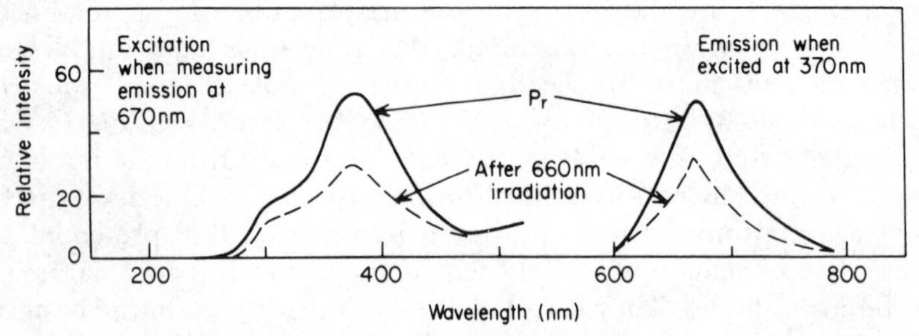

FIG 7. Fluorescence excitation and emission spectra of P_r. (Correll *et al.*, 1968b.)

F CIRCULAR DICHROISM

Circular dichroism (c.d.) measures the optical activity of absorption bands of both protein and chromophore, as well as asymmetric interaction between chromophore and protein. Analysis of these spectra is therefore complex and the individual contribution of the chromophore and protein to such spectra have not been settled. Kroes (1968, 1970) first showed c.d. spectra of P_r and P_{fr} in the region 300–800 nm. Essentially similar results have been described by Anderson *et al.* (1970), Hopkins and Butler (1970) and Burke *et al.* (1972). Above 300 nm the negative and positive Cotton effects seen in c.d. spectra (Fig. 8) for P_r and P_{fr} in the red and far-red regions of the spectrum are interpreted as being due to the optical activity of the chromophore and asymmetry of its environment. Circular dichroism spectra below 300 nm are assumed to represent optical activity of the protein only, although the activity due to the chromophore could possibly be involved. Differences in c.d. spectra in this region observed by some workers (Anderson *et al.*, 1970; Hopkins and Butler, 1970) have not been found with large rye phytochrome by Tobin and Briggs (1973), who attributed these

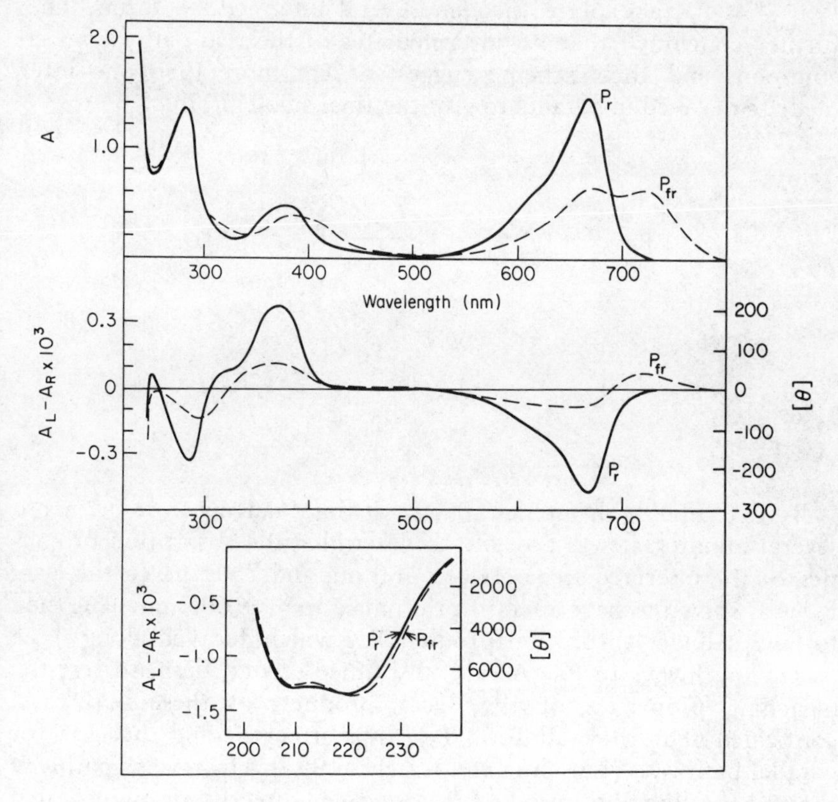

FIG. 8. Circular dichroism spectra and absorption spectra of purified oat phytochrome. The c.d. spectra below 300 nm were obtained by diluting the sample by one half. The measured values of the c.d. spectra ($A_L - A_R$) are given in the left-hand scale; on the right the values are converted to decimolar ellipticities (θ). (Hopkins and Butler, 1970.)

differences to partial denaturation of the samples used by the earlier workers. The significance of c.d. spectra in relation to the conformational changes of phytochrome on phototransformation is discussed in detail in Section IXD.

V Intermediates in phototransformation

A FLASH PHOTOLYSIS

A detailed study using the technique of flash photolysis (Linschitz *et al.*, 1966) with purified oat phytochrome of 60 000 mol. wt, revealed that the apparently simple photoreactions $P_r \xrightarrow{h\nu} P_{fr}$ and $P_{fr} \xrightarrow{h\nu} P_r$ take place through several intermediate forms. They further concluded that no intermediates of the two pathways were common, and their scheme suggested that more than one intermediate decayed in parallel to give the final stable product

$$P_r \xrightarrow{h\nu} P_r^* \longrightarrow P_{r_1} \rightarrow \{P_{r_2}, P_{r_3}, P_{r_4}\} \rightarrow P_{fr}$$

$$P_{fr} \xrightarrow{h\nu} P_{fr}^* \rightarrow \{P_{fr_1}, P_{fr_2}\} \rightarrow P_r$$

By determining difference spectra at times ranging from 0·2 ms to several minutes, it was possible to determine the absorption properties of the intermediates between 560 nm and 750 nm. In this way P_r^* was shown to have an absorption maximum at 695 nm. This then formed a lower absorption product P_{r_1} which decayed along three parallel pathways to P_{fr}. A carefully timed second flash, to test the bleaching properties of the decay products of these pathways, confirmed that they all form P_{fr}, thus strengthening the case for parallel pathways (Linschitz and Kasche, 1967). The reverse pathway showed that the photoproduct P_{fr}^* decayed directly, via two parallel pathways to P_r. Parallel pathways have been deduced on the basis of a kinetic analysis that assumes all the reactions are first order, and that the phytochrome used is homogeneous. Flash photolysis studies on large rye phytochrome are eagerly awaited.

B LOW TEMPERATURE STUDIES

Spruit (1966a,b,c,d) demonstrated phytochrome intermediates *in vivo* using etiolated pea plumules and isolated corn phytochrome by means of low-temperature techniques. At the temperature of liquid nitrogen (–196°C) it was found that irradiation of P_r with red light produced a stable intermediate with an absorbance maximum at 698 nm (called P_{698}). This reaction was at least partially photoreversible. Similarly, irradiation of P_{fr} with far-red light at this temperature gave a stable intermediate with a maximum absorbance at 650 nm (called P_{650}), a reaction which was also photoreversible. Cross *et al.* (1968), Pratt and Butler (1968, 1970) and Kroes (1970) used low temperatures to investigate intermediates in highly purified samples of phytochrome. These experiments demonstrated a stable photoproduct of P_r having a maximum absorbance at 695 nm, which could be photoconverted back to P_r. Presumably this intermediate is analogous to P_{698} observed *in vivo*. On warming, this intermediate was shown to decay. At higher temperatures Cross *et al.* (1968) obtained evidence for two other intermediates between P_{698} and P_{fr}, one of which had an absorbance maximum at 710 nm (P_{710}), the other having a very low absorbance (called P_{bl}). Pratt and Butler (1968) examined the $P_{fr} \longrightarrow P_r$ reaction *in vitro* and at temperatures below –150°C a stable intermediate with an absorbance maximum at 660 nm was formed on irradiating P_{fr}. On warming, this product decayed to a lower extinction form before forming P_r. These low-temperature studies confirmed the conclusion of the flash photolysis experiments, that the two pathways have no common intermediates.

Experiments *in vivo* using etiolated pea epicotyl tissue have demonstrated that, on warming P_{698}, produced at –196°C, little if any P_{fr} is produced (Spruit and Kendrick, 1973). Similarly, warming P_{650} produced very little P_r. This was explained on the basis of the dark reversion of P_{698} to P_r and P_{650} to P_{fr}. *In vivo* these reactions occur preferentially to the reactions along the pathways towards P_{fr} and P_r, and have been interpreted as events restricted to the phytochrome chromophore. Warming of the photoreaction products formed at –70°C and above *in vivo*, has provided evidence for the presence of P_{710} and P_{bl} in the pathway from $P_r \longrightarrow P_{fr}$. Similar experiments have shown two intermediates, P_{690} (having a maximum absorbance at 690 nm) and P_x (having a very low absorbance) present in the $P_{fr} \longrightarrow P_r$ pathway (Kendrick and Spruit, 1973b; Spruit and Kendrick, 1973). Recently a more convenient technique has been used to study intermediates *in vivo* in freeze-dried pea epicotyl tissue (Kendrick, 1974) where complete photoconversion is prevented (Tobin and Briggs, 1969; Tobin *et al.*, 1973; Balangé, 1974). However, in samples freeze-dried as P_r, red light clearly shows the formation of P_{698}, which decays back to P_r in darkness at 0°C. Samples freeze dried after red light show a reversible photoreaction on exposure to alternate red and far-red light. This reaction is interpreted as the photoreversible reaction $P_{fr} \underset{\text{red}}{\overset{\text{far-red}}{\rightleftharpoons}} P_{650}$. In darkness P_{650} shows a slow, partial reversion to P_{fr}, a reaction which may be of significance in relation to the inverse dark reversion reaction observed in dehydrated seeds and discussed in Section VIE.

C INTERMEDIATES AND PIGMENT CYCLING

If phytochrome is irradiated with mixed red and far-red light which simultaneously excites both P_r and P_{fr}, the pigment cycles. Depending on the irradiance an amount of intermediate accumulates under steady state conditions predominantly behind the slowest of the thermal dark reactions. Measurement of intermediates under these conditions raises the technical problem of separating the light used for pigment excitation from the measuring beams as seen by the photodetector. Briggs and Fork (1969a,b), Everett and Briggs (1970) and Gardner and Briggs (1974) demonstrated, under such conditions both *in vitro* and *in vivo*, the presence of intermediates that underwent reactions in darkness. This was done by restricting measurements to the blue region of the spectrum, separating the excitation source from the photodetector by means of cut-off filters. This method enabled intermediate difference spectra in the blue region to be investigated and compared with the complete phototransformation difference spectrum. With this technique, however, it was not possible to measure in the red and far-red regions of the spectrum. Using a quasicontinuous spectrophotometer (see Chapter 23, Sections IIB, IIE) Kendrick and Spruit (1972a,b, 1973a,c) obtained a complete difference spectrum for the dark reactions of intermediates that accumulate under conditions of pigment cycling. It was concluded that they represent predominantly P_{bl} forming P_{fr} at 0°C (Fig. 9). At lower temperatures (–20°C) no P_{fr} is formed; the intermediates that accumulate have relatively low absorbance, and form P_r and P_{710} in darkness. This experiment demonstrates that P_{710} can be photoconverted back to P_r without forming P_{fr}, a fact confirmed in other experiments (Spruit and Kendrick, 1973). A scheme for phytochrome photoconversion *in vivo* (Fig. 10) that

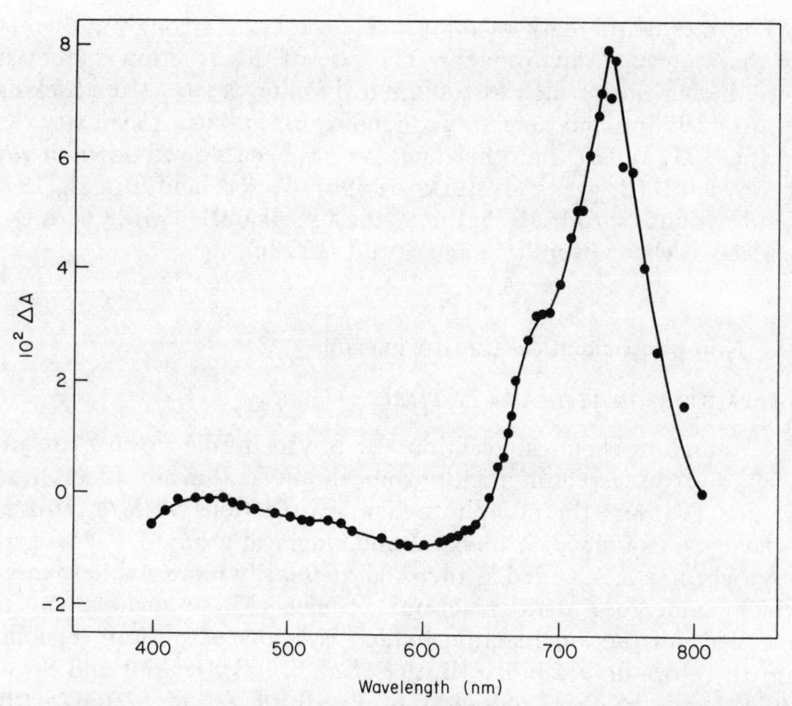

FIG. 9. Difference spectrum for the dark reactions of intermediates maintained by mixed red/far-red light. Sample was partially purified from oats and was in glycerol : buffer (3 : 1). (Kendrick and Spruit, 1973c.)

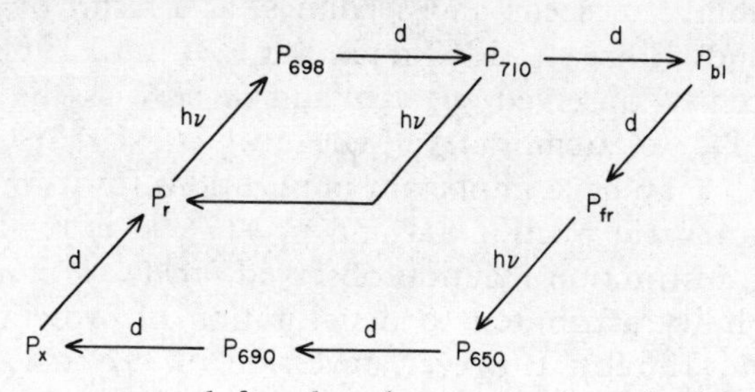

FIG. 10. Scheme proposed for the photoconversion of phytochrome. hν: photoreaction; d: dark reaction. P_{698}, P_{710}, P_{650} and P_{690} are intermediates having peak absorption at 698, 710, 650 and 690 nm respectively. P_{bl} and P_x are relatively weakly absorbing intermediates. (Kendrick and Spruit, 1973c.)

takes into account most of our knowledge *in vitro* has been presented (Kendrick and Spruit, 1973b,c). The scheme gives a sequential pathway for the phototransformation in both directions and points out that P_{710} may be photoconverted to P_r via an as yet unknown pathway.

The rate of the dark reaction $P_{bl} \longrightarrow P_{fr}$ is strongly influenced by the molecular environment. The rate of this reaction is increased by reducing agents such as sodium dithionite, *in vitro* (Kendrick and Spruit, 1973c) and anaerobic conditions *in vivo* (Kendrick and Spruit, 1973a). On the other hand, it has been slowed down *in vitro* by glycerol (Briggs and Fork, 1969b; Everett and Briggs, 1970; Kendrick and Spruit, 1973c), and the use of buffer made with D_2O in place of water (Kendrick and Spruit, 1973c).

VI Non-photochemical transformations

A REACTIONS *IN VIVO* AND *IN VITRO*

The non-photochemical reactions of phytochrome *in vivo* are outlined and discussed in relationship to physiology in Chapter 15, Section IV. Here the non-photochemical reactions both *in vitro* and *in vivo* are examined in physical and chemical terms. The P_r form of phytochrome is regarded as thermodynamically more stable than P_{fr}, which undergoes dark reactions *in vivo*. These include (a) the so-called destruction reaction, which is a loss of photoreversibility and therefore detectability (Butler *et al.*, 1963; De Lint and Spruit, 1963), and (b) dark reversion to P_r (Butler *et al.*, 1963). Other reactions which have been observed *in vivo* are inverse dark reversion of P_r to P_{fr} in seeds (Boisard *et al.*, 1968) and increases in total photoreversibility in seeds and seedlings, as a result of P_r synthesis (Clarkson and Hillman, 1967a; Boisard *et al.*, 1968). Another reaction recently observed *in vivo* and *in vitro* is the preferential binding of P_{fr} to membranes (Marme *et al.*, 1973; Quail *et al.*, 1973a) which may have important implications for its mechanism of action (Chapter 15, Section III and p. 417). *In vitro,* P_{fr} does not undergo the destruction reaction observed *in vivo*, although a loss of photoreversibility attributed to denaturation of protein is observed (Butler *et al.*, 1964b). However, reversion of P_{fr} to P_r does occur (Mumford, 1966).

B DENATURATION

There is a temperature-dependent loss of photoreversibility *in vitro* which is attributed to protein denaturation. This process can be distinguished from the destruction reaction observed *in vivo* because it can take place with phytochrome in the P_r as well as the P_{fr} form. P_{fr} is, however, the more susceptible form in the presence of proteolytic enzymes, urea and sulphydryl reagents (Butler *et al.*, 1964b) although Roux and Hillman (1969) have demonstrated P_r to be more susceptible than P_{fr} to denaturation with glutaraldehyde. A partly denatured form of phytochrome retains photoreversibility but has an A_{725}/A_{665} ratio less than unity. Furuya and Hillman (1966) obtained a small molecular weight substance from peas which was not fully characterized (called P_{fr} "killer") but which rapidly brought about denaturation of P_{fr} *in vitro*. Lisansky (1973) has investigated *in vitro* loss of photoreversibility and has identified at least one process which is metal ion-dependent. A substance was extracted from the apical portions of light-grown pea seedlings which retards metal-dependent loss of photoreversibility *in vitro*. The molecular weight of this substance is around 465 daltons and it is thought to act by metal ion chelation.

C DESTRUCTION

The *in vivo* destruction of P_{fr} was first reported simultaneously by Butler *et al.* (1963) and De Lint and Spruit (1963) in etiolated seedlings of corn. Subsequent work has indicated that this reaction is ubiquitous in etiolated seedlings, is oxygen dependent, being inhibited by respiratory inhibitors (Butler and Lane, 1965), requires a metallic ion (Furuya *et al.*, 1965), has a high Q_{10} (between 2·7 and 3·5 in corn (Butler and Lane, 1965) and 4·3 in *Amaranthus* (Kendrick and Frankland, 1969b)), and is probably enzymic (Furuya *et al.*, 1965). Detailed analysis of this reaction has been conducted in several species of etiolated seedlings (Hillman, 1967; Frankland, 1972). The destruction rate in several dicotyledons depends on the concentration of P_{fr} present and is a first-order reaction. Under continuous irradiation the destruction of total phytochrome is proportional to the photostationary equilibrium maintained (Fig. 11) (Kendrick and Frankland, 1968).

Thus

$$\log_e \frac{P}{P_0} = -k\phi t$$

where P/P_0 is the proportion of the initial phytochrome at time (t); $\phi = (P_{fr}/P_{total})$; k = decay constant of P_{fr}.

The destruction reaction is irradiance independent once the photoequilibrium characteristic of the light used is established (Kendrick and Frankland, 1969b; Marmé, 1969).

In monocotyledons the destruction kinetics appear to be independent of P_{fr} concentration and follow zero-order kinetics (Butler

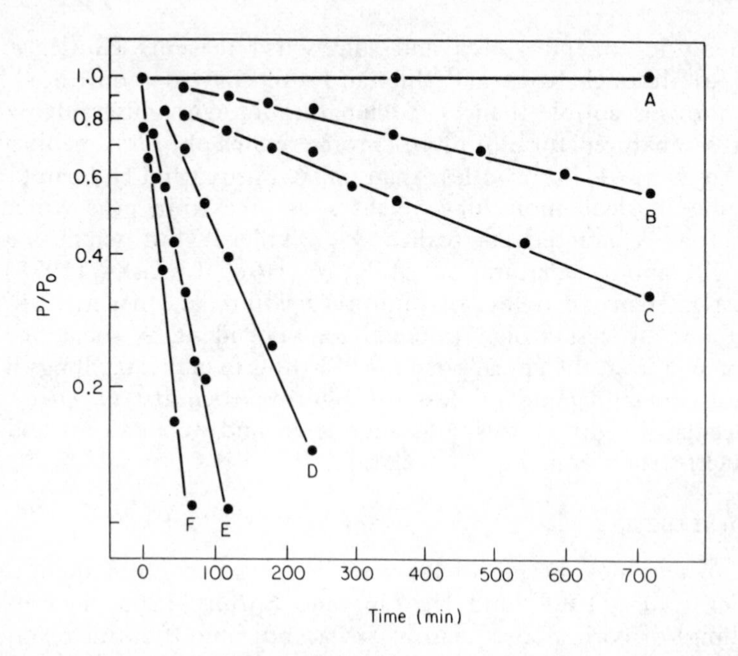

FIG. 11. Decay of total phytochrome in *Amaranthus* under continuous irradiation at 25° P/P_0 plotted on logarithmic scale. A: darkness; B: far-red $\phi = 0{\cdot}02$; C: far-red $\phi = 0{\cdot}05$; D: blue $\phi = 0{\cdot}22$; E: mixed red/far-red $\phi = 0{\cdot}49$; F: decay of P_{fr} in dark following irradiation with red $\phi = 0{\cdot}80$ (Kendrick and Frankland, 1969b.)

et al., 1963; De Lint and Spruit, 1963; Pratt and Briggs, 1966; Dooskin and Mancinelli, 1968; Kendrick, 1972). Also a lag phase has been observed before destruction proceeds. The facts indicate that some factor necessary for the destruction is limiting and is perhaps induced by the $P_r \longrightarrow P_{fr}$ conversion. Support for this idea has recently been obtained by Kidd and Pratt (1973), who demonstrated that the lag phase was extended in the presence of protein synthesis inhibitors. The exact mechanism of destruction is not known. Spruit (1972) has proposed a model to account for phytochrome destruction on the basis of its translocation to small sites in the cell, where it would contribute less optically to the absorbance difference on phototransformation. However, Coleman (1973) demonstrated a loss of phytochrome by an immunological technique which agrees well with the kinetics of destruction as measured spectrophotometrically, suggesting that a major disorganization of the protein occurs during destruction.

D REVERSION

Dark reversion of P_{fr} to P_r had been predicted on the basis of physiological experiments (Borthwick *et al.*, 1954), but was first measured spectrophotometrically in cauliflower florets (Butler *et al.*, 1963; Butler and Lane, 1965). Subsequently it has been demonstrated in etiolated seedlings of dicotyledons, members of the Centrospermae so far studied being an exception (Kendrick and Frankland, 1968, 1969b; Kendrick and Hillman, 1971). Also the monocotyledons do not appear to show this reaction *in vivo* (Hillman, 1964; Frankland, 1972). Since in most species P_{fr} undergoes both reversion and destruction simultaneously, kinetic analysis is difficult. However in tissues which only show reversion, e.g. cauliflower florets, kinetic analysis reveals a fast and slow component. The amount and rate of reversion in mustard seedlings has been shown to be dependent on the physiological age (Marme *et al.*, 1971), photostationary equilibrium established (Schmidt and Schäfer, 1974), and temperature (Butler and Lane, 1965; Schäfer and Schmidt, 1974). In peas reversion is particularly prominent at an early stage of development (McArthur and Briggs, 1971). The amount of reversion is increased by metabolic inhibitors which inhibit the destruction process (Furuya *et al.*, 1965). This was interpreted as evidence for competition between the destruction and reversion processes for P_{fr} molecules.

Dark reversion of P_{fr} to P_r *in vitro* has been observed in extracts from light (Taylor, 1968) and dark-grown plants (Mumford, 1966; Anderson *et al.*, 1969; Mumford and Jenner, 1971; Pike and Briggs, 1972). Two-component kinetics have been observed in most cases and the contribution of the fast and slow components vary in large and small rye phytochrome (Pike and Briggs, 1972). It is interesting to point out that although reversion is seen *in vitro* in phytochrome from monocotyledons, it does not occur *in vivo*. Anderson *et al.* (1969) demonstrated that reversion was stimulated by low pH. Under these conditions a protonated form of P_{fr} exists (P_{frH}) which can form P_r and P_{fr}. A low oxidation-reduction potential has been shown to stimulate dark reversion *in vitro* (Fig. 12) (Mumford and Jenner, 1971; Pike and Briggs, 1972; Kendrick and Spruit, 1973c). Clearly the molecular environment is very important for reversion to occur and it was suggested that *in vivo* this reaction only occurs where the necessary conditions prevail (Mumford and Jenner, 1971; Kendrick and Spruit, 1973c). This implies that the destiny of a P_{fr} molecule is predetermined and that the molecules undergoing

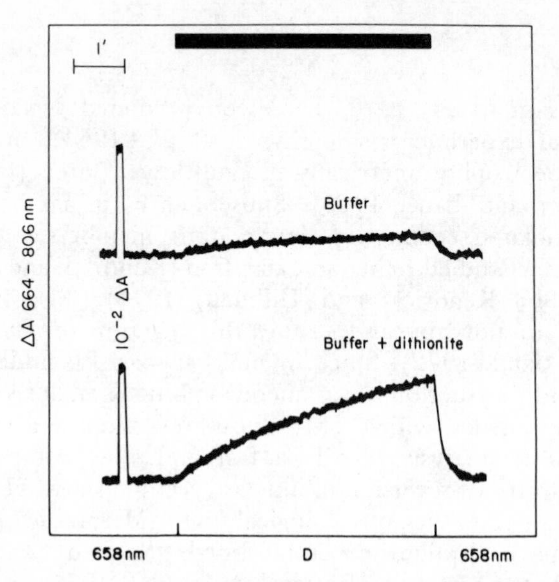

FIG. 12. Recordings of ΔA between 664 nm and 806 nm. They demonstrate the enhancement of dark reversion of P_{fr} to P_r after 658 nm light by 5 mM sodium dithionite. Sample of partially purified oat phytochrome at 0°C. (Kendrick and Spruit, 1973c.)

destruction and reversion are spatially separated within the cell. The restriction of P_{fr} molecules to the later pool could account for the photostable phytochrome in cauliflower and other tissues grown in the light (Hillman, 1964).

E INVERSE REVERSION

Inverse dark reversion is a reaction observed *in vivo* in dehydrated or partly hydrated seeds and was initially interpreted as a production of P_{fr} from P_r (Boisard *et al.*, 1968; Kendrick *et al.*, 1969; Spruit and Mancinelli, 1969; Malcost *et al.*, 1970). *In vivo* (Tobin and Briggs, 1969; Tobin *et al.*, 1973; Kendrick, 1974) and *in vitro* (Tobin *et al.*, 1973; Balange, 1974) under conditions of dehydration, complete photoconversion of phytochrome is not possible. In freeze-dried pea epicotyl tissue P_r undergoes a transient reaction on exposure to red light:

$$P_r \underset{d}{\overset{h\nu}{\rightleftharpoons}} P_{698}$$

This reaction is probably restricted to the chromophore and represents the production and decay of the intermediate P_{698} (Kendrick, 1974). On the other hand, in pea epicotyl tissues freeze-dried after red light treatment, phytochrome in the P_{fr} form undergoes a stable photoreversible reaction on subsequent exposure to alternate red and far-red light. Under these conditions P_{fr} produces a low absorption product with maximum absorbance in the red region of the spectrum which is interpreted as being similar to the intermediate P_{650}.

$$P_{fr} \underset{\text{red}}{\overset{\text{far-red}}{\rightleftharpoons}} P_{650}$$

The similarity of the difference spectrum of this reaction to that of phytochrome in dry cucumber seeds (Spruit and Mancinelli, 1969) is striking. The demonstration that P_{650} in freeze-dried pea epicotyl tissue can slowly revert to P_{fr}, a reaction of P_{650} also seen at low temperatures (Kendrick and Spruit, 1973b), offers a more reasonable explanation of inverse dark reversion (Kendrick and Spruit, 1974). The production of P_{fr} in the dark is therefore not from P_r, but from P_{650} present after exposure of P_{fr} to far-red light under the extreme molecular environment of dehydration. This can adequately explain the observed data and account for the physiological behaviour of seeds which require continuous far-red light to inhibit their germination (Kendrick and Frankland, 1969a; Rollin, 1972).

F SYNTHESIS

The phytochrome content of seeds is low and rapidly increases on addition of water. This rehydration makes complete photoreversibility and therefore detectability possible. After this initial increase a lag phase occurs before a second phase of phytochrome increase associated with seed germination and seedling development (Fig. 13). This second pool appears in the P_r form and is capable of destruction. In peas this second increase in phytochrome can be inhibited by protein synthesis inhibitors (McArthur and Briggs, 1971), but in rye (Correll *et al.*, 1968a) no evidence for incorporation of labelled amino acids or δ-amino-laevulinic acid was found. In this case it was concluded that the protein and chromophore components were already synthesized in the seed before maturation. After destruction, synthesis is apparently initiated, resulting in an increase in darkness of detectable phytochrome (Clarkson and Hillman, 1967a,b; Kendrick and Frankland, 1969b; Marmé, 1969). Under continuous

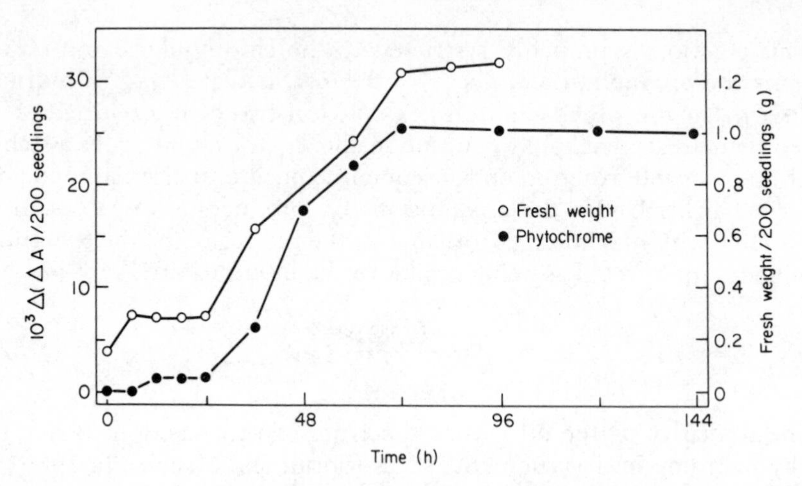

FIG. 13. Time course of fresh weight and phytochrome changes in *Amaranthus* seedlings from the time of sowing at 25°C. (Kendrick and Frankland, 1969b.)

irradiation a balance occurs between synthesis and destruction (Clarkson and Hillman, 1968). Density labelling experiments suggest that the increase in phytochrome associated with seedling development and apparent synthesis after destruction in *Cucurbita* is a result of *de novo* synthesis (Quail *et al.*, 1973b,c). Also, the same authors suggest that the plateau level eventually reached in darkness involves a gradual turnover of phytochrome molecules, as a result of synthesis and degradation of the P_r form.

G REACTIONS AND THE MOLECULAR ENVIRONMENT

In Fig. 14 a scheme is presented as a working hypothesis to understand the dark reactions of the phytochrome system *in vivo*. The synthesis of phytochrome occurs in the P_r form, which is thermodynamically stable. On exposure to red light unstable P_{fr} is formed which in darkness can undergo destruction or reversion to P_r. The reversion reaction is favoured by low pH and reducing agents by formation of the protonated form of P_{fr} (P_{frH}). It is proposed that P_{frH} is similar if not identical to P_{650}, the first intermediate produced by exposure of P_{fr} to far-red light. Under "normal" (hydrated) conditions this proceeds to P_r via a series of dark reactions. The occurrence of dark reversion will depend strongly on the molecular environment of the phytochrome in terms of oxidation-reduction potential and its steric interaction with membranes. One or both of these factors could account for the lack of reversion

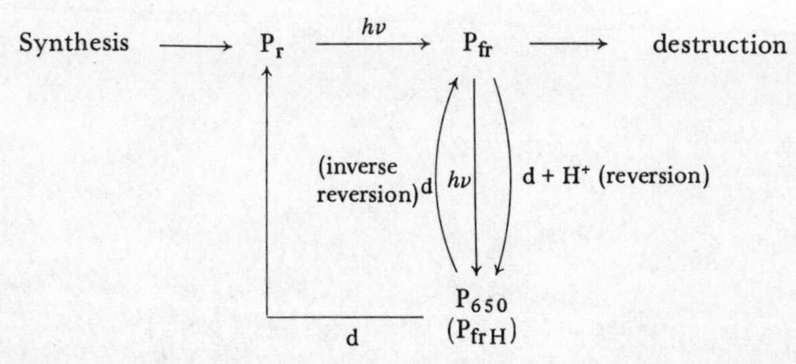

FIG. 14. Scheme proposed to account for the dark reactions of the phytochrome system.

seen in monocotyledons *in vivo* and for the existence of photostable phytochrome in light-grown plants. Another extreme molecular environment is dehydration which restricts protein and chromophore movement. Under these conditions P_{fr} is photoconverted into P_{650} which is restricted from forming P_r, but can undergo a dark reversion to P_{fr}. This is the inverse dark reversion reaction observed under conditions of dehydration in seeds, initially thought to be the dark transformation of P_r to P_{fr}. In this way the molecular environment is all important in determining the destiny of P_{fr} and an understanding of these reactions should help in determining their physiological significance.

VII Chemical and conformational changes upon photoconversion

A CHROMOPHORE ISOMERIZATION

The exact nature of the chemical changes which occur during photoconversion of phytochrome are as yet unknown, although several suggestions have been made. These include *cis–trans* isomerization (Siegelman and Butler, 1965; Suzuki and Hamanaka, 1969; Kroes, 1970; Burke *et al.*, 1972), lactim–lactam interconversion (Siegelman and Butler, 1965) and intramolecular proton or electron migration (Siegelman and Butler, 1965; Crespi *et al.*, 1968; Siegelman *et al.*, 1968; Kroes, 1970; Rüdiger, 1972).

The strongest argument in favour of the involvement of *cis–trans* isomerization comes from circular dichroism measurements. Working at low temperatures, Burke *et al.* (1972) obtained large Cotton effects for P_r, P_{bl} and a further intermediate named A. P_{bl} showed a particularly large Cotton effect, suggesting that the tetrapyrrole was folded in some way. Figure 15 shows two possible models of the interconversions based on these results. However, the interpretation

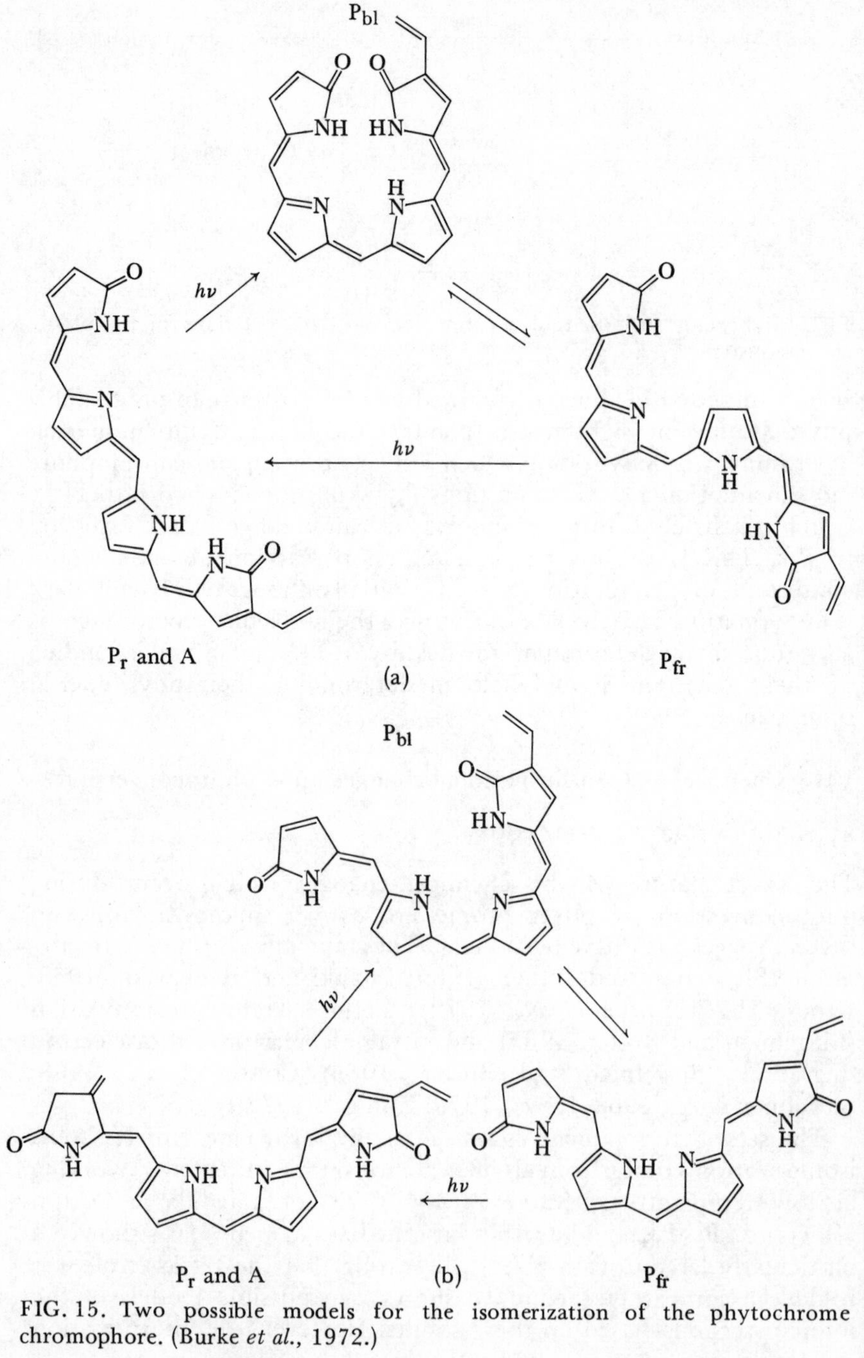

P_{bl}

P_r and A

(a)

P_{fr}

P_{bl}

P_r and A

(b)

P_{fr}

FIG. 15. Two possible models for the isomerization of the phytochrome chromophore. (Burke *et al.*, 1972.)

of this work is somewhat of a problem since all measurements were made in 66% glycerol, in which phytochrome would be considerably dehydrated, with possible effects on the intermediates produced (see Section VIE above).

The lactim–lactam conversion is thought to be unlikely. Infrared spectra of phytochrome reported by Kroes (1970) suggest that the lactam configuration is preferred and provide no evidence for interconversion between the lactim and lactam configurations.

Siegelman *et al.* (1968) were first to propose a full model for the interconversion based on intramolecular electron migration. This scheme is given in Fig. 1 and is dependent on a P_r structure having ethylidene substituents on both terminal rings. Rüdiger's (1972) chemical analysis of the chromophore, on the other hand, is not compatible with the presence of a second ethylidene substituent. In addition, as Kroes (1970) points out, although the proposed $P_r \longrightarrow P_{fr}$ isomerization appears to be energetically possible, the reintroduction of double bonds into rings 1 and 4 in the $P_{fr} \longrightarrow P_r$ transformation seems most unlikely, even for a photochemical reaction.

Rüdiger has proposed another protonation mechanism (Rüdiger and Correll, 1969; Rüdiger, 1972). In this model, shown in Fig. 3, it is assumed that a conformational change of the protein takes place such that proton transfer from the chromophore to the protein, and vice versa, can occur. Recently Suzuki *et al.* (1973) have shown, by theoretical argument only, that none of the models so far advanced can account for the unique optical properties of phytochrome, as long as their π electron systems are assumed to have no ionic structure.

B CHANGES IN THE PROTEIN

In 1964, Butler *et al.* showed that the absorption spectrum of P_{fr} was more affected than that of P_r by treatment with 5 M urea, 5×10^{-3} M *p*-chloromercuribenzoate and 5×10^{-3} M *N*-ethylmaleimide. Trypsin and pronase, on the other hand, altered the P_r spectrum but only affected that of P_{fr} if present during the $P_r \longrightarrow P_{fr}$ photoconversion. These results were taken to indicate a change in conformation of the protein during phototransformation.

Briggs *et al.* (1968), however, were subsequently unable to find any differences between P_r and P_{fr} in velocity sedimentation in sucrose gradients, electrophoretic mobility, behaviour on molecular sieve chromatography, and binding to and elution from calcium phosphate (brushite) gels. On the other hand, they did report that P_{fr} was more labile than P_r during ammonium sulphate precipitation. On spectral grounds, Roux and Hillman (1969) found that P_r was more labile than P_{fr} in 0·5% glutaraldehyde. Subsequently Roux (1972) performed amino acid analyses on highly purified oat phytochrome which had been reacted with glutaraldehyde either as P_r or P_{fr}, or whilst cycling between P_r and P_{fr}. Of 27 lysine residues (per 60 000 mol. wt), 13 reacted with glutaraldehyde in the P_r form or whilst cycling, whereas only 11 reacted in the P_{fr} form. These data were interpreted as indicating that a conformational change occurs during the $P_r \longrightarrow P_{fr}$ transformation in which two lysine residues become inaccessible to the glutaraldehyde.

Hopkins and Butler (1970) observed minor differences between P_r and P_{fr} in ultraviolet difference spectra, in microcomplement fixation and in far-ultraviolet circular dichroism. On the other hand, Anderson *et al.* (1970) found no difference in circular dichroism spectra below 240 nm. Hopkins (1971) showed that P_r had a larger sedimentation coefficient than P_{fr} by about 0·1 *s* ($s^{Pr}_{20,\omega} = 5{\cdot}1$; $s^{Pr}_{20,\omega} = 5{\cdot}0$). Thus, although all the data are not consistent, there would seem to be considerable evidence for a protein conformational change during the photoconversions.

The above reports, however, all relate to the degraded, small phytochrome of molecular weight *c.* 60 000 daltons. Recently, attempts have been made to obtain direct information on the possibility of protein conformation changes occurring in native, 120 000 mol. wt phytochrome, with results that in general contradict the view obtained from small phytochrome.

Tobin and Briggs (1973) have carried out detailed studies of the fluorescence and circular dichroism spectra of native rye phytochrome. The fluorescence spectra of both P_r and P_{fr} showed a peak for excitation at 286 nm, and for emission at 331 nm. The position of the fluorescence emission maximum is thought to be related to the exposure of tryptophan residues to the aqueous medium. The failure to observe any change in the emission spectra on photoconversion indicates that there is no change in the environment of tryptophan residues in the molecule, thus ruling out major changes in conformation. Tobin and Briggs (1973) suggest that the position of the emission maximum (i.e. near 330 nm) indicates that the tryptophan residues are in a relatively non-polar environment, i.e. buried within the molecule. If this is so, however, a certain degree of conformational change of the surface of the molecule may be possible, without altering the environment of the buried tryptophan residues.

The circular dichroism spectra of native rye P_r and P_{fr} below 240 nm were also identical, again indicating the lack of substantial conformational changes during the photoconversion. On the other hand, the ultraviolet difference spectra reported do indicate some differences in the protein moieties of P_r and P_{fr} (Fig. 16). Tobin and Briggs (1973) point out that significant (23–38%) chromophore absorption occurs at these wavelengths, making interpretation difficult; however, the following possibilities were offered.

The 318 nm peak is probably chromophore absorption, since

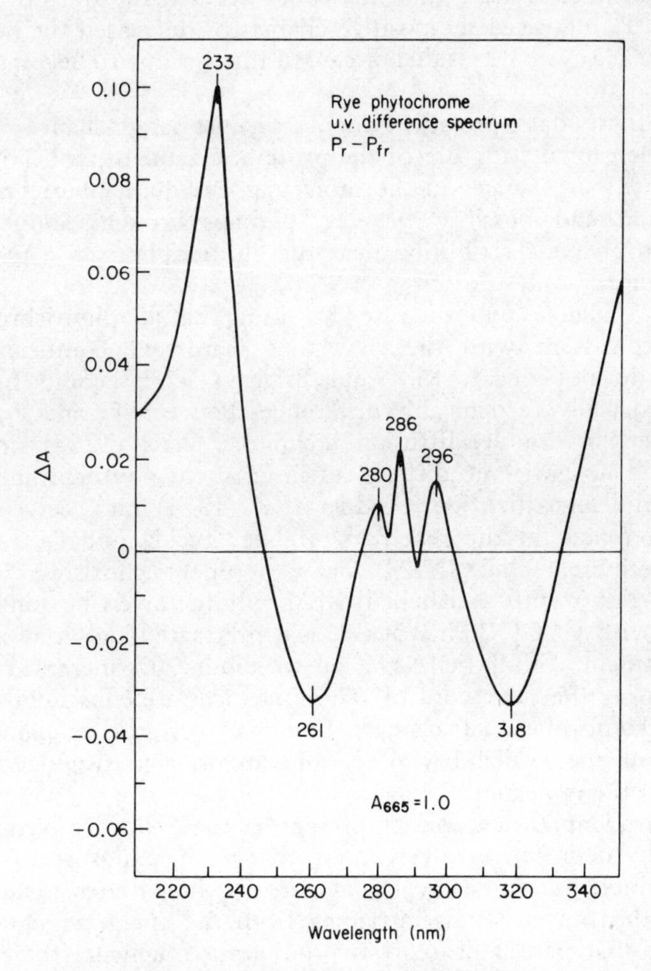

FIG. 16. Ultraviolet difference spectrum of purified high molecular weight rye phytochrome. (Tobin and Briggs, 1973.)

proteins do not absorb significantly above 300 nm. The 261 nm peak is also thought to be due to the chromophore, since similar peaks are not seen with other proteins. The region between 275 and 310 nm is suggested as being similar to the perturbation difference spectra of tyrosine or tryptophan produced by placing them in a relatively non-polar solvent such as ethylene glycol. This may be considered as being similar to moving them from the surface to the interior of a protein molecule or vice versa. One of several possibilities for the 233 nm peak is a change in the degree of ionization of a cysteine residue, and the authors indicate that this could occur if the transformation of P_{fr} to P_r increased the positive charge (or decreased the negative charge) near a cysteine residue or caused that residue to become more exposed.

Thus these data provide evidence against large changes in the three-dimensional structure of the protein. On the other hand, there is evidence for changes in the molecular environment of tyrosine, tryptophan and possibly cysteine, perhaps brought about by a change in charge distribution near the chromophore as a result of chromophore isomerization.

Other evidence published so far using native phytochrome is mainly consistent with the view that significant conformational changes do not occur. Rice and Briggs (1973b) could find no immunologically recognizable differences between P_r and P_{fr} from rye using the double diffusion technique. Pratt (1973) similarly could find no antigenic differences in large oat phytochrome using microcomplement fixation. Gardner *et al.* (1974) did observe significant differences in the reactivity of large rye P_r and P_{fr} towards $[^{14}C]$*N*-ethylmaleimide (NEM). The experiments consisted of reacting P_r or P_{fr} with unlabelled NEM, photoconverting, and then reacting with $[^{14}C]$NEM. When P_r was pretreated and converted to P_{fr}, treatment with $[^{14}C]$NEM led to about 70% increased radioactivity over the dark control. These changes are consistent with a protein conformational change, but may also be explained by changes in the availability of certain amino acid residues at the surface of the molecule.

The prudent conclusion at present is that photoconversion of native phytochrome involves minor changes in exposure of amino acid residues, but not large changes in the three-dimensional structure of the protein. It is intriguing from the energetic viewpoint, however, that partial proteolysis produces a fragment (the 60 000 mol. wt phytochrome) which appears to exhibit considerable conformational changes upon photoconversion.

VIII The mode of action of phytochrome

Phytochrome is clearly an extremely interesting molecule with a unique set of properties, and it is to be expected that the molecular mechanism of action of phytochrome will reflect those properties. At present, the mode of action of phytochrome is mainly a topic for speculation, but there are indications that progress may soon be made through the investigation of phytochrome action in cell-free preparations.

As described in detail in Chapter 15 there is considerable evidence, mainly circumstantial at present, that at least some of the cellular phytochrome is located either in or near various membranes and several hypotheses of phytochrome action have been developed around this concept. Unfortunately, although phytochrome has now been isolated as a membrane component, it has not yet been demonstrated that membrane-bound phytochrome controls any biochemical or biophysical processes *in vitro*. The isolation of membrane-bound phytochrome is discussed in detail in Chapters 15 and 23 and it is not necessary to repeat that coverage here. Certain aspects of the work are worth discussing, however.

The isolation of membrane-bound phytochrome depends on the pH and Mg^{2+} concentration of the extracting medium and on the nature of any pre-extraction irradiation treatments given to the plants (Marmé *et al.*, 1973, 1974; Quail *et al.*, 1973; Boisard *et al.*, 1974). In all this work, the criterion for membrane-bound phytochrome is whether or not the spectral photoreversibility is located in the pellet after a high-speed centrifugation. Quail (personal communication) has found that phytochrome present in a 100 000 $\times g$ supernatant may become pelletable at 100 000 $\times g$ subsequent to irradiation *in vitro*. Thus soluble phytochrome, in the absence of any membrane components, becomes pelletable merely by being photoconverted. This pelletability in the absence of membrane material is highly dependent on salt concentration, as is the binding in the presence of membrane fragments (Marmé *et al.*, 1973, 1974; Boisard *et al.*, 1974). Thus the use of pelletability at high g forces as a criterion for binding of phytochrome to membranes may be questionable.

Boisard *et al.* (1974) investigated the effects of pretreatment with red and far-red light on pelletability of phytochrome at 17 000 $\times g$ using Zucchini squash seedlings. They found that red light treatment led to approximately 40% of the total phytochrome being present in the 17 000 $\times g$ pellet, whereas dark controls had only 2–4% in this fraction. After red/far-red treatment, about 12% was pelletable, which decreased to dark control levels within two hours of darkness. These data were interpreted as indicating that P_{fr} could bind to specific membrane sites whereas P_r could not; in addition it was suggested that phytochrome was a ligand which interacted allosterically with the membrane binding sites. It was further postulated that such molecular interactions at the membrane level could be responsible for functional changes including ion movements in *Albizzia* leaflets (Satter *et al.*, 1970), rapid electrical potential changes in oat coleoptiles (Newman and Briggs, 1972) and suppression of lipoxygenase activity (Oelze-Karow *et al.*, 1970).

This hypothesis could certainly account for the observed data, not least because allosteric interactions with membranes can, in a formal sense, account for practically every problem in biology. On the other hand, it seems to these reviewers that the evidence for specific binding of phytochrome to membranes is not yet conclusive. For example, the published data appear to be equally consistent with phytochrome being a normal membrane component which is differentially released from the membrane during extraction depending on whether or not it is in the P_{fr} state. Nevertheless, irrespective of the speculative models engendered, the demonstration of phytochrome in membrane fractions is a notable step which should provide many fruitful lines to follow.

An intriguing aspect of phytochrome and membranes is the insertion of phytochrome into artificial "black lipid" membranes. Roux and Yguerabide (1973) incorporated purified oat phytochrome (est. 110 000 mol. wt on a Bio-Gel P-150 column) into a membrane made from oxidized cholesterol. Figure 17 shows that the membrane resistance could be reversibly altered by red and far-red light. Unfortunately, the oxidized cholesterol membranes were too unstable in the presence of phytochrome to allow the longer term experiments which would be necessary to determine whether or not the conductance changes involve selective permeability to ions. Since that paper, S. J. Roux (personal communication) has confirmed photoreversible conductance changes with oxidized cholesterol membranes, but has found that phytochrome does not incorporate into lecithin, phosphatidylcholine, or phosphatidylethanolamine model membranes. On the other hand, there appears to be some hope that defined mixtures of other lipids might prove more successful.

There is also evidence from studies on cell-free preparations for phytochrome being an active component of other cell organelles. Manabe and Furuya (1973, 1974) have demonstrated the presence of phytochrome in purified mitochondria and in membranes prepared

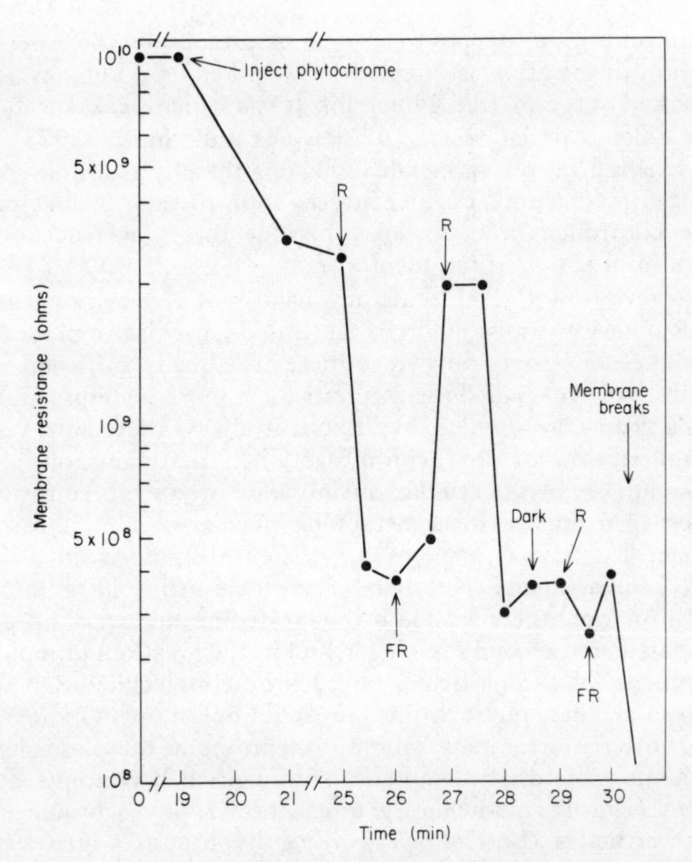

FIG. 17. Photoreversible conductance changes observed in an artificial black-lipid membrane in the presence of purified high molecular weight oat phytochrome. R = red light; FR = far-red light. (Roux and Yguerabide, 1973.)

from the mitochondria. In intact mitochondria, red and far-red light controlled the rate of NADPH formation, but only when exogenous $NADP^+$ was supplied. In the absence of exogenous $NADP^+$, very little NADPH was formed with or without red light. After lysis with Triton X-100, $NADP^+$ reduction rose substantially but was no longer controlled by phytochrome. These results are most readily interpreted in terms of phytochrome controlling the transport of $NADP^+$ across the mitochondrial membrane.

Evans and Smith (1975) have recently demonstrated the existence of phytochrome in purified etioplast preparations from barley leaves. In this case, a considerable enrichment of phytochrome occurs during the purification of the etioplasts, suggesting that a relatively high concentration of phytochrome is present in the etioplasts. Irradiation of the etioplast preparation with red light initiates a rapid and marked surge in free gibberellin levels which peaks at about 5 min. (Cooke and Saunders, 1975; Evans and Smith, 1975). This response is red/far-red reversible indicating the phytochrome control of an *in vitro* reaction. Further investigation of such *in vitro* phytochrome controlled processes may provide direct evidence on the mechanism of action of phytochrome.

When reviewing the molecular mechanism of phytochrome action, it is customary to finish with an illustrated, speculative model. This we shall refrain from doing since there are already sufficient in the literature. All we shall do is to plead for a more widespread use of Occam's razor—the simplest hypothesis is always preferable. Perhaps the simplest view of phytochrome action is that the molecule is a component of plant cellular membranes where it controls the transport of a single critical metabolite (which we may call a second messenger by formal analogy to cyclic-AMP), between different cellular compartments. The specificity of the manifold responses to phytochrome photoconversion is thus vested in the special properties of the postulated second messenger, and in the location, or molecular environment, of the particular phytochrome molecule; i.e. it would be expected that phytochrome in an etioplast membrane would control different reactions from phytochrome in the plasmalemma, or in a mitochondrial membrane. Although such concepts do not provide a rigorous, mathematical description of phytochrome action, they nevertheless constitute a working hypothesis which suggests several lines of investigation.

ACKNOWLEDGEMENTS

The authors are grateful to the following who kindly provided hitherto unpublished material, or made helpful comments on the design of the chapter: A. P. Balangé, W. R. Briggs, S. G. Lisansky, D. Marmé, L. H. Pratt, P. H. Quail, S. J. Roux, W. Rüdiger, and C. J. P. Spruit.

REFERENCES

Anderson, G. R., Jenner, E. L. and Mumford, F. E. (1969). *Biochemistry, N.Y.* 8, 1182.

Anderson, G. R., Jenner, E. L. and Mumford, F. E. (1970). *Biochim. biophys. Acta* 221, 69.

Balangé, A. P. (1974). *Physiol. Veg.* 12, 95.

Boisard, J., Spruit, C. J. P. and Rollin, P. (1968). *Meded. LandbHoogesch. Wageningen* 68-1.

Boisard, J., Marmé, D. and Schäfer, E. (1971). *Planta* 99, 302.
Boisard, J., Marmé, D. and Briggs, W. R. (1974). *Pl. Physiol, Lancaster* 54, 272.
Borthwick, H. A., Parker, W. M. and Hendricks, S. B. (1950). *Am. soc. Nat.* 84, 117.
Borthwick, H. A., Hendricks, S. B., Took, E. H. and Took, V. K. (1954). *Bot. Gaz.* 115, 205.
Briggs, W. R. (1975). *Anais Acad. bras. Cienc.* (in press).
Briggs, W. R. and Fork, D. C. (1969a). *Pl. Physiol., Lancaster* 44, 1081.
Briggs, W. R. and Fork, D. C. (1969b). *Pl. Physiol., Lancaster* 44, 1089.
Briggs, W. R. and Rice, H. V. (1972). *A. Rev. Pl. Physiol.* 23, 293.
Briggs, W. R., Zollinger, W. D. and Platz, B. B. (1968). *Pl. Physiol., Lancaster* 43, 1239.
Burke, M. J., Pratt, D. C. and Moscowitz, A. (1972). *Biochemistry, N.Y.* 11, 4025.
Butler, W. L. (1972). *In* "Phytochrome" (K. Mitrakos and W. Shropshire Jr, eds), p. 185. Academic Press, London and New York.
Butler, W. L. and Lane, H. C. (1965). *Pl. Physiol., Lancaster* 40, 13.
Butler, W. L., Norris, K. H., Siegelman, H. W. and Hendricks, S. B. (1959). *Proc. natn. Acad. Sci. U.S.A.* 45, 1703.
Butler, W. L., Lane, H. C. and Siegelman, H. W. (1963). *Pl. Physiol., Lancaster* 38, 514.
Butler, W. L., Hendricks, S. B. and Siegelman, H. W. (1964a). *Phytochem. Photobiol.* 3, 521.
Butler, W. L., Siegelman, H. W. and Miller, C. O. (1964b). *Biochemistry, N.Y.* 3, 851.
Clarkson, D. T. and Hillman, W. S. (1967a). *Nature, Lond.* 213, 468.
Clarkson, D. T. and Hillman, W. S. (1967b). *Pl. Physiol., Lancaster* 42, 933.
Clarkson, D. T. and Hillman, S. W. (1968). *Pl. Physiol., Lancaster* 43, 88.
Coleman, R. A. (1973). Ph.D. Thesis, Vanderbilt University, Nashville, Tennessee.
Cope, B. T., Smith, U., Crespi, H. L. and Katz, J. J. (1967). *Biochim. biophys. Acta* 133, 446.
Cooke, R. J. and Saunders, P. F. (1975). *Planta* 123, 399.
Correll, D. L., Edwards, J. L., Klein, W. H. and Shropshire, W. Jr (1968a). *Biochim. biophys. Acta* 168, 36.
Correll, D. L., Steers, E. Jr, Towe, K. M. and Shropshire, W. Jr (1968b). *Biochim. biophys. Acta* 168, 46.
Crespi, H. L., Smith, U. and Katz, J. J. (1968). *Biochemistry, N.Y.* 7, 2232.
Cross, D. R., Linschitz, H., Kasche, V. and Tenenbaum, J. (1968). *Proc. natn Acad. Sci. U.S.A.* 61, 1095.
Cundiff, S. C. and Pratt, L. H. (1973). *Pl. Physiol, Lancaster* 51, 210.
De Lint, P. J. A. L. and Spruit, C. J. P. (1963). *Meded. LandbHoogesch. Wageningen,* 63-1.
Dooskin, R. H. and Mancinelli, A. L. (1968). *Bull. Torrey bot. Club* 95, 474.
Dubois, M., Giles, K. A., Hamilton, J. K., Rebers, P. A. and Smith, F. (1956). *Analyt. Chem.* 28, 350.
Evans, A. E. and Smith, H. (1975). *In* "Light and Plant Development" (H. Smith, ed.), Proc. 22nd Easter School, University of Nottingham. Butterworths, London.
Everett, M. S. and Briggs, W. R. (1970). *Pl. Physiol., Lancaster* 45, 679.

Everett, M. S., Briggs, W. R. and Purves, W. K. (1970). *Pl. Physiol., Lancaster* **45**, 805.
Frankland, B. (1972). *In* "Phytochrome" (K. Mitrakos and W. Shropshire Jr, eds), p. 196. Academic Press, London and New York.
Fry, K. T. and Mumford, F. E. (1971). *Biochem. biophys. Res. Commun.* **45**, 1466.
Furuya, M. and Hillman, W. S. (1966). *Pl. Physiol., Lancaster* **41**, 1241.
Furuya, M., Hopkins, W. G. and Hillman, W. S. (1965). *Archs Biochem. Biophys.* **112**, 180.
Gardner, G. M. (1972). Ph.D. Thesis, Harvard University, Cambridge, Mass.
Gardner, G. M. and Briggs, W. R. (1974). *Photochem. Phytobiol.* **19**, 367.
Gardner, G. M., Pike, C. S., Rice, H. V. and Briggs, W. R. (1971). *Pl. Physiol., Lancaster* **48**, 686.
Gardner, G. M., Thompson, W. F. and Briggs, W. R. (1974). *Planta* **117**, 367.
Greenfield, N. and Fasman, G. D. (1969). *Biochemistry, N.Y.* **8**, 4108.
Grill, R. and Spruit, C. J. P. (1972). *Planta* **108**, 203.
Hanke, J., Hartmann, K. M. and Mohr, H. (1969) *Planta* **86**, 235.
Hartmann, K. M. (1966). *Photochem. Photobiol.* **5**, 349.
Hendricks, S. B., Butler, W. L. and Siegelman, H. W. (1962). *J. phys. Chem.* **66**, 2550.
Hillman, W. S. (1964). *Am. J. Bot.* **51** 1102.
Hillman, W. S. (1967). *A. Rev. Pl. Physiol.* **18**, 301.
Hopkins, D. W. (1971). Ph.D. Thesis, Harvard University, Cambridge, Mass.
Hopkins, D. W. and Butler, W. L. (1970). *Pl. Physiol., Lancaster* **45**, 567.
Kidd, G. H. and Pratt, L. H. (1973). *Pl. Physiol., Lancaster* **52**, 309.
Kendrick, R. E. (1972). *Planta* **102**, 286.
Kendrick, R. E. (1974). *Nature, Lond.* **250**, 159.
Kendrick, R. E. and Frankland, B. (1968). *Planta* **82**, 317.
Kendrick, R. E. and Frankland, B. (1969a). *Planta* **85**, 326.
Kendrick, R. E. and Frankland, B. (1969b). *Planta* **86**, 21.
Kendrick, R. E. and Hillman, W. S. (1971). *Am. J. Bot.* **58**, 424.
Kendrick, R. E. and Spruit, C. J. P. (1972a). *Nature, Lond.* (New Biol.) **234**, 28.
Kendrick, R. E. and Spruit, C. J. P. (1972b). *Planta* **107**, 341.
Kendrick, R. E. and Spruit, C. J. P. (1973a). *Photochem. Photobiol.* **18**, 139.
Kendrick, R. E. and Spruit, C. J. P. (1973b). *Photochem. Photobiol.* **18**, 153.
Kendrick, R. E. and Spruit, C. J. P. (1973c). *Pl. Physiol., Lancaster* **52**, 327.
Kendrick, R. E. and Spruit, C. J. P. (1974). *Planta* **120**, 265.
Kendrick, R. E., Spruit, C. J. P. and Frankland, B. (1969). *Planta* **88**, 293.
Kroes, H. H. (1968). *Biochem. biophys. Res. Commun.* **31**, 877.
Kroes, H. H. (1970). *Meded. LandbHoogesch., Wageningen.* 70–81.
Linschitz, H. and Kasche, V. (1967). *Proc. natn. Acad. Sci. U.S.A.* **58**, 1059.
Linschitz, H., Kasche, V., Butler, W. L. and Siegelman, H. W. (1966). *J. biol. Chem.* **10**, 98.
Lisansky, (1973). Ph.D. Thesis, Yale University, New Haven, Conn.
Malcost, R., Boisard, J., Rollin, P. and Spruit, C. J. P. (1970). *Meded. LandbHoogesch., Wageningen* 70–16.
Marmé, D. (1969). *Planta* **88**, 43.
Marmé, D., Marchal, B. and Schäfer, E. (1971). *Planta* **100**, 331.
Marmé, D., Boisard, J. and Briggs, W. R. (1973). *Proc. natn. Acad. Sci. U.S.A.* **73**, 3861.

Marmé, D., Mackenzie, J. M. Jr, Boisard, J. and Briggs, W. R. (1974). *Pl. Physiol., Lancaster* 54, 263.

Manabe, K. and Furuya, M. (1973). *Pl. Physiol., Lancaster* 51, 982.

Manabe, K. and Furuya, M. (1974). *Pl. Physiol., Lancaster* 53, 343.

McArthur, J. A. and Briggs, W. R. (1971). *Pl. Physiol., Lancaster* 48, 46.

Mitrakos, K. and Shropshire, W. Jr, (eds), (1972). "Phytochrome". Academic Press, London and New York.

Mumford, F. E. (1966). *Biochemistry, N.Y.* 5, 522.

Mumford, F. E. and Jenner, E. L. (1966). *Biochemistry, N.Y.* 5, 3657.

Mumford, F. E. and Jenner, E. L. (1971). *Biochemistry, N.Y.* 10, 98.

Newman, I. A. and Briggs, W. R. (1972). *Pl. Physiol., Lancaster* 50, 687.

Oelze-Karow, H., Schopfer, P. and Mohr, H. (1970). *Proc. natn Acad. Sci. U.S.A.* 65, 51.

Ó hEocha, C. (1965). *A. Rev. Pl. Physiol.* 16, 415.

Parker, M. W., Hendricks, S. B. and Borthwick, H. A. (1950). *Bot. Gaz.* 111, 242.

Pike, C. S. and Briggs, W. R. (1972). *Pl. Physiol., Lancaster* 49, 514.

Pratt, L. H. (1973). *Pl. Physiol., Lancaster* 51, 203.

Pratt, L. H. and Briggs, W. R. (1966). *Pl. Physiol., Lancaster* 41, 1189.

Pratt, L. H. and Butler, W. L. (1968). *Photochem. Photobiol.* 8, 477.

Pratt, L. H. and Butler, W. L. (1970a). *Photochem. Photobiol.* 11, 361.

Pratt, L. H. and Butler, W. L. (1970b). *Photochem. Photobiol.* 11, 503.

Pratt, L. H. and Coleman, R. A. (1971). *Proc. natn. Acad. Sci. U.S.A.* 60, 2431.

Pringle, J. R. (1970). *Biochem. biophys. Res. Commun.* 39, 46.

Purves, W. K. and Briggs, W. R. (1968). *Pl. Physiol., Lancaster* 43, 1259.

Quail, P. H., Marmé, D. and Schäfer, E. (1973a) *Nature, Lond.* (New Biol.) 39, 11.

Quail, P. H., Schäfer, E. and Marmé, D. (1973b). *Pl. Physiol., Lancaster* 52, 124.

Quail, P. H., Schäfer, E. and Marmé, D. (1973c). *Pl. Physiol., Lancaster* 52, 128.

Rice, H. V. and Briggs, W. R. (1973a). *Pl. Physiol., Lancaster* 51, 927.

Rice, H. V. and Briggs, W. R. (1973b). *Pl. Physiol., Lancaster* 51, 939.

Rice, H. V., Briggs, W. R. and Jackson-White, C. J. (1973). *Pl. Physiol., Lancaster* 51, 917.

Rollin, P. (1972). *In* "Phytochrome" (K. Mitrakos and W. Shropshire, Jr, eds), p. 229. Academic Press, London and New York.

Roux, S. J. (1971). Ph.D. Thesis, Yale University, New Haven, Conn.

Roux, S. J. (1972). *Biochemistry, N.Y.* 11, 1930.

Roux, S. J. and Hillman, W. S. (1969). *Archs Biochem. Biophys.* 131, 423.

Roux, S. J. and Yguerabide, J. (1973). *Proc. natn. Acad. Sci. U.S.A.* 70, 762.

Rüdiger, W. (1972). *In* "Phytochrome" (K. Mitrakos and W. Shropshire, Jr, eds), p. 129. Academic Press, London and New York.

Rüdiger, W. and Correll, D. L. (1969). *Justus Liebig's Annln Chem.* 723, 208.

Sanger, F. (1945). *Biochem. J.* 39, 507.

Satter, R. L., Marinoff, P. and Galston, A. W. (1970). *Am. J. Bot.* 57, 916.

Schäfer, E., Marchal, B. and Marmé, D. (1971). *Planta* 101, 265.

Schäfer, E. and Schmidt, W. (1974). *Planta* 116, 257.

Schmidt, W. and Schäfer, E. (1974). *Planta* 116, 267.

Schmidt, W., Marmé, D., Quail, P. and Schäfer, E. (1973). *Planta* 111, 329.

Shropshire, W. Jr. (1972). *In* "Photophysiology" (A. C. Giese, ed.), vol. 7, p. 33. Academic Press, London and New York.

Siegelman, H. W. and Butler, W. L. (1965). *A. Rev. Pl. Physiol.* **16**, 383.

Siegelman, H. W. and Firer, E. M. (1964). *Biochemistry, N.Y.* **3**, 418.

Siegelman, H. W., Chapman, D. J. and Cole, W. J. (1968). *In* "Porphyrin and Related Compounds" (T. W. Goodwin, ed.), p. 107. Academic Press, London and New York.

Siegelman, H. W., Turner, B. C. and Hendricks, S. B. (1966). *Pl. Physiol., Lancaster* **41**, 1289.

Smith, H. (1974). "Phytochrome and Photomorphogenesis". McGraw-Hill, London and New York.

Song, P.-S., Chae, Q., Lightner, D. A., Briggs, W. R. and Hopkins, D. (1973). *J. Am. chem. Soc.* **95**, 7892.

Spruit, C. J. P. (1966a). *In* "Currents in Photosynthesis" (J. B. Thomas and J. C. Goedheer, eds), p. 67. Douker, Rotterdam.

Spruit, C. J. P. (1966b). *Biochim. biophys. Acta* **112** 186.

Spruit, C. J. P. (1966c). *Biochim. biophys. Acta* **120**, 454.

Spruit, C. J. P. (1966d). *Meded. LandbHoogesch., Wageningen* 66-15.

Spruit, C. J. P. (1972). *In* "Phytochrome" (K. Mitrakos and W. Shropshire, Jr, eds), p. 78. Academic Press, London and New York.

Spruit, C. J. P. and Kendrick, R. E. (1972). *Planta* **103**, 319.

Spruit, C. J. P. and Kendrick, R. E. (1973). *Photochem. Photobiol.* **18**, 145.

Spruit, C. J. P. and Mancinelli, A. L. (1969). *Planta* **88**, 303.

Stark, G. R. and Smyth, D. G. (1963). *J. biol. Chem.* **238**, 214.

Suzuki, H. and Hamanaka, T. (1969). *J. physiol. Soc. Japan* **26**, 1462.

Suzuki, H., Sugimoto, T. and Nakachi, K. (1973). *J. physiol. Soc. Japan* **34**, 1045.

Taylor, A. O. (1968). *Pl. Physiol., Lancaster* **43**, 767.

Taylor, A. O. and Bonner, B. A. (1967). *Pl. Physiol., Lancaster* **42**, 762.

Tobin, E. M. and Briggs, W. R. (1969). *Pl. Physiol., Lancaster* **44**, 148.

Tobin, E. M. and Briggs, W. R. (1973). *Phytochem. Photobiol.* **18**, 487.

Tobin, E. M., Briggs, W. R. and Brown, P. K. (1973). *Photochem. Photobiol.* **18**, 497.

Walker, T. S. and Bailey, J. L. (1968). *Biochem. J.* **107**, 603.

Walker, T. S. and Bailey, J. L. (1970a). *Biochem. J.* **120**, 607.

Walker, T. S. and Bailey, J. L. (1970b). *Biochem. J.* **120**, 613.

Weber, K. and Osborn, M. (1969). *J. biol. Chem.* **244**, 4406.

Chapter 8

Nature and Properties of Flavonoids*

T. SWAIN

ARC Laboratory of Biochemical Systematics, Royal Botanic Gardens, Kew, Richmond, Surrey, England

I Introduction

Nature as it surrounds us in the height of summer is predominantly green. It is not surprising, therefore, that plants or parts of plants which are in strong contrast to this overwhelming greenness have always attracted man and other denizens of the animal kingdom. Although the various shades of yellow are pleasing to the majority of people, the most fascinating colours are those which make the deepest contrast to green, that is, various shades of red and blue. Almost all the brilliant colours of this type which are found in flowers and fruit are due to the presence of one or more of the groups of flavonoid compounds known as anthocyanins (e.g. **I**,

Fig. 1). Other classes of flavonoid compound are responsible for the yellow colour of certain flowers, although more usually this is due to the presence of carotenoid pigments. Yet other flavonoids account for the actual whiteness in most white flowers, which without them would perhaps appear almost translucent. Finally some of the brown and black pigments found in nature are due either to the products of oxidation of flavonoid and related phenolic compounds, or to their chelates with metals.

II Classes of flavonoid compounds

The term flavonoid was first applied twenty years or so ago by Geissman and Hinreiner (1952) (Harborne *et al.*, 1975) to embrace all those compounds whose structure is based on that of flavone (2-phenylchromone, **II**) (L. *flavus*, yellow). Occasionally the term is misspelt as flavanoid which might be a better one to use as the parent skeleton of the group is really flavan (**III**, 2-phenylchroman) in which the heterocyclic ring is fully reduced. A number of other terms have been, and still are sometimes generically used, often in a rather loose way, to cover various groups of flavonoid compounds; for example, anthoxanthin (Gr. *anthos*, flower; *xanthos*, yellow); anthochlor (Gr. *chloros*, pale green); chymochrome (Gr. *chymos*, juice; *chroma*, colour); and, as mentioned above, anthocyanin (Gr. *kyanos*, blue). This last term was the first to be introduced to describe a class of flavonoid pigments (as anthocyan) by Marquart in 1835.

It can be seen that flavone (**II**, Fig. 1) consists of two benzene rings (A and B) joined together by a three-carbon link which is formed into a γ-pyrone ring. The various classes of true flavonoid compounds differ one from another only by the state of oxidation of this 3-C link. There is a limitation to the number of structures commonly found in nature, which vary in their state of oxidation from flavan-3-ols (catechins, **IV**) to flavonols (3-hydroxyflavones, **V**) and anthocyanins (**I**). Also included in the flavonoids are the flavanones (**VI**), flavanonols or dihydroflavonols (**VII**) and the flavan-3,4-diols (proanthocyanidins) (**VIII**). It should be noted that there are also five classes of compounds (the dihydrochalkones, 3-phenylpropiophenones (**IX**); the chalkones, phenylstyryl ketones (**X**; Gr. *chalcos*, copper); the isoflavones, 3-phenylchromones (**XI**); the neoflavones, 4-phenylcoumarins (**XII**) and the aurones, 2-benzylidine-3-coumaranones (**XIII**; L. *aurum,* gold)) which do not actually possess the basic 2-phenylchroman skeleton (**II**), but are so closely related both chemically and biosynthetically to the other flavonoid types that they are always included in the flavonoid group.

The individual compounds within each class are distinguished mainly by the number and orientation of hydroxyl, methoxyl and other groups substituted in the two benzene rings (A and B, **II**, Fig. 1). These groups are usually arranged in restricted patterns in the flavonoid molecule, reflecting the different biosynthetic origins of the two aromatic nuclei. Thus in the A ring (**II**) of the majority of flavonoid compounds, hydroxyl groups are substituted at either both C-5 and C-7, or only at C-7, and generally are unmethylated. This pattern of hydroxylation follows from the acetate or malonate origin of this ring (see Chapter 20). The B ring (**II**) of the flavonoids on the other hand is usually substituted by either one, two, or three hydroxyl or methoxyl groups. The first, which is rarely methylated, is substituted in the position *para* to the point of attachment of this ring to the rest of the molecule (C-4′, in **II**), with the second and third groups *ortho* to it at C-3′ and C-5′, these latter two groups often being methylated. The hydroxylation pattern of the B ring thus resembles that found in the commonly occurring cinnamic acids (**XIV**) and coumarins and reflects their common biosynthetic origin from prephenic acid and its congeners.

Flavonoid compounds usually exist in the plant, except in non-living woody tissues, as glycosides; that is, one or more of their hydroxyl groups is joined by a semiacetal link to C-1 of a sugar. Here again, as will be discussed later, there is a restriction on the hydroxyl groups which are so affected. The sugar-free compounds are referred to as *aglycones*, and although their presence has often been reported in non-woody tissues, it is probable that in most cases they are formed as artefacts during the course of extraction, since most living tissues contain very active glycosidases which work even in the presence of high concentrations of organic solvents. Glycosidation not only confers sap solubility to the generally somewhat insoluble flavonoid aglycones, but also confers stability, especially for the anthocyanidins and more highly hydroxylated compounds. For example whereas quercetin and myricetin (**V**, Fig. 1) are susceptible to oxidation catalysed by phenolase, the corresponding 3-*O*-glycosides are stable (Roberts, 1960).

The sugars which have been found in flavonoid glycosides include simple hexoses and pentoses (monosides), and di- and trisaccharides (biosides and triosides) always combined through the oxygen at the C-1 usually by a β link. In many cases, more than one phenolic hydroxyl group in the flavonoid molecule may be glycosylated, giving rise to di-monosides and so on (e.g. cyanin, cyanidin-3,5-di-β-D-glucoside, **I**). D-Glucose, occurring either alone or as part of a

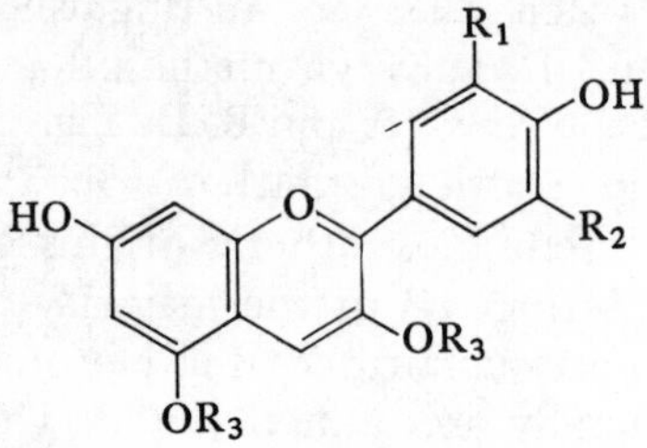

(I)

Anthocyanidins (R_3 = H)
R_1R_2 = H pelargonidin
R_1 = OH, R_2 = H cyanidin
R_1R_2 = OH delphinidin
Anthocyanins (R_3 = sugar)
R_1 = OH, R_2 = H, R_3 = glucose cyanin

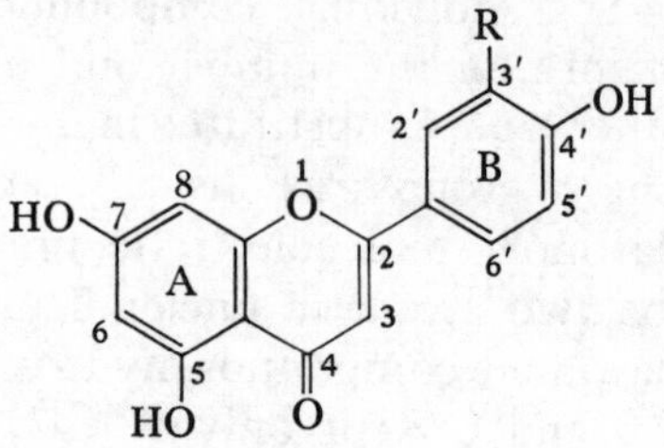

(II)

Flavones
R = H apigenin
R = OH luteolin
Common glycosides have the sugar at C-7 hydroxyl group

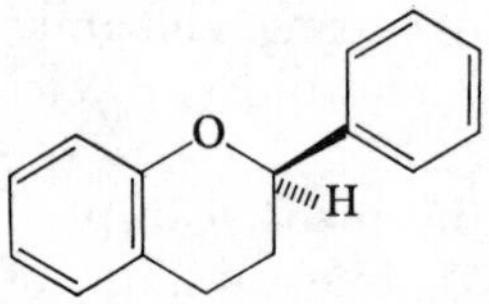

(III)

Flavan
Note that the 2*S* form is shown (phenyl group above plane of heterocyclic ring)

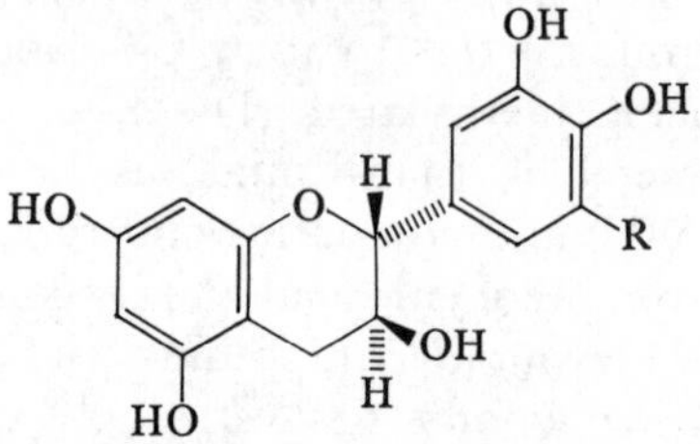

(IV)

Catechins
(2*R*, 3*S* as shown)
R = H (+)-catechin
R = OH (+)-gallocatechin
(2*R*, 3*R*: opposite configuration at C-3)
R = H (−)-epicatechin
R = OH (−)-epigallocatechin

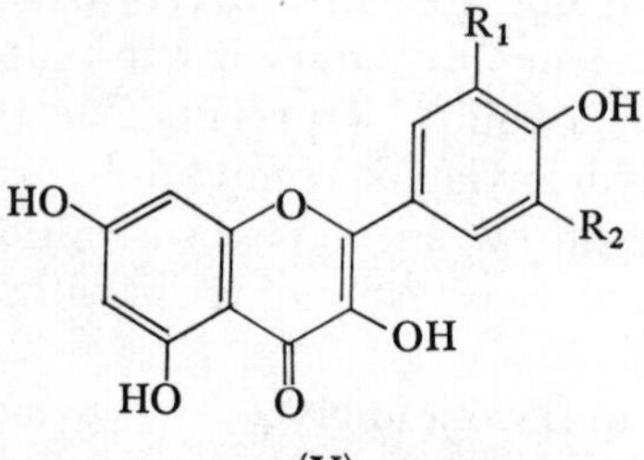

(V)

Flavonols (3-hydroxyflavones)
R_1R_2 = H kaempferol
R_1 = OH, R_2 = H quercetin
R_1R_2 = OH myricetin
Common glycosides have the sugar(s) at C-3 and/or C-7

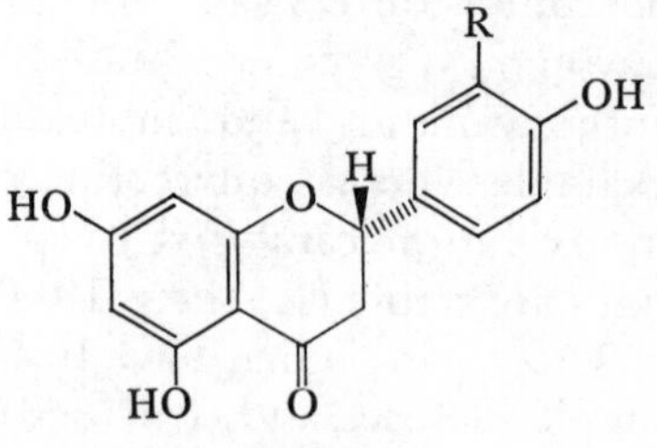

(VI)

Flavanones
Note that stereochemistry at C-2 (2*R*) is the same as the catechins
R = H naringenin
R = OH eriodictyol

FIG. 1. Structures of flavonoid aglycones.

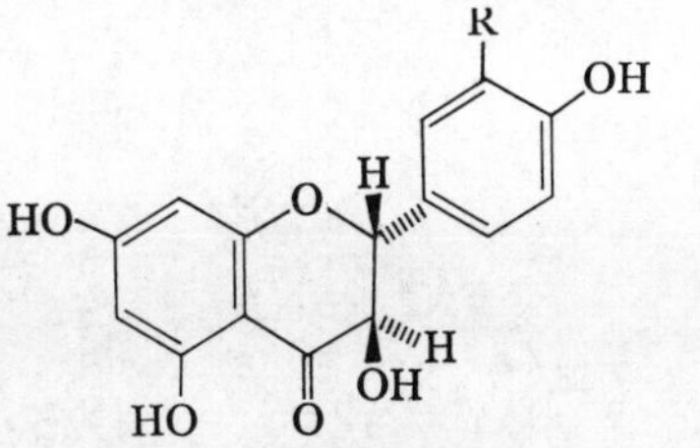

(VII)

Dihydroflavonols (flavanones)
Note that the stereochemistry at C-2 C-3 (2*R*-3*S*) is the same as (−)-epicatechin, IV
R = H, dihydrokaempferol
R = OH dihydroquercetin (taxifolin)

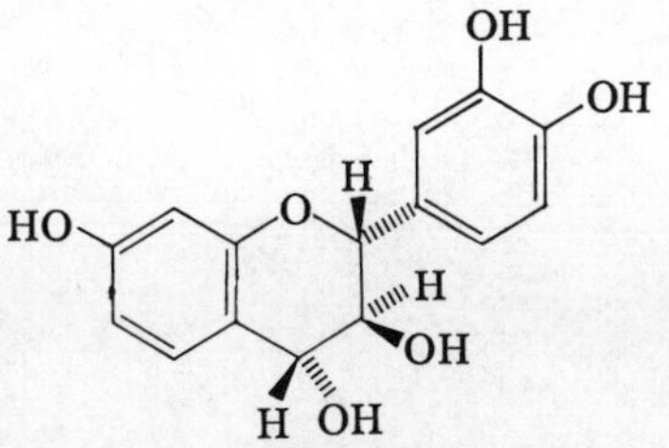

(VIII)

Flavan-3,4-diols or proanthocyanidins
Mollisacacidin (2*R*, 3*S*, 4*R*)
(+)-Leucofisetinidin has opposite configuration (2*S*, 3*R*, 4*S*)
Flaven-3,4-diols with a 5 hydroxy group do not appear to exist in nature except as dimers (see p. 459)

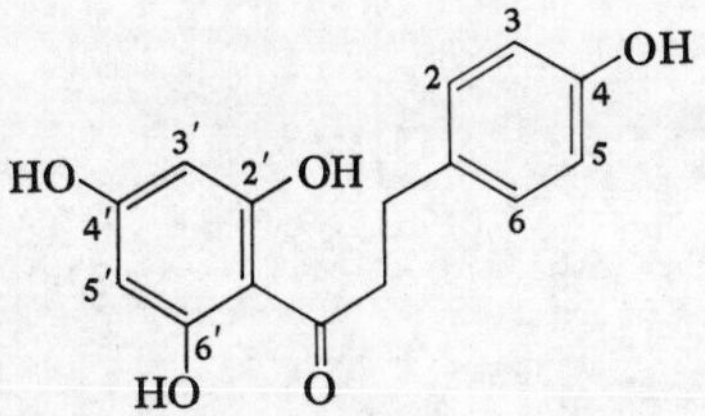

(IX)

Dihydrochalkones (note numbering)
phloretin

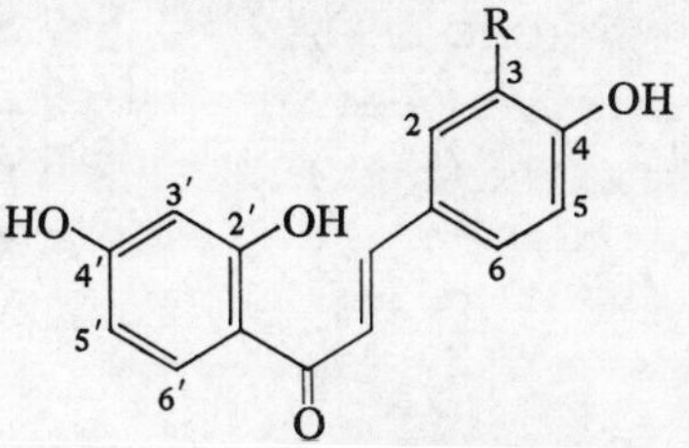

(X)

Chalkones (note numbering)
R = H isoliquiritigenin
R = OH butein

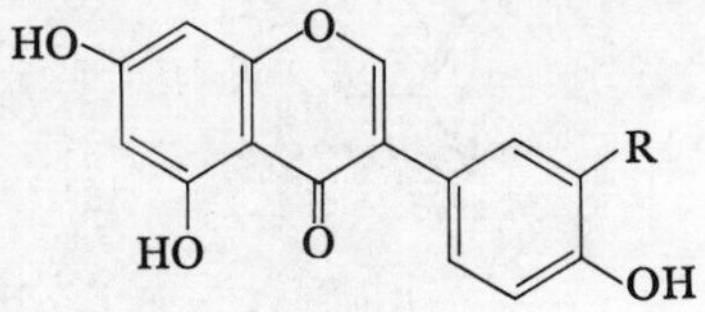

(XI)

Isoflavones (numbering same as flavones)
R = H genistein
R = OH orobol

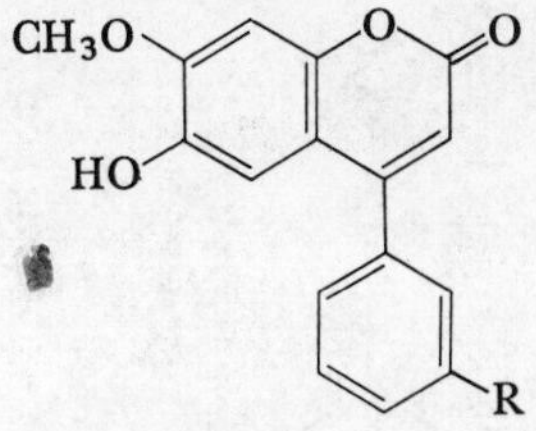

(XII)

Neoflavones (dalbergins or 4-phenylcoumarins)
R = H dalbergin
R = OH stevenin

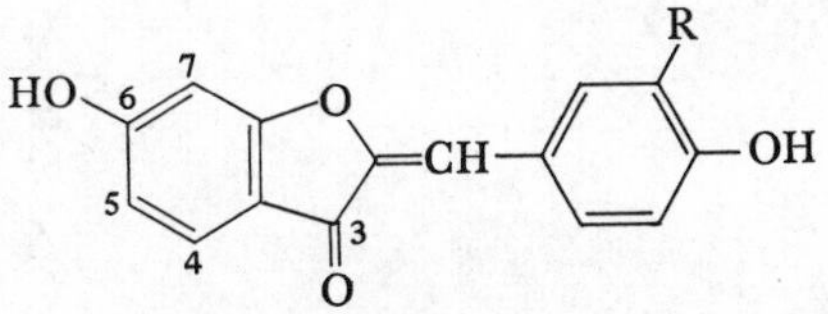

(XIII)

Aurones (note numbering)
R = H hispidol
R = OH sulphuretin

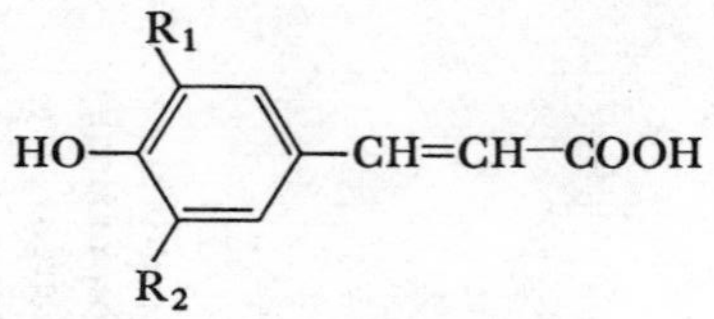

HO—C6H2(R1)(R2)—CH=CH—COOH

(XIV)

Cinnamic acids
R_1R_2 = H, *p*-coumaric acid
R_1 = OH, R_2 = H caffeic acid
R_1 = OMe, R_2 = H ferulic acid
R_1R_2 = OMe sinapic acid

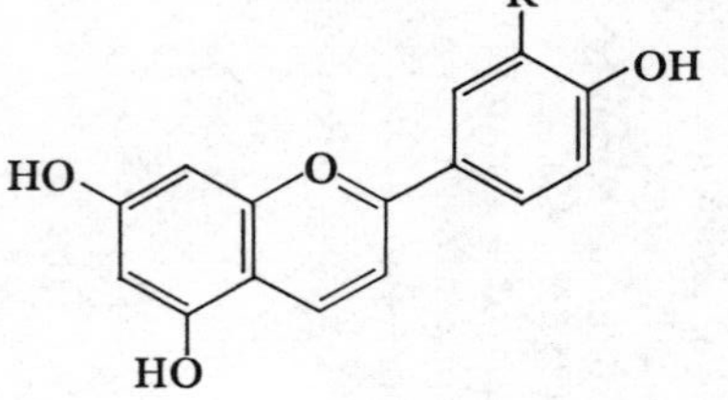

(XV)

3-Deoxyanthocyanidins
R = H apigeninidin
R = OH luteolinidin

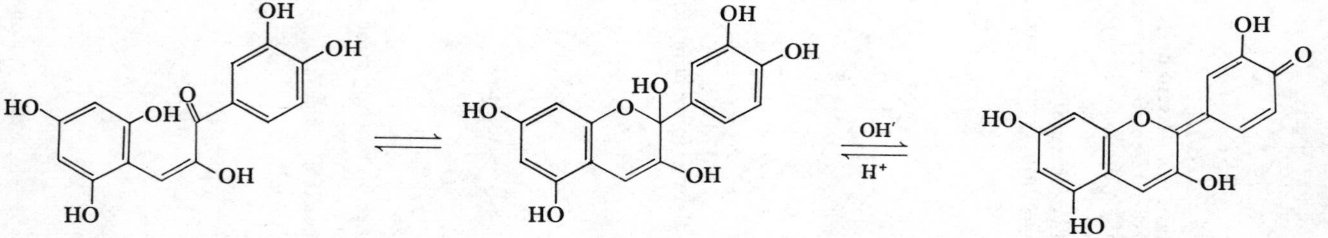

$\rightleftharpoons$ $\underset{H^+}{\overset{OH'}{\rightleftharpoons}}$

(XVII)
Chalkone form of pseudobase of cyanidin

(XVI)
Pseudo base of cyanidin (pH 7·0)

(XVIII)
One tautomeric form of the anhydro base of cyanidin (pH > 9·0)

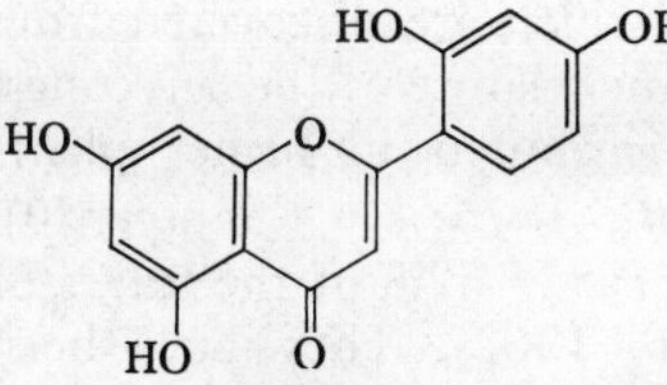

(XIX) Morin

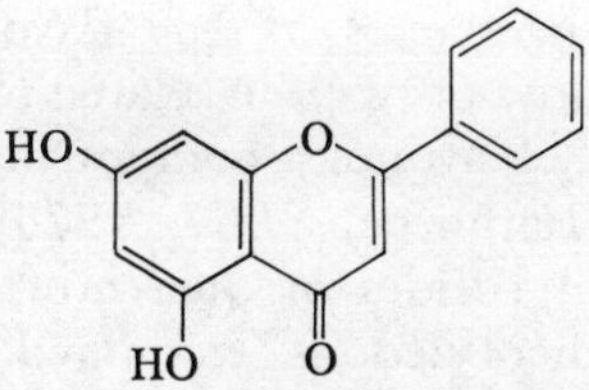

(XX) Chrysin

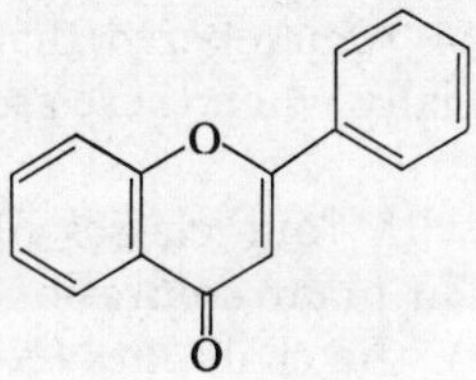

(XXI) Flavone

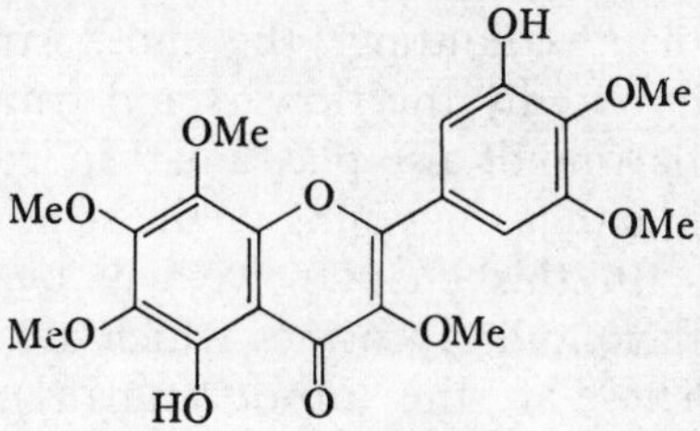

(XXII) Digicitrin

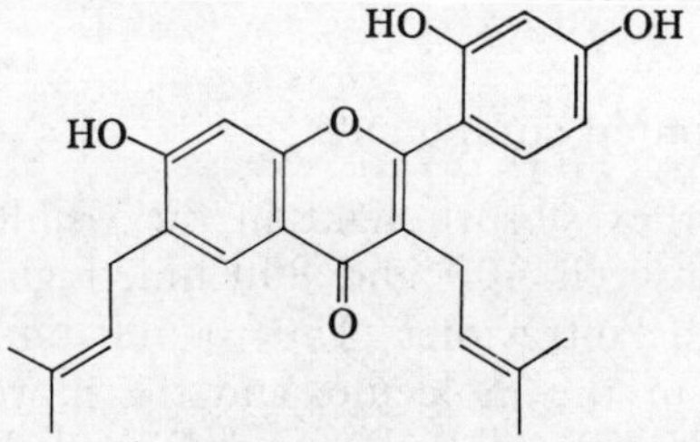

(XXIII) Artocarpin

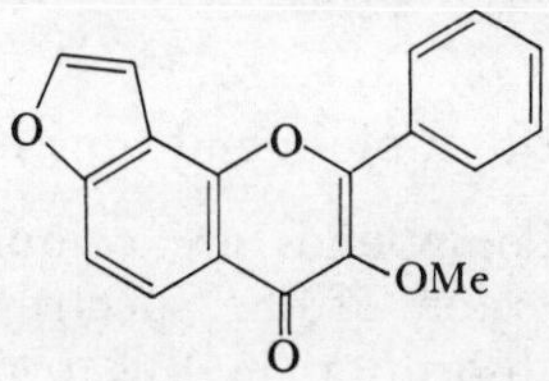

(XXIV) Karanjin

disaccharide, is the most common sugar in glycosides; D-galactose and L-rhamnose are less frequent; and L-arabinose, D-glucuronic acid and D-xylose are rather rare.

Until the last ten years or so, most of the flavonoid glycosides were each given a trivial name, often confusing and usually derived from the species or genus of plant from which they were first isolated. For example, the 3-β-L-rhamnoside of quercetin (**VI**) is called quercitrin, and that from kaempferol (**V**) afzelin. This multiplicity of names makes it very difficult even for the expert to search the literature for compounds belonging to a single class, or derivatives of a single compound. In some cases different names are still being used for the same substance! It is to be hoped, however, that the names of any flavonoid glycosides which are discovered from now on will be related to the present nomenclature of the aglycones.

Flavonoid compounds are widely distributed in higher plants (Harborne, 1967, 1973; Harborne *et al.*, 1975; see Chapter 20); glycosides of quercetin, for example, occur in 62% of the leaves of dicotyledons examined by Bate-Smith (1962). They have been isolated from all the different parts of plants, although there are usually variations in the types of compound found in the various anatomical tissues of any one plant (cf. Griffiths, 1958). In terms of plant colouring, the most striking of such plant organs are undoubtedly the flowers and fruits, but it should be remembered that flavonoids also play a part in leaf colouring, especially when these are in either the young or the senescent state.

In this chapter we will be dealing mainly with four classes of flavonoid glycosides which are involved directly in plant coloration. These are the anthocyanins (**I**), the flavonols (**V**), the chalkones (**X**) and the aurones (**XIII**). Before discussing the nature and properties of these classes in detail, brief consideration must be given to the relationship between their colour and their constitution.

III Colour and constitution of flavonoid compounds

Compounds are coloured because they absorb light in the visible region of the spectrum, that is between 400 and 800 nm. Light absorption in this region and in the ultraviolet (150 to 400 nm) causes the excitation of electrons in the molecule, and the more firmly such electrons are bound the higher will be the energy needed, that is, the shorter will be the wavelength at which the light is absorbed. Thus, saturated hydrocarbons show no absorption above 160 nm, but compounds containing double or triple bonds, or hetero atoms with non-bonded electrons show absorption at much higher wavelengths.

Graebe and Liebermann (1868) were the first to recognize the relationship between colour and unsaturation. Some years later Witt (1888) coined the terms chromophore (Gr. *phoros*, bearing) for organic groups which produced colour when present in a molecule and auxochrome for other groups which deepened such colours. The relationship between unsaturation, the structure of the chromophores and colours did not become apparent, however, until the electronic theories of organic chemistry had been fully worked out.

It is now generally recognized that the presence of conjugation, chromophores and auxochromes in a molecule all act in promoting greater ease of electron transition to higher energy levels by stabilization of the resulting excited structure. In any given type of molecule the longer the conjugated chain, or the greater the number of chromophores and auxochromes, the easier are such transitions, and the longer the wavelength of absorption.

These facts are well illustrated in the flavonoid field (see Chapter 20, Table IV). The hydroxy-substituted flavans (catechins, **IV**, and proanthocyanins, **VIII**) in which no exocyclic conjugation of the two benzene rings occurs, have single-banded absorption spectra like that of the simple phenols with λ_{max} 275–280 nm (Fig. 2). Similarly, the flavanones (**VI**) and isoflavones (**XI**) have spectra like those of the corresponding hydroxyacetophenones, since here only the hydroxy groups in the A ring are in conjugation with the carbonyl group at C-4 (Fig. 3). However, the conjugation between the hydroxy groups of the B ring and the carbonyl groups at C-4 becomes important in the flavones, and the resulting spectra consist of two well-separated bands of high absorptivity ($\log \epsilon > 4{\cdot}0$). The one at longer wavelength (band I) being associated with B-ring conjugation, and

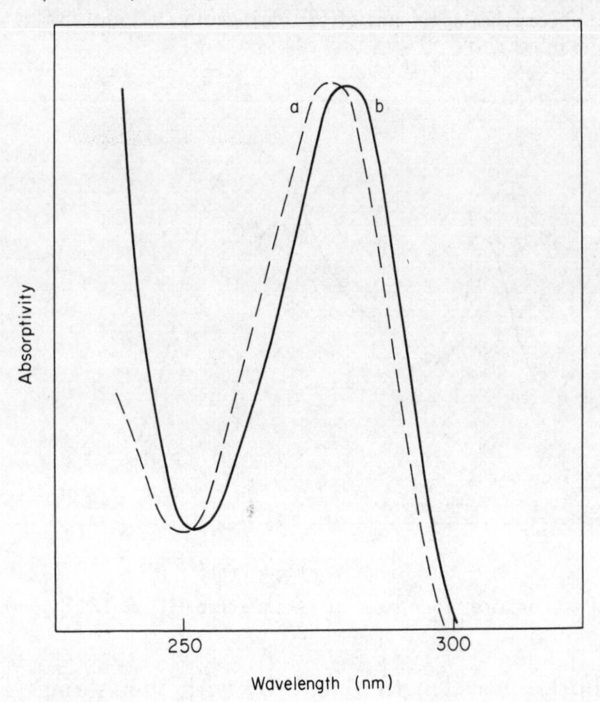

FIG. 2. The absorption spectrum of (a) (+)-catechin (**IV**) and (b) catechol (1,2-dihydroxybenzene) in ethanol.

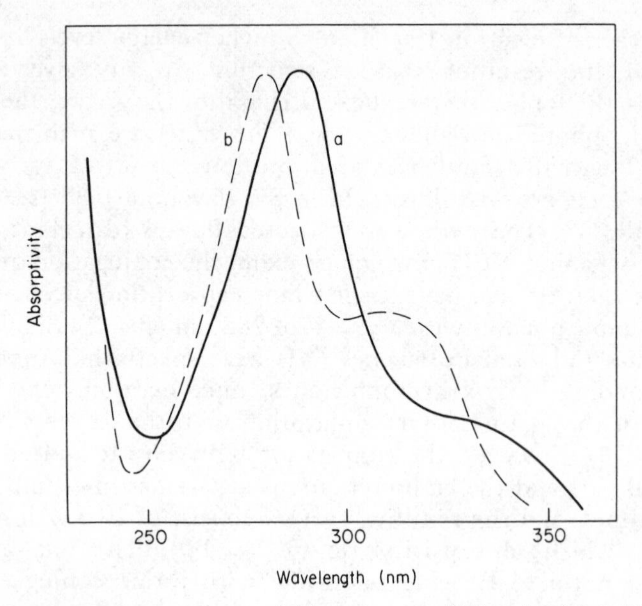

FIG. 3. The absorption spectrum of (a) naringenin (**VI**) and (b) 2,4-dihydroxyacetophenone in ethanol.

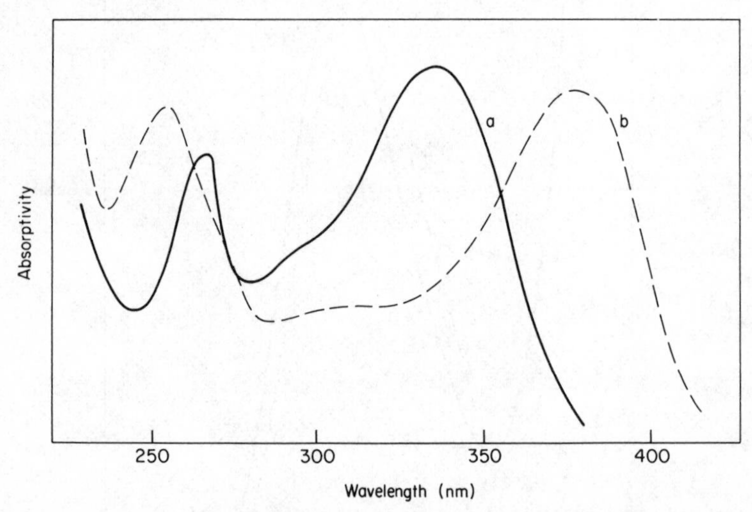

FIG. 4. The absorption spectrum of (a) apigenin (**II**) and (b) quercetin (**V**) in ethanol.

that at shorter wavelength (band II) with the A ring (Fig. 4). In polyhydroxy-substituted compounds there is a great deal of overlap of individual absorption bands which is due to the contributions made by the various canonical forms, and many weaker bands (mainly arising from transitions of the non-bonded oxygen electrons) are masked or combined to give broad maxima. In the flavonols (**V**), band I moves to longer wavelengths than in the flavones owing to the contribution of the vinylic hydroxyl group at C-3 (Fig. 4). At sufficiently high concentration, solutions of flavonols absorb sufficiently in the blue region of the visible spectrum to appear yellow. Chalkones (**X**) are deeper yellow because they absorb a slightly higher wavelength range than the flavonols. In aurones (**XIII**), band I is at an even longer wavelength probably owing to the planarity of the five-membered heterocyclic ring and consequent greater ease of resonance (Jurd, 1962a).

The anthocyanin salts (**I**) in acid solution, in which the electrons of the heterocyclic oxygen atom are involved in π bond formation, absorb at the longest wavelength of any class of flavonoid compounds (Fig. 5). It should be noted, however, that in neutral or basic solution the anthocyanins form unstable tautomeric structures from the pseudobase or phenolate ions of I and therefore have different absorption spectra (see p. 438).

It can be seen, therefore, that as conjugation increases, flavonoid compounds absorb light at longer wavelengths. In any one class of compounds, increasing hydroxylation has the same effect, owing to the increase in the non-bonding electrons supplied by the hydroxyl group. Naturally there is a difference in the contribution made by each hydroxyl group depending on its position of substitution (cf. Harborne, 1967), generally in accordance with the expectation from

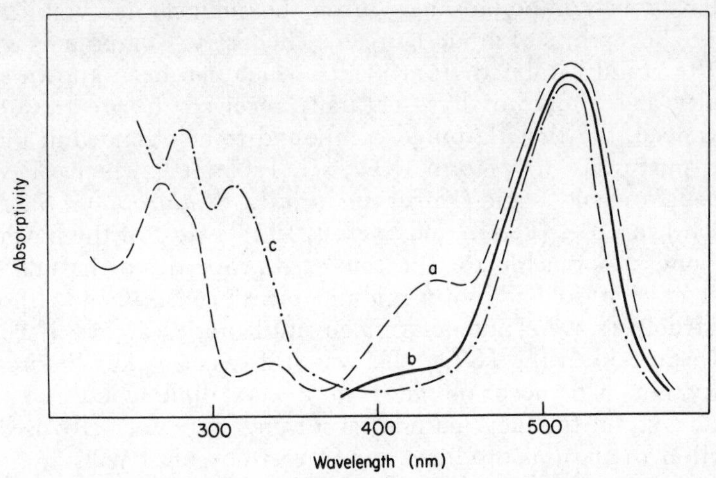

FIG. 5. The absorption spectrum of (a) pelargonidin-3-glucoside, (b) pelargonidin-3,5-diglucoside and (c) pelargonidin-3-*p*-coumarylglucoside-5-glucoside in 0·01% methanolic hydrochloric acid.

organic chemical theory. Acylation or glycosylation of all the hydroxyl groups in flavonoid compounds results in a hypsochromic (to lower wavelength) shift, usually resulting in a spectrum like that of the parent nucleus. Here the non-bonded electrons of the hydroxyl oxygens are involved in resonance with the acyl group itself. Methylation or glycoside formation has a much lesser hypsochromic effect, except where the vinylic hydroxyl group of flavonols is involved.

Finally one should remember that the spectra of all groups of flavonoid compounds are markedly altered when the phenolic hydroxyl groups are ionized (i.e. above pH 9) or involved in copigmentation or in chelate formation with metals (Mabry *et al.*, 1970). Indeed, the formation of deeper colours when flavonoids are treated with iron alum and similar mordants, which also makes the pigments substantive, formed the basis of the old dyestuffs industry. For example, dyers' weld (*Reseda lutea* L.), which is the oldest known European dyestuff (Perkin and Everest, 1918) contains luteolin (**II**), which is itself hardly coloured (λ_{max} 350 nm) but which with aluminium gives an attractive yellow colour (λ_{max} 390 nm).

IV Anthocyanins

A INTRODUCTION

The flavonoid compounds have been the subject of investigations since the beginnings of modern organic chemistry. Compounds which could be readily isolated from plants which had been known since antiquity as a source of dyes naturally received the greatest attention. Indeed, the first flavonoid compound to be obtained in a more or less pure state was morin (**XIX**, see Table III), which Chevreul obtained from old fustic (*Morus tinctoria* L. = *Chlorophora tinctoria* Gaudich.) in 1814 (Perkin and Everest, 1918). Most of the flavonoid compounds responsible for the tinctorial properties of natural dyestuffs had been isolated by the mid-nineteenth century, and although their structures were not determined until the researches of Perkin and Kostanecki in the 1890s, this was still twenty years before any anthocyanin had been obtained in a crystalline condition. The difficulty lay in the fact that neutral solvents were normally used for extraction of natural products and it was not until Willstätter and Everest (1913) recognized the oxonium nature of anthocyanins and used acids for their extraction that rapid progress was made. These two workers isolated cyanin (**I**) from cornflower and recognized that it was a diglucoside. They also proposed the terms *anthocyanin* and *anthocyanidin* for the glycosides and aglycones respectively, and introduced methods for distinguishing them based on differences in their distribution between amyl alcohol and aqueous acid. The flavylium salt structure of the anthocyanidins (**I**) was proposed one year later by Everest (1914) and was rapidly confirmed by the brilliant synthetic work of Willstätter and his collaborators (summarized by Dean, 1963). Later Robinson and Karrer and their co-workers extended such syntheses to include all the then known anthocyanins.

B ANTHOCYANIDINS

A total of twenty-two anthocyanidins are known, four of which are found only as products from proanthocyanidins (Timberlake and Bridle, 1975). They range from apigeninidin (**XV**, gesneridin), first isolated as a 5-glucoside from *Gesneria fulgens* (Robinson and Robinson, 1932), to hirsutidin, first obtained from *Primula hirsuta* (Karrer and Widmer, 1927b) (Table I). The variation in the pattern of substitution is more limited; twelve of the compounds being based on the structure given in (**I**) with the 3,5,7-trihydroxy substitution pattern most expected on biosynthetic grounds. The glycosides of three of these anthocyanidins, cyanidin, delphinidin and pelargonidin (**I**), are the most common in nature; derivatives being present in 80% of pigmented leaves, 69% of fruits and 50% of flowers. By contrast, anthocyanidins lacking a hydroxyl group at C-3 (**XV**) are relatively rare in higher plants.

As might be expected from what was said previously, the visible colour of the anthocyanidins is dependent on the number and orientation of hydroxyl and methoxyl groups in the molecule and varies from orange yellow (apigenidin, λ_{max} 476 nm) to blue (delphinidin, λ_{max} 546 nm) (Harborne, 1967). The colour is, however, dependent on the solvent and is bluer (i.e. at lower wavelengths, usually 5–10 nm) in alcoholic acid than in aqueous acid. Owing to the ionic character of the anthocyanidins the intensity of colour is also markedly dependent on pH, and as this is raised to near neutrality more and more of the pigment is transformed into the colourless pseudobase form (2-hydroxy chromene) (**XVI**). According to the data of Swain and Hillis (1959) cyanin (**I**) has a p*K* of approximately 2, but the values for other flavylium salts may be somewhat higher (Albert, 1959). It should be noted that only band I is affected, the intensity of band II (λ_{max} 280 nm)

TABLE I

Some naturally occurring anthocyanidins. Compounds are arranged first by number and arrangement of hydroxyl and other groups in the A ring and then by number of substituents in the B ring

Trivial name	Structure[a]					Source[b]
	A ring		B ring			
	Hydroxylation	Other	3′	4′	5′	
3-Deoxyanthocyanidins (e.g. **XV**)						
Apigenidin	5,7-diOH	—	—	OH	—	*Rechsteineria*
Luteolinidin	5,7-diOH	—	OH	OH	—	*Rechsteineria*
Tricetinidin	5,7-diOH	—	OH	OH	OH	*Camelia*
Columnidin[c]	5,7-diOH?	—			—	*Columnea*
Carexidin[c]	5,7-diOH?					*Carex*
Anthocyanidins (e.g. **I**)						
None[d]	7-OH	—	—	OH	—	*Acacia*
Fisetinidin[d]	7-OH	—	OH	OH	—	*Acacia*
Robinetinidin[d]	7-OH	—	OH	OH	OH	*Robinia*
Pelargonidin	5,7-diOH	—	—	OH	—	*Pelargonium*
Margicassidin	5,7-diOH	6 (or 8) alkenyl	—	OH	—	—
Cyanidin	5,7-diOH	—	OH	OH	—	*Centaurea*
Peonidin	5,7-diOH	—	OMe	OH	—	*Paeonia*
Rosinidin	5,7-diOH	7-OMe	OMe	OH	—	*Primula*
Delphinidin	5,7-diOH	—	OH	OH	OH	*Delphinium*
Petunidin	5,7-diOH	—	OMe	OH	OH	*Petunia*
Malvidin	5,7-diOH	—	OMe	OH	OMe	*Malva*
Pulchellidin	5,7-diOH	5-OMe	OH	OH	OH	*Plumbago*
Europinidin	5,7-diOH	5-OMe	OMe	OH	OH	*Plumbago*
Capensinidin	5,7-diOH	5-OMe	OMe	OH	OMe	*Plumbago*
Hirsutidin	5,7-diOH	7-OMe	OMe	OH	OMe	*Primula*
None[d]	7,8-diOH	—	OH	OH	—	*Acacia*
Aurantinidin	5,6,7-triOH	—	—	OH	—	*Impatiens*

[a] See (**I**) and (**II**) for numbering; OMe = *O*-Methyl. [b] Primary source given.

[c] Structures unknown: Columnidin was originally thought to be 5,6 (or 8) 7,3′4′-pentahydroxy flavylium salt (Harborne, 1967) but comparison with synthetic products shows this is not so (J. B. Harborne, 1974, personal communication); Carexidin is a luteolinidin-like compound (Clifford

hardly changing when the pH is raised from 1·0 to 6·0. The pseudobase form can undergo ring opening to form the corresponding chalkone (**XVII**) (Jurd, 1962b), and other ill-understood changes usually ensue, some of which are light catalysed and non-reversible.

At pH values above 7·0, the anthocyanidins form anhydro bases (**XVIII**) all of which have bluer shades, but which are unstable in the presence of water unless the 3-hydroxyl group is absent (**XV**). Further raising of the pH induces ionization of the phenolic hydroxyl groups, but under such basic conditions the anthocyanidins are very unstable, undergoing both cleavage and aerial oxidation, and reacidification of such solution usually leads to poor recoveries.

C GLYCOSIDIC FORMS

All the common anthocyanidins exist in nature, mainly as glycosides (anthocyanins), and often more than one derivative of a single aglycone is present in a given plant. Indeed, there are only three reports of free anthocyanidins in plants (Timberlake and Bridle, 1975). Nearly forty different types of anthocyanin glycosides are known (Harborne, 1967; Timberlake and Bridle, 1975). The relative order of frequency of occurrence of the various monosaccharides in the anthocyanins is as stated previously (p. 430). So far L-arabinose has only been found in a few plant genera, in each case linked with an α link. Recently both xylose and fructose have been reported in anthocyanidin monosides, and an oligosaccharide derivative has also been reported (Timberlake and Bridle, 1975). Although glucose is still the only sugar reported attached to the C-5 hydroxyl group, there is now good evidence that galactose, arabinose and xylose can, along with glucose, occur in combination with the C-7 hydroxyl. Only a few of the wider number of possible disaccharides have been found attached to anthocyanidins. Thus only one rhamnosylglucoside (rutinose, $\alpha 1 \rightarrow 6$), one xylosylglucoside (sambubiose, $\beta 1 \rightarrow 2$) and two glucosylglucosides (sophorose, $\beta 1 \rightarrow 2$; gentiobiose, $\beta 1 \rightarrow 6$) have been observed, the first three being relatively common. There are also a number of rare or even unique biosides found in anthocyanins including arabinosylglucosides, galactosylglucosides and rhamnosylgalactosides; triosides are rare and include 2^G-xylosyl- and 2^G-glucosylrutinose as well as a linear glucosylrhamnosylglucose (Timberlake and Bridle, 1975). There are other unusual combinations of 3-biosides with 5- or 7-glucosides and a number of derivatives with glucose on the B-ring hydroxyls (Harborne, 1967; Timberlake and Bridle, 1975).

The λ_{max} of band I in the spectra of the glycosides of the common anthocyanins shows in all cases a hypsochromic shift from that of the aglycone. The two common classes, the 3- and 3,5-diglycosides have similar spectral maxima, but the former all have a pronounced shoulder in the 440–460 nm region (Fig. 5) which can be used for identification purposes (Harborne, 1967; see Chapter 20).

Besides the glycosides proper, acylated anthocyanins have long been known although their structures were only elucidated in the last ten years (Harborne, 1967). In each case investigated the sugar attached to the 3-hydroxyl group is acylated. The main acyl groups so far found in these compounds are those derived from the four common cinnamic acids, *p*-coumaric acid, caffeic acid, ferulic and sinapic acid (**XIV**): an earlier report on the occurrence of *p*-hydroxybenzoic acid (Willstätter and Mieg, 1915) in such compounds has not been substantiated (Harborne, 1967). The presence of cinnamic acids can be readily demonstrated in these compounds by spectroscopic means, since the absorption spectra of the acid residues are superimposed on those of the unacylated anthocyanins (Harborne, 1967) (see Fig. 5). The methods by which their mode of linkage was determined is described below. Besides cinnamic acid derivatives, malonyl (Karrer and Widmer, 1928; Bloom and Geissman, 1973) and acetyl (Anderson *et al.*, 1970) esters are known, and it seems probable that more of these will be found when more rigorous searches are carried out.

D CHEMISTRY OF ANTHOCYANINS

1 *Synthesis*

The first synthesis of an anthocyanidin was carried out by Willstätter and Mallison (1914), who reduced quercetin (V) in poor yield to cyanidin (I). In the same year Willstätter and Zechmeister (1914) carried out a total synthesis of pelargonidin (I) from 3,5,7-trimethoxycoumarin and *p*-methoxyphenyl magnesium bromide. However, the most convenient methods of synthesis of anthocyanidins and the only ones used with success for the glycosides were devised by Robinson and his co-workers (Robinson, 1934). In these syntheses, a series of 2-*O*-benzoyl- (or *O*-tetraacetylglucosyl)-phloroglucinaldehydes were condensed with a variety of ω-hydroxy- (or *O*-tetraacetylglucosyl)-acetophenones to produce all the common anthocyanidins and their glycosides as illustrated in the synthesis of hirsutin (Fig. 6) (Robinson and Todd, 1932).

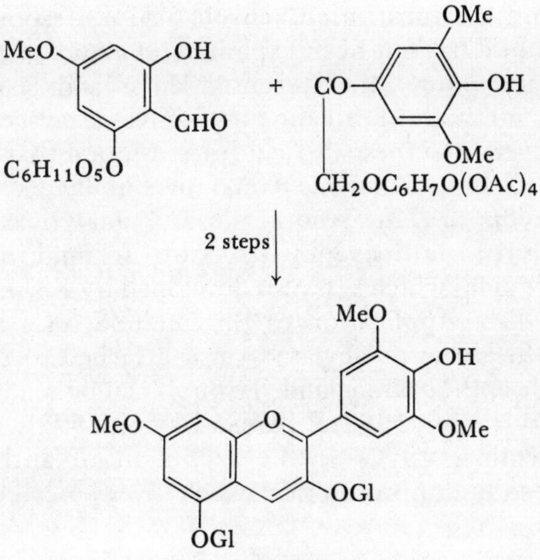

MeO, OH, CHO, $C_6H_{11}O_5O$ + CO–, OMe, OH, OMe, $CH_2OC_6H_7O(OAc)_4$

2 steps

MeO, OH, MeO, O, OMe, OGl, OGl

Hirsutin
(Gl = glucosyl)

FIG. 6. Synthesis of hirsutin.

2 *Reactions*

Since there are several excellent textbooks available (e.g. Dean, 1963; Harborne *et al.*, 1975) the chemistry of these and other flavonoid compounds will not be dealt with in any detail. However, there are several reactions which deserve attention as they are of importance in determining the structure of anthocyanins on a microscale. Obviously the first reaction that needs to be used for this purpose is hydrolysis. When this is carried out under mild acid conditions and the products examined by paper chromatography a number of structural features can be elucidated (Harborne, 1967; Ribereau-Gayon, 1972). For example, Harborne (1958) detected seven readily separable products from the controlled hydrolysis of pelargonidin-5-glucoside-3-sophoroside; the known 3,5-diglucoside, the 3-sophoroside, the 3- and 5-glucosides, pelargonidin itself and the two sugars, glucose and sophorose. From these data the structure of the compound was easily and unequivocally determined.

Mild alkali degradation under hydrogen using 10% barium hydroxide was used by Karrer to determine the structure of the anthocyanidins (Karrer and Widmer, 1927a). Under these conditions the compounds break down to give phloroglucinol (or other A-ring fragments) and the substituted benzoic acid corresponding to the B ring. This method has not been exploited on a microscale because the yields are very poor. More recently Hurst and Harborne (1967) developed a method of alkaline reductive cleavage with sodium amalgam to overcome these difficulties and found that good yields of both A and B ring fragments could be obtained on a microscale (Ribereau-Gayon, 1972). Another useful analytical tool for the determination of anthocyanin structure is mild oxidation with hydrogen peroxide, which was also developed by Karrer to determine the sugar at C-3. Application of this method on a microscale has shown that the sugar or acylated sugar attached to the 3-hydroxyl group of both anthocyanins and flavonols can be readily determined (Chandler and Harper, 1962; Ribereau-Gayon, 1972). Another useful method (Hrazdina, 1972) is the reduction of anthocyanidins in ethanol with sodium borohydride which gives *epi*catechin racemates (**IV** 2*R* 3*R*).

V Flavones and flavonols

A AGLYCONES

Although, as mentioned above, morin (**XIX**) was isolated as early as 1814, the structure of these compounds was not properly elucidated until 1891 when Hertzig (1891) showed that quercetin had the molecular formula shown in (**V**; Fig. 1). Shortly afterwards Kostanecki (1893) determined the structure of chrysin (**XX**), and within ten years he and his group had synthesized most of the then known flavones and flavonols (see Perkin and Everest, 1918).

Today, if glycosides and other derivatives are excluded, over 280 flavones and flavonols have been isolated (Tables II and III), the latter class making nearly two-thirds of the total (Harborne *et al.*, 1975). Undoubtedly many more remain to be discovered. The compounds range in hydroxylation pattern from flavone (**XXI**) itself, which occurs as a mealy deposit on the leaves of *Primula pulverulenta* (Müller, 1915) and other species, to digicitrin (**XXII**, Table III) from the red foxglove *Digitalis purpurea* (Meier and Furst, 1962). Besides compounds containing a simple flavone nucleus, several *C*-methyl and *C*-prenyl derivatives have been isolated. Artocarpin (**XXIII**) from the heartwood of *Artocarpus integrefolia* (Bringi and Dave, 1956) is distinguished as being the only flavone substituted with a carbon side chain at C-3. It is accompanied in the wood by its 2′-*O*-isopentenyloxy-ether (Dave *et al.*, 1962). Several furanoflavones like karanjin (**XXIV**) from *Pongamia glabra* (Limaye, 1936) and biflavonyls such as ginkgetin (**XXV**) from *Ginkgo biloba* are also known (Harborne and Williams, 1975). Finally there is a number of glycoflavones which contain a 6- or 8- (or both) substituted *C*-glycosyl residue as in vitexin (**XXVI**) from *Vitex littoralis* (Perkin, 1898).

Although the hydroxylation patterns of the simple flavones and flavonols are more varied than those of the anthocyanidins (cf. Tables I–III), over 90% of the glycosides so far isolated contain hydroxyl or methoxyl groups at C-5 and at C-7. Nearly 80% are substituted at C-4′ and about half at C-3′, as expected from the mode of biosynthesis. Of the less usual positions of hydroxy or methoxy substitution, C-6 is most favoured (over one-third of glycosides containing "extra" hydroxyls), C-8 next (about a quarter) with C-2′ and C-5′ relatively rare (approximately 10% each). Except for hydroxyl groups at C-5 (*c*. 15%) methylation occurs in nearly half of the hydroxyl groups at the expected positions (7, 3′, 4′, 5′), but is much more common (70–80% at C-6 and C-8 (see Dean, 1963; Harborne *et al*., 1975).

Many of the compounds listed in Tables II and III have so far only been isolated (either as glycosides or aglycones) from a single species but the three most commonly occurring flavonols, kaempferol, quercetin and myricetin (**V**, Table III), are much more widely distributed than the corresponding anthocyanidins (**I**, Table I). Indeed, in some groups of plants they can almost be regarded as common metabolites; for example, 90% of the leaves of the woody dicotyledons contain glycosides of one or more of these three flavonols (Harborne, 1967).

B GLYCOSIDES

As with other flavonoid compounds, the hydroxy flavones and flavonols mainly occur as glycosides usually with a β-linked sugar. Many of the aglycones listed in Tables II and III give rise to at least five or six glycosides, quercetin, for example, having over seventy (Harborne and Williams, 1975). Most plants contain more than one glycoside of any aglycone, huckleberry (*Vaccinium myrtillis*), for example, having five glycosides of quercetin (Ice and Wender, 1953). Only a few generalizations will, therefore, be given here.

The most common sugar in flavone and flavonol glycosides is again glucose followed by galactose, rhamnose, arabinose and xylose; glucuronides are also found. L-Arabinose occurs both α- and β-linked to the 3-hydroxyl group of quercetin, and in the first case furanoside

TABLE II

Some naturally occurring flavones. Compounds are arranged first by number and arrangement of hydroxyl and other groups in the A ring and then by number of substituents in the B ring

Trivial name	Substitution pattern[a]						Source[b]
	A ring		B ring				
	Hydroxylation	Other	2′	3′	4′	5′	
Flavone	—	—	—	—	—	—	*Primula*
None	5,6-diOH	—	OH	—	—	—	*Casimiroa*
Zapotinin	5,6-diOH	6,6′-di*O*Me	*O*Me	—	—	OH	*Casimiroa*
Cerrosillin	5,6-diOH	5,6-di*O*Me	—	OMe	—	OMe	*Sargentia*
Chrysin	5,7-diOH	—	—	—	—	—	*Populus*
Apigenin	5,7-diOH	—	—	—	OH	—	widespread
Acacetin	5,7-diOH	—	—	—	OMe	—	*Robinia*
Genkwanin	5,7-diOH	7-*O*Me	—	—	OH	—	*Daphne*
Luteolin	5,7-diOH	—	—	OH	OH	—	widespread
Chrysoeriol	5,7-diOH	—	—	OMe	OH	—	*Eriodictyon*
Diosmetin	5,7-diOH	—	—	OH	OMe	—	*Diosma*
Tricetin	5,7-diOH	—	—	OH	OH	OH	*Lathyrus*
Tricin	5,7-diOH	—	—	OMe	OH	OMe	*Triticum*
Baicalein	5,6,7-triOH	—	—	—	—	—	*Scutellaria*
Scutellarein	5,6,7-triOH	—	—	—	OH	—	*Scutellaria*
Hispidulin	5,6,7-triOH	6-*O*Me	—	—	OH	—	*Ambrosia*
Pectolinarigenin	5,6,7-triOH	6-*O*Me	—	—	OMe	—	*Linaria*
6-Hydroxyluteolin	5,6,7-triOH	—	—	OH	OH	—	*Catalpa*
Nepetin	5,6,7-triOH	6-*O*Me	—	OH	OH	—	*Nepeta*
Jaceosidin	5,6,7-triOH	6-*O*Me	—	OMe	OH	—	*Digitalis*
Eupatilin	5,6,7-triOH	6-*O*Me	—	OMe	OMe	—	*Eupatorium*

Eupatorin	5,6,7-triOH	6,7-di*O*Me	—	OH	OMe	—	*Eupatorium*
Norwogonin	5,7,8-triOH	—	—	—	—	—	*Scutellaria*
Isoscutellarein	5,7,8-triOH	—	—	—	OH	—	*Pinguicula*
Hypolaetin	5,7,8-triOH	—	—	OH	OH	—	*Hypolaena*
Wightin	5,7,8-triOH	7,8-di*O*Me	OMe	OH	—	—	*Andrographis*
Alnetin	5,6,7,8-tetraOH	6,7,8-tri*O*Me	—	—	—	—	*Alnus*
Nevadensin	5,6,7,8-tetraOH	6,8-di*O*Me	—	—	OMe	—	*Iva*
Tangeretin	5,6,7,8-tetraOH	5,6,7,8-tetra*O*Me	—	—	OMe	—	*Citrus*
Sudachitin	5,6,7,8-tetraOH	6,8-di*O*Me	—	OMe	OH	—	*Citrus*
Acerosin	5,6,7,8-tetraOH	6,8-di*O*Me	—	OH	OMe	—	*Iva*
Hymenoxia	5,6,7,8-tetraOH	6,8-di*O*Me	—	OMe	OMe	—	*Hymenoxia*
Gardenin D	5,6,7,8-tetraOH	6,7,8-tri*O*Me	—	OH	OMe	—	*Gardenia*
Nobiletin	5,6,7,8-tetraOH	5,6,7,8-tetra*O*Me	—	OMe	OMe	—	*Citrus*
Gardenin A	5,6,7,8-tetraOH	6,7,8-tri*O*Me	—	OMe	OMe	OMe	*Gardenia*

[a] See (**II**); OMe = *O*-Methyl. [b] See Table I.

TABLE III

Some naturally occurring flavonol aglycones. Compounds are arranged first by number and arrangement of hydroxyl and other groups in the A ring and then by number of substituents in the B ring

Trivial name	Substitution pattern[a]						Source[b]
	A ring		B ring				
	Hydroxylation	Other	2	3	4	5	
None	7-OH	—	—	—	OH	—	*Schinopsis*
Fisetin	7-OH	—	—	OH	OH	—	*Rhus*
Robinetin	7-OH	—	—	OH	OH	OH	*Robinia*
Kanugin	7-OH	—	—	OH	—O—CH_2—O		*Pongamia*
Galangin	5,7 di-OH	—	—	—	—	—	*Alpinia*
Datiscetin	5,7 di-OH	—	OH	—	—	—	*Datisca*
Kaempferol	5,7 di-OH	—	—	—	OH	—	widespread
Rhamnocitrin	5,7 di-OH	7-*O*Me	—	—	OH	—	*Rhamnus*
Kaempferide	5,7 di-OH	—	—	—	OMe	—	*Alpina*
Morin	5,7 di-OH	—	OH	—	OH	—	*Morus*
Quercetin	5,7 di-OH	—	—	OH	OH	—	widespread
Azaleatin	5,7 di-OH	5-*O*Me	—	OH	OH	—	*Rhododendron*
Rhamnetin	5,7 di-OH	7-*O*Me	—	OH	OH	—	*Rhamnus*
Rhamnazin	5,7 di-OH	7-*O*Me	—	OMe	OH	—	*Rhamnus*
Isorhamnetin	5,7 di-OH	—	—	OMe	OH	—	widespread
Persicarin	5,7 di-OH	3-OSO_3K	—	OMe	OH	—	*Polygonum*
Tamarixetin	5,7 di-OH	—	—	OH	OMe	—	*Tamarix*
Caryatin	5,7 di-OH	3,5-di*O*Me	—	OH	OH	—	*Carya*
Oxyayanin A	5,7 di-OH	3,7-di*O*Me	OH	OH	OMe	—	*Apulia*
Myricetin	5,7 di-OH	—	—	OH	OH	OH	widespread
Europetin	5,7 di-OH	7-*O*Me	—	OH	OH	OH	*Plumbago*
Syringetin	5,7 di-OH	—	—	OMe	OH	OMe	*Lathyrus*

Galetin	5,6,7-triOH	—	—	—	OH	—	*Galega*
Eupalitin	5,6,7-triOH	6,7-di*O*Me	—	—	OH	—	*Eupatorium*
Penduletin	5,6,7-triOH	3,6,7-tri*O*Me	—	—	OH	—	*Briquelia*
Quercetagetin	5,6,7-triOH	—	—	OH	OH	—	common
Patuletin	5,6,7-triOH	6-*O*Me	—	OH	OH	—	common
Jaceldin	5,6,7-triOH	3,6-di*O*Me	—	OMe	OH	—	*Centaurea*
Oxyayanin B	5,6,7-triOH	3,7-di*O*Me	—	OH	OMe	—	*Distemonanthus*
Eupatoretin	5,6,7-triOH	5,6,7-tri*O*Me	—	OH	OMe	—	*Eupatorium*
Melisimplin	5,6,7-triOH	3,6,7-tri*O*Me	—	O—CH_2—O		—	*Melicope*
Melisimplexin	5,6,7-triOH	3,5,6,7-tetra*O*Me	—	O—CH_2—O		—	*Melicope*
Apulein	5,6,7-triOH	3,5,6,7-tetra*O*Me	OH	—	OMe	OH	*Apulia*
None	5,6,7-triOH	3,6-di*O*Me	—	OH	OH	OH	*Centaurea*
None	5,6,7-triOH	3,5,6,7-tetra*O*Me	—	OMe	OMe	OMe	*Murraya*
Alnustinol	5,7,8-triOH	7-*O*Me	—	—	—	—	*Alnus*
Graphalin	5,7,8-triOH	3,8-di*O*Me	—	—	—	—	*Graphalium*
Herbacetin	5,7,8-triOH	—	—	—	OH	—	widespread
Tambuletin	5,7,8-triOH	8-*O*Me	—	—	OH	—	*Xanthoxylum*
Prudometin	5,7,8-triOH	8-*O*Me	—	—	OMe	—	*Prunus*
Tambulin	5,7,8-triOH	7,8-di*O*Me	—	—	OMe	—	*Xanthoxylum*
Gossypetin	5,7,8-triOH	—	—	OH	OH	—	widespread
Corniculatusin	5,7,8-triOH	8-*O*Me	—	OH	OH	—	*Lotus*
Limocitrin	5,7,8-triOH	8-*O*Me	—	OMe	OH	—	*Citrus*
Wharangin	5,7,8-triOH	7,8-OCH_2O 3-*O*Me	—	OH	OH	—	*Melicope*
Ternatin	5,7,8-triOH	3,7,8-tri*O*Me	—	OMe	OH	—	*Melicope*
Meliternin	5,7,8-triOH	3,5,7,8-tetra*O*Me	—	O—CH_2—O		—	*Melicope*
Hibiscetin	5,7,8-triOH	—	—	OH	OH	OH	*Hibiscus*
None	5,6,7,8-tetraOH	6,8-di*O*Me	—	—	—	—	*Gnaphalium*
Calycoptarin	5,6,7,8-tetraOH	6,7,8-tri*O*Me	—	—	OH	—	*Calycopteris*
Limocitrol	5,6,7,8-tetraOH	6,8-di*O*Me	—	OMe	OH	—	*Citrus*
Natsudaidain	5,6,7,8-tetraOH	5,6,7,8-tetra*O*Me	—	OMe	OMe	—	*Citrus*
Digicitrin	5,6,7,8-tetraOH	3,6,7,8-tetra*O*Me	—	OH	OMe	OMe	*Digitalis*

[a] See (**V**) and (**II**); OMe = *O*-Methyl; O—CH_2—O = methylenedioxy. [b] See Table I.

and pyranoside forms are present, giving three known isomeric compounds in all. There are numerous compounds containing either sugars on more than one hydroxyl group or di- and trisaccharides. One kaempferol derivative is even reported to contain five sugar residues (Beckmann and Geiger, 1963). The structure of the complex sugars in these compounds has not been examined in many cases, probably owing to the difficulty in obtaining the higher glycosides in a pure state. In flavones the commonest position for sugar substitution is the hydroxyl group at C-7 (80%), followed by those at C-5 and C-4′. In flavonols the C-3 hydroxyl is most favoured (70%) followed by C-7 (20%), and C-4′ (5%). As with the anthocyanins, several acylated compounds, e.g. tiliroside, the 3-*p*-coumarylglucoside of kaempferol (Harborne, 1967), have been isolated and it is to be expected that many more will be found in the future.

Obviously, the structures of only a few of the 280 flavone and flavonol aglycones can be given here. All the common members of both groups which are found in flowers and fruits are listed, together with a selection of the more highly oxygenated and hence more highly coloured compounds (Tables II and III). It should be remembered, however, that even those flavones and flavonols which are barely coloured to the human eye may be important in flowers as insect guides (since insects can detect light absorption in the near ultraviolet (*c.* 350 nm) (Burkhardt, 1964; Thompson *et al.*, 1972) or as co-pigments (Asen *et al.*, 1972) or, in some cases, even impart a visible colour because of metal chelation.

C CHEMISTRY OF FLAVONES AND FLAVONOLS

1 *Synthesis*

Kostanecki's synthesis of flavones by a Claisen condensation between suitably substituted phloracetophenones and benzoic esters (Emilewicz *et al.,* 1899) has been superseded by the Baker-Venkataraman (Baker, 1933) and the Allan–Robinson syntheses (Allan and Robinson, 1924). All these syntheses involve the formation and subsequent cyclization of a suitable *o*-hydroxydibenzoylmethane as shown in the synthesis of quercetin (V), by the condensation of ω-methoxyphloracetophenone with veratric anhydride (Allan and Robinson, 1924) which goes via the intermediate shown in Fig. 7. Flavanones and chalkones which may be readily synthesized can be converted into flavones and flavonols in a variety of ways (Harborne *et al.*, 1974). For example, naringenin (VI) can be

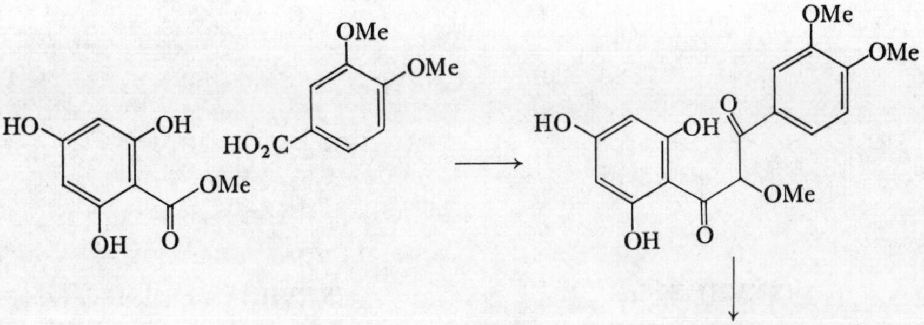

Quercetin V

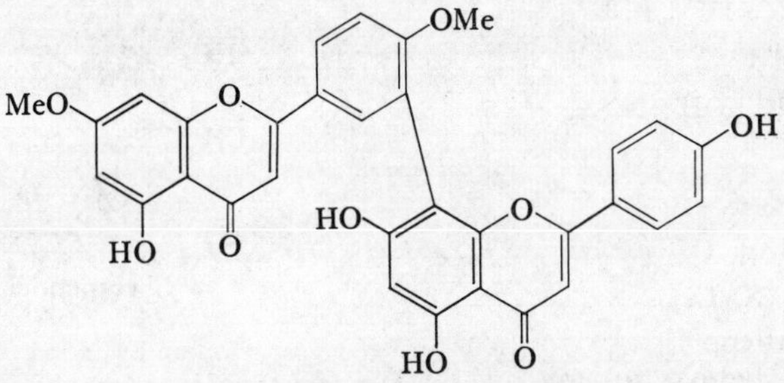

(XXV) Ginkgetin

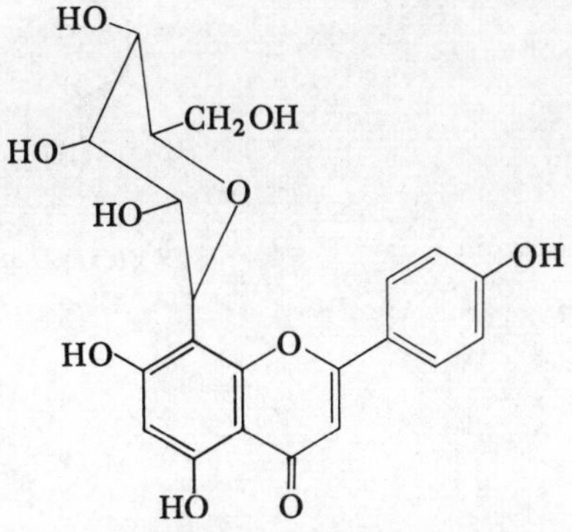

(XXVI) Vitexin

FIG. 7. Synthesis of quercetin.

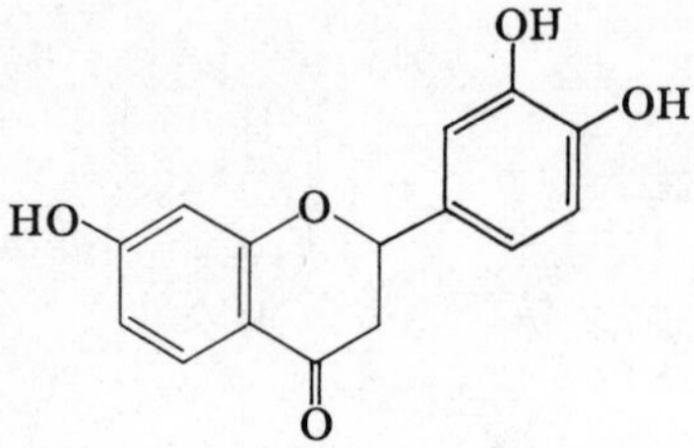

(XXVII) Butin

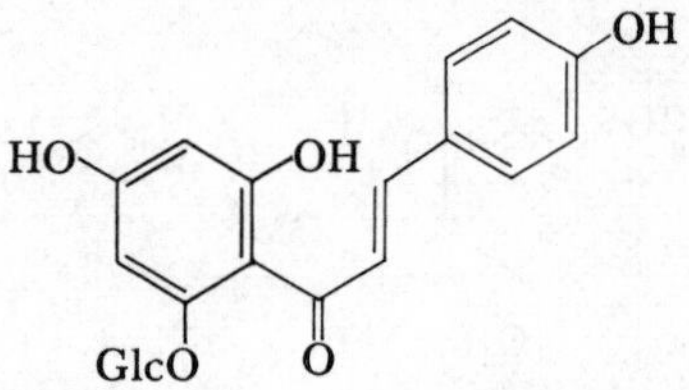

(XXVIII) Isosalipurpuroside

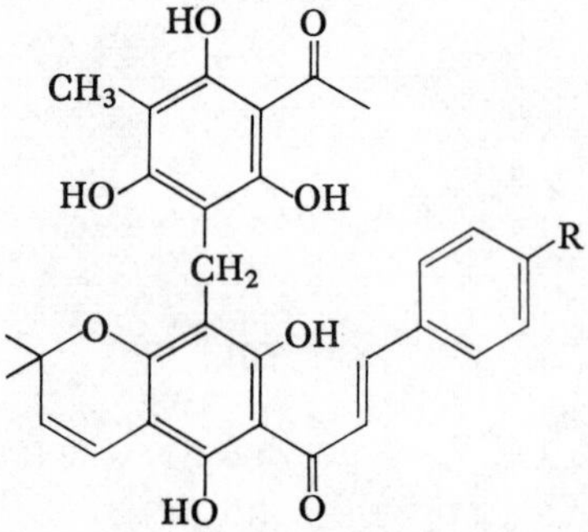

(XXIX)
R = H Rottlerin
R = OH 4-Hydroxyrottlerin

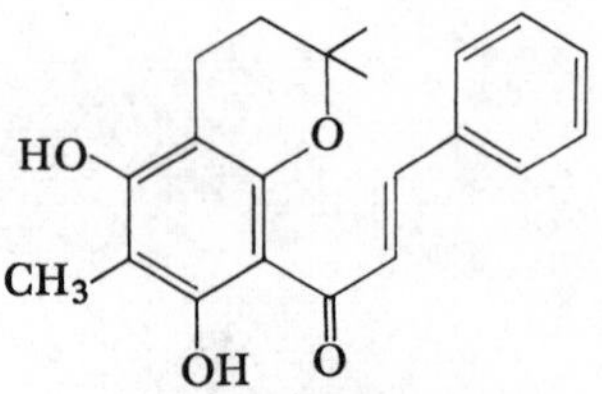

(XXX)
Chromenochalkone

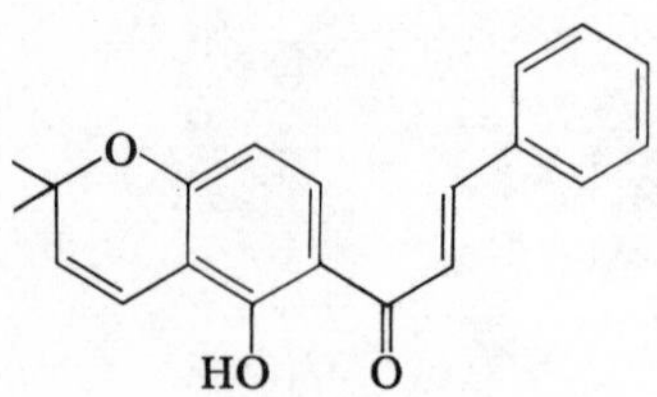

(XXXI) Lonchocarpin

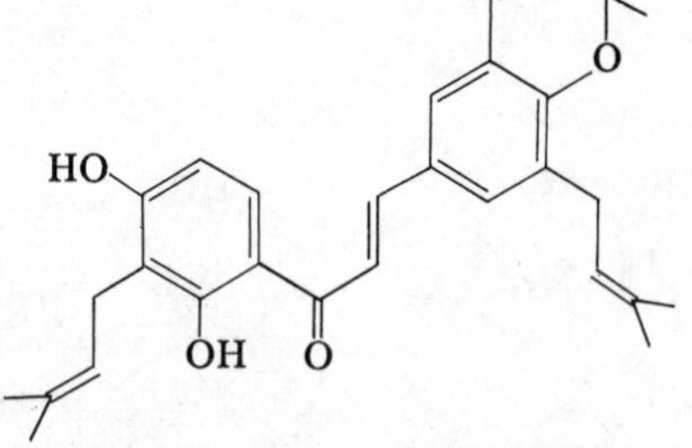

(XXXII) Sophoradochromene

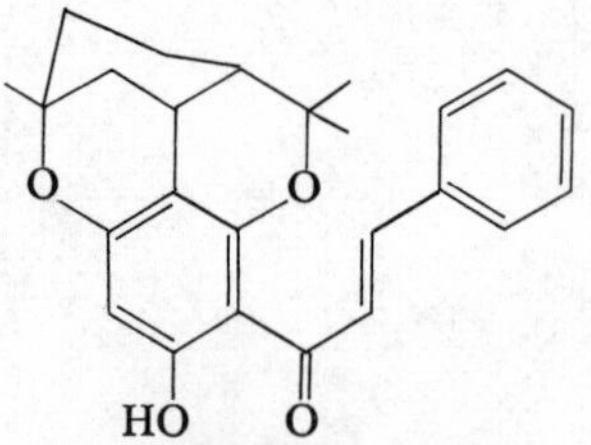

(XXXIII) Rubranine

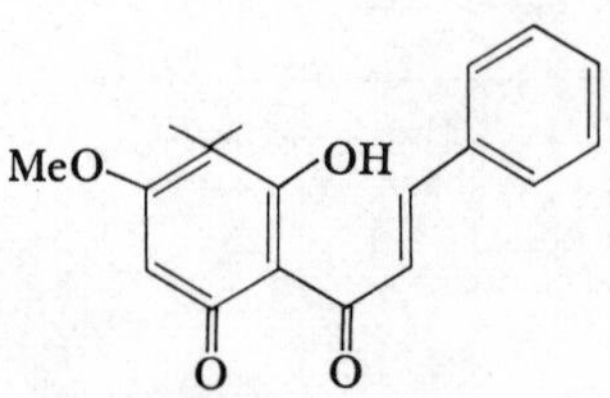

(XXXIV) Ceroptene

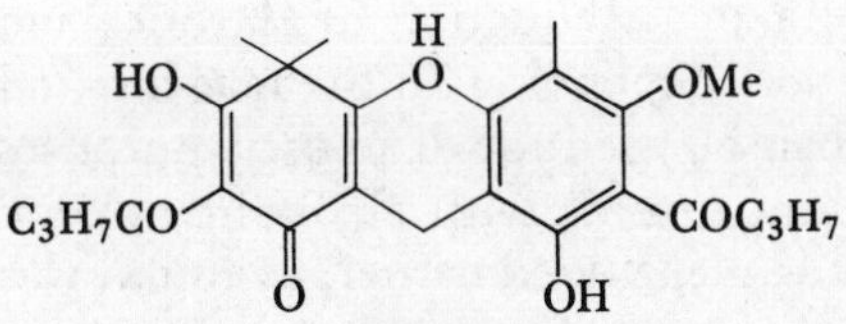

(XXXV)
Aspidin

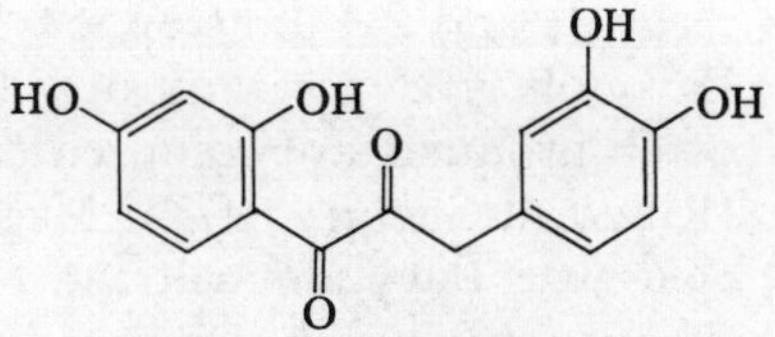

(XXXVI)
Keto form of α-hydroxychalkone from *Trachylobium verracosum*

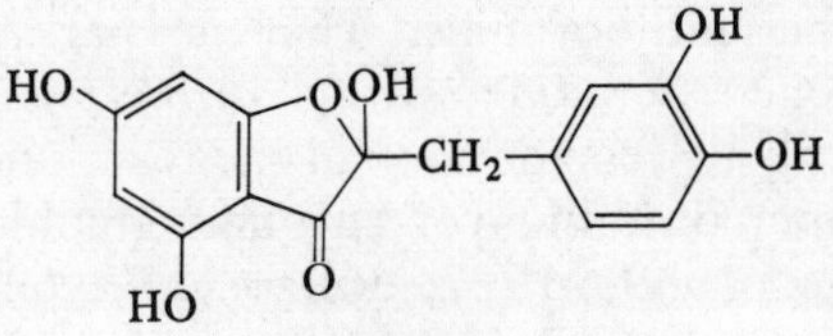

(XXXVII)
Alphitonin

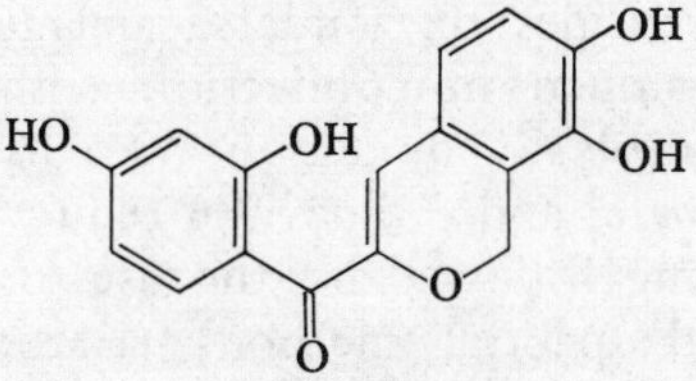

(XXXVIII)
Chromenochalkone from *Goniorrhachis marginata*

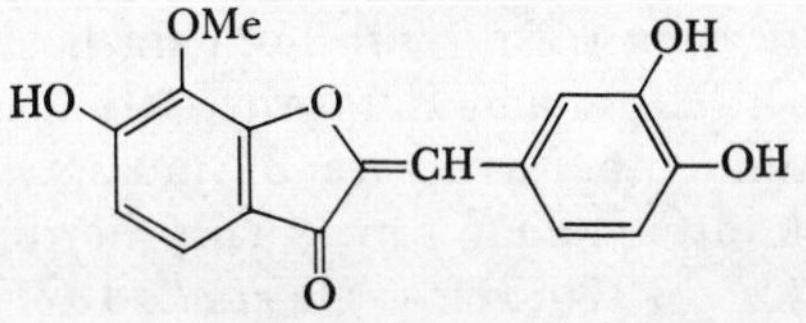

(XXXIX)
Leptsosidin

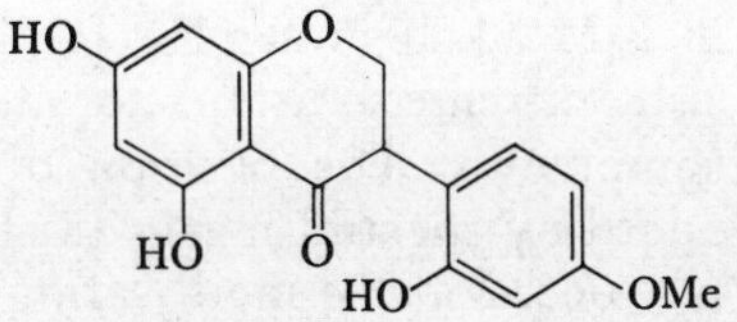

(XL)
Ferreirin

converted into apigenin (**II**) by dehydrogenation of the trimethyl ether with selenium dioxide and subsequent demethylation; tetra-*O*-methyleriodictyol (**VI**) yields 5,7,3′,4′-tetra-*O*-methylquercetin (**I**) on treatment with amyl nitrate to introduce an isonitroso group at C-3 followed by subsequent hydrolysis with mineral acid. A large number of flavone and flavonol glycosides have been synthesized either by direct coupling of the acylated sugar bromide or by mild versions of the above syntheses. These have recently been reviewed by Bognár (1973) and Farkas and Major (1973).

2 *Reactions*

The carbonyl group in flavones and flavonols does not show the general reactions of such groups with keto reagents except Grignard reagents and 2,4-dinitrophenylhydrazine. Reduction of flavones and flavonols gives a variety of substances depending on the reagent, and often products are coloured and can be used for diagnostic purposes (Ribereau-Gayon, 1972). Magnesium or zinc with alcoholic mineral acid gives flavylium salts; in fact, as mentioned earlier, cyanidin was first synthesized in this way from quercetin (Willstätter and Mallison, 1914). Catalytic reduction, on the other hand, gives flavan-4-ols or flavan-3,4-diols (King and Clark-Lewis, 1954).

The determination of the structure of flavone and flavonol glycosides on a microscale makes use of similar methods to those described in connection with the anthocyanins. That is, partial hydrolysis, oxidation with hydrogen peroxide, ozone or permanganate, and ordinary or reductive alkali degradation, all the reactions being followed by the use of paper of thin-layer chromatographic and spectral methods (Harborne, 1959, 1967; Ribereau-Gayon, 1972). As the flavones and flavonols are somewhat more stable than the anthocyanins ordinary degradation with alkali is usually carried out by fusion with solid potassium hydroxide for 1–2 minutes, or by heating with 50% aqueous sodium hydroxide under N_2 (Doporto *et al.*, 1955). Reductive cleavage, by heating with 2% sodium amalgam in 2 M alkali under N_2 for 1–1·5 h (Hurst and Harborne, 1967) is also extremely useful for determining the structure of unknown compounds. The position of sugar attachment can usually be deduced by spectral means (Mabry *et al.*, 1970; Harborne *et al.*, 1975) but occasionally methylation followed by hydrolysis is also useful. (See Chapter 20.)

VI Chalkones

A AGLYCONES

Although chalkones (**X**) had been prepared as intermediates in the synthesis of flavones before the turn of the century, the first compound of proven structure of this class was not isolated until 1939 when butein (**X**, Table IV) was obtained from *Dahlia variabilis* by Price (1939). Perkin and Hummel (1904) had actually obtained this compound during the isolation of the corresponding flavanone butin (**XXVII**) from *Butea frondosa*.

In 1921, Klein introduced the term *anthochlor* for some pigments in certain yellow flowers which turned red on treatment with ammonia, and later Gertz (1938) showed that pigments of this class were especially widespread in the subtribe Coreopsidinae of the Compositae. Starting in 1941, Geissman and his co-workers

(Shimokoriyama, 1962) showed that these colour changes were due to the presence of chalkones and aurones (the first of which, leptosidin, was isolated from *Coreopsis grandiflora* (Geissman and Heaton, 1943)) and their researches considerably extended the number of known chalkones and their glycosides.

As can be seen, chalkones (**X**) do not contain a heterocyclic ring. However, if they possess a free 2′(or 6′)-hydroxyl group they can be relatively readily isomerized to give the corresponding flavanone (e.g. butein **X** → butin **XXVII**). Since all known flavanones are colourless (λ_{max} 275–295 nm with a shoulder at *c.* 327 nm), and this transformation occurs *in vivo* catalysed by chalcone–flavanone isomerase (Grisebach, 1972; Hahlbrock and Grisebach, 1974; see Chapter 20), it is obvious that the contribution which the deep yellow chalkones make to petal colour may be regulated relatively easily. This could account, in part, for the observed changes in the depth of colour of some chalkone-containing flowers as the bud opens.

The chalkones are, of course, important as precursors of all other flavonoids (Chapter 20) and hence can be regarded as ubiquitous compounds in the plant kingdom. However, as is true for several other classes of compound (e.g. terpenoids), most of the presumed biosynthetic intermediates of the flavonoids are undetectable except under special circumstances (e.g. using ^{14}C-labelled precursors). It is only when the chalkones (or other intermediates) accumulate, usually in glycosidic form, that they become important as colouring matters.

In the past few years many more new chalkones have been isolated from plants and there are now well over fifty known aglycones, forty of which are shown in Table IV. Most of these, as might be expected from their mode of biosynthesis, have A-ring oxygenation patterns based on resorcinol (e.g. **X**) or phloroglucinol (e.g. **XXVIII**) with B rings commonly substituted in a similar way to those found in other groups of flavonoids (see Tables I–III). Many contain extra hydroxy or methoxy groups and *C*-prenyl groups in the A ring. *O*-Alkylation (or glycosidation) is common especially in those chalkones having a 2′,6′-oxygen substitution pattern, since these isomerize to flavanones more readily than those compounds which have only one 2′-hydroxy group. This is due to the fact that a single hydroxyl group in the *ortho* position (2′ in **X**) to the carbonyl group of the three-carbon link is hydrogen bonded, and is then less likely to be involved in heterocyclic ring formation.

The simplest chalkone structurally is the 2′,4′-dihydroxy derivative isolated from *Flemingia chappax* (Adityachaudhury *et al.*,

TABLE IV

Some naturally occurring chalkone aglycones. Compounds are arranged first by number and arrangement of hydroxyl and other groups in the A ring and then by number of substituents in the B ring

Trivial name	Substitution pattern[a]						Source[b]
	A ring		B ring				
	Hydroxylation	Other	2	3	4	5	
Echinatin	4′-OH	—	*O*Me	—	OH	—	*Glycyrrhiza*
None	2′,4′-diOH	—	—	—	—	—	*Flemingia*
Derricidin	2′,4′-diOH	4′-*O*P	—	—	—	—	*Derris*
Isoliquiritigenin	2′,4′-diOH	—	—	—	OH	—	*Robinia*
None	2′,4′-diOH	4′-*O*Me	—	—	OH	—	*Xanthorrhoea*
Butein	2′,4′-diOH	—	—	OH	OH	—	*Butea*
Homobutein	2′,4′-diOH	—	—	*O*Me	OH	—	*Acacia*
Robtein	2′,4′-diOH	—	—	OH	OH	OH	*Robinia*
Derricin	2′,4′-diOH	3′-P,4′-*O*Me	—	—	—	—	*Derris*
Isobarachalcone	2′,4′-diOH	3′-P	—	—	OH	—	*Psoralea*
Sophoridin	2′,4′-diOH	3′-P	—	P	OH	P	*Sophora*
Barachalcone	2′,4′-diOH	5′-P,4′-*O*Me	—	—	OH	—	*Psotalea*
None	2′,3′,4′-triOH	—	—	—	OH	—	*Acacia*
Okanin	2′,3′,4′-triOH	—	—	OH	OH	—	*Acacia*
Lanceoletin	2′,3′,4′-triOH	3′-*O*Me	—	OH	OH	—	*Coreopsis*
Flemichapparin	2′,4′,5′-triOH	5′-*O*Me	—	—	—	—	*Flemingia*
Stillopsidin	2′,4′,5′-triOH	—	—	OH	OH	—	*Coreopsis*
5-Deoxyhomo-flemingin	2′,4′,5′-triOH	3′-G, 5′-*O*Me	OH	—	—	—	*Flemingia*
Homoflemingin	2′,4′,5′-triOH	3′-G, 5′-*O*Me	OH	—	—	OH	*Flemingia*
None	2′,4′,6′-triOH	4′-*O*Me	—	—	—	—	*Lindera*
None	2′,4′,6′-triOH	2′-*O*Me	—	—	—	—	*Piper*

Excelsin	2′,4′,6′-triOH	4′,6′-di*O*Me	—	—	—	—	*Pinus*
Isosalipurposide	2′,4′,6′-triOH	2′-*O*Gl	—	—	OH	—	*Salix*
Neosakuranin	2′,4′,6′-triOH	2′-*O*Gl, 4′-*O*Me	—	—	OH	—	*Prunus*
None	2′,4′,6′-triOH	4′-*O*Me	—	—	OMe	—	*Pityrogramma*
None	2′,4′,6′-triOH	2′,4′-di*O*Me	—	—	OMe	—	*Xanthorrhoea*
None	2′,4′,6′-triOH	—	—	OH	OH	—	*Limonium*
None	2′,4′,6′-triOH	2′,4′-di*O*Me	—	OH	OMe	—	*Merrillia*
None	2′,4′,6′-triOH	2′-*O*Gl	—	OH	OH	OH	*Helichrysum*
None	2′,4′,6′-triOH	2′,4′-di*O*Me	—	OMe	OMe	OMe	*Merrillia*
None	2′,4′,6′-triOH	5′-Me	—	—	—	—	*Daemonorops*
Xanthumol	2′,4′,6′-triOH	3′-P, 6′-*O*Me	—	—	OH	—	*Humulus*
Kuraridin	2′,4′,6′-triOH	3′-iG, 6′-*O*Me	OH	—	OH	—	*Sophora*
Carthamin	2′,3′,4′,6′-tetraOH	6′-*O*Gl	—	—	OH	—	*Carthamus*
Carthamone	2′,3′,4′,6′-tetraOH	2′-*O*Gl, 3′,6′-Q	—	—	OH	—	*Carthamus*
Pedicinin	2′,3′,4′,5′,6′-tetraOH	4′-*O*Me, 3′,6′-Q	—	—	—	—	*Didymocarpus*
Methylpedicin	2′,3′,4′,5′,6′-tetraOH	4′,5′-di*O*Me, 3′,6′-Q	—	—	—	—	*Didymocarpus*
Pedicin	2′,3′,4′,5′,6′-tetraOH	3′,4′,6′-tri*O*Me	—	—	—	—	*Didymocarpus*
Pedicellin	2′,3′,4′,5′,6′-tetraOH	2′,3′,4′,5′,6′-penta *O*Me	—	—	—	—	*Didymocarpus*

[a] See (**X**); OMe = *O*-Methyl; P = prenyl $[(CH_3)_2C{=}CHCH_2-]$; G = geranyl; Gl = glucosyl; Q = quinone; iG = isogeranyl $[(CH_3)_2C{=}CH-CH_2-CH(C(CH_3){=}CH_2)-CH_2-]$.

[b] Only primary source given. Including glycosides, chalkones are reported from 22 species of Leguminosae, 9 Compositae, 3 Gesneriaceae and 1 member of 17 other plant families.

1971). Rottlerin (XXIX) and its hydroxy congeners (XXIX) from *Mallotus phillipinensis* (*Rottlera tinctoria*) (Merlini, 1968) are the most complex compounds containing the basic chalkone skeleton, and they are accompanied in this plant by a simpler chromeno-chalkone (XXX) (Crombie *et al.*, 1968). Several other chromeno-chalkones containing an additional pyran ring originating by reaction between a prenyl group and an *ortho* hydroxyl are known, including lonchocarpin (XXXI) from *Derris sericea* (Nascimento and Mors, 1970) and sophoradochromene (XXXII) from *Sophora subprostata* (Komatsu *et al.*, 1970); the latter being obviously formed from sophoridin (Table V). An equally complex chalkone is rubraine (XXXIII) from *Aniba resaeodora* (Combes *et al.*, 1970) in which a C_{10} terpenoid moiety is attached to the simple α',4',6'-trihydroxy-chalkone skeleton. Ceroptene (XXXIV), which is found as a mealy deposit on the fronds of the fern *Pityrogramma triangularis* (Nilsson, 1959), has a quinone methide structure, suggesting that it may be related to the acylphloroglucinols (XXXV) found in all species of the fern genus *Dryopteris*.

In the last two years, three α-hydroxychalkones have been found (Roux and Ferreira, 1974). These exist in the diketone form (e.g. XXXVI) and are therefore colourless. Roux and Ferreira (1974) have suggested that these compounds are possibly widespread in nature and may be intermediates in the biosynthesis of flavonols, via the *trans* dihydroflavonols, and of the peltogynols. The tautomers of these compounds, the 2-hydroxy-2-benzylcoumaranones, have long been known, alphitonin (XXXVII) being obtained in 1922 by Smith and Read. However, only six are known (Bohm, 1975) although because of their lack of colour several others may have been overlooked in recent investigations.

B GLYCOSIDES

Like other flavonoids, many of the simple chalkones exist in the plant as glycosides, but, except in two cases (an apioglucoside and a rhamnoglucoside of isoliquiritigenin from *Glycyrrhiza glabra*; Bohm 1975) glucose is the only sugar, either alone or in biosides, which has been found. For example, Harborne (1963) isolated four different glycosides of 4,2',4'-trihydroxychalkone from *Ulex europaeus* all of which contained only glucose (see Chapter 20). Often the compounds co-occur with the glucosides of the corresponding flava-none and aurone. This is not surprising in view of the relative ease of

TABLE V

Naturally occurring aurone aglycones. Compounds are arranged first by number and arrangement of hydroxyl and other groups in the A ring and then by number of substituents in the B ring

Trivial name	Substitution pattern[a]					Source[b]
	A ring		B ring			
	Hydroxylation	Other	3′	4′	5′	
Hispidol	6-OH	—	—	OH	—	*Glycine*
Sulfuretin	6-OH	—	OH	OH	—	*Cosmos*
None	4,6-diOH	—	—	OH	—	*Limonium*
Aureusidin	4,6-diOH	—	OH	OH	—	*Antirrhinum*
Rengasin	4,6-diOH	4-*O*Me	OH	OH	—	*Melanorrhoea*
Bracteatin	4,6-diOH	—	OH	OH	OH	*Helichrysum*
Maritimetin	6,7-diOH	—	OH	OH	—	*Bidens*
Leptosidin	6,7-diOH	7-OMe	OH	OH	—	*Coreopsis*

[a] See (**XIII**); Me = methyl; Gl = glucose.
[b] Only primary source given. Including glycosides, aurones are found in 8 species of Cyperaceae, 7 Compositae, 4 Gesneriaceae and in 1–3 species of 6 other families.

enzymic interconversion of chalkones and flavanones, and of another enzyme which catalyses the oxidation of chalkones to aurones (Shimokoriyama and Hattori, 1953; Wong and Wilson, 1972).

C CHEMISTRY OF CHALKONES

The synthesis of chalkones is relatively easy, since they are obtained by the condensation of suitably substituted acetophenones (including glucosyloxy derivatives) with benzaldehydes in the presence of base (Farkas and Major, 1973). The yields, however, are often rather poor. Chalkones are readily cleaved by strong alkali but, as mentioned earlier, on treatment with dilute alkali or acid (usually the latter) they undergo partial or complete ring closure to the corresponding flavanones. The spectral changes which accompany this isomerization (cf. Chapter 20) are a good indication of the presence of chalkones. This conversion does, however, make the identification of chalkone glycosides somewhat difficult, and removal of sugars is best carried out enzymically (Harborne, 1963). Methylation, other than with diazomethane and demethylation, often gives poor yields of product, since the vinyl ketone structure leads to ready polymerization. The position of sugar attachment is therefore probably best determined by u.v. spectra (Harborne, 1963; Mabry *et al.*, 1970). In the more complex cases, for example the chalkone (**XXXVIII**) from *Goniorrhachis marginata*, recourse to p.m.r. and m.s. measurements, and determination of the ORD curve of the corresponding flavanone has proved most useful (Gottlieb and de Sousa, 1972).

VII Aurones

A STRUCTURE OF AURONES

As mentioned previously, Geissman and Heaton (1943) isolated the first aurone (**XIII**) (the name suggested by Bate-Smith and Geissman (1954) for the class hitherto known as benzylidene-3-coumaranones) from *Coreopsis grandiflora* and called it leptosidin (**XXXIX**). Since that time only seven other aurones have been found, their substitution pattern being similar to the other flavonoids all containing a hydroxyl group at C-6 (Table V).

Unlike the chalkones and other flavonoids there are no prenylated or chromeno aurones, the substitution pattern being (with the exception of maritimetin and leptosidin, Table V) that of the common flavonols. The only sugar found in glycosidic combination is glucose. Since aurones are readily identified on paper and thin-layer chromatograms by their intense yellow to orange fluorescence in u.v. light and by their characteristic u.v. spectra (Mabry *et al.*, 1970; Saleh and Bohm, 1972), it seems highly unlikely that they have been missed in modern surveys. It must be concluded, therefore, that they are rather rare compounds, quite restricted in their distribution.

B CHEMISTRY OF AURONES

The aurones are readily synthesized by condensing suitable aldehydes with β-coumarones, themselves made by cyclization of ω-chloro-2-hydroxyacetophenones. The ease of condensation enables glucosides to be synthesized without difficulty (Farkas and Major, 1973).

Like the other flavonoids discussed previously, the aurones are degraded by alkali to yield A- and B-ring fragments. They also react with cyanides to give the isomeric flavones and may be oxidized by alkaline hydrogen peroxide under suitable conditions to give flavonols.

Determination of the structure of aurones on a microscale follows the conventional patterns described earlier (Harborne, 1962; Mabry *et al.*, 1970) and as they are more stable than chalkones, methylation and other manipulations present little difficulty (Nordstrom and Swain, 1956; Farkas and Major, 1973).

VIII Other flavonoid compounds

In general the flavanones (**VI**) and their 3-hydroxy derivatives, flavanonols or dihydroflavonols (**VII**), isoflavones (**XI**) and the rare isoflavanones (**XL**), are probably not involved directly in plant coloration except as mentioned (p. 426) in white flowers. Flavanones and 3-hydroxyflavanones, especially those having vicinal hydroxy groups (e.g. taxifolin) however, are, susceptible to oxidation and yield brown pigments which in some cases are probably responsible for the browning of flowers as they wither. Nevertheless most of the brown pigments produced from flavonoids appear to be formed by the oxidation of catechins (flavan-3-ols, **IV**) and proanthocyanin oligomers and polymers (flavolans) produced from the flavan-3,4-diols (**VIII**). Indeed, these polymeric forms, which are often called condensed tannins, are themselves coloured. The structure of many such compounds is now known (Jacques *et al.*, 1973). Enzymic oxidation of these oligomers leads to the formation of polyquinones.

Such quinones are brown in colour and are relatively stable since they are not reduced by either ascorbic acid or potassium ferrocyanide but only by borohydride (Swain, 1963).

The proanthocyanins are extremely widespread in plants (Bate-Smith, 1974) especially the woody dicotyledons, 60% of which have these compounds in their leaves. They occur predominantly in the polymeric form, and are present in specially high concentration in heartwoods (Roux, 1962). In fact, the distinctive dark colour by which most heartwoods are recognized is almost always due to the presence of flavolans; species which do not contain these compounds (e.g. the box *Buxus sempervirens*) have no distinctive heartwood. Flavolans and their oxidation products are also responsible for the brown colour of autumn leaves. Senescent green leaves, when freed from chlorophyll by extraction with organic solvents, show the presence of large amounts of polymeric insoluble brown proanthocyanin (cf. Hillis and Swain, 1959). This apparently is the partial cause of the dull olive colour of these mature leaves. When such leaves are dried slowly, the chlorophyll disappears, and the flavolans are further oxidized to give the deep brown appearance characteristic of many autumn leaves. The proanthocyanins, besides giving rise to brown pigments, occasionally may impart pink colours to wood owing to their partial conversion into anthocyanidins.

The browning described above may or may not be due to enzyme-catalysed oxidations, but the discoloration which ensues when fresh parts of plants are cut certainly is. This is generally most noticeable in the case of fruits, such as the apple, or other fleshy organs like the potato tuber, but it does occur in all plant tissues having the requisite enzyme system and a suitable substrate. In the majority of cases browning is due to the oxidation of hydroxy flavans; indeed, in tea and chocolate the desirable colour of the beverage is due to the products of this reaction (Swain, 1962). However, other non-flavonoid phenolic compounds may be involved. In some cases darkening may also be caused by the formation of metal chelates but, except for modifications in flower petal colour, there are no well-authenticated cases other than in foods (Swain, 1962; Harborne, 1967).

IX Conclusions

It can be seen that although the flavonoids represent a relatively restricted range of structural types, they certainly offer a greater variation in colour than any of the other classes of compounds involved in plant pigmentation. Variation in the degree of conjugation shifts the λ_{max} of the main near u.v. and visible absorption maxima from *c*. 370 nm (pale yellow) from the common flavonols, through deep yellow (chalkones, hydroxyflavonols and aurones), orange (aurones and deoxyanthocyanins) to reds and blues (anthocyanins). Coupled with co-pigmentation effects and the formation of metal chelates this gives the plant kingdom an extremely wide range of possible shades of colour. Since the flavonoids probably evolved in non-flowering plants for quite other reasons (Swain, 1974, 1975) their utilization in gymnosperms and angiosperms as long-range attractants for insect and animal pollinators and seed dispersal agents is noteworthy and shows the plasticity of the evolutionary process.

ACKNOWLEDGEMENTS

I am grateful to Dr J. B. Harborne for help in preparing this chapter and to Mrs S. Saunders for preparing the bibliography.

REFERENCES

Adityachaudhury, N., Kirtaniya, C. L. and Mukherjee, B. (1971). *Tetrahedron* 27, 2111-2117.
Albert, A. (1959). "Heterocyclic Chemistry". Athlone Press, London.
Allan, J. and Robinson, R. (1924). *J. chem. Soc.* 125, 2192.
Anderson, D. W., Gueffroy, D. E., Webb, A. D. and Kepner, R. E. (1970). *Phytochemistry* 9, 1579.
Asen, S., Norris, K. H. and Stewart, R. N. (1972). *Phytochemistry* 11, 2739.
Baker, W. (1933). *J. chem. Soc.* 1381.
Bate-Smith, E. C. (1962). *J. Linn. Soc.* (*Bot*) 58, 95.
Bate-Smith, E. C. (1974). "Chemistry in Botanical Classification" (G. Bendz and J. Santesson, eds). Nobel Foundation, Stockholm.
Bate-Smith, E. C. and Geissman, T. A. (1954). *Nature, Lond.* 167, 788.
Beckmann, S. and Geiger, H. (1963). *Phytochemistry* 2, 281.
Bloom, M. and Geissman, T. A. (1973). *Phytochemistry* 12, 2005.
Bognár, R. (1973). *In* "Recent Flavonoid Research" (R. Bognár *et al.*, eds). Akademiai Kiadó, Budapest.
Bohm, B. (1975). *In* "The Flavonoids" (J. B. Harborne *et al.*, eds). Chapman and Hall, London.
Bringi, N. V. and Dave, K. G. (1956). *Scient. Proc. R. Dubl. Soc.* 27, 93.
Burkhardt, D. (1964). *Adv. Insect Physiol.* 2, 131.
Chandler, B. V. and Harper, K. A. (1962). *Aust. J. Chem.* 15, 114.
Clifford, H. T. and Harborne, J. B. (1969). *Phytochemistry* 8, 123.
Combes, G., Vassort, P. and Winternitz, F. (1970). *Tetrahedron* 26, 5981-5992.
Crombie, L., Green, C. L., Tuck, B. and Whiting, D. A. (1968). *J. chem. Soc. C* 2625-2630.
Dave, K. G., Telang, S. A. and Venkataraman, K. (1962). *Tetrahedron Lett.* 1, 9.
Dean, F. M. (1963). "The Naturally Occurring Oxygen Ring Compounds". Butterworths, London.

Doporto, M. L., Gallagher, K. M., Gowan, J. E., Hughes, A. C., Philbin, E. M., Swain, T. and Wheeler, T. S. (1955). *J. chem. Soc.* 4249.
Emiliwcz, T., Kostanecki, St. v. and Tambor, J. (1899). *Ber. dt. chem. Ges.* 32, 2448.
Everest, A. E. (1914). *Proc. R. Soc. Ser. B* **87**, 449.
Farkas, L. and Major, A. (1973). *In* "Recent Flavonoid Research" (R. Bognár *et al.*, eds). Akademiai Kiadó, Budapest.
Geissman, T. A. and Heaton, C. D. (1943). *J. Am. chem. Soc.* **65**, 677.
Geissman, T. A. and Hinreiner, E. (1952). *Bot. Rev.* **18**, 77.
Gertz, O. (1938). *K. physiograf. Sällskop Lund Forh.* **8**, 62.
Gottlieb, O. R. and de Sousa, J. R. (1972). *Phytochemistry* **11**, 2841.
Graebe, J. and Liebermann, F. (1868). *Ber. dt. chem. Ges.* **1**, 106.
Griffiths, L. A. (1958). *Biochem. J.* **70**, 120.
Grisebach, H. (1972). *In* "Chemistry in Evolution" (T. Swain, ed.), p. 487. Butterworths, London.
Hahlbrock, K. and Grisebach, H. (1975). *In* "The Flavonoids" (J. B. Harborne *et al.*, eds). Chapman and Hall, London.
Harborne, J. B. (1958). *J. Chromat.* **1**, 473.
Harborne, J. B. (1959). *J. Chromat.* **2**, 581.
Harborne, J. B. (1962). *Fortschr. Chem. org. NatStoffe* **20**, 165.
Harborne, J. B. (1963). *Phytochemistry* **2**, 85.
Harborne, J. B. (1967). "Comparative Biochemistry of the Flavonoids". Academic Press, London and New York.
Harborne, J. B. (1973). "Phytochemical Methods". Chapman and Hall, London.
Harborne, J. B. and Williams, C. (1975). *In* "The Flavonoids" (J. B. Harborne *et al.*, eds). Chapman and Hall, London.
Harborne, J. B., Mabry, T. M. and Mabry, H. (eds). (1975). "The Flavonoids". Chapman and Hall, London.
Hertzig, J. (1891). *Monatsh.* **12**, 172.
Hillis, W. E. and Swain, T. (1959). *J. Sci. Fd Agric.* **10**, 135.
Hrazdina, G. (1972). *Phytochemistry* **11**, 3491.
Hurst, H. M. and Harborne, J. B. (1967). *Phytochemistry* **6**, 1111.
Ice, C. H. and Wender, S. H. (1953). *J. Am. chem. Soc.* **75**, 50.
Jacques, D., Haslam, E., Bedford, G. R. and Greatbanks, D. (1973). *Chem. Commun.* 518.
Jurd, L. (1962a). *In* "The Chemistry of Flavonoid Compounds" (T. A. Geissman, ed.), pp. 107–155. Pergamon Press, New York and Oxford.
Jurd, L. (1962b). *Chemy Ind. Rev.* 1197.
Karrer, P. and Widmer, R. (1927a). *Helv. chim. Acta* **10**, 5.
Karrer, P. and Widmer, R. (1927b). *Helv. chim. Acta* **10**, 758.
Karrer, P. and Widmer, R. (1928). *Helv. chim. Acta* **11**, 837.
King, F. E. and Clark-Lewis, J. W. (1954). *J. chem. Soc.* 1399.
Klein, G. (1921). *Sber. Akad. Wiss. Wien* **130**, 247.
Komatsu, M., Temomori, T., Hatayama, K. and Mikuriya, N. (1970). *Yakugaku Zasshi* 90, 463–468.
Kostanecki, St. v. (1893). *Ber. dt. chem. Ges.* 26, 2901.
Limaye, D. B. (1936). *Rasayanam* **1**, 1.
Mabry, T. J., Markham, K. R. and Thomas, M. B. (1970). "The Systematic Identification of Flavonoids". Springer Verlag, Berlin.

Marquart, L. C. (1835). "Eine chemisch-physiol. Abhandlung". Bonn.
Meier, W. and Furst, A. (1962). *Helv. chim. Acta* 45, 232.
Merlini, L. (1968). *Corsi Semin. Chim.* 11, 97-98; *Chem. Abstr.* 1970. 72, 21565.
Müller, H. (1915). *J. chem. Soc.* 107, 87S.
Nascimento, Do. M. C. and Mors, W. B. (1970). *Annais Acad. Bras. Cienc.* 42 (suppl.), 87-92 (*Chem. Abstr.* 1971, 75, 72477g.).
Nilsson, M. (1959). *Acta chem. scand.* 13, 750-757.
Nordstrom, C. G. and Swain, T. (1956). *Archs Biochem. Biophys.* 60, 329.
Perkin, A. G. (1898). *J. chem. Soc.* 73, 1019.
Perkin, A. G. and Everest, A. E. (1918). "The Natural Organic Colouring Matters". Longmans, Green, London.
Perkin, A. G. and Hummel, I. J. (1904). *J. chem. Soc.* 85, 459.
Price, J. R. (1939). *J. chem. Soc.* 1017.
Ribereau-Gayon, P. (1972). "Plant Phenolics". Oliver and Boyd, Edinburgh.
Roberts, E. A. H. (1960). *Nature, Lond.* 185, 536.
Robinson, G. M. and Robinson, R. (1932). *Biochem. J.* 26, 1647.
Robinson, R. (1934). *Ber. dt. chem. Ges.* 67A, 85-98.
Robinson, R. and Todd, A. R. (1932). *J. chem. Soc.* 2299.
Roux, D. G. (1962). *Chemy Ind. Rev.* 278.
Roux, D. G. and Ferreira, D. (1974). *Phytochemistry* 13, 2039.
Saleh, N. A. M. and Bohm, B. A. (1972). *J. Chromat.* 57, 166.
Shimokoriyama, M. (1962). *In* "The Chemistry of Flavonoid Compounds" (T. A. Geissman, ed.), pp. 286-316. Pergamon Press, New York and Oxford.
Shimokoriyama, M. and Hattori, S. (1953). *J. Am. chem. Soc.* 75, 2277.
Smith, H. G. and Read, J. (1922). *J. Proc. R. Soc. N.S.W.* 56, 253-259.
Swain, T. (1962). *In* "The Chemistry of Flavonoid Compounds" (T. A. Geissman, ed.). Pergamon Press, New York and Oxford.
Swain, T. (1963). *Int. Congr. pure appl. Chem.* London.
Swain, T. (1974). *In* "Chemistry in Botanical Classification" (G. Bendz and J. Santesson, eds). Nobel Foundation, Stockholm.
Swain, T. (1975). *In* "The Flavonoids" (J. B. Harborne, T. M. Mabry and H. Mabry, eds). Chapman and Hall, London.
Swain, T. and Hillis, W. E. (1959). *J. Sci. Fd Agric.* 10, 54.
Thompson, W. R. Meinwald, J., Aneshansley, D. and Eisner, T. (1972). *Science, N.Y.* 177, 528.
Timberlake, C. F. and Bridle, P. (1975). *In* "The Flavonoids" (J. B. Harborne, T. M. Mabry and H. Mabry, eds). Chapman and Hall, London.
Willstätter, R. and Everest, A. E. (1913): *Justus Liebig's Annln Chem.* 401, 189.
Willstätter, R. and Mallison, G. (1914). *Sber. preuss. Akad. Wiss.* 769.
Willstätter, R. and Mieg, W. (1915). *Justus Liebig's* Annln Chem. 408, 61.
Willstätter, R. and Zechmeister, L. (1914). *Sber. preuss. Akad. Wiss.* 886.
Witt, E. (1888). *Ber. dt. chem. Ges.* 21, 325.
Wong, E. and Wilson, J. M. (1972). *Phytochemistry* 11, 875.

Chapter 9

Biosynthesis of Flavonoids

EDMON WONG

Applied Biochemistry Division, D.S.I.R., Palmerston North, New Zealar

I Introduction

The biosynthesis of flavonoid compounds has been estimated to account for the consumption of as much as one-sixtieth of the total carbon fixed by photosynthesis in plants (Smith, 1972). The broad outline of the pathway of biosynthesis to this quantitatively important group of natural products has been understood for some time but knowledge of the details of the pathway has not been precise. Research in recent years has been concerned with the elucidation of individual steps and with the isolation and characterization of enzymes of the pathway.

Flavonoid biosynthesis can conveniently be considered in three stages. The first concerns the formation of the basic $C_6C_3C_6$ skeleton from a combination of the acetate–malonate and shikimic acid pathways to aromatic compounds. The precise nature of the phenylpropanoid intermediate involved here has long been a question of interest. The next stage is concerned with the ways by which the different classes of flavonoids are synthesized by a combination of sequential and parallel routes from a C_{15} flavonoid prototype. This is the major area of interest in flavonoid biosynthesis, and in spite of a wealth of studies, cannot yet be said to be well understood. Finally, consideration must be directed to the elaboration of individual compounds within each flavonoid class, involving steps such as hydroxylation, glycosylation etc. which in the main need not be specifically part of the flavonoid pathway. The interest here centres mainly on the stage of the pathway at which these reactions take place.

Insights into all three aspects of the problem of flavonoid biosynthesis have come in the past from studies of "comparative anatomy" (Geissman and Hinreiner, 1952; Birch and Donovan, 1953; Robinson, 1955; Whalley, 1961), and chemical genetic studies (Alston, 1964; Harborne, 1967), and in more recent times from feeding experiments with radioactive tracers. The bulk of the latter studies has been carried out by Grisebach and his co-workers and the results have regularly been reviewed (Grisebach, 1961, 1965, 1968; Grisebach and Barz, 1969). More recent research has centred on detailed studies of individual steps of flavonoid biosynthesis and much progress has lately been achieved in the isolation and characterization of enzymes of the pathway. Progress has been made possible largely by a better understanding of the physiology of flavonoid biosynthesis, particularly of the effect of light. The use of young plant tissues and cell suspension cultures as source materials has also greatly facilitated the study of flavonoid biosynthesis at the enzymic level.

This chapter will touch upon all of the aspects of flavonoid biosynthesis mentioned above but attention will be concentrated mainly on the results of newer tracer and enzymic studies.

II Formation of the basic $C_6C_3C_6$ skeleton

By analogy with fatty acid biosynthesis, the basic $C_6C_3C_6$ skeleton of flavonoids has long been postulated to arise from the condensation of a C_6C_3 unit with three malonate–acetate units, all being activated as CoA thio esters (Birch and Donovan, 1953; Grisebach, 1962). The C_6C_3 unit is presumably a cinnamic acid, thus giving rise to a chalcone as the first C_{15} flavonoid prototype (Fig. 1). All available evidence is consistent with this cinnamic acid-chalcone hypothesis and direct proof has recently been provided by the isolation of the chalcone synthetase enzyme (Section IID).

The involvement of cinnamic acids in flavonoid biosynthesis links this pathway to that of lignin biosynthesis. Phenylalanine is the immediate biogenetic precursor of cinnamic acid, the enzyme responsible for the conversion being phenylalanine ammonia lyase (PAL, Section IIA). This aromatic amino acid thus serves as the branching point from primary metabolism to these two major areas of secondary plant metabolism.

With few exceptions, all flavonoid compounds possess oxygen at position C-4′ of the B ring. *Para*-Oxygenation is also a hallmark of phenylpropanoid compounds. This strongly suggests that *p*-coumaric acid, rather than cinnamic acid itself, is the direct phenylpropanoid intermediate to the great majority of flavonoids. The enzyme catalyzing the formation of *p*-coumaric acid from cinnamic acid, cinnamic acid 4-hydroxylase (CAH), is thus another key enzyme in the pathway to both flavonoids and lignin. The biosynthesis of

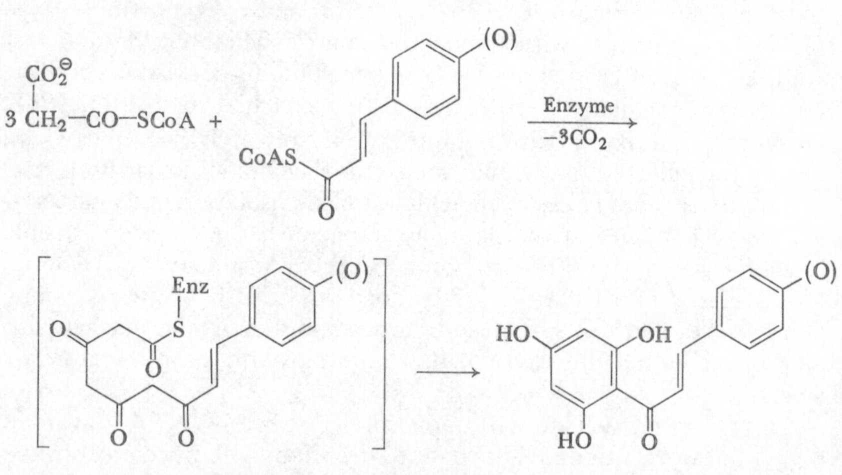

FIG. 1. Probable origin of the $C_6C_3C_6$ skeleton of flavonoid compounds.

p-coumaric acid is also possible via tyrosine (Fig. 2), but this route is apparently of significance in plants of the Gramineae family only.

The ability to combine the acetate–malonate and shikimic acid pathways to produce flavonoid compounds has long been thought to be characteristic of higher plants only, and sporadic reports in the past of the presence of flavonoid compounds in micro-organisms have not been conclusive. The compounds chloroflavonin (I) (Bird and Marshall, 1969) and flavonin (II) (Marchelli and Vining, 1973), however, have recently been isolated from cultures of *Aspergillus candidus* and their status as fungal metabolites has been proved by their *de novo* synthesis by *Aspergillus* from administered [^{14}C]phenylalanine (Marchelli and Vining, 1973). This study also indicates that the pathway to flavonoids in micro-organisms is apparently the same as in higher plants, unlike the situation existing for some other groups of phenolic natural products (Zenk and Leistner, 1968; Carpenter *et al.*, 1969).

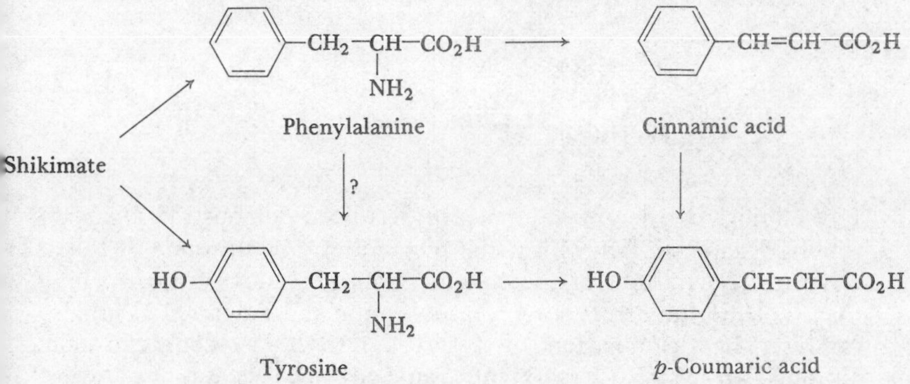

Shikimate → Phenylalanine (C_6H_5–CH_2–CH(NH_2)–CO_2H) → Cinnamic acid (C_6H_5–CH=CH–CO_2H)

Shikimate → Tyrosine (HO–C_6H_4–CH_2–CH(NH_2)–CO_2H) → *p*-Coumaric acid (HO–C_6H_4–CH=CH–CO_2H)

Phenylalanine ↓ ? Tyrosine; Cinnamic acid ↓ *p*-Coumaric acid

FIG. 2. Possible routes to *p*-coumaric acid.

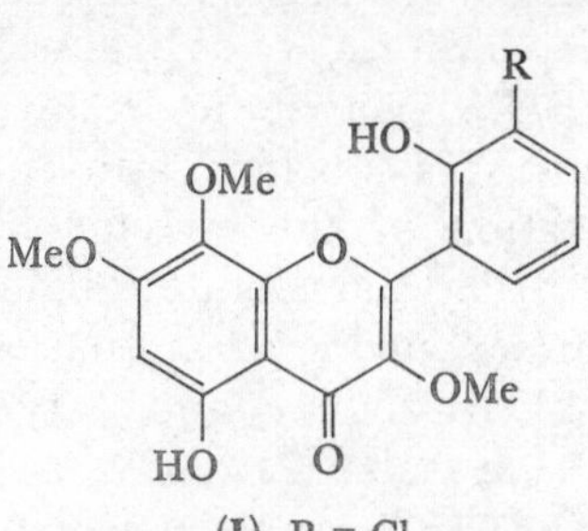

(I) R = Cl
(II) R = H

A PHENYLALANINE AMMONIA LYASE (PAL)

Phenylalanine ammonia lyase (PAL) (EC 4.1.1.5) is without doubt the key enzyme in phenylpropanoid (C_6C_3) metabolism. Since its first report by Koukol and Conn (1961), numerous studies on various aspects of this enzyme have been made, and it has been the subject of several recent reviews (Hanson and Havir, 1972a,b; Camm and Towers, 1973a). In this section only those properties more relevant to flavonoid biogenesis will be summarized. External and intrinsic factors affecting PAL levels in plants will be dealt with in Section V.

The enzyme has a wide distribution (Young *et al.*, 1966; Towers and Subba Rao, 1972) and has been isolated and purified from numerous plant sources and from certain fungi (Camm and Towers, 1973a). In fungi the enzyme appears to have a role in catabolism (Towers, 1969).

The reaction catalysed by PAL is the antiperiplanar deamination of L-phenylalanine to yield *trans*-cinnamic acid (Hanson and Havir, 1972a):

$$C_6H_5CH_2CH(NH_3^{\oplus})CO_2^{\ominus} \xrightleftharpoons{\text{PAL}} C_6H_5CH{=}CHCO_2^{\ominus} + NH_4^{\oplus} \qquad (1)$$

Reversibility of the enzyme reaction has been demonstrated *in vitro* (Subba Rao *et al.*, 1967) but the biosynthetic significance of this is obscure. Preparations of the enzyme from plants of the grass family (Neish, 1961) and from some yeasts (Camm and Towers, 1973a) also catalyse the deamination of L-tyrosine to *trans-p*-coumaric acid, although always to a less extent. Some enzyme preparations are also effective with dopa (Young and Neish, 1966). Tyrosine ammonia lyase (TAL) is sometimes considered to be a separate enzyme (Neish, 1961). In barley preparations, two peaks of activity with very different ratios of PAL/TAL activity are resolved on Sephadex G-200 (Kindl, 1970). In contrast, a constant ratio of PAL/TAL activity was found in the purification of the enzyme from the yeast *Sporobolomyces pararoseus* (Parkhurst and Hodgins, 1971). Havir *et al.* (1971) have produced convincing additional evidence for a common catalytic site for L-phenylalanine and L-tyrosine in the enzyme isolated and purified from maize (*Zea mays*).

Multiple forms of PAL have been isolated by DEAE column chromatography of preparations from sweet potatoes (Minamikawa and Uritani, 1965b) and from oak leaves (*Quercus pedunculata*) (Boudet *et al.*, 1971). In the latter study the two forms showed different sensitivities to benzoic acids and cinnamic acids which are inhibitors of PAL *in vitro*. This difference in sensitivities to inhibitors was also exhibited by the two isoenzymes of sweet potato. These studies indicate that pathways of biosynthesis leading to different end products (e.g. benzoic acid and cinnamic acid derivatives) may involve different PAL isoenzymes. It should be noted, however, that attempts to obtain PAL isoenzymes in many other systems have been unsuccessful (Camm and Towers, 1973a).

In addition to benzoic and cinnamic acids, end-product inhibitions of PAL have also been observed with flavonoid compounds such as kaempferol and quercetin (Attridge *et al.*, 1971).

B CINNAMIC ACID 4-HYDROXYLASE (CAH)

Cinnamic acid 4-hydroxylase (CAH) is the second key enzyme in the pathway of biosynthesis of phenylpropanoid/flavonoid compounds from phenylalanine. Its isolation from spinach was first reported by Nair and Vining (1965) but attempts by other workers to demonstrate enzyme activity under the conditions reported have been unsuccessful (Russell, 1971). CAH activity has since been isolated from pea seedlings by Russell and Conn (1967) and many of its properties have been studied in detail (Russell, 1971).

The CAH of pea seedlings is a mixed function oxidase which requires molecular oxygen and NADPH for activity;

$$C_6H_5\text{—CH=CH—CO}_2H + O_2 \xrightarrow{\text{NADPH} + H^{\oplus} \;\to\; \text{NADP}^{\oplus}} HO\text{—}C_6H_4\text{—CH=CH—CO}_2H + H_2O \qquad (2)$$

NADH will not substitute for NADPH and tetrahydrofolate is not required. 2-Mercaptoethanol is needed for optimal activity; it does not serve as an external reductant for the hydroxylation reaction but is needed for functional integrity of the enzyme.

CAH appears to be specific for cinnamic acid. Phenylalanine and *p*-coumaric acid were not hydroxylated by the enzyme. Kinetic studies showed that CAH has high affinity for cinnamic acid ($K_m = 1{\cdot}7 \times 10^{-5}$ M) and that it is inhibited by low concentration of *p*-coumaric acid. The high degree of inhibition indicates that the concentration of *p*-coumarate in the tissues is under stringent self-regulation and suggests that regulation of CAH activity is an important control point in phenolic biosynthesis (Russell, 1971).

It has been shown that when [*p*-^{3}H]cinnamic acid is used as the substrate, as much as 93% of the tritium is retained in the *meta* position of the hydroxylated product (Russell *et al.*, 1968). This is an example of the NIH shift accompanying hydroxylation by a plant enzyme. The mechanism proposed for oxygenation and rearrangement of tritium is given in Fig. 3 (Reed *et al.*, 1973).

Since the report of Russell and Conn (1967), CAH activity has been isolated from buckwheat seedlings (Amrhein and Zenk, 1968, 1971), parsley cell suspension cultures (Hahlbrock *et al.*, 1971a), garbanzo bean (Reed *et al.*, 1973) and potato tubers (Camm and Towers, 1973b). CAH activity has also been detected in bamboo and asparagus shoots and gingko and cedar seedlings (Shimada *et al.*, 1970), and in grapefruit tissue slices (Hasegawa and Maier, 1972). Enzymic activity in all cases has been found in the microsomal fraction. Camm and Towers (1973b), however, recently reported that in dormant potato tuber tissues CAH is mostly in a soluble form, becoming associated with microsomal fraction in slices on ageing in light.

As with PAL, activity of CAH in plants has been found to be affected by light, wounding, ethylene treatment and ontogeny (Section V).

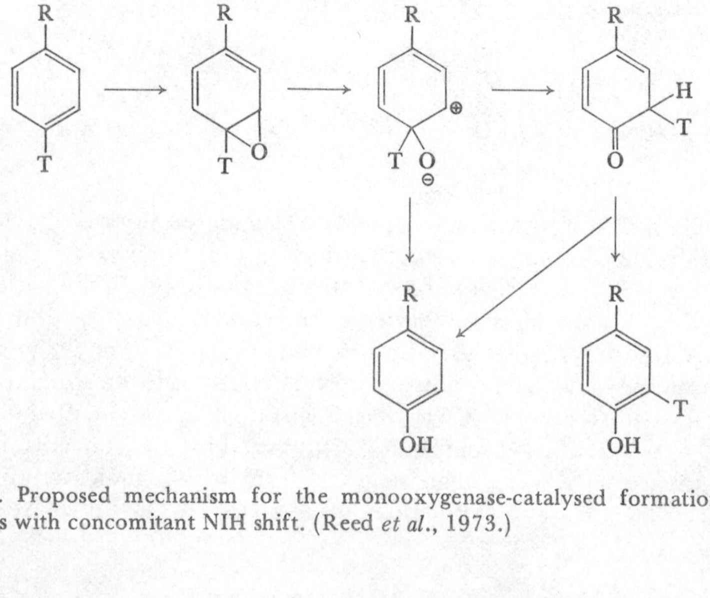

R
T
R
T
O
R
T
O
⊕
⊖
R
H
T
O
R
OH
R
T
OH

FIG. 3. Proposed mechanism for the monooxygenase-catalysed formation of phenols with concomitant NIH shift. (Reed *et al.*, 1973.)

C FORMATION OF CoA ESTERS OF CINNAMIC ACIDS

Early attempts to demonstrate the activation of cinnamic acids with cell-free preparations from parsley and garbanzo bean were inconclusive (Grisebach *et al.*, 1966). Only very weak activation of cinnamic acids was obtained, and it was considered possible that this activation was due to a non-specific acid:CoA ligase not related to aromatic metabolism.

Walton and Butt (1970, 1971) have prepared extracts from leaves of spinach beet (*Beta vulgaris*) which activated cinnamic acid in the presence of CoA. A much higher rate of cinnamate activation relative to acetate was found, and the cinnamate activation was distinguished by its sensitivity to oxidizing conditions and its greater lability on storage. The enzyme activity was assayed by direct spectrophotometric analysis of cinnamoyl CoA; CoA-dependent formation of hydroxamate; and incorporation of pyrophosphate into nucleotide phosphate. By analogy with acetyl-CoA synthetase (acetate:CoA ligase (AMP), EC 6.2.1.1), the reaction can thus be assumed to proceed in two stages.

$$\text{Cinnamate} + \text{ATP} \rightleftharpoons \text{Cinnamoyl-AMP} + \text{PP}_i$$

$$\text{Cinnamoyl-AMP} + \text{CoASH} \rightleftharpoons \text{Cinnamoyl-SCoA} + \text{AMP}$$

The activation system was reported to be virtually specific for the unsubstituted cinnamic acid although the negative results found for *p*-coumaric, caffeic, ferulic and sinapic acid were not conclusive. Measurable activity with cinnamic acid was also found in leaves from young plants of pea, runner bean and spinach.

The formation of CoA esters of cinnamic, *p*-coumaric, *p*-methoxycinnamic, and ferulic acids by cell suspension cultures of parsley was demonstrated by Hahlbrock and Grisebach (1970). Of these acids *p*-coumaric acid served as the most efficient substrate. Acetate and malonate were also activated but whereas the formation of acetyl-CoA is not influenced by light, enzyme activity for the activation of *p*-coumaric acid is markedly increased after illumination, in a manner very similar to that in which the activities of a number of enzymes involved in flavone biosynthesis in parsley are stimulated by light. It was concluded from these results that the formation of *p*-coumaroyl-CoA was catalysed by a specific enzyme (*p*-coumarate:CoA ligase) directly related to the biosynthesis of flavonoids. The activation of *p*-coumaric acid has also been demonstrated with cell-free preparations from young parsley leaves.

A similar enzyme exists in cell suspension cultures of soyabean and has been partly purified (Lindl *et al.*, 1973). The enzyme is very unstable under the conditions of the purification ($MnCl_2$ and $(NH_4)_2SO_4$ precipitations and Sephadex G-100 gel column chromatography) but can be stored frozen for short periods. The reaction catalysed required Mg^{2+} and ATP at a molar ratio of 1:1. This enzyme also exhibits a high degree of specificity for *p*-coumaric acid (relative percentage conversion: *p*-coumaric 100, ferulic 49, caffeic 14, *p*-methoxycinnamic 12, cinnamic 10) and is also considered to be a specific enzyme of aromatic metabolism.

A substituted-cinnamic acid CoA ligase, possibly involved in the biosynthesis of lignin in swede roots (*Brassica napobrassica*) has recently been described by Rhodes and Wooltorton (1973). This enzyme differs interestingly in substrate specificity from that of the enzymes already described. It has an apparent specificity for the presence of a free phenolic-OH group in the substrate. *p*-Coumaric and ferulic acids are the two most active ($K_m = 1{\cdot}4 \times 10^{-5}$, $3{\cdot}1 \times 10^{-5}$ M respectively), followed by caffeic and isoferulic acids. Cinnamic acid and methoxylated cinnamic acids on the other hand are inactive. The activity of this enzyme in swede root tissues rises during ageing and with ethylene treatment, in parallel to the increased formation of cinnamic acids and lignin under the same conditions.

D CHALCONE SYNTHETASE

The first demonstration of the cell-free formation of a flavonoid from *p*-coumaroyl-CoA and malonyl-CoA was recently reported by Kreuzaler and Hahlbrock (1972). Success was made possible by using as source material for enzyme isolation parsley cell suspension cultures 24 h after illumination with light. Based on the highly co-ordinated changes in the activities of the enzymes related to flavone-glycoside biosynthesis previously found in this system (Hahlbrock *et al.*, 1971a), it was predicted that the "chalcone synthetase" should also be most active at about that stage.

After incubation of a protein extract from illuminated parsley cell cultures with malonyl-CoA and *p*-coumaroyl-CoA, one or the other substrate being ^{14}C-labelled in separate experiments, [^{14}C]naringenin (5,7,4′-trihydroxyflavanone) was isolated as the product. Evidence for the specific incorporation of malonyl-CoA into ring A was obtained by degradative studies. The probable reactions catalysed by the enzyme are as shown in Fig. 1. The isolation in this work of the flavanone rather than the chalcone as the reaction product is not surprising, since spontaneous or enzyme-catalysed isomerization of chalcone to flavanone in the crude enzyme system is to be expected (see Section IIIA). Taking results at face value would mean that flavanone is the true product of the enzymic reaction, which seems most unlikely.

III Biosynthesis of classes of flavonoids

The central role of chalcone as the common C_{15} biogenetic intermediate of the different classes of flavonoids is now well established. The earlier evidence has been reviewed (Grisebach, 1965, 1968) and results need only be presented here in summary form (Fig. 4). Reference to the more recently studied transformations will be made in the sections following dealing with the individual classes of flavonoids. It should be pointed out that in many of the studies summarized in Fig. 6, the flavanone naringenin (V) was actually used as the precursor in feeding experiments. The corresponding chalcone (III), however, can be taken as being definitely implicated, representing in either case the immediate precursor or the immediate transformation product of the flavanone (see Section IIIB).

Having established that all classes of flavonoids are chalcone derived, the problem then becomes one of elucidation of class relationships beyond the chalcone stage. Existence of a direct relationship between flavonoid classes indicates their synthesis via sequential steps and absence of relationship indicates synthesis by parallel steps from the chalcone. Biochemical genetic evidence available strongly suggests that a combination of parallel and sequential steps is operative in the biosynthesis of flavonoids.

The conversion of chalcone/flavanone to the other classes of flavonoids requires in essence oxidation and reduction reactions and many schemes based on good chemical analogies have been proposed for the transformations (for a summary, see Pelter *et al.*, 1971). From the biochemical viewpoint, these reactions are presumably mediated by enzymes of the mono-oxygenase and dehydrogenase types. Catalysis by each of the two types of enzymes can in principle lead to either a net gain of oxygen or a net loss of two hydrogens in the flavonoid substrate, if hydration and dehydration are allowed as ancillary processes. This situation can best be summarized by the two stoicheiometric equations below:

$$-2[H] = +[0] - H_2O$$
$$+[0] = -2[H] + H_2O$$

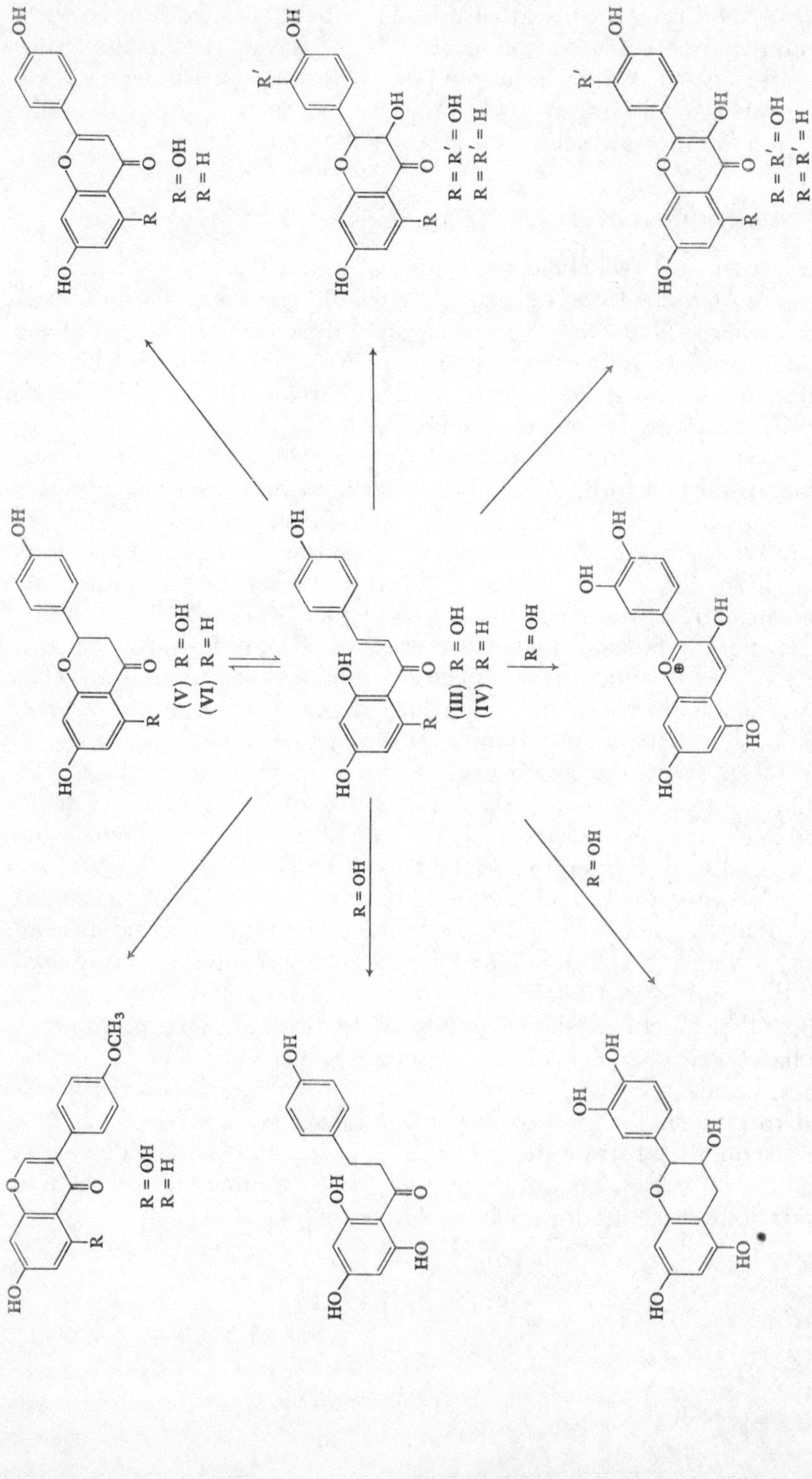

FIG. 4. Incorporation of chalcones into other flavonoid compounds.

Alternative routes to certain of the flavonoid classes can be envisaged on this principle and examples will be met with in the sections following.

A CHALCONE–FLAVANONE ISOMERIZATION

Although isomerization of chalcones to the corresponding flavanones occurs spontaneously in solution, in nature the reaction is catalysed by an enzyme, chalcone–flavonone isomerase (EC 5.5.1.6).

The isolation and purification of chalcone–flavanone isomerase from soyabean seed was first reported by Wong and Moustafa (1966; Moustafa and Wong, 1967). The enzyme catalyses the cyclization of 4,2′,4′-trihydroxychalcone (VII) to (–)-4,7-dihydroxyflavanone (IX) having the *S* configuration at C-2. The activity of the enzyme is not dependent on co-enzymes or activators but is strongly inhibited by *p*-hydroxymercuribenzoate.

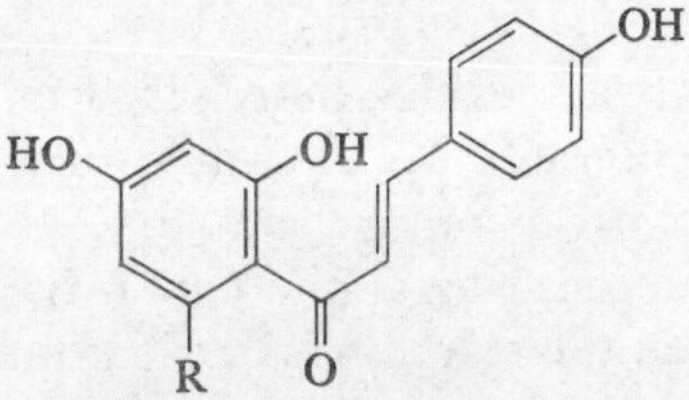

(VII) Isoliquiritigenin, R = H
(VIII) Chalconaringenin, R = OH

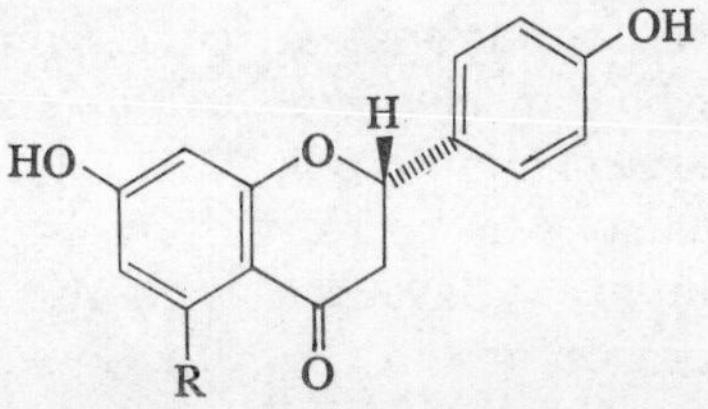

(IX) (–)-Liquiritigenin, R = H
(X) (–)-Naringenin, R = OH

Substrate specificity was studied with chalcones having different substitution patterns in rings A and B. Hydroxylation in ring B is not a specific requirement of the enzyme and the 4,2′,4′,6′-tetrahydroxychalcone (VIII) was found to be as good a substrate as (VII). This broad specificity of the enzyme with respect to the substitution pattern of ring A is considered to have biogenetic significance, since chalcones (VII) and (VIII) are expected to be separate progenitors of two parallel series of flavonoid compounds (see Section IVA).

Chalcone–flavanone isomerase has since been isolated from a number of other sources. In all plants studied the enzyme exists in a varying number of multiple forms. By means of DEAE and gel filtration chromatography and preparative electrophoresis, isomerase enzymes and isoenzymes have been isolated from mung bean (*Phaseolus aureus*), garbanzo bean (*Cicer arietinum*) and parsley (*Petroselinum hortense*) and their properties compared (Hahlbrock *et al.*, 1970a). These heteroenzymes and isoenzymes differ appreciably in regard to the pH optimum and substrate specificity (Table I). The two isoenzymes from mung bean have overlapping specificities towards the chalcones (**VII**) and (**VIII**). Since both normal (5-hydroxy) and 5-deoxy flavonoids occur in mung bean, isomerase specificity for chalcones of the two series would be expected. From the K_m values (Table I) it is tempting to speculate that mung bean isoenzyme I might *in vivo* be the enzyme responsible for the normal whilst isoenzyme II is responsible for the 5-deoxy series.

In contrast, the isoenzyme purified from parsley is inactive with chalcones belonging to the 5-deoxy series (Table I) in line with the fact that in this plant only flavonoids of the normal series occur. This specificity for chalcones with normal phloroglucinol type A ring has been found also for the enzymes from *Datisca cannabina* and *Tulipa* (Grambow and Grisebach, 1971; Wiermann, 1972b). The results thus indicate that there is a correlation between substrate specificity of chalcone–flavanone isomerase and substitution pattern of flavonoid compounds in a particular plant.

The mechanism of the enzyme catalysed cyclization has been studied (Hahlbrock *et al.*, 1970b). α-Deuterated 4,2′,4′-trihydroxychalcone is converted by the isomerase from mung bean to 7,4′-dihydroxyflavanone with the deuterium appearing at the equatorial position at C-3 (Fig. 5). Conversely, when the reaction is carried out

TABLE I

K_m Values (10^{-5}M) and pH optima for chalcone–flavanone isomerase enzymes and substrates. (Data from Hahlbrock *et al.*, 1970a)

	Enzyme from			
	Mung bean isoenzyme I	Mung bean isoenzyme II	Garbanzo bean[a]	Parsley[b]
Chalcone[c]				
4,2′,4′-Trihydroxy (**VII**)	5·7	1·4	1·6	–
4,2′,4′,6′-Tetrahydroxy (**VIII**)	1·8	4·4	4·4	1·6
2′,4′-Dihydroxy	26	4·1	1·8	–
4,2′-Dihydroxy	19	3·7	0·7	–
pH optimum	7·5	8·4	8·5	8·4

[a] One of two electrophoretically separable forms isolated.
[b] One of five electrophoretically separable forms isolated.
[c] The 4′-glucoside of chalcone (**VII**) and 4′- and 6′-glucosides of chalcone (**VIII**) show no activity with any of the four enzymes.

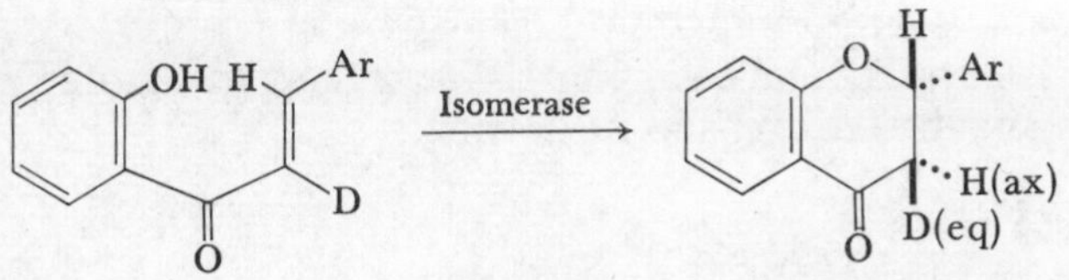

FIG. 5. Stereochemical course of the reaction catalysed by chalcone–flavanone isomerase.

in D_2O, the labelling introduced was found predominantly at the axial position of C-3. The cyclization thus represents formally a *cis* addition to the double bond.

B THE ROLE OF FLAVANONES AS INTERMEDIATES

Since flavanone is the thermodynamically more stable isomer of chalcone and since the majority of flavonoid compounds contain a 6-membered heterocyclic ring in common with flavanone, it might reasonably be assumed that flavanone represents the primary *heterocyclic* intermediate in the pathway to other classes of flavonoids. Earlier tracer studies have abundantly shown that flavanones are excellent precursors of many types of flavonoids, including flavones, flavonols, anthocyanins and isoflavonoids (Grisebach, 1968). That flavanones are involved directly as intermediates in the biosynthesis of these compounds, and not via chemical back-conversion to chalcones, found apparent experimental support when Patschke *et al.* (1966b) showed that the natural (–)-enantiomer of naringenin (X) was incorporated into the anthocyanidin, cyanidin, and the isoflavone biochanin A, at a much higher rate than the corresponding (+)-enantiomer.

The possibility that flavanones may not, after all, be more direct precursors of other flavonoids than chalcones has been raised by Wong (1968). Parallel competitive feeding experiments were carried out in which either (a) ^{14}C-labelled chalcone (isoliquiritigenin) (**VII**) diluted with an equal amount of (–)-flavanone (liquiritigenin) (**IX**) or (b) ^{14}C-labelled (–)-flavanone (liquiritigenin) diluted similarly with chalcone (isoliquiritigenin) was fed to subterranean clover seedlings. The radioactive products, 7,4′-dihydroxyflavone (**XI**), 7,4′-dihydroxyisoflavone (daidzein) (**XIII**) and 7-hydroxy-4′-methoxyisoflavone (formononetin) (**XIV**) were isolated and their specific activities determined. The results (Table II) show that the specific activities for the products from experiment (a) are higher than those from experiment (b). This was interpreted as indicating that the chalcone, and not the flavanone, is the more immediate precursor for the flavone and isoflavone.

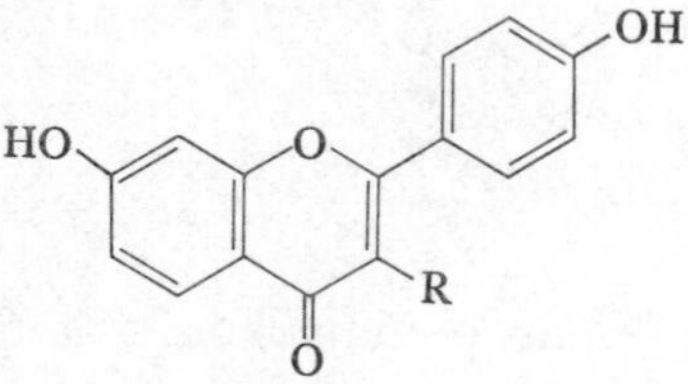

(XI) R = H
(XII) R = OH

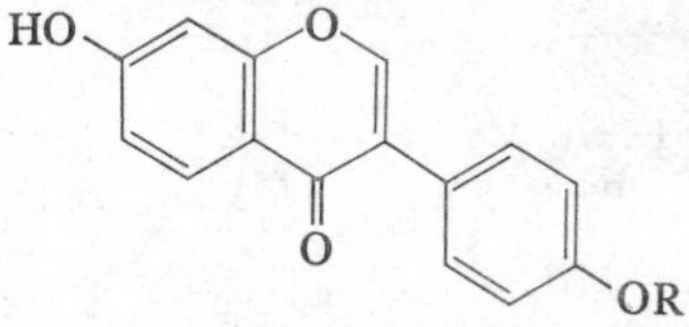

(XIII) Daidzein, R = H
(XIV) Formononetin, R = Me

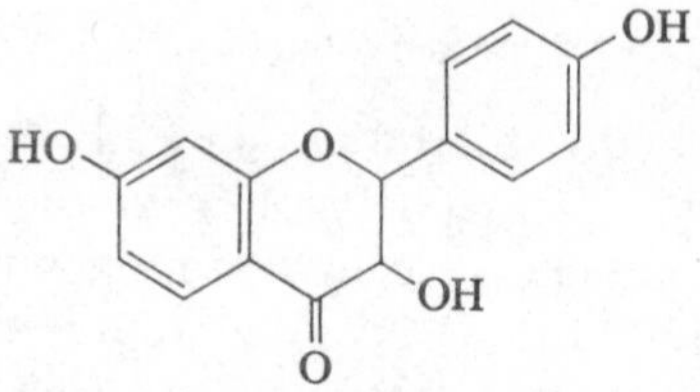

(XV)
Garbanzol

TABLE II

Specific activities (c.p.m./μmole x 10^{-3}) of flavonoid compounds from clover seedlings fed [^{14}C] chalcone or [^{14}C] flavanone diluted with an equal amount of the appropriate unlabelled tautomer. (Data from Wong, 1968)

Precursor[a b]	Flavanone	Chalcone	7,4′-Dihydroxy flavone	7,4′-Dihydroxy isoflavone	Formononetin
Experiment					
(a) [^{14}C] Chalcone +(–)-flavanone	74	144	135	106	28
(b) [^{14}C] (–)-Flavanone + chalcone	134	54	66	51	13

[a] Chalcone = isoliquiritigenin (**VII**). Flavanone = liquiritigenin (**IX**).
[b] Sp. act. 199 c.p.m./μmol x 10^{-3}.

Further support for this conclusion was obtained from double-labelling experiments using a second plant system (Wong and Grisebach, 1969). [^{14}C] Chalcone (**VII**) and [^{3}H] -(–)-flavanone (**IX**) were fed to seeds of garbanzo bean. $^3H/^{14}C$ ratios of the radioactive products, 7,4′-dihydroxyflavone (**XI**), 7,4′-dihydroxyflavonol (**XII**), formononetin (**XIV**) and 7,4′-dihydroxyflavanonol (garbanzol) (**XV**), were determined at various times after feeding. From the $^3H/^{14}C$ values, precursor composition in terms of the ratio of original [^{14}C] chalcone and [^3H] flavanone incorporated (ch/fl or fl/ch) could be calculated. Results (Fig. 6) again strongly suggest that all of the products are derived directly from chalcone.

The results from both studies also show conclusively that *in vivo* the (—)-flavanone and chalcone are biochemically interconvertible. *In vitro* catalysis of the reverse reaction, flavanone to chalcone, by purified chalcone–flavanone isomerase has also been studied (Boland and Wong, 1975).

It would seem from these results that the reaction chalcone → flavanone catalysed by isomerase is not directly involved in the pathway from chalcones to flavones, isoflavones, flavanonols and flavonols. Flavanone formation is then a parallel rather than an intermediate step in the biosynthesis of these compounds (Fig. 7). The stereo-

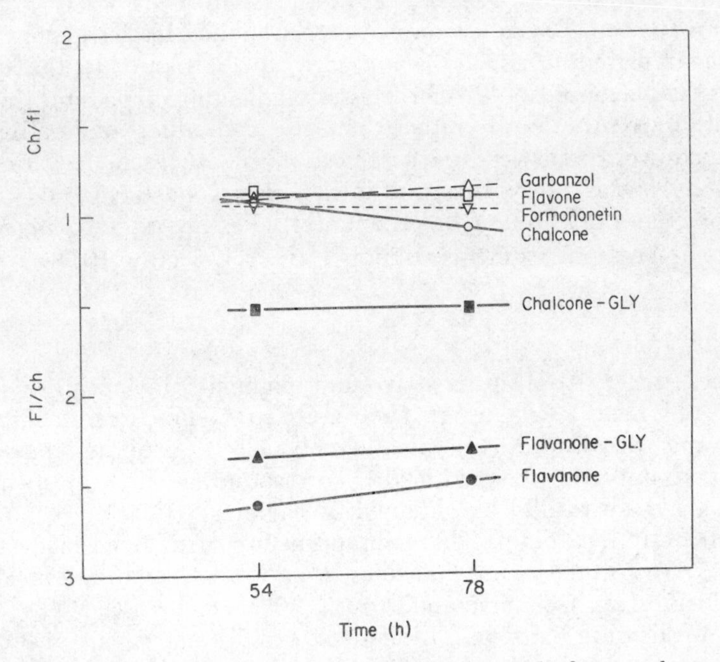

FIG. 6. Precursor composition of flavonoid compounds from garbanzo bean after feeding [^{14}C] chalcone and [^3H] flavanone. (From Wong and Grisebach, 1969.)

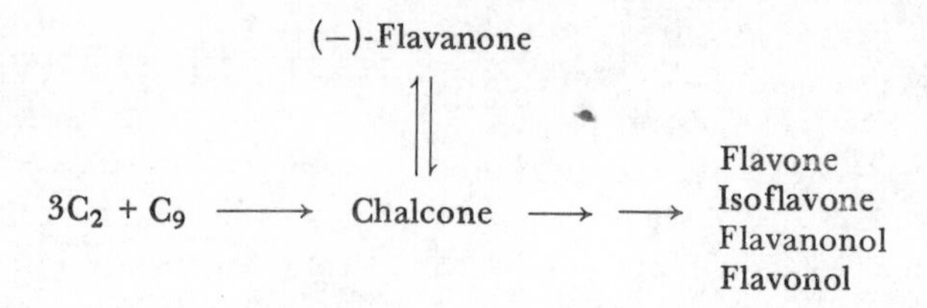

FIG. 7. The probable biogenetic relationship of flavanone and other classes of flavonoids.

specific incorporation results of Patschke *et al.* (1966b) are also readily explicable in this scheme since only natural (−)-flavanones would be expected to undergo enzymic isomerization to chalcones.

It should be noted, however, that much indirect evidence exists to support the alternative view that the chalcone ⟶ flavanone step is of important significance in flavonoid biosynthesis. Isomerase activity has been correlated in several plant systems with stage of development and differentiation when intensive biosynthesis of flavonoid compounds occurs (Section V). Correlation of isomerase specificity with the substitution pattern of flavonoids in a plant has already been mentioned. Except in rare cases, flavanones, like chalcones, are not accumulated in metabolically active plant tissues to the same extent as flavonoids of other classes, indicating a possible intermediate biosynthetic role rather than as end products of a pathway. An unequivocal answer to the question of whether flavanone is involved in the biosynthesis of other classes of flavonoids from chalcone can only come from enzymic studies in systems where the chalcone–flavanone isomerase is inoperative (cf. Section IIID).

C AURONES

Aurones differ from other flavonoid pigments in having a five-membered heterocyclic ring. They are also of much more limited taxonomic distribution. In plants in which they occur, they are usually found together with the corresponding chalcones, from which they can readily be obtained by chemical oxidation (e.g. with oxygen or ferricyanide). The assumption has often been made that aurones are formed from chalcones in nature by a simple oxidative ring closure (e.g. Geissman and Crout, 1969). Mechanistically, such a facile cyclization can be rationalized readily in terms of initial oxidation (loss of one or two electrons) at the 4-OH group of ring B (Dean and Podimuang, 1965; Pelter *et al.*, 1971). Figure 8 presents the mechanism postulated in terms of one-electron steps.

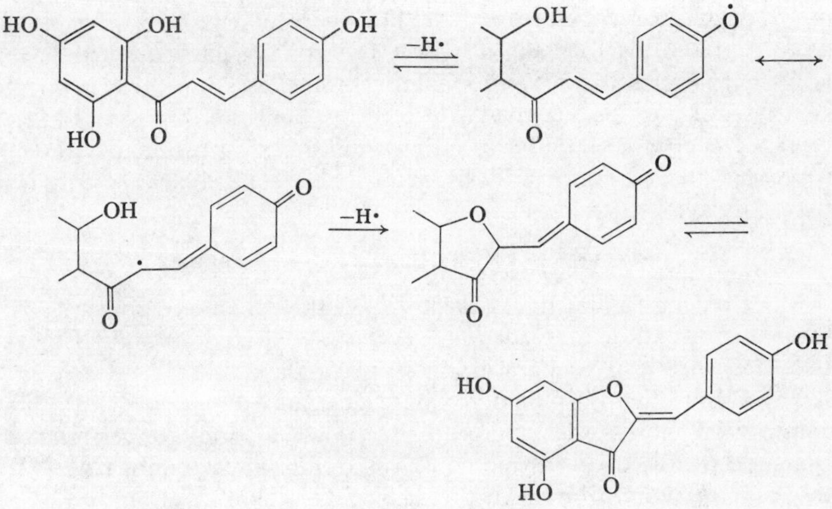

FIG. 8. Postulated mechanism for the oxidation of chalcone to aurone.

Enzyme preparations capable of converting chalcones into aurones have been studied. Crude cell-free extracts of soyabean converted isoliquiritigenin-4′-glucoside (**XVI**) to the corresponding aurone, hispidol-6-glucoside (**XVIII**) (Wong, 1966). When the free chalcone (**XVII**) was used as substrate with this enzyme preparation (Wong, 1967), or with cell-free extracts of *Cicer arietinum* (Wong, 1965), two intermediate products were obtained which readily converted to the aurone hispidol (**XIX**). These products have been identified as

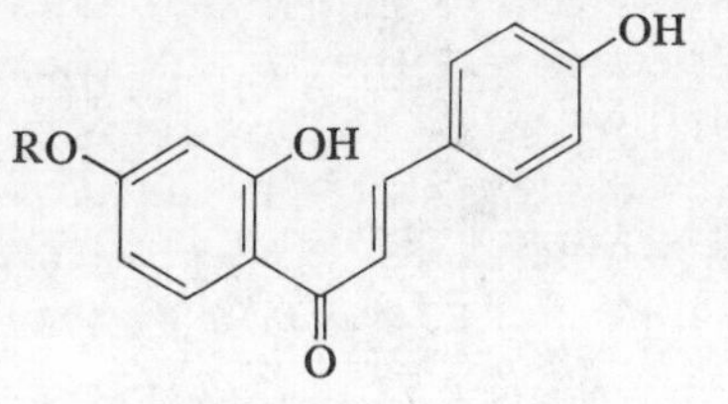

(**XVI**) R = Glucosyl
(**XVII**) R = H

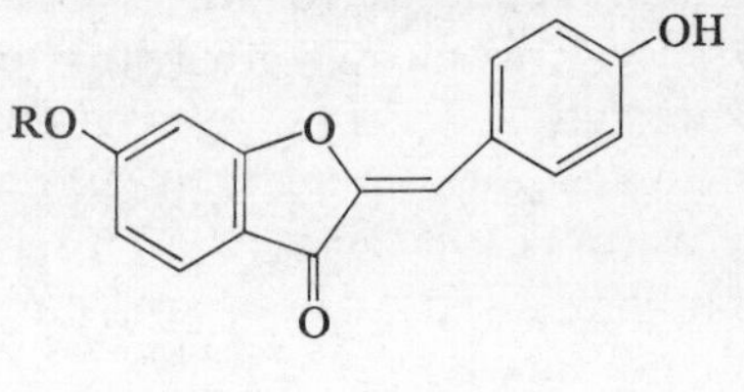

(**XVIII**) R = Glucosyl
(**XIX**) Hispidol, R = H

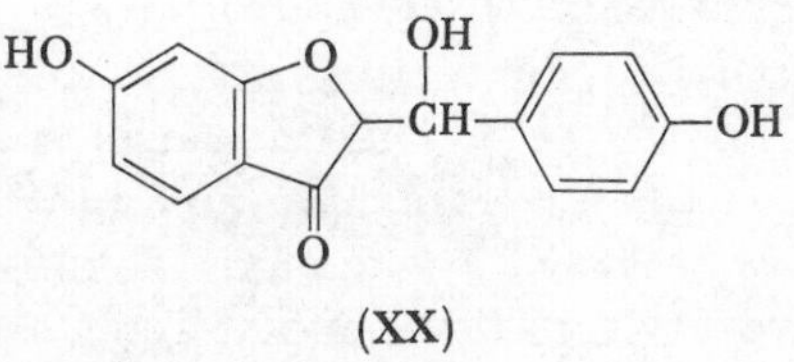

(**XX**)

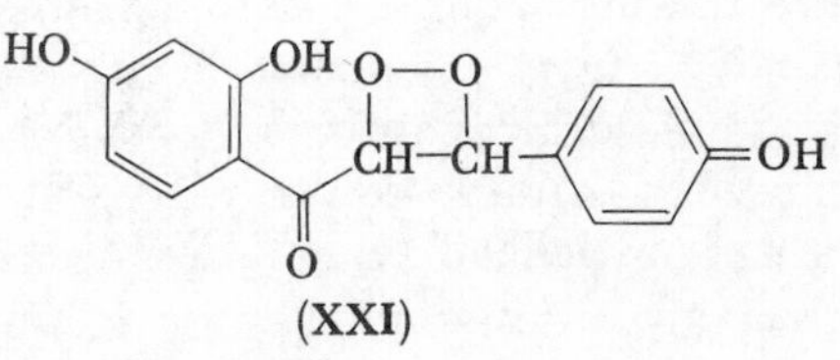

(**XXI**)

the diastereoisomeric forms of the 2(α-hydroxybenzyl)coumaranone (XX) representing formally a hydrated aurone. Such structures have long been postulated as possible intermediates in the chemical synthesis of aurones via the AFO reaction, and the discovery in this work of such a compound as an enzymic reaction product strongly suggested that the biosynthesis of aurones from chalcones follows the pathway:

$$\text{Chalcone} \longrightarrow \text{Hydrated Aurone} \longrightarrow \text{Aurone}$$

More recent studies have revealed that the enzyme responsible for the transformation of chalcone (**XVII**) to hydrated aurone (**XX**) in the *Cicer* cell-free preparation is a peroxidase (Wong and Wilson, 1972). With the purified peroxidase enzyme from *Cicer* and with commercial horseradish peroxidase, however, the conversion of chalcone to hydrated aurone constituted only a very minor reaction. Instead, the main product of the enzymic reaction was found to be a very unstable compound, the structure of which is not yet unequivocally established but is most probably that of the novel cyclic peroxide derivative (**XXI**) (E. Wong and J. M. Wilson, unpublished results). With the crude cell-free enzyme preparation the conversion of chalcone (**XVII**) into this product was not observed, probably owing to the presence of some specific inhibitor(s) in the system. Under these conditions the reaction leading to the hydrated aurones became manifest, enabling the discovery of this enzymic reaction to be made in the first place.

The multiplicity of reactions and products obtainable with the *in vitro* chalcone–peroxidase system studied (see also Sections IIIE and IIIF), although of great interest in providing hitherto unknown examples of the type of enzymic oxidative reactions possible with chalcones, also renders it extremely difficult to assess the *in vivo* significance of these reactions. The biosynthetic significance of the pathway to aurone via the aurone hydrate must remain at present uncertain.

D FLAVONES

The work of Grisebach and Bilhuber (1967) has demonstrated that the flavones apigenin (**XXII**) and chrysoeriol (**XXIII**) are formed in parsley from chalconaringenin (**III**). Concomitant feeding of [^{3}H]dihydrokaempferol (**XXIV**) with [^{14}C]chalconaringenin showed a much lower incorporation of the former compound, confirming that flavones are more likely formed by the oxidation of chalcones–flavanones and not by dehydration of flavanonols.

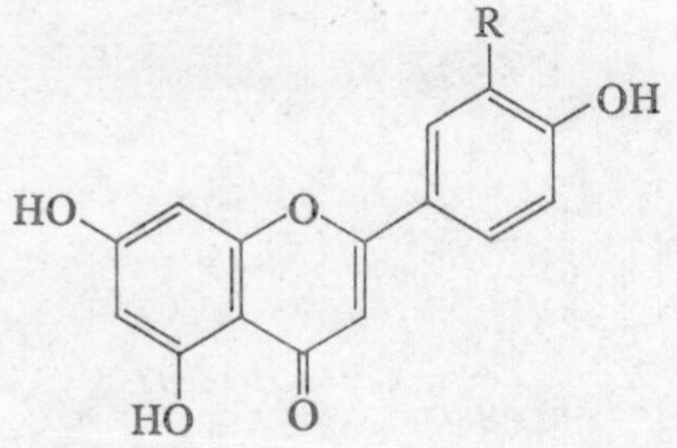

(XXII) Apigenin, R = H
(XXIII) Chryseriol, R = OMe

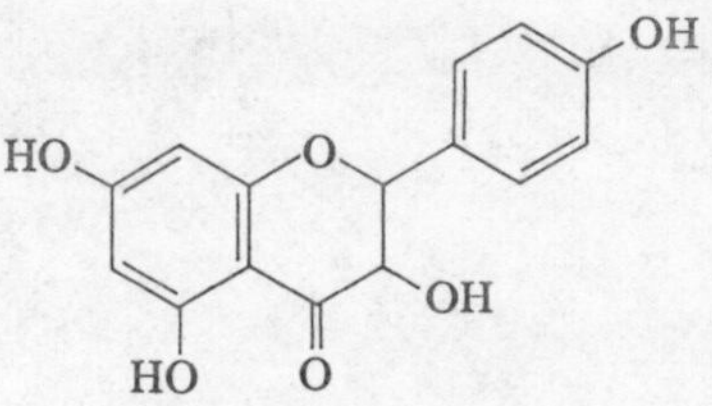

(XXIV)
Dihydrokaempferol

Attempts to isolate this flavone-forming enzyme from illuminated parsley cell culture by Grisebach's group have not led to reproducible results, but its presence in young parsley leaves has very recently been demonstrated by these workers (Sutter, 1972). The enzyme requires molecular oxygen, ferrous or ferric iron, and a heat-labile co-factor (as yet unidentified) which can be removed by dialysis or gel filtration through Sephadex G-25.

The 'substrate' used for the *in vitro* study was the flavanone naringenin (V), but since chalcone-flavanone isomerase was also active in the crude extracts, the possibility cannot be excluded that the flavanone added has first been converted into the corresponding chalcone which then serves as the true substrate. From the information available, it would seem that the enzyme is more likely an oxygenase of the mixed function oxidase type, rather than a dehydrogenase. This distinction is of interest from the viewpoint of deducing whether chalcone or flavanone is the more likely intermediate to the flavone.

If chalcone is the intermediate, the more likely enzymic reactions would be oxygenation followed by dehydration, viz.

$$\text{Chalcone} + [\text{O}] - \text{H}_2\text{O} = \text{Flavone}$$

Dehydrogenation mechanisms for the conversion of chalcone to flavone can, however, also be postulated (Pelter *et al.*, 1971). On the other hand, if flavanone is the true substrate for the enzymic reaction, direct dehydrogenation catalysed by a dehydrogenase would seem the most likely. Further information regarding the precise nature of the enzyme from parsley will be awaited with interest.

A sequence of reactions concordant with the oxygenation-dehydration mode of flavone biosynthesis is shown in Fig. 9. Two lines of circumstantial evidence can be cited in support of this

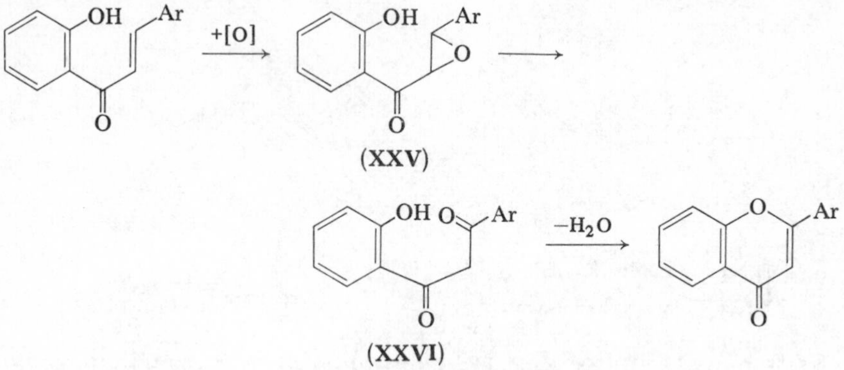

FIG. 9. Hypothetical scheme for the biosynthesis of flavone via oxygenation.

hypothetical scheme: dibenzoylmethanes of type (**XXVI**) have in recent years been isolated as natural products and have been shown to be labile chemical precursors of flavones (Williams, 1967); and photochemical analogy is available for the conversion of a chalcone epoxide (**XXV**) into dibenzoylmethane (**XXVI**) (Ramakrishnan and Kagan, 1970).

E FLAVANONOLS

The studies of Wong and Grisebach (1969), described earlier (Section IIIB), have indicated that flavanonols are derived directly from

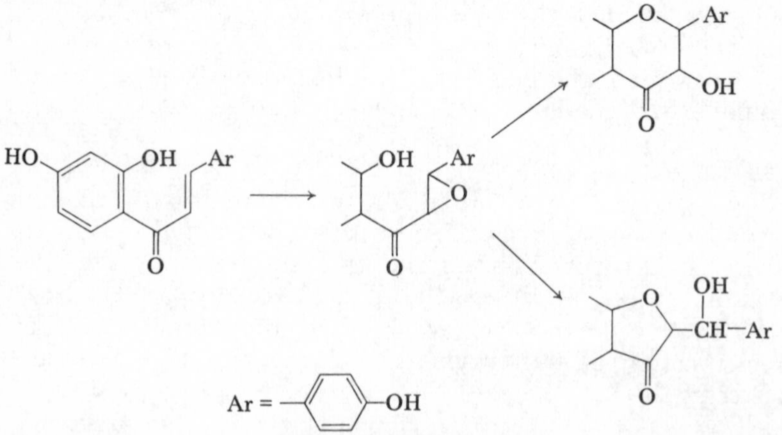

FIG. 10. Possible origin of the products formed in the peroxidase-catalysed oxidation. (Wong and Wilson, 1972.)

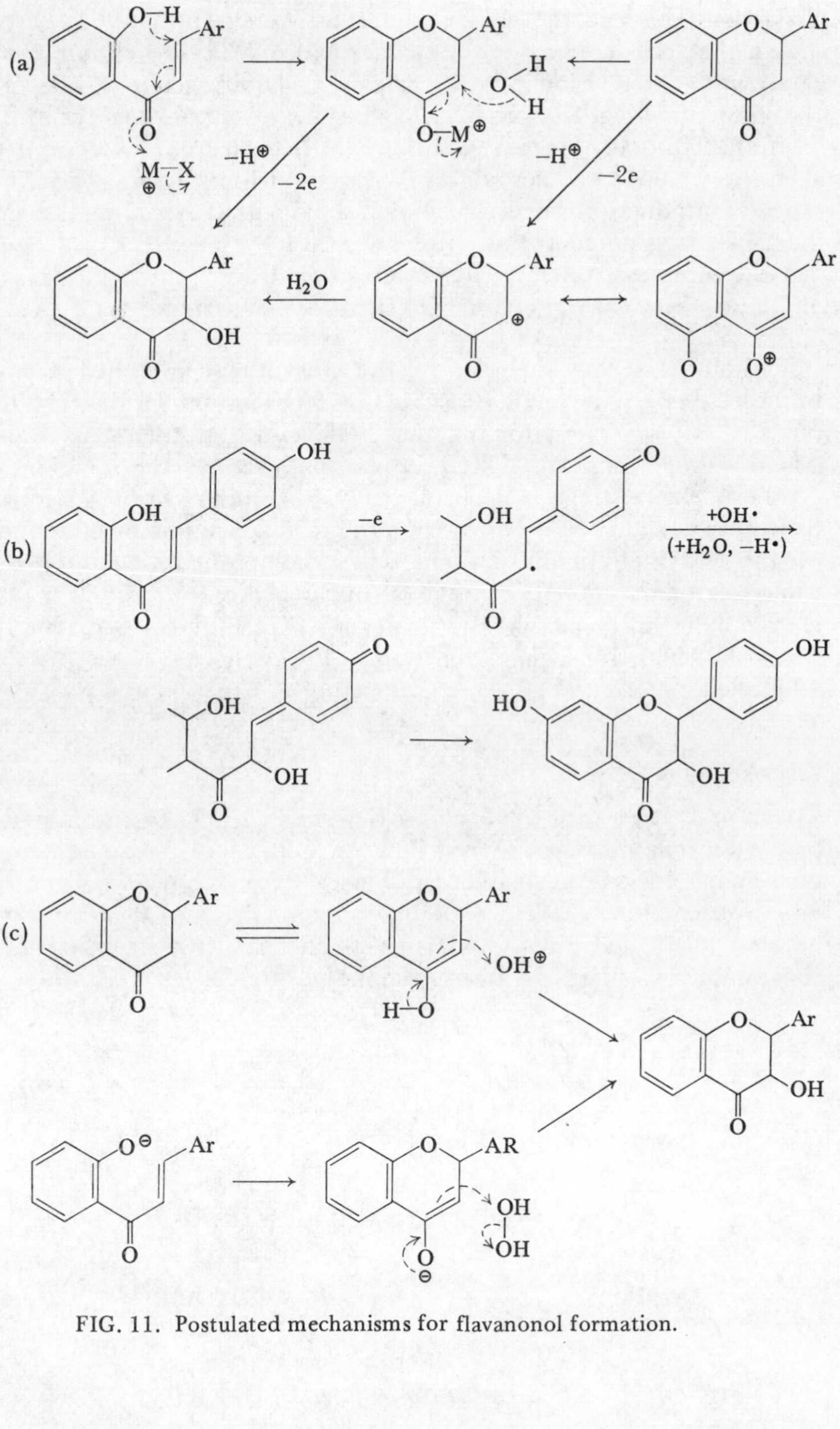

(a)

O–H Ar O $M^{\oplus}$–X → O Ar H O H $O–M^{\oplus}$ ← O Ar O

$-H^{\oplus}$ $-2e$ $-H^{\oplus}$ $-2e$

O Ar OH O ← H_2O O Ar $\oplus$ O ⟷ O Ar O $O^{\oplus}$

(b)

OH OH O $\xrightarrow{-e}$ OH O O $\xrightarrow[(+H_2O,\ -H\cdot)]{+OH\cdot}$

OH O OH O → HO O OH OH O

(c)

O Ar O ⇌ O Ar $OH^{\oplus}$ H–O

O Ar OH O

$O^{\ominus}$ Ar O → O AR OH OH O $\ominus$

FIG. 11. Postulated mechanisms for flavanonol formation.

chalcones. This and the earlier studies by Wong (1964, 1965) have shown that cell-free enzyme preparations from *Cicer arietinum* can catalyse the conversion of the chalcone isoliquiritigenin (**XVII**) to the flavanonol garbanzol (**XV**). The enzyme responsible for this reaction, like that of hydrated aurone (**XX**) formation, has recently been shown also to be peroxidase (Wong and Wilson, 1972). The two types of products formed, flavanonol and hydrated aurone, can in theory be accounted for as different manifestations of a chalcone epoxide intermediate (Fig. 10). As already indicated, the significance of these reactions catalysed *in vitro* by peroxidase is as yet uncertain.

Flavanonol seems likely to be the biosynthetic intermediate of other classes of flavonoids having the 3-OH group in their basic nucleus. The introduction of the 3-OH group at this stage thus constitutes an important step in the pathway of flavonoid biosynthesis. Interestingly, mechanistically plausible reaction schemes have been formulated by different authors in which the introduction of the 3-OH group in flavanonols is accomplished via attack of chalcones or flavanone by either (a) nucleophilic, (b) radical, or (c) electrophilic oxygenating species (Fig. 11) (Birch, 1957, 1963; Bu'Lock, 1965; Dean and Podimuang, 1965; Geissman and Crout, 1969; Pelter *et al.*, 1971).

F FLAVONOLS

Flavanonols are easily oxidized to flavonols chemically and a close biogenetic relationship between these two classes of flavonoid compounds has long been postulated. Tracer studies showed that both dihydrokaempferol (**XXIV**) and dihydroquercetin (**XXVII**) are incorporated into quercetin (**XXVIII**) in buckwheat with an efficiency comparable to that of phenylalanine (Patschke *et al.*, 1966a;

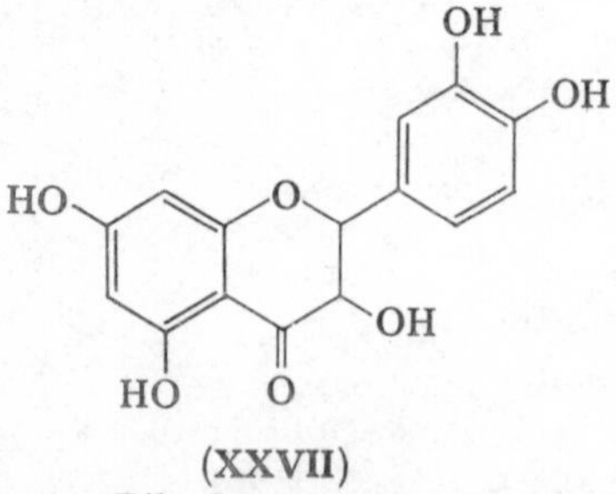

(**XXVII**)
Dihydroquercetin

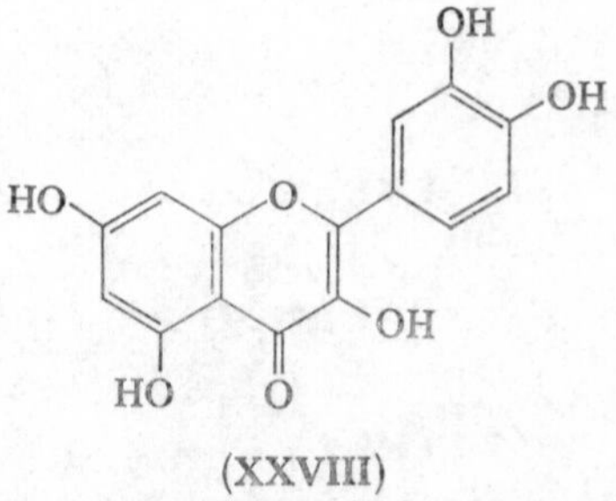

(**XXVIII**)
Quercetin

Patschke and Grisebach, 1968). Enzymic evidence for this apparently facile dehydrogenation reaction, however, is as yet lacking. The main product of the reaction of the chalcone isoliquiritigenin (**XVII**) with purified peroxidase (Wong and Wilson, 1972), already referred to in Section IIIC, has been found to be readily convertible into flavonol (**XII**), particularly on treatment with alkali (0·1 M). As with the formation of the aurone hydrate (**XX**) and flavanonol (**XV**) by the same enzyme, the biosynthetic significance of these interesting transformations is again obscure.

Rathmell and Bendall (1972) have independently studied the reaction of the same chalcone (**XVII**) with horseradish peroxidase and with a partly purified enzyme extract of *Phaseolus vulgaris*. These workers isolated the same aurone and flavonol products (**XIX** and **XII**) from the reaction mixture but were unaware of the existence of the labile intermediate compounds found by Wong and Wilson (Section IIIC). They postulated instead that the immediate product of the enzymic reaction leading to the flavonol was the flavanonol (**XV**). These workers argued for a possible physiological role of peroxidase in flavonol biosynthesis. The interesting point was made that the high correlation between the occurrence of lignin and flavonols in the plant kingdom (Bate-Smith, 1963) could be rationalized in terms of catalysis of both reactions by the same enzyme, namely peroxidase.

G ANTHOCYANIDINS

The *in vivo* conversion of flavanonols into anthocyanidins has been demonstrated in buckwheat seedlings and in pea (Patschke *et al.*, 1966a; Patschke and Grisebach, 1968). These earlier results have more recently been confirmed by feeding experiments with cell suspension cultures of *Haplopappus gracilis* (Fritsch *et al.*, 1971). Dihydrokaempferol (**XXIV**) was incorporated into cyanidin (**XXIX**) at a much higher efficiency (up to 12%) and with a much lower dilution value (20) than the corresponding chalcone (**VIII**) or L-phenylalanine in this system.

The mechanism for the conversion of a flavanonol into an anthocyanidin is at present a matter only of conjecture. Since the two classes of compounds are formally of the same oxidation level, reaction schemes involving non-redox steps, such as that shown in Fig. 12, have been proposed for their transformation (Geissman and Crout, 1969). The possibility of a flavanonol first undergoing reduction to a level equivalent to that of flavan-3,4-diol, which then

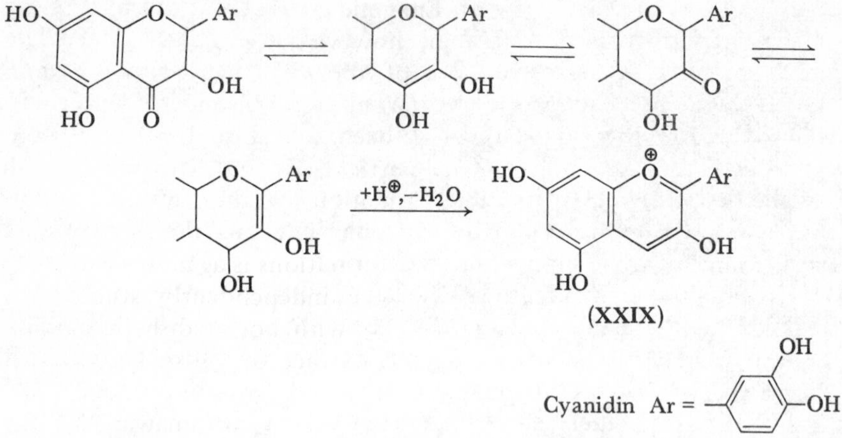

FIG. 12. Hypothetical non-redox route to anthocyanidins.

becomes oxidized to anthocyanidin, has now largely been discounted, although no firm experimental evidence one way or the other is available.

The occurrence in nature of 3-deoxyanthocyanidins such as apigenidin (**XXX**) and luteolidin (**XXXI**) in *Sorghum* poses an intriguing biosynthetic problem. By analogy with anthocyanidin biogenesis, flavanones would be expected to be the progenitors of these flavylium compounds. It is possible that in plants such as *Sorghum*, enzymes involved in the normal sequence, flavanonol to anthocyanidin, could have substrate specificities which allow them to accept 3-deoxy analogues of the normal substrates. Alternatively, specific redox reactions involving chalcone/flavanone only could be operative in these plants and be responsible for the biosynthesis of compounds of this class. In this connection it is interesting to note that the unusual flavan-4-ol (luteoforol) (**XXXII**) has recently been

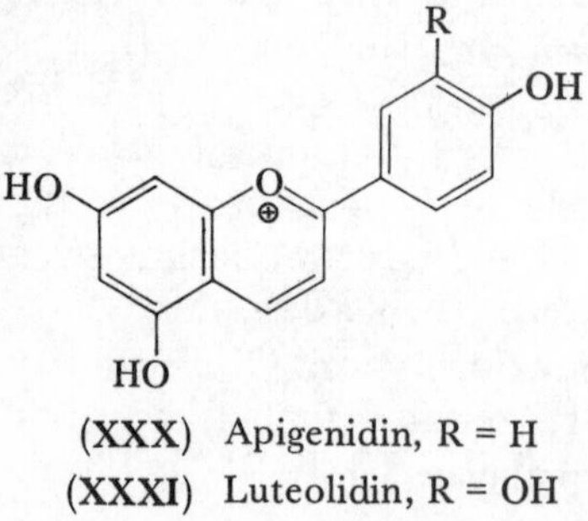

(**XXX**) Apigenidin, R = H
(**XXXI**) Luteolidin, R = OH

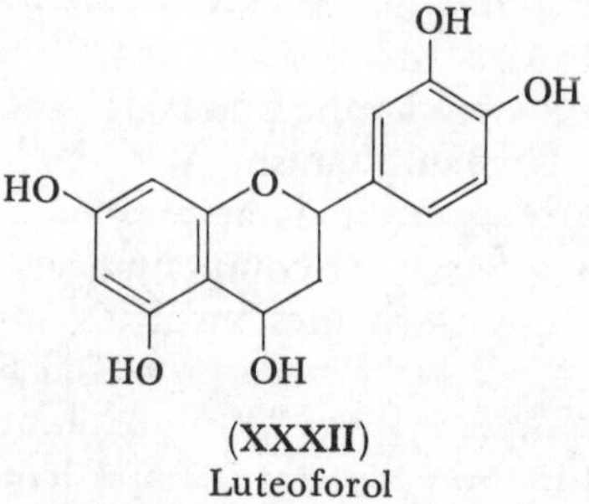

(**XXXII**)
Luteoforol

found to occur in *Sorghum* (Bate-Smith, 1969). This compound, as expected, readily yields luteolidin on treatment with acids. Whether it represents a biosynthetic intermediate in the pathway of the 3-deoxyanthocyanidin remains to be defined experimentally.

H CATECHINS, FLAVAN-3,4-DIOLS, PROANTHOCYANIDINS

In this section we are dealing with classes of flavonoids having a lower state of oxidation than chalcone. These are the flavan derivatives—catechins (flavan-3-ols) and flavan-3,4-diols, and condensation products of these two types of flavonoids. From a practical viewpoint, these compounds are important as sources of discoloration or colour formation rather than as pigments.

Flavan-3,4-diols (leucoanthocyanidins), on heating in acid in the presence of air, are converted into anthocyanidins. Certain polymeric flavan derivatives also exhibit this property and such substances are collectively designated as proanthocyanidins.* The proanthocyanidins that have been chemically well characterized mostly can be represented by the type structure (**XXXVII**). Formally these may be regarded as polymerization products of a catechin (**XXXVI**) with one or more molecules of a flavan-3,4-diol (**XXXIII**), or some derivative thereof at the same oxidation level, such as (**XXXIV**) or (**XXXV**) (Fig. 13). Biogenesis of these proanthocyanidins is therefore likely to be intimately connected with the biogenesis of the monomeric flavan units.

Patschke and Grisebach (1965) first demonstrated the incorporation of chalconaringenin (**VIII**) into epicatechin (**XXXIX**) in tea leaves. More recently, the incorporation of dihydrokaempferol (**XXIV**) into catechins in the same plant was shown. Incorporation efficiency of the flavanonol, however, was only about the same as that of shikimic acid and L-phenylalanine (Zaprometov and Grisebach, 1973). Little experimental work has yet been reported on the biosynthesis of flavan-3,4-diols. Thompson *et al.* (1972) have postulated that the moiety equivalent to flavan-3,4-diol in the B type procyanidin dimers, represented for example by the top flavan unit in (**XXXVII**) ($n = 1$), is formed via oxidation of catechins.

If one bears in mind that, like all other flavonoid types, catechins

* The nomenclature followed here are those of Weinges *et al.* (1969) and Thompson *et al.* (1972). The older term leucoanthocyanidin (or even leucoanthocyanin) is frequently still used to designate compounds of the proanthocyanidin type as well as for flavan-3,4-diols, creating confusion. Use of the term leucoanthocyanidin should now be restricted to the monomeric flavan-3,4-diols only.

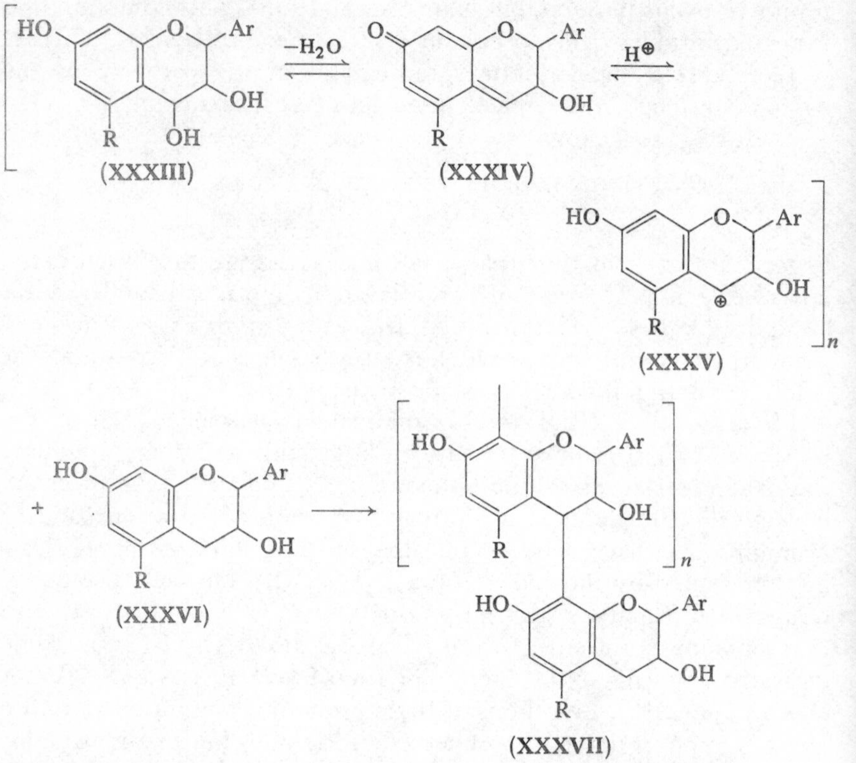

HO O Ar
OH
R OH
(XXXIII)

$\xrightleftharpoons{-H_2O}$

O O Ar
OH
R
(XXXIV)

$\xrightleftharpoons{H^{\oplus}}$

HO O Ar
OH
⊕
R
(XXXV)
n

+

HO O Ar
OH
R
(XXXVI)

⟶

HO O Ar
OH
R
n
HO O Ar
OH
R
(XXXVII)

FIG. 13. Possible biogenetic relationship of proanthocyanidins, flavan-3,4-diols and catechins.

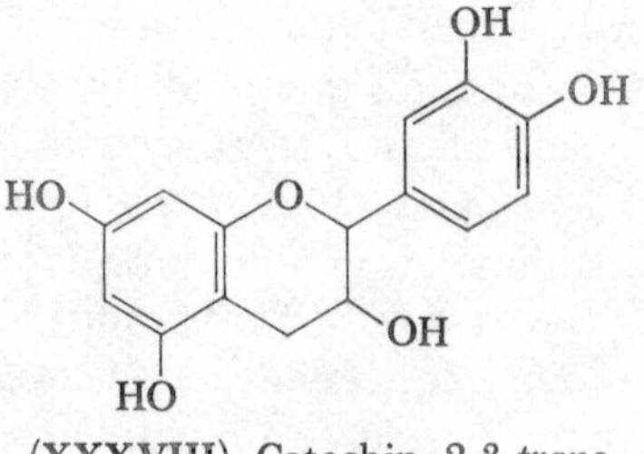

OH
OH
HO O
OH
HO

(XXXVIII) Catechin, 2,3-*trans*
(XXXIX) Epicatechin 2,3-*cis*

and flavan-3,4-diols are derived initially from chalcones, then the involvement of flavanonol in the postulated (or implied) sequence:

$$\text{Chalcone} \xrightarrow{-2e} \text{Flavanonol} \xrightarrow{+4e} \text{Catechin} \xrightarrow{-2e} \text{Flavan-3,4-diol}$$

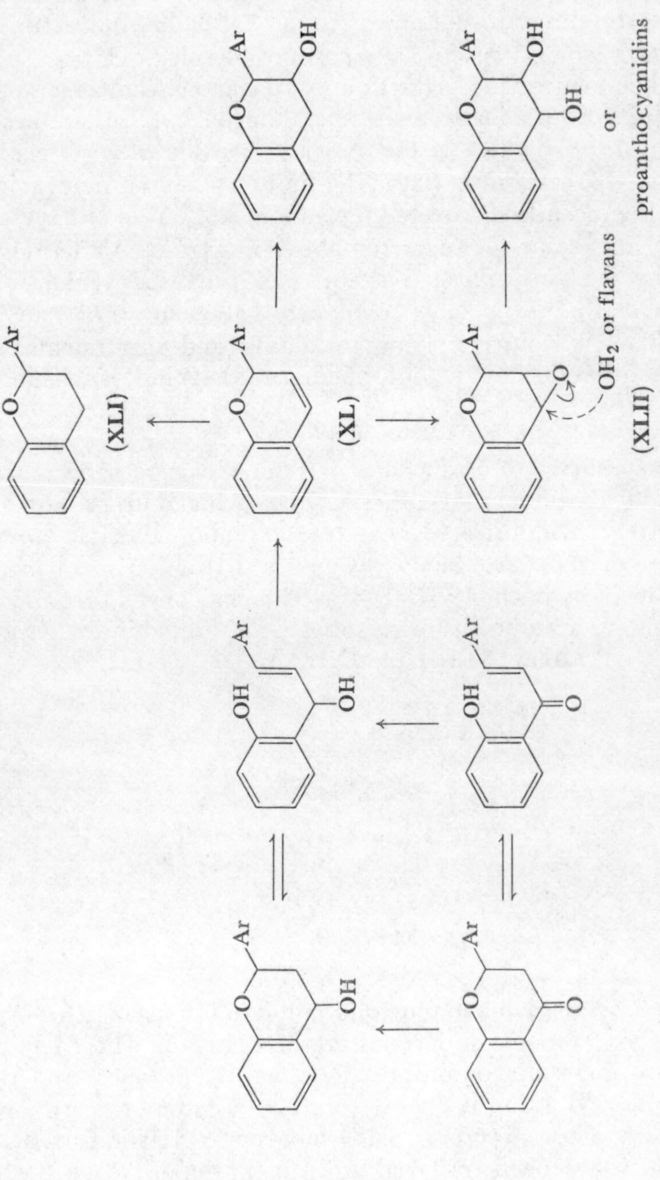

FIG. 14. Hypothetical scheme for the role of flavene in the biosynthesis of flavonoids of lower oxidation levels.

seems rather roundabout and unlikely. A factor which must also be taken into account in considering flavan-3,4-diol biogenesis is the existence in nature of various stereoisomeric forms of these compounds. The direct formation of flavan-3,4-diols from natural *trans*-flavanonols would therefore also seem to be ruled out.

A scheme involving reduction of flavanone/chalcone to flavene (**XL**) (Jurd, 1967; Clark-Lewis and Skingle, 1967; Clark-Lewis and Jamison, 1968), which is theoretically capable of interrelating the formation of catechins, flavan-3,4-diols, and proanthocyanidin (as well as the recently discovered flavan-4-ol and flavan (**XLI**) classes of reduced flavonoids), is suggested above in Fig. 14. Variations in the stereochemistry of flavan-3-4-diols and proanthocyanidins can be accounted for to a large extent by allowing variations in the stereochemical course of ring formation and ring opening of the postulated flavene epoxide intermediate (**XLII**) in this scheme.

I ISOFLAVONES AND ISOFLAVONOIDS

The biosynthesis of isoflavones has long been a subject of study along with flavonoid biosynthesis. Grisebach and his co-workers have conclusively demonstrated that these compounds arise from intermediates of the flavonoid pathway via a reaction involving B ring migration (Grisebach, 1961, 1968). The chalcones (**VII**) and (**VIII**), for example, have been shown to be incorporated intact respectively into the isoflavones (**XLIII**) and (**XLIV**).

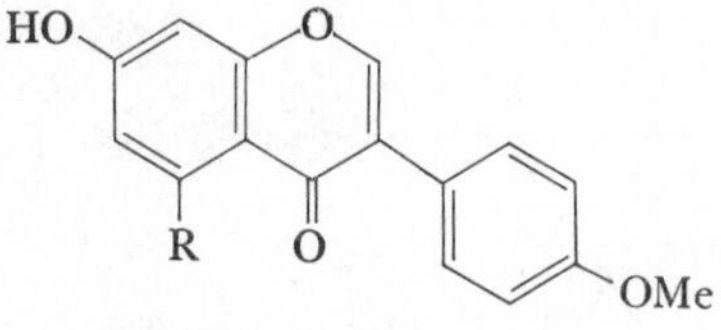

HO, O, R, O, OMe

(**XLIII**) R = H
(**XLIV**) R = OH

Many classes of plant phenolic compounds having the branched $C_6C_3C_6$ skeleton of isoflavones are now known. These compounds constitute the isoflavonoid group of natural products and this 1,2 shift of an aryl group is almost certainly a common feature of their biogenesis. The structural and biogenetic relationships of isoflavonoids have been reviewed recently in detail (Wong, 1970) and reference is made here only to two recent incorporation studies in this area, that of coumestan (coumestrol) (**XLV**) biosynthesis in this area, that of coumestan (coumestrol) (**XLV**) biosynthesis in *Phaseolus aureus* (Dewick *et al.*, 1970; Berlin *et al.*, 1972) and rotenoid (amorphigenin) (see Fig. 24) biosynthesis in *Amorpha fructicosa* seedlings (Crombie *et al.*, 1973). Valuable experimental evidence for the intimate biogenetic relationship existing between isoflavanones and coumestans, and between 2′-methoxyisoflavones and rotenoids (Fig. 24) were obtained in these studies. Knowledge of the enzymology of isoflavonoid biosynthesis is as yet completely lacking.

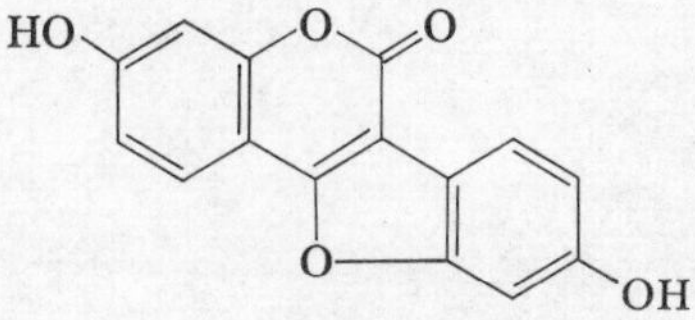

(XLV)
Coumestrol

J SUMMARY OF BIOGENETIC INTERRELATIONSHIPS

It can be seen from the foregoing discussions that, with few exceptions, knowledge of the mechanisms of formation of the different classes of flavonoid compounds, both at the structural and the enzymic level, is still at a rudimentary stage. Tracer studies have led to qualitative statements of the probable presence or absence of sequential or parallel biogenetic relationships among the different classes of flavonoids. In some cases "quantitative" statements as to the relative degree of biogenetic relationship have also been made. However, detailed understanding of the oxidative and reductive biochemical processes leading to the different flavonoid types cannot be made by tracer studies alone, but can be arrived at only by isolation and purification of individual enzymes catalysing discrete steps in the biosynthetic pathways. We have also seen that in some cases even where enzyme-catalysed conversions of different classes of flavonoid can be demonstrated *in vitro*, the biogenetic significance of the reactions *in vivo* can still remain a problem. It is in this area of enzymic and biochemical study of formation of the different classes of flavonoids that further research efforts would best be directed.

Figure 15 summarizes the broad biogenetic relationships of the flavonoid classes as indicated from tracer studies. Such a scheme is consistent with much of the data from biochemical genetic studies (Geissman *et al.*, 1954; Harborne, 1967; Wong and Francis, 1968a). An arrow joining two classes of compounds in the pathway presented

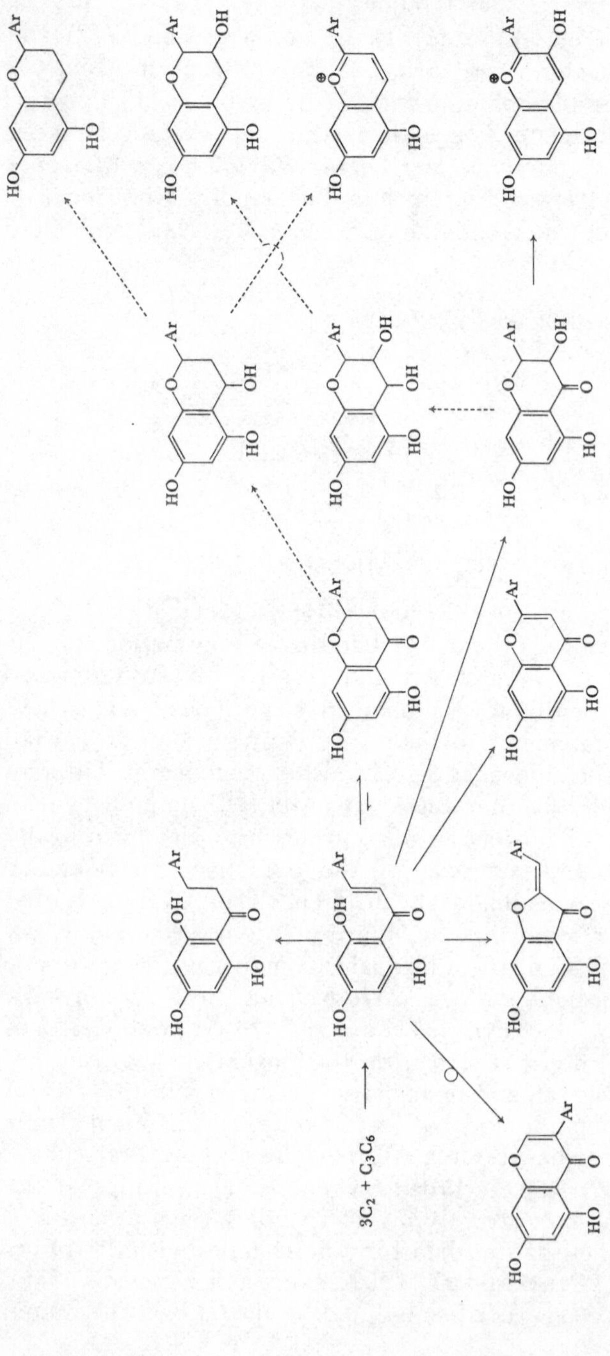

$3C_2 + C_3C_6$

FIG. 15. Possible biogenetic relationships among the different classes of flavonoids. (Dotted arrows indicate experimental proof lacking.)

should not be interpreted, however, as indicating that the reaction is understood in a biochemical sense, but should serve only as a useful starting point for further studies at the enzymic level. Until more information of this type is forthcoming, this or any other scheme for flavonoid biogenesis should be regarded as only tentative.

IV Biosynthesis of secondary features

The myriad of individual compounds within each flavonoid class are distinguished one from another by the number and orientation of hydroxyl groups and by the degree these groups are modified by methyl, isoprenyl, and glycosyl substituents. In addition, substitution by these groups may also take place at the carbon atoms of the ring giving rise to the *C*-substituted flavonoid compounds. Many of the processes leading to these secondary flavonoid features are of the type common to other areas of plant metabolism, and need not necessarily be biogenetically part of the flavonoid pathway. Progress in the understanding of some of these reactions at the enzymic level has again come from studies with the parsley cell suspension system by Grisebach and his colleagues. Many of the enzymes catalysing such reactions with flavonoid compounds have been found to have surprisingly high degrees of specificity.

It should be noted that certain hydroxylation patterns in flavonoid compounds are part and parcel of the main biosynthetic pathway. Chalconaringenin (**VIII**) is the expected first C_{15} product of the normal operation of the pathway, whilst isoliquiritigenin (**VII**) constitutes the corresponding prototype for the 5-deoxyflavonoid compounds, the removal of the C-5 oxygen having taken place at the pre-aromatization stage (see Section **IVA**). The use of these two chalcones in tracer experiments for conversion to other flavonoids having the same or modified oxygenation pattern has already been met with many times in this chapter. Two other theoretical prototype chalcones, (**XLVI**) and (**XLVII**), would appear not to be of

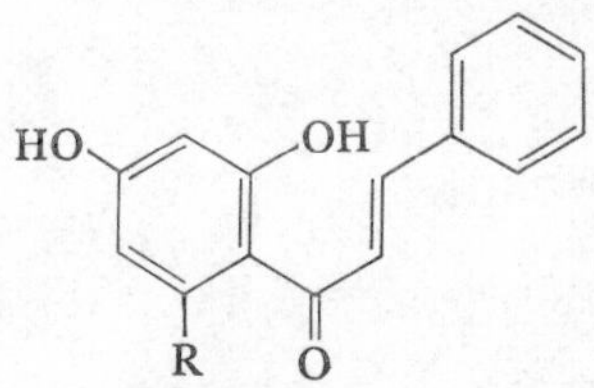

HO OH R O

(**XLVI**) R = OH
(**XLVII**) R = H

much biosynthetic importance since flavonoid natural products lacking the 4′-OH group are very rare.

A RING A HYDROXYLATION

That the absence of the 5-OH group in ring A, common to many flavonoids of the Leguminosae, is determined at a pre-chalcone stage of biosynthesis can be inferred from three lines of reasoning. (1) Both chemical and biochemical precedence would suggest that removal of oxygen at the polyketide stage rather than the phenolic stage would be energetically more favourable (Birch, 1957). (2) Available chemical genetic evidence indicates that genes affecting presence or absence of the 5-OH group act at very early stages of the pathway, not later than the chalcone stage, since all flavonoid types are affected together by such genes (Wong and Francis, 1968a; Birch, 1973). (3) Tracer studies have demonstrated the specificities of chalcones (**III**) and (**IV**) as precursors of the normal and 5-deoxy series of flavonoid products respectively (Fig. 4).

Flavonoid compounds with extra oxygenation at C-6 or C-8 are quite common. Biochemically both types represent products of *o*-hydroxylation in ring A. Little is known of the enzymology involved in plants, but enzymic evidence for such reactions in micro-organisms is available (Jeffrey *et al.*, 1972a). Taxifolin (**XLVIII**) is *o*-hydroxylated by cell-free extracts of a soil *Pseudomonas* in the presence of NADPH and O_2 to both 2,3-dihydrogossyptin (**XLIX**) and 2,3-dihydroquercetagetin (**L**). The enzyme was partly purified and shown to be a flavoprotein. The hydroxylation reactions serve as a prelude to oxidative fissions of the A ring by the micro-organism (Jeffery *et al.*, 1972b).

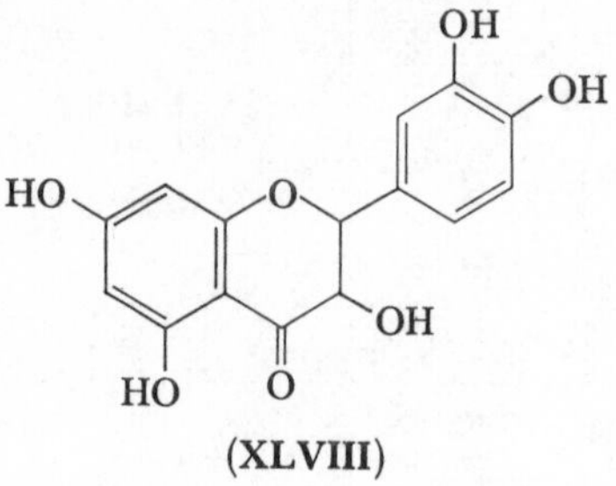

(**XLVIII**)

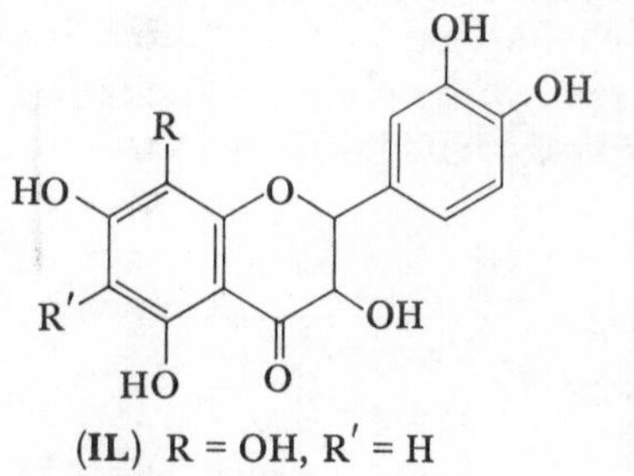

(**IL**) R = OH, R′ = H
(**L**) R = H, R′ = OH

B RING B HYDROXYLATION

The hallmark of compounds of phenylpropanoid origin is the

presence of 4-, 3,4-, and 3,4,5-hydroxy groupings in the aromatic ring and this characteristic is reflected in ring B of flavonoid compounds.* That the 4′-hydroxy function present in the vast majority of flavonoid compounds is most probably introduced at the C_6C_3 level, via cinnamic acid hydroxylase action, has already been discussed (Section II). The question as to what stage of the biosynthetic pathway introduction of the second or third hydroxyl group takes place is a more complex one.

Incorporations of various substituted cinnamic acids with differing efficiencies into flavonoid compounds have been cited as evidence for the view that the substitution pattern of ring B of flavonoids is determined at the cinnamic acid stage (Meier and Zenk, 1965; Hess, 1968). On the other hand, much evidence from tracer studies with C_{15} precursors support the alternative view that B ring hydroxylation takes place at the chalcone/flavanone stage or later. Thus naringenin (**LI**) is a good precursor of *O*- and *C*-glucosides of luteolin (**LII**) in *Spirodela polyrhizo* (Wallace and Grisebach, 1973); and dihydrokaempferol (**LIII**) and dihydroquercetin (**LIV**) but not kaempferol (**LV**) are good precursors of quercetin (**LVI**) (Patschke *et al.*, 1966a; Patschke and Grisebach, 1968). Another example

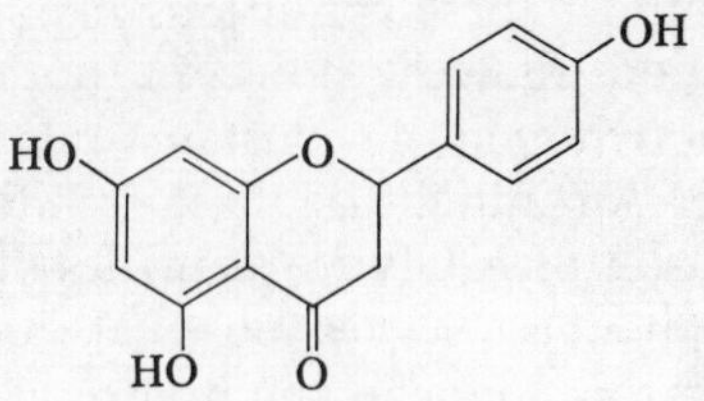

(**LI**)

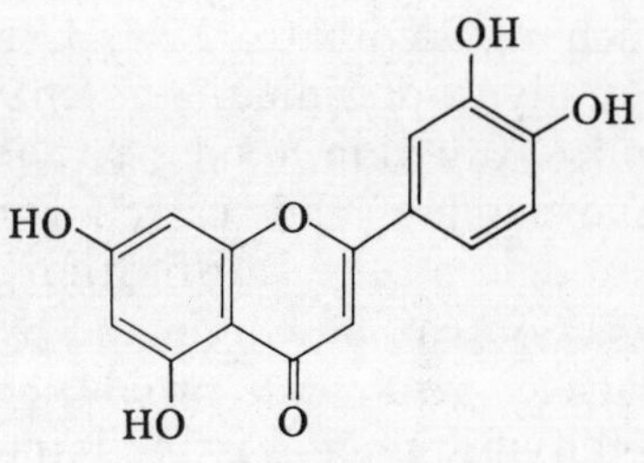

(**LII**)

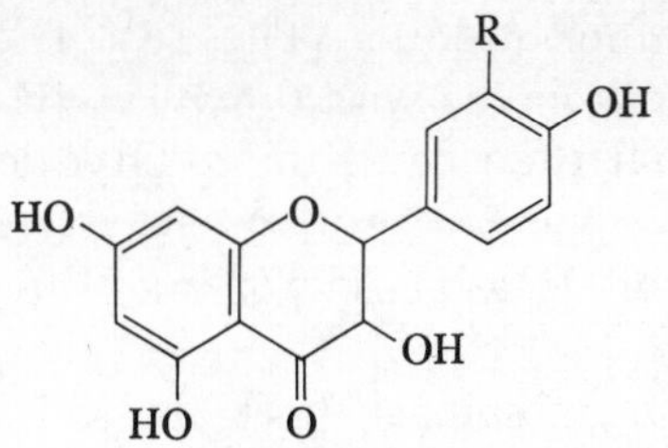

(**LIII**) R = H
(**LIV**) R = OH

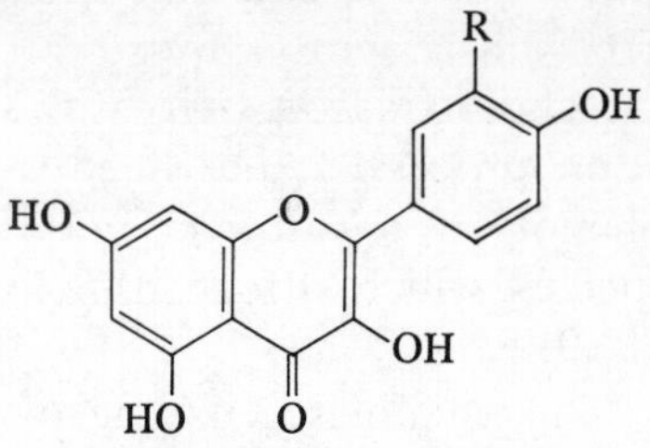

(**LV**) Kaempferol, R = H
(**LVI**) Quercetin, R = OH

* Corresponding numbering in flavonoids 4′, 3′4′ etc. Account is not taken here of modification of the –OH group by methylation and glycosylation.

involves the flavonol datiscetin (**LVII**) with the unusual 2′-OH in ring B. Dihydrogalangin (**LVIII**) has been found to be a much more efficient precursor of this compound than *o*-coumaric acid (Grisebach and Grambow, 1968; Grambow and Grisebach, 1971) (for other examples, see Fig. 4).

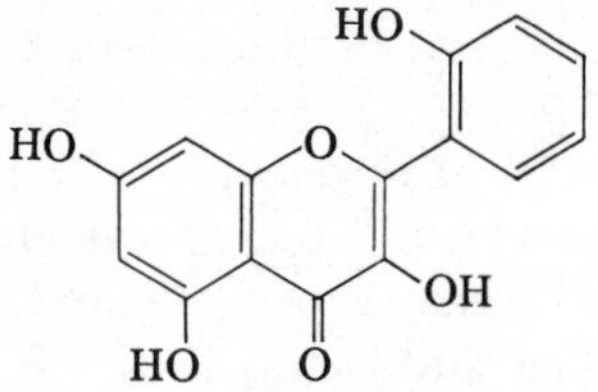

HO HO O OH HO O

(**LVII**)
Datiscetin

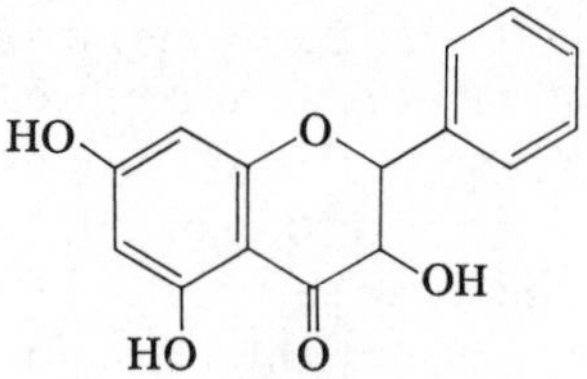

HO O OH HO O

(**LVIII**)
Dihydrogalingin

Enzymes of the phenolase* type capable of hydroxylating *p*-coumaric acid and to some extent 4′-hydroxy flavonoid compounds have in recent years been studied. Vaughan and Butt (1969, 1970) have purified an enzyme from the leaves of spinach beet (*Beta vulgaris*) which catalyses the hydroxylation of *p*-coumaric acid to caffeic acid in the presence of O_2, and an external electron donor such as ascorbate, NADH, NADPH, tetrahydrofolate or dimethyltetrahydropteridine. The enzyme acts under these conditions like a mixed-function oxidase. The preparation showed, however, both hydroxylase and catechol oxidase activity in a constant ratio throughout the purification procedure. An initial lag phase in the reaction can be eliminated by the addition of catalytic quantities of caffeic acid and other *o*-dihydric phenols, suggesting that an *o*-dihydric phenol is the immediate electron donor in the hydroxylation reaction, and that the added electron donor such as ascorbate acts subsequently to reduce the *o*-quinone product (Fig. 16). The catechol oxidase activity is thus obligatorily linked with the hydroxylation process. It has been suggested that the role of the *o*-dihydric phenols is to interact with enzymic copper to produce an active species which effects the hydroxylation (Mason, 1957; Hamilton, 1969).

The phenolase enzyme preparation from spinach beet also catalyses the *o*-hydroxylation of various 4′-hydroxy flavonoids, producing their 3′,4′-dihydroxy analogues (Vaughan *et al.*, 1969). Further

* Other names: tyrosinase, catechol oxidase, monophenol oxidase, polyphenol oxidase, monophenol mono-oxygenase, EC 1.14.18.1.

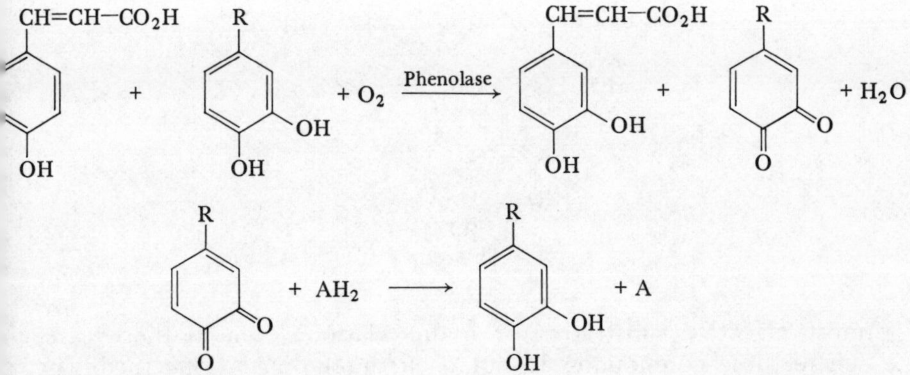

FIG. 16. Suggested mechanism for the hydroxylation of *p*-coumaric acid catalysed by phenolase.

studies have given a measure of the relative effectiveness of flavanones, flavanonols and flavonols as substrates for hydroxylation. Thus K_m values of 1·6, 2·2 and 52·0 x 10^{-4} M were obtained for naringenin (**LI**), kaempferol (**LV**) and dihydrokaempferol (**LIII**) respectively, and maximal velocities for these substrates were 8, 9, and 125 nmol product/min/m unit enzyme (Roberts and Vaughan, 1971). The spinach beet phenolase has been found to be associated mainly with the chloroplast (Barlett *et al.*, 1972). Sato (1966) has previously reported that the chloroplasts of several plants, especially woody angiosperms, are able to oxidize *p*-coumaric acid to caffeic acid.

An enzyme apparently specific for the hydroxylation of *p*-coumaric acid to caffeic acid, present in the first internodes of *Sorghum vulgare,* has been reported (Stafford and Dresler, 1972). The reaction requires molecular oxygen and a suitable electron donor was obligatory. In contrast to the two phenolases from green shoots of the same plant, this third enzyme apparently lacks catechol oxidase activity. While phenolase preparations which lack or have attenuated hydroxylase ability are common, preparations in which the hydroxylase activity predominates are in general unknown. This, together with the finding that *o*-dihydric cinnamic acid derivatives occur in the first internodes of *Sorghum,* suggests that this enzyme could be a specific *p*-coumaric acid hydroxylase.

A comparative study of phenolase specificity in *Malus, Pyrus* and other plants towards *p*-coumaric acid derivatives and flavonoid compounds has been made by Challice and Williams (1970). The dihydrochalcone phloretin (**LIX**) and its glycosides appear to be the

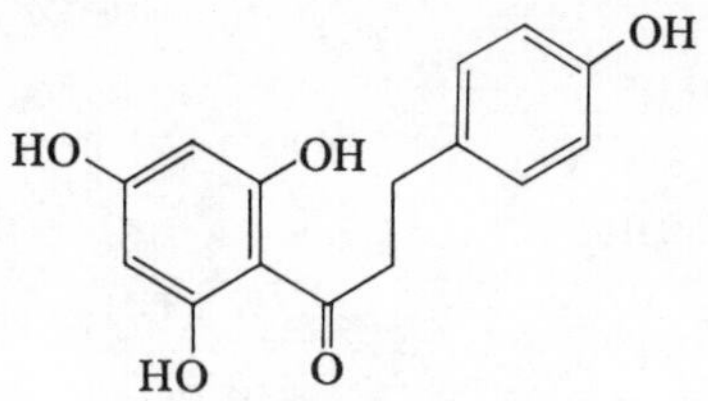

HO OH OH HO O

(LIX)
Phloretin

most effective substrates for hydroxylation, even in those cases where these compounds do not occur in the plant concerned. The flavanone naringenin (**LI**) is a moderately good substrate only with the phenolase from *Pyrus* leaves. Other flavonoids of the flavone, flavonol and flavanonol classes are not transformed to any appreciable extent as compared with the dihydro chalcones. *p*-Coumaric and 3-*O*-*p*-coumaroylquinic acids, on the other hand, are good substrates. Their relative and absolute effectiveness as substrates was found to be dependent upon the nature of the tissue within the plant from which the enzyme was extracted.

Whilst the *in vitro* studies with enzymes of the phenolase type clearly demonstrated that phenolase could catalyse the hydroxylation of flavonoid compounds under controlled conditions, the *in vivo* role of phenolase in flavonoid biosynthesis is by no means certain. Hess (1967) has found the same isoenzyme pattern for phenolase in flower buds of nine pure lines of *Petunia hybrida,* even though these lines contain the genes of B ring "hydroxylation" in various combinations.

Irrespective of the role of phenolase, it is most probable that other hydroxylase enzymes are also involved in the production of 3′,4′-hydroxy flavonoid compounds *in vivo*. Chemical genetic evidence is available to indicate that genes, and therefore presumably enzymes, controlling hydroxylation patterns of ring B are specific to different classes of flavonoid compounds. Hydroxylation of flavonols and anthocyanins in particular are controlled separately from other flavonoids in many genotypes (Harborne, 1967; Wong and Francis, 1968b). The gene M in *Antirrhinum* is known to affect the hydroxylation pattern of flavones (but not aurones) in addition to those of flavonols and anthocyanins (Geissman *et al.,* 1954). Hydroxylation of the B ring of flavonoid compounds must therefore be deemed possible at many points of the biosynthetic pathway. The synthesis of a particular 3′,4′-hydroxy compound thus can follow more than one sequence of steps in different plants, and in some cases, even in different tissues and at different stages of development of the same plant. The relative importance of the different pathways can be expected to be determined by kinetic factors such as relative concentrations of substrates and enzymes, and by different relative specificities of the enzyme involved. A hypothetical scheme consistent with experimental evidence and utilizing this metabolic grid concept to 3′,4′-dihydroxy flavonoids is given in Fig. 17. Little is yet known of enzymes catalysing the introduction of 5′- and 2′-hydroxyl groups in ring B of flavonoid compounds.

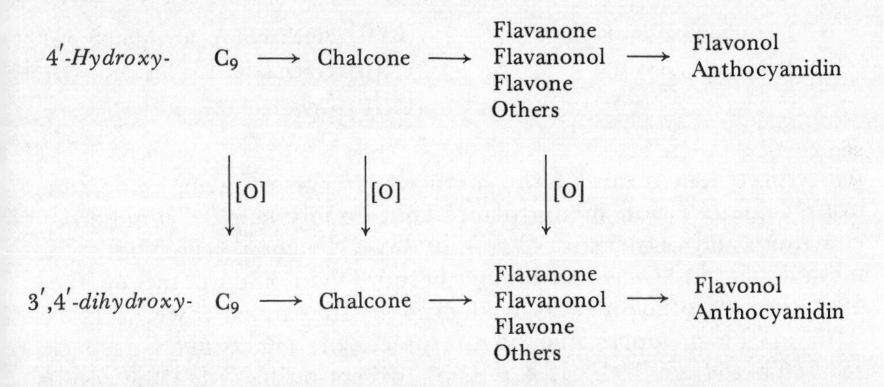

FIG. 17. Possible multiple routes to 3′,4′-dihydroxy flavonoids.

C *O*-METHYLATION

Chemical-genetic evidence and studies of "comparative anatomy" indicate that methylation constitutes a late stage in flavonoid biosynthesis. The biological processes responsible often appear to be highly specific, both with respect to the class of compound methylated and to the position of methylation. Thus, for example, in subterranean clover (Geraldton variety) it has been found that the methylated isoflavone constitutents are all 4′-methoxy derivatives, including formononetin (**LX**), biochanin A (**LXII**) and pratensein (**LXIII**). On the other hand, chalcone, flavone and flavonol constituents, if methylated, all have the 3′-methoxy-4′-hydroxy grouping (Wong and Francis, 1968a). In the mutant A258 of this clover the methylated isoflavones (**LX**), (**LXII**) and (**LXIII**) are almost completely absent, being replaced by greatly increased amounts of the free phenols, daidzein (**LXI**) and genistein (**LXIV**) while the other methylated flavonoids are little changed (Wong and Francis, 1968b).

Results from tracer studies also clearly indicate that methylation of flavonoid compounds can take place at the C_{15} level (cf. Fig. 4). Earlier results indicating an alternative route to methylated flavonoids via methylated cinnamic acids can in most cases be now

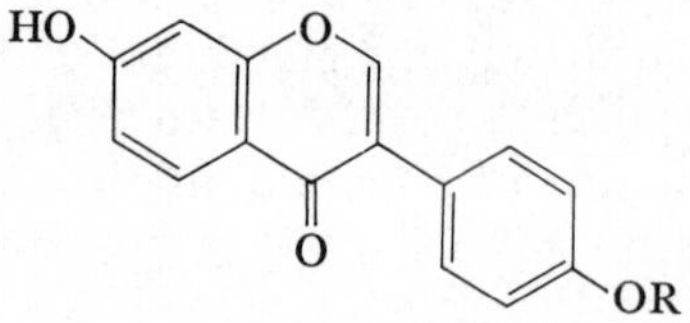

(LX) Formononetin, R = Me
(LXI) Daidzein, R = H

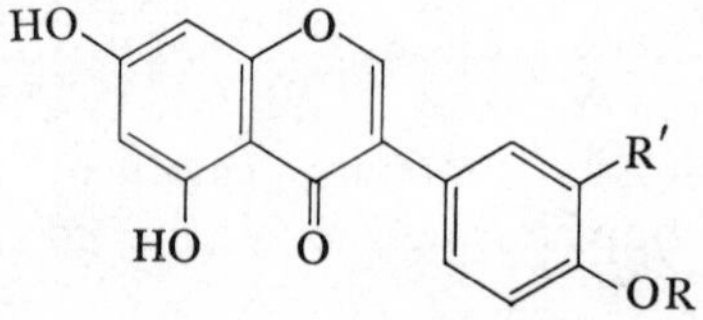

(LXII) Biochanin A, R = Me, R′ = H
(LXIII) Pratensein, R = Me, R′ = OH
(LXIV) Genistein, R = R′ = H

reinterpreted in terms of the methoxy precursor having undergone prior demethylation in the plant. Thus doubly labelled *p*-methoxycinnamic and 4-methoxy-2′,4′-dihydroxychalcone were both demethylated in *Cicer* seedlings before their incorporation into 4′-methoxy isoflavones, as evidenced by the much lower $^{3}H/^{14}C$ ratio in the products than in the precursors (Barz and Grisebach, 1967; Ebel *et al.*, 1970a). A possible exception has been the results reported by Ebel *et al.* (1970b), who found that under short-term feeding conditions *p*-methoxy cinnamic was apparently incorporated intact into the flavone acacetin (**LXV**) in the young leaves of *Robinia pseudoacacia.*

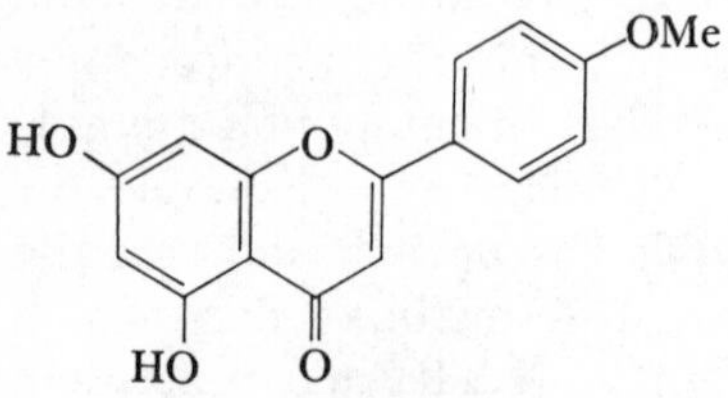

(LXV)
Acacetin

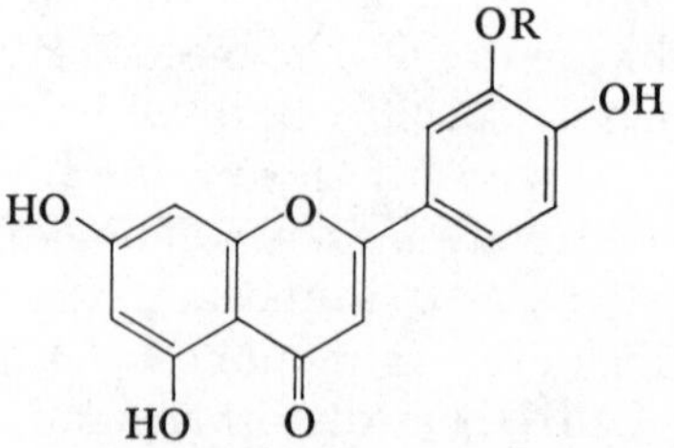

(LXVI) Chryseriol, R = Me
(LXVII) Luteolin R = H

An *O*-methyltransferase apparently involved in flavonoid biosynthesis has been isolated from illuminated cell-suspension culture

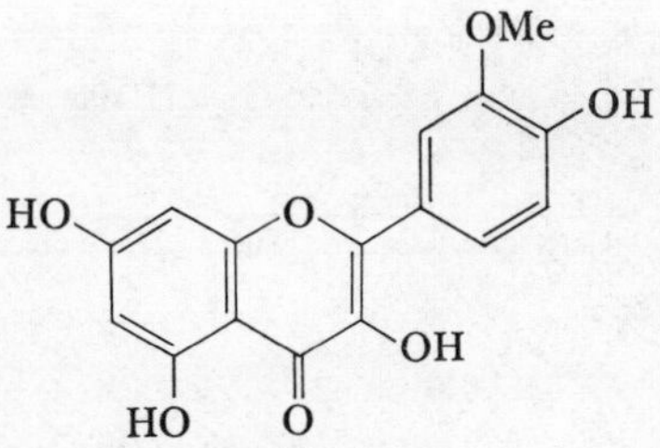

(LXVIII)
Isorhamnetin

of parsley (Ebel *et al.*, 1972). The enzyme catalyses the transfer of the methyl group of *S*-adenosyl-L-methionine to 3′,4′-dihydroxy flavonoids exclusively in the 3′ position (Fig. 18) in agreement with the occurrence of a number of glycosides of chrysoeriol (**LXVI**) and isorhamnetin (**LXVIII**) in these cells. The substrates include luteolin (**LXVII**) and luteolin 7-*O*-glucosides, and although ferulic acid is also readily formed from caffeic acid in the presence of the methyl transferase, the affinity of the enzyme for caffeic acid (K_m $1{\cdot}6 \times 10^{-3}$ M) is considerably lower than that for luteolin ($4{\cdot}6 \times 10^{-5}$ M) and luteolin 7-*O*-glycoside ($3{\cdot}1 \times 10^{-5}$ M).

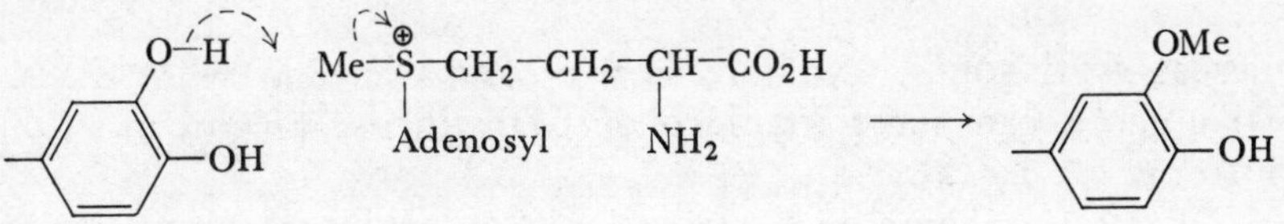

FIG. 18. Reaction catalysed by *O*-methyltransferase from parsley.

D *O*-GLYCOSYLATION

The flavonoid glycosides of parsley cell-suspension culture consists of a number of malonylated and non-acylated flavone 7-*O*-glucosides and 7-*O*-apiosylglucosides, and flavonol 7-*O*-glucosides and 3,7-*O*-diglucosides. The sequence of enzymic steps involved in the formation of apigenin 7-*O*-apiosylglucoside malonate from apigenin is shown in Fig. 19. All of the enzymes indicated have recently been isolated from this plant source.

The first enzyme, UDP-glucose:flavonoid 7-*O*-glycosyl transferase, was shown to be specific for the 7-*O* position of various flavanones, flavones and flavonols (but not flavonol-3-*O*-glycosides) as acceptors,

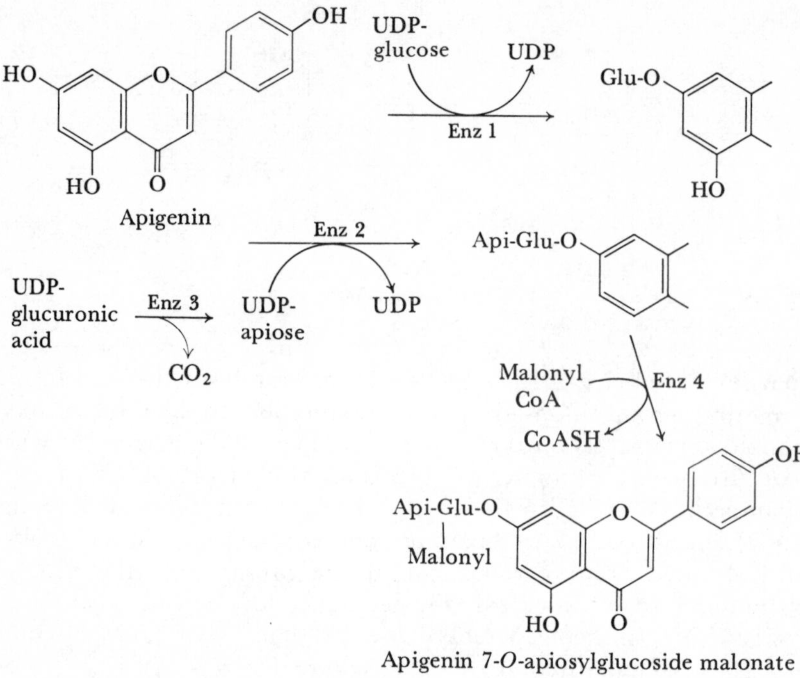

FIG. 19. Enzymic steps in the formation of acylated flavone bioside.

whereas isoflavones, cyanidin and *p*-coumaric acid were inactive. TDP-glucose can serve in place of UDP-glucose as glucosyl donor (Sutter *et al.*, 1972).

UDP-apiosyl:flavonoid 7-*O*-glucoside apiosyl transferase (Enz 2, Fig. 19) is highly specific for UDP-D-apiose as glycosyl donor. By contrast, 7-*O*-glucosides of a large variety of flavones, flavanones and isoflavones, and glucosides of *p*-substituted phenols can serve as acceptors. No reaction takes place, however, with flavonol 7-*O*-glucosides (Ortmann *et al.*, 1970, 1972).

The probable enzymic sequence for flavonol 3,7-diglucoside formation in the parsley system is illustrated in Fig. 20 (Sutter and Grisebach, 1973). The UDP-glucose:flavonol 3-*O*-glucosyl transferase was isolated and separated completely from the 7-*O*-glucosyl transferase mentioned above. While the 3-*O*-glucosyl transferase is specific for the position of glucosylation, several flavonols, including quercetin 7-*O*-glucoside (**LXIX**) can serve as substrates. The flavanonol, dihydroquercetin, was however not glucosylated.

These findings from direct enzymic studies provide elegant con-

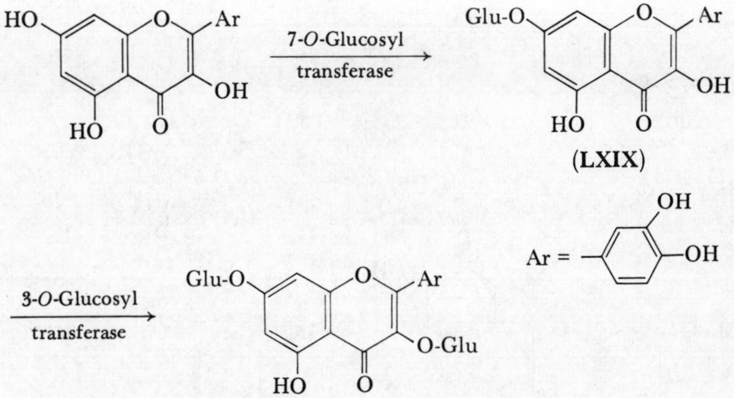

HO, Ar, OH, HO, O → 7-*O*-Glucosyl transferase → Glu-O, Ar, OH, HO, O (**LXIX**)

→ 3-*O*-Glucosyl transferase → Glu-O, Ar, O-Glu, HO, O

Ar = 3,4-dihydroxyphenyl (OH, OH)

FIG. 20. Enzymic sequence of flavonol 3,7-diglycoside formation.

firmation of the stepwise mode of complex flavonoid glycoside biosynthesis long indicated by chemical genetic data (Harborne, 1967).

Besides that described for the parsley system, the formation of flavonol-3-glycosides from quercetin and other flavonols and sugar nucleoside diphosphates (UDP or TDP) has been reported to be catalysed by cell-free extracts from *Phaseolus vulgaris* (Marsh, 1960), *Phaseolus aureus* (Barber, 1962), *Leucaena glauca* (Barber and Chang, 1968), *Impatiens balsamina* (Miles and Hagen, 1968) and *Zea mays* (Larson, 1971).

E *C*-GLYCOSYLATION

C-Glycosyl derivatives are found predominantly among the flavones but are known also in other classes of flavonoids (Alston, 1968). The 8-*C*-glucosides of apigenin (vitexin) (**LXX**) and luteolin (orientin) (**LXXI**) are two widely occurring examples of this class.

Addition of ^{14}C-labelled apigenin and luteolin to the growth media of axenically cultured *Spirodela* and *Lemna* plants (duckweed) which synthesize a number of *O*- and *C*-glycosides containing the apigenin and luteolin oxygenation patterns resulted in the incorporation of labelling only in the *O*-glycosylated but not the *C*-glycosylated flavones (Wallace *et al.*, 1969). It was concluded therefore that *C*-glycosylation occurs at an earlier biosynthetic step. In a more recent study, the incorporation of [^{14}C] naringenin into both *C*- and *O*-glycosides of apigenin and luteolin was demonstrated (Fig. 21) (Wallace and Grisebach, 1973). These results, together with the earlier finding that *C*-glycosyl flavones can also undergo *O*-glycosyl-

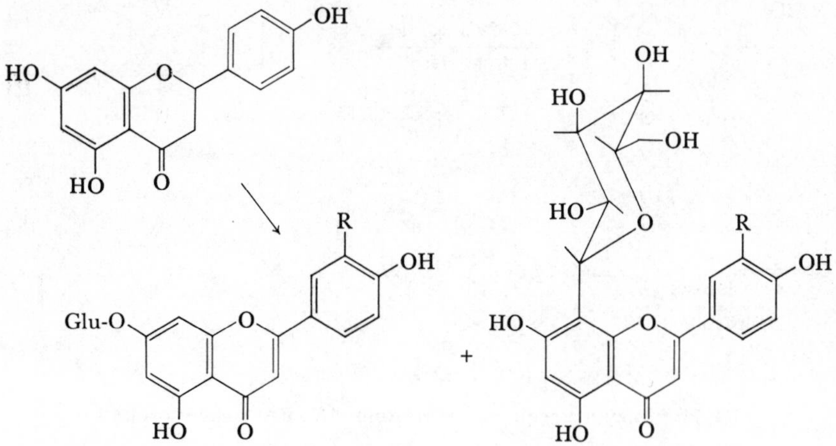

Apigenin 7-*O*-glucoside, R = H
Luteolin 7-*O*-glucoside, R = OH

(LXX) Vitexin, R = H
(LXXI) Orientin, R = OH

FIG. 21. Incorporation of flavanone into flavone *O*- and *C*-glycosides in *Spirodela*.

Chalcone/Flavanone —Oxidation→ Flavone —*O*-Glycosylation→ *O*-Glycosyl flavone

↓ *C*-Glycosylation

C-Glycosyl chalcone/flavanone —Oxidation→ *C*-Glycosyl flavone —*O*-Glycosylation→ *O*-Glycosyl-*C*-glycosyl flavone

FIG. 22. Probable biogenetic relationships among the various types of flavone glycosides.

ation, suggest that the relationship among the various types of flavone glycosides are as summarized in Fig. 22 (Wallace and Grisebach, 1973).

Little is known of the mechanism of *C*-glycosylation. The anion **(LXXII)** of the phenolic aglycone by interaction with a 1-phosphorylated sugar in the form of UDP- or TDP-sugar will presumably furnish the *C*-glycosyl derivative in a manner analogous to *O*-glycosylation.

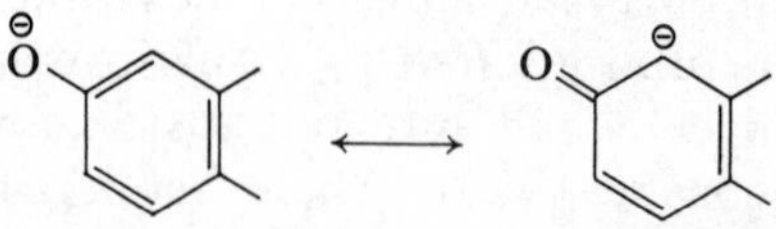

(LXXII)

F *C*-ALKYLATION

Occurrences of *C*-methylated flavonoid compounds are far fewer than *O*-methylated ones. Methylation via electrophilic attack of the methyl cation from methionine on both oxygen or carbon is a mechanistically acceptable process and was predicted on this basis (Birch *et al.*, 1958). The introduction of isoprenoid units by alkylation on carbon, on the other hand, is much more common compared with *O*-alkylation. Birch (1973) has suggested that since the biochemically fundamental process of terpene elaboration requires alkylation on carbon, other processes such as alkylation of aromatic rings may reflect deviation via modification of the *C*-alkylating enzymes of normal terpenoid biosynthesis. Both *C*-methylation and *C*-isoprenylation are more commonly found in ring A of flavonoids presumably owing to the greater nucleophilic character of this ring in the species undergoing *C*-alkylation. At what stage or stages of the flavonoid biosynthetic pathway the alkylations occur is not clear. Notably many chalcones have been found in nature to be highly alkylated (Merlini, 1973), indicating that *C*-isoprenylation can take place at an early stage.

Isoprenoid substituents in flavonoids are very frequently manifest as the 2,2-dimethylchromene (**LXXIII**) or the 2-isopropenyl-

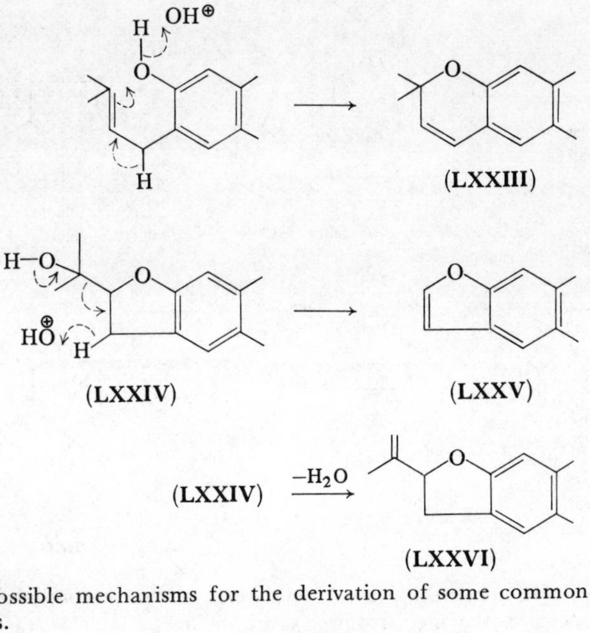

(LXXIII)

(LXXIV) **(LXXV)**

(LXXIV) $\xrightarrow{-H_2O}$ **(LXXVI)**

FIG. 23. Possible mechanisms for the derivation of some common isoprenoid substituents.

coumaran (**LXXVI**) and benzofuran (**LXXV**) ring systems. The formation of (**LXXIII**) and (**LXXV**) via oxidative mechanisms can be variously formulated, as for example in Fig. 23. Genesis of the 2-hydroxyisopropyl coumaran ring system (**LXXIV**) is illustrated in the proposed scheme for the late steps of the pathway to the rotenoid amorphigenin (Fig. 24) (Crombie *et al.*, 1973). The broad pathway to rotenoids, based on incorporation studies with carefully

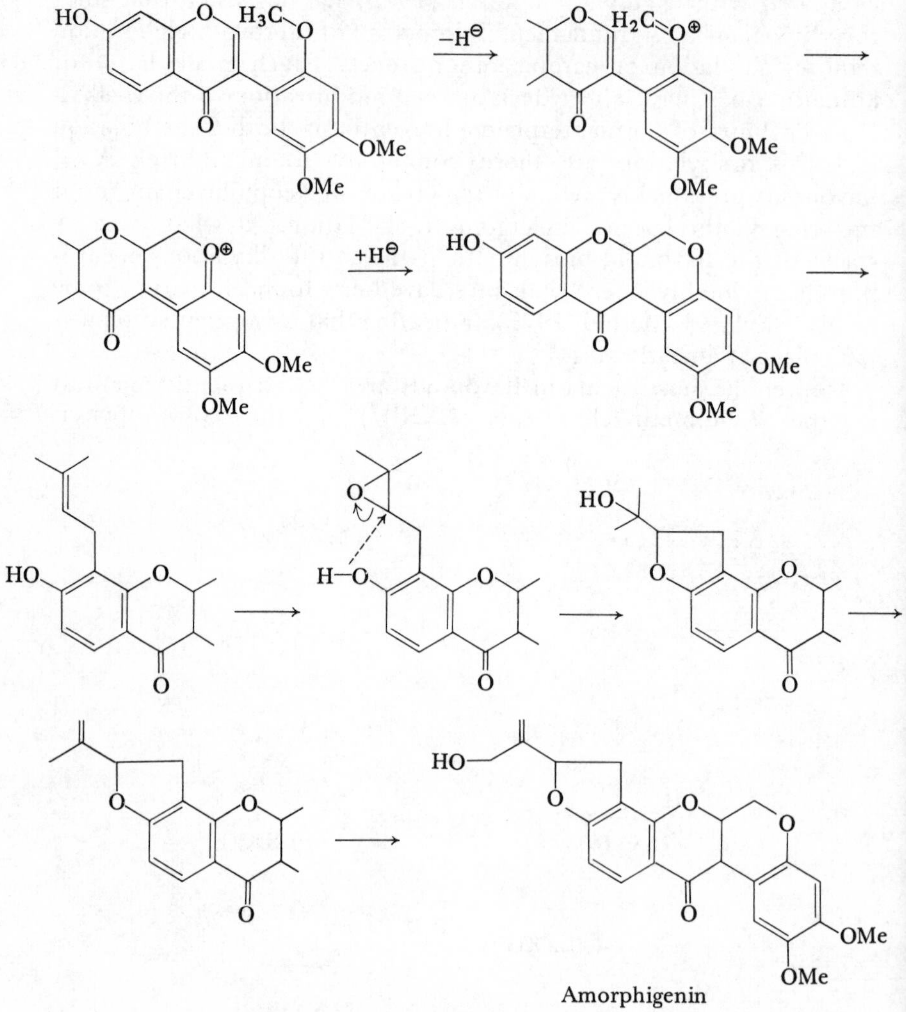

FIG. 24. Proposed scheme for the biosynthesis of the rotenoid amorphigenin from a 2′-methoxyisoflavone. (Crombie *et al.*, 1973.)

selected precursors, and on the natural occurrence of suspected intermediates in the plant, provides another example of alkylation processes possible in these compounds in the formation of the extra pyran ring via oxidation-reduction steps involving the 2′-methoxy group (Fig. 24).

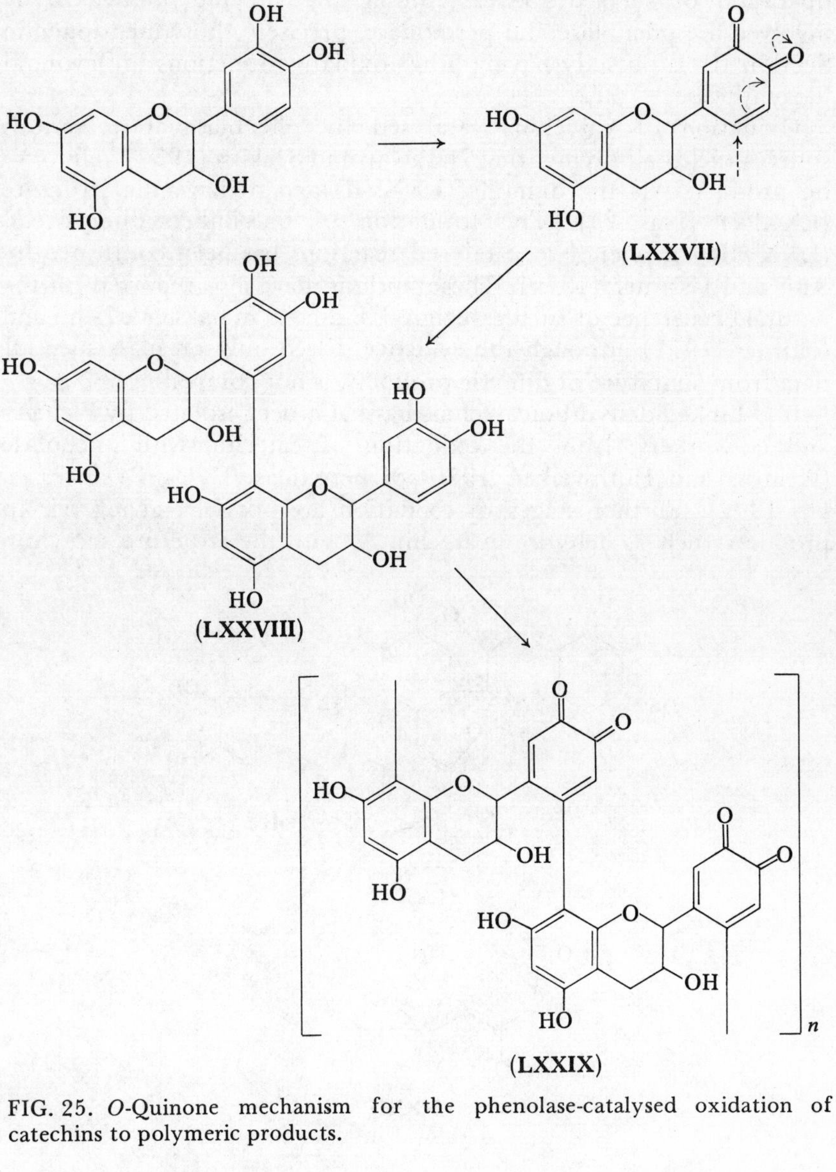

(LXXVII)

(LXXVIII)

(LXXIX)

FIG. 25. *O*-Quinone mechanism for the phenolase-catalysed oxidation of catechins to polymeric products.

G OXIDATIVE POLYMERIZATION OF FLAVONOIDS

Colourless flavonoid compounds often contribute heavily to colour formation by oxidation and polymerization and the changes undergone in such reactions are generally complex. The rapid browning of cut or damaged plant tissues provides commonplace examples of the operation of such processes. Among the enzymes known to be involved are phenolase and peroxidase, precisely those mentioned in Section III as possibly having other oxidative functions in flavonoid biosynthesis.

Oxidation of catechins catalysed by phenolase from various sources yields a polymer that Hathway and Seakins (1957) believe to be produced via the quinone (**LXXVII**) and to have the structure (**LXXIX**) (Fig. 25). The formation of 6′,8-linked dimers e.g. (**LXXVIII**) in phenolase catalysed reactions has been confirmed by Ahn and Gstirner (1970). These authors have also reported on the natural occurrence of such catechin 6′,8-dimers in oak bark (Ahn and Gstirner, 1971) although the evidence, based only on mass spectral data from a mixture of dimeric products, is not compelling.

6′,8-Linked dehydrodicatechins have also been isolated by Weinges and co-workers, from the oxidation of catechin with phenolase (Weinges and Huthwelker, 1970) or peroxidase/H_2O_2 (Weinges *et al.*, 1972). Further stages of oxidation are possible, giving rise to products such as dehydrodicatechin A, with the structure shown in

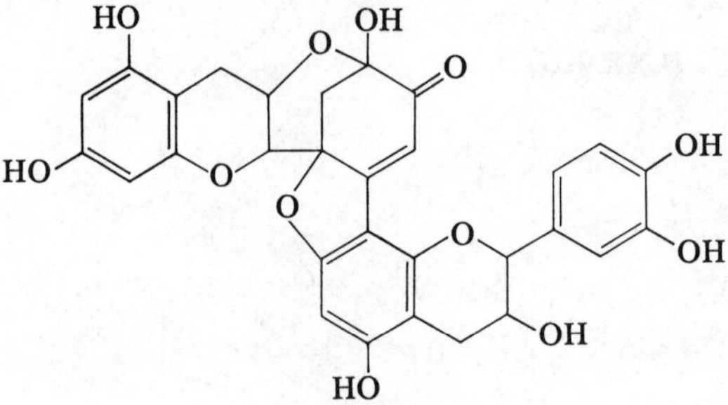

(LXXX)

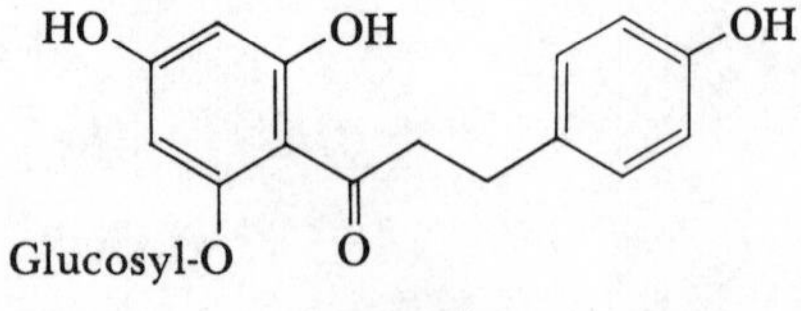

(LXXXI)
Phloridzin

(**LXXX**) (Weinges *et al.,* 1971). Phenolase catalysed oxidation and polymerization of the dihydrochalcone phloridzin (**LXXXI**) in apple tree tissues have also been studied in some detail (Sarapu, 1971).

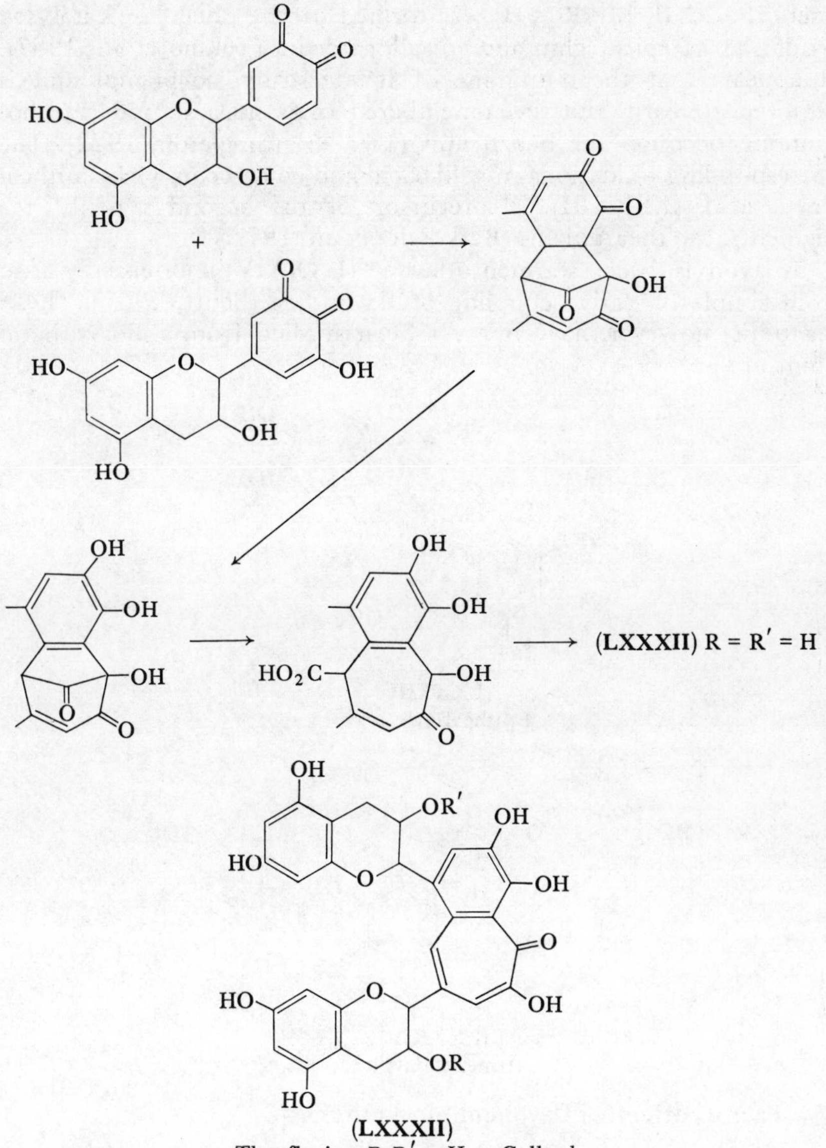

→ (**LXXXII**) R = R′ = H

(**LXXXII**)
Theaflavins, R,R′ = H or Galloyl

FIG. 26. Possible mechanism for the formation of theaflavin.

The enzymic oxidation of *vic*-trihydric phenols to benzotropolone structures plays an important part in the formation of the pigments in black tea. Five theaflavin pigments of the general formula (**LXXXII**) have been reported (Bryce *et al.*, 1972), and theaflavin itself (**LXXXII**, R = R′ = H) was formed by the phenolase catalysed oxidation of epicatechin and epigallocatechin (Takino *et al.*, 1967). It appears that the *o*-quinone of the *vic*-trihydric phenol unit is required to form the seven-membered ring and a second catechol quinone becomes the benzo unit (Fig. 26) (Singleton, 1972). The corresponding oxidations of gallic acid and epicatechin yield epitheaflavic acid (**LXXXIII**), a precursor of the second type of tea pigments, the thearubigins (Berkowitz *et al.*, 1971).

Biflavonyls, such as armentoflavone (**LXXXIV**) undoubtedly arise from simple oxidative coupling of flavonoids (Scott, 1967). These reactions, however, have not yet been studied from a biosynthetic point of view.

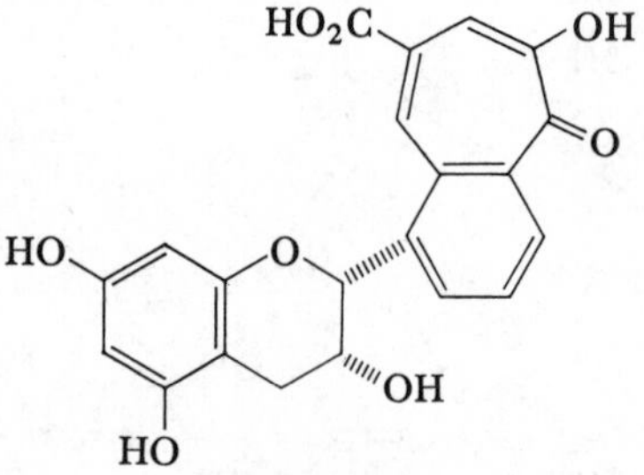

(**LXXXIII**)
Epitheaflavic acid

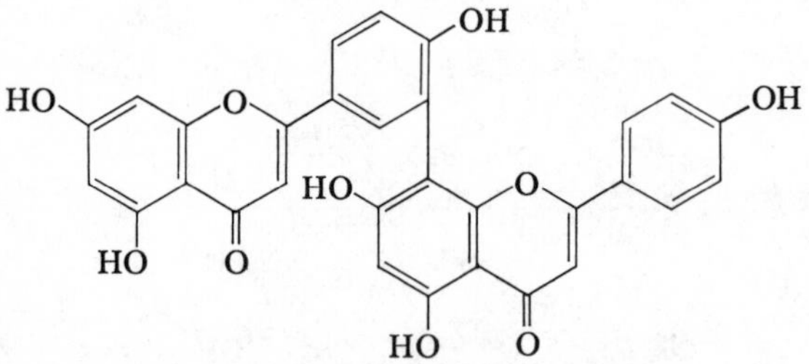

(**LXXXIV**)
Armentoflavone

V Factors affecting flavonoid biosynthesis

Flavonoid biosynthesis of plants is regulated by a wide variety of both internal and environmental factors. Endogenous control by substrates, enzymes, end products and hormonal substances is greatly influenced by exogenous factors such as light, infection and stress. The wealth of factors known to affect flavonoid biosynthesis in one way or another has been responsible for a minor explosion of research in recent years. Much of this interest has hitherto been centred on physiological rather than biochemical aspects, but with a better understanding of the enzymology of flavonoid biosynthesis in the last few years biochemical mechanisms underlying factors controlling flavonoid biosynthesis have begun to receive more attention.

When it is considered that phenylalanine lies at the hub of a wide range of biosynthetic reactions and is the branching point of flavonoid biosynthesis from the major pathway of protein synthesis, it may be expected that the enzyme responsible for its conversion to cinnamic acid, PAL, has regulatory properties. In general, PAL seems to be extraordinarily sensitive to the physiological stage of the plant. Levels change as the plant germinates and develops, or changes may follow wounding, infection or activation by light and growth modifiers. Concomitant increases in levels of PAL and flavonoid compounds have been demonstrated in many plants and plant tissues, including flavanones in grapefruit (Maier and Hasegawa, 1970), flavanols in strawberry (Hyodo, 1971), anthocyanins in buckwheat (Scherf and Zenk, 1967) and isoflavonoids (Hadwiger and Schwochau, 1971) and flavonols (Smith and Attridge, 1970) in pea. On the other hand, it has been noted (Camm and Towers, 1973a,b) that there are many examples where there does not appear to be a direct correlation between the level of PAL and the production of a specific phenolic compound. Such results would suggest that, in these situations, the PAL level may not be rate-limiting (Creasy, 1971).

Indeed, much evidence from recent enzymic studies (to be discussed in the sections following) indicates that effects such as light do not act by regulating only one pacemaker enzyme, but each operates in a co-ordinant fashion to regulate a whole series of enzymes in the pathway. The study of all enzymes along the pathway to a particular class of flavonoid compounds is obviously necessary before assessments of the level of control by individual enzymes can be made. As indicated in the earlier sections, most of these enzymes remain to be discovered. It is possible that PAL functions as a primary control in phenylpropanoid/flavonoid biosynthesis and that other enzymes concerned with subsequent steps display secondary control (Camm and Towers, 1973a,b). The fact that PAL was the first enzyme specific to this area of secondary metabolism to be discovered and the ease with which it can be assayed could also be cited as reasons for the emphasis which this enzyme has recently received.

A EFFECT OF LIGHT

Light is by far the most extensively studied of the environmental factors affecting flavonoid production. This topic has recently been reviewed by Smith (1972, 1973). Much of the earlier work has been directed towards defining the nature and mode of action of the photoreceptors involved. At least three photoresponse systems active in regulating flavonoid synthesis have now been recognized in plants (Smith, 1972). Phytochrome involvement was demonstrated in mustard (Mohr, 1958), turnip (Siegelman and Hendricks, 1957), gherkin seedlings (Engelsma and Meizer, 1965), artichoke tubers and pea leaves (Bottomley *et al.*, 1966), and a host of other plant tissues (Smith, 1972).

In addition to exhibiting red–far-red sensitivity, many of the systems are also sensitive to blue light (Nitsch and Nitsch, 1966; Engelsma, 1968b), the photoreceptor responsible being probably a flavin/flavoprotein (Smith, 1972; Zucker, 1972). A far-red absorbing pigment, which may or may not be identical with phytochrome, is possibly a third photoreceptor system responsible for the stimulatory effect of continuous far-red illumination observed in some plants (Smith, 1972; Camm and Towers, 1973a,b).

As a result of Zucker's discovery (Zucker, 1965) of the increased activity of PAL in potato slices incubated in light, a number of investigations have been carried out on the effect of light on PAL in other tissues (Zucker, 1972) and on the extractable activities of other known enzymes of flavonoid biosynthesis. The results of these studies are summarized in Table III.

It is evident from these investigations that the photocontrol of flavonoid synthesis is exerted only on the B-ring pathway. In the two plants in which an enzyme of A-ring biosynthesis (acetate:CoA ligase) has been studied, light had apparently no effect on the extractable activity. This negative finding with respect to the A-ring pathway is also in accord with results from precursor incorporation experiments (Harper *et al.*, 1970).

The stimulatory effect of light on PAL can be seen to be quite a general phenomenon (Table III). In almost all cases, the increase in PAL levels brought about by light treatment does not occur immediately upon irradiation and is only transient in character. Much recent effort has been directed towards the elucidation of the molecular mechanisms underlying the light-mediated rise and fall in PAL levels. An increased level of PAL activity may in principle be due to *de novo* synthesis (induction) or conversion of inactive into active form. Incorporation of isotopic labelling during light stimulation in mustard (Schopfer and Hock, 1971) and *Xanthium* disc (Zucker, 1969) has been cited as evidence for *de novo* synthesis of enzyme. The fall in PAL levels after the light-mediated peak has been reached has been attributed to the action of a specific proteinaceous inactivator (Zucker, 1968; Engelsma, 1970b). The formation of the inactivator appears to be in some way correlated with the level of PAL present and Engelsma (1968a) has shown that the products of PAL action, i.e. cinnamic acid and *p*-coumaric acid, are probably the agents which bring about the synthesis of the inactivator. Attridge and Smith (1973) have recently presented evidence for the existence of a pool of inactive PAL in gherkin seedlings. The presumed PAL-inactivator complex may be dissociated *in vivo* by low temperature treatment (Engelsma, 1970b), releasing the active enzyme.

The investigations of Hahlbrock *et al.* (1971a), carried out with parsley cell suspension cultures, have revealed a more complex picture of photoregulation at the enzymic level. Activities of all enzymes responsible for the C_6C_3 stage of flavonoid biosynthesis, i.e. PAL, CAH and *p*-coumarate:CoA ligase, are all increased by light treatment, reaching their maximum about 12 hours after illumination and then decreased. A second set of enzymes, all involved specifically at the C_{15} stage of biosynthesis, is also induced by the same light treatment. These increase in activity more slowly and reach a maximum much later in time. Regulation of the two sets of enzymes thus appear to be clearly different. The concomitant induction of PAL and CAH during illumination of excised buckwheat hypocotyls has earlier been studied by Amrhein and Zenk (1970).

B EFFECT OF WOUNDING AND INFECTION

An increased level of PAL upon wounding occurs in pea seedlings, citrus fruit peel, bean, sweet potato, swedes, gherkin and buckwheat (Minamikawa and Uritani, 1965b; Engelsma, 1968a; Walton and Sondheimer, 1968; Riov *et al.*, 1969; Amrhein and Zenk, 1971; Hyodo and Yang, 1971a; Rhodes and Wooltorton, 1971). In the latter plant, wounding also causes cinnamic acid hydroxylase activity to increase (Amrhein and Zenk, 1968).

Infection by a pathogenic organism following wounding has been

TABLE III

Effect of light on enzymes of flavonoid biosynthesis

Region of pathway	Enzyme	Plant	Effect	References
Shikimic acid pathway	5-Dehydroquinase	Mung bean (*Phaseolus aureus*)	None	Ahmed and Swain, 1970
		Pea	Stimulation	Ahmed and Swain, 1970
	Shikimate : NADP oxidoreductase	Mung bean	None	Ahmed and Swain, 1970
		Pea	Stimulation	Ahmed and Swain, 1970
		Pea	None	Attridge and Smith, 1967
		Buckwheat	None	Amrhein and Zenk, 1971
C_6C_3 stage	Phenylalanine ammonia lyase (PAL)	Potato	Stimulation	Zucker, 1965
		Artichoke (*Helianthus tuberosus*)	Stimulation	Nitsch and Nitsch, 1966
		Mustard	Stimulation	Durst and Mohr, 1966
		Gherkin	Stimulation	Engelsma, 1967
		Buckwheat	Stimulation	Scherf and Zenk, 1967
		Pea	Stimulation	Attridge and Smith, 1967
		Strawberry	Stimulation	Creasy, 1968
		Cocklebur (*Xanthium stramatium*)	Stimulation	Zucker, 1969
		Mung bean	Stimulation	Ahmed and Swain, 1970

		Red cabbage	Stimulation	Engelsma, 1970a
		Radish (*Raphanus sativus*)	Stimulation	Bellini and van Poucke, 1970
		Parsley (*Petroselinum hortense*)	Stimulation	Hahlbrock *et al.*, 1971a
		Soyabean	Stimulation	Hahlbrock *et al.*, 1971b
	Cinnamic acid 4-hydroxylase (CAH)	Pea	Stimulation	Russell and Conn, 1967
		Buckwheat	Stimulation	Amrhein and Zenk, 1968
		Parsley[a]	Stimulation	Hahlbrock *et al.*, 1971a
	p-Coumarate : CoA ligase	Parsley[a]	Stimulation	Hahlbrock *et al.*, 1971a
		Soyabean[a]	Stimulation	Hahlbrock *et al.*, 1971b
A-ring synthesis	Acetate : CoA ligase	Parsley[a]	None	Hahlbrock and Grisebach, 1970
		Soyabean[a]	None	Hahlbrock *et al.*, 1971b
$C_6C_3C_6$ stage	Chalcone–flavanone isomerase	Buckwheat	None	Amrhein and Zenk, 1971
		Parsley[a]	Stimulation	Hahlbrock *et al.*, 1971a
Secondary features	SAM : 3,4-dihydroxyphenol-3-*O*-methyltransferase	Parsley[a]	Stimulation	Hahlbrock *et al.*, 1971a
	Glucosyl transferase	Parsley	Stimulation	Hahlbrock *et al.*, 1971a
	UDP-apiose synthetase	Parsley	Stimulation	Hahlbrock *et al.*, 1971a
	Apiosyl transferase	Parsley	Stimulation	Hahlbrock *et al.*, 1971a

[a] Cell suspension culture.

found to stimulate greatly PAL production in bean (Farkas and Szirmai, 1969), tobacco (Pegg and Sequeira, 1968; Paynot *et al.*, 1971), sweet potato (Minamikawa and Uritani, 1965a) and soyabean (Biehn *et al.*, 1968). In leguminous plants PAL synthesis has been correlated with the production of phytoalexins (antimicrobial compounds produced by plants in response to infection by pathogens) of the isoflavonoid class, such as pisatin (**LXXXV**) and phaseollin (**LXXXVI**), providing an interesting functional relationship

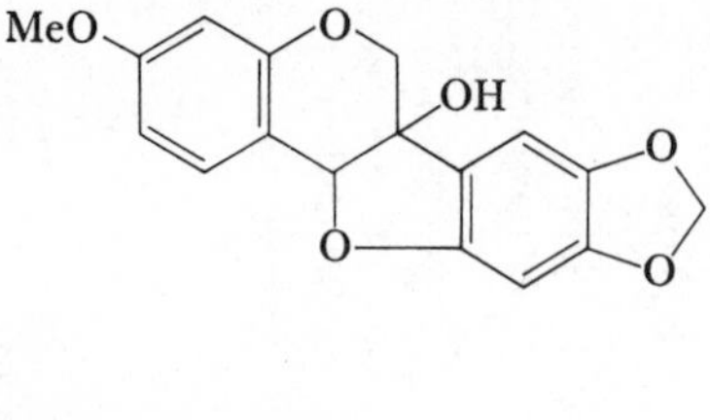

MeO OH O O O O

(**LXXXV**)
Pisatin

HO O O O

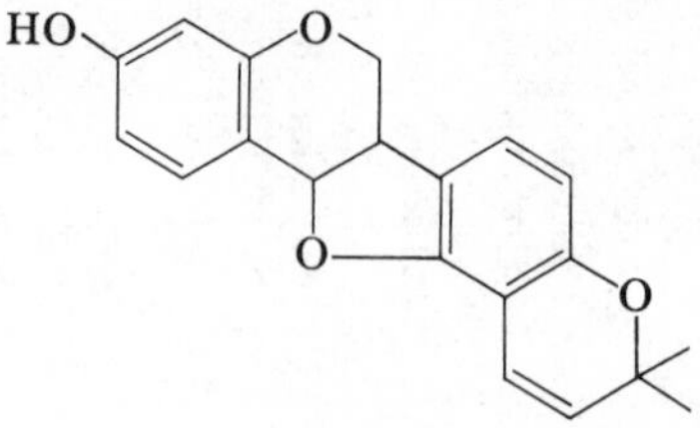

(**LXXXVI**)
Phaseollin

(Hadwiger, 1968; Hess *et al.*, 1971). Production of other isoflavonoid and flavonoid compounds has also been found to be increased in infected tissues of some leguminous plants. Coumestans, flavones and flavonols, for example, are greatly increased in concentration in fungus-infected alfalfa and white clover leaves (Bickoff *et al.*, 1967; Wong and Latch, 1971). The general increases of flavonoid compounds in these cases could merely reflect by-product formation arising from the increased activity of a common biosynthetic pathway which is primarily intended for the production of the isoflavonoid phytoalexins responsible for disease resistance in these plants.

C EFFECT OF HORMONAL SUBSTANCES

Ethylene treatment enhances PAL production in many plant tissues (Imaseki *et al.*, 1968; Riov *et al.*, 1969; Engelsma and van Bruggen, 1971; Hyodo and Yang, 1971a; Rhodes and Wooltorton, 1971). Cinnamic acid hydroxylase activity in peas (Hyodo and Yang, 1971b) and *p*-coumarate:CoA ligase activity in swedes (Rhodes and Wooltorton, 1973) have also been found to be affected similarly. Wounding of a tissue stimulates the endogenous production of ethylene, which is also produced by plants following infection by pathogenic organisms. Many of the PAL enhancing effects resulting from wounding and infection quoted in the last section could well be ascribed to the production of ethylene under such conditions (Camm and Towers, 1973a,b).

Influence on PAL induction activity by other hormonal substances has been demonstrated in a few tissues. The effective hormones include gibberellin (Cheng and Marsh, 1968; Reid and Marsh, 1969), abscisin (Walton and Sondheimer, 1968) and kinetin (Rubery and Fosket, 1969).

D INTRINSIC FACTORS

The ability to synthesize flavonoids in a given plant is often specifically restricted to certain tissues, cell layers, or even to individual cells. Anthocyanin synthesis in mustard hypocotyls, for example, is restricted to the hypodermal layer (Oelze-Karrow and Mohr, 1970). Age and stage of development of the tissue is generally also an important factor. Young tissues typically produce higher concentrations of flavonoids than mature plants. Thus synthesis and accumulation of flavonoid compounds can generally be expected to vary greatly with plant growth and differentiation. Such physiological factors are of great practical importance to those studying biosynthetic pathways in higher plants. Difficulties involved in the isolation of enzymes, owing to their relative low concentrations and variation in their activities during most stages of plant growth, have been circumvented somewhat in recent years by the use of plant cell suspension cultures (Street, 1973). The recent successes of Hahlbrock and co-workers with parsley and soyabean cell culture systems offer excellent examples of this approach which holds promise of leading to further progress in the study of flavonoid biosynthesis.

An example of changes observed with plant growth and differentiation is found in the study of Hahlbrock *et al.* (1971c). The activities of five enzymes related to the biosynthesis of flavone glycosides (PAL, chalcone-flavanone isomerase, glucosyl transferase, apiosyl transferase and apiose synthetase), as well as the accumulation of these flavone glycosides were studied during the growth of young parsley plants. On a specific fresh weight basis, the glycoside content as well as the activities of all five enzymes was found to be higher in very young cotyledons or leaves, decreasing to a lower level as the organs develop. When allowance is made for the change in weight of the organs with development, the curves shown in Fig. 27 for enzyme activity or flavone glycoside content per organ *vs.* growth

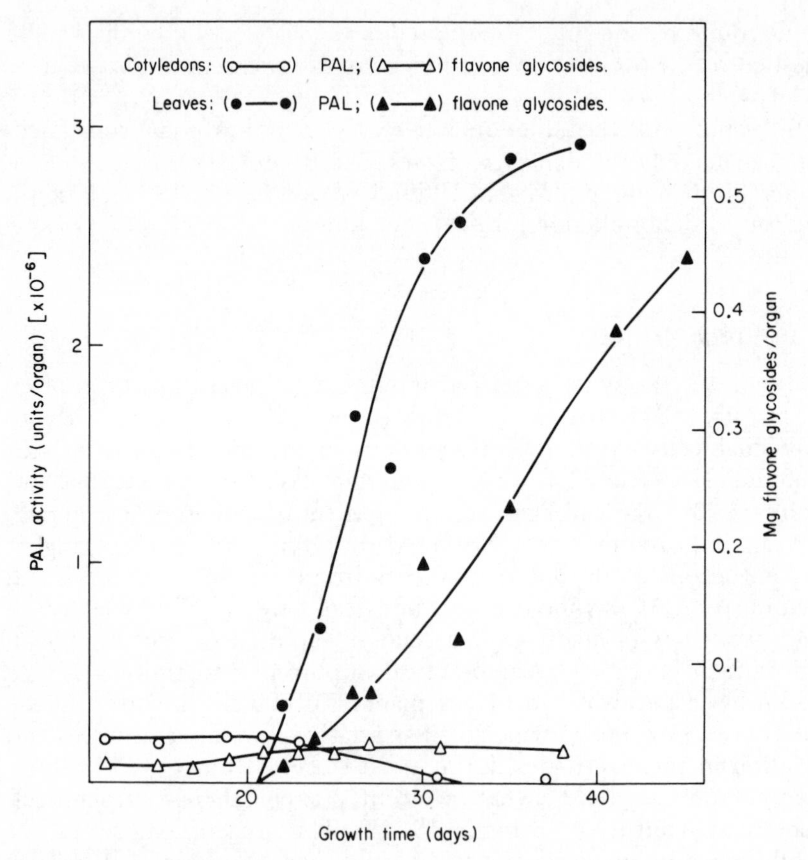

FIG. 27. PAL activity and flavone glycoside content of cotyledons and leaves of parsley (Hahlbrock *et al.*, 1971c).

were obtained. The striking similarity among the activity/growth curves for the five enzymes suggest that these enzymes are regulated interdependently, at least under the conditions of growth and during the period of organ development investigated. Such results also indicate that successful isolation of other enzymes in this biochemical pathway would be most likely also in the same initial stages of cotyledon or leaf development.

The association of highest enzymic activity with young developing tissues has also been reported for cinnamic acid hydroxylase in pea seedlings (Russell, 1971) and cinnamate:CoA ligase in leaves of spinach beet (Walton and Butt, 1970).

Sequential stages of flavonoid synthesis during pollen development in *Narcissus*, *Lilium* and *Tulipa* have been detected by Wiermann

(1970). In these species, the pigmentation process is initiated by the synthesis of cinnamic acids coinciding with maximal PAL activity during meiosis (Wiermann, 1972a). An intense synthesis of flavonol glycosides takes place in *Narcissus* during the separation of the tetrad. In *Tulipa*, however, chalcones are synthesized in an intermediate phase before flavonols and anthocyanins are produced in significant amounts. Maximal chalcone-flavanone activity appears at a time when chalcone concentration is decreasing and flavonol concentration increasing (Wiermann, 1972b). PAL activity seems specifically localized in the tapetum fraction of the developing *Tulipa* anther (Wiermann, 1972a) and the final stages in the synthesis of flavonol glycosides and anthocyanins apparently take place on the pollen wall (Wiermann, 1970).

Maier and Hasegawa (1970) studied the changes of PAL activity in developing grape fruit under normal growth concentrations (without wounding or abnormal light treatment) and reported a high degree of correlation with the rate of naringenin glycoside accumulation. Production of the glycoside occurs mainly during the period of fruit growth associated with extensive cell division, as opposed to the period of cell enlargement, suggesting that naringenin glycoside synthesis occurs mainly early in the life cycle of each group of new cells.

REFERENCES

Ahmed, S. I. and Swain, T. (1970). *Phytochemistry* **9**, 2287–2290.
Ahn, B. Z. and Gstirner, F. (1970). *Arch. Pharm.* **303**, 925–932.
Ahn, B. Z. and Gstirner, F. (1971). *Arch. Pharm.* **304**, 666–673.
Alston, R. E. (1964). *In* "Biochemistry of Phenolic Compounds" (J. B. Harborne, ed.), pp. 171–204. Academic Press, London and New York.
Alston, R. E. (1968). *Recent Adv. Phytochem.* **1**, 305–327.
Amrhein, N. and Zenk, M. H. (1968). *Naturwissenschaften* **55**, 394.
Amrhein, N. and Zenk, M. H. (1970). *Naturwissenschaften* **57**, 312.
Amrhein, H. and Zenk, M. H. (1971). *Z. PflPhysiol.* **64**, 145–168.
Attridge, T. H. and Smith, H. (1967). *Biochim. biophys. Acta* **148**, 805–807.
Attridge, T. H. and Smith, H. (1973). *Phytochemistry* **12**, 1569–1574.
Attridge, T. H., Stewart, G. R. and Smith, H. (1971). *FEBS Lett.* **17**, 84–86.
Barber, G. A. (1962). *Biochemistry, N.Y.* **1**, 463–468.
Barber, G. A. and Chang, M. T. Y. (1968). *Phytochemistry* **7**, 35–39.
Barlett, D. J., Poulton, J. E. and Butt, V. S. (1972). *FEBS Lett.* **23**, 265–267.
Barz, W. and Grisebach, H. (1967). *Z. Naturf.* **22b**, 627–633.
Bate-Smith, E. C. (1963). *In* "Chemical Plant Taxonomy" (T. Swain, ed.), pp. 127–139. Academic Press, London and New York.
Bate-Smith, E. C. (1969). *Phytochemistry* **8**, 1803–1810.
Bellini, E. and van Poucke, M. (1970). *Planta* **93**, 60–70.
Berkowitz, J. E., Coggon, R. and Sanderson, G. W. (1971). *Phytochemistry* **10**, 2271–2278.

Berlin, J., Dewick, P. M., Barz, W. and Grisebach, H. (1972). *Phytochemistry* **11**, 1689-1693.

Bickoff, E. M., Loper, G. M., Hanson, C. H., Graham, J. H., Witt, S. C. and Spencer, R. R. (1967). *Crop Sci.* 7, 259-261.

Biehn, W. L., Kuc, J. and Williams, E. B. (1968). *Phytopathology* **58**, 1255-1260.

Birch, A. J. (1957). *Fortschr. chem. Org. NatStoffe* **14**, 186-216.

Birch, A. J. (1963). *In* "Chemical Plant Taxonomy" (T. Swain, ed.), pp. 141-166. Academic Press, London.

Birch, A. J. (1973). *Pure appl. Chem.* **33**, 17-38.

Birch, A. J. and Donovan, F. W. (1953). *Aust. J. Chem.* 6, 360-368.

Birch, A. J., English, R. E., Massy-Westropp, R. A., Slaytor, M. and Smith, H. (1958). *J. chem. Soc.* 365-368.

Bird, A. E. and Marshall, A. C. (1969). *J. chem. Soc. C* 2418-2420.

Boland, M. J. and Wong, E. (1975). *Eur. J. Biochem.* **50**, 383-389.

Bottomley, W., Smith, H. and Galston, A. W. (1966). *Phytochemistry* **5**, 117-123.

Boudet, A., Ranjeva, R. and Gadal, P. (1971). *Phytochemistry* **10**, 997-1005.

Bryce, T., Collier, P. D., Mallows, R., Thomas, P. E., Frost, D. J. and Wilkins, C. K. (1972). *Tetrahedron Lett.* 463-466.

Bu'Lock, D. (1965). "Biosynthesis of Natural Products", p. 90. MacGraw-Hill, New York.

Camm, E. L. and Towers, G. H. N. (1973a). *Phytochemistry* **12**, 961-973.

Camm, E. L. and Towers, G. H. N. (1973b). *Phytochemistry* **12**, 1575-1580.

Carpenter, I., Locksley, H. D. and Scheinmann, F. (1969). *Phytochemistry* **8**, 2013-2025.

Challice, J. S. and Williams, A. H. (1970). *Phytochemistry* **9**, 1261-1269.

Cheng, C. K.-C. and Marsh, H. V. (1968). *Pl. Physiol., Lancaster* **43**, 1755-1759.

Clark-Lewis, J. W. and Jamison, R. W. (1968). *Aust. J. Chem.* **21**, 2247-2254.

Clark-Lewis, J. W. and Skingle, D. C. (1967). *Aust. J. Chem.* **20**, 2169-2190.

Creasy, L. L. (1968). *Phytochemistry* **7**, 441-446.

Creasy, L. L. (1971). *Phytochemistry* **10**, 2705-2711.

Crombie, L., Dewick, P. M. and Whiting, D. A. (1973). *J.C.S. Perkin I* 1285-1294.

Dean, F. M. and Podimuang, V. (1965). *J. chem. Soc.* 3978-3987.

Dewick, P. M., Barz, W. and Grisebach, H. (1970). *Phytochemistry* **9**, 775-783.

Durst, F. and Mohr, H. (1966). *Naturwissenschaften* **53**, 707.

Ebel, J., Achenbach, H., Barz, W. and Grisebach, H. (1970a). *Biochim. biophys. Acta* **215**, 203-205.

Ebel, J., Barz, W. and Grisebach, H. (1970b). *Phytochemistry* **9**, 1529-1534.

Ebel, J., Hahlbrock, K. and Grisebach, H. (1972). *Biochim. biophys. Acta* **269**, 313-326.

Engelsma, G. (1967). *Planta* **75**, 207-219.

Engelsma, G. (1968a). *Planta* **82**, 355-368.

Engelsma, G. (1968b). *Acta Bot. Neerl.* **17**, 85-89.

Engelsma, G. (1970a). *Acta bot. Neerl.* **19**, 403-414.

Engelsma, G. (1970b). *Planta* **91**, 246-254.

Engelsma, G. and van Bruggen, J. M. H. (1971). *Pl. Physiol., Lancaster* **48**, 94-96.

Engelsma, G. and Meizer, G. (1965). *Acta bot. neerl.* **14**, 54-72.

Farkas, G. L. and Szirmai, J. (1969). *Neth. J. pl. Path.* **75**, 82–85.
Fritsch, H. J., Hahlbrock, K. and Grisebach, H. (1971). *Z. Naturf.* **26b**, 581–585.
Geissman, T. A. and Crout, D. H. G. (1969). "Organic Chemistry of Secondary Plant Metabolism." Freeman, Cooper, San Francisco.
Geissman, T. A. and Hinreiner, E. (1952). *Bot. Rev.* **18**, 77–244.
Geissman, T. A., Jorgensen, E. C. and Johnson, B. L. (1954). *Archs Biochem. Biophys.* **49**, 368–388.
Grambow, H. J. and Grisebach, H. (1971). *Phytochemistry* **10**, 789–796.
Grisebach, H. (1961). *In* "Chemistry of Natural Phenolic Compounds" (W. D. Ollis, ed.), pp. 59–73. Pergamon Press, Oxford.
Grisebach, H. (1962). *Planta med.* **10**, 385–397.
Grisebach, H. (1965). This volume 1st edition, pp. 279–308.
Grisebach, H. (1968). *Recent Adv. Phytochem.* **1**, 379–406.
Grisebach, H. and Barz, W. (1969). *Naturwissenschaften* **56**, 538–544.
Grisebach, H. and Bilhuber, W. (1967). *Z. Naturf.* **22b**, 746–751.
Grisebach, H. and Grambow, H. J. (1968). *Phytochemistry* **7**, 51–56.
Grisebach, H., Barz, W., Hahlbrock, K., Kellner, S. and Patschke, L. (1966). *In* "Biosynthesis of Aromatic Compounds" (G. Billek ed.), vol. 3, pp. 25–36. Pergamon Press, Oxford.
Hadwiger, L. A. (1968). *Neth. J. pl. Path.* **74** (*suppl.*), 163–169.
Hadwiger, L. A. and Schwochau, M. E. (1971). *Pl. Physiol., Lancaster* **47**, 746–751.
Hahlbrock, K. and Grisebach, H. (1970). *FEBS Lett.* **11**, 62–64.
Hahlbrock, K., Wong, E., Schill, L. and Grisebach, H. (1970a). *Phytochemistry* **9**, 949–958.
Hahlbrock, K., Zilg, H. and Grisebach, H. (1970b). *Eur. J. Biochem.* **15**, 13–18.
Hahlbrock, K., Ebel, J., Ortmann, R., Sutter, A., Wellmann, I. and Grisebach, H. (1971a). *Biochim. biophys. Acta* **244**, 7–15.
Hahlbrock, K., Kuhlen, E. and Lindl, T. (1971b). *Planta* **99**, 311–318.
Hahlbrock, K., Sutter, A., Wellman, E., Ortmann, R. and Grisebach, H. (1971c). *Phytochemistry* **10**, 109–116.
Hamilton, G. A. (1969). *Adv. Enzymol.* **32**, 55–96.
Hanson, K. R. and Havir, E. A. (1972a). *Recent Adv. Phytochem.* **4**, 45–85.
Hanson, K. R. and Havir, E. A. (1972b). *In* "The Enzymes" (P. D. Boyer, ed.), 3rd edn, vol. 7, pp. 75–166. Academic Press, London and New York.
Harborne, J. B. (1967). "Comparative Biochemistry of the Flavonoids". Academic Press, London and New York.
Harper, D. B., Austin, D. J. and Smith, H. (1970). *Phytochemistry* **9**, 497–505.
Hasegawa, S. and Maier, V. P. (1972). *Phytochemistry* **11**, 1365–1370.
Hathway, D. E. and Seakins, J. W. T. (1957). *Biochem. J.* **67**, 239–245.
Havir, E. A., Reid, P. D. and Marsh, H. V. (1971). *Pl. Physiol., Lancaster* **48**, 130–136.
Hess, D. (1967). *Z. PflPhysiol.* **56**, 295–298.
Hess, D. (1968). "Biochemische Genetik." Springer-Verlag, Berlin.
Hess, S. L., Hadwiger, L. A. and Schwochau, M. (1971). *Phytopathology* **61**, 79–82.
Hyodo, H. (1971). *Plant Cell Physiol.* **12**, 989–991.
Hyodo, H. and Yang, S. F. (1971a). *Pl. Physiol., Lancaster* **47**, 765–770.
Hyodo, H. and Yang, S. F. (1971b). *Archs Biochim. Biophys.* **143**, 338–339.

Imaseki, H., Uchiyama, M. and Uritani, I. (1968). *Agric. Biol. Chem.* **32**, 387-389.

Jeffery, A. M., Knight, M. and Evans, W. C. (1972a). *Biochem. J.* **130**, 373-381.

Jeffery, A. M., Jerina, D. M., Self, R. and Evans, W. C. (1972b). *Biochem. J.* **130**, 383-390.

Jurd, L. (1967). *Tetrahedron* **23**, 1057-1064.

Kindl, H. (1970). *Hoppe-Seyler's Z. physiol. Chem.* **351**, 792-798.

Koukol, J. and Conn, E. E. (1961). *J. biol. Chem.* **236**, 2692-2698.

Kreuzaler, F. and Hahlbrock, K. (1972). *FEBS Lett.* **28**, 69-72.

Larson, R. L. (1971). *Phytochemistry* **10**, 3073-3076.

Lindl, T., Kreuzaler, F. and Hahlbrock, K. (1973). *Biochim. biophys. Acta* **302**, 457-464.

Maier, V. P. and Hasegawa, S. (1970). *Phytochemistry* **9**, 139-144.

Marchelli, R. and Vining, L. C. (1973). *Chem. Commun.* 555-556.

Marsh, C. A. (1960). *Biochim. Biophys. Acta* **44**, 359-361.

Mason, H. S. (1957). *Adv. Enzymol.* **19**, 79-233.

Meier, H. and Zenk, M. H. (1965). *Z. PflPhysiol.* **53**, 415-421.

Merlini, L. (1973). *Phytochemistry* **12**, 669-670.

Miles, C. D. and Hagen, C. W. (1968). *Pl. Physiol., Lancaster* **43**, 1347-1354.

Minamikawa, T. and Uritani, I. (1965a). *J. Biochem., Tokyo* **57**, 678-688.

Minamikawa, T. and Uritani, I. (1965b). *J. Biochem., Tokyo* **58**, 53-59.

Mohr, H. (1958). *Naturwissenschaften* **45**, 448-449.

Moustafa, E. and Wong, E. (1967). *Phytochemistry* **6**, 625-632.

Nair, P. M. and Vining, L. C. (1965). *Phytochemistry* **4**, 161-168.

Neish, A. C. (1961). *Phytochemistry* **1**, 1-24.

Nitsch, C. and Nitsch, J. P. (1966). *C. R. Acad. Sci.* **262D**, 1102-1105.

Oelze-Karow, H. and Mohr, H. (1970). *Z. Naturf.* **25b**, 1282.

Ortmann, R., Sandermann, H. and Grisebach, H. (1970). *FEBS Lett.* **7**, 164-166.

Ortmann, R., Sutter, A. and Grisebach, H. (1972). *Biochim. biophys. Acta* **289**, 293-302.

Parkhurst, J. R. and Hodgins, D. S. (1971). *Phytochemistry* **10**, 2997-3000.

Patschke, L. and Grisebach, H. (1965). *Z. Naturf.* **20b**, 399.

Patschke, L. and Grisebach, H. (1968). *Phytochemistry* **7**, 235-237.

Patschke, L., Barz, W. and Grisebach, H. (1966a). *Z. Naturf.* **21b**, 45-47.

Patschke, L., Barz, W. and Grisebach, H. (1966b). *Z. Naturf.* **21b**, 201-205.

Paynot, M., Martin, C. and Giraud, M. (1971). *C.r. hebd. Séanc. Acad. Sci., Paris* **273D**, 537-539.

Pegg, G. F. and Sequeira, L. (1968). *Phytopathology* **58**, 476-483.

Pelter, A., Bradshaw, J. and Warren, R. F. (1971). *Phytochemistry* **10**, 835-850.

Ramakrishnan, V. T. and Kagan, J. (1970). *J. org. Chem.* **35**, 2898-2900.

Rathmell, W. G. and Bendall, D. S. (1972). *Biochem. J.* **127**, 125-132.

Reed, D. J., Vimmerstedt, J., Jerina, D. M. and Daly, J. W. (1973). *Archs Biochem. Biophys.* **154**, 642-647.

Reid, P. D. and Marsh, H. V. (1969). *Z. PflPhysiol.* **61**, 170-172.

Rhodes, M. J. C. and Wooltorton, L. S. C. (1971). *Phytochemistry* **10**, 1989-1997.

Rhodes, M. J. C. and Wooltorton, L. S. C. (1973). *Phytochemistry* **12**, 2381-2387.

Riov, J., Monselise, S. P. and Kahan, R. S. (1969). *Pl. Physiol., Lancaster* **44**, 631-635.

Roberts, R. J. and Vaughan, P. F. T. (1971). *Phytochemistry* **10**, 2649-2652.

Robinson, R. (1955). "The Structural Relations of Natural Products." Clarendon, Oxford.
Rubery, P. H. and Fosket, D. E. (1969). *Planta* **87**, 54-62.
Russell, D. W. (1971). *J. biol. Chem.* **246**, 3870-3878.
Russell, D. W. and Conn, E. E. (1967). *Archs Biochem. Biophys.* **122**, 256-258.
Russell, D. W., Conn, E. E., Sutter, A. and Grisebach, H. (1968). *Biochim. biophys. Acta* **170**, 210-213.
Sarapu, L. (1971). *Biokhimiya* **36**, 343-353.
Sato, M. (1966). *Phytochemistry* **5**, 385-389.
Scherf, H. and Zenk, M. (1967). *Z. PflPhysiol.* **57**, 401-418.
Schopfer, P. and Hock, B. (1971). *Planta* **96**, 248-253.
Scott, A. I. (1967). *In* "Oxidative Coupling of Phenols" (W. I. Taylor and A. R. Battersby, eds), pp. 95-117. Arnold, London.
Shimada, M., Yamazaki, T. and Higuchi, T. (1970). *Phytochemistry* **9**, 1-4.
Siegelman, H. W. and Hendricks, S. B. (1957). *Pl. Physiol., Lancaster* **32**, 393-398.
Singleton, V. L. (1972). *In* "The Chemistry of Plant Pigments." (C. O. Chichester, ed.), pp. 143-191. Academic Press, New York and London.
Smith, H. (1972). *In* "Phytochrome" (K. Mitrakos and W. Shropshire, eds), pp. 433-481. Academic Press, New York and London.
Smith, H. (1973). *In* "Biosynthesis and Its Control in Plants" (B. V. Milborrow, ed.), pp. 303-321. Academic Press, London and New York.
Smith, H. and Attridge, T. H. (1970). *Phytochemistry* **9**, 487-495.
Stafford, H. A. and Dresler, S. (1972). *Pl. Physiol., Lancaster* **49**, 590-595.
Street, H. E. (1973). *In* "Biosynthesis and Its Control in Plants" (B. V. Milborrow, ed.), Chap. 5. Academic Press, London and New York.
Subba Rao, P. V., Moore, K. and Towers, G. H. N. (1967). *Can. J. Biochem.* **45**, 1863-1872.
Sutter, A. (1972). Ph.D. Thesis, University of Freiburg/Br., Germany.
Sutter, A. and Grisebach, H. (1973). *Biochim. biophys. Acta* **309**, 289-295.
Sutter, A., Ortmann, R. and Grisebach, H. (1972). *Biochim. biophys. Acta* **258**, 71-87.
Takino, Y., Ferritti, A., Flanagan, V., Gianturco, M. A. and Vogel, M. (1967). *Can. J. Chem.* **45**, 1949-1956.
Thompson, R. S., Jacques, D. and Haslam, E. (1972). *J.C.S. Perkin I* 1387-1399.
Towers, G. H. N. (1969). *In* "Perspectives in Biochemistry" (J. B. Harborne, and T. Swain, eds), pp. 179-91. Academic Press, New York and London.
Towers, G. H. N. and Subba Rao, P. V. (1972). *Recent Adv. Phytochem.* **4**, 1-43.
Vaughan, P. F. T. and Butt, V. S. (1969). *Biochem. J.* **113**, 109-115.
Vaughan, P. F. T. and Butt, V. S. (1970). *Biochem. J.* **119**, 89-94.
Vaughan, P. F. T., Butt, V. S., Grisebach, H. and Schill, L. (1969). *Phytochemistry* **8**, 1373-1378.
Wallace, J. W. and Grisebach, H. (1973). *Biochim. biophys. Acta* **304**, 837-841.
Wallace, J. W., Mabry, T. J. and Alston, R. E. (1969). *Phytochemistry* **8**, 93-99.
Walton, D. C. and Sondheimer, E. (1968). *Pl. Physiol., Lancaster* **43**, 467-469.
Walton, E. and Butt, V. S. (1970). *J. exp. Bot.* **21**, 887-891.
Walton, E. and Butt, V. S. (1971). *Phytochemistry* **10**, 295-304.
Weinges, K. and Huthwelker, D. (1970). *Justus Liebig's Annln Chem.* **731**, 161-170.

Weinges, K., Baehr, W., Ebert, W., Goertz, K. and Marx, H. D. (1969). *Fortschr. Chem. org. NatStoffe* 27, 158–260.
Weinges, K., Mattauch, H., Wilkins, C. and Frost, D. (1971). *Justus Liebig's Annln Chem.* **754**, 124–136.
Weinges, K., Kloss, P. and Jaggy, H. (1972). *Arzneimittel-Forsch.* 22, 166–168.
Whalley, W. B. (1961). *In* "Chemistry of Natural Phenolic Compounds" (W. D. Ollis, ed.), pp. 20–58. Pergamon Press, Oxford.
Wiermann, R. (1970). *Planta* **95**, 133–145.
Wiermann, R. (1972a). *Z. PflPhysiol.* **66**, 215–221.
Wiermann, R. (1972b). *Planta* **102**, 55–60.
Williams, H. H. (1967). *Chemy Ind.* 1526–1527.
Wong, E. (1964). *Chemy Ind.* 1985.
Wong, E. (1965). *Biochim. biophys. Acta* **111**, 358–363.
Wong, E. (1966). *Phytochemistry* 5, 463–467.
Wong, E. (1967). *Phytochemistry* 6, 1227–1233.
Wong, E. (1968). *Phytochemistry* 7, 1751–1758.
Wong, E. (1970). *Fortschr. Chem. org. NatStoffe* **28**, 1–73.
Wong, E. and Francis, C. M. (1968a). *Phytochemistry* 7, 2131–2137.
Wong, E. and Francis, C. M. (1968b). *Phytochemistry* 7, 2139–2142.
Wong, E. and Grisebach, H. (1969). *Phytochemistry* **8**, 1419–1426.
Wong, E. and Latch, G. C. M. (1971a). *N.Z. Jl agric. Res.* 14, 633–638.
Wong, E. and Moustafa, E. (1966). *Tetrahedron Lett.* 3021–3022.
Wong, E. and Wilson, J. M. (1972). *Phytochemistry* **11**, 875.
Young, M. R. and Neish, A. C. (1966). *Phytochemistry* **5**, 1121–1132.
Young, M. R., Towers, G. H. N. and Neish, A. C. (1966). *Can. J. Bot.* **44**, 341–349.
Zaprometrov, M. N. and Grisebach, H. (1973). *Z. Naturf.* in press.
Zenk, M. H. and Leistner, E. (1968). *Lloydia* **31**, 275–292.
Zucker, M. (1965). *Pl. Physiol., Lancaster* **40**, 779–784.
Zucker, M. (1968). *Pl. Physiol., Lancaster* **43**, 365–374.
Zucker, M. (1969). *Pl. Physiol., Lancaster* **44**, 912–922.
Zucker, M. (1972). *A. Rev. Pl. Physiol.* 23, 133–156.

Chapter 10
Quinones: Nature, Distribution and Biosynthesis

R. H. THOMSON

Department of Chemistry, University of Aberdeen, Scotland

I Introduction

Although they exist in large numbers and are widely distributed, quinones make relative little contribution to natural colouring in the plant kingdom. Mostly they are yellow to red, the range of colour being from pale yellow to almost black. Blue and green quinones are extremely rare. They have been observed in all phyla except the ferns and mosses, the majority occurring in flowering plants and micro-organisms. A large proportion are present in roots, wood or bark, and consequently are not readily observed. For example, madder *(Rubia tinctorum)* contains nearly twenty anthraquinones and was formerly of great importance as a natural dyestuff, but the pigments are confined entirely to the roots and are not visible above ground. Occasionally the quantity of pigment produced is substantial, amounting to over 17% of the dry weight of the bark of *Coprosma australis* (Briggs and Dacre, 1948), and 30% of the dried mycelium of *Pyrenophora graminea* (= *Helminthosporium gramineum*) (Charles *et al.*, 1933; cf. Birch *et al.*, 1958). A few higher fungi and lichens are brightly coloured by quinones, and these, together with sea urchins and sea lilies provide the most obvious manifestation of these pigments. Quinones are seldom found in flowers or leaves,* and in any case are usually masked by other pigments, and a few occur normally in a reduced, colourless form. These pigments have also been extracted from soil and in one remarkable example the quinone content is sufficient to colour the ground green (Butler *et al.*, 1964).

Natural quinones are derived from malonate, mevalonate and shikimate, and from combinations of these, so that in structure they are more diverse than any other group of plant pigments. The benzoquinones, naphthoquinones and anthraquinones are the most numerous, and within each group there are substantial variations. The existence of benzoquinones of type (I), (II) and (III) within one family (Cyperaceae) provides a good illustration. Smaller groups of polycyclic quinones are based on the phenanthrene, naphthacene, 1,2-benzanthracene, pentaphene, perylene and naphthodianthrene systems, and a number contain *N*-heterocyclic rings. Others belong biogenetically to other classes of compounds; terpenoid quinones are quite numerous and (III) is obviously a flavanoid. There is every reason to suppose that the natural quinone family will continue to grow and diversify still further.†

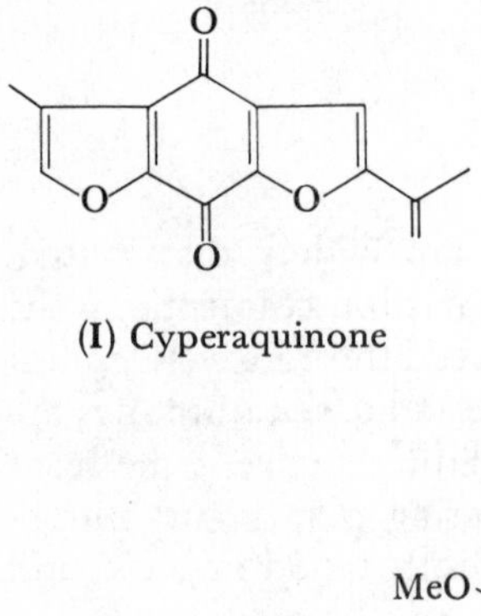

(I) Cyperaquinone

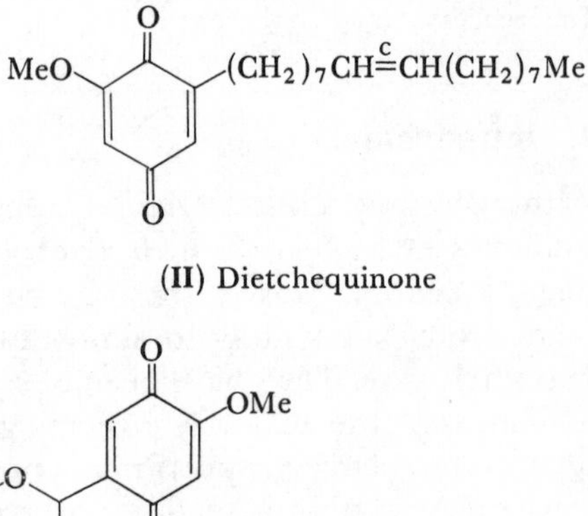

(II) Dietchequinone

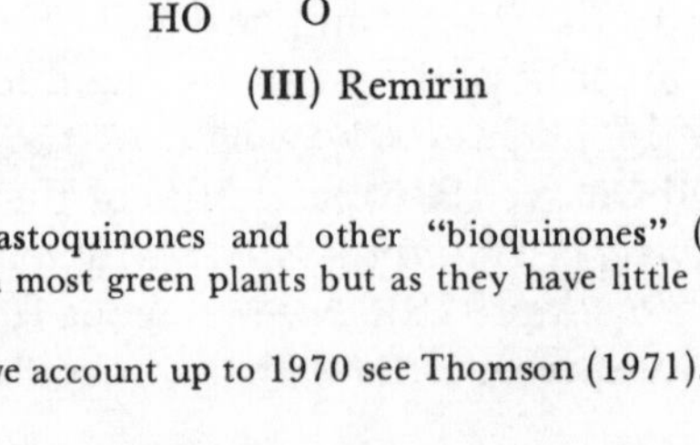

(III) Remirin

* The important plastoquinones and other "bioquinones" (ubiquinones and menaquinones) are present in most green plants but as they have little interest as pigments they are not considered here.

† For a comprehensive account up to 1970 see Thomson (1971).

II Benzoquinones

The sixty or so benzoquinones can be subdivided into several groups, each of which has a limited distribution. A number of simple hydroxy/methoxy derivatives of benzoquinone and toluquinone are elaborated by lower fungi (especially *Penicillium* and *Aspergillus* spp.); examples are coprinin (**IV**) (yellow), spinulosin (**V**) (dark purple) and the dimer, oosporein (**VI**) (bronze). On the other hand 2,6-dimethoxybenzoquinone is a common heartwood constituent (*inter alia* elm, oak, beech), and has been found in poplar bark, maple sap, in a bamboo (Yamagishi *et al.*, 1972), and elsewhere, and is probably derived from a lignin precursor, or from lignin itself (Young and Steelink, 1973).

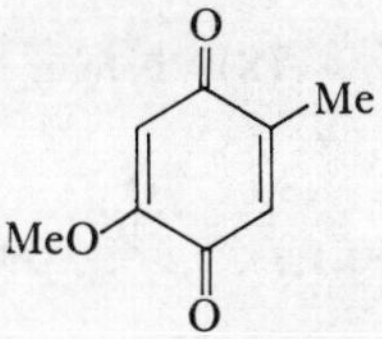

(**IV**) Coprinin 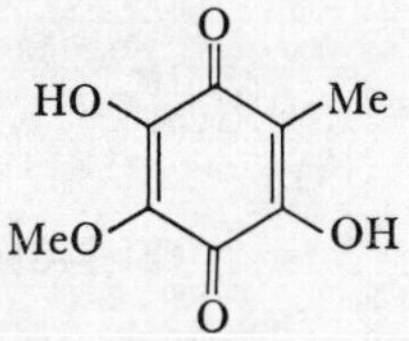 (**V**) Spinulosin

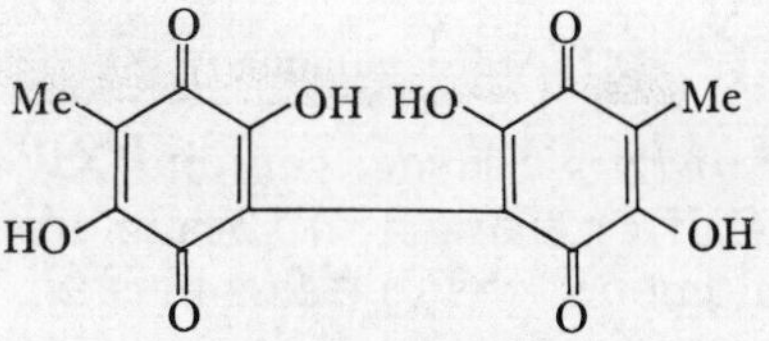

(**VI**) Oosporein

Sarcodontic acid (**VII**), produced by the wood-rotting fungus *Sarcodontia setosa* found on old apple trees, is the only fungal benzoquinone with a long side chain but several examples are found in higher plants. Primin (**VIII**), the dermatitic principle on the leaves of *Primula obconica*, is the simplest while embelin (**IX**) and vilangin (**X**), found in the fruit of *Embelia ribes*, are typical quinones from the Myrsinaceae. Although a useful taxonomic character for this family, quinones of the same type have been found in Liliaceae (Yoshihira and Natori, 1966) and in Cyperaceae (sedges) (Dunlop, 1971; R. J. Wells and J. K. MacLeod, personal communication); compare (**II**; R = OH) (*Cyperus dietricheae*) and ardisiaquinone (**XI**) (Myrsinaceae).

In addition to the ubiquitous plastoquinones and ubiquinones several other prenylated benzoquinones occur in higher and lower

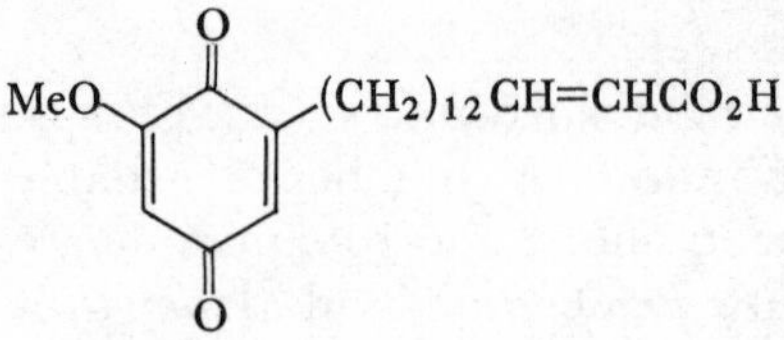

(VII) Sarcodontic acid

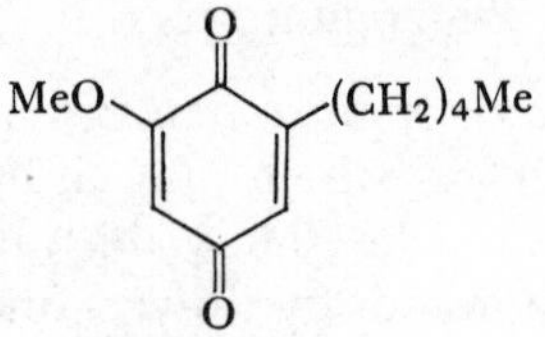

(VIII) Primin

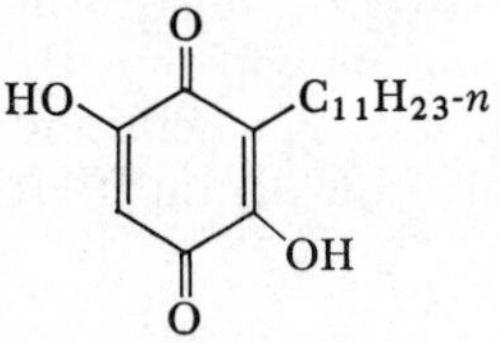

(IX) Embelin

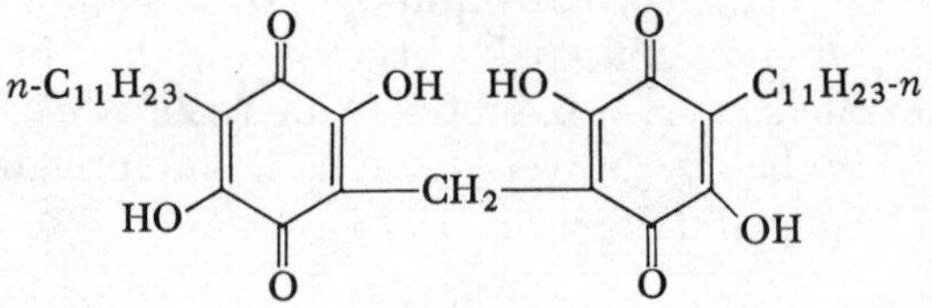

(X) Vilangin

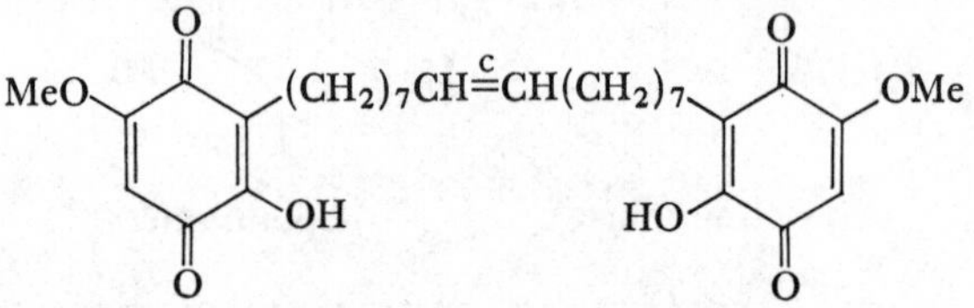

(XI) Ardisiaquinone A

plants. The leaves of *Pyrola media* contain (**XII**; $n = 1$ and 2) while the boviquinones (**XIII**; $n = 3$ and 4) (Beaumont and Edwards, 1969 and 1971) and amitenone* (**XIV**) (Minami *et al.*, 1968), from higher fungi, are analogous to embelin (**IX**) and vilangin (**X**). Tauranin (**XIVa**; $R = HO$, $R' = CH_2OH$) is a fungal metabolite (*Oospora aurantia*) while the related metabolite zonarone (**XIVa**; $R = R' = H$) occurs in the seaweed *Dictyopteris zonaroides*, mainly in the quinol form (Fenical *et al.*, 1973). Prenylation is also an important biosynthetic process in the Cyperaceae; breviquinone (**XV**) (Allan *et al.*, 1973) is an obvious example but in most of the pigments the

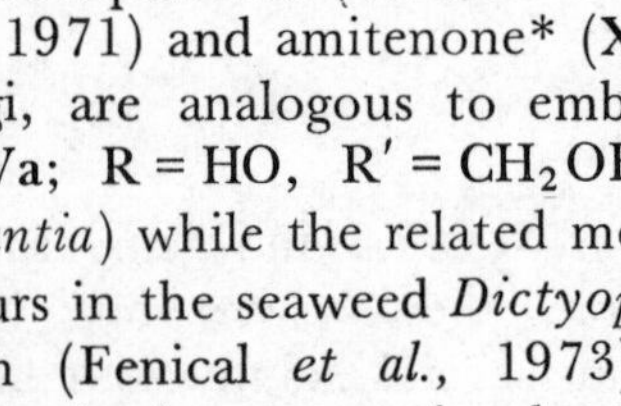

(XII)

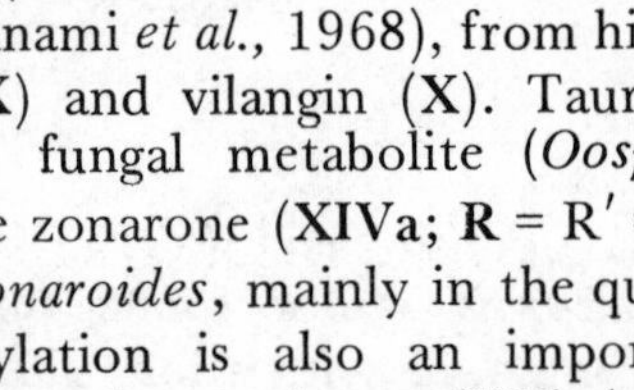

(XIII) Boviquinones

* Amitenone is possibly an artefact (Beaumont and Edwards, 1969).

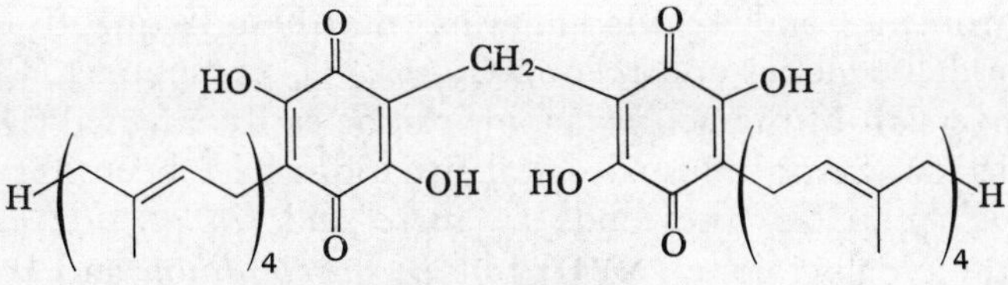

(XIV) Amitenone

C_5 side chain has undergone cyclization to form a furan ring as in cyperaquinone (**I**) (Allan *et al.*, 1969). The co-existence of scabequinone (**XVI**) with breviquinone (**XV**) suggests that the unusual isopropylchroman ring in the former is also derived from mevalonate (Allan *et al.*, 1973). With respect to benzoquinones the Cyperaceae family is unique; so far it has yielded eleven quinones of the cyperaquinone (**I**) type (including five dimers), two breviquinones of type (**XV**), two scabequinones of type (**XVI**), and two dietchequinones of type (**II**). As the pigments are very easily detected they are of considerable taxonomic value in this family where classical methods of identification are difficult (Dunlop, 1971; Allan, 1972; R. J. Wells and J. K. MacLeod, personal communication).

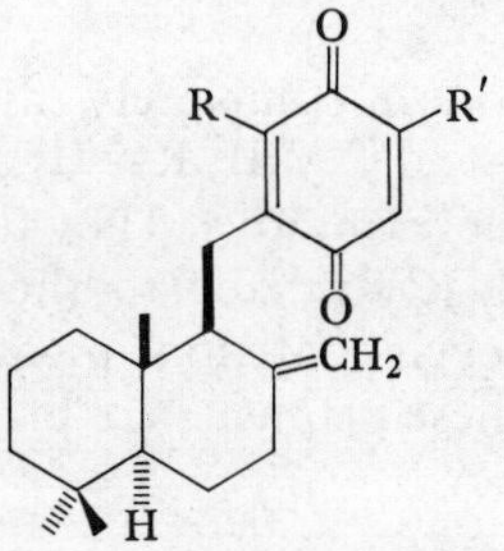

(XIVa)

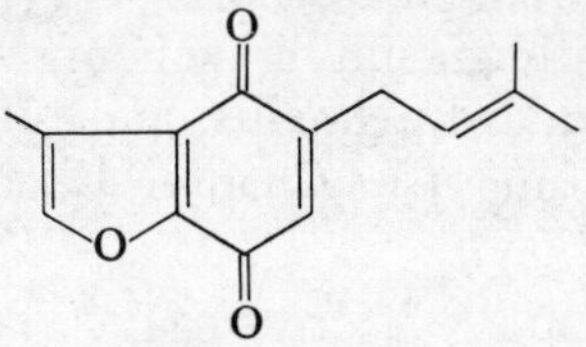

(XV) Breviquinone

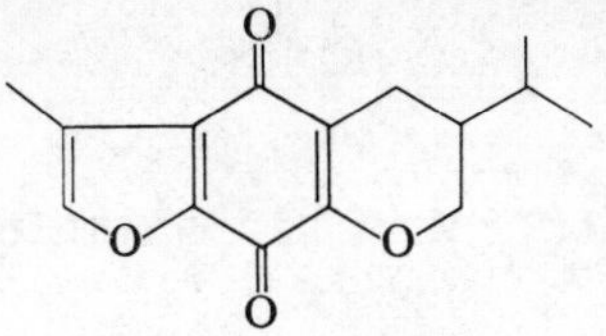

(XVI) Scabequinone

The dalbergiones (e.g. **XVII**) are a small group of benzoquinones restricted mainly to *Dalbergia* and *Machaerium* heartwoods (Leguminosae) (Braga de Oliveira *et al.*, 1971) although also found in the related *Goniorrhachis* genus (and possibly *Peltogyne*) (Rêgo de Sousa *et al.*, 1967). They are neoflavanoids, a group which includes the

4-arylchromenes and 4-arylcoumarins, and these frequently co-exist with the dalbergiones and related $C_6-C_3-C_6$ compounds. Chemical interconversion of the neoflavanoids can be easily effected (Donnelly *et al.*, 1973). Some unusual chiral situations have been observed in *Dalbergia* spp. The rosewoods *D. nigra* and *D. latifolia* elaborate *R*-4-methoxydalbergione (**XVII**) but *D. miscolobium* and *D. baroni*

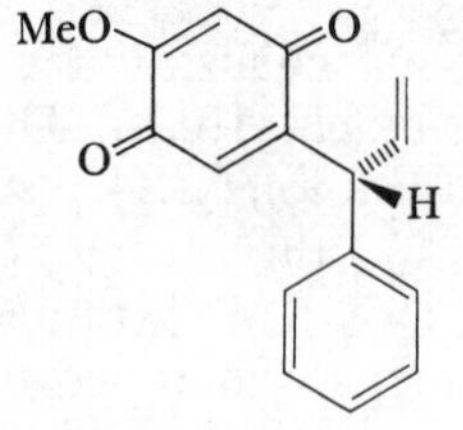

(**XVII**) (*R*)-4-Methoxydalbergione

produce the *S* form. Moreover, two other dalbergiones have been isolated from *D. nigra* which have the *S* configuration, and in addition a quasiracemate of (**XVII**) and *S*-4,4′-dimethoxydalbergione. The blue black pigment in *D. retusa* is the quinhydrone of (**XVII**) (Jurd *et al.*, 1972).

Another small group of benzoquinones, completely different again, are terphenyl derivatives. Polyporic acid (**XVIII**; R = H) and the almost black thelephoric acid (**XIX**) are representative. They occur in higher fungi and in lichens, and are closely related to the lichen tetronic acid pigments, and to the diarylcyclopentenone pigments in higher fungi (see Chapter 12). Some of these pigments are present in

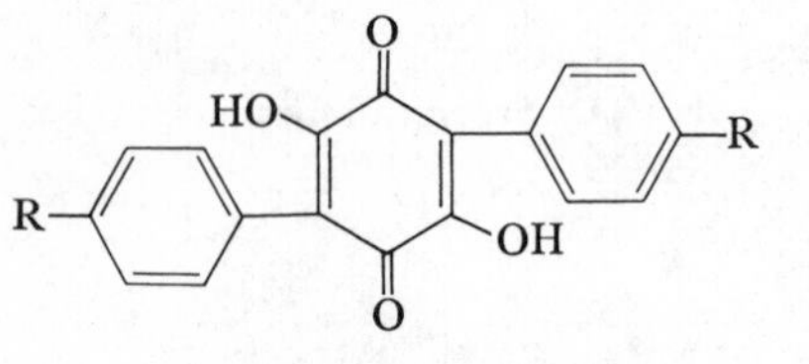

(**XVIII**) Polyporic acid (R = H) 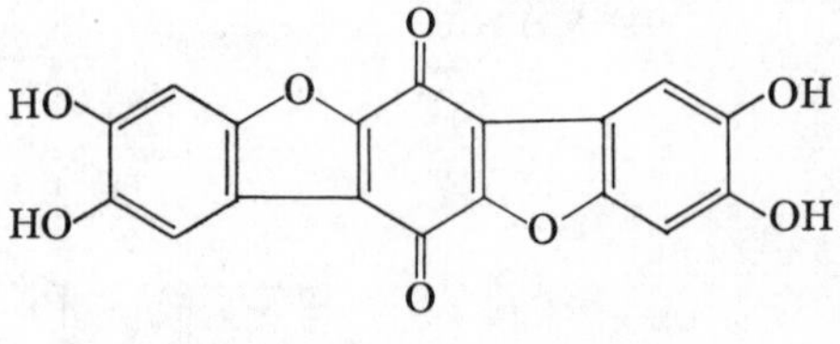 (**XIX**) Thelephoric acid

the colourless quinol form, and leuco esters and leuco ethers have been isolated. The two terphenyl quinones which occur in the lower fungi are different from the others; volucrisporin (**XX**) has an unusual hydroxylation pattern and phlebiarubrone (**XXI**) is a rare *o*-benzoquinone. On the other hand the terphenyl derivative (**XXII**) from cultures of *Aspergillus candidus* (Marchelli and Vining, 1973) is closer to the usual pattern.

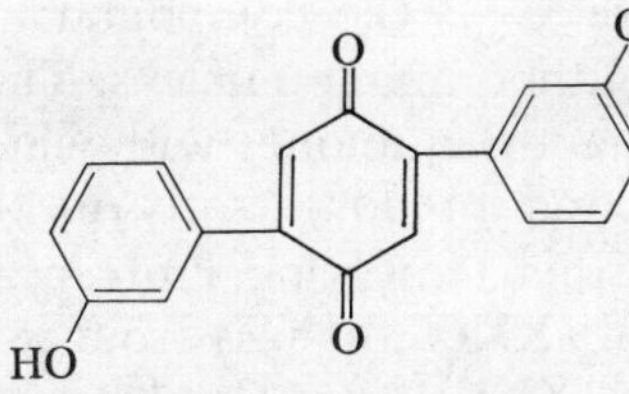

(**XX**) Volucrisporin 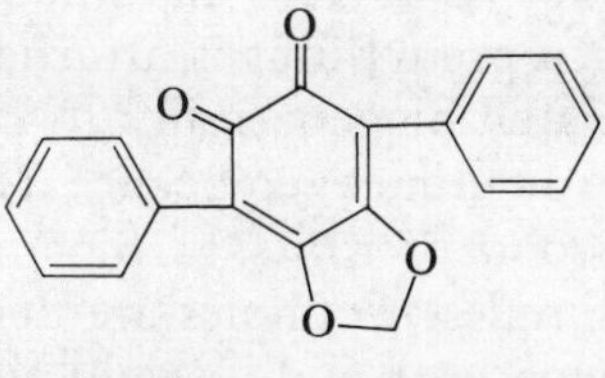 (**XXI**) Phlebiaquinone

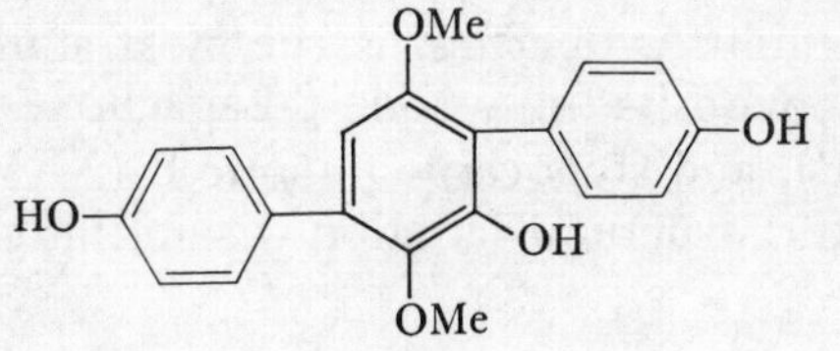

(**XXII**)

III Naphthoquinones

Naphthoquinones are more numerous than benzoquinones, and the majority are found in flowering plants, usually in heartwood. They may also occur in bark, roots, leaves and seeds. Lawsone (2-hydroxy-1,4-naphthoquinone), the dyeing principle of henna, is found in the leaves of *Lawsonia* and *Impatiens* spp., mainly in a reduced form, but, exceptionally, it also occurs in the flowers of *I. balsamina*. Lawsone has not been found outside the Balsaminaceae and this restricted distribution is typical of most naphthoquinones. The 5-hydroxy isomer, juglone, is characteristic of the Juglandaceae, and is present as the quinol-4-glucoside, but recently it has appeared in *Lomatia* (Proteaceae) together with the reduced form (**XXIII**) (Moir and Thomson, 1973a).

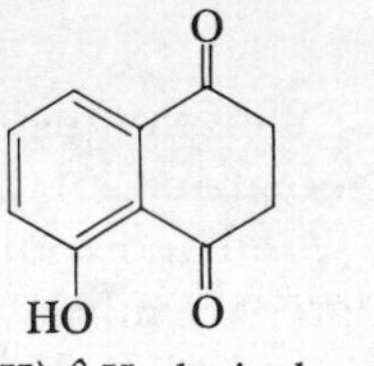

(**XXIII**) β-Hydrojuglone 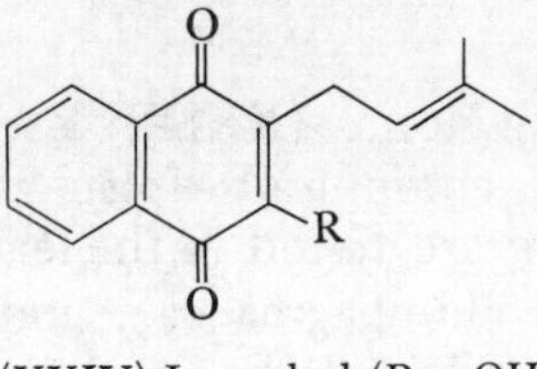 (**XXIV**) Lapachol (R = OH)

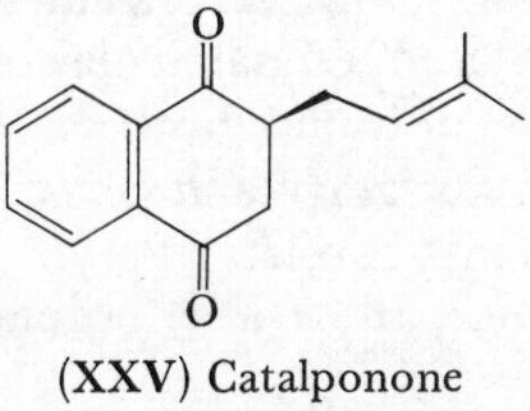

(**XXV**) Catalponone

About a dozen naphthoquinones are C_{10}–C_5 compounds in which a prenyl unit is attached to a naphthoquinone nucleus. They are related biogenetically to certain C_{15} anthraquinones with which they sometimes co-occur. They are found predominantly in the heartwood or roots of Bignoniaceae, Verbenaceae, and Proteaceae. The simplest examples are deoxylapachol (**XXIV**; R = H) from teak (Verbenaceae) and its reduction product (**XXV**) from *Catalpa orata* (Bignoniaceae) (Inouye *et al.*, 1971). Lapachol (**XXIV**; R = OH), which has antitumour properties, is the most abundant, sometimes forming yellow deposits in wood. β-Lapachone (**XXVI**), α-caryopterone (**XXVII**), and stenocarpoquinone B (**XXVIII**) (Mock *et al.*, 1973) illustrate the structural variation within this group.

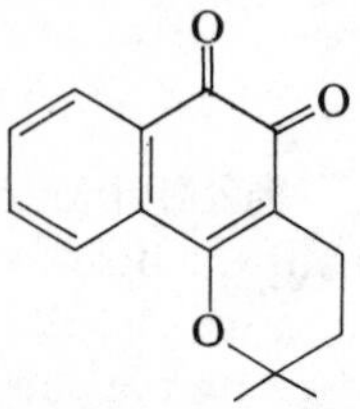

(**XXVI**) β-Lapachone

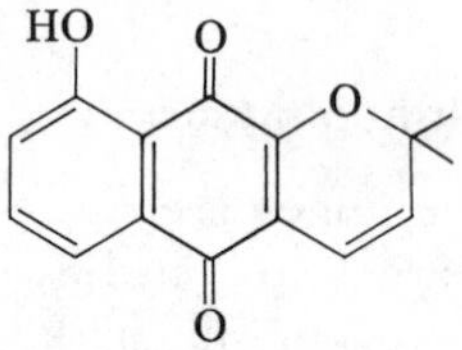

(**XXVII**) α-Caryopterone

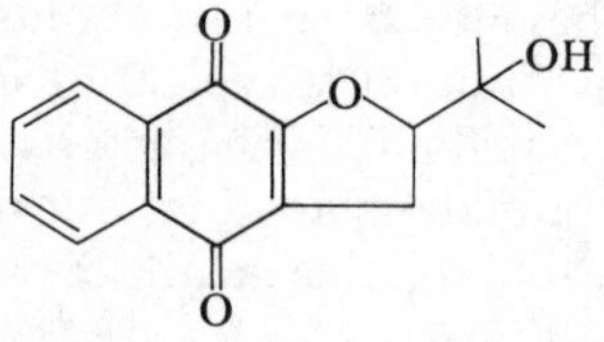

(**XXVIII**) Stenocarpoquinone B

7-Methyljuglone (**XXIX**; R = H), 2-methyljuglone (plumbagin) and related compounds form another group of about thirty pigments. They are found in the leaves, wood and roots of Droseraceae, Ebenaceae, Plumbaginaceae, and related families. *Diospyros* heartwoods are a particularly rich source from which about twenty dimers (e.g. **XXX**), two trimers, and a tetramer derived from 7-methyljuglone, have been isolated. Of particular interest is the blue quinone (**XXXI**) which has been found both in *Diospyros* and in *Euclea* (Ferreira *et al.*, 1973; Musgrave and Skoyles, 1974). In the Liliaceae a small group of pigments consists of stypandrone (**XXIX**; R = Ac), a dimer, dianellinone, and the trimer, trianellinone (**XXXII**) (Cooke and Down, 1971).

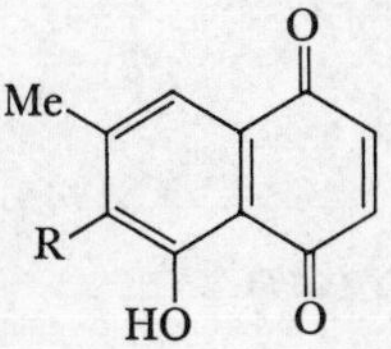

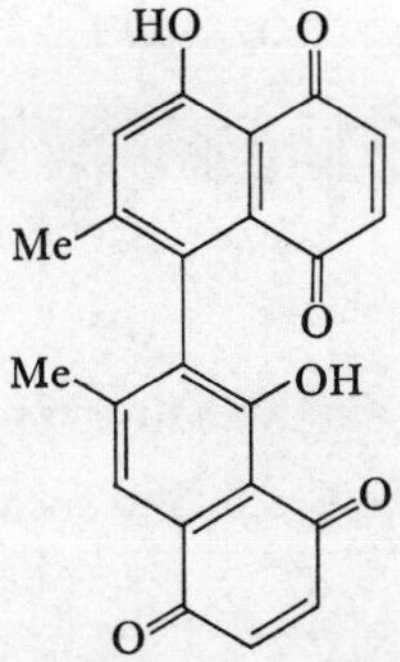

Me, R, HO, O, O

(**XXIX**) 7-Methyljuglone (R=H)

HO, O, Me, O, Me, OH, O, O

(**XXX**) Isodiospyrin

MeO, HO, O, Me, Me, O, OH, OMe

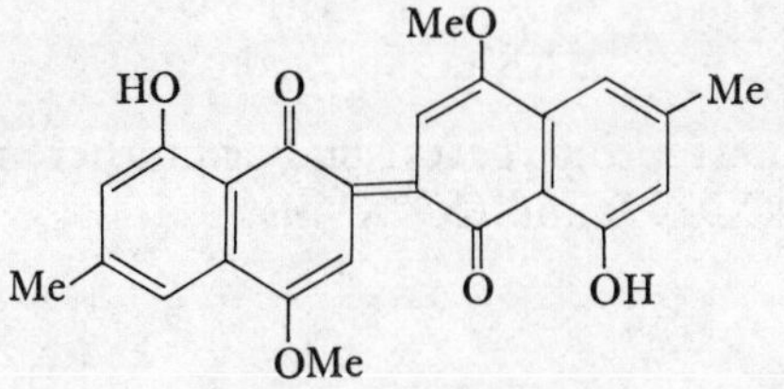

(**XXXI**) Diosindigo

Me, HO, Ac, O, O, Ac, Me, HO, O, O, O, O, OH, Me, Ac

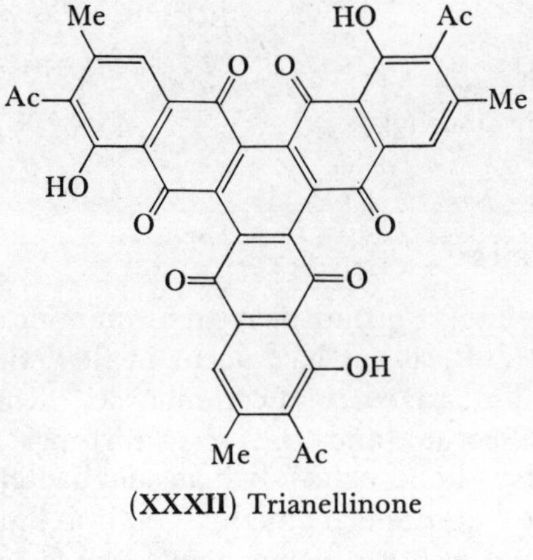

(**XXXII**) Trianellinone

Naphthoquinones do not seem to occur in the higher fungi but more than twenty have been found in moulds. Several of these are elaborated by *Fusarium* spp. (e.g. **XXXIII**). In general the fungal pigments tend to be more complex in structure and *O*-heterocyclic rings are common; pigments (**XXXIV**) and (**XXXV**) are illustrative. The same remark also applies to the few naphthoquinones isolated

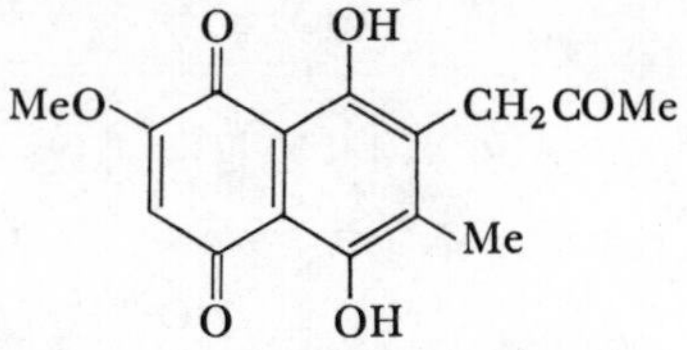

(**XXXIII**) Javanicin

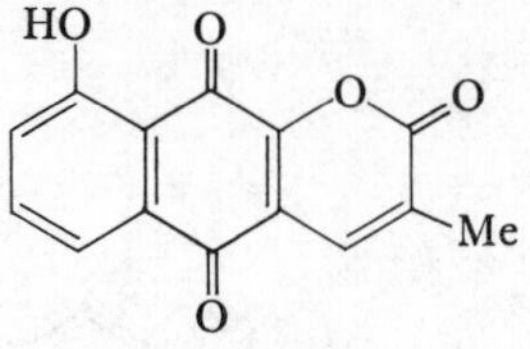

(**XXXIV**) Lambertellin

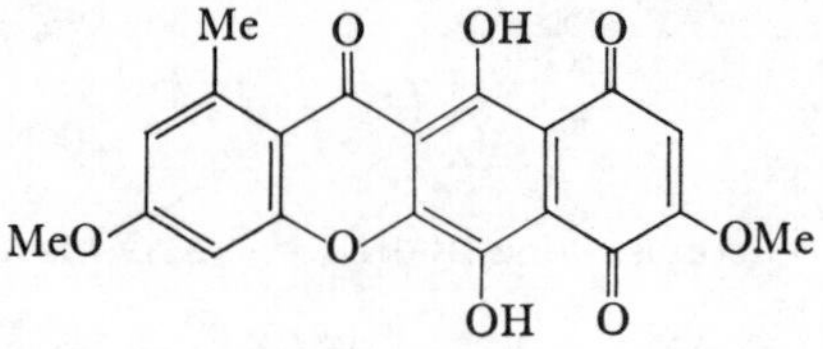

(**XXXV**) Bikaverin

from bacteria (all streptomycetes) and again there is a wide variation as indicated by (**XXXVI**) and (**XXXVII**).

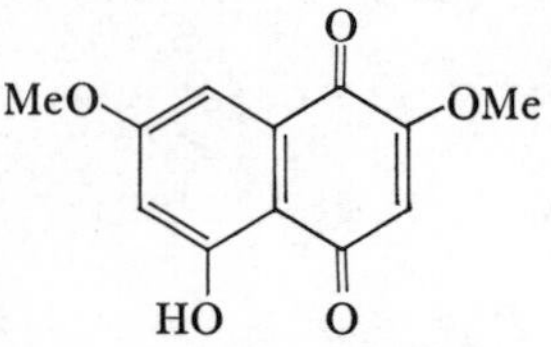

(**XXXVI**) Flaviolin dimethyl ether

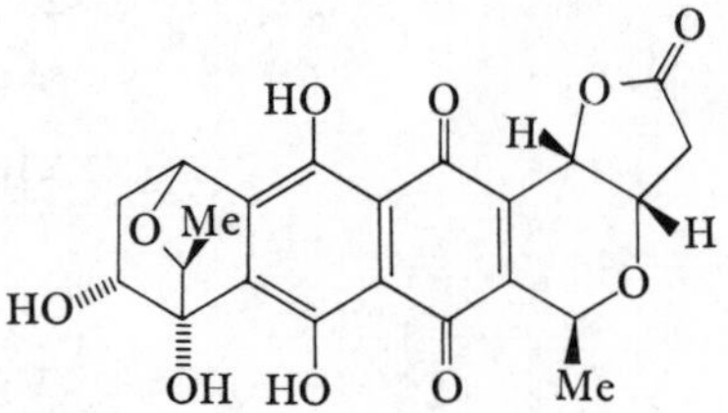

(**XXXVII**) Granaticin

IV Anthraquinones

This is much the largest group of natural quinones. Of those in the plant kingdom (~200) about half occur in flowering plants (chiefly Rubiaceae, Scrophulariaceae, Leguminosae, Rhamnaceae, Polygonaceae and Liliaceae) and the rest in fungi (deuteromycetes, ascomycetes and basidiomycetes), lichens and bacteria. In contrast to the other types of quinone pigment, it is not unusual to find the same anthraquinone in both higher plants and fungi, and frequently anthraquinones exist as glycosides in flowering plants and occasionally in higher fungi (Steglich and Lösel, 1972). The quinones are sometimes accompanied by the corresponding anthrones, and may occur in all parts of the plant (except flowers) but they are present most in wood and roots. Exceptionally, the *Digitalis* (Scrophulariaceae) quinones occur almost exclusively in the leaves.

Most anthraquinones are simply polyhydroxy/methoxy derivatives with or without a one-carbon side chain (CH_3, CH_2OH, CHO, CO_2H) which invariably occupies a β position. Dimers are quite common, and several chloro derivatives are known, chiefly in lichens. Perhaps the most remarkable compound is anthraquinone itself which is present in the cuticular wax of perennial rye grass (*Lolium perenne*) (Allebone *et al.*, 1971).

Some forty anthraquinones have been found in the Rubiaceae, and most of these are substituted only in one ring (e.g. **XXXVIII**). A similar situation exists among the leaf pigments of *Digitalis* (more than twenty quinones) but a higher proportion are substituted in two rings, and certain orientations are very unusual (e.g. (**XXXIX**) and (**XL**) (Brew and Thomson, 1971; Imre *et al.* 1974)). The most

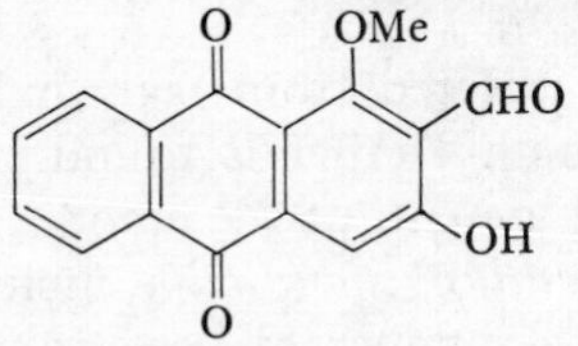

O OMe CHO OH O

(**XXXVIII**) Damnacanthal

HO O OH Me O

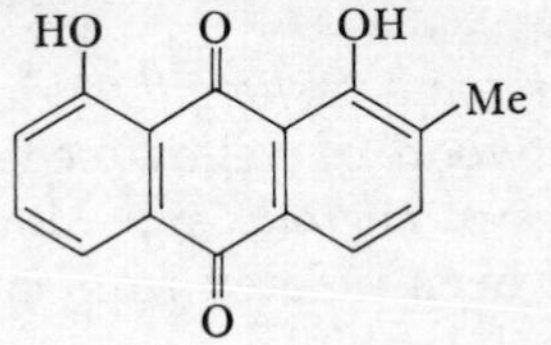

(**XXXIX**) Isochrysophanol

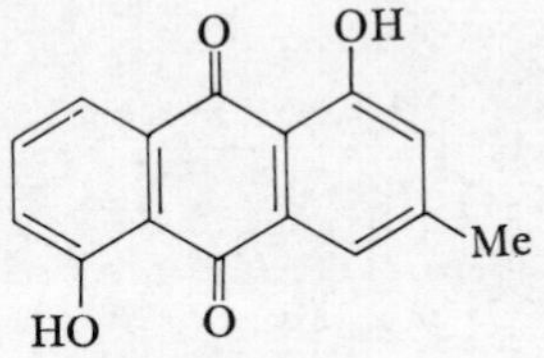

O OH Me HO O

(**XL**) Ziganein

frequent substitution pattern is that of emodin (**XLI**). About forty pigments are derived from this structure by methylation, side-chain oxidation, chlorination, oxidative dimerization etc., and they occur widely in fungi and lichens, and in higher plants (especially Leguminosae, Polygonaceae, and Rhamnaceae). Madagascin, the 3-($\gamma\gamma$-dimethylallyl) ether of emodin, from the bark of *Harungana madagascariensis*, is the only example of prenylation in the anthraquinone series. Other variations of (**XLI**) arise by further hydroxylation (e.g. **XLII**), and the absence of one or more hydroxyl groups. Chrysophanol (1,8-dihydroxy-3-methylanthraquinone) and its derivatives frequently co-exist with emodin and related metabolites. Pachybasin (**XLIII**) is of more limited distribution but nevertheless it is elaborated by several micro-organisms (*Penicillium, Phoma, Tricho-*

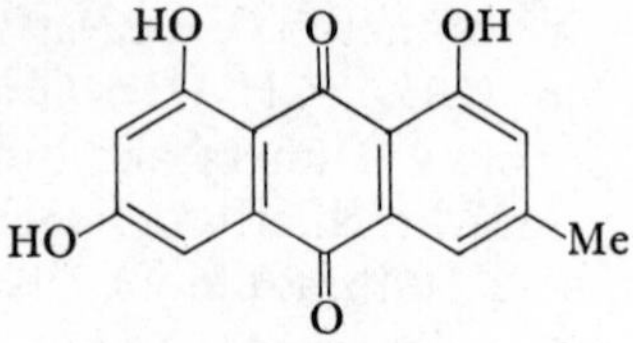

(XLI) Emodin

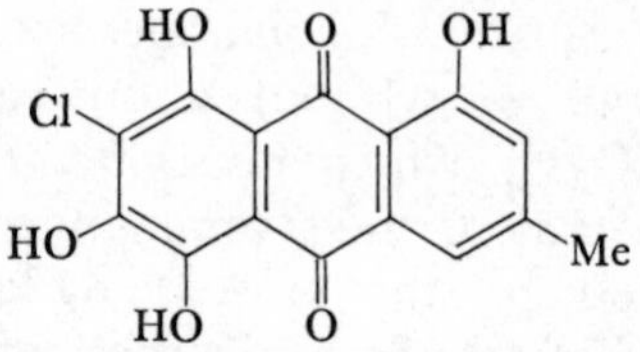

(XLII) Valsarin

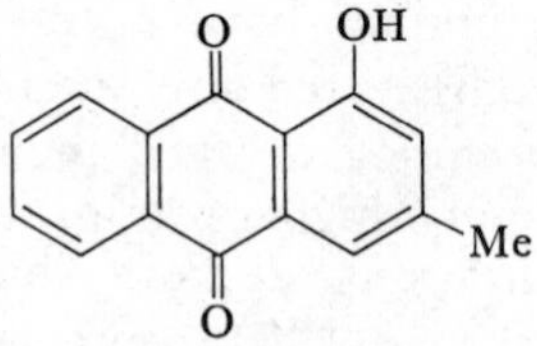

(XLIII) Pachybasin

derma and *Aspergillus*), and has been isolated from teak and the heartwood of yellow cedar, while its methyl ether is found in the leaves of *Digitalis* spp. More than thirty pigments of the emodin type have been isolated from cultures of *Penicillium islandicum*, including several simple dimers (e.g. **XLIV**) and a number of very complex dimeric compounds such as the toxic luteoskyrin (**XLV**) (Takeda *et al.*, 1973).

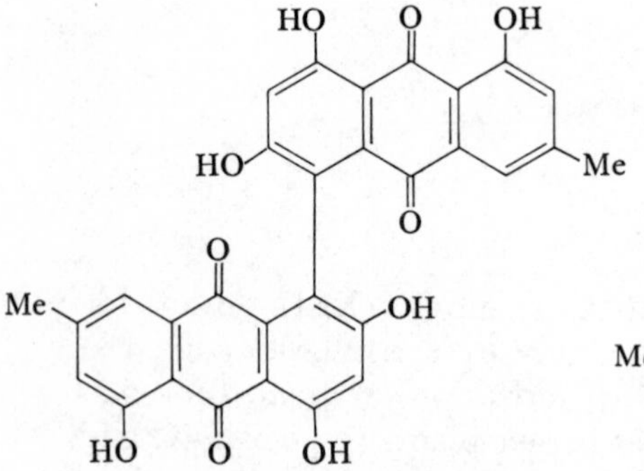

(XLIV) (+)-Skyrin

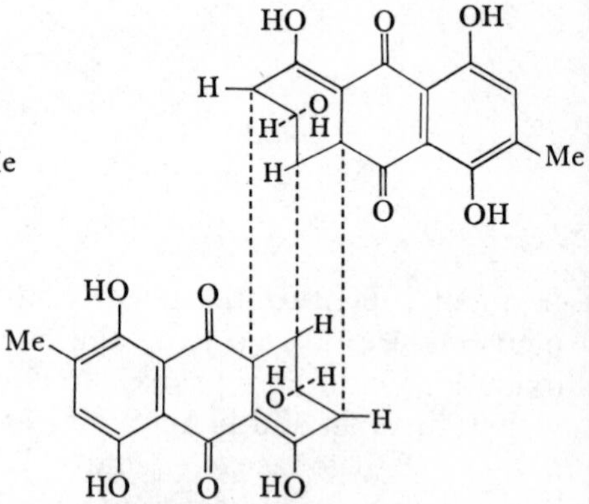

(XLV) (–)-Luteoskyrin

Aspergillus versicolor elaborates a dozen anthraquinones of a different type, one or two of which also occur in lichens. These are C_{18} or C_{20} compounds having a C_4 or C_6 side chain attached in a β position to 1,3,6,8-tetrahydroxyanthraquinone. The simplest example is averythrin (**XLVI**) but usually the side chain is oxygenated and exists in cyclic forms such as (**XLVII**) and (**XLVIII**). Dothistromin, a furanofuran pigment of type (**XLVIII**) from the fungus *Dothistroma pini*, has three α-hydroxyl groups (Gallagher and Hodges, 1972).

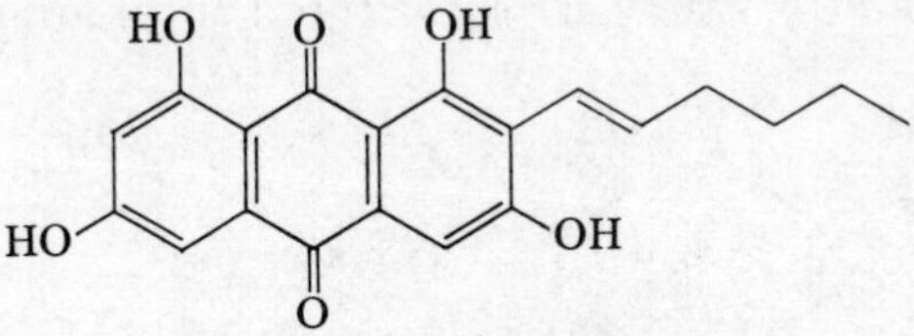

(**XLVI**) Averythrin

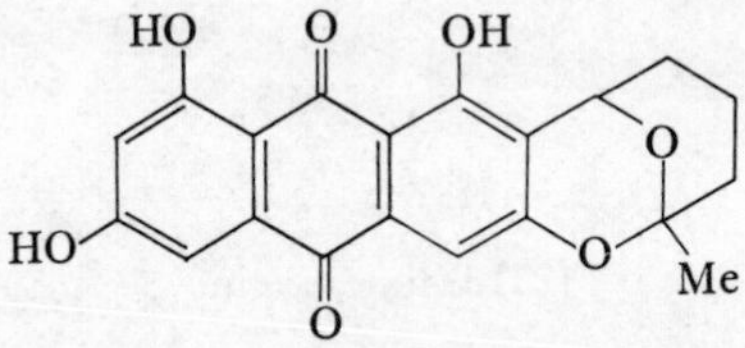

(**XLVII**) Averufin

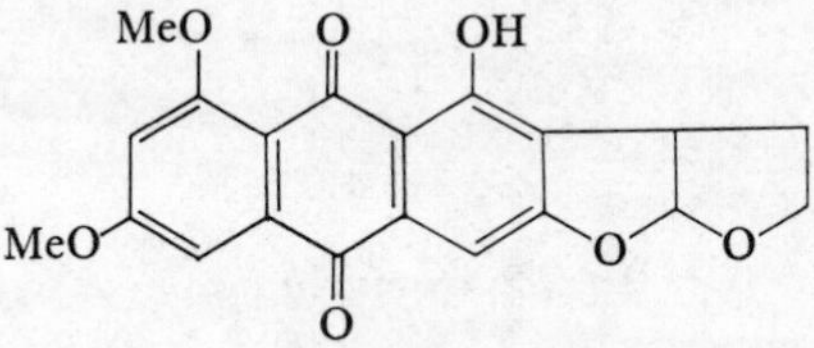

(**XLVIII**) Aversin

Finally the bacterial anthraquinones should be mentioned. So far these have been found only in cultures of *Streptomyces shiodaensis* which produces about twenty pigments known as julichromes (Tsuji and Nagashima, 1970). They all possess the same skeleton (See **XLIX**) and differ in the degree of oxidation. Exceptionally they have a carbon side chain at C-1 in addition to the normal β-methyl group, a feature otherwise confined to anthraquinones in the animal kingdom.

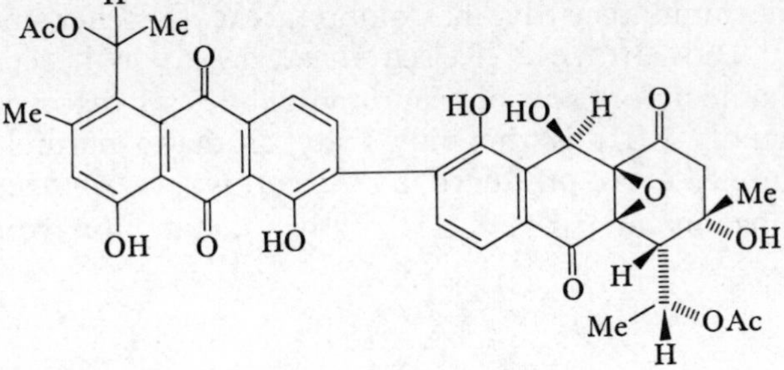

(**XLIX**) Julichrome $Q_{2.3}$

V Other polycyclic quinones

In this section the anthracyclinones, found exclusively in bacteria (streptomycetes), are the most numerous (~30) (Brockmann, 1963).

They have the same tetracyclic skeleton as the tetracyclines, and occur both free and as glycosides (anthracyclines) in combination with various sugars, including aminosugars, frequently rhodosamine. Only a few are fully aromatic (e.g. **L**) and in most cases ring A is reduced when in effect they are substituted anthraquinones (e.g. **LI**).

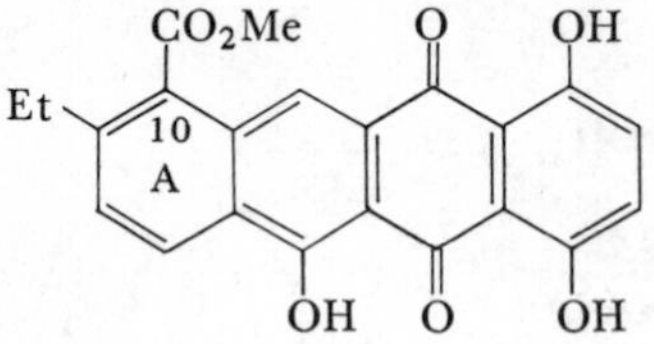

(**L**) η-Pyrromycinone

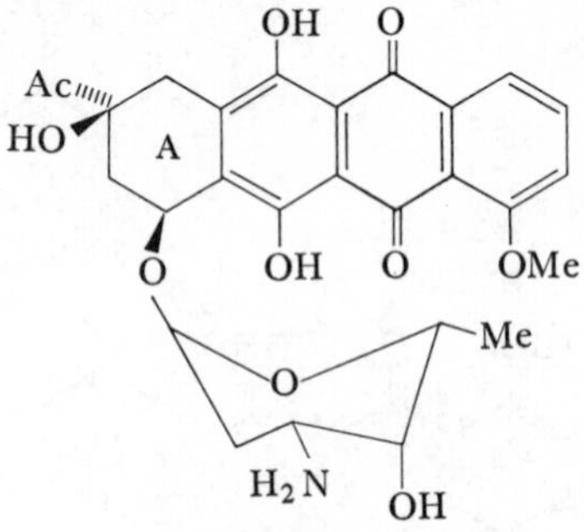

(**LI**) Daunomycin

They differ in the number of α-hydroxyl groups they possess and in the structure of ring A. In particular the ester side chain at C-10 in (**L**) may be absent or replaced by a hydroxyl group. The stereochemistry is almost constant throughout the series. The anthraquinones have antibiotic properties, and daunomycin (**LI**) shows considerable antitumour activity.

Piloquinone (**LII**), and its hydroxy derivative, are the only natural phenanthraquinones known so far apart from the mevalonate-derived examples in Section VI. It is also a product of streptomycete metabolism (Polonsky *et al.*, 1963). However, several phenanthrols have been found recently in Combretaceae (Letcher and Nhamo, 1972) and Dioscoreaceae (Reisch *et al.*, 1973) so it seems only a matter of time before new phenanthraquinones are discovered.

Tetrangulol (**LIII**) is the only fully aromatic natural 1,2-benzanthraquinone. It is produced by *Streptomyces rimosus* together with tetrangomycin (**LIV**; R = H). Other metabolites from strepto-

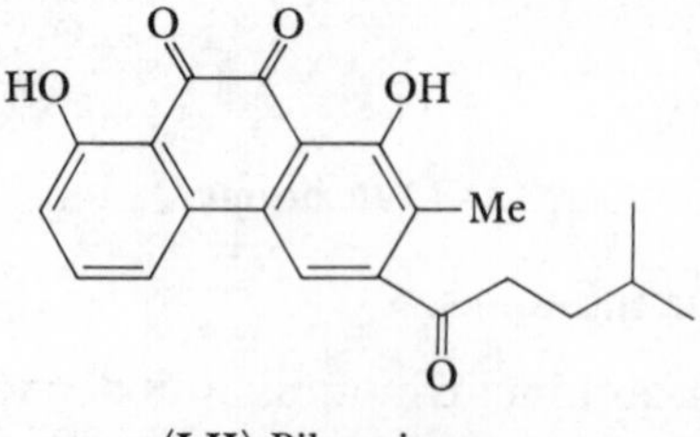

(**LII**) Piloquinone

mycetes are rabelomycin (**LIV**; R = OH) (Liu *et al.*, 1970) and aquayamycin (**LV**) (Sezaki *et al.*, 1970).

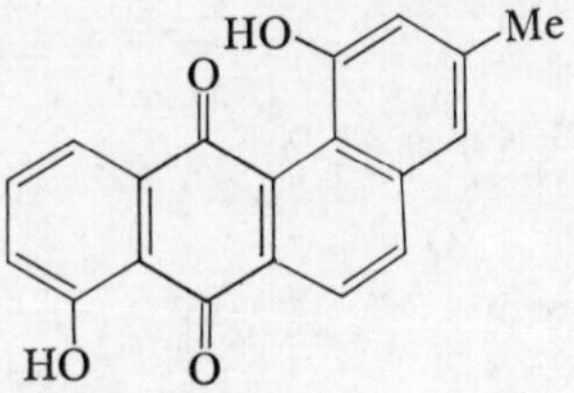

(**LIII**) Tetrangulol

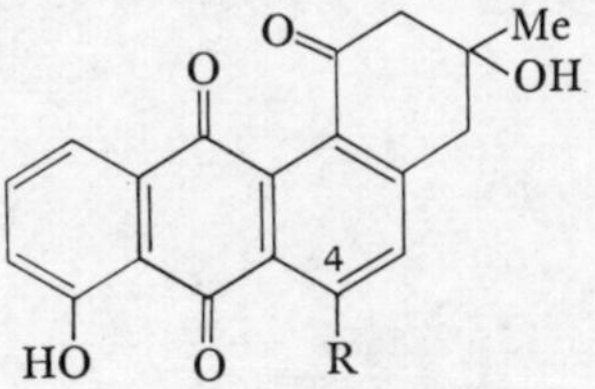

(**LIV**) Tetrangomycin

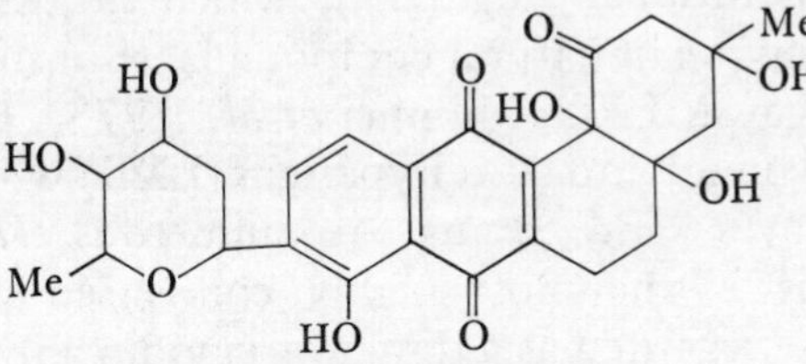

(**LV**) Aquayamycin

Streptomyces coelicolor, which elaborates the binaphthoquinone, actinorhodin (**LVI**), also produces the related pigment phenocyclinone (**LVII**) (Brockmann and Christiansen, 1970). The heterocyclic moieties of the latter are also present in granaticin (**XXXVII**) and in kalafungin (**LVIII**) both of which are streptomycete metabolites.

Another small group of polycyclic quinones is related to 4,9-dihydroxyperylene-3,10-quinone (**LIX**). They are all fungal metabolites, the parent (**LIX**), which is almost black, occurring in a large

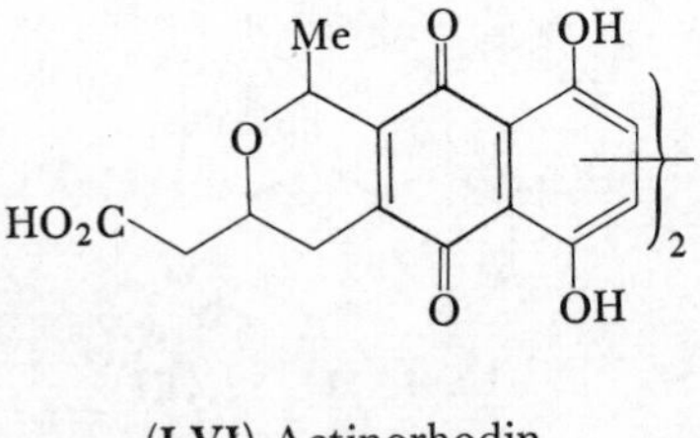

(**LVI**) Actinorhodin

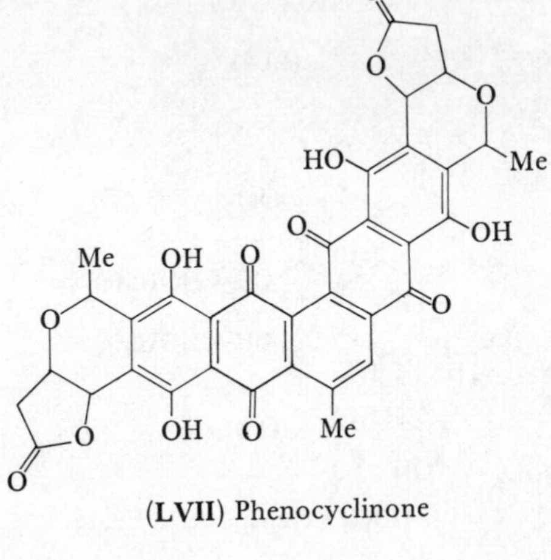

(**LVII**) Phenocyclinone

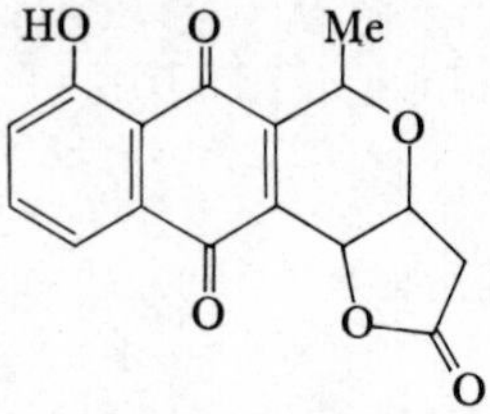

(**LVIII**) Kalafungin

ascomycete (*Daldinia concentrica*). The others are mould products and possess an additional ring (**LX**), which in the case of cercosporin (**LXI**) is unique in natural compounds. (Lousberg *et al.*, 1971; Yamazaki and Ogawa, 1972; Mumma *et al.*, 1973). Cercosporin is a photodynamic pigment and so is hypericin (**LXII**) which is present in the flowers, leaves and stems of numerous *Hypericum* spp. (Guttiferae). This is the most highly condensed natural quinone known and can be regarded as a dimer of emodin anthrone. A related pigment occurs in the flowers of buckwheat (Polygonaceae).

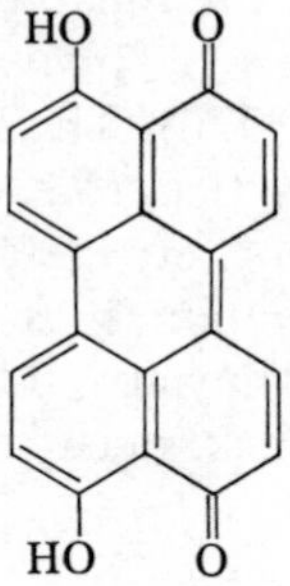

(**LIX**)

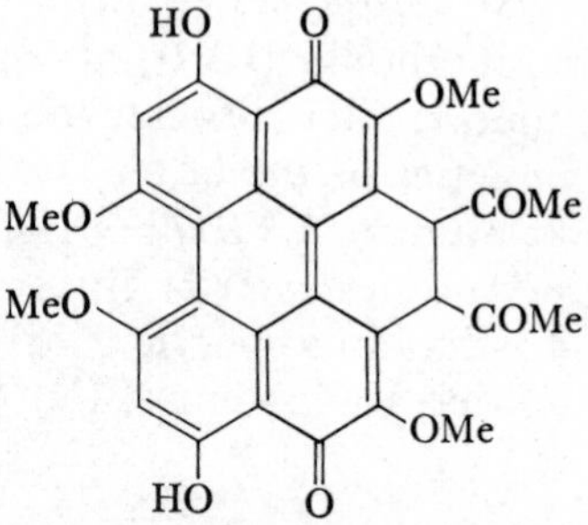

(**LX**) Elsinochrome A

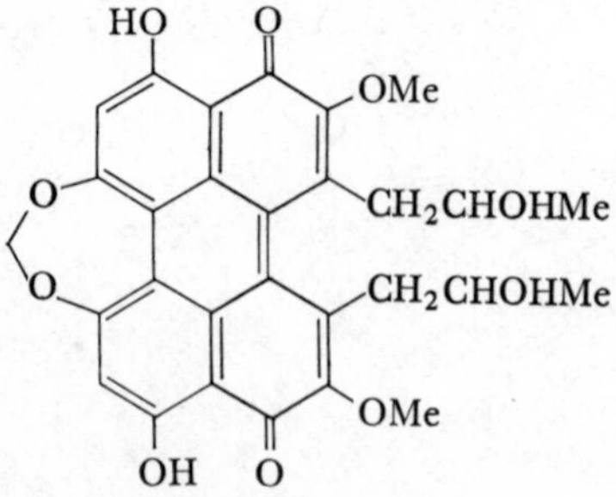

(**LXI**) Cercosporin

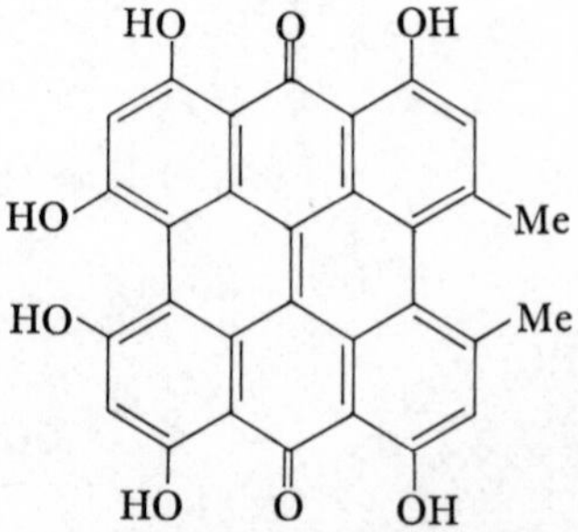

(**LXII**) Hypericin

VI Quinones belonging to other classes

As quinones are formed both *in vitro* and *in vivo* by the oxidation of phenols, any class of natural compound which includes phenolic derivatives is potentially a source of quinones. Indeed it is surprising how few quinones in this category have been discovered. There are no coumarin quinones*, no lignan quinones, and until very recently virtually no flavonoid quinones. In fact the largest number of such quinones are terpenes.

Quinones which are biosynthesized by attaching a prenyl side chain to an aromatic system have already been mentioned, but in the present context terpenoid quinones are those in which the entire carbon skeleton is derived from mevalonate. They are, in fact, highly oxidized terpenes and there appear to be as many terpenoid quinones as terpenoid phenols. With few exceptions they are found in higher plants. Among the monoterpenes thymoquinone, its hydroxy derivatives, and the trimeric compound (LXIII) occur in the coniferae. The sesquiterpenes provide perezone (LXIV) (Compositae) and helicobasidin (LXV) (from cultures of *Helicobasidium mompa* which

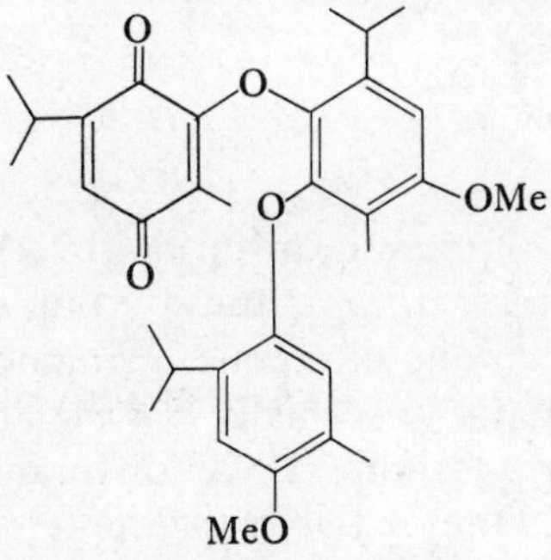

(LXIII) Libocedroxythymoquinone

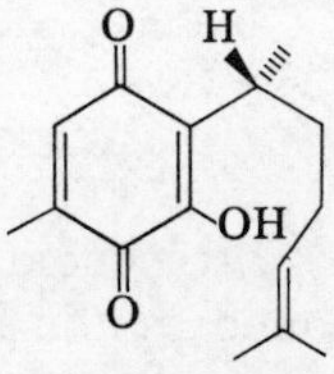

(LXIV) Perezone

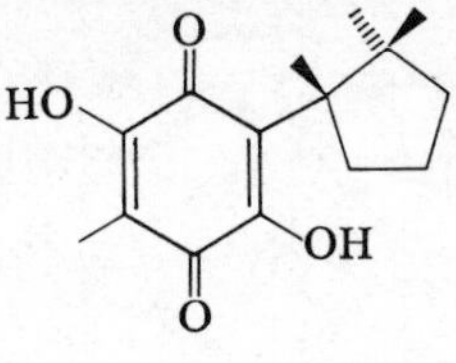

(LXV) Helicobasidin

also synthesizes a polyhydroxynaphthoquinone), and numerous naphthoquinones. These include the bombaxquinones (LXVI) from

*Added in proof: one has been discovered (Joshi *et al.*, 1974).

Bombax malabaricum (Seshadri *et al.*, 1971), and the mansonone group (e.g. **LXVII**) from the heartwood of *Mansonia altissima*; mansonones have also been found in *Ulmus* along with related cadalene derivatives (Chen *et al.*, 1972; Rowe *et al.*, 1972; Nishikawa *et al.*, 1973). Maturinone (C_{14}) (**LXVIII**) is one of a group of aromatic sesquiterpenes in *Cacalia decomposita* (Compositae).

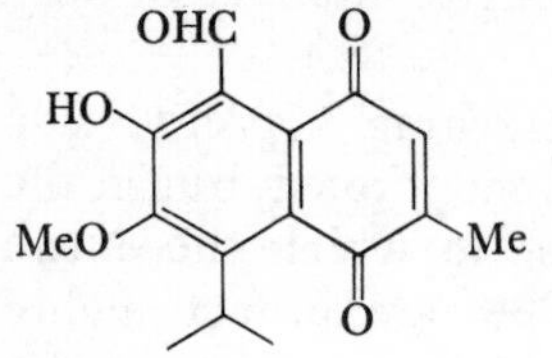

(**LXVI**) Bombaxquinone A

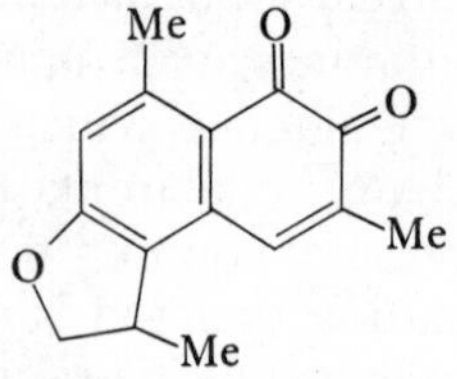

(**LXVII**) Mansonone D

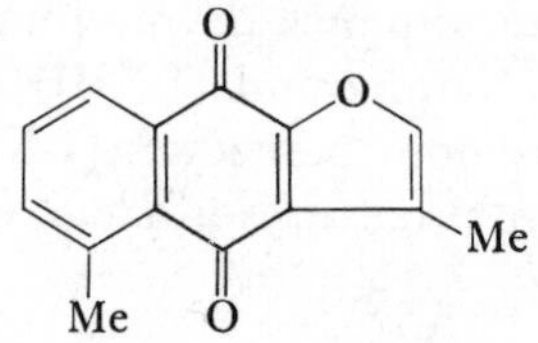

(**LXVIII**) Maturinone

Diterpenoid quinones include the *o*-quinone biflorin (**LXIX**) found in the leaves and flowers of *Capraria biflora*, and a group of some fifteen pigments related to the phenolic diterpene ferruginol. These are found principally in the Labiatae (especially *Salvia*) and are usually either hydroxyquinones like royleanone (**LXX**) or furanoquinones like cryptotanshinone (**LXXI**). One or two are completely aromarized C_{18} compounds (**LXXII**). Triterpenoid quinones are not known although there are several quinone methides in that series (Chapter 12).

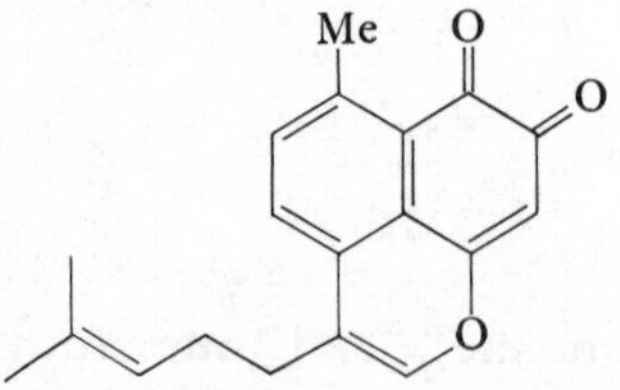

(**LXIX**) Biflorin

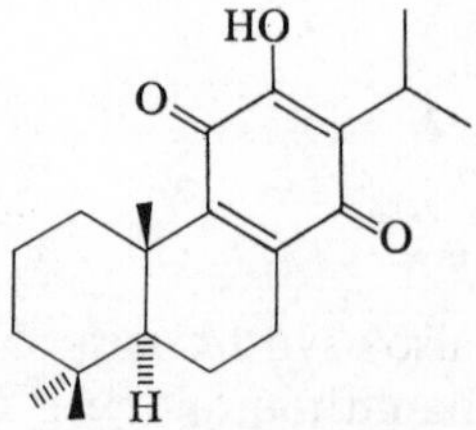

(**LXX**) Royleanone

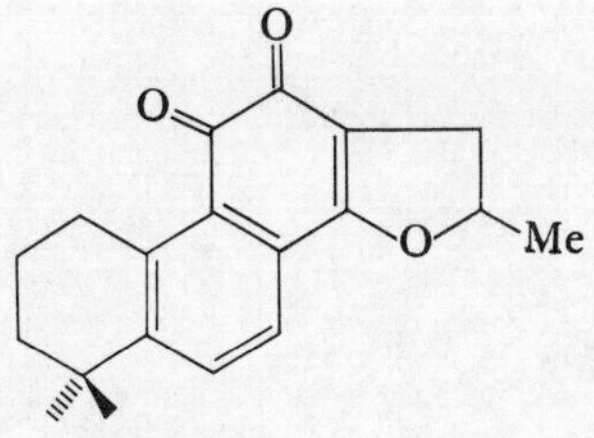

(**LXXI**) Cryptotanshinone

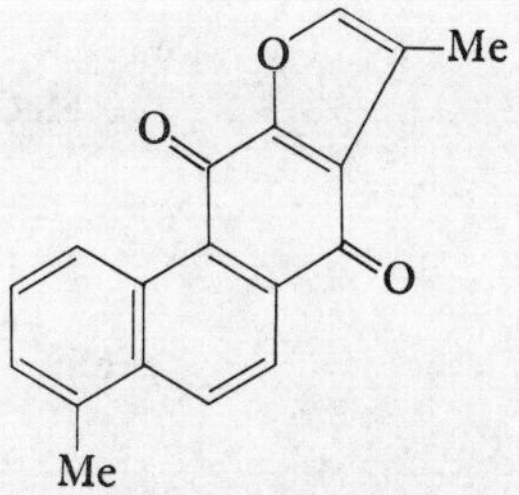

(**LXXII**) Isotanshinone I

There are surprisingly few flavonoid quinones. The mixture of flavonoids in safflower (*Carthamus tinctorius*) includes carthamone, a chalcone-quinone glucoside, and the pedicinins (**LXXIII**) are present in the leaves of *Didymocarpus pedicellata.* Three flavanone quinones (e.g. **III**) have been reported in the Cyperaceae (Allan and Wells, 1973), there is an isoflavan quinone (**LXXIV**; R = OMe) in the wood of *Machaerium mucronulatum* and also in that of *Cyclobium clausseni* together with (**LXXIV**; R = H) (Braga de Oliveira, 1971), and the isoflavone quinone bowdichione (**LXXV**) is one of a complex mixture of flavanoids in the wood of *Bowdichia nitida* (Brown *et al.*, 1974).

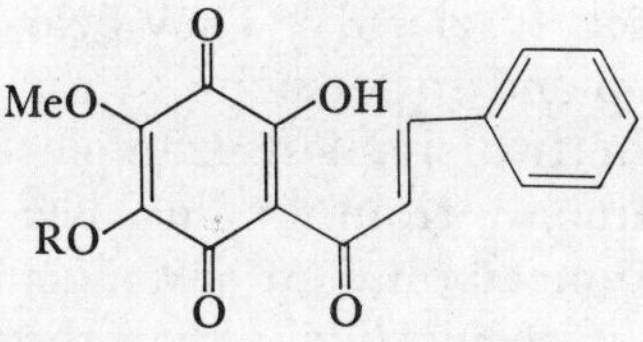

(**LXXIII**) Pedicinin (R=H)

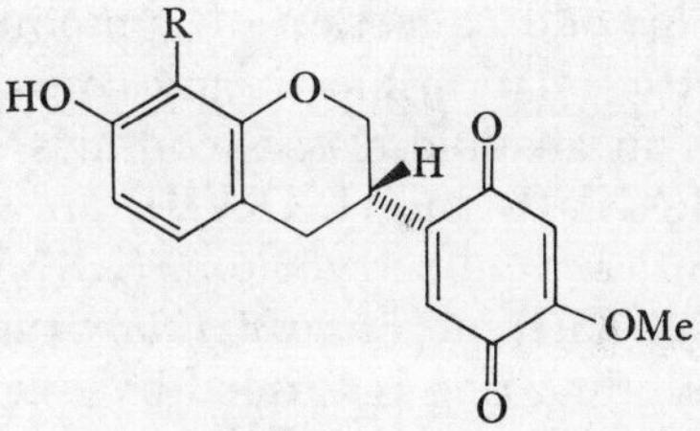

(**LXXIV**) Mucroquinone (R=OMe)
Claussequinone (R=H)

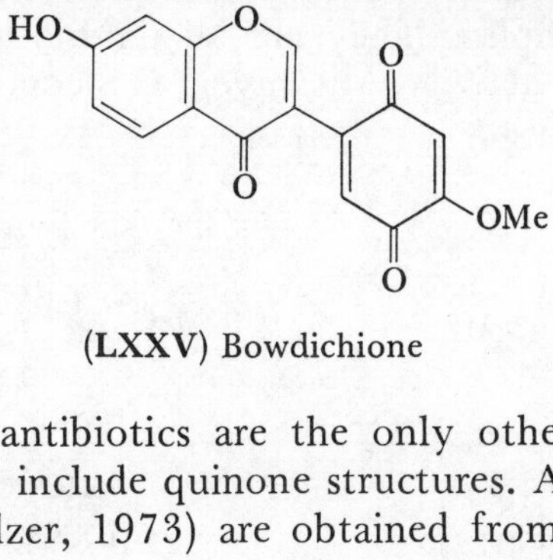

(**LXXV**) Bowdichione

The macrolide antibiotics are the only other group of natural compounds which include quinone structures. All these ansamycins (Prelog and Oppolzer, 1973) are obtained from streptomycetes. In

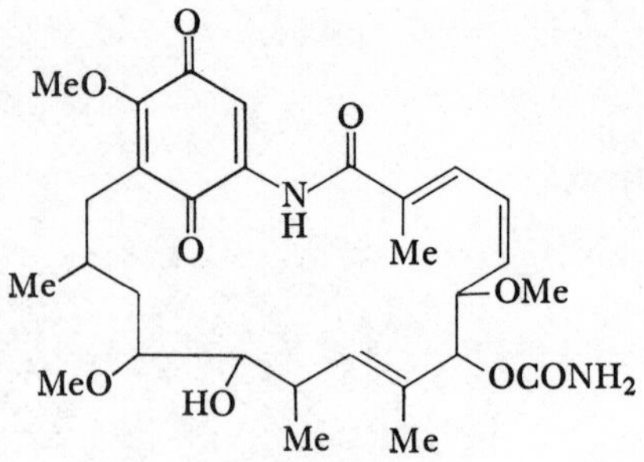

(**LXXVI**) Geldamycin

geldamycin (**LXXVI**), the simplest (Sasaki *et al.*, 1970), a benzoquinone unit is part of the large ring structure, and this is replaced by a naphthoquinone unit in the rifamycins, streptovaricins, and tolpomycins. In the axenomycins the naphthoquinone chromophore is not part of the macrolide structure (Arcamone *et al.*, 1973a,b).

VII *N*-Heterocyclic quinones

This odd collection of compounds is difficult to classify. They bear little relationship to any conventional group and so far none of the main alkaloid classes contains quinones. The two simplest examples (**LXXVII**) and (**LXXVIII**) are aza-anthraquinones from moulds. The antibiotic bostrycoidin (**LXXVII**) is a co-metabolite of javanicin (**XXXIII**) in *Fusarium bostrycoides*, and it seems likely that the pyridine ring is formed by oxidation of the ring methyl group to an aldehyde function, followed by condensation with ammonia. Streptonigrin (**LXXIX**), the mitomycins (e.g. **LXXX**), mitiromycin (**LXXXI**), and the kanamycins (e.g. **LXXXII**) (Omura *et al.*, 1971) are much more complex. They are all antibiotics and some have marked anti tumour activity. Mitomycin C is commercially available in Japan for clinical use.

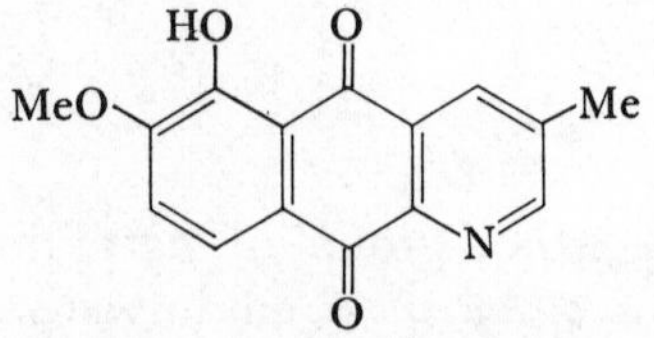

(**LXXVII**) Bostrycoidin

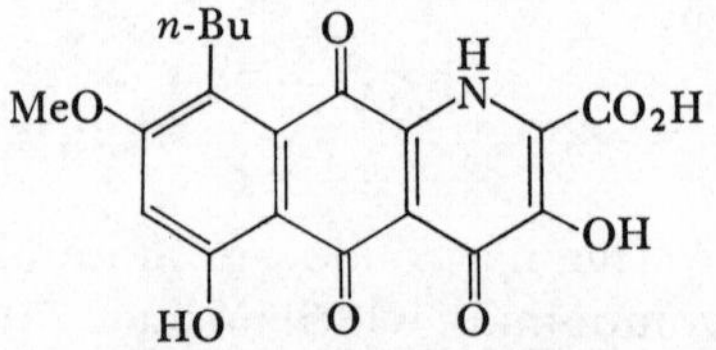

(**LXXVIII**) Phomazarin

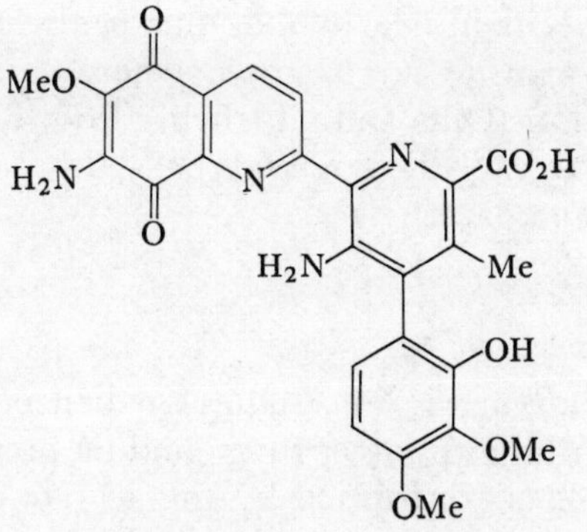

(LXXIX) Streptonigrin

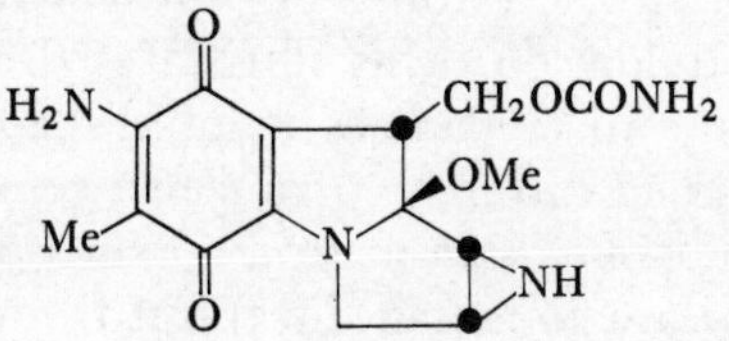

(LXXX) Mitomycin C

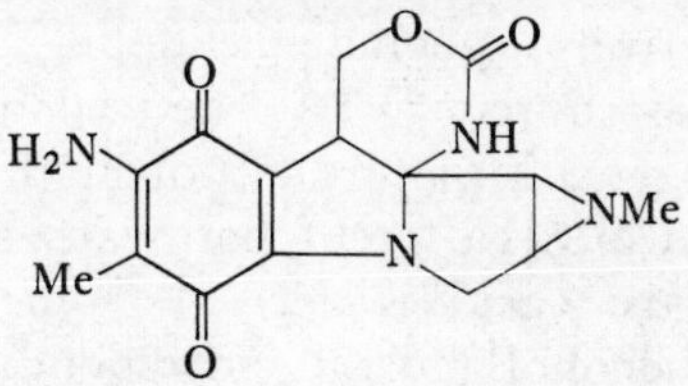

(LXXXI) Mitiromycin

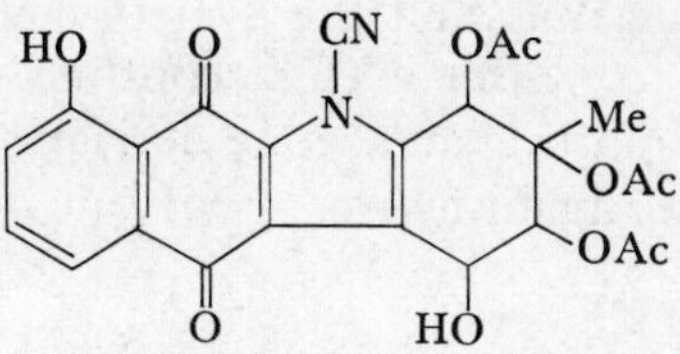

(LXXXII) Kinamycin A

VIII Biosynthesis

To a large extent the biogenesis of quinones is an extension of the biogenesis of aromatic compounds, oxidation of a phenol being one of the final steps. It is therefore no surprise to find that most quinone rings originate from acetate-malonate or from shikimate. although a number of less obvious pathways have been discovered. Broadly speaking, the formation *in vivo* of the main types of quinone is now understood, and the pathway leading to most of them can be inferred although the actual number of experiments remains fairly small. The biosynthesis of several subgroups has yet to be studied.

Where quinones occur in heartwoods this is obviously difficult, and the dalbergiones provide an example where there has been much plausible speculation (Ollis and Gottlieb, 1968; Braga de Oliveira *et al.*, 1971) supported by biogenetic-type synthetic work, but direct experimental data are lacking.

A BENZOQUINONES

For obvious reasons nearly all studies on benzoquinone biogenesis have been carried out with lower fungi, and on present evidence most fungal benzoquinones are formed by the acetate-malonate pathway (Turner, 1971). For example, spinulosin (**V**) arises from a tetraketide by way of orsellinic acid and orcinol which then undergoes further hydroxylation, oxidation to a quinone, and *O*-methylation. Alternatively, phenolic coupling of orcinol gives a dimer from which oosporein (**VI**) is derived by further oxidation. This illustrates the general picture although there are some discrepancies in published work. Different pathways may operate in different organisms; gentisic acid is derived from acetate in *Penicillium urticae* (gentisyl alcohol and the corresponding quinone are co-metabolites) but from shikimate in *Polyporus tumulosus*. Packter (1969) found that tyrosine was effectively incorporated into coprinin (**IV**), the benzene ring providing the quinone ring, and the *C*-methyl group arising from the β-carbon atom of tyrosine. This conflicts with earlier work (Pettersson, 1966) reporting an acetate derivation for this metabolite (possibly a different organism was used) but it is clear that simple,

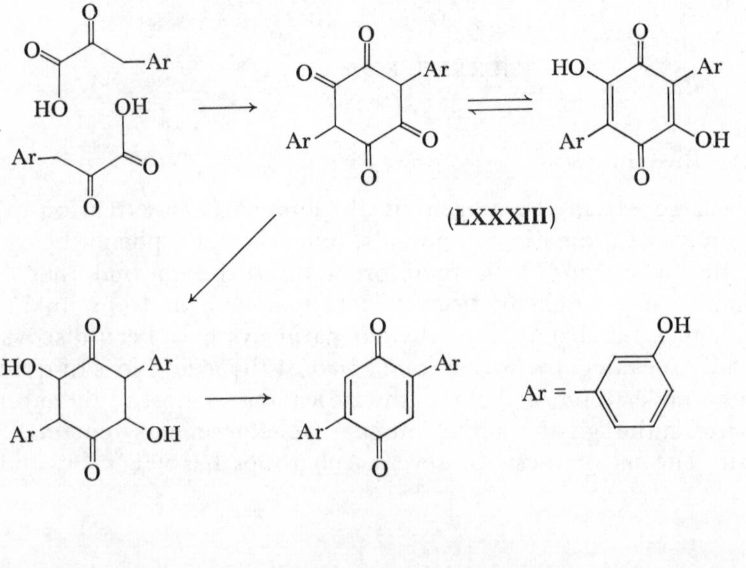

(LXXXIII)

structurally related benzoquinones, may be biosynthesized by different routes. It also appears that in many cases the quinol is the true natural product, the quinone being an artefact of the isolation procedure.

The terphenylquinones are essentially C_6-C_3 dimers derived from shikimic acid as is suggested by the hydroxylation pattern observed in the phenyl substituents. Phlebiarubrone (**XXI**) is derived from two molecules of phenylalanine, methionine providing the methylene carbon (Anchel *et al.*, 1970). An earlier study of volucrisporin (**XX**) (Chandra *et al.*, 1966) showed that shikimic acid, phenylalanine, phenyllactic acid and *m*-hydroxytyrosine were incorporated, the labelling pattern suggesting the biosynthetic route via phenylpyruvic acid shown on page 548. The *m*-hydroxylation step is not yet understood. It will be observed that the intermediate (**LXXXIII**) is at the oxidation level of the other terphenylquinones (e.g. **XVIII**).

B NAPHTHOQUINONES

Among fungal naphthoquinones it has been established that javanicin (**XXXIII**) and mollisin (**LXXXIV**) are acetate derived. Two polyketide chains are involved in the formation of the latter (Bentley and Gatenbeck, 1965a; Seto *et al.*, 1973), while in the former the ring methyl arises by reduction of a terminal carboxyl group which is unusual (Bentley and Gatenbeck, 1965b). Inspection of formulae suggests that most fungal naphthoquinones are also derived from

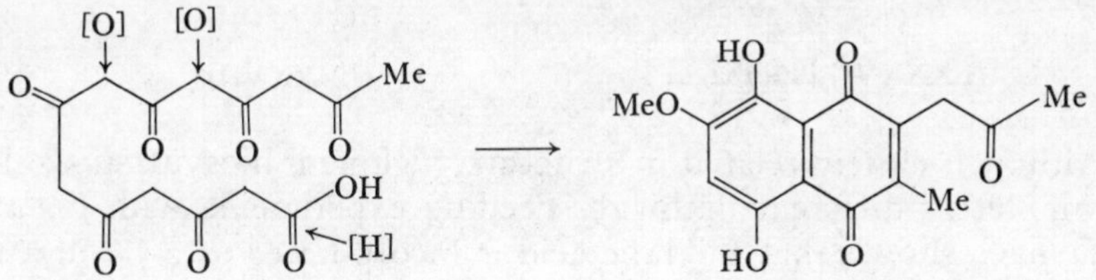

[O] [O] Me OH [H] → HO O MeO Me O Me HO O

(**XXXIII**) Javanicin

Cl_2→Me O CO_2H HO_2C O O Me → Cl_2HC O O Me Me HO O

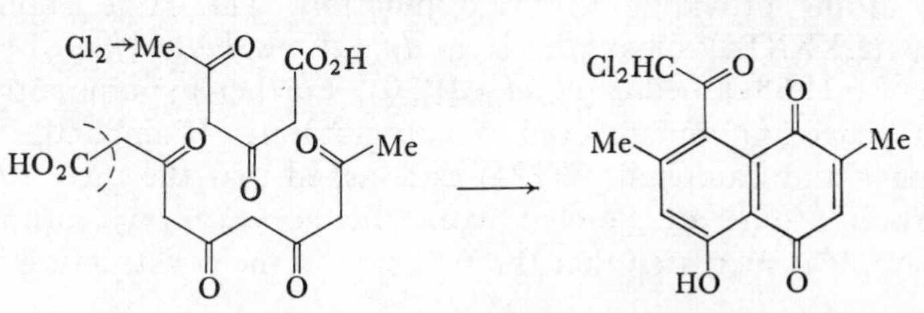

(**LXXXIV**) Mollisin

polyketides but the simplest compound (**LXXXV**) (from *Marasmius graminum*) is probably an exception.

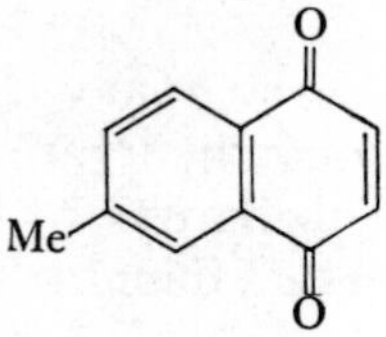

(**LXXXV**)

Most naphthoquinones are found in higher plants and happily some of these have been studied in recent years. Several pathways have been discovered, one being the polyacetate–malonate route, as in fungi, which operates in *Plumbago* and *Drosera* spp. to produce plumbagin (**LXXXVI**) (Durand and Zenk, 1971). 7-Methyljuglone (**XXIX**; R = H) is formed in the same way in *Drosera* (Durand and Zenk, 1971)* and if this is true also in *Diospyros* then all the binaphthoquinones in this family can be accounted for by phenolic coupling of an acetate-derived naphthalene derivative such as (**LXXXVII**).

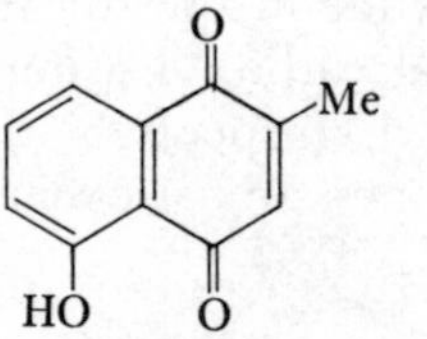

(**LXXXVI**) Plumbagin

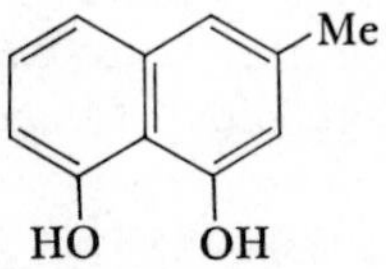

(**LXXXVII**)

Although closely related in structure, juglone is biosynthesized by a completely different pathway. Feeding experiments with *Juglans regia* have shown that shikimic acid is incorporated as a C_7 unit to form the benzenoid ring and one of the carbonyl groups, the vinyl carbon atoms providing the ring junction. The same applies to lawsone (**LXXXIX**) biosynthesis in *Impatiens balsamina* (Leistner and Zenk, 1968a; Leduc *et al.*, 1970). Further experiments with both labelled glutamate and 2-ketoglutarate (Campbell, 1969; Grotzinger and Campbell, 1972a) established that the three remaining carbon atoms are derived from the non-carboxyl carbons of glutamate. It is suggested that the first step in the construction of the

* Experimental data not yet established.

quinone ring is the addition of succinoyl semialdehyde as its thiamine pyrophosphate complex to shikimic acid (see Scheme below) (Campbell, 1969; Grotzinger and Campbell, 1972b) leading to the formation of *o*-succinoylbenzoic acid (**LXXXVIII**) which Dansette and Azerad (1970) have shown to be efficiently incorporated into both juglone and lawsone. The subsequent steps by which (**LXXXVIII**) is converted into lawsone and juglone have still to be elucidated. When *I. balsamina* is fed with [2-^{14}C] acetate the label is found predominantly at C-2 in lawsone, which means that no symmetrical intermediate is involved between (**LXXXVIII**) and (**LXXXIX**) (Grotzinger and Campbell, 1972b) but there is evidence that juglone biosynthesis does proceed by way of a symmetrical intermediate (Scharf *et al.*, 1971). For example, when *J. regia* was fed with (6*R*)-[7-^{14}C,6-^{3}H]-shikimate only 1·5% of the tritium was retained in the resulting juglone whereas 47·6% was incorporated from the *pro*-6*S* position. This suggests an equal distribution of tritium from the *pro*-6*S* position of shikimate between H-5 and H-8 of an unhydroxylated juglone precursor which might be 1,4-naphtho-quinone or -quinol. Subsequent hydroxylation would

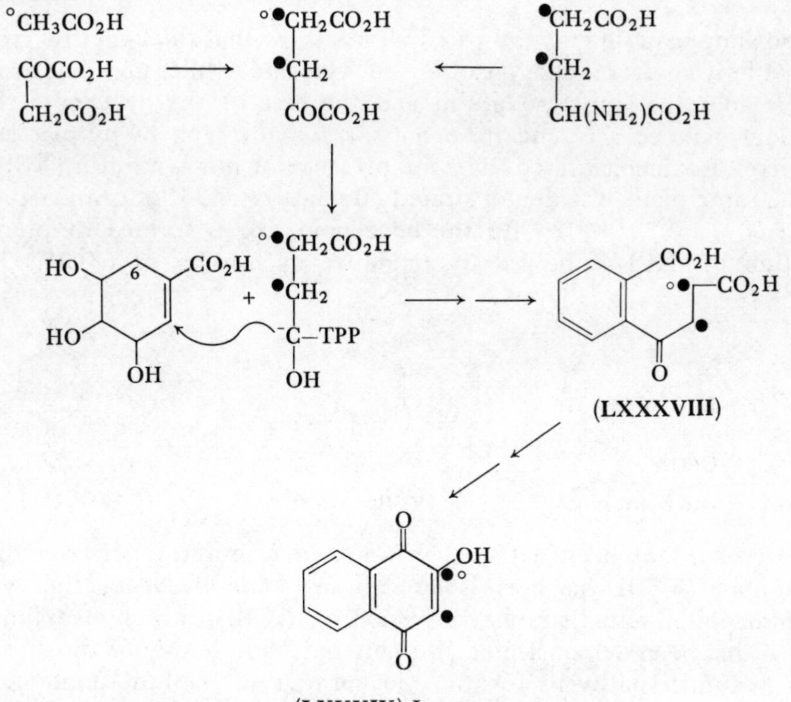

then eliminate half the tritium. These observations also support the idea that chorismic acid is involved in naphthoquinone biosynthesis as the formation of chorismic acid also proceeds with loss of the *pro*-6*R* and retention of the *pro*-6*S* hydrogen of shikimate. This was first proposed by Cox and Gibson (1966), and Dansette and Azerad (1970) have suggested a pathway for the formation of juglone (and other quinones) from chorismate, but so far incorporation of this precursor has not been achieved.

Yet another route to naphthoquinones operates in *Chimaphila umbellata* (Pyrolaceae). Shikimic acid provides the quinone ring of chimaphilin (**XC**) by the homogentisic acid pathway, and the

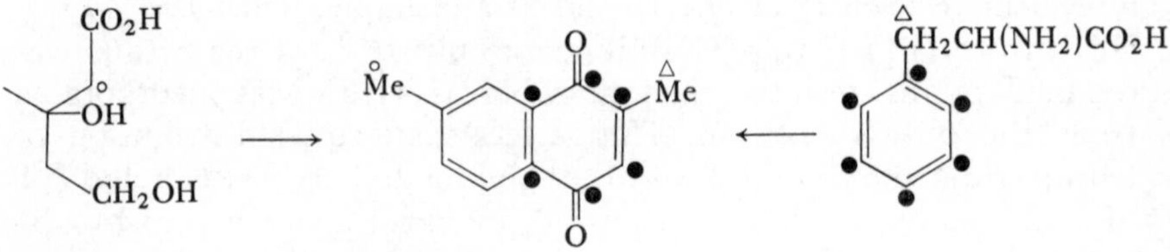

(**XC**) Chimaphilin

labelling experiments indicated above, show that the 2-methyl group is derived from the β-carbon of tyrosine while mevalonic acid provides the 7-methyl group and the rest of the benzenoid ring. Homogentisic acid and its precursor, *p*-hydroxyphenylpyruvic acid, were also incorporated, and the presence of homo-arbutin (**XCI**) in the same plant was demonstrated (Bolkart *et al.*, 1968; Bolkart and Zenk, 1968, 1969). Thus the benzenoid ring is formed by prenylation of (**XCI**) followed by oxidative cyclization of (**XCII**). The

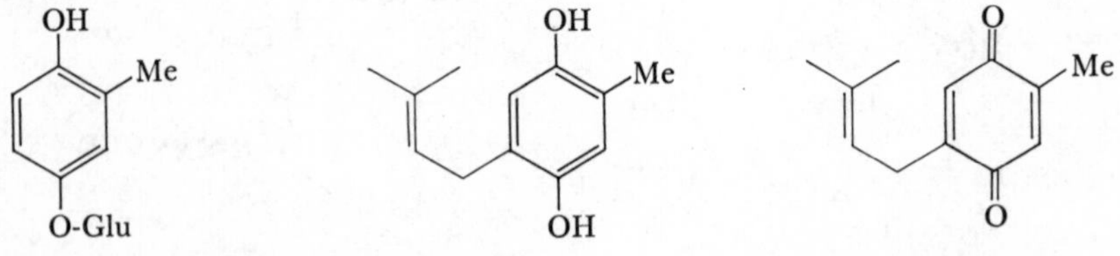

(**XCI**) Homo-arbutin (**XCII**) (**XCIII**)

latter has not been detected in *Chimaphila* but the corresponding quinone (**XCIII**) has been isolated from *Pyrola media* together with chimaphilin, and a simple conversion of (**XCIII**) into chimaphilin *in vitro* has been demonstrated (Burnett and Thomson, 1968a).

A fourth pathway for the biosynthesis of naphthoquinones in higher plants has been observed by Schmid and Zenk (1971) in

Plagiobothrys arizonicus. Like many other Boraginaceae this plant elaborates alkannin (**XCV**), and feeding experiments established that the aromatic ring (A) is derived from *p*-hydroxybenzoic acid while two molecules of mevalonic acid provided the remaining ten carbon atoms. Phenylalanine and cinnamic acid were also incorporated but, surprisingly, shikimic acid and tyrosine were not. The biosynthesis of alkannin thus appears to be similar to that of chimaphilin, and involves essentially prenylation of *p*-hydroxybenzoic acid (or a related phenol) with geranylpyrophosphate to give (**XCIV**) followed by oxidative cyclization and introduction of oxygen.

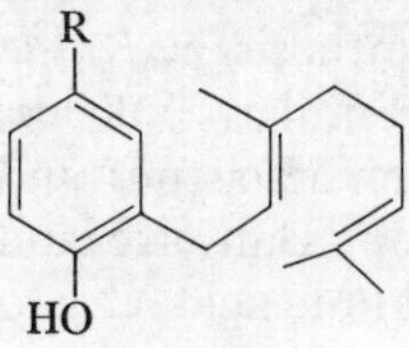

(**XCIV**) 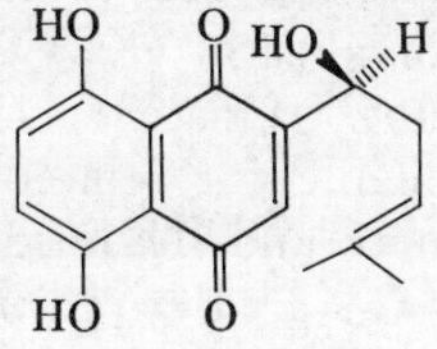 (**XCV**) Alkannin 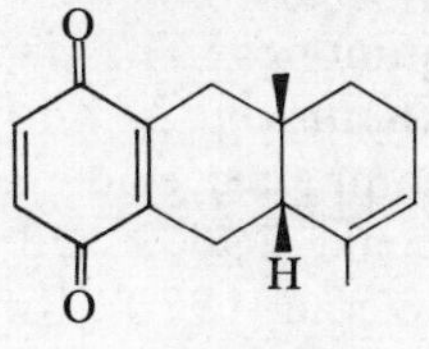 (**XCVI**) Cordiachrome A

The cordiachromes (e.g. **XCVI**) also occur in Boraginaceae and probably have a similar mode of biogenesis in which prenylation of a phenolic precursor is followed by oxidative cyclization onto a terminal carbon atom of a geranyl side chain, and a subsequent transannular reaction (Moir and Thomson, 1973b).

C ANTHRAQUINONES

In the original structure analyses by Birch and Donovan (1953, 1955) it was noted that many natural anthraquinones of the emodin (**XLI**) type had structures which were in accord with the "acetate hypothesis" but a substantial number (e.g. alizarin) did not fit the theory. This immediately suggested that at least two biosynthetic routes were utilized by plants in the formation of anthraquinones. Subsequently, this was established by experiments. As predicted, it was found by Birch and others that many fungal anthraquinones, which are of the emodin type, are, in fact, acetate derived and this is probably true for all anthraquinones of fungal and lichen origin (Thomson, 1971). The early experiments were naturally carried out with moulds but more recently Steglich *et al.* (1972) have used higher fungi in field experiments. *Dermocybe* toadstools elaborate numerous anthraquinones of the emodin (**XLI**) type and it was shown by feeding labelled emodin and endocrocin (emodin-2-

carboxylic acid) to the young sporophores of *Dermocybe* spp. that these quinones were precursors of the more highly hydroxylated derivatives which are normally present. Endocrocin was incorporated into the anthraquinone acids dermolutein and dermorubin but not into emodin and the other non-carboxylated pigments. As mentioned earlier the same type of pigment is also found in higher plants, and it has been established that chrysophanol and emodin are derived entirely from acetate in *Rhamnus frangula* and *Rumex alpinus* (Leistner, 1971).

The second pathway leading to the formation of anthraquinones in higher plants is related to the routes in operation for naphthoquinones. The pathway for alizarin (**XCVII**) biosynthesis in *Rubia tinctorum* is outlined below (Leistner, 1973a). Similar, but less complete, evidence is available for related quinones (Leistner and Zenk, 1967, 1968b; Burnett and Thomson, 1968b; Dansette and Azerad, 1970). Ring A and C-9 are derived from shikimic acid, C-1 to

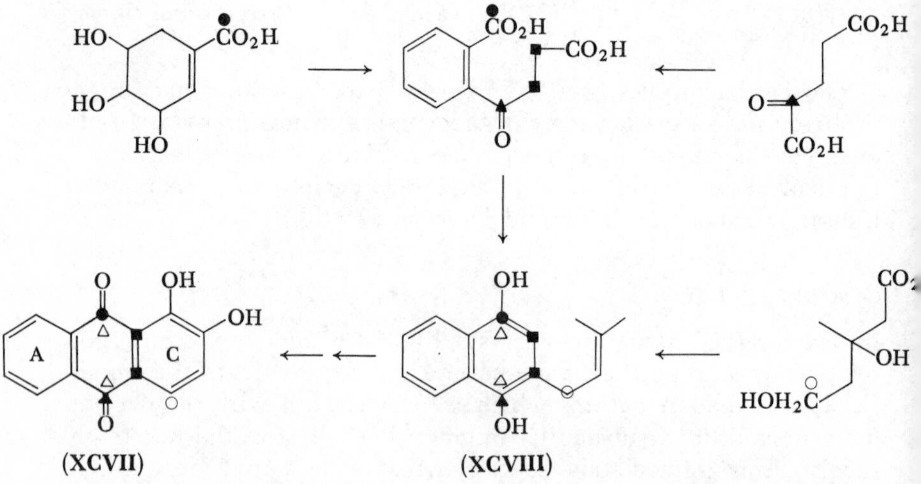

(XCVII) (XCVIII)

● – Shikimic acid; ▲ – α-Ketoglutaric acid; ■ – Succinoylbenzoic acid;
○ – Mevalonic acid; △ – Deoxylapachol quinol

C-4 come from mevalonic acid, and the three remaining carbon atoms are provided by glutamic (≡α-ketoglutaric) acid. *o*-Succinoylbenzoic acid was specifically incorporated so that the formation of rings A and B parallels the biosynthesis of lawsone. The precise structure of the first naphthalene derivative is not yet known. 1,4-Dihydroxynaphthalene-2-carboxylic acid has been suggested but has not been identified in the plant. Nevertheless it is clear that ring

C must be formed by prenylation of a naphthol precursor followed by oxidative cyclization.

Deoxylapachol (**XXIV**; R = H) can be cyclized *in vitro* to 2-methylanthraquinone, and these compounds and related anthraquinones co-occur in Verbenaceae and Bignoniaceae heartwoods (Thomson, 1971). The corresponding quinol (**XCVIII**) was incorporated into alizarin by *Rubia tinctorum* although it may have been non-specific. In alizarin formation one methyl group of mevalonic acid is incorporated into ring C and the other is removed by oxidation. In pseudopurpurin the latter appears as a carboxyl group, and in related pigments it survives as a methyl, hydroxymethyl, or aldehyde substituent.

A few anthraquinones in Rubiaceae and Scrophulariaceae contain substituents in both benzenoid rings, although the orientation is different from that of the emodin group. An example is morindone (**XCIX**) the principal anthraquinone pigment in *Morinda citrifolia*. The other quinones in this plant, including alizarin, are substituted in ring C only except for a minor constituent soranjidiol (**C**). Leistner

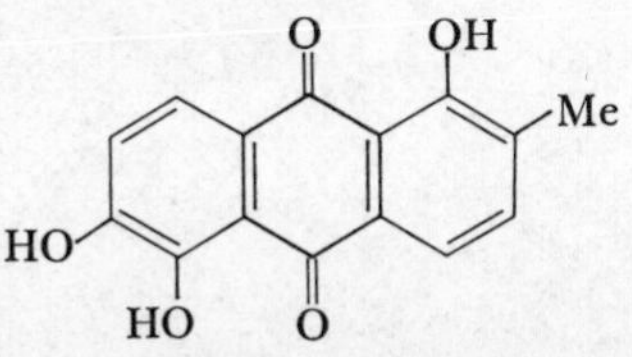

(**XCIX**) Morindone

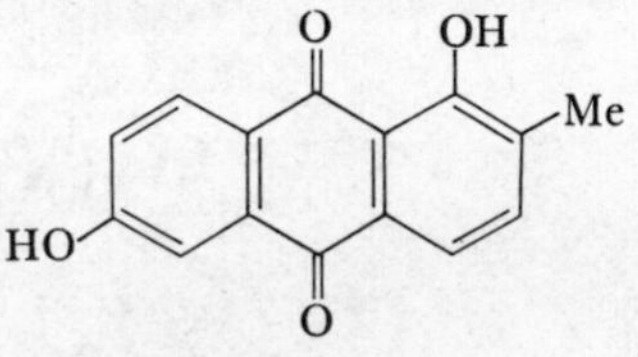

(**C**) Soranjidiol

(1973b) has shown that morindone is also derived from shikimic acid by way of *o*-succinoylbenzoic acid although curiously there was no incorporation of mevalonic acid. The specific incorporation of *o*-succinoylbenzoic acid shows that the ring A hydroxyl groups in morindone are not derived from the hydroxyl groups of shikimic acid.

In view of the similarity of the *o*-succinoylbenzoic acid pathway to naphthoquinones and anthraquinones it might be expected that both types of quinone would be found together. This does occur in Verbenaceae and Bignoniaceae, and a new example was discovered recently in Gesneriaceae (Stöckigt *et al*., 1973). Leaves and roots of *Streptocarpus dunnii* contain both dunnione (**CI**) and the anthraquinone (**CII**), and the latter was found to incorporate shikimic acid, *o*-succinoylbenzoic acid, and mevalonic acid. Both compounds are presumably derived from a common naphthalenic precursor by an appropriate prenylation and cyclization mechanism.

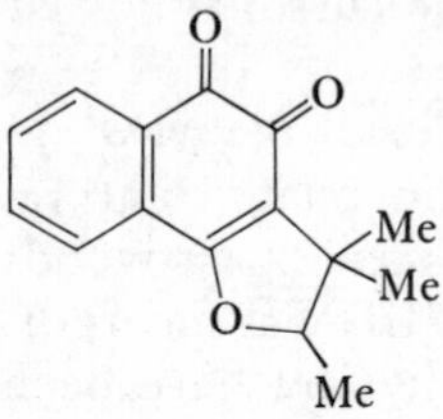

(**CI**) Dunninone

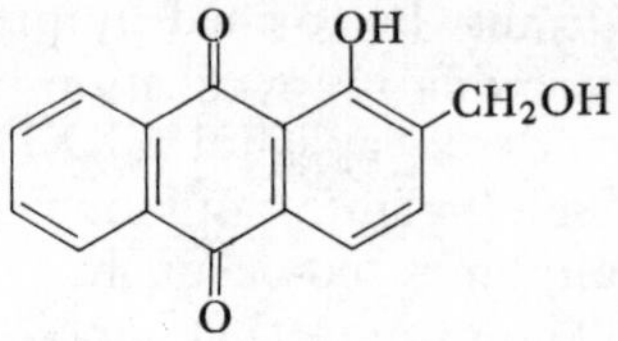

(**CII**)

The most intriguing question in anthraquinone biosynthesis which now remains is the origin of anthraquinone itself in *Lolium perenne.*

D OTHER QUINONES

The biogenesis of three of the larger polycyclic quinones has been studied, and all were found to be of polyketide origin (Thomson, 1971). These are the bacterial pigments piloquinone (**LII**) and ϵ-pyrromycinone (**CIII**), and the fungal metabolite elsinochrome A

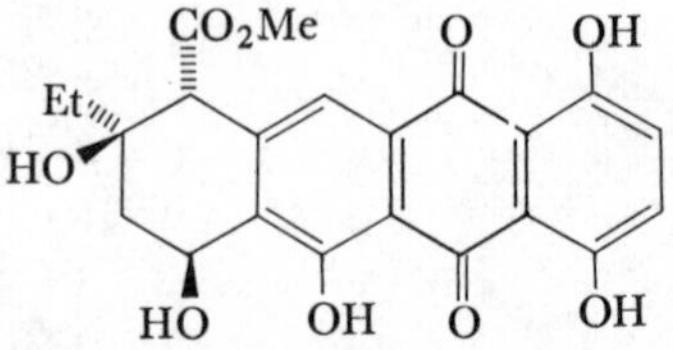

(**CIII**) ϵ-Pyrromycinone

(**LX**). The starter unit in piloquinone is isobutyrate (fed as such or as valine), and in ϵ-pyrromycinone it is propionate. The labelling pattern in elsinochrome A shows that there is a dimerization step in the biosynthetic pathway resulting in the coupling of two monomeric naphthalene units. An unusual feature is the formation of an additional ring by the linkage of two benzylic groups.

REFERENCES

Allan, R. D. (1972). Ph.D. Thesis, James Cook University, Townsville, Queensland, Australia.

Allan, R. D. and Wells, R. J. (1973). *Tetrahedron Lett.* 7-8.

Allan, R. D., Correll, R. L. and Wells, R. J. (1969). *Tetrahedron Lett.* 4669-4672.

Allan, R. D., Dunlop, R. W., Kendall, M. J. and Wells, R. J. (1973). *Tetrahedron Lett.* 3-5.

Allebone, J. E., Hamilton, R. J., Bryce, T. A. and Kelly, W. (1971). *Experientia* 27, 13-14.

Anchel, M., Bose, A. K., Khanchandani, S. K. and Funke, P. T. (1970). *Phytochemistry* 9, 2335-2338.

Arcamone, F., Barbieri, W., Francheschi, G., Penco, S. and Vigevani, A. (1973a). *J. Am. chem. Soc.* 95, 2008-2009.

Arcamone, F., Francheschi, G., Gioia, B., Penco, S. and Vigevani, A. (1973b). *J. Am. chem. Soc.* 95, 2009-2011.

Beaumont, P. C. and Edwards, R. L. (1969). *J. chem. Soc. C* 2398-2403.

Beaumont, P. C. and Edwards, R. L. (1971). *J. chem. Soc. C* 2582-2585.

Bentley, R. and Gatenbeck, S. (1965a). *Biochemistry, N.Y.* 4, 1150-1156.

Bentley, R. and Gatenbeck, S. (1965b). *Biochem. J.* 94, 478-481.

Birch, A. J. and Donovan, F. W. (1953). *Aust. J. Chem.* 6, 360-368.

Birch, A. J. and Donovon, F. W. (1955). *Aust. J. Chem.* 8, 529-533.

Birch, A. J., Ryan, A. J. and Smith, H. (1958). *J. chem. Soc.* 4773-4774.

Bolkart, K. H. and Zenk, M. H. (1968). *Naturwissenschaften* 55, 444-445.

Bolkart, K. H. and Zenk, M. H. (1969). *Z. PflPhysiol.* 61, 356-359.

Bolkart, K. H., Knoblock, M. and Zenk, M. H. (1968). *Naturwissenschaften* 55, 445.

Braga de Oliveira, A., Gottlieb, O. R. and Ollis, W. D. (1971). *Tetrahedron* 10, 1863-1876.

Braga de Oliveira, A., Gottlieb, O. R., Machado Goncalves, T. M. and Ollis, W. D. (1971). *Anais Acad. bras. Cienc.* 43, 129-130.

Brew, E. J. C. and Thomson, R. H. (1971). *J. chem. Soc. C* 2007-2010.

Briggs, L. H. and Dacre, J. C. (1948). *J. chem. Soc.* 564-568.

Brockmann, H. (1963). *Prog. Chem. Org. Nat. Prods.* 21, 121-182.

Brockmann, H. and Christiansen, P. (1970). *Chem. Ber.* 103, 708-717.

Brown, P. M., Thomson, R. H., Hausen, B. M. and Simatupang, M. H. (1974). *Annalen* 1295-1300.

Burnett, A. R. and Thomson, R. H. (1968a). *J. chem. Soc.* 857-860.

Burnett, A. R. and Thomson, R. H. (1968b). *J. chem. Soc.* 2437-2441.

Butler, J. H. A., Downing, D. T. and Swaby, R. J. (1964). *Aust. J. Chem.* 17, 817-819.

Campbell, I. M. (1969). *Tetrahedron Lett.* 4777-4780.

Chandra, P., Read, G. and Vining, L. C. (1966). *Can J. Biochem.* 44, 403-413.

Charles, J. H. V., Raistrick, H., Robinson, R. and Todd, A. R. (1933). *Biochem. J.* 27, 499-511.

Chen, F., Liu, Y.-M. and Chen, A.-H. (1972). *Abstr. 8th IUPAC Int. Symp. Chem. Nat. Prods.* 203.

Cooke, R. G. and Down, J. G. (1971). *Aust. J. Chem.* 24, 1257-1265.

Cox, G. B. and Gibson, F. (1966). *Biochem. J.* 100, 1-6.

Dansette, P. and Azerad, R. (1970). *Biochem. biophys. Res. Commun.* 40, 1090-1095.

Donnelly, D. M. X., Kavanagh, P. J., Kunesch, G. and Polonsky, J. (1973). *J.C.S. Perkin I* 965-967.

Dunlop, R. W. (1971). B.Sc. Thesis, James Cook University, Townsville, Queensland, Australia.

Durand, R. and Zenk, M. H. (1971). *Tetrahedron Lett.* 3009-3012.

Fenical, W., Sims J. J., Squatrito, D., Wing, R. M. and Radlock, P. (1973). *J. org. Chem.* 38, 2383-2386.

Ferreira, M. A., Costa, M. A. C., Correia Alves, A. and Lopes, M. H. (1973). *Phytochemistry*, **12**, 433–435.
Gallagher, R. T. and Hodges, R. (1972). *Aust. J. Chem.* **25**, 2399–2407.
Grotzinger, E. and Campbell, I. M. (1972a). *Tetrahedron Lett.* 4685–4686.
Grotzinger, E. and Cämpbell, I. M. (1972b). *Phytochemistry*, **11**, 675–679.
Imre, S., Oztunç, A. and Büyüktimkin, N. (1974). *Phytochemistry*, **13**, 681–682.
Inouye, H., Okuda, T. and Hayashi, T. (1971). *Tetrahedron Lett.* 3615–3618.
Joshi, B. S., Kamat, V. N. and Gawad, D. H. (1974). *J.C.S. Perkin I* 1561–1564.
Jurd, L., Stevens, K. and Manners, G. (1972). *Phytochemistry*, **11**, 3287–3292.
Leduc, M. M., Dansette, P. M. and Azerad, R. G. (1970). *Eur. J. Biochem.* **15**, 428–435.
Leistner, E. (1971). *Phytochemistry*, **10**, 3015–3020.
Leistner, E. (1973a). *Phytochemistry*, **12**, 337–345.
Leistner, E. (1973b). *Phytochemistry*, **12**, 1669–1674.
Leistner, E. and Zenk, M. H. (1967). *Z. Naturf.* **22b**, 865–868.
Leistner, E. and Zenk, M. H. (1968a). *Z. Naturf.* **23b**, 259–268.
Leistner, E. and Zenk, M. H. (1968b). *Tetrahedron Lett.* 1395–1396.
Letcher, R. M. and Nhamo, L. R. M. (1972). *J.C.S. Perkin I* 2941–2946 and previous papers.
Liu, W.-C., Parker, W. L., Slusarchyk, D. S., Greenwood, G. L., Graham, S. F. and Meyer, E. (1970). *J. Antibiot., Tokyo* 23, 437–441.
Lousberg, R. J. J. Ch., Weiss, U., Salemink, C. A., Arnone, A., Merlini, L. and Nasini, G. (1971). *Chem. Commun.* 1463–1464.
Marchelli, R. and Vining, L. C. (1973). *Chem. Commun.* 555–556.
Minami, K., Aswawa, K. and Sawada, M. (1968). *Tetrahedron Lett.* 5067–5070.
Mock, J., Murphy, S. T., Ritchie, E. and Taylor, W. C. (1973). *Aust. J. Chem.* **26**, 1121–1130.
Moir, M. and Thomson, R. H. (1973a). *Phytochemistry*, **12**, 1351–1353.
Moir, M. and Thomson, R. H. (1973b). *J.C.S. Perkin I* 1352–1357, 1556–1561.
Mumma, R. O., Lukezic, F. L. and Kelly, M. G. (1973). *Phytochemistry*, **12**, 917–922.
Musgrave, O. C. and Skoyles, D. (1974). *J.C.S. Perkin I* 1128–1131.
Nishikawa, K., Yasuda, S. and Hanzawa, M. (1973). *Mokuzai Gakkaishi* **18**, 623.
Ollis, W. D. and Gottlieb, O. R. (1968). *Chem. Commun.* 1396–1397.
Omura, S., Nakagawa, A., Yamada, H., Hata, T., Furusaki, A. and Watanabe, T. (1971). *Chem. Pharm. Bull., Tokyo* 19, 2428–2430.
Packter, N. M. (1969). *Biochem. J.* **114**, 369–371.
Pettersson, G. (1966). *Acta chem. scand.* **20**, 151–158.
Polonsky, J., Johnson, B. C., Cohen, P. and Lederer, E. (1963). *Bull. Soc. chim. Fr.* 1909–1917.
Prelog, V. and Oppolzer, W. (1973). *Helv. chim. Acta* **56**, 2279–2287 et seq.
Rêgo de Sousa, J., Gottlieb, O. R. and Taveira Magelhães (1967). *Anais Acad. bras. Cienc.* 39, 227–231.
Reisch, J., Báthory, M., Szendrei, K., Novák, K. and Minker, E. (1973). *Phytochemistry* 12, 228 and previous papers.
Rowe, J. W., Seikel, M. K., Roy, D. N. and Jorgensen, E. (1972). *Phytochemistry* 11, 2513–2517.
Sasaki, K., Rinehart, K. L., Slomp, G., Grostic, M. F. and Olson, E. C. (1970). *J. Am. chem. Soc.* 92, 7591–7593.
Scharf, K.-H., Zenk, M. H., Onderka, D. K., Carroll, M. and Floss, H. G. (1971). *Chem. Commun.* 576–577 and references therein.

Schmid, H. V. and Zenk, M. H. (1971). *Tetrahedron Lett.* 4151–4155.
Seshadri, V., Batta, A. K. and Rangaswami, S. (1971). *Curr. Sci.* **40**, 630–631.
Seto, H., Cary, L. W. and Tanabe, M. (1973). *Chem. Commun.* 867–868.
Sezaki, M., Kondo, S., Maeda, K., Umezawa, H., and Ohno, M. (1970). *Tetrahedron* **26**, 5171–5190.
Steglich, W. and Lösel, W. (1972). *Chem. Ber.* **105**, 2928–2932.
Steglich, W., Arnold, E., Lösel, W. and Reininger, W. (1972). *Chem. Commun.* 102–103.
Stöckigt, J., Srocka, U. and Zenk, M. H. (1973). *Phytochemistry* **12**, 2389–2391.
Takeda, N., Seo, S., Ogihara, Y., Sankawa, U., Iitaka, I., Kitagawa, I. and Shibata, S. (1973). *Tetrahedron* **29**, 3703–3719.
Tsuji, N. and Nagashima, K. (1970). *Tetrahedron* **26**, 5201–5213.
Thomson, R. H. (1971). "Naturally Occurring Quinones", 2nd edn. Academic Press, London and New York.
Turner, W. B. (1971). "Fungal Metabolites". Academic Press, London and New York.
Yamazaki, S. and Ogawa, T. (1972). *Agric. Biol. Chem., Tokyo* **36**, 1707–1718.
Yamagishi, K., Yasuda, S. and Hanzawa, M. (1972). *Mokuzai Gakkaishi* **18**, 131–135.
Yoshihira, K. and Natori, S. (1966). *Chem. Pharm. Bull., Tokyo* **14**, 1052–1053.
Young, M. and Steelink, C. (1973). *Phytochemistry* **12**, 2851–2861.

Chapter 11

Betalains

MARIO PIATTELLI

Institute of Organic Chemistry, University of Catania, Catania, Italy

I Introduction

The term betalain has been recently coined (Mabry and Dreiding, 1968) to describe collectively two groups of water-soluble vacuolar plant pigments of restricted distribution, closely related both chemically and biogenetically: the red-violet betacyanins and the yellow betaxanthins. Originally called "Caryophyllinenroth" (Bischoff, 1876) and successively renamed "Rübenroth" (Weigert, 1894), "Betacyane" (Schudel, 1918) and, confusingly, "nitrogenous anthocyanins" (Lawrence *et al.*, 1939), the red pigments were finally termed betacyanins (Dreiding, 1961). Although their chemistry was investigated by numerous workers, little of value was obtained until 1957, when the conspicuous red-violet glucoside from *Beta vulgaris* root, betanin, was isolated in crystalline form (Wyler and Dreiding,

1957; Schmidt and Schönleben, 1957); subsequent progress was relatively rapid and the structure of the relevant aglycone, betanidin, was elucidated a few years later (Wyler *et al.*, 1963). Betaxanthins, formerly referred to as "flavocyanins" (Price *et al.*, 1939), are a more recently discovered group. No chemistry of any significance was carried out until 1964, when the yellow pigment from *Opuntia ficus-indica* fruits, indicaxanthin, was isolated and its structure fully clarified (Piattelli *et al.*, 1964c). Since then, notable advances have been made not only in the chemistry, but also in the biochemistry and physiology of these pigments.

II Betacyanins

A DETECTION AND ISOLATION

A number of colour reactions, mostly based on changes in pH and not exempt from uncertainty, have been proposed as a means of distinguishing between betacyanins and anthocyanins (Bischoff, 1876; Haverland, 1892; Weigert, 1894). More recently (Reznik, 1955), paper chromatography has also been used; the chief drawback which precluded widespread adoption of the method is that the R_F values are very low. Paper electrophoresis, however, is the most reliable method for detecting the presence of betacyanins in plant material (Lindstedt, 1956; Schmidt and Schönleben, 1956; Reznik, 1957), since they migrate as anions at pHs as low as 2·4, whereas anthocyanins show no appreciable migration.

Attempts to isolate betacyanins were made by several workers (*inter alia* Bischoff, 1876; Haverland, 1892; Schudel, 1918; Ainley and Robinson, 1937; Pucher *et al.*, 1938; Aronoff and Aronoff, 1948) but success was achieved only in 1957, when crystalline betanin was obtained by Wyler and Dreiding (1957) using preparative electrophoresis, and independently by Schmidt and Schönleben (1957), who resorted to a combination of chromatographic and electrophoretic techniques.

In the author's experience, the best method for use in both small and large-scale isolation is a preliminary separation of the total betacyanin fraction from the crude aqueous plant extract by non-ionic absorption on strongly acid resin and subsequent chromatography on a polyamide column, using increasing concentrations of methanol in aqueous citric acid as the developing solvent (Piattelli and Minale, 1964a,b). The resolved bands are freed from citric acid by resin treatment, concentrated *in vacuo* and allowed to crystallize.

If crystallization is not achieved (this is usually the case in small-scale separations), the individual betacyanins can be handled in solution for characterization by chromatography and electrophoresis. As many as sixteen betacyanins from a single plant source have been cleanly separated by this procedure (Piattelli and Minale, 1964b).

B STRUCTURE OF THE AGLYCONES

Since at present it clearly appears that all betacyanins are based on two isomeric aglycones, betanidin and isobetanidin, a brief outline of the salient features of their chemistry is in order.

Acid hydrolysis of betanin gave, along with glucose, a mixture of both aglycones (Wyler and Dreiding, 1959), whereas under milder conditions (Schmidt *et al.*, 1960) or in enzyme-catalysed hydrolysis (Piattelli and Minale, 1964a) betanidin only was obtained. In alkaline media, starting from either betanidin or isobetanidin, an equilibrium was attained with 30% of betanidin and 70% of the isomer (Wyler and Dreiding, 1959).

Upon alkali fusion, betanidin was split into 4-methylpyridine-2,6-dicarboxylic acid (**I**), 5,6-dihydroxy-2,3-dihydroindole-2-carboxylic acid (leucodopachrome or cyclodopa, **II**) and formic acid (Wyler and Dreiding, 1959, 1962) (Fig. 1); together, these fragments account for all the carbon atoms of betanidin. The main outlines of the molecule were revealed by the structural analysis of derivatives of neobetanidin (14,15-dehydrobetanidin) (Mabry *et al.*, 1962, 1967). Esterification of betanidin with methanolic hydrochloric acid (Fig. 2) gave a trimethylester hydrochloride and this on acetylation afforded 5,6-di-*O*-acetylneobetanidin trimethylester (**III**), a yellow compound which displayed, like all the neobetanidin derivatives, a large bathochromic shift (100 nm) from neutral to acidic solution, characteristic of cyanine dyes. The structure of this substance, which maintained most of the distinctive features of the parent pigment but proved to

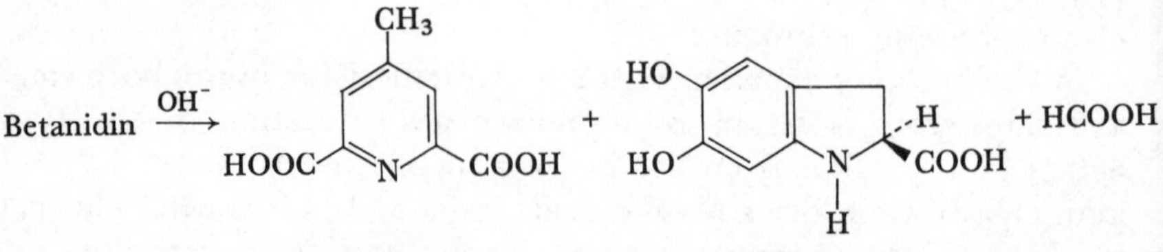

Betanidin $\xrightarrow{OH^-}$ (I) + (II) + HCOOH

FIG. 1. Alkaline degradation of betanidin.

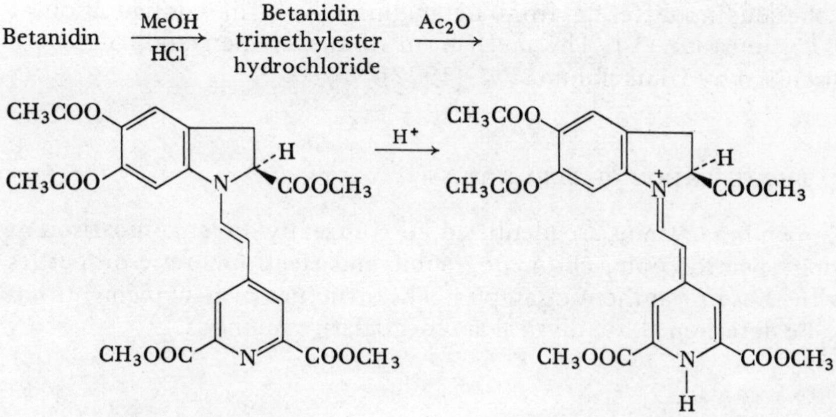

Betanidin $\xrightarrow[HCl]{MeOH}$ Betanidin trimethylester hydrochloride $\xrightarrow{Ac_2O}$

(III; λ_{max} 383 nm) $\xrightarrow{H^+}$ (λ_{max} 483 nm)

FIG. 2. Conversion of betanidin into a neobetanidin derivative.

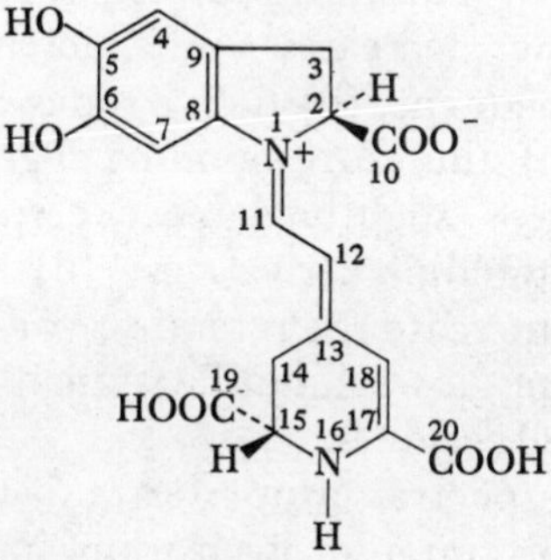

(IV) 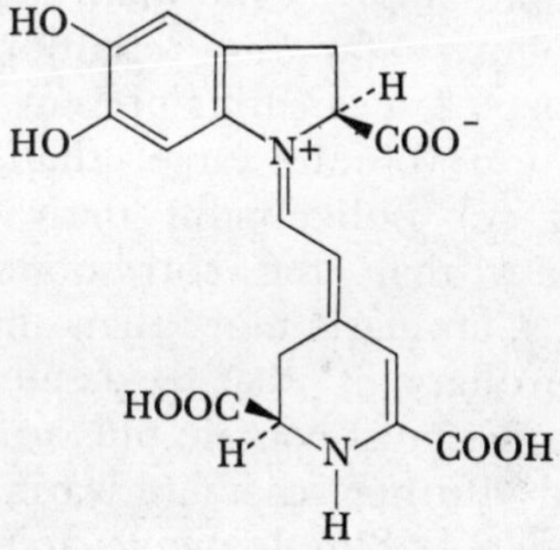 (V)

be much more amenable to investigation, was deduced largely from a comparison of its n.m.r. spectrum with those of model compounds. At this point it was finally possible to infer the gross structure of betanidin from the n.m.r. spectrum of 5,6-di-*O*-acetylbetanidin (Wyler *et al.*, 1963). The isolation of 5,6-dihydroxy-2,3-dihydro-2*S*-indole-2-carboxylic acid from the alkaline degradation products of betanidin permitted the absolute stereochemical assignment at C-2. The absolute configuration of the only other chiral centre in betanidin, C-15, was determined by Wilcox *et al.* (1965b) by a correlation with indicaxanthin, which was known to possess the *S* configuration at the corresponding carbon atom (C-11). In this way it was proved that the structure of betanidin is (IV), while that of isobetanidin, differing from betanidin only in its configuration at C-15, must be (V). The mechanism of the epimerization has been discussed by Dunkelblum *et al.* (1972).

C IDENTIFICATION OF BETACYANINS

Known betacyanins are identified, of course, by direct comparison of their spectroscopic, chromatographic and electrophoretic properties with those of authentic samples. The structure of new pigments has to be determined by physical and degradative methods.

1 *Physical properties*

Preliminary information useful for the identification of an individual betacyanin can be deduced from its behaviour on polyamide column chromatography. The main facts relating mobilities to structure are as follows: (a) the retention volume decreases with increasing number of sugar units present in the pigment; (b) 6-glycosides have retention volumes larger than those of the corresponding 5-glycosides; (c) isobetanidin derivatives have slightly larger retention volumes than the corresponding betanidin derivatives; (d) acyl groups, aromatic more than aliphatic, increase the retention volume. A corollary is that a pigment having lower mobility than the aglycones must contain one or more acyl residues.

For further characterization, the spectral properties must be considered. Simple betacyanins exhibit intense absorption in the visible region between 534 and 554 nm. Acylated betacyanins containing at least one aromatic (cinnamic) acyl residue have strong absorption in the ultraviolet region between 280 and 320 nm, where the absorption of simple betacyanins is weak. As an acylated betacyanin can be considered, with good approximation, spectroscopically equivalent to a mixture of the deacylated parent compound and the pertinent aromatic acid(s), the number of cinnamic acid units that are present in the pigment can be estimated from the intensity of the u.v. peak when compared with that of the colour maximum.

Nuclear magnetic resonance spectroscopy can also afford valuable structural information, but unfortunately the usefulness of this method is severely circumscribed by the low solubility of betacyanins.

2 *Chemical methods*

To begin with, it is to be noted that usually in a given plant source betacyanins occur in pairs of C-15 diastereoisomers, the betanidin derivatives being present as a rule in larger amounts than the corresponding isobetanidin ones. Such pairs can be readily identified since in suitable conditions each member is transformed into an equilibrium mixture of both epimers (Piattelli and Minale, 1964a,b). Furthermore, it has been observed that upon controlled acid hydrolysis the isobetanidin glycosides yield only isobetanidin, while the betanidin glycosides give a mixture of betanidin and isobetanidin, as a result of the partial isomerization of the aglycone. On the basis of these facts, it is possible to establish whether two betacyanins of unknown structure are C-15 epimers and, if it is so, which of them is a derivative of betanidin and which of isobetanidin. It is worth noting that, when preliminary experiments establish that two pigments bear an epimeric relationship to each other, degradative work to elucidate their structure can be conveniently carried out using a mixture of both, which is more easily obtained than the individual compounds.

Identification of the sugar moiety (there is no known example of a betacyanin in which both the hydroxyl groups of the aglycone have sugar residues attached to them) can be achieved by acid hydrolysis. In carefully selected conditions it is often possible to obtain a stepwise liberation of the sugar residues from a diglycoside; the intermediate monoglycoside can be identified by electrophoresis and chromatography. On the other hand, dilute acetic acid hydrolyses a polyoside to give a mixture of the relevant polyose and its products of partial and total hydrolysis, whose components can be identified by standard procedures (Piattelli and Imperato, 1970a,b). Sugars may also be obtained from betacyanins by oxidative destruction of the aglycone with alkaline hydrogen peroxide (Piattelli and Minale, 1966). Enzymic hydrolysis has also been used successfully in degradative studies; amaranthin, for instance, is hydrolysed by β-glucuronidase to glucuronic acid and betanin, which is in turn attacked by almond emulsin yielding glucose and betanidin (Piattelli and Minale, 1966). The sugar residue can be located as follows: treatment with excess diazomethane converts a betacyanin into an *O*-methylneobetanidin trimethylester glycoside, in which the originally free phenolic hydroxyl group of the aglycone has been methylated and thus labelled for identification; subsequent alkali fusion affords an indole carboxylic acid (5-hydroxy-6-methoxyindole-2-carboxylic acid or 5-methoxy-6-hydroxyindole-2-carboxylic acid) which is identified by comparison with authentic samples (Piattelli *et al.*, 1964a).

Acyl residues can be easily removed from complex betacyanins by cold alkaline hydrolysis under nitrogen, isolated and identified by chromatography and spectroscopy. In order to determine the exact position of the acyl group in the molecule (as yet, the structure of no betacyanin carrying two or more acyl residues has been completely elucidated), two methods have been used: (a) periodate oxidation in very mild conditions to minimize the possibility of acyl migration, followed by borohydride reduction, mild acid hydrolysis, a second borohydride reduction and chromatographic identification of the obtained polyols; (b) methylation of the pigment and subsequent identification of the products obtained by hydrolysis of the permethylated compound (Minale *et al.*, 1966).

D NATURALLY OCCURRING BETACYANINS

Betacyanins range in structure from the parent aglycones to various glycosides (Fig. 3), which can also occur as acyl derivatives. Two pairs of isomeric monosides have been described: betanin (**VI**) and isobetanin, 5-*O*-β-D-glucopyranosides of betanidin and isobetanidin, respectively (Piattelli *et al.*, 1964a; Wilcox *et al.*, 1965a), and gomphrenin-I (**VII**) and -II, obtained from inflorescences of globe amaranth, *Gomphrena globosa*, and shown to be 6-*O*-β-D-glucopyranosides of the isomeric aglycones (Minale *et al.*, 1967).

Three pairs of biosides are known. Amaranthin (**VIII**), 5-*O*-[2-*O*-(β-D-glucopyranosyluronic acid)-β-D-glucopyranoside] of betanidin, and its epimer isoamaranthin have been isolated from leaves of *Amaranthus tricolor* (Piattelli *et al.*, 1964b; Piattelli and Minale, 1966). From purple bracts of a horticultural variety of *Bougainvillea* ("Mrs Butt") two epimeric betacyanins have been isolated, bougainvillein-r-I (**IX**) and isobougainvillein-r-I, and proved to be 5-*O*-β-sophorosides of betanidin and isobetanidin (Piattelli and Imperato, 1970a). The 6-*O*-β-sophoroside (**X**) of betanidin and its epimer have been obtained by deacylation of the pigments, bougainvillein-v's, from bracts of *Bougainvillea glabra* var. *sanderiana* (Piattelli and Imperato, 1970b). The only known triosides, an incompletely characterized 6-*O*-rhamnosylsophoroside of betanidin and its C-15 epimer, have been obtained from the same source (Piattelli and Imperato, 1970b).

A considerable number of acylated betacyanins have also been reported and are listed in Table I. It has been established that in every case the acyl groups are linked to the sugar moiety. For most of the pigments, the ratio of acid(s) to deacylated pigment has been

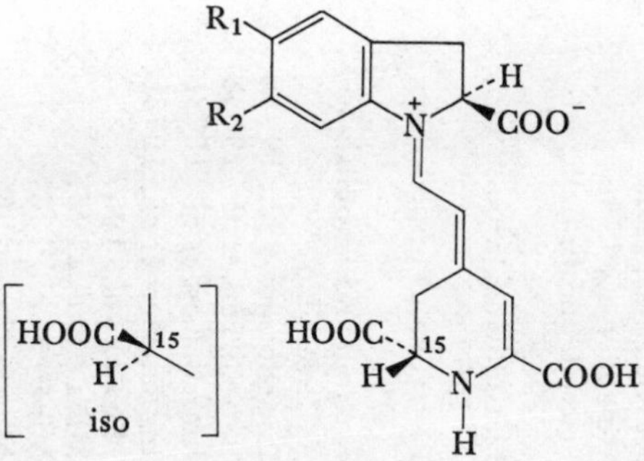

R_1

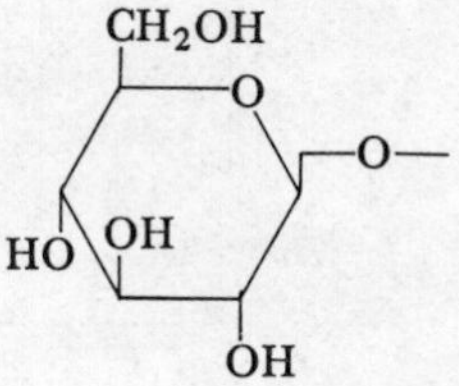

R_2 $\overset{+}{N}$ H COO^-

[HOOC 15 H iso] HOOC 15 H N COOH H

CH_2OH O O—

(VI) R_1 = HO OH OH ; R_2 = OH

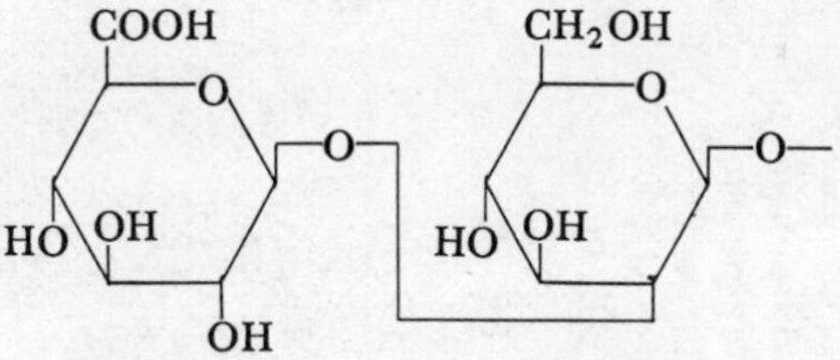

CH_2OH O O—

(VII) R_1 = OH; R_2 = HO OH OH

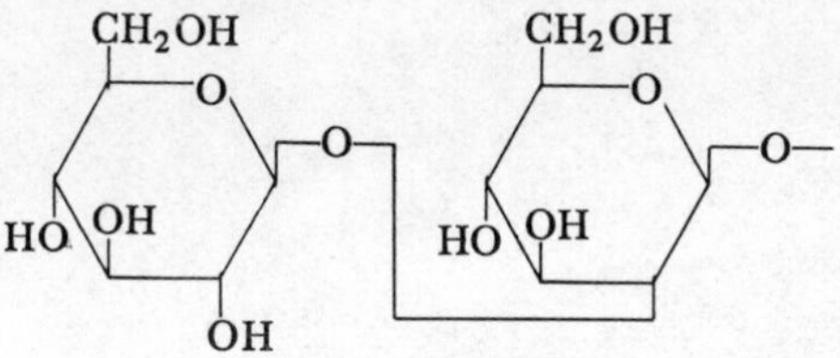

COOH O O— CH_2OH O O—

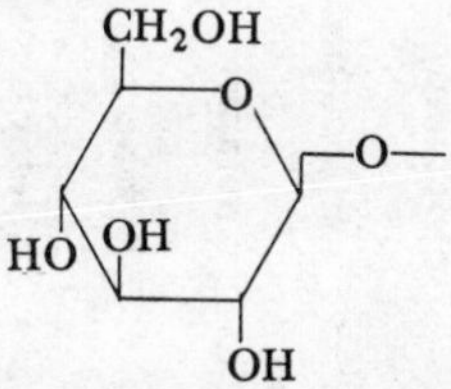

(VIII) R_1 = HO OH OH HO OH ; R_2 = OH

CH_2OH O O— CH_2OH O O—

(IX) R_1 = HO OH OH HO OH ; R_2 = OH

CH_2OH O O— CH_2OH O O—

(X) R_1 = OH; R_2 = HO OH OH HO OH

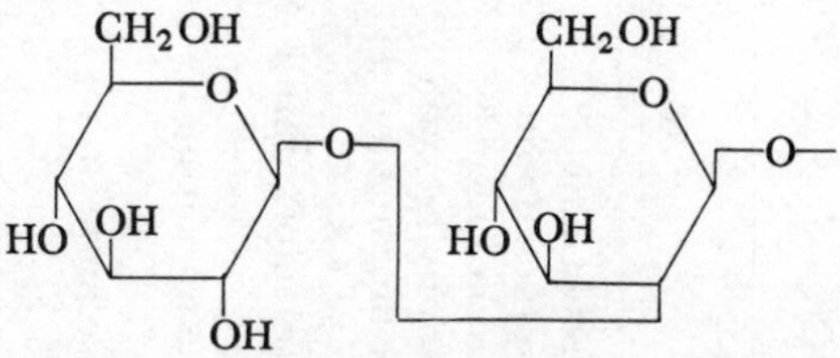

FIG. 3. Naturally occurring glycosides of betanidin and isobetanidin.

TABLE I

Acylated betacyanins[a]

Name	Acid(s)	Deacylated pigment	References
Prebetanin	Sulphuric	Betanin	Wyler *et al.*, 1967
Isoprebetanin	Sulphuric	Isobetanin	Wyler *et al.*, 1967
Phyllocactin	MA	Betanin	Minale *et al.*, 1966
Isophyllocactin	MA	Isobetanin	Minale *et al.*, 1966
Lampranthin-I	FE, PC	Betanin	Piattelli and Impellizzeri, 1969
Isolampranthin-I	FE, PC	Isobetanin	Piattelli and Impellizzeri, 1969
Lampranthin-II	FE, PC	Betanin	Piattelli and Impellizzeri, 1969
Isolampranthin-II	FE, PC	Isobetanin	Piattelli and Impellizzeri, 1969
Drosanthemin-Ia	CA, FE	Betanin	Impellizzeri *et al.*, 1973b
Drosanthemin-Ib	CA, FE	Isobetanin	Impellizzeri *et al.*, 1973b
Drosanthemin-IIa	CA	Betanin	Impellizzeri *et al.*, 1973b
Drosanthemin-IIb	CA	Isobetanin	Impellizzeri *et al.*, 1973b
Gomphrenin-III	PC	Gomphrenin-I	Minale *et al.*, 1967
Gomphrenin-V	FE	Gomphrenin-I	Minale *et al.*, 1967
Gomphrenin-VI	FE	Gomphrenin-II	Minale *et al.*, 1967
Gomphrenin-VII	FE	Gomphrenin-I	Minale *et al.*, 1967
Gomphrenin-VIII	PC	Gomphrenin-I	Minale *et al.*, 1967
Iresinin-I	HMG	Amaranthin	Minale *et al.*, 1966
Isoiresinin-I	HMG	Isoamaranthin	Minale *et al.*, 1966

Iresinin-III	PC, FE, SA, CA	Amaranthin	Minale *et al.*, 1966
Iresinin-IV	FE, SA	Amaranthin	Minale *et al.*, 1966
Celosianin	FE, PC	Amaranthin	Minale *et al.*, 1966
Isocelosianin	FE, PC	Isoamaranthin	Minale *et al.*, 1966
Suaedin	CA, CI, PC	Amaranthin	Piattelli and Imperato, 1971
Bougainvillein-r-II	CA, PC	Betanidin or isobetanidin 5-sophorosides	Piattelli and Imperato, 1970a
Isobougainvillein-r-II	CA, PC		
Bougainvillein-r-III	PC		
Bougainvillein-r-IV	PC		
Bougainvillein-r-V	PC		
Bougainvillein-v-A	PC	Betanidin or isobetanidin 6-sophorosides	Piattelli and Imperato, 1970b
Bougainvillein-v-B	PC		
Bougainvillein-v-C	PC		
Bougainvillein-v-D1	PC		
Bougainvillein-v-D2	PC		
Bougainvillein-v-E	PC		
Bougainvillein-v-F	PC		
Bougainvillein-v-G	PC	Betanidin or isobetanidin 6-rhamnosylsophorosides	
Bougainvillein-v-H1	CA		
Bougainvillein-v-H2	CA, PC	Betanidin or isobetanidin 6-sophorosides	
Bougainvillein-v-I	CA, PC		

[a] Abbreviations: CA = caffeic acid, CI = citric acid, FE = ferulic acid, HMG = 3-hydroxy-3-methylglutaric acid, MA = malonic acid, PC = *p*-coumaric acid, SA = sinapic acid.

determined, whereas the exact location of the acyl group has been ascertained, and always found to be the 6 position of the glucose unit, for the following monoacyl derivatives: prebetanin, a minor pigment from *Beta vulgaris*, identified as sulphuric acid half-ester of betanin (Wyler *et al.*, 1967); phyllocactin from *Phyllocactus hybridus*, malonic acid half-ester of betanin (Minale *et al.*, 1966); iresinin, the major pigment from leaves of *Iresine herbstii*, 3-hydroxy-3-methylglutaric acid half-ester of amaranthin (Minale *et al.*, 1966); gomphrenin-III and -V, from *Gomphrena globosa*, *p*-coumaroyl and, respectively, feruloyl esters of gomphrenin-I (Minale *et al.*, 1967). The occurrence of C-15 epimers of these pigments has also been reported.

Finally, it is worth mentioning the natural occurrence of 2-decarboxybetanidin (**XI**), isolated from flowers of *Carpobrotus acinaciformis* (Piattelli and Impellizzeri, 1970), since it provides the only exception in containing a modified aglycone moiety.

III Betaxanthins

A DETECTION AND ISOLATION

There are no special colour tests for betaxanthins. As they possess electrophoretic properties similar to those of betacyanins, paper electrophoresis is the best method for their detection.

The first crystalline betaxanthin was obtained from fruits of *Opuntia ficus-indica* and named indicaxanthin (Piattelli *et al.*, 1964c). The isolation procedure was similar to that used for betacyanins: from the total betalain fraction separated by absorption on strongly acid ion-exchange resin, indicaxanthin was isolated by polyamide chromatography. Other betaxanthins have similarly been isolated; it must be remarked, however, that polyamide chromatography is not in general so satisfactory as for betacyanins. Preparative paper electrophoresis and chromatography have also been used. As a rule, a judicious combination of chromatographic and electrophoretic techniques gives the best results and allows individual pigments to be separated from co-occurring betacyanins and from other plant constitutents.

B DETERMINATION OF THE STRUCTURES

The structure of indicaxanthin has been established by physical and degradative studies. Alkali fusion of the pigment yielded 4-methylpyridine-2,6-dicarboxylic acid and proline (racemic), while acid degradation gave the same amino acid as the L form. Hydrogen peroxide oxidation afforded pyridine-2,4,6-tricarboxylic acid and

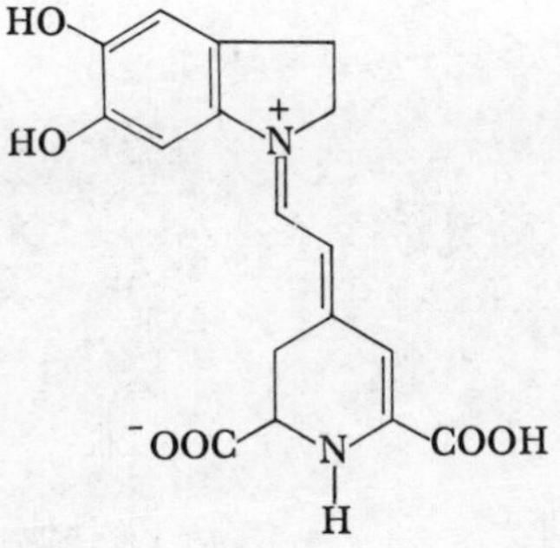

(XI)

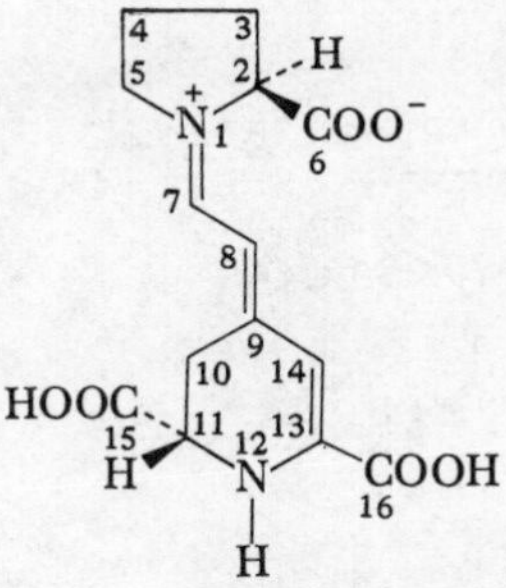

(XII)

minor amounts of L-aspartic acid. These data, combined with those from n.m.r. spectrum, allowed the absolute structure (**XII**) to be assigned to indicaxanthin (Piattelli *et al.*, 1964c).

Other betaxanthins have been characterized on the basis of the following observations. All the pigments of this group have an absorption maximum near 480 nm. On diazomethane methylation they give derivatives in which the dihydropyridine ring has been oxidized to a pyridine ring. These compounds, neobetaxanthins, exhibit absorption maxima between 340 and 360 nm, which in acidic solution show a bathochromic shift of 80–100 nm (Piattelli *et al.*, 1965c). Alkali fusion of betaxanthins yields 4-methylpyridine-2,6-dicarboxylic acid, whereas acid hydrolysis and subsequent chromatography allow the identification of the amino acid or amine linked to the dihydropyridine system.

C NATURALLY OCCURRING BETAXANTHINS

In addition to indicaxanthin, the following betaxanthins have been so far identified as naturally occurring pigments (Fig. 4). Portulaxanthin (**XIII**) was isolated from flowers of *Portulaca grandiflora* (Piattelli *et al.*, 1965a), vulgaxanthin-I (**XIV**) and vulgaxanthin-II (**XV**) were obtained from *Beta vulgaris* root (Piattelli *et al.*, 1965b), four miraxanthins (**XVI–XIX**)—and in addition two incompletely characterized betaxanthins—from flowers of *Mirabilis jalapa* (Piattelli *et al.*, 1965c), and dopaxanthin (**XX**) from flowers of *Glottiphyllum longum* (Impellizzeri *et al.*, 1973a). The structures of all these pigments have been substantiated by synthesis.

It seems pertinent to mention here that nudicaulin, the water-soluble yellow pigment isolated by Price *et al.* (1939) from petals of *Papaver nudicaule* and *Meconopsis cambrica* (Papaveraceae), has been supposed to be related to "nitrogenous anthocyanins", but the

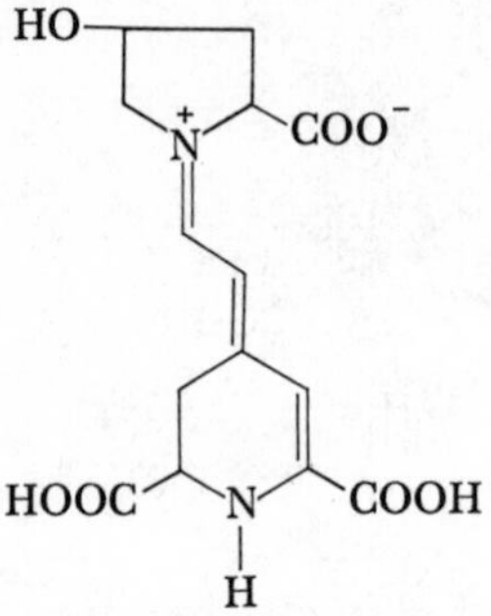

(XIII)

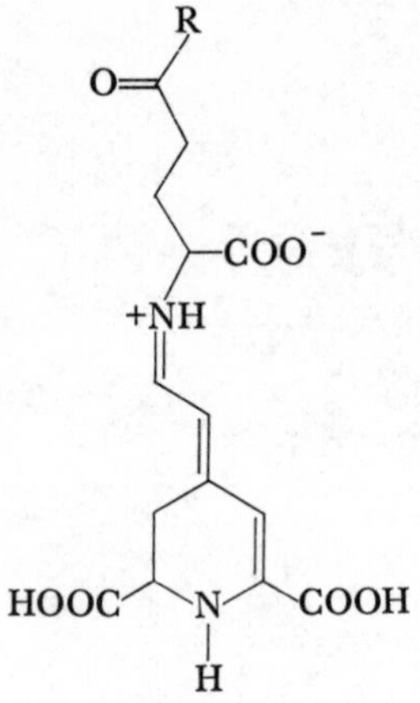

(XIV) R = NH_2
(XV) R = OH

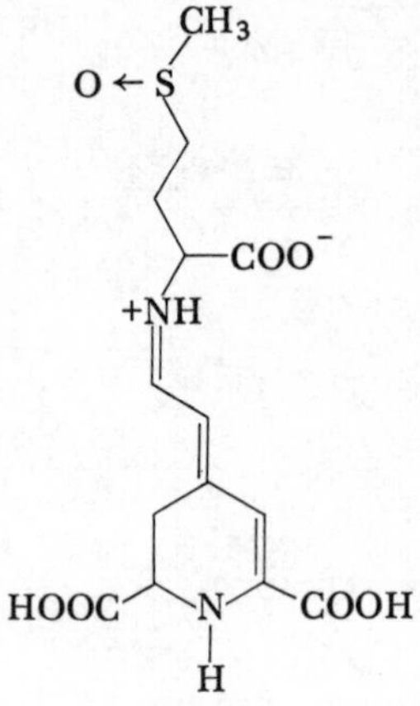

(XVI)

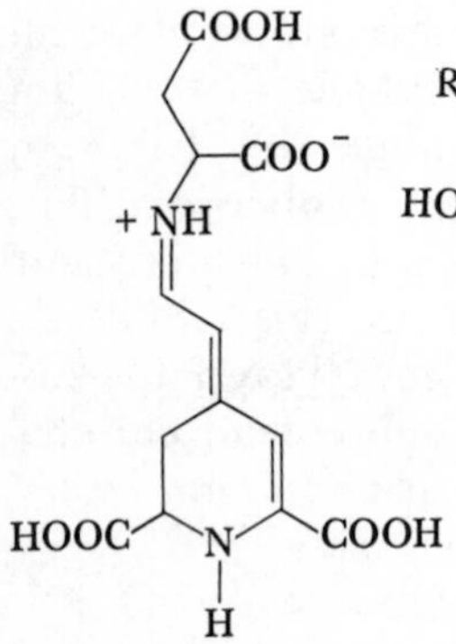

(XVII)

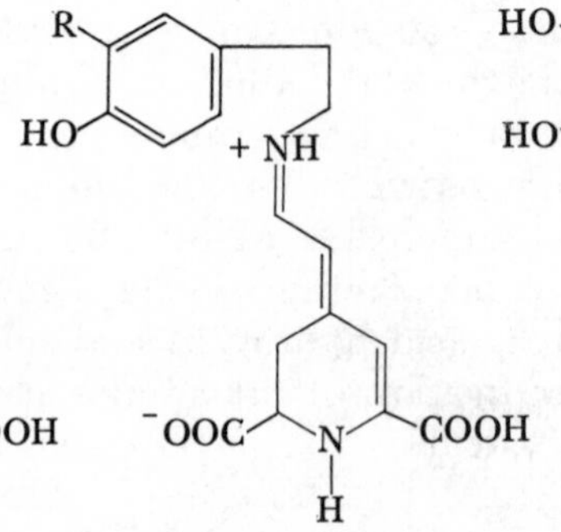

(XVIII) R = H
(XIX) R = OH

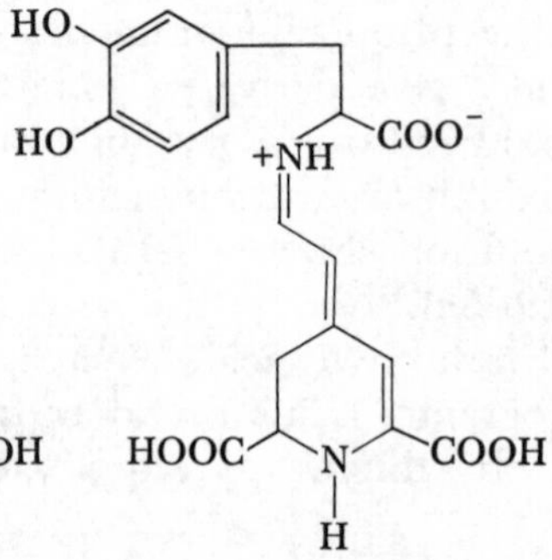

(XX)

FIG. 4. Naturally occurring betaxanthins.

possibility that it is a betaxanthin was ruled out by Harborne (1965) on the basis of spectral and chemical evidences.

IV Synthesis of betalains

Whereas the synthesis of a neobetanidin derivative (5,6-di-*O*-methyl-neobetanidin trimethylester) has been carried out successfully by Badgett *et al.* (1970), the total synthesis of a betalain has not been so far achieved.

A partial synthesis of indicaxanthin has been effected by treatment of betanin with excess proline in alkaline medium. Conversely, indicaxanthin was transformed into betanidin by a base exchange with cyclodopa (Wyler *et al.*, 1965). At about the same time, Piattelli *et al.* (1965c) obtained indicaxanthin and other betaxanthins (vulgaxanthin-I and various miraxanthins) by a slightly different method; betanin was dissolved in water, the solution saturated with sulphur dioxide and, after bleaching, concentrated *in vacuo*. Addition of the requisite amino acid or amine resulted in the formation of the expected betaxanthin.

The method described by Wyler has also been used for the synthesis of vulgaxanthins (Nilsson, 1970) and dopaxanthin (Impellizzeri *et al.*, 1973a).

Probably both the base-exchange procedures involve the formation of betalamic acid (**XXII**); this compound has been in fact isolated from the degradation products of betanin (Kimler *et al.*, 1971) or betanidin (Sciuto *et al.*, 1972) in alkaline media and shown to react readily with amino acids or amines to give betalains.

V Biosynthesis of betalains

Before any experimental evidence was available, it has been pointed out (Wyler *et al.*, 1963; Mabry, 1964) that betanidin might possibly be formed *in vivo* from two molecules of 3,4-dihydroxyphenylalanine (dopa, **XXI**) (Fig. 5). Oxidative opening of the aromatic ring and subsequent closure to a dihydropyridine system would yield betalamic acid (**XXII**), which could give betanidin on condensation with cyclodopa. This could derive from a second dopa molecule by oxidative cyclization, presumably by way of dopaquinone, as postulated for the early stages of melanogenesis (Raper, 1927, 1928; Nicolaus *et al.*, 1964; Harley-Mason, 1965). Reaction of betalamic acid with amino or imino compounds other than cyclodopa could by analogy yield a range of betalains. In fact, feeding experiments have shown that labelled dopa is incorporated into the betanin occurring in the root and hypocotyl of *Beta vulgaris* (Hörhammer *et al.*, 1964) and into amaranthin in *Amaranthus* seedlings (Garay and Towers, 1966). Minale *et al.* (1965) have isolated indicaxanthin and betanin from fruits of *Opuntia ficus-indica* after incubation with labelled dopa; from measurements of the distribution of the radioactivity in the pigments and degradation experiments it was demonstrated conclusively that dopa indeed acts as precursor of the dihydropyridine moiety of betalains. Subsequently Miller *et al.* (1968) have found by radioactive feeding

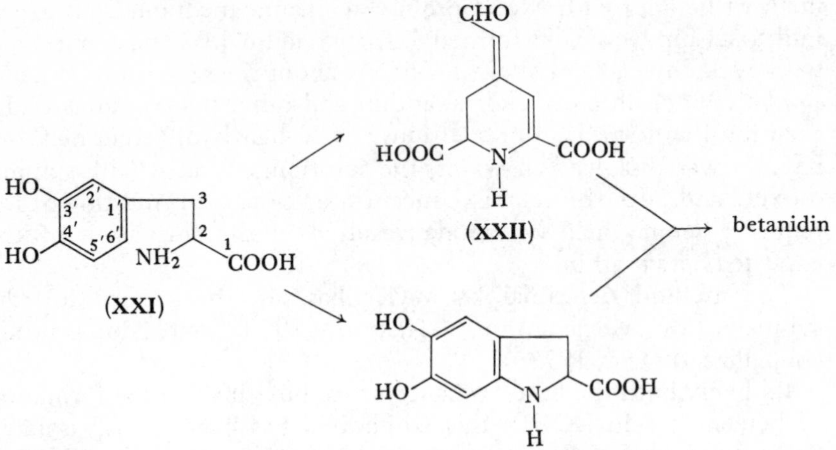

FIG. 5. Hypothetical biogenesis of betanidin.

experiments on the incorporation of dopa into betanin in the fruits of *Opuntia* species that a minor proportion of the radioactivity appears in the dihydroindole unit and a major proportion in the dihydropyridine moiety. This result has been confirmed by feeding experiments in beets (Liebisch *et al.*, 1969) with $[1\text{-}^{14}C,^{15}N]$ tyrosine or $[2\text{-}^{14}C,^{15}N]$ tyrosine which showed incorporation of the amino acid, including the amino nitrogen atom, into both the heterocyclic units of betanin.

In theory, the C_9N skeleton of the dihydropyridine system of betalains could be derived from dopa in either of two ways: (a) opening of the aromatic ring between the two hydroxyl groups ("intradiol" cleavage) followed by bonding of nitrogen to carbon 5′; or (b) ring opening between carbons 4′ and 5′ ("extradiol" cleavage) and subsequent closure by bonding of nitrogen to carbon 3′. The occurrence in plants of oxygenases which catalyse either of these cleavages of dihydroxyaromatic compounds has in fact been reported (Fujisawa and Hayaishi, 1968; Nozaki *et al.*, 1968; Hayaishi and Nozaki, 1969). To differentiate between these two possibilities, the incorporation of doubly labelled L-$[1\text{-}^{14}C,3',5'\text{-}^{3}H]$-tyrosine into betanin in the fruit of *Opuntia decumbens* (Fischer and Dreiding, 1972) and into indicaxanthin in the fruits of *Opuntia ficus-indica* (Impellizzeri and Piattelli, 1972) has been studied. Since the results were consistent with the pathway involving an extradiol cleavage of the aromatic ring of dopa distal with respect to the side chain, it was possible to derive the following tentative biogenetic pathway to betanidin and betaxanthins (Fig. 6). The alternative

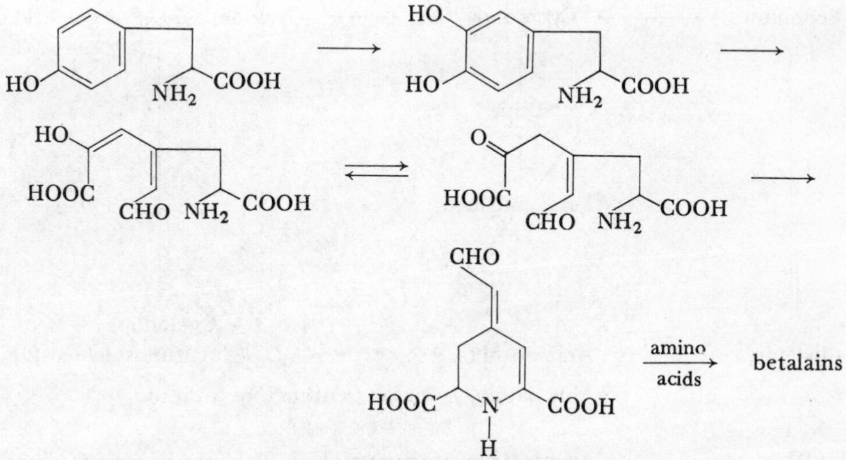

FIG. 6. Tentative pathway for the biosynthesis of betalains.

possibility, that the closure of the dihydropyridine ring occurs after the step of condensation with amino acids or amines, must also be taken into consideration. However, the recent detection of free betalamic acid as a natural constituent of a number of Centrospermae (Kimler *et al.*, 1971) lends support to the view that this compound is the key intermediate in the biosynthesis of all betalains.

In an attempt to determine at what stage glycosylation occurs in the biosynthesis of betacyanins, labelled betanidin has been fed to fruit of *Opuntia dillenii* (Sciuto *et al.*, 1972) and found to be efficiently incorporated into betanin, a result that seems to indicate that glucosylation occurs at the aglycone level. However, more recent results (Sciuto *et al.*, 1974) on the biosynthesis of amaranthin militate against the view that glycosylation is a late or terminal step and raise the doubt that the observed incorporation of betanidin into betanin is merely a reaction the plant is able to carry out. In these experiments, non-labelled potential precursors (cyclodopa, cyclodopa 5-*O*-β-D-glucoside, betanidin and betanin) were administered to seedlings of a betaxanthin-producing yellow variety of *Celosia plumosa* ("Golden Feather") which possess the biosynthetic potential for the production of the betacyanin normally synthesized by red varieties of the same species, that is, amaranthin. From the relative rates of incorporation into amaranthin (the best precursors are cyclodopa and its 5-*O*-β-D-glucoside; betanidin and betanin are rather less good) it appears that the route cyclodopa → cyclodopa 5-*O*-β-D-glucoside → cyclodopa 5-*O*-β-glucuronosylglucoside → amaranthin (Fig. 7a) constitutes the major pathway for amaranthin

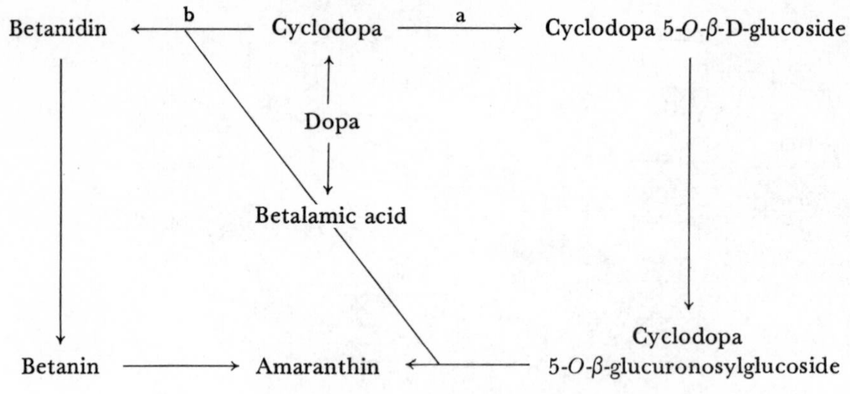

FIG. 7. Possible pathways for amaranthin biosynthesis.

synthesis, while the alternative pathway (Fig. 7b) involving aglycone formation as an intermediate step seems to have far less importance. Observation of this kind, however, cannot be easily interpreted until it is known whether or not the exogenous precursors reach the site of biosynthesis with approximately the same efficiency.

One last point deserves a brief comment. It has been mentioned previously that generally in a plant source betacyanins occur in pairs of C-15 diastereoisomers. However, it has been observed that in a few plants some betanidin glycosides occur unaccompanied by their corresponding isomers (Piattelli and Minale, 1964b), whereas the opposite was never observed. Furthermore, it has been reported (Piattelli *et al.*, 1969) that seedlings of *Amaranthus tricolor* contain amaranthin only, while in mature leaves of the same plant this pigment is accompanied by its isomer. These facts can be taken as circumstantial evidence that the betanidin pigments are the primary products of metabolism, the isobetanidin derivatives being formed by spontaneous epimerization.

VI Factors governing betalain biosynthesis

The most investigated environmental factor controlling betalain biosynthesis seems to be illumination. Light is an absolute requirement for pigment synthesis in certain species, while in others betalain accumulation also takes place in darkness and is increased by irradiation with white light (Wohlpart and Mabry, 1968b). Far-red reversibility (phytochrome control) of the light-induced betalain synthesis has been demonstrated in seedlings of *Amaranthus tricolor*

(Piattelli *et al.*, 1969), *A. caudatus* (Koehler, 1972c), *Chenopodium rubrum* (Wagner and Cumming, 1970)—a circadian rhythm in betacyanin accumulation in this species has been reported by Wagner and Frosch (1971)—*Celosia plumosa* (Giudici de Nicola *et al.*, 1973a) and *C. cristata* (Giudici de Nicola *et al.*, 1974), while no phytochrome intervention could be demonstrated in *A. salicifolius* (Heath and Vince, 1962). Measurements of phytochrome in *A. caudatus* (Kendrick and Frankland, 1969) have shown that its concentration increases from the time of germination until 72 h from sowing and then levels off over a 48 h period; the sensitivity of the seedlings to illumination develops in parallel and the pigment formation in reponse to light stimulation is maximal between 72 and 96 h (Koehler, 1972c).

Kinetin (6-furfurylaminopurine) is also able to promote betalain synthesis (Bamberger and Mayer, 1960; Koehler, 1967, 1972c; Piattelli *et al.*, 1971; Giudici de Nicola *et al.*, 1972a, 1973a). The results of experiments with this hormone are difficult to interpret, as it is known to affect almost every aspect of plant growth. Nevertheless, it appears that its effect on the pigment accumulation in darkness is strong in plants with a compulsory light requirement, and moderate in species capable of synthesis in the dark.

Since the effects of both light and kinetin have been ascribed, on the basis of the depressing action of inhibitors of nucleic acid and protein biosynthesis (Birnbaum and Koehler, 1970; Koehler and Birnbaum, 1970b; Piattelli *et al.*, 1970a,b, 1971; Giudici de Nicola *et al.*, 1972a; Koehler, 1972b), to the activation of genes coding for the enzymes involved in pigment synthesis, this differential response to kinetin is possibly due to the fact that in etiolated seedlings of plants which do not have an absolute light requirement the genes for betalain synthesis are already "open" (Giudici de Nicola *et al.*, 1973a,b).

From his extensive researches on light-induced anthocyanin synthesis, Mohr (1966a,b) has put forward the hypothesis that gene activation is controlled by the low-energy phytochrome system. Evidence has been reported (Giudici de Nicola *et al.*, 1972a), however, that the action of light on the genes governing betalain formation is mediated by photoreceptor(s) other than phytochrome. Furthermore, the inhibitory effect of salicylaldoxime and 2,4-dinitrophenol on the pigment biosynthesis stimulated by short-term irradiation or kinetin points to the implication of oxidative phosphorylation in the regulation of betalain production (Giudici de Nicola *et al.*, 1972a,b, 1973a).

Under continuous illumination with white light, in conditions of "high energy reaction" (h.e.r.), the photosynthetic system seems to be the most important factor governing pigment synthesis. Indeed, the inhibition of chlorophyll formation following administration of laevulinic acid is paralleled by a strong depression of pigment accumulation. The fact that the pigment formation stimulated by continuous white light is also reduced in the presence of 2,4-dinitrophenol and 3-(3,4-dichlorophenyl)-1,1-dimethylurea is evidence for the contribution of both cyclic and non-cyclic photophosphorylation in the biosynthesis of betalains (Giudici de Nicola *et al.*, 1972b, 1973a, 1974).

The above results suggest that both light and kinetin act on betalain synthesis at two different levels, activation of genes and availability of energy-rich compounds. The effect of light on the gene system appears to be mediated by photoreceptor(s) other than phytochrome, whereas that on the availability of the energy-rich compounds seems to be mediated by phytochrome in the early hours of irradiation and successively by the photosynthetic system.

Also the far-red h.e.r. associated with betalain synthesis in *C. plumosa, C. cristata* and *A. caudatus* (Giudici de Nicola *et al.*, 1973b, 1974), species able to synthesize in darkness, is dependent on the photosynthetic system, essentially via the cyclic pathway, while continuous far-red lacks any effect on *A. tricolor*, which has an absolute light requirement for pigment production. Assuming that in etiolated seedlings of this species the genes involved in the pigment formation are inactive and taking into account that continuous far-red is known to maintain a low but constant level of active P_{fr} form (Hartmann, 1966), this result agrees with the view that gene activation is not the primary action of phytochrome. Moreover, the fact that in *A. tricolor* pigment synthesis is strongly stimulated by continuous white light whereas far-red is completely ineffective suggests that photoactivation of the genes involved in betacyanin formation requires wavelengths other than 720 nm (Giudici de Nicola *et al.*, 1973b).

It has been recently reported (Rast *et al.*, 1973) that in etiolated seedlings of *A. paniculatus* adenosine 3′,5′-cyclic monophosphate (cAMP) and its dibutyryl derivative can satisfy the light requirement for the synthesis of betacyanin. This result agrees with the suggestion that cAMP could act as a "second messenger" in some phytochrome-controlled responses (Mohr, 1972).

The effects of added precursors on betalain synthesis have also been studied. Administration of dopa or tyrosine increases the pigment accumulation in seedlings of *Amaranthus* (Garay and Towers, 1966; Koehler, 1967), but no such increase is observed for the betanin synthesis in callus cultures (Constabel and Nassif-Makki, 1971) and in excised leaf discs (Wohlpart and Black, 1973) of *Beta vulgaris*; it has been suggested that the developmental stage of the tissue rather than the availability of exogenous precursors is the limiting factor in pigment formation (Constabel and Nassif-Makki, 1971).

Betacyanin accumulation is enhanced in leaf discs of red beet plants by added sucrose (Wohlpart and Black, 1973) and in *A. caudatus* seedlings by potassium nitrate (Koehler and Birnbaum, 1970a).

VII Chemical taxonomy of betalains

Even before the structure of betalains was clarified, the potential importance of these pigments to plant taxonomy had been evaluated by several authors. At first found in eight families—Amaranthaceae, Aizoaceae (excluding the Molluginaceae, which elaborate anthocyanins), Basellaceae, Chenopodiaceae, Cactaceae, Nyctaginaceae, Phytolaccaceae and Portulacaceae (Reznik, 1955, 1957)—they were successively observed by Rauh and Reznik (1961) in Didieraceae, a small family from Madagascar, while Mabry *et al.* (1963) found them in *Stegnosperma halimifolium* of the family Stegnospermaceae. This taxon, formerly considered as a subfamily (Stegnospermatoidiae) of the Phytolaccaceae, was elevated to the rank of family by Hutchinson (1959) and considered as belonging to the order Pittosporales, although most taxonomists include it in the Phytolaccaceae. Among the families of the Englerian Centrospermae (Engler, 1936), only the non-pigmented Cynocrambaceae (syn. Thelygonaceae) and the anthocyanin-containing Caryophyllaceae apparently lack betalains. On the basis of this striking correlation between chemical and morphological characters, Mabry (1964) has suggested that the order Centrospermae, including the Cactaceae—treated as a separate order, Opuntiales, by Engler but included in Centrospermae by other taxonomists, e.g. Wettstein (1935)—be reserved for the betalain-containing families, and that the anthocyanin-containing Caryophyllaceae be considered as a phyletic group related but not belonging to the betalain-producing order.

The totally different chemical structure and biosynthetic pathways of betalains and anthocyanins, the fact that there is no known case of the coexistence of the two types of pigment in the same plant

TABLE II

Distribution of betalains[a-c]

Family and species	Betacyanins	Betaxanthins	References
Aizoaceae: 130 (2500)			
Aptenia cordifolia	p	a	Reznik, 1955
Bergeranthus multiceps	a	p	Reznik, 1957
Carpobrotus acinaciformis	Bd, Bt, Lm II, Db	—	Piattelli and Impellizzeri, 1970
Conophytum albescens	a	p	Reznik, 1957
C. augustum	a	p	Reznik, 1957
C. bifidum	a	p	Reznik, 1957
C. citrinum	a	p	Reznik, 1957
C. conradi	a	p	Reznik, 1957
C. corculum	a	p	Reznik, 1957
C. ectypum	a	p	Reznik, 1957
C. elishae	a	p	Reznik, 1957
C. ernianum	p	a	Reznik, 1957
C. exsertum	a	p	Reznik, 1957
C. flavum	a	p	Reznik, 1957
C. globuliforme	a	p	Reznik, 1957
C. impolitum	a	p	Reznik, 1957
C. laetum	a	p	Reznik, 1957
C. limbatum	p	a	Reznik, 1957
C. luteolum	a	p	Reznik, 1957
C. luteum	p	p	Reznik, 1957
C. marginatum	p	a	Reznik, 1957
C. marnierianum	p	p	Reznik, 1957
C. maximum	p	p	Reznik, 1957
C. meyeri	a	p	Reznik, 1957
C. minutum	p	a	Reznik, 1957
C. mirabile	Bd	p	Wyler and Dreiding, 1961b

C. nelianum	a	p	Reznik, 1957
C. ornatum	p	p	Reznik, 1957
C. pearsonii	p	a	Reznik, 1957
C. piluliforme	p	a	Reznik, 1957
C. plenum	a	p	Reznik, 1957
C. poellnitzianum	p	—	Dreiding, 1961
C. praegratum	p	a	Reznik, 1957
C. sellatum	p	a	Reznik, 1957
C. simplum	a	p	Reznik, 1957
C. subacutum	a	p	Reznik, 1957
C. taylorianum	p	a	Reznik, 1957
C. tetracarpum	a	p	Reznik, 1957
C. tischeri	p	a	Reznik, 1955, 1957
C. truncatum	p	—	Lawrence *et al.*, 1939
C. tubatum	p	a	Reznik, 1957
C. turrigerum	p	a	Reznik, 1957
C. wettsteinii	p	a	Reznik, 1955, 1957
Dorotheanthus gramineus	p	a	Reznik, 1957
Drosanthemum floribundum	Bd, Bt, Dr I–II	—	Impellizzeri *et al.*, 1973b
Faucaria albidens	a	p	Reznik, 1957
F. duncani	a	p	Reznik, 1955, 1957
F. kingiae	a	p	Reznik, 1957
Fenestraria aurantiaca	p	p	Reznik, 1957
Gibbeum gibbosum	p	a	Reznik, 1957
G. velutinum	p	a	Reznik, 1955, 1957
Glottiphyllum erectum	a	p	Reznik, 1957
G. difforme	a	p	Reznik, 1957
G. longum	a	Dp	Impellizzeri *et al.*, 1973a
G. nelii	a	p	Reznik, 1957
G. platycarpum	a	p	Reznik, 1955
G. salmi	a	p	Reznik, 1957
Lampranthus brownii	a	p	Reznik, 1957
Lampranthus sp.	Bt, Lm I–II	—	Piattelli and Impellizzeri, 1969

TABLE II—*continued*

Family and species	Betacyanins	Betaxanthins	References
L. zeyheri	p	p	Reznik, 1957
Lithops kuibisensis	p	p	Reznik, 1957
Malephora mollis	p	a	Reznik, 1957
Mesembryanthemum conspicuum	Bt	—	Piattelli and Minale, 1964b
M. edule	Bd, Bt	—	Piattelli and Minale, 1964b
M. nodiflorum	p	—	Gertz, 1906; Lawrence *et al.*, 1939
M. roseum	Bd, Bt	—	Wyler and Dreiding, 1961b
Pleiospilos bolusii	p	—	Dreiding, 1961
P. dekanahi	a	p	Reznik, 1957
P. herrei	a	p	Reznik, 1957
P. simulans	p	p	Reznik, 1957
Rabiaea lesliei	a	p	Reznik, 1957
Rhombophyllum dolabriforme	a	p	Reznik, 1957
Sesuvium portulacastrum	p	—	Taylor, 1940
Tetragonia crystallina	p	—	Gertz, 1906; Lawrence *et al.*, 1939
Trianthema portulacastrum	p	—	Mabry *et al.*, 1963
Trichodiadema bulbosum	p	a	Reznik, 1957
T. densum	p	a	Reznik, 1957
Amaranthaceae: 60 (900)			
Achyrantes verschaffeltii	p	—	Gertz, 1906
A. aspera	p	—	Koehler, 1972a
Aerva sanguinolenta	p	—	Reznik, 1955, 1957
Alternanthera amoena	p	p	Reznik, 1957
A. ficoidea	p	—	Taylor, 1940
A. halimifolia	p	—	Taylor, 1940
A. macrophylla	p	—	Taylor, 1940
A. petsikiana	Am	—	Dreiding, 1961
A. versicolor	Am, Bt	—	Piattelli and Minale, 1964b

Amaranthus acanthocarpa	p	—	Mabry *et al.*, 1963
A. caruru	p	—	Koehler, 1972a
A. caudatus	Am	—	Piattelli and Minale, 1964b
A. gangeticus	p	—	Koehler, 1972a
A. gracilis	p	—	Taylor, 1940
A. graecizans	Am, Bt	—	Piattelli and Minale, 1964b
A. hybridus	Am	—	Piattelli and Minale, 1964b
A. hypocondriacus	Am	—	Piattelli and Minale, 1964b
A. palmeri	p	—	Mabry *et al.*, 1963
A. paniculatus	Am	—	Wyler and Dreiding, 1961b
A. pringlei	p	—	Mabry *et al.*, 1963
A. quitensis	p	—	Taylor, 1940
A. retroflexus	Am, Bt	—	Piattelli and Minale, 1964b
A. salicifolius	p	—	Bischoff, 1876
A. tricolor	Am	—	Piattelli and Minale, 1964b
Celosia argentea	p	—	Taylor, 1940
C. cristata	Am, Ce	p	Reznik, 1955; Minale *et al.*, 1966
C. plumosa	Am, Bt	p	Piattelli and Minale, 1964b; Giudici de Nicola *et al.*, 1973a
C. thompsonii	p	—	Reznik, 1957
Froelichia drummondii	p	—	Mabry *et al.*, 1963
Gomphrena decumbens	p	—	Mabry *et al.*, 1963
G. globosa	Gm I–VIII, Am, Ce	—	Minale *et al.*, 1967
G. nealleyi	p	—	Mabry *et al.*, 1963
G. sonorae	p	—	Mabry *et al.*, 1963
Iresine herbstii	Am, Ir I–IV	—	Minale *et al.*, 1966
I. lindenii	Am, Bt, Ir I–II	—	Piattelli and Minale, 1964b
Mogiphanes brasiliensis	p	—	Gertz, 1906
Tidestromia lanuginosa	p	—	Mabry *et al.*, 1963
Basellaceae: 4–5 (20)			
Anredera vesicaria	p	—	Wohlpart and Mabry, 1968a
Basella alba	p	a	Reznik, 1955
B. rubra	p	a	Reznik, 1955

TABLE II—*continued*

Family and species	Betacyanins	Betaxanthins	References
Cactaceae: 200 (2000)			
Ariocarpus kotschubeyanus	p	a	Reznik, 1957
Aylostera pseudodeminuta	p	a	Reznik, 1957
Aporocactus flagelliformis	Bt, Ph	—	Piattelli and Imperato, 1969
Borzicactus sepium	Bt, Ph	—	Piattelli and Imperato, 1969
Cereus comarapanus	Bt, Ph	—	Piattelli and Imperato, 1969
C. peruvianus	p	—	Dreiding, 1961
C. speciosus	p	—	Lawrence *et al.*, 1939
C. stenogonus	Bt, Ph	—	Piattelli and Imperato, 1969
C. thouarsii	p	—	Taylor, 1940
Chamaecereus silvestris	p	p	Reznik, 1957
Cleistocactus jujuyensis	Bt, Ph	—	Piattelli and Imperato, 1969
C. parviflorus	Bt, Bd, Ph	—	Piattelli and Imperato, 1969
C. smaragdiflorus	Bt, Ph	—	Piattelli and Imperato, 1969
C. strausii	Bt, Ph	—	Piattelli and Imperato, 1969
Epiphyllum truncatum	p	p	Reznik, 1955, 1957
Eriocereus guelichii	Bt, Ph	—	Piattelli and Imperato, 1969
Gymnocalycium andreae	p	a	Reznik, 1957
G. baldianum	p	a	Reznik, 1957
G. mihanovichii	p	—	Dreiding, 1961
Haageocereus acranthus	Bt, Ph	—	Piattelli and Imperato, 1969
Hariota salicornioides	p	p	Reznik, 1955
Hylocereus undatus	p	—	Dreiding, 1961
Lobivia chlorogona	p	p	Reznik, 1957
L. famatimensis	p	p	Reznik, 1957
Mammillaria centricirrha	Bt, Ph	—	Piattelli and Imperato, 1969
M. hidalgensis	p	—	Dreiding, 1961
M. heyderi	p	—	Wohlpart and Mabry, 1968a
M. magnimamma	Bt, Ph	—	Piattelli and Imperato, 1969

M. neumanniana	Bt, Ph	—	Piattelli and Imperato, 1969
M. pusilla	p	p	Reznik, 1957
M. rhodantha	p	a	Reznik, 1957
M. seitziana	Bt, Ph	—	Piattelli and Imperato, 1969
M. setigera	p	a	Reznik, 1957
M. surculosa	p	a	Reznik, 1957
M. woodsii	p	a	Reznik, 1957
M. zeilmanniana	p	a	Reznik, 1957
M. zuccariniana	Bt, Ph	—	Piattelli and Imperato, 1969
Melocactus peruvianus	p	a	Reznik, 1957
Monvillea spegazzini	Bt	p	Wyler and Dreiding, 1961b
Neoporteria ebenacantha	p	p	Reznik, 1957
Nopalea dejecta	Bt, Ph	—	Piattelli and Imperato, 1969
Notocactus mammulosus	p	a	Reznik, 1957
N. ottonis	p	a	Reznik, 1957
Opuntia azurea	a	p	Reznik, 1957
O. bergeriana	Bt, Bd, Ph	In, Vu I–II, Mr II	Piattelli *et al.*, 1965a; Piattelli and Imperato, 1969
O. decumbens	Bt	—	Fischer and Dreiding, 1972
O. dillenii	Bt, Bd	—	Sciuto *et al.*, 1972
O. engelmannii	Bt, Ph	—	Piattelli and Imperato, 1969
O. ficus-indica	Bt	In, Vu I–II, Mr II	Piattelli *et al.*, 1964c; Piattelli *et al.*, 1965a
O. guatemalensis	Bt	—	Piattelli and Imperato, 1969
O. leptocaulis	p	—	Mabry *et al.*, 1963
O. lindheimeri	p	—	Mabry *et al.*, 1963
O. macrocentra	p	—	Wohlpart and Mabry, 1968a
O. monacantha	Bd, Pt	—	Piattelli and Imperato, 1969
O. paraguayensis	Bt, Ph	—	Piattelli and Imperato, 1969
O. polyacantha	Bd, Bt	—	Piattelli and Imperato, 1969
O. ritteri	Bt, Ph	—	Piattelli and Imperato, 1969
O. robusta	Bt	—	Erkut, 1962
O. soehrensi	p	a	Reznik, 1957
O. streptacantha	Bt, Ph	—	Piattelli and Imperato, 1969

TABLE II—*continued*

Family and species	Betacyanins	Betaxanthins	References
O. tomentella	Bt, Ph	—	Piattelli and Imperato, 1969
O. tomentosa	Bt, Ph	—	Piattelli and Imperato, 1969
O. vulgaris	Bt	—	Piattelli and Minale, 1964b
Parodia mutabilis	a	p	Reznik, 1957
P. sanguiniflora	p	p	Reznik, 1957
P. stuemeri var. *tilcarensis*	Am	p	Wyler and Dreiding, 1961b
Pereskia aculeata	p	p	Reznik, 1955
Phyllocactus hybridus	Bt, Ph	In, Vu I–II, Mr II	Piattelli *et al.*, 1965a; Minale *et al.*, 1966
P. phyllanthoides	p	—	Kryz, 1919, 1920
Pilosocereus glaucescens	Bt, Ph	—	Piattelli and Imperato, 1969
P. nobilis	Bt, Ph	—	Piattelli and Imperato, 1969
P. leucocephalus	Bd, Bt, Ph	—	Piattelli and Imperato, 1969
Rebutia krainziana	p	p	Reznik, 1957
R. marsoneri	p	p	Reznik, 1957
R. minuscola	p	p	Reznik, 1957
R. senilis	p	p	Reznik, 1957
Rhodocactus grandifolius	Bt, Ph	—	Piattelli and Imperato, 1969
Selinocereus grandiflorus	p	—	Lawrence *et al.*, 1939
Soehrensia bruchii	Bt, Ph	—	Piattelli and Imperato, 1969
Stenocereus stellatus	Bt, Ph	—	Piattelli and Imperato, 1969
Thelocactus bicolor	p	—	Dreiding, 1961
Chenopodiaceae: 100 (1500)			
Atriplex barclayana	p	—	Mabry *et al.*, 1963
A. hastata	p	—	Gertz, 1906
A. hortensis	Am, Ce	—	Piattelli and Minale, 1964b
A. hymenotheca	p	—	Wohlpart and Mabry, 1968a
A. lentiformis	p	—	Mabry *et al.*, 1963
A. littoralis	p	—	Gertz, 1906; Lawrence *et al.*, 1939

A. portulacoides	Am, Ce	—	Piattelli and Imperato, 1971
Beta vulgaris	Bt, Pb	Vu I–II	Piattelli and Minale, 1964b; Piattelli *et al.*, 1965b
Chenopodium album	Am, Bt, Ce	—	Piattelli and Minale, 1964b
C. amaranticolor	Am, Bt, Ce	—	Piattelli and Minale, 1964b
C. ambrosioides	p	—	Mabry *et al.*, 1963
C. berlandieri	p	—	Mabry *et al.*, 1963
C. glaucum	p	—	Mabry *et al.*, 1963
C. nuttalliae	p	—	Wohlpart and Mabry, 1968a,b
C. quinoa	p	—	Bischoff, 1876
C. rubrum	p	—	Wagner and Cumming, 1970
C. urbicum	Am, Bt	—	Piattelli and Imperato, 1971
C. virgatum	p	—	Gertz, 1906
Corispermum canescens	p	—	Gertz, 1906
C. nitidum	p	—	Mabry *et al.*, 1963
Cycloma atriplicifolium	p	—	Mabry *et al.*, 1963
Enchilaena tomentosa	p	—	Wohlpart and Mabry, 1968a
Kochia pentatropis	p	—	Wohlpart and Mabry, 1968a
K. scoparia	Bt, Ph	—	Piattelli and Minale, 1964b
K. tomentosa	p	—	Wohlpart and Mabry, 1968a
K. tricophylla	p	—	Lawrence *et al.*, 1939
Rhagodia nutans	p	—	Wohlpart and Mabry, 1968a
R. parabolica	p	—	Wohlpart and Mabry, 1968a
Salicornia fruticosa	Am, Ce	—	Piattelli and Imperato, 1971
S. perennis	p	—	Mabry *et al.*, 1963
Salsola soda	Am, Ce	—	Piattelli and Minale, 1964b
Spinacia oleracea	Am, Bt, Ph, Ce	—	Piattelli and Minale, 1964b
Suaeda fruticosa	Am, Ce, Su	—	Piattelli and Imperato, 1971
S. linearis	p	—	Mabry *et al.*, 1963
S. maritima	p	—	Lawrence *et al.*, 1939
Didieraceae: 4–5 (11)			
Alluaudia ascendens	p	a	Rauh and Reznik, 1961
A. humberti	p	a	Rauh and Reznik, 1961

TABLE II—*continued*

Family and species	Betacyanins	Betaxanthins	References
A. procera	p	a	Rauh and Reznik, 1961
Alluaudiopsis fiherenensis	p	a	Rauh and Reznik, 1961
A. marnieriana	p	a	Rauh and Reznik, 1961
Decarya madagascariensis	p	a	Rauh and Reznik, 1961
Didierea madagascariensis	p	a	Rauh and Reznik, 1961
D. trollii	p	a	Rauh and Reznik, 1961
Nyctaginaceae: 30 (300)			
Abronia amaliae	p	—	Mabry *et al.*, 1963
A. cycloptera	p	—	Mabry *et al.*, 1963
A. fragrans	p	—	Mabry *et al.*, 1963
A. villosa	p	—	Mabry *et al.*, 1963
Allionia incarnata	p	—	Mabry *et al.*, 1963
Anulocaulis gypsogenus	p	—	Wohlpart and Mabry, 1968a
Boerhaavia coccinea	p	—	Taylor, 1940
B. erecta	p	—	Mabry *et al.*, 1963
B. intermedia	p	—	Mabry *et al.*, 1963
B. scandens	p	—	Taylor, 1940
B. spicata	p	—	Mabry *et al.*, 1963
Bougainvillea fastuosa	p	—	Piattelli and Minale, 1964b
B. glabra var. *sanderiana*	11 Bg-v's	—	Piattelli and Imperato, 1970b
Bougainvillea sp. cv. "Mrs Butt"	Bt, 5 Bg-r's	—	Piattelli and Imperato, 1970a
B. spectabilis	p	p	Reznik, 1955; Wyler and Dreiding, 1961b
Cryptocarpus pyriformis	p	—	Taylor, 1940
Cyphomeris gypsophiloides	p	—	Mabry *et al.*, 1963
Mirabilis himalaica	p	—	Acheson, 1956
M. jalapa	Bt	In, Vu, I, Mr I–VI	Piattelli *et al.*, 1965c
M. lindheimeri	p	—	Mabry *et al.*, 1963
Nyctaginia capitata	p	—	Mabry *et al.*, 1963

Oxybaphus nyctagineus	p	a	Reznik, 1957
Phytolaccaceae: 17 (120)			
Phytolacca americana	Bt, Pb	—	Wyler and Dreiding, 1961a
P. australis	p	—	Taylor, 1940
Rivinia americana	p	p	Reznik, 1955, 1957
R. humilis	Bt	p	Reznik, 1955; Piattelli and Minale, 1964b
Trichostigma peruvianum	p	a	Reznik, 1955, 1957
Portulacaceae: 19 (500)			
Anacampseros rufescens	p	a	Reznik, 1955, 1957
Calandrinia grandiflora	p	a	Reznik, 1957
Claytonia linearis	p	—	Mabry *et al.*, 1963
C. megarrhiza	p	—	Mabry *et al.*, 1963
C. virginica	p	—	Mabry *et al.*, 1963
Montia perfoliata	p	—	Mabry *et al.*, 1963
Portulaca grandiflora	Bd, Bt	In, Vu I–II, Mr II, Px	Piattelli and Minale, 1964b; Piattelli *et al.*, 1965a
P. intraterranea	—	p	Adachi and Katayama, 1968
P. marginata	—	p	Adachi and Katayama, 1968
P. oleracea	p	p	Piattelli and Minale, 1964b; Adachi and Katayama, 1968
P. pilosa	p	p	Mabry *et al.*, 1963; Adachi and Katayama, 1968
Spraguea umbellata	p	—	Mabry *et al.*, 1963
Stegnospermaceae: 1 (3)			
Stegnosperma halimifolium	p	—	Mabry *et al.*, 1963

[a] The size of the family is indicated by the estimated number of genera (species).

[b] Abbreviations: –, no report; a, absent; p, present; Am, amaranthin; Bd, betanidin; Bg, bougainvillein; Bt, betanin; Ce, celosianin; Db, decarboxybetanidin; Dp, dopaxanthin; Dr, drosanthemin; Gm, gomphrenin; In, indicaxanthin; Ir, iresinin; Lm, lampranthin; Mr, miraxanthin; Pb, prebetanin; Ph, phyllocactin; Px, portulaxanthin; Su, suaedin; Vu, vulgaxanthin.

[c] Betanidin derivatives are almost invariably accompanied by the corresponding isobetanidin derivatives; therefore, the occurrence of the latter pigments is not shown in the table.

or even in the same family (other classes of flavonoids are, however, common in betalain families) and the restricted distribution of betalains are good arguments in favour of the outstanding taxonomic significance ascribed by Mabry to these pigments. However, of an estimated number of approximately 8000 species in the Centrospermae, only about 300 have been as yet examined for betalains (Table II), so the coverage is less than 4%. More extensive surveys are therefore badly needed to corroborate the present view on the phylogenetical correlation among the betalain-containing plants. Possibly a more thorough investigation, not limited simply to ascertaining the presence or absence of betalains would allow their systematic significance to be extended to disclose relationships between distribution patterns and botanical classification of plants of the Centrospermae at lower taxonomic level. To this end, plant material collected in the wild should be chosen for examination, excluding that obtainable from botanical gardens which is often of unknown origin or has been hybridized in cultivation. In this connection, it is unfortunate that a number of the data recorded in Table II, in particular those referring to more detailed examinations, have been obtained from cultivated plants (the aim of the work was in general the identification and characterization of new pigments rather than a deliberate phytochemical survey) and therefore cannot be confidently used to relate chemical data to classification. Anyway, on the basis of present knowledge, betalain distribution in Centrospermae seems not to be completely erratic. Amaranthin, for example, appears to be restricted to Amaranthaceae and Chenopodiaceae (the presence of this betacyanin in *Parodia stuemeri*, of the family Cataceae, deserves confirmation), and phyllocactin to Cactaceae and Chenopodiaceae.

Lastly, it is to be mentioned that recently (Döpp and Musso, 1973a,b; Musso 1973) a violet pigment, muscapurpurin (λ_{max} 540 nm), and seven yellow ones, musca-aurins I–VII (λ_{max} 475 nm), have been isolated from the poisonous mushroom *Amanita muscaria* (fly agaric) and identified as betalains on the basis of u.v. spectra and degradation to betalamic acid. One of these compounds, musca-aurin I, could be allocated structure (**XXIII**), which was confirmed by partial synthesis from betalamic acid and ibotenic acid. The occurrence of pigments of the same type in otherwise unrelated plants is likely to be due to chemical convergence under evolutionary pressure; clearly, it should be of great interest to establish whether or not the biosynthetical route to the fungal pigments is the same as that found in the family Centrospermae.

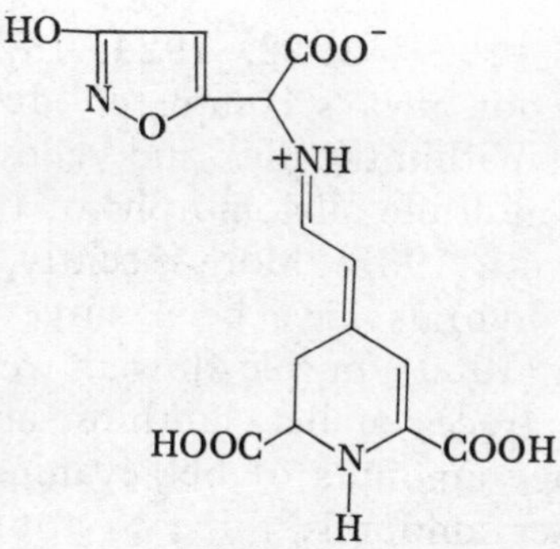

HO COO⁻ N O +NH HOOC N COOH H

(XXIII)

VIII Function of betalains

The physiological function of these pigments in the plant is unknown, except that when present in flowers or fruits they may have a role, like the anthocyanins, in insect or bird pollination and in seed dispersal by animals, respectively. The occurrence in other plant parts (leaves, stem, root) may be devoid of immediate function. Also the transient coloration of many seedlings of the Centrospermae and the autumnal reddening of the leaves of certain plants of this order, e.g. *Kochia scoparia*, have no obvious physiological significance and are perhaps related to the accumulation of carbohydrates. Betalains may also form in injured tissues, normally not pigmented, of plants possessing the specific factors necessary for their synthesis, possibly as a defence mechanism against viral infection. These pigments, in fact, are known to inhibit viral reproduction (Sosnova, 1970).

IX Pharmacology of betalains

Although structurally related to alkaloids—betacyanins have been described as "chromoalkaloids" (Mabry *et al.*, 1962)—betalains are non-toxic, as can be deduced from the fact that they are present in considerably large concentrations in certain foodstuffs such as beetroot or prickly pear. After ingestion of beetroot, betanin occasionally appears in the urine; the aetiology and the mechanism of this disorder (beeturia, or betaninuria) are still controversial (Watson, 1964; Geldmacher-Mallinckrodt and Mendner, 1965; Forrai *et al.*, 1968; Adam *et al.*, 1969).

X Inheritance of betalains

The mode of flower colour inheritance within the species *Portulaca grandiflora* has been extensively studied by Japanese geneticists

(Yasui, 1920; Ikeno, 1921, 1922, 1924, 1928; Enamoto, 1923, 1927). White forms, not always completely devoid of pigment, are recessive to coloured. Within the pink and yellow series, three colour shades expressed by multiple allelomorphs of the C gene have been distinguished (Yamaguti, 1935). More recently, a number of colour forms and their F_1 hybrids have been surveyed for betalains by Ootani and Hagiwara (1969); purple flowers were shown to contain five betacyanins and traces of betaxanthins, and yellow ones three betaxanthins and trace amounts of betacyanins. Four to five betacyanins and, in lesser amounts, three to four betaxanthins were detected in red flowers. Intermediate colour forms are due to varying proportion of betacyanins and betaxanthins. Betacyanins are, in general, dominant to betaxanthins and, in F_1 hybrids, the higher content of betacyanin in purple form is dominant to the lower content in pink form, while the higher content of betaxanthin in yellow orange form is recessive to the lower content in pink form.

A genetical analysis of betalain formation in petals of *P. grandiflora* in sterile culture has also been carried out (Adachi and Katayama, 1970).

XI Conclusion

After nearly a century of frustrating efforts, in the last decade or so the fundamentals of betalain chemistry have been soundly based, primarily as the result of the introduction of new sophisticated techniques of isolation and structural analysis of organic compounds. The biosynthesis and the physiology of these pigments have also been actively investigated. Much work, however, requires to be done. Some aspects of the biosynthesis are still obscure, and nothing at all is known about the enzymology of the biosynthesis and the metabolic pathways by which the pigments are degraded *in vivo*. Finally, it is to be hoped that the control mechanisms of betalain biosynthesis will become more clear with the acquisition of further data.

REFERENCES

Acheson, R. M. (1956). *Proc. R. Soc. Ser. B* 145, 549–553.

Adachi, T. and Katayama, Y. (1968). *Bull. Fac. Agri. Univ. Miyazaki* 15, 239–247.

Adachi, T. and Katayama, Y. (1970). *Bull. Fac. Agri. Univ. Miyazaki* 16, 137–145.

Adam, E., Farriaux, J. P. and Fontaine, G. (1969). *Acta paediat. belg.* 23, 209–227.

Ainley, A. D. and Robinson, R. (1937). *J. chem. Soc.* 446–449.
Aronoff, F. and Aronoff, E. M. (1948). *Fd Res.* 13, 59–65.
Badgett, B., Parikh, I. and Dreiding, A. S. (1970). *Helv. chim. Acta* **53**, 433–448.
Bamberger, E. and Mayer, A. M. (1960). *Science, N.Y.* **131**, 1094–1095.
Birnbaum, D. and Koehler, H.-K. (1970). *Biochem. Physiol. Pfl.* **161**, 521–531.
Bischoff, H. (1876). Inaugural Dissertation, Tübingen.
Constabel, F. and Nassif-Makki, H. (1971). *Ber. dt. bot. Ges.* **84**, 629–636.
Döpp, H. and Musso, H. (1973a). *Naturwissenschaften* **60**, 477–478.
Döpp, H. and Musso, H. (1973b). *Chem. Ber.* **106**, 3473–3482.
Dreiding, A. S. (1961). *In* "Recent Developments in the Chemistry of Natural Phenolic Compounds" (W. O. Ollis, ed.), pp. 194–211. Pergamon Press, Oxford.
Dunkelblum, E., Miller, H. E. and Dreiding, A. S. (1972). *Helv. chim. Acta* **55**, 642–648.
Enamoto, N. (1923). *Jap. J. Bot.* 1, 137–151.
Enamoto, N. (1927) *Jap. J. Bot.* 3, 267–288.
Engler, A. (1936). "Syllabus der Pflanzenfamilien", 11th edn. (L. Diels, ed.). Borntraeger, Berlin.
Erkut, H. (1962). *Istanb. Univ. Fen. Fak. Mecm. C* 27, 67–73.
Fischer, N. and Dreiding, A. S. (1972). *Helv. chim. Acta* **55**, 649–658.
Forrai, G., Vágújfalvi, D. and Bölcskey, P. (1968). *Acta paediat. hung.* 9, 43–51.
Fujisawa, H. and Hayaishi, O. (1968). *J. biol. Chem.* **243**, 2673–2681.
Garay, A. S. and Towers, G. H. N. (1966). *Can. J. Bot.* **44**, 231–236.
Geldmacher-Mallinckrodt, M. and Mendner, K. (1965). *Dt. Z. ges. gericht. Med.* 56, 287–298.
Gertz, O. (1906) Dissertation, Lund.
Giudici de Nicola, M., Piattelli, M., Castrogiovanni, V. and Molina, C. (1972a). *Phytochemistry* 11, 1005–1010.
Giudici de Nicola, M., Piattelli, M., Gastrogiovanni, V. and Amico, V. (1972b). *Phytochemistry* 11, 1011–1017.
Giudici de Nicola, M., Piattelli, M. and Amico, V. (1973a). *Phytochemistry* **12**, 353–357.
Giudici de Nicola, M., Piattelli, M. and Amico, V. (1973b). *Phytochemistry* **12**, 2163–2166.
Giudici de Nicola, M., Amico, V. and Piattelli, M. (1974). *Phytochemistry* 13, 439–442.
Harborne, J. B. (1965). *Phytochemistry* **4**, 647–657.
Harley-Mason, J. (1965). *In* "Comprehensive Biochemistry" (M. Florkin and E. H. Stotz, eds), vol. 6, pp. 254–257. Elsevier, Amsterdam.
Hartmann, K. M. (1966). *Photochem. Photobiol.* 5, 349–366.
Haverland, F. (1892). Inaugural Dissertation, Erlangen.
Hayaishi, O. and Nozaki, M. (1969). *Science, N.Y.* 164, 389–396.
Heath, O. V. S. and Vince, D. (1962). *Symp. Soc. exp. Biol.* **16**, 114–137.
Hörhammer, L., Wagner, H. and Fritzsche, W. (1964). *Biochem. Z.* 339, 398–400.
Hutchinson, J. (1959). "The Families of Flowering Plants", vol. 1. Clarendon Press, Oxford.
Ikeno, S. (1921). *J. Coll. Agric. imp. Univ. Tokyo* **8**, 93–133.
Ikeno, S. (1922). *Z. indukt. Abstamm.-u. VererbLehre.* **29**, 122–135.
Ikeno, S. (1924). *Jap. J. Bot.* 2, 45–62.

Ikeno, S. (1928). *Jap. J. Bot.* **4**, 189–218.
Impellizzeri, G. and Piattelli, M. (1972). *Phytochemistry* **11**, 2499–2502.
Impellizzeri, G., Piattelli, M. and Sciuto, S. (1973a). *Phytochemistry* **12**, 2293–2294.
Impellizzeri, G., Piattelli, M. and Sciuto, S. (1973b). *Phytochemistry* **12**, 2295–2296.
Kendrick, R. E. and Frankland, B. (1969). *Planta* **86**, 21–32.
Kimler, L., Larson, R. A., Messenger, R., Moore, J. B. and Mabry, T. J. (1971). *Chem. Commun.* 1329–1340.
Koehler, K.-H. (1967). *Ber. dt. bot. Ges.* **80**, 403–415.
Koehler, K.-H. (1972a). *Biol. Rundschau* **10**, 273–274.
Koehler, K.-H. (1972b). *Phytochemistry* **11**, 127–131.
Koehler, K.-H. (1972c). *Phytochemistry* **11**, 133–137.
Koehler, K.-H. and Birnbaum, D. (1970a). *Biol. Zbl.* **89**, 201–211.
Koehler, K.-H. and Birnbaum, D. (1970b). *Biochem. Physiol. Pfl.* **161**, 511–520.
Kryz, F. (1919). *Z. Nahr. Genussm.* **38**, 364–365.
Kryz, F. (1920). *Ost. Chem. Ztg* **23**, 55–56.
Lawrence, W. J. C., Price, J. R., Robinson, G. M. and Robinson, R. (1939). *Phil. Trans. R. Soc. Ser. B* **230**, 149–178.
Liebisch, H. W., Matschiner, B. and Schuette, H. R. (1969). *Z. PflPhysiol.* **61**, 269–278.
Lindstedt, G. (1956). *Acta chem. scand.* **10**, 698–699.
Mabry, T. J. (1964). *In* "Taxonomic Biochemistry and Serology" (C. A. Leone, ed.), pp. 239–254. Ronald Press, New York.
Mabry, T. J. and Dreiding, A. S. (1968). *In* "Recent Advances in Phytochemistry" (T. J. Mabry, R. E. Alston and V. C. Runeckles, eds), vol. 1, pp. 145–160. Appleton-Century-Crofts, New York.
Mabry, T. J., Wyler, H., Sassu, G., Mercier, M., Parikh, I. and Dreiding, A. S. (1962). *Helv. chim. Acta* **45**, 640–647.
Mabry, T. J., Taylor, A. and Turner, B. L. (1963). *Phytochemistry* **2**, 61–64.
Mabry, T. J., Wyler, H., Parikh, I. and Dreiding, A. S. (1967). *Tetrahedron* **23**, 3111–3127.
Miller, H. E., Rösler, H., Wohlpart, A., Wyler, H., Wilcox, M. E., Frohofer, H., Mabry, T. J. and Dreiding, A. S. (1968). *Helv. chim. Acta* **51**, 1470–1474.
Minale, L., Piattelli, M. and Nicolaus, R. A. (1965). *Phytochemistry* **4**, 593–597.
Minale, L., Piattelli, M., De Stefano, S. and Nicolaus, R. A. (1966). *Phytochemistry* **5**, 1037–1052.
Minale, L., Piattelli, M. and De Stefano, S. (1967). *Phytochemistry* **6**, 703–709.
Mohr, H. (1966a). *Z. PflPhysiol.* **54**, 63–83.
Mohr, H. (1966b). *Photochem. Photobiol.* **5**, 469–483.
Mohr, H. (1972). "Lectures on Photomorphogenesis". Springer-Verlag, Berlin.
Musso, H. (1973). *Chimia* **27**, 659.
Nicolaus, R. A., Piattelli, M. and Fattorusso, E. (1964). *Tetrahedron* **20**, 1163–1172.
Nilsson, T. (1970). *LantbrHögsk. Annl* **36**, 179–219.
Nozaki, M., Ono, K., Nakazawa, T., Kotani, S. and Hayaishi, O. (1968). *J. biol. Chem.* **243**, 2682–2690.
Ootani, S. and Hagiwara, T. (1969). *Jap. J. Genet.* **44**, 65–79.
Piattelli, M. and Impellizzeri, G. (1969). *Phytochemistry*, **8**, 1595–1596.
Piattelli, M. and Impellizzeri, G. (1970). *Phytochemistry* **9**, 2553–2556.

Piattelli, M. and Imperato, F. (1969). *Phytochemistry* **8**, 1503-1507.
Piattelli, M. and Imperato, F. (1970a). *Phytochemistry* **9**, 455-458.
Piattelli, M. and Imperato, F. (1970b). *Phytochemistry* **9**, 2557-2560.
Piattelli, M. and Imperato, F. (1971). *Phytochemistry* **10**, 3133-3134.
Piattelli, M. and Minale, L. (1964a). *Phytochemistry* **3**, 307-311.
Piattelli, M. and Minale, L. (1964b). *Phytochemistry* **3**, 547-557.
Piattelli, M. and Minale, L. (1966). *Annali Chim.* **56**, 1060-1064.
Piattelli, M., Minale, L. and Prota, G. (1964a). *Annali Chim.* **54**, 955-962.
Piattelli, M., Minale, L. and Prota, G. (1964b). *Annali Chim.* **54**, 963-968.
Piattelli, M., Minale, L. and Prota, G. (1964c). *Tetrahedron* **20**, 2325-2329.
Piattelli, M., Minale, L. and Nicolaus, R. A. (1965a). *Rc. Accad. Sci. fis. mat., Napoli* 32, 55-56.
Piattelli, M., Minale, L. and Prota, G. (1965b). *Phytochemistry* **4**, 121-125.
Piattelli, M., Minale, L. and Nicolaus, R. A. (1965c). *Phytochemistry* **4**, 817-823.
Piattelli, M., Giudici de Nicola, M. and Castrogiovanni, V. (1969). *Phytochemistry* **8**, 731-736.
Piattelli, M., Giudici de Nicola, M. and Castrogiovanni, V. (1970a). *Atti Accad. naz. Lincei Rc.* **VIII**, **48**, 255-260.
Piattelli, M., Giudici de Nicola, M. and Castrogiovanni, V. (1970b). *Phytochemistry* **9**, 785-789.
Piattelli, M., Giudici de Nicola, M. and Castrogiovanni, V. (1971). *Phytochemistry* **10**, 289-293.
Price, J. R., Robinson, R. and Scott-Moncrieff, R. (1939). *J. chem. Soc.* 1465-1468.
Pucher, G. W., Curtis, L. C. and Vickery, H. B. (1938). *J. biol. Chem.* **123**, 61-70.
Raper, H. S. (1927). *Biochem. J.* **21**, 89-96.
Raper, H. S. (1928). *Physiol. Rev.* **8**, 245-282.
Rast, D., Skřivanová, R. and Bachofen, R. (1973). *Phytochemistry* **12**, 2669-2672.
Rauh, W. and Reznik, H. (1961). *Bot. Jb.* **81**, 94-105.
Reznik, H. (1955). *Z. Bot.* **43**, 499-530.
Reznik, H. (1957). *Planta* **49**, 406-434.
Schmidt, O. T. and Schönleben, W. (1956). *Naturwissenschaften* **43**, 159.
Schmidt, O. T. and Schönleben, W. (1957). *Z. Naturf.* **12b**, 262-263.
Schmidt, O. T., Becher, P. and Hübner, M. (1960). *Chem. Ber.* **93**, 1296-1304.
Schudel, G. (1918). Dissertation, Zürich.
Sciuto, S., Oriente, G. and Piattelli, M. (1972). *Phytochemistry* **11**, 2259-2262.
Sciuto, S., Oriente, G., Piattelli, M., Impellizzeri, G. and Amico, V. (1974). *Phytochemistry* **13**, 947-951.
Sosnova, V. (1970). *Biol. Plant.* **12**, 425-427.
Taylor, T. W. J. (1940). *Proc. R. Soc. Ser. B* **129**, 230-237.
Wagner, E. and Cumming, B. G. (1970). *Can. J. Bot.* **48**, 1-18.
Wagner, E. and Frosch, S. (1971). *Can. J. Bot.* **49**, 1981-1985.
Watson, W. C. (1964). *Biochem. J.* **90**, 3.
Weigert, L. (1894). "Jahresbericht und Programm der k. k. önologischen und pomologischen Lehranstalt in Klosterneuburg". Wien.
Wettstein, R. von (1935). "Handbuch der systematischen Botanik", 4th edn. Deuticke, Leipzig-Wien.

Wilcox, M. E., Wyler, H., Mabry, T. J. and Dreiding, A. S. (1965a). *Helv. chim. Acta* **48**, 252–258.
Wilcox, M. E., Wyler, H. and Dreiding, A. S. (1965b). *Helv. chim. Acta* **48**, 1134–1147.
Wohlpart, A. and Black, S. M. (1973). *Phytochemistry* **12**, 1325–1329.
Wohlpart, A. and Mabry, T. J. (1968a). *Taxon* **17**, 148–152.
Wohlpart, A. and Mabry, T. J. (1968b). *Pl. Physiol., Lancaster* **43**, 457–459.
Wyler, H. and Dreiding, A. S. (1957). *Helv. chim. Acta* **40**, 191–192.
Wyler, H. and Dreiding, A. S. (1959). *Helv. chim. Acta* **42**, 1699–1702.
Wyler, H. and Dreiding, A. S. (1961a). *Helv. chim. Acta* **44**, 249–257.
Wyler, H. and Dreiding, A. S. (1961b) *Experientia* **17**, 23–25.
Wyler, H. and Dreiding, A. S. (1962). *Helv. chim. Acta* **45**, 638–640.
Wyler, H., Mabry, T. J. and Dreiding, A. S. (1963). *Helv. chim. Acta* **46**, 1745–1748.
Wyler, H. Wilcox, M. E. and Dreiding, A. S. (1965). *Helv. chim. Acta* **48**, 361–366.
Wyler, H., Rösler, H., Mercier, M. and Dreiding, A. S. (1967). *Helv. chim. Acta* **50**, 545–561.
Yamaguti, Y. (1935). *Jap. J. Genet.* **11**, 109–112.
Yasui, K. (1920). *Bot. Mag., Tokyo* **34**, 55–65.

Chapter 12

Miscellaneous Pigments

R. H. THOMSON

Department of Chemistry, University of Aberdeen, Scotland

I Introduction

The major groups of plant pigments, discussed in the preceding chapters, by no means exhaust the range of chemical structures found in natural colouring matters. It is important that investigators should be aware of this, and the purpose of the present chapter is to draw attention to the existence of a number of smaller groups. The significance of these pigments is not clear, and some have a very restricted distribution although the actual number of compounds can be substantial. More than thirty phenazines and well over a hundred xanthones have been described, and the numbers continue to increase, but most groups are much smaller and in some cases only individual compounds can be cited.

A collection of miscellaneous pigments naturally includes a number of different chromophores but the great majority contain a conjugated carbonyl system which is modified in various ways. Quinone methides and phenoxazones are obviously related to the true quinones, and the γ-pyrone pigments share a ring system with the flavones. Many yellow pigments are simply phenolic ketones but a few, like erythroskyrin* (I) (*Penicillium islandicum*) contain a

* References to most of the fungal pigments cited are given by W. B. Turner (1971).

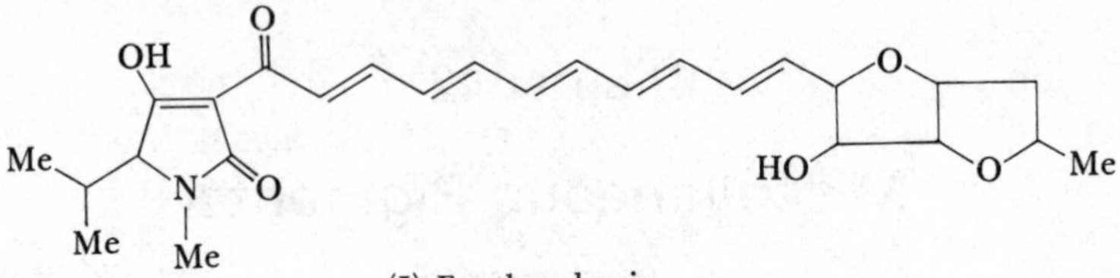

(I) Erythroskyrin

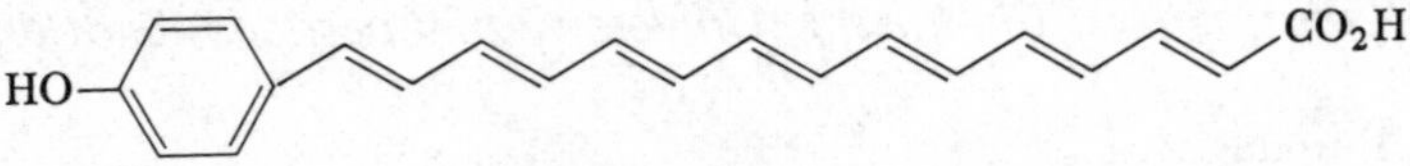

(II) Cortisalin

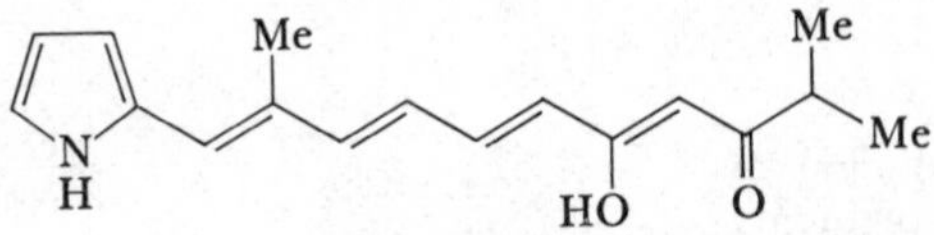

(III) Wallemia A

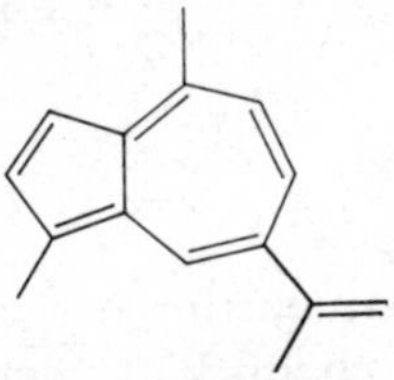

(IV) Lactarazulene

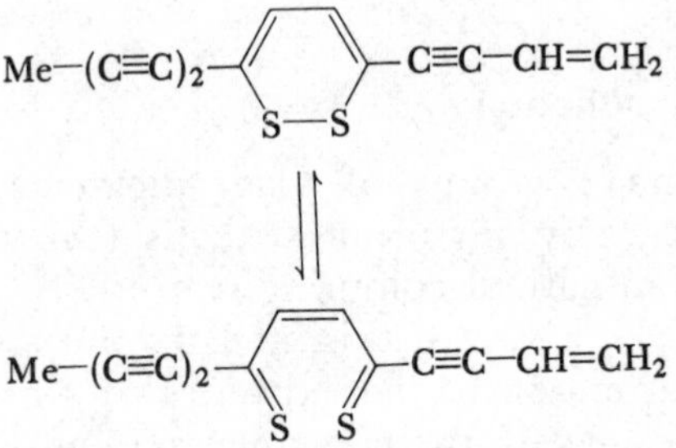

$Me-(C\equiv C)_2-$ ... $-C\equiv C-CH=CH_2$

(V)

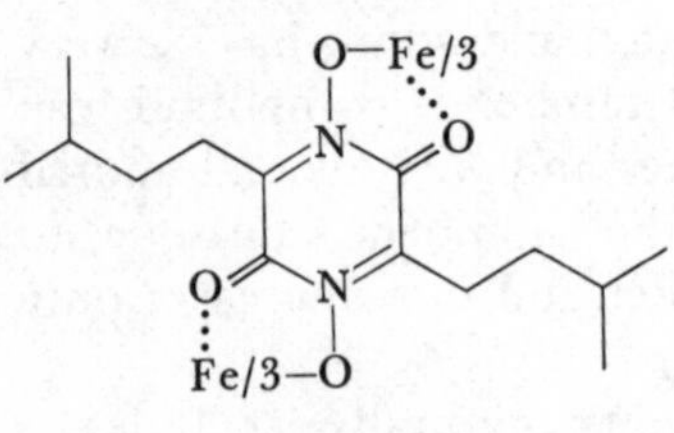

(VI) Pulcherrimin

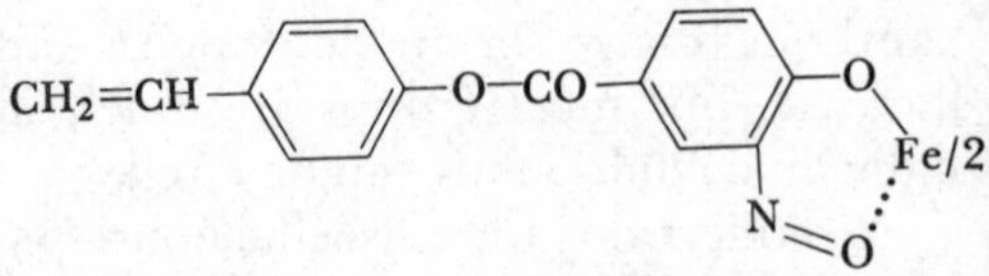

(VII) Ferroverdin

polyene system. Cortisalin (**II**), the violet-red pigment in the fruit bodies of *Corticium salicinum*, is reminiscent of the carotenoids, and so is wallemia A (**III**) the major pigment in the bright orange fungus *Wallemia sebi* (Badar *et al.*, 1973).

Pigments lacking a carbonyl group are the exception. They include one or two azulenes, for example lactarazulene (**IV**), a blue pigment in the toadstool *Lactarius deliciosus*, and a few red oils of type (**V**), present in various Compositae, which could be described as thiocarbonyl compounds (Bohlmann *et al.*, 1973). Some pigments owe their colour, at least in part, to a metal atom, and there may be no carbonyl group in the chromophoric system. The red pulcherrimin (**VI**) (*Candida pulcherrima*) and the green ferroverdin (**VII**) (*Streptomyces* spp.) (Ballio *et al.*, 1963) are examples containing iron. Amavadine, a blue pigment from the toadstool *Amanita muscaria*, is a unique vanadium complex (Kneifel and Bayer, 1973).

II *o*-Acylphenols

The appearance of colour is often incidental, arising from a minor modification of a chromophore which normally absorbs only in the ultra-violet region. Aromatic ketones provide many examples; the parent compounds and their monohydroxy derivatives are normally colourless but the introduction of additional hydroxyl groups may shift the long-wave absorption into the visible region. Thus a simple ketone like (**VIII**) (*Daldinia concentrica*) is yellow, and so is pyoluteorin (**IX**) (*Pseudomonas aeruginosa*) and the heartwood constituent (**X**) (*Maesopsis eminii*), sorbicillin (**XI**) (*Penicillium notatum*) is orange, and the glyoxylic acid (**XII**) (*Polyporus tumulosus*) is red.

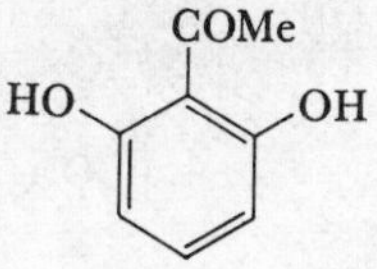

(**VIII**)

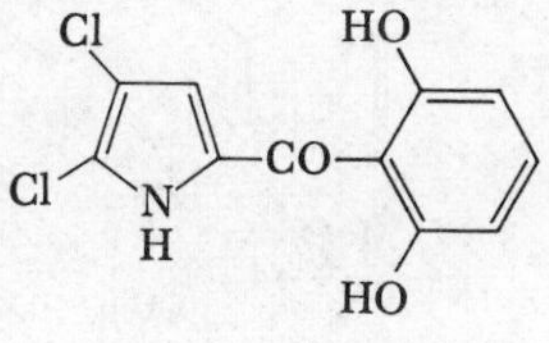

(**IX**) Pyoluteorin

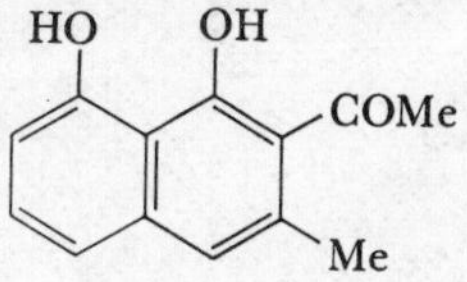

(**X**) Musizin

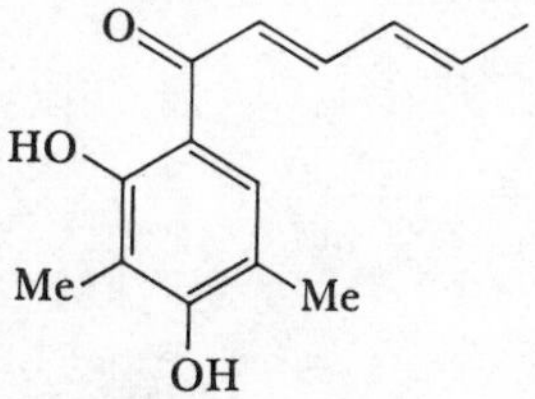

(**XI**) Sorbicillin

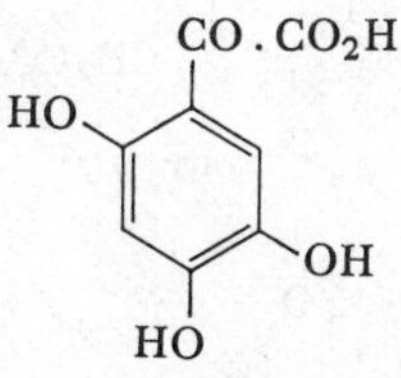

(**XII**)

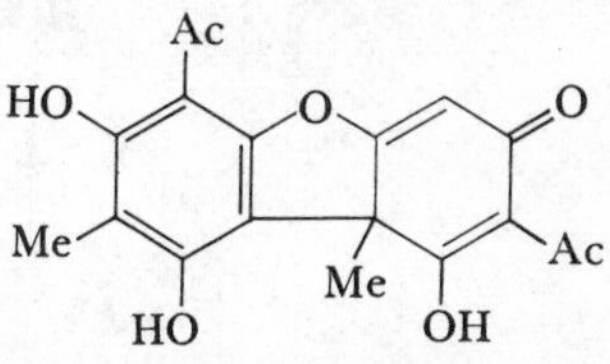

(**XIII**) Usnic acid

A more complex example is usnic acid (**XIII**), a yellow pigment widely distributed in lichens, occurring in both (+) and (–) forms. Feeding experiments with several species (Taguchi *et al.*, 1969) established the polyketide origin of usnic acid, and methylphloroacetophenone from which it is derived by phenolic coupling (Barton *et al.*, 1956), was incorporated. The monomer was also detected in the lichens.

Among higher fungi some *Cortinarius, Dermocybe* and *Tricholoma* spp. are intensely yellow owing to the presence of pigments of the flavomannin (**XIV**) and phlegmacin (**XV**) type (Steglich *et al.*, 1972b, 1973; Steglich and Töpfer-Petersen, 1972, 1973). These tetrahydroanthracene derivatives are easily converted into quinones by oxidation and dehydration, and they frequently co-occur with mono- and bianthraquinones. Related bianthrones occur in other fungi, and lichens. Chromophorically the pigments of type (**XIV**) and (**XV**) are similar to musizin (**X**) and biogenetically they belong to the emodin group of anthraquinones (p. 537).

Most of the pigments in this section are derived from polyketides but the coleones are a group of diterpenoid colouring matters, related to the royleanone and tanshinone quinones, found on the leaves of *Coleus* spp. (Labiatae). Some of these pigments are quinone methides and some are keto quinols. An example of the second type

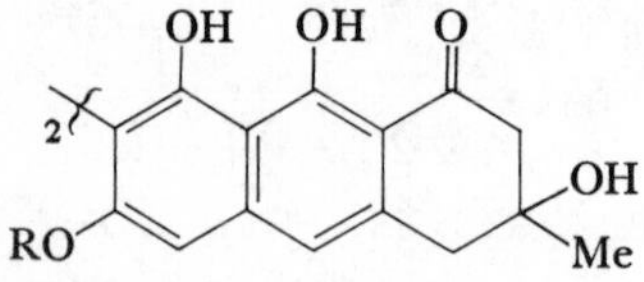

OH OH O
RO
OH
Me
2

(**XIV**) Flavomannin

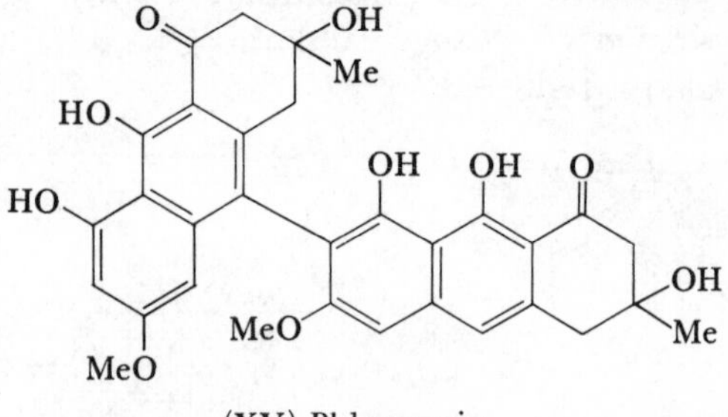

O
OH
Me
HO
HO
MeO
OH OH O
MeO
OH
Me

(**XV**) Phlegmacin

OH CH_2OAc
HO
CH_2OAc
OH
HO
O
O

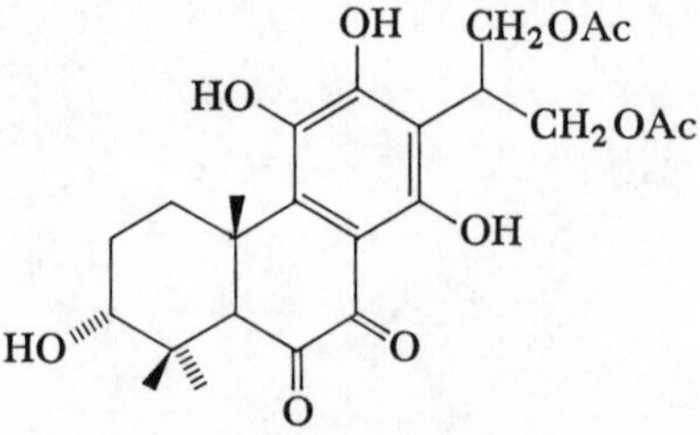

(**XVI**) Coleone K

is the orange coleone K (**XVI**) (Moir *et al.*, 1973) whose chromophore is very similar to that of the glyoxylic acid (**XII**). A closely related yellow pigment, lycoxanthol, has been found in the clubmoss *Lycopodium lucidulum* (Burnell *et al.*, 1972).

III Quinone methides

This scattered group of pigments has certain features in common with true quinones, and they cover a range of structural types (A. B. Turner, 1966). Like quinones they are easily reduced, the resulting leuco compounds containing one, and not two, additional phenolic groups. Examples from fungi are the yellow pulvilloric acid (**XVII**) and the orange fuscin (**XVIII**). The red pigment, obtusaquinone (**XIX**), has a more extended conjugated system and occurs in the heartwood of *Dalbergia obtusa* along with dalbergiones and other C_6-C_3-C_6 compounds (Gregson *et al.*, 1968). A small group of quinone methides is elaborated by the wood-rotting fungus *Peniophora sanguinea*. Two of these are xylerythrin (**XX**) and peniosanguin (**XXI**) both of which form almost black crystals (Gripenberg

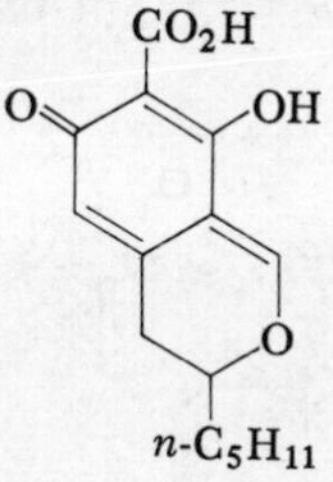

(**XVII**) Pulvilloric acid

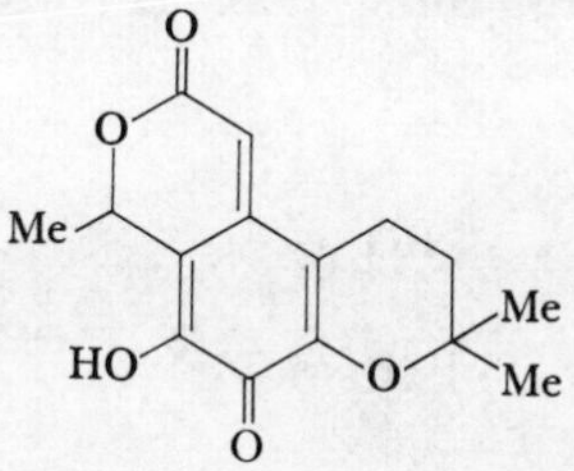

(**XVIII**) Fuscin

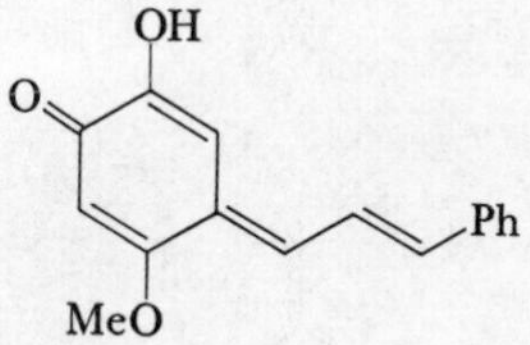

(**XIX**) Obtusaquinone

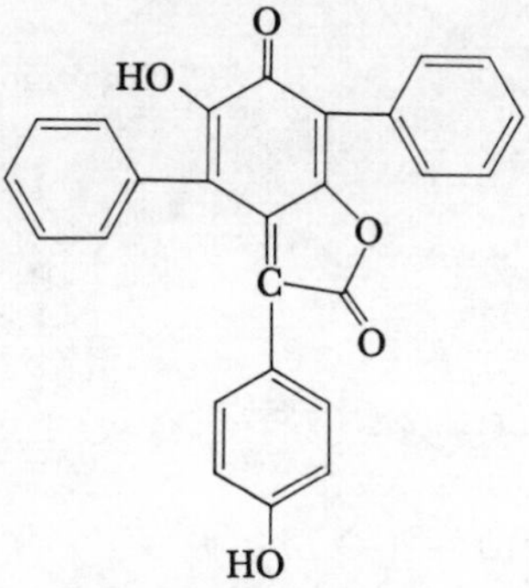

(**XX**) Xylerythrin

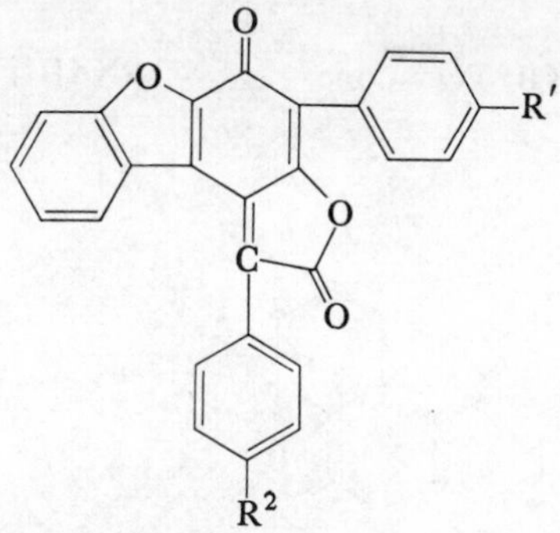

(**XXI**) Peniosanguin ($R' = HO$, $R^2 = H$ or $R' = H$, $R^2 = HO$)

and Martikkala, 1969; Gripenberg, 1971). They are closely related to the terphenylquinones, and formally xylerythrin can be derived by condensation of *p*-hydroxyphenylacetic acid and polyporic acid; the latter has been found in *Peniophora filamentosa.*

There are several diterpene and triterpene quinone methides some of which were at first mistaken for true quinones. Fuerstione (**XXII**) is the red pigment in the leaf glands of *Fuerstia africana* (Labiatae) (Karanatsios *et al.*, 1966). The coleone leaf gland pigments have already been mentioned and coleone E (**XXIII**) is one of the quinone methides; the conjugation extends through three rings (Rüedi and Eugster, 1972). Taxodione (**XXIV**) and taxodone, the antitumour pigments from the seeds of *Taxodium distichum* (Taxodiaceae), are closely related to fuerstione (Kupchan *et al.* , 1969). The root-bark of *Maytenus dispermus* (Celastraceae) contains both maytenoquinone, a diterpenoid pigment with the same chromophore as taxodione (Martin, 1973a), and pristimerin and dispermoquinone which are triterpenoid (Martin, 1973b). All the known triterpenoid quinone methides are Celastraceae root pigments. With the exception of dispermoquinone (**XXV**), the chromophore in rings A and B is like that of fuerstione, e.g. tingenone (**XXVI**) (Brown *et al.*, 1973; Delle

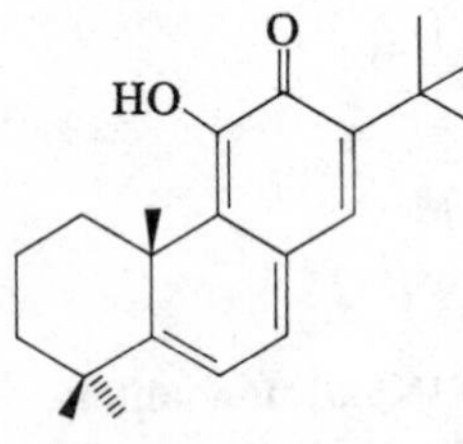

(**XXII**) Fuerstione

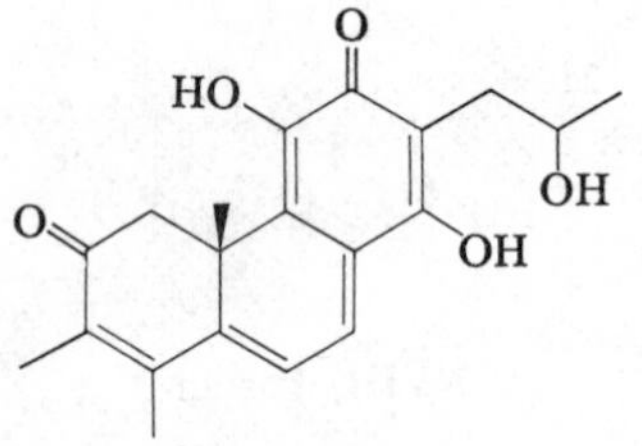

(**XXIII**) Coleone E

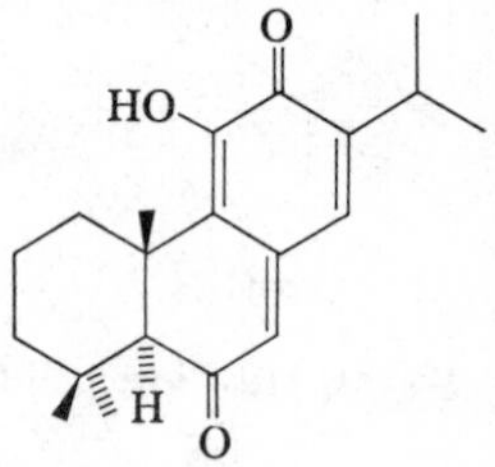

(**XXIX**) Taxodione

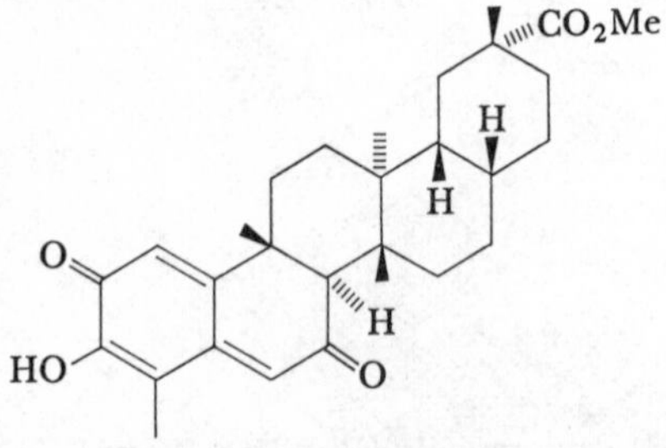

(**XXV**) Dispermoquinone

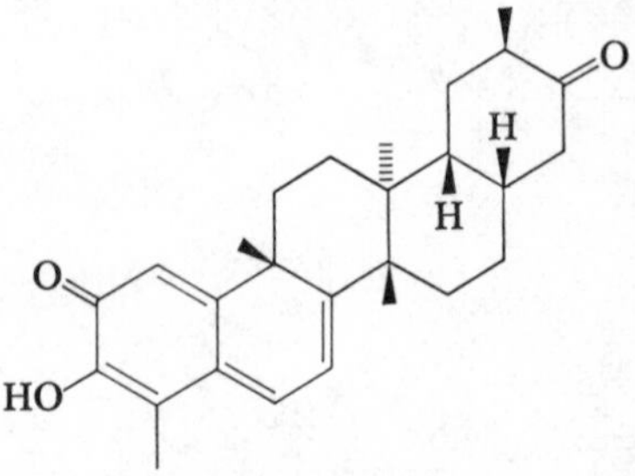

(**XXVI**) Tingenone

Monache *et al.*, 1973; Nakanishi *et al.*, 1973). The saptarangi quinones in the roots of *Salacia macrosperma* are probably also triterpenoid (Krishnan and Rangaswami, 1971).

A few quinone methides form a subsection of the flavanoid pigments (Dean, 1963). These red compounds, of limited distribution, are "anhydrobases" corresponding to flavylium salts of which carajurin (**XXVII**), from the leaves of *Bignonia chica*, is a simple example. They are stable only in the absence of a substituent at position 3, and while they readily form flavylium salts (**XXVIII**) in acid solution conversion of the latter into pseudo bases is relatively difficult. As anthocyanins are normally isolated from plant material by acid extraction it is difficult to exclude the possibility that some of these may exist *in vivo* in the anhydro base form. Dracorubin (**XXIX**) is one of the complex mixture of pigments found in "dragon's blood", a dark red resin exuded by certain *Dracaena* (Liliaceae) and *Deamonorops* (Palmaceae) trees. Dracorubin is a "bisflavanoid" clearly derived from a monomer of the carajurin type, but santalin A (**XXX**) is a more complex dimer of a new type apparently formed by coupling of an isoflavanoid monomer with an open chain C_6-C_3-C_6 unit (Mathieson *et al.*, 1973; Arnone *et al.*,

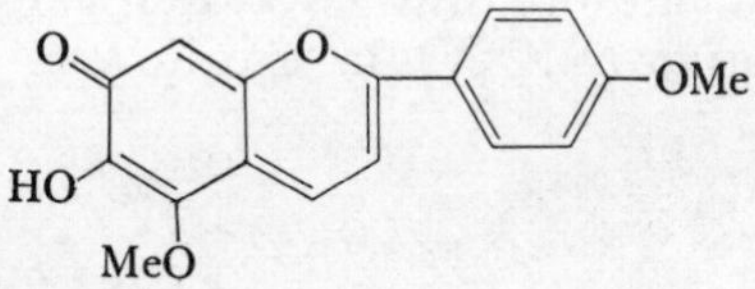

(XXVII) Carajurin

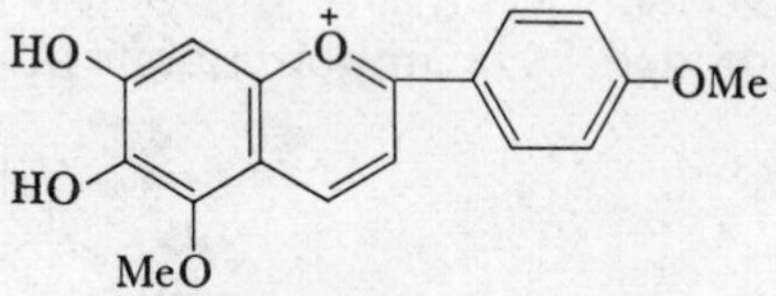

(XXVIII)

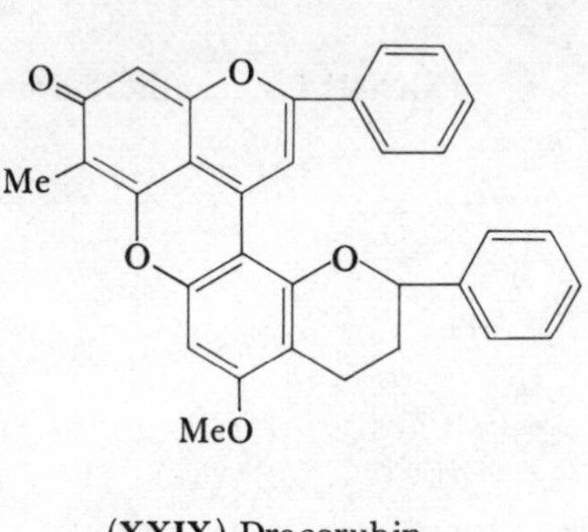

(XXIX) Dracorubin

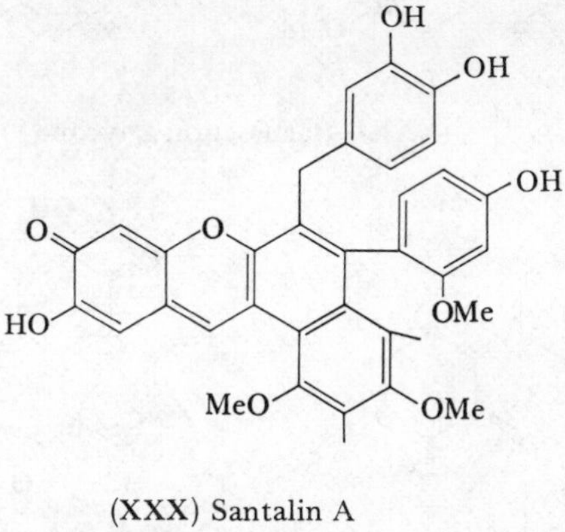

(XXX) Santalin A

1975). Santalin is one of the pigments isolated from the "insoluble red" woods, camwood (= barwood) (*Baphia nitida,* Leguminosae) and red sandalwood (*Pterocarpus santalinus,* Leguminosae) formerly used for dyeing purposes (Robertson and Whalley, 1954).

IV Phenalones

These pigments have been found only in cultures of a few *Penicillium*, and in the Haemodoraceae, especially in their highly coloured root sytems (Thomas, 1973). Haemocorin, a red glycoside from the bulbous roots of *Haemodorum corymbasum*, was the first to be identified (Cooke and Segal, 1955; Cooke *et al.*, 1958) (see **XXXI**), and closely related pigments have been found in *H. distichophyllum, Lachnanthes tinctoria* (red root), and other species (Bick and Blackman, 1973; Edwards and Weiss, 1974). All the natural phenalones are hydroxylated and can exist in tautomeric forms. They undergo oxidative ring cleavage readily, and related naphthalides and naphthalic anhydrides co-exist with the phenalone pigments. The pigments of *L. tinctoria* include several modified 9-phenylphenalones; the dark purple lachnanthofluorone (**XXXII**) and the yellow naphthalide (**XXXIII**) occur in the roots (Edwards and Weiss, 1974), and the yellow lachnanthopyrone (**XXXIV**) and the orange lachnanthopyridone derivative (**XXXV**) in the flowers

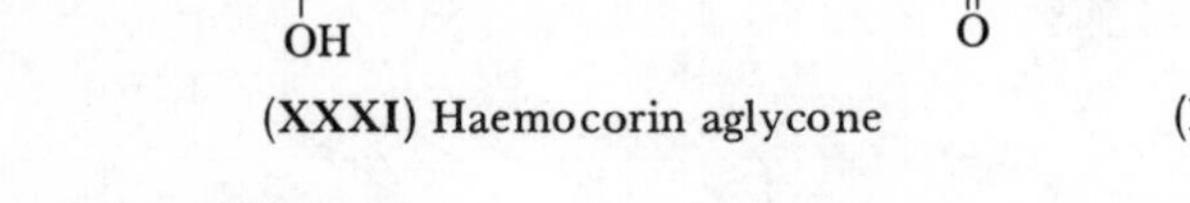

(XXXI) Haemocorin aglycone

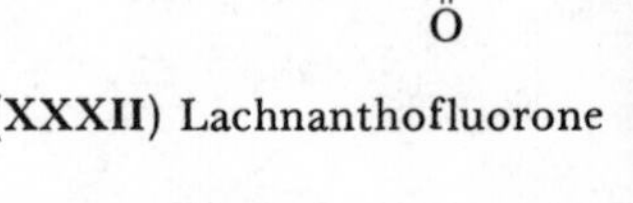

(XXXII) Lachnanthofluorone

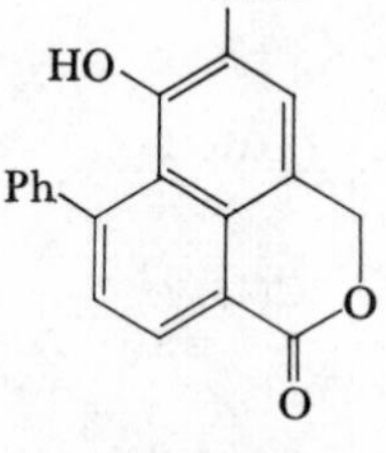

(XXXIII)

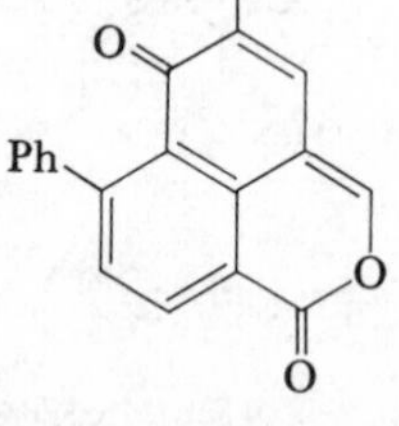

(XXXIV) Lachnanthopyrone

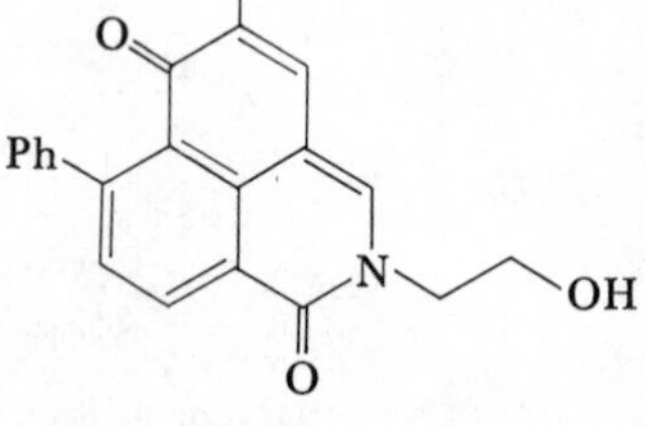

(XXXV)

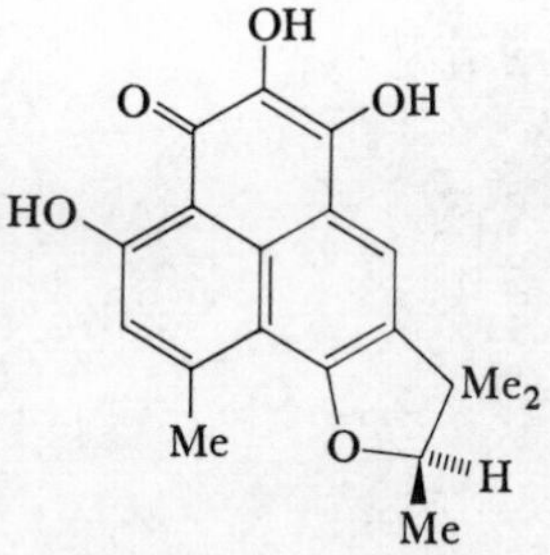

(XXXVI) Atrovenetin 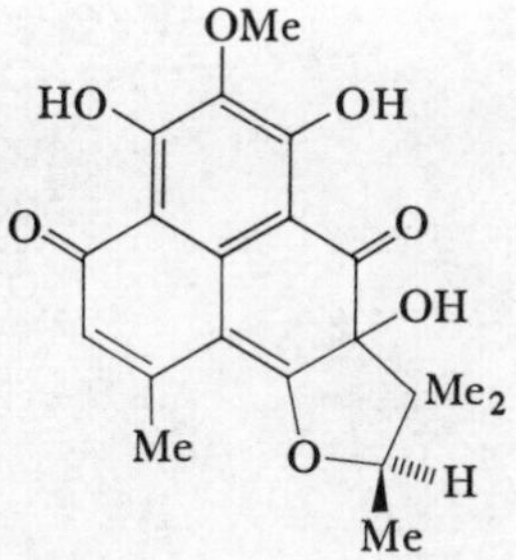 (XXXVII) Herqueinone 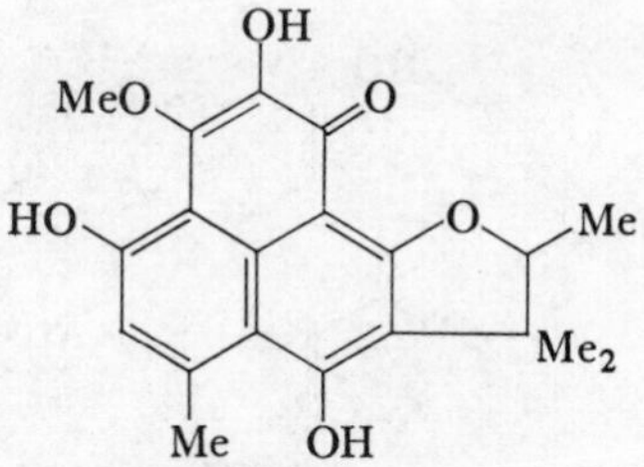 (XXXVIII) Herqueichrysin

(Edwards and Weiss, 1972). Presumably the pyridone (**XXXV**) arises by reaction of ethanolamine (or an equivalent) with the lactone (**XXXIV**) which in turn is formed by cleavage of the phenalone system and loss of C-5.

The fungal phenalones are represented by atrovenetin (**XXXVI**), herqueinone (**XXXVII**), herqueichrysin (**XXXVIII**) and closely related compounds from *P. herquei* (A. B. Turner, 1966; Narasimhachari and Vining, 1972) and the complex dimeric duclauxin group from *P. duclauxi* (Shibata, 1967).

In the fungal metabolites the phenalone ring system is acetate derived, and mevalonate provides the dihydrofuran ring, but the Haemodoraceae pigments are biosynthesized by a completely different pathway (Thomas, 1973). Labelling experiments with haemocorin and lachnanthoside (a closely related glycoside in *L. tinctoria*) showed that phenylalanine, tyrosine and acetate were incorporated. Label from [2-^{14}C]tyrosine appeared exclusively at C-5, and from the relative incorporations of 1- and [2-^{14}C]acetate it was concluded that C-6a was derived from C-2 of acetate (Thomas, 1971; Edwards *et al.*, 1972). The results suggest that biosynthesis proceeds by way of an intermediate such as (**XXXIX**), the amino acids providing two C_6-C_3 units, although an alternative (**XL**), in which half the molecule derives from a polyketide, is not excluded on the available evidence (Roughley and Whiting, 1973). The hypothetical structure (**XXXIX**) is related to several natural 1,7-diarylheptanoids of which the yellow pigment curcumin (**XLa**), from *Curcuma* spp. (Zingiberaceae, which includes Haemodoraceae), is the best known. Here again biosynthetic studies (Roughley and Whiting, 1973) do not distinguish decisively between two possible pathways analogous to those discussed for the phenylphenalones, although the formation of curcumin from an unsymmetrical precursor would be surprising.

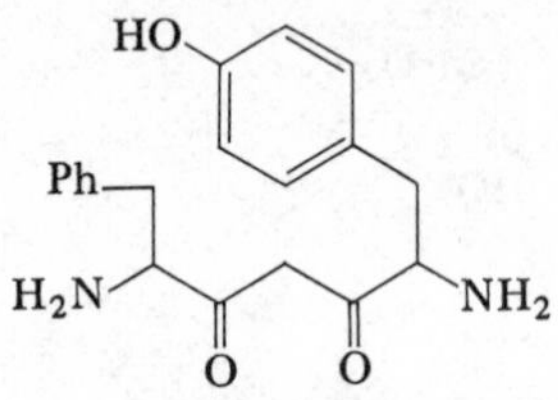

(XXXIX)

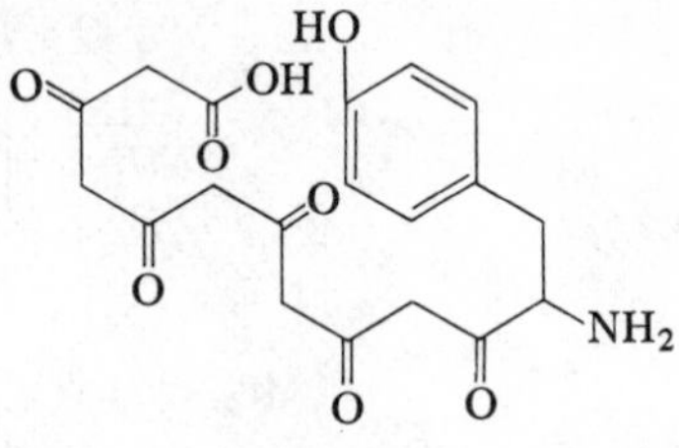

(XL)

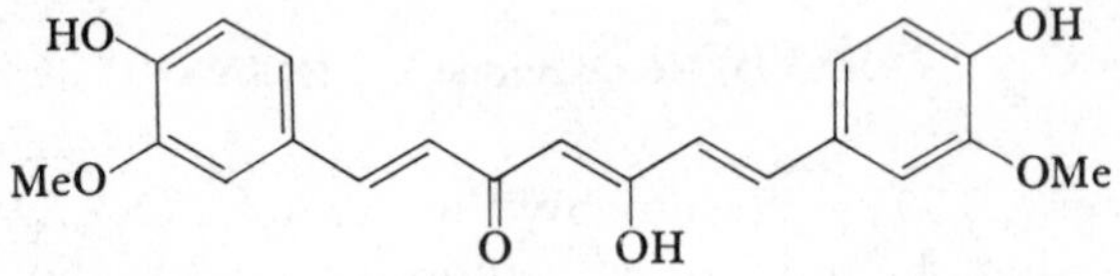

(XLa) Curcumin

V Pyrones

The majority of pigments containing a pyrone ring are, of course, flavanoid but in addition there are a substantial number of xanthones, and a few naphthopyrones and simple pyrones. α-Pyrone pigments are relatively few, the pyrone ring being conjugated with a polyene or an aromatic system. The orange pigment hispidin (**XLI**), from the wood-rotting fungus *Polyporus hispidus*, is representative of a small group of styryl α-pyrones found mainly in flowering plants (Dean, 1963). It is biosynthesized, presumably, from caffeic acid and two C_2 units, and there is evidence that the C_6-C_3 moiety is obtained from lignin. Mature fruit bodies of *P. hispidus* become increasingly tough and darker in colour, attributed to oxidative polymerization of the pigment to form a "fungus lignin" (Bu'Lock *et al.*, 1962). Citreoviridin (**XLII**) (*Penicillium* spp.) and the orange fulvoplumierin (**XLIII**), from the rind of *Plumiera acutifolia* (Albers-Schönberg *et al.*, 1962), are diverse examples of polyene α-pyrones.

γ-Pyrones include the yellow fungal pigments citromycetin (**XLIV**) and fulvic acid (**XLV**), and the orange-red naphthopyrones rubrofusarin (**XLVI**) and ustilaginoidin A (**XLVII**). A few related naphthopyrones are found in higher plants but the xanthones are much more numerous, occurring particularly in Guttiferae and Gentianaceae. The majority possess a hydroxy(methoxyl) group at C-1 and a phloroglucinol nucleus (ring A), and structural variation lies mainly in the hydroxylation pattern of ring B, and the position of prenyl side chains; gentisin (**XLVIII**) and mangostin (**XLIX**) are typical examples. However, xanthones from fungi and lichens usually

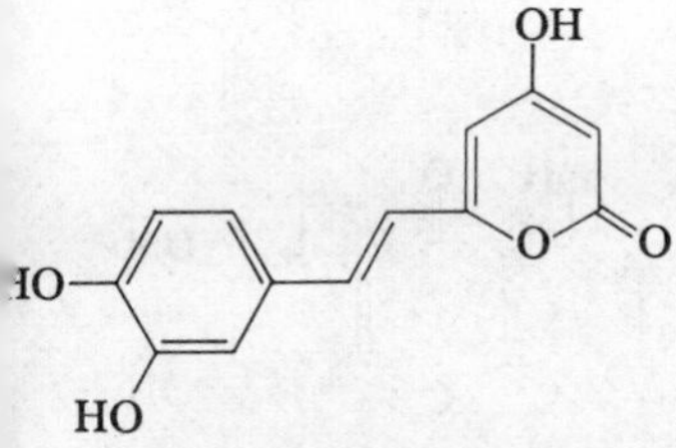

(XLI) Hispidin

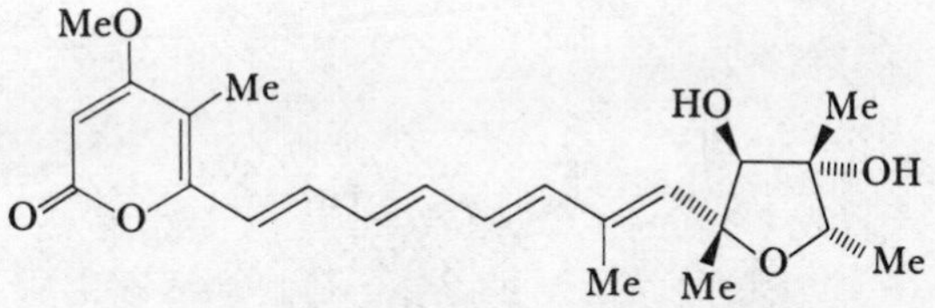

(XLII) Citreoviridin

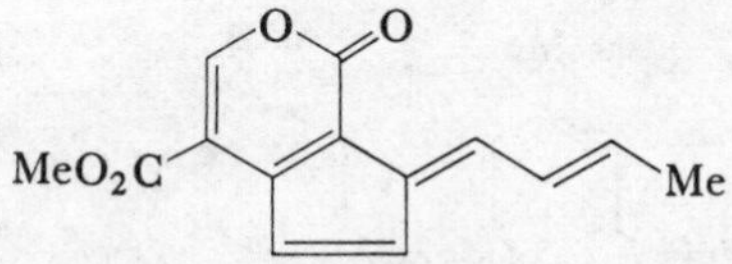

(XLIII) Fulvoplumierin

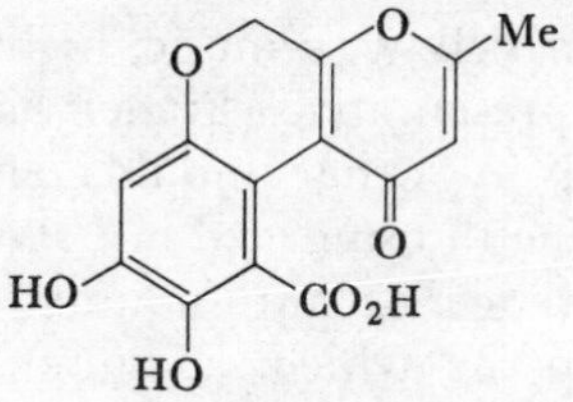

(XLIV) Citromycetin

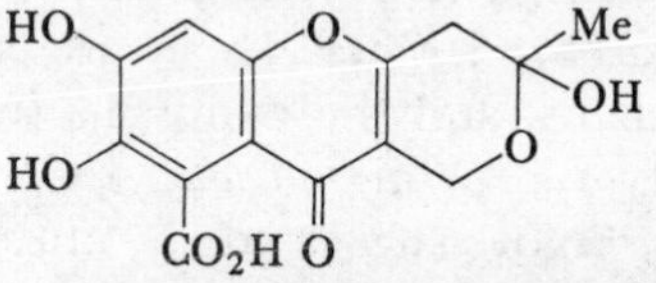

(XLV) Fulvic acid

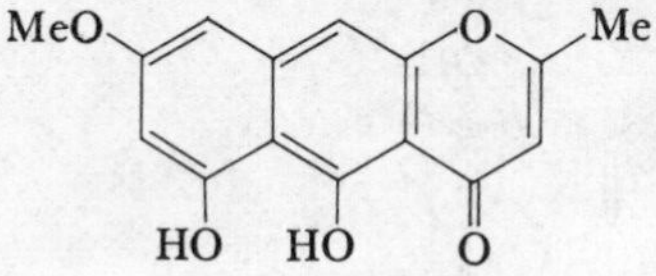

(XLVI) Rubrofusarin

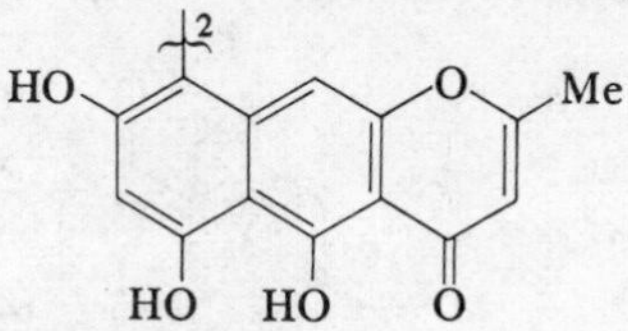

(XLVII) Ustilaginoidin A

carry a methyl side chain and have a different oxygenation pattern (see L) (Carpenter *et al.*, 1969). Although not yet established, probably all the fungal xanthones are of wholly polyketide origin. On the other hand, experiments with *Gentiana lutea* show that in flowering plants xanthones are derived partly from acetate (ring A) and partly from shikimic acid (ring B) leading to the initial formation of a hydroxylated benzophenone, frequently co-existent in the plant, which is finally cyclized by oxidative coupling (Gupta and Lewis, 1971).

Claviceps purpurea (ergot), the well-known parasite of rye, produces a complex mixture of pigments including the ergochromes

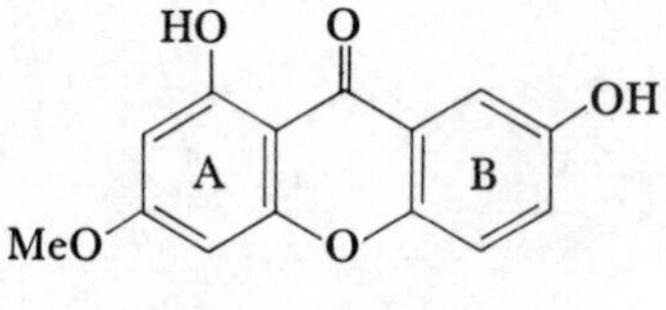

(XLVIII) Gentisin

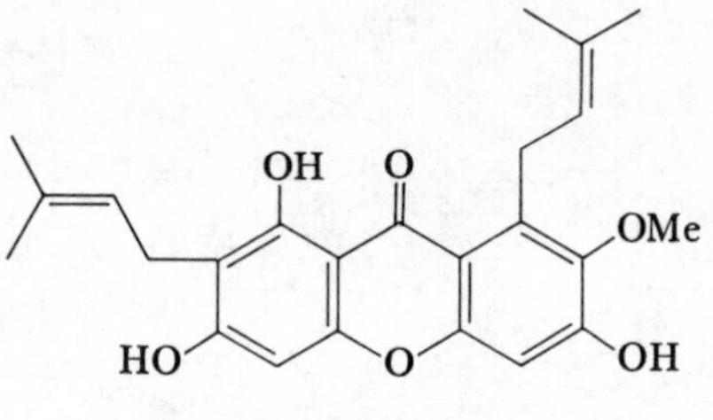

(XLIX) Mangostin

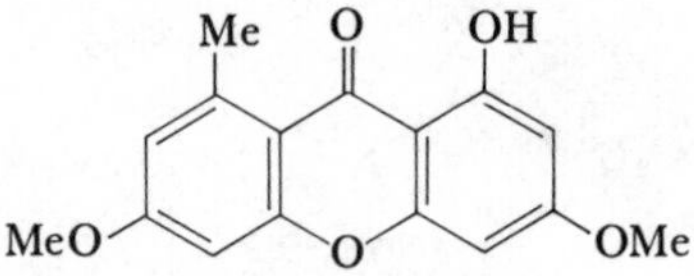

(L) Lichexanthone

(secalonic acids), a remarkable group of ten yellow, dimeric, highly modified xanthones; (**LI**) and (**LII**) are representative (Franck and Flasch, 1973). They are also elaborated by other moulds and lichens, and the eumitrins (e.g. **LIII**) occur with usnic acid in *Usnea bayleyi* (Yang *et al.*, 1973). The co-existence of anthraquinones in ergot suggested a mutual relationship in which a quinone functions as precursor to the ergochromes, the key step being an oxidative cleavage analogous to a Baeyer-Villiger oxidation (see **LIV**).

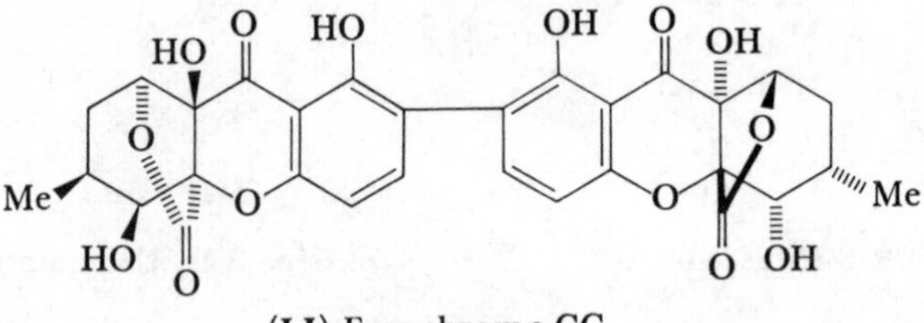

(LI) Ergochrome CC
(ergoflavin)

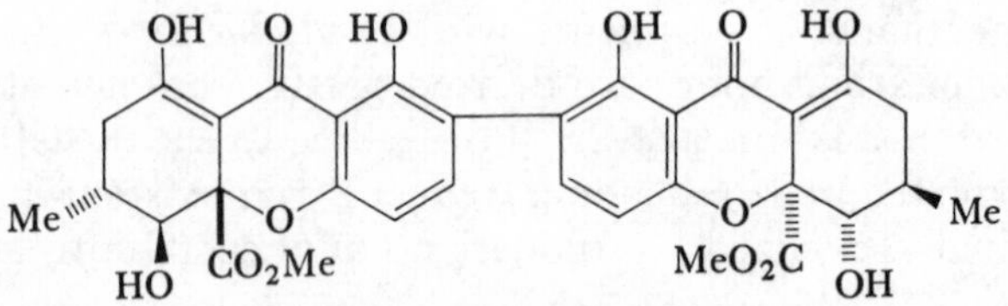

(LII) Ergochrome AB
(secalonic acid C)

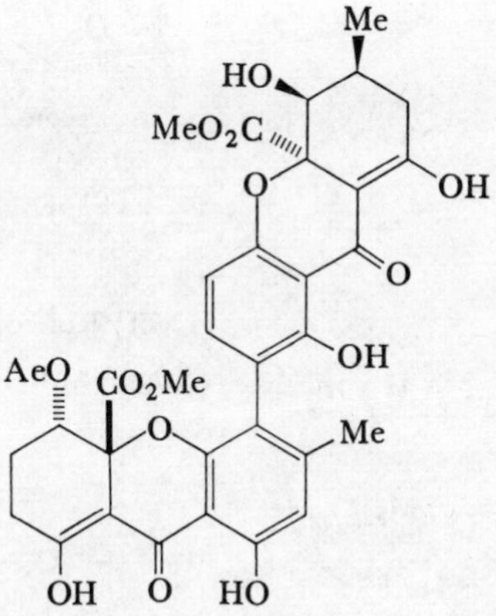

(LIII) Eumitrin A_1

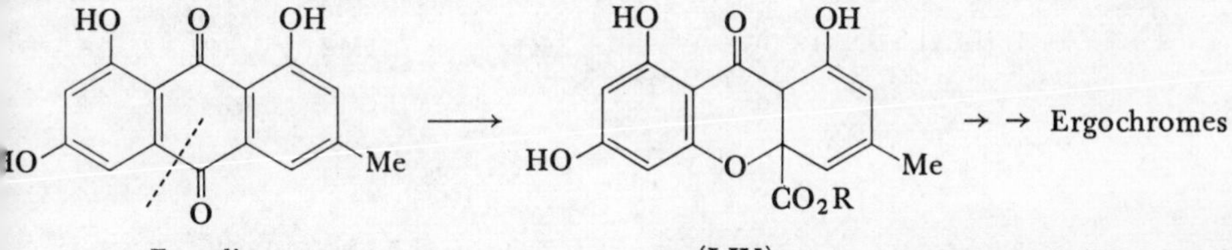

Emodin → (LIV) → → Ergochromes

This has been amply supported by acetate-labelling experiments, and by the incorporation of labelled emodin into the ergochromes (Franck *et al.*, 1968; Gröger *et al.*, 1968).

VI Sclerotiorins

This small group of yellow to red mould pigments (e.g. **LV–LX**) is produced by a limited number of *Monascus* and *Penicillium* spp. (Whalley, 1963). The most recent example, ankaflavin (**LVI**) (Manchand *et al.*, 1973), is the pigment from *M. purpureus* which is responsible for the colour of the red rice used in the preparation of certain foods and drinks in eastern countries. They share some structural features in common with quinone methides and pyrones; biogenetically they are polyketides. Extensive biosynthetic studies (W. B. Turner, 1971) have shown that the chromophore is derived normally from acetate–malonate but enolization of (**LXI**) (to form a quinone methide) is prohibited by the introduction of both a methyl and an ester function onto the methylene group. In the biosynthesis

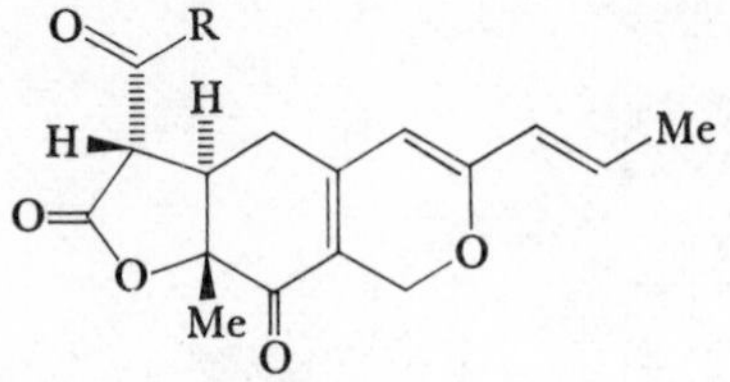

(**LV**) Monascin (R = n-C_5H_{11})
(**LVI**) Ankaflavin (R = n-C_7H_{15})

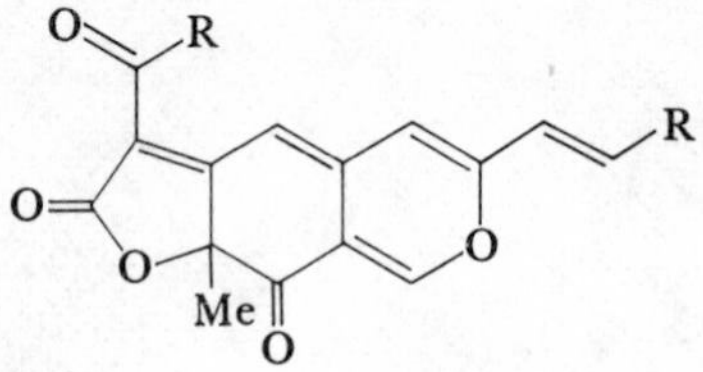

(**LVII**) Rubropunctatin (R = n-C_5H_{11})
(**LVIII**) Monascorubrin (R = n-C_7H_{15})

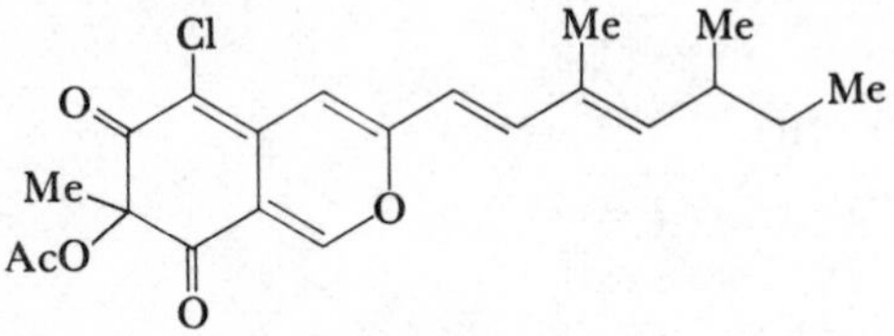

(**LIX**) Sclerotiorin

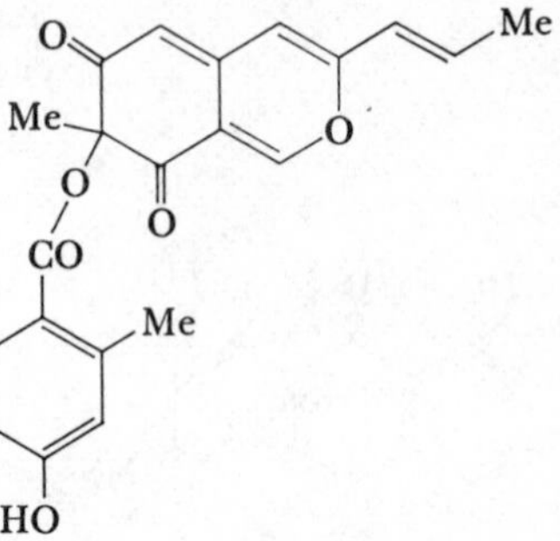

(**LX**) Mitorubrin

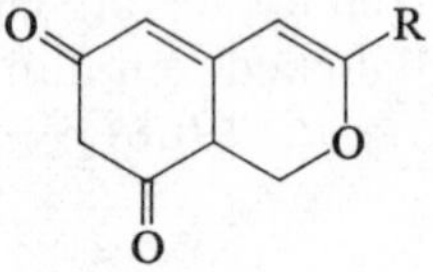

(**LXI**)

of (**LV**) to (**LVIII**) this function is a β keto ester which subsequently condenses to form the lactone. Surprisingly, in rotiorin (**LVII**; R=Me, side chain as for **LIX**) the requisite acetoacetyl unit is derived solely from acetate, and malonate is not involved (Holker *et al*., 1964).

VII Vulpinic acid pigments

This group of mainly yellow pigments is quite widely distributed in lichens; they are sometimes referred to as lichen tetronic acids (Dean, 1963). Vulpinic acid (**LXII**), from *Letharia* (*Cetraria*) *vulpina*, was

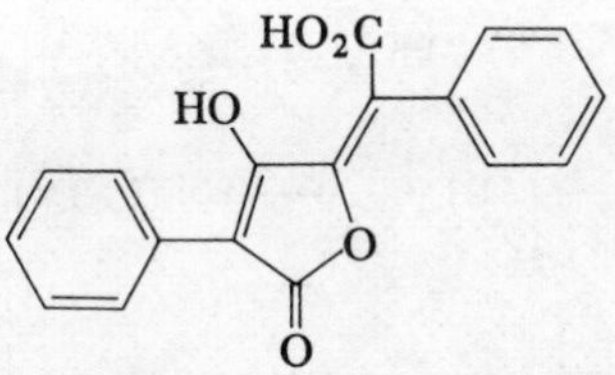

(**LXII**) Vulpinic acid (R = Me)
(**LXIII**) Pulvinic acid (R = H)

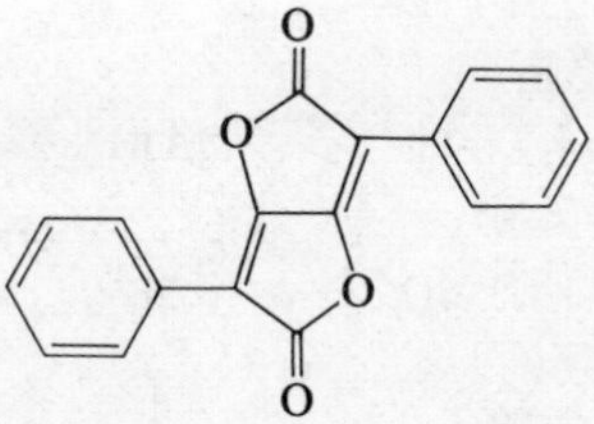

(**LXIV**) Pulvinic anhydride

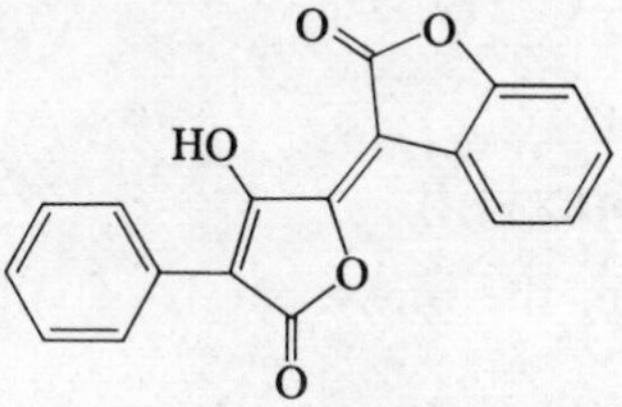

(**LXV**) Calycin

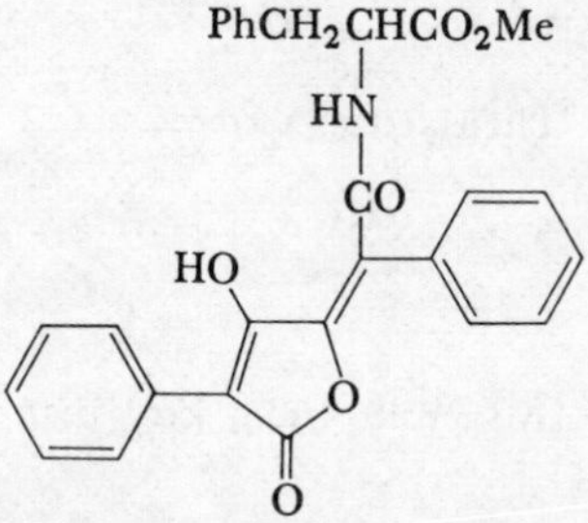

(**LXVI**) Rhizocarpic acid

the first member of the group to be identified. It is easily hydrolysed to pulvinic acid (**LXIII**) which gives pulvinic anhydride (**LXIV**) on heating. The dilactone (**LXIV**) behaves like an anhyride and reverts to vulpinic acid on heating in methanol. All these compounds are found in lichens, and indeed all three, together with calycin (**LXV**) are elaborated by the fungal mycobiont of *Candariella vitellina* (Mosbach, 1967). Further structural variation in these lichen pigments is confined to hydroxylation of the benzene rings, and conjugation with amino acids (**LXVI**).

Biogenetically these compounds are C_6-C_3 dimers closely related to the terphenylbenzoquinones. Pulvinic acid co-exists with polyporic acid (**LXVII**) in *Sticta* spp. In fact polyporic acid can be easily oxidized *in vitro* to pulvinic anhydride and the biogenetic pathway to the vulpinic acids is similar. The key step is an oxidative ring cleavage. In support, Mosbach (1964) has shown that the lichen *Evernia vulpina* incorporates [1-^{14}C]phenylalanine into vulpinic acid which has a labelling pattern (**LXVIII**) consistent with its formation by way of polyporic acid. It has also been demonstrated (Maass and Neish, 1967) that polyporic acid is incorporated into pulvinic anhydride and calycin when fed to *Pseudocyphellaria crocata*; pulvinamide is also present in this lichen and it is suggested (Maass, 1970)

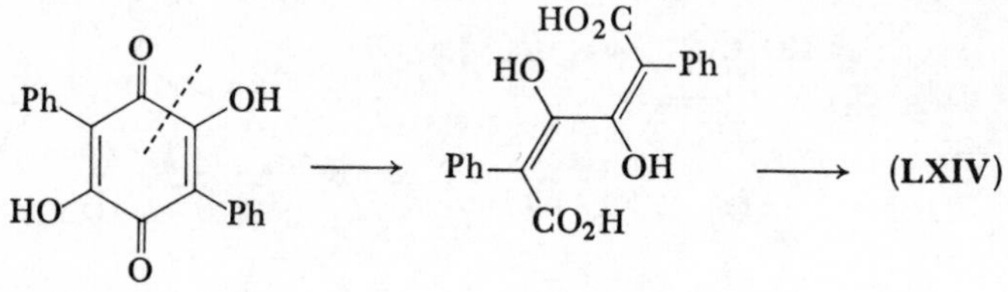

→ (LXIV)

(LXVII) Polyporic acid

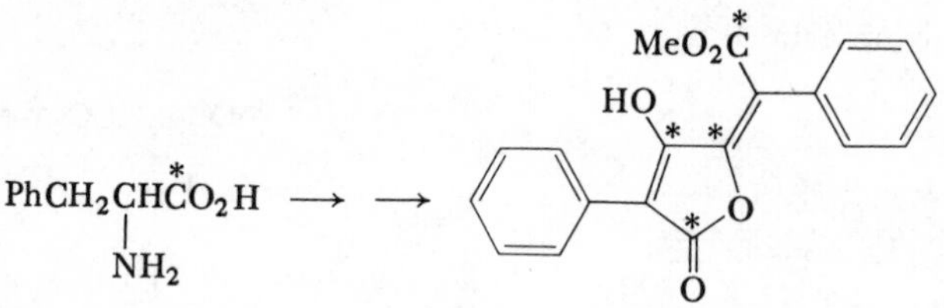

$PhCH_2CH(NH_2)CO_2H$ → →

(LXVIII)

that this may be a key intermediate in the biosynthesis of vulpinic acids.

In the last few years additional vulpinic acids have been found as pigments in higher fungi (Boletaceae and related genera) (Beaumont *et al.*, 1968; Steglich *et al.* 1968, 1969, 1970). These are hydroxylated derivatives, mostly yellow to red, the substitution patterns being those expected of shikimate-derived compounds, for example variegatic acid (**LXIX**), atromentic acid (4,4′-dihydroxyvulpinic acid), and the brown-violet dilactone variegatorubin (3,3′,4,4′-tetrahydroxycalycin). Atromentic acid anhydride is easily obtained by oxidation of atromentin (4′,4″-dihydroxypolyporic acid) (Wikholm and Moore, 1972), and it is of interest that *Clitocybe illudens* produces atromentic acid in surface cultures whereas the sporophores contain atromentin (Singh and Anchel, 1971). The flesh of certain *Boletus* spp. turns blue when damaged, and this was formerly attributed to "boletol", a mythical anthraquinone. It is now known that the colour arises from vulpinic acid derivatives which contain a catechol system in ring A, and the colour is attributed to the formation of a quinone methide anion (e.g. **LXX** from **LXIX**) by enzymic oxidation on exposure to air (Beaumont *et al.*, 1968; Steglich *et al.*, 1968).

A related type of blueing principle is gyrophorin (**LXXI**), a lemon-yellow diarylcyclopentenetrione from the sporophores of *Gyroporus cyanescens*; the blue colour obtained on oxidation is attributed to the mesomeric anion (**LXXII**) (Besl *et al.*, 1973). Another structural variation is seen in the grevillins (**LXXIII**) which, together with related pigments, colour the yellow sporophores of

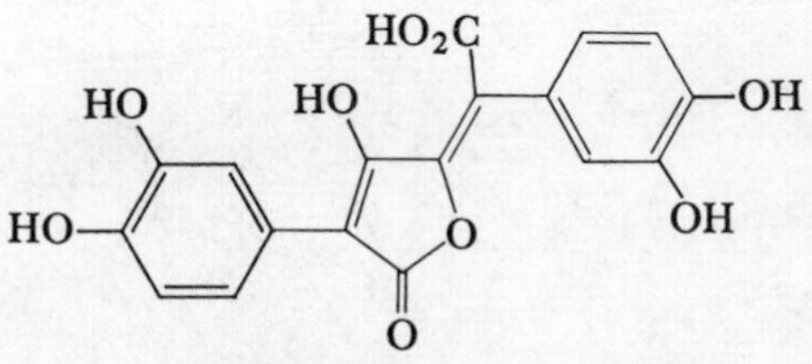

(LXIX) Variegatic acid

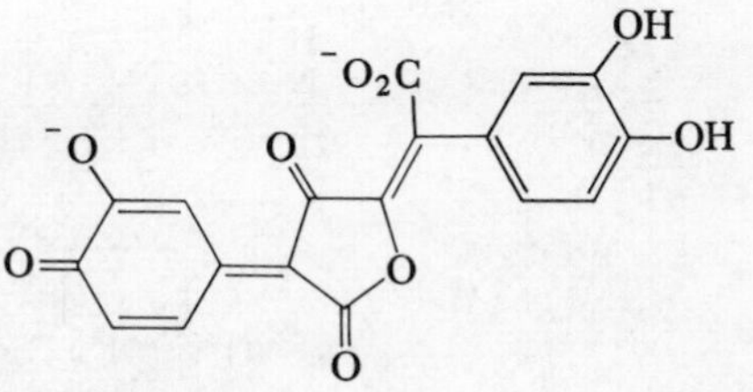

(LXX)

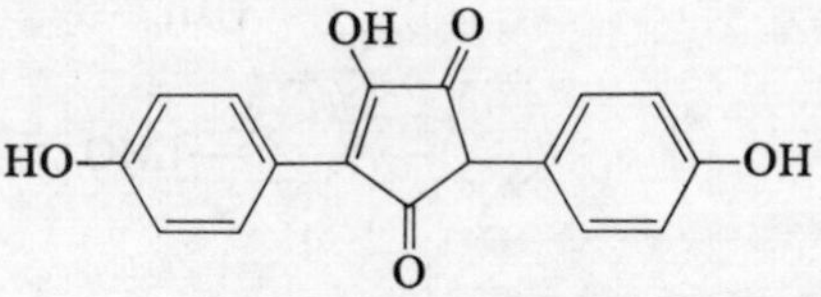

(LXXI) Gyrophorin

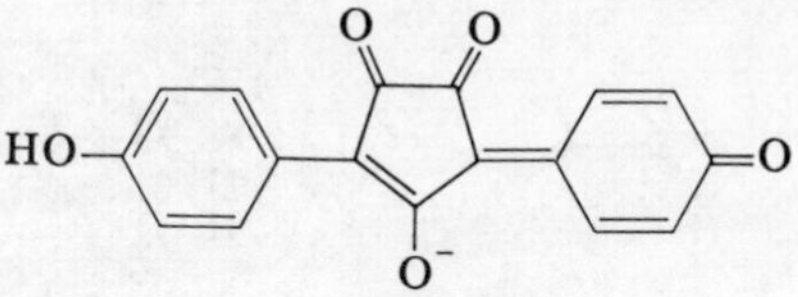

(LXXII)

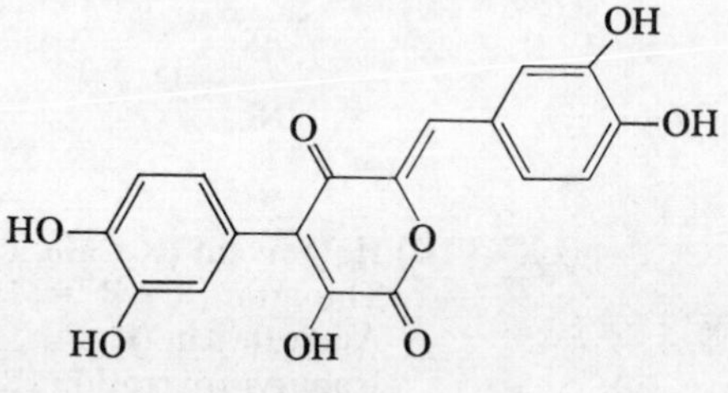

(LXXIII) Grevillin C

Suillus grevillei (= *Boletus elegans*) (Steglich *et al.*, 1972a; Edwards and Gill, 1973).

VIII Nitrogenous pigments

Apart from chlorophyll, remarkably few nitrogenous pigments are found in flowering plants, the betalains (Chapter 11) being the only significant group. A few alkaloids are coloured but most nitrogenous pigments relevant to this chapter are found in micro-organisms, especially streptomycetes. Mostly they occur in very small groups and some are difficult to classify as can be seen from the following examples.

Violacein (LXXIV), the dark violet pigment from *Chromobacterium violaceum*, is derived solely from tryptophan, the central ring probably arising from the side chains of two precursor molecules

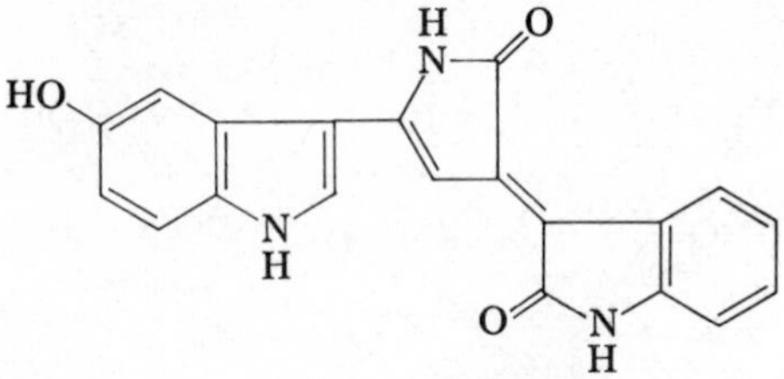

(LXXIV) Violacein

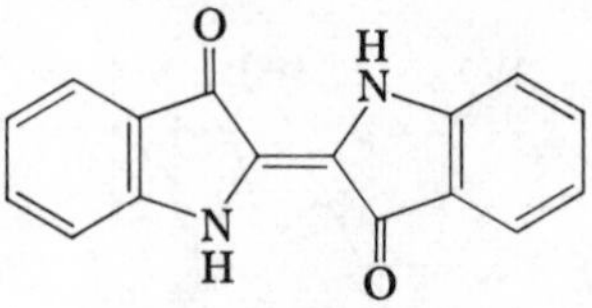

(LXXV) Indigo

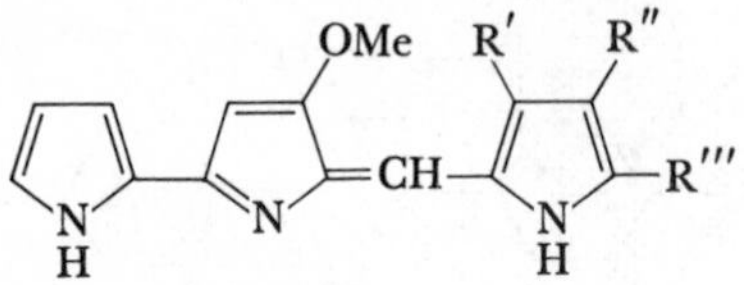

(LXXVI) Prodigiosin

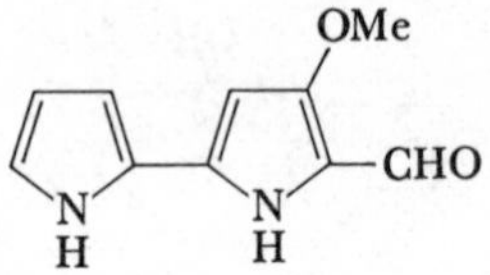

(LXXVII)

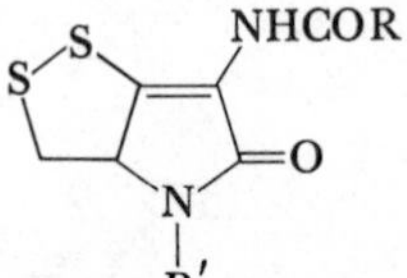

(LXXVIII) Holomycin (R = Me, R′ = H)
Thiolutin (R = R′ = Me)
Aureothricin (R = Et, R′ = Me)
Isobutyropyrrothin (R = i-Pr, R′ = Me)

(DeMoss and Evans, 1960). Lyophilized washed cells of the same bacterium will also metabolize tryptophan to indigo (**LXXV**) (Sebek and Jäger, 1962), a pigment also produced by *Ps. indoloxidans* and a mutant of *Schizophyllum commune*. (Indigo from higher plants is an artefact.) *Serratia marcescens* and other bacteria elaborate prodigiosins (**LXXVI**; R′, R″, R‴ = H or alkyl), a small group of violet-red polypyrrole pigments (Williams and Hearn, 1967). There is also a blue *Serratia* pigment containing two bipyrrole residues (Wassermann *et al.*, 1968), and in one remarkable case, metacycloprodigiosin, from a species of *Streptomyces*, the *meta* positions C-2 and C-4 are bridged by an alkyl chain (Wassermann *et al.*, 1969). In the biosynthesis of prodigiosins the bipyrryl system is formed first, the final step being condensation of the aldehyde (**LXXVI**) with an alkylated pyrrole (Bogorad and Troxler, 1967). Several streptomycetes produce yellow antibiotics of type (**LXXVIII**) which may derive from glycine and cysteine (Miller, 1961).

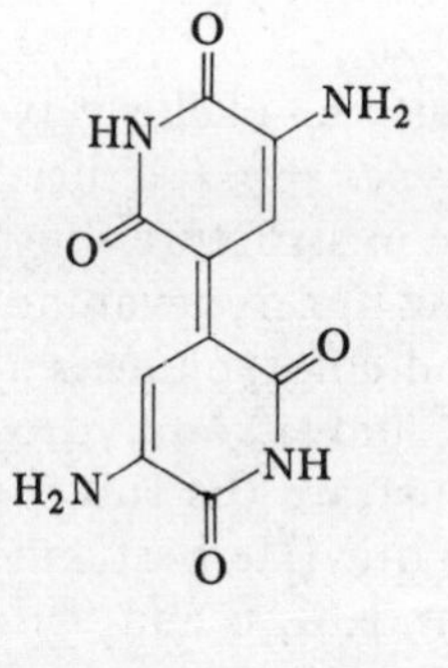

(LXXIX) Indigoidine

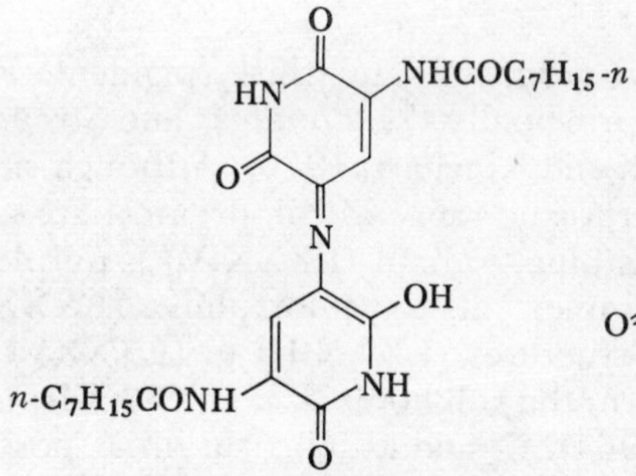

(LXXX)

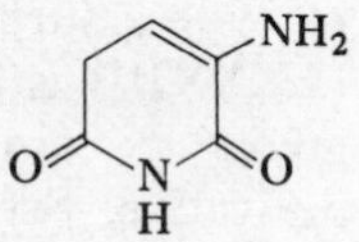

(LXXXI)

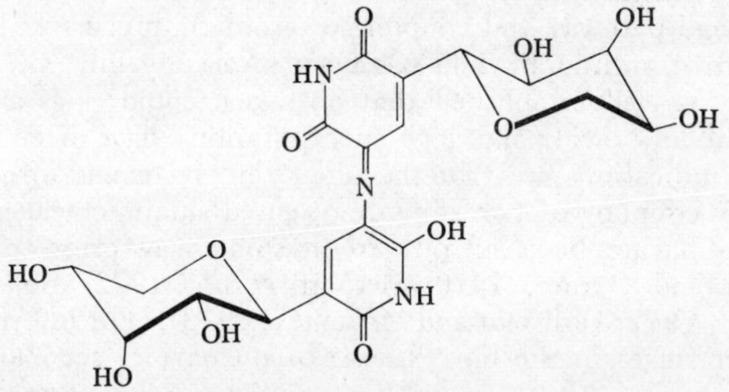

(LXXXII) Indochrome A

"Bacterial indigo" is not the same as true indigo (LXXV) already mentioned. This pigment, indigoidine (LXXIX), produced by *Pseudomonas indigofera, Arthrobacter atrocyaneus*, and other species, is a diaza-*o*-diphenoquinone while *Ps. lemonnieri* forms the diazaindophenol derivative (LXXX) (Kuhn *et al.*, 1965; Knackmuss, 1967). Along with indigoidine, *A. polychromogenes* produces water-soluble blue pigments, the indochromes, which are diazaindophenol *C*-ribosides (LXXXII) (Knackmuss *et al.,* 1969). The pigments (LXXIX) and (LXXX) have been synthesized by oxidative coupling of suitable pyridine derivatives, and the biosynthetic pathway is similar. 2,6-Dihydroxypyridine appears to be a common precursor for this small group of pigments but the earlier stages are uncertain. Indigoidine is readily formed by autoxidation of aminoglutaconimide (LXXXI) which may be derived from glutamine by way of amino-glutarimide (Knackmuss, 1973).

A PHENAZINES

This compact group of some thirty pigments is confined exclusively to bacteria, principally *Pseudomonas* and *Streptomyces* spp. (Gerber, 1969; Tipton and Rinehart, 1970). Although simple in structure they provide a surprising range of colour; most are yellow but pyocyanine (**LXXXIII**) is blue, iodinin (**LXXXIV**) is purple, and chloraphine is a green π complex of oxychloraphine (**LXXXV**) and its dihydro derivative. Structures (**LXXXIII**) to (**LXXXVI**) illustrate the substitution pattern; the sulphonic acid (**LXXXVII**) is unique (Herbert and Holliman, 1964). *C*- and *O*-substitution at positions 1, 4, 6 and 9 is predominant with a tendency for the same substituent to appear at C-1 and C-5, or C-4 and C-9. This symmetrical trend suggested that the biosynthesis of these compounds might proceed by the aromatic amino acid pathway and coupling of aromatic precursors but incorporation of anthranilic acid has been invariably low. On the other hand, it is well established that both benzenoid rings are derived from shikimic acid, and high incorporations have been achieved. Present indications are that the phenazine system is formed essentially by coupling of two suitably oriented shikimic acid molecules, nitrogen having been introduced in some way prior to coupling (Podojil and Gerber, 1970; Herbert *et al.*, 1972; Hollstein and Marshall, 1972; Hollstein and McCamey, 1973). (For information on the later stages in the biosynthesis of phenazines see Flood *et al.*, 1972.)

Caulerpin (**LXXXVIII**), a red compound from *Caulerpa* spp. (green algae), is unique (Aguilar-Santos, 1970). Chemically it is a dibenzodihydrophenazine but its biogenesis is a matter for conjecture.

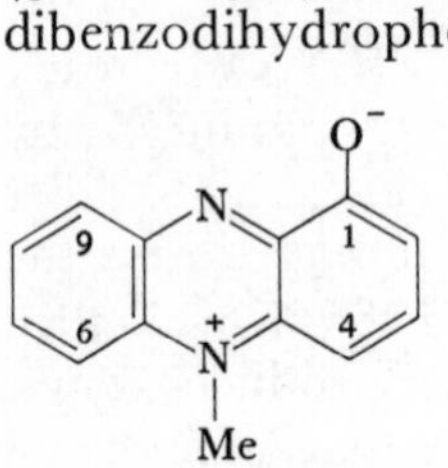

(**LXXXIII**) Pyocyanine

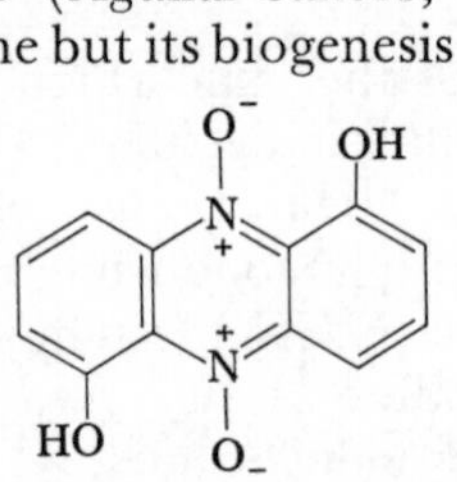

(**LXXXIV**) Iodinin

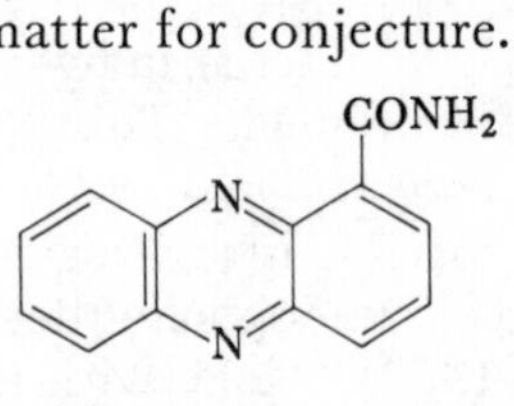

(**LXXXV**) Oxychlororaphine

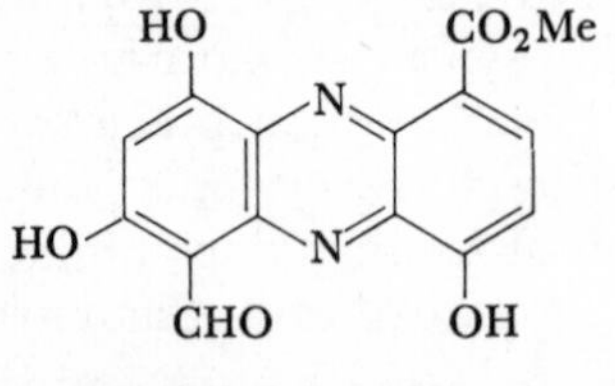

(**LXXXVI**) Lomofungin

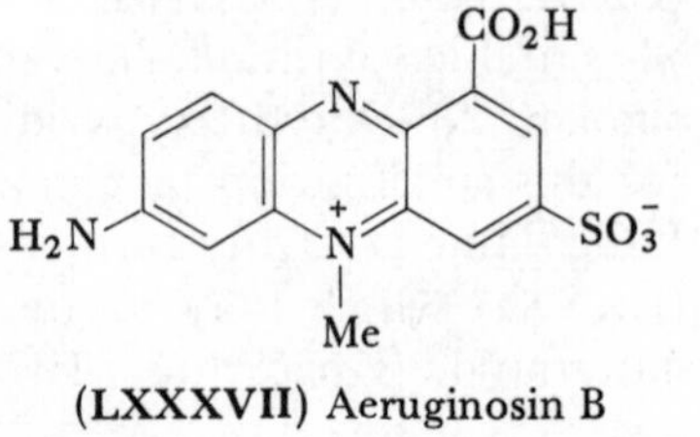

(**LXXXVII**) Aeruginosin B

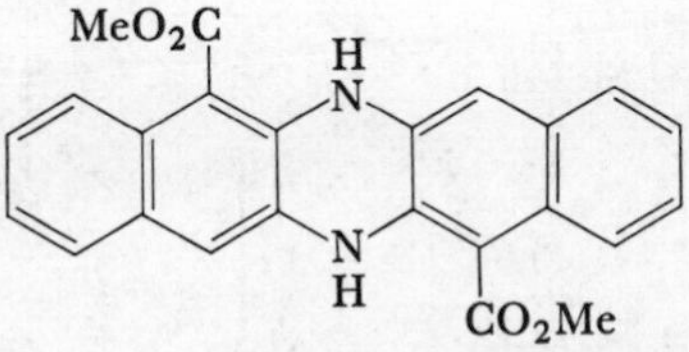

(**LXXXVIII**) Caulerpin

B PHENOXAZINONES

The phenoxazinone chromophore is present in the actinomycins (**LXXXIX**), a group of bright red antibiotics produced by *Streptomyces* (Brockmann, 1960), in the red cinnabarin pigments (**XC**) isolated from the wood-rotting fungus *Polyporus cinnabarinus*, and in a few simple bacterial metabolites related to (**XC**; R = R′ = H) (Gerber, 1967a). The 2-amino derivative (**XC**; R = R′ = H) is also produced by *Penicillium notatum* (Baer *et al*., 1971). It is of interest that litmus and the orcein pigments obtained from certain lichens by aerial oxidation in the presence of ammonia, also contain the phenoxazinone system (Musso, 1960). The actinomycins are chromopeptides which differ only in the nature of the peptide chains (R and R′ in **LXXXIX**) each of which consists of five amino acid residues (e.g. **XCI**).

Phenoxazinones can be synthesized *in vitro* by oxidative coupling of 3-hydroxyanthranilic acids (Gerber, 1967b), and actinomycin D (**XCI**) has been synthesized by the same route (Brockmann and Lackner, 1968; Meienhofer, 1970). A similar process takes place in the biosynthesis of the insect phenoxazinones (ommochromes), and almost certainly this is true for the actinomycins. This is suggested by the co-occurrence of *o*-aminophenol and questiomycin A (**XC**; R = R′ = H) in a culture of *Streptomyces* (Anzai *et al*., 1960), by the incorporation of labelled tryptophan into the actinomycin chromophore, presumably via 3-hydroxyanthranilic acid (Sivak *et al*., 1962),

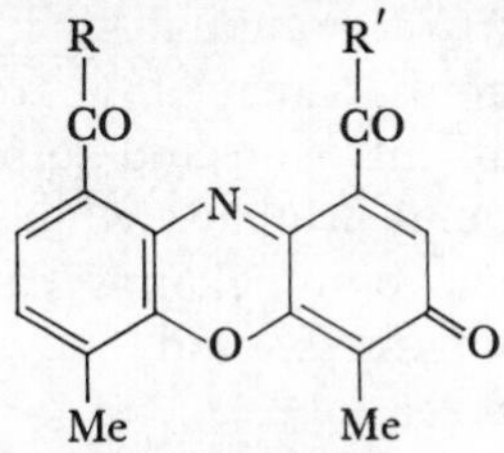

(**LXXXIX**) Actinomycins

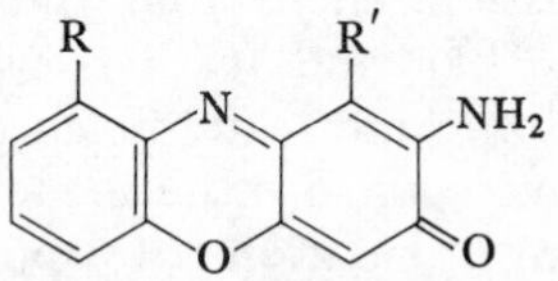

(**XC**) Cinnabarin (R = CH_2OH, R′ = CO_2H)
Cinnabarinic acid (R = R′ = CO_2H)
Tramesanguin (R = CO_2H, R′ = CHO)

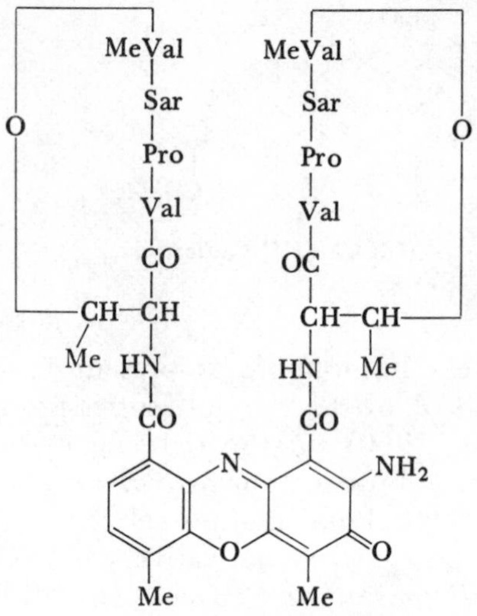

(XCI) Actinomycin D

and by the conversion of 3-hydroxy-4-methylanthranilic acid into actinocin (**LXXXIX**; R = R′ = H) using a cell-free enzyme system obtained from *S. antibioticus*.

C MELANINS

The term melanin is often used loosely to describe any natural dark brown or black pigment. A less sweeping generalization refers to melanins as products of high molecular weight formed by enzymic oxidation of phenols, but this is still a very broad definition. As originally used the term was applied to a nitrogenous black pigment derived from tyrosine. Such pigments are now classified as eumelanins (Nicolaus, 1968), being nitrogenous polymers formed by the action of oxygen and tyrosinase on tyrosine or closely related compounds. The eumelanins belong mainly to the animal kingdom, and are distinguished from the phaeomelanins, the yellow to red pigments in hair and feathers which contain nitrogen and sulphur, and the allomelanins, the black pigments found in plants which are devoid of nitrogen. Synthetic melanins, similar to eumelanins, can be prepared by oxidation of appropriate substrates, and they are referred to as tyrosine melanin, dopa melanin etc.

The initial stages in the biogenesis of eumelanin leading from tyrosine to 5,6-dihydroxyindole are well documented (Thomson,

1962; Nicolaus, 1968; Swan, 1973) and the final stage is essentially an oxidative polymerization of indole-5,6-quinone. This, however, is a gross simplification as other monomers are involved, there is cross linking between the main chains, some indole units are degraded to pyrrole units, and the melanin macromolecule is conjugated with protein. Consequently melanins have a highly irregular structure and are resistant to chemical degradation and difficult to study by physical methods. All work in this field has been carried out with eumelanins from animals and with synthetic materials, and the evidence that eumelanins occur in plants is based on the co-existence of a tyrosine–tyrosinase or related system. It should be noted that black pigmentation may arise during normal development or it may appear only after injury, death or decay. It seems likely that the black pigments in *Vicia faba, Cytisus nigricans* and *Sarothamnus scoparius* are eumelanins (Thomas, 1955), and also those produced when potatoes, bananas, and mushrooms are damaged, but the evidence is weak. The black pigments produced by *Bacillus salmonicida* and by the vibrio *Microspira tyrosinatica* may also qualify as eumelanins but none of these plant products have been studied chemically.

Of the great majority of dark pigments found as markings on petals, in the spores of higher fungi, in senescent leaves and seedpods, and in the dead cells of bark and pericarps, virtually nothing is known (Thomson, 1962). However, the black pigments from the seeds of *Helianthus annuus, Citrullus vulgaris, Luffa cylindrica* and others and from the spores of *Ustilago maydis* give catechol on alkali fusion and are known as catechol melanins. (Nicolaus *et al.*, 1964).

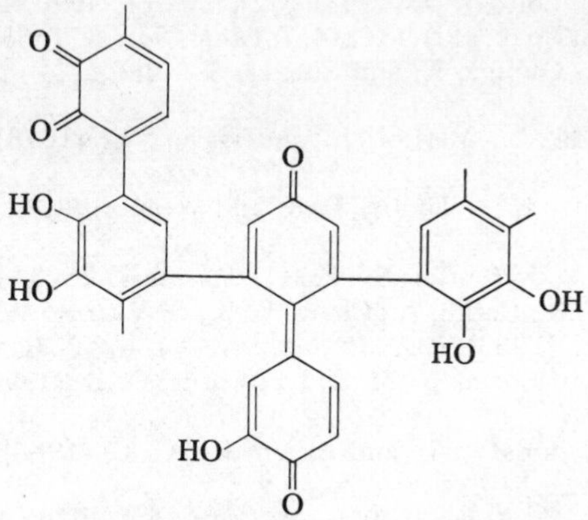

(XCII)

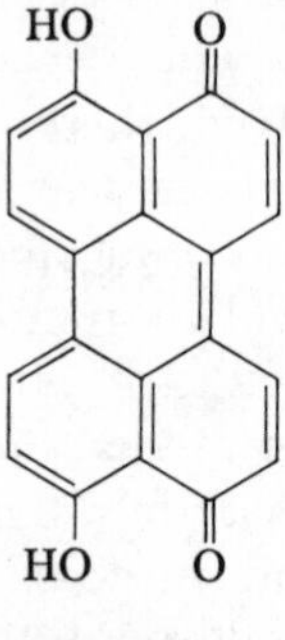

(XCIII) 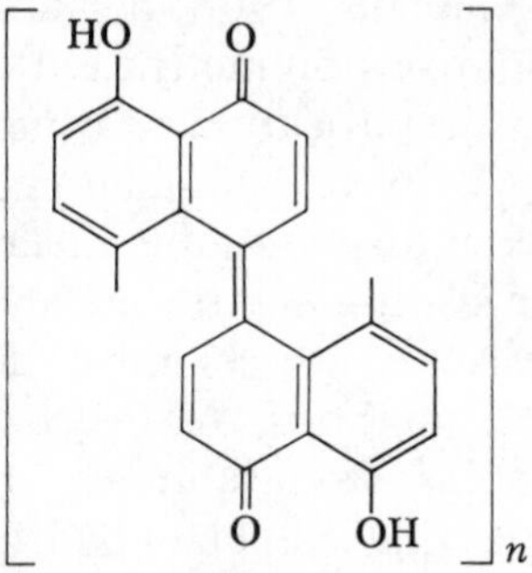 (XCIV)

Dark polymers can be obtained *in vitro* by oxidation of catechol and may be represented by part structures such as (**XCII**). The black pigment from the ascomycete *Daldinia concentrica* gave catechol and 1,8-dihydroxynaphthalene on alkali fusion. In laboratory culture this fungus elaborates simple derivatives of 1,8-dihydroxynaphthalene while in the wild state it produces the almost black quinone (**XCIII**), as well as polymer. It seems likely that *Daldinia* melanin is a polymer of type (**XCIV**) formed by oxidative polymerization of 1,8-dihydroxynaphthalene (Allport and Bu'Lock, 1958, 1960). The black pigment aspergillin, produced by *Aspergillus niger*, appears to be related (Nicolaus, 1968).

REFERENCES

Aguilar-Santos, G. (1970). *J. chem. Soc. C* 842-843.

Albers-Schönberg, G., von Philipsborn, W., Jackman, L. M. and Schmid, H. (1962). *Helv. chim. Acta.* 45, 1406-1408.

Allport, D. C. and Bu'Lock, J. D. (1958). *J. chem. Soc.* 4090-4094.

Allport, D. C. and Bu'Lock, J. D. (1960). *J. chem. Soc.* 654-662.

Anzai, K., Isono, K., Okuma, K. and Suzuki, S. (1960). *J. Antibiot., Tokyo* 13, 125-132.

Arnone, A., Camarda, L., Merlini, L. and Nasini, G. (1975). *J.C.S. Perkin I* 186-194.

Badar, Y., Lockley, W. J. S., Toube, T. R. and Weedon, B. C. L. (1973). *J.C.S. Perkin I* 1416-1424.

Baer, H., Zarnack, J. and Pfleiffer, S. (1971). *Pharmazie* 26, 314.

Ballio, A., Bertholdt, H., Carilli, A., Chain, E. B., Di Vittorio, V., Tonolo, A. and Vero-Barcellona, L. (1963). *Proc. R. Soc. Ser. B* 158, 43-70.

Barton, D. H. R., De Florin, A. M. and Edwards, O. E. (1956). *J. chem. Soc.* 530-534.

Beaumont, P. C., Edwards, R. L. and Elsworthy, G. C. (1968). *J. chem. Soc. C* 2968-2974.

Besl, H., Bresinsky, A., Steglich, W. and Zipfel, K. (1973). *Chem. Ber.* 106, 3223-3229.

Bick, I. R. C. and Blackman, A. J. (1973). *Aust. J. Chem.* 26, 1377-1380.
Bogorad, L. and Troxler, R. F. (1967). *In* "Biogenesis of Natural Compounds" (P. Bernfeld, ed.), 2nd edn, p. 301. Pergamon Press, Oxford.
Bohlmann, F., Burkhardt, R. and Zdero, C. (1973). "Naturally Occurring Acetylenes". Academic Press, London and New York.
Brockmann, H. (1960). *Prog. Chem. Org. Nat. Prods.* 18, 1-54.
Brockmann, H. and Lackner, H. (1968). *Chem. Ber.* **101**, 1312-1340.
Brown, P. M., Moir, M., Thomson, R. H., King, T. J., Krishnamoorthy, V. and Seshadri, T. R. (1973). *J.C.S. Perkin I* 2721-2725.
Bu'Lock, J. D., Leeming, P. R. and Smith, H. G. (1962). *J. chem. Soc.* 2085-2089.
Burnell, R. H., Mo, L. and Moinas, M. (1972). *Phytochemistry* 11, 2815-2820.
Carpenter, I., Locksley, H. D. and Scheinmann, F. (1969). *Phytochemistry* **8**, 2013-2025.
Cooke, R. G. and Segal, W. (1955). *Aust. J. Chem.* **8**, 107-113.
Cooke, R. G., Johnson, B. L. and Segal, W. (1958). *Aust. J. Chem.* **11**, 230-235.
Dean, F. M. (1963). "Naturally Occurring Oxygen Ring Compounds". Butterworths, London.
Delle Monache, F., Marini-Bettòlo, G. B., Conçalves de Lima, O., d'Albuquerque, I. L. and de Barros Coêlho, J. S. (1973). *J.C.S. Perkin I* 2725-2728.
DeMoss, R. D. and Evans, N. R. (1960). *J. Bact.* 79, 729-733.
Edwards, J. M. and Weiss, U. (1972). *Tetrahedron Lett.* 1631-1634.
Edwards, J. M. and Weiss, U. (1974). *Phytochemistry* 13, 1597-1602.
Edwards, J. M., Schmitt, R. C. and Weiss, U. (1972). *Phytochemistry* 11, 1717-1720.
Edwards, R. L. and Gill, M. (1973). *J.C.S. Perkin I* 1921-1929.
Flood, M. E., Herbert, R. B. and Hollimann, F. G. (1972). *J.C.S. Perkin I* 622-626 and refs. therein.
Franck, B. and Flasch, H. (1973). *Prog. Chem. Org. Nat. Prods.* 30, 151-206.
Franck, B., Hüper, F., Gröger, D. and Erge, D. (1968). *Chem. Ber.* **101**, 1954-1969.
Frank, R. L., Clark, G. R. and Coker, J. N. (1950). *J. Am. Chem. Soc.* 72, 1824-1826.
Gerber, N. N. (1967a). *J. org. Chem.* 32, 4055-4057.
Gerber, N. N. (1967b). *Can. J. Chem.* 46, 790-792.
Gerber, N. N. (1969). *J. heterocyclic Chem.* 6, 297-300 and refs. therein.
Gregson, M., Ollis, W. D., Redman, B. T. and Sutherland, I. O. (1968). *Chem. Commun.* 1395-1396.
Gripenberg, J. (1971). *Acta chem. scand.* 25, 2999-3005.
Gripenberg, J. and Martikkala, J. (1969). *Acta chem. scand.* 23, 2583-2588.
Gröger, D., Erge, D., Franck, B., Ohnsorge, U., Flasch, H. and Hüper, F. (1968). *Chem. Ber.* **101**, 1970-1978.
Gupta, P. and Lewis, J. R. (1971). *J. chem. Soc. C* 629-631.
Herbert, R. B. and Hollimann, F. G. (1964). *Proc. chem. Soc.* 19.
Herbert, R. B., Hollimann, F. G. and Ibberson, D. N. (1972). *Chem. Commun.* 355-356.
Holker, J. S. E., Staunton, J. and Whalley, W. B. (1964). *J. chem. Soc.* 16-22.
Hollstein, U. and McCamey, D. A. (1973). *J. org. Chem.* 38, 3415-3417.
Hollstein, U. and Marshall, L. G. (1972). *J. org. Chem.* 37, 3510-3514.

Karanatsios, D., Scarpa, J. S. and Eugster, C. H. (1966). *Helv. chim. Acta* **49**, 1151–1172.

Knackmuss, H.-J. (1967). *Chem. Ber.* **100**, 2537–2545.

Knackmuss, H.-J. (1973). *Angew. Chem.* 12, 139–145.

Knackmuss, H.-J., Cosens, G. and Starr, M. P. (1969). *Eur. J. Biochem.* **10**, 90–95.

Kneifel, H. and Bayer, E. (1973). *Angew. Chem.* 12, 508.

Krishnan, V. and Rangaswami, S. (1971). *Indian J. Chem.* 9, 117–120.

Kuhn, R., Bauer, H. and Knackmuss, H.-J. (1965). *Chem. Ber.* **98**, 2139–2153.

Kupchan, S. M., Karim, A. and Marcks, C. (1969). *J. org. Chem.* **34**, 3912–3918.

Maass, W. S. G. (1970). *Phytochemistry* 9, 2477–2481.

Maass, W. S. G. and Neish, A. C. (1967). *Can. J. Bot.* **45**, 59–72.

Manchand, P. S., Whalley, W. B. and Chen, F.-C. (1973). *Phytochemistry* **12**, 2531–2532.

Martin, J. D. (1973a). *Tetrahedron* **29**, 2553–2559.

Martin, J. D. (1973b). *Tetrahedron* **29**, 2997–3000.

Mathieson, D. W., Millard, B. J., Powell, J. W. and Whalley, W. B. (1973). *J.C.S. Perkin I* 184–188.

Meienhofer, J. (1970). *J. Am. chem. Soc.* **92**, 3771–3777.

Miller, M. W. (1961). "The Pfizer Handbook of Microbial Metabolites". McGraw-Hill, New York.

Moir, M. Rüedi, P. and Eugster, C. H. (1973). *Helv. chim. Acta* **56**, 2534–2539.

Mosbach, K. (1964). *Biochem. biophys. Res. Commun.* **17**, 363–367.

Mosbach, K. (1967). *Acta chem. scand.* **21**, 2331–2334.

Musso, H. (1960). *Planta med.* **8**, 432–446.

Nakanishi, K., Gullo, V. P., Miura, I., Govindachari, T. R. and Viswanathan, N. (1973). *J. Am. chem. Soc.* **95**, 6473–6475.

Narasimhachari, N. and Vining, L. C. (1972). *J. Antibiot., Tokyo* **25**, 155–162.

Nicolaus, R. A. (1968). "Melanins". Hermann, Paris.

Nicolaus, R. A., Piatelli, M. and Fattorusso, E. (1964). *Tetrahedron* **20**, 1163–1172.

Podojil, M. and Gerber, N. N. (1970). *Biochemistry* **9**, 4616–4618.

Robertson, A. and Whalley, W. B. (1954). *J. chem. Soc.* 2794–2801.

Roughley, P. J. and Whiting, D. A. (1973). *J.C.S. Perkin I* 2379–2388.

Rüedi, P. and Eugster, C. H. (1972). *Helv. chim. Acta* **45**, 1994–2014.

Sebek, O. K. and Jäger, H. (1962). *Nature, Lond.* **196**, 793–795.

Shibata, S. (1967). *Chem. Brit.* 3, 110–121.

Singh, P. and Anchel, M. (1971). *Phytochemistry* **10**, 3259–3262.

Sivak, A., Meloni, M. L., Nobili, F. and Katz, E. (1962). *Biochim. biophys. Acta* **57**, 283–289.

Steglich, W. and Töpfer-Petersen, E. (1972). *Z. Naturf.* **27b**, 1286–1288.

Steglich, W. and Töpfer-Petersen, E. (1973). *Z. Naturf.* **28c**, 255–259.

Steglich, W., Furtner, W. and Prox, A. (1968). *Z. Naturf.* **23b**, 1044–1050.

Steglich, W., Furtner, W. and Prox. A. (1969). *Z. Naturf.* **24b**, 941–942.

Steglich, W., Furtner, W. and Prox. A. (1970). *Z. Naturf.* **25b**, 557–558.

Steglich, W., Besl, H. and Prox. A. (1972a). *Tetrahedron Lett.* 4895–4898.

Steglich, W., Töpfer-Petersen, E., Reininger, W., Gluchoff, K. and Arpin, N. (1972b). *Phytochemistry* **11**, 3299–3304.

Steglich, W., Töpfer-Petersen, E. and Pils, I. (1973). *Z. Naturf.* **28c**, 354–355.

Swan, G. A. (1974). *Prog. Chem. Org. Nat. Prods.* 31, 522–582.
Taguchi, H., Sankawa, U. and Shibata, S. (1969). *Chem. Pharm. Bull., Tokyo* 17, 2054–2060.
Thomas, M. (1955). *In* "Modern Methods of Plant Analysis" (K. Paech and M. W. Tracey eds), vol. 4. Springer-Verlag, Berlin.
Thomas, R. (1971). *Chem. Commun.* 739–740.
Thomas, R. (1973). *Pure appl. Chem.* 34, 515–528.
Thomson, R. H. (1962). *In* "Comparative Biochemistry" (M. Florkin and H. S. Mason, eds), vol. 3A. Academic Press, London and New York.
Tipton, C. D. and Rinehart, K. L. (1970). *J. Am. chem. Soc.* 92, 1425–1426.
Turner, A. B. (1966). *Prog. Chem. Org. Nat. Prods.* 24, 288–328.
Turner, W. B. (1971). "Fungal Metabolites". Academic Press, London and New York.
Wassermann, H. H., Friedland, D. J. and Morrison, D. A. (1968). *Tetrahedron Lett.* 641–644.
Wassermann, H. H., Rodgers, G. C. and Keith, D. D. (1969). *J. Am. chem. Soc.* 91, 1263–1264.
Whalley, W. B. (1963). *Pure appl. Chem.* 7, 565–587.
Wikholm, R. J. and Moore, H. W. (1972). *J. Am. Chem. Soc.* 94, 6152–6158.
Williams, R. P. and Hearn, W. R. (1967). *In* "Antibiotics" (D. Gottlieb and P. D. Shaw, eds). Springer-Verlag, Berlin.
Yang, D.-M., Takeda, N., Iitaka, Y., Sankawa, U. and Shibata, S. (1973). *Tetrahedron* 29, 519–528.

Part II. FUNCTION

Chapter 13

Function in Photosynthesis

C. P. WHITTINGHAM

Botany Department, Rothamsted Experimental Station, Harpenden, Hertfordshire, England

I Introduction

Photosynthesis involves the photoreduction of carbon dioxide to an organic form such as carbohydrate. In green plants there is a concomitant liberation of oxygen from water.

$$CO_2 + H_2O \rightarrow [CHOH] + O_2 (\Delta F + 112 \text{ kcal})$$

In photosynthetic bacteria in place of water much stronger reducing substances are used such as H_2, H_2S and H_2R where R is an organic residue.

$$CO_2 + 2H_2R \rightarrow CH_2O + H_2O + 2R$$

In this case the energy required for the reaction is considerably less.

The pigment which photosensitizes the reaction has been known since the time of Dutrochet to be chlorophyll in the case of green plants; in the bacteria it is bacteriochlorophyll. Chlorophyll in organic solution has two absorption maxima in the visible region of the spectrum both at wavelengths shorter than 720 nm. The absorption of bacteriochlorophyll *in vitro* extends into the infrared to 800 nm. Excitation of either the blue or red absorption band of chlorophyll produces, after initial excitation to a short-lived state, the same common fluorescent state. This may return to the ground state either by the emission of fluorescence by dissipating energy as heat, or by chemical reaction. Prior to any of these there is the possibility of transfer of energy by resonance from one pigment molecule to another. *In vitro* a non-fluorescent long-lived excited triplet stage has also been demonstrated (Livingstone, 1954). While in principle energy migration can also take place from this metastable state, it is likely to be small in comparison with migration from the fluorescent state.

Evidence from homogeneous energy migration can be obtained from observations of the depolarization of fluorescence. Unless all molecules in a pigment aggregate are arranged in parallel, energy transfer results in a change in the direction of the oscillating electric dipole responsible for the emission of fluorescence. After several transfers, parallelism between the electric vectors of the absorbed and of the emitted light will be lost. If excitation is produced by plane-polarized light, the light emitted in fluorescence will be polarized also; but if energy migration occurs, fluorescence will be more or less depolarized. This has been observed by Goodheer (1957) in the fluorescence of phycobilins, even when observed in molecular dispersion; it can be interpreted as evidence that a quantum absorbed by one of the pigment molecules attached to a given protein molecule in the chromoprotein migrates through other pigment molecules before it is emitted as fluorescence. A similar process with excitation quanta absorbed *in vivo* by chlorophyll occurs although it may preferentially take place through different forms of chlorophyll which exist *in vivo* (see Seely, 1971).

Energy migration can lead not only to depolarization, but also to quenching of fluorescence if the energy quantum encounters centres in which the excitation energy is dissipated, either by conversion into heat, or by utilization in a photochemical reaction. The fluorescence yield of chlorophyll *a* in solution may be of the order of 30%. In the living cell it is 2–3%.

In addition to the different forms of excited molecule there are chemically different forms of chlorophyll in the ground state (Krasnovsky, 1959). In organic solution chlorophyll is reversibly bleached when illuminated in complete absence of oxygen. Chlorophyll dissolved in pyridine can be reduced in light by ascorbate or phenylhydrazine (Krasnovsky reaction).

$$\text{Chl } a + \text{ascorbate} \xrightarrow{h\nu} \text{Chl } a.\text{H}_2(\text{eosinophyll}) + \text{dehydroascorbic acid}$$

The reduced form of chlorophyll (eosinophyll) is a pink compound with absorption bands at 518 and 585 nm. It exhibits no paramagnetic resonance and is not therefore a free radical. It has a strong negative redox potential since it can reduce riboflavin or other oxidants, e.g. saffranin, with redox potentials down to −0·30 V. The photoreduction of riboflavin by ascorbate which is photosensitized by chlorophyll involves a free energy change of +9 kcal/mol. Such a system is a simple model of a photosystem in which the pigment chlorophyll photosensitizes the transfer of an electron from an oxidant to a reductant.

II Photochemistry of pigments *in vivo*

Microscopic studies of the photosynthetic green plant show that the pigments are concentrated in the chloroplast particles with an average concentration of 10^{-1} M corresponding to an average distance between molecules in the grana of the order of 50 Å. Theoretical studies of resonance transfer suggest that the probability of energy transfer within the lifetime of the fluorescent state should be high (Förster, 1959). This calculation assumes that chlorophyll is uniformly distributed throughout the granum of the chloroplast. Detailed studies (to be discussed later) of the absorption band in the red for chloroplasts or living cells show a complexity involving two or three separate maxima. On extraction of the pigment a single absorption band is obtained so that the complexity *in vivo* may be due either to dipole–dipole interaction or the existence of a single chromatophore in different molecular configurations. In addition the chloroplast contains accessory pigments which are predominantly carotenoids in the green plant and phycobilins in the blue-green and red algae. The carotenoids are relatively lipophilic and the phycobilins hydrophilic compared with chlorophyll. This may indicate the existence of at least a two-phase system. In spite of this complexity it is probable that the mean separation of molecules of different pigments does not exceed the distance through which resonance transfer is likely to take place.

Studies with the electron microscope and particularly of freeze-etched preparations show the thylakoids which make up the chloroplast to have characteristic patterns on both the inner and outer faces of the membrane but it is still not possible to state what is the minimum structural unit necessary for photosynthetic activity. The two primary photochemical reaction products must be in different phases and undergo enzymic stabilization, ultimately leading to the formation of carbohydrate and molecular oxygen, without danger of immediate recombination. Calvin (1959) suggested that a separation of products could be based on an electron migration similar to that occurring in crystalline photoelectric conductors although the absorption curve *in vivo* gives little evidence that the bulk of the chlorophyll *in vivo* is crystalline. But it is extremely probable that pigment molecules are systematically organized within basic subunits.

In vitro neither carotenoids nor phycobilins show fluorescence. *In vivo* they are able to sensitize the fluorescence of chlorophyll. In the diatom *Nitzchia* absorption of light quanta predominantly by the carotenoid fucoxanthin produced fluorescence of chlorophyll *a* with approximately the same yield as light absorbed by chlorophyll itself (Dutton *et al.*, 1943). Similarly the fluorescence of chlorophyll *a* may be sensitized by absorption by chlorophyll *b* in *Chlorella*, and in the red and blue-green algae phycobilins have been shown to transfer their energy to chlorophyll. The transfer of energy takes place with high efficiency, but can occur only from a pigment absorbing at a shorter wavelength to one absorbing at a longer wavelength and not vice versa. Furthermore, in the red algae, efficiency of energy transfer from phycoerythrin to chlorophyll *a* depends on treatment (Brody, 1958). If the cells are pre-illuminated by light absorbed by phycoerythrin the efficiency of transfer from phycoerythrin to

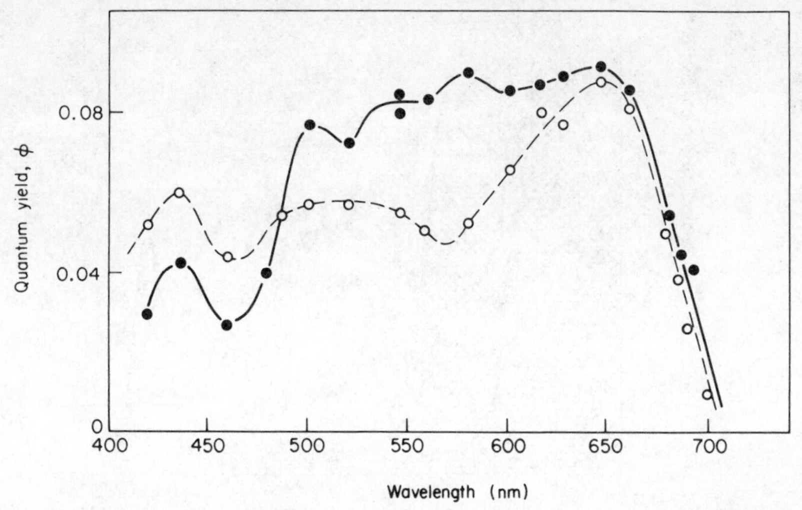

FIG. 1. Action spectra of photosynthesis in the red alga, *Porphyridium* after pre-illumination with green (●——●) or blue light (○ - - - ○). (After Brody and Emerson, 1959.)

chlorophyll is improved; whereas pre-illumination with light absorbed by chlorophyll decreases the apparent efficiency (compare Fig. 1).

The efficiency of excitation of photosynthesis by absorption by different pigments can also be seen from action spectra. The action spectrum shows the quantum yield of photosynthesis as a function of the wavelength of monochromatic light. The action spectrum can be compared with the absorption spectra for individual pigments isolated from the plant and the photosynthetic activity compared with the proportion of light at each wavelength absorbed by individual pigments (Fig. 2). Such comparisons have shown that, taken as a whole, the carotenoids of higher plants and green algae are only half as efficient as chlorophyll in producing photosynthesis (Emerson and Lewis, 1943). By contrast the carotenoid fucoxanthin in the brown algae and diatoms is almost as efficient as chlorophyll itself (Tanada, 1951). The phycobilin pigments are highly efficient in

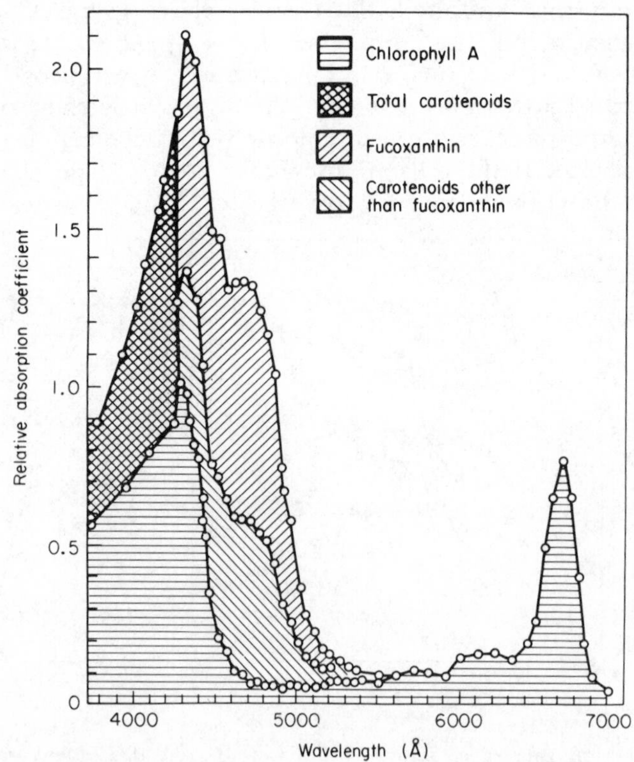

FIG. 2. The contribution of different pigments to the absorption spectrum of the diatom *Nitzchia closterium*. (After Dutton *et al.*, 1943.)

photosynthesis in the red and blue-green algae (Emerson and Lewis, 1942). One possible hypothesis is that all these pigments sensitize photosynthesis directly. A more attractive hypothesis, consistent with the observations of fluorescence discussed previously, is that light absorbed by pigments other than chlorophyll *a* is transferred to this pigment which then alone sensitizes photosynthesis. In general terms we may postulate a transfer by resonance from various pigments to that pigment which shows absorption at the longest wavelength. In the early studies this was regarded as being chlorophyll *a*.

III "Enhancement" effects

A detailed study of the action spectrum in the far-red region of the visible spectrum gave evidence of further complexity. Emerson and Lewis (1943) observed that the quantum efficiency of photosynthesis in the green alga *Chlorella* decreased very rapidly in the region between 680 and 700 nm where there was still appreciable absorption. The effect was even more striking in certain red algae where the quantum yield declined beyond 650 nm although there was appreciable absorption at 680 nm (Fig. 3). This suggested that a form of chlorophyll which absorbed in the far red was ineffective in photosynthesis.

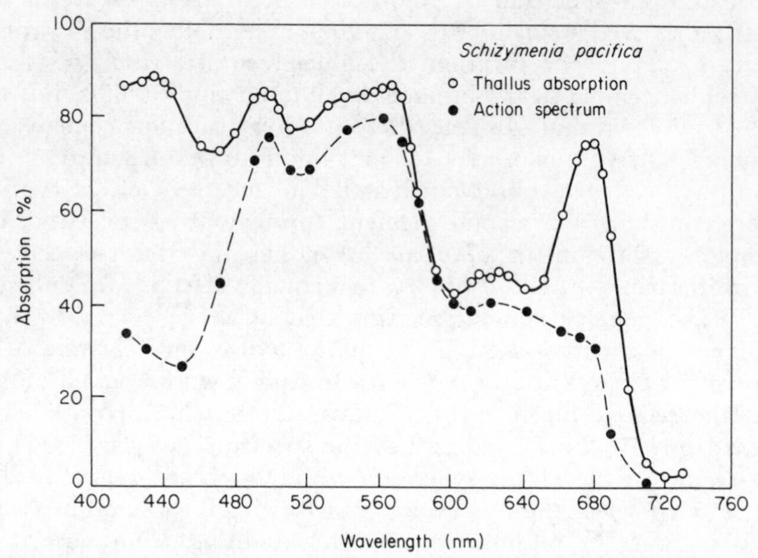

FIG. 3. Action and absorption spectrum of a red alga, *Schizymenia pacifica*. Absorption ○——○, Photosynthesis ● - - - ● (After Haxo and Blinks, 1953.)

Emerson and co-workers showed that far-red absorption could be made effective if it was supplemented by simultaneous absorption at a shorter wavelength. The action spectrum for the increased rate of photosynthesis resulting from a second wavelength superimposed on a beam of light of 697 nm showed two characteristic peaks, one at 650 nm and one at 670 nm. This suggested that the photosynthetic rate in light of 697 nm is limited in some way whereas that at 650 nm is not. At 650 nm illumination results in excitation of both chlorophyll *a* and *b*, whereas at 697 nm the absorption is largely due to chlorophyll *a*. Emerson concluded that the simultaneous excitation of chlorophyll *b* must improve the photosynthetic efficiency of the light absorbed in the far red by chlorophyll *a* (Emerson *et al.*, 1957). At that time Emerson made no comment on the second peak in the enhancement action spectrum at 670 nm.

An increased photosynthetic activity (enhancement effect) due to simultaneous illumination by two different wavelengths has now been found in a large number of organisms (Haxo, 1960). In the red alga *Porphyridium*, Brody and Emerson (1959) showed a marked increase in activity of light absorbed in the far red when this was supplemented by light absorbed by phycocyanin. In *Chlorella* the action spectrum for the effectiveness of the second light showed two peaks, one at 480 nm and one at 658 nm, giving a curve resembling the absorption spectrum of chlorophyll *b*; in *Porphyra* the action spectrum showed a single peak at 550 nm resembling the absorption spectrum of phycoerythrin. In the blue-green alga *Anacystis* maximum enhancement was obtained by illumination at 600 nm; this implied phycocyanin. In the diatom *Navicula* enhancement was obtained at 540 nm and 645 nm implicating fucoxanthin and chlorophyll *c*. It was observed that the enhancement effect resulted from excitation of a second pigment throughout its spectrum. For example, excitation in *Chlorella* of either the blue or the red absorption bands of chlorophyll *b* was equally effective for enhancement. The general conclusion was that it is necessary to have simultaneous excitation of both chlorophyll *a* and of some other pigment for efficient photosynthesis. In studies with monochromatic light, the relative inefficiency of absorption by chlorophyll *a* alone appears only at the far-red end of the spectrum because this is the only region in the visible where chlorophyll *a* is the sole absorbing pigment. It is clear there is evidence suggesting the operation of two light reactions in photosynthesis, one resulting from absorption by chlorophyll *a* (reaction I), the other from absorption by the accessory pigments (reaction II).

Independent evidence of a different type supporting this general conclusion came from studies by Blinks (1960) on the short-term changes in rate consequent upon a change in the wavelength of incident light. Even although the intensity at two different wavelengths was adjusted to give equal steady state photosynthetic rates, a marked change in rate was observed immediately upon a change from one wavelength to another. For example, in the green alga *Ulva* on changing the wavelength of light from 688 to 640 nm an abrupt increase in oxygen evolution was observed. On return to 688 nm there was a corresponding decrease. When in *Porphyridium* the activity resulting from light of wavelength 702 nm was taken as a fixed reference rate, the action spectrum for the size of transient due to alternate illumination with light of a variable wavelength corresponded to the absorption spectrum of phycoerythrin. In *Ulva* an action spectrum obtained in a similar way corresponded to the absorption spectrum of chlorophyll *b*. Blinks chose to interpret these transients as due in part to differing changes in respiratory rate produced by light of different wavelengths. French and Fork (1961) have shown that a stimulation of respiration does take place as a consequence of illumination at certain wavelengths. They determined the action spectrum both for the effect on respiration immediately following illumination and for the enhancement effect on photosynthesis. The former gave an action spectrum similar to absorption of chlorophyll in the far red, whereas the latter resulted from absorption by accessory pigments.

Myers and French (1960) have shown that the enhancement effect observed with *Chlorella* illuminated with two wavelengths does not require that the two wavelengths be given simultaneously, but that these may be given alternatively in periods of several seconds duration. This suggested the formation of an intermediate common to the two reactions with a life of several seconds.

Emerson made no comment on the second peak at 670 nm in the enhancement action spectrum observed by him. The peak (Govindjee and Rabinowitch, 1960), or shoulder (Myers and French, 1960), at 670 nm in action spectra for the Emerson effect in *Chlorella*, was interpreted by the first-mentioned authors as an indication of the participation in the enhancement effect of a chlorophyll *a* type, with maximum absorption at 670 nm. Emerson and Rabinowitch (1960) (cf. also Franck, 1958) proposed the following hypothesis. Two photochemical reactions occur in photosynthesis: one of these is caused by a non- or weakly-fluorescent chlorophyll *a*, the second by a fluorescent chlorophyll *a*. Excitation of the non-fluorescent chlorophyll *a* alone does not lead to photosynthesis. However, if both chlorophylls are excited simultaneously, the two reactions can co-operate, and lead to an enhanced photosynthesis. The so-called accessory pigments, e.g. chlorophyll *b* in *Chlorella*, and the phycobilins in red and blue-green algae, effect photosynthesis by transferring their excitation energy by resonance transfer to fluorescent chlorophyll *a* but not to the non-fluorescent form.

Franck (1958) suggested that the two kinds of chlorophyll involved in photosynthesis are a non-fluorescent form, in which the excited molecules in the S* singlet state are instantaneously converted, via the nπ state, into metastable molecules T; and a fluorescent kind, which permits the S* state to survive long enough either to fluoresce, or to transfer its energy by resonance, through a sequence of chlorophyll molecules, to a molecule already in the T state, raising it into an excited triplet state, T*. The latter was postulated to be able to permit direct sensitization of an electron transfer from H_2O to CO_2 (as R . COOH), i.e. through 1·2 V.

IV Existence of pigment forms *in vivo*

Brown and French (1959) made a detailed study of the absorption spectra *in vivo* of living plants by the use of a differential spectrophotometer. This instrument measures the first derivative of optical extinction as a function of wavelength. The observed spectra are then matched by adding together derivatives of normal probability curves showing a single absorption maximum. The absorption in the red region of the spectrum for the higher plant could be analysed in terms of the presence of at least three distinct forms of chlorophyll. These are characterized by the position of their absorption maxima and are referred to as Ca_{673}, Ca_{683}, Ca_{695}. On extracting the chlorophyll from these organisms the extract shows only a single peak due to chlorophyll *a*. The presence of chlorophyll *b* also results in a shoulder on the absorption curve *in vivo* at 650 nm. The proportion of the different forms of pigment varies from organism to organism and may indeed vary according to the cultural conditions in any one organism. For example, the spectrum of *Euglena* is characterized by three components having maxima at 673, 683 and 695 nm; as the culture ages, or if a culture is grown at lower light intensity, the 695 nm component becomes more and more pronounced (French and Elliot, 1958). Again after exposure of chloroplast suspensions to high intensities of illumination the absorption spectrum shifts, suggesting that the 673 component is preferentially bleached compared with 683. Three forms of phycoerythrin have been observed in different organisms, and phycocyanin also shows differences in different organisms (Ó hEocha, 1960).

In the purple bacteria, three forms of bacteriochlorophyll (with absorption bands at 800, 850 and 890 nm) appear *in vivo* but give rise to only one band (at about 770 nm) upon extraction into methanol. The wider separation of these bands is in agreement with the generally greater influence of solvent on the position of the absorption bands of bacteriochlorophyll compared with those of chlorophyll *a*. It is still unclear whether distinct photochemical functions must be attributed to these different forms in bacteria as well as in algae.

The two photochemical reactions required for efficient photosynthesis have been called by French the long-wave chlorophyll reaction and the accessory pigment reaction. French and Fork (1961) have suggested that chlorophyll a_{683} and chlorophyll a_{695} are both capable of a single photochemical step, referred to as the long wavelength chlorophyll reaction. The accessory pigment reaction can be effected by absorption by chlorophyll a_{673} and chlorophyll *b* in *Chlorella* and in the higher plant, by chlorophyll *c* and fucoxanthin in the brown algae, and by phycoerythrin and phycocyanin in the red algae.

As pointed out by Duysens it should not be assumed that the action spectrum for enhancement corresponds to the action spectrum of a photochemical system. The quantum yield is not necessarily maximal at shorter wavelengths where the accessory pigments absorb. The action spectrum for one system should be observed as the action spectrum of photosynthesis against a background of intense light activating the other system and vice versa. French *et al.* (1960) measured the action spectrum for *Chlorella* against strong background light of 700 nm and of 650 nm. The two resulting action spectra crossed at 683 nm, the wavelength which presumably results in equal excitation of both systems. Beyond 683 nm the short wavelength system is relatively less excited; at shorter wavelengths the accessory system is less excited.

V Photochemical reactions of isolated chloroplasts

The study of the photochemical reactions of photosynthesis received a great impetus when R. Hill showed that it was possible to demonstrate photochemical activity of chloroplasts outside the living plant. He isolated chloroplasts from higher plants and showed that if ferric salts were added and the system illuminated the ferric ions were reduced to ferrous ions and oxygen was evolved. It was suggested that the essential photochemical step was a "splitting" or photolysis of water and molecules. The oxidized radical produced from water [OH] * gave rise to the evolution of oxygen and the reduced radical [H] * reacted with the ferric salt.

The view that the splitting of water was the primary event in the photochemical reaction of photosynthesis had been suggested by Van Niel on the basis of his studies of the photosynthetic bacteria. Van Niel distinguished two large groups, the Thiorhodaceae, which require not only light energy and carbon dioxide to grow but in addition a reduced sulphur compound such as H_2S, and the Athiorhodaceae, which require in place of the sulphur compound an organic substance such as an alcohol. In no case in the photosynthetic bacteria was oxygen produced. Van Niel suggested that in these organisms the oxidized radical produced from the splitting of water did not give rise to oxygen as in the green plant but was disposed of necessarily by reaction with the added substance which thus became oxidized. This was in agreement with the observation that in the Thiorhodaceae sulphur was a product of photosynthesis, and in the Athiorhodaceae a ketone.

A number of hydrogen acceptors are capable of stimulating oxygen production by illuminated chloroplasts. These include quinones which are reduced to hydroquinones and dyes such as dichlorphenolindophenol. At first it was not realized that during the preparation and isolation of the chloroplasts a soluble protein ferredoxin† was lost from the chloroplasts. When this was added back it was found that chloroplasts could reduce NADP in light with the simultaneous production of oxygen (Davenport, 1959).

Arnon and his colleagues showed later that if adenosine diphosphate, phosphate and magnesium ions are supplied, chloroplasts produce ATP in the light. For this phosphorylative activity, light is essential and oxygen is not; hence the process has been called photophosphorylation to distinguish it from the analogous reactions of mitochondria which require oxygen (oxidative phosphorylation). Arnon showed that when hydrogen acceptor and phosphate acceptors were added together the rate of both phosphorylation and of oxygen evolution increased above their separate independent individual rates. Under certain conditions the amount of reduced product (NADPH) was stoicheiometrically equivalent to the amount

* [ ] square brackets indicate the radicals are probably in a combined form.

† Ferredoxin of chloroplasts is the same substance as the methemoglobin-reducing factor of Davenport *et al.* (1952), the NADP-reducing factor of Arnon *et al.* (1957) and the photosynthetic pyridine nucleotide reductase (PPNR) of San-Pietro and Lang (1958).

of ATP produced. This suggested a "coupling" between the phosphorylative and the oxidoreductive reactions.

In addition, a further phosphorylative activity of isolated chloroplasts was found when such substances as vitamin K or FMN (flavin mononucleotide) were added to the chloroplasts without any added hydrogen acceptor. Under these conditions the phosphorylative activity was greatly increased but the oxidoreductive activity diminished, demonstrating an alternative sequence of reactions resulting in the production of ATP as the sole product without any accumulation of reduced product. This was called cyclic photophosphorylation.

If the essential feature of photosynthesis is the photolysis of water, then during the transfer of hydrogen to an intermediate hydrogen acceptor phosphorylation must take place. The hydrogen acceptor normally reacts ultimately with carbon dioxide in the green plant or in isolated chloroplasts with added substances, e.g. ferric ions. However, in the presence of vitamin K it must be reoxidized reforming a water molecule and thus reversing the initial photolysis. There would then be no net oxidation-reduction reaction, but if the reoxidation was coupled to phosphorylation ATP could result as the sole product.

Later Arnon (1961) compared the photochemical activity of chloroplasts isolated from higher plants and chromatophores isolated from bacteria. Not all the photochemical activities exhibited by chloroplasts isolated from green plants are shown by isolated bacterial chromatophores. For example, addition of NADP was found to have little effect on the photophosphorylative activity of chromatophores. One reaction shown to be common to both types of particle was the formation of ATP in the presence of vitamin K but absence of an external hydrogen acceptor–cyclic phosphosphorylation. Arnon was then led to suggest that it is not necessary to regard the photochemical process as universally involving water splitting. He suggested that during excitation by light, chlorophyll loses an electron which is expelled with a high potential energy. This electron is returned to the chlorophyll molecule by a series of carriers of which vitamin K is one, and during this sequence of reactions a phosphate bond is formed.

Photosynthetic bacteria when supplied with hydrogen gas do not require light energy for the reduction of NADP, but if succinate or thiosulphate is the hydrogen donor additional energy must be provided by light for the reduction. In this case only some of the electrons ejected by light could return to chlorophyll and result in

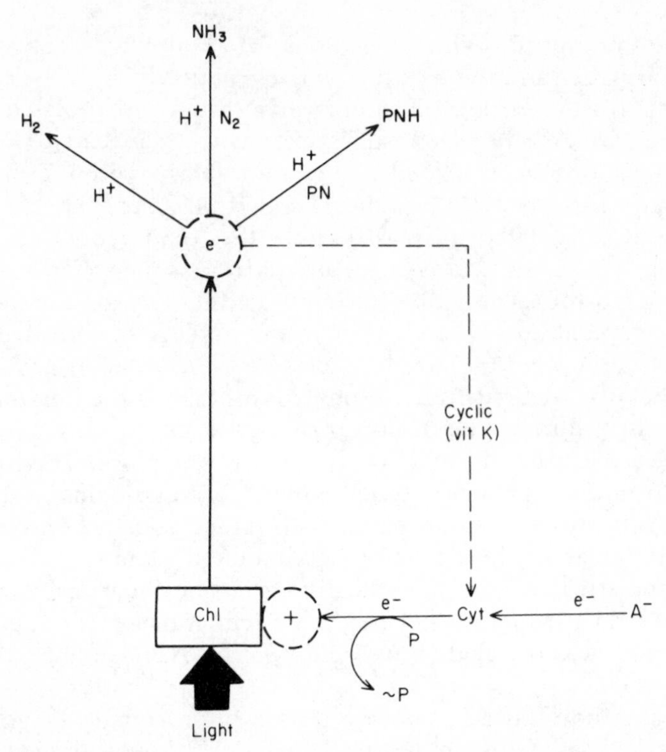

FIG. 4. Electron transport system in photosynthetic bacteria, e.g. *Chromatium*. (After Arnon, 1961.)

photophosphorylation, the remainder being used to reduce NADP. In the latter case some electrons must be restored to chlorophyll from the external hydrogen donor. This process in the case of thiosulphate and succinate is catalysed by cytochromes and is thought to be coupled to phosphorylation. Light energy is therefore used to give an electron from thiosulphate a potential which will enable it to reduce NADP. Alternatively the strongly reducing electron produced can be used in the photoproduction of hydrogen gas or the photofixation of nitrogen gas. In agreement with these hypotheses both these processes are stimulated by addition of thiosulphate (Fig. 4).

VI Light-induced absorption changes: the primary photochemical reactants

The difference spectrum of photosynthesis is measured as the change in the absorption spectrum of photosynthesizing cells consequent upon light activation. The first studies were those of Duysens (1952), who observed the absorption changes resulting from steady-state illumination. Later Witt (1955) and Kok (1957) used flash excitation first of millisecond duration and most recently with laser pulses of 10 ns duration. A repetitive flash technique with a device for storing and averaging signals was designed to decrease the noise relative to the signal.

Using steady-state illumination Duysens first measured the difference spectrum resulting from steady-state photosynthesis in the red alga *Porphyridium*. The difference spectrum obtained from chemical reduction of cytochrome *f* *in vitro* corresponded quite closely to the difference spectrum resulting from illumination in *Porphyridium* (Fig. 6). It follows that cytochrome *f* probably undergoes changes in oxidoreduction during the course of photosynthesis. Subsequent work showed that *Porphyridium* has the simplest difference spectrum of all organisms so far studied. With *Chlorella*, instead of a relatively simple difference spectrum at least five regions of significant changes were observed in the visible part of the spectrum. It is still not possible to interpret completely this complex difference spectrum unequivocally in terms of known biochemical intermediates.

Kok (1961) used intense flash illumination both with green algae and isolated chloroplasts and showed a significant absorption change in the red part of the spectrum at 700 nm associated with a smaller change at 430 nm. This could not be related to any known substance. He therefore called the substance responsible for this change in difference spectrum P700. Kok considered P700 to be a form of chlorophyll *a* occurring *in vivo* which is reversibly bleached by light. By addition of redox agents of different potential and observation of their effect on the difference band at 700 nm, it was shown that the bleaching of P700 related to its oxidation and the potential was determined as about $E_0' = +0{\cdot}400$ V. It is a one-electron redox carrier. The oxidation in light can be observed at −150°C when the dark reduction is significantly slowed down. Its probable concentration is very small, perhaps as little as one molecule for each 200 to 1000 molecules of chlorophyll *a*, assuming a molar extinction coefficient for the reduced form the same as that of chlorophyll. Many workers consider that P700 is a special part of the chlorophyll *a* present and may be that part which gives an absorption maximum at 692 nm. The rise time of the oxidation in light is less than 29 ns; the dark reduction time is of the order of 10^{-2} s. If chloroplasts are extracted with an acetone 70–72% (v/v) water mixture a great deal of the chlorophyll *a* is extracted but less of the P700. After such treatment the absorption changes due to P700 remain although changes attributable to cytochrome *f* are removed.

In a similar manner if chromatophores, isolated from photosynthetic bacteria, are treated with detergent most of the bacteriochlorophyll can be removed and only a special part called P890 remains (Clayton, 1963; Vredenberg and Duysens, 1963). This pigment is strongly bleached by light due to its oxidation; accompanying this change there is a shift of the absorption band at 800 nm to shorter wavelengths. By the use of laser beams Parsons (1967) showed that this change takes place within 10^{-6} s at 37°K. Both P700 and P890 may be the first stable substances to be oxidized by the excited chlorophyll molecule. The term reaction centre has been used to define the particular molecular grouping at which excitation energy is first converted to chemical bond energy.

By using flashes of different duration and particularly in recent work with the use of giant laser flashes Witt (1971) has determined the half-time of rise and decay of a number of differential absorption peaks.

An increase in absorption at 520 nm accompanied by smaller decreases at 430, 460 and 490 nm has a rapid rise time of less than 20 ns and a decay of 3 μs. Witt proposes that this is due to the formation of a metastable form of carotenoid sensitized by a triplet state of chlorophyll which is likely to exist under intensities of illumination in excess of those required to saturate photosynthesis. The triplet state of chlorophyll is unlikely to be populated significantly at normal light intensities and direct evidence for its occurrence *in vivo* is absent except in chloroplasts treated with detergent or in mutants with low chlorophyll content.

A photoactive pigment whose lifetime is one hundredth that of P700 has been observed by Döring *et al.* (1969) and called P680 since it is characterized by a decrease in absorption at 682 and 435 nm with a smaller band at 640 nm. Its redox potential is as yet not known but it is considered by Witt and colleagues to be the reaction centre pigment of system II as P700 may be for system I. Both the absorption changes at 700 and at 680 nm disappear in the presence of DCMU but upon addition of PMS the change at 700 nm reappears but not that at 680 nm.

A relatively slow change involves an increase in absorption at 515 nm accompanied by a decrease at 475 nm. This was first shown by Duysens who could not attribute it to the oxidoreduction of any known electron carrier. Some attributed the changes to carotenoids (de-epoxidation) and others showed it was contributed to by changes in light scattering arising from conformational changes in illuminated chloroplasts. Junge and Witt (1968) have analysed changes at these same wavelengths but with a rise time of less than 10 μs and a decay of 20 ms. With inhibitors which did not affect oxygen evolution but did inhibit photophosphorylation the changes were unaffected. Witt considered that they indicate the presence of a high energy state which is capable of conversion to phosphate bond energy; they may be due to electromagnetic effects resulting from the development of an electrical field across the thylakoid membrane. The absorption changes are induced to similar extents by both photochemical systems I and II. Ionophores such as gramicidin D or uncouplers such as DPIP greatly accelerate the decay time of these absorption changes. The complete difference spectrum is similar (although not identical) to the change induced in the absorption spectrum of a chlorophyll carotenoid multilayer *in vitro* by an electrical field of 10^6 V/cm. Similar changes have been observed in photosynthetic bacteria (Jackson and Crofts, 1971) and in that case the same absorption change could be induced in the dark if a gradient of ion concentration were established across the membrane.

The existence of two possible light reactions can be most clearly demonstrated if their activation is separated in time. Duysens *et al.* (1961) showed that if far-red light (absorbed by pigment system I) was given to red algae, cytochrome *f* became oxidized, e.g.

$$\text{Chl cyt} \rightarrow \text{Chl}^{*}\ \text{cyt} \rightarrow \text{Chl}^{+}\ \text{cyt}^{-}$$

Cyt^{-} indicates cytochrome which has gained an electron and become oxidized. The Chl^{+} must regain its electron ultimately from water.

If subsequently shorter wavelength light absorbed by system II was given, the return of the oxidized cytochrome to the reduced state was accelerated (see Fig. 6). Again, the Emerson enhancement effect can be observed also if the two light beams are given not simultaneously but successively. Spectroscopy can identify intermediates which are situated between the two light reactions. P700 can be shown to be placed between the two photosystems. Actinic light of wavelengths less than 700 nm results in rapid oxidation followed by a slower reduction as shown by a decrease in absorption but far-red light beyond 700 nm can result only in oxidation.

In green algae and leaves there is evidence from photosynthetic enhancement studies that the reaction centres associated with photoreaction I contains P700 and forms of chlorophyll *a* absorbing at longer wavelengths, identified as chlorophyll a_{690} and probably some of the chlorophyll a_{680} component also. The pigments concerned with photoreaction II in system II include certainly the largest part, if not all, of chlorophyll a_{670}, all the accessory pigments (including chlorophyll *b*), and possibly also some chlorophyll a_{680}. As the wavelength of exciting light varies, the relative amount of energy absorbed by the two systems varies; beyond 680 nm "excess" energy is absorbed by system I relative to that absorbed by system II. The excess is wasted and the "red drop" in photosynthesis occurs. When the reverse situation occurs, i.e. the excess is in system II, some of the excess energy could be transferred theoretically from system II to I. (The reverse would be thermodynamically impossible.) Such a "spill-over" hypothesis was proposed by Myers (1971) as an alternative to the completely independent operation of the two systems as formulated in the "separate package" hypothesis. Experimental data are still inadequate to establish one view rather than the other. Some evidence suggests that pre-illumination with system II light for 2–3 min may increase the proportion of pigment absorbing in system I and vice versa; this suggests a movement of pigment molecules between the photochemical units rather than a transfer of energy after absorption. By studying the action spectra for the various absorption changes of the difference spectrum it can be shown that the change at 433 nm is greatest with stimulation of system I and that of 515 nm with stimulation of excess system II.

In *Porphyridium* and in blue-green algae the phycobilins are the predominant pigment of system II. System I must contain most of the chlorophyll. Thus in these groups of organisms the "red drop" of photosynthesis extends over a wider range of wavelengths so that absorption by chlorophyll *a* alone is relatively inefficient both with respect to photosynthesis and fluorescence.

In photosynthetic bacteria a small special part of the bacteriochlorophyll is considered to be a reaction centre, either P890 or P870 according to species and to the wavelength at which maximum change in absorption takes place upon illumination. Other changes in absorption result from cytochrome components, probably of the *c* type and a ubiquinone; these constituents form the reaction centre of bacteria. Cytochromes of the *b* type are also present. Transfer of electrons from pigment through ubiquinone and cytochrome forms a cyclic reaction path which can convert light energy to phosphate energy. There is the possibility of a second light reaction coupled to the oxidation of substrate and an additional cytochrome component; this is also dependent on light excitation by bacteriochlorophyll.

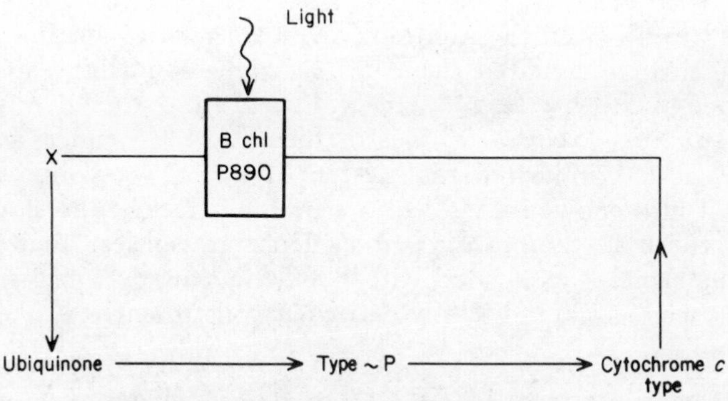

FIG. 5. Cyclic electron transport in bacteria.

In general light energy is absorbed by or transferred to a chlorophyll molecule and thence to a reaction centre molecule P. P is part of a molecular complex ZPQ with an electron donor Z and an electron acceptor Q. The excitation energy is used to transfer an electron from Z to Q:

$$ZPQ + h\nu \rightarrow Z^{+}PQ^{-}$$

In green algae and higher plants in system I, P is P700; in system II it is probably P680. The nature of the substance Z for the two systems will be discussed in the following section.

VII Electron paramagnetic resonance studies

Electron paramagnetic resonance (e.p.r.) spectroscopy can be used to detect the formation of free radicals or other types of paramagnetic centres. Suspensions of algal cells and of isolated chloroplasts give a characteristic e.p.r. signal upon illumination. Detailed examination of the signal in *Chlorella* has shown it to consist of two components; (1) a signal which is excited by far-red absorption and rapidly decays when the light is turned off and (2) a signal which can also be excited with shorter wavelengths and which persists for a longer time after darkening. Kok and Beinert (1962) prepared a particulate preparation from red algae from which two-thirds of the chlorophyll had been removed by acetone but most of the P700 remained. This preparation after dispersion by sonication showed a reversible absorption change in the light and a free radical signal of the same type as that resulting from illumination of whole chloroplasts. The signal was affected by light, ferricyanide and PMS in a similar manner to the effect on changes in absorption at 700 nm. Thus it was suggested that the light-dependent short-lived e.p.r. signal in photosynthetic material came from the photo-oxidized form of P700. In *Anacystis* 713 nm light stimulated the e.p.r. signal to a much greater extent than 635 nm light. Levine and Smillie (1962) studied mutants of *Chlamydomonas reinhardii*. Measurements of biochemical activity in mutant ac-141 indicated that system II was blocked but system I active; it also showed a fast e.p.r. signal, but lacked the slow signal. Whilst there is therefore good evidence to suggest that the fast decaying signal is associated with P700 the slow decaying signal has not yet been attributed to any particular constituent.

VIII Electron carriers between the two photochemical systems

Two photochemical reactions are required for green plant photosynthesis. Photochemical process I sensitizes the transfer of electrons from some intermediate to a potential more negative than that of pyridine nucleotides. Process II produces this intermediate from water with the simultaneous production of oxygen. Hill and Bendall (1960) proposed on theoretical grounds that cytochromes might act as intermediates between the two systems. They proposed that cytochrome b_6 could be reduced with the simultaneous oxidation of water in process II and then transfer an electron to cytochrome *f*, which is then oxidized in process I with the ultimate reduction of NADP. The standard oxidation reduction potential of cytochrome b_6 (cytochrome$_{563}$) is 0 V and of cytochrome *f* (cytochrome$_{553}$) + 0.34 V. They proposed that in a dark reaction between photochemical reactions I and II the electron transfer between the two cytochromes could be coupled to phosphorylation (Fig. 6).

Cytochrome oxidoreduction is characterized by absorption

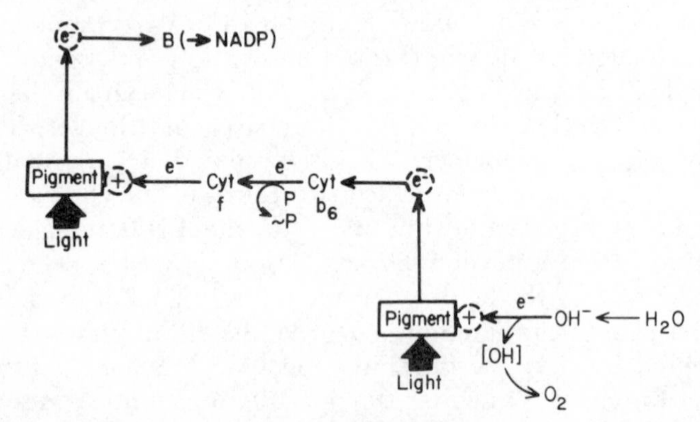

FIG. 6. Electron transport system in green plant photosynthesis.

changes near 405, 430 and 555 nm. In green plants such changes have not been readily observed under normal conditions. Duysens attributed a small change at 420 nm to a cytochrome and assumed that the changes at 405 and 555 nm which should have accompanied it were obscured by larger changes resulting from changes in state of other substances. By exciting chloroplasts at low temperature (−150°C) or by using 720 nm (procedures which might be expected to slow down the reduction of cytochrome and allow the oxidation to be more readily seen) changes in cytochrome absorption could be observed. Chance reported observations with whole leaves at 77°K, both of Swiss chard and spinach, where changes at 555 nm were observed which he attributed to a cytochrome. Chance preferred to consider cytochrome as the photoreagent for process I and P700 to be a secondary change induced by process I and not itself in the main electron chain. Bonner and Hill (1963) observed that plastids isolated from etiolated mung bean leaves show an absorption change at 557 nm which becomes less apparent when the leaf is green.

Duysens first showed changes characteristic of cytochromes in the red algae *Porphyridium*. In *Anacystis*, Duysens and Amesz (1962) observed light oxidation of a cytochrome 552 nm; later, Olson and Smillie (1963) found the same changes in *Euglena* chloroplast fragments. Far-red light resulted in an oxidation of the cytochrome, whereas shorter wavelengths caused an initial oxidation followed by a slower reduction to a final steady state. Cytochrome oxidation was more effective when chlorophyll *a* was excited rather than phycocyanin. When DCMU is added to stop reduction of the cytochrome, it is found that excitation of a pigment absorbing at 705 nm is more effective than absorption by chlorophyll *a* at 690 nm. This suggests that P700 might be primarily concerned in a cytochrome oxidation. Olson and Smillie showed that washed chloroplast fragments from *Euglena* lost much of the cytochrome 552. When such fragments were illuminated in tris buffer strong light caused gradual reduction of cytochrome b_6, but if the light intensity were appreciably lower a light-induced oxidation was found.

Rumberg, using spinach chloroplasts, observed reduction of cytochrome b_6 (cytochrome b_{563}) to system II light and a rapid re-excitation in the dark or far-red light. He first suggested that re-excitation resulted from the transfer of electrons from b_6 to cytochrome *f* but later he found that cytochrome *f* became reduced ten times faster than b_6 was oxidized. He therefore proposed that b_6 was not on the dark path of electron transfer between systems I and II. Hind and Olson investigated the effect of various inhibitors with spinach chloroplasts and showed that with DCMU the oxidoreduction of cytochrome b_6 was less sensitive and that of cytochrome *f* more sensitive than photophosphorylation. Similar effects were observed with antimycin A. They proposed that there were two parallel paths of electron transport between systems I and II, b_6 occurring in one path and *f* in the other and that the oxidoreduction of cytochrome *f* was coupled with phosphorylation. An additional cytochrome component, cytochrome b_{559}, has been observed in mature bean leaves although in the etiolated leaf only b_6 and *f* occur. Cytochrome b_{559} has also been observed in spinach chloroplasts and remains bound to the membrane system after b_6 and *f* have been extracted with Triton X100. There is some doubt as to whether some absorption changes attributed by earlier workers to cytochrome b_6 were in fact due to b_{559}. Cramer and Butler (1967), Ben-Hayyim and Avron (1970) but not Hind (1968) nor Knaff and Arnon (1969) nor Hiller *et al.* (1971), observed light-induced absorption changes of cytochrome b_{559} in untreated chloroplasts; all agree that cytochrome b_{559} is reduced by system II but there is disagreement as to its mechanism of photo-oxidation. Now it has been shown that this cytochrome can exist in two distinct forms which differ in potential by 300 mV, i.e. 0·065 V and 0·37 V. Several authors have considered that b_{559} plays a part in some side reaction relating to the oxidized state of system II, presumably in its high potential forms whilst recognising that it may (in its low potential form) be oxidized by system I. Which of these oxidations is important *in vivo* has not been established.

Knaff and Arnon (1969) proposed a scheme of three light reactions consisting of (i) photosystem II, which contains two short wavelength light reactions (2A and 2B) in series joined by an electron transport system containing cytochrome b_{559} and (ii) a photosystem I in parallel with II but concerned only with cyclic electron transport; they consider cytochrome *f* and b_6 are components of this electron flow. This view is not consistent with the observed antagonistic effect of red and far-red light on the oxidoreduction of cytochrome *f*. Cytochrome *f* probably plays some role although not necessarily the main role in the transfer of electrons between systems I and II. The function of cytochrome$_{559}$ and b_6 is less clear.

Plastocyanin

In 1960 Katoh (1960) isolated from *Chlorella* and from leaves of higher plants a copper-containing protein, plastocyanin (mol. wt 21 000) in an abundance of approximately 1 mole for each 400 mol chlorophyll; it contained two copper atoms per molecule. After treatment with detergents chloroplasts were no longer able to reduce coenzymes utilizing water; when the plastocyanin was added back this ability was restored. The same result was obtained from the photosensitized reduction of coenzymes by reduced indophenol effected by system I alone, and for cyclic photophosphorylation. Hence plastocyanin must react in some part of the electron path common to systems I and II. The potential of plastocyanin is +0·40 V. The reduced form is colourless and not oxidized by air; the oxidized blue form (absorption bands at 460, 597 and 770 nm) is rapidly reduced by chloroplast preparations in the light. The two components, plastocyanin and cytochrome *f* are present in about the same abundance. Some authors consider that there may be two parallel electron paths between systems I and II; some electrons reacting through cytochrome *f* and others through plastocyanin. Selected mutants of *Chlamydomonas* were studied by Levine (1969) some of which were devoid of cytochrome *f* (ac-206) and others devoid of plastocyanin (ac-208). The absence of plastocyanin reduced cyclic electron flow but the loss of cytochrome *f* had little effect, suggesting that cytochrome *f* may serve as an additional reservoir for the storage of electrons but is not an essential constituent.

Mutants of the green algae *Chlamydomonas reinhardii, Euglena gracilis* and *Scenedesmus obliquus* have been obtained in large numbers by Levine. Each mutant strain has a reduced rate of NADP reduction with water as the electron donor and this can be correlated with the relative absence of certain electron carriers. A mutant of *C. reinhardii* ac-80_a lacks P700 as shown by the absence of the characteristic difference spectrum; it can carry out the Hill reaction but only with DPIP and not with NADP as acceptor. Also the light-induced oxidation of cytochrome_{559} and cytochrome_{553} are not observed in this organism. Cyclic phosphorylation catalysed by PMS also cannot be demonstrated. This suggests that this mutant has a functional photosystem II but not photosystem I.

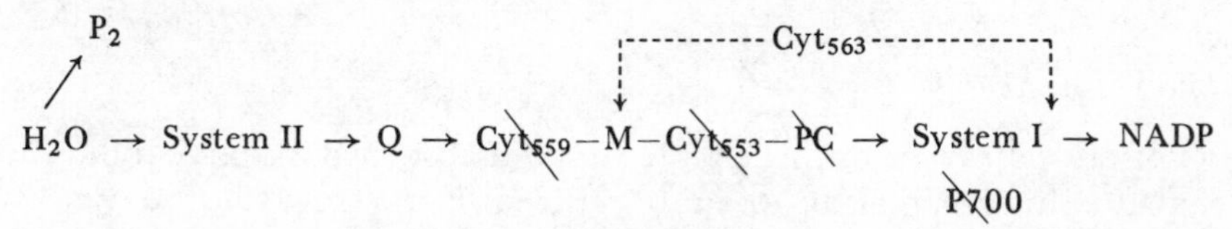

FIG. 7. Electron transport chain in *Chlamydomonas reinhardii* according to Levine, with sites indicated (\) when mutant blocks have been observed and also showing cyclic flow (- - -).

Plastoquinone

Plastoquinone was originally obtained from leaves by Kofler and isolated from chloroplasts by Lester and Crane (1965). It occurs as a number of isomers and is present in an overall ratio of one mole for each seven moles chlorophyll. The structure is similar to that of ubiquinones. After plastoquinone is extracted from chloroplasts by petroleum ether the ability to oxidize water is lost; on restoring plastoquinone the chemical activity is restored. Plastoquinone is probably oxidized and reduced in the course of photosynthesis, reduction being accompanied by a decrease in optical density at 263 nm; excess of system I light results in oxidation and excess of system II reduction. Photoreduction is inhibited in the presence of DCMU. The oxidation-reduction potentials are between +0·00 V and +0·10 V so that plastoquinones must be assumed to operate at the other end of the thermal reaction step from plastocyanin. Only one in ten plastoquinone molecules are oxidized in a short flash but all in a long flash; hence plastoquinone is considered to exist in two "pools". Extraction of plastoquinone has rather less effect on cyclic phosphorylation than on non-cyclic photophosphorylation, suggesting that electrons from cyclic flow must react somewhere between plastoquinone and the electron donor to P700. Witt and co-workers believe that the small plastoquinone pool is the primary acceptor of photosystem II; Amesz believes that compound Q (see next section) fills this role.

Ferredoxin has been frequently proposed as the primary electron acceptor for photosystem I. Suggestions have also been made that other more reducing compounds act as primary acceptor, with potentials as low as –0·6 V, ferredoxin then being secondary. Fuller and Nugent (1969) proposed pteridines, but the low potential claimed was strongly questioned by Archer and Scrimgeour (1970). Yocum and San Pietro (1970) discovered a ferredoxin-reducing substance (FRS), which it was essential to add to sonicated chloroplasts to retain the ability to reduce NADP.

The substance Q

As discussed earlier in this chapter quanta absorbed by phycobilins in red and blue-green algae are more efficient in exciting fluorescence of chlorophyll *a* than excitation of chlorophyll *a* itself. From this observation it was concluded that energy was transferred preferentially from the phycobilins to some part of chlorophyll *a*, say chlorophyll a^2, and not to another part of chlorophyll, say chlorophyll a^1, which is only weakly fluorescent or does not fluoresce at all. In *Porphyridium cruentum* actinic light of 680 and 430 nm results in oxidation of cytochrome and hence must activate photochemical process I more than process II. Since it gives relatively little fluorescence the weakly fluorescent form chlorophyll a^1 is to be associated with process I. Addition of 560 nm light (which excites fluorescence with high efficiency) to illumination with 680 nm light causes reduction of cytochrome, suggesting that the fluorescent form of chlorophyll a^2 is associated with process II.

Govindjee *et al.* (1960) investigated the effect of fluorescence in *Chlorella* of the addition of far-red light to illumination with shorter wavelengths. They found that the total fluorescence from combined illumination with a far-red and a red beam given together was smaller than the sum of the fluorescence intensities excited by the beams given separately. This is consistent with the mechanism just discussed. They claim that the fluorescence yield at both wavelengths was independent of intensity, but Duysens and Sweers (1963) showed that the steady-state fluorescence yield increased with an increase in light intensity for light preferentially activating process II. The latter authors observed marked changes in fluorescence intensity upon changing from a light which primarily activated process II to one primarily activating process I and vice versa. They postulated that the decrease in fluorescence resulting from addition of light activating process I, indicated that some reagent for process I called Q must quench the fluorescence of chlorophyll a^2 when oxidized but not when reduced. In *Porphyridium* absorption of only one quantum per 100 chlorophyll a^2 molecules restored fluorescence so that Q is present in concentration of about one hundredth that of chlorophyll. In the presence of DCMU, even upon illumination with weak light, chlorophyll a^2 fluorescence is not decreased by addition of light-activating system I. Hence DCMU must inhibit the reoxidation of QH.

$$H_2O \longrightarrow \text{(fluorescent chlorophyll } a^2 \text{ (system II))} \longrightarrow Q \longrightarrow$$
$$\text{plastoquinone} \longrightarrow \text{(non-fluorescent chlorophyll } a^1 \text{ (system I))} \longrightarrow \text{NADP}$$

Duysens and Sweers suggest further that Q may be converted by a dark reaction into a form Q^1. This also is postulated to quench fluorescence but to be non-photoactive. The reverse reaction $Q^1 \rightarrow Q$ which takes place in light II (since Q is then converted to QH and hence Q^1 to Q) is considered to be slow. This indeed may be the slow reaction which has been observed in a number of induction phenomena, and would mean that no slow reaction need be postulated in the main electron transport generally. Other possible slow dark reactions include a back reaction between NADPH and Q.

The integrated electron chain

The abundance of each intermediate carrier relative to chlorophyll has been determined. One molecule of cytochrome *f* is present for each 350 molecules of chlorophyll *a*. Plastocyanin is in about the same proportion as cytochrome *f* and the proportion of P700 is also the same. Hence a system I particle may be visualized as containing one plastocyanin, one cytochrome *f*, one P700 and something like 300 chlorophyll *a* molecules. The absorption of light can take place in any one molecule; energy is then transferred to a special molecule (e.g. P700) where electron transfer takes place. This is the reaction centre. Energy "harvesting" by the pigment "antennae" is used to describe the absorption by the bulk of the pigment molecules which are not themselves directly involved in initiation of a chemical process. There are relatively twice as many cytochrome b_6 molecules as cytochrome *f*, and there are three molecules of plastoquinone for each molecule of b_6. Hence for each molecule of plastoquinone there are approximately 50 chlorophyll molecules. Again several pigment molecules must be associated with one centre of reaction for photochemical system II.

In 1932 Emerson and Arnold had, on the basis of physiological studies with flashing light upon photosynthesis in algal cells, postulated a photosynthetic unit. The exciting illumination was sufficiently intense to excite every chlorophyll molecule, but with a sufficiently brief flash no molecule could be excited twice within the duration of a single flash. The average maximum photosynthetic yield obtained from such an intense but very brief flash was a measure of the absolute quantity of some yield-limiting intermediate between the chlorophyll and the ultimate biochemical product, in this case measured as oxygen. With a flash of 10^{-6} s duration so bright that further increase in energy did not increase the photosynthetic yield, Emerson and Arnold found a limiting maximum yield per flash of 1 molecule of oxygen for every 2000 moles of chlorophyll; they measured the production of oxygen with a manometer so that the yield per flash was obtained as an average from a sequence of several hundred flashes. Since one molecule of oxygen represents the transfer of four electrons each electron transfer centre must relate to 2000/4 or 500 chlorophyll molecules. Thus within the intermediates linked together to form the electron chain, there must be some component present in approximately this relative quantity. A similar limiting factor was indicated from the results of experiments with DCMU.

In recent studies Joliot (1968) has measured the oxygen produced per flash from chloroplasts or algae using a sensitive polarographic method. The flashes were intense but short (10^{-5} s) and separated by dark intervals of seconds. If such a sequence of flashes follows a prolonged dark period, the first flash, no matter how intense, produces no oxygen, the second very little, the third excess and the fourth approximately the same as the average of many subsequent flashes. The same relative magnitudes are observed with flashes 5–8 but the effect is smaller and so on in subsequent cycles. It is proposed that four electron transfers (corresponding to four states, Z, Z^{+}, Z^{2+}, Z^{3+}) are required for the release of one molecule of O_2; the accumulation of individual changes takes place in the first two flashes and only in the third and fourth are units of four available. The experiments clearly indicate that the reaction centres must act independently in the accumulation of four separate electrons.

IX Physical separation of the two photosystems

The possibility exists that the two photosystems occur as physically discrete entities within the chloroplast *in vivo*. To try and separate them chloroplasts have been treated with detergents like digitonin or Triton X100 or other dispersing agents. Using digitonin followed by differential centrifugation, Anderson and Boardman (1966) obtained a less heavy and a heavy fraction. The first striking difference between them was the ratio of chlorophyll *a* to chlorophyll *b*; the light fraction had a ratio of anything between four and seven to one, the heavier system a ratio of two to one, whereas the ratio for the whole tissue was about three to one. Hence the light fraction was relatively enriched in chlorophyll *a* and heavier in chlorophyll *b*. The ratio of cytochrome *f* to the total chlorophyll present for the whole tissue is normally about 1/350 but the light fraction was found to have relatively more. On the contrary, cytochrome b_6 was more abundant in the heavy fraction. Cytochrome b_{559} was also unequally distributed between the particles. The results suggest that the light particles are relatively richer with respect to system I and the heavy with respect to system II. The results are consistent with the view that separation of two particles containing the separate photochemical systems has been achieved. The ability to utilize reduced indophenol to reduce coenzyme is shown equally by the heavy particles and the light, but the heavy particles more actively catalyse reactions which require photochemical system II. It cannot be established whether the two types of particle actually exist in the chloroplast before treatment, or whether the detergent has broken up a system of membranes into two distinct forms by selectively destroying certain components. The general evidence favours the view that the particles occur *in vivo* and are liberated by suitable detergent treatment with their biochemical integrity preserved. Other methods of preparation have included sonication treatment (Jacobi and Lehmann, 1968) and exposure of chloroplasts to very high pressures (Michel and Michel-Wolwertz, 1970); by means of these different methods, preparations of similar particle types can be obtained.

X Phosphorylation

Phosphorylation was first considered in terms of a reaction between an oxidoreduction pair. The reduced product was believed to react with a hydrogen acceptor C transferring part of the energy of the BH_2-I linkage to that between I and B.

$$AH_2 + B \longrightarrow BH_2 + A$$
$$BH_2 + I \longrightarrow BH_2 - I$$
$$BH_2 - I + C \longrightarrow CH_2 + B \sim I$$
$$B \sim I + ADP + P \longrightarrow I + B + ATP$$

The "high-energy compound" I ~ B could then react with ADP and phosphate to give ATP. The mechanism postulates that there must be a high-energy complex formed in the system prior to phosphorylation. Hind and Jagendorf (1963) sought to demonstrate such a high-energy intermediate by illuminating chloroplasts in the absence of phosphorylating reagents. When subsequently phosphorylation reagents were added in the dark, phosphorylation took place, thus separating in time the formation of a high-energy complex in the light from its subsequent use in phosphorylation in the dark. Thus they showed that a state generated by light in the absence of phosphate was capable of subsequently effecting phosphorylation; it persisted for some seconds in the dark. They were unable to isolate chemically any specific intermediate.

An alternative mechanism of phosphorylation suggested by Mitchell (1966) proposed that the high-energy state was not within a chemical compound but present as a physical state capable of generating electromotive force. For example, if during illumination an electrical potential difference is established between the inside and outside of the chloroplast, the electrical energy thus stored could subsequently be utilized for phosphorylation. Mitchell supposed that the chloroplast membrane was such that during illumination protons moved selectively across it to establish a difference in hydrogen ion concentration between the two sides. It follows that chloroplasts should accumulate protons as a result of illumination. Hind and Jagendorf were indeed able to demonstrate that light did induce pH changes in an unbuffered suspension of chloroplasts in the absence of phosphorylating agents, the medium becoming alkaline. Spectroscopic measurements of pH change have been made after the introduction into the chloroplasts of bromthymol blue or umbelliferone, the latter indicating a pH change by a change in its fluorescence. The time course of pH rise in chloroplasts in the light closely but not precisely antiparallels the absorption change at 515 nm previously mentioned as an indicator of a change in electrical field force. Together with the change in pH, movement of other ions, e.g. potassium, out of the chloroplast have been observed upon illumination (Dilley and Vernon, 1965).

Other methods have been used to study the formation of a high-energy state. For example, the average size of the chloroplast particles has been shown to increase during illumination (Packer and Siegenthaler, 1965). Under other conditions shrinkage has been observed. The type of response is determined by the nature of the anions present. Shrinkage is reversible in the dark but swelling is not. Also by studying the optical properties of a chloroplast suspension it has been shown that light is scattered to a different extent during illumination of the suspension suggesting physical changes which may be correlated with the formation of a high-energy state. By treating chloroplasts with a chelating agent such as EDTA, a specific factor, the coupling factor, can be extracted. After such treatment chloroplasts still show changes in pH and in light scattering and in swelling upon illumination, but they cannot catalyse phosphorylation. When the coupling factor is added back, phosphorylation is restored.

According to the hypothesis of Mitchell it follows that generation within the chloroplasts of an electrochemical potential difference by any means other than light should result in a state potentially capable of generating ATP. In agreement with this prediction Hind and Jagendorf (1965) found that if chloroplasts were equilibrated at an acid pH in the dark with ADP and inorganic phosphate and the pH was then rapidly raised, phosphorylation resulted. Moreover, the maximum synthesis of ATP observed was stoicheiometrically related to the number of protons passing through the membrane during the equilibrium of hydrogen ions. Later it was confirmed that a change in pH was more important than the absolute value of the initial or final pH. The production of ATP can be so large as to make it extremely unlikely that the phosphorylation can be due to the presence of any chemical intermediate within the chloroplast whose state is changed by the change in pH. The phosphorylation resulting from an acid-base transfer in the dark is specifically inhibited by a serum prepared from the coupling factor mentioned previously. The dark phosphorylation is also uncoupled by these agents believed to act as proton-conducting reagents, e.g. nitrophenols or carbonyl cyanide *m*-chlorophenylhydrazide (CCC).

When spinach chloroplasts are rapidly changed from an acidic to a basic suspension medium, Mayne and Clayton (1966) observed that they emit for a brief time chlorophyll fluorescence. It is considered that chlorophyll has been excited to the singlet state at the expense of some high-energy state, a phenomenon demonstrated by Strehler and Arnold (1951) with algae some years earlier.

REFERENCES

Anderson, J. M. and Boardman, N. K. (1966). *Biochim. biophys. Acta* **112**, 403.

Arnon, D. I. (1961). *Nature, Lond.* **190**, 601.

Arnon, D. I., Whatley, F. R. and Allen, M. B. (1957). *Nature, Lond.* **180**, 182.

Archer, M. C. and Scrimgeour, K. G. (1970). *Can. J. Biochem.* **48**, 526.

Ben Hayyim, G. and Avron, M. (1970). *Eur. J. Biochem.* **14**, 205.

Blinks, L. R. (1960). *In* "Comparative Biochemistry of Photoreactive Systems" (M. B. Allen, ed.), chap. 22. Academic Press, New York and London.

Bonner, N. and Hill, R. (1963). *In* "Photosynthetic Mechanism of Green Plants", pp. 82–90. Nat. Acad. Sci. Washington, D.C.

Brody, M. (1958). *Science, N.Y.* **128**, 838.

Brody, M. and Emerson, R. (1959). *J. gen. Physiol.* **43**, 251.

Brown, J. S. and French, C. S. (1959). *Pl. Physiol., Lancaster* **34**, 305.

Calvin, M. (1959). *Brookhaven Symp. Biol.* p. 160.

Clayton, R. K. (1963). *Biochim. biophys. Acta* **75**, 312.

Cramer, W. A. and Butler, W. L. (1967). *Biochim. biophys. Acta* **143**, 332.

Davenport, H. E. (1959). *Biochem. J.* **73**, 45P.

Davenport, H. E., Hill, R. and Whatley, F. R. (1952). *Proc. R. Soc. B* **139**, 346.

Dilley, R. A. and Vernon, L. P. (1965). *Archs Biochem. Biophys.* **111**, 365.

Döring, G., Renger, G., Vater, J. and Witt, H. T. (1969). *Z. Naturf.* **24b**, 1139.

Dutton, H. J., Manning, W. M. and Duggar, B. M. (1943). *J. phys. Chem.* **47**, 308.

Duysens, L. N. M. (1952). Ph.D. Thesis, Univ. Utrecht.

Duysens, L. N. M. and Amesz, J. (1962). *Biochim. biophys. Acta* **64**, 243.
Duysens, L. N. M. and Sweers, H. E. (1963). *In* "Microalgae and Photosynthetic Bacteria", a special edition of *Plant Cell Physiol.* 353.
Duysens, L. N. M., Amesz, J. and Kamp, B. M. (1961). *Nature, Lond.* **190**, 510.
Emerson, R. and Arnold, W. (1932). *J. gen. Physiol.* **16**, 191.
Emerson, R. and Lewis, C. S. (1942). *J. gen. Physiol.* **25**, 579.
Emerson, R. and Lewis, C. S. (1943). *Am. J. Bot.* **30**, 165.
Emerson, R. and Rabinowitch, E. (1960). *Pl. Physiol., Lancaster* **35**, 477.
Emerson, R., Chalmers, R. F. and Cederstrand, C. (1957). *Proc. natn. Acad. Sci. U.S.A.* **43**, 133.
Förster, T. (1959). *Disc. Faraday Soc.* **27**, 7.
Franck, J. (1958). *Proc. natn. Acad. Sci. U.S.A.* **44**, 941.
French, C. S. and Elliott, R. F. (1958). *Yearb. Carneg. Instn* **57**, 278.
French, C. S. and Fork, D. C. (1961). *Int. Congr. Biochem.*, Moscow.
French, C. S., Myers, J. and McLeod, G. C. (1960). *In* "Symposium on Comparative Biochemistry". vol. 7. Academic Press, New York and London.
Fuller, R. C. and Nugent, N. A. (1969). *Proc. natn. Acad. Sci., U.S.A.* **63**, 1311.
Goodheer, J. C. (1957). "Optical properties and *in vivo* orientation of photosynthetic pigments". Ph.D. Thesis, Univ. Utrecht.
Govindjee, R. and Rabinowitch, E. I. (1960). *Science, N.Y.* **132**, 355.
Govindjee, R., Ichimura, S., Cederstrand, C., and Rabinowitch, E. I. (1960). *Archs Biochem. Biophys.* **89**, 321.
Haxo, F. T. (1960). *In* "Comparative Biochemistry of Photoreactive Systems" (M. B. Allen, ed.), chap. 21. Academic Press, New York and London.
Haxo, F. T. and Blinks, L. R. (1953). *J. gen. Physiol.* **73**, 389.
Hill, R. and Bendall, F. (1960). *Nature, Lond.* **186**, 136.
Hiller, R. G., Anderson, J. M. and Boardman, N. K. (1971). *In* "Proceedings of the Second International Congress on Photosynthesis Research" (G. Forti, M. Avron and A. Melandri, eds), vol. 1, p. 547. Dr W. Junk, N.V., The Hague.
Hind, G. (1968). *Photochem. Photobiol.* **7**, 369.
Hind, G. and Jagendorf, A. T. (1963). *Proc. natn. Acad. Sci., U.S.A.* **49**, 715.
Hind, G. and Jagendorf, A. T. (1965). *J. biol. Chem.* **240**, 3195.
Jackson, J. B. and Crofts, A. R. (1971). *Eur. J. Biochem.* **18**, 120.
Jacobi, G. and Lehmann, H. (1968). *Z. PflPhysiol.* **59**, 457.
Joliot, P. (1968). *Photochem. Photobiol.* **8**, 451.
Junge, W. and Witt, H. T. (1968). *Z. Naturf.* **23b**, 244.
Katoh, S. (1960). *Nature, Lond.* **186**, 533.
Knaff, D. B. and Arnon, D. I. (1969). *Proc. natn. Acad. Sci. U.S.A.* **63**, 956.
Kok, B. (1957). *Acta bot. neerl.* **6**, 316.
Kok, B. (1961). *Biochim. biophys. Acta* **48**, 527.
Kok, B. and Beinert, H. (1962). *Biochem. biophys. Res. Commun.* **9**, 349.
Krasnovsky, A. A. (1959). *In* "Progress in Photosynthetic Research", vol. 2, p. 709. IUBS.
Lester, R. J. and Crane, F. L. (1965). *J. biol. Chem.* **234**, 2169.
Levine, R. P. (1969). *A. Rev. Pl. Physiol.* **20**, 523.
Levine, R. P. and Smillie, R. M. (1962). *Proc. natn. Acad. Sci. U.S.A.* **48**, 417.
Livingstone, R. (1954). *Nature, Lond.* **173**, 485.
Mayne, B. C. and Clayton, R. (1966). *Proc. natn. Acad. Sci. U.S.A.* **55**, 494.
Michel, J. M. and Michel-Wolwertz, M. R. (1970). *Photosynthetica* **4**, 146.
Mitchell, P. (1966). *Biol. Rev.* **41**, 445.

Myers, J. (1971). *A. Rev. Pl. Physiol.* 43, 723
Myers, J. and French, C. S. (1960). *J. gen. Physiol.* 43, 723.
Ó h Eocha, C. (1960). *In* "Comparative Biochemistry of Photoreactive Systems" (M. B. Allen, ed.), chap. 12. Academic Press, New York and London.
Olson, J. M. and Smillie, R. M. (1963). *In* "Photosynthetic mechanism of green plants", p. 56. Natn. acad. Sci.–Natn. Res. Council Publn no. 1145.
Packer, L. and Siegenthaler, P. A. (1965). *Pl. Physiol., Lancaster* 40, 1080.
Parsons, W. W. (1967). *Biochim. biophys. Acta* 131, 154.
San-Pietro, A. and Lang, H. M. (1958). *J. biol. Chem.* 231, 211.
Seely, G. R. (1971). *In* "Proceedings of the Second International Congress on Photosynthesis Research" (G. Forti, M. Avron and A. Melandri, eds), vol. 1, p. 341. Dr W. Junk, N. V., The Hague.
Strehler, B. L. and Arnold, W. (1951). *J. gen. Physiol.* 34, 809.
Tanada, T. (1951). *Am. J. Bot.* 38, 276.
Vredenberg, W. J. and Duysens, L. N. M. (1963). *Nature, Lond.* 197, 355.
Witt, H. T. (1955). *Naturwissenschaften* 42, 72.
Witt. H. T. (1971). *Q. Rev. Biophys.* 4, 4.
Yocum, C. F. and San-Pietro, A. (1970). *Archs Biochem. Biophys.* 140, 152.

Chapter 14
Functions of Carotenoids Other Than in Photosynthesis

J. H. BURNETT

Department of Agricultural Science, Oxford University, England

I Introduction

There has been a tendency to seek a universal funtion for carotenoids in all plants, whether chlorophyllous or not. If one exists it has not yet been ascertained with certainty. The most plausible proposal for a universal function is that carotenoids protect cells from photo-oxidative damage caused by the incidental absorption of visible light. This is a relatively recent proposal first put forward by Stanier and his co-workers in relation to bacteria and certain algae (Stanier and Cohen-Bazire, 1957). A somewhat older hypothesis is that the carotenoids act as photoreceptors involved in the phototropism of green plants and fungi, or in the phototaxis of motile algae and some bacteria. This hypothesis, although not disproved, seems increasingly less probable. A third hypothesis implicated carotenoids in reproduction, especially in fungi. It is now clear that this is indeed the case although the connection is secondary and somewhat tenuous. This is certainly not a universal function for carotenoids.

Whatever the cellular, metabolic function(s) of carotenoids may be, it is clear that the presence of oxygenated carotenoids in flowers, fruit and, to some extent, even in leaves, stems and roots of higher plants subserves important, if incidental, roles in attracting or repelling animals during pollination and dispersal (whether of fruits or vegetative fragments). Once carotenoids have been formed in any organ of a plant they may affect visual stimulation in animals and, thereafter, selection may operate to perpetuate and increase the numbers of individuals forming such carotenoids, or vice versa. In terrestrial higher plants, at least, this may explain the apparently bewildering range of species-specific xanthophylls which are so frequently found in petals and fruits. An implication of this view is that such carotenoids are initially functionless, metabolic by-products whose perpetuation has been maintained through visual selection by animals. Such a view is a slightly more sophisticated version of that expressed forty years ago by Frey-Wyssling (1935).

It is difficult to suppose that this kind of explanation accounts for the widespread occurrence of the plastid carotenoids of terrestrial green plants, or the carotenoids of algae, fungi and bacteria, whether or not they are located in these organisms in specific organelles, e.g. plastids, eye spots or submicroscopic chromatophores. The functional role of carotenoids in plastids and chromatophores in photosynthesis has been dealt with in Chapter 13. Other possible functions will now be discussed.

II Photoprotection

This topic has been recently reviewed in some detail by Krinsky (1968). Here, therefore, the fundamental observations will be described, subsequent work outlined and the present status of the protective function of carotenoids assessed.

A PHOTOPROTECTION IN BACTERIA

In 1951 Swart-Füchtbauer and Rippel-Baldes showed that sunlight in the spectral region 366–405 nm had a marked bactericidal effect. This suggested some form of pigment-sensitized photo-oxidation, operative in the visible spectrum. The first evidence for such an effect was described for a mutant of the photosynthetic purple bacterium *Rhodopseudomonas spheroides* in 1955 and the situation has now been investigated in considerable detail in the mutant (Griffiths *et al.*, 1955; Griffiths and Stanier, 1956; Sistrom *et al.*,

1956; Dworkin, 1958) and in normal cells in which carotenoid formation has been inhibited (Cohen-Bazire and Stanier, 1958; Fuller and Anderson, 1958; Dworkin, 1959).

In the mutant bacterial cell the normal coloured carotenes are replaced by the more highly saturated, colourless polyene phytoene; bacteriochlorophyll is also formed but in lesser amounts than in the normal cell. The cells can photosynthesize normally under the usual strictly anaerobic conditions, but if exposed to light under aerobic conditions the cells are rapidly killed and their bacteriochlorophyll destroyed. Normal cells are not killed by exposure to these conditions nor are mutant cells killed by exposure to aerobic conditions alone; indeed in both cases growth continues for at least 10–12 h and bacteriochlorophyll is synthesized by the former and not lost in the latter. Thus the death of the mutant is due to exposure both to light and oxygen. The bacteriochlorophyll absorbs light *in vivo* between 820 and 920 nm. The mutant was exposed to light passed through a filter completely opaque to all wavelengths shorter than 800 nm and air introduced, once again cells were killed and the bacteriochlorophyll destroyed. There seems little doubt, therefore, that in this mutant the bacteriochlorophyll is the photosensitizing agent and, in the non-mutant cell, that protection is afforded by the normal carotenoids.

The protective action of the carotenoids in normal cells was demonstrated in *R. spheroides, Rhodospirillum rubrum* and *Chromatium* by interfering with carotenoid synthesis. Diphenylamine was added to cultures and as the carotenoids were depleted the characteristic lethal symptoms developed in light and aerobic conditions; removal of the diphenylamine restored synthesis of carotenoids and reduced the photosensitivity of the cells. It was also noted that the effect of diphenylamine was to block carotenoid synthesis in *R. rubrum* at neurosporene which has nine conjugated double bonds. Subsequent work with several other mutants of *R. rubrum* by Crounse *et al.* (1963) revealed that neurosporene is, indeed, the least unsaturated carotenoid capable of providing some protection against photo-oxidation.

Comparable behaviour was demonstrated with normal and diphenylamine-treated clones of *Chromatium* but here it was achieved using only isolated chromatophores. Chromatophores from the mutant had their photophosphorylative ability totally destroyed by an exposure of 10 min to light and air while those of normal cells retained 63% of their initial activity under the same conditions, and even after 30 min exposure still showed 42% activity. It is clear, therefore, that the phenomenon within the cell is intimately connected with the chromatophores in which both carotenoids and bacteriochlorophylls are located. However, the phenomenon does not depend upon the kind of energy transfer from carotenoid to bacteriochlorophyll involved in photosynthetic phosphorylation since *Chromatium* chromatophores isolated in 0·4 M glucose failed to achieve this, although photoprotection was maintained.

Investigations were extended to non-photosynthetic bacteria such as *Corynebacterium poinsettiae* (Kunisawa and Stanier, 1958), *Sarcina lutea* (Mathews and Sistrom, 1959; Mathews, 1964) and *Mycobacterium* spp. (Mathews, 1963; Wright and Rilling, 1963). In all these cases it has been shown that the absence of carotenoids is associated with a lack of protection against lethal photodynamic action in the presence of a photosensitizer. In *Corynebacterium* lacking carotenoids, whether diphenylamine-treated normal cells or pigmentless mutants, cells were killed only in air and on exposure to light in the presence of the exogenous, photosensitizing dye, toluidene blue. Similar photosensitized killing was known from *Serratia marescens* (Kaplan, 1956) and occurs in *Sarcina*. However, in this latter organism 99% of caroteneless mutant cells are killed in 2 h on exposing them to sunlight in the presence of oxygen without the addition of an exogenous photosensitizing pigment. Normal cells were unaffected by light whether in atmospheres of oxygen or nitrogen, and mutant cells were unaffected by light in an atmosphere of nitrogen. It has also been shown that the carotenoids in normal *Sarcina* cells are probably in the cell envelope, i.e. interposed between the light and the internal, unknown photosensitizing pigment (Mathews and Sistrom, 1960). This is also true of *Mycobacterium marinum* but in this organism light and oxygen together stimulate carotenoid production: it is photochromogenic. Despite studies in this and other organisms, the nature of the cellular photosensitizing pigment(s) in non-photosynthetic bacteria is still unknown, although Wright and Rilling (1963) suggested that the photosensitizer absorbed in much the same spectral region as the carotenoids in their caroteneless *Mycobacterium* sp. If this is so then the system differs from that in photosynthetic bacteria or in toluidine blue-mediated photodynamic killing, where the light is absorbed at a different wavelength from that absorbed by the carotenoids. It could be that in such a system the light is partitioned between different pigments, the carotenoids functioning as screening pigments.

The best evidence for an endogenous photosensitizing pigment has come from more recent work by Burchard and Dworkin (1966) with *Mycococcus xanthus*. This bacterium only forms carotenoids after it has entered the stationary phase in aerobic conditions in the light. Dark-grown cells exposed to light, equivalent to daylight, lysed. On the other hand, a mutant formed carotenoids both in the dark and in the light, in the stationary phase; it only became photosensitive when treated with diphenylamine, i.e. when carotenoid synthesis was inhibited. A porphyrin similar to protoporphyrin IX was isolated from normal dark-grown cells and shown to render them photosensitive. Moreover, the action spectrum for photolysis of *M. xanthus* resembled that of this porphyrin (Burchard *et al.*, 1966). In contrast the protected mutant was found to contain only about one-sixteenth of the amount of this porphyrin present in normal, sensitive cells.

Two other observations seem to be common to many of the bacteria studied. Firstly, carotenoids either afford no protection against light at low temperature, e.g. *R. rubrum* and *R. spheroides* at 1°C (Dworkin, 1959), or only protect with reduced efficiency, e.g. *S. lutea* at 4°C with, or without an exogenous photosensitizer (Mathews, 1964). Secondly, the Q_{10} of the lethal effect was about 1, suggesting either that the effect was purely photochemical, or that the photosensitive and photo-protective molecules are so close that energy transfer between them was exceedingly efficient (Dworkin, 1958, 1959; Mathews and Sistrom, 1960).

B PHOTOPROTECTION IN ALGAE AND FUNGI

Claes (1954, 1956, 1957, 1958) isolated mutants of *Chlorella vulgaris* which lacked the primary carotenoids but possessed phytoene, phytofluene etc., a situation analogous to that in the purple bacteria. Mutants 5/871, 5/515 and 9a are all killed by light under aerobic conditions, but unfortunately these mutants are said to show greatly reduced chlorophyll synthesis and, in the case of 9a, red light is ineffective in destroying the pigments and killing the cells (Kandler and Schotz, 1956). It is known from studies on the bleaching of chlorophyll in the related *C. pyrenoidosa* that the speed and effectiveness of the breakdown is in part conditional on the chlorophyll content, notably the chlorophyll *b* fraction which has a stabilizing function (Allen, 1958). Thus the situation is not so clear-cut in the *Chlorella* mutants as in the purple bacteria and it seems possible that a different interpretation is required, even although

Claes and Nakayama (1959) have shown that the aerobic destruction of chlorophyll *in vitro* can be alleviated only by carotenoids with a conjugated chain length at least equivalent to that in neurosporene, i.e. longer than in the carotenoids of the light-sensitive mutants. Similar difficulties arise in the case of caroteneless mutants of *Chlamydomonas* such as that studied by Sager and Zalokar (1958). This *pale-green* mutant had less than 0·5% of the total carotenoid content of the normal plant and it lacked phytoene and phytofluene. The chlorophyll content was only one-fifteenth that of normal cells although, because of the reduction in carotenoids, the chlorophyll/carotenoid ratio was higher than in normal cells. The mutant can photosynthesize but dies in the light whether grown photosynthetically or heterotrophically. It is tempting to compare this situation with that in the purple bacteria, but the greater biochemical differences in this algal mutant and the structural modifications described in the lamellar membranes of the chloroplast render comparisons almost meaningless. To date, therefore, data from algae have been disappointing and difficult to interpret.

Many caroteneless mutants of fungi are available, they have frequently been used to study the effects of inhibitors of carotenoid synthesis and many of them are photochromogenic. Yet few experiments bearing upon photoprotection have been reported using fungi. A striking exception is provided by the investigations of Goldstrohm and Lilly (1965) on the photochromogenic *Dacryopinax spathularia*. They exposed aerobic cultures of both dark-grown and light-grown cells to sunlight and found that 89% of all the cells from the former culture were killed after 2 hours, whereas the latter were unaffected after 4 hours. The phenomenon was truly photodynamic for in nitrogen there was virtually no lethal effect in the light. They also made a petroleum extract of β-carotene (which has been tentatively identified in light-grown *D. spathularia* cultures) from *Phycomyces blakesleeanus* and interspersed this as a filter between the light source and the *Dacryopinax* cultures. After one hour dark-grown cultures had only shown 24% lethality but similar cultures where the filter was replaced by petroleum alone showed 74% death. They concluded that the photosensitive sites were associated with a yellow pigment. Alternatively, endogenous carotene could have been acting as a filter protecting the photosensitive areas in the fungus.

Studies with *Rhodotorula* and *Saccharomyces* spp. have provided some evidence for photoprotection by carotenoids provided that the cells have been artificially photosensitized with toluidine blue. The results and their interpretations are ambiguous, however, and, as the phenomenon is far from being understood, they will not be considered further (MacMillan *et al.*, 1966; Maxwell *et al.*, 1966).

C PHOTOPROTECTION IN HIGHER GREEN PLANTS

Although a number of mutants are known in which pigment synthesis, including that of carotenoids, has been disturbed, they have not proved easy to use in studies of photoprotection. As long ago as 1902, Kohl suggested that carotenoids acted as protective agents against the photodestruction of chlorophyll, and Willstätter and Stoll (1918) thought that yellow carotenes prevented chlorophyll from bleaching. The best evidence for such effects has come

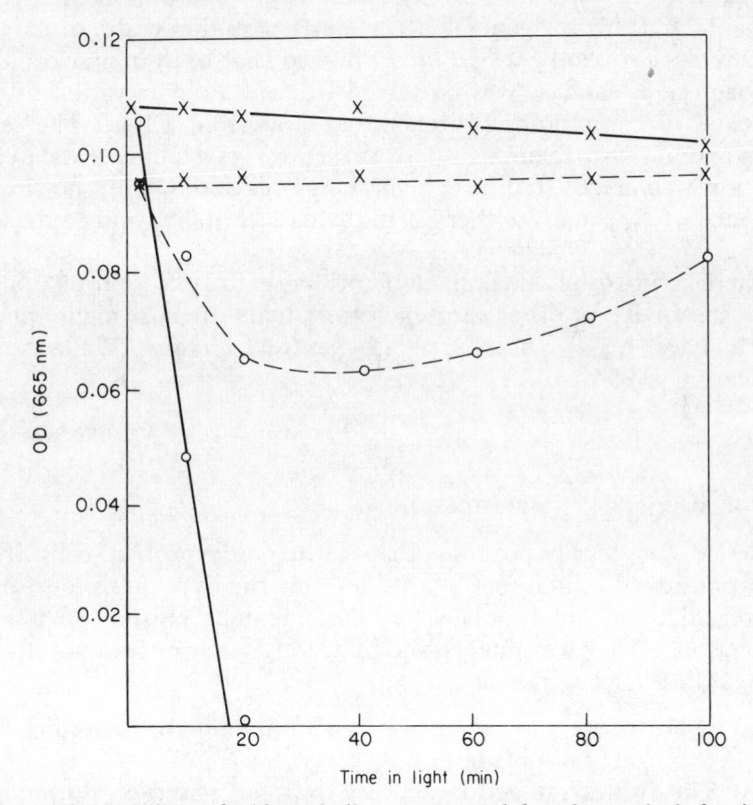

FIG. 1. The stability of chlorophyll, as measured by the optical density at 665 nm, of: (a) Normal seedlings in air (○---○) and in nitrogen (X---X); (b) Mutant (*white-3*) seedlings in air (○—○) and in nitrogen (X—X), as affected by daylight. (From Anderson and Robertson, 1960.)

from studies with the *white-3* mutant of maize (*Zea mays*) (Anderson and Robertson, 1960, 1961). In faint light (*c*. 5·4 lux) both normal and mutant plants developed chlorophyll. The latter only formed phytoene, phytofluene and ζ-carotene. In bright light in air, chlorophyll was rapidly destroyed in the mutant which only developed phytoene. Chlorophyll was only partially destroyed in the normal plants which showed full carotene development. In nitrogen no chlorophyll destruction occurred either in light or in darkness, in normal or in mutant plants (Fig. 1).

Anderson and Robertson also made the interesting suggestion that carotenoids might also protect porphyrin compounds other than chlorophyll. Catalase was destroyed in the mutant maize and persisted in normal plants when exposed to light and oxygen. This is comparable to the results obtained with chlorophyll. It seems probable that this is a general effect. Further studies with normal and mutant *Z. mays* and *Sarcina lutea* showed that both in maize and in the bacterium catalase was protected from photoinactivation by the presence of carotenoids (Mitchell and Anderson, 1965). The same investigators also demonstrated that pure crystalline catalase in solution was inactivated in light in the presence of O_2 but not in the presence of N_2, nor in either gas in the dark (Mitchell and Anderson, 1965).

There is no other unambiguous evidence for photoprotection or photosensitivity in other caroteneless mutants of those higher plants which have been studied, e.g. *Helianthus annuus* (Wallace and Schwarting, 1954).

D THE MECHANISM OF PROTECTION

Evidence for the hypothesis that carotenoids protect cells from photosensitized oxidations is scattered but the hypothesis must now be regarded as highly plausible. Possible mechanisms of photoprotection are far less understood. The only features common to all, or most of the examples, are:

(a) A protective carotenoid requires a minimum of nine to eleven conjugated double bonds.
(b) The protective action appears to be exercised on porphyrins which may, or may not, be photosynthetic pigments.
(c) In micro-organisms carotenoids afford no protection, or less efficient protection at low temperatures. (The effect of

temperature on photoprotection in higher green plants is not known.)

(d) The photoprotective and photosensitive compounds are probably in close proximity as shown either by their location or from Q_{10} values (this too has only been investigated in micro-organisms).

In addition to these observations with organisms, attention should be directed to the protective action of carotenoids in experiments *in vitro*.

As early as 1925 Noack observed that carotenoids protected the photobleaching of chlorophyll in solution; the hypothesis propounded by Willstätter and Stoll (1918) for the situation *in vivo*. Aronoff and Mackinney (1943) repeated Noack's observation but noted that protection was associated with the destruction of the carotenoids. There is a suggestion from later experiments with β-carotene/chlorophyll mixtures that the former was oxidized but it is not known whether this is related to its protective action (Mackinney and Chichester, 1960). The most detailed studies have been made by Claes and her co-workers (Claes and Nakayama, 1959; Claes, 1960, 1961). Employing carotenoids and chlorophyll *a* in petrol ether solutions they have shown that:

(a) Only carotenoids with more than seven conjugated double bonds are effective as photoprotectants.

(b) This is the only important structural feature of the carotenoids in respect of their role as protectants.

(c) Protection is a function of the carotene/chlorophyll ratio, increasing to a saturation point as the ratio increases.

(d) Under anaerobic conditions carotenoids also protect against the photoreduction of chlorophyll by light.

Similar observations have been made by other investigators. For example, Krasnovsky *et al.* (1960) demonstrated a protective effect against photo-oxidation of chlorophyll in acetone solution by β-carotene as well as its inhibition of photoreduction. They confirmed the minimal, double-bond requirement.

Thus the results of experiments *in vitro* support those made *in vivo* in respect of the structural requirements for carotenoids and the fact that a porphyrin molecule is protected. Moreover, the inhibition of the photoreduction of chlorophyll under anaerobic conditions by carotenoids *in vitro* is not in conflict with observations made *in vivo*. However, nothing is known of the temperature dependence or Q_{10} of the photoprotection reactions *in vitro*.

Despite these similarities, an understanding of the mechanism of photoprotection is not thereby much improved. Attention has been concentrated on the ability of carotenes to quench lethal triplet-state chlorophyll since the original observation of this effect by Fujimori and Livingston (1957). An alternative hypothesis has been that there is a direct energy transfer from singlet-state oxygen to β-carotene yielding triplet-state carotene and ground-state triplet oxygen (Foote and Denny, 1968). In 1970, Foote *et al.* combined these views with earlier ideas in the following scheme:

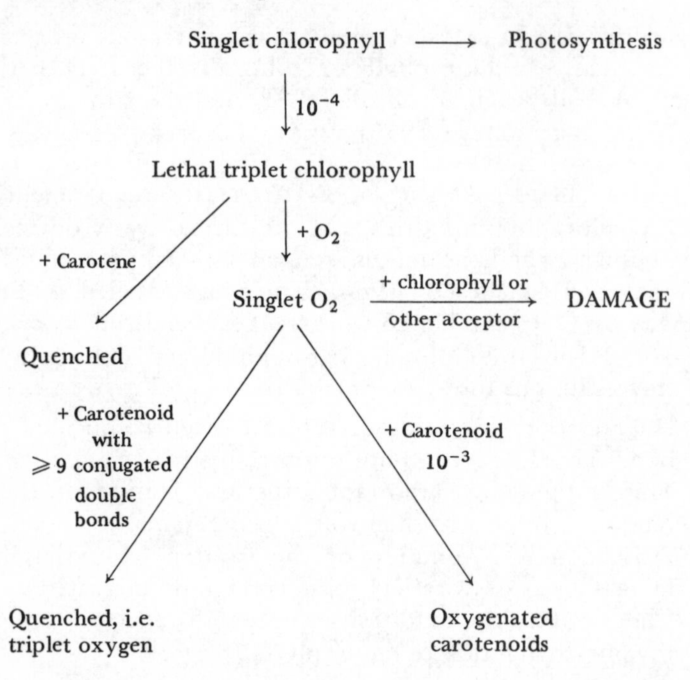

This scheme accounts for the structural requirement in carotenoids which afford photoprotection. It accounts for the favourable *in vitro* effect of a high carotene/chlorophyll ratio since the quenching of triplet chlorophyll by carotenes or oxygen compete at the same rate unless the local concentration of carotenoids exceeds that of oxygen. It can also be seen that damage results if singlet oxygen reacts with chlorophyll or a similar acceptor: hence the implication of porphyrins in the system. Finally, the scheme involves not only a photosensitizing molecule but also the participation of oxygen, as found *in vivo*. However, this explanation suffers from three drawbacks. Firstly, some of the *in vivo* data are not accounted for, notably the Q_{10} and temperature-dependence effects. Secondly, it relates almost exclusively to β-carotene and chlorophyll, so leaving the situation in non-photosynthetic organisms obscure. Lastly, the significance *in vivo* of triplet state chlorophyll is in any case questionable.

There can then be little doubt that carotenoids can protect a range of micro-organisms and plants from photosensitized oxidations but the mechanism and overall significance of such protection is not yet clear.

III Phototropism and phototaxis

A GENERAL OBSERVATIONS

The experimental evidence that carotenoids play some role in photoresponses, such as in phototropism and phototaxis, is inconclusive at present. The evidence is derived almost entirely from the comparison of the action spectrum of the photoresponse with the absorption spectrum of, usually, the major carotenoid component extracted from the organism, or organ showing the response. If these spectra are, in the opinion of the investigator, reasonably congruent and provided that no other compounds exist with absorption spectra which are at least as congruent, then it has been concluded that the carotenoid is the effective photoreceptive pigment involved in the response. It will readily be appreciated that this procedure is liable to a variety of experimental errors and that quite small errors may have a profound effect on interpretations made when the action and absorption spectra are compared.

Two kinds of experimental difficulties lie behind all attempts to investigate the role of carotenoids in photoresponses. The first is to obtain a sufficiently accurate and comprehensive action spectrum of the response. French (1959) has clearly described the difficulties inherent in the accurate measurement of action spectra in plants. It has been stated that: "An action spectrum measured to a precision of 12% at 10 mμ wavelength intervals with a half-band width of 5 to 10 mμ is considered good work. The precision now ordinarily obtained in absorption spectroscopy is far greater than in action spectroscopy". (Smith and French, 1963.) Very few measurements of the action spectra for phototropic responses have achieved this accuracy and none for phototaxis. The second problem is to measure, or assess, the effective absorption spectrum of the carotenoid(s) in the intact organism. It is so difficult to deal with in experiments that most measurements have been made, in fact, on extracted carotenoids. Measurements made on more-or-less intact cells are difficult to interpret. Such absorption spectra suffer interference by other pigments or absorbing systems in the cells, by scattering losses due to the nature of the material examined and possibly by distortion due to the many fluorescent substances which can occur at least in higher plants (R. H. Goodwin, 1953). A further source of error is that it may well be that in the living plant the carotenoids occur as carotenoid–protein complexes. This is suggested by the fact that, in higher plants, carotenoids are readily extracted after treatment with a polar solvent (which denatures proteins) but not after treatment with non-polar solvents and by the fact that absorption maxima *in vivo* are somewhat higher than in solvent extracts. Moreover, although it appears to be a special case at present, a unique β-carotene protein has been identified with reasonable certainty in spinach leaves (Nishimura and Takamatsu, 1957). Shibata (1958) has also drawn attention to the fact that alterations can be brought about in absorption spectra if the carotenoids are present in the cells in a crystalline form, e.g. a peak at 515 nm in the spectrum of β-carotene in carrot tissue which is not present when extracted in a solvent such as benzene.

It will now be convenient to consider these problems in more detail for phototropism in higher plants and fungi and for phototaxis in algae and flagellates.

B PHOTOTROPISM

There have been several relatively recent reviews of phototropism in higher plants (Briggs, 1963, 1964; Ball, 1969) and in fungi (Page, 1968; Carlile, 1970). These detail, to a greater or lesser extent, the pros and cons for carotenoids as the photoreceptor molecules mediating phototropic responses. On balance, opinion is generally against carotenoids and in favour of riboflavin. The arguments are summarized below:

(a) There are several well-defined phototropic systems (Briggs, 1964) as well as the "light-growth" reactions first described by Blaauw (1914, 1915, 1918). Carotenoids do not seem to be implicated in all of these.

(b) Even in system I—the positive curvature shown by graminaceous coleoptiles to a light source, the perceptive region being the coleoptile tip—there was a discrepancy between a peak at 370 nm in the action spectrum which was replaced by a hollow in the absorption spectrum of the extracted carotenoids (Shropshire and Withrow, 1958, Thimann and Curry, 1960). It has been claimed that riboflavin, especially if associated with lipid material, gives a better fit between action and absorption spectra (Galston, 1964).

(c) Similar difficulties arise in comparisons between the action spectra for both phototropism and the light-grown response in *Phycomyces blakesleeanus* and the absorption spectra of extracted carotenoids (Curry and Gruen, 1959; Delbrück and Shropshire, 1960). Moreover, the absorption spectrum *in vivo* of the conspicuous carotenoid in *Pilobolus kleinii* includes a strongly absorbing peak at 500 nm; a wavelength that induces little phototropic response (Page and Curry, 1966).

(d) Riboflavin or flavins are more likely to be the appropriate photoreceptor molecules because of their involvement in electron-transfer paths, their *in vitro* photosensitivity and their possible involvement with a growth hormone, indole acetic acid (Galston, 1959; Carlile, 1970) although this last claim has been seriously challenged (Briggs *et al.,* 1957; Reisener, 1958; Shropshire and Withrow, 1958; Briggs, 1963).

It seems improbable that claims and counter-claims based on such comparisons are ever likely to resolve the issue. However, it has recently become clear that evidence of an entirely different kind renders such comparisons virtually worthless. In 1963 Shropshire predicted that the photoreceptor molecule for phototropism need only be present in extremely low concentration. This view has been borne out by the work of Meissner and Delbrück (1968), who employed mutants of *P. blakesleeanus* differing in their carotenoid content. By making a number of reasonable assumptions, they calculated that the necessary amount of photoreceptor pigment was so small that there was probably at least ten times more β-carotene in the sensitive zone than was required to give the response. Indeed, the necessary absorption would be less than that detectable by about a factor of thirty. It is clear, therefore, that at the present time, and indeed in the foreseeable future, the sensitivity of spectral methods will be quite unsuitable for the detection of any photoreceptor which mediates phototropism.

C PHOTOTAXIS

The nature of the photoreceptive site in mobile algal, fungal or flagellate cells which show directional responses to light in their movement has been debated. Wager (1900) described a paracrystalline body, the paraflagellar body at the base of the major flagellum of *Euglena gracilis*, and identified this as the photosensitive site. Many motile cells possess a pigmented, red or orange, organelle, the stigma or so-called eye spot, which has been thought to be the photosensitive site. This now seems improbable. As long ago as 1878 Strasburger described phototaxis in motile algae and fungal zoospores, including several which lacked a stigma. Many dinoflagellates lack a stigma yet show normal positive responses. Gössel (1957) obtained a *Euglena* which lacked an eyespot, by treatment with pyribenzamine but it showed normal response. It has been claimed that this mutant does possess a paraflagellar body, however, although details have not been published (Kivic and Vesk, 1972). Cells which lack a stigma are often less sensitive in their response to light but, if they lack both a stigma and a paraflagellar body, they become indifferent to light (Tchakhotine, 1936). Mast (1927) suggested that the pigments in a stigma screen the primary photoreceptor and that organisms respond in such a way as to minimize such screening. In this case the action spectrum for phototaxis would be compounded from the absorptive effects of stigma and photoreceptor. Halldal (1958) demonstrated that, if this were so, there could be serious distortion of the action spectrum in colourless motile cells but not in coloured cells, such as those of algae.

A wide range of cells has been studied with some precision: green algae such as *Chlamydomonas, Platymonas, Eudorina, Volvox*; gametes or zoospores, e.g. *Ulva*; pigmented and unpigmented euglenoids, especially *Euglena* spp; dinoflagellates like *Peridinium, Gonyaulax* and *Prorocentrum*; colourless flagellates, e.g. *Chilomonas*; and a fungal zoospore of an *Allomyces* sp. Some of the data obtained are presented in Table I.

It can be seen that *Euglena* spp. have received intensive study and it now seems probable that, in their case at least, the "screening hypothesis" of Mast is applicable. Kivic and Vesk (1972) have interpreted the action and absorption spectra obtained by Wolken and his co-workers (Wolken and Shin, 1958; Strother and Wolken, 1961; Wolken, 1967) on this hypothesis. Wolken and Shin (1958) provided an action spectrum for photokinesis, i.e. random movement under constant illumination, of *E. gracilis*. This is assumed by Kivic and Vesk to approximate to the absorption spectrum of the primary photoreceptor of *Euglena*. They then calculated a composite "difference" spectrum from the sum of the absorption curves of the stigma and chloroplast, less the action spectrum for photokinesis, all in appropriate comparable units. This they compared with the action

TABLE I
Phototaxis action spectra shown by algae and a fungus

Organisms	Major peaks (nm)	Minor peaks or shoulder(s) (nm)	Authors
ALGAE			
Chlorophyta			
Eudorina, *Volvox*	492		Lunz, 1931
Ulva gametes	485	435(s)	Halldal, 1958
Euglenophyta			
Euglena	495 475	450 425	Bünning and Schneiderhöhn, 1956
	490	420	Wolken and Shin, 1958
Pyrrophyta			
Dunaliella, *Stephanoptera*	493		Halldal, 1958
Platymonas	495 335 275	435	Halldal, 1961
Peridinium, *Gonyaulax*	475		Halldal, 1958
Prorocentrum	570		Halldal, 1958
Cryptophyta			
Chilomonas (colourless)	366		Lunz, 1931
FUNGI			
Phycomycetes			
Allomyces	525/515 480/470	560/570 430/420	Robertson, 1972

spectrum for phototaxis obtained by Wolken and demonstrated a remarkable correlation between them (Figs. 2 and 3).

The peaks in the action spectrum for photokinesis do not suggest that a carotenoid is involved. Thus the most plausible working hypothesis for *Euglena* is that the primary photoreceptor is not carotenoid, although the carotenoids of the stigma and chloroplast play some role as they act, with chlorophylls, as screening pigments.

It can be seen from Table 1 that most action spectra for phototaxis in other organisms, save for *Chilomonas* and *Allomyces,* are similar to that obtained for *Euglena*. In the other algae, excepting the dinoflagellates, an explanation similar to that for *Euglena* could account for their action spectra. It could well be in the dinoflagellates that only the chloroplasts act as screening structures. It is clear, however, that action spectra for phototaxis will have to be reinterpreted in the light of Kivic and Vesk's (1972) claim. At present, it

FIG. 2.

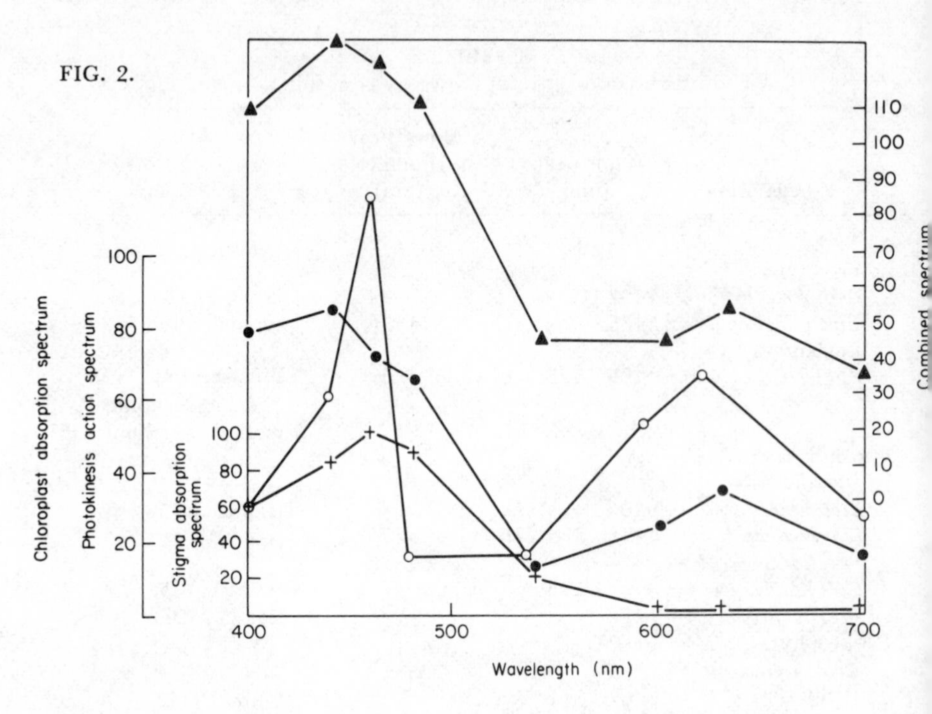
Chloroplast absorption spectrum
Photokinesis action spectrum
Stigma absorption spectrum
Combined spectrum
Wavelength (nm)

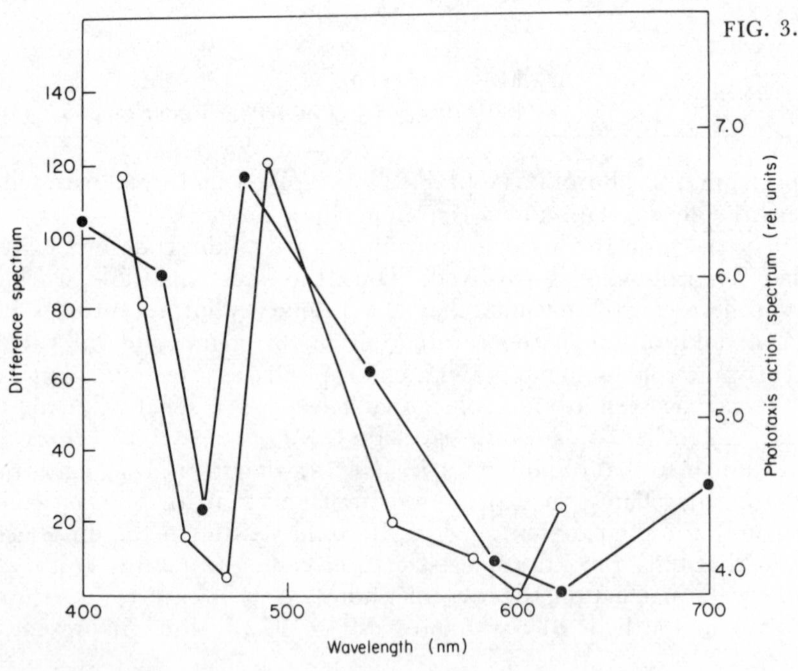
Difference spectrum
Phototaxis action spectrum (rel. units)
Wavelength (nm)

FIG. 3.

seems unlikely that carotenoids act as primary photoreceptor molecules for this type of response.

IV Reproduction

A GENERAL OBSERVATIONS

Goodwin (1950) has drawn attention to the widespread occurrence and accumulation of carotenoids in reproductive structures of plants and animals. It has been argued already, in the introduction to this chapter, that the carotenoids subserve the passive function of making flowers, fruits and even perhaps certain vegetative parts of higher plants attractive to animals, thereby promoting processes such as pollination and dispersal. There is no firm evidence in support of any metabolic role for carotenoids in the reproduction of higher plants, or indeed among terrestrial lower plants. The situation in the algae and fungi is at last becoming clear. The claims of Moewus (Kuhn and Moewus, 1940) for the functional role of crocetin esters in *Chlamydomonas* species are now effectively disproved on theoretical (Philip and Haldane, 1939; Thimann, 1940) and practical grounds (Ryan, 1955; Hartmann, 1955; Renner, 1958) and are best forgotten. It is clear that light is nearly always involved in the initiation of the mating reaction, but the action spectrum appears to implicate photosynthesis and it need not be considered further here.

In the fungi, the notion that carotenoids are involved in reproduction must be attributed to Chodat and Schopfer (1927), who drew attention to the differential accumulation of "carotene" in the + and – mating types of *Mucor hiemalis* and concluded that it was an important sexual difference (Blakeslee, 1904, had noted this difference but drew no conclusions from it). Since then mycologists have sporadically noted this and similar phenomena, and the idea has grown up that carotene and sexuality are in some way related. There are certainly fungi which possess no carotenoids and many

FIG. 2. The absorption spectrum of the stigma (+—+), of the chloroplast (●—●) and their sum, the combined spectrum (▲—▲), all as percentage absorption in relation to wavelength. The action spectrum for photokinesis (○—○) determined as the mean rate of swimming in mm/3 min is also shown. (After Kivic and Vesk, 1972.)

FIG. 3. A comparison of the action spectrum for phototaxis (○—○), shown as "relative" spectral sensitivity, with the difference spectrum, i.e. combined spectrum (Fig. 2) minus photokinesis action spectrum. (From Kivic and Vesk, 1972.)

which possess predominantly only one, usually β-carotene or lycopene (Goodwin, 1952). Carotenoids therefore cannot play any universal role in fungal reproduction. Moreover, Carlile (1956) has shown that in *Fusarium macrosporum* it is possible to obtain macrospores in the absence of carotenoids and vice versa, although both processes are the result of a photostimulus; similarly in *Pyronema confluens* (Carlile and Friend, 1956) sexual reproduction was shown to be a light-stimulated process which could still take place in a mutant lacking carotenoids. It is, however, amongst the Phycomycetes that the supposed connection has been most strongly pressed, and they alone will be considered.

B REPRODUCTION IN PHYCOMYCETES

Historically, carotenoids have been implicated in two kinds of process: sexual or mating-type dimorphism and the processes involved in sexual reproduction. Their role has never been clear and correlations between these activities and carotene synthesis or accumulation have never been wholly satisfactory. Now that it is known that carotenoids are only involved indirectly in such processes much of the confusion has been eliminated.

In 1956 Barnett *et al.* reported that mixed + and – cultures of *Choanephora cucurbitarum* (Mucorales) not only formed zygospores when grown together in liquid medium in shake cultures but also increased the production of β-carotene. Hesseltine and Anderson (1957) extended this observation to both interspecific and intraspecific crosses in the Choanephoraceae, e.g. *C. cucurbitarum* + and –; *Blakeslea trispora* + and –; *C. cucurbitarum* + and *B. trispora* –. Intensive studies were made to investigate the causes of the enhanced carotenoid synthesis for which *B. trispora* was found to be particularly valuable. These studies culminated in the discovery by Caglioti *et al.* (1964, 1966) of the trisporic acids in *B. trispora*. At that time they were thought to be, primarily, carotenogenic agents. However, Van den Ende (1967, 1968), studying the diffusable sex factors from culture filtrates of several mucorales, e.g. *B. trispora* + and –, identified these as trisporic acids B and C. This identity was elegantly and unambiguously confirmed by Austin *et al.* (1969), who obtained trisporic acids from *B. trispora* and compared them with the zygophore-inducing hormone isolated from mated mycelia of *Mucor mucedo* (Gooday, 1968a,b). This was shown to be indistinguishable from trisporic acid C. This identity has formed the basis of an assay for the trisporic acids. Zygophore formation by the – strain of

M. mucedo is sensitive to about 0·01 μg of trisporic acid C (Bu'lock *et al.*, 1972).

Subsequent studies have shown that the trisporic acids are themselves probably derived from β-carotene by cleavage to give retinol, the C_{18}-ketone followed by a series of oxidations. The evidence for this suggestion can be summarized briefly:

(a) In *Phycomyces blakesleeanus* mutants incapable of β-carotene synthesis are known, some at least blocked at one β-cyclase step (Heisenberg and Cerdà-Olmedo, 1968; Bergman *et al.*, 1969; Sutter in Bu'lock, 1973).
(b) Diphenylamine added to mated shake cultures of *B. trispora* prevents trisporic acid formation (Austin *et al.*, 1969).
(c) Significant amounts of radioactive carbon or tritium are contributed to trisporic acids when [^{14}C]-β-carotene, [^{14}C]-labelled retinol, β-C_{18}-ketone, 4-hydroxy-β-C_{18}-ketone or [11, 12-^{3}H]retinyl acetate are supplied to mated shake cultures of *B. trispora.*

Thus in fungi producing trisporic acid the association of β-carotene with the process of sexual reproduction lies in its cyclical relationship to these acids (Fig. 4).

To date, trisporic acids have been detected in *M. mucedo, P. blakesleeanus, B. trispora* and mixed *B. trispora* and *Zygorrhynchus moelleri* cultures (Gooday, 1968b; Van den Ende, 1968; Austin *et al.*, 1969; Sutter, 1970; Sutter *et al.*, 1972). The last of these, *Z. moelleri*, differs in two ways from the others. Firstly, as in many other species of mucoraceous fungi, it lacks β-carotene and, secondly, it is self-fertile. The course of the mating reaction in all mucoraceous fungi is very similar both in interspecific and intraspecific crosses, although in the latter the reactions are nearly always incomplete or somewhat abnormal. This has provided good circumstantial evidence that the course of the reaction is mediated by similar diffusable hormones in all mucoraceous fungi (Burgeff, 1924; Blakeslee and Cartledge, 1927; Satina and Blakeslee 1930; Burnett, 1953a,b; Plempel, 1957). The results with mixed cultures of *B. trispora* and *Z. moelleri* suggest strongly that the trisporic acids are, indeed, the common denominator in the mating reactions of both cross- and self-fertilizing mucorales. In those fungi which lack β-carotene, however, the origin of the trisporic acids is not yet known; presumably they are derived from some other compound in the carotene biosynthetic pathway. In this sense, therefore, in these fungi there is a relationship between carotenoids and sexuality.

One other, more direct association has been demonstrated (Fig. 4). When zygospores, the end product of the sexual reaction, of *M. mucedo* are subjected to exhaustive alkaline and acid extraction the residue is chemically very similar to the sporopollenin of the exine of pollen in higher plants (Gooday *et al.*, 1973). Sporopollenin is

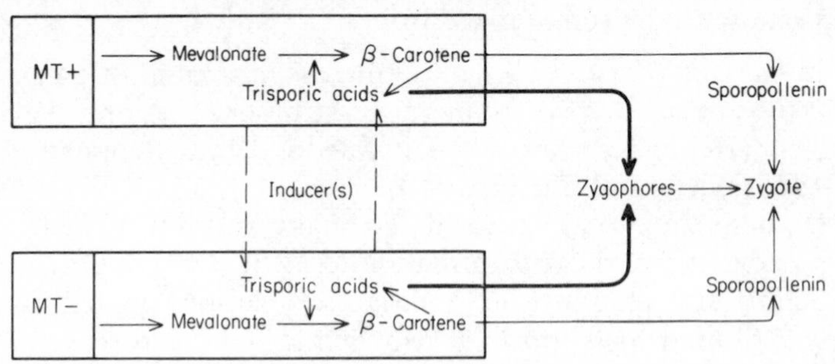

FIG. 4. A diagram to illustrate how β-carotene is involved in sexual reproduction in a fungus such as *Mucor mucedo*. It not only provides the precursor for the trisporic acids, which act as zygophore inducers, but also gives rise to sporopollenin in the zygospore wall.

derived by the oxidative polymerization of carotenoids and carotenoid esters and, indeed, Shaw (1970, 1971) has prepared a sporopollenin from β-carotene by oxidative polymerization in the presence of BF_3 as catalyst. A comparison of the properties of *M. mucedo* sporopollenin and such a synthetic compound is shown in Table II.

These analyses were supported by the finding that radioactivity derived from D,L-[2-^{14}C]mevalonic acid, [G-^{14}C]-β-carotene, or [15, 15′-^{3}H]-β-carotene was incorporated in *M. mucedo* sporo-

TABLE II
Comparisons between *Mucor* and synthetic sporopollenins

A. Percentage composition of major dicarboxylic acids produced by ozonolysis	No. of carbon atoms			
	3	4	5	6
Mucor sporopollenin	3·8	4·1	5·4	70·1
Synthetic sporopollenin	0·8	6·0	7·9	79·2
B. Empirical formula (in terms of C_{90} units)				
Mucor sporopollenin	C_{90}	H_{130}	O_{33}	
Synthetic sporopollenin*	C_{90}	$H_{130-132}$	O_{27-32}	

* The analysis varies with the conditions: more dilute solutions of β-carotene give values nearer to those from *Mucor*.

pollenin. There is good circumstantial evidence that the polymer is located in the outer part of the zygospore wall. Similar studies with comparable zygospores of *Rhizopus sexualis* failed to detect any sporopollenin and this may be correlated with Hocking's finding that *R. sexualis* lacks β-carotene (Hocking, 1963).

Thus it seems that the localized accumulation of β-carotene in the zygophores of *M. mucedo* has been exploited to provide a compound in the zygote wall which increases its resistance both to chemical and biological degradation. This suggests a possible explanation for the accumulation of carotene in the gametangia of other fungi. Bu'lock (1973), for example, has drawn attention to the interesting observation (Cantino, 1967) that γ-carotene synthesis is most marked in *Blastocladiella emersonii* when it develops to form a thick-coated resistant sporongium. However, it is not certain that accumulation of carotenes is always related to either sexual reproduction or sporopollenin production. For example, the characteristic accumulation of γ-carotene in the male gametangium and gametes of *Allomyces* spp. (Emerson and Fox, 1940) or the occurrence of carotene in male plants of *Blastocladiella variabilis* but not in females (Harder and Sörgel, 1938) still defy explanation. On the other hand, the identification of hormone A, the hormone which initiates the sexual reaction in *Achlya* spp. as a 24-ethyl steroid, antheridiol, suggests that in ways as yet unclear, the isoprene pathway is involved in morphogenetic processes leading to reproduction. In this sense there may well be associations between carotenoid synthesis and reproduction but they are likely to be indirect rather than causal associatons.

V Conclusions

Comparison with the first edition of this book will show that in the last decade considerable changes have taken place in our understanding of the place of carotenoids in the life of plants, other than in photosynthesis. In particular, it is now evident that the most probable role of general applicability is that of protection against damage by light in aerobic and, possibly, in anaerobic conditions. Earlier, and originally more popular, hypotheses concerning a role for carotenoids in photoreception related to tropistic and taxic responses, or in reproduction seem less probable.

Whether or not further and wider studies will demonstrate that photoprotection is indeed a universal function of carotenoids is still uncertain. That they do seem to have this function in a wide range of micro-organisms suggests that photoprotection is probably a very ancient role for carotenoids. The evolution of porphyrins and photosynthesis perhaps conferred danger as well as benefit upon autotrophic organisms!

It is never safe to speculate about the future of scientific investigations. However, it would be surprising if the mechanism of photoprotection by carotenoids was not to be elucidated more clearly in the next decade. Yet even if this were found to be a universal function of carotenoids, it seems not improbable that other, unexpected, roles will be found for these ancient, ubiquitous and relatively abundant compounds in plants.

REFERENCES

Allen, M. B. (1958). *Brookhaven Symp. Biol.* 11, 339-342.

Anderson, I. C. and Robertson, D. S. (1960). *Pl. Physiol. Lancaster* 35, 531-534.

Anderson, I. C. and Robertson, D. S. (1961). *Third Int. Congr. Photobiol.*, Copenhagen, pp. 477-479. Elsevier, Amsterdam.

Aronoff, S. and Mackinney, G. (1943). *J. Am. chem. Soc.* 65, 956-958.

Austin, D. J., Bu'lock, J. D. and Gooday, G. W. (1969). *Nature, Lond.* 223, 1178-1179.

Ball, N. G. (1969). *In* "Plant Physiology VA" (F. C. Steward, ed.), pp. 119-228. Academic Press, New York and London.

Barnett, H. L., Lilly, V. G. and Krause, R. F. (1956). *Science, N.Y.* 123, 141.

Bergman, K., Burke, P. V., Credà-Olmedo, E., David, C. N., Delbrück, M., Foster, K. W., Goodell, E. W., Heisenberg, M., Meissner, G., Zalokar, M., Dennison, D. S. and Shropshire, W. (1969). *Bact. Rev.* 33, 99-157.

Blaauw, A. H. (1914). *Z. Bot.* 6, 641-703.

Blaauw, A. H. (1915). *Z. Bot.* 7, 465-532.

Blaauw, A. H. (1918). *Meded. LandbHoogesch. Wageningen* 15, 89-203.

Blakeslee, A. F. (1904). *Proc. Am. Acad. Arts Sci.* 40, 205-319.

Blakeslee, A. F. and Cartledge, J. L. (1972). *Bot. Gaz.* 84, 51-58.

Briggs, W. R. (1963). *Rev. Pl. Physiol.* 14, 311-352.

Briggs, W. R. (1964). *In* "Photophysiology" (A. C. Giese, ed.), pp. 223-271. Academic Press, New York and London.

Briggs, W. R., Tocher, R. D. and Wilson, J. P. (1957). *Science, N.Y.* 126, 210-212.

Bu'lock, J. D. (1973). *Pure appl. Chem.* 34, 435-461.

Bu'lock, J. D., Drake, D. and Winstanley, D. J. (1972). *Phytochemistry* 11, 2011-2018.

Bünning, E. and Schneiderhöhn, G. (1956). *Arch. Mikrobiol.* 24, 80-90.

Burchard, R. P. and Dworkin, M. (1966). *J. Bact.* 91, 535-545.

Burchard, R. P., Gordon, S. A. and Dworkin, M. (1966). *J. Bact.* 91, 896-897.

Burgeff, H. (1924). *Bot. Abh., K. Goebel* 4, 1-135.

Burnett, J. H. (1953a). *New Phytol.* 52, 58-64.

Burnett, J. H. (1953b). *New Phytol.* 52, 86-88.

Caglioti, L., Cainelli, G., Camerino, B., Mondelli, R., Prieto, A., Quilico, A., Salvatori, T. and Selva, A. (1964). *Chimica Ind., Milano* 46, 961-966.

Caglioti, L., Cainelli, G., Camerino, B., Mondelli, R., Prieto, A., Quilico, A., Salvatori, T. and Selva, A. (1966). *Tetrahedron* 7 (suppl.), 175-187.
Cantino, E. C. (1967). *Symp. Soc. gen. Microbiol.* 11, 243-271.
Carlile, M. J. (1956). *J. Gen. Microbiol.* 14, 643-654.
Carlile, M. J. (1970). *In* "Photobiology of Micro-organisms" (P. Halldal, ed.), pp. 309-344. Wiley, London and New York.
Carlile, M. J. and Friend, J. S. (1956). *Nature, Lond.* 178, 369.
Chodat, R. and Schopfer, W. H. (1972). *C.r. Soc. Phys. hist. nat., Genève* **44**, 176-179.
Claes, H. (1954). *Z. Naturf.* **9b**, 461-469.
Claes, H. (1956). *Z. Naturf.* **11b**, 260-266.
Claes, H. (1957). *Z. Naturf.* **12b**, 401-407.
Claes, H. (1958). *Z. Naturf.* **13b**, 222-224.
Claes, H. (1960). *Biochem. biophys. Res. Commun.* 3, 585-590.
Claes, H. (1961). *Z. Naturf.* **16b**, 445-454.
Claes, H. and Nakayama, T. O. M. (1959). *Z. Naturf.* 14b, 746-747.
Cohen-Bazire, G. and Stanier, R. Y. (1958). *Nature, Lond.* **181**, 250-252.
Crounse, J., Feldman, R. P. and Clayton, R. K. (1963). *Nature, Lond.* **198**, 1227-1228.
Curry, G. M. and Gruen, H. E. (1959). *Proc. natn. Acad. Sci. U.S.A.* 45, 797-804.
Delbrück, M. and Shropshire, W. (1960). *Pl. Physiol., Lancaster.* 35, 194-204.
Dworkin, M. (1958). *J. gen. Physiol.* 41, 1099-1112.
Dworkin, M. (1959). *Nature, Lond.* 184, 1891-1892.
Emerson, R. and Fox, D. L. (1940). *Proc. R. Soc. B* 128, 275-293.
Foote, C. and Denny, R. W. (1968). *J. Am. chem. Soc.* 90, 6233-6235.
Foote, C. S., Denny, R. W., Weaver, L., Chang, Y. and Peters, J. (1970). *Ann. N.Y. Acad. Sci.* 171, 139-148.
French, C. S. (1959). *In* "Photoperiodism and Related Phenomena in Plants and Animals", p. 15. A.A.A.S., Washington, D.C.
Frey-Wyssling, A. (1935). 'Die Stoffausscheidungen der höheren Pflanzen." Springer-Verlag, Berlin.
Fuller, R. C. and Anderson, I. C. (1958). *Nature, Lond.* **181**, 252-254.
Fujimori, E. and Livingston, R. (1957). *Nature, Lond.* 180, 1036-1038.
Galston, A. W. (1959). *In* "Encyclopedia of Plant Physiology" (W. Ruhland, ed.), vol. 17(1), pp. 492-529. Springer-Verlag, Berlin.
Galston, A. W. (1964). *Abstr. tenth int. bot. Congr.* Edinburgh, pp. 185-186.
Goldstrohm, D. D. and Lilly, V. G. (1965). *Mycologia* 57, 612-623.
Gooday, G. W. (1968a). *New Phytol.* 67, 815-821.
Gooday, G. W. (1968b). *Phytochemistry* 7, 2103-2105.
Gooday, G. W., Fawcett, P., Green, D. and Shaw, G. (1973). *J. gen. Microbiol.* 74, 233-239.
Goodwin, R. H. (1953). *A. Rev. Pl. Physiol.* 4, 283-304.
Goodwin, T. W. (1950). *Biol. Rev.* 25, 391-413.
Goodwin, T. W. (1952). *Bot. Rev.* 18, 291-316.
Gössel, I. (1957). *Arch. Mikrobiol.* 27, 288-305.
Griffiths, M. and Stanier, R. Y. (1956). *J. gen. Microbiol.* 14, 698-715.
Griffiths, M., Sistrom, W. R., Cohen-Bazire, G. and Stanier, R. Y. (1955). *Nature, Lond.* 176, 1211-1214.
Halldal, P. (1958). *Physiologia Pl.* 11, 118-153.
Halldal, P. (1961). *Physiologia Pl.* 14, 133-139.

Harder, R. and Sörgel, G. (1938). *Nachr. Ges. Wiss. Göttingen, Phys. Kl., N.F.* **6**, *Biol.* 3, 119-127.
Hartmann, M. (1955). *Am. Nat.* **89**, 321-346.
Heisenberg, M. and Cerdà-Olmedo, E. (1968). *Molec. gen. Genet.* **102**, 187-195.
Hesseltine, C. W. and Anderson, R. F. (1957). *Mycologia* **49**, 449-452.
Hocking, D. (1963). *Nature, Lond.* **197**, 404.
Kandler, O. and Schotz, F. (1956). *Z. Naturf.* **11b**, 708-718.
Kaplan, R. W. (1956). *Arch. Mikrobiol.* **24**, 60-79.
Kivic, P. A. and Vesk, M. (1972). *J. exp. Bot.* **23**, 1070-1075.
Kohl, F. G. (1902). "Untersuchungen über das Carotin und seine physiologische Bedeutung in der Pflanze". Borntraeger, Leipiz.
Krasnovksy, A. A., Prozdova, N. N. and Pakshina, E. V. (1960). *Biochemistry, N.Y.* **25**, 217-222. (Trans. from *Biokhimiya* **25**, 288-295.)
Krinsky, N. L. (1968). *In* "Photophysiology: Current Topics" (A. C. Giese, ed.), pp. 123-195. Academic Press, New York and London.
Kuhn, R. and Moewus, F. (1940). *Ber. dt. chem. Ges.* **73**, 547-559.
Kunisawa, R. and Stanier, R. Y. (1958). *Arch. Mikrobiol.* **31**, 146-156.
Lunz, A. (1931). *Z. vergl. Physiol.* **14**, 68-92.
MacKinney, G. and Chichester, C. O. (1960). *In* "Comparative Biochemistry of Photoreactive Systems" (M. B. Allen, ed.), pp. 205-214. Academic Press, New York and London.
MacMillan, J. D., Maxwell, W. A. and Chichester, C. O. (1966). *Photochem. Photobiol.* **5**, 555-565.
Mast, S. O. (1927). *Z. vergl. Physiol.* **5**, 730-738.
Mathews, M. M. (1963). *Photochem. Photobiol.* **2**, 1-8.
Mathews, M. M. (1964). *Photochem. Photobiol.* **3**, 75-77.
Mathews, M. M. and Sistrom, W. R. (1959). *Nature, Lond.* **184**, 1892-1893.
Mathews, M. M. and Sistrom, W. R. (1960). *Arch. Mikrobiol.* **35**, 139-146.
Maxwell, W. A., MacMillan, J. P. and Chichester, C. O. (1966). *Photochem. Photobiol.* **5**, 567-577.
Meissner, G. and Delbrück, M. (1968). *Pl. Physiol., Lancaster* **43**, 1279-1283.
Mitchell, R. L. and Anderson, I. C. (1965). *Science, N.Y.* **150**, 74.
Noack, K. (1925). *Z. Bot.* **17**, 481-548.
Nishimura, M. and Takamatsu, K. (1957). *Nature, Lond.* **180**, 699-700.
Page, R. M. (1968). *In* "Photophysiology: Current Topics" (A. C. Giese, ed.), pp. 65-90. Academic Press, New York and London.
Page, R. M. and Curry, G. M. (1966). *Photochem. Photobiol.* **5**, 31-40.
Philip, U. and Haldane, J. B. S. (1939). *Nature, Lond.* **143**, 334.
Plempel, M. (1957). *Arch. Mikrobiol.* **26**, 151-174.
Reisener, H.-J. (1958). *Z. Bot.* **46**, 474-505.
Robertson, J. A. (1972). *Arch. Mikrobiol.* **85**, 259-266.
Renner, O. (1958). *Z. Naturf.* **13b**, 399-403.
Ryan, F. J. (1955). *Science, N.Y.* **122**, 470.
Sager, R. and Zalokar, M. (1958). *Nature, Lond.* **182**, 98-100.
Satina, S. and Blakeslee, A. F. (1930). *Bot. Gaz.* **90**, 299-311.
Shaw, G. (1970). *In* "Phytochemical Phylogeny" (J. B. Harborne, ed.), pp. 31-58. Academic Press, London and New York.
Shaw, G. (1971). *In* "Sporopollenin" (J. Brooks, P. R. Grant, M. D. Muir, P. van Gijzel and G. Shaw, eds), pp. 305-350. Academic Press, London and New York.

Shibata, K. (1958). *J. Biochem., Tokyo* **45**, 599-623.
Shropshire, W. (1963). *Physiol. Rev.* **43**, 38-67.
Shropshire, W. and Withrow, R. B. (1958). *Pl. Physiol., Lancaster* **33**, 360-365.
Sistrom, W. R., Griffiths, M. and Stanier, R. Y. (1956). *J. cell. comp. Physiol.* **48**, 459-472.
Smith, J. H. and French, C. S. (1963). *A. Rev. pl. Physiol.* **14**, 181-224.
Stanier, R. Y. and Cohen-Bazire, G. (1957). *In* "Microbial Ecology" (R. E. O. Williams and C. C. Spicer, eds), pp. 56-89. Cambridge University Press, London.
Strasburger, E. (1878). *Jena. Z. Naturw.* **12**, 551-625.
Strother, G. K., and Wolken, J. J. (1961). *J. Protozool.* **8**, 261-265.
Sutter, R. P. (1970). *Science, N.Y.* **168**, 1590-1592.
Sutter, R. P., Dehaven, R. N. and Whitaker, J. P. (1972). *Abstr. a. Meet. Am. Soc. Microbiol.* G227, 68.
Swart-Füchtbauer, H. and Rippel-Baldes, A. (1951). *Arch. Mikrobiol.* **16**, 358-362.
Tchakhotine, S. (1936). *C.r. Séanc. Soc. Biol.* **121**, 1162.
Thimann, K. V. (1940). *Chronica bot.* **6**, 31-32.
Thimann, K. V. and Curry, G. M. (1960). *In* "Comparative Biochemistry" (M. Florkin and H. S. Mason, eds), vol. 1, pp. 243-309. Academic Press, New York and London.
Van den Ende, H. (1967). *Nature, Lond.* **215**, 211-212.
Van den Ende, H. (1968). *J. Bact.* **96**, 1298-1303.
Wager, H. (1900). *Rep. Par. Ass. Advmt Sci.* **70**, 931.
Wallace, R. H. and Schwarting, A. E. (1954). *Pl. Physiol., Lancaster* **29**, 431-436.
Willstätter, R. and Stoll, A. (1918). "Untersuchungen über die Assimilation der Kohlensaure". Springer-Verlag, Berlin.
Wolken, J. J. (1967). "*Euglena*: An Experimental Organism for Biochemical and Biophysical Studies," 2nd edn. Meredith, New York.
Wolken, J. J. and Shin, E. (1958). *J. Protozool.* **5**, 39-46.
Wright, C. J. and Rilling, H. C. (1963). *Photochem. Photobiol.* **2**, 339-342.

Chapter 15

The Physiological Functions of Phytochrome*

RUTH L. SATTER and ARTHUR W. GALSTON

Biology Department, Yale University, New Haven, Connecticut, U.S.A.

* This chapter is dedicated to the memory of Harry A. Borthwick, outstanding scholar and warm human being. Together with Dr Sterling B. Hendricks, Dr Borthwick designed and executed the experiments that led to the discovery of phytochrome, described in the first few pages of this review.

I Introduction

The two principal interconvertible forms of the plant chromoprotein phytochrome act as the poles of a switch that controls metabolic reactions in response to light signals from the environment. Red light (action peak at *c.* 660 nm) converts P_r (red-absorbing) phytochrome to P_{fr} (far-red absorbing) and far-red light (action peak at *c.* 730 nm) converts P_{fr} to P_r. Although P_r is the longer lived of the two forms, P_{fr} often persists during several hours of darkness following red irradiation; thus the ratio of red to far-red energy just prior to darkness can influence metabolism during many subsequent hours. Phytochrome regulates all major phases of growth and development from seed germination to senescence and is found in all eukaryotic plants, with the possible exception of the fungi. Phytochrome mediates similar responses in lower and higher plants, as shown in Table I.

Phytochrome has been purified from homogenates of various green tissues including an alga, a liverwort (Taylor and Bonner, 1967), and two mosses (Giles and von Maltzahn, 1968) and from etiolated seedlings of several angiosperms (Butler *et al.*, 1959). *In vivo* absorption spectra have been measured in *Ginkgo*, pine (Grill and Spruit, 1972) and several angiosperms (Furuya and Hillman, 1964; Briggs and Siegelman, 1965; Correll and Shropshire, 1968; Correll *et al.*, 1968; McArthur and Briggs, 1970). Both *in vivo* and *in vitro* light absorption differ with taxonomic group, leading to speculation about evolutionary changes in the structure of the molecule. However, absorption spectra in green tissue result in part from other pigments; since these also vary somewhat among plant groups, corrections need to be made before comparative analyses will be meaningful (Grill, 1972).

A photoreversible pigment with absorption peaks at 520 and 650 nm was recently extracted from the blue green alga *Tolypothrix* (Scheibe, 1972), and brief red and green irradiations regulate pigment synthesis in this organism (Fujita and Hattori, 1962) and growth habit in another blue green alga, *Nostoc* (Lazaroff, 1966). Thus it seems likely that morphogenesis in prokaryotes is also regulated by a photoreversible pigment.

This review emphasizes phytochrome action *in vivo*; *in vitro* properties are described in Chapter 23 and pigment isolation and purification in Chapter 7. Other recent reviews have been written by Rollin (1970), Briggs and Rice (1972), Mohr (1972), Shropshire (1972) and Smith (1973); in addition, a symposial volume has recently been published (Mitrakos and Shropshire, 1972).

TABLE I
Phytochrome-mediated responses in lower and higher plants

Response	Plant group	Selected reference
Spore germination	Algae	Takatori and Imahori, 1971
	Ferns	Mohr *et al.*, 1964
		Raghavan, 1973
		Sugai and Furuya, 1968
Propagule germination	Mosses	Larpent-Gourgaud *et al.*, 1972
Seed germination	Angiosperms	Rollin, 1972
De-etiolation	Liverworts	Fredericq and DeGreef, 1972
	Angiosperms	Borthwick *et al.*, 1951
Chlorophyll synthesis	Liverworts	Fredericq and DeGreef, 1972
	Angiosperms	Virgin, 1972
Growth habit	Algae	Rethy, 1968
	Liverworts	Fredericq and DeGreef, 1972
	Mosses	Larpent and Jacques, 1971
	Ferns	Miller and Miller, 1963
		Laetsch and Briggs, 1962
		Etzold, 1965
	Angiosperms	Downs *et al.*, 1957
Rhizoid production	Algae	Valio and Schwabe, 1969
Root branching	Angiosperms	Jaffe, 1970
Senescence	Liverworts	DeGreef *et al.*, 1971
	Angiosperms	Dodge *et al.*, 1970
Dormancy	Algae	Wilson and Schwabe, 1964
	Angiosperms	Williams *et al.*, 1972
Phototaxis	Dinoflagellate	Forward, 1970
Leaf movement	Angiosperms	Fondéville *et al.*, 1966

II History

A DISCOVERY OF PHYTOCHROME

The discovery of phytochrome resulted from unexpected congruities in the action spectra for several different phenomena, some related and some seemingly disparate. They included: inhibition of flowering in the short-day plants *Biloxi* soyabeans (Parker *et al.*, 1945) and *Xanthium* (Parker *et al.*, 1946), promotion of flowering in the long-day plants *Hordeum* (Borthwick *et al.*, 1948) and *Hyoscyamus* (Parker *et al.*, 1950), inhibition of stem growth in dark-grown peas (Parker *et al.*, 1949) and barley (Borthwick *et al.*, 1951), and germination of photosensitive lettuce seed (Borthwick *et al.*, 1952b).

All these spectra peaked in the red region (R) at or near 660 nm and events potentiated by R light could be prevented by subsequent far-red light (FR) at or near 730 nm (Fig. 1). The energies required for these several responses were also similar. When R and FR irradiations were repeated several times, the last light treatment before the dark period always determined the physiological response (Fig. 2); for example, the germination of lettuce seed, subjected to

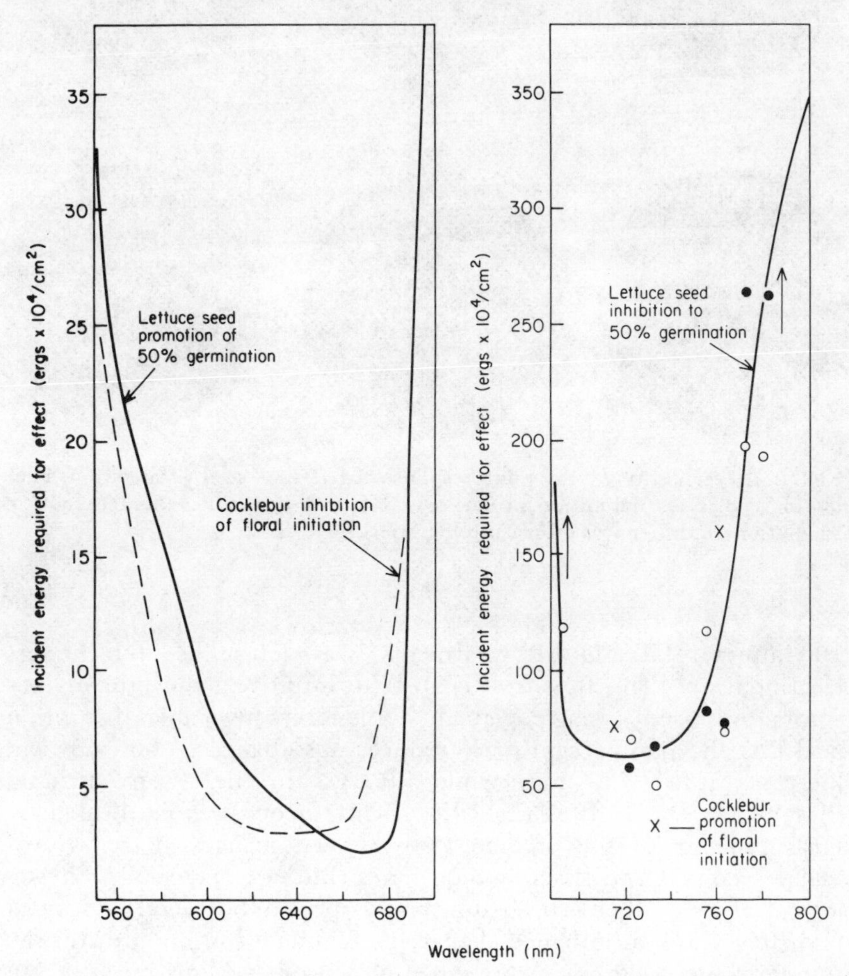

FIG. 1. Action spectra for promotion (left) and inhibition (right) of lettuce seed (*Lactuca sativa* var. Grand Rapids) germination and inhibition (left, broken line) and promotion (right) of flowering of cocklebur (*Xanthium saccharatum*). The solid and open circles are from two experiments with lettuce seed. (From Borthwick *et al.*, 1952a, 1954.)

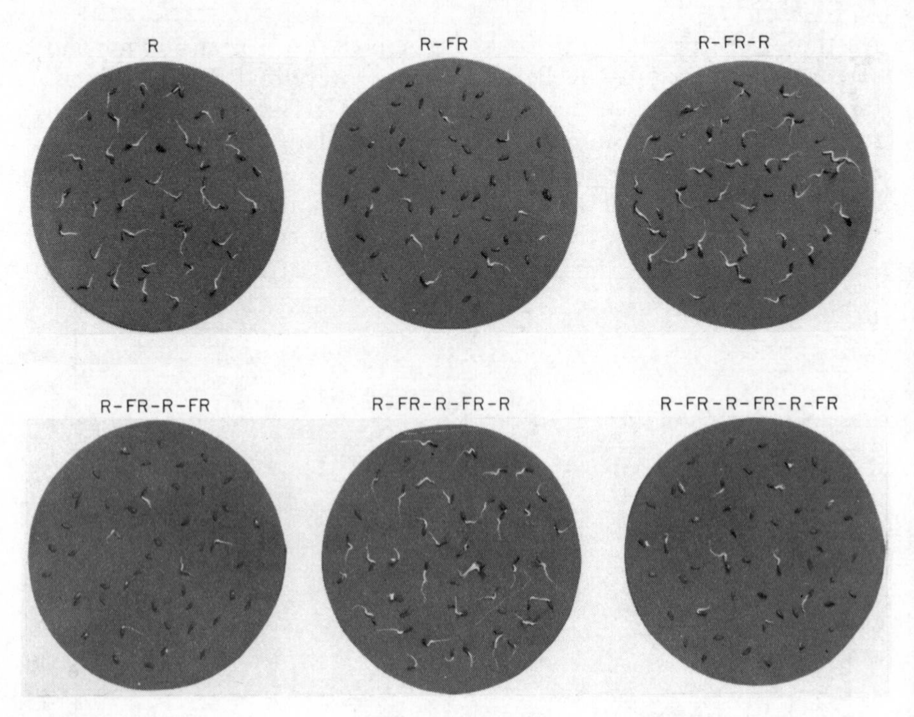

FIG. 2. Reversibility of the photoresponse for lettuce seed germination. Each lot of seed, after imbibition in darkness, received the indicated succession of irradiations and then was returned to darkness for 2 days.

100 alternate R and FR treatments, was determined by the last treatment only (Borthwick, 1972). With intuitive genius, the investigators proposed a single pigment with interconvertible absorption peaks as the photoreceptor: R reduces absorbance at 660 nm and increases it at 730 nm, while FR acts in the reverse manner (Borthwick *et al.*, 1952a,b, 1954). This theory was confirmed by measurements of the predicted absorbance changes in dark-grown maize seedlings and tissue homogenates (Butler *et al.*, 1959). Physiological assay of R-FR reversibility and spectrophotometric assay of photoreversible absorbance changes at 660 and 730 nm are still the standard methods for detecting phytochrome. More recently, an immunochemical technique for detecting the protein moiety has also been developed (Pratt and Coleman, 1971). The dual wavelength spectrophotometer used for measuring P_r, P_{fr} and P (total), and the immunoassay, are both described in Chapter 23.

B P_{fr} AS THE PHYSIOLOGICALLY ACTIVE FORM

Phytochrome in completely etiolated tissue is in the P_r form. Irradiation with 10^{-12} Einsteins/cm^2 of energy in the 660 nm region converts only 0·001% of the phytochrome to P_{fr}, yet has a measurable effect on the enlargement of etiolated pea leaves (Parker *et al.*, 1949). However, after irradiation with saturating R energy, subsequent 10^{-10} Einsteins/cm^2 in the 730 nm region has little effect on leaf size. Thus the early investigators proposed that P_{fr} is the physiologically active form (Borthwick and Hendricks, 1961), and this view has not been seriously challenged by any subsequent data. Irradiation of etiolated tissue with far-red light (*c.* 717 nm), which converts a small fraction of the phytochrome molecules to P_{fr}, is standard procedure in several laboratories for studying phytochrome-mediated effects (Mohr, 1972).

Reactions potentiated by P_{fr} escape from photochemical control (i.e. can no longer be reversed by FR) in periods varying from 60 seconds (leaf unrolling in grasses (Wagné, 1964)) to several hours (lettuce seed germination (Borthwick *et al.*, 1954)) after R irradiation.

C PARADOXES AND EVOLUTION OF NEW CONCEPTS

Spectrophotometric and physiological assays of P_r and P_{fr} have given equivalent results in some, but not all, cases. The situation in which biological and spectrophotometric data do not agree have been called paradoxes. Hillman (1967a) describes the paradoxes in detail and groups them into two categories, one typified by *Pisum* and the other by *Zea*. In the *Pisum* paradox, etiolated pea stem segments respond to different combinations of R and FR light as though 20% of the phytochrome is P_{fr}, although spectrophotometric measurements indicate that all the phytochrome is P_r (Hillman, 1965a). In the *Zea* paradox, a physiological response to R is saturated by a spectrophotometrically undetectable amount of P_{fr}, but reversed by FR light that increases P_{fr} to a detectable level (Briggs and Chon, 1966). The concepts of "bulk" (spectrophotometrically detectable but physiologically inactive) and "active" phytochrome were proposed to rationalize the paradoxes. Subsequent investigations have provided data to support this interpretation. Cell fractionation studies (see Section IIIC) reveal that phytochrome is in the soluble fraction of etiolated seedlings but R irradiation promotes its attachment to a sedimentable particle (Quail *et al.*, 1973a; Boisard *et al.*, 1974; Marmé, 1974). However, there are fewer receptor sites than phytochrome molecules, so that only a portion of the phytochrome is membrane-bound. Soluble and bound phytochrome have different rates of decay (Boisard *et al.*, 1971, 1974) and might be the "inactive" and "active" species predicted by Hillman and Briggs and Chon.

Other interpretations have also been presented. The *Zea* paradox involves change in phototropic sensitivity to blue irradiation one hour after R or FR. Identification of the photoreceptor for phototropism and investigation of its interaction with phytochrome might resolve this paradox, and progress in this direction has already been made. A yellow flavoprotein whose absorption spectrum resembles the action spectrum for phototropism has been extracted from *Cucurbita*, and is in the same particulate fraction (plasmalemma) as phytochrome (W. R. Briggs and D. Marmé, personal communication).

Attempts to correlate spectrophotometric measurements with physiological responses are often futile because spectrophotometric assays measure the average properties of phytochrome molecules in several different cells, while the phytochrome controlling a specific response is localized in a particular environment (Kendrick and Spruit, 1973a). Studies of seed germination that support this viewpoint are discussed in Section IVA.

Immunochemical assays of phytochrome (Pratt and Coleman, 1971, 1974) reveal specific distribution patterns (described in Section III) and suggest that phytochrome molecules in specific locations have unique physiological functions.

III Localization

A DISTRIBUTION WITHIN ETIOLATED SEEDLINGS

An immunocytochemical technique (Pratt and Coleman, 1971, 1974) has been used to locate and assay phytochrome in several monocotyledonous plants. The distribution in *Zea* shoots is relatively uniform, but in oats, rye and rice it is highly localized with respect to organs and cell types. In 3 day old Garry oats, for example, phytochrome is absent from the extreme coleoptile tip, localized in parenchyma cells 0·1 to 0·5 mm below the apex and in inner and outer epidermal cells 0·5 mm to 1·5 mm below the apex. It is barely detectable just above the node but heavily concentrated in the node. It is also found in the vascular tissue of the shoot, and is abundant in root cap cells but less concentrated in or absent from other parts of the root (Fig. 3).

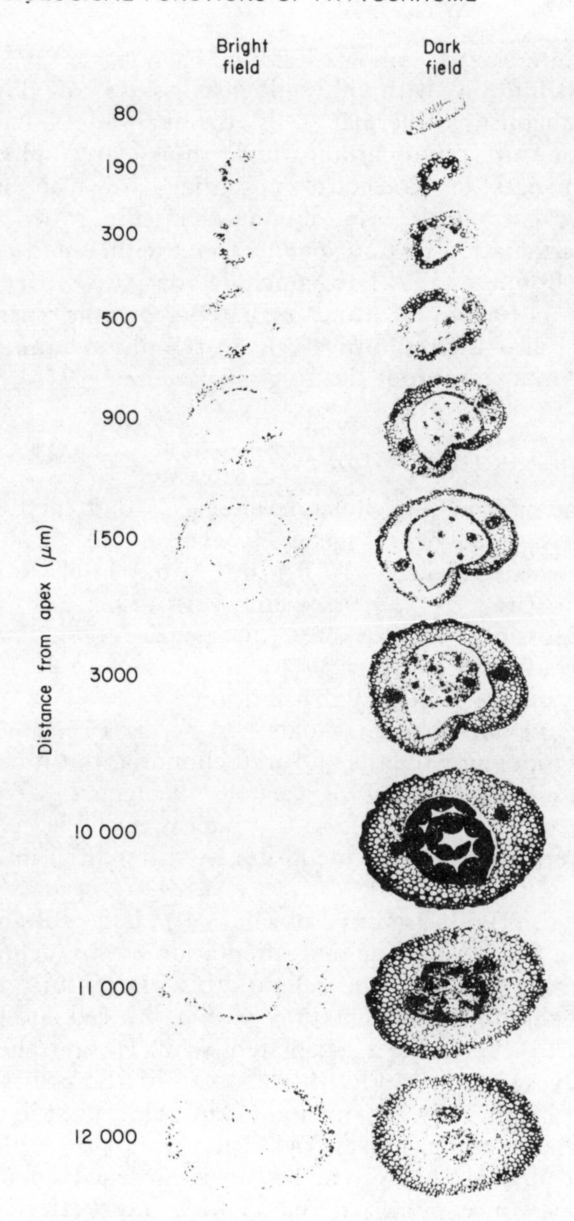

FIG. 3. Immunochemical pattern of phytochrome distribution in cross sections of 3 day old oat shoot. The numbers indicate distance in μm behind the coleoptile apex. The dark-field micrographs (right column) indicate structural details and the dark areas in the bright field micrographs (left column) represent the presence of phytochrome. (From Pratt and Coleman, 1971.)

It is of interest to compare immunochemical assays of phytochrome distribution with physiological assays of phytochrome activity. Subapical epidermal cells of oat shoots have a high concentration of phytochrome and show large phytochrome-mediated changes in bioelectric potential (Newman and Briggs, 1972). Phytochrome is also abundant in the root cap region; microbeam irradiation of *Convolvulus* roots with R and FR light led Tepfer and Bonnett (1972) to conclude that the cap cells are the photoreceptors for phytochrome-mediated geotropic sensitivity. The cap cells are also the responsive cells in the phytochrome-mediated adhesion of mung bean root tips to glass (Racusen, 1973).

B INTRACELLULAR LOCALIZATION

Immunoassay at the subcellular level reveals that phytochrome in etiolated grass seedlings is generally distributed throughout the cytoplasm when localized as P_r, but after brief R irradiation phytochrome (now as P_{fr}) associates with discrete regions of the cytoplasm, possibly receptor sites (MacKenzie *et al.*, 1974). Subsequent FR irradiation converts phytochrome back to P_r and it slowly relaxes and resumes its general distribution.

Phytochrome in dark-grown oats and rye is associated with the nuclear envelope, amyloplasts and mitochondria, but it has not been detected inside the nucleus or vacuoles (Coleman, 1973). Phytochrome was also detected close to the nuclear envelope in lyophilized sections of etiolated *Avena* coleoptiles by differential microspectrophotometry (Galston, 1968).

Haupt *et al.* (1969) used microbeams of polarized R and FR light to determine the localization and orientation of phytochrome in the alga *Mougeotia* (Fig. 4). The cells of this filamentous alga have a single large chloroplast which turns within the cell and exposes its broad face to the light if irradiated with R, and this effect is prevented by subsequent FR. If only part of the cell is irradiated with a microbeam, only the portion of the chloroplast in or adjacent to the irradiated region moves. Detailed scans of the cell led Haupt (1970) to conclude that phytochrome is localized on or near the plasmalemma. Since polarized red light is most effective when it vibrates parallel to the cell wall, but polarized FR light is most effective when it vibrates perpendicular to the cell wall, Haupt concluded that each phytochrome molecule has a strong dichroic orientation that changes 90° upon transformation of P_r to P_{fr} or vice versa. Phytochrome molecules in the fern *Dryopteris* are also

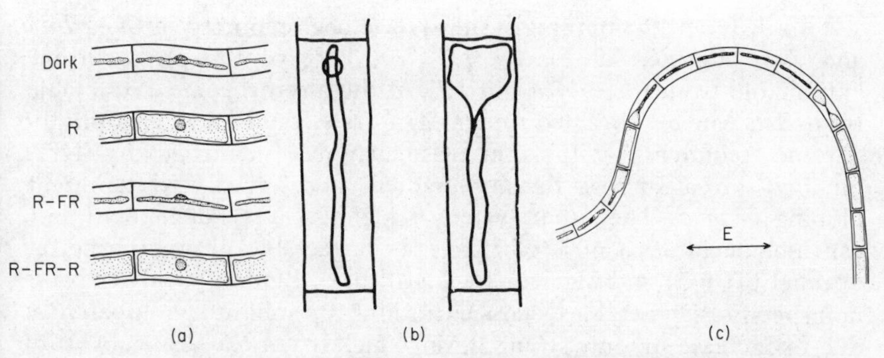

FIG. 4. (a) *Mougeotia* with its giant chloroplast orienting to the light. (b) The upper part of the cell is irradiated with a microbeam of R. (c) Irradiation with polarized R. The arrow indicates the electrical vector. (From Haupt, 1970.)

strongly dichroic (Etzold, 1965). More recently, Marmé and Schäfer (1972) reported that the amount of phytochrome converted from P_r to P_{fr} and vice versa following irradiation of corn coleoptiles with polarized R and FR light depends upon the plane of polarization. Thus fixed orientation of phytochrome in a layer like the plasmalemma would appear to be the rule.

Wellburn and Wellburn (1973) separated etioplasts from *Avena* shoots by a Sephadex column technique, then irradiated the plastids with R or FR light. The subsequent structural development of the etioplast was regulated by the light treatments, implying that phytochrome is inside the plastid or is bound to its outer envelope.

C ISOLATION IN SUBCELLULAR FRACTIONS (See Addendum, p. 727)

Differential centrifugation of homogenates of etiolated tissue yields fractions containing phytochrome (Rubinstein *et al.*, 1969), and brief R or FR treatments determine the percentage pelletable (Marmé *et al.*, 1973; Quail *et al.*, 1973a; Quail and Schäfer, 1974). A fraction which binds naphthylphthalamic acid (NPA), presumably a marker for the plasmalemma (Lembi *et al.*, 1971), preferentially binds P_{fr} under specified conditions (Marmé *et al.*, 1971a).

Isolation of membrane-bound phytochrome involves a slow speed spin (500*g*) followed by centrifugation of the supernatant fraction at 17 000 *g* or higher. Only 4% to 7% of the phytochrome in darkened tissue is in the high speed pellet, but R irradiation prior to extraction or R irradiation of an extract prior to high speed centrifugation increases particulate phytochrome severalfold.

Two different P_{fr}-binding systems have been reported in *Cucurbita* and *Zea* (Marmé, 1974). At pH 7·0, 20% of the phytochrome extractable from *Cucurbita* and 13% of the phytochrome extractable from *Zea* can be obtained under Mg^{2+}-free conditions. This phytochrome sediments with the plasmalemma (identified by NPA binding) on a sucrose density gradient and appears to represent binding *in vivo*. The other system is Mg^{2+}-(or Ca^{2+})-dependent and can include 50% to 60% of the total extractable phytochrome, at optimal pH 6·5) and Mg^{2+} or Ca^{2+} (10 mM). Binding at lower pH is non-specific. Higher Mg^{2+} concentrations are inhibitory, apparently due to increase in ionic strength, since high K^+ or Na^+ can substitute for Mg^{2+} in this regard. The membrane fraction associated with the Mg^{2+}-dependent system has not been identified, nor has the physiological significance of the two systems been assessed. Both contribute to measured values of bound phytochrome in *Cucurbita* hooks and *Zea* coleoptiles, reported by Quail *et al.* (1973a), Quail and Schäfer (1974) and Boisard *et al.* (1974). In a more recent study of phytochrome pelletability, Williamson and Morré (1974 abstract) reported that phytochrome from soyabean hypocotyls sediments with rough endoplasmic reticulum when the extraction buffer contains Mg^{2+}.

1 *Irradiation of tissue extracts*

The binding properties of phytochrome extracted from darkened *Cucurbita* tissue and irradiated *in vitro*, relate primarily to the Mg^{2+}-dependent system (Marmé et al, 1973, 1974). Irradiation of either the 500 s (supernatant fraction of 500 *g* centrifugation) prior to centrifugation at 17 000 *g*, or the 17 000 s prior to centrifugation at 50 000 *g* promotes binding, if the Mg^{2+} or Ca^{2+} concentration in the extraction buffer is appropriate.

Binding can be reversed by FR if Mg^{2+} is absent from the incubation medium during *in vitro* irradiation, but once the vesicles are reconstituted by Mg^{2+}, R-inducible binding is no longer FR reversible.

2 *Irradiation* in vivo

Marmé (1974) proposed that the receptor to which P_{fr} binds is soluble prior to R irradiation. This conclusion was deduced from the following experiment: Phytochrome was extracted from dark-grown *Cucurbita* seedlings under Mg^{2+}-free conditions, and was centrifuged at 50 000 *g* for 30 min to remove cell organelles and membrane vesicles. The supernatant fraction was divided into two parts; one was exposed briefly to R, and the other kept in the dark. Both were then layered on top of sucrose density gradients and centrifuged at 80 000 *g* for 30 min. The R-treated fraction formed a second peak that was completely absent from the dark fraction. Marmé concluded that photo-conversion of P_r to P_{fr} promotes the binding of phytochrome to a partially solubilized vesicle system in the cytoplasm, and the complex is then incorporated into the plasmalemma. This concept has been challenged by Quail (1975) as described in the Addendum on page 727.

The several investigators who studied the effect of *in vivo* phytochrome photoconversions on binding agree that FR without prior R does not raise the binding significantly above the dark control level in *Cucurbita* or *Zea* but FR after R is more effective than FR alone, i.e. conversion to P_{fr} changes the affinity of the receptor for P_r (Quail *et al.*, 1973a; Boisard *et al.*, 1974; Quail and Schäfer, 1974). Once again, the effect varies with experimental conditions, particularly the Mg^{2+} concentration of the extraction buffer (Marmé, 1974). FR reverses the effect of R by 50% (Quail *et al.*, 1973a) or 75% (Boisard *et al.*, 1974) in *Cucurbita*, but enhances R-promoted binding in *Zea* (Quail *et al.*, 1973a).

Further studies are clearly required before relating these results to *in vivo* systems. It is quite possible that during homogenization of tissue for *in vitro* experiments a bond between phytochrome and its membrane receptor is more readily ruptured when phytochrome is in one form rather than the other (Galston, 1974). Haupt's (1970) physiological experiments imply that both P_r and P_{fr} are attached to the plasmalemma in *Mougeotia.*

Quail and Schäfer (1974) studied the effect of irradiation with wavelengths between 660 nm and 730 nm that establish intermediate P_{fr} levels. Wavelengths *c.* 697 nm are 50% more effective than 660 nm in inducing binding, although they are less than half as effective in converting phytochrome to P_{fr}. However, light sources that induce intermediate P_{fr} levels do not enhance binding unless the extraction buffer contains Mg^{2+}; once again, the effects of the two binding systems need to be disentangled before conclusions regarding *in vivo* behaviour can be drawn.

These investigators also examined the effect of light intensity on 708 nm-induced binding. Increase in intensity has a large promotive effect on binding when irradiations continue for 10–15 min after P_{fr} has attained a photo-stationary level. Thus the system responds to the total number of P_{fr} molecules integrated over time rather than simply the steady state concentration. These data might have important consequences in interpreting HIR (discussed in Section IX).

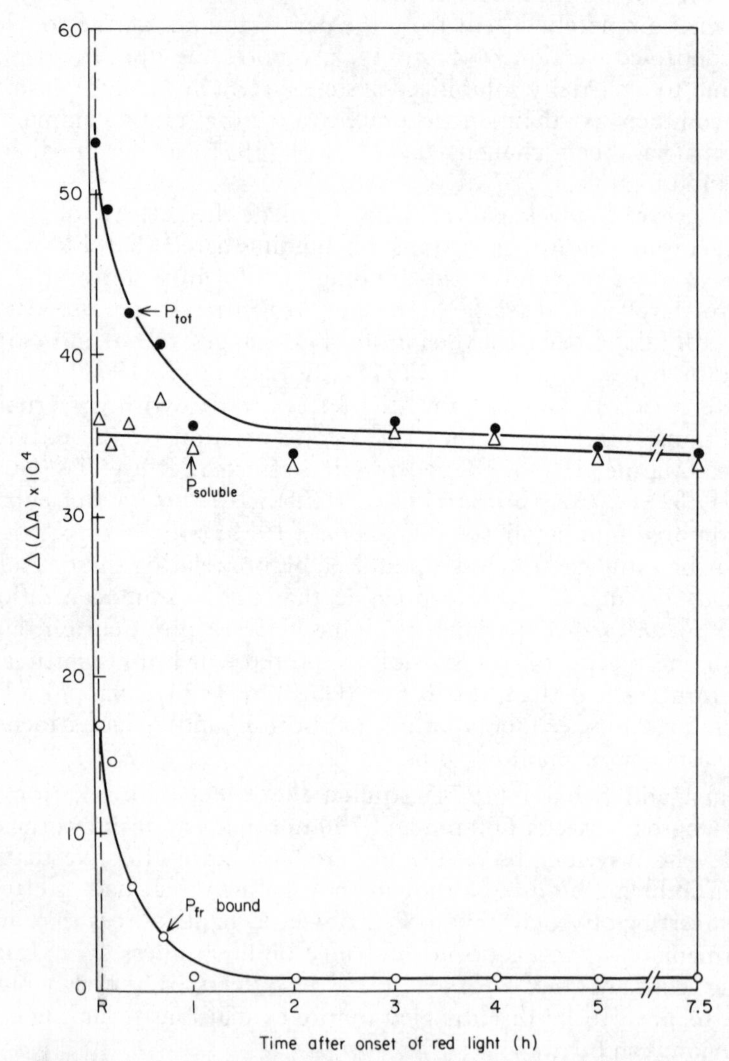

FIG. 5. Kinetics of destruction of the bound and soluble fractions of phytochrome extracted from *Cucurbita*. Seedlings were irradiated with R, then kept in the dark at 25° for 0–7·5 h prior to extraction. Open circles = bound phytochrome; triangles = soluble phytochrome; solid circles = total phytochrome. (From Boisard *et al.*, 1974.)

3 *Release of bound phytochrome*

When R-irradiated *Cucurbita* seedlings are stored in the dark at 25°C for 0–60 minutes prior to extraction, the amount of pelletable phytochrome decreases as the length of dark incubation increases, while the soluble fraction remains constant, i.e. only membrane-bound phytochrome is labile (Fig. 5). Another R irradiation one hour after the first increases binding only slightly, indicating that most of the binding sites are either blocked or altered. The binding capacity is restored slowly; after 24 h of darkness, it is almost normal again (Boisard *et al.*, 1974). This could be due either to synthesis of new binding sites or to restoration of binding capacity in old sites.

Behaviour is quite different when *Zea* (Quail *et al.*, 1973a) or *Cucurbita* (Boisard *et al.*, 1974) seedlings irradiated with R followed by FR are kept in the dark at 25°C for 0–100 min prior to extraction. Pelletable phytochrome is released quantitatively to the soluble pool in the P_r form ($t_{1/2}$ = 50 min) and a second irradiation after two hours of darkness restores pelletability to the original level. Phytochrome lability following brief R irradiation and stability following brief FR irradiation were first detected spectrophotometrically by Butler *et al.* (1963).

Although the soluble fraction of phytochrome remains constant following brief irradiation of *Cucurbita* seedlings, it decreases rapidly if R irradiation persists for more than 30 min; after 60 min, phytochrome is barely detectable (Boisard *et al.*, 1974). It appears that etiolated tissues have efficient mechanisms for inactivating phytochrome after it has been converted to P_{fr}.

4 *Incorporation into lipid bilayers*

Support for the concept that phytochrome acts by control of membrane function is derived from experiments with model "black-lipid" membranes. When partially oxidized cholesterol is spread as a thin film over a pore in a Teflon cup, the reasonably stable membrane has a high initial resistance which declines gradually and is unaffected by light. If phytochrome is added to the solution, some of it apparently makes its way into the membrane, for R irradiation now decreases membrane resistance while FR reverses the R effect. R–FR reversals can be carried out for a few cycles before the membranes collapse (Roux and Yguerabide, 1973). While the chemical basis for such an effect is not known, it could involve the P_r *v.* P_{fr} difference in free lysine residues noted by Roux (1972) and/or surface conformational differences noted by Hopkins and Butler (1970), and reinterpreted by Tobin and Briggs (1973).

D IMPLICATIONS OF PHYTOCHROME-MEMBRANE INTERACTION

Both *in vivo* and *in vitro* experiments provide convincing evidence that phytochrome acts at a membrane locale. Interactions between phytochrome and a membrane-localized receptor could lead to changes in membrane properties, leading in turn to ion flux, change in bioelectric potential and enzyme synthesis. Several questions remain unresolved. It is not known whether phytochrome binding *in vivo* is enhanced by conversion of P_r to P_{fr}, as experiments on pelletable phytochrome suggest, or whether this is an artefact of the extraction procedure. Neither is it clear which membranes bind phytochrome and whether pH and ionic gradients play a role in determining phytochrome distribution *in vivo*. Phytochrome has been localized at several different membranes: (1) the plasmalemma, by microbeam irradiation (Haupt *et al.*, 1969) and cell fractionation (Quail *et al.*, 1973a; Boisard *et al.*, 1974); (2) plastids, by experiments with isolated organelles (Wellburn and Wellburn, 1973); (3) the endoplasmic reticulum, by cell fractionation (Williamson and Morré, 1974); (4) the nuclear envelope, by microspectrophotometry (Galston, 1968). Phytochrome has also been detected immunochemically near all these membranes (Coleman, 1973). Phytochrome localization at all or most membranes would be consistent with its regulation of a broad array of rapid and developmental responses, enumerated in Sections V–VII.

IV Non-photochemical $P_r \rightleftharpoons P_{fr}$ transformations

A HYDRATION AND SYNTHESIS

Phytochrome has been detected in dry seeds of cucumber (Spruit and Mancinelli, 1969), *Cucurbita* (Zouaghi *et al.*, 1972) and pine (Tobin and Briggs, 1969). Since photoreversibility in embryos from pine seeds decreases following lyophilization and is restored by a subsequent rehydration, it appears that measured values are more indicative of seed desiccation than of its phytochrome content.

Absorption spectrum differences between phytochrome in dry cucumber seeds and in etiolated seedlings led to the proposal (Spruit and Mancinelli, 1969) of distinct phytochrome species. It seems more likely that these spectral differences are due to differences in hydration, since dehydration-rehydration reversibly alters the absorption spectrum of phytochrome in etiolated peas (Kendrick and Spruit, 1974) and in an extract from rye (Tobin *et al.*, 1973).

The phytochrome content of imbibing seeds increases in two distinct steps. Imbibition at 0°C or on cycloheximide does not prevent the appearance of the early pool (Tobin and Briggs, 1969) but does prevent the later one (McArthur and Briggs, 1970; Zouaghi *et al.*, 1972), suggesting that pre-existing phytochrome is hydrated during the early phase while new phytochrome is synthesized during the later phase. The size of the early pool in *Amaranthus* (Kendrick *et al.*, 1969) and pine (Tobin and Briggs, 1969) is not affected by light treatments, indicating that this phytochrome is photostable, but the size of the second pool decreases rapidly following R irradiation.

Continuous FR has a relatively minor effect on the size of the second pool in *Amaranthus*, while R reduces it below detectable level, yet FR prevents and R permits seed germination. Kendrick *et al.* (1969) and Kendrick and Frankland (1969a,b) concluded that the two phytochrome pools have different physiological functions: molecules in the first pool are the photoreceptors for light effects on germination, and molecules in the second pool are photoreceptors for de-etiolation. This conclusion also applies to cucumber (Yaniv *et al.*, 1967). In each of these seeds, some phytochrome in the first pool appears in the P_{fr} form during imbibition, and a FR pulse inhibits subsequent dark germination. All the phytochrome in the second pool, by contrast, appears as P_r. In some seeds, such as Alaska pea, germination is light insensitive; such seeds do not have early-appearing cycloheximide-insensitive phytochrome (McArthur and Briggs, 1970).

B INVERSE DARK REVERSION: FACT OR MYTH?

The appearance of P_{fr} during imbibition is usually gradual and is not prevented by FR prior to or early in the imbibition period, and this increase has been reported to be correlated with the disappearance of P_r. In addition, prolonged or intermittent FR irradiation is required to prevent germination in some seeds. These data have led to the suggestion that phytochrome is hydrated as P_r, then reverts to P_{fr} (called "inverse dark reversion" since reversion in the opposite direction is predicted on thermodynamic grounds) (Boisard *et al.*, 1968; Kendrick *et al.*, 1969; Spruit and Mancinelli, 1969; Malcoste *et al.*, 1970; Rollin, 1972; Zouaghi *et al.*, 1972). An alternative explanation is that seeds contain a mixture of P_r and P_{fr} which becomes detectable during slow imbibitional hydration (Mancinelli *et al.*, 1966; Briggs and Rice, 1972; Taylorson and Hendricks, 1972b; Tobin *et al.*, 1973). Recent studies support the latter view. Experiments of McCullough and Shropshire (1970), Vidaver and Hsiao (1972), Shropshire (1973) and Hayes and Klein (1974) imply that phytochrome can be stored in seed as P_r or P_{fr} and rehydrated in the stored form (see Section VIA). Kendrick and Spruit's (1974) study suggests that the reported spontaneous disappearance of P_r during imbibition is an artefact. They detected P_{650}, an intermediate in $P_{fr} \rightarrow P_r$ photoconversion, in lyophilized etiolated pea hooks and reported that P_{650} rather than P_r reverts to P_{fr} in the dark. Despite the lack of evidence of "inverse dark reversion" during imbibition, data obtained from studies of floral initiation in photoperiod-sensitive plants are difficult to interpret without proposing "inverse dark reversion" or *de novo* synthesis of P_{fr} (Evans and King, 1969; King and Cumming, 1972). These data are discussed in Section VIC.

C DESTRUCTION

The terms "destruction" and "decay" are used interchangeably to describe the gradual loss of photoreversibility following conversion of P_r to P_{fr} in etiolated tissue (Butler *et al.*, 1963; DeLint and Spruit, 1963). Coleman (1973) reported loss of immunodetectability of phytochrome in etiolated oat tissue exposed to brief R, and the time course for this loss paralleled the time course for loss of photoreversibility. These results imply that the protein is degraded rather than merely bleached or compartmentalized, as had been suggested earlier (Clarkson and Hillman, 1967).

Destruction usually begins immediately after R irradiation, but a 10 to 60 min lag has been noted in very young seedlings of *Avena* (Kidd and Pratt, 1973), *Pisum* (McArthur and Briggs, 1971) and *Sinapis* (Marmé *et al.*, 1971b). Inhibitors of protein synthesis prevented this light-mediated induction of the destructive mechanism in *Avena*, but did not prevent its operation after induction (Kidd and Pratt, 1973).

Furuya *et al.* (1965) proposed that destruction is an enzymic process. It is also an oxidative, energy-requiring, metal-dependent process. However, it is not clear whether these conditions are required for destruction *per se*, or for binding to membranes which in turn leads to destruction. Destruction is inhibited by anaerobiosis, azide and CO (Butler and Lane, 1965) and has a high Q_{10} (between 2·7 and 3·5 in *Zea* (Pratt and Briggs, 1966) and 4·3 in *Amaranthus* (Kendrick and Frankland, 1969b)). Metal-complexing agents and sulfhydryl compounds inhibit destruction *in vivo* in *Avena* and *Pisum* (Furuya *et al.*, 1965; Hillman, 1967a) and Fe^{2+}, Fe^{3+} and Zn^{2+} overcome the inhibitions in *Avena* (Manabe and Furuya, 1971). Heavy metals, especially Hg^{2+}, Cd^{2+}, Cu^{2+} and Zn^{2+}, also promote decay *in vitro* (Lisansky and Galston, 1974). Cu^{2+} also interferes with photoperiodic time measurement in *Lemna gibba* (Hillman, 1962) and *L. perpusilla* (Hillman, 1965b). Cu^{2+}-treated plants respond as though the days are short regardless of the actual daylength. This might be due to promotion of decay, which would reduce the amount of P_{fr} (Lisansky and Galston, 1974).

Decay in *Cucurbita* consists of a rapid phase with first-order kinetics and a slower second phase. The time course of decay during the rapid phase parallels that of the bound fraction (Boisard *et al.*, 1971, 1974). In other dicotyledons, decay is first order (Butler and Lane, 1965), leading one to question whether soluble and bound phytochrome decay at the same rate in these seedlings or whether all their P_{fr} is bound. Decay in monocotyledons is zero order (Butler *et al.*, 1963; DeLint and Spruit, 1963; Chorney and Gordon, 1966; Pratt and Briggs, 1966; Dooskin and Mancinelli, 1968). Phytochrome in etiolated tissue is barely detectable after a few hours of continuous R irradiation (Clarkson and Hillman, 1967; Boisard *et al.*, 1974) but phytochrome in light-grown tissue appears to behave quite differently. Photostable phytochrome has been detected spectrophotometrically in hydrated seed, cauliflower florets (Butler and Lane, 1965), parsnip roots and artichoke receptacle tissue (Hillman, 1964). Although chlorophyll screening has interfered with spectrophotometric measurements of phytochrome in green tissues, the large number of phytochrome-mediated responses in light-grown plants imply pigment stability.

Since environmental conditions such as pH, concentration of metals, and sulphydryl compounds influence phytochrome stability *in vivo* and *in vitro*, differences in molecular environment in dark-grown and light-grown tissue might promote lability under the former conditions and stability under the latter. A reasonable possibility is that differences in membrane composition are responsible for differences in phytochrome stability, e.g. light might promote the synthesis of molecules that protect phytochrome in its membrane environment, just as it promotes the synthesis of carotenoids that protect chlorophyll molecules in thylakoid membranes from degradative attack by molecular oxygen (Stanier, 1960).

D DESTRUCTION AND SYNTHESIS: ARE THEY INTERRELATED?

Phytochrome levels in etiolated peas decreased to a very low level following repeated R light treatments, then started to increase. Clarkson and Hillman (1967) postulated that destruction to a level below critical stimulates "apparent" synthesis. Kendrick and Frankland (1969b) came to a similar conclusion based on *Amaranthus* data. Since synthesis and destruction occur simultaneously and both processes alter phytochrome levels, these interpretations were equivocal. To resolve the question, Quail *et al.* (1973b) combined kinetic spectrophotometric measurements with a density labelling technique that enabled them to separate newly synthesized phytochrome molecules from pre-existing ones. They found no evidence that the rate of synthesis is related to destruction. Their experiments revealed continuous synthesis of phytochrome in the P_r form with a zero rate order constant. They concluded that destruction and recovery are both transitions between steady state levels, with degradation predominating in the former case and synthesis in the latter.

E REVERSION

Reversion is detected in the dark by an increased absorbance at 660 nm correlated with a decreased absorbance at 730 nm. It was predicted soon after the pigment had been discovered, on the basis of experiments on lettuce seed germination. Incubation at high temperature prevents R-promoted germination if the high temperature treatment follows R, but not if it precedes the light treatment; Borthwick *et al.* (1954) concluded that P_{fr} reverts to P_r by a thermal process.

Reversion has been detected spectrophotometrically in seeds of *Pinus palustris* during the early phase of imbibition (Tobin and Briggs, 1969) and in etiolated seedlings of dicotyledons (Butler *et al.*, 1963; Hillman, 1967) with the exception of *Centrospermae* (Kendrick and Frankland, 1968; Kendrick and Hillman, 1971) but not in any monocotyledons (Butler and Lane, 1965; Furuya *et al.*, 1965; Pratt and Briggs, 1966; Pike and Briggs, 1972). Particularly high rates of reversion have been reported in carrot tissue culture (Wetherell and Koukkari, 1970). Kinetic analysis of reversion is complicated by the fact that destruction usually occurs simultaneously.

Reversion *in vivo* is very sensitive to temperature ($Q_{10} = 8$, in

Amaranthus retroflexus; Taylorson and Hendricks, 1969) but not to oxygen (Butler and Lane, 1965), metal-complexing agents (Hillman, 1967) or respiratory inhibitors. *Pisum* is an exception; its reversion is promoted by azide (Furuya *et al.*, 1965). Although reversion in rye has not been detected *in vivo*, it is very rapid *in vitro* (Pike and Briggs, 1972) with the rate dependent upon pH (Anderson *et al.*, 1969) and oxidation-reduction state (Mumford and Jenner, 1971; Kendrick and Spruit, 1973a).

Reversion is usually more rapid in etiolated tissue than in light-grown tissue. For example, reversion in etiolated peas at 24°C is virtually complete 4 h after R irradiation (Hopkins and Hillman, 1965) but one-third of the phytochrome in light-grown cauliflower heads is still in the P_{fr} form 17 h after R light treatment (Butler and Lane, 1965). In etiolated *Sinapis* seedlings, reversion is confined to the lag phase of decay and is completed in 10 min (Marmé *et al.*, 1971b). Reversion in photoperiodically-sensitive green plants has been assayed physiologically by the "null method" described in Section VIC.

V Rapid responses to R and FR light

Since phytochrome is a quantitatively small component of plant cells, and even low actinic irradiances absorbed by the pigment suffice to trigger large physiological changes in the organism, some amplification mechanism must connect initial photon absorption with ultimate response. Among candidate mechanisms are control of gene activity, control of enzyme activity, control of membrane permeability and control of the level of oligodynamic substances such as hormones, cyclic nucleotides and neurohumoral effectors.

The first two theories received widespread support during a 10–15 year period following discovery of the pigment. Research during this period focused on developmental responses that were often not detectable until several hours or even days after absorption of stimulus energy. For example, Mohr and co-workers studied anthocyanin formation and photomorphogenesis in the mustard seedling and gathered considerable evidence that phytochrome-mediated changes in enzyme activity are involved in photomorphogenesis. This led Mohr (1966a,b) to propose that differential gene activation and repression are the primary regulatory processes that follow photoreception.

A EVIDENCE THAT PHYTOCHROME ALTERS MEMBRANE PROPERTIES

This theory was challenged by the finding that nyctinastic leaflet movement in *Mimosa pudica* is controlled by phytochrome status with effects detectable 5 min after the light treatments (Fondéville *et al.*, 1966). Since Haupt (1959, 1965) had demonstrated that chloroplast orientation in *Mougetia* is regulated by phytochrome within 10 min of R and FR, and some phytochrome-mediated responses escape photochemical control 60 s after brief exposure to R (Fredericq, 1964, 1965; Wagné, 1964), it became apparent that P_{fr} performs its physiological function far too rapidly for so cumbersome a mechanism as gene-activated enzyme synthesis. In 1967, Hendricks and Borthwick proposed that phytochrome alters membrane permeability; an abundance of physiological evidence presented during the next few years supports this view.

1 *Leaf movements and* K^+*/flux*

Phytochrome regulates leaflet movement in *Albizzia julibrissin* (Hillman and Koukkari, 1967; Jaffe and Galston, 1967) and *Samanea saman* (Sweet and Hillman, 1969) as well as in *Mimosa pudica*, and leaflet movement, which results from turgor changes in pulvinal motor cells, is regulated by massive K^+ flux through motor cell membranes (Toriyama, 1955; Allen, 1969; Satter *et al.*, 1970, 1974b,c; Satter and Galston, 1971a,b, 1973), detectable 10 min after R and FR light treatments. The photoreceptors are in the pulvinus (Koukkari and Hillman, 1968; Satter *et al.*, 1974b,c) and the effect is ATP dependent, leading to the conclusion that P_{fr} mediates the activity of ion pumps or channels in pulvinal motor cells (Satter *et al.*, 1970, 1974c). However, it is not known whether phytochrome has a direct effect on K^+ movement or whether K^+ flux is a secondary manifestation of the flux of protons or other ions. Phytochrome also regulates changes in transmembrane potential of *Samanea* motor cells, detectable 90 s after the beginning of the R and FR light treatments (Racusen and Satter, 1975). In addition, changes in the surface charge of motor cells, determined by a cell electrophoresis technique, are under phytochrome control. Similar phytochrome-mediated changes in electrical surface charge have been detected in root cap cells of etiolated plants and are described below.

2 *The "Tanada" effect*

Barley (Tanada, 1967) and mung bean (Jaffe, 1968; Tanada, 1968; Yunghans and Jaffe, 1970) root tips incubated in an appropriate medium attach to negatively charged glass or to a negative electrode (Racusen and Miller, 1972) if pre-irradiated with R, and detach when exposed to FR. These effects are detectable 15 s after the light treatments. Microscopic observation reveals that only root cap cells are involved in the adhesion phenomenon. H^+ or Ca^{2+} can substitute for R, suggesting that the cap cells carry fixed ionic charges; the net charge can be reversed by light, H^+ or Ca^{2+} (Racusen and Etherton, 1973). Experiments with an anilinonaphthosulphonate fluorescent dye (DANS) revealed R-induced changes in the number of charged binding sites, and Racusen (1973) proposed that phytochrome controls changes in the conformation of a membrane protein in the cap cells. This interpretation supports a model of Boisard *et al.* (1974) which proposes that P_{fr} binds to the plasmalemma and changes its conformation.

3 *Changes in bioelectric potential*

Jaffe (1968) reported phytochrome-mediated changes in the bioelectric potential between the apex and cut end of mung bean roots. R makes the apex 1 mV more positive than the cut end, and FR reverses the polarity.

R and FR also alter the electrical potentials measured between the apical coleoptile tissue of *Avena* and the solution bathing the base of

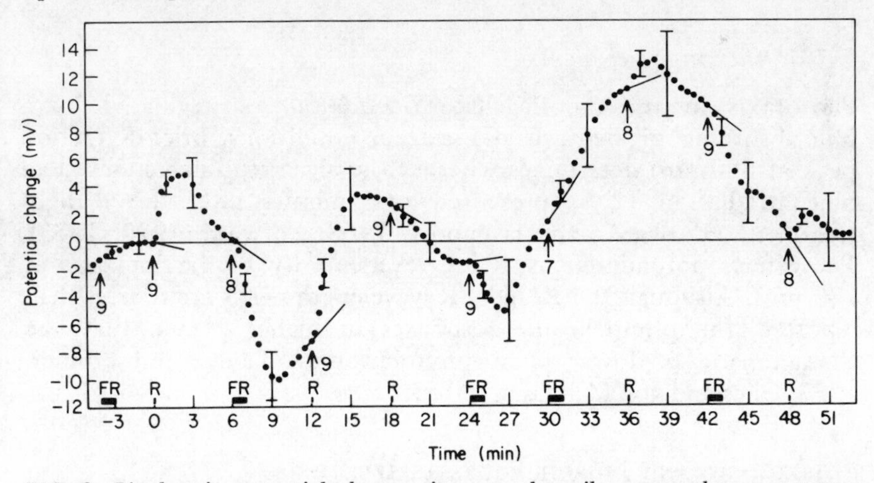

FIG. 6. Bioelectric potential changes in oat coleoptiles exposed to a sequence of five FR and five R light treatments. The start of each treatment is indicated by an arrow, and its duration is indicated by a bar on the bottom of the figure. Numbers against the arrows refer to sample size. The changes were measured using a flowing solution of 10 mM KCl to contact the surface of the coleoptile, 3 to 5 mm below the apex. (From Newman and Briggs, 1972.)

the organ (Newman and Briggs, 1972). Intense light sources saturated phytochrome photoconversions with 15 s R and 65 s FR, and changes in potential were detected 10 to 15 s after the beginning of the light treatments. R increased the potential 5 to 10 mV and FR partly reversed this effect; these effects could be repeated for several consecutive R–FR cycles (Fig. 6). Since photoreversibility was incomplete, Newman and Briggs postulated that R initiates two independent processes, one reversible by FR and the other irreversible. They proposed that R lowers membrane permeability and turns off an ion pump; subsequent FR irradiation turns the pump on again, but does not alter the R-induced permeability changes.

4 *Permeability to water*

Phytochrome regulates permeability to water in *Mougeotia* (Weisenseel and Smeibidl, 1973), epidermal cells of *Taraxacum* (Carceller and Sanchez, 1972) and lettuce seed (Nabors and Lang, 1971). This effect was detected only 15 s after R or FR irradiation of *Mougeotia*. Cells plasmolyse more rapidly upon incubation in mannitol and deplasmolyse more rapidly when mannitol is removed, if they are irradiated with R rather than FR. One would expect this to have especially important physiological consequences for aquatic organisms.

5 *Motility in lower plants*

Phototaxis in the dinoflagellate *Gyrodinium* is regulated by a blue-absorbing pigment. Under certain conditions, brief R (action peak at 620 nm) determines whether the algae stop in response to a blue irradiation (2 s) presented one minute after the R light treatment (Forward and Davenport, 1968; Forward, 1970, 1973). The effects potentiated by R are reversible by FR (action peak at 720 nm). Although the R and FR wavelength peaks are shorter than reported for phytochrome responses in higher plants, they are characteristic of lower plant phytochromes (Taylor and Bonner, 1967; Giles and von Maltzahn, 1968).

B HORMONES AND NEUROHUMOURS AS EFFECTORS

Like phytochrome, plant hormones occur in minute concentrations, and it is conceivable that the effects of phytochrome might be interpreted in terms of hormone synthesis or destruction, or, alternatively, hormone release or binding. Either of these types of phenomena could probably occur rapidly enough to satisfy even the virtually instantaneous time requirement for phytochrome action. Synthesis or destruction might be effected either by direct action of P_{fr}, acting enzymically, or through phytochrome control of activity of synthetic or destructive enzymes, existing as proenzymes prior to irradiation. Binding or release of hormones, either from an active locus or a sequestration site, could similarly be effected through modification of the site by P_{fr}.

1 *IAA and ABA*

Tanada (1973a,b) demonstrated the remarkable antagonism of IAA and ABA, at 10nM or lower concentration, in the phytochrome-mediated attachment of barley and mung bean root tips to phosphate-charged glass. As previously explained, such attachment is promoted by R and negated by FR, provided that the incubation medium is appropriate. ABA induces attachment and IAA detachment in mung beans, and vice versa in barley, and Tanada proposed that phytochrome regulates the endogenous levels of both hormones which in turn regulate changes in surface charge at the plasmalemma. This bespeaks a membrane locale for hormonal as well as phytochrome action.

2 *Acetylcholine*

Jaffe (1970, 1971) reported that mung bean root tips have a high titre of acetylcholine (ACh) with levels regulated by R and FR. Furthermore, ACh substitutes for R in several phytochrome-mediated phenomena: root tip attachment, H^+ efflux, changes in respiratory metabolism (Yunghans and Jaffe, 1972) and secondary root formation. Atropine, an inhibitor of ACh action, reduces the effect, as does cholinesterase, an ACh-degrading enzyme; eserine, an inhibitor of the enzyme, inhibits FR action. Jaffe proposed that change in the titre of ACh is the primary event that follows absorption of stimulus energy by phytochrome and couples photoreception to both rapid and developmental responses.

Subsequent experiments have both supported (Gressel *et al.*, 1971; Hartmann, 1971; Kandeler, 1972) and negated (Kasemir and Mohr, 1972; Satter *et al.*, 1972; Tanada, 1972; White and Pike, 1974) this suggestion. It is clear that ACh (or compounds acting similarly on clam hearts) exists in plants, that its titre may depend on irradiation and on phytochrome, and that ACh can mimic certain aspects of R-induced photomorphogenesis. But action of external ACh in, for example, the root adhesion system is apparently trivial, and easily negated by increasing the K^+ level. Also, ACh changes do not always follow R or FR irradiation. More work is needed to clarify this subject.

VI Developmental responses

A SEED GERMINATION

1 Dark germination related to P_{fr}

The germination of many seeds is sensitive to light; light usually promotes but may also inhibit. These effects, whenever investigated in detail, reveal a requirement for P_{fr} (Borthwick and Hendricks, 1961). A similar requirement has been demonstrated for spore germination in algae (Takatori and Imahori, 1971), mosses (Bauer and Mohr, 1959) and ferns (Mohr, 1956; Sugai and Furuya, 1967; Raghavan, 1973). When light promotes, the action spectrum peaks near 660 nm, as shown for *Lactuca sativa* var. Grand Rapids (Fig. 1). When light inhibits, as in *Cucumis, Nemophila, Nigella* and *Phacelia,* the effects are mainly due to blue and FR wavelengths (Isikawa, 1954; Black and Wareing, 1960; Yaniv *et al*., 1967).

Phytochrome is synthesized during seed development, stored in dry seed as dehydrated P_r or P_{fr}, and rehydrated early in the imbibition period. Experiments with *Arabidopsis* reveal that the light regime during seed development (or more accurately, light conditions immediately preceding the dehydration phase of seed development) affects the subsequent light requirement for germination (McCullough and Shropshire, 1970; Shropshire, 1973; Hayes and Klein, 1974) (Fig. 7). This plant produces completely viable seed when grown in light containing ratios of R/FR between 0·6 and 7·0. When the developing seed was exposed to a light quality different from that illuminating the rest of the plant, the light regime of the parent plant was unimportant but seed exposed to the highest R/FR ratio prior to dehydration showed the highest dark germination. Lettuce seed behaves similarly; Vidaver and Hsiao (1972) exposed hydrated seed to R or FR, then lyophilized the seed and stored it in the dark for more than a year. When rehydrated, it germinated in response to the light treatment preceding desiccation. Temperature during seed development (Koller, 1962) and relative humidity during

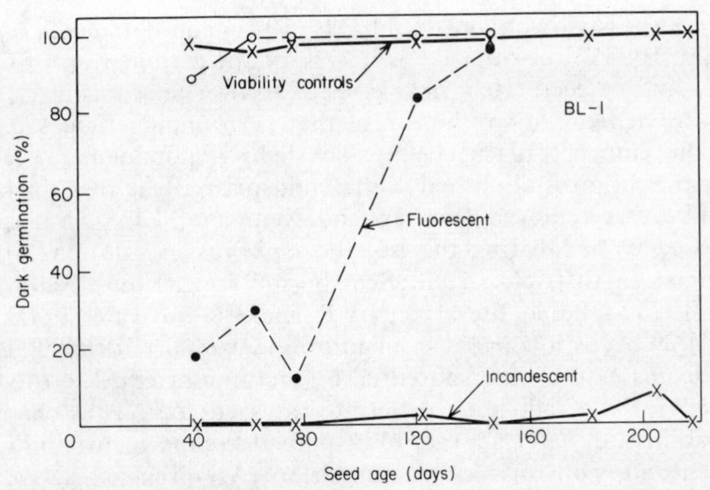

FIG. 7. Dark germination of *Arabidopsis* seed harvested from plants in which floral primordia were irradiated with cool white fluorescent (R:FR = 7·0) or incandescent (R:FR = 0·6) light. The viability controls are the germination percentages obtained if seed are in white light during imbibition. (From Shropshire, 1973.)

storage also alter dark germination patterns. P_{fr} is stable in dry seed, but reverts to P_r if the seed is stored in a moist atmosphere (Hsiao and Vidaver, 1973).

Temperature and phytochrome interact to control the germination of many types of seeds. For example, *Amaranthus retroflexus* seeds incubated at 35°C require R for germination, but prechilling at 20°C or lower for several days prior to the 35°C incubation eliminates the light requirement (Taylorson and Hendricks, 1969). Dark germination in other seeds such as *Rumex crispus* (Taylorson and Hendricks, 1972b) and lettuce (Scheibe and Lang, 1965) is also promoted by a temperature shift of this type. An interpretation proposed by Taylorson and Hendricks (1969, 1972a) is that pre-existing P_{fr} is hydrated during the prechilling period and remains stable at 20°C but reverts to P_r at higher temperature. In *Amaranthus*, for example, P_{fr} to P_r reversion is four times as rapid at 25° as at 20°C. Thus requirements for specific temperature and moisture regimes during seed stratification can be understood in terms of equilibria between hydrated and unhydrated phytochrome, and between P_r and P_{fr}.

2 *Analysis of mechanism*

The P_{fr} level required for germination varies considerably. For example, 1 to 2% P_{fr} potentiates germination in *Amaranthus retroflexus* (Taylorson and Hendricks, 1971) and *Wittrockia superba* (Downs, 1974), but 50% P_{fr} is necessary for germination of Grand Rapids lettuce seed (Hendricks *et al.*, 1956). The requisite P_{fr} level seems to depend upon the stress that surrounding tissues impose upon the embryo (Black, 1969). The light requirement for lettuce seed germination is abolished if the endosperm tissue that surrounds the embryo is removed (Evenari and Neumann, 1953), but can be reimposed by incubating the isolated embryos in solutions of high osmotic strength (Kahn, 1960; Scheibe and Lang, 1965). Nabors and Lang (1971) studied the effect of R and FR on water uptake by isolated embryos incubated in mannitol, and reported that R irradiation decreases the water potential by an amount equal to the force required for the radicle to penetrate the seed coat. Thus change in permeability to water appears to be part of the mechanism by which phytochrome controls seed germination. As discussed above (see Section V), phytochrome regulates permeability to water in mature plants also.

R irradiation increases the level of extractable gibberellic acid (GA) in imbibing lettuce seeds (Koller, 1966) and exogeneous GA can abolish the light requirement for germination of many seeds (Negbi *et al.*, 1966); some investigators have proposed that the R-promoted increase in gibberellin synthesis underlies light promotion of seed germination. Data by Bewley *et al.* (1968) do not support this view. When lettuce seed are irradiated by R followed by FR 5 min later, sensitivity to subsequent externally supplied GA increases over dark controls. This strong synergism suggests that both GA and P_{fr} are necessary for germination (Black, 1969).

To recapitulate: The light sensitivity of many kinds of seeds is related to the requirement for P_{fr}. Seeds that require R for germination lack sufficient endogenous P_{fr}, while those that are inhibited by light already have an adequate P_{fr} level. Variation in the $P_{fr}:P_r$ ratio of phytochrome stored in the seed varies greatly for different species and varieties, and depends in part upon the light regime during seed dehydration, and temperature and humidity during storage. A decrease in water potential appears to be part of the mechanism by which P_{fr} promotes germination.

B DE-ETIOLATION

Dark-grown seedlings usually have long internodes and small leaves lacking chlorophyll. Dicotyledons often form a plumular hook and leaves of monocotyledons are rolled. Brief irradiation with R causes profound changes in these growth patterns. The rate of stem elongation is reduced and leaves expand; the plumular hook opens (Fig. 8) and grass leaves flatten; the tissue greens (Parker *et al.*, 1949; Withrow *et al.*, 1957; Mohr and Noblé, 1960; Downs, 1964; Wagné, 1964). These morphological changes are correlated with changes in structural development, respiratory metabolism, hormone levels,

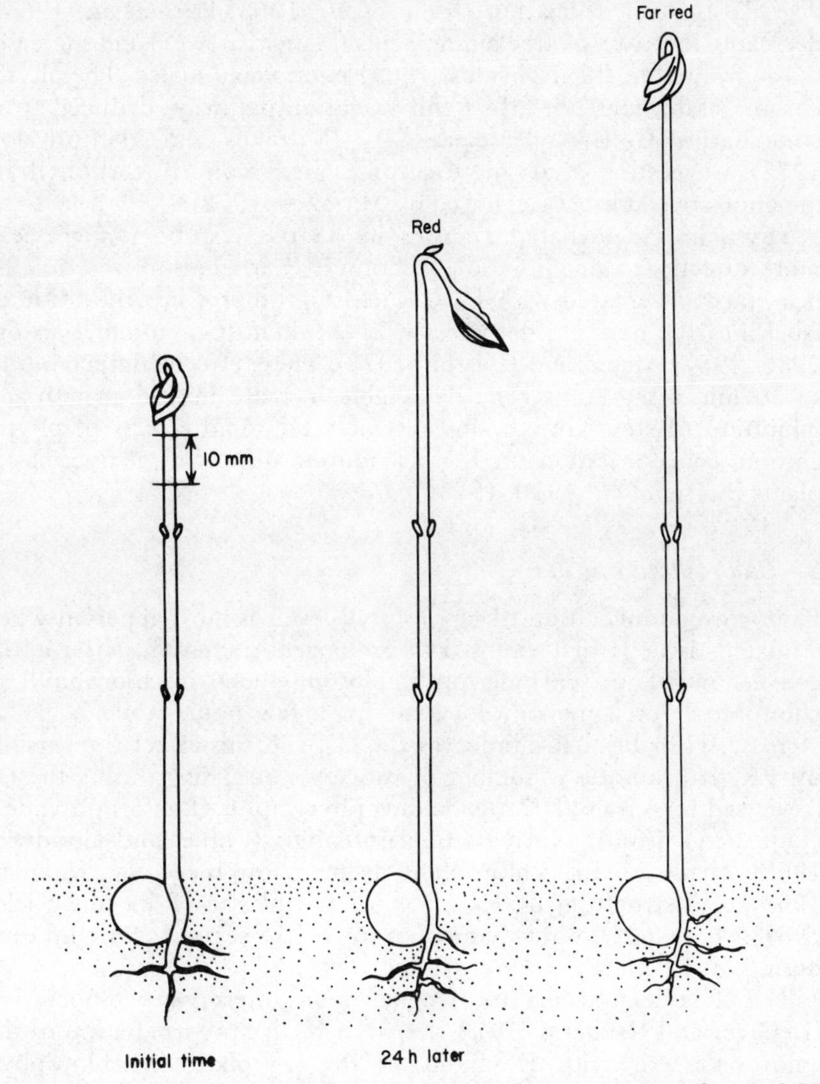

FIG. 8. Six to 7 day old dark-grown *Pisum sativum* var. Alaska before and 24 h after R and FR.

enzyme synthesis and activity, and transport of inorganic and organic compounds.

1 *Carbohydrate metabolism and transport*

Red light, acting through conversion of P_r to P_{fr}, stimulates starch degradation and utilization (Klein *et al.*, 1963; Price *et al.*, 1964), decreases the titre of free amino acids (Margaris, 1972) and increases CO_2 production (Mantouvalos, 1969) in etiolated maize. The phytochrome state does not affect the oxidation pathway, deduced from comparative C_6:C_1 release as CO_2 (Mitrakos and Mantouvalos, 1972), suggesting that phytochrome mediation of carbohydrate metabolism is indirect (reviewed by Mitrakos, 1972).

Phytochrome-mediated changes in sucrose transport might lead more directly to de-etiolation responses. R irradiation of etiolated pea epicotyls treated with [^{14}C]sucrose promotes incorporation of label into the bud and decreases it in the stem (Goren and Galston, 1966, 1967; Anand and Galston, 1972). These effects, detectable 15 to 30 min after R, precede detectable increase in bud growth and inhibition of stem growth, and resemble the rapid effects of phytochrome transformation on K^+ distribution in pulvini of nyctinastic plants (Satter *et al.*, 1970, 1974c).

2 *Chlorophyll synthesis*

Phytochrome mediation of chlorophyll levels is most apparent when etiolated tissue is first exposed to prolonged irradiation. After initial conversion of pre-existing protochlorophyllide to chlorophyll *a*, chlorophyll levels remain constant for a few hours, then begin to increase. Prior brief R eliminates the lag, and this effect is reversible by FR. R promotes a number of processes occurring during the lag (reviewed by Virgin, 1972) including phytylation of chlorophyllide a (Liljenberg, 1966), synthesis of carotenoids (Cohen and Goodwin, 1962) that protect chlorophyll from photo-bleaching (Stanier, 1960), and structural development of the plastid (Price and Klein, 1961). It is not known which of these processes is rate-limiting during the lag phase.

The R effect seems to depend upon interorgan co-operation (DeGreef and Caubergs, 1972, 1973). Selective pre-irradiation of the embryonic axis with R eliminates the lag phase in chlorophyll synthesis and enhances respiration in dark-grown bean leaves, indicating that a stimulus is translocated from the axis to the leaves.

R irradiation also eliminates the lag in chlorophyll synthesis in excised bean leaves, but only if they are floated on sucrose. R promotion of leaf growth in excised pea epicotyls also requires exogenous sucrose (Bertsch and Hillman, 1961). Thus metabolic changes during de-etiolation are closely related to changes in energy production and distribution (DeGreef and Caubergs, 1973).

3 *Enzyme synthesis and activity*

There can be no doubt that phytochrome status controls the ultimate expression of gene action, as seen in the activities of various enzymes. Recall that mustard seedlings are pale yellow when grown in the dark and synthesize a considerable amount of anthocyanin after irradiation that converts P_r to P_{fr} (Lange *et al.*, 1971). Clearly phytochrome regulates the activity of the enzymes controlling anthocyanin synthesis, either through control of transcription, translation or activity of the already-formed enzyme. One such crucial enzyme is phenylalanine ammonia lyase, or PAL (EC 4.3.1.5), which converts phenylalanine into cinnamic acid and ammonia. The cinnamate so formed furnishes the C_6–C_3 portion of flavonoid molecules including the B ring and the bridge connecting A and B rings.

In some etiolated plants, such as pea seedlings, the PAL activity of terminal buds is greatly increased following R irradiation, and the effects of R are annulled by FR administered soon thereafter (Attridge and Smith, 1967). Following brief R, there is a lag period (30–60 min), followed by a rise to maximal enzyme activity at 6–7 h and a sharp decline, although not down to the pre-irradiation level. In mustard seedlings, continuous FR irradiation which maintains a P_{fr} level of a few percent produces similar results (Durst and Mohr, 1966), albeit with different kinetics.

The appearance of increased PAL activity in these various systems is partially prevented by applying the usual inhibitors of RNA and protein synthesis soon after irradiation (Engelsma, 1967; Zucker, 1969; Amrhein and Zenk, 1971). This and other evidence led Mohr (1966a, 1970) to propose gene activation as a basis for the action of P_{fr}. But the use of inhibitors is not enough to establish the validity of the theory, both because of possible side effects of the inhibitors themselves and the possibility that one rapidly metabolized protein can directly affect the activation of a second-pre-existing protein. Proof of control of gene activation would require the direct demonstration of increased rates of enzyme synthesis through labelling with ^{14}C amino acids or D_2O, or possibly through the use of specific immunological techniques. In both *Xanthium* leaf discs (Zucker, 1970) and potato tubers (Sacher *et al.*, 1972) PAL synthesis as deduced from labelling data occurs at elevated rates following irradiation, but also occurs in completely etiolated tissue. Thus total gene activation seems untenable as a hypothesis in these instances and the most one could advocate is control of *rate* of synthesis. Also it is not clear whether P_{fr} acts directly on genetic systems or whether its action is indirect, i.e. changing permeability to K^+, which could result in gene activation (Kroeger and Lezzi, 1966).

There is evidence that light may indirectly activate PAL in seedlings of gherkin (Attridge and Smith, 1973). Treatment with either cycloheximide or low temperature increases PAL activity, suggesting that a temperature-sensitive process, presumably synthesis, leads to the appearance of a protein which masks PAL activity. This repression by light of the synthesis of the masking protein could cause "activation" of PAL.

Phytochrome also controls the activity of other enzymes, including lipoxygenase (Oelze-Karow *et al.*, 1970; Oelze-Karow and Mohr, 1970), nitrate reductase (Jones and Sheard, 1972), amylase (Drumm *et al.*, 1971) and glutathione reductase (Drumm and Mohr, 1973). In no instance has the mechanism of activation been analysed in such terms as to permit a clear answer concerning gene activation *v.* direct activation of a pre-existing enzyme precursor.

4 *Hormonal metabolism*

There are many examples in the literature of the control by phytochrome of hormonal levels in plants. The most dramatic examples deal with gibberellin and ethylene. Irradiation of etiolated cereal seedlings (Reid *et al.*, 1968; Beevers *et al.*, 1970) or their homogenates (Reid *et al.*, 1972) leads to a dramatic rise in their assayable gibberellin titre within *c.* 20 min. Although such rapid effects would suggest release rather than synthesis, inhibitors of RNA and protein synthesis (Reid *et al.*, 1968) and of gibberellin synthesis (Loveys and Wareing, 1971) partly prevented the effect of light.

Ethylene production by various etiolated seedlings is inhibited by red light (Goeschl *et al.*, 1967; Kang *et al.*, 1967; Imaseki *et al.*, 1971), and the resulting decrease in ethylene can account for some if not all of the growth responses triggered by the light. In certain instances, such as hook opening, geotropic stem curvature, and carotenoid synthesis in pea seedlings, the R-mediated response can be prevented by the application of ethylene (Kang *et al.*, 1967; Kang and Burg, 1972a,b,c). With other de-etiolation responses, such as the inhibition of epicotyl or hypocotyl elongation, the response is prevented by application of gibberellin (Lockhart, 1956). It is instructive that both gibberellin (Wood and Paleg, 1972) and ethylene (Abeles, 1973) have been implicated in membrane-associated processes. Conceivably, if phytochrome acts by allosteric alteration of a membrane site dealing with transport, either ethylene or gibberellin could mimic or reverse this effect.

C PHOTOPERIODISM AND FLOWERING

1 *Translocation of the floral stimulus*

Photoperiod-sensitive plants remain vegetative indefinitely if grown on unfavourable photoperiods, but flowering in some short-day plants such as *Pharbitis* (Imamura, 1967) and *Xanthium* (Hamner and Bonner, 1938) can be initiated by a single inductive cycle. Some types continue to flower indefinitely even after minimal induction, while others revert to the vegetative habit when grown on non-inductive photoperiods.

In some plants, such as *Xanthium*, the whole plant need not be exposed to an inductive cycle to initiate flowering (Hamner and

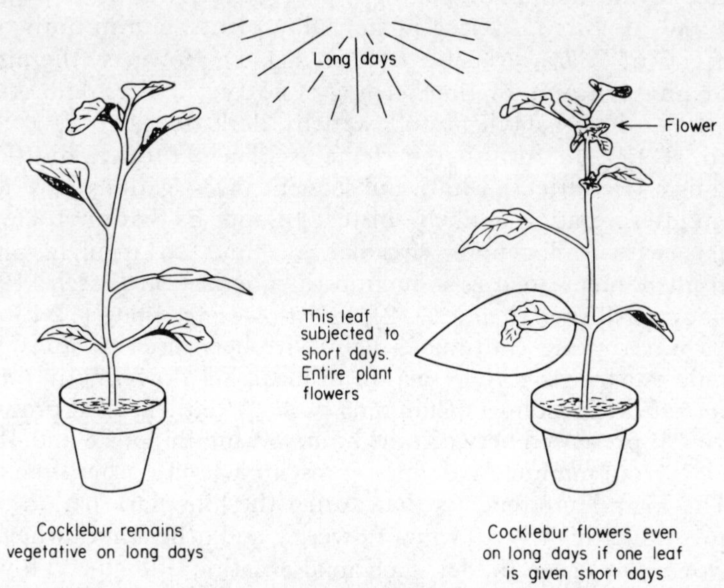

FIG. 9. Induction of flowering in cocklebur when only one leaf is exposed to short days. (After Hamner and Bonner, 1938.)

Bonner, 1938). One induced leaf suffices, and the floral stimulus (florigen) then travels from the induced leaf and causes initiation of reproductive buds in other parts of the plant (Fig. 9). A leaf from an induced plant grafted onto a non-induced one will even cause the latter to flower (Imamura, 1967), but girdling the stem to kill the phloem blocks transport of florigen (Withrow and Withrow, 1943). It seems clear that some influence originating in the induced leaf travels through the phloem to the bud, converting it from vegetative to floral growth. Yet all attempts to obtain the stimulus in cell-free extracts have failed. Both gibberellins and auxins have been extracted from induced leaves, and both can alter flowering in certain plants (Salisbury, 1955; Lona and Fioretti, 1962); thus it is possible that the postulated florigen is a combination of these two hormones and perhaps other substances as well.

2 *Time measurement*

Long-day plants initiate floral buds when the dark period is shorter than a critical length, and short-day plants initiate when the dark period exceeds a critical length. It is clear that phytochrome mediates the response in both long- and short-day plants, since a short R break in the middle of the night promotes flowering in long-day plants and inhibits flowering in short-day plants, and in both types the effects of R are reversible by FR (Fig. 1). However, the precise role of phytochrome in timekeeping is not yet understood. A high level of P_{fr} during daylight followed by dark decay to P_r at night seemed at first to furnish the elements of a photoperiodic timing mechanism (Hendricks, 1960). Subsequent investigations have made such an interpretation unlikely. Instead, it appears that phytochrome interacts with endogenous circadian rhythms to regulate photoperiod phenomena such as induction of tuberization (Esashi, 1962), dormancy (Williams *et al.*, 1972), and flowering. Long (72 h) dark periods interrupting continuous light have been used to study joint rhythmic and phytochrome action in initiating flowering in *Chenopodium rubrum* seedlings (Cumming *et al.*, 1965). Brief R promotes flowering if presented between the hours 10 and 26, or 38 and 48, or 66 and 72, and inhibits flowering if presented at any other time (Fig. 10). The logical inference is that during the long dark period there are times when high P_{fr} favours flowering and other times when low P_{fr} favours flowering. In fact, such an alternation of high and low P_{fr} states may be mandatory for the inception of flowering. This is true for both long- and short-day plants, although sometimes the low and

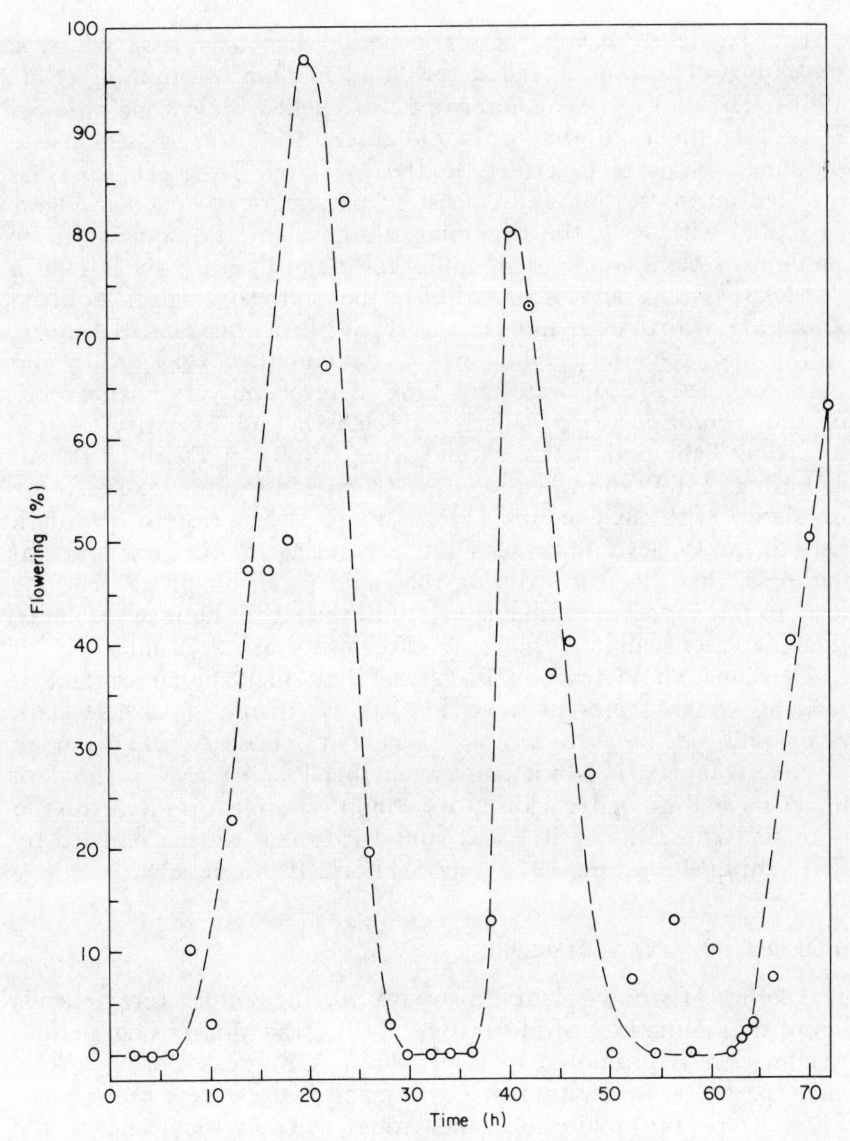

FIG. 10. Flowering of *Chenopodium rubrum* seedlings exposed to R for 2 min at one of the indicated times during a 72 h dark period. (After Cumming *et al.*, 1965.)

high P_{fr} reactions in long-day plants occur simultaneously rather than sequentially. The difference between long- and short-day plants may thus simply depend on whether low P_{fr} or high P_{fr} reactions are more limiting (Evans and King, 1969).

The P_{fr} level during the dark period has been assayed by a physiological technique called the "null method" (Cumming *et al.*, 1965). The dark period is interrupted by light with varying ratios of R to FR; the ratio that does not alter the flowering response is presumed to have no effect on the P_{fr} level. This technique has revealed an interesting and consistent pattern of reversion in plants irradiated with R at the beginning of the night: P_{fr} appears to be stable for several hours, then rapidly reverts to P_r at the sixth hour in *Pharbitis* (Evans and King, 1969), the sixth or eighth hour in *Xanthium* (Borthwick and Downs, 1964; Papenfuss and Salisbury, 1967) and the third hour in *Chenopodium rubrum* (King and Cumming, 1972). The rate and time of reversion vary with experimental conditions, particularly the length and intensity of the preceding light period (Evans and King, 1969). If *Pharbitis* (Evans and King, 1969) and *Chenopodium* (King and Cumming, 1972) are irradiated with BCJ lamps (FR:R energy = 5) prior to the dark period, the P_{fr} level increases gradually during the first few hours of darkness, then decreases during the rapid reversion phase. The P_{fr} level in *Chenopodium* remains low until the eighth hour of darkness, then increases again, i.e. the P_{fr} level seems to vary rhythmically.

The apparent increase in P_{fr} in the dark might be an artefact; if real, it could represent reversion of P_r to P_{fr} ("inverse dark reversion"), *de novo* synthesis of P_{fr}, or release of P_{fr} from an inactive complex. It is not known whether P_{fr} increases in the dark under natural as well as laboratory conditions, but it is interesting to note that the ratio of R:FR at sunset is similar to that emitted by BCJ lamps (Shropshire, 1973) (see Section XIC for details).

D DORMANCY AND SENESCENCE

The ability of woody plants to survive freezing temperature depends upon the induction of dormancy during the prefreezing period. Acclimation is promoted by short days. A R break during a long night prevents acclimation in *Cornus* and *Weigela*, and subsequent FR relieves the inhibition, indicating that phytochrome is the photoreceptor (Williams *et al.*, 1972). The photoreceptors are in the leaves and the stimulus then moves to other parts of the plant (Irving and Lanphear, 1967a,b; Steponkus and Lanphear, 1967; Fuchigami *et al.*, 1971), indicating the importance of translocation in this as in other phytochrome-mediated developmental responses.

Phytochrome may also serve as the photoreceptor for induction of senescence. In the liverwort *Marchantia*, FR at the beginning of the night decreases the chlorophyll content and promotes a rapid decay of the ultrastructural organization (DeGreef *et al.*, 1971). These senescence events are reversible by R, and resemble changes that occur in leaves of angiosperms under natural conditions at the end of a growing season (Dodge, 1970).

VII Linkage to respiratory systems

Since many of the reactions mediated by phytochrome utilize respiratory energy, it is relevant to inquire whether phytochrome status affects energy transduction. A report of P_{fr}-promoted increase in NAD kinase activity (Yamamoto and Tezuka, 1972) was reinterpreted by Hopkins and Briggs (1973), who reported that the postulated R-promoted increase in activity vanished with increased purification of the enzyme. Investigating a possibly related phenomenon, Manabe and Furuya (1973) measured NADP reduction in a low speed (1000–7000 *g*) pellet from etiolated bean tissue irradiated *in vitro*. The rate of NADPH formation during the first 2 min after R was twice as rapid as after FR, although differential effects of the light treatments had disappeared 20 min later.

Yunghans and Jaffe (1972) reported that R irradiation of mung bean root tips caused an immediate twelvefold decrease in ATP, a 15-fold increase in P_i and a 2·5-fold increase in O_2 uptake; subsequent FR partially reversed these effects. White and Pike (1974) reported a decline in ATP levels in bean buds 15 s after R, followed by a rapid increase that peaks at 1 min, then returns to the dark level. The promotive effect of R at 1 min was reversed by FR.

These studies do not reveal a consistent pattern. One problem is that the phytochrome state affects both ATP synthesis and reactions that utilize high energy compounds. Another problem is discussed by White and Pike; the light sources used in these investigations were not sufficiently intense for rapid photoconversion of P_r to P_{fr}, essential for accurate characterization of a rapid response.

VIII Interaction of phytochrome and endogenous rhythms

The activity of enzymes involved in energy transduction in *Chenopodium rubrum* seedlings is endogenously rhythmic during a long dark period at constant temperature (Frosch *et al.*, 1973). If seedlings are grown in the light on alternating temperature cycles prior to the dark period, the temperature cycles set the phase of the rhythm. Phytochrome phototransformations modulate the rhythms in NADP-dependent glyceraldehyde-3-phosphate dehydrogenase and adenylate kinase activity if the seedlings are transferred to darkness when the former is increasing and the latter decreasing, but R and FR light treatments have no effect if seedlings are transferred to darkness during the opposite phases of the rhythms (Frosch and Wagner, 1973a,b). Since endogenous circadian rhythms underlie metabolic activities in most if not all eukaryotic plants, it is not surprising that phytochrome mediation of energy transduction is subject to rhythmic control.

Endogenous rhythms interact with phytochrome to regulate both rapid responses such as ion fluxes and changes in respiration rate, and slower developmental responses such as induction of flowering and dormancy (Cumming and Wagner, 1968). The developmental responses are usually not detectable for days or even weeks after the light treatments, and thus reflect past rather than current light treatments. The rapid responses have been studied in the hope they might reveal the molecular basis for phytochrome–rhythmic interaction.

1 CO_2 output in Lemna

Lemna perpusilla is an obligate short-day plant. Flowering and CO_2 output are both regulated by phytochrome–rhythmic interaction, and the same rhythmic oscillator appears to be involved in both processes (Hillman 1964, 1970, 1971a,b, 1972, 1974). CO_2 output changes rapidly following R and FR light signals, and Hillman regards CO_2 output as a "real time" indicator of processes underlying photoperiodism. *Lemna* can complete its life cycle heterotrophically on a defined medium, permitting investigation of the effects of nutrition on rhythmic and phytochrome-controlled phenomena.

When plants are grown in darkness punctuated by brief R and FR light breaks, the CO_2 output pattern is regulated by the light treatments. However, the plant cannot discriminate between R and FR if the medium lacks nitrogen. This is reminiscent of the requirement for sucrose for manifestation of differential R and FR effects on de-etiolation in peas (Bertsch and Hillman, 1961) and beans (DeGreef and Caubergs, 1972) (discussed in Section VIB), and suggests that P_{fr} is associated with synthetic activities in all three cases.

When darkness is interrupted by a 15 min light break every eight hours, CO_2 output oscillates with a circadian rhythm if two of the breaks are R and the other FR or vice versa, but CO_2 output is essentially constant if all the light treatments are alike. Hillman concludes that a daily change in the P_{fr} level is required for circadian oscillations in *Lemna*. Since phytochrome binds to membranes (Marmé *et al.*, 1973; Quail *et al.*, 1973a; Boisard *et al.*, 1974) and phototransformations change membrane permeability or transport (Satter *et al.*, 1970; see below) or surface charge (Tanada, 1967), we may infer that changes in membrane properties are part of the biological clock mechanism.

2 K^+ *and* Cl^- *flux in nyctinastic plants*

Studies of leaf movement, Cl^- and K^+ flux in *Albizzia* and *Samanea* also imply that phytochrome and endogenous rhythms interact at a membrane locale (Satter and Galston, 1971a,b, 1973; Galston and Satter, 1972; Satter *et al.*, 1973, 1974a,b,c, 1976; Applewhite *et al.*, 1973; Racusen and Satter, 1975). Leaflets open and close with an endogenous circadian rhythm during prolonged darkness and such movement is a consequence of ion redistribution in pulvinal motor tissue (Fig. 11). Rhythmic ion flux in key motor cells involves alternating predominance of ion pumps and leakage through diffusion channels, each lasting *c.* 12 h. Thus the rhythm consists of oscillations in active transport, membrane permeability, or both.

Brief R and FR light treatments alter leaflet angle and K^+ distribution when leaflets are open and active transport predominates, but have little or no effect on closed leaflets. Time-dependence also characterizes phytochrome action in photoperiod phenomena.

IX High irradiance responses (HIR)

Some responses potentiated by brief R and prevented by subsequent brief FR, such as increase in lipoxygenase activity in *Sinapis* (Oelze-Karow *et al.*, 1970), are promoted most effectively by prolonged FR irradiation, action peak *c.* 720 nm. Most investigators agree that phytochrome is a photoreceptor for such duration-dependent responses. Although 720 nm light is only 15% as effective as 660 nm light in converting phytochrome to P_{fr} (Hartmann and Unser, 1973), it maintains P_{fr} for a longer period since the rate of destruction is lower (Hartman, 1966; Schäfer *et al.*, 1972) (see Section IVC). However, some of the duration-dependent reactions such as increase in activity of phenylalanine ammonia lyase (Dittes *et al.*, 1971) and ascorbate oxidase (Drumm *et al.*, 1972) are intensity dependent even after photostationary P_{fr}/P_r ratio has been established, and whether phytochrome is the *sole* photoreceptor for these HIR is still controversial.

Schopfer and Oelze-Karow (1971), Schopfer and Mohr (1972) and Mohr (1972) claim that "excited P_{fr}" is required for the intensity-dependent reactions while other reactions are promoted by P_{fr} in the ground state, but they have presented no biophysical evidence for "excited P_{fr}". This theory, initially proposed by Hartmann (1966), was subsequently rejected by Hartmann and Unser (1973), who attribute the high irradiance effects to the increase in the rate of $P_{fr} \rightleftharpoons P_r$ interconversions, also called pigment cycling. They claim

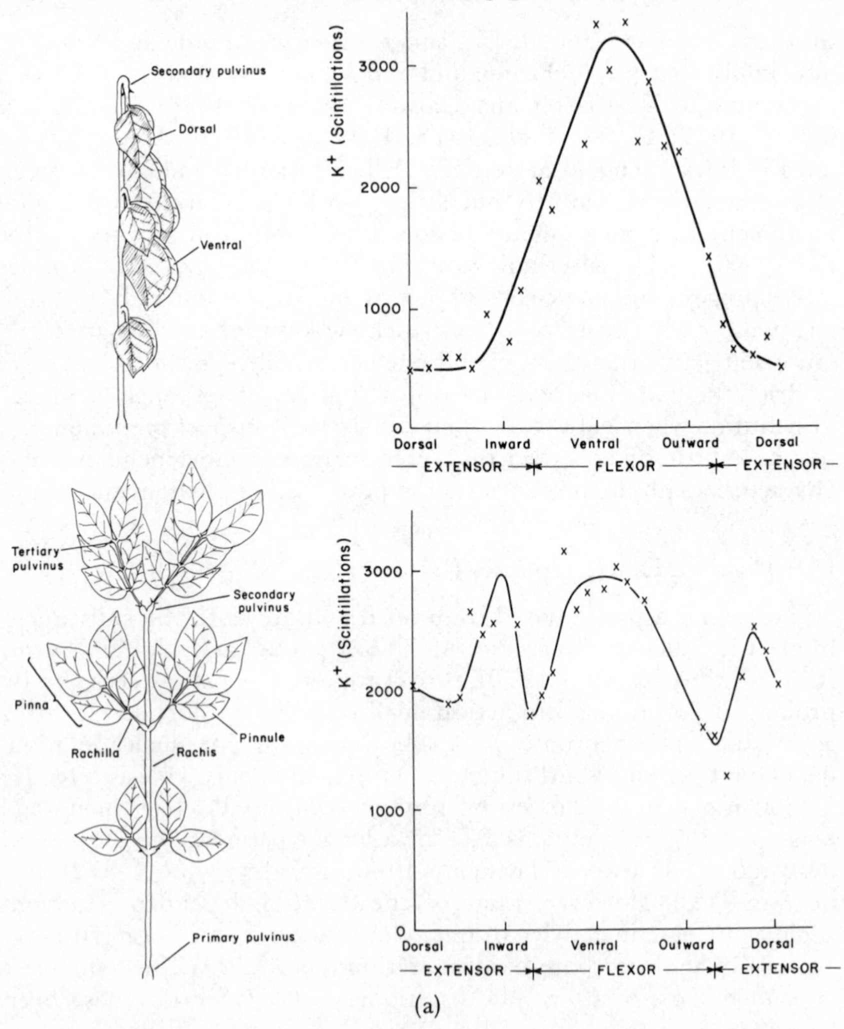

(a)

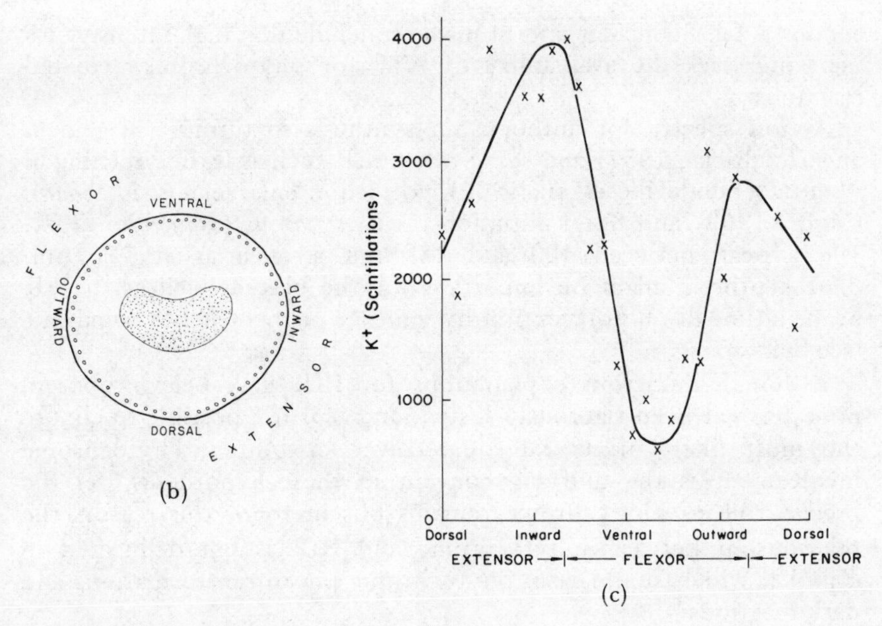

FIG. 11. (a) *Samanea saman* pinnae in the closed (above) and open (below) positions. (b) Transverse section of the secondary pulvinus. Extensor cells expand during opening and contract during closure, and flexor cells change in the reverse direction. ▒ Vascular core. ○○○ Motor cell region. (c) Rhythmic and P_{fr}-controlled changes in K^+ in motor cells of transverse sections, analysed with the electron microprobe. Abscissa: distance from a middorsal reference point. Upper: Closed (0°) pulvinus, h 5 of dark preceded by white (high P_{fr}). Middle: Partially open (75°) pulvinus, h 13 of dark preceded by white (high P_{fr}). Lower: Open (120°) pulvinus, h 13 of dark preceded by FR (low P_{fr}). After Satter *et al.*, 1974b,c.)

that "newly formed P_{fr}" is more effective than "old P_{fr}". This concept is incorporated into Schäfer's (1975) model.

Schneider and Stimson (1971, 1972) present a different interpretation for HIR, based on their study of anthocyanin synthesis in turnip seedlings. They propose that two photoreceptors are involved: phytochrome and chlorophyll *a*, operating via cyclic photophosphorylation. They report that the chlorophyll content of turnip seedlings exposed to FR for 72 h is fifty times as high as that of dark controls. Anthocyanin synthesis is inhibited by antimycin A and 2,4-dinitrophenol, inhibitors of cyclic phosphorylation, by NH_4^+, an uncoupler of photophosphorylation, and by laevulinic acid, an inhibitor of chlorophyll synthesis. In addition, it is promoted by compounds that inhibit non-cyclic phosphorylation and by exogenous ATP. Schneider and Stimson conclude that high intensity FR light increases the availability of ATP for phytochrome-mediated reactions.

Action spectra for anthocyanin synthesis in turnips (Siegelman and Hendricks, 1957) and some other HIR such as leaflet opening in *Mimosa* (Fondéville *et al.*, 1967), cotyledon enlargement in *Sinapis* (Mohr, 1959) and floral initiation in *Hyoscyamus* (Schneider *et al.*, 1967) peak between 420 and 480 nm as well as at 720 nm. Photosynthesis might be important for the blue as well as the FR peak, although photoreception by another pigment is not ruled out (see below).

Although numerous explanations for HIR have been proposed, none has yet been rigorously tested, and it is not possible to choose the most likely of several competitive possibilities. Phytochrome involvement is the unifying concept in these hypotheses, yet the precise role of phytochrome remains obscure. For this reason, the relationship between phytochrome and HIR is not delineated in Fig. 12, which summarizes phytochrome phototransformations and dark reactions.

X Interaction of R, FR and blue light

Blue light interacts with brief R and FR irradiations to regulate many growth responses. Both P_r and P_{fr} have absorption peaks in the blue region (Fig. 4 in Chap. 7) and some effects of blue light, such as alteration of CO_2 metabolism and floral induction in *Lemna* (Hillman, 1967b, 1971b) can be explained by the *c.* 35% P_{fr} level (Butler *et al.*, 1964) induced by the light source. However, many of the interactions are exceedingly complex. For example, spore germination in *Cheilanthes* (Raghavan, 1973) is promoted by R, but blue light is required before or after the R irradiation to induce FR photoreversibility of the R effect. Raghavan proposed that a blue-absorbing pigment interferes with $P_r \leftrightarrow P_{fr}$ photoconversions. In the dinoflagellate *Gyrodinium* (Forward and Davenport, 1968; Forward 1970, 1973) the phytochrome system mediates the phototactic response to blue light. R and FR light treatments determine the number of algae that stop in response to prior blue light and also the wavelength peak of the blue action spectrum. Forward proposed that the blue-absorbing pigment is a carotenoprotein. In *Mougeotia* (Haupt, 1971), the effect of polarized R light on chloroplast orientation depends upon the plane of polarization, but subsequent

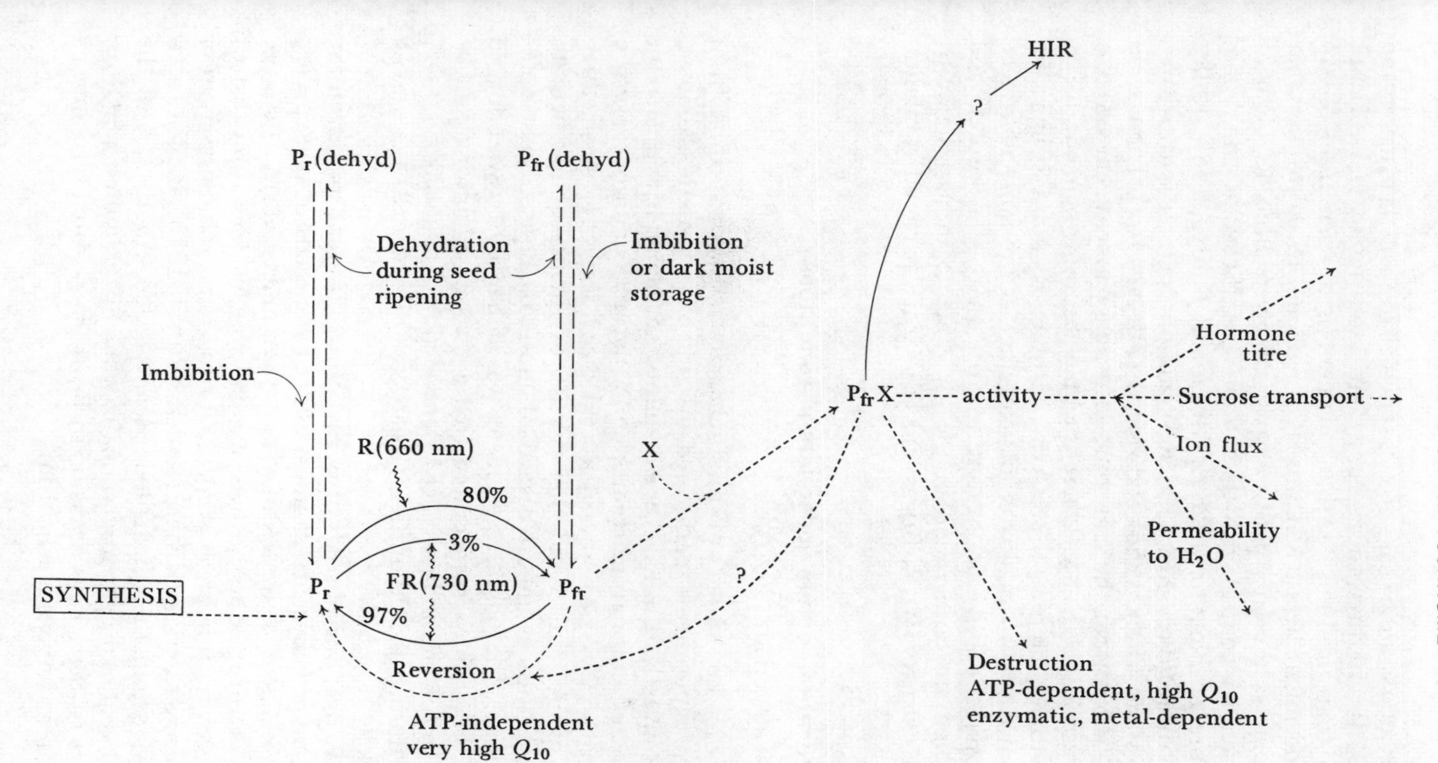

FIG. 12. A model for the photo and dark transformations that lead to P_{fr}-promoted and high irradiance responses (HIR). Part of the model was developed by Hartmann (1966). Symbols: — phototransformations; - - - - dark transformations; $P_{r\ (dehyd)}$ and $P_{fr\ (dehyd)}$ are not interconvertible. X = membrane receptor.

blue light alters this sensitivity. Haupt (1971) and Hartmann and Unser (1973) interpret these same data in different ways. Haupt concludes that the blue-absorbing pigment is distinct from phytochrome and might be a flavin, but Hartmann and Unser propose that phytochrome is the only photoreceptor involved. Blue, R, FR light also interact to control growth in *Lunularia* (Schwabe and Valio, 1970), spore germination in *Pteris* (Sugai and Furuya, 1967, 1968), stem growth in *Pisum* (Bertsch, 1963), and phototropic sensitivity in *Zea* (Briggs and Chon, 1966) and *Avena* (Briggs, 1963). There is no reason to believe that the same blue-absorbing pigment is involved in all these responses. In cases where the blue-absorbing pigment differs from phytochrome but the effects of blue, R and FR light are interrelated, the two pigments may interact at the molecular level. In this connection it is interesting to note that a flavoprotein and phytochrome are both found in the same membranous fraction of *Cucurbita* extracts (W. R. Briggs and D. Marmé, personal communication).

XI Phytochrome action under natural conditions

A SEED GERMINATION

Most of our knowledge of phytochrome action has come from laboratory experiments performed in controlled growth chambers, but some investigators have drawn important generalizations about phytochrome action from studies under natural conditions. Wesson and Wareing (1969a,b) studied buried weed seeds of twenty-three different species and found that twenty of them required light to germinate, although unburied seed of most of these species is not light-sensitive. The light requirement was enhanced by waterlogged soil, and the authors conjectured that a gaseous inhibitor might be responsible for the induced light dependency. Enhanced P_{fr} to P_r reversion during high hydration (Hsiao and Vidaver, 1973) seems an equally likely possibility.

In 1879, Beal buried seeds of twenty-three different species on the Michigan State University campus; 20% of the seeds of one species, *Verbascum blattaria*, were still viable after ninety years, and were grown to maturity. Their growth was indistinguishable from that of controls grown from fresh seed, indicating remarkable preservation of respiratory reserves (Kivilaan and Bandurski, 1973). Progeny seed were planted and their light dependency tested. About 30% of the fresh seed required light to germinate, while 80% required light after dark moist storage. Thus reversion of P_{fr} to P_r in stored seed appears to be a potent survival mechanism.

Taylorson and Borthwick (1969) studied the effect of light filtration by a foliar canopy on seed germination in six weed species. A single layer of tobacco, corn or soyabean leaves effectively filters R wavelengths from sunlight but has little effect on FR intensity; it thus reduces the R:FR ratio from 1·1 to 0·18. Seed germination is inhibited significantly by such filtered radiation, particularly in species such as lamb's quarters, which require a high P_{fr} level for germination. This differential filtering depends upon absorption by chlorophyll, and has obvious adaptive value, since it hinders germination after a foliar canopy has developed and newly emerging seedlings would have difficulty competing with established plants. Seeds of many species do not germinate without prolonged low temperature treatment during stratification, and this depends upon a requirement for P_{fr} (see Section VIA). Phytochrome is thus an important and effective regulator of seasonal germination patterns in nature.

B GROWTH UNDER A FOREST CANOPY

Genetically identical plants become morphologically distinct when grown in different environments, such as field and forest. Kasperbauer (1971, 1973) and Kasperbauer *et al.* (1970) investigated the role of phytochrome in this phenomenon by comparing the growth of tobacco plants in the forest with that of plants in a controlled chamber that were exposed to FR at the end of each day. Plants on the outside edges of the canopy, and plants in controlled chambers exposed to end of day R irradiation served as controls. As contrasted with these controls, plants in the forest and those exposed to terminal FR developed long internodes, thin leaves and an apical growth habit. Differences in the P_{fr} level affected the rate of CO_2 fixation and concentrations of free sugar, organic acids and amino acids. Thus differences in growth patterns in a sunny field compared to a forest are due in large measure to differences in light quality, acting through the ratio of $P_{fr}:P_r$.

C SPECTRAL QUALITY OF SUNLIGHT

Recording equipment on the roof of the Smithsonian Institute in Washington, D.C. monitors the energy of sunlight in discrete portions of the spectrum. These monitors reveal that the ratio of R:FR energy is *c.* 3 during most of the day, but falls to 0·7 for *c.* 30 min at dusk and at dawn (Shropshire, 1973) (Fig. 13). The rate of stomatal opening in tobacco and sunflower is increased if prolonged low

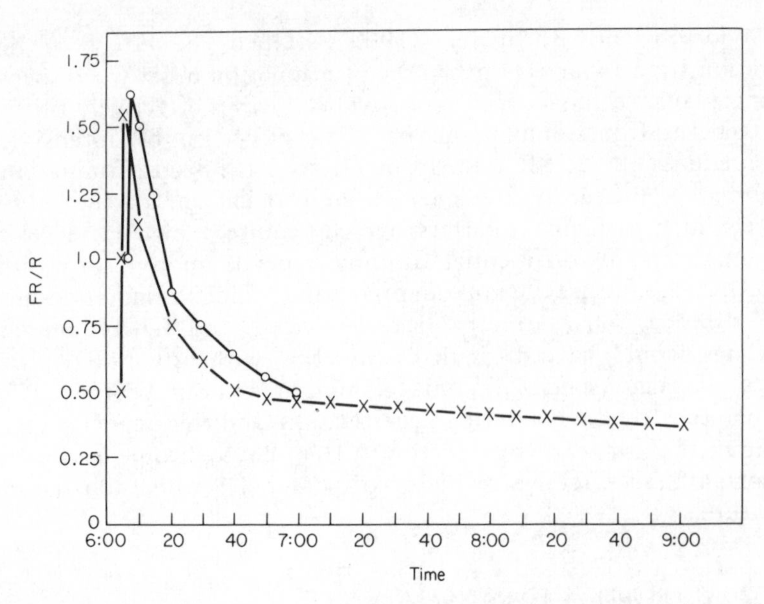

FIG. 13. Ratio of FR (730 nm ± 4 nm) to R (660 nm ± 4 nm) irradiance, measured at 1 min intervals after sunrise for two clear days in Washington, D.C. The measurements were made at the Smithsonian North Plaza Tower on 31 May and 1 June 1963. (From Shropshire, 1973.)

energy FR precedes R (Habermann, 1973), indicating that stomatal guard cells have an adaptive mechanism for utilizing the FR energy available when the sun is low on the horizon.

A R:FR ratio of 0·7 at the beginning of the night is lower than that used in many investigations of light effects on floral initiation. Most of these studies have been conducted in controlled chambers illuminated by fluorescent lights with a small incandescent supplement. The rate of pigment cycling is much lower under these conditions than in sunlight, where a large percentage of the phytochrome might be in intermediate form at steady state (Briggs and Fork, 1969; Kendrick and Spruit 1973b,c; Spruit and Kendrick, 1973). Correlated studies of light effects on floral initiation in plants grown under field and controlled conditions are clearly required to evaluate phytochrome action under natural conditions.

XII The mode of phytochrome action

Is it possible to attribute all the varied manifestations of phytochrome transformation to a single reaction or event in which P_{fr} participates? Such a unitary hypothesis of phytochrome action should be proposed if it is possible to do so, both for its heuristic value and for its logical parsimony. Let us recall that manifestations of phytochrome action range from virtually immediate (seconds or less) to leisurely (weeks or months); from the physical (changed surface charge) to the chemical (altered enzyme activity); from reactions not involving protein synthesis (leaf movements) to those which do (enhanced leaf expansion); from those involving no cell division (plumular hook opening) to those which do (seed germination); from the completely R/FR reversible (chloroplast movements) to the poorly reversible (photocontrol of lipoxygenase); and from purely localized effects (leaf movements) to transmitted effects (floral induction).

Recognizing that such a model might be a current fad, one must concede that allosteric alteration of phytochrome in membranes (Boisard *et al.*, 1974) could be expected to control virtually any membrane-mediated process, and in turn virtually all processes in eukaryotic cells. Certainly altered transport of ions through membranes accounts for phytochrome control of leaflet movement (Satter *et al.*, 1970) and together with control of permeability to water (Weisenseel and Smeibidl, 1973) could account for hook opening. It could also account for altered enzyme activity and gene activation. These in turn, together with altered patterns of sugar transport (Goren and Galston, 1966, 1967; Anand and Galston, 1972) could lead to changes in protein synthesis, cell division and morphogenesis.

Are changes in permeability and ion transport sufficient to co-ordinate all the activities within a cell that are controlled by phytochrome? Phytochrome mediated chloroplast movements in *Mougeotia* illustrate logical problems with respect to such a scheme. Photoreception in this organism occurs in an oriented, phytochrome-containing lamella, presumably the plasmalemma. Within several minutes of phytochrome photoconversion, the chloroplast rotates, and each region of the chloroplast reacts locally upon localized irradiation of the membrane near it. Presumably such chloroplast movements involve the contraction of cytoplasmic microfibrils attached to the chloroplasts (Wagner *et al.*, 1972), much as chromosome movements result from contraction of microtubule assemblies. Could such contractions result from alteration of membrane permeability? Even if one envisions a K^+-activated ATPase on microfibrils responding to altered levels of the cation resulting from phytochrome-controlled alteration of nearby membrane permeability, it is still difficult to account for the precise localizations inherent in the system. Would not K^+ be expected to diffuse around and affect a larger area than it appears to do?

Phytochrome-controlled changes in the electrical field around the plasmalemma could stimulate a contractile ATPase and produce movements in localized regions of the chloroplast. Such changes in surface charge at the plasmalemma are predicted by phytochrome mediation of the electrophoretic movement of mung bean root cap cells (Racusen, 1973; Racusen and Etherton, 1973) and *Samanea* motor cells (R. H. Racusen and R. L. Satter, unpublished results) and by the effect of the light treatments on the ultraviolet difference spectrum of a phytochrome extract from rye (Tobin and Briggs, 1973). Phytochrome-mediated changes in the electrical field in the region near membranes could provide intracellular communication, and synchronize the many activities within a cell that are regulated by R and FR light.

How does one explain the effects of phytochrome photoconversion that are propagated through the plant? It is useful to comment here about *florigen*, the hypothetical flowering hormone of plants (Chailakhian, 1958). Its occurrence was proposed almost 40 years ago on convincing physiological grounds, yet all attempts at its isolation have failed. We know only that some stimulus originates in leaves under photoperiods whose physiological effectiveness is determined by phytochrome status. This stimulus is propagated through living cells only from leaf to bud, probably via phloem (Withrow and Withrow, 1943). Once at the bud, it exerts a morphogenetic effect, converting vegetative growing points to the reproductive habit. In many plants the induced state persists and can be graft transmitted. Clearly some effective principle is present. Why then has it not been amenable to extraction and reintroduction into receptor plants?

Two possibilities present themselves. One is that the effective entity is complex and particulate, possibly a self-replicating nucleoprotein entity whose very size, complexity and lability militate against *in vitro* manipulations. Somehow the replicative process of such an entity is triggered by phytochrome transformation, much as lysogenic bacteriophage is activated by ultraviolet irradiation. The other possibility is the cascade-like release of minute quantities of some membrane active material, like acetylcholine (Jaffe, 1971) or a cyclic nucleotide (Amrhein and Filner, 1973). This material might never accumulate in sufficient quantity to be extracted and reintroduced, since, in effect, it would be consumed in the action which promotes the appearance of another molecule further down the membrane.

Whatever the merit of these hypotheses, they highlight one remaining unexplored mystery concerning phytochrome effects. Why are many consequences of phytochrome conversion localized, both within the cell (*Mougeotia*) or within a tissue (*Albizzia, Samanea*), while developmental responses such as de-etiolation and induction of flowering and dormancy depend upon a stimulus that is released by phytochrome and travels throughout the plant? An answer to this question might tell us much about the central problem of phytochrome action.

XIII Addendum

Recent investigations of phytochrome physiology were described at the Easter School Conference at the University of Nottingham in April 1975 (Smith, 1975, 1976).

The identity of the presumably membranous phytochrome-containing subcellular fractions of *Cucurbita* extracts was reinvestigated in detail by Quail (1975a,b; see also Smith, 1975, 1976). Quail's data suggest that phytochrome in *Cucurbita* binds non-specifically to ribosomal material. When the extraction buffer contains Mg^{2+}, phytochrome binds to intact ribosomes on the rough endoplasmic reticulum; however, if the buffer is Mg^{2+}-free, ribonucleoproteins separate from the endoplasmic reticulum, but bind phytochrome nevertheless. Quail proposes that the material described as "partially solubilized" membrane receptor by Marmé (1974), and Marmé *et al.* (1974), is actually degraded ribosomal material. In addition, he suggests that the bond between phytochrome and ribosomes is electrostatic in nature, and does not necessarily indicate *in vivo* binding. At the Easter School Conference, doubt was also expressed that NPA is a specific marker for the plasmalemma in tissue homogenates. This further challenges the identity of subcellular fractions previously identified as plasmalemma.

ACKNOWLEDGEMENTS

Much of the organization of the chapter and some of the figures are borrowed from the original version by Dr S. B. Hendricks, whom we thank, and the late Dr H. A. Borthwick. Many colleagues sent information in advance of publication; for this kindness we thank J. Boisard, W. R. Briggs, R. G. Hayes, W. S. Hillman, W. H. Klein, D. Marmé, L. H. Pratt, P. H. Quail, E. Schäfer, M. J. Schneider, W. Shropshire and M. H. Weisenseel. We are especially grateful to Dr Dieter Marmé, who read an early draft and offered valuable suggestions, and to the National Science Foundation, for continued support.

REFERENCES

Abeles, F. B. (1973). "Ethylene in Plant Biology". Academic Press, New York and London.

Allen, R. D. (1969). *Pl. Physiol., Lancaster* **44**, 1101-1107.

Amrhein, N. and Filner, P., (1973). *Proc. natn. Acad. Sci. U.S.A.* **70**, 1099-1103.

Amrhein, N. and Zenk, M. H. (1971). *Z. PflPhysiol.* **64**, 145-168.

Anand, R. and Galston, A. W. (1972). *Am. J. Bot.* **59**, 327-336.

Anderson, G. R., Jenner, E. L. and Mumford, F. E. (1969). *Biochemistry, N.Y.* **8**, 1182-1187.

Applewhite, P. B., Satter, R. L. and Galston, A. W. (1973). *J. gen. Physiol.* **62**, 707-713.

Attridge, T. H. and Smith, H. (1967). *Biochim biophys. Acta* **148**, 805-807.

Attridge, T. H. and Smith, H. (1973). *Phytochemistry* **12**, 1569-1574.

Bauer, L. and Mohr, H. (1959). *Planta* **54**, 68-73.

Beevers, L., Loveys, B. R., Pearson, J. A. and Wareing, P. F. (1970). *Planta* **90**, 286-294.

Bertsch, W. F. (1963). *Am. J. Bot.* **50**, 754-760.

Bertsch, W. F. and Hillman, W. S. (1961). *Am. J. Bot.* **48**, 504-511.

Bewley, J. D., Negbi, M. and Black, M. (1968). *Planta* **78**, 351-357.

Black, M. (1969). *Proc. Soc. exp. Biol. Med.* **23**, 193-217.

Black, M. and Wareing, P. F. (1960). *J. exp. Bot.* **11**, 28-39.

Boisard, J., Spruit, C. J. P. and Rollin, P. (1968). *Meded. LandbHoogesch. Wageningen* **68**,17, 1-5.

Boisard, J., Marmé, D. and Schäfer, E. (1971). *Planta* **99**, 302-310.

Boisard, J., Marmé, D. and Briggs, W. R. (1974). *Pl. Physiol., Lancaster* **54**, 272-276.

Borthwick, H. A. (1972). *In* "Phytochrome" (K. Mitrakos and W. Shropshire, Jr., eds), pp. 3-44. Academic Press, London and New York.

Borthwick, H. A. and Downs, R. J. (1964). *Bot. Gaz.* **125**, 227-231.

Borthwick, H. A. and Hendricks, S. B. (1961). *In* "Handbuch der Pflanzenphysiologie" (W. Ruhland, ed.), vol. 16, pp. 299-330. Springer-Verlag, Berlin.

Borthwick, H. A., Hendricks, S. B. and Parker, M. W. (1948). *Bot. Gaz.* **110**, 103-118.

Borthwick, H. A., Hendricks, S. B. and Parker, M. W. (1951). *Bot. Gaz.* **113**, 95-105.

Borthwick, H. A., Hendricks, S. B. and Parker, M. W. (1952a). *Proc. natn. Acad. Sci. U.S.A.* **38**, 929-934.

Borthwick, H. A., Hendricks, S. B., Parker, M. W., Toole, E. H. and Toole, V. K. (1952b). *Proc. natn. Acad. Sci. U.S.A.* **38**, 662-666.

Borthwick, H. A., Hendricks, S. B., Toole, E. H. and Toole, V. K. (1954). *Bot. Gaz.* **115**, 205-215.

Briggs, W. R. (1963). *Am. J. Bot.* **50**, 196-207.

Briggs, W. R. and Chon, H. P. (1966). *Pl. Physiol., Lancaster* **41**, 1159-1166.

Briggs, W. R. and Fork, D. C. (1969). *Pl. Physiol., Lancaster* **44**, 1089-1094.

Briggs, W. R. and Rice, H. V. (1972). *A. Rev. Pl. Physiol.* **23**, 293-334.

Briggs, W. R. and Siegelman, H. W. (1965). *Pl. Physiol., Lancaster* **40**, 934-941.

Butler, W. L. and Lane, H. C. (1965). *Pl. Physiol., Lancaster* **40**, 13-17.

Butler, W. L., Hendricks, S. B. and Siegelman, H. W. (1964). *Photochem. Photobiol.* **3**, 521-528.

Butler, W. L., Norris, K. H., Siegelman, H. W. and Hendricks, S. B. (1959). *Proc. natn. Acad. Sci. U.S.A.* 45, 1703-1708.

Butler, W. L., Lane, H. C. and Siegelman, H. W. (1963), *Pl. Physiol., Lancaster* **38**, 514-519.

Carceller, M. S. and Sanchez, R. A. (1972). *Experientia* 28, 364.

Chailakhian, M. Kh. (1958). *Biol. Zbl.* 77, 641-662.

Chorney, W. and Gordon, S. A. (1966). *Pl. Physiol., Lancaster* **41**, 891-896.

Clarkson, D. T. and Hillman, W. S. (1967). *Nature, Lond.* 213, 468-470.

Cohen, R. and Goodwin, T. W. (1962). *Phytochemistry* 1, 67-72.

Coleman, R. A. (1973). Ph.D. Thesis, Vanderbilt University, Nashville, Tennessee.

Correll, D. L. and Shropshire, W. Jr. (1968). *Planta* 79, 275-283.

Correll, D. L., Edwards, J. L. and Medina, V. J. (1968). *Planta* 79, 284-291.

Cumming, B. G. and Wagner, E. (1968). *A. Rev. Pl. Physiol.* 19, 381-416.

Cumming, B. G., Hendricks, S. B. and Borthwick, H. A. (1965). *Can. J. Bot.* 43, 825-853.

DeGreef, J. A. and Caubergs, R. (1972). *Physiologia Pl.* 26, 157-165.

DeGreef, J. A. and Caubergs, R. (1973). *Physiologia Pl.* 28, 71-76.

DeGreef, J. A., Butler, W. L., Roth, T. F. and Fredericq, H. (1971). *Pl. Physiol., Lancaster* **48**, 407-412.

DeLint, P. J. A. L. and Spruit, C. J. P. (1963). *Meded. LandbHoogesch. Wageningen* 63, 1-7.

Dittes, L., Rissland, I. and Mohr, H. (1971). *Z. Naturf.* **26b**, 1175-1180.

Dodge, J. D. (1970). *Ann. Bot.* 34, 817-824.

Dooskin, R. H. and Mancinelli, A. L. (1968). *Bull. Torrey bot. Club* 95, 474-487.

Downs, R. J. (1964). *Phyton, B. Aires* 21, 1-6.

Downs, R. J., Hendricks, S. B. and Borthwick, H. A. (1957). *Bot. Gaz.* 118, 199-208.

Drumm, H. and Mohr, H. (1973). *Z. Naturf.* **28c**, 559-563.

Drumm, H., Elchinger, J., Moller, J., Peter, K. and Mohr, H. (1971). *Planta* 99, 265-274.

Drumm, H., Brüning, K. and Mohr, H. (1972). *Planta* 106, 259-267.

Durst, F. and Mohr, H. (1966a). *Naturwissenschaften* 53, 531-532.

Engelsma, G. (1967). *Planta* 75, 207-219.

Esashi, Y. (1962). *Plant Cell Physiol.* 3, 67-82.

Etzold, H. (1965). *Planta* 64, 254-280.

Evans, L. T. and King, R. W. (1969). *Z. PflPhysiol.* 60, 277-288.

Evenari, M. and Neumann, G. (1953). *Palest. J. Bot. Jerusalem Ser.* 6, 96.

Fondéville, J. C., Borthwick, H. A. and Hendricks, S. B. (1966). *Planta* 69, 357-364.

Fondéville, J. C., Schneider, M. J., Borthwick, H. A. and Hendricks, S. B. (1967). *Planta* 75, 228-238.

Forward, R. B. Jr. (1970). *Planta* 92, 248-258.

Forward, R. B. Jr. (1973). *Planta* 111, 167-178.

Forward, R. B. Jr. and Davenport, D. (1968). *Science, N.Y.* 161, 1028-1029.

Fredericq, H. (1964). *Pl. Physiol., Lancaster* 39, 812-816.

Fredericq, H. (1965). *Biol. Jaarb.* 33, 66.

Fredericq, H. and DeGreef, J. A. (1972). *In* "Phytochrome" (K. Mitrakos and W. Shropshire, Jr., eds), pp. 319-346. Academic Press, London and New York.

Frosch, S. and Wagner, E. (1973a). *Can. J. Bot.* 51, 1521-1528.
Frosch, S. and Wagner, E. (1973b). *Can. J. Bot.* 51, 1529-1535.
Frosch, S., Wagner, E. and Cumming, B. G. (1973). *Can. J. Bot.* 51, 1355-1367.
Fuchigami, L. H., Evert, D. R. and Weiser, C. J. (1971). *Pl. Physiol., Lancaster* 47, 164-167.
Fujita, Y. and Hattori, A. (1962). *Plant Cell Physiol.* 3, 209-220.
Furuya, M. and Hillman, W. S. (1964). *Planta* 63, 31-42.
Furuya, M., Hopkins, W. G. and Hillman, W. S. (1965). *Archs Biochem. Biophys.* 112, 180-186.
Galston, A. W. (1968). *Proc. natn. Acad. Sci. U.S.A.* 61, 454-460.
Galston, A. W. (1974). *Pl. Physiol., Lancaster.* 54, 427-436.
Galston, A. W. and Satter, R. L. (1972). *In* "Structural and Functional Aspects of Phytochemistry" (V. C. Runeckles and T. C. Tso, eds), pp. 51-79. Academic Press, London and New York.
Giles, K. L. and Maltzahn, K. E. von (1968). *Can. J. Bot.* 46, 305-306.
Goeschl, J. D., Pratt, H. K. and Bonner, B. A. (1967). *Pl. Physiol., Lancaster* 42, 1077-1080.
Goren, R. and Galston, A. W. (1966). *Pl. Physiol., Lancaster* 41, 1055-1064.
Goren, R. and Galston, A. W. (1967). *Pl. Physiol., Lancaster* 42, 1087-1090.
Gressel, J., Galun, E. and Strausbauch, L. (1971). *Nature, Lond.* 232, 648-649.
Grill, R. (1972). *Planta* 108, 185-202.
Grill, R. and Spruit, C. J. P. (1972). *Planta* 108, 203-213.
Habermann, H. M. (1973). *Pl. Physiol., Lancaster* 51, 543-548.
Hamner, K. C. and Bonner J. (1938). *Bot. Gaz.* 100, 388-431.
Hartmann, E. (1971). *Planta* 101, 159-165.
Hartmann, K. M. (1966). *Photochem. Photobiol.* 5, 349-366.
Hartmann, K. M. and Unser, I. C. (1973). *Z. PflPhysiol.* 69, 109-124.
Haupt, W. (1959). *Planta* 53, 484-501.
Haupt, W. (1965). *A. Rev. Pl. Physiol.* 16, 267-290.
Haupt, W. (1970). *Physiol. Vég.* 8, 551-563.
Haupt, W. (1971). *Z. PflPhysiol.* 65, 248-265.
Haupt, W., Mortel, G. and Winkelnkemper, I. (1969). *Planta* 88, 183-186.
Hayes, R. G. and Klein, W. H. (1974). *Plant Cell Physiol.* 15, 643-653.
Hendricks, S. B. (1960). *Cold Spring Harb. Symp. quant. Biol.* 25, 245-248.
Hendricks, S. B. and Borthwick, H. A. (1967). *Proc. natn. Acad. Sci. U.S.A.* 58, 2125-2130.
Hendricks, S. B., Borthwick, H. A. and Downs, R. J. (1956). *Proc. natn. Acad. Sci. U.S.A.* 42, 19-26.
Hillman, W. S. (1962). *Am. J. Bot.* 49, 892-897.
Hillman, W. S. (1964). *Am. J. Bot.* 51, 1102-1107.
Hillman, W. S. (1965a). *Physiologia Pl.* 18, 346-358.
Hillman, W. S. (1965b). *Plant Cell Physiol.* 6, 499-506.
Hillman, W. S. (1967a). *A. Rev. Pl. Physiol.* 18, 301-324.
Hillman, W. S. (1967b). *Plant Cell Physiol.* 8, 467-473.
Hillman, W. S. (1970). *Pl. Physiol., Lancaster* 45, 273-279.
Hillman, W. S. (1971a). *Pl. Physiol., Lancaster* 47, 431-434.
Hillman, W. S. (1971b). *Pl. Physiol., Lancaster* 48, 770-774.
Hillman, W. S. (1972). *Pl. Physiol., Lancaster* 49, 907-911.
Hillman, W. S. (1975). *Photochem. Photobiol.* 21, 39-48.
Hillman, W. S. and Koukkari, W. L. (1967). *Pl. Physiol., Lancaster* 42, 1413-1418.

Hopkins, D. W. and Briggs, W. R. (1973). *Pl. Physiol., Lancaster* **51** (suppl.), 52.
Hopkins, D. W. and Butler, W. L. (1970). *Pl. Physiol., Lancaster* **46**, 567-570.
Hopkins, W. G. and Hillman, W. S. (1965). *Am. J. Bot.* **52**, 427-432.
Hsiao, A. I. and Vidaver, W. (1973). *Pl. Physiol., Lancaster* **51**, 459-463.
Imamura, S. (1967). *In* "Physiology of Flowering of *Pharbitis nil*" (S. Imamura, ed.), pp. 15-28. Jap. Soc. pl. Physiol., Tokyo.
Imaseki, H., Pjon, C. J. and Furuya, M. (1971). *Pl. Physiol., Lancaster* **48**, 241-244.
Irving, R. M. and Lanphear, F. O. (1967a). *Pl. Physiol., Lancaster* **42**, 1191-1196.
Irving, R. M. and Lanphear, F. O. (1967b). *Pl. Physiol., Lancaster* **42**, 1384-1388.
Isikawa, S. (1954). *Bot. Mag.* **67**, 51.
Jaffe, M. J. (1968). *Science, N.Y.* **162**, 1016-1017.
Jaffe, M. J. (1970). *Pl. Physiol., Lancaster* **46**, 768-777.
Jaffe, M. J. (1971). *In* "Structural and Functional Aspects of Phytochemistry". (V. C. Runeckles and T. C. Tso, eds), pp. 81-105, Academic Press, London and New York.
Jaffe, M. J. and Galston, A. W. (1967). *Planta* **77**, 135-141.
Jones, R. W. and Sheard, R. W. (1972). *Nature, Lond.* **238**, 221-222.
Kahn, A. (1960). *Pl. Physiol., Lancaster* **35**, 333-339.
Kandeler, R. (1972). *Z. PflPhysiol.* **67**, 86-92.
Kang, B. G. and Burg, S. B. (1972a). *Pl. Physiol., Lancaster* **49**, 631-633.
Kang, B. G., and Burg, S. B. (1972b). *Pl. Physiol., Lancaster* **50**, 132-135.
Kang, B. G. and Burg, S. B. (1972c). *Planta* **104**, 275-281.
Kang, B. G., Yocum, C. S., Burg, S. B. and Ray, P. M. (1967). *Science, N.Y.* **156**, 958-959.
Kasemir, H. and Mohr, H. (1972). *Pl. Physiol., Lancaster* **49**, 453-454.
Kasperbauer, M. J. (1971). *Pl. Physiol., Lancaster* **47**, 775-778.
Kasperbauer, M. J. (1973). *Abstr. Sixth int. Congr. Photobiol.* 170.
Kasperbauer, M. J., Tso, T. C. and Sorokin, T. P. (1970). *Phytochemistry* **9**, 2091-2095.
Kendrick, R. E. and Frankland, B. (1968). *Planta* **82**, 317-320.
Kendrick, R. E. and Frankland, B. (1969a). *Planta* **85**, 326-339.
Kendrick, R. E. and Frankland, B. (1969b). *Planta* **86**, 21-32.
Kendrick, R. E. and Hillman, W. S. (1971). *Am. J. Bot.* **58**, 424-428.
Kendrick, R. E. and Spruit, C. J. P. (1973a). *Pl. Physiol., Lancaster* **52**, 327-331.
Kendrick, R. E. and Spruit, C. J. P. (1973b). *Photochem. Photobiol.* **18**, 139-144.
Kendrick, R. E. and Spruit, C. J. P. (1973c). *Photochem. Photobiol.* **18**, 153-160.
Kendrick, R. E. and Spruit, C. J. P. (1974). *Planta* **120**, 265-272.
Kendrick, R. E., Spruit, C. J. P. and Frankland, B. (1969). *Planta* **88**, 293-302.
Kidd, G. H. and Pratt, L. H. (1973). *Pl. Physiol., Lancaster* **52**, 309-311.
King, R. W. and Cumming, B. G. (1972). *Planta* **108**, 39-57.
Kivilaan, A. and Bandurski, R. S. (1973). *Am. J. Bot.* **60**, 140-145.
Klein, W. H., Price, L. and Mitrakos, K. (1963). *Photochem. Photobiol.* **2**, 233-240.
Koller, D. (1962). *Am. J. Bot.* **49**, 841-844.

Koller, D. (1966). *Planta* 70, 42-45.
Koukkari, W. L. and Hillman, W. S. (1968). *Pl. Physiol., Lancaster* 43, 698-704.
Kroeger, H. and Lezzi, M. (1966). *A. Rev. Ent.* 11, 1-22.
Laetsch, W. M. and Briggs, W. R. (1962). *Pl. Physiol., Lancaster* 37, 142-148.
Larpent, M. and Jacques, R. (1971). *C.r. hebd. Séanc. Acad. Sci., Paris, Ser. D* 273, 162-164.
Larpent-Gourgard, M., Larpent, J. P. and Jacques, R. (1972). *Physiol. Vég.* 10, 553-558.
Lange, H., Shropshire, W. and Mohr, H. (1971). *Pl. Physiol., Lancaster* 47, 649-654.
Lazaroff, N. (1966). *J. Phycol.* 2, 7-17.
Lembi, C. A., Morré, D. J., Thomson, K. S. and Hertel, R. (1971). *Planta* 99, 37-45.
Liljenberg, C. (1966). *Physiologia Pl.* 19, 848-853.
Lisansky, S. A. and Galston, A. W. (1974). *Pl. Physiol., Lancaster* 53, 352-359.
Lockhart, J. A. (1956). *Proc. natn. Acad. Sci. U.S.A.* 42, 841-848.
Lona, F. and Fioretti, L. (1962). *Nuovo G. bot. ital.* 64, 236.
Loveys, B. R. and Wareing, P. F. (1971). *Planta* 98, 109-116.
MacKenzie, J. M., Jr., Coleman, R. A. and Pratt, L. H. (1974). *Pl. Physiol., Lancaster* (suppl.) 2.
Malcoste, R., Boisard, J., Spruit, C. J. P. and Rollin, P. (1970). *Meded. LandbHoogesch. Wageningen* 70-16, 1-16.
Manabe, K. and Furuya, M. (1971). *Bot. Mag., Tokyo* 84, 417-423.
Manabe, K. and Furuya, M. (1973). *Pl. Physiol., Lancaster* 51, 982-983.
Mancinelli, A. L., Borthwick, H. A. and Hendricks, S. B. (1966). *Bot. Gaz.* 127, 1-5.
Mantouvalos, G. (1969). Ph.D. Thesis, Univ. of Athens.
Margaris, N. (1972). Ph.D. Thesis, Univ. of Athens.
Marmé, D. (1974). *J. supramolec. Struct.* 2, 751-768.
Marmé, D. and Schäfer, E. (1972). *Z. PflPhysiol.* 67, 192-194.
Marmé, D., Schäfer, E., Trillmich, F. and Hertel, R. (1971a). *Eur. A. Symp. Pl. Photomorphogenesis*, Eretria Abstract: 36.
Marmé, D., Marchal, B. and Schäfer, E. (1971b). *Planta* 100, 331-336.
Marmé, D., Boisard, J. and Briggs, W. R. (1973). *Proc. natn. Acad. Sci. U.S.A.* 70, 3861-3865.
Marmé, D., Mackenzie, J. M., Boisard, J. and Briggs, W. R. (1974). *Pl. Physiol., Lancaster* 54, 263-271.
McArthur, J. A. and Briggs, W. R. (1970). *Planta* 91, 146-154.
McArthur, J. A. and Briggs, W. R. (1971). *Pl. Physiol., Lancaster* 48, 46-49.
McCullough, J. M. and Shropshire, W., Jr. (1970). *Plant Cell Physiol.* 11, 139-148.
Miller, J. H. and Miller, P. M. (1963). *Plant Cell Physiol.* 4, 65-72.
Mitrakos, K. (1972). *In* "Phytochrome" (K. Mitrakos and W. Shropshire, Jr. eds), pp. 587-612. Academic Press, London and New York.
Mitrakos, K. and Mantouvalos, G. (1972). *Z. PflPhysiol.* 67, 97-104.
Mitrakos, K. and Shropshire, W., Jr. (1972). "Phytochrome." Academic Press, London and New York.
Mohr, H. (1956). *Planta* 46, 534-551.
Mohr, H. (1959). *Planta* 53, 109-124.
Mohr, H. (1966a). *Photochem. Photobiol.* 5, 469-483.
Mohr, H. (1966b). *Z. PflPhysiol.* 54, 63-83.

Mohr, H. (1970). *Biol., Rundschau* 23, 187-194.
Mohr, H. (1972). "Lectures on Photomorphogenesis." Springer-Verlag, New York, Heidelberg, Berlin.
Mohr, H. and Noblé, A. (1960). *Planta* **55**, 327-340.
Mohr, H., Meyer, U. and Hartmann, K. (1964). *Planta* **60**, 483-496.
Mumford, F. E. and Jenner, E. L. (1971). *Biochemistry, N.Y.* **10**, 98-101.
Nabors, M. W. and Lang, A. (1971). *Planta* **101**, 1-25.
Negbi, M., Rushkin, E. and Koller, D. (1966). *Plant Cell Physiol.* 7, 363-376.
Newman, I. A. and Briggs, W. R. (1972). *Pl. Physiol., Lancaster* **50**, 687-693.
Oelze-Karow, H. and Mohr, H. (1970). *Z. Naturf.* **25b**, 1282-1286.
Oelze-Karow, H., Schopfer, P. and Mohr, H. (1970). *Proc. natn. Acad. Sci. U.S.A.* **65**, 51-57.
Papenfuss, H. D. and Salisbury, F. B. (1967). *Pl. Physiol., Lancaster* **42**, 1562-1568.
Parker, M. W., Hendricks, S. B., Borthwick, H. A. and Scully, N. J. (1945). *Science, N.Y.* **102**, 152-155.
Parker, M. W., Hendricks, S. B., Borthwick, H. A. and Scully, N. J. (1946). *Bot. Gaz.* **108**, 1-26.
Parker, M. W., Hendricks, S. B., Borthwick, H. A. and Went, F. W. (1949). *Am. J. Bot.* 36, 194-204.
Parker, M. W., Hendricks, S. B. and Borthwick, H. A. (1950). *Bot. Gaz.* **111**, 242-252.
Pike, C. S. and Briggs, W. R. (1972). *Pl. Physiol., Lancaster* **49**, 514-520.
Pratt, L. H. and Briggs, W. R. (1966). *Pl. Physiol., Lancaster* **41**, 467-474.
Pratt, L. H. and Coleman, R. A. (1971). *Proc. natn. Acad. Sci. U.S.A.* **68**, 2431-2435.
Pratt, L. H. and Coleman, R. A. (1974). *Am. J. Bot.* **61**, 195-202.
Price, L. and Klein, W. H. (1961). *Pl. Physiol., Lancaster* 36, 733-735.
Price, L., Mitrakos, K. and Klein, W. H. (1964). *Q. Rev. Biol.* 39, 11-18.
Quail, P. H. (1975a). *Planta* **123**, 223-234.
Quail, P. H. (1975b). *Planta* **123**, 235-246.
Quail, P. H. and Schäfer, E. (1974). *J. memb. Biol.* **15**, 393-404.
Quail, P. H., Marmé, D. and Schäfer, E. (1973a). *Nature, Lond.* **245**, 189-190.
Quail, P. H., Schäfer, E. and Marmé, D. (1973b). *Pl. Physiol., Lancaster* **52**, 128-131.
Racusen, R. H. (1973). *Pl. Physiol., Lancaster* **51** (suppl.), 51.
Racusen, R. H. and Etherton, B. (1973). *Pl. Physiol., Lancaster* **51** (suppl.), 51.
Racusen, R. H. and Miller, K. (1972). *Pl. Physiol., Lancaster* **49**, 654-655.
Racusen, R. H. and Satter, R. L. (1975). *Nature, Lond.* **255**, 408-410.
Raghavan, V. (1973). *Pl. Physiol., Lancaster* **51**, 306-311.
Reid, D. M., Clements, J. B. and Carr, D. J. (1968). *Nature, Lond.* **217**, 580-582.
Reid, D. M., Tuing, M. S., Durley, R. C. and Railton, I. D. (1972). *Planta* **108**, 67-75.
Rethy, R. (1968). *Z. PflPhysiol.* 59, 100-102.
Rollin, P. (1970). "Phytochrome, Photomorphogénèse, et Photoperiodisme". Masson et Cie, Paris.
Rollin, P. (1972). *In* "Phytochrome" (K. Mitrakos and W. Shropshire, Jr., eds), pp. 229-254. Academic Press, London and New York.
Roux, S. (1972). *Biochemistry, N.Y.* 11, 1930-1936.
Roux, S. and Yguerabide, J. (1973). *Proc. natn. Acad. Sci. U.S.A.* 70, 762-764.

Rubinstein, B., Drury, K. S. and Park, R. B. (1969). *Pl. Physiol., Lancaster* **44**, 105-109.
Sacher, J. A., Towers, G. H. N. and Davies, D. D. (1972). *Phytochemistry* **11**, 2383-2391.
Salisbury, F. B. (1955). *Pl. Physiol., Lancaster* **30**, 327-334.
Satter, R. L. and Galston, A. W. (1971a). *Science, N.Y.* **174**, 518-520.
Satter, R. L. and Galston, A. W. (1971b). *Pl. Physiol., Lancaster* **48**, 740-746.
Satter, R. L. and Galston, A. W. (1973). *Bio. Sci.* **23**, 407-416.
Satter, R. L., Applewhite, P. B. and Galston, A. W. (1972). *Pl. Physiol., Lancaster* **50**, 523-525.
Satter, R. L., Marinoff, P. and Galston, A. W. (1970). *Am. J. Bot.* **57**, 916-926.
Satter, R. L., Applewhite, P. B., Kreis, D. J. and Galston, A. W. (1973). *Pl. Physiol., Lancaster* **52**, 202-207.
Satter, R. L., Applewhite, P. B. and Galston, A. W. (1974a). *Pl. Physiol., Lancaster* **59**, 280-285.
Satter, R. L., Geballe, G. T., Applewhite, P. B. and Galston, A. W. (1974b). *J. gen. Physiol.* **64**, 413-430.
Satter, R. L., Geballe, G. T. and Galston, A. W. (1974c). *J. gen. Physiol.* **64**, 431-442.
Satter, R. L., Schrempf, M. and Galston, A. W. (1976). *Chronobiol.* In press.
Schäfer, E. (1975). *J. math Biol.* In Press.
Schäfer, E., Marchal, B. and Marmé, D. (1972). *Photochem. Photobiol.* **15**, 457-464.
Scheibe, J. (1972). *Science, N.Y.* **176**, 1037-1039.
Scheibe, J. and Lang, A. (1965). *Pl. Physiol., Lancaster* **40**, 485-492.
Scheibe, J. and Lang, A. (1969). *Photochem. Photobiol.* **9**, 143-150.
Schneider, M. J. and Stimson, W. (1971). *Pl. Physiol., Lancaster* **48**, 312-315.
Schneider, M. J. and Stimson, W. (1972). *Proc. natn. Acad. Sci. U.S.A.* **69**, 2150-2154.
Schneider, M. J., Borthwick, H. A. and Hendricks, S. B. (1967). *Am. J. Bot.* **54**, 1241-1249.
Schopfer, P. and Mohr, H. (1972). *Pl. Physiol., Lancaster* **49**, 8-10.
Schopfer, P. and Oelze-Karow, H. (1971). *Planta* **100**, 167-180.
Schwabe, W. W. and Valio, I. F. M. (1970). *J. exp. Bot.* **21**, 122-137.
Shropshire, W. Jr. (1972). *In* "Photophysiology" (A. C. Glese, ed.), vol. 7, pp. 33-72. Academic Press, London and New York.
Shropshire, W. Jr. (1973). *Sol. Energy* **15**, 99.
Siegelman, H. W. and Hendricks, S. B. (1957). *Pl. Physiol., Lancaster* **32**, 393-398.
Smith, H. (1973). *In* "International Review of Science," Medical and Technical Publ. Co.
Smith, H. (1975). "Twenty-Second Easter School in Agricultural Sciences: Light and Plant Development." Abstracts. School of Agriculture, Univ. of Nottingham, Sutton Bonington, Loughborough, Leicestershire.
Smith, H. (1976). "Twenty-Second Easter School in Agricultural Sciences: Light and Plant Development." Conference proceedings. Butterworth, London. In press.
Spruit, C. J. P. and Kendrick, R. E. (1973). *Photochem. Photobiol.* **18**, 145-152.
Spruit, C. J. P. and Mancinelli, A. L. (1969). *Planta* **88**, 303-310.
Stanier, R. Y. (1960). *Harvey Lect.* **54**, 219.
Steponkus, P. L. and Lanphear, F. O. (1967). *Pl. Physiol., Lancaster* **42**, 1673-1679.

Sugai, M. and Furuya, M. (1967). *Plant Cell Physiol.* **8**, 737-748.

Sugai, M. and Furuya, M. (1960). *Plant Cell Physiol.* **9**, 671-680.

Sweet, H. G. and Hillman, W. S. (1969). *Physiologia Pl.* **22**, 776-786.

Takatori, S. and Imahori, K. (1971). *Phycologia* **10**, 221-228.

Tanada, T. (1967). *Proc. natn. Acad. Sci. U.S.A.* **59**, 376-380.

Tanada, T. (1968). *Pl. Physiol., Lancaster* **43**, 2070-2071.

Tanada, T. (1972). *Pl. Physiol., Lancaster* **49**, 860-861.

Tanada, T. (1973a). *Pl. Physiol., Lancaster* **51**, 150-153.

Tanada, T. (1973b). *Pl. Physiol., Lancaster* **51**, 154-157.

Taylor, A. O. and Bonner, B. A. (1967). *Pl. Physiol.* **42**, 762-766.

Taylorson, R. B. and Borthwick, H. A. (1969). *Weed Sci.* **17**, 48-51.

Taylorson, R. B. and Hendricks, S. B. (1969). *Pl. Physiol., Lancaster* **44**, 821-825.

Taylorson, R. B. and Hendricks, S. B. (1971). *Pl. Physiol., Lancaster* **47**, 619-622.

Taylorson, R. B. and Hendricks, S. B. (1972a). *Pl. Physiol., Lancaster* **49**, 663-665.

Taylorson, R. B. and Hendricks, S. B. (1972b). *Pl. Physiol., Lancaster* **50**, 645-648.

Tepfer, D. A. and Bonnett, H. T. (1972). *Planta* **106**, 311-324.

Tobin, E. M. and Briggs, W. R. (1969). *Pl. Physiol., Lancaster* **44**, 148-150.

Tobin, E. M. and Briggs, W. R. (1973). *Photochem. Photobiol.* **18**, 487-495.

Tobin, E. M., Briggs, W. R. and Brown, P. K. (1973). *Photochem. Photobiol.* **18**, 497-503.

Toriyama, H. (1955). *Cytologia* **20**, 367-377.

Valio, I. F. M. and Schwabe, W. W. (1969). *J. exp. Bot.* **20**, 615-628.

Vidaver, W, and Hsiao, A. I. (1972). *Can. J. Bot.* **50**, 687-689.

Virgin, H. I. (1972). *In* "Phytochrome" (K. Mitrakos and W. Shropshire, Jr., eds), pp. 371-406. Academic Press, London and New York.

Wagné, C. (1964). *Physiologia Pl.* **17**, 751-756.

Wagner, G., Haupt, W. and Laux, A. (1972). *Science, N.Y.* **176**, 808-809.

Weisenseel, M. H. and Smeibidl, E. (1973). *Z. PflPhysiol.* **70**, 420-431.

Wellburn, F. A. and Wellburn, A. R. (1973). *New Phytol.* **72**, 55-60.

Wesson, G. and Wareing, P. F. (1969a). *J. exp. Bot.* **20**, 402-413.

Wesson, G. and Wareing, P. F. (1969b). *J. exp. Bot.* **20**, 414-425.

Wetherell, D. F. and Koukkari, W. L. (1970). *Pl. Physiol., Lancaster* **46**, 350-351.

White, J. M. and Pike, C. S. (1974). *Pl. Physiol., Lancaster* **53**, 76-79.

Williams, B. J. Jr., Pellett, N. E. and Klein, R. M. (1972). *Pl. Physiol., Lancaster* **50**, 262-285.

Williamson, F. A. and Morré, D. J. (1974). *Pl. Physiol., Lancaster* **46** (suppl.).

Wilson, J. R. and Schwabe, W. W. (1964). *J. exp. Bot.* **15**, 368-380.

Withrow, A. P. and Withrow, R. B. (1943). *Bot. Gaz.* **104**, 409-416.

Withrow, R. B., Klein, W. H. and Elstad, V. (1957). *Pl. Physiol., Lancaster* **32**, 453-462.

Wood, A. and Paleg, G. A. (1972). *Pl. Physiol., Lancaster* **50**, 103-108.

Yamamoto, Y. and Tezuka, T. (1972). *In* "Phytochrome" (K. Mitrakos and W. Shropshire, Jr., eds), pp. 408-429. Academic Press, London and New York.

Yaniv, Z., Mancinelli, A. L. and Smith, P. (1967). *Pl. Physiol. Lancaster* **42**, 1479-1482.

Yunghans, H. and Jaffe, M. J. (1970). *Physiologia Pl.* **23**, 1004-1016.

Yunghans, H. and Jaffe, M. J. (1972). *Pl. Physiol., Lancaster* **49**, 1-7.

Zouaghi, M., Malcoste, R. and Rollin, P. (1972). *Planta* **106**, 30-43.

Zucker, M. (1969). *Pl. Physiol., Lancaster* **44**, 912-922.

Zucker, M. (1970). *Biochim. biophys. Acta* **208**, 331-333.

Chapter 16

Functions of Flavonoids in Plants

J. B. HARBORNE

Plant Science Laboratories, The University, Reading, England

I Introduction

Water-soluble pigments of the flavonoid group (i.e. anthocyanins, flavonols and flavones) are very widely distributed in nature. The intensely coloured anthocyanins are most abundantly present in petals and fruits, but also occur in leaves, roots and tubers. In leaves of many plants their presence is presumably masked by the ubiquitous chlorophyll, but purple-red anthocyanin colour sometimes appears either in young or in dying leaves. Flavonols and flavones are also very common leaf constituents, but their pale yellow or cream colours are again masked by chlorophyll. In contrast to chlorophylls, carotenoids and quinones, flavonoid pigments are characteristic only of higher plants, being virtually absent from lower phyla.

The most significant function of the sap-soluble flavonoids is their ability to impart colour to the plants in which they occur. They are responsible for most orange, scarlet, crimson, mauve, violet and blue colours, as well as contributing much to yellow, ivory and cream flowers. The only other considerable groups of colouring matters in higher plants are the lipid-soluble chlorophylls and carotenoids. Chlorophylls *a* and *b* provide the prevailing green of plant leaves, and the carotenoids are the most important sources of yellow colours in flowers and fruits. These lipid pigments are considered in other chapters in this volume, but the relative contributions of carotenoids and flavonoids to yellow colour will be considered in a later section in this chapter.

Flavonoids are important to plants because flower colour, together with flower shape, scent and nectar, attracts bees, butterflies and other animals to higher plants to ensure fertilization. Indeed, many plant species are self-incompatible and are thus completely dependent on animal vectors for this purpose. Even in self-pollinated plants (e.g. the sweet pea) seed set is usually higher because of the activities of insect visitors, and this explains why such plants retain their attractive cyanic petal colours even after generations of inbreeding.

The association between bee activity and flower colour has long been studied (Sprengel, 1793; Darwin, 1876; see also Manning, 1956a). Bees can discriminate four basic colours and natural evolution of plants towards producing blue flowers must be associated with the bees' preference for this colour (von Frisch, 1950). By contrast, humming birds, which are significant pollinators in tropical and subtropical climates, are attracted by bright orange and red colours and this can be correlated with selection for anthocyanidins (pelargonidin and luteolinidin) which give just these colours in many tropical plant species. The range of known pollinating vectors and their preferences in terms of flower colour are indicated in Table I. The role of flower colour in pollination ecology is a subtle and complex one and is further discussed by Faegri and van der Pijl (1971) and by Baker and Hurd (1968).

The importance of anthocyanin colour in fruits such as the strawberry, cherry, blackcurrant and so on as an aid to seed dispersal

TABLE I
Pollinating vectors and their colour preferences

Animal	Flower colour preferences
Carrion and dung flies	Purple-brown or green
Bees	Blue, yellow, "ultraviolet" (blind to red)
Moths	Usually white or pale colours
Butterflies	Pale colours (often pink)
Humming birds	Orange, scarlet and bright red
Beetles	White or dull colours
Bats	White or off-white

Blue, red and pink colours = anthocyanins.
Yellow = carotenoids (occasionally yellow flavonols).
Cream and white = flavones and flavonols.

by animals is self-evident. Both fruit and flower colour give immense aesthetic pleasure to man and conscious selection for colour varieties among garden plants and horticultural crops has been practised for a very long time. For example, it is possible that colour varieties of the potato tuber were prized by the Peruvian peasants even in pre-Inca times (Salaman, 1949); and colour "breaks" in the tulip were known and preserved in European gardens as far back as the sixteenth century (McKay and Warner, 1933).

Serious study of the water-soluble plant pigments may be dated from the time of Robert Boyle, who showed in 1664 that the purple pigment of *Viola tricolor* was a natural indicator, becoming green in alkaline and red in acid solution. The term "anthocyanin" was coined by Marquart in 1835 and the basic chemistry of flavones and anthocyanins was worked out at the turn of the nineteenth century by von Kostanecki and Willstätter in Germany and Perkin in this country. The isolation and determination of structure of cyanin, the principal anthocyanin of the cornflower, by Willstätter and Mallison in 1915 remain a landmark in the study of flavonoid pigments. During the period 1930 to 1940, the chemistry of flower colour variation was intensively studied by geneticists and biochemists at the John Innes Institute. In active collaboration with the Robinsons and their school at Oxford, who were the first to synthesize the natural anthocyanins, these workers laid the foundation for our present knowledge of the subject. The valuable contributions of Karrer and his co-workers in Zurich and of Hayashi in Japan must also be mentioned.

In recent years with the development of more accurate and refined analytical techniques, determinations of the structures of all the more important water-soluble pigments have been completed (Geissman 1962; Harborne, 1962a; Dean, 1963; Harborne *et al.* 1975). Work has also gone forward on the contributions of co-pigmentation, metal-complexing and pH to the blueing of flower colours, and many of the factors responsible for subtle differences in shade and tone in flowers have now been defined.

In this chapter, our present knowledge of the chemical factors controlling plant colour will be reviewed and some of the gaps remaining in our understanding of the subject will be mentioned. Earlier reviews dealing *inter alia* with flavonoids and flower colour are those of Scott-Moncrieff (1936), Beale (1941), Blank (1947, 1958), Lawrence (1950), Paech (1954), Reznik (1956) and Harborne (1967). For an account of the properties and nature of flavonoid pigments, the reader is referred to Chapter 8 by T. Swain.

II Contribution to flower colour

The contribution of flavonoids to flower colour and their importance in relation to other pigments are summarized in Table II. The

TABLE II
Contribution of flavonoids to flower colour

Colour[a]	Pigments[b]	Examples
Ivory and cream	Flavones (e.g. apigenin) and/or flavonols (e.g. quercetin)	Ivory *Antirrhinum majus* or *Dahlia variabilis*
Yellow	(a) Carotenoid alone	Yellow *Rosa*
	(b) Flavonol alone (i.e. gossypetin)	*Primula vulgaris*
	(c) Aurone alone	Yellow *Antirrhinum majus*
	(d) Carotenoid and flavonol or chalkone	*Lotus corniculatus, Ulex europeaus*
Orange	(a) Carotenoid alone	*Lilium regale*
	(b) Pelargonidin and aurone	*Antirrhinum majus*
Scarlet	(a) Pure pelargonidin	*Pelargonium, Salvia splendens*
	(b) Cyanidin and carotenoid	*Tulipa* cultivar
	(c) Cyanidin and flavonoid	*Chasmanthe* and *Lapeyrousa*
Brown	Cyanidin on carotenoid background	*Cheiranthus cheiri, Rosa* cv. "Cafe", *Primula polyanthus*
Magenta or crimson	Pure cyanidin	Red *Camellia hortense*, red *Begonia* cultivars
Pink	Pure peonidin	*Paeonia, Rosa rugosa*
Mauve or violet	Pure delphinidin	*Verbena, Brunfelsia calycina*
Blue	(a) Cyanidin and co-pigment	*Meconopsis betonicifolia*
	(b) Cyanidin as metal complex	*Centaurea cyanus*
	(c) Delphinidin and co-pigment	*Plumbago capensis*
	(d) Delphinidin as metal complex	*Commelina communis*
	(e) Delphinidin[c] at "alkaline" pH	*Primula sinensis*
Black (purple black)	Delphinidin (at high concentration)	*Tulipa* cv. "Queen of the Night", *Viola* x *wittrockiana* (pansy)

[a] Green flowers (e.g. of *Helleborus foetidus*) are presumably pigmented by chlorophyll. Chlorophyll appears in some plants (*Tulipa*) in immature petals, only to disappear as the flower matures.

[b] For brevity, the pigment aglycones only are given, but it should be remembered that flavonoids practically always occur in flowers in glycosidic form.

[c] Present here in a partly methylated form.

descriptions of colour here are rather broad; for more precise notes on the colours of individual flowers, colour charts, such as those of Wilson (1938) and Wanscher (1953), should be consulted. The colours in Table II refer to that of the corolla or petal; it should, perhaps, be pointed out that flavonoids also contribute to pigmentation in other parts of the flower, be it sepal, bract, stamen, style or pollen. Anthocyanins have, for example, been isolated from style and stamen of *Anemone*, from sepals of many tuber-bearing *Solanum*, from bracts (floral leaves) of *Poinsettia* and *Musa*, and from pollen of *Fuchsia*.

The various points about flower colour shown in Table II will be amplified in the following sections.

A THE ANTHOCYANINS

1 *Hydroxylation and colour*

The most important flower pigments are undoubtedly the anthocyanins. For example, of 832 dicotyledonous species in the British flora surveyed (cf. Beale *et al.*, 1941), no less than 49% have anthocyanins in the wild type or in a naturally occurring variety. The contribution of anthocyanins to flower colour is basically simple.

TABLE III
Colour of the three major anthocyanidins

Name (abbreviation) $\lambda_{max}^{MeOH-HCl}$ (nm) and colour	Chemical structure
Pelargonidin (Pg) 520 orange-red	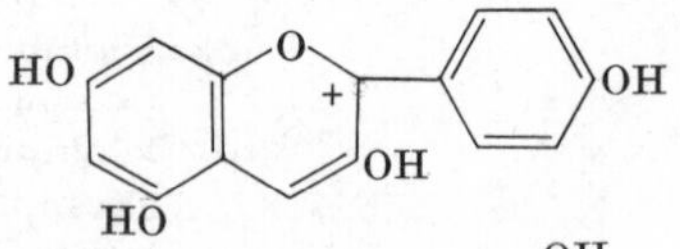 HO, HO, OH, OH (flavylium cation, +)
Cyanidin (Cy) 535 magenta	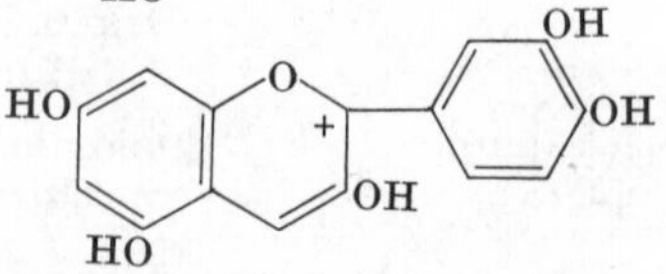 HO, HO, OH, OH, OH (flavylium cation, +)
Delphinidin (Dp) 545 mauve	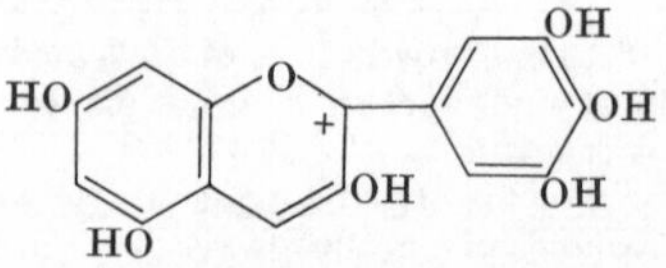 HO, HO, OH, OH, OH, OH (flavylium cation, +)

There are three main pigments: pelargonidin (Pg), cyanidin (Cy) and delphinidin (Dp), which differ in structure only by the number of their hydroxyl groups (Table III). Neglecting for the present complications of glycosylation, methylation and co-pigmentation, these three pigments, either singly or as mixtures, provide the whole range of flower colour from pink, orange and scarlet to mauve, violet and blue. Broadly speaking, all pink, scarlet and orange-red flowers have pelargonidin, first isolated from the garden *Pelargonium*, all crimson and magenta flowers have cyanidin, which occurs typically in the crimson rose, and mauve and blue flowers have delphinidin, a pigment named after *Delphinium*. In considering the relative contributions in more detail of pelargonidin, cyanidin and delphinidin, their general occurrence in nature will be discussed first, before their occurrence in colour varieties of garden plants are considered. The two aspects are interrelated, since both depend on the fact that selection for particular colours (especially blue and scarlet) has occurred during the evolution of higher plants.

The many surveys of anthocyanins that have been carried out show that the distribution of Pg, Cy and Dp types is strongly correlated with flower colour. Some typical results, obtained by Forsyth and Simmonds (1954) in a study of 247 Trinidad plants, are shown in Table IV. In a survey of Australian flora, Gascoigne *et al.* (1948) obtained similar results, observing, for example, that Dp was present in 90% of the blue-flowered species. In the remaining 10%, blueing is presumably produced by strong co-pigmentation of cyanidin (see later).

While there is clearly a relationship between flower colour and geographical distribution, earlier suggestions by Beale *et al.* (1941) that anthocyanidin type is simply related to climate now seem to be unsatisfactory. Thus the view that high pelargonidin and low delphinidin frequencies are characteristic of tropical floras appears to be

TABLE IV
Relation between colour and anthocyanidin types in Trinidad plants

	No. of species containing					
Flower colour	Dp	Dp + Cy	Cy	Cy + Pg	Pg	Totals
Yellow-red	0	0	6	3	6	15
Red	7	6	57	14	8	92
Bluish-red	21	8	19	1	0	49
Reddish-blue	19	5	4	0	0	28
Blue	11	1	2	0	0	14

TABLE V
Occurrence of anthocyanidin types in flower colour mutants of garden plants[a]

Plant	Colour forms	No. of varieties examined	Percentage occurrence of			Accompanying flavonols[c]
			Pg	Cy	Dp[b]	
Lathyrus odoratus (sweet pea)	Cerise, pink[d] and salmon	7	100	0	0	Km
	Crimson and carmine	5	0	100	0	Km, Qu
	Mauve and blue	9	0	0	100	Km, Qu, My
Verbena hybrida (verbena)	Pale pink	2	100	0	0	
	Scarlet-magenta	1	95	5	0	
	Pink	1	80	20	0	Km, Qu, Ap
	Scarlet	2	85	10	5	
	Maroon	1	40	30	30	
	Purple-blue	2	0	15	85	My, Qu, Km, Lu, Ap
	Purple	2	0	10	90	
	White	1	0	0	0	Lu, Ap
Hyacinthus orientalis (hyacinth)	Deep red "Scarlet O'Hara"	—	90	10	0	
	Pink "Pink Perfection"	—	60	40	0	
	Mauve "Lord Balfour"	—	20	80	0	Ap, Km
	Mauve "Mauve Queen"	—	0	100	0	
	Blue "Delft Blue"	—	0	10	90	
	Pale blue "Springtime"	—	0	0	100	

Streptocarpus hybrida[e] (cape primrose)	Pink	—	100	0	0	Ap, Km
	Salmon	—	80	20	0	Ap
	Rose and magenta	—	0	100	0	Ap, Lu
	Mauve and blue	—	0	0	100	Ap, Lu
Primula sinensis[e] (Chinese primrose)	Orange and coral	—	90	10	0	Km
	Maroon, mauve and blue	—	0	0	100	Km, Qu, My
Lupinus polyphyllus (lupin)	Pink and red	3	40	60	0	Qu, Km, Lu, Ap
	Purple, mauve and blue	3	0	20	80	Lu, Ap
Tulipa (tulip)	Red and orange	48	46	48	6	Km, Qu
	Pink, crimson and deep red	38	36	56	7	
	Black, purple and violet	21	6	32	61	Km, Qu, My

[a] Except in the case of *Tulipa* (Shibata and Ishikura, 1960), data are from Harborne (1962a, 1963a, 1967).
[b] Dp = delphinidin; Cy = cyanidin; Pg = pelargonidin. Pigments present in *Lathyrus, Primula* and *Streptocarpus* are mainly methylated (i.e. peonidin, petunidin and malvidin are present).
[c] Km = kaempferol; Qu = quercetin; My = myricetin; Lu = luteolin; Ap = apigenin.
[d] Two pink shades also contain traces of cyanidin.
[e] Forms of known genotype were examined in these cases.

incorrect: Forsyth and Simmonds (1954) found that the Trinidad plants contained fewer Pg and Dp types (17% and 41%) and more Cy types (67%) than temperate plants. Delphinidin predominates in alpine floras because blue-flowered plants are so common here; thus Acheson (1956), in a survey of Himalayan plants, found 28 Dp-, 17 Cy- and 1 Pg-containing species.

When wild plants are brought into cultivation, mutations occur in the direction: Dp → Cy → Pg (Beale, 1941). Mutant forms have been preserved in many garden plants and among these flowers there is again excellent correlation between colour and anthocyanidin type (Table V). In a few plants (e.g. *Lathyrus odoratus*) genetic factors controlling hydroxylation are involved in epistasy and so are absolute in their action: pure Dp, Cy and Pg types predominate. In others (e.g. *Primula sinensis*), intermediate cyanidin types are absent and the main mutation affecting hydroxylation, K → k, is from Dp to Pg; only traces of Cy are present in the kk mutant.

In most garden plants, varieties containing mixtures of anthocyanidin types are quite as common as those having single pigments. Flowers with mixtures are of the expected intermediate colour shades. Good examples here are *Verbena* and the garden hyacinth (Table V). At the other extreme, there are plants (e.g. *Tulipa*) in which forms having single anthocyanidins in their petals are rare and the colour range is due to the presence of varying proportions of the three main types. Shibata and Ishikura (1960), who analysed 107 tulip varieties, found that the majority had mixtures of all three pigments; nevertheless, there is still a good correlation between the predominating type and colour (Table V).

In some garden plants, pelargonidin forms have not yet been isolated. This is true, for example, of the cultivated *Cyclamen*, in which Dp and Cy types only are known and the colour range extends from deep purple to purplish-red (van Bragt, 1962). Pelargonidin is not found among the flower colour forms of the tuber-bearing *Solanum* for a different reason, i.e. because selection for colour has taken place, unusually, in the tuber and not in the flower (Dodds and Long, 1955).

By contrast with the plants mentioned above and in Table V, there are plants which are not able to synthesize delphinidin and in which colour variation is more restricted. One example is the garden rose, most varieties of which contain cyanidin and in which mutations to pelargonidin are of rather rare occurrence. The orange-red varieties available today are derived from the dwarf polyantha "Paul Crampel", which was introduced about 1930. This orange-flowered variety is unusual in that it back-mutates somatically to produce crimson-flowered offshoots, which are pigmented by cyanidin. A second example is *Dahlia variabilis*, a plant in which cyanic colour is due to cyanidin or pelargonidin or their mixture. A third example is *Antirrhinum majus*, magenta forms of which have cyanidin and pink forms pelargonidin. Two wild *Antirrhinum* species (*A. cornutum* and *A. nuttallianum*) have blue delphinidin-containing flowers (Harborne, 1963b), so that there is a reasonable chance that a blue snapdragon will be produced one day.

A rare change that affects the hydroxylation pattern of anthocyanidins, and hence colour, is the loss of the 3-hydroxyl group. Pigments lacking this 3-hydroxyl, analogous otherwise to delphinidin, cyanidin and pelargonidin, are much nearer the yellow end of the colour spectrum; thus tricetinidin has $\lambda_{max}^{MeOH-HCl}$ 520, luteolinidin $\lambda_{max}^{MeOH-HCl}$ 493 and apigeninidin $\lambda_{max}^{MeOH-HCl}$ 476 nm. Plants containing these pigments have orange-red (*Gesneria cuneifolia, Rechsteineria cardinalis*) or orange-yellow flowers (*Kohleria eriantha*). The striking russet-orange flowers of *Columnea* x *banksii*, *C. stavengeri* and related species contain a pigment of a similar nature, called columnin. This pigment was originally formulated (Harborne, 1966) as being either 5,6,7,3′,4′- or 5,7,8,3′,4′-pentahydroxyflavylium but recent synthetic studies (J. B. Harborne, unpublished results) indicate that neither of these structures is correct; further work on its chemistry is in progress.

Substitution of a hydroxyl group in the 6-position also has a hypsochromic effect on colour and the 6-hydroxy derivatives of pelargonidin, cyanidin and delphinidin have been synthesized (Charlesworth and Robinson, 1934). Of these, only 6-hydroxypelargonidin $\lambda_{max}^{MeOH-HCl}$ 497 nm has been found as a flower pigment. It has been isolated from the tangerine-coloured petals of *Impatiens aurantiaca* (Clevenger, 1964) and identified by comparison with synthetic material (Jurd and Harborne, 1968). A survey of anthocyanidins in other members of the genus *Impatiens* has indicated that it is unique to this species (Clevenger 1971).

2 *Methylation of anthocyanins*

Methylation of some of the hydroxyl groups of the anthocyanidin molecule has a small reddening effect on colour. This is apparent from a consideration of the absorption spectra of the known methylated pigments (Table VI). Few examples of this reddening of flower colour can be quoted, because the effect of methylation is frequently obscured by other factors, particularly by co-pigmen-

TABLE VI
Long-wave absorption maxima of methylated anthocyanidins

	$\lambda_{max}^{MeOH-HCl}$ (nm)	$\Delta\lambda$
Delphinidin derivatives		
Parent compound (Dp)	546	–
3′-*O*-Methyl ether (petunidin) (Pt)	546	0
3′,5′-di-*O*-Methyl ether (malvidin) (Mv)	542	4
7,3′,5′-tri-*O*-Methyl ether (hirsutidin) (Hs)	536	10
5,3′,5′-tri-*O*-Methyl ether(?) (capensinidin) (Cp)	538	8
Cyanidin derivatives		
Parent compound (Cy)	535	0
3′-*O*-Methyl ether (peonidin) (Pn)	532	3
7,3′-di-*O*-Methyl ether (rosinidin) (Rs)	524	11

tation. However, the relative amounts of the various methylated pigments have been measured in several mutants of *Primula sinensis*, all of which are recessive for the co-pigment gene B. The results (Table VII) support the thesis that methylation has a reddening effect (Harborne and Sherratt, 1961). Another example is *Rosa*, in which cyanidin-peonidin mixtures are found almost exclusively in pinker varieties (*Rosa rugosa* and derived hybrids) whereas crimson and deeper red varieties have only cyanidin (Harborne, 1961).

While pigments based on malvidin, peonidin and petunidin are quite common, 5- or 7-*O*-methylated anthocyanins are very rare; in this case, the hypsochromic effect of methylation is more pronounced (8 to 11 nm). However, this is not reflected to any extent in flower colour, because of other modifying factors. Thus the flower colour of the capensinidin-containing *Plumbago capensis* is sky-blue, the result of strong co-pigmentation of the anthocyanin by the flavonol, azalein (5-*O*-methylquercetin 3-rhamnoside) (Harborne,

TABLE VII
Methylated pigments in *Primula sinensis*

Variety	Flower colour	Percentage composition		
		Dp	Pt	Mv
"Duchess Fern Leaf"	Blue	100	0	0
"Reading Pink"	Pink	37	51	13
"Oak Tongue"	Maroon	5	29	65

1962b). Again, although *Primula rosea* (containing rosinidin) has distinctive rose-pink petals, *Primula* species having hirsutidin (e.g. *Primula denticulata, P. rubra*) cannot be distinguished by their flower colour from those having malvidin (e.g. *P. obconica, P. malacoides*) (Harborne, 1968).

3 *Glycosylation of anthocyanidins*

Although glycosylation of the 3-hydroxyl group of anthocyanidins has a relatively large hypsochromic effect (−15 nm) in the visible region of the spectrum, glycosylation is not an important factor in flower colour, because anthocyanidins practically always occur in flowers with at least one sugar attached to the 3-hydroxyl group. It is true that there have been reports from time to time of anthocyanidins being isolated in the free state from flowers: e.g. from *Begonia* spp. (Bopp, 1957), *Camellia japonica* (Hayashi and Abe, 1953) and *Lathyrus hirsutus* (Pecket, 1960). The isolation of free anthocyanidins from these flowers does not necessarily mean that they occur as such *in vivo,* since some 3-glycosides, and notably 3-pentosides, are labile and invariably undergo partial hydrolysis during extraction and isolation with acid-containing solvents. Thus re-examination in this laboratory of flower pigments of the plants mentioned above failed to disclose the presence of any free aglycones; significantly however, a cyanidin 3-pentoside was detected in one *Begonia* species and a malvidin 3-rhamnoside in *Lathyrus hirsutus.* Obviously, free anthocyanidin may have a transient existence in flower petals where relatively rapid changes in pigmentation are occurring. Indeed, free cyanidin has been detected unequivocally in the pink basal blotch of petals of *Hibiscus mutabilis* but the pigment remains present for only a few hours (Lowry, 1971).

From the point of view of flower colour, the nature of the 3-substituted sugar is immaterial; all 3-glycosides of a particular anthocyanidin have the same visible spectra. Although 3-glycosides and 3,5-diglycosides (the two major classes of anthocyanins) have almost identical visible maxima, the substituent of a sugar in the 5-position does have a small effect on colour. Thus, the 3,5-diglucosides of pelargonidin, peonidin and malvidin differ from the corresponding 3-glycosides by being fluorescent in solution. This is probably related to the intensity of coloration in the flowers in which they occur; petals of *Pelargonium* and *Punica granatum*, containing pelargonidin 3,5-diglucoside, do have an intensely orange appearance. Acylation of the sugar residues of such glycosides with *p*-coumaric or caffeic acid partly quenches this fluorescence; thus acylated pelargonidin 3,5-diglucosides occur in the duller scarlet blooms of *Salvia splendens* and *Monarda didyma.*

Anthocyanins with sugars attached to the 3- and 7-positions, instead of the more usual 3- and 5-positions, are rare in nature. One such pigment, pelargonidin 3-sophoroside-7-glucoside (Harborne, 1962a), has a visible maximum ($\lambda_{max}^{MeOH-HCl}$ 498 nm) different from pelargonidin 3-glucoside ($\lambda_{max}^{MeOH-HCl}$ 507 nm) and the 3,5-diglucoside ($\lambda_{max}^{MeOH-HCl}$ 504 nm). The flowers which contain it, *Papaver orientalis* and some forms of *P. nudicaule,* are distinctly more orange-yellow in colour than the scarlet *Papaver rhoeas,* which contains pelargonidin and cyanidin 3-sophoroside, so that, in this instance, a change in glycosylation pattern has affected flower colour.

4 *Quantitative effects*

Variations in the amounts of anthocyanin in the petal have profound effects on colour and large discontinuous differences in anthocyanin content have been noted in the flowers of some plant varieties. At one end of the scale, low pigment concentrations give flowers with a faint pinkish blush (e.g. the rose "Madame Butterfly") and at the other, high concentrations are found in the deep purple-black petals of the tulip "Queen of the Night" or of the pansy "Jet Black".

Rather few quantitative measurements have been made, mainly because reliable extinction values for most anthocyanins have only been readily available in recent years. In making such measurements, allowance must be made for variations due to the environment, age of the petal and so on. Values obtained by the early workers (quoted by Blank (1958)) are, for cyanin in cornflower, 0·05 to 0·07% of the dry weight in normal varieties and 13 to 14% in some dark-purple forms. Similarly, flowers of *Pelargonium peltatum* were found to have 1% and *P. zonale* 6 to 14% of the dry weight as pigment.

More recently, Harborne and Sherratt (1961) have found that the orange "Dazzler" mutant of *Primula sinensis* has three times as much pelargonidin glucoside as the coral form. This difference in anthocyanin concentration (3·2 as compared with 1·1% dry weight) is under monogenic control. A single gene also appears to separate deep and pale mauve colour forms of *Solanum iopetalum*, which differ in anthocyanin concentration by a factor of four (J. B. Harborne, unpublished data). Deep purple forms of *Torenia fournieri* have as much as nine to ten times as much pigment as pale forms. Anthocyanin concentration is here controlled by two complementary genes, and hybrids between the two forms have intermediate quantities and colours (Endo, 1962). A gene controlling anthocyanin concentration in *Pisum* increases pigment in the petal wing more than in the petal standard; pigment ratios in intense and pale phenotypes are 1:5·5 for the wings and 1:2·5 in the standard (Harborne, 1964a).

Pigment concentrations in petals of *Dianthus caryophyllus* and *Antirrhinum majus* have been measured by Geissman and his co-workers in connection with genetical studies. In red carnations, Geissman and Mehlquist (1947) found pelargonidin 3-glucoside to occur in the ratios of 4:2:1 in petals of different intensity; magenta forms contained 2·4% of their dry weight as cyanidin 3-glucoside. In *Antirrhinum*, Jorgensen and Geissman (1955) found that the concentration of cyanidin 3-rutinoside in magenta forms varied (0·3 to 1·4% dry weight) inversely with the aurone concentration (see also below, p. 754).

To summarize, it is clear that anthocyanin concentration in the flower varies within the range from 0·01 to 15% dry weight and is controlled in a very precise manner by genetic factors in certain plants.

5 *Co-pigmentation*

The phenomenon of co-pigmentation, the blueing of anthocyanin colour *in vivo* by flavones and related substances, was discovered independently by Robinson and Robinson (1931) and by Lawrence (1932). It is easily demonstrated *in vitro*. At room temperature aqueous acid extracts of co-pigmented flowers are bluer in tone than those of unco-pigmented petals. On heating, the loose co-pigment–pigment complex is dissociated and there is a reddening in colour; on cooling, the colour reverts to the original blue shade. While the co-pigment in most flowers which show this phenomenon is probably a flavone glycoside or a hydrolysable tannin and can be removed by ethyl acetate extraction, some unrelated substances will co-pigment with anthocyanins in the test tube; examples are 2-hydroxyxanthone, narcotine and papaverine. Co-pigmentation can be demonstrated in *Primula sinensis*, in which plant it is under simple genetic control (Scott-Moncrieff, 1936). The spectral shift shown by the malvidin 3-glucoside of maroon flowers (bb types) when co-pigmented with flavone to give mauve flowers (BB types) is 5 nm (the λ_{max} in aqueous acid changes from 516 to 521 nm). The two flavonols responsible, i.e. kaempferol 3-gentiobioside and 3-gentiotrioside, occur in both genotypes although in amounts varying with the genotype. Measurements in two different families of colour forms showed that co-pigmented forms had three and five times more flavonol than the unco-pigmented varieties (Harborne and Sherratt, 1961).

Co-pigmentation is also a factor controlling flower colour in the genus *Rosa*. A blue rose has long been searched for; the rather unsatisfactory mauve and purple varieties (e.g. "Reine de Violette") so far available contain the cyanidin 3,5-diglucoside of crimson roses co-pigmented with large amounts of gallotannin. The spectral shift in rose is from 507 to 512 nm (Harborne, 1961). A more unusual co-pigment, the *C*-glucosidic xanthone mangiferin (Bate-Smith and Harborne, 1963) is present in *Iris* cultivars and produces, by interaction with delphinidin 3-(*p*-coumarylrutinoside)-5-glucoside, a range of colours through red and mauve to blue. Blueness in flowers of *Lathyrus* is due to co-pigmentation of delphinidin, petunidin and malvidin 3-rhamnoside-5-glucosides. (Harborne, 1963c) with kaempferol and quercetin glucosides. Pecket and Selim (1962) have demonstrated this by hybridizing a species with red flowers, which lack flavonols, with one having cream flowers, containing flavonols. The resulting hybrid, *L. clymenum* x *ochrus*, had the expected mauve-coloured flowers. In addition, colour differences between wing and standard in cyanic flowers of *Lathyrus, Pisum* and *Vicia* are probably a result of variations in amount of these co-pigments.

The role of co-pigmentation in flower colour has been the subject of some controversy and as recently as 1962, Hayashi, in reviewing the factors controlling blueness in flowers, tended to dismiss co-pigmentation as a minor factor compared to metal complexing (see below). However, a recent series of experiments involving both *in vivo* and *in vitro* studies, carried out by Asen and his co-workers (1972), leaves no doubt that co-pigmentation is one of the most important factors controlling petal colour *in vivo*. Indeed, these authors (Asen *et al.* 1972b) indicate that co-pigmentation is not only involved in producing mauve and blue shades but is also essential for the full expression of cyanic colour in many higher plants.

Asen *et al.* (1972) point out that at the normal pH of the cell sap (between 3·0 and 5·0) aqueous solutions of pure anthocyanins are not as strongly coloured as at pH 2 and below and, furthermore, they absorb only in the red region. However, at pH 3·0 to 5·0, addition of flavonol co-pigment, at a similar concentration as the anthocyanin, causes a bathochromic shift and also a large increase in absorbance. This effect can be observed with the common anthocyanidin 3,5-diglucosides (see Table VIII) and complexes are formed with both red flavylium salts and purple anhydro bases. The response is different in the case of the corresponding 3-glycosides, since addition of flavonol to delphinidin 3-glucoside had no effect on absorbance at physiological pHs (Table VIII); presumably here the anhydro base is so stable that co-pigmentation is not required for further stabilization. While practically all flavones, flavonols and related phenolics can co-pigment, presumably by hydrogen bonding, with anthocyanins at physiological pHs, their effectiveness varies with structure, so that both their nature and concentration determines the degree of co-pigmentation *in vivo.* In fact, all colours between red and blue can be achieved *in vitro* by varying (a) the nature and concentration of anthocyanin, (b) the molar ratio of anthocyanin to flavonol co-pigment, and (c) the pH. That co-pigmentation occurs *in vivo* has been confirmed by successful matching of the absorption spectra of co-pigmented anthocyanin solutions against the spectral curves of cyanic colours measured in intact petals for a range of plants (Stewart *et al.* 1969).

Further confirmation of the *in vivo* importance of co-pigmentation has been provided by the isolation from the flowers of a blue *Iris* cultivar, "Prof. Blaauw", of a stable blue non-metallic co-pigment complex. The material contained the anthocyanin delphanin, as the anhydrobase, several *C*-glycosylflavones (as co-pigments) and some pectin (Asen *et al.*, 1970). A related complex has also been obtained from larkspur, *Delphinium ajacis* (Bayer *et al.*, 1966).

Finally, a dramatic example of the effect of lack of co-pigmentation on flower colour may be mentioned. This is the case of an orange petal "sport" of the *Azalea* cultivar "Red Wing". The hypsochromic shift (11 nm) in colour caused by cytoplasmic mutation is entirely due to the suppression of flavonol glycoside synthesis in the sport since the same anthocyanin, cyanidin 3,5-diglycoside, is present in both forms (Asen *et al.*, 1971b).

As a result of these recent studies, the phenomenon of co-pigmentation may have to be considered under two heads: first, the usual situation where sufficient flavone is present to allow full expression of the natural colour of the anthocyanin; and second, the less common situation where a significant increase in the flavone:anthocyanin ratio produces additional co-pigmentation and hence bathochromic shifts from red and purple to mauve and blue shades. At least these new results explain why flavones (or flavonols), besides

TABLE VIII
Co-pigment effect of adding quercitrin to anthocyanin solutions at pH 3·32

Anthocyanin (2×10^{-3} M)	Co-pigmented solution[a]		
	Spectral max. (nm)	Bathochromic shift (nm)	Increase in absorbance (%)
Pelargonidin 3,5-diglucoside	510	15	226
Cyanidin 3,5-diglucoside	527	19	218
Delphinidin 3-glucoside	530	20	16
Peonidin 3,5-diglucoside	533	22	230
Petunidin 3,5-diglucoside	538	23	188
Malvidin 3,5-diglucoside	549	30	560

[a] Quercitrin (quercetin 3-rhamnoside) added at a concn. of 6×10^{-3} M. Similar effects were observed in the pH range 2·12 and 5·10 and were less pronounced at lower co-pigment concentrations. Data from Asen *et al.* (1972b).

occurring in white and cream flowers (see below), are widespread, indeed universal, in their occurrence in cyanic petals.

6 *Metal complexing*

The recent isolation of blue-coloured metal-chelated anthocyanins from several plant species has suggested that such chelation is a factor in the blueing of flower colour. However, the fact that such pigment complexes always contain flavone as well raises some doubts as to whether the presence of metal is actually essential for blueing since this can be so effectively accomplished by co-pigmentation alone (see above). No doubt, the metal could be important in stabilizing anthocyanin–flavone complexes *in situ* but this has yet to be proved. A significant restriction on such chelation is that it is confined to anthocyanins with a free catechol nucleus, i.e. to cyanidin, delphinidin and petunidin glycosides. Thus, in blue flowered plants which have malvidin as the major anthocyanidin, some other mechanism must be invoked to explain blueness.

Another restriction on metal chelation as a factor in flower colour is the availability in the tissue of suitable chelating metals, such as aluminium or iron. This can be seen in the indirect nature of the control of colour in the sepal of *Hydrangea macrophylla*, where the complexing of delphinidin 3-glucoside by aluminium and molybdenum is affected by the internal mineral balance in the plant. Thus the colour change from red to blue is controlled by the availabilities of nitrogen, phosphorus and potassium salts to the plant, which in turn have a large effect on the accumulation of the chelating metals (Asen *et al.*, 1959).

Success in isolating chelated anthocyanins from blue flowers has been achieved by directly expressing the petal sap, followed by immediate precipitation with alcohol or ether. The complexes are unstable *in vitro* and are readily dissociated by acid or by ion-exchange columns. They can, however, be successfully purified by dialysis or by chromatography on cellulose columns. Two such pigments have been studied in great detail: commelinin from *Commelina communis*, and protocyanin from the blue cornflower *Centaurea cyanus*. These pigments have been analysed by several groups of workers, in some cases with conflicting results. However, it now appears (Asen and Jurd, 1967) that the two pigments have essentially the same type of structure, although their anthocyanins and metal ions are different.

In commelinin, the anthocyanin delphanin [delphinidin 3-(*p*-coumaryl)rutinoside-5-glucoside] and the flavone, 6-*C*-glucosylapigenin 7-methyl ether 4′-glucoside, are complexed with magnesium in the ratio 2:2:1 (Takeda *et al.* 1966; see also Hayashi and Takeda, 1970). Protocyanin, on the other hand, is based on cyanidin 3,5-diglucoside (cyanin) complexed with iron (Bayer *et al.*, 1960; Saito and Hayashi, 1965) and the flavone, apigenin 7-glucuronide-4′-glucoside (Asen and Horowitz, 1974). The crystalline pigment in this case is an iron complex with four molecules of cyanin to three of the flavone (Asen and Jurd, 1967).

If, indeed, metals are important in some plants for producing blue colours, then isolation of "anthocyanin complexes" from some corresponding red flowers should yield metal-free pigments. This is, in fact, the case and so-called "genuine red anthocyanins" have been obtained, using mild methods of extraction at neutral pHs, from red cornflower, red dahlia and a red rose (Saito *et al.*, 1964). A violet pigment, violanin, similarly extracted from the pansy *Viola* x *wittrockiana*, was originally reported to contain some potassium (Hayashi and Takeda, 1962) but further purification has provided a metal-free material (Takeda and Hayashi, 1965). Electron spin resonance (e.s.r.) measurements on these red pigments, which are assumed to have a quinonoid structure, have indicated that they can exist as free radicals (Takeda *et al.*, 1968).

7 *Effect of pH*

Solutions of anthocyanins are red in acid, blue in alkali and the pH of the cell sap was once considered to be an important factor in flower colour. However, a survey by Shibata *et al.* (1949) of the cell

TABLE IX
Factors controlling cyanic flower colour in higher plants[a]

1. Nature of the anthocyanin (especially degree of hydroxylation)
2. Concentration of anthocyanin
3. Molar ratio of anthocyanin to flavone co-pigment
4. pH
5. Chelation with metals, such as magnesium and iron
6. Mixtures of anthocyanins
7. Background effects of chlorophyll or carotenoids
8. Binding of anthocyanin to macromolecules (e.g. pectins)
9. Anatomical modifications

[a] Arranged in approximate order of importance.

sap of 200 plants showed that flowers were all acidic (pH about 5·5), irrespective of colour. Nevertheless, small changes of pH within the acid range appear to have some effect on flower colour. Thus Scott-Moncrieff (1936) noted differences of 0·5 and 1·0 pH units between red and mauve colour forms of *Primula sinensis*, *P. acaule* and *Papaver rhoeas*. In *P. sinensis*, the change of pH from 5·4 to 6·2 has a blueing effect and is controlled by the gene R, which is independent of factors controlling co-pigmentation and methylation. The difference in pH of petal sap between the two phenotypes of *P. sinensis* has been confirmed using modern instrumentation (J. R. S. Fincham and J. B. Harborne, unpublished results).

More recently, shifts in flower colour in the rose cultivar "Better Times" (Asen *et al.*, 1971a) and the garden statice *Limonium* (Asen *et al.* 1973) have been attributed to small changes in pH. *In vitro* experiments with anthocyanin–metal–co-pigment complexes (Jurd and Asen, 1966) have also confirmed that increasing pH in the range 3·0 to 6·0 has a blueing effect on colour. Compared with other modifying factors, however, it seems that pH is only of minor importance in controlling flower colour. For convenience, a summary of the various factors concerned in cyanic flower production in higher plants is given in Table IX.

8 *Anthocyanins on a yellow background*

Many colour effects are produced by interaction of anthocyanins with other pigments. The co-pigment effects of the cream flavones have already been discussed but the effects of co-occurrence with yellow pigments have not yet been mentioned. A good example here is the garden snapdragon, which contains yellow aurone pigments but no carotenoids. Mixtures of cyanidin with these aurones give orange-red colours, while mixtures of pelargonidin and aurone give orange-yellow shades. High anthocyanin concentration is here related to low aurone concentration and vice versa; the interaction of factors controlling pigment synthesis has a considerable effect on flower colour (Jorgensen and Geissman, 1955). The general effect of anthocyanins occurring with water-soluble yellow pigments is such that cyanidin-containing flowers appear as if they have pelargonidin. Examples of flowers having cyanidin and water-soluble, but unidentified, yellow pigments are the yellow-red petals of *Lapeyrousia cruenta, Chasmanthe* and *Crocosma masonorum* (Harborne, 1963a) and those of *Tecomaria capensis* and *Holmskioldia sanguinea* (Forsyth and Simmonds, 1954).

When anthocyanins and lipid-soluble yellow pigments co-occur, the resulting flower colour is more often brown than orange. This may be because, whereas water-soluble yellow pigments are in the cell vacuole with the anthocyanins, carotenoids are located in the plastids. Brown colours formed by magenta cyanidin on a yellow carotenoid background can be seen in the wallflower *Cheiranthus cheiri*, in *Primula polyanthus* and in rose varieties, e.g. the coffee-coloured "Café". Brown colours are not confined to the flower; brown anthers of the flowers of some *Solanum* plants are coloured by the petunidin glycoside, petanin, on a carotenoid background.

9 *Flower colour changes*

Besides colour variation between mutants of garden plants which has formed the basis of the discussion up to this point, there are also the alterations in colour undergone by individual flowers as they develop, mature, fade and die. These changes are very pronounced in a few plants (e.g. from yellow to red in *Cheiranthus mutabilis*) and have therefore attracted the attention of pigment chemists. The assumption that such changes are due to the interconversion of one kind of pigment into another is most unlikely, however attractive the hypothesis that a colour change from yellow to red is due to the *in vivo* reduction of flavonol to anthocyanidin. Thus tracer and enzymatic studies show that flavonoid and carotenoid pigments, although often described as "end products" of metabolism, are regularly "turned over" in the plant. Furthermore, both Reznik (1961) and Hess (1963) have shown that the initiation and rate of flavonol synthesis in petals of *Primula obconica* and *Petunia* are quite independent of those of anthocyanin synthesis. More probably, explanations for colour changes during the life of a flower lie in differential rates of pigment synthesis, and in alterations of pH and of availability of metal ions. These factors operating singly or together will inevitably produce effects on flower colour.

Three examples may be mentioned. First, the popular rose "Masquerade" is yellow in bud, orange-yellow when freshly open and deep red before fading. It is clear that yellow carotenoid is produced at an early stage of development, whereas the synthesis of the anthocyanin, cyanin, is delayed until maturity. Significantly, the undersides of red petals have yellow patches, indicating that anthocyanin synthesis in this variety is particularly light dependent. Second, the flower of *Hibiscus mutabilis* changes during the course of one day from yellow in the morning to pink in the evening. It was originally suggested that the pigment of the yellow morning flowers, quercetin 7-glucoside, was the precursor of cyanidin 3,5-diglucoside in the pink evening flowers (Subramanian and Swamy, 1964) but, more recently, it has been found that the change is due to a threefold increase in anthocyanin synthesis during the day, with the flavonol levels remaining unchanged (Subramanian and Nair, 1970). The flavonol glycosides present are all derivatives of quercetin: the 3-, 7- and 4′-glucosides, the 3-galactoside and the 3-rutinoside. According to Subramanian and Nair (1970) there are two anthocyanins, the 3,5-diglucoside and 3-rutinoside-5-glucoside of cyanidin, while Lowry (1971) claims that there is only one pigment, namely cyanidin 3-sambubioside, in the flowers of *Hibiscus mutabilis*. A third example is the garden pea, *Pisum sativum* and related legume species where notable colour changes occur in both the wing and keel of the flower during ontogeny. Statham and Crowden (1974) have found that there is a change in the type of anthocyanin synthesized during flower development and the older flowers have pigments that are less methylated and there is also a change in the glycosylation pattern. Such changes in pigment synthesis could be responsible, in part, for the alterations in flower colour observed during floral ontogeny.

10 Flower patterns and virus breaks

It is not possible to discuss the chemistry of flower colour without a brief mention of flower patterns. These are quite elaborate in such plants as the foxglove, violet, orchid, iris and primrose. Patterns are usually due to local increase in pigment production in some areas of the petal (e.g. the foxglove). Alternatively, a second pigment is locally superimposed on the main colouring matter (e.g. the dark purple spot in some poppy flowers is due to cyanidin on a pelargonidin background). Patterning is under complex genetic control and occasionally genes controlling patterns are mutable and their effects vary with the environment. Some patterns are related to the needs of insect vectors; the "honey lines" of *Streptocarpus* and many other flowers direct the bees towards the nectar (Manning, 1956b).

A more complicated partitioning of pigments has been observed by Bloom and Vickery (1973) in the *Mimulus luteus* complex, the flowers of which have red spots on a yellow background. Some, but not all, of the flavonoids are differentially distributed between the spots and the background. For example, quercetin 3-glucoside, which is structurally related to the red anthocyanin, co-occurs in the spot with cyanidin 3-glucoside, whereas quercetin 7-glucoside does not. Again, one of the two classes of yellow pigment present, namely carotenoid, is not partitioned, whereas the other yellow flavonol (in this case herbacetin 7-glucoside) is exclusive to the background. The significance of flower patterning produced by the presence of flavonoids which absorb strongly in the ultraviolet will be discussed in more detail in the next section.

Some colour patterns have quite a different cause: that is, virus infection. Virus breaks in tulips were first reported in the literature in 1568. The effect of virus is to inhibit anthocyanin in some areas of the tulip petals; the variegated forms, once so popular, are now rapidly being replaced by healthy self-coloured varieties. Virus infection has a mottling effect on flower colour in the garden stock *Matthiola incana*, and may either increase anthocyanin colour or inhibit it (Feenstra *et al.,* 1963). The concentration of flavonol in virus-infected "white" flowers is higher than that in healthy white flowers, but there is no simple relationship between inhibition of anthocyanin and increase in flavonol. Precisely how the virus inhibits or stimulates pigment synthesis in these flowers is still obscure.

B CHALKONES, AURONES AND FLAVONOLS

1 *Yellow flower colour*

In considering the contribution of flavonoids to yellow flower colour, it is necessary to point out at the outset that the commonly occurring flavonols and flavones, which are often incorrectly stated to be yellow pigments, do not contribute *per se* to yellow flower colour. It is true that the common flavonols, myricetin, quercetin and kaempferol, are pale yellow in the solid state. Nevertheless, their

3-glycosides (and they always occur in petals in glycosidic form) are either colourless or pale buff in colour. Furthermore, the spectral maxima of the majority of flavonol glycosides lie at about 340–360 nm and there is very little absorption above 380 nm. If they do contribute to yellow colour, it is only if they occur in some unusual glycosidic or other combination or if they act as co-pigments to true yellow flavonoid pigments (see below).

Flavonoids which do contribute to yellow colour can be considered under the following four headings.

(1) The *anthochlor pigments* (chalkones and aurones) are deep yellow in colour and characteristically change to red when the petal is treated with ammonia vapour (Gertz, 1938). Chalkones have absorption maxima at 370 to 390 nm, aurones at 390 to 420 nm. A typical chalkone is coreopsin, which occurs in many *Coreopsis* species, and a typical aurone is aureusin, which is the principal colouring matter of the yellow snapdragon (*Antirrhinum majus*) (for formulae, see Fig. 1). Chalkones and aurones have a restricted distribution in nature, having been identified as flower pigments in about nine plant families (Table X). Although they do occasionally provide the only source of yellow colour in the petal (e.g. in *Dahlia*), more frequently they co-occur with carotenoids (e.g. in *Coreopsis*)

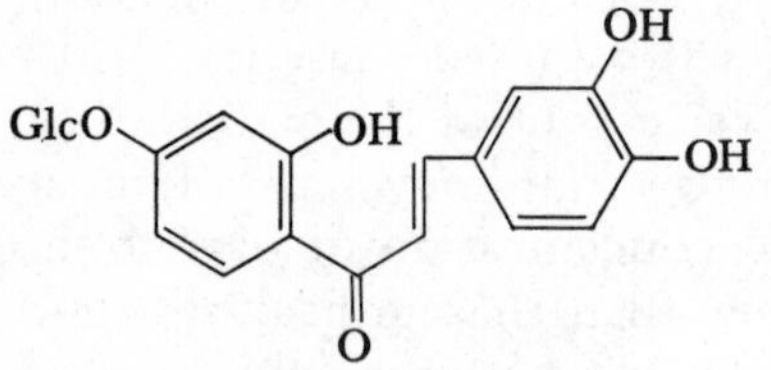

Coreopsin, a chalkone of *Coreopsis tinctoria*

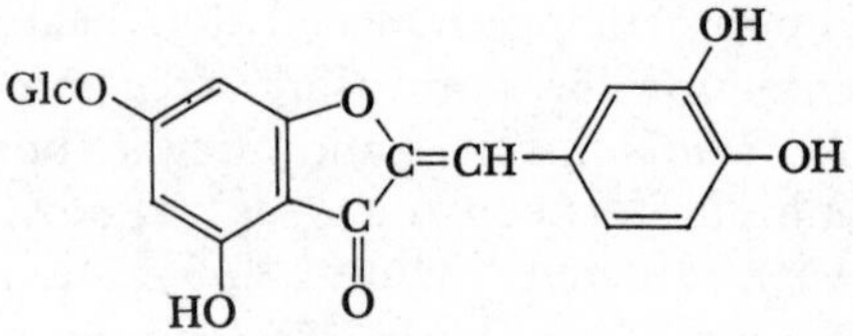

Aureusin, an aurone of *Antirrhinum majus*

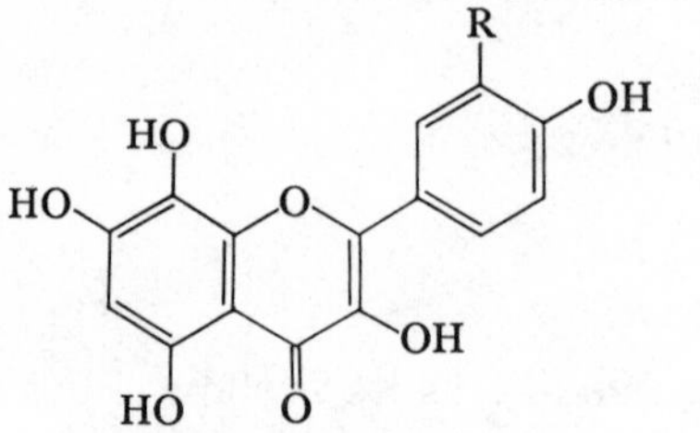

Herbacetin R=H from *Meconopsis paniculata*
Gossypetin R=OH from *Gossypium herbaceum*

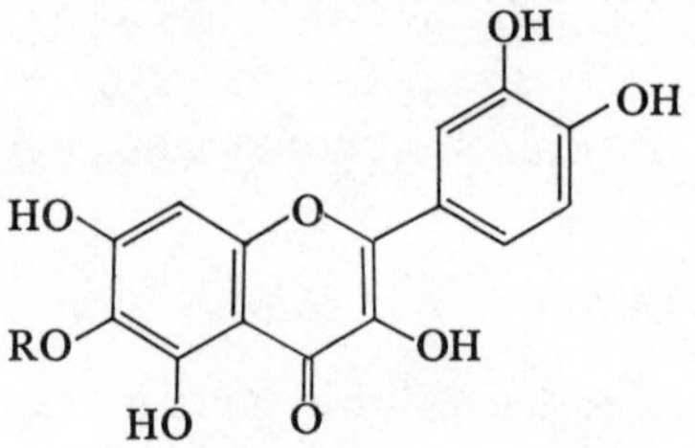

Quercetagetin R=H from *Tagetes erecta*
Patuletin R=Me from *Tagetes patula*

FIG. 1. Some yellow flavonoids.

TABLE X
Plant families with anthochlor flower pigments

Family[a]	Genus and species	Chalcone(s) present	Aurone(s) present
Acanthaceae	*Asystasia gangetica*	Isosalipurposide	—
	Rungia repens		
Caryophyllaceae	*Dianthus caryophyllus* (yellow carnation)	Isosalipurposide	—
Compositae	Many species in the Heliantheae, e.g.		
	Coreopsis tinctoria	Coreopsin	Sulphurein
	Dahlia variabilis	Butein, coreopsin	Sulphurein
Gesneriaceae	Many species in the Cyrtandrioideae, e.g. *Aeschynanthus ellipticus*	Isosalipurposide	—
	Chirita micromusa	—	Cernuoside
Leguminosae	*Butea frondosa*	Coreopsin	
	Ulex europeaus	Isoliquiritigenin Glucosides	—
Oxalidaceae	*Oxalis cernua*	—	Cernuoside
Plumbaginaceae	*Limonium* cv.	2′,4′,6′,3,4-Pentahydroxy-chalcone	4,6,4′-Tri-hydroxyaurone
Ranunculaceae	*Paeonia trollioides*	Isosalipurposide	—
Scrophulariaceae	*Antirrhinum* and *Linaria* spp.	2′,4′,6′,3,4-Pentahydroxy-chalcone	Aureusin, bractein

[a] This list is not exhaustive: unidentified chalkones have been detected in flowers in other families (e.g. *Corylopsis*, Hamemalidaceae). For references and further details, see Swain, this volume, Chapter 8, and also Harborne *et al.* (1975).

and thus share with lipid pigments the task of producing yellow coloration.

(2) *Gossypetin* and related compounds impart a distinctive pale yellow colour to plant petals where they occur. Gossypetin (8-hydroxyquercetin) itself was isolated some time ago as the principal yellow pigment of cotton flowers, *Gossypium* (Perkin, 1899). More recently, it has been recognized as the main pigment, together with the related herbacetin (8-hydroxykaempferol), in the primrose *Primula vulgaris*, and occurs, together with some carotenoid, in many other yellow-flowered *Primula* species (e.g. the

cowslip *P. veris* and the oxlip *P. elatior*). Gossypetin or its derivatives also contribute to yellow flower colour in *Rhododendron, Meconopsis, Papaver* and a range of other genera (see Table XI). In an earlier report (Harborne, 1965) the above plants were thought to be pigmented by a closely related isomer of gossypetin, namely 6-hydroxyquercetin or quercetagetin (see Fig. 1 for formula). Improved methods of analysis, however, have provided a clear means of distinguishing the two closely similar yellow flavonols (Harborne, 1969) and it is now apparent that quercetagetin is much rarer than gossypetin and in general has a different distribution pattern. As a flower pigment, quercetagetin was first isolated from the African marigold *Tagetes erecta* (Latour and Magnier de la Source, 1877) and it accompanies the related 6-methyl ether, patuletin, in flowers of

TABLE XI
Plant families with 6- or 8-hydroxyflavonols in the flowers

Family[a]	Genus and species	Yellow pigment(s) present
Ericaceae	*Rhododendron* spp. (6)	Gossypetin 3-galactoside
Compositae	*Anthemis, Chrysanthemum, Dendranthema, Lepidophorum* and *Tanacetum* spp.	Quercetagetin and patuletin 7-glucosides (gossypetin 7-glucoside in *C. segetum*)
Leguminosae	*Coronilla valentina* subsp. *glauca*	Gossypetin or quercetagetin
	Leucaena glauca	Quercetagetin and patuletin
	Lotus corniculatus	Gossypetin and its 7- and 8-methyl ethers
	Prosopis spicigera	Patuletin 7-glucoside
Papaveraceae	*Meconopsis paniculata*	Herbacetin 3-glucoside
	Papaver nudicaule	Gossypetin 7-glucoside
Primulaceae	*Primula* spp. (7) *Dionysia* spp. (3) and *Vitaliana primulifera*	Gossypetin, mainly as the 3-gentiotrioside
Malvaceae	*Abutilon, Gossypium* and *Hibiscus* spp.	Gossypetin and herbacetin (variously as the 7-, 8- or 3-glucoside)
Sterculiaceae	*Fremontia californica*	Gossypetin
Scrophulariaceae	*Mimulus luteus*	Herbacetin 7-glucoside
Ranunculaceae	*Ranunculus* spp. (8)	Gossypetin 7-glucoside and a gossypetin methyl ether

[a] For references, see Harborne (1968, 1969); also Harborne *et al.* (1975).

Tagetes patula. These two compounds seem to be almost completely restricted to the Compositae as flower pigments, since their only other major sources are the yellow rays of various members of the *Chrysanthemum* complex. Indeed, the sole case where quercetagetin and gossypetin co-occur seems to be in the petals of the corn marigold *Chrysanthemum segetum* (Harborne *et al.*, 1970). The only unrelated occurrences of quercetagetin and patuletin seem to be in the petals of *Prosopis spicifera* and *Leucaena glauca* (both Leguminosae) (Nair and Subramanian, 1962).

As with the anthochlor pigments, the yellow flavonols seem to occur fairly often (and especially in the Compositae) with yellow carotenoids so that their presence in a sense seems superfluous. That they have a raison d'être in such instances is apparent from the recent discovery of Thompson *et al.* (1972) that they can act as nectar guides in flowers in which they occur. In particular, these authors identified glycosides of quercetagetin, patuletin and patuletin 7-methyl ether in the petals of the composite "Black-eyed Susan", *Rudbeckia hirta*, and demonstrated that, unlike the co-occurring carotenoids, their distribution was completely restricted to the petal bases. Their intense u.v. absorption between 340 and 380 nm means that they are visible to pollinating insects, e.g. bees, which can "see" in this range of the spectrum, and they provide a means for the insects to locate the centre of the flower, where the pollen and nectar are. Flavonoids may well fulfil such a role in a range of different plants (Horowitz and Cohen, 1972; Eisner *et al.* 1973).

(3) The *common flavonols* probably contribute to yellow flower colour when (a) they are methylated or (b) they are present in certain unusual glycosidic forms. Thus a myricetin dimethyl ether (syringetin) contributes to yellow colour in the meadow pea *Lathyrus pratensis*, and isorhamnetin (quercetin 3′-methyl ether) may do the same in the common marigold *Calendula officinalis*. Quercetin 7- and 4′-glucosides have absorption spectra similar to quercetin itself and may therefore provide some yellow in gorse *Ulex europeaus*, in *Rosa foetida* and in other petals in which they occur.

The common flavones absorb at shorter wavelengths than the flavonols and probably never directly contribute to yellow flower colour. However, they may well have an indirect role as co-pigments in intensifying the colour of chalkones, aurones or yellow flavonols in petals. At least, Asen *et al.* (1972a) have been able to demonstrate that the flavone apigenin 7-glucuronide co-pigments with the aurone, aureusin in petals of the yellow snapdragon cultivar "Yellow

Rocket", producing a bathochromic shift from 395 to 440 nm. These authors were also able to show that the u.v. and visible spectrum of the co-pigmented aurone solution precisely matched that observed by optical measurements in the fresh petals. Another possibility—that flavonols and flavones contribute to yellow flower colour when complexed with metals such as aluminium (compare the anthocyanin-metal complexes, p. 752)—has yet to be demonstrated *in vivo*.

(4) *Anthocyanins with unusual hydroxylation patterns* (e.g. apigeninidin, λ_{max} 476 nm) and their presence in orange-yellow flowers have already been discussed in Section IIA3.

Finally in this section, the contribution of flavonoids to yellow petal colour *relative to* other classes of yellow pigment needs some discussion. The range of other types may be considerable, although most emphasis has been placed on the widespread carotenoids (Goodwin, 1952). Little is known, by contrast, of the distribution in flowers of yellow alkaloids. Under this heading come not only the amino acid-based betaxanthins of the Centrospermae (see Chapter 11) but also more conventional alkaloids such as the isoquinoline berberine from *Berberis* and the as yet incompletely characterized alkaloid glycoside nudicaulin of the Iceland poppy. Whether quinones, another chemical class which could contribute to yellow colour, occur at all in flower petals as colouring matters is also not generally known.

With regard to flavonoids vis-à-vis carotenoids, a number of broad surveys (see e.g. Seybold, 1954) have indicated that most yellow-flowered angiosperm species have carotenoids as the major yellow pigments. Certainly, most bright yellow petal colours in nature, as in the buttercup, are characteristically carotenoid based. Most surveys have been based on simple colour and solubility tests so that there may well be a degree of error in the results. For example, the possibility of carotenoids being accompanied by yellow flavonoids has not often been considered, so that there are certain gaps in our information on this point.

In fact, if the distribution of yellow pigments within particular groups of plant (see Table XII) is considered, one finds that there is apparently a considerable number of different ways that yellow colour is achieved. At the extremes, there are plants which have only yellow flavonoid (primrose, cotton flower etc.) and there are those, probably the considerable majority, which have yellow carotenoids. In between, there are cases where both vacuolar and plastid

TABLE XII
Some typical patterns of flower pigmentation among angiosperm plants

Plant Group(s)	Pigments and distributions
Centrospermae	(1) Alkaloidal betaxanthins in most yellow-petalled plants
	(2) Yellow flavonoids in the Caryophyllaceae (e.g. isosalipurposide in the carnation)
Tribe Heliantheae of the Compositae	(1) Yellow anthochlors + carotenoids in many species (e.g. *Coreopsis*)
	(2) Carotenoids alone in many other species
Crocus, *Lilium* and *Tulipa*	Yellow carotenoids (violaxanthin, carotenes) the only pigments; colourless flavonols also present
Papaveraceae	(1) Alkaloid nudicaulin in *Meconopsis cambrica*
	(2) Nudicaulin + gossypetin in *Papaver nudicaule*
	(3) Herbacetin alone in *Meconopsis paniculata*
	(4) Carotenoids alone in *Eschscholtzia californica*
Primula (Primulaceae)	(1) Gossypetin main flower pigment in sections Vernales and Sikkimenses
	(2) Carotenoids main pigments in sections Candelabra, Nivales, Farinosae and Verticillata
Scrophulariaceae	(1) Chalkones and aurones only pigments in *Antirrhinum* and *Linaria*
	(2) Unidentified novel aurone in *Calceolaria*
	(3) Herbacetin + carotenoid in *Mimulus luteus*
	(4) Water-soluble carotenoids in *Nemesia* and *Verbascum*
	(5) Carotenoids in *Mimulus longiflorus*

pigments co-occur and other cases where yellow alkaloids are involved. Even within the same genus, there may be different patterns. This is true in *Primula,* where yellow flavonols predominate in some sections, replacing carotenoids which are the main yellow pigments in other sections (see Table XII). Similar cases of such variations at the generic level are found in *Paeonia* and *Rhododendron* and, at the family level, in the Gesneriaceae (Harborne, 1967). Such differences may reflect differences in geographical habitat at the generic or family levels and be related to differential visual sensitivities of pollinating insects to carotenoid- or to flavonoid-yellow. One could argue that the angiosperms, having at hand in the carotenoids a perfectly satisfactory method of producing yellow colour, should not require any other source of yellow. The fact that they have evolved a variety of other ways of synthesizing yellow colours requires some explanation and the full answer will only be given when both biochemical and ecological techniques have been applied together to the phenomenon.

2 *Cream, ivory and white flower colour*

A large proportion of higher plants have white, ivory or cream flowers; in addition, cyanic-flowered plants in cultivation not infrequently produce acyanic (white) mutants. The vast majority of such flowers have "colourless" flavones or flavonols. Surveys of white-flowered wild plant species by Roller (1956) and Reznik (1956) showed that 86% of the sample contained kaempferol and 17% quercetin; luteolin and apigenin also commonly occur. Dihydroflavonols, leucoanthocyanidins and flavanones are of less frequent occurrence in such flowers.

Although not visible to the human eye, these pigments absorb strongly in the ultraviolet and can be "seen" by bees and, presumably, other insects; thus they may serve some purpose in the plant. The presence of these flavonoids also adds "body" to the flower petal to give a cream or ivory appearance. Thus the albino mutant in *Antirrhinum majus* (a rare form which lacks flavones altogether) is readily distinguished in its appearance from ivory flavone-containing flowers of the same plant. There are also several different white forms of *Dahlia variabilis:* some have the flavones (luteolin, apigenin) present in cyanic forms, others have the flavonols kaempferol and quercetin, and yet others (e.g. "Clare White") have flavanones (Nordström and Swain, 1958), but no true albino is known. In fact, white petals which completely lack flavonoids of any kind are very rare. Besides the albino snapdragon, the only other examples known to the author are certain strains of *Petunia* and *Pisum.* Such plants are of great interest from the point of view of flavonoid biogenesis (cf. Geissman and Harborne, 1955).

The biogenetic relationship of the ivory flavones to the anthocyanins is now well established (see e.g. Harborne, 1962c) and some examples of the co-occurrence of structurally-related anthocyanins and flavones are given in Table V. For a complete discussion of flavonoid biosynthesis, the reader is referred to Chapter 9.

III Leaf colour

Anthocyanins are practically the only flavonoids which are visible in leaves and their contribution may be considered under three headings: (1) red pigment formed transiently in young leaves but disappearing as the leaves mature; (2) permanent leaf pigmentation; and (3) autumnal colouring. One general point about leaf anthocyanins is that they are simpler in structure than those of flowers. Indeed, cyanidin (as the 3-glucoside) is the characteristic pigment in the leaf in all stages of its growth and other anthocyanidins are much less frequent (Lawrence *et al.*, 1939)

A TRANSIENT LEAF COLOURS

Young leaves of many higher plants, at an early stage in development, produce a flush of anthocyanin which disappears rapidly as the leaves mature. Price and Sturgess (1938) surveyed 200 species from 110 genera and identified the anthocyanin in 93% of this sample as a cyanidin 3-glycoside. In a narrower survey, Reznik (1956) found cyanidin in nine species, delphinidin in four and peonidin in three.

The cause of anthocyanin formation at this stage in leaf development is probably that sugar accumulates in the tissues in amounts in excess of the immediate requirements for growth. Indeed, anthocyanin synthesis in leaves seems to be rather closely related to carbohydrate metabolism. Experiments with tomato seedlings show that under any environmental or cultural stress (i.e. high light intensity, nitrogen starvation, feeding of sucrose, removal of "carbohydrate sinks" or low temperatures), quantities of petunidin glycoside accumulate (Hussey, 1963). If subjected to both a high light intensity and conditions which stop growth, the tomato plant becomes deep purple and the chlorophyll green is completely masked.

Other conditions leading to anthocyanin accumulation in leaves include viral or fungal infection, treatment with growth regulators or wounding. For example, purple tinges appear on leaves of leaf roll-infected potatoes as a consequence of the disturbed translocation caused by this virus. With regard to wounded tissue, Bopp (1959) surveyed 191 species and found anthocyanin accumulation around the wound in 20% of this sample.

The transient anthocyanin soon disappears as the plant grows away, but it is not known whether the pigment produced in the juvenile stage is enzymically destroyed by anthocyanase (Huang, 1955) or by an oxidase (Bopp, 1957) or is "diluted out" as the cells divide and multiply.

B PERMANENT LEAF COLOUR

While anthocyanin colour is rather widespread in young and autumnal leaves, it is rather uncommon as a permanent leaf feature. While cyanidin is again the most usual pigment (present in twenty-eight out of thirty-six types surveyed by Reznik (1956)), delphinidin types are known and pelargonidin also occurs.

Many of the plants that show leaf coloration are of considerable ornamental value. The copper beech, a "sport" from the normal green *Fagus sylvatica*, is pigmented with the 3-galactosides of cyanidin and pelargonidin. Cyanidin glycosides also provide permanent colours in *Begonia* (*B. rex* has a deep purple-red leaf), in *Coleus* (notable for the striking anthocyanin patterning), in *Rosa*, in *Acer* and in *Rubus*. Intense leaf anthocyanin colour is sometimes correlated with deep colours in the flowers; the dark magenta form of *Antirrhinum* is one example. One of the most intensely pigmented of all plants is the red cabbage *Brassica oleracea*.

Primula and *Solanum* plants are unusual in having delphinidin derivatives in their leaves. In *Primula sinensis*, leaf pigmentation is correlated with that of the flower; thus K types have delphinidin in both plant parts, whereas kk types have mainly pelargonidin in the flower and mainly cyanidin and peonidin in leaves. Leaf pigments are commonly less methylated than those of the flower (Harborne and Sherratt, 1961). In *Solanum* plants and in the tomato, the prevailing leaf pigment is petunidin (Harborne, 1960) and one wild potato species, *S. microdontum*, has a remarkably intensely pigmented purple-black stem, together with normal green leaves.

Anthocyanins differing from the usual type occur in the leaves of members of the Gesneriaceae. This is not surprising since apigeninidin and luteolinidin have been detected in flowers of plants of this family. Unusual "pigments" are present in leaves of aquatic plants, which, according to Reznik and Neuhausel (1959), have anthocyanins present in a colourless leucobase form. Treatment of the leaves with cold acid liberates anthocyanin, so these compounds are not leucoanthocyanidins. In passing, it may be mentioned that colourless leucobase forms of anthocyanin have also been found in flowers, e.g. in white blooms of *Lespedeza hortensis* (Hayashi and Abe, 1953) and in some, but not all, white *Iris* varieties (Werckmeister, 1955; J. B. Harborne, unpublished data).

Anthocyanin colour in leaves is often localized in its distribution. It sometimes occurs only on the undersurface (e.g. *Hoffmannia ghiesbreghtii*), or in the leaf hairs (e.g. *Gynura aurantiaca*). Permanent anthocyanin in leaf appears to lack function; many plants with red leaves are variants of a more normal green form. However, the presence of anthocyanin does not appear to be a handicap; the plants are able to photosynthesize normally.

Other classes of flavonoid besides anthocyanin occasionally provide permanent colour in plants in the form of coloured deposits on the undersides of leaves. One of the best known examples is the gesnerad *Didymocarpus pedicellata*, which produces a highly coloured dust under the leaves consisting chemically of a range of methylated chalcones which are present in the quinonoid form (Seshadri, 1951; see also Chapter 10, by R. H. Thomson). A number of ferns are distinguished by having white, yellow or other coloured deposits on the undersurfaces of the fronds. In *Pityrogramma*, these are based on either methylated chalcones (Nilsson, 1967) or on kaempferol in methylated form (Smith *et al.*, 1971).

C AUTUMNAL COLOURING

Autumnal colouring is well known to be provided by both anthocyanins and carotenoids. Anthocyanin coloration is particularly striking in certain genera (*Acer* and *Pyrus*, for example) and is dependent on climatic factors for its full development. The pigment of autumnal leaves is almost always the simplest anthocyanin, cyanidin 3-glucoside. Hayashi and Abe (1953) analysed seventy-four plants from twenty-five families and found cyanidin 3-glucoside in all of them. Again, Reznik (1956) found cyanidin in forty-seven out of forty-nine plants; autumnal leaves of forty-five species had a quercetin glycoside as well. While it is assumed that anthocyanin formation in autumn is connected with the liberation of sugar (from starch degradation) in dying leaves, no quantitative measurements correlating these two factors have yet been made.

IV Fruit colour

As in the leaves and flowers, anthocyanin is a major contributor to colour in the fruits of higher plants. Not unnaturally, most attention has been given to the pigments of edible fruits. One or other of the six common anthocyanidins provides colour here (Table XIII); the glycosidic patterns of the pigments of most of these fruits are known (see Harborne, 1964b) but are not given since glycosylation is not immediately relevant to colour production. Ornamental fruits have also been studied; for example, cyanidin and peonidin 3-galactosides are known to pigment the red berries of *Ardisia crispa* and a petunidin glycoside is present in the purple-black berries of the deadly nightshade *Atropa belladonna* (Harborne, 1963b).

As indicated in Table XIII, anthocyanin colour is sometimes present throughout the fruit (e.g. raspberry), but in other cases is

TABLE XIII
Anthocyanidins of edible fruit

Anthocyanidin	Fruit
Pelargonidin	Passion fruit, strawberry[a]
Cyanidin	Apple (skin), pear (flesh), mulberry, blackberry, raspberry, cherry, plum, peach, sloe, redcurrant, elderberry and cranberry[b]
Cyanidin and delphinidin	Blackcurrant, "blood" orange (juice), "red" banana (skin)[c]
Delphinidin	Pomegranate (juice), aubergine (skin), whortleberry and grape[d]

[a] Wild strawberry has a 1:1 mixture of Pg and Cy, whereas cultivated strawberry has Pg, with only traces of Cy (Sondheimer and Karash, 1956).

[b] Also contains peonidin.

[c] Pigments are methylated (peonidin and malvidin); the pigment of the more common yellow-skinned cultivars is carotenoid.

[d] Also contains petunidin and malvidin.

confined to the juice (e.g. blood orange) or the skin (e.g. apple). Among the legumes, anthocyanin may colour the pod of some species (e.g. strains of *Pisum sativum*) or the seed coat of others (e.g. *P. nepalensis, Phaseolus vulgaris, Phaseolus multiflorus*). The seed coat pigments of the various colour forms of the broad bean have been studied in great detail by Feenstra (1960), who has isolated from them all six common anthocyanidins, as well as several flavonol glycosides and some leucoanthocyanidins.

Many of the factors modifying anthocyanin colour in flowers presumably also operate in the fruit, but little work has been done on this aspect of fruit colour. Differences in pigment concentration are mainly responsible for the distinctive appearance of the two main groups of cherry varieties, the "black" (actually purple) and the "white" (actually pale red) cherry. The colours in the skin and flesh of the cherry are controlled by different genes, and yellow cherry cultivars, e.g. "Stark's Gold", lack anthocyanin in the flesh but have a yellow carotenoid in the skin. Metal complexing rather than co-pigmentation is presumably responsible for most of the blue colours in plant berries; thus Chenery (1948) found that 87% of 154 blue-fruited species were strong aluminium accumulators.

Flavonols and flavones occur widely in fruits but their contribution, if any, to fruit colour has not been extensively studied. Other classes of flavonoid undoubtedly contribute occasionally to yellow, orange, red or brown colours in fruits. Thus the yellow isoprenoid isoflavones, osajin and pomiferin, contribute colour to the osaje orange, *Maclura pomifera* (Wolfrom and Mahan, 1942). Again, three methylated chalcones have been isolated from the fruit of *Merrillia caloxylon* (Rutaceae) by Fraser and Lewis (1972). Finally, the fruit of the cyperaceous plant *Kyllingia brevifolia* has yielded the bright yellow chalcone okanin and that of the related *Gahnia clarkei* the yellow-brown aurone, aureusidin (Clifford and Harborne, 1969).

V Functions of flavonoids in leaves

A PROTECTIVE ROLES

Although the most important function of flavonoids in plants appears to be their contribution to colour, this does not tally with their natural distribution within the plant. Indeed, in quantitative terms, the bulk of the flavonoid occurs hidden by chlorophyll within the leaf and thus cannot play any direct part in the visual attraction of animals to plants. It is, however, well established that insects can "see" in the u.v. region of the spectrum. Thus flavonoids, because of their strong adsorption band at 330–380 nm, in a region where chlorophylls and carotenoids do not absorb strongly, may be detectable by insects in leaves. Alternatively, their presence may be elicited by taste during insect feeding. There is, thus, the possibility that leaf flavonoids have a protective role in plant–animal interactions and some of the recent evidence in favour of this hypothesis will be discussed below.

An even more important protective role for leaf flavonoids that has been proposed (see Caldwell, 1971; Swain, 1975) is as a light screen against damaging u.v. radiation. Some early experiments of Barber (1965) on natural stands of *Eucalyptus urnigera* in Tasmania are relevant here. He found that increasing anthocyanin content in the young leaves was correlated with increasing altitude and suggested that more anthocyanin was produced at the higher altitudes because the u.v. radiation from the outer atmosphere was more intense there.

Flavone and flavonol glycosides, because of their strong u.v. adsorption both at 250–270 nm and at 330–370 nm, would seem to offer equally good protection against such radiation. Their presence in leaves, for example, would prevent mutagenesis of thymine (λ_{max} 260 nm) in DNA and also possible photodestruction of such coenzymes as NAD and NADP (λ_{max} 340 nm). The recent discovery of flavonoids actually occurring within leaf chloroplasts has brought this role further into the limelight since this would be one possible explanation for their presence in this vital organelle. A brief account of the recent work on flavonoids in chloroplasts is therefore given below.

B FEEDING DETERRENTS

The ecological role of flavonoids, together with other secondary constituents, has received considerable attention in recent years (e.g. Harborne, 1971) and there is growing evidence to suggest that some of these compounds are involved in the feeding responses of insects and other animals in natural habitats. The chemical interaction between a food plant and the animals parasitic on it is a complex and subtle one and has been reviewed at length (Schoonhoven, 1968; Dethier, 1970). Two aspects are particularly relevant in the case of the flavonoids.

Firstly, it appears that a chemical deterrent in the leaf is apparent to the animal not only through its toxicity but also because there is an associated "warning signal" indicating its presence. In the case of alkaloids or glucosinolates, this signal is given by the bitter taste. For the flavonoids, it may be either taste (e.g. the well-known astringency of the flavolans or proanthocyanidins) or colour, perceived as yellow or as strong u.v. absorption. The second point is that it is not important, in an evolutionary context, whether a particular compound is detected as being an attractant or a deterrent. All such compounds appear to have been originally synthesized by the plant as deterrents and only the metabolic flexibility of insects has allowed them to avoid the toxic consequences of this activity and to turn deterrents into attractants. Thus any compound which appears as an attractant to a particular insect predator must have originally been a deterrent and is still presumably a deterrent to the majority of other insect feeders.

Examples where flavonoids have definitely been implicated in insect feeding responses are given in Table XIV and Fig. 2. The work involved in establishing such relationships is considerable, so that it is not surprising that only a few cases of such interactions are at present known. There is some circumstantial evidence suggesting that this is a fairly general phenomenon among the angiosperms. In particular, it is likely that many 6- or 8-hydroxyflavonols and flavones, besides the one mentioned in Table XIV, are involved as specific deterrents. The cytotoxicity of many of these substances, especially when occurring in *O*-methylated form, has been determined by Kupchan and his coworkers (1969a,b) and by Stout *et al.*

TABLE XIV
Examples of specific leaf flavonoids involved in insect feeding responses

Class and compound	Plant species	Insect species	Reference
FLAVANOLS			
Catechin 7-xyloside[a]	Elm, *Ulmus americana*	Elm-bark beetle *Scolytus multistriatus*	Doskotch *et al.*, 1973
Condensed flavolan[b]	Oak, *Quercus robur*	Winter moth *Operophthera brumata*	Feeny, 1969
FLAVONOLS			
Quercetin 3-glucoside[a]	Mulberry *Morus alba*	Silkworm *Bombyx mori*	Hamamura *et al.*, 1962; Ishikawa, 1966
Quercetin 3-rhamnoside[b]			
Morin[a]			
Quercetin glycosides[b]	Tobacco *Nicotiana tabacum*	Tobacco budworm *Heliothis virescens*	Shaver and Lukefahr, 1969
FLAVONE			
6-Methoxyluteolin 7-rhamnoside[a]	Alligatorweed *Alternanthera phylloxeroides*	Beetle *Agasicles* sp.	Zielske *et al.*, 1972

[a] Attractant. [b] Repellent (or toxin).

(1964). Their fairly wide distribution in leaves has also been established (Harborne, 1969; Harborne and Williams, 1971a). Such a protective function, for example, would explain why in *Rhododendron* gossypetin occurs in the flowers (as a yellow pigment) in only 6% of the species but is present in the leaves of the majority (76%) of species (Harborne and Williams, 1971b). Again, it would explain why the same pigment is never present in the flowers of the related family Empetraceae but is universal in the leaves (Moore *et al.*, 1970). Also, many examples are accumulating where leaf flavonoid patterns, within the same species or genus, change with geography, suggesting that plants modify their flavonoid synthesis according to the animal predators in their particular environment (Harborne, 1975). While it would be surprising if the common leaf flavones and flavonols were regularly involved as feeding deterrents, it is at least a reasonable hypothesis that many of the more unusual flavonoid structures (such as 6-methoxyluteolin, gossypetin or morin) found in the leaf were originally produced as a protection against overgrazing.

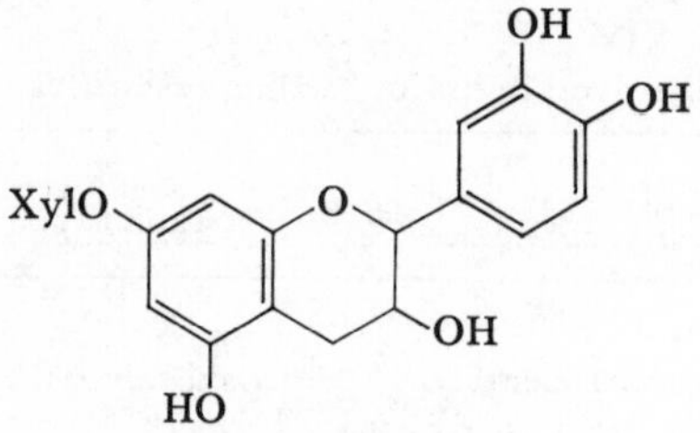

Catechin 7-xyloside
attractant to the elm-bark beetle

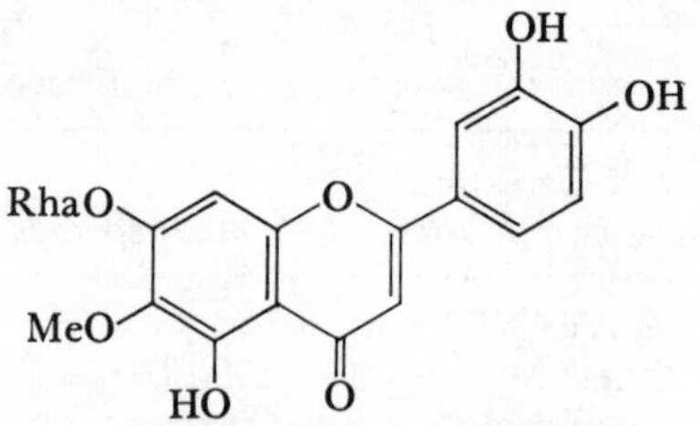

6-Methoxyluteolin 7-rhamnoside
attractant to the *Agasicles* beetle

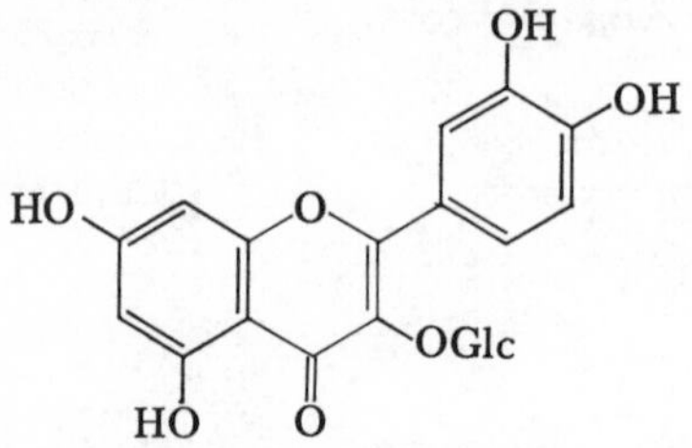

Quercetin 3-glucoside

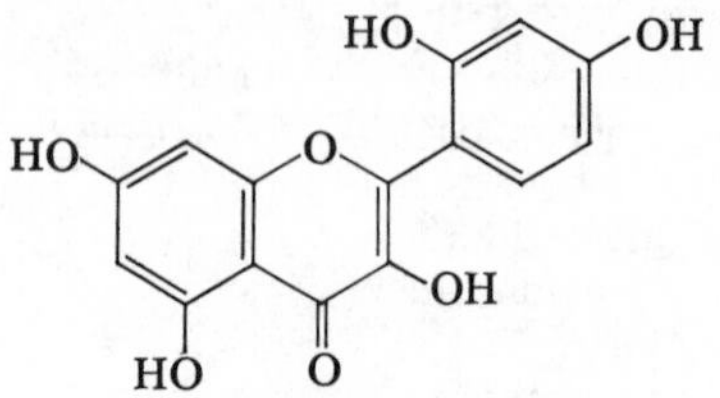

Morin

attractants to the silkworm *Bombyx mori*

FIG. 2. Some flavonoids as insect-feeding guides.

C FLAVONOIDS IN CHLOROPLASTS

The localization of flavonoids within the plant cell has generally been little studied. The coloured anthocyanins can be observed with a light microscope to accumulate in the cell vacuole and it has usually been assumed, on little actual evidence, that the structurally related flavone and flavonol glycosides, also being water-soluble, are similarly located. Occasionally, flavonoids have been detected in lipid fractions. In such cases, they are not present as *O*-glycosides but in *O*-methylated form and are assumed to be cytoplasmic or, more likely, to occur outside the cell on the surface or in a bud excretion (Wollenweber and Egger, 1971).

The possibility of water-soluble flavonoids occurring in other organelles within the cell was not seriously considered, until quite recently various enzymes of phenolic biosynthesis were detected in chloroplast fractions (see e.g. Stafford, 1969). This has been rapidly followed by the independent discovery by four groups of workers (Table XV) of flavonoids within the chloroplasts of a range of plants. This has aroused not only much interest from the functional viewpoint but also some scepticism. The problem of proving that these flavonoids actually exist *in vivo* in the chloroplast is a difficult

TABLE XV
Flavonoids isolated from chloroplasts

Flavonoid	Source	Isolation techniques	Other components	Reference
Saponarin	Barley, *Hordeum vulgare*	Aqueous/ non-aqueous	Enzyme PAL	Saunders and McClure, 1972
Quercetagetin (in partly methylated form)	Spinach, *Spinacia oleracea*	In antigen fraction, following ether treatment	*p*-Coumaryl tartaric acid	Oettmeier and Heupel, 1972
Kaempferol 3-arabinoside (and other glycosides)[a]	*Impatiens balsamina*	Aqueous/ non-aqueous	Hydroxy-cinnamic acids	Weissenböck *et al.*, 1971
Glycoflavones[b]	Wheat, *Triticum dicoccum*[c]	—	—	Monties, 1969

[a] Flavonoid yield 1·5–2·0% dry wt.
[b] Flavonoid yield 0·5%.
[c] Eight other angiosperms also gave flavonoid-containing chloroplasts.

one, since it is always possible that flavonoids in the vacuole might transfer to the chloroplast surface during cell disruption. However, considerable care has been taken to avoid this possibility. For example, in recording the flavonoid saponarin (see Fig. 3) in barley plastids, Sanders and McClure (1972) thoroughly washed etioplasts or chloroplasts, following isolation by several aqueous or non-aqueous techniques, and still found the flavone within the chloroplast. Also, aqueous solutions of the flavonol rutin were added artificially at various stages during the isolation procedures, but it was never found to contaminate the final purified plastid preparation.

It now appears from a survey of over a hundred angiosperms (J. W. McClure, personal communication) that all plant species have flavonoids in the chloroplasts, the compounds in general being the same as those in other parts of the leaf. If this is so, then it seems that these compounds must have a general function, which is not affected by small changes in chemical structure (i.e. variations in hydroxylation or glycosylation). The possibility that flavonoids provide a protective u.v. screen has been mentioned above, in VA; this could be especially

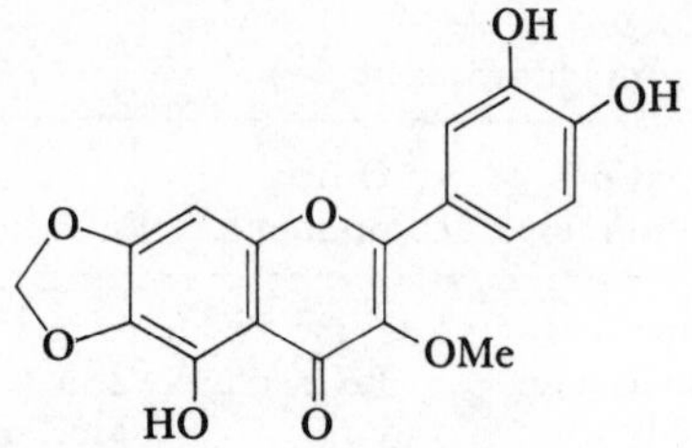

Methylated quercetagetin (from spinach)

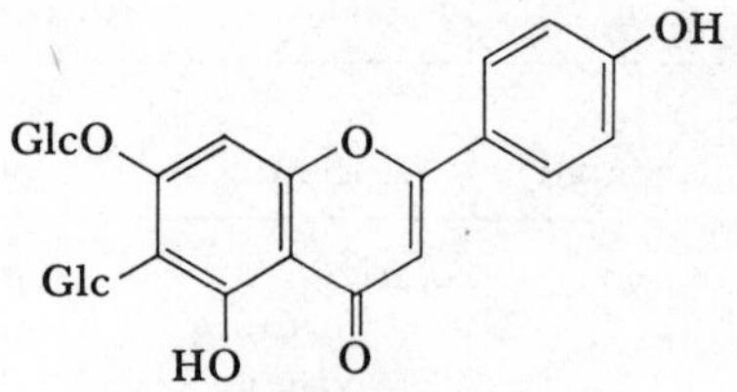

Saponarin (from barley)

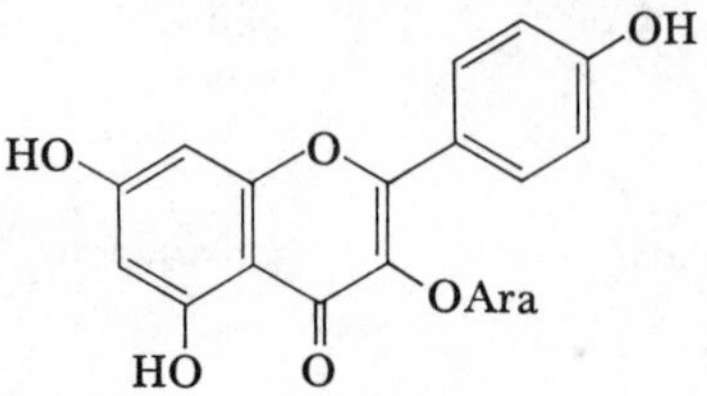

Kaempferol 3-arabinoside (from *Impatiens*)

FIG. 3. Flavonoids identified in plant chloroplasts.

important in such a vital organelle as the chloroplast. Another possibility (Oettmeier and Heupel, 1972) is that flavonoids actually play a part in photosynthesis. They may be involved, at least in spinach, as part of the primary acceptor complex for photosystem I of photosynthetic electron transport. Clearly, much more work is needed to see if this is so or not.

REFERENCES

Acheson, R. M. (1956). *Proc. R. Soc. B* **145**, 549.

Asen, S. and Horowitz, R. M. (1974). *Phytochemistry* 13, 1219.

Asen, S. and Jurd, L. (1967). *Phytochemistry* **6**, 577.

Asen, S., Stuart, N. W. and Siegelman, H. W. (1959). *Proc. Am. Soc. hort. Sci.* 73, 495.

Asen, S., Stewart, R. N., Norris, K. H. and Massie, D. R. (1970). *Phytochemistry* 9, 619.

Asen, S., Norris, K. H. and Stewart, R. N. (1971a). *J. Am. Soc. hort. Sci.* **96**, 770.

Asen, S., Stewart, R. N. and Norris, K. H. (1971b). *Phytochemistry* **10**, 171.

Asen, S., Norris, K. H. and Stewart, R. N. (1972a). *Phytochemistry* **11**, 2739.

Asen, S., Stewart, R. N. and Norris, K. H. (1972b). *Phytochemistry* **11**, 1134.

Asen, S., Norris, K. H., Stewart, R. N. and Semenink, P. (1973). *J. Am. Soc. hort. Sci.* **98**, 174.

Baker, H. G. and Hurd, P. D. (1968). *Ann. Rev. Ent.* 13, 385.

Barber, H. N. (1965). *Heredity, Lond.* **20**, 551.

Bate-Smith, E. C. and Harborne, J. B. (1963). *Nature, Lond.* **198**, 1307.

Bayer, E., Nether, K. and Egeter, H. (1960). *Chem. Ber.* 93, 2871.
Bayer, E., Egeter, H., Fink, A., Nether, K. and Weginann, K. (1966). *Angew. Chem. Int. Ed. Engl.* 5, 791.
Beale, G. H. (1941). *J. Genet.* 42, 197.
Beale, G. H., Price, J. R. and Sturgess, V. C. (1941). *Proc. R. Soc. B* 130, 113.
Blank, F. (1947). *Bot. Rev.* 13, 241.
Blank, F. (1958). *In* "Encyclopedia of Plant Physiology" (K. Paech and M. V. Tracey, eds), vol. 10, p. 300. Springer-Verlag, Berlin.
Bloom, M. and Vickery, R. K. (1973). *Phytochemistry* 12, 165.
Bopp, M. (1957). *Planta* 48, 631.
Bopp, M. (1959). *Z. Bot.* 47, 197.
Boyle, R. (1664). "Experiments and Considerations Touching Colours." London.
Bragt, J. van (1962). *Meded. LangbHoogesch. Wageningen* 62, 1.
Caldwell, M. M. (1971). *In* "Photophysiology" (Glese, A. C. ed.), vol. 6, p. 131. Academic Press, New York and London.
Charlesworth, E. H. and Robinson, R. (1934). *J. chem. Soc.* 1619.
Chenery, E. M. (1948). *Ann. Bot.* 12, 121.
Clevenger, S. (1964). *Can. J. Biochem.* 42, 154.
Clevenger, S. (1971). *Evolution, Lancaster, Pa.* 25, 669.
Clifford, H. T. and Harborne, J. B. (1969). *Phytochemistry* 8, 123.
Darwin, C. (1876). "The Effects of Cross and Self-fertilization in the Vegetable Kingdom." Murray, London.
Dean, F. (1963). "Naturally Occurring Oxygen Ring Compounds." Butterworth, London.
Dethier, V. G. (1970). *In* "Chemical Ecology" (E. Sondheimer and J. B. Simeone, eds), pp. 83–102. Academic Press, New York and London.
Dodds, K. S. and Long, D. H. (1955). *J. Genet.* 53, 136.
Doskotch, R. W., Mikhail, A. A. and Chatterji, S. K. (1973). *Phytochemistry* 12, 1153.
Eisner, J., Eisner, M. and Aneshansley, D. (1973). *Proc. natn. Acad. Sci. U.S.A.* 70, 1002.
Endo, T. (1962). *Jap. J. Genet.* 37, 284.
Faegri, K. and van der Pijl, L. (1971). "The Principles of Pollination Ecology," 2nd edn. Pergamon Press, Oxford.
Feenstra, W. J. (1960). *Meded. LandbHoogesch. Wageningen* 60, 1.
Feenstra, W. J., Johnson, B. L., Ribereau-Gayon, P. and Geissman, T. A. (1963). *Phytochemistry* 2, 273.
Feeny, P. (1969). *Phytochemistry* 8, 2119.
Forsyth, W. G. C. and Simmonds, N. W. (1954). *Proc. R. Soc. B* 142, 549.
Fraser, A. W. and Lewis, J. R. (1972). *Phytochemistry* 11, 868.
Frisch, K. von (1950). "Bees, Their Vision, Chemical Senses and Language." Cornell, Ithaca, New York.
Gascoigne, R. M., Ritchie, E. and White, D. R. (1948). *J. Proc. R. Soc. N.S.W.* 82, 44.
Geissman, T. A. (ed.) (1962). "Chemistry of the Flavonoid Compounds." Pergamon Press, Oxford.
Geissman, T. A. and Harborne, J. B. (1955). *Archs Biochem. Biophys.* 55, 447.
Geissman, T. A. and Mehlquist, G. A. L. (1947). *Genetics, Princeton* 32, 410.
Gertz, O. (1938). *K. physiograf. Sällskop Lund Forh.* 8, 62.

Goodwin, T. W. (1952). "Comparative Biochemistry of the Carotenoids." Chapman and Hall, London.
Hamamura, Y. K., Hayashiya, K., Naite, K., Matsuura, K. and Nishida, J. (1962). *Nature, Lond.* 194, 754.
Harborne, J. B. (1960). *Biochem. J.* 74, 262.
Harborne, J. B. (1961). *Experientia* 17, 72.
Harborne, J. B. (1962a). *Fortschr. Chem. org. NatStoffe.* 20, 165.
Harborne, J. B. (1962b). *Archs Biochem. Biophys.* 96, 171.
Harborne, J. B. (1962c). *In* "Chemistry of the Flavonoid Compounds" (T. A. Geissman, ed.), pp. 598-617. Pergamon Press, Oxford.
Harborne, J. B. (1963a). *In* "Chemical Plant Taxonomy" (T. Swain, ed.), pp. 359-388. Academic Press, London and New York.
Harborne, J. B. (1963b). *Phytochemistry* 2, 85.
Harborne, J. B. (1963c). *Phytochemistry* 2, 327.
Harborne, J. B. (1964a). *Rep. John Innes hort. Instn* 45.
Harborne, J. B. (1964b). *In* "Biochemistry of Phenolic Compounds" (J. B. Harborne, ed.), pp. 129-169. Academic Press, London and New York.
Harborne, J. B. (1965). *Phytochemistry* 4, 647.
Harborne, J. B. (1966). *Phytochemistry* 5, 589.
Harborne, J. B. (1967). "Comparative Biochemistry of the Flavonoids." Academic Press, London and New York.
Harborne, J. B. (1968). *Phytochemistry* 7, 1215.
Harborne, J. B. (1971). (ed.) "Phytochemical Ecology." Academic Press, London and New York.
Harborne, J. B. (1975). *In* "The Flavonoids" (J. B. Harborne, T. J. Mabry and H. Mabry, eds), pp. 1056-1095. Chapman and Hall, London.
Harborne, J. B. and Sherratt, H. S. A. (1961). *Biochem. J.* 78, 298.
Harborne, J. B. and Williams, C. A. (1971a). *Phytochemistry* 10, 367.
Harborne, J. B. and Williams, C. A. (1971b). *Phytochemistry* 10, 2727.
Harborne, J. B., Heywood, V. H. and Saleh, N. A. M. (1970). *Phytochemistry* 9, 2011.
Harborne, J. B., Mabry, T. J. and Mabry, H. (1975). "The Flavonoids." Chapman and Hall, London.
Hayashi, K. (1962). *In* "Chemistry of Flavonoid Compounds" (T. A. Geissman, ed.), pp. 248-285. Pergamon Press, Oxford.
Hayashi, K. and Abe, Y. (1953). *Misc. Rep. Res. Inst. nat. Resour., Tokyo* 29, 1.
Hayashi, K. and Takeda, K. (1962). *Proc. imp. Acad. Japan* 38, 161.
Hayashi, K. and Takeda, K. (1970). *Proc. Japan Acad.* 46, 535.
Hess, D. (1963). *Planta* 59, 567.
Horowitz, A. and Cohen, Y. (1972). *Am. J. Bot.* 59, 706.
Huang, H. T. (1955). *J. agric. Fd Chem.* 3, 141.
Hussey, G. C. (1963). *J. exp. Bot.* 14, 326.
Ishikawa, S. (1966). *J. cell. comp. Physiol.* 67, 1.
Jorgensen, E. C. and Geissman, T. A. (1955). *Archs Biochem. Biophys.* 55, 389.
Jurd, L. and Asen, S. (1966). *Phytochemistry* 5, 1263.
Jurd, L. and Harborne, J. B. (1968). *Phytochemistry* 7, 1209.
Kupchan, S. M., Sigel, C. W., Hemingway, R. J., Knox, J. R. Udayamurthy, M. D. (1969a). *Tetrahedron* 25, 1603.
Kupchan, S. M., Sigel, C. W., Knox, J. R. and Udayamurthy, M. D. (1969b). *J. org. Chem.* 34, 1460.

Latour and Magnier de la Source (1877). *Bull. Soc. chim. Fr.* 228, 337.
Lawrence, W. J. C. (1932). *Nature, Lond.* 129, 834.
Lawrence, W. J. C. (1950). *Symp. Biochem. Soc.* 4, 3.
Lawrence, W. J. C., Price, J. R., Robinson, R. and Robinson, G. M. (1939). *Phil. Trans. R. Soc. Ser. B* 230, 149.
Lowry, J. B. (1971). *Phytochemistry* 10, 673.
Manning, A. (1956a). *Natur., New Biol.* 21, 59.
Manning, A. (1956b). *Behaviour* 9, 114.
Marquart, L. Cl. (1835). "Die Farben der Bluthen, Eine chemischphysiologische Abhandlung." Bonn.
McKay, M. B. and Warner, M. F. (1933). *Natn. hort. Mag.* 178.
Monties, P. (1969). *Physiologie Veg.* 15, 29.
Moore, D. M., Harborne, J. B. and Williams, C. A. (1970). *J. Linn. Soc.* 63, 277.
Nair, A. G. R. and Subramanian, S. C. (1962). *Curr. Sci.* 31, 155, 504.
Nilsson, E. (1967). *Acta chem. scand.* 21, 1942.
Nordström, C. G. and Swain, T. (1958). *Archs Biochem. Biophys.* 73, 220.
Oettmeier, W. and Heupel, A. (1972). *Z. Naturforsch.* 27B, 177.
Paech, K. (1954). *A. Rev. Pl. Physiol.* 6, 273.
Pecket, R. C. (1960). *New Phytol.* 59, 138.
Pecket, R. C. and Selim, A. R. A. A. (1972). *Nature, Lond.* 195, 620.
Perkin, W. H. (1899). *J. chem. Soc.* 75, 825.
Price, J. R. and Sturgess, V. C. (1938). *Biochem. J.* 32, 1658.
Reznik, H. (1956). *Sber. heidelb. Akad. Wiss.* 125.
Reznik, H. (1961). *Flora* 150, 454.
Reznik, H. and Neuhausel, R. (1959). *Z. Bot.* 47, 471.
Robinson, G. M. and Robinson, R. (1931). *Biochem. J.* 25, 1687.
Roller, K. (1956). *Z. Bot.* 44, 477.
Saito, N. and Hayashi, K. (1965). *Scient Rep. Tokyo Kyoiku Daigaku* 12B, 39.
Saito, N., Hirate, K., Hotta, R. and Hayashi, K. (1964). *Proc. Japan Acad.* 40, 516.
Salaman, R. N. (1949). "The History and Social Influence of the Potato." Cambridge University Press.
Saunders, J. A. and McClure, J. W. (1972). *Am. J. Bot.* 59, 673.
Schoonhoven, L. M. (1968). *Ann. Rev. Ent.* 13, 115.
Scott-Moncrieff, R. (1936). *J. Genet.* 32, 117.
Seshadri, T. R. (1951). *Rev. pure appl. Chem.* 1, 186.
Seybold, A. (1953–4). *Sber. heidelb. Akad. Wiss.* 2, 31.
Shaver, T. N. and Lukefahr, M. J. (1969). *H. econ. Ent.* 62, 643.
Shibata, M. and Ishikura, N. (1960). *Jap. J. Bot.* 17, 230.
Shibata, K., Hayashi, K. and Isaka, T. (1949). *Acta phytochim., Tokyo* 15, 17.
Smith, D. M., Craig, S. P. and Santarosa, J. (1971). *Am. J. Bot.* 58, 292.
Sondheimer, E. and Karash, C. B. (1956). *Nature, Lond.* 178, 648.
Sprengel, C. K. (1793). "Das entdeckte Geheimnis der Natur im Bau und in der Befruchtung der Blumen."
Stafford, H. A. (1969). *Phytochemistry* 8, 743.
Statham, C. M. and Crowden, R. K. (1974). *Phytochemistry* 13, 1835.
Stewart, R. N., Asen, S., Norris, K. H. and Massie, D. R. (1969). *Am. J. Bot.* 56, 227.
Stout, M. G., Reich, H. and Huffinan, M. N. (1964). *Cancer Chemother. Rep.* 36, 23.

Swain, T. (1975). *In* "The Flavonoids" (J. B. Harborne, T. J. Mabry and H. Mabry, eds), pp. 1096–1129. Chapman and Hall, London.
Subramanian, S. S. and Nair, A. G. R. (1970). *Curr. Sci.* 39, 323.
Subramanian, S. S. and Swamy, M. N. (1964). *Curr. Sci.* 33, 112.
Takeda, K. and Hayashi, K. (1965). *Proc. Japan Acad.* 41, 449.
Takeda, K., Mitsui, S. and Hayashi, K. (1966). *Bot. Mag., Tokyo* 79, 578.
Takeda, K., Saito, N. and Hayashi, K. (1968). *Proc. Japan Acad.* 44, 352.
Thompson, W. R., Meinwald, J., Aneshansley, D. and Eisner, T. (1972). 177, 528.
Wanscher, J. H. (1953). *K. Vet Hojsk Aarsskr.* 91.
Weissenböck, G., Terini, M. and Reznik, H. (1971). *Z. PflPhysiol.* 64, 274.
Werckmeister, P. (1955). *Züchter* 25, 315.
Willstätter, R. and Mallison, H. (1915). *Justus Liebig's Annln Chem.* 408, 147.
Wilson, R. F. (1938). "Horticultural Colour Chart."
Wolfrom, M. L. and Mahan, J. (1942). *J. Am. chem. Soc.* 64, 308.
Wollenweber, E. and Egger, K. (1971). *Z. PflPhysiol* 65, 427.
Zielske, A. G., Simons, J. N. and Silverstein, R. M. (1972). *Phytochemistry* 11, 393.

Part III. METABOLISM IN SENESCENT AND STORED TISSUES

Chapter 17

Metabolism in Senescent and Stored Tissues

K. L. SIMPSON, TUNG-CHING LEE, DELIA B. RODRIGUEZ and C. O. CHICHESTER

Department of Food and Resource Chemistry, University of Rhode Island, Kingston, Rhode Island, U.S.A.

I Introduction

Since the review of Chichester and Nakayama (1965) on the "pigment changes in senescent and stored tissues" our understanding of these changes has broadened considerably. It is hoped that the present review will bring together from many fields what is known about the senescence of pigments. It is also the hope of the reviewers that the obvious gaps in our knowledge will be exposed and thus direct basic research into these areas.

It is not an easy task to define senescence as it applies to pigments in plants. One can certainly describe a sequence of events in the ripening of fruit in which growth is combined with changes in flavour, texture, colour etc. At a given stage the fruit is "ripe" and we recognize that some loss of pectin is important to this process. Further loss of pectin results in overripe or senescent tissue. However, long before this senescent state many fruits have become devoid of chlorophyll and chloroplasts have been replaced by chromoplasts. Thus in a series of complicated events degradation of chlorophyll occurs during the biosynthesis of carotenoids and/or anthocyanins or betalains. Likewise destruction of violaxanthin and neoxanthin may lead to the biosynthesis of abscisic acid-like compounds. We might agree that a rotten apple is an example of senescent tissue but even here we must recognize that the liberation of seeds is the ultimate point of the whole exercise.

For sake of simplicity we have chosen to separate the various pigments and to treat their biosynthesis and metabolism as separate events. We realize that in some cases what we treat as metabolism may overlap with what others treat as biosynthesis. We have not considered the formation of bixin, the well-known pigment in annatto seeds, as a senescent change even though it can be regarded as a diapocarotenoid and the formation of apocarotenoids are treated as a possible mechanism in pigment degradation.

We have considered degradation of pigments in stored tissues from a broad point of view, realizing that it is difficult and perhaps not worthwhile to separate metabolic changes in pigments from the chemical changes which ensue as a result of metabolic changes. As many fruits approach ripening there is a stage in which a number of biochemical events are initiated by the autocatalytic production of ethylene. This increased respiration marks the change between growth and senescence. This climacteric has been described in terms of biochemical mechanisms of control; however, Bain and Mercer (1964) present observations showing that "organization resistance" or the more commonly used term, compartmentalization, could be an important control factor in the metabolism in plant cells. Bain and Mercer (1964) followed the protoplasts of immature and mature preclimacteric fruit and mature climacteric fruit. They found that just prior to the climacteric rise in respiration there was an almost complete disorganization of the protoplast and loss of structure of the chloroplasts and cytoplasm together with the colour change in the skin. With the continued loss of organization in senescence it is likely that enzymes, acids etc. would no longer be physically separated from pigments and other substrates.

Harris and Spurr (1969) present pictures of mature chloroplasts from three distinct tomato lines (low pigment, normal pigment and high beta) in which the chloroplasts exhibit different shapes, and in which the biosynthetic pigment formation is best explained in terms of compartmentalization. It remains to be seen what relationship exists between the various levels of cellular control and senescent changes.

The object of the storage of plant tissues is generally to maintain the tissue with the least amount of change. Changes here are equated with degradation and methods of dehydration, canning, freezing, chemical preservation etc. are used to accomplish this purpose. We have reported the results of a number of model systems which have been devised to show *in vitro* what is taking place *in vivo*. An enormous amount of work remains in interpreting the results of model systems in terms of the vastly more complex *in vivo* systems.

II Chlorophyll

All plant materials containing chlorophyll can be characterized as undergoing a loss in colour during senescence, processing, or storage. While the pathways for chlorophyll biosynthesis are known in some detail (Bogorad, 1966; Ellsworth, 1972; Chapter 2), knowledge of the intermediate steps of chlorophyll degradation, one of the most widely observed phenomena in nature, is still limited. The *in vivo* degradation of chlorophyll occurs very quickly. Park *et al.* (1973) reported that within a few weeks during the autumn months the chlorophyll content of ginko leaves fell from *c.* 1·35 mg/g to 0·05. Thus intermediates do not normally accumulate in detectable amounts for identification. In addition, artefacts are easily formed during extraction and isolation procedures (Bacon, 1966; Schanderl and Co, 1966). A number of papers have been published on the degradation of chlorophyll during processing and storage; however, there are a limited number of comprehensive reviews on this subject

(Chichester and Nakayama, 1965; Chichester and McFeeters, 1971).

A number of chlorophyll pigments are known to exist (Seely, 1966; Ellsworth, 1972; Chapter 2), but from a chemical and technological standpoint chlorophylls *a* and *b* are the most important. The majority of work on the changes of chlorophyll concentration has been concerned with these pigments, although chlorophylls *c* and *d* are of importance in marine flora and bacterial chlorophyll in micro-organisms.

The ratio of chlorophylls *a* : *b* in higher plants is usually between 2·5 and 3·5 : 1. Limited changes in this ratio have been related to the intensity of illumination to which individual plants have been exposed during growth and to which the species as a whole has become adapted (Goodwin and Mercer, 1972). Their behaviour during senescence has been reported by many investigators (Seybold, 1943; Wolf, 1956; Goodwin, 1958; Laval-Martin, 1969; Gribanovski-Sassu *et al.*, 1969; Sanger, 1971; Whitfield and Rowan, 1974). Chlorophyll *a* invariably decreases more rapidly than chlorophyll *b*. Figure 1 shows changes in chlorophylls *a* and *b* and the ratio of chlorophylls *a* : *b* in the course of one year in leaves of *Lycium europaeum* (Gribanovski-Sassu *et al.*, 1969). Schanderl *et al.* (1963) also demonstrated that during the *in vitro* conversion of chlorophylls *a* and *b* into their respective pheophytins, the rate of conversion of *a* exceeds that of *b* by 5 : 1.

A GENERAL REACTIONS

In general, the principal steps in the decomposition of chlorophyll involve the loss of phytol to give chlorophyllide, the loss of magnesium and phytol to give pheophorbide, oxidation and isomerization of chlorophylls to give their corresponding reaction products, and finally the decomposition of chlorophyll and its derivatives to give colourless, low molecular weight compounds after the destruction of the porphyrin ring. Figure 2 indicates schematically several degradation pathways of chlorophyll *a*.

When leaves and other plant materials are heated, such as in blanching, isomeric chlorophyll *a′* and chlorophyll *b′* are formed (Strain, 1958). The isomers are separable from the parent compounds on sugar columns although they give similar phase tests. *In vitro* the isomerization reactions occur at room temperature in a few hours, or more rapidly if alkali is added. The isomerized products are reconvertible to an equilibrium mixture of the parent compounds by treatment with additional alkali or by heating in *n*-propanol.

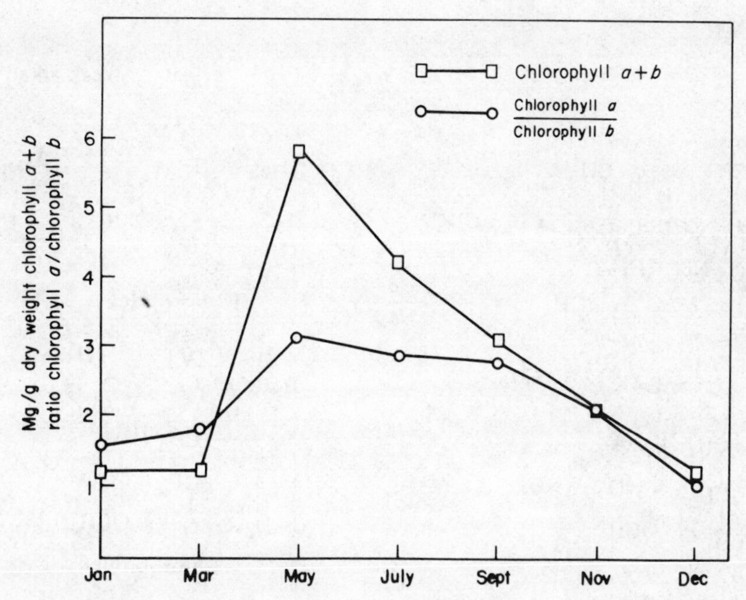

FIG 1. Changes in chlorophyll *a* and *b* in ratio Chlorophyll *a*/Chlorophyll *b* in the course of one year in leaves of *Lycium europaeum*. (Reproduced with permission from Gribanovski-Sassu *et al.*, 1969.)

It is also known that pheophytin is formed when green leaves are boiled (Bacon and Holden, 1967); the rate of formation increases proportionately with increasing temperature. The rate of conversion of chlorophyll *a* into pheophytin *a* is much greater than chlorophyll *b* to pheophytin *b*. As with other immediate effects, pH and duration of heating are also important. If the pheophytins *a* and *b* are heated in propanol, they are converted into pheophytins *a′* and *b′*, which are less strongly absorbed than the initial pheophytins on sugar columns.

Acid treatment of chlorophyll solutions or green leaves produces pheophytin. Egle (1944), from his studies on silage, concluded that acids arising during fermentation constituted an important factor in chlorophyll destruction with chlorophyll *a* less resistant than *b*. Similar observations on drying plant material suggested to him that acids normally present in leaves are very important in the destruction of chlorophyll.

Allomerization occurs when chlorophyll is dissolved in inorganic solvents containing oxygen and allowed to stand in the dark. The irreversible process was called allomerization by Willstätter because the properties of the chlorophyll were altered while the appearance and chemical composition seemed to undergo no change. It was

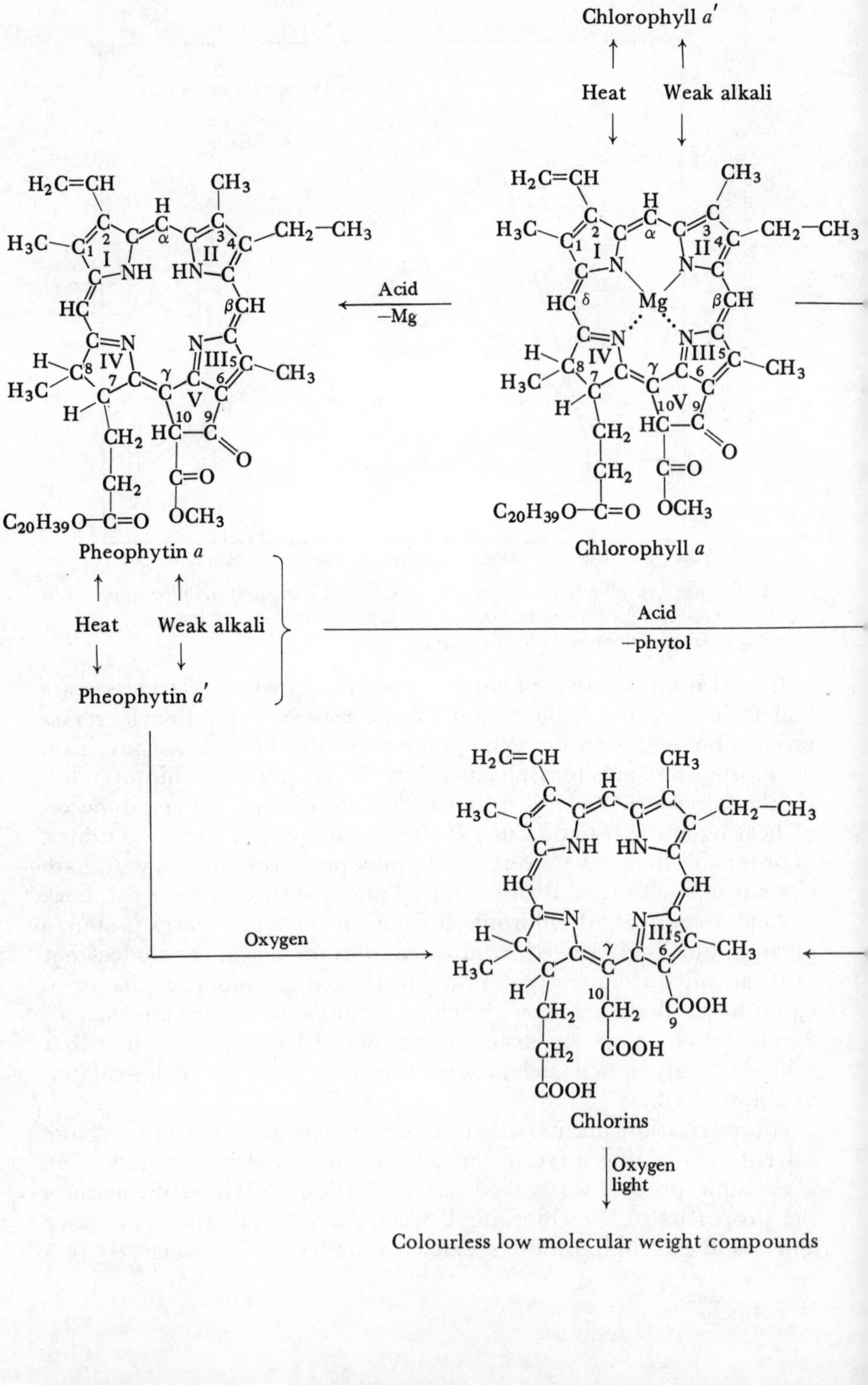

Chlorophyll *a*′

Heat Weak alkali

Pheophytin *a*

Acid
−Mg

Chlorophyll *a*

Heat Weak alkali

Pheophytin *a*′

Acid
−phytol

Oxygen

Chlorins

Oxygen
light

Colourless low molecular weight compounds

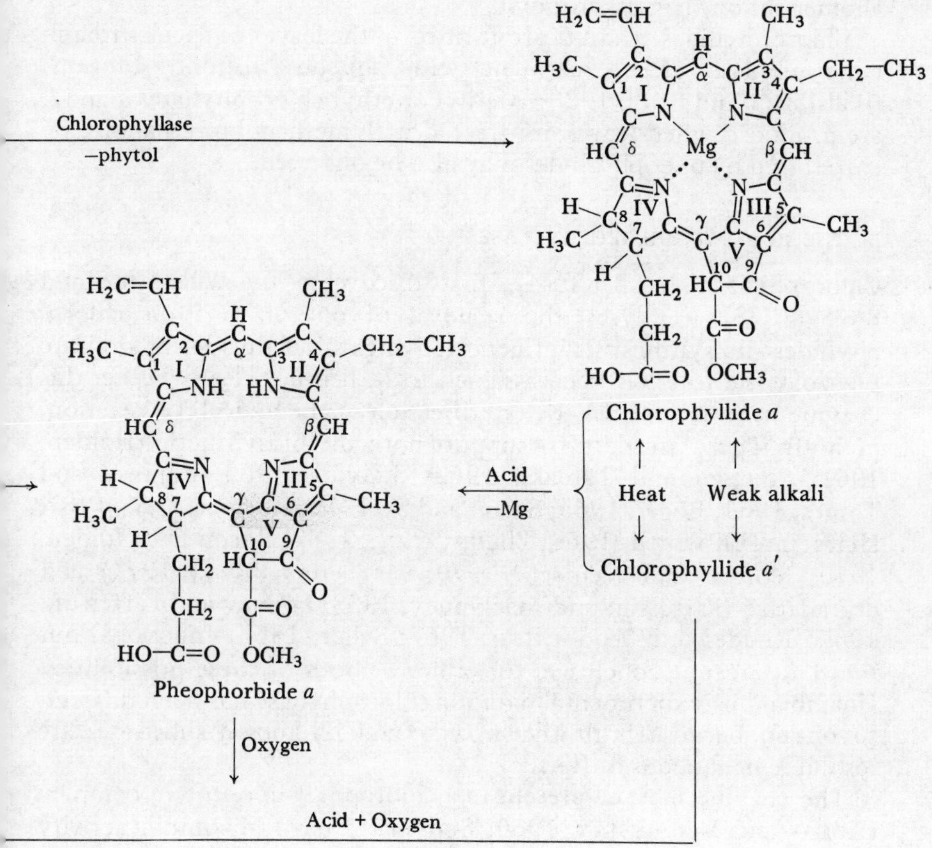

Chlorophyllase
−phytol

Chlorophyllide *a*

Heat

Weak alkali

Chlorophyllide *a*′

Acid
−Mg

Pheophorbide *a*

Oxygen

Acid + Oxygen

FIG. 2. Pathways for the degradation of chlorophyll *a*.

found subsequently that allomerization was due to a slow process of oxidation involving the removal of the hydrogen at C-10 and its replacement by an –OH or by an alkyloxy group when alcohols were used as solvents. The presence of magnesium in the molecule seems to favour formation of the enol. Pheophytin is less susceptible to allomerization than chlorophyll.

Chlorophyllides *a* and *b* are formed in the leaves of species rich in chlorophyllase after treatment with aqueous acetone solutions (Willstätter and Stoll, 1928). Methyl or ethyl chlorophyllides *a* and *b* are produced when leaves are treated with methanol or ethanol. The corresponding pheophorbides may also be observed.

B THE ROLE OF CHLOROPHYLLASE

Chlorophyllase (EC 3.1.1.14), first discovered by Willstätter and Stoll (1928), catalyses the removal of phytol to form chlorophyllides. Its synthesis is influenced by light, but thus far no definite physiological role has been assigned to it. It is not clear whether this enzyme *in vivo* catalyses the esterification or the hydrolytic reaction, or both. There are data to support both the biosynthetic (Holden, 1961; Shimizu and Tamaki, 1962; Sudyina, 1963; Ramirez and Tomes, 1964; Böger, 1965; Strain and Svec, 1966; Chiba *et al.*, 1967; Hines and Ellsworth, 1969; Phillips *et al.*, 1969; Bacon and Holden, 1970; Moreth and Yentsch, 1970; McFeeters *et al.*, 1971) and degradative (Patterson and Mackinney, 1938; Looney and Patterson, 1967; Rhodes and Wooltorton, 1967; Seller, 1968) functions, but the data are not conclusive for either or both of these possibilities. Unambiguous experimental data for chlorophyllase has been difficult to obtain because both the enzyme and its known substrates are insoluble in aqueous buffers.

The enzyme may be present in a chlorophyll–lipoprotein complex (Ardao and Vennesland, 1960; Sudyina, 1963). Maximum activity can be obtained when the enzyme is solubilized. Many procedures (Ardao and Vennesland, 1960; Böger, 1965; McFeeters *et al.*, 1971) have been developed for solubilization of chlorophyllase from a number of plant tissues. Solubilization of substrate has been accomplished in nearly all cases by using buffers containing 40 to 80% acetone. Interpretation of enzyme kinetics for reactions run under these conditions is difficult. Another problem is that substrates are only partially soluble even at high acetone concentration (Bacon and Holden, 1970), making reproducibility of results very difficult and quantitative assays of enzymic activity subject to large errors. Most of the quantitative data for chlorophyllase in the literature and their interpretations are of doubtful validity (McFeeters *et al.*, 1971).

More information about the physical, chemical and catalytic properties of chlorophyllase is needed to elucidate the definite physiological function of the enzyme.

C ENZYMIC OXIDATION

Strain (1941) first observed that chlorophylls *a* and *b* were oxidized to colourless substances in the presence of oxygen in a system consisting of an aqueous extract from soyabeans and a lipid. Mapson and Moustafa (1955) found that an enzyme preparation from peas catalysed the oxidation of various substances, including chlorophylls, when unsaturated fatty acids such as linoleic or linolenic were added. Wagenknecht *et al.* (1952) found that chlorophyll was destroyed in frozen green peas which had not previously been blanched. It was later determined that the degradation obtained was mainly caused by lipoxygenase (Wagenknecht and Lee, 1956, 1958; Lee and Wagenknecht, 1958). Walker (1964) showed that the peroxidase system in frozen green beans also degraded chlorophyll. These systems were similar to a lipoxygenase–linoleate system responsible for the oxidation of carotene (Sumner, 1941; Blain *et al.*, 1953; and Tookey *et al.*, 1958). Holden (1965) found that legume seeds and legume extracts bleached chlorophyll in the presence of long-chain fatty acids. The most effective acids were the known substrates for lipoxygenase. The bleaching effect was inhibited by commercial antioxidants. Chlorophyll was bleached by extracts which had lipoxygenase activity but not by purified lipoxygenase preparations. Addition of linoleic acid to seed extracts stimulated bleaching two to three times that of the control whereas oleic acid had little effect. Chlorophyll thus appears to be bleached as a secondary substrate during a chain reaction involving peroxidation of fatty acids and the breakdown of hydroperoxide by a heat-labile factor in that system.

The bleaching of chlorophyll was extremely rapid as pheophytin and pheophorbide were not detected. No attempt has been made by Holden, however, to identify the colourless breakdown products formed in the chlorophyll bleaching reaction. While investigating the formation of "changed" chlorophyll in leaves, Bacon and Holden (1967) observed that most of the chlorophyll disappeared from chopped barley leaves suspended in 50% aqueous acetone at room temperature overnight in the dark. It is probable that a similar chlorophyll bleaching system might be involved in the bleaching of leaf tissue in other aqueous organic solvents.

Holden (1970) investigated in detail the relationship between lipoxygenase activity and chlorophyll using cereal seedlings and a wide range of other species. Leaves that bleached in aqueous acetone were all rich in lipoxygenase although some with high enzyme activity did not bleach significantly. Although leaves with low lipoxygenase activity did not bleach, pheophytin and pheophorbide were sometimes formed. In the intact leaf tissue the enzyme and its substrates are probably separated and the role of the acetone may be to bring enzyme and substrate together. Holden (1970) points out, however, that the way in which chlorophyll is bleached in leaves suspended in aqueous acetone may be unrelated to normal chlorophyll breakdown in senescing tissues. For example, oat leaves show negligible bleaching when kept in aqueous acetone overnight, but the loss of chlorophyll under moist dark conditions is faster than most other leaves.

D PHOTODEGRADATION

A possible non-biochemical mechanism for chlorophyll decomposition is the process of photodecomposition. Moreth and Yentsch (1970) studied the role of chlorophyllase and light in the decomposition of chlorophyll from marine phytoplankton. Their experiments show that a similar photo-oxidation process exists in fractured cells and in the faeces of marine herbivores, although the rate of photobleaching is slower *in vivo* than *in vitro*. They suggest that photobleaching might be a principal mechanism for pigment decomposition because of its rapidity and high efficiency.

Goldwaite and Laetsch (1967) found that high intensity light increased the senescence and the degradation of both chlorophyll and protein in bean leaf discs. Edelman and Schoolar (1969) also showed that light is a major factor in chlorophyll destruction in sugar cane leaf tissue. However, Khudairi (1970) demonstrated that the photodestruction of chlorophyll in the presence of sugar was also temperature sensitive, suggesting a combination of photo and enzymic effects.

A chlorophyll solution is irreversibly bleached when exposed to light in the presence of oxygen (Jen and Mackinney, 1970a,b; Morris *et al.*, 1973). The action spectra parallel the absorption spectra for both chlorophylls *a* and *b*; the reaction is second order. Jen and Mackinney (1970a) noted that whereas in most chemical reactions chlorophyll *a* is lost at a substantially higher rate than chlorophyll *b*, no such difference is found for the photochemical reaction. The greater stability of pheophytin or pheophorbide over chlorophyll and chlorophyllide implies the necessity of Mg^{2+} for the excited triplet state of the molecule. No pheophytin has been detected in the reaction products of the irradiated chlorophyll solution in the absence of water. The photodecomposition of chlorophyll in solution first produces transient pink or an intermediate red colour before becoming colourless (Aronoff and Mackinney, 1943; Dilung, 1958; Jen and Mackinney, 1970a,b). The red intermediate compounds can be precipitated by the use of highly non-polar solvents (Jen and Mackinney, 1970b). The precipitates are more highly oxygenated but retain phytol and a N–Mg ratio of 4 : 1. The n.m.r. spectra indicate that the methine bridge and vinyl protons have been lost. Methyl ethyl maleimide was a product of chromic acid oxidation of the red precipitate. It was also found in the supernatant solution and in the fully bleached end product.

Chlorophyll irradiated with light will yield, in addition to red pigmented intermediates, a colourless blue-fluorescing petroleum ether extract and a colourless and more strongly fluorescing aqueous extract. The former contains a phytyl ester and 10 to 15% of the original nitrogen, whereas the latter contains a substantial fraction of the original nitrogen and magnesium (Morris *et al.*, 1973).

E EFFECT OF GAMMA IRRADIATION

The effect of gamma rays on the chlorophyll content in peas (*Pisum sativum*) was recently reported by Snauwaert *et al.* (1973). The stability of chlorophyll was studied *in situ* (in pea varieties Onyx, Finette, and Cobri) and in solution. Gamma irradiation with doses up to 2 Mrad produced no visible changes in colour. Irradiation had a marked effect on the extractability of chlorophylls. The corresponding pheophytin and colourless derivatives of the chlorophylls were isolated after irradiation. As with other pigments, the higher the state of purity the more sensitive it is to gamma irradiation.

F DEGRADATION IN SENESCENT LEAVES

Since the early work of Willstätter and Stoll (1918) showed that chlorophyll values were lower in yellow leaves than in green leaves, the subject of autumn pigment degradation has been of particular scientific interest. Goodwin (1958) reviewed many of the early investigations with special emphasis on their errors and inconsistencies.

Sanger (1971) followed the pigment formation and loss in hazel (*Coylus americana*), aspen (*Populus tremaloides*) and pin oak

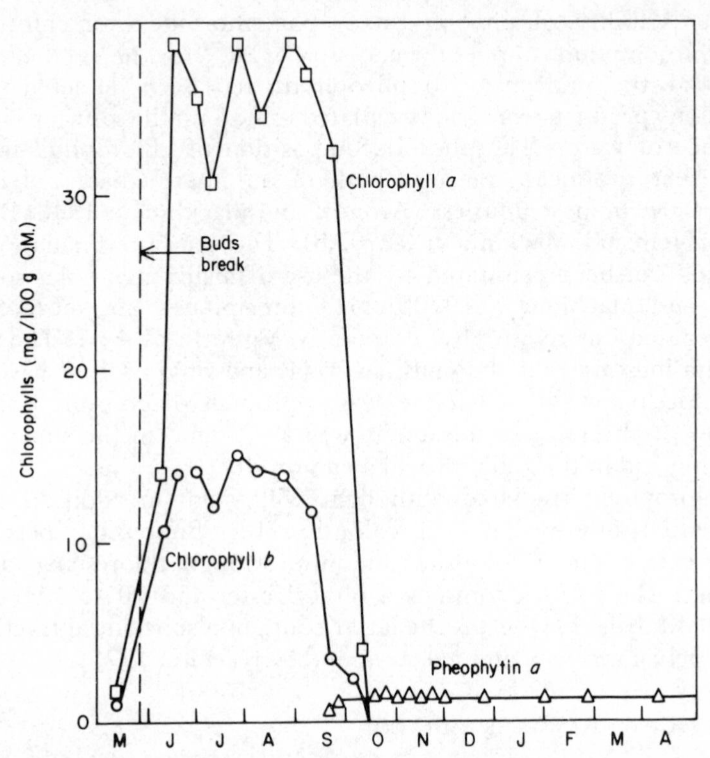

FIG. 3. Temporal changes in chlorophyll *a* and *b* per unit organic matter (oak). (Reproduced with permission from Sanger, 1971.)

(*Quercus ellipsoidlis*) from their inception in buds, development of a summer maximum, through the autumn coloration period and to the decomposition in dry, falling leaves. Leaves contained generally high but varying concentrations of chlorophyll and carotenoid through the summer months (Fig. 3). During the autumn senescent (coloration) period, preceding leaf desiccation and fall, chlorophyll decays rapidly, producing low levels of pheophytin with only occasional faint traces of pheophorbide and chlorophyllide. The levels of carotenoids begin declining at the same time as chlorophyll, but at a much lower rate. Oak leaves, whether dropped to the ground in autumn or held on the tree throughout the winter, still contain measurable amounts of lutein and β-carotene and low concentrations of pheophytin *a*. At leaf-fall, aspen and hazel leaves are devoid of all pigments.

Park *et al.* (1973) found that the phytol content of green leaves coincides with that calculated on the basis of the chlorophyll present. During senescence, 95% of the chlorophyll may disappear, leaving no trace of coloured breakdown products as the leaf turns yellow, brown or red. In all the cases studied, the phytol ester linkage was highly stable during yellowing. The phytol was often recovered in amounts equivalent to those found in green leaves. The bulk of the nitrogen in petroleum ether extracts from green leaves was associated with the chlorophyll, i.e. the *N*-phytol mole ratio was close to 4 : 0. In yellow leaves, this ratio varied from 0·79 to 0·13—higher in fresh leaves, lower in many older ones. The bulk of the chlorophyll nitrogen at this stage was in water or alcohol-soluble forms. Both light and oxygen were required for the bleaching of chlorophyll solutions *in vitro*. Although both are assumed to be normally required in the disappearance of chlorophyll, light is not an invariable factor *in vivo*.

Park *et al.* (1973) further commented that an abundance of source material is a prerequisite for tracing the fate of chlorophyll nitrogen. This is admittedly a difficult undertaking, as the chlorophyll nitrogen is less than 2% of the total nitrogen in ginkgo and hemlock (Lu, 1970) and there was a rapid and extensive cleavage of the chlorophyll molecule with the formation of colourless, low molecular weight compound (Seybold, 1943).

Lu (1970) attempted a systematic fractionation of non-protein nitrogenous constituents of green and yellow beans. If substances such as haematinic acid or methyl ethyl maleimide were to be found in yellow leaves, it would be strong presumptive evidence that it had been derived from chlorophyll. Although Lu concluded that it was possibly unrealistic to expect that relatively small fractions of such material would not be masked by more abundant "contaminants", two imide fractions were found to be of special interest: one from the neutral fraction of yellow ginko leaves (m/e 173) and the second from the acidic fraction of both yellow hemlock and ginko with a strong blue fluorescence. The latter chromatographed with a blue-fluorescing compound obtained by photodecomposition of chlorophyll *in vitro* (Jen, 1969). The discovery that phytol can be demonstrated in yellow leaves in quantities comparable with those in green leaves only after saponification simplified the problem considerably. There must exist a phytol-containing fragment (or fragments) in the yellow leaf, derived from ring IV, and it seems probable that some of its nitrogen has been retained in the imide form. The phytol, bound in its ester linkage, is a useful marker as if ring IV had been uniquely labelled with an isotope.

The ultimate fate of chlorophyll during senescence still remains unclear. It appears that a rapid decomposition of chlorophylls into small fragments must occur, because no large, obviously derived, compounds are observed (Seybold, 1943; Park *et al.*, 1973).

McFeeters and Schanderl (1968) developed an *in vivo* chlorophyll degradation system based on the incorporation of radioactive chlorophyll into bell peppers (*Capsicum frutescens*). ^{14}C-labelled chlorophyll was prepared and injected into ripening carpels for different periods of time. The distribution of radioactive compounds into three fractions was established. Preliminary chromatography did not permit isolation and identification of degradation products. If further work confirms that this degradation is physiologically normal, this system might provide a system for further studies of biological degradation of chlorophyll. ^{14}C- and or ^{15}N-labelling compounds will facilitate the isolation, identification and establishment of the origin and identity of trace amounts of breakdown products.

G CHLOROPHYLL DEGRADATION IN FOODS

The importance of chlorophylls in food technology derives not from their photosynthetic role, but chiefly from their part in the green coloration of vegetables and fruits. Since colour is an important attribute of food quality, the colour changes that take place during the processing or storage of foods has been the subject of many studies. It is generally agreed that the principal cause of the discoloration of green vegetables during processing or storage is the conversion of chlorophylls to pheophytins (Mackinney and Joslyn, 1941; Schanderl *et al.*, 1963). In some canned vegetables, like peas and spinach, it was apparent that the discoloured pigments were almost entirely pheophytin (Tan and Francis, 1962; Schanderl *et al.*, 1963). It is known that during this reaction hydrogen ions can transform chlorophylls into corresponding pheophytins by replacing the Mg ion in the porphyrin ring. Pheophytinization of chlorophylls occurs under many processing and storage conditions Clydesdale and Francis, 1968; Buckle and Edwards, 1970; Clydesdale *et al.*, 1970; Sweeney, 1970). Schanderl *et al.* (1965) showed that the regreening of processed vegetables can occur on storage in the presence of copper or zinc. The pigments responsible for the regreening are stable complexes of pheophytins *a* and *b* with copper or zinc. The reaction takes place under normal storage conditions.

Kinetic studies of the acid-catalysed conversion of chlorophylls *a* and *b* to pheophytins *a* and *b* in an acetone-aqueous buffer medium were conducted by Cho (1966). The rate of the reaction is second order with respect to hydrogen ion concentration and first order with respect to chlorophyll concentration. The greater stability of chlorophyll *b* to pheophytinization compared with chlorophyll *a* was explained by the following mechanism:

$$\text{CHl} + 2\text{H}^+ \underset{}{\overset{\text{Keg}}{\rightleftharpoons}} [(\text{Chl H}_2)^{2+}] \xrightarrow{\text{K}_2} \text{Pheophytin} + \text{Mg}^{2+}$$

in which the rate is determined by Keq. The Keq of chlorophyll *b* is about ten times smaller than for chlorophyll *a* because of the contribution of resonance structures.

Mechanisms other than pheophytinization can cause chlorophyll degradation in foods. The action of chlorophyllase and acid can lead to pheophorbide formation (White *et al.*, 1963). Allomerized chlorophylls can be formed by oxidation at C-10 on the isocyclic ring. Compounds without the characteristic chlorophyll spectra may be formed by lipoxygenase and lipid oxidation (Holden, 1965; Buckle and Edwards, 1970). Walker (1964) showed that under anaerobic conditions pheophytinization of chlorophyll can also be related to lipid oxidation in beans.

LaJollo *et al.* (1971) studied chlorophyll degradation as a function of water activity (A_w) in freeze-dried, blanched spinach puree and in model systems containing cellulose–citrate buffer and chlorophyll *a*. At 37°C and A_w values higher than 0·32, the most important mechanism of chlorophyll degradation was conversion into pheophytin. At A_w values lower than 0·32, the rate of pheophytin formation in spinach was low. Some of the modified chlorophylls were isolated and appeared to have spectral characteristics similar to those products isolated from processed pea puree by Buckle and Edwards (1969) and from legume seeds (Holden, 1965), but sufficient material could not be prepared for further characterization. The possibility of removal of the Mg ion from the porphyrin ring of chlorophyll *a* was shown in model systems even at A_w values below Brunnauer-Emmet-Teller (BET) monolayer coverage.

Gupte and Francis (1964) studied the effect of pH adjustment and high-temperature short-time (h.t.s.t.) processing on colour and pigment retention in spinach puree. Spinach processed by h.t.s.t. methods retained more pigment than a conventional retort process, the improvement in colour due primarily to decreased degradation of chlorophyll *a*. Adjustment to neutral or slightly alkaline pHs prevent the acid hydrolysis of chlorophylls and thus increase pigment retention in retort processing.

H ALTERATION OF CHLOROPHYLL IN SOIL

Chlorophyll-type compounds in soil have been studied by Hoyt (1966a, b). He first conducted a series of experiments to investigate the decomposition of chlorophyll in excised leaves. Conditions of high humidity and normal temperatures resulted in rapid chlorophyll loss. Freezing, boiling, desiccation and waterlogging all stopped chlorophyll loss. These data are consistent with an enzymic mechanism of degradation.

Hoyt further reported that tissue enzyme rapidly decomposed chlorophyll in chopped plant material mixed with soil. He stated that the slow decomposition was caused by micro-organisms which seem to be the main cause of chlorophyll degradation in soil. Micro-organisms decomposed both chlorophylls *a* and *b* in two to four months in field soils—chlorophyll $a > b$. Of the porphyrin ring compounds, pheophytin was the most resistant, and chlorophyllide and pheophorbide were rarely found in soil and never in large amounts. Microbiological decomposition increased with increased moisture content and temperature ($5° \rightarrow 25°$) of soil. Decomposition slowed with increasing soil acidity and was very slow at pHs below 4.0. Neither the species nor the quantity of plant material had much effect on the rate of decomposition. Although the mechanism of microbiological degradation of chlorophyll is not well understood, the decomposition of chlorophyll-type compounds by tissue enzymes, by micro-organisms and direct chemical action may contribute a significant amount of nitrogen to the soil.

I EFFECT OF CHEMICALS ON CHLOROPHYLL METABOLISM

Changes in chlorophylls resulting from various chemical and physical treatment of leaves and leaf extracts were studied by Bacon and Holden (1967). Chlorophylls *a* and *b* are easily altered chemically when leaves or leaf extracts are exposed to treatment which includes heating and the action of organic solvents such as acetone, ethanol and methanol. Treatment with such solvents leads to the formation of pheophytin chlorophyllides and a' and b' isomers, and the eventual loss of colour. Chlorophylls *a* and *b* can each be converted into two other derivatives with absorption spectra almost identical with those of parent compounds.

The results of the experiments by Bacon and Holden (1967) and of other workers show that certain extraction processes and treatments of leaves can lead to chemical alteration of the chlorophylls. Holden (1972) further reported that degradation of chlorophyll was inhibited in the dark in detached leaves floating on neutral EDTA (ethylene diamine tetraacetic acid) solutions, but stimulated in the light. Salts of inorganic acids, organic acids, surface active agents (Triton 100, Manoxol OT, cetyl trimethylammonium bromide) and herbicides (Diquat, Ioxynil, DCMU) likewise prevented breakdown of chlorophyll in the dark and caused it to be photo-oxidized in the light. Kinins inhibited degradation in the dark but did not promote bleaching in illuminated leaves. Srivastava (1968) reported the acceleration of senescence in excised barley leaves by abscisic acid and its reversal by kinetin. Mishra and Mishra (1968) studied the effect of various growth regulators such as auxins, kinins, gibberellins and growth-retarding chemicals on the prevention of chlorophyll and starch degradation of detached leaves undergoing senescence. They found that: (a) benzimidazole was most effective for rice leaves in preventing the chlorophyll degradation, (b) aminotriazole can effectively preserve the chlorophyll and starch content of groundnut leaves, (c) phytokinins were very effective on a wide range of crop plants, (d) gibberellins and growth-retarding chemicals were least effective in preventing the degradation of chlorophyll.

Retardation of chlorophyll degradation in *Zea mays* by coumarin, Phosfon D (2,4-dichlorobenzyltributylphosphonium chloride), and CCC (2-chloroethyl-trimethylammonium chloride) was reported by Knypl (1967). The effect of 2-chloroethylphosphonic acid (CEPA) and vanillin on chlorophyll, protein and RNA synthesis in detached cucumber cotyledons, and chlorophyll degradation in senescing leaf discs of kale was further studied by Knypl (1971). Kefford *et al.* (1973) tested approximately 900 urea derivatives and related compounds (urea cytokinins) for ability to retard leaf senescence as measured by chlorophyll retention in the radish (*Raphanus sativus*) leaf discs. Ninety compounds were found to be active. There was a high correlation between ability to promote chlorophyll retention and initiation of cell division. They further suggested from their results that the receptor site of cytokinin activity is the same as for senescence retardation and cell division initiation. Osborne (1965) reported that the hormonal retardation of leaf senescence was associated with the synthesis of RNA and protein. This was confirmed by Knypl (1971), who suggested that CEPA acts primarily on RNA metabolism whereas vanillin acts on protein turnover.

Holden (1972) stated that chlorophyll breakdown is clearly linked with the degradation of protein and probably also of lipids. In detached leaves (floated on water in the light), photosynthesis and protein synthesis can be maintained for a limited period with the chlorophyll level being retained or even increased in amount. When photosynthesis was inhibited, the chloroplasts became nonfunctional and catabolic reactions associated with senescence commenced. In the dark, even without inhibitors present, some protein and chlorophyll were degraded.

Martin and Thimann (1972) studied the senescence of oak leaves after detachment and darkening by following the loss of chlorophyll and protein and the formation of a proteolytic enzyme (Fig. 4).

Draper (1969) found that the final stages of the senescent phase in cucumber cotyledons were marked by a loss of all classes of lipids. Newman *et al.* (1973) recently studied further the lipid transformations in greening and senescing leaf tissue and demonstrated that senescent tissues showed a decline in chlorophyll and fatty acids. These results further support Holden's comments.

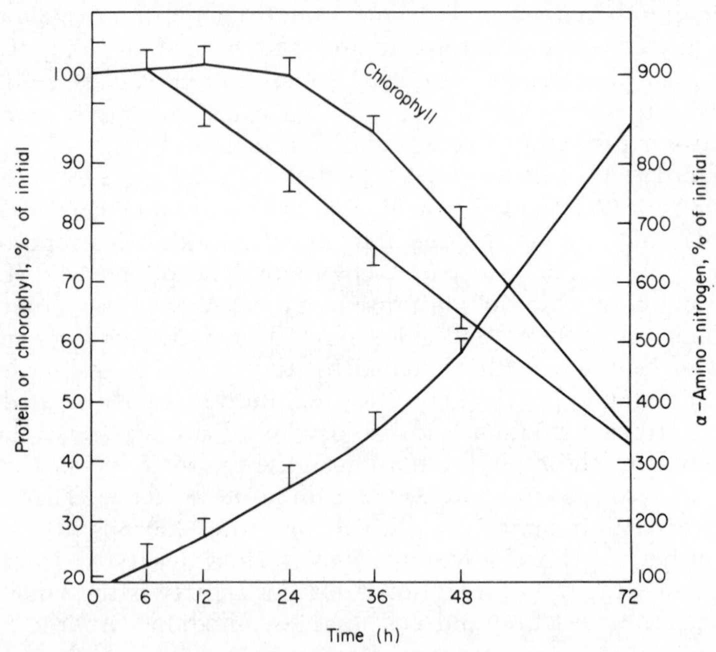

FIG. 4. Changes in chlorophyll, α-amino nitrogen, and protein in 7-day-old *Avena* first leaves, with time after detachment and darkening. The vertical bars represent one-half the 95% confidence limits. (Reproduced by permission from Martin and Thimann, 1971.)

It appears that ultimately chlorophyll will be oxidized. The oxidative destruction may be caused by enzymes, or exposure to heat, light, or other factors. Strain and his co-workers (Strain, 1954, 1958; Pennington *et al.*, 1967) reported the formation of "oxidized chlorophylls" in leaves exposed to methanol, acetone, ethanol and diethyl ether. These compounds were isolated by chromatography on powdered sugar columns. Bacon and Holden (1967) isolated "changed chlorophylls *a*-1, *a*-2, *b*-1 and *b*-2" produced by alkaline treatment of heated leaves or by the action of aqueous acetone or methanol on unheated leaves. Buckle and Edwards (1969) found three new chlorophyll degradation products from processed pea puree. These three products were similar in many respects to chlorophyll degradation products reported recently in ripening peppers, banana peel and cucumber peel (Schanderl and Co, 1966; Co and Schanderl, 1967a,b), to "changed chlorophylls" (Bacon and Holden, 1967), to artefacts from chromatography of purified pigments on certain adsorbents (Bacon, 1966) and to pigments in extracts of *Chlorella* (Michel-Wolwertz and Sironval, 1965). From these investigations it would appear that there exists a number of chlorophyll-like products which are separable from chlorophylls and are shown to have chemical spectral properties similar to various alteration products of parent chlorophylls. Their structures have not yet been completely clarified. Additional investigation is needed to determine their relation to the parent compounds and their position in any oxidation scheme, either *in vivo* or *in vitro*.

Beyond the formation of the oxidized chlorophylls, the non-enzymic breakdown of these substrates is probably concerned with oxygen attack on the isocyclic carbon C-10. The process probably involves the oxidation of C-10 to a hydroxyl group, followed by a breaking of the ring to form a variety of purpurins and chlorins. These materials are found in moderate quantities in dried plant materials (Aronoff, 1953). The exact pathway of oxidation is not yet clearly known *in vivo* as well as *in vitro*. At present, the formation of pheophorbides and/or purpurins causes a considerable change of the colour of plant materials. The further oxidation of these derivatives occurs through complete scission of one isocyclic ring, followed by oxidation of the tetrapyrrole (Aronoff, 1953). Seybold (1943) suggested that in plant materials the time required for the decomposition of chlorophylls to materials of low molecular weight is comparatively rapid. Recent work by Jen and Mackinney (1970a,b), Sanger (1971), Park *et al.* (1973) and Morris *et al.* (1973) all confirm this (see detailed description in this section).

It is difficult to identify these colourless, low-molecular weight compounds by spectroscopic and chromatographic means. However, recent investigation on the chlorophyll decomposition in *in vitro* systems (Jen and Mackinney, 1970a,b; Fuhrhop and Manzerall, 1971; Morris *et al.*, 1973) and *in vivo* systems (Lu, 1970; Park *et al.*, 1973) appear to start to reveal the identity of the ultimate breakdown products of chlorophyll decomposition.

III Carotenoids

Pure solutions of carotenoids are altered or destroyed by acids and, in some cases, alkalies, by specific enzymes, by oxygen and by light. A number of model systems have been developed to test various conditions and to determine the mechanisms that are involved in the changes. The extrapolation of the results of model systems to the vastly more complex plant systems is a difficult step. Nevertheless, the progress obtained in delineating the mechanism of the changes caused by various agents has come largely from model systems.

It can be stated generally that disruption of tissues or purification invariably results in a loss of carotenoids. In some tissues the carotenoids are relatively stable while in others they are formed and destroyed in a few days. In macerated green leaves, half of the carotenoids were lost in 20 min (Friend and Nakayama, 1959). The formation and loss of carotenoids occurs in only a few days in the corona of the narcissus (Booth, 1957, 1963).

Maize (corn) stored at 7°C for 3 years was found to retain 50% of its carotenoid content (Quackenbush, 1963). However, a purified extract of maize pigments stored for a week in the dark in petroleum ether was found in the authors' experience to deteriorate substantially (cf. Simpson and Goodwin, 1965).

A great deal of the work reported concerns the loss of β-carotene in senescent and stored tissues because of the value of this compound as a pigment and as a provitamin. Stahl *et al.* (1957) studied the loss of β-carotene under various lighting conditions of drying hay (Table I) and the results show that light as well as other factors are involved in β-carotene loss.

A number of papers purport to show the loss of β-carotene by measuring the decrease of absorption at 450 nm. However, a number of carotenes as well as metabolized or degraded products such as xanthophylls, epoxides, *cis*-isomers etc. also absorb light at that

TABLE I
β-Carotene losses in drying hay[a]

Lighting condition	Time (h)	Percentage loss
Intense sunlight	48	70
Shade	48	36
Diffuse sunlight indoors (through glass)	48	44
Diffuse sunlight after steaming	—	40
Ultraviolet light after steaming	48	40

[a] Reproduced by permission from Stahl *et al.* (1957).

wavelength. The carotenoids in the shell of the pineapple were studied (Gortner, 1965) and it was found that during development there was a decrease in carotenoids followed by an increase during senescence However, no attempt was made to distinguish between carotenes, xanthophylls, esters or epoxides. Work which considers the mechanism of the oxidation and degradation in general of carotenoids should distinguish between the various types of carotenoids.

A FORMATION OF EPOXIDES

It has been postulated that the initial step in the breakdown of carotenoids occurs through the formation of epoxides (Zechmeister *et al.*, 1943; Fishwick, 1962) and some studies support this hypothesis. However, a number of 5,6-epoxides are found in green plants, fruits and algae. The common epoxides are either structural derivatives of zeaxanthin (antheraxanthin 5,6-epoxide; violaxanthin 5,6,5′,6′-diepoxide) or the allenic monoepoxide neoxanthin. In the higher plants and algae, which are O_2 evolving photosynthetic organisms, light-induced changes in violaxanthin concentration have been described as the violaxanthin cycle:

$$\text{Violaxanthin} \rightleftharpoons \text{Antheraxanthin} \rightleftharpoons \text{Zeaxanthin}$$

Although the function of the violaxanthin cycle has been the subject of some disagreement, it seems clear from the results reported by Yamamoto and Takeguchi (1971) that the cycle is a pathway for photosynthetic uptake of O_2. Lee and Yamamoto (1968) suggest that the cycle is not the O_2 uptake pathway in photosystem 1 but a pathway for the consumption of excess photoproducts or for the conversion of these products into other forms of energy. In apricots and peaches violaxanthin is formed simultaneously with the formation of carotenes and xanthophylls (Katayama *et al.*, 1971) and at maximum maturity actually decreases in concentration (Chichester and Nakayama, 1965).

The epoxide cycle reported to be present in *Euglena gracilis* (Bamji and Krinsky, 1965) was shown to be composed of the monoepoxide diadinoxanthin (I) and the non-epoxide diatoxanthin (II) (Aitzetmüller *et al.*, 1968). Krinsky (1968) proposed that the function of this cycle is to protect the cells against lethal photo-oxidations.

While the involvement of epoxide cycles in senescent or stored tissue is not clear, evidence of the formation of epoxides in these

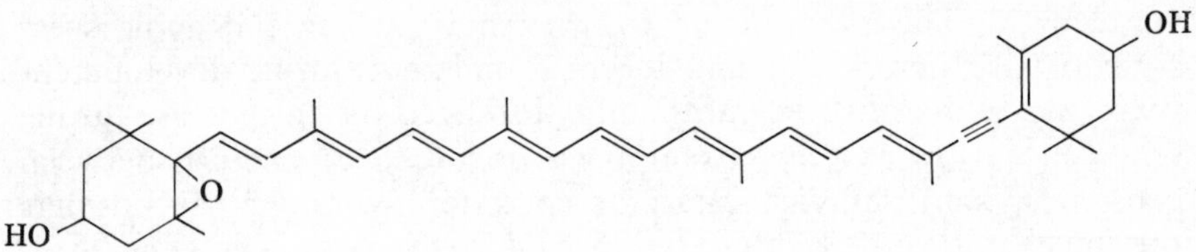

Diadinoxanthin (I)

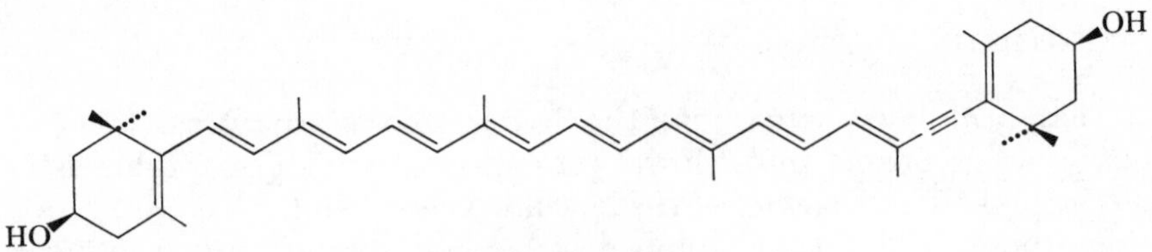

Diatoxanthin (II)

tissues has been presented. In addition, the data clearly show the formation of epoxides in model systems.

Glover and Redfearn (1953) showed that epoxides of β-carotene were formed in leaves in the light and that these epoxides were not directly reduced to β-carotene again but are probably degraded further. Goodwin (1958) detected lutein 5-6 epoxide and traces of β-carotene epoxide in autumn leaves.

Britton and Goodwin (1969) and Ben-Aziz *et al.* (1973) have detected a number of acyclic and cyclic epoxides in tomatoes. These authors suggest that the epoxides may be the result of tissue injury but also suggest that they may be early products in oxidative degradation in senescing tissues. The latter suggestion was based on unpublished observations that the epoxides may be present in greater quantities in overripe tomatoes.

The juice of the Shamouti orange (Gross *et al.*, 1971) was found to contain over 50 carotenoids, many of which were epoxides. The epoxides do not appear to be artefacts of isolation, although such epoxides as mutatochrome (III), mutatoxanthin (IV), luteoxanthin (V) etc. are associated with carotenoid degradations.

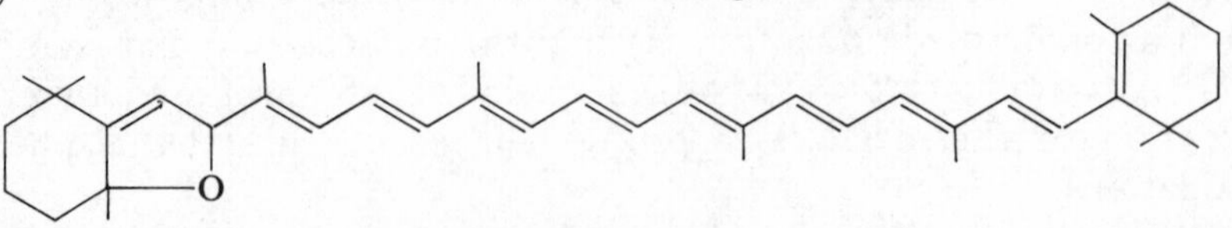

Mutatochrome (III)

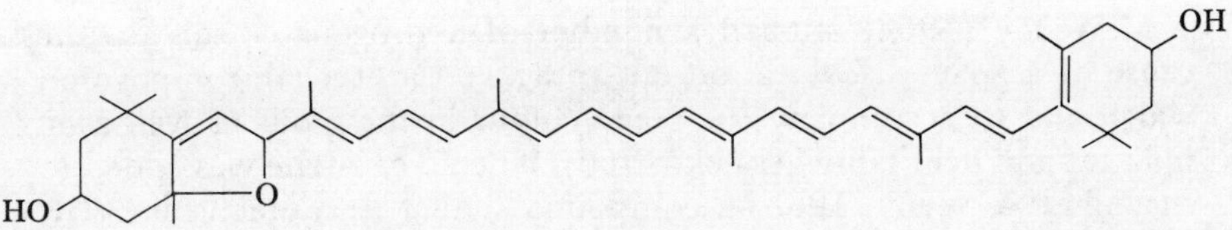

OH

HO O

Mutatoxanthin (IV)

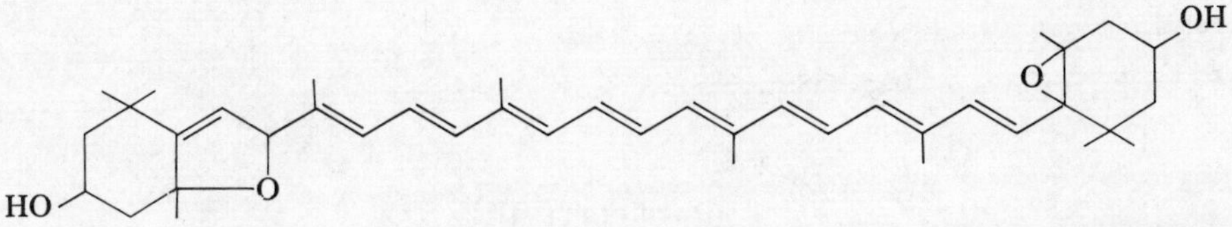

OH

O

HO O

Luteoxanthin (V)

Carotenoid formation was followed in *Triphasia trifolia* (Yokoyama and White, 1970) from the early green stage to the fully ripe stage. Epoxides were not reported although there was an accumulation of secocarotenoids (semi-β-carotenone (VI), β-carotenone (VII) etc.) at the expense of carotenes.

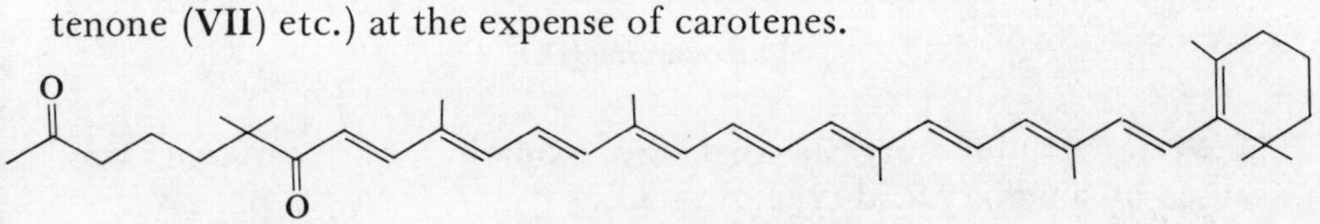

O

O

Semi-β-carotenone (VI)

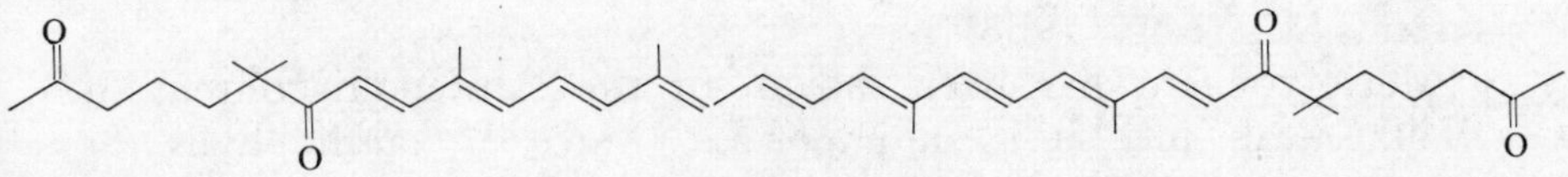

O O

O O

β-Carotene (VII)

Thus neither epoxide nor secocarotenoid formation should be considered as obligate senescent compounds in citrus.

The significance of epoxide occurrence is further complicated by the finding of Bodea (1969) which showed that epoxide formation increases and carotene formation decreases with elevation from sea level within the same plant.

We suggest that if tissues other than the tomato, orange etc. were examined for epoxides in the same manner as those above, similar results would be obtained.

McWeeny (1968) studied a number of hydrogenated oils for the cause of a green discoloration on storage. β-Carotene decomposition was found to accompany the green colour in these oils as well as in milk fat and beef tallow (Luck, 1966). When β-carotene was added to the oils (McWeeny, 1968) a compound similar to mutachrome (**III**) (5,8-epoxy-β-carotene) and termed pseudomutatochrome was isolated. The presence of the diepoxides aurochrome (**VIII**) and possibly luteochrome (**IX**) was indicated in this study.

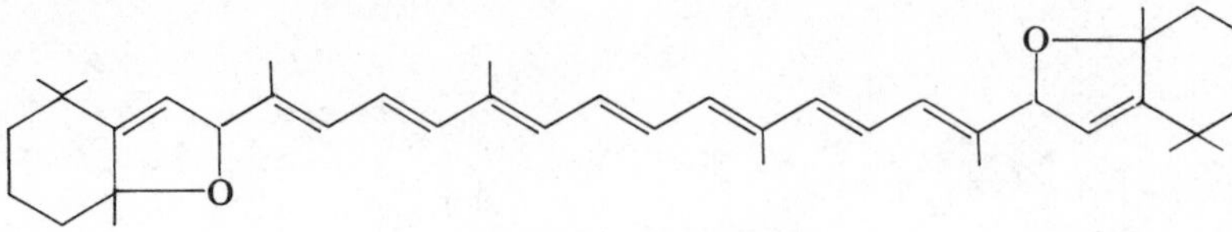

Aurochrome (**VIII**)

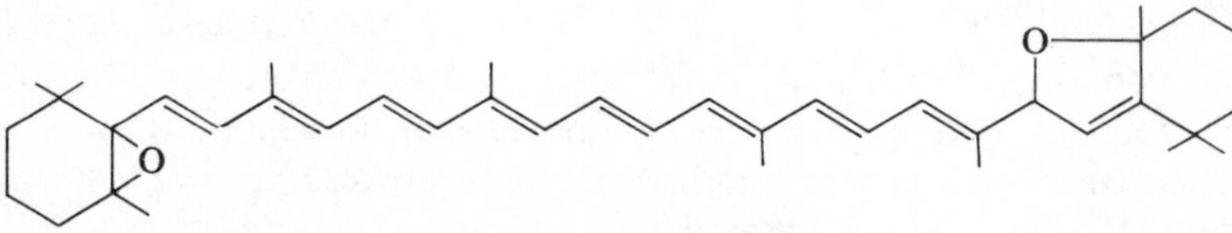

Luteochrome (**IX**)

McWeeny (1968) suggests that the oxidation of β-carotene was caused by a peroxy acid type of agent.

Figure 5 gives the suggested general reaction mechanism leading to the formation of green discoloration in β-carotene hydrogenated fat systems (McWeeny, 1968).

Much concern has been shown for the carotenoid content of alfalfa meal since it is an important source of xanthophylls for broiler and egg pigmentation (cf. Taylor *et al.*, 1968). Epoxide xanthophylls are not effective pigmenters for either broilers (Livingston *et al.*, 1969) or eggs (Marusich *et al.*, 1960). Lutein was found during pilot and industrial scale processing to be more stable than either violaxanthin or neoxanthin. Neoxanthin was found to be more stable than the other epoxide–violaxanthin (Livingston *et al.*, 1968). Freeze-dried lucerne (alfalfa) meals, while more stable than commercially dehydrated meals, lost 83% of the original violaxanthin content. Neoxanthin and lutein were found to be much less labile (Knowles *et al.*, 1968).

El-Tinay and Chichester (1970) studied the oxidation of solutions of β-carotene in toluene at various temperatures. It was found that

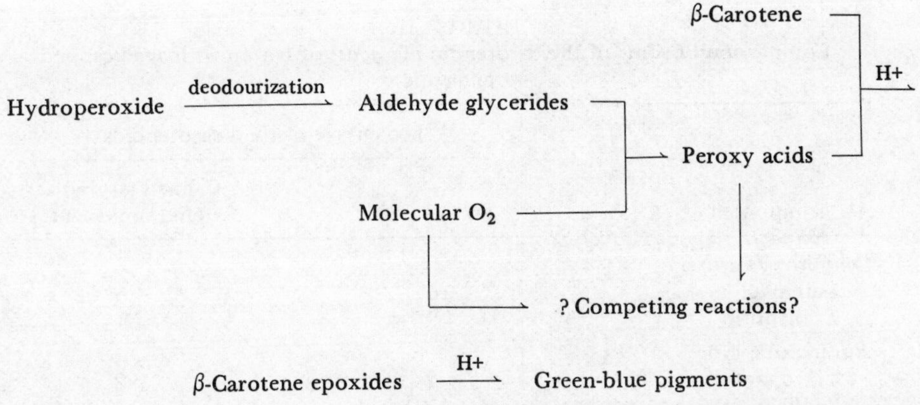

FIG. 5. Suggested general reaction mechanism leading to the formation of green discoloration in β-carotene hydrogenated fat systems (McWeeny, 1968).

the rate of β-carotene disappearances indicated a zero order reaction in the presence of excess oxygen. The site of the initial attack under these conditions was the terminal double bonds (5,6 and 5′,6′) and a series of epoxides of β-carotene were formed (5,6-monoepoxide; 5,6-monoepoxide isomer; 5,6,5′,6′-diepoxide; 5,8-monoepoxide; 5,8-monoepoxide isomer; 5,8,5′,8′-diepoxide). When a free radical initiator (*N*-bromosuccinimide) was added to the system the products of the reaction were echinenone (4-keto-β-carotene) and canthaxanthin (4,4′-diketo-β-carotene) (cf. Petracek and Zechmeister, 1956). These authors found that the addition of diphenylamine, a known free-radical inhibitor, stopped the loss of β-carotene.

Nicoara *et al.* (1970) oxidized canthaxanthin with perbenzoic acid and found, as predicted by Bodendorf (1930), that the 5,6-epoxides were not formed. These authors isolated instead the 8,10- and 13,14-epoxy canthaxanthins. The in-chain epoxides are more labile than the 5,6-epoxides which might explain their lack of isolation *in vivo* systems (Osianu *et al.*, 1969).

Tsukida *et al.* (1966) irradiated β-carotene with visual light in various solvents and reported the formation of xanthophylls, retrodehydrocarotene and carotene epoxides.

Chemical oxidations of β-carotene are known to yield a number of epoxide products (Karrer and Jucker, 1945; Hunter and Krakenberger, 1947; Seely and Meyer, 1971). Hasegawa *et al.* (1969) irradiated β-carotene solutions with a continuous-wave gas laser in the presence of the photosensitizing dye toluidine blue and found thirteen compounds composed of *cis* isomers and epoxides. Seely and Meyer (1971) studied the photosensitized oxidation of

TABLE II
Comparison of some of the carotenoid pigments of frozen with aged canned orange juice[a]

Component	Percentage of total carotenoids	
	Fresh frozen	Canned (stored at room temperature)
Dihydroxy carotenes		
Lutein	6	5
Zeaxanthin	9	13
Furanoxide type		
Flavoxanthin	1	2
Trollein	15	14
Mutatoxanthin	7	19
Trollichrome-like	12	36
Auroxanthin	22	19
Epoxide type		
Trollixanthin-like	26	0

[a] Reproduced by permission from Curl and Bailey (1955, 1956) as quoted by Blundstone *et al.* (1971).

β-carotene by the photodynamic agent hypericin. Mutatochrome was the first product formed in this photo-oxidation system, whereas in the chemical oxidation the 5,6-monoepoxide was formed first and later rearranged to the 5,8-monoepoxide. These authors found that when the 5,6-monoepoxide was subjected to photo-oxidation no mutachrome was isolated after 3 h. It is significant that the violaxanthin cycle epoxides antheraxanthin and violaxanthin were not isolated from the photo-oxidation of zeaxanthin. In this case a compound exhibiting properties similar to auroxanthin (zeaxanthin 5,8,5′,8′-diepoxide) was detected. The data of Seely and Meyer (1971) suggests that mutachrome and auroxanthin may have a photochemical origin at some stage of development including senescence but that a different mechanism is probably involved in the violaxanthin cycle.

Tsukida *et al.* (1966) irradiated β-carotene with visual light in the presence of chlorophyll and oxygen. These authors found, in addition to epoxy carotenoids, hydroxy, acetylenic and allenic pigments. This latter paper illustrates the difficulties in moving from *in vitro* to *in vivo* systems.

Plant carotenoids stored for periods of time under less than ideal conditions will be partly converted into epoxide derivatives. De La

Mar and Francis (1969) studied the effect of bleaching paprika under accelerated storage conditions. Nearly 96% of the original β-carotene was lost and many new pigments were formed. Most of the pigments in bleached paprika were oxygenated derivatives possessing hydroxyl, keto or epoxy groups.

Curl and Bailey (1955, 1956) studied the effect of storage of canned orange juice. Under the low pH conditions of storage there was a significant isomerization of the 5,6- to the 5-8-furanoxide-type xanthophyll epoxides (Table II).

Singleton *et al.* (1961) and Gortner and Singleton (1961) found that handling and processing of pineapple fruit released fruit acids causing a rapid isomerization resulting in the formation of *cis–trans* carotenoids and 5,8-epoxide carotenoids.

B FORMATION AND ESTERIFICATION OF XANTHOPHYLLS

The usual pigment content of photosynthetic tissue consists of chlorophylls, β-carotene as the main carotene and lutein as the main xanthophyll. Other xanthophylls invariably present include the violaxanthin cycle pigments and neoxanthin. As green tissue ripens, as in the case of fruit, or ages, as in the case of autumn leaves, there is a drop in the level of chlorophyll and in many cases carotene and an increase in xanthophylls—commonly esterified xanthophylls.

It is not surprising that the beautiful autumn leaf colours should be some of the first pigments to be studied. The pigments of autumn leaves were termed "autumn carotenes" by Berzelius (1837a,b) owing to their non-polar nature. Kuhn and Brockmann (1932) subsequently showed that the non-polar nature of the "autumn carotenes" was due to the esterification of xanthophylls during senescence. It was thought that the carotenes were converted into xanthophylls which remained as the chlorophyll decreased. However, there is a decrease in total carotenoids with the carotenes apparently being more labile than the xanthophylls. This stability may be due in part to the fact that the xanthophylls are esterified as they are liberated into the cytoplasm on disruption of the chloroplasts (Goodwin, 1958; Grob and Eichenberger, 1962). As Sanger (1971) points out, the loss of pigments in tree leaves is very dependent on the species and the climatic conditions. Some leaves lost all pigments whereas others still had measurable amounts of lutein and β-carotene with violaxanthin and neoxanthin disappearing most rapidly. The relative loss in the carotenoids or their esterification may either precede, occur simultaneously, or follow the decrease in chlorophyll, depending upon the species. Clearly then, there is no direct relationship between chlorophyll and xanthophyll changes (Chichester and Nakayama, 1965; Reid *et al.*, 1970).

The carotenoids of fruit have been extensively studied (cf. Goodwin and Goad, 1970) but the changes in individual pigments during ripening and senescent stages are not generally well understood. The tomato (Raymundo *et al.*, 1967; Ben-Aziz *et al.*, 1973), apricot and peach (Katayama *et al.*, 1971) are examples of fruit in which individual pigments have been studied. In the latter study [^{14}C]-mevalonic acid (MVA) was readily incorporated into carotene and xanthophylls in ripening fruit. As fruit approached maturity the incorporation of [^{14}C]MVA into xanthophylls became less than the carotenes. Katayama *et al.* concluded that the β-carotene formed in ripe or overripe fruit was not converted to xanthophylls but was degraded, thus accounting for the change in ratio. At the ripe stage most of the xanthophylls were esterified, which is a general characteristic of fruit (Tsumaki *et al.*, 1954; Galler and Mackinney, 1965). Carotenoids accumulate in oranges after the "colour break" (Lewis and Coggins, 1964; Eilati *et al.*, 1969, 1972) and decrease in regreening valencia oranges (Lee *et al.*, 1971). When valencia oranges were left on the tree for an additional year there was a decrease in the total xanthophylls with a proportional decrease in the concentration of esterified xanthophylls. It would appear that the xanthophyll esters in the lipid globules of the chromoplasts of regreening oranges are metabolically available in contrast to crystalline carotenes in the carrot root and in tomatoes.

The β-citraurin ester has been isolated from Marsh seedless grapefruit and characterized as β-citraurin myristate (Philip, 1973). Palmitic acid has been isolated from other xanthophylls (Karrer and Schlientz, 1934; Karrer *et al.*, 1948).

C *CIS–TRANS* ISOMERIZATION

The majority of the natural carotenoids occur in the generally more stable all-*trans* form. There are, however, fruits such as the tangerine tomato (Raymundo and Simpson, 1972) and *Pyracantha angustifolia* (Zechmeister and Pinckard, 1947) where significant amounts of poly-*cis* carotenes exist. In the tangerine tomato poly-*cis* isomers of ζ-carotene, γ-carotene, neurosporene and lycopene have been isolated (Raymundo and Simpson, 1972; R. W. Glass and K. L. Simpson, unpublished observation). Poly-*cis* carotenes from tangerine tomato fruit develop in the dark in contrast to the chloroplastidic pigments

(Raymundo *et al.*, 1974). The poly-*cis* pigments are not formed from the all-*trans* in vitro (Weedon, 1971).

Cis isomers occur in a number of plants (Weedon, 1971) but they also are formed in isolation procedures and under storage conditions. The rate of formation of *cis–trans* isomers from all-*trans* carotenoids is directly proportional to the intensity of light, particularly light at wavelengths of main absorption bands, increase in temperature and presence of catalysts such as acid or iodine. A single *cis* bond may shift the absorption bands to shorter wavelengths (2–5 nm). The intensity of the absorption may also be decreased.

Gortner and Singleton (1961) and Singleton (1963) showed that the processing of pineapple released sufficient fruit acids to cause the formation of *cis–trans* isomers from the natural carotenoids present. In practical terms the isomerization resulted in a lightening of fruit colour.

Although there are few specific data on the random formation of *cis*-isomers in senescent or stored tissue, one would expect that any processing or storage condition involving elevated temperatures, low pH and light would produce these isomers (cf. Seely and Meyer, 1971). The chromatographic observation (Bickoff *et al.*, 1954) for forty coloured bands from dehydrated lucerne (alfalfa) meal in contrast to the twelve carotenoids isolated from fresh lucerne would be consistent with what one would predict based on model systems. While there has been very little work done on the biological significance of *cis* isomers, the introduction of a *cis* band can greatly reduce the potency of β-carotene as a vitamin A precursor (Ames, 1958).

D ENZYMIC CHANGES

In Table I it could be seen that carotenoids were lost owing to a number of factors and that some protection was afforded by blanching the product. These results are suggestive of an enzymic destruction of the pigments.

The carotenoids are always present in photosynthetic tissue and these systems nearly always contain a carotene-destroying enzyme. One exception to this "rule" is the iris leaf, but here the enzyme may be present but inhibited by natural antioxidants (Booth, 1960). Friend and Mayer (1960) used the water-soluble carotenoid glycoside crocin as a model substrate to study the destruction of carotenoids by sugar beet chloroplasts. It was found that the oxidation was aerobic, inhibited by cyanide and sodium diethyl dithiocarbamate and caused by a haematin-type enzyme. Results with catalase and peroxidase showed that the destruction of carotenoids by chloroplasts was not a coupled reaction involving hydrogen peroxide nor was the enzyme a lipoxygenase. The pH optimum for this enzyme was found to lie between 4·0 and 5·0 in lucerne extracts (Walsh and Hauge, 1953) and at 7·3 in chloroplasts from sugar beet leaves (Friend and Mayer, 1960). The chloroplast enzyme was found not to act through a coupled oxidation of a fatty acid (Friend and Mayer, 1960; Friend and Dicks, 1967; Dicks and Friend, 1968). Enzymic action was stimulated by oxidized linoleate but not fresh linoleate (Friend and Mayer, 1960).

Blain *et al.* (1968) have characterized a haematin–type carotene oxidase system from tomato extracts that also required oxidized linoleate and was not specific to the pentadiene system. A similar factor has been isolated from soya extracts (Blain and Styles, 1959). The very rapid destruction of carotene in field-cured lucerne is partly caused by the chloroplast or haematin enzyme (Walsh and Hauge, 1953).

The major work on enzymically catalysed oxidation of the carotenes has been concerned with the lipoxygenase system primarily from soya. In these systems the carotenoids act as antioxidants or secondary substrates to the fat oxidizing systems. Friend (1958) isolated the products from a coupled oxidation of β-carotene by a linoleate–lipoxygenase system. In addition to unchanged β-carotene and its *cis* isomer, the following compounds were isolated: β-carotene monoepoxides, conjugated polyene ketones, aurochrome (**VIII**) and β-C_{18}-ketone (**X**). The isolation of long-chain carbonyl and apocarotenoids would suggest a form of β-oxidation.

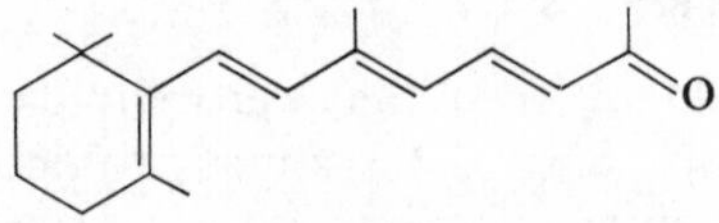

O

β-C_{18}-ketone (**X**)

Blain and Shearer (1963) developed a model system for the auto-oxidation of β-carotene in agar gels. When pelargonic acid, a liquid-saturated fat, was added, the stability of β-carotene was improved both with and without antioxidant. When the unsaturated ester methyl linoleate was added, the oxidation of carotene was accelerated but the antioxidants were also more effective.

The oxidation of unprotected carotene was greatly accelerated with the addition of either haemoglobin or soya lipoxygenase. It is interesting that of the antioxidants used, nordihydroguaiaretic acid (NDGA) was superior to santoquin (ethoxyquin = 6-ethoxy-1,2-dihydro-2,2,4-trimethylquinoline) for the lipoxygenase catalysis whereas the reverse was true for haemoglobin.

The isolation of a series of β-apocarotenals and apolycopenals (Winterstein *et al.*, 1960) from natural sources led Isler *et al.* (1962)

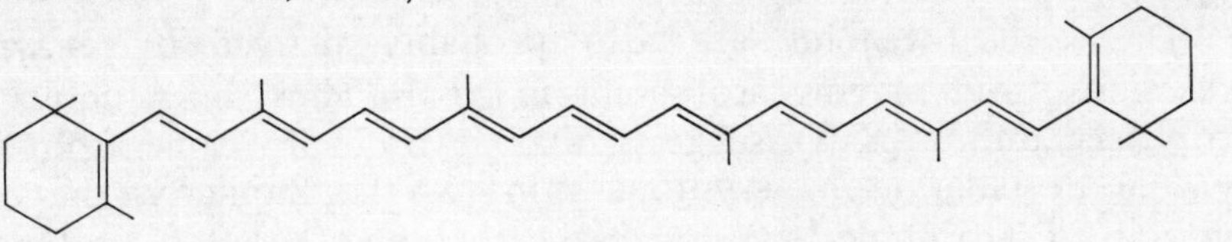

β-Carotene (XI)

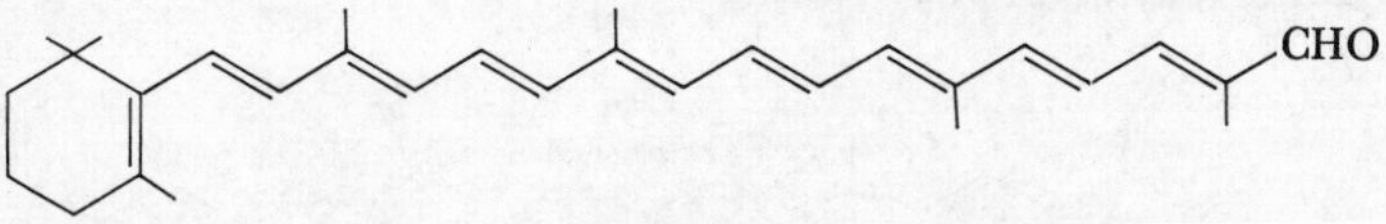

CHO

β-Apo-8′-carotenal (XII)

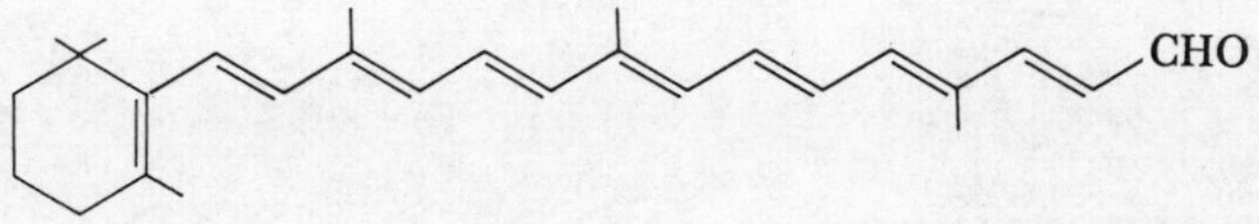

CHO

β-Apo-10′-carotenal (XIII)

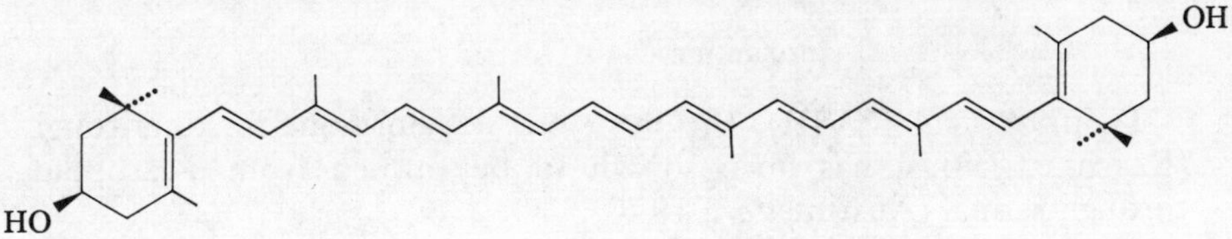

OH

HO

Zeaxanthin (XIV)

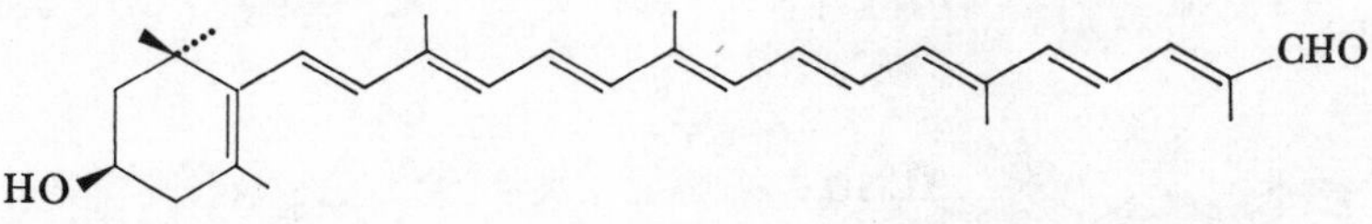

CHO

HO

β-Citraurin (XV)

SCHEME 1. Suggested degradative transformations (Yokoyama and White, 1966).

to postulate a β-oxidation mechanism as one pathway for the destruction of carotenoids *in vivo* (cf. Glover, 1960).

Weedon (1971) discusses a number of apocarotenoids under a section on degraded carotenoids. Yokoyama and White (1966) suggested that β-apo-10′-carotenal and β-apo-8′-carotenal are natural degradation products of β-carotene, and β-citraurin the product of zeaxanthin.

The secocarotenoids are also probably derived by enzymic reactions from parent carotenoids in citrus. Thus the structure of triphasiaxanthin (**XVI**) suggests that it probably came from the partial degradation of cryptoxanthin (**XVII**). Similarly, the conversion of β-carotene (**XI**) → semi-β-carotenone (**VI**) → β-carotenone (**VII**) and α-carotene → semi-α-carotenone has been suggested on the basis of structural similarities and ripening data (Table III; Yokoyama and White, 1970).

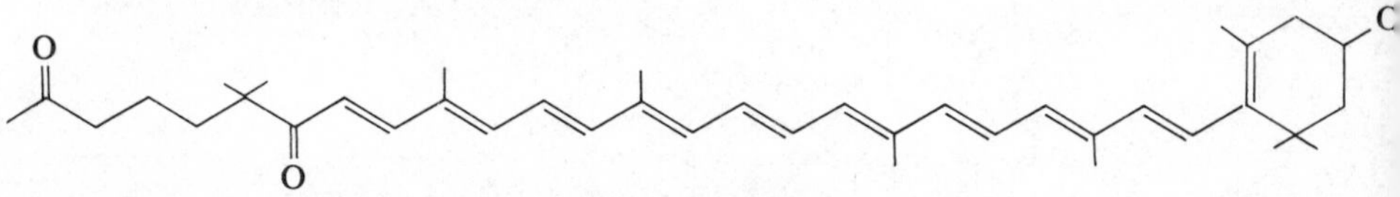

Triphasiaxanthin (**XVI**)

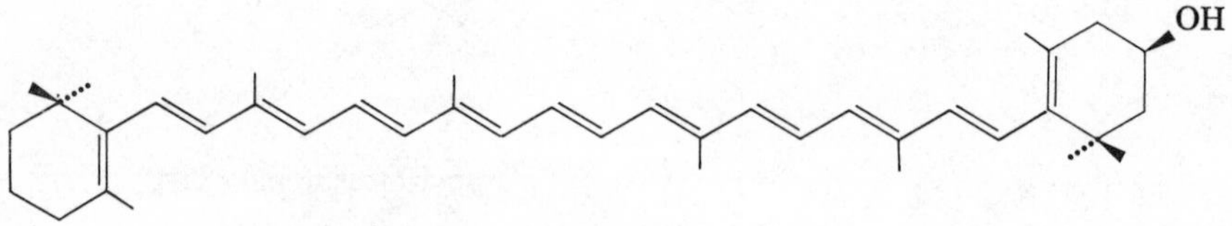

Cryptoxanthin (**XVII**)

Trisporic acid (**XVIII**), the principal sex hormone in some fungi (Reschke, 1969), has been shown to be formed from β-carotene through retinal (Austin *et al.*, 1970).

Structural similarities between the naturally occurring growth inhibitor abscisic acid (**XIX**) and some xanthophyll epoxides led to the suggestion that abscisic acid was formed by degradation of carotenoids.

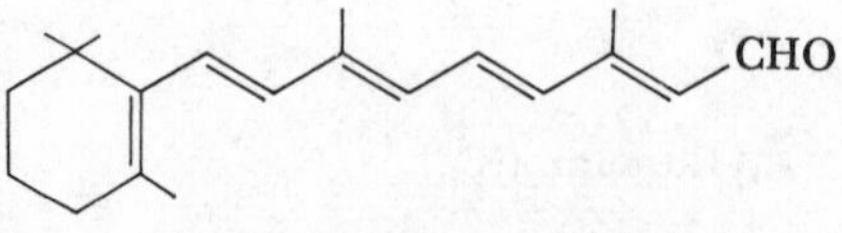

Trisporic acid (**XVIII**)

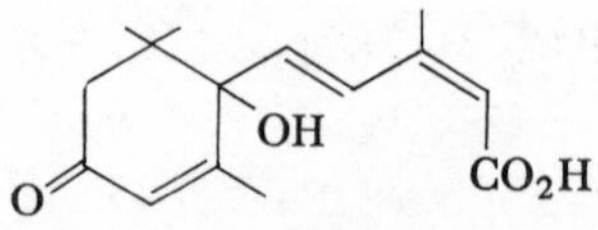

Abscisic acid (**XIX**)

TABLE III

Composition of some carotenoids in the ripening fruit of *Triphasia trifolia*[a]

Carotenoids	In % of total carotenoids[b]			
	Early green (dark green) colour	Mature green (light green colour with blotches of orange)	Early season (deep orange) colour	Fully ripe (deep crimson) colour
α-Carotene	1	4	2	—
β-Carotene	6	21	8	0·5
Semi-α-carotenone	—	—	2	4
Semi-β-carotenone	—	17	46	63
Triphasiaxanthin	—	—	3	6
β-Carotenone	—	—	7	21
Cryptoxanthin	2	6	3	—

[a] Reproduced by permission from Yokoyama and White (1970).

[b] Total carotenoids content (mg/100 g wet wt): Early green, *c.* 10; mature green, *c.* 90; early season, *c.* 130; fully ripe, *c.* 300.

Taylor and Burden (1970) showed that the photo-oxidation of violaxanthin and neoxanthin produced growth inhibitors with biological activities similar to abscisic acid. Incorporations of [2-^{14}C]MVA into abscisic acid by intact tomato fruit was obtained (Noddle and Robinson, 1969) and the stereochemical aspects of the formation of double bonds in abscisic acid did not rule out violaxanthin as its precursor (Milborrow, 1972).

It should be pointed out that the seco- or apocarotenoids and related terpenes may be "biosynthetic" compounds even though they might be formed by a "degradation" step from parent carotenoids. Some of these compounds produced by the photo-oxidation of violaxanthin and neoxanthin may be formed in senescent or stored tissue or may be similar to compounds formed in these tissues.

E EFFECT OF IONIZING RADIATION

Early studies by Lukton and Mackinney (1956) showed that solid films of β-carotene in the absence of oxygen were relatively stable to ionizing radiation. Carotenoids were found to be much more labile in the various solvents tested. In the latter case the damage may be either a direct effect or an indirect effect with the pigment being attacked at a latter stage of a solvent-borne free radical chain. In biological materials the stability of carotenoids is much greater owing to natural free radical scavengers.

Sawant *et al.* (1970) irradiated orange juice and mango pulp with 1·0 and 2·0 Mrad and found that the more dense mango pulp was more resistant to radiation. While some new carotenoid fractions were detected, the spectral curves do not show the gross qualitative changes obtained with sun lamp irradiation of carotenes.

Villegas *et al.* (1972) found that gamma irradiation delayed carotene development in tomato fruits. These authors found that softening of the fruit and mould infection greatly limited the usefulness of the method. It would seem highly unlikely on the basis of toxicological problems that ionizing radiation will be applied to food. The data available show that the carotenoids are stable and are not the limiting factors in the use of ionizing radiation treatment of fruit.

IV Anthocyanins

The water-soluble anthocyanin pigments are usually dissolved in the cell sap rather than associated with chloroplast and other lipoid bodies. They differ from chlorophylls and carotenoids in stability, chemical reactions and time of biosynthesis.

The development of anthocyanins in leaves, fruits and flowers is characterized by a slow increase at the early stages of organ growth, followed by an abrupt and substantial rise to a maximum and, finally, by a sharp or slow decrease until the death of the plant cells (Nagornaya, 1968; Vasquez *et al.*, 1970; Sanger, 1971; Stickland, 1972; Woodward, 1972). Maximum anthocyanin accumulation takes place late in maturity when plastid pigments are undergoing rapid decomposition (Goodwin, 1958; Sanger, 1971; Stickland, 1972; Woodward, 1972; Trojan and Gol'yan, 1973). In strawberries, for example, anthocyanin synthesis begins 28 to 35 days after petal fall when net synthesis of chlorophyll and carotenoid pigments has long ceased and their concentrations are down to very low levels (Fig. 6).

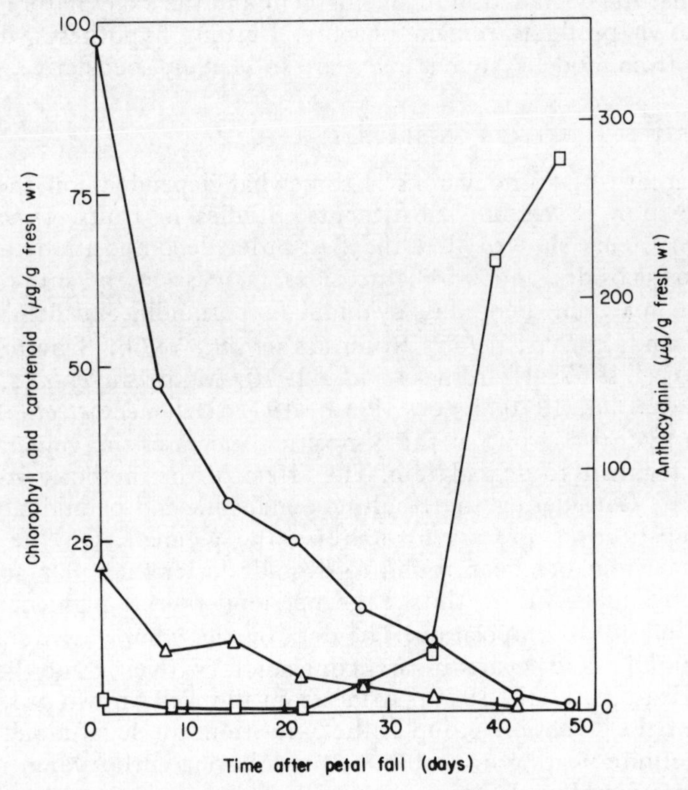

FIG. 6. Changes in pigment concentration per unit fresh weight in developing strawberry fruits. ○ = Chlorophyll. △ = Carotenoid. □ = Anthocyanin. (Reproduced by permission from Woodward, 1972.)

Similarly, in autumn leaves anthocyanins reach their highest level when chlorophyll and carotenoids have declined to less than 40% in oak leaves and chlorophyll has dropped to 20% in hazel leaves (Sanger, 1971).

After the peak of autumn coloration, hazel leaves fall to the ground devoid of all pigments. On the other hand, oak leaves which overwinter on the tree lose their anthocyanins at a low rate until they fall in April with trace amounts of the pigments (Sanger, 1971). Qualitatively, the anthocyanins of autumn leaves may differ from those of the young leaves (Ishikura, 1972).

Anthocyanin content is reduced considerably during storage of processed and unprocessed plant materials. Much attention has been given to the rate and extent of degradation and to the factors which accelerate or inhibit such destruction. The precise reactions involved in the natural degradation of the pigment and the exact nature of the breakdown products remain obscure. Current hypotheses, derived chiefly from model system studies, are still largely speculative.

A SUBSTITUENT EFFECTS ON STABILITY

The stability of anthocyanins is somewhat dependent on the type and position of certain substituents. Studies in fruits, wines and model systems showed that the first-order decomposition rates of 3-monoglucosides and 3,5-diglucosides increase in the order pelargonidin, malvidin, peonidin, cyanidin or petunidin and delphinidin (Keith and Powers, 1965; Robinson *et al.*, 1966; Saberrov and Ul'yanova, 1967; Hzardina *et al.*, 1970; Montreau *et al.*, 1970; Wrolstad *et al.*, 1970; Estevez-Pinto, 1972; Dzheneev *et al.*, 1973). Thus a hydroxyl group at the 3′-position increases the vulnerability of the pigment to degradation. The effect of the methoxy group is less clear. Considering the results on malvidin and peonidin, methoxyl substitution appears to stabilize the pigment. On the other hand, malvidin has been found to degrade faster than pelargonidin, indicating that the methoxyl group renders the pigment more susceptible to decomposition. The data on the comparative stability of petunidin and cyanidin are contradictory (Keith and Powers, 1965; Hzardina *et al.*, 1970). Aside from the well known protective effect of the glycosidic group at the 3-position, the sugar moiety also has a definite bearing on stability. Wines having anthocyanin monoglucosides discolour faster on exposure to light and heat than those containing diglucosides (Van Buren *et al.*, 1968). Acylated diglucosides resist the harmful effect of light better than the corresponding non-acylated diglucoside. Arabinoside pigments degrade faster than the galactosides (Starr and Francis, 1968).

B ENZYMIC DECOLORATION

Enzymes capable of decolourizing anthocyanins have been prepared from fungi, fruits, flowers and other plant organs (Huang, 1955, 1956; Forsyth and Quesnel, 1957; Wagenknecht *et al.*, 1960; Peng and Markakis, 1963; Goodman and Markakis, 1965; Harborne, 1965; Sakamura *et al.*, 1966; Boylen *et al.*, 1969; Casoli *et al.*, 1969a,b; Dall'aglio and Leoni, 1969; Proctor and Creasy, 1969; Segal and Segal, 1969; Uchiyama, 1969; Dall'aglio *et al.*, 1970; Schmid, 1971; Medvedeva, 1973; Skalski and Sistrunk, 1973). *In vitro* experiments have shown these enzymes to be effective towards pigment extracts of a variety of fruits and other plant sources. They fall into two general types: the glucosidases (anthocyanases) and the polyphenol oxidases (phenolases).

The glycosidase-mediated discoloration occurs in two steps: (1) enzymic hydrolysis of the anthocyanin to anthocyanidin and sugar; (2) spontaneous decomposition of the aglycone (anthocyanidin), which has been destabilized by the removal of the glycosidic group, resulting in irreversible loss of colour. The glycosidases differ in their substrate specificity. The enzyme from the petals of *Impatiens balsamina* has a strict requirement for an aromatic aglycone and a mono-β-like glycoside, but shows both glucosidase and galactosidase activity (Boylen *et al.*, 1969). The commercial fungal anthocyanase examined by Harborne (1965) has a wider specificity range. Anthocyanidin 3-galactosides, 3-glucosides, 3,5-diglucosides and 3-(diglycosides) are rapidly hydrolysed and the rate of hydrolysis is not appreciably affected by the structure of the aglycone moiety. The enzyme is much less active towards the α-glycosides—3-rhamnosides, 3-arabinosides and 3-rutinosides—and inactive towards the acylated glycosides. The cacao anthocyanase, on the other hand, works on 3-galactosides and 3-arabinosides but not on 3-glucosides and 3-xylosides.

Anthocyanin-decolourizing polyphenol oxidases require the presence of catechol or other *o*-dihydroxyphenols for activation (Wagenknecht *et al.*, 1960; Peng and Markakis, 1963; Goodman and Markakis, 1965; Sakamura *et al.*, 1966; Proctor and Creasy, 1969; Segal and Segal, 1969; Skalski and Sistrunk, 1973). Maximum activity is generally exhibited in the pH range 6–7 (Wagenknecht *et al.*, 1960; Peng and Markakis, 1963; Sakamura *et al.*, 1970; Proctor

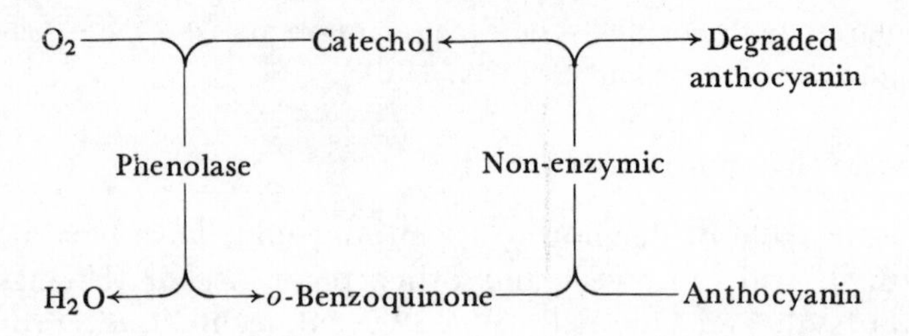

and Creasy, 1969), in contrast to the glycosidases which work best in the pH range 3–5 (Huang, 1955, 1956; Boylen *et al.*, 1969). To account for the participation of catechol, Peng and Markakis (1963) proposed that the phenolase–catechol–anthocyanin reaction proceeds first by the enzymic oxidation of catechol to *o*-quinone, followed by the non-enzymic oxidation of anthocyanin by the *o*-quinone to colourless products. *o*-Quinone by itself has not been shown to oxidize anthocyanins, however. Phenolases also exhibit substrate selectivity. The enzymes from chicory florets (Proctor and Creasy, 1969) and from eggplant (Sakamura and Obata, 1961) decolourize delphinidin glycosides more readily than other anthocyanins.

The degradation of anthocyanins *in vivo* has not been linked directly to the activity of glycosidases and polyphenol oxidases. One may anticipate the enzyme activity to parallel anthocyanin degeneration and thus increase with senescence and storage. Wagenknecht *et al.* (1960) measured at several stages the amount of an enzyme they called anthocyanase, although it was more a polyphenol oxidase in character. Except for blueberries and blackberries, enzyme level was highest at the green unripe stage and diminished with increasing maturity (Table IV). Obviously, the interplay of other factors such as physical barrier and endogenous inhibitors has to be considered here. No clear-cut relationship is seen between colour intensity and enzyme content. The heavily pigmented cherries Windsor, Bing, Royal Duke and English Morello as well as the lightly pigmented Yellow Glass show very high activity. In contrast, Medvedeva (1973) found polyphenolase activity in *Petunia hybrida* to be consistently higher in coloured flowers than in white ones. In chicory florets the activity of the enzyme and anthocyanin content decreased simultaneously during the day with the breakdown of cellular structure (Proctor and Creasy, 1969). Proctor and Creasy suggested that the loss of membrane integrity allows contact between enzyme and vacuolar substrates, resulting in discoloration and subsequent inactivation of the enzyme by noxious substances.

TABLE IV
Changes in anthocyanase activities of several fruits with increasing maturity[a]

Sample	Anthocyanase activity, units/100 mg dry wt			
	Green	Ripe	Post-harvest	Pre-harvest
Sweet cherries (*Prunus arium*), var.				
Windsor	448	196	–	286
Bing	263	158	–	224
Yellow Glass	320	208	–	250
Sour cherries (*Prunus cerasus*), var.				
Montmorency	290	140	120	220
English Morello	274	124	146	262
Cherries (*Prunus arium, Prunus cerasus*), var.				
Royal Duke	354	172	–	258
Black raspberries (*Rubus occidentalis* L.), var.				
Dundee	48	0	0	0
Blackberries (*Rubus* spp.)				
Mixed varieties	96	306	186	32
Blueberries (*Vaccinium corymbosum* I.), var.				
Rubel	236	284	–	262

[a] Reproduced by permission from Wagenknecht *et al.* (1960).

C CHANGES IN MOLECULAR STRUCTURE WITH pH

Anthocyanins and especially anthocyanidins are extremely sensitive to fluctuations in pH—a point which is made more relevant to this discussion by the observed increase in pH on fruit ripening and storage (Ulrich, 1970). pH is the one single factor which shows the greatest effect on the colour quality of stored frozen strawberry fruits and puree, the colour loss being greater at higher pH (Sistrunk and Cash, 1970; Wrolstad *et al.*, 1970). Wrolstad *et al.* (1970) calculated that strawberries should have a pH of 3·51 or lower to have acceptable colour after freezing and storage. The rate of anthocyanin destruction in the presence of oxygen is pH dependent and proportional to the percentage of the pigment in the pseudobase form (Lukton *et al.*, 1956).

The modifications in molecular structure induced by pH changes have been verified by recent spectroscopic and polarographic studies

(Kuhn and Sperling, 1960; Jurd, 1963, 1964a; Jurd and Geissman, 1963; Timberlake and Bridle, 1965, 1966, 1967a; Harper and Chandler, 1967a,b; Nilsson, 1967; Harper, 1968; Pyysalo and Makitie, 1973). Since the pH of most plant extracts fall in the range 3–6 (Hayashi, 1962), only the transformations at acidic to neutral pH will be considered.

Anthocyanidins undergo more drastic alterations in acid solutions than the corresponding anthocyanins. The combined action of glycosidases and increased pH presents a reasonable explanation for at least part of the *in vivo* destruction of anthocyanins. Harper (1968) has described the changes in the structure of pelargonidin with pH (Scheme 2); these are based on his spectral and polarographic data. The pigment exists as the deep red flavylium ion (**XX**) up to pH 3. At higher pH it is hydrated to the colourless pseudobase (**XXI**), which establishes an equilibrium with the keto pseudobase (**XXII**).

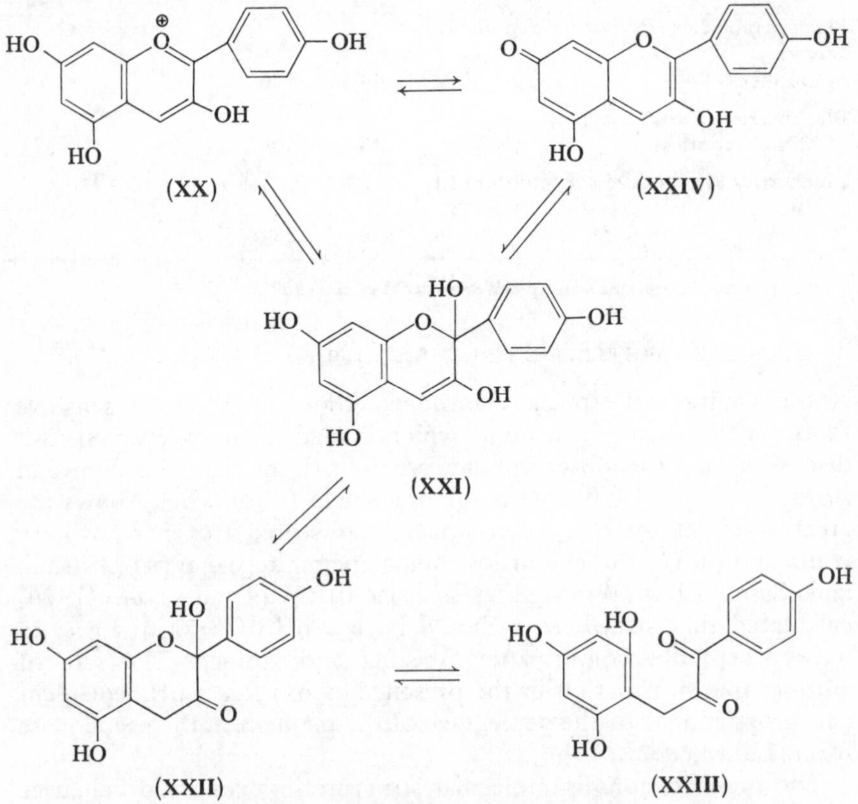

SCHEME 2. Transformation of pelargonidin with pH (Harper, 1968).

The keto form undergoes ring fission to yield the α-diketone (**XXXIII**), which appears in the solution at pH 3 to 7. The anhydrobase (**XXIV**) is formed in appreciable amounts at pH 6, giving the solution a purple colour. The discoloration that occurs with time up to pH 7 is attributed to the formation of the pseudobase and the α-diketone.

Jurd (1972) noted that cyanidin in aqueous solutions at pH 2–5 decomposes irreversibly on standing, liberating 3,4-dihydroxybenzoic acid from the B ring. With Harper's demonstration of α-diketone formation and considering the well-known observation that α-diketones are easily hydrolysed to acids, Jurd pointed out that the observed irreversible decomposition of anthocyanin can occur through the formation of α-diketones which then hydrolyse to phenolic benzoic acids (see below).

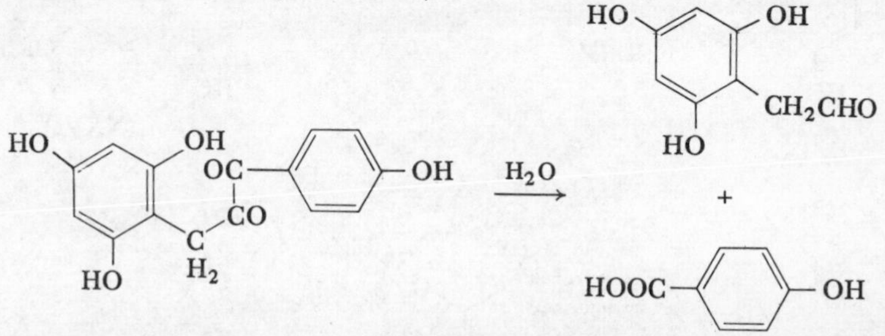

The existence of an equilibrium between the flavylium salt and the corresponding open-ringed 2-hydroxy chalcone in acidic aqueous solutions is well established for synthetic flavylium salts unsubstituted in the 3-position (Jurd, 1963, 1964a; Timberlake and Bridle, 1966, 1967a; Harper and Chandler, 1967a,b). The spectra of 3′-methoxy-4′,7-dihydroxy flavylium chloride show a marked decrease in the absorption at 468 nm (λ_{max} of the cation) with increasing pH. The absorption at 373 nm increases simultaneously, reflecting the formation of the 2-hydroxychalcone (Jurd, 1963). At pH 4 the flavylium salt exists mainly in the chalcone form (90%). Further confirmatory evidence for the flavylium–chalcone equilibrium has been derived from extensive spectral, polarographic and kinetic studies. The position of the equilibrium is markedly affected by light which favours flavylium cation formation (Timberlake and Bridle, 1965; Jurd, 1969). The rare natural anthocyanidin luteolinidin, which lacks a 3-substituent, is transformed in a manner analogous to the model flavylium compounds (Nilsson, 1967).

Appreciable quantities of the coloured cationic form of anthocyanins remain under conditions where the anthocyanidins are rapidly decolourized. The anthocyanins in acid solutions exist as equilibrium mixtures of the coloured flavylium salt (**XXV**) and the colourless carbinol or chromenol (**XXVI**) (Sondheimer and Kertesz, 1953; Swain, 1962; Dean, 1963; Timberlake and Bridle, 1966). The chromenols are fairly stable at room temperature and can be readily transformed back to the cation on acidification. Views are conflicting on whether the flavylium cation loses a proton first to form the coloured quinoidal base which is then hydrated to the colourless chromenol, e.g. (**XXV**) → (**XXVI**) → (**XXVII**) (Jurd and Asen, 1966;

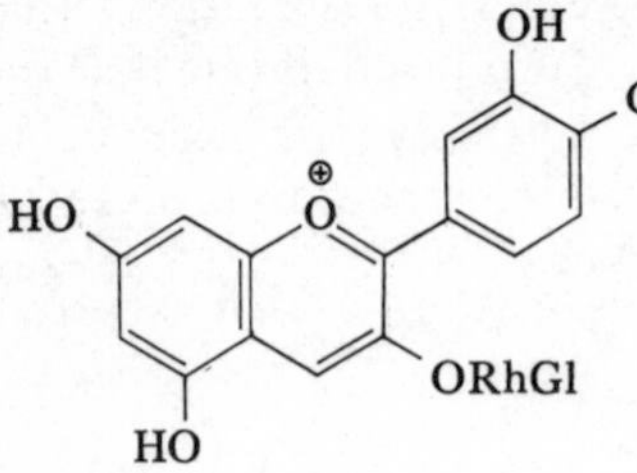

(**XXV**) 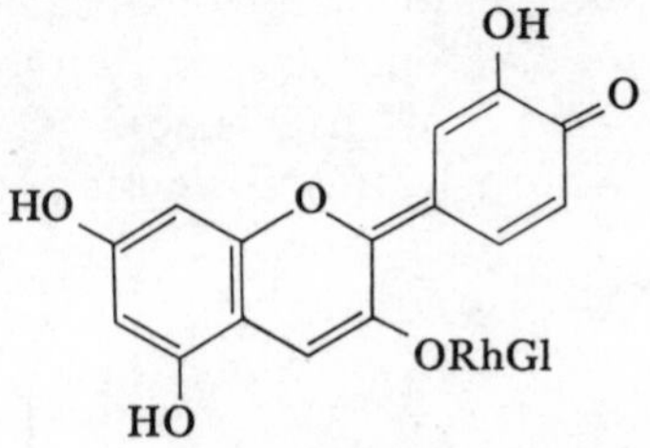 (**XXVI**)

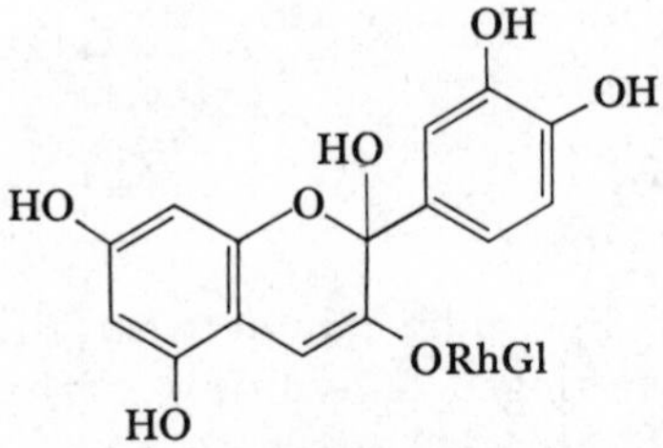

(**XXVII**)

Jurd, 1972), or the anhydrobase is formed from the cation through the chromenol, e.g. (**XXV**) → (**XXVII**) → (**XXVI**) (Timberlake and Bridle, 1967a; Pyysalo and Makitie, 1973). Jurd and Asen observed that solutions of cyanidin 3-glucoside were red up to pH 4. Above pH 4, the λ_{max} shifted from 510 nm, the λ_{max} of the flavylium salt, to 538, the λ_{max} of the anhydrobase. The solutions at the pH range 4–5 are virtually colourless, leading Jurd and Asen to conclude that the anthocyanin is immediately and almost quantitatively converted into its colourless carbinol base via the unstable purple anhydrobase. The anhydrobase is most stable above pH 5, but even at this pH it loses almost all colour in an hour. On the other hand, Pyysalo and Makitie contend that neither the acid dissociation reaction nor the disappearance of the red colour of anthocyanin proceeds directly from the flavylium ion to the anhydrobase but through an intermediate proposed to be the carbinol base.

Spectral examination of anthocyanins shows no indication of chalcone formation in acid to neutral solutions (Jurd, 1964a). The absorption of cyanidin-3-rhamnoside at 510 nm decreases as the pH is raised without an increase in the absorption at 300–400 nm. Anthocyanins form ionized chalcone in strongly alkaline solutions.

The A ring of natural anthocyanidins are invariably hydroxylated at the 5- and 7-positions. Studies on synthetic flavylium salts with or without hydroxyl substituents at the 5- and 7-positions revealed that hydroxylation at these positions increases the stability of the coloured cationic form (Timberlake and Bridle, 1967a).

D ASCORBIC ACID RELATED DEGRADATION

The electron deficient anthocyanin molecule is susceptible to the nucleophilic attack of compounds which may exist naturally in plant and food materials. Foremost among these nucleophiles is ascorbic acid, which has been shown to accelerate anthocyanin breakdown in fruits, wines and model systems (Beattie *et al.*, 1943; Mechter, 1953; Sondheimer and Kertesz, 1953; Pratt *et al.*, 1954; Markakis *et al.*, 1957; Timberlake, 1960; Tinsley and Bockian, 1960; Ashmawy *et al.*, 1967; Starr and Francis, 1968; Valuiko and Germanova, 1968; Lovric *et al.*, 1969; Kuznetsova, 1971; Lewicki, 1973), the rate being greater with increasing levels of oxygen and ascorbic acid (Markakis *et al.*, 1957; Starr and Francis, 1968). An exception to the general observation was reported by Wrolstad *et al.* (1970) and Sistrunk and Cash (1970), who found no significant correlation between ascorbic acid and colour quality. Conversely, anthocyanin hastens the oxidation of ascorbic acid except in the presence of copper ions, when it shows some protective effect toward ascorbic acid (Clegg and Morton, 1968; Harper *et al.*, 1969; Lewicki, 1973). Two mechanisms have been proposed for the participation of ascorbic acid in anthocyanin degradation: (1) oxidation of anthocyanin by hydrogen peroxide formed by the autoxidation of ascorbic acid (Sondheimer and Kertesz, 1952; Harper *et al.*, 1969), and (2) actual condensation of anthocyanin and ascorbic acid (Jurd, 1972).

The role of hydrogen peroxide as the possible reactive species in the ascorbic acid related decomposition of anthocyanin is still an open question. Whatever the case, it is known that hydrogen peroxide is capable of oxidizing anthocyanins and related flavylium salts in model systems (Karrer and de Meuron, 1932; Sondheimer and Kertesz, 1952; Jurd, 1964b, 1966, 1968; Valuiko and Germanova, 1968; Hzardina and Franzese, 1974). In aqueous acidic solution, the flavylium molecule is attacked by the nucleophilic hydrogen peroxide at the 2-position. The heterocyclic ring is then cleaved between the 2- and 3-positions to form *o*-benzoyloxyphenylacetic acid esters of the malvone type. The structure of the malvone, first reported by Karrer, has now been established as (**XXIX**) (Hzardina, 1970).

Jurd (1966) proposed a Baeyer-Villiger type of oxidation as the mechanism for the reaction of hydrogen peroxide with alkylflavylium salts in aqueous solutions (Scheme 3). A similar pathway is suggested by Hzardina (1970) as the reaction mechanism for the oxidation of malvidin 3,5-diglucoside (**XXVIII**). In this case, the

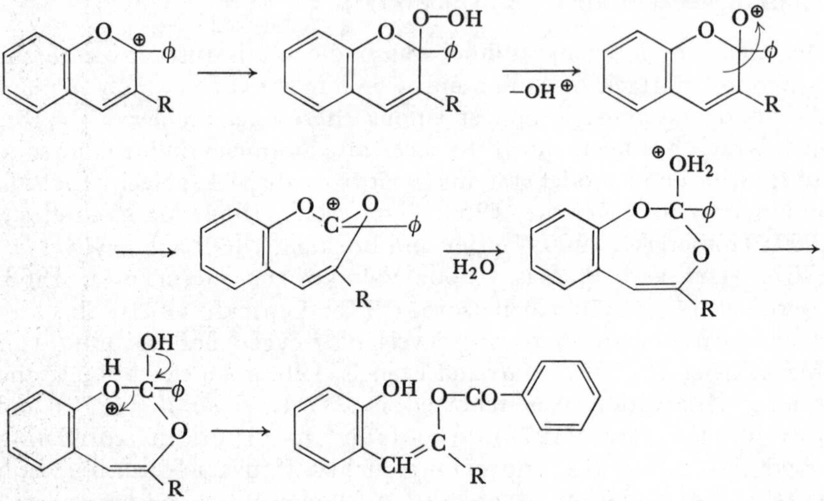

SCHEME 3. Proposed mechanism for the H_2O_2 oxidation of alkylflavylium salts (Jurd, 1966).

nucleophilic attack of water on the carbonium ion is followed by the migration of one of the hydrogen atoms to the oxygen in the 3-position of the seven-membered ring, resulting in the formation of the malvone (**XXIX**).

Acylated anthocyanidin 3,5-diglucosides are oxidized by hydrogen peroxide under acidic conditions to acylated *o*-benzoyloxyphenylacetic acid esters (Hzardina and Franzese, 1974). The fading of the solution was considerably slower (4 h) than the corresponding, non-acylated diglucoside (15 min), probably owing to a lowered activity of the C-2 position, to steric hindrance, or to both. When the oxidation of the acylated anthocyanin was carried out under neutral conditions, the product formed was 3-*o*-arylglucosyl-5-*o*-glucosyl-7-hydroxy coumarin.

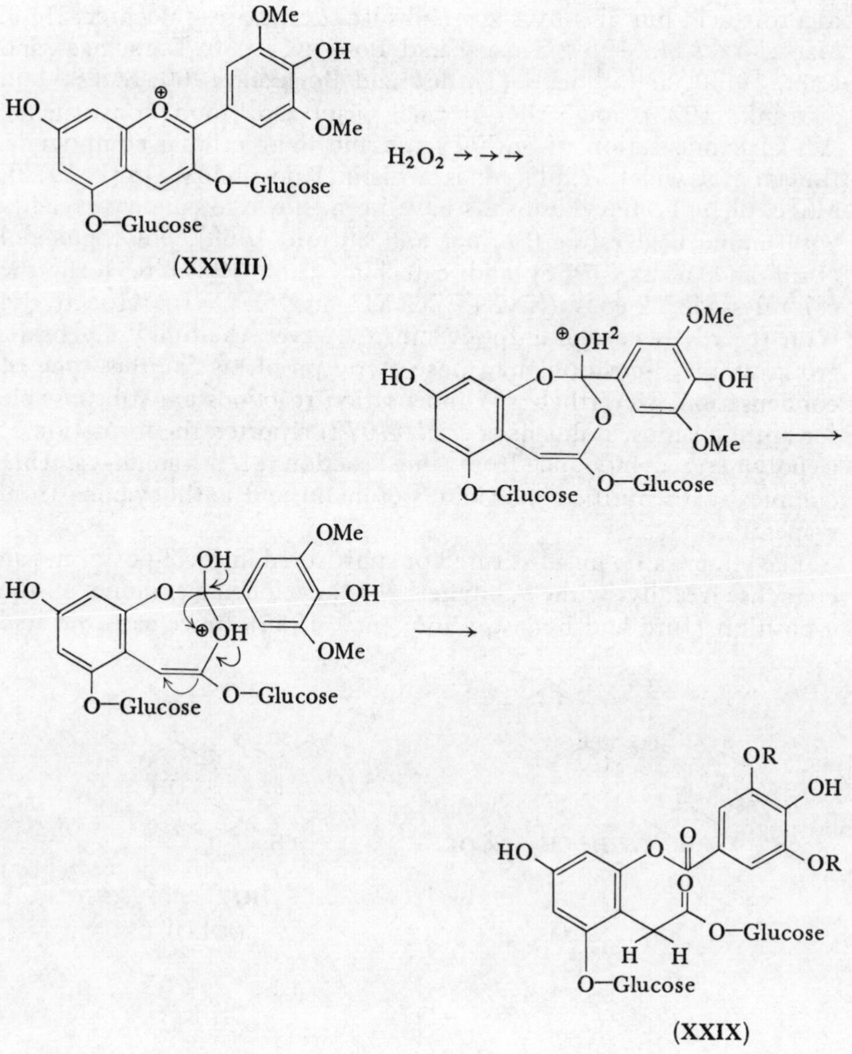

(XXVIII) $H_2O_2 \rightarrow \rightarrow \rightarrow$

(XXIX)

SCHEME 4. Proposed mechanism for the H_2O_2 oxidation of malvidin-3,5-diglucoside (Hzardina, 1970).

The involvement of hydrogen peroxide in the oxidation of anthocyanins in natural systems is somewhat difficult to assess. The formation of hydrogen peroxide from ascorbic acid and other substances is no guarantee that the amount formed is sufficient and selectively utilized for anthocyanin oxidation.

The discoloration of anthocyanin is intensified not only by

ascorbic acid but also by sugar and sugar derivatives (Mechter, 1953; Markakis *et al.*, 1957; Tinsley and Bockian, 1960; Daravingas and Cain, 1968), amino acids (Tinsley and Bockian, 1960; Skalski and Sistrunk, 1973) and other normal plant and food constituents. Actual condensation of anthocyanin and these cellular compounds, though not widely explored, is a distinct possibility (Jurd, 1972). Model phenolic flavylium salts have been shown to condense readily with amino acid esters (Shriner and Sutton, 1963), phloroglucinol (Jurd and Waiss, 1965) and catechin (Jurd, 1967) to form the colourless flav-2-enes (**XXX**), (**XXXI**) and (**XXXII**), respectively. With regard to natural anthocyanins, however, the bulky glycosidic group at the 3-position may pose steric problems for this type of condensation. Nevertheless, condensation reactions are still possible for anthocyanins. Bokuchava *et al.* (1972) reported the formation of a polymeric compound from the reaction of a tannin–catechin complex with purified malvidin, petunidin and anthocyanins from grapes.

Flavylium salts, unsubstituted or substituted in the 3-position, can condense readily with 5,5-dimethyl-1,3-cyclohexane dione at the 4-position (Jurd and Bergot, 1965; Jurd, 1965). Since ascorbic acid

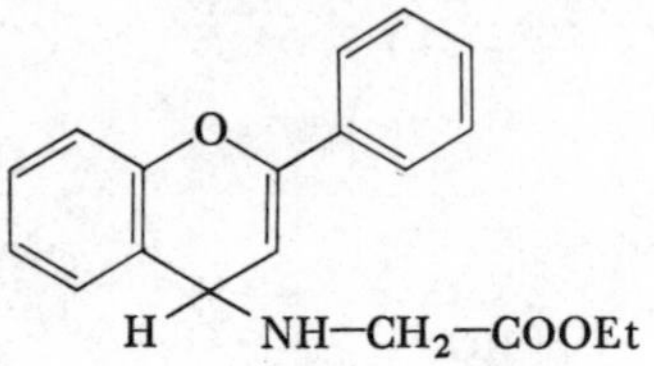

H NH—CH_2—COOEt

(XXX)

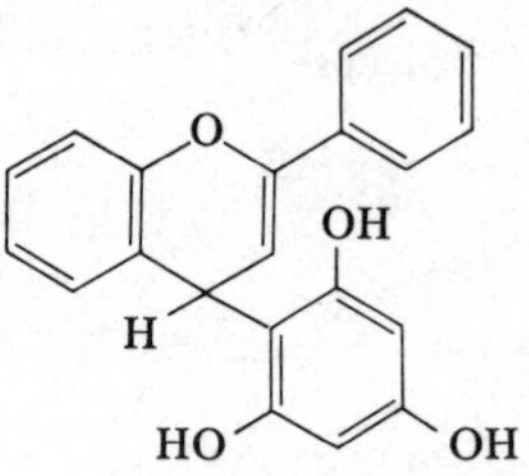

OH H HO OH

(XXXI)

OH OH HO OH HO

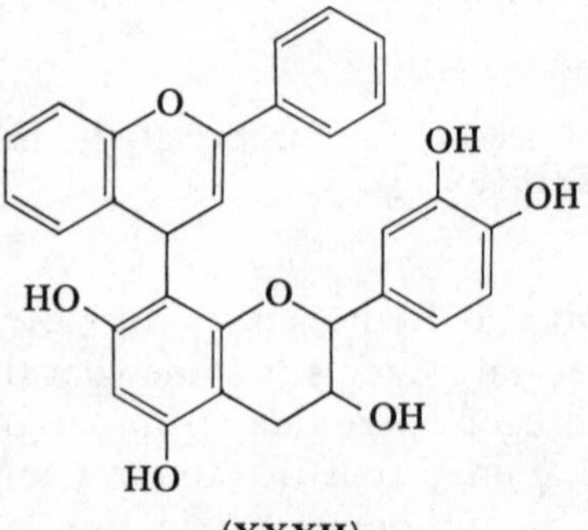

(XXXII)

resembles dimedone structurally, Jurd (1972) postulated that it may act in the same manner (Scheme 5). This hypothesis is strengthened by the ability of ascorbic acid to react with benzyl carbonium ions to form a *C*-benzyl product (Buncel *et al.*, 1965).

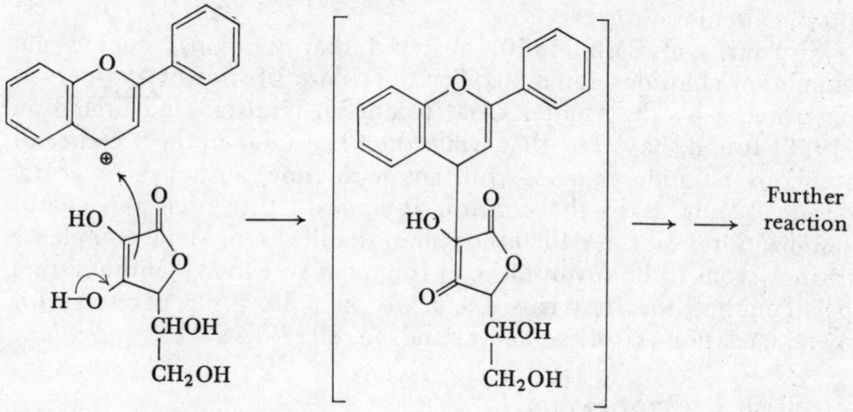

SCHEME 5. Proposed condensation reaction of flavylium salts and ascorbic acid (Jurd, 1972).

E METAL COMPLEX FORMATION

Anthocyanins that contain *o*-dihydroxyl systems can chelate metals, producing colours distinctly different from that of the free pigment. Some of the coloured metal complexes, in contrast to the free anhydrobases, are remarkably stable at pH 4–6 (Jurd and Asen, 1966; Chandler and Clegg, 1970; Jurd, 1972). Maximum complex formation between cyanidin 3-glucoside and aluminum occurs at pH 5.5. The cerise complex formed at this pH is stable on long standing in diffused light (Jurd and Asen, 1966). Similarly, ferrous and ferric salts yield stable blue complexes, but maximum complex formation in this case takes place above pH 5·5.

The colour of a solution of cyanidin 3-glucoside complexed with aluminum salts changes from red to blue-violet as the pH increases, the change being most pronounced in the pH range 3·0–3·5 (Asen *et al.*, 1969). The sharp change in colour of the aluminum–anhydrobase complex with small pH changes was offered by Asen *et al.* (1969) as a possible explanation for the blueing of some red flowers with age. The pH of flowers increases with senescence and the free anhydrobase is considered too unstable to account for the blue coloration.

Anthocyanin complexes with stannous ions also produce a blue colour. Tin (II) complexes with cyanidin and delphinidin glycosides are responsible for the appearance of a strong blue colour in berry juices (Pyysalo and Kevusi, 1973). Luh *et al.* (1962) have also reported that canned peaches with high anthocyanin content react with stannous ions under processing conditions to cause blue or purple discoloration.

Sistrunk and Cash (1970) observed that stannous, stannic and aluminum chlorides could stabilize the colour of strawberry puree as determined by the Hunter CDM (redness). Wrolstad and Erlandson (1973) found, however, that while the CDM value in the presence of stannous chloride remains constant with time, anthocyanin degradation occurs as in the control. It appears then that the colour stabilization is not due to anthocyanin stabilization. Metal complexes do not seem to be involved here; the main strawberry anthocyanin, pelargonidin, does not have the *ortho* phenolic group necessary for metal chelation (Wrolstad and Erlandson, 1973).

F SULPHITE DECOLORATION

Sulphur dioxide in moderate concentrations is often utilized in commercial fruit processing to prevent browning and to inhibit microbiological spoilage. Unfortunately, the treatment creates another discoloration problem–the bleaching of anthocyanins. Sulphite decoloration was originally thought to arise from actual reduction of the pigments. This assumption was discounted by the observation that bleached anthocyanins could be rapidly regenerated on acidification (Ribereau-Gayon, 1959). Considering that the rate of decoloration by sulphur dioxide was pH dependent, Ribereau-Gayon suggested that the decoloration could be due to the formation of a chalcone-bisulphite addition product. Unable to find spectral evidence for the formation of the chalcone from anthocyanin, Jurd (1964a) tested Ribereau-Gayon's hypothesis using 4′,7-dihydroxy-3-methoxyflavylium chloride, a compound known to form chalcone. The spectrum of this flavylium salt in aqueous solution at pH 3·8 shows the absorptions of both the flavylium cation and the chalcone. As the solution is instantly decolourized on addition of sodium bisulphite, the flavylium cation absorption disappears immediately and completely whereas the chalcone peak remains unaffected. The cation, not the chalcone, is therefore the reactive form in sulphite decoloration. The product is believed to be a colourless chrom-4(or 2)-sulphonic acid (**XXXIII**). Kinetic studies by Timberlake and Bridle (1967b) support the formation of the complex between sulphur dioxide and the flavylium cation at a mole ratio of 1 : 1. Synthetic flavylium compounds with a methyl or phenyl group at the 4-position are not affected by sulphur dioxide (Timberlake and Bridle, 1968).

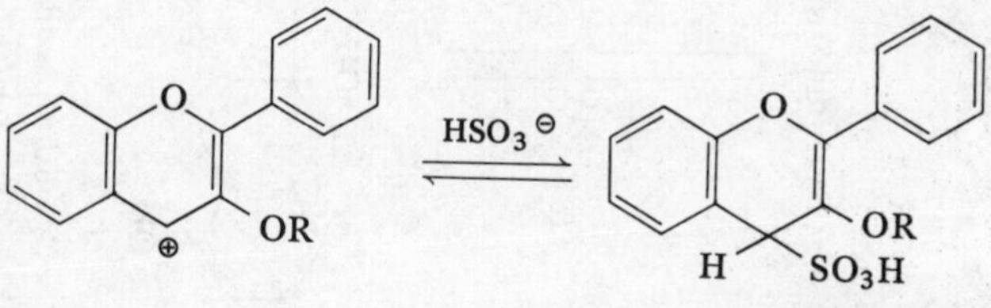

$$\text{flavylium cation (3-OR)} \underset{}{\overset{HSO_3^{\ominus}}{\rightleftharpoons}} \text{4-H, 4-}SO_3H\text{ adduct (3-OR)}$$

(XXXIII)

G THERMAL DEGRADATION

Degradation of anthocyanins during heat processing of foodstuffs is a well-known phenomenon. The kinetics of thermal degradation are generally first order up to 110°C (Lukton *et al.*, 1956; Markakis *et al.*, 1957; Keith and Powers, 1965; Daravingas and Cain, 1968; Ruskov and Tanchev, 1971, 1973; Tanchev, 1971, 1972a,b; Tanchev and Ioncheva, 1972) and influenced by temperature, plant variety, pH, presence of oxygen and sugar content (Mechter, 1953; Lukton *et al.*, 1956; Daravingas and Cain, 1968; Ruskov and Tanchev, 1971, 1973; Tanchev, 1972a,b; Tanchev and Ioncheva, 1972). To prevent the thermal destruction of anthocyanins in grapes, El-Gindy *et al.* (1972) recommended that the temperature must not exceed 63°C (145°F). This agrees basically with the earlier findings of Decareau *et al.* (1956) that 60°C (140°F) is the highest possible temperature that can be used without appreciable loss in strawberry pigments. Thermal decomposition is also inhibited by a decrease in pH and the exclusion of oxygen (Markakis *et al.*, 1957; Daravingas and Cain, 1968).

Two hypotheses have been presented to explain the thermal degradation of anthocyanins: (1) hydrolytic opening of the pyrylium ring to form a substituted chalcone which degrades to a brown, insoluble polyphenolic compound, and (2) hydrolysis of the 3-glycosidic linkage. Their failure to detect the aglycone from pure pelargonidin 3-monoglucoside led Markakis *et al.* (1957) to propose the first mechanism. This is corroborated by the earlier report of Lukton *et al.* (1956) that aglycone solutions heated at 45°C for 4 days do not produce the red-brown precipitate formed as the principal breakdown product of pelargonidin 3-monoglucoside. Evidence supporting both mechanisms has been obtained from the same model system (Adams, 1973). The rate of sugar liberation from acidic solutions of cyanidin 3-glycosides at 100°C parallels the rate of red

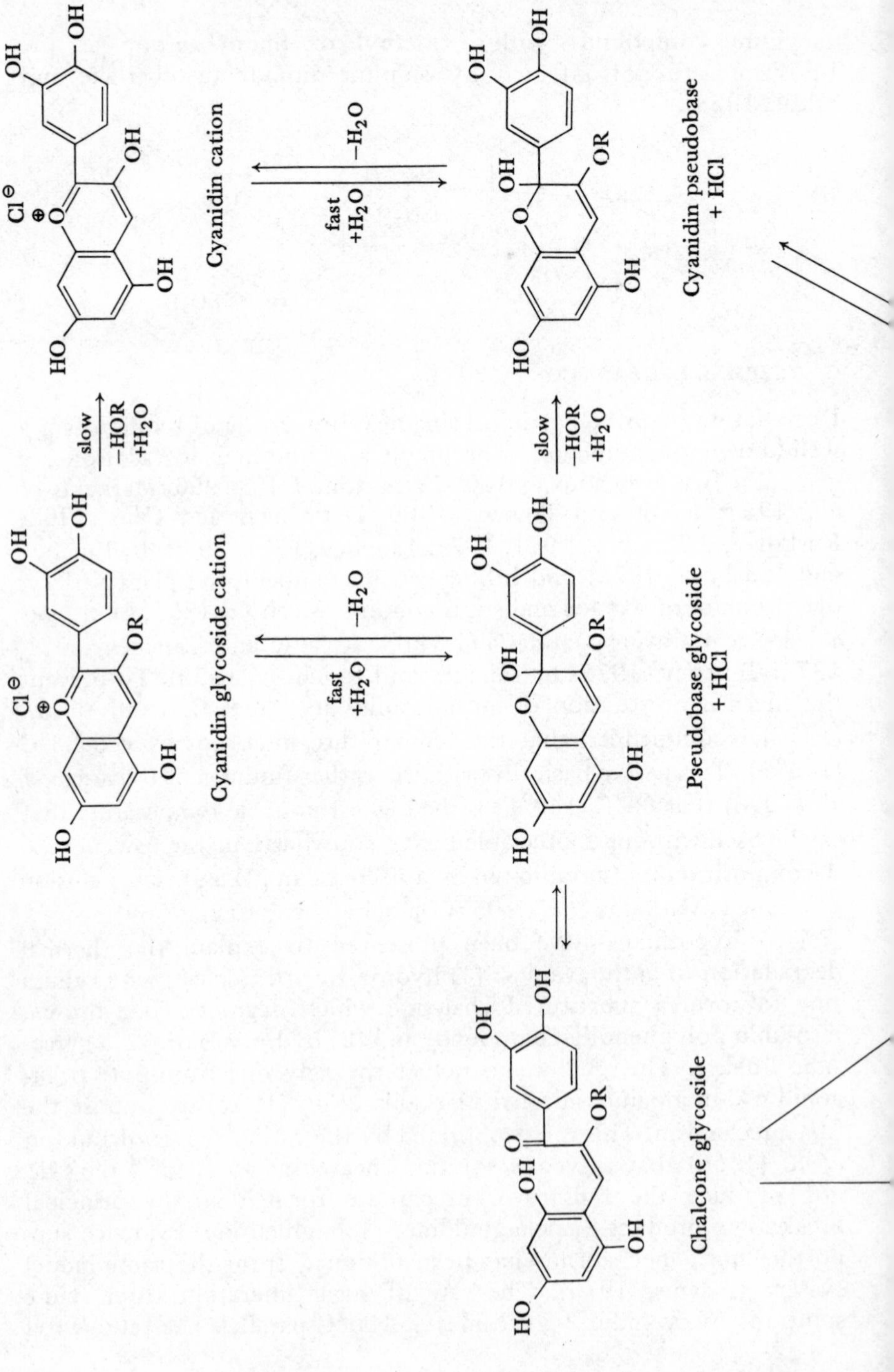

Cyanidin glycoside cation

slow, $-HOR$, $+H_2O$ →

Cyanidin cation

fast, $+H_2O$ ⇌ $-H_2O$

Pseudobase glycoside
+ HCl

slow, $-HOR$, $+H_2O$ →

Cyanidin pseudobase
+ HCl

Chalcone glycoside

SCHEME 6. Proposed mechanism for the anaerobic thermal degradation of cyanidin glycoside (Adams, 1973).

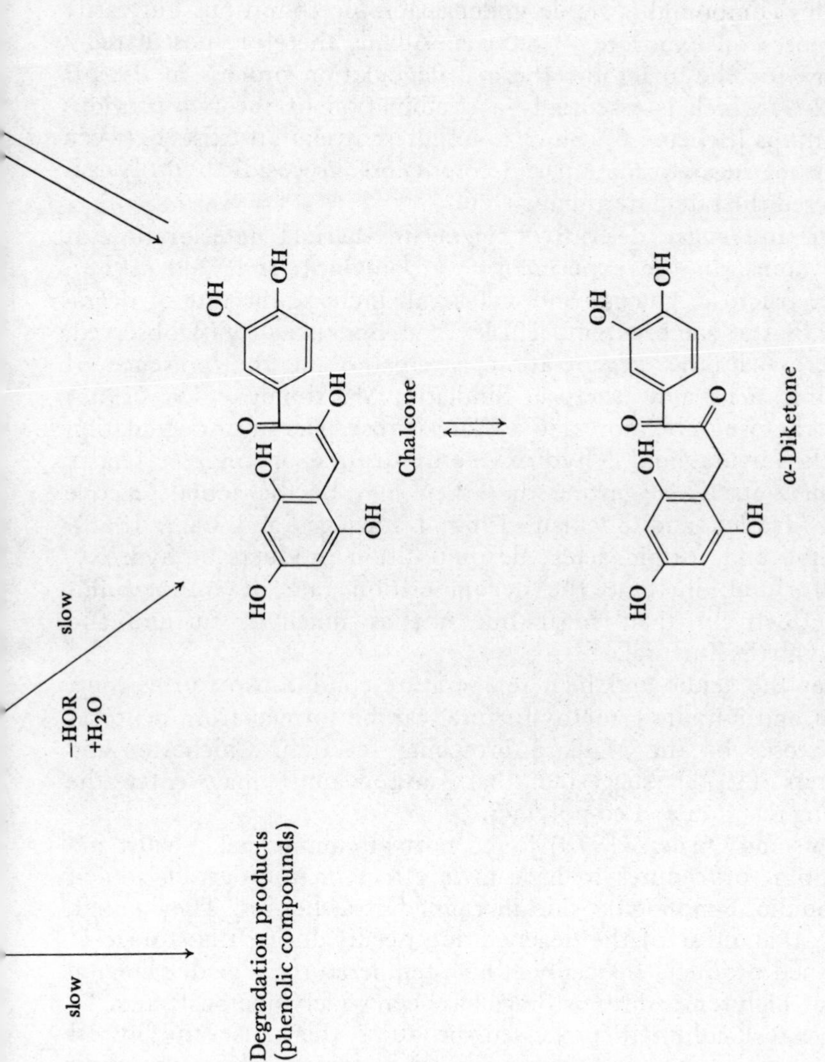

colour loss and the rate of anthocyanin disappearance, indicating that glycosidic hydrolysis is a principal reaction in the discoloration process. Phenolic aglycones and glucosides of undetermined nature appear in the solution apparently as high temperature breakdown products. In addition, an almost colourless phenolic product, tentatively identified as chalcone or α-diketone, is detected during the degradation of all the cyanidin glycosides examined in the pH range 1–4. This compound is stable under anaerobic conditions but easily decomposes on exposure to oxygen. Adams therefore postulated a pathway for the anaerobic thermal degradation process in the pH range 2–4, which is essentially a combination of the two previous propositions (Scheme 6). Since a definite correlation exists between the rates of sugar formation and colour loss, glycosidic hydrolysis is considered the rate-determining step.

Sugar and sugar derivatives aggravate thermal deterioration of anthocyanins. In the experiments of Daravingas and Cain (1968), sucrose, fructose, glucose and xylose all increase the rate of degradation to the same extent. Tinsley and Bockian (1960) observed, however, that the degeneration is greatest in the presence of glucuronic acid and fructose. Similarly, Mackinney *et al.* (1955) found fructose to be more active than sucrose. The sugar degradation products furfural and 5-hydroxymethylfurfural are more effective than sugar itself, suggesting that they may be the actual reactive species (Tinsley and Bockian, 1960; Daravingas and Cain, 1968). Laevulinic and formic acids, decomposition products of hydroxymethylfurfural, increase the decomposition rate of anthocyanins more effectively than sugar but not as much as furfural and hydroxymethylfurfural.

Under the acidic and high temperature conditions of processing, furfural and 5-hydroxymethylfurfural can be formed from pentoses and hexoses or the Maillard browning reaction. Chichester and McFeeters (1971) suggested that anthocyanins may enter the browning reaction as a co-polymer.

Adams and Ongley (1973) found normal commercial exhaust and sterilization procedures to have little effect on the degradation of perargonidin 3-monoglucoside in canned strawberries. They noted, instead, that most of the degeneration occurs during the storage of the canned product, especially at high temperatures. The detrimental effect of high temperature storage has been widely demonstrated. To produce well-coloured processed products then, use the lowest possible processing and storage temperatures, select the right plant varieties, exclude oxygen, decrease the pH, prevent metal contamination as have been recommended (Markakis *et al.*, 1957; Dalal and Salunkhe, 1964; Papov *et al.*, 1969a,b; Kuznetsova, 1971; El-Gindy and Raouf, 1972; Adams and Ongley, 1973).

V Betalains

The betacyanins and betaxanthins are the red-violet and yellow pigments that are found in the ten families of flowering plants belonging to the order Centrospermae (see Chapter 11). The initial interest and continuing economic interest lies in the fact that the betalains are the pigments in the red beet root *Beta vulgaris*. These pigments are limited to the Centrospermae and the anthocyanins are excluded from this order (Piattelli and Minale, 1964; Mabry, 1970).

Studies on the fate of these pigments in senescence or storage are mainly limited to the red beet, the only food plant in the order, and the information available is much more limited than with the other plant pigments. Recent safety investigations have led to the restriction in the use of FDC red no. 2 food colour and prompted investigators to study the use of approved food colours such as the beet pigment (Von Elbe *et al.*, 1972; Aurstad and Dahle, 1973; Von Elbe and Maing, 1973; Von Elbe *et al.*, 1974b).

A EFFECT OF pH

Habib and Brown (1956) found that beets processed at pHs of 3·5 or 5·5 retained most of their natural colour. However, at pH values of 7·5 and 8·5 discoloration occurred. The stability of betanin was found by Von Elbe *et al.* (1974a) to lie between pH values of 4·0 and 5·0.

Storage at pH values of less than 3·0 and greater than 7·0 resulted in a decrease in the intensity of the absorbance and some shift in the wavelength maxima. Table V shows the effect of storing betanin solutions at different pH values.

The reaction with mild acids or β-glucosidase results in the conversion of betanin or isobetanin to the aglucones betanidin and isobetanidin (Mabry, 1966). Betanin and isobetanin are isomeric compounds with inversion of configuration at C-15 (Scheme 7).

Treatment of betanidin with alkali results in the formation of 2,6-dihydroxy-2,3-dihydroindole-2-carboxylic acid, formic acid and 4-methylpyridine-2,6-dicarboxylic acid.

The extent to which the reactions above proceed in senescent or stored beet root tissue is not clear. Presumably some conversion of

TABLE V
Absorbances at absorption maxima (A_{max}) of betanin solutions[a] at different pH value and stored at 4°C[b]

pH Value (± 0·05)	Storage time: 0 days		Storage time: 7 days	
	Absorbance ± 0·01	A_{max} ± (nm)	Absorbance ± 0·01	A_{max} ± (nm)
2·0	0·47	535	0·34	534
3·0	0·51	535	0·47	534
4·0	0·53	537	0·52	537
5·0	0·54	537	0·53	537
6·0	0·53	537	0·51	537
7·0	0·52	537	0·47	537
8·0	0·49	538	0·37	538
9·0	0·49	544	0·34	544

[a] Betanin concentration: 0·45 mg/100 ml.
[b] Reproduced with permission from Von Elbe *et al.* (1974a).

betanin to the aglucone and isomeric compounds occurs especially with heat processing.

B EFFECT OF HEATING

When solutions of betanin pigment or beet tissue are heated, the colour fades and the red pigment is replaced by brown pigments (Aronoff and Aronoff, 1948). Von Elbe *et al.* (1974a) studied the degradation rates for betanin solution and beet juice heated at various pH values (Table VI). Both solutions were most stable at a pH of 5·0; however, the beet juice was more stable than the pure solution.

Aurstad and Dahle (1973) eluted both the red and yellow pigments with distilled water from electrophoretograms and determined the effect of heating on the visual spectra. At the end of 12 min at 90°C most of the characteristic absorption of both of the pigments was lost.

C EFFECT OF LIGHT AND/OR AIR

In common with other pigments the betalains are destroyed by light, air or a combination of the two. It is a common experience that the

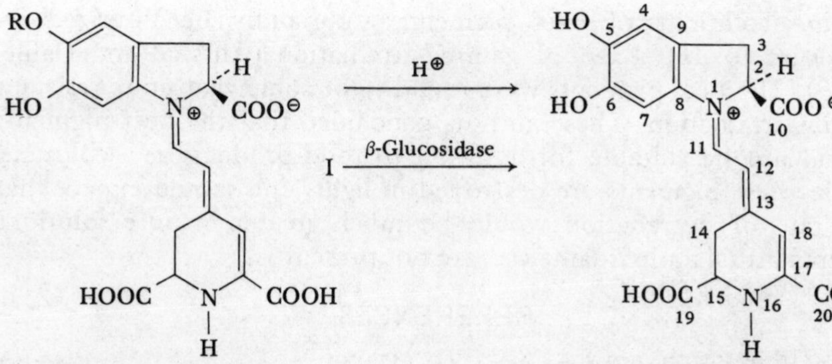

I. Betanin, Isobetanin, R = β-D-Glucosyl

$\xrightarrow{H^{\oplus}}$

I $\xrightarrow{\beta\text{-Glucosidase}}$

III. Betanidin, Isobetanidin

SCHEME 7. Hydrolytic and degradative reactions given for several betacyanin pigments (Mabry, 166).

colour of processed beet roots packed in clear glass jars fade with time, especially with exposure to direct light.

Vilece *et al.* (1955) and Habib and Brown (1956) observed that a small amount of air in the head space of a can of beet products caused the surface to turn brown. Von Elbe *et al.* (1974a) stored betanin solutions at 7·0 under air or nitrogen in sealed clear vials with or without light. The presence of air increased the rate of degradation by 14·6% and light by 15·6%. The combination of air and light caused a degradation rate of 28.6%.

TABLE VI
Degradation rates[a] for betanin solutions and beet juice as a function of pH at 100°C ± 1[b]

pH ± 0·05	$K(\text{min})^{-1} \times 10^{-3}$	$T_{1/2}$ (min) ± CE[c]
3·0	94	7·4 ± 1
4·0	51	13·6 ± 2
5·0	48	14·5 ± 2
6·0	79	8·1 ± 1
7·0	118	5·9 ± 1
Beet juice		
3·0	79	8·8 ± 1
5·0	24	28·6 ± 3
7·0	135	5·1 ± 1

[a] Average of triplicate determinations.
[b] Reproduced with permission from Von Elbe *et al.* (1974a).
[c] CE = maximum calculated error.

Pure solutions of beet pigments are rapidly decolourized by exposure to 100 Krad of gamma irradiation (Aurstad and Dahle, 1973). The beet pigments were found to be somewhat more resistant to u.v. irradiation. These authors concluded that the beet pigments would not be suitable for addition to food products as a colorant. While these pigments are destroyed in light, one should expect that the rate of degradation would be much greater in pure solutions where natural antioxidants etc. are not present.

REFERENCES

Adams, J. B. (1973). *J. Sci. Fd Agric.* 24, 747–762.

Adams, J. B. and Ongley, M. H. (1973). *J. Fd Technol.* 8 139–145.

Aitzetmüller, K., Svec, W. A., Katz, J. J. and Strain, H. H. (1968). *Chem. Commun.* 1, 32–33.

Ames, S. (1958). *In* "Annual Review of Biochemistry" (J. M. Luck, ed.), p. 375. Stanford University Press, Stanford, California.

Ardao, C. and Vennesland, B. (1960). *Pl. Physiol., Lancaster* 35, 368–371.

Aronoff, S. (1953). *Adv. Fd Res.* 4, 133–184.

Aronoff, S. and Aronoff, E. M. (1948). *Fd Res.* 13, 59–65.

Aronoff, S. and Mackinney, G. (1943). *J. Am. chem. Soc.* 65, 956–958.

Asen, S., Norris, K. H. and Stewart, R. N. (1969). *Phytochemistry* 8, 653–659.

Ashmawy, H., El-Gindy, M. M. and Elmanawaty, H. (1967). *Agric. Res. Rev. Cairo* 45, 48–58.

Aurstad, K. and Dahle, H. K. (1973). *Z. Lebensmittelunters, u.-Forsch.* 151, 171–174.

Austin, D. J., Bu'Lock, J. D. and Drake, D. (1970). *Experientia* 26, 348–349.

Bacon, M. F. (1966). *Biochem. J.* 101, 34C–36C.

Bacon, M. F. and Holden, M. (1967). *Phytochemistry* 6, 193–210.

Bacon, M. F. and Holden, M. (1970). *Phytochemistry* 9, 115–125.

Brain, J. M. and Mercer, F. V. (1964). *Aust. J. biol. Sci.* 17, 78–85.

Bamji, M. S. and Krinsky, N. I. (1965). *J. biol. Chem.* 240, 467–470.

Beattie, H. G., Wheeler, K. A. and Pederson, C. S. (1943). *Fd Res.* 8, 395–404.

Ben-Aziz, A., Britton, G. and Goodwin, T. W. (1973). *Phytochemistry* 12, 2759–2764.

Berzelius, J. J. (1837a). *Justus Liebig's Annln Chem.* 21, 257–267.

Berzelius, J. J. (1837b). *Justus Liebig's Annln Chem.* 22, 69–70.

Bickoff, E. M., Livingston, A. L., Bailey, G. F. and Thompson, C. R. (1954). *J. agric. Fd Chem.* 2, 563–567.

Blain, J. A. and Shearer, G. (1963). *Proc. Nutr. Soc.* 22, 162–171.

Blain, J. A. and Styles, E. C. C. (1959). *Nature, Lond.* 184, 1141.

Blain, J. A., Hawthorn, J. and Todd, J. P. (1953). *J. Sci. Fd Agric.* 4, 580–587.

Blain, J. A., Patterson, J. D. E. and Pearce, M. (1968). *J. Sci. Fd Agric.* 19, 713–717.

Blundstone, H. A. W., Woodman, J. S. and Adams, J. B. (1971). *In* "The Biochemistry of Fruits and Their Products" (A. C. Hulme, ed.), vol. 2, p. 547. Academic Press, London and New York.

Bodea, C. (1969). *Pure appl. Chem.* 20, 517–530.

Bodendorf, K. (1930). *Arch. Pharm.* 286, 491–499.

Böger, P. (1965). *Phytochemistry* 4, 435.
Bogorad, L. (1966). *In* "The Chlorophylls" (Vernon, L. P. and Seely, G. R., eds), pp. 481-510. Academic Press, London and New York.
Bokuchava, M. A., Solnyshkin, V. I., Valviko, G. G., Sturra, Z. S. and Siashvili, A. I. (1972). *Soobshch. Akad. Nauk Gruz. SSR* 68, 437-440. (*Chem. Abstr.* 78, 54577 X.)
Booth, V. H. (1957). *Biochem. J.* 65, 660-663.
Booth, V. H. (1960). *J. Sci. Fd Agric.* 11, 8-13.
Booth, V. H. (1963). *Biochem. J.* 87, 238-239.
Boylen, C. W., Hagen, C. W. and Manzell, R. L. (1969). *Phytochemistry* 8, 2311-2315.
Britton, G. and Goodwin, T. W. (1969). *Phytochemistry* 8, 2257-2258.
Buckle, K. A. and Edwards, R. A. (1969). *Phytochemistry* 8, 1901-1906.
Buckle, K. A. and Edwards, R. A. (1970). *J. Sci. Fd Agric.* 21, 307-312.
Buncel, E., Jackson, K. G. A. and Jones, J. K. N. (1965). *Chemy Ind.* 89.
Casoli, U., Dall'aglio, G. and Leoni, C. (1969a). *Industria Conserve* 44, 102-106.
Casoli, U., Dall'aglio, G. and Leoni, C. (1969b). *Industria Conserve* 44, 193-198.
Chandler, B. V. and Clegg, K. M. (1970). *J. Sci. Fd Agric.* 21, 315-319.
Chiba, Y., Aiga, I., Idemori, M., Satoh, Y., Matsushita, K. and Sasa, T. (1967). *Plant Cell Physiol.* 8, 623-635.
Chichester, C. O. and McFeeters, R. F. (1971). *In* "The Biochemistry of Fruits and Their Products" (A. C. Hulme, ed.), vol. 2, pp. 707-719. Academic Press, London and New York.
Chichester, C. O. and Nakayama, T. O. M. (1965). *In* "Chemistry and Biochemistry of Plant Pigments" (T. W. Goodwin, ed.), 439-457. Academic Press, London and New York.
Cho, D. H. (1966). Ph.D. Thesis, University of California, Davis, California.
Clegg, K. M. and Morton, A. D. (1968). *J. Fd Technol.* 3, 277-284.
Clydesdale, F. M., Fleishman, D. L. and Francis, F. J. (1970). *Food Prod. Dev.* 4, 127, 130, 134, 136, 138.
Clydesdale, F. M. and Francis, F. J. (1968). *Fd Technol.* 22, 135-138.
Co, D. Y. C. L. and Schanderl, S. H. (1967a). *Phytochemistry* 6, 145-148.
Co, D. Y. C. L. and Schanderl, S. H. (1967b). *J. Chromat.* 26, 442-448.
Curl, A. L. and Bailey, G. F. (1955). *Fd Res.* 20, 371-376.
Curl, A. L. and Bailey, G. F. (1956). *J. Agric. Fd Chem.* 4, 156-159.
Dalal, K. B. and Salunkhe, D. K. (1964). *Fd Technol.* 18, 1198-1200.
Dall'aglio, G. and Leoni, C. (1969). *Industria Conserve* 44, 102-106.
Dall'aglio, G., Balestrazzi, A. and Gherardi, S. (1970). *Ind. Conserve* 45, 301-306.
Daravingas, G. and Cain, R. F. (1968). *J. Fd Sci.* 33, 138-142.
Dean, F. M. (1963). "Naturally Occurring Oxygen Ring Compounds." Butterworth, London.
Decareau, R. V., Livingston, G. E. and Fellers, C. R. (1956). *Fd Technol.* 10, 125-128.
De La Mar, R. R. and Francis, F. J. (1969). *J. Fd. Sci.* 34, 287-290.
Dicks, J. W. and Friend, J. (1968). *Phytochemistry* 7, 1933-1947.
Dilung, I. I. (1958). *Ukr. khim. Zh.* 24, 202-207.
Draper, S. R. (1969). *Phytochemistry* 8, 1641-1647.
Dzheneev, S. Y., Tyutyunnik, V. I. and Sivetsev, M. V. (1973). *Izv. Vyssh. Ucheb. Zaved., Pishch. Tekhnol.* 6, 158-159. (*Chem. Abstr.* 80, 131762 z.)

Edelman, J. and Schoolar, A. I. (1969). *Z. PflPhysiol.* 60, 470–471.
Egle, K. (1944). *Bot. Arch.* 45, 93–148.
Eilati, S. K. Monselise, S. P. and Budowski, P. (1969). *J. Proc. Am. hort. Soc.* 94, 346–348.
Eilati, S. K. Budowski, P. and Monselise, S. P. (1972). *Plant Cell Physiol.* 13, 741–746.
Elbe, J. H. von, Sy, S. H., Maing, I.-Y. and Gabelman, W. H. (1972). *J. Fd Sci.* 37, 932–934.
Elbe, J. H. von and Maing, I.-Y. (1973). *Cereal Sci. Today* 18, 263–264, 316–317.
Elbe, J. H. von, Klemment, J. T., Amundson, C. H., Cassens, R. G. and Lindsay, R. C. (1974a). *J. Fd Sci.* 39, 128–132.
Elbe, J. H. von, Maing, I.-Y. and Amundson, C. H. (1974b). *J. Fd Sci.* 39, 334–337.
El-Gindy, M. M. and Raouf, M. S. (1972). *Agric. Res. Rev.* 50, 291–300.
El-Gindy, M. M., Raouf, M. S. and El-Manawaty, H. (1972). *Agric. Res. Rev.* 50, 281–290.
Ellsworth, R. K. (1972). *In* "The Chemistry of Plant Pigments" (C. O. Chichester, ed.), pp. 85–100. Academic Press, New York and London.
El-Tinay, A. H. and Chichester, C. O. (1970). *J. org. Chem.* 35, 2290–2293.
Estevez-Pinto, M. J. R. (1972). *Vinea et Vino Portugalie Documento, Series II: Enologia* 5, 1–63. (*Fd Sci. Technol. Abstr.* 5, 1H 147.)
Fishwick, M. J. (1962). *Abstr. First Int. Congr. Fd Sci. Technol.* 369–379. Warsaw, Poland.
Forsyth, W. G. C. and Quesnel, V. C. (1957). *Biochem. J.* 65, 177–179.
Friend, J. (1958). *Chemy Ind.* 597–598.
Friend, J. and Dicks, J. W. (1967). *Phytochemistry* 6, 1193–1202.
Friend, J. and Mayer, A. M. (1960). *Biochim. biophys. Acta* 41, 422–429.
Friend, J. S. and Nakayama, T. O. M. (1959). *Nature, Lond.* 184, 66–67.
Fuhrhop, J. and Manzerall, D. (1971). *Photochem. Photobiol.* 13, 453–458.
Galler, M. and Mackinney, G. (1965). *J. Fd Sci.* 30, 393–395.
Glover, J. (1960). *Vitams Horm.* 18, 371–386.
Glover, J. and Redfearn, E. R. (1953). *Biochem. J.* 54, viii.
Goldwaite, J. J. and Laetsch, W. M. (1967). *Pl. Physiol., Lancaster* 42, 1757–1762.
Goodman, L. P. and Markakis, P. (1965). *J. Fd Sci.* 30, 135–137.
Goodwin, T. W. (1952). "The Comparative Biochemistry of Carotenoids", p. 31. Chapman and Hall, London.
Goodwin, T. W. (1958). *Biochem. J.* 68, 503–511.
Goodwin, T. W. and Goad, L. J. (1970). *In* "The Biochemistry of Fruits and Their Products" (A. C. Hulme, ed.), vol. 1, pp. 305–368. Academic Press, London and New York.
Goodwin, T. W. and Mercer, E. T. (1972). "Introduction to Plant Biochemistry". p. 41. Pergamon Press, Oxford.
Gortner, W. A. (1955). *J. Fd. Sci.* 30, 30–32.
Gortner, W. A. and Singleton, V. L. (1961). *J. Fd. Sci.* 26, 53–55.
Gribanovski-Sassu, O., Pellicciari, R. and Hinghez, C. C. (1969). *Annali Ist. sup. Sanita* 5, 51–53.
Grob, E. and Eichenberger, W. (1962). *Biochem. J.* 85, 11P.
Gross, J., Gabai, M. and Lifshitz, A. (1971). *J. Fd Sci.* 36, 466–473.

Gupte, S. M. and Francis, F. J. (1964). *Fd Technol.* 18, 141–144.
Habib, A. T. and Brown, H. D. (1956). *J. Proc. Am. hort. Soc.* 68, 482–490.
Harborne, J. B. (1965). *Phytochemistry* 4, 107–120.
Harper, K. A. (1968). *Aust. J. Chem.* 21, 221–227.
Harper, K. A. and Chandler, B. V. (1967a). *Aust. J. Chem.* 20, 731–744.
Harper, K. A. and Chandler, B. V. (1967b). *Aust. J. Chem.* 20, 745–756.
Harper, K. A., Morton, A. D. and Rolfe, E. J. (1969). *J. Fd Technol.* 4, 255–267.
Harris, W. H. and Spurr, A. R. (1969). *Am. J. Bot.* 56, 380–389.
Hasegawa, K., Macmillan, J. D., Maxwell, W. A. and Chichester, C. O. (1969). *Photochem. Photobiol.* 9, 165–169.
Hayashi, K. (1962). *In* "The Chemistry of Flavonoid Compounds" (T. A. Geissman, ed.), pp. 248–285. Macmillan, New York.
Hines, G. D. and Ellsworth, R. K. (1969). *Pl. Physiol., Lancaster* 44, 1742–1744.
Holden, M. (1961). *Biochem. J.* 78, 359–364.
Holden, M. (1965). *J. Sci. Fd Agric.* 16, 312–325.
Holden, M. (1970). *Phytochemistry* 9, 1771–1777.
Holden, M. (1972). *Phytochemistry* 11 2393–2402.
Hoyt, P. B. (1966a). *Pl. Soil* 25, 167–180.
Hoyt, P. B. (1966b). *Pl. Soil* 25, 313–327.
Huang, H. T. (1955). *J. Agric. Fd Chem.* 3, 141–146.
Huang, H. T. (1956). *J. Am. chem. Soc.* 78, 2390–2393.
Hunter, R. F. and Krakenberger, R. M. (1947). *J. Am. chem. Soc.* 1, 1–4.
Hzardina, G. (1970). *Phytochemistry* 9, 1647–1652.
Hzardina, G. and Franzese, A. J. (1974). *Phytochemistry* 13, 231–234.
Hzardina, G., Borzell, A. J. and Robinson, W. (1970). *Am. J. Enol. Vitic.* 21, 201–204.
Ishikura, N. (1972). *Phytochemistry* 11, 2555–2558.
Isler, O., Rüegg, R. and Schudel, P. (1962). *In* "Recent Progress in the Chemistry of Natural and Synthetic Colouring Matters and Related Fields" (T. S. Gore, B. S. Joshi, S. V. Sunthankar and B. D. Tilak, eds), pp. 39–57. Academic Press, New York and London.
Jen, J. J. (1969). Ph.D. Thesis, University of California, Berkeley, California.
Jen, J. J. and Mackinney, G. (1970a). *Photochem. Photobiol.* 11, 297–302.
Jen, J. J. and Mackinney, G. (1970b). *Photochem. Photobiol.* 11, 303–308.
Jurd, L. (1963). *J. org. Chem.* 28, 987–991.
Jurd, L. (1964a). *J. Fd Sci.* 9, 16–19.
Jurd, L. (1964b). *J. org. Chem.* 29, 2602–2605.
Jurd, L. (1965). *Tetrahedron* 21, 3707–3714.
Jurd, L. (1966). *Tetrahedron* 22, 2913–2921.
Jurd, L. (1967). *Tetrahedron* 23, 1057–1064.
Jurd, L. (1968). *Tetrahedron* 24, 4449–4457.
Jurd, L. (1969). *Tetrahedron* 25, 2367–2380.
Jurd, L. (1972). *In* "The Chemistry of Plant Pigments" (C. O. Chichester, ed.), pp. 123–142. Academic Press, New York and London.
Jurd, L. and Asen, S. (1966). *Phytochemistry* 5, 1263–1271.
Jurd, L. and Bergot, B. J. (1965). *Tetrahedron* 21, 3697–3705.
Jurd, L. and Geissman, T. A. (1963). *J. org. Chem.* 28, 2394–2397.
Jurd, L. and Waiss, A. C. (1965). *Tetrahedron* 21, 1471–1483.
Karrer, P. and de Meuron, G. (1932). *Helv. chim. Acta* 15, 505–512.

Karrer, P. and Jucker, E. (1945). *Helv. chim. Acta* 28, 427–436.
Karrer, P. and Schlientz, W. (1934). *Helv. chim. Acta* 17, 55–57.
Karrer, P., Jucker, E. and Steinlin, K. (1948). *Helv. chim. Acta* 31, 113.
Katayama, T., Nakayama, T. O. M., Lee, T. H. and Chichester, C. O. (1971). *J. Fd Sci.* 36, 804–806.
Kefford, N. P., Bruce, M. I. and Zwar, J. A. (1973). *Phytochemistry* 12, 995–1003.
Keith, E. S. and Powers, J. J. (1965). *J. Agric. Fd Chem.* 13, 577–579.
Khudairi, A. K. (1970). *Physiologie Pl.* 23, 613–622.
Knowles, R. E., Livingston, A. L., Nelson, J. W. and Kohler, G. O. (1968). *J. Agric. Fd Chem.* 16, 654–658.
Knypl, J. S. (1967). *Naturwissenschaften* 54, 146.
Knypl, J. S. (1971). *Acta Soc. Bot. Pol.* 15, 257–274.
Krinksy, N. I. (1968). *Photophysiology* 3, 123–195.
Kuhn, R. and Brockmann, H. (1932). *Hoppe-Seyler's Z. physiol. Chem.* 206, 41–64.
Kuhn, H. and Sperling, W. (1960). *Experientia* 16, 237–244.
Kuznetsova, N. A. (1971). *Nauch. Tr. Nauchnoizsled Inst. Konserv. Prom. Plovdiv.* 8, 147–157. (*Chem. Abstr.* 77, 125015 e.)
La Jollo, F., Tannenbaum, S. R. and Labuza, T. P. (1971). *J. Fd Sci.* 36, 850–853.
Laval-Martin, D. (196). *Physiol. Vég.* 7, 251–259.
Lee, F. A. and Wagenknecht, A. C. (1958). *Fd Res.* 23, 584–590.
Lee, K. H. and Yamamoto, H. Y. (1968). *Photochem. Photobiol.* 7, 101–107.
Lee, T. H., Erickson, L. C. and Chichester, C. O. (1971). *Hortscience* 6, 231–232.
Lewicki, P. P. (1973). *Zesz. Nauk. Akad. Roln. Warszawie. Technol. Rolno. Spozyn.* 8, 53–63. (*Chem. Abstr.* 80, 133733 q.)
Lewis, L. N. and Coggins, C. W. (1964). *J. Proc. Am. Hort. Soc.* 84, 177–180.
Livingston, A. L., Knowles, R. E., Nelson, J. W. and Kohler, G. O. (1968). *J. Agric. Fd Chem.* 16, 84–87.
Livingston, A. L., Nelson, J. W. and Kohler, G. O. (1969). *J. Ass. off. anal. Chem.* 52, 617–622.
Looney, N. E. and Patterson, M. E. (1967). *Nature, Lond.* 214, 1245–1246.
Lovric, T., Debicki, J. and Zadrevec, V. (1969). *Prehrambeno-Tehnol. Revija* 7, 4–8. (*Chem. Abstr.* 74, 30873 g.)
Lu, S. L. (1970). M.S. Thesis, University of California, Berkeley, California.
Luck, H. (1966). *J. Dairy Res.* 33, 25–29.
Luh, B. S., Chichester, C. O. and Leonard, S. J. (1962). *Abstr. First int. Congr. Fd Sci. Technol.* London.
Lukton, A. and Mackinney, G. (1956). *Fd Technol.* 10, 630–632.
Lukton, A., Chichester, C. O. and Mackinney, G. (1956). *Fd Technol.* 10, 427–432.
Mabry, T. J. (1966). *In* "Comparative Phytochemistry" (T. Swain, ed.), pp. 231–244. Academic Press, London and New York.
Mabry, T. J. (1970). *In* "Chemistry of the Alkaloids" (S. W. Pelletier, ed.), pp. 367–384. Van Nostrand Reinhold, New York.
McFeeters, R. F. and Schanderl, S. H. (1968). *J. Fd Sci.* 33, 547–553.
McFeeters, R. F., Chichester, C. O. and Whitaker, J. R. (1971). *Pl. Physiol., Lancaster* 47, 609–618.
Mackinney, G. and Joslyn, M. A. (1941). *J. Am. chem. Soc.* 63, 2530–2531.

Mackinney, G., Lukton, A. and Chichester, C. O. (1955). *Fd Technol.* 9, 324-326.
McWeeny, D. J. (1968). *J. Sci. Food Agric.* 19, 250-253, 254-258, 259-265.
Mapson, L. W. and Moustafa, E. M. (1955). *Biochem. J.* 60, 71-80.
Markakis, P., Livingston, G. E. and Fellers, C. R. (1957). *Fd Res.* 22, 117-130.
Martin, C. and Thimann, K. V. (1972). *Pl. Physiol., Lancaster* 49, 64-71.
Marusich, W., DeRitter, E. and Bauernfeind, J. C. (1960). *Poult. Sci.* 39, 1338-1345.
Mechter, E. E. (1953). *J. Agric. Fd Chem.* 1, 574-579.
Medvedeva, E. A. (1973). *Izv. Akad. Nauk. Tadzh. SSR, Otd. Biol. Nauk.* 2, 22-27. (*Chem. Abstr.* 79, 102749 z.)
Michel-Wolwertz, M.-R. and Sironval, C. (1965). *Biochim. biophys. Acta* 94, 330-343.
Milborrow, B. V. (1972). *Biochem. J.* 128, 1135-1146.
Mishra, D. and Misra, B. (1968). *Z. PflPhysiol* 58, 207-211.
Montreau, F. R., Latter, A. and Margulis, H. (1970). *C. r. hebd. Séanc. Acad. Sci., Serie D. Science Naturelles* 270, 1178-1181. (*Fd Sci. Technol. Abstr.* 2, 7J730.)
Moreth, C. M. and Yentsch, C. S. (1970). *J. exp. mar. Biol. Ecol.* 4, 238-249.
Morris, M. M., Park, Y. and Mackinney, G. (1973). *J. Agric. Fd Chem.* 21, 277-279.
Nagornaya, R. V. (1968). *Fiziol. Biokhim. Osn. Pitan. Rast.* 4, 203-209. (*Chem. Abstr.* 71, 46735 u.)
Newman, D. W., Rowell, B. W. and Byrd, K. (1973). *Pl. Physiol., Lancaster* 51, 229-233.
Nicoară, E., Oșianu, D. and Bodea, C. (1970). *Rev. roum. Chim.* 15, 965-971.
Nilsson, E. (1967). *Acta chem. scand.* 21, 1942-1951.
Noddle, R. C. and Robinson, D. R. (1969). *Biochem. J.* 112, 547-548.
Osborne, D. J. (1965). *J. Sci. Fd Agric.* 16, 1-13.
Oșianu, D., Nicoară, E. and Bodea, C. (1969). *Rev. roum. Chim.* 16, 765-767.
Papov, K., Kolev, D. and Vladimirov, G. (1969a). *Nauch. Tr. Nauchnoizsled Inst. Konserv. Prom., Plovdiv.* 6, 65-70. (*Chem. Abstr.* 72, 88983 t.)
Papov, K., Richev, G. and Vladimirov, G. (1969b). *Nauch. Tr. Nauchnoizsled Inst. Konserv. Prom., Plovidiv.* 6, 71-74. (*Chem. Abstr.* 72, 88984 u.)
Park, Y., Morris, M. M. and Mackinney, G. (1973). *J. Agric. Fd Chem.* 21, 279-281.
Petracek, F. J. and Zechmeister, L. (1956). *J. Am. chem. Soc.* 78, 1427-1434.
Patterson, P. D. and Mackinney, H. H. (1938). *Phytopathology* 28, 329-342.
Peng, C. Y. and Markakis, P. (1963). *Nature, Lond.* 199, 597-598.
Pennington, F. C., Strain, H. H., Svec, W. A. and Katz, J. J. (1967). *J. Am. chem. Soc.* 89, 3875-3880.
Philip, T. (1973). *J. Agric. Fd Chem.* 21, 693-694.
Phillips, D. R., Horton, R. F. and Fletcher, R. A. (1969). *Physiologia Pl.* 22, 1050-1054.
Piattelli, M. and Minale, L. (1964). *Phytochemistry* 3, 547-557.
Pratt, D. E., Balkcom, C. M., Powers, J. J. and Mills, L. W. (1954). *J. Agric. Fd Chem.* 2, 367-372.
Proctor, J. T. A. and Creasy, L. L. (1969). *Phytochemistry* 8, 1401-1403.
Pyysalo, H. and Kevusi, T. (1973). *Z. Lebensm.-Unters. Forsch.* 153, 224-233. (*Fd Sci. Technol. Abstr.* 6, 3H287.)

Pyysalo, H. and Makitie, O. (1973). *Acta chem. scand.* 27, 2681-2682.
Quackenbush, F. W. (1963). *Cereal Chem.* 40, 266-269.
Ramirez, D. A. and Tomes, M. L. (1964). *Bot. Gaz.* 125, 221-226.
Raymundo, L. C. and Simpson, K. L. (1972). *Phytochemistry* 11, 397-400.
Raymundo, L. C., Chichester, C. O. and Simpson, K. L. (1974). *J. Agric. Fd Chem.* In press.
Raymundo, L. C., Griffiths, A. E. and Simpson, K. L. (1967). *Phytochemistry* 6, 1527-1532.
Reid, M. S., Lee, T. H., Pratt, H. K. and Chichester, C. O. (1970). *J. Proc. Am. hort. Soc.* 95, 814-815.
Reschke, T. (1969). *Tetrahedron Lett.* 39, 3435-3439.
Rhodes, M. J. C. and Wooltorton, L. S. C. (1967). *Phytochemistry* 6, 1-12.
Ribereau-Gayon, P. R. (1959). *In* "Recherches sur les Anthocyannes des Vegetaux Application au Genre Vitis". Librairie Generale de l'Enseignement, Paris.
Robinson, W. B., Weirs, L. O., Bertino, J. J. and Mattick, L. R. (1966). *Am. J. Enol. Vitic.* 17, 178-184.
Ruskov, P. Z. and Tanchev. S. (1971). *Nauch. Tr. Vissh Inst. Khranit. Vkusova Prom., Plovdiv.* 18, 215-220. (*Chem. Abstr.* 77, 60288 z.)
Ruskov, P. Z. and Tanchev. S. S. (1973). *Prikl. Biokhim. Mikrobiol.* 9, 907-910. (*Chem. Abstr.* 80, 46625 u.)
Saberrov, N. V. and Ul'yanova, D. A. (1967). *Dakl. TSKHA (Timiryasez Sel'skokhoz, Akad.)* 132, 259-264. (*Chem. Abstr.* 69, 75707 u.)
Sakamura, S. and Obata, Y. (1961). *Agric. Biol. Chem.* 25, 750-756.
Sakamura, S., Shibusa, S. and Obata, Y. (1966). *J. Fd Sci.* 31, 317-319.
Sanger, J. E. (1971). *Ecology* 52, 1075-1089.
Sawant, P. L., Ramakrishnan, T. V. and Kumta, U. S. (1970). *Radiat. Bot.* 10, 169-174.
Schanderl, S. H. and Co, D. Y. C. L. (1966). *J. Fd Sci.* 31, 141-145.
Schanderl, S. H., Chichester, C. O. and Marsh, G. (1963). *J. org. Chem.* 27, 3865-3868.
Schanderl, S., Marsh, G. L. and Chichester, C. O. (1965). *J. Fd Sci.* 30, 312-316.
Schmid, P. (1971). *Z. Lebensm.-Unters. Forsch.* 146, 198-205. (*Fd Sci. Technol. Abstr.* 3, 12J1501.)
Seely, G. R. (1966). *In* "The Chlorophylls" (L. P. Vernon and G. R. Seely, eds), pp. 67-103. Academic Press, London and New York.
Seely, G. R. and Meyer, T. H. (1971). *Photochem. Photobiol.* 13, 27-32.
Segel, B. and Segel, R. M. (1969). *Rev. Ferment. Ind. Aliment.* 24, 22-24. (*Chem. Abstr.* 71, 79741 q.)
Seller, J. P. (1968). Ph.D. Thesis, Universität Bern, Germany.
Seybold, A. (1943). *Bot. Arch.* 43, 71-77.
Shimizu, S. and Tamaki, E. (1962). *Bot. Mag.* 75, 426-427.
Shriner, R. L. and Sutton, R. (1963). *J. Am. chem. Soc.* 85, 3989-3991.
Simpson, K. L. and Goodwin, T. W. (1965). *Phytochemistry* 4, 193-195.
Singleton, V. L. (1963). *Fd Technol.* 17, 112-115.
Singleton, V. L., Gortner, W. A. and Young, H. Y. (1961). *J. Fd Sci.* 26, 49-52.
Sistrunk, W. A. and Cash, J. N. (1970). *Fd Technol.* 24, 473-477.
Skalski, C. and Sistrunk, W. A. (1973). *J. Fd Sci.* 38, 1060-1062.
Snauwaert, F., Tobback, P. P., Verhees, J. and Maes, E. (1973). *In* "Radiation Preservation Food, Proc. Symp." IAEA, Vienna, Austria.

Sondheimer, E. and Kertesz, Z. I. (1952). *Fd Res.* 17, 288-298.
Sondheimer, E. and Kertesz, Z. I. (1953). *Fd Res.* 18, 475-479.
Srivastava, B. I. S. (1968). *Biochim. biophys. Acta* 169, 534-536.
Stahl, W., Steger, H., Kasdorff, K. and Pueschel, F. (1957). *Z. Tierphysiol. Tierenahr. Futtermittelk.* 12, 333-339.
Starr, M. S. and Francis, F. J. (1968). *Fd Technol.* 22, 1293-1295.
Stickland, R. G. (1972). *Ann. Bot.* 36, 459-469.
Strain, H. H. (1941). *J. Am. chem. Soc.* 63, 3542.
Strain, H. H. (1954). *J. Agric. Fd Chem.* 2, 1222-1225.
Strain, H. H. (1958). *In* "Chloroplast Pigments and Chromatographic Analysis" 32nd Annual Priestley Lectures, The Pennsylvania State University, University Park, Pa.
Strain, H. H. and Svec., W. A. (1966). *In* "The Chlorophylls" (L. P. Vernon and G. R. Seely, eds.), pp. 22-61. Academic Press, London and New York.
Sudyina, E. G. (1963). *Photochem. Photobiol.* 2, 181-190.
Summer, R. J. (1942). *J. biol. Chem.* 146, 215-218.
Swain, T. (1962). *In* "The Chemistry of Flavonoid Compounds" (T. A. Geissman, ed.), pp. 513-552. Macmillan, New York.
Sweeney, J. P. (1970). *Fd Technol.* 23, 186-189.
Tan, C. T. and Francis, F. J. (1962). *J. Fd. Sci.* 27, 232-241.
Tanchev, S. S. (1971). *Nauch. Tr. Vissh Inst. Khranit. Vkusova Prom., Plovdiv.* 18, 383-390. (*Chem. Abstr.* 79, 41144 z.)
Tanchev, S. S. (1972a). *Ind. Obst. Germuerserwert* 57, 315-317. (*Chem. Abstr.* 8, 14555 g.)
Tanchev, S. S. (1972b). *Z. Lebensmittelunters. u.-Forsch.* 150, 28-30.
Tanchev, S. and Ioncheva, N. I. (1972). *Izv. Vyssh, Ucheb. Zaved. Pishch. Tecknol.* 4, 58-60. (*Chem. Abstr.* 78, 41764 f.)
Taylor, H. F. and Burden, R. S. (1970). *Phytochemistry* 9, 2217-2223.
Taylor, R. D., Kohler, G. O., Maddy, K. H. and Enochian, R. V. (1968). "Alfalfa Meals in Poultry Feeds—An Economic Evaluation Using Parametric Linear Programming". U.S. Economic Research Ser., U.S.D.A., Agricultural Econ. Report no. 130.
Timberlake, C. F. (1960). *J. Sci. Fd Agric.* 11, 268-273.
Timberlake, C. F. and Bridle, P. (1965). *Chemy Ind.* 1520-1521.
Timberlake, C. F. and Bridle, P. (1966). *Nature, Lond.* 212, 158-159.
Timberlake, C. F. and Bridle, P. (1967a). *J. Sci. Fd Agric.* 18, 473-478.
Timberlake, C. F. and Bridle, P. (1967b). *J. Sci. Fd Agric.* 18, 479-485.
Timberlake, C. F. and Bridle, P. (1968). *Chemy Ind.* 1489.
Tinsley, I. J. and Bockian, A. H. (1960). *Fd Res.* 25, 161-173.
Tookey, H. L., Wilson, R. G., Lohmar, R. L. and Dutton, H. J. (1958). *J. biol. Chem.* 230, 65-72.
Trojan, A. V. and Gol'yan, D. S. (1973). *Izv. Vyssh. Ucheb. Zaved, Pishch. Tekhnol.* 1, 22-24. (*Chem. Abstr.* 79, 2805 w.)
Tsukida, K., Yokota, M. and Ikeuchi, K. (1966). *Vitamin* 33, 174-178.
Tsumaki, T., Yamaguchi, M. and Hori, F. (1954). *Scient. Rep. Fac. Sci. Kyushu Univ.* 2, 35.
Uchiyama, Y. (1969). *Agric. Biol. Chem.* 33, 1342-1345.
Ulrich, R. (1970). *In* "The Biochemistry of Fruits and Their Products" (A. C. Hulme, ed.), vol. 1, pp. 89-118. Academic Press, London and New York.

Valuiko, G. G. and Germanova, L. M. (1968). *Prikl. Biokhim. Mikrobiol.* 4, 464-467. (*Chem. Abstr.* 69, 95075 q.)

Van Buren, J. P., Bertina, J. J. and Robinson, W. B. (1968). *Am. J. Enol. Vitic.* 19, 147-154.

Vasquez, R. A., Maestro, D. R. and Valle, M. L. J. del (1970). *Grasas Aceites* 21, 337-341. (*Fd Sci. Technol. Abstr.* 3, 9J1166.)

Vilece, R. J., Fagerson, I. S. and Esselen, W. B. (1955). *J. Agric. Fd Chem.* 3, 433-435.

Villegas, C. N., Chichester, C. O., Raymundo, L. C. and Simpson, K. L. (1972). *Pl. Physiol., Lancaster* 50, 694-697.

Wagenknecht, A. C. and Lee, F. A. (1956). *Fd Res.* 21, 605-610.

Wagenknecht, A. C. and Lee, F. A. (1958). *Fd Res.* 23, 25-31.

Wagenknecht, A. C., Lee, F. A. and Boyle, F. P. (1952). *Fd Res.* 17, 343-350.

Wagenknecht, A. C., Scheiner, D. M. and Van Buren, J. P. (1960). *Fd Technol.* 14, 47-49.

Walker, G. C. (1964). *J. Fd. Sci.* 29, 383-388.

Walsh, K. A. and Hauge, S. M. (1953). *J. Agric. Fd Chem.* 1, 1001-1004.

Weedon, B. C. L. (1971). *In* "Carotenoids" (O. Isler, ed.), pp. 267-323. Birkhauser, Basle.

White, R. C., Jones, I. D. and Gibbs, E. (1963). *J. Fd Sci.* 28, 431-436.

Whitfield, D. M. and Rowan, K. S. (1974). *Phytochemistry* 13, 77-83.

Willstätter, R. and Stoll, A. (1918). *In* "Untersuchungen über die Assimilation der Kohlensäure" p. 448. Springer-Verlag, Berlin.

Willstätter, R. and Stoll, A. (1928). *In* "Investigations on Chlorophyll" (F. M. Schertz and A. R. Merg, trans.), p. 385. Science Press, Lancaster, Pa.

Winterstein, A., Studer, A. and Ruëgg, R. (1960). *Chem. Ber.* 93, 2951-2965.

Wolf, F. T. (1956). *Am. J. Bot.* 43, 714-718.

Woodward, J. R. (1972). *J. Sci. Food Agric.* 23, 465-473.

Wrolstad, R. E. and Erlandson, J. A. (1973). *J. Fd Sci.* 38 460-463.

Wrolstad, R. E., Putnam, T. P. and Varseveld, G. W. (1970). *J. Fd. Sci.* 35, 448-452.

Yamamoto, H. Y. and Takeguchi, C. A. (1971). *Second Int. Congr. Photosynthesis* 621-627.

Yokoyama, H. and White, M. J. (1966). *Phytochemistry* 5, 1159-1173.

Yokoyama, H. and White, M. J. (1970). *Phytochemistry* 9, 1795-1797.

Zechmeister, L. and Pinckard, J. H. (1947). *J. Am. chem. Soc.* 69, 1930-1935.

Zechmeister, L., LeRosen, A. L., Schroeder, W. A., Polgar, A. and Pauling, L. (1943). *J. Am. Chem. Soc.* 65, 1940-1951.

Specific Name Index

D

Subject Index

D

E

V

W

X

Z